AF323503

BIOTECHNOLOGY

CYTOGENETICS, BIOTECHNOLOGY AND BIOINFORMATICS SERIES

BIOTECHNOLOGY

Darbeshwar Roy

Alpha Science International Ltd.
Oxford, U.K.

Biotechnology

980 pgs. | 124 figs. | 58 tbls.

Darbeshwar Roy
Department of Genetics and Plant Breeding
College of Agriculture
Govind Ballabh Pant University of Agriculture and Technology
Pantnagar, India

Copyright © 2010

ALPHA SCIENCE INTERNATIONAL LTD.
7200 The Quorum, Oxford Business Park North
Garsington Road, Oxford OX4 2JZ, U.K.

www.alphasci.com

All rights reserved. No part of this publication may be reproduced, stored in a retrieval system or transmitted in any form or by any means, electronic, mechanical, photocopying, recording or otherwise, without prior written permission of the publisher.

ISBN 978-1-84265-538-2

Printed in India

To
Shishir and Cynthia with love

Preface

For understanding genes of different organisms and their manipulations for human benefit, it is essential to understand the structure and function of their genetic material at the molecular level. Further, in case of plants it is essential to protect the crop varieties against biotic and abiotic stresses whereas in case of humans it is necessary to protect them from many killer diseases. These problems can be solved only by understanding the genes and the genomes of these species and their related wild and weedy species. This book describes the genetic material(DNA, RNA), its structure and function. Gene, its structure and regulatory elements- promoter, enhancer and silencer are described. Genetic code and other types of codes, RNA editing and DNA imprinting are described. Structural genomics and functional genomics technologies are described to understand the structure and function of genes at individual gene level and at genome-wide level. Study of genes' products- the proteins are as important as the study of genes' immediate products- the transcriptome and so proteins and proteomics are discussed in detail. Further, genes do not work in isolation rather they work together as pathways and networks. Then there are different types of interactions such as DNA x protein, RNA x protein, DNA x RNA and protein x protein and techniques for studying various types of interactions are discussed. It is also important to understand the cell function at the molecular level using systems biology approach. Sequencing is the best way to characterize a gene, its regulatory elements and genome whereas chemical synthesis of DNA is essential for designing gene, regulatory elements, probes and primers and both can help in tailoring an organism. Isolation of gene, gene cloning and *in vitro* mutagenesis are important for studying gene structure and function. Gene silencing through antisense RNA and RNAi can help in understanding the gene function and has practical implications. Finally, various molecular cytogenetics techniques and genetic engineering techniques described in this book will help in setting up experiments. This book provides the latest methods of analyses and technologies for studying structure and function of genes and genomes. Finally, applications of bioinformatics tools in sequence assembly and in structural and functional genomics and proteomics have been described.

I would like to thank Mr. N. K. Mehra, M/s Narosa Publishing House for conceiving to publish my manuscript in a series of books (Cytogenetics, Biotechnology and Bioinformatics) and also for his sincere interest and attention. Further, I would like to thank Mrs. Anupama Jauhry, Asstt. General Manager (Production) for taking care of my manuscript.

Finally, I would like to thank Veena, Shishir and Cynthia for their encouragement during preparation of the book.

Darbeshwar Roy

Contents

17. Systems Biology 17.1

1

Nucleic Acid and Genome Organization

Nucleic acid was demonstrated to be the carrier of genetic information first in 1944 by Avery, Mc Leod and Mc Carty's transformation experiment that showed that DNA isolated from one virulent strain of bacterium (donor) can enter and transform the cells of another bacterium (recipient) strain endowing it with some heritable characteristics of the donor. Further experiment by Hershy and Chase in 1952 showed that it was DNA of the bacteriophage (bacterial virus, T2) and not its protein coat, which carried the genetic information for replication of the virus in the host cell of bacterium, E. coli. Later experiment by Fraenkel-Conart in 1955 with tobacco mosaic virus causing yellow molting of the leaf showed that the nucleic acid was RNA to be the bearer of genetic information. Thus there are two types of nucleic acid- DNA (deoxy ribonucleic acid) and RNA (ribonucleic acid).

1.1 BIOCHEMISTRY OF NUCLEIC ACID

Nucleic acids are polymers of nucleotides. A **nucleotide** consists of nitrogenous base, sugar and phosphate group. The nitrogenous bases are of two types- purines and pyrimidines. Adenine and guanine are purines whereas primidines are thymine, cytosine and uracil. The pentose sugars are of two types- 2'deoxyribose and ribose. The DNA differs from RNA in that the former contains 2' deoxyribose sugar (2' Oxygen is absent) whereas RNA contains ribose sugar. Further, in the DNA the pyrimidine bases are thymine and cytosine whereas RNA contains cytosine and uracil. There is one methyl group in thymidine whereas there is a hydrogen atom in uracil. Thus there are four types of nucleotides and that part of nucleotide which consists of nitrogenous base plus sugar is called **nucleoside**. The four nucleosides in DNA are called deoxy adenosine (dA), deoxy cytidine (dC), deoxy guanosine (dG) and thymidine (T) whereas in case of RNA it is adenosine (A), uridine (U), cytidine (C) and guanosine (G). The four nucleotides are deoxyadenosine monophosphate (AMP), UMP, GMP and TMP. The precursors of DNA are dATP, dCTP, dGTP and TTP and for RNA ATP, CTP, GTP and UTP. The nucleotides are not always found in the form of nucleoside monophosphate. They may contain either diphosphate or triphosphate groups. The cell machinery for DNA and RNA synthesis only acts on nucleotides containing a triphosphate group. Further deoxyribonucleotides are made from ribonucleotides. Pyramidine nucleotides are made from PRPP, aspartate and carbamyl phosphate (PRPP (5-phosphoribosyl 1-pyrophosphate), Aspartate and CPhosphate $\rightarrow$ UMP $\rightarrow$ UTP $\rightarrow$ CTP). Purine nucleotides are made from PRPP, glutamine, glycine, folate and aspartate (PRPP $\rightarrow$ IMP and XMP and from XMP to GMP). The nitrogenous base is joined covalently (at N-1 of pyrimidines and N-9 of purines) in an N-B- glycosyl bond at the C-1' of the pentose and the phosphate is esterified to C-5' of pentose (see figure 1.2). The linkage between nucleotides is of the ester type and the C-3' And C-5' of the two different sugars forms a double ester with phosphoric acid. This is called **phospho-diester bond**. The internucleotide linkage in one strand is 3'-5' while in the other strand it is 5'-3'

1.2   DNA MODEL

The components of nucleic acid does not say anything about the physical structure of the DNA. The discovery of composition of DNA came from Chargaff in 1947. The chemical analysis of DNA obtained from different species showed that the base equivalences, i.e. for each nucleotide of adenine there was one of thymine (A= T) and for each nucleotide of guanine there was one of cytosine (G= C) in the DNA. Further the ratio of A/T and G/C was constant in all DNAs, i.e. purines/pyrimidines ratio was constant but (A+T)/(G+C) ratio varied in different species, i.e. base composition generally varied from one species to another. The physical structure of DNA became clear from the X-ray diffraction study of DNA fiber by Franklin and Wilkins in 1953 who observed the characteristic X-ray diffraction pattern. DNA molecule was helical in structure with two **periodicities** along their long axis- a primary one of 3.4 Å and a secondary one of 34 Å and DNA from different organisms produced very definite and almost identical patterns. This indicated that the nucleotides were oriented spatially in the same fashion in different DNA macromolecules. On the basis of these discoveries Watson and Crick (1953) postulated a model of DNA structure (Figure 1.1). They hypothesized that the DNA consists of two anti-parallel strands in a right handed double helical form. The two strands are held together by complementary base pairs formed by hydrogen bonding, the two bonds between A and T (A=T) and three bonds between G and C (G≡C) stacked within the double helix and the hydrophilic sugar phosphate backbones are located outside.

1.2.1   Identity of Nucleotide Strands

The ends of each nucleotide strand have distinct chemical identities indicated by the symbol 5' and 3'. The two strands differ in polarity. One nucleotide strand runs in the direction of 5'-3' whereas the other strand runs in the direction of 3' to 5'. The 5'-3' has all its bases pointing down whereas the second strand has all its bases pointing up. Further, the direction of 3' to 5' sugar-phosphate linkage in 3' to 5' is opposite to that of in 5'-3' strand. The directionality is thus due to the specific internucleotide phosphodiester linkage between the O3' and O5' atoms from the neighboring deoxyriboses. Now if a nucleotide were represented as a right arrow, ($\rightarrow$) then the geometry of the DNA would require that the base available for pairing be attached to the shank of the arrow from below and similarly, if the nucleotide were represents as a left arrow, ($\leftarrow$), then the geometry would require that the base available for pairing be attached to the shank of the arrow from above. This model of DNA structure satisfies the requirements of any model to explain self-replication and protein synthesis which are the two important functions of DNA. The base pairs are 3.4 Å (0.34nm) apart and there are 10.5 base pairs (36 Å or 3.6nm) in each complete turn of the double helix, i.e. a complete turn occurs at every 36 Å. The diameter of double helix is 20 Å (2nm). It was earlier postulated that there were only 10 base pairs per complete turn and the complete turn occurred at every 34 Å.

1.2.2 Characteristics of DNA

The three characteristics of the **secondary structure** of DNA are:

    I. Right handed helical structure $\rightarrow$

    II. Diameter of DNA double helix and

    III. Number of bases per complete turn.

Variations in the above three parameters have been observed. The two helices are right handed, i.e., they turn in a clockwise direction as viewed from the rear end or turn in the direction of fingers of the right hand when the thumb indicated the line of sight. Left handed structures or DNA structure with diameter and number of bases per complete turn more or less than Watson and Crick's DNAs have been found. The negative charge of the DNA is associated with the phosphate group which can be neutralized by the positive charges of metal ions (Na, K, Mg), polyamines or proteins. DNA exists predominantly as an antiparallel double stranded helix. The two strands are helically coiled which maximizes the exposure of the negatively charged sugar-phosphate backbone to water and shields the hydrophobic aromatic bases in the middle from water. There is close interaction between the stacked bases in a nucleic acid. The offset pairing between the two strands creates **major** and **minor groove** on the surface of the duplex. The stacked bases form a hydrophobic core but the amino and keto groups point into the grooves and allow interaction with the solvent. The backbone of both DNA and RNA are hydrophilic with one negative charge per phosphate. **Regulatory protein** has been found to be associated with major groove whereas water molecules are associated with minor groove.

1.3 NUCLEIC ACID STRUCTURE

Nucleic acid- Nucleic acids show primary, secondary and tertiary structures. The **primary structure** refers to its covalent structure and nucleotide sequence. Any regular or stable structure taken up by some or all of the nucleotide in a nucleic acid is referred to as **secondary structure**. The double helix in case of DNA as described above is its secondary structure. A complete folding of large chromosomes within eukaryotic chromatin and bacterial nucleoids is generally considered as **tertiary structure** (see Roy, 2009). Tertiary structure refers to a specific conformation of a double stranded DNA molecule containing extra twists in the double helix as a result of either overwinding (+ supercoil) or underwinding (- supercoil) causing it to coil upon itself.

1.4 DYNAMIC NATURE OF DNA

DNA is a remarkably flexible molecule. The sugar-phosphate polynucleotide chain may consist of many single bonds around which the attached atoms may rotate. The deoxyribose ring is non-plannar and adopts two preferred 'pluckers' namely, C2'-endo and C3'-endo conformations. Two parameters, the sugar pseudorotation phase angle and its maximum torson angle, i.e. pseudorotation amplitude, are normally used to describe the pluckering modes available in terms of sugar torson angles. Two preferred orientations (*anti* or *syn*) of the base with respect to sugar as defined by the torson angle about the 4 C1'-N glycosyl bond x are found. Thermal fluctuations can cause bending, stretching and melting (unwinding) of the strands.

1.5 DNA'S OTHER STRUCTURAL FORMS

The DNA can exist in several **structural forms**. The B form of DNA or the B-DNA is the most stable structure which Watson and Crick postulated. A and Z forms have been characterized in DNA crystal structures. A-form double helices differ from B-DNA in the conformation of the sugar and the displacement of the bases from the helical axis. The A-form, a right handed helix is shorter and of greater diameter (~ 26 Å) than a B-form helix with the same sequence whereas the Z-form is a left handed helix of lesser diameter (~18 Å). In comparison to 10.5 bps/turn in B-DNA, the number of base pairs per complete turn in A and Z forms is 11 and 12, respectively. B, A and Z forms thus differ in the **pitch** (distance between successive repeating units, the helical distance) and the number of bases per turn and in A and Z forms bases are not flat but tilted. The A –form is formed under condition of water stress, major grooves deepen and minor grooves become shallower. Under condition of high salt concentration the DNA tends to convert to Z-form in which the major groove barely appears but the minor groove is narrow and deep. As the DNA is made up of A-T and G-C base pairs, these two types of base pairs could be present in either equal or in different proportions in the DNA and the higher the $G\equiv C/A=T$ ratio the separation of paired DNA stands is more difficult and such DNAs will be susceptible to thermal collapse. Other pairings of bases (other than A-T-rich DNAs) will tend to destabilize the double helical structure. All higher plants and animals have an excess of A+T over G+C whereas among bacteria and lower plants thee is much more variation and both A+T-rich and G-C-rich species occur. Watson and Crick model assumes unspecified (random) sequences of bases but when there is an alternating purine and pyrimidine sequences in the DNA, e.g. poly (d_A-d_T) then this poly (d_A-d_T) tends to form a left handed double helix known as Z-DNA. These alternating purine and pyrimidine bases occur about once every 3, 000 base pairs in human chromosomes. The question is then what Z-DNA is doing within the B-DNA that makes up chromosomes. The Z-DNA is assumed to provide a regulatory element to control gene expression. If Z-DNA does indeed have a biological role in gene regulation and signaling then such sequence should readily adopt a left handed form. The DNA polymer d $(GC)_n.d (GC)_n$ can go under a transition from the usual right handed 10.4 base pairs per turn B form to a left handed 12bp per turn Z form in response to altered environmental conditions (Haniford and Pulleyblank, 1983). Although other sequences poly (d_G-d_T): poly (dA.dC) could equally adopt a left handed Z structure in a semicrystalline fibre sample but only in

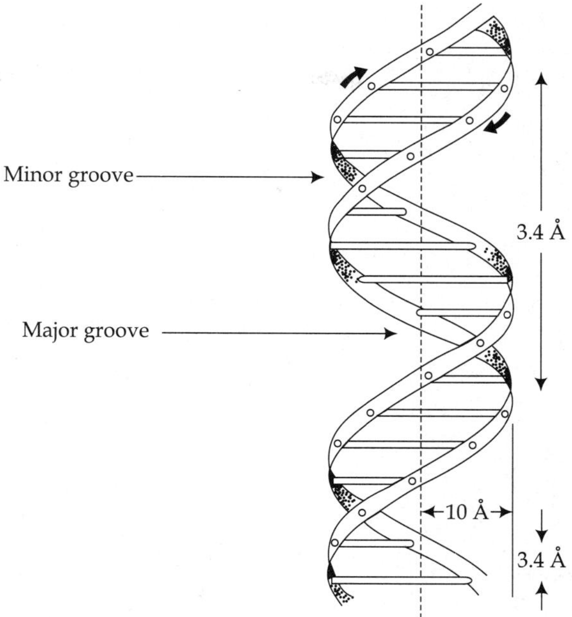

FIGURE 1.1 DNA double helix model (B-DNA).

extremely environmental conditions. Further, the adjacent bases are related to each other by roll, slide and twist (or angle between bases). The helical **twist angle** is highly dependent on the particular sequence and varies from 27.7^0 (ApG) to 40.0^0 (for GpC). The base tilt in case of A-DNA is 20^0, in B-DNA is 6^0 and in Z-DNA is 7^0 and there are different ways in which the adjacent bases could move relative to one another. Thus the forms of DNA depending on the particular sequence of bases will have effect on the structure of DNA. Left handed Z-DNA is a higher energy form of double helix, stabilized by negative supercoiling generated by transcription or unwinding nucleosomes. Regions near the transcription start site frequently contain motifs favoring formation of Z-DNA and formation of Z-DNA near the promoter region stimulates transcription. Z-DNA is also stabilized by specific protein binding. There are proteins which bind specifically to Z-DNA and so it is a biologically functional DNA component. Z-DNA occurs in a dynamic state, forming as a result of physiological processes then relaxing to the right handed B-DNA. Each time a DNA segment turns into Z-DNA, two B-Z junctions are formed. Ha et al. (2005) presented the X-ray crystal structure of a junction between B- and Z-DNA. The structure of Z-DNA structure was given by Wang et al. in 1979. They showed that a single base pair is broken and flipped out of the DNA double helix, thereby facilitating the juxtaposition of a left handed Z-DNA helix with the right handed B-DNA. Z-DNA is inherently unstable. Further, in comparison to B-DNA, the hydrogen bonded pairs are turned upside down in Z-DNA (Sinden, 2005). The extruded bases on each side may be sites for DNA modifications.

The other forms of DNA are **C-DNA** (right handed with 9.33 residues per turn and 19 Å in diameter), **E-DNA** (7.5 residues per turn, contains no guanine bases), **G-DNA** (four stranded

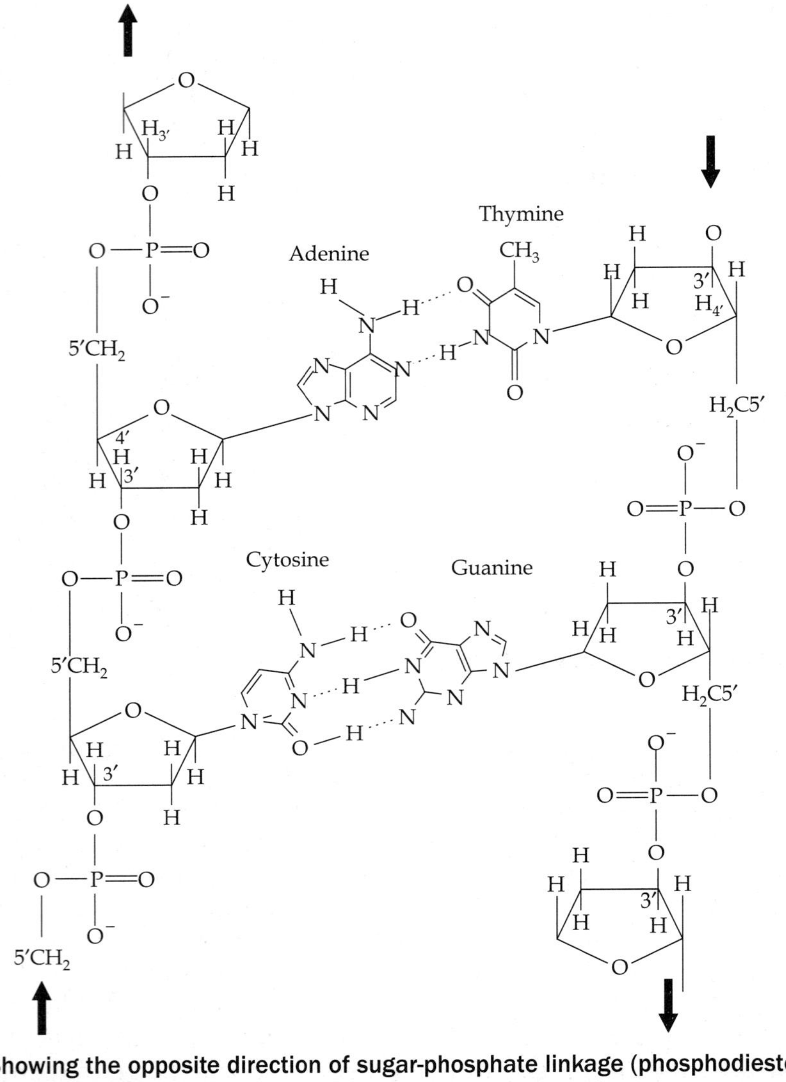

FIGURE 1.2 Showing the opposite direction of sugar-phosphate linkage (phosphodiester bond) in the two strands. One strands. One strand running in the direction $5' \rightarrow 3'$ and another in $3' \rightarrow 5'$.

DNA with G-rich strands), **G4-DNA** (all four strands run in parallel orientation and are linked by Hoogsteen bonded guanine quartets), **H (hinged)-DNA** (contain both single and triple stranded regions), M-DNA, P-DNA, V-DNA and ε DNA.

Because the level of DNA superhelicity varies with the cellular energy charge, it can change rapidly in response to a wide variety of altered nutritional and environmental conditions. This is a global alteration affecting the entire chromosome and the expression levels of all operons whose promoters are sensitive to superhelicity.

1.6 CHROMOSOMAL DNAS

Chromosomal DNAs are many orders of magnitude longer than the cell or viruses that contain them. All DNAs from all organisms contain the same sugar and phosphate and the same four nucleotides but the difference in DNA lies in the sequence of their four bases. The genetic information of a DNA molecule resides in the linear sequence of the four bases.

1.6.1 Types of Base Sequence in DNA

There are three types of base sequences of DNA:

    I. Unique (or low copy) sequences DNA

    II. Moderately repetitive sequences DNA

    III. Highly repetitive sequences DNA.

I. Unique sequences DNA—The unique sequences are those sequences that are repeated only a few times, i.e. they have got one or a few copies. These sequences are thought to consist of sequences coding for most enzyme function. These are also called **coding sequences** or genes. In human cells it has a Cot ½ value of about 1000 moles of nucleotides seconds per litre. Viral and bacterial genomes consist of pre-dominantly unique (single copy) DNA with little repetitive sequence. Unique sequences constitute less than 0.5% of the total chromosomal DNA in wheat. Thus these unique sequences are buried in a vast excess of nucleotide sequences whose function is yet to be known.

II. Repetitive DNA—It refers to any nucleotide sequence that is present in multiple copies per genome. Repetitive DNA can be of two types:

    1. **Moderately repetitive DNA** and

    2. **Highly repetitive DNA.**

Repetitive sequences often occur clustered in specific chromosomal regions.

1. Moderately repetitive DNA—DNA sequences that are repeated at least 1000 times, such DNAs are called moderately repetitive. The DNA sequences consist of up to 100 to 500 bps. Such sequences are scattered throughout the genome. Moderately repeated sequences have Cot value of between 100 and 10,000. These sequences are thought to be those coding for major structural proteins of cells such as histones and thus moderately repetitive sequences constitute functional genes. Genes encoding five classes of histones occur in clusters, one gene of each kind in each cluster, not necessarily oriented in the same direction. The clusters themselves tend to be clustered. The genes for rRNA and tRNA also fall into this category. rRNA genes (rDNA) are clustered in tandem arrays at one or a few chromosome loci and identifiable microscopically as nucleolus organizers. Most species have a few hundred rRNA gene copies, 200 (human and Drosophila) and 140 (yeast). The other moderately repetitive DNA is tRNA genes. Intermediately repetitive DNA is more heterogeneous than highly repetitive DNA. Some of the smaller multigene families are highly clustered whereas the larger families tend to be dispersed throughout the genome. All eukaryotes have multiple genes specifying an additional small (5S) rRNA. In yeast these genes form part of the major rDNA cluster whereas in case of multi-cellular eukaryotes they form separate clusters.

2. Highly repetitive sequences—Highly repeated sequences or simple sequence DNAs are repeated over millions of times and they are not transcribed into RNA and thus are called

junk DNA that has no function. The length of such sequences is less than 5-100 bps. The proportion of a genome that is classified as repeated or single copy depends on the method of determination.

1.7 ORGANIZATION OF THE REPEATED DNA SEQUENCES

Repeated sequences are organized as
1. **Tandem repeats** and
2. **Dispersed repeats**.

Plant's chromosomes carry large number of dispersed (often retrotransposons) and tandemly repeated DNA sequences.

1.7.1 Tandem Repeats

On class of tandemly repeated sequences consists of rDNA, the gene encoding the ribosomal RNA and they are thus functional genes in repetitive arrays and discussed above. Tandem arrays of rDNA are found at one or a few places in a given plant genome and in most cases these clusters are associated with the NORs.

Satellite DNA — A second major class of tandemly repeated sequences are satellite DNAs. The simple sequence DNA is also called satellite DNA which is usually quite short and are arranged in tandem arrays and are clustered together and when isolated by caesium chloride buoyant density gradients of sheared DNA it forms a satellite band due to differing content of adenine and thymine residues. The term satellite tends now to be used to describe all tandemly repetitive high copy number DNA sequences even though they are not separable by density. Such sequences account for the highly repetitive non-dispersed DNA fraction identified by Cot analysis. Satellite is made up of the simple sequence, (GAA)m (CAC)n in barley and mainly found in the pericentromeric regions of all seven chromosomes. Satellite DNA comprises about 5-10% of the total DNA in Drosophila, mice and humans. Satellite DNA is a major component of constitutive heterochromatin (those parts of chromatin that remain condensed and deeply-stained through out the cell cycle.

1.7. 2 Interspersed DNA Elements

To understand the architecture of genome, there is a need to define all interspersed DNA elements, the families of sequences repeated throughout the genome which make up the majority of the genome. These repeat sequences are arbitrarily classified into CpG island, SINES, LINES, minsatellite and microsatellites.

A. Dispersed simple sequence (Minisatellite and microsatellite) — The micro-satellite and mini-satellite are based on short sequences that are repeated in tandem at one or more places in the genome. In particular, micro-satellite repeats may be present at many different locations in the genome and each is flanked by restriction sites whose distance from the core repeat differs from one location to the next. The micro-satellite has a very short core repeating unit of 2 to 9 base pair, for example, (5'-CAT-3') n where n is the number of repeats in the micro-satellite whereas in case of mini-satellite the core unit is typically of 10 to 60 base pairs. It has a Cot value as low as 10^{-3} moles of nucleotides per litre. In humans minisatellites occur scattered over the chromosomes in small arrays of tandem repeats, the numbers in each array often varying from one individual to another. Satellite DNA has no identified function for the organism. The repetitive sequences vary greatly between different species within the same genus but within a species the same satellite sequences often recur on every chromosome. The non-coding sequences influence the spacing between the genes and probably the recombination between them. They affect the amount of DNA per chromosome arm and therefore the chromosomal organization in the nucleus, rates of cell cycle and plant development by modifying the total DNA mass. DNA repeats such as the mammalian alpha-satellite interacts with kinetochore proteins and other conserved sequences are thought to have required function in chromosome disjunction and segregation.

B. Dispersed repeats —Dispersed repeats or interspersed repetitive DNA are sequences that differ from the satellite in that they are not clustered together but are dispersed throughout the genome. These include inverted repeats and repeated sequences interspersed among one another or among unique sequences. Interspersion, the alternating sequence of repetitive and single copy DNA in eukaryotes, shows basically two patterns. 1. Where short interspersed elements (SINES) of less tan 0.5kb alternate with unique sequences of about 0.5 to 2kb in length and 2. where long interspersed elements (LINES) of 5-7kb alternate with long DNA sequences of about10kb.The interspersion patterns differ from organism to organism. Such types of repeats have been formed in the chromosomal DNA as a result of invasion by parasitic or selfish DNA or RNA or mobile genetic elements which have the capacity for proliferation and dispersal and which perpetuate themselves without greatly reducing the fitness of the organism. The dispersed repeated sequences include retrotransposons, transposons (autonomous type) and MITES (non-autonomous type).

Mobile elements—These mobile elements have been classified into 1. Retrotransposons 2. Transposons. Transposition can occur by reverse transcription of an RNA intermediate or by excision and reintegration of DNA itself.

1. Retrotransposons (RTs)—Retrotransposons are the most abundant class of dispersed repeats. Some dispersed repeats may have evolved by infection with an RNA sequence called retrotransposons. Retrotransposons are mobile genetic elements that transpose through reverse transcriptase of an RNA intermediate. RTs are ubiquitous in plants and plant TRs are structurally and functionally similar to retrotransposons and retroviruses found in other eukaryotes. They are often high copy in numbers, extensively heterogeneous populations and their chromosomal dispersal pattern and are the major constituents of plant genome. These retrotransposons are heavily methylated in comparison with genes (Bennetzen et al., 1994). RTs consist of LTR and nonLTR retrotransposons which terminate with poly (A) sequences. RTs are found between genes. They constitute at least 50% of the maize genome.

LTR (long terminal repeat) transposons—LTR RTs are classified into Ty1-copia and Ty3-gypsy groups. They differ from each other in both degree of sequence similarity and the order of encoded gene products. Ty1-copia TRs are present throughout the plant kingdom. Ty3-gypsy RTs are also present in both angiosperm and gymnosperm. Both groups of RTs are commonly found in high copy number (up to a few millions copies per haploid nucleus). Both are ubiquitous in cereal genomes. LTR-RTs have long terminal repeats in direct orientation at each end. Within LTRs are U3, R, and U5 regions that contains signal for initiation and termination of transcription. The genes within the RTs encode capsid like protein (CP), endonuclease (EN), integrase (INT), protease (PR) and reverse transcriptase (RT) and RNAase-H. The envelope (*env*) gene like sequence in the position of ORF 3 where a functional *env* gene is present in the animal retroviruses has been found in both Ty1-copia and Ty4-gypsy groups. The function of the *env* gene like sequence in plant RTs is unknown.

Non LTR RTs—The nonLTR RTs comprising **LINES** and **SINES** can also be found in high copy number (up to 250,000). LINES are present throughout the plant kingdom. In many cases RTs comprise over 50% of the nuclear DNA content. LTR TRs range from a few kb up to 15kb. LINES range from less than 1kb to may be 8kb whereas SINES range from 100bp to 300bp. LINES possess ORF that codes for protein with homology to reverse transcriptase. In *Arabidopsis* about 4% of the genome consists of RTs. In maize RTs make up 50-80% of the DNA contributing only 10% or less of the mRNA in most tissues. In mammalian genome LINES and SINES (L1 and ALU elements) are major mobile sequences, constituting about 35% of the total DNA up to 100,000 copies of each element type. RTs sequences are present either within, or close to many genes. In many cases Ty1-copia, Ty3-gypsy, LINES and SINES RTs are dispersed widely through plant chromosomes. However, each RT group also has families that are clustered or deficient in specific regions of chromosomes. For example, some Ty3-gypsy RTs are clustered in the centromeric regions in maize, rye and wheat whereas some Ty1-copia RTs are present in the terminal heterochromatic regions of *Allium* or absent from the same region in rye. RT sequences have not been found in the NORs in plants. Many RTs are preferentially found in euchromatic regions where most genes are located. Sequence analysis of large genomic clones in maize has revealed that euchromatic RTs are mostly found in intergeneric regions. RTs move to new chromosomal locations via an RNA intermediate that is converted into extrachromosomal double stranded DNA by the encoded reverse transcriptase/RNaseH enzymes prior to insertion into genome.This replicative mode of transposition can rapidly increase in copy numbers of elements and can thereby greatly increase plant genome size. The DNA transposable elements (e.g. Ac, Tam1 and EN/Spm) transpose by an excision/repair mechanism and usually do not greatly increase in plant genome size. Retrotransposons differ from transposons in being flanked by long terminal direct repeats (LTRs). They move not by excision but by reverse transcription and chromosomal integration. After infection into the host cell the RNA sequence is reverse transcribed into DNA copies by reverse transcriptase. The reverse transcript is made double stranded and integrated into the host genome and thus present as disperse repeats. Thus they are inserted at new loci without removal from their previous sites of integration. Retrotransposons resemble chromosomally integrated retroviral genomes (proviruses) in all these characteristics. Retrotransposons result in the production of non-functional '**processed pseudogenes**'. There are two classes of dispersed repeats: 1. SINES-short interspersed nuclear elements consisting of repeats of less than 500bp and repeated over 300,000 times and 2. LINES (long interspersed nuclear elements) consisting of 6-7Kbp and repeated up to 10,000 times. SINES are found in Alu element of human cells and LINES in human L1 (Kpn elements) element and are called **retrotransposons.** The mouse B1 and human Alu SINES are unique among SINES in being derived from 7SL RNA. Some repeats of the highly repetitive sequences are related to (or may be) transposons that move about the genome. Thus repeated DNA sequences are most likely to result from amplification and activity of different types of mobile genetic elements. RTs like the DNA TEs can generate mutations by inserting within or near gene. Moreover RT induced mutations are relatively stable because they transpose via replication and the sequence at the insertion site is retained. RTs are parasitic or selfish DNAs. As they are expressed in the genome they can be subjected to the same rules of inheritance as are the genes of their genomic host. A RT can be co-opted for use in biological process that benefits organism. For instance, the preferential insertion of some RTs in Drosophila at telomeric locations has removed the need for a telomerase function in these insects. Tos17 is a retrotransposon of rice and its transposition is activated through tissue culture.

2. Transposons — Transposable elements are the second major class of dispersed repeats. The presence of transposable elements in plants is characterized by terminal inverted repeats with some elements in different species sharing the same or similar inverted repeats and replicate through a DNA intermediate. Transposable elements contain genes that facilitate their excision from and re-insertion into the chromosome. The process of insertion results in the duplication of a short sequence at the site if insertion(see chapter 22). The duplicated copies lie on either side of the TE. The eukaryotic DNA TEs were identified by Barbara McClintock in maize and different TEs are Ac/Ds, En/Spm, or Mu. Specific TEs have been found close to an α-amylase gene and within a glutenin gene in wheat. Most transposable elements are lying inactive and the inactivation could occur due to mutation within its gene (s), by suppression of transcription of its gene by methylation of cytosine or other nucleotides. Transposons have not been reported in human or other mammalian genomes. Further, there is frequent translocation of certain pieces of DNA in bacterial, yeast and Drosophila genomes. Sequence transposition is playing an important role in the evolution of these genomes and in generating new genetic variation and linkage relationship.

Many eukaryotic transposons are related to retroviruses and their mechanism of transposition includes an RNA intermediate.

MITES (Miniature invert repeat transposable elements) — MITES are characterized by short sequence, terminal or subterminal inverted repeats flanked by short direct repeats and have no coding potential. They are distributed ubiquitously and seem to originate from DNA transposons. MITES are non-autonomous TE of about 0.3kb, flanked by 14bp terminal inverted repeats and occurs preferably in 3' untranslated regions of the genes. They initiate antisense transcription when they are activated in methylation mutants. MITES are found in worms, insects, mammals and plants. MITES are associated with the genes of many cereals. MITES attain a very high copy numbers in most genomes (from 1000 to over 15000). They do not possess coding capacity and insert preferentially into noncoding regions of a single or low copy sequences and exceptionally function as part of transcription initiation or polyadenylation sites. MITES do not recognize any specific target site but rather distinct secondary structure of the target DNA. MITES have a tendency to be dispersed along the chromosome whereas the retrotransposons and other autonomous type DNA- mediated transposable elements are clustered in the pericentromeric region.

For more on mobile elements see chapter 22.

1.8 SPECIAL FUNCTION REPETITIVE SEQUENCE

The eukaryotic chromosomes have two specific functions repetitive DNA sequences.

1. Centromere

Repeated DNA sequences have a role in the organization and function of higher eukaryotic centromere.

The centromere (kinetochore) of eukaryotes generally contains simple sequence DNA. It consists of thousands of tandem copies of one or a few short sequences (~5 – 10 bps sequences). The centromere DNA is organized in three domains (CDEI, II and III). The domain I and III flank an internal segment and are binding sites for centromeric proteins. The centromeres are frequently flanked by heterochromatic A-T rich, repetitive sequences of about 100-120 bp which prevents crossingover. The centromeric core required for accurate chromosome segregation is about 100bp in length. In yeast the CEN-DNA is composed of three conserved sequence elements: a 14bp left terminus (CDEI), a central A-T-rich segment 80-90bp in length (CDEII) and an 11bp right terminus (CDEIII). Mammalian centromeres are not defined by a consensus DNA sequence. The centromere DNA mediates chromosome attachment to the meiotic and mitotic spindles and often forms dense heterochromatin. *Arabidopsis* centromeres like those of many higher eukaryotes contain numerous repetitive elements including retroelements, transposons, microsatellites and middle repetitive DNA. The function of higher eukaryotic centromeres may be specified by proteins that bind to centromere DNA, by epigenetic modifications or by secondary or higher order structures. The centromere DNA is noncoding DNA, does not contain any ORF and so not transcribed. α-**satellite**- It refers to 171bp sequence elements of the human genome which is tandemly reiterated many fold to form a higher repeat units which are arranged in arrays of 1.2×10^6bp at the centromere of each chromosome. Within each α-satellite a specific 17bp-CENPB-box functions as address site for a centromeric specific CENP-B binding protein involved in interations with mictotubule-associated proteins. This alphoid DNA is transcriptionally inactive but nevertheless constitutes up to 5% of the human genome.

Knobes — Knobs are composed primarily of a 180bp element distributed in tandem arrays. This type of organization (tandem repeats of about 180bp) is typical of other known centromeric elements. The evidence of knob/centromere homology is a fact that knobs can function as meiotic centromeres when a variant of chromosome 10, known as Abnormal 10 (Ab 10) is present in the cell (see Roy, 2009).

2. Telomere

The telomeres are sequences found at the ends of the eukaryotic chromosomes that help in stabilizing the chromosomes. The best characterized yeast telomere consists of about 100 bps of imprecisely repeated sequences of the form,

5′ (TxGy) n

3′ (CyAx) nÀ

where n can range from 20 to 100 and x and y can lie between 1 and 4. *Arabidopsis* telomeres are composed of CCCTAAA repeats and average about 2-3 kb. Repeated telomeric sequences are added to the ends of eukaryotic chromosomes by enzyme called telomerase and thus telomeres are not replicated through the cell machinery.

The end of a eukaryotic chromosome consists of a terminal, densely stained **protelomere** with repetitive DNA sequences and a subterminal of weakly stained **eutelomere**. The protelomere consists of short tandemly repeated GC rich sequences of up to 64bp in length with a core consensus sequence as follows.

TTTAGGG (Arabidopsis)

TGG (G) (Yeast)

TTAGGG (Human)

Telomere length of the same chromosome varies in different cellular stages (decreases in human fibroblasts during aging, increases by 7-10bp per generation in trypanosomes and can range from less than 50bp in Euplotes to more than 100kb in mouse).

1.9   PLANT GENOME ORGANIZATION

The DNA sequence in chromosome contains unique sequences (coding sequences) and repeated sequences (70-80% in barley). The repeated sequences includes inverted repeats (6% in barley), tandemly arranged repeated sequences (long tandem arrays of repeats, small tandem arrays of repeats, 10-20%), repeated sequences interspersed among one another or among unique nucleotide sequences (50-60%) and highly repeated sequences or simple sequences (satellite). Further, nucleotide sequences could be functional or non-functional. Repetitive DNA constitutes 42-25% of the rice genome The MITES contributes about 4% and RTs- the most numerous large repeats contributes more than 15% to rice genome. Other transposable elements such as Ac in maize possess inverted terminal repeats. The long tandem arrays of identical sequences are found in the subtelomeric and pericentric heterochromatin regions and array of rRNA genes with the NORs. There is a close correlation between the site of C and N bands and the presence of long tandem arrays of short, repeating sequences in the chromosomal DNA.

Duplication of chromosome segments reflected in low copy sequence duplication— Mapping of loci has highlighted the areas of the genome that are duplicated. Study of cDNA clones showed about 12-15% of *A. thaliana* loci to be duplicated. In maize, 28.6% of informative clones testes detected duplicated sequences arranged in a non-random order. Chromsomes 2 and 7 share 13 pairs of duplicated loci covering well over 50cM; chromosomes 3 and 8 share 10 duplicated loci and chromosomes 6 and 8 share 6, none of which are present on chromosome. A small number of loci appear to be duplicated in *Sorghum bicolor*. The origin of duplicated sequences in *Zea mays* and *S. bicolor* may involve either polyploidization of maize genomes after the divergence of maize and sorghum or ancestral duplication followed by sequence divergence in different genomic regions.

Variation at DNA sequence level—Plant genomes vary widely in size ranging from 0.7pg per nucleus in *Arabidopsis thaliana* to 17 pg in *Triticum aestivum* (Lapitan. 1992) (see Roy, 2009). Some of the differences can be due to the difference in the level of ploidy but a considerable portion of variation is due to variation in the amount of repetitive DNA. Plant species with relatively low genome size are characterized by a low number of repetitive sequence. For example, Arabidopsis has only 14% repeated sequence whereas grasses contain as much as 60 to 80% repetitive DNA. Variation due to repetitive DNA can result in difference of several orders of magnitude even among species within a single family (Bennette et al., 1982). The repetitive DNA ranges from simple duplications of protein coding loci (i.e. gene families) to highly repeated tandem arrays of ribosomal gene (rDNA) to dispersed repeats of simple sequences (VNTRS). Duplicated gene loci constitute a relatively small amount of total repetitive DNA. The next larger fraction of repetitive DNA is the high copy number nuclear ribosomal DNA (nrDNA). Ribosomal repeats are normally similar if not identical sequences in all copies (Dvorak, 1988). Of the remainder of repetitive DNA, the majority is present as tandem repeats of short repeated sequences referred to as satellite or variable number tandem repeats (VNTRs).

Smaller repeat units (> 10 bp) are called the *minisatellites* whereas very short repeat units (<10 bp) are called *microsatelites* (see chapter 15 for detail). The tandem repeats are present throughout the genome. Although much of the plant genome is composed of repetitive DNA, a significant portion consists of low or single copy nuclear DNA (scn DNA).

The molecular studies conducted so far indicate that the base pair changes are more frequent than large rearrangements, heterogeneity is not restricted by coding regions and polymorphisms at the DNA level are much more frequent than are charge changes in protein (Burr, 1994). Differences at the DNA level vary from 1 to 2 base pairs/1000 in humans to more than 40bp/1000 in maize. Maize has been found to be exceptionally polymorphic because of the presence of transposable element; Transposable elements not only create rearrangements but also generate small sequence changes. Transposable elements when excised from a locus leave behind a 'foot print' most commonly a variant of the short sequence duplicated upon insertion and thus generate diversity. Sometimes transposable elements when inserted in exons, alter the activity (Schwarz-Sommer et al. 1985). Some regions of the genome are significantly more polymorphic than the single copy sequence. Tandem repeats exhibit is a high degree polymorphism. Polymorphism of tandem repeats is thought to be generated by unequal crossingover or by slippage during DNA polymerization. Understanding genetic variation in the DNA is important because the types of genetic variation found in the DNA reflects the detection methods employed.

1.10 HUMAN GENOME ORGANIZATION

The human genome consists of various types of sequences, coding sequences translated into protein (1.1 %-1.4%), coding sequences transcribed into RNA (25%), simple sequence repeats or microsatellites (3%), transposons (45%), large duplications (5%) and other intergeneric DNA (20.7%). Permanent changes in the composition and size of the genome is caused by genome contraction by deletions and genome expansion through amplification and transposition of retrotransposons.

1.11 OTHER TYPES OF DNA SEQUENCES

DNA cruciforms, triplexes, quadruplexes and even unwound structure have all been characterized structurally and their biological role have been clearly identified (Sinden, 1994).

A. Palindromic sequences—DNA strands with self-complementary **inverted repeats** are called palindromes. In other words, DNA strands can have base sequences which when read forwards and backwards from a point of symmetry give the same sense. In the figure 1.3— the palindrome shows symmetry around the axis ACTC and the asymmetric sequence TAGCGT can pair with the complementary sequence, ATCGCA of the same strand and forms a hair pin structure. In case of double stranded DNA this term is used for DNA sequences which read alike on both strands if read in the same direction (e.g. 5′ to 3′ or 3′ to 5′) and they form a 'cruciform' structure. 3-6% of the eukaryotic sequences are palindromic. The base pair length ranges from 300 to 1200 and the molecules have a Cot1/2 value of < 10^{-5} mol. of nucleotide seconds per litre. Palindromic sequences are found in all classes of DNA and are widely distributed throughout the chromosomes. Such DNA sequence has the ability to renature instaneously. Inverted repeats are thus constituents of palindromes and are target sites for various DNA-affine proteins. Several regulatory sites contain varying lengths of inverted repeat sequences. There is presence of cruciform structure in supercoiled DNA. Cruciforms have been favorite hypothetical recognition sites on DNA. There can be formation of internal DNA cruciform structures with resultant influence on transcription or formation of secondary structures at the mRNA level with a resultant influence on translation.

B. Fold-back DNA — It refers to a special structure within a DNA duplex molecule generated by Watson-Crick base pairing of complementary inverted repeat sequences on the same strand (intra strand annealing). A fold-back structure consists of a base-paired stem and a loop formed by a variable number of unpaired bases (up to several hundred). Fold-back structures are also observed in RNA molecules, e.g. tRNA. A DNA sequence motif which forms such a structure is for example the oriA (origin of assembly in TMV).

C. Mirror repeats or reverse repeats — When inverted repeats occur within each individual strand of the DNA the sequences are called mirror repeats as shown in Figure 1.3. Mirror repeats do not have complementary sequences within the same strand and thus they can not form cruciform or hair pin structure.

D. Terminal inverted repeats (TIRs) — TIRs are sequence motifs that flank transposons and are identical or partly identical and present in inverse orientations. TIRs function as recognition sites for the excision of the transposons.

E. Direct repeat — Short direct repeats of chromosomal DNA are generated by the integration of 'insertion sequences (ISs).

F. Triplex or quadruplex DNA — During replication, recombination and transcription three or four DNA strands can pair to form triplex DNA or quadruplex DNA through hydrogen bonding. H-DNA- When there is along stretch of strand made up of sequences with alternating T and T residues and thus looking like mirror repeat centered around T or C (Figure 1.3). It is found in the polypyrimidine or polypurine tracts which also incorporates a mirror repeat. It looks like triple stranded form. Two of the three strands in H-DNA triple helix contain pyrimidines and the third contains purines. **Relics DNA** – High molecular weight DNA which consists of highly repeated DNA but lacking the recognition sites for the enzyme used. It is also called **'junk DNA'** which is not expressed and whose function is not fully understood. Now it is becoming clear that many regulatory sequences reside within this DNA.

1.12 CAUSES OF VARIATION IN DNA STRUCTURE

The structural variations can occur in sugar, phosphate group and bases. The structural variations can arise because of different possible conformations of the deoxyribose, rotation about the contiguous bonds that make up the phosphodeoxyribose backbone (rotation is

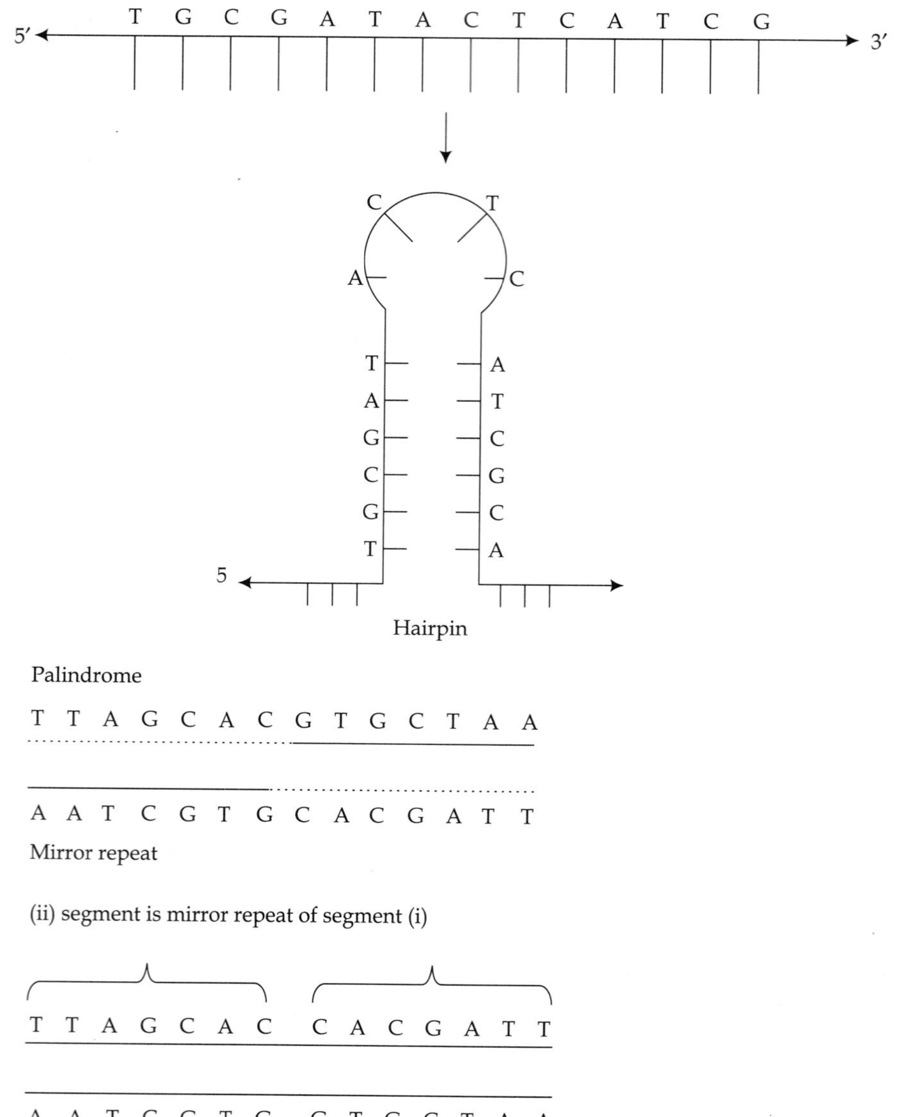

FIGURE 1.3 **Showing palindromic sequence and the formation of a hairpin and mirror repeat.**

possible around a number of bonds in the sugar-phosphate backbone) and free rotation about C-1'-N- glycosyl bond. Purines in purine nucleotides are restricted to two stable conformations with respect to deoxyribose, called *syn* and *auto* and pyramidines are restricted to the anti-conformation.

Sugar — The pentose sugar exists in two forms-aldehyde (straight chain) and ring (β-furanose) in solution and are in equilibrium. RNA contains only β-D-ribofuranose and DNA exists solely as β-2'-deoxy-D-ribofuranose. The Ribofuranose rings exist in four different conformations. In all the cases 4 of 5 C atoms are in a single plane and the 5th (either C-2′ or C-3′) is on either the same (endo) or opposite (exo) side of the plane relative to C-3′ atom. There is potential for the 2′-OH of the ribose to participate in hydrogen bonding to form non-helical structure which is not possible with DNA. This OH group not only limits the range of possible secondary structure available to RNA but also makes it more susceptible to chemical and enzymatic degradation.

Phosphate—The phosphate group is negatively charged and ionized at pH=7. Negative charges are neutralized by ionic interactions with the positive charge on proteins, metal ions and polyamines. Nucleotides with phosphate groups can be in positions other than on the 5′ carbon. RNA on hydrolysis by ribonucleases produces ribonucleoside 2′, 3′- cyclic monophosphates (intermediate) which gives rise ribonucleoside 3′ monophosphate, the final product. 3′,5′- cyclic monophosphate (cAMP), present in animal kingdom and guanosine 3′,5′- cyclic monophosphate (cGMP) have regulatory function. ATP gives rise to cAMP in the presence of adenylate cyclase. ppGpp (guanosine tetraphosphate) is produced in bacteria.

Bases—Bases can exist in two tautomeric forms, keto- and enol-. Modifications in bases are of two types. 1. Methylated form, hydroxyl methylated or glucosylated is most common in viral DNA. 2. Altered or unusual bases involved in regulating or protecting the genetic information.

1.13 METHYLATED BASES OCCURRING IN DNA

Methylated bases can be incorporated, e.g. Thymine and hydroxyl methyl cytosine in T_4 DNA or methylation occurs after DNA synthesis. 5-methyl cytosine occurs in higher animals and plants (eukaryotes), N^6-methyl adenosine and 4N-methyl cytosine occur in lower organisms like bacteria (prokaryotes). In plant a third of the cytosine present as 5-methylated.Thus adenine and cytosine are methylated more often than guanosine and thymine. Methylation is most common at CpG sequences. In plants the DNA sequence recognized and methylated by methyl transferase is 5′-CNG-GNC-

Base methylation will affect DNA-protein interaction thus determining the potential of a gene for future activity. In methylation one will have to study

1. the level of methylation
2. site of methylation and
3. heritability of methylation.

Methylation is catalyzed by specific enzyme- the DNA methylases (methyl tranferases). Methyl group comes from the amino acid methionine which exists in the form of S-adenosyl-L-methionine. S-adnosyl-L-methionine + Base (in DNA or RNA) → S-adenosyl-homocysteine, methylated base in DNA or RNA. The level of methylation of cytosine in barley is 23% whereas in adenine is 0.2%.

1.13.1 Methylation and Other Small NuclearRNA (snRNA) Nucleotide Modification

Minor bases of tRNAs are methylated as a result of post-transcriptional modification. Inosine (ribonucleoside from hypoxanthin is called inosine) contains the base hypoxanthin (6-hydroxy purine). Pseudouridine contains uracil and in case of pseudouridine (ψ) the base uracil is removed and attached to the sugar through a different N atom. The other modified bases besides inosine and pseudouridine are, methyl guanosine (m^1G), 4-Thiouridune (S^4U), N^6-Isopentenyl adenosine (i^6A), Ribothymidine (T), and Dihydrouridine (D). These bases are modified by methylation, deamination or reduction. rRNA is also methylated and contains some pseudouridine. mRNA and its precursors are methylated at their 5′ cap structures. Mature mRNA is formed from heterogeneous nuclear RNA (hnRNA).

1.13.2 Function of Methylation

They play a role in mismatch repair, recombination and the initiation of DNA replication. One of the funct ions of DNA methylation is to define which is the parental strand during DNA replication. The nascent or daughter strand is unmethylated and could be subject to mismatch repair. Repairing of mismatched base pairs formed occasionally during DNA replication is done by Dam (DNA adenine methylation methylase. It is also a kind of defense mechanism. It distinguishes its DNA from foreign DNA by making its own DNA with methylated groups and destroying foreign DNA without methyl group. It is an example of restriction modification system. Methylation of cytosine increases the tendency to segment of DNA to assume Z form which is a left handed and more stretched form of DNA and is involved in the regulation of expression of some genes or in genetic recombination. Methylation suppresses the migration of segments of DNA called transposons.

Time of methylation—Marked changes in pattern of methylation occurs during gametogenesis and in the very early vertebrate embryo indicate that methylation may be important in determining the potential for future activity.

Change of pattern of methylation—In somatic cells DNA methyl transferase serves only to maintain the pattern of methylation in newly synthesized DNA. In the absence of methylation there is loss of methyl group from cystosine. Methylation of unmethylated-CG-sequence occurs in embryo. Changes in methylation are associated with changes in gene expression in eukaryotes.

Place of methylation—Methylation is confined to certain sequences or regions of the DNA. In eukaryotes the pattern of methylation is tissue specific and not all the recognition sequences are methylated. 1. **Degree of methylation** at specific sites does vary with gene expression. Gene (s) methylated in specific tissue or stages of development is not expressed and the same gene (s), globin gene and genes for egg white protein undermethylated (some but not all of the CCGG sequences are undermethylated) in specific tissues are expressed. 2. **Loss of methylayion** has been observed in case of immunoglobin light and heavy chain genes. 3. **Relative hypomethylation**- Hypomethylation of genes (metallothionein genes is induced by cadmium or glucocorticoid). Correlation is found between gene expression and hypomethylation of α-foetoprotein genes. Under methylation of gene (s) occurs in 5′-flanking sequences. Hypomethylated **'CpG' island** regions have a higher probability of association with genes that are transcriptionally active in several tissues but whether such association exists in plants is not yet known. There are some 45000 short 1-2kb long regions in vertebrate genomes which are characterized by a high GC content, lack of cytosine methylation, absence or under representation of histone H1 and presence of hypoacetylated histones H3 and H4. CpG islands are nucleosome free regions and are associated with the 5′

domains of all known genes (house keeping genes and some tissue specific genes). In about 60% of human genes CpG islands cover the promoter region and first one or two exons. It may be involved in transcriptional regulation. About 2% of mammalian genomes contain 5'-CpG-3' motif once every 10bp in which C-residue is consistently non-methylated. In major fraction (98%) 5'-CpG-3' occurs on an average once every 50-100bp and the cytosine is methylated. Cluster of methylation sites are found in either 5' or 3'-flanking sequences of the genes. C_PG dinucleotides the sites of methylation in mammals are underrepresented in DNA. Clusters of C_PGd called C_PG islands are often found in association with genes, most often in the promoters and first exons but also in regions more toward the 3'-end. C_PG island is unmethylated in the germ line (and indeed most somatic tissues) and thus ensuring its continued existence in the face of strong mutagenic pressure of 5-methyl cytosine deamination. C_PG islands often function as strong promoters and have also been proposed to function as replication origins. But there are examples such as DNA of Drosophila and other insects in which there is almost total lack of methylation. Methylation of structural gene (a gene coding for a protein or RNA molecule, as distinct from a regulatory gene) had no inhibitory effect on transcription. Considering the results it can be concluded that specific demethylation is involved in gene expression and it appears that methylation is a pre-requisite for gene expression but not enough to ensure it. DNA methylation especially of bases with the promoter regions of the genes can modulate (and usually diminish) the transcription of the adjacent genes whereas demethylation of such bases (induced by incorporation of the modified base-azacytidine) which does not allow the transfer of methyl group onto its c^5 position usually leads to activation of the previously silent genes. Possible the binding of DNA- affine proteins (transcription factors) is prevented or less specific if bases within the recognition sequence are methylated. On the other hand, methylation of bases in CG-rich regions may induce conformational changes in DNA from B- to Z-DNA.

Plants and filamentous fungi share with mammals enzymes responsible for DNA methylation. DNA methylation is associated with gene silencing and transposon control. Plants and fungi differ from mammals in the genomic distribution, sequence specificity and heritability of methylation. DNA sequences as short as 30bps can be targets for methylation which occurs at all Cs, including those not present in symmetrical C_PG and in plants C_PN_PG nucleotide groups. **Hemimethylation**- It refers to the presence of methylated nucleotides, for example, 5methyl cytosine or N6–methyl adenine in only one strand of a DNA duplex molecule resulting from semiconservative replication. Usually the newly synthesized strand is methylated by methyl transferase so that both strands of a DNA are methylated at comparative sites.

The variation in DNA structure can arise spontaneously due to deamination of certain bases, hydrolysis of base-sugar N-glycosyl bonds, radiation induced changes leading to formation of pyrimidine dimmers and oxidative changes although occurring at very low rates. For example, loss of exocyclic amino groups (deamination) from cytosine will result in the production of uracil which will lead to decrease in G≡C base pairs but increase in A=U base pairs although it is immediately recognized and removed by repair system. The hydrolysis of base –sugar N-glycosyl bonds leads to loss of purines. In the radiation induced changes U.V. light induces condensation of two ethylene groups of the adjacent pyrimidine bases leading to formation of cyclobutane pyrimidine dimmers. Ionizing radiations such as X-ray, Gamma ray can cause structural changes by way of ring opening, fragmentation of bases and breaks in covalent backbone of nucleic acid. The chemical mutagens such as nitrous acid, HNO_2,

Bisulfite (deaminating agents) and alkylating agents such as DMS can also cause changes in base structure. Deaminating agents cause deamination of bases whereas alkylating agents methylate a guanine to O^6-methyl guanine which can not base pair with cytosine. HNO2 reacts with amino groups and converts them into hydroxyls. Cytsosine upon deamination produces uracil, adenine produces hypxanthin and guanine produces xanthin The hydrogen peroxide, hydroxyl radicals and superoxide radicals which arise during irradiation or as a product of aerobic metabolism can cause oxidative damage in DNA by way of oxidation of deoxyribose and base moieties or even break in the DNA strand. Methylation of bases- Methylation acts like a brake. It can turn gene expression up or down, on or off depending upon how much of it is around and what part of the genetic machinery it affects. 5-Methyl cytidine occurs in the DNA of plants and animals and N^6- methyl adenosine in bacterial DNA.

1.14 NONCODING DNA

The noncoding DNA neither encodes a polypeptide nor an RNA and it includes, introns, spacers, pseudogenes, centromeres and most repetitive DNA (satellite, microsatellite, minisatellite, retrotransposions).

Mitochondrial and chloroplast DNA—Mitochondrial DNAs are far smaller than nuclear DNA. The size of DNA ranges from 15 to 20kb in animals and from 200-250kb in flowering plants.

1.15 DNA SEQUENCES WITH UNIQUE CHARACTERISTICS

Certain DNA sequences have abnormal mobilities in gel electrophoresis. DNA fragments having repeats of $(A)_n$ nucleotides with n equal to or greater than 4 separated by another 4 to 5 nucleotides and phased with the helical repeat form **smooth banding** and thus migrate more slowly than those with having mixed sequences, for example, 434 repressor DNA complex. **Sharp kink** is caused and stabilized by DNA-binding protein. The DNA is kinked sharply at the C2pG3 steps in the Sac7D-DNA complex. Kink is caused by the intercalation of the side chains of amino acid residues Val26 and Met29 of the Sac7D protein into DNA base pairs from the minor groove direction. Sharp DNA kink has been observed in the complexes of TATA-box binding protein (TBP) and two HMG-box containing protein, LEF-1 and SRY with the cognate specific DNA sequences.

1.16 TRIPLET REPEAT SEQUENCES

A number of diseases have been associated with expansions of triplet repeats of the DNA sequence $(CNG)_n$. $(CGC)_n$ repeat in X-chromsome is associated with fragile syndrome. $(CAG)_n$ repeat is associated with Huntington's disease and spinobulbar muscular dystrophy whereas $(CTG)_n$ is associated with myotonic dystrophy.

1.17 DNA OF ORGANELLES

It was a long established rule in classical genetics of plants and animals that with the exception of sex linkage male and female parents contributed equally to the genotype of the next generation. It was based on the realization that male and female gametes though very different in cytoplasmic mass contributed nuclei equally. In other words, they showed equal results from reciprocal crosses. But then exception to this rule of equal results from reciprocal crosses has long been known in plants. Plant embryos usually acquire their chloroplasts along with their DNA from the egg and in some plants such as Pelargonium there is some transmission through the pollen tube as well. The cell organelles such as mitochondria and chloroplasts contain closed-loop duplex DNA molecules. Both contain genes determining a number of traits. The genomes of nucleus, plastids (chlroroplasts) and mitochondria differ markedly in gene number, organization and stability. Plastid genes are densely packed in an order highly conserved in all plants (Sato, 1999) whereas mitochondrial are widely dispersed and subjected to extensive recombination (Unseld et al.). Further, plastids are derived from the cyanobacterial lineage and mitochondria from the α-Proteobacteria.

All plant organelle genomes are highly autopolyploid. There are 20-500 plastids per leaf cell and within each plastid there are hundreds of identical plastid genomes(plastomes) depending on species, light levels and stage of development(Baumgartner et al., 1989). Similarly, there are high numbers of mitochondria per cell, 260 in pea and 200-300 in watermelon (Lampa and Bendich, 1984) but only few estimates of genome copy number per mitochondria are available.

Mitochondrial DNA (mtDNA) — mtDNA encodes essential proteins for the oxidative phosphorylation pathways. They are inherited maternally in human and show a rapid evolutionary rate. The mitochondrial genome in human is divided into two distinct regions: (i) The DNA –loop (or control region), an ~1100bp noncoding region and (ii) a functional region which encodes 13 proteins, 12S and 16S rRNAs and 22 tRNAs. Mitochondria have their own special translation machinery with rRNAs and a complete set of tRNAs all distinct from the corresponding components of the cytoplasmic apparatus and these are transcribed from the mtDNA. The genes for mt ribosomal proteins are chromosomal. Other mtDNA genes encode components of respiratory pigments and enzymes- cytochrome b and a number of the subunits of the complex enzymes such as NADH, cytochrome c oxidase and ATPase. Mitochondrial circular DNA is found in about 5-10 copies per organelle. mtDNAs are 15-20kb in animals (about 16kb in *Drosophila* and 16.5kb in human), 75kb in budding yeast, 63kb in Neurospora, 94kb in *Podospora anserine*, a filamentous ascomycete and 200-2500kb in flowering plants. There are a number of examples of mitochondrial inherited traits in humans. A series of mutations in mtDNA due to deletions cause human disease such as myopathies and encephalomyopathies, (LHON, MEREF, RSS, etc). These diseases show maternal inheritance as transmission of mitochondria is exclusively through egg. As the patient contains some normal mitochondria along with the defective (mutant) one and so it is not lethal and the mixed population of mitochondria is maintained and transmitted maternally from generation to generation. The different mitochondrial lineages (a group of closely related and therefore mostly homologous mitochondrial DNA molecules) do not segregate out during cell division as is found in *S. cerevisiae*. Since the proportion of mitochondria with mutations at specific sites in mtDNA varies after repeated cycles of cell division the mitochondrial diseases are highly variable. Animal mtDNAs are organized in an extremely compact fashion with almost no spaces between the coding sequences and nearly all the boundaries between genes are marked by tRNA genes. The mtDNA yields a single RNA transcript from which the tRNA molecules are excised and the remaining fragments left are mRNAs for single protein –coding genes. The mitochondrial genome A comparison between Human and Drosophila mtDNAs show high degree of similarity between the two with respect to size, gene content and even to a large extent in gene order. mtDNAs from fungi contain nearly the same set of genes as animal mtDNAs, or perhaps fewer in yeast as NADH dehydrogenase genes have not been identified as yet. The extra length of fungal mtDNA is accounted for partly by the presence of more or less abundant self-splicing mitochondrial introns. Most spacer sequence in the yeast mtDNA consists of A/T rich repetitive sequences with shorter G/C-rich repeats. The mtDNAs of plants are different in sense that they consist of greater than 500kb of unique sequence but not necessarily in the same molecule and sometimes at least distributed between a number of different DNA circles. It appears to harbour more or less the same complement of genes as animal mtDNA but with a great deal of redundant or spacer sequence. In case of maize (*Zea mays*) the variants of mtDNA have been shown to cause male sterility (pollen sterility). Although mitochondria are the organs of respiration in all eukaryotes, *S. cerevisiae*, common yeast can dispense with mitochondrial function and is able to live by alcoholic fermentation in the absence of oxygen.

The closed loop DNA structure, lack of nucleosomes, sizes of the rRNAs and susceptibility of the ribosomes to antibacterial antibiotics all these features of mitochondria resemble the bacterial genome but mt genes in vertebrate mitochondria differ from both eukaryotes and bacteria (which share 'universal genetic code', see chapter 3). Mitochondrial genes at least in fungi differ from chromosomal genes in that they have self-splicing introns and interaction between different parts of the intron sequence is often stabilized by a protein encoded in the intron itself. Another unique feature associated with yeast and various filamentous fungi is that the mutants mtDNA molecules originated as a result of deletions and thus lost essential functions are often more efficient in replication in comparison to their wild-type.

Yeast mitochondrial recombination — Mitochondrial recombination and segregation occur in the diploid phase immediately following karyogamy and not specifically at meiosis and so mitochondrial markers differ in their mode of inheritance from chromosomal markers. When haploid cells of different mating types but of similar size fuse, they contribute mitochondria to the zygote. A diploid formed by sexual fusion of haploids differing in their mitochondria will initially contain a mixture of both mitochondrial types but after a few diploid budding cycles the mitochondria lineages become segregated and each cell comes to contain a single type of mitochondria. When the two haploid parents differ with respect to more than one mitochondrial trait their diploid segregants tend to contain not only the two parental types mitochondria but also the recombinant type mitochondria. Thus knowing the frequencies of different recombinant classes one can construct a linkage map but the problem comes from the occurrence of very high frequency of recombination occurring over an extended period of time instead of being limited to one precise moment as in meiosis. So linkage can be demonstrated only for relatively closely placed markers. The linkage map of mitochondrial DNA has been constructed using deletion mapping procedure.

Chloroplast DNA (cpDNA) — The chloroplast genome generally consists of several copies of closed circular DNA. The genome size varies from 85-2000kb in green algae to 130-160kb in higher plants. The number of cpDNA per organelle also varies from an average of 2-20 molecules per nucleoide to about 10-20 nucleoids per chloroplast and depends on the developmental stage of the leaf. Two inverted repeats of 20-30kb each separate a small single copy region (sscr) from a large single copy region (lscr) which contain most of the genes (some

100-150 genes). Chloroplast DNA has a high degree of homology to prokaryotic DNA and functional chloroplast requires the products of about 1000 nuclear genes. Choloplast contains genes for the special apparatus of chloroplast protein synthesis- chloroplast specific rRNA and tRNAs (unlike the mitochondrial situation) for chloroplast ribosomal proteins and electron transport chain. Further, a number of the components of the photosynthetic apparatus including the larger of the two polypeptide subunits of the enzyme –ribulose biphosphate carboxylase are encoded by chloroplast genome. Figure 1.4 shows three forms (linear, open circle and supercoiled) that the plasmid DNA can take.

RNA (RIBONUCLEIC ACID)

1.18   RNA (RIBONUCLEIC ACID)

The functions of RNA molecules range from the enzyme-like activity of ribozymes (self-cleaving reaction of ribozymes), protein synthesis by rRNAs, recognition of aminoacyl-tRNA synthetase by tRNAs to storage of genetic information in RNA viruses. RNA thus has structural, catalytic and even regulatory properties. Ribonucleic acid is a single stranded structure consisting of nucleotides which in turn is made up of ribose sugar, nitrogenous bases- adenine, cytosine, guanine and uracil and phosphoric acid. RNA has a covalent structure very similar to that of DNA, the only differences being the change from a 2-deoxyribose sugar in DNA to a ribose sugar in RNA and from a methyl group in thymidine to a hydrogen atom in uracil. For RNA the ratio, A/U= G/C= 1, does not hold for most RNAs. RNAs of reovirus and wound tumor virus are double stranded. As A can pair with U and G can pair with C so even though RNA is a single stranded structure it is found in the form of paired helical structure. The different types of RNA differ in length and secondary and tertiary structure. The secondary and tertiary structure of RNA is more complex than that of DNA with single RNA strands often fold into hairpins, double stranded regions and complex loops. There are equivalences of base pairs and so many bases on the same chain are hydrogen bonded into hairpin type turns found in tRNA. The joint presence of single stranded and double helical regions gives individual tRNA molecule an irregular 3-D shapes.

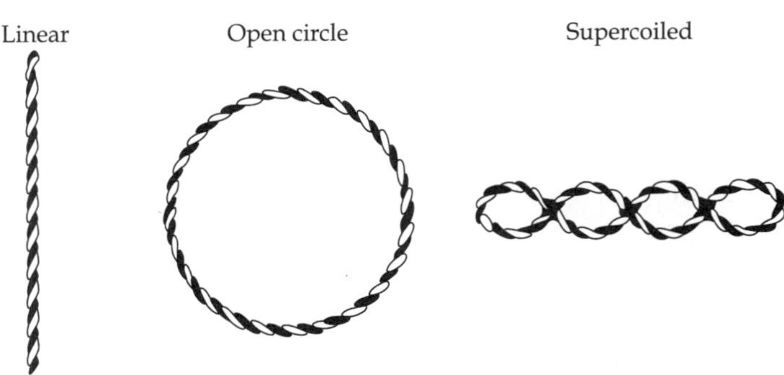

FIGURE 1.4   Showing molecular forms of plasmid DNA.

1.18.1 Types of RNA

There are three types of RNA:

1. Ribosomal RNA (rRNA)
2. Transfer RNA (tRNA) and
3. Messenger RNA (mRNA).

Single stranded mRNA is an intermediate in gene expression whereas rRNA and tRNA have structural, catalytic and information-decoding roles in the process of protein synthesis.

1.19 SYNTHESIS OF RNA

RNA is synthesized from the double stranded DNA. The two complementary strands are defined by their function in transcription. Only one strand of the DNA duplex is usually transcribed. The antisense strand (**anticoding strand, minus strand**), 3′-5′ strand of DNA serves as the **template** for RNA synthesis and has a base sequence complementary to the transcript. The other strand, 5′-3′ strand (the nontranscribed strand) is referred to as the **coding strand (plus** or **sense strand**). RNA is synthesized on the template strand (3′-5′) and is identical in sequence with U in place of T to the non-templete (5′-3′) or coding strand. The base thymine in the DNA strand is replaced by uracil in the RNA. In some cases both strands of the same site are transcribed but from opposite direction.

5′-ATGCTAAAATCG-3′ (Coding or plus strand)

3′-TACGATTTTAGC-5 (Minus or antisense strand)

↓

5′-AUGCUAAAAUCG-3′

To deduce the amino acid sequence from the transcribed strand using the standard codon tables (see chapter 3), one must convert it into its complement: A to U, T to A, G to C and C to G. The amino acid sequence can be read directly from the nontranscribed strand, the coding strand.

The RNA synthesis is in the direction of 5′-3′ and requires the enzyme DNA dependent RNA polymerase. RNA polymerase binds to specific sequences in the DNA called promoters which directs the transcription of adjacent segments of DNA (genes). The promoters sequences have been described in the chapter 5 on regulation. **RNA polymerases** are of four types.

1. RNA polymerase I
2. RNA polymerase II
3. RNA polymerase III
4. RNA polymerase IV

In addition to the three RNA polymerases plants also contain a fourth RNA polymerase which is involved in transcribing the small inhibitory RNAs involved in transcriptional repression(see chapter 23 on Gene silencing)

RNA polymerase I (located in nucleoli and transcription of the gene for rRNA occurs in nucleoli) is used in the synthesis of rRNA whereas RNA polymerase II (occurs in nucleoplasm) is involved in the synthesis of mRNA and some other specialized RNAs and RNA polymerase III is used in the synthesis of tRNAs, 5S rRNA and some other specialized RNAs (snRNA and 7SRNA). In prokaryotes all these functions are performed by two types of RNA polymerase. One type synthesizing RNA primer (RNA primase) necessary for DNA replication and another type transcribing structural, ribosomal and tRNA genes. The newly synthesized RNA is called a **primary transcript** which is mono-cistronic in eukaryotes. As we will see that gene consists of exons and introns sequences so the primary transcript contains all exons and introns. The introns are finally removed (spliced) from the primary transcript through a process involving a number of steps called **RNA processing** which leads to the formation of a mature, fully functional RNA which contains no introns and all the exons are joined together. Although RNA polymerase II catalyzes DNA-dependent RNA synthesis during gene transcription, there is ,however, evidence that it also posseses RNA-dependent RNA polymerase (RdRP) activity. Polymerase II can use a homopolymeric RNA template and can extend RNA by several nucleotides in the absence of DNA and has been implicated in the replication of the RNA genomes of hepatitis delta virus and plant viroids (Lehman et al., 2007).

Belotserkovskaya et al. (2003) proposed the mechanism by which pol II can transcribe the chromatin without disrupting its epigenetic status. Transcription by RNA Pol II is accompanied by the alteration of chromatin structure in transcribed regions of the genome. The FACT (facilitates chromatin transcription) complex is required for transcription elongation through nucleosomes by RNA polII in vitro. FACT facilitates RNA polII driven transcription by destabilizing nucleosomal structure so that one histone H2A-H2B dimmer is removed during enzyme passage. FACT also possesses intrinsic histone chaperone activity and can deposit core histones into DNA. In another mechanism, enzyme DNA topoisomerase IIβ, associated with DNA-repair machinery activates transcription by generating a break in double stranded DNA within a nucleosome and allows chromatin to relax which is required to drive gene expression (Ju et al., 2006).

RNAs made by eukaryotic RNA polymerase II undergoes a series of post-transcriptional processing events to produce mature mRNA which are then exported to the cytoplasm. These modifications generally include i. methylation of the 2'-hydroxyl group of the ribose sugar (s) near the co-transcriptionally added cap, splicing to remove introns and cleavage with subsequent polyadenylation (addition of ~200 adenosines) at the 3' end. Internal base editing and methylation of the 6 position of adenosines also occur in some mRNAs.

1.20   RIBOSOMAL RNA (rRNA)

Ribosomal RNA is formed by the nucleolar reorganizer region (NOR) of the chromosome. Nucleolus, a densely staining body and found in the nucleus of eukaryotic cell is involved in rRNA synthesized ribosome formation. In human cells rDNA genes are clustered in the NORs of five pairs of acrocentric chromosomes (13,14,15,21 and 22). Ribosomal RNA in association with proteins forms a cellular body called the **ribosome** which is the site of protein synthesis. 85 to 90% of the total cellular RNA is in the form of ribosomal RNA. Ribosomal RNAs from different organisms differ in their sedimentation constants. The RNA polymerase I produces a single long pre-rRNA that contains three species of rRNA, 18S, 5.8S and 28S rRNAs joined to each other by the transcripts of transcribed spacers (as the ribosomal transcription unit contains the genes for 18S, 5.8S and 28S separated by internal spacers (ITS) and flanked on its 5 side by an external transcribed spacer (ETS). The fourth species of rRNA, 5S rRNA of most eukaryotes is made completely separately as the genes for 5S rRNA are located in a separate part of chromosome and transcribed by RNA polymerase III. The different steps in the production of rRNA are:

1. Transcription and processing of pre-rRNA

2. Splicing of pre-rRNA

3. Post-transcriptional modification of rRNA

4. Precursor ribonucleoprotein in the formation of ribosomes.

Excision of the group I introns found in some rRNAs requires a guanosine co-factor. Some group I and group II introns are capable of self-splicing, i.e., no protein enzymes are required. Assembly of new ribosomes occurs predominantly in the nucleolus in case of eukaryotes. Eukaryotic ribosomes are larger than prokaryotic ribosomes and have diameter of 23 nm. It is composed of two subunits, 60S and 40S and which in turn form unit of 80S. The eukaryotic 60S ribosome consists of 5S rRNA (120 nucleotide residues), 5.8S rRNA (160 nucleotide residues) and 28S rRNA (4700 nucleotide residues) and 40S ribosome consists of 5.8S rRNA and 18S rRNA (1900 nucleotide residues).Ribosomal RNA is transcribed from an upstream promoter by polII. The ratio of protein to RNA is 1:1. Ribosomes vary in size among different species. It contains over 80 different types of proteins in eukaryotes. In bacteria (prokaryote) three species of rRNA, 16S, 23S and 5S rRNA (and some tRNAs) arise from a single 30S RNA precursor of about 6500 nucleotides. 5S rRNA appears to be a ubiquitous component of all prokaryotic and eukaryotic ribosomes. However, it has not been found in mitochondrial ribosomes of some fungi and animals. Bacterial ribosome contains 65% rRNA and 35% protein and has diameter of 18nm. It is composed of two unequal subunits, 30S and 50S and which in turn form units of 70S which in turn may form units of 100S. In prokaryotes and organelles, 5S rRNAs are synthesized as parts of a single long transcript together with 16S and 23S rRNAs. The three individual components are separated in a maturation process. In eukaryotes, 5S rRNAs are coded by separate genes arranged in tandem arrays of repeating units. Their number varies significantly up to several thousands in vertebrates and plants. These genes are transcribed by pol III. The synthesis of 5S rRNA in eukaryotes depends on binding of a 40kDA protein, transcription factor IIIA (TFIIIA). The TFIIIA specifically binds to the 5S rRNA gene and gene product with high affinity and specificity although three-dimensional structures of RNA and DNA are distinctly different (Szymanski et al., 1998). 5S rRNA is the only known rRNA species which binds ribosomal proteins before it is incorporated into the ribosome in both prokaryotes and eukaryotes. In eukaryotes, the 5S rRNA molecule binds only ribosomal protein L5 whereas in bacteria it interacts with three ribosomal proteins L5, L18 and L25.

The nucleotide sequences of 5S rRNAs are strongly conserved and thus can be used in evolutionary analyses for the reconstruction of phylogenetic relationship between distant taxa (see Roy, 2009 for more on phylogenetic relationship).

1.21   TRANSFER RNA (tRNA)

It is also called adapter RNA and as the name implies it transports amino acid to the site of protein synthesis. There are at least 20 different tRNAs and all of them have similar structure and all have molecular weight of about 25000 to 30,000. Each tRNA has a specific attraction for a specific amino acid and to hydrogen bond with at least one nucleotide triplet of mRNA. In other words, a tRNA molecule has two recognition sites, one for its particular amino acid and the other for a specific codon in an mRNA. The tRNA decodes the genetic information of mRNA into a sequence of amino acids constituting protein. Transfer RNA is transcribed from an internal promoter by polIII. tRNA is derived from an entirely separate sequence of DNA molecules. The exact location of DNA template for tRNA has not been found yet. The sedimentation constants of all the tRNAs appear to be around 4S and they comprise some 5% of the total cellular RNA. In eukaryotes the tRNA genes are transcribed by DNA dependent RNA polymerase III and are present in multiple copies with an average frequencies for each tRNA gene ranges from 5, in yeast, 10 in Drosophila and 15 in humans to over 200 in Xenopus. The various steps in tRNA synthesis in eukaryotes are as follows:

1. 5′- end processing (cleavage)
2. 3′- end processing (cleavage)
3. Addition of CCA to 3′-end
4. Modification of bases
5. Splicing (removal of introns).

A fourth class of introns is found in some tRNAs and it is the only class of introns known to be spliced by protein enzymes, ribozymes which include self splicing introns and the RNA component of RNase P (the enzyme that cleaves the 5′ end of tRNA precursors). The primary transcript RNA undergoes modification and 5′- and 3′- ends are processed (there is excision from precursor and 5′ and 3′ trimming) and have got extra sequences at 5′- and 3′-ends. A tRNA consists of about 80 nucleotides and it contains high content of methylated bases and it is shown to be composed of 4 loops forming a clover leaf structure, called the secondary structure (Figure 1.5). All tRNAs have the same following end groups- 5′ G-C-C-A 3′. The 3′end consists of ACC plus a variable fourth nucleotide. Amino acid is always attached to the 3′ terminal A and is called Amino acid arm. Away from the 3′ end is first loop made up of 7 unpaired bases and always contains the sequence- 5′T pseudoUCG-3′ and it is called T ψ C arm. The second loop, a highly variable in size ,often called lump is not present in all tRNAs. The third loop consists of 7 unpaired bases and contains the anticodon which binds to the codon on the mRNA. The anticodon arm is bracketed on its 3′ end by an often alkylated base purine and on its 5′ end by pyrimidine uracil. The third base (called **wobble**) in the anticodon (5′end) has the ability to hydrogen bond with any of the two or three bases at 3′ end of codon and a single tRNA can recognize several different codons. The fourth arm is called D-arm which contains 2 to 3 D (dihydrouridine) residues depending upon the tRNA. The fourth loop contains 8-12 unpaired bases. It is thought to be involved in binding to specific amino acid activating enzyme, amino acyl synthetase. Many of the unusual bases differ from the normal base by the presence of one or more methyl ($-CH_2$) groups. These unusual bases do not form conventional base pair and the exposed free amino and keto groups can then form secondary bonds to template RNA, to a ribosome or to the enzymes needed to attach a specific

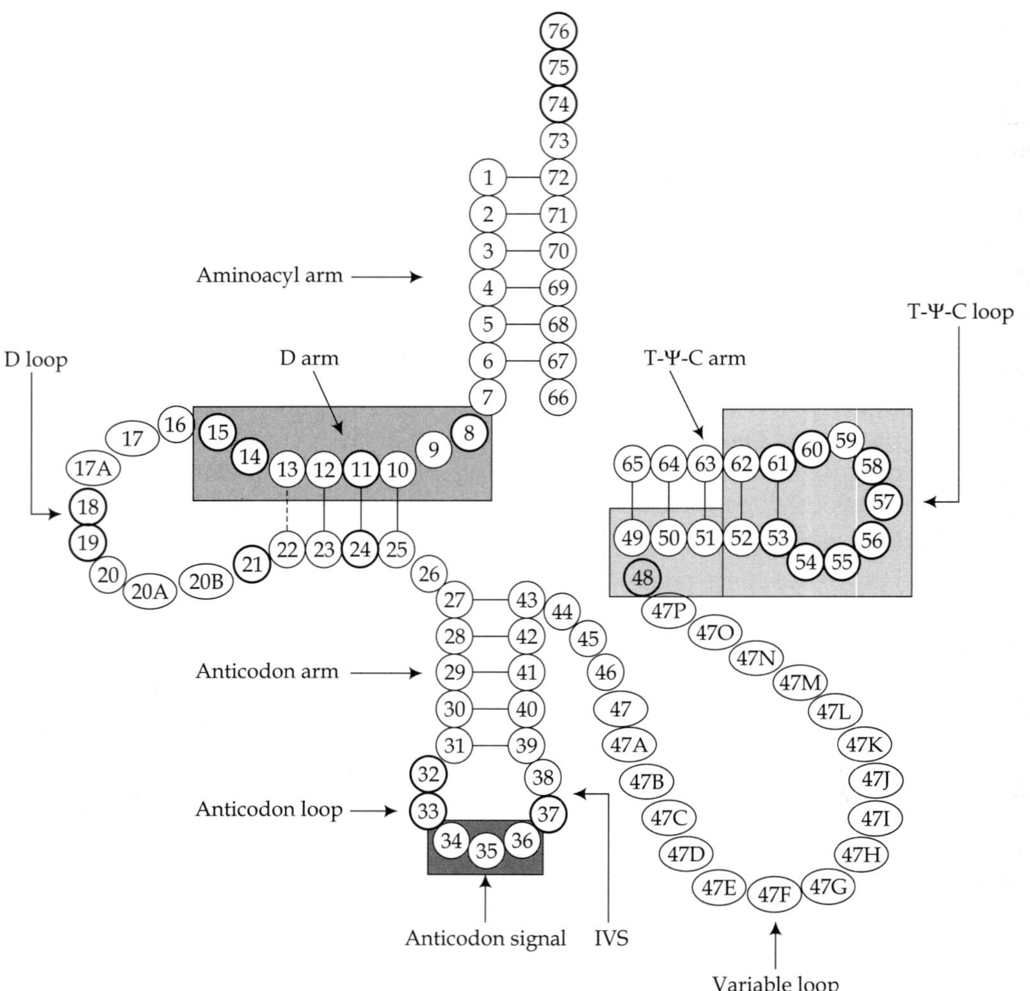

FIGURE 1.5   Cloverleaf secondary structure of a tRNA sequence. The standard system of numbering tRNA sequences is given (Sprinzl *et al.*, 1989). Circles represent nucleotides that are always present; among them, the thick-edged circles denote invariant or semi-invariant nucleotides. Ovals represent nucleotides that are not present in each tRNA sequence.

amino acid to its specific tRNA. It is as a result of post-transcriptional modifications or base transposition (replacement). These methylated bases are: ψ pseudouridine (The presence of ψ depends on the presence of an intron in the gene), I-Inosine, T-ribothymidine, D,5,6 dihydrouridine, m′1,1 methylinosine, m′1G,1- methylguanosine and m^{2}G,N^2-dimethylguanosine. Uracil is sometimes joined in an unusual way to the ribose moity to form pseudouridine instead of uridine and these various modifications do nor affect base-pairing. In the tertiary structure many of the interactions are between invariant or semivariant bases. It results from a large number of hydrophobic stacking interaction in the augmented helices plus the additional hydrogen bonding interactions between nucleotide residues that are often widely separated in the secondary clover leaf structure. Some of the interactions involve interaction among three bases. The unusual bases tRNAs vary in length from 73 to 93 nucleotides. The bacterial tRNAs consist of 73 nucleotides whereas the eukaryotic tRNAs consist of 93 nucleotide residues. The tRNAs thus range in length from 73 to 94 nucleotides and are characterized by a relatively large proportion of modified or non-standard nucleosides. The tRNAs also have primary, secondary and tertiary structure. Mitochondrial and chloroplast tRNAs are smaller than these tRNAs. Mitochondrial DNAs (mtDNAs) encode 13 proteins, two rRNAs and 22 tRNAs and all these 22 tRNAs decode all 64 possible triplet codon and thus not all the 32 tRNAs required for the normal code are needed.

1.21.1 Isoacceptor tRNA

It refers to any one of a group of tRNAs which accepts the same amino acid but differs in their primary sequence (e.g. the anticodon). Codon usage preference occurs in the selection of synonymous codons by different tRNA species charged with the same amino acid. As expected in case of strongly expressed genes, there is a correlation between the relative amounts of different isoaccepting tRNA species and the use of corresponding codons (Ikemura, 1981). In other words, the higher the concentration of a particular isoaccepting tRNA, the more often the corresponding codons appears in the sequence of the strongly expressed gene. But then this indicates that translation may be modulated and controlled by rare codons for which the corresponding tRNAs occur in trace amounts only (Winnacker, 1987) Different isoacceptor tRNAs are encoded by different genes.

1.22 MESSENGER RNA (mRNA)

The mRNAs act as blueprints for protein construction. The mRNA constitutes about 5% of the total cellular RNA. In some specialized cells a high percentage of total mRNA codes for one or a few abundant proteins. The mRNA is involved in carrying the genetic information from the DNA to the site of protein synthesis, the ribosome. Since the DNAs from different species differ only in the sequences of bases, mRNA must reflect this difference in base sequence. The complete DNA sequence between transcription initiation (cap site) and transcription termination site recognized by DNA dependent RNA polymerase is referred to as the **'transcription unit'**. Messenger RNA is transcribed from an upstream promoter by polII.

The pre-mRNA is subject to 5′ end capping, splicing, 3′ end cleavage and polyadenylation. In other words, the different steps involved in mRNA processing are as follows.

1. 5′ capping—5′ end of the primary transcript is capped by addition of methylated guanosine cap (7-methyl guanosine) and 5′-leader sequence. The 5′ terminus of most eukaryotic mRNAs consist of a 7-methyl guanosine residue linked in a 5′ to 3′ orientation to the mRNA via 3 phosphate groups m^7G (5′) ppp (5′) Nmp… where N stands for the first nucleotide encoded by the DNA. The capping is catalyzed by a guanylyl transferase. After addition of the cap a specific guanine-7-methyl transferase methylates the guanosine (cap 0) found in unicellular eukaryotes. Further methylation may involve 2′-0-methylation of ribose of the first or first two nucleotides of the transcript thus forming cap '1' (intranuclear) and cap'2' (cytoplasmic), respectively. Between the cap and the translational initiation codon (AUG) a length of non-translated RNA, known as **leader sequence,** is added. 5′ cap is added before the completion of primary transcript. Capping is necessary for the correct splicing of the primary transcript. It protects mRNA from exonucleolytic attack and 7-methyl guanosine residue serves as a key signal for translation initiation.

2. 3′ polyadenylation—3′ end is cleaved and about 80 to 250 adenylate residues are added 3′ end of most but not all pre-mRNA are modified by addition of poly (A) tail. Poly (A) tail provides stability to the mRNA. 3′ terminus is often not generated by simple termination but termination followed by processing. Polyadenylation may also require a sequence downstream of the coding sequence. Between translational termination codon and polyadenylation site eukaryotic mRNAs often have substantial segment of non-coding sequence (30 to 637bp in length). The transcript thus terminates well beyond the apparent end of the gene. There are also examples of mRNAs which are not polyadenylated and further some mRNA species are bimorphic in that they are having poly (A) $^{+}$ and poly (A) $^{-}$ forms. mRNAs which do not carry poly (A) tail at its 3′end are called poly (A) $^{-}$ mRNA, for example, rRNA, tRNA and also histone mRNA. Eukaryotic mRNAs have a distinctive 3′ end structure. Almost all mRNAs can be identified by 8 nucleotide sequence and the distance of this sequence from the poly (A) tail. Poly (A) addition is prompted by a polyadenylation signal present at the 3′ end of the transcript. It is a hexanucleotide consensus sequence,5′-AATAAAA-3′ (in animals), 5′-AATAAAN-3′ (in plants) and generally, 5′-AATAA-3′ sequence is found close to the 3′ end of most of the eukaryotic genes transcribed by RNA polymerase and a consensus sequence 5′-AAUAAA-3′ in an mRNA molecule directs the cleavage of the message 10-30 bases 3′ of the element. This works with a downstream U/GU-rich element which is believed to regulate the complex of proteins necessary to complete 3′ processing (Pauws et al., 2000). The specific site of cleavage of pre-mRNA is located between these regulatory elements and is determined by the nucleotide composition of the cleavage region with the following nucleotide preference A>U>C>>G. Pauws et al. (2001) found that 44% of human regulatory genes used more than one cleavage site and thus resulted in generation of slightly different mRNA species. **Alternative polyadenylation**—It is a variant of conventional post-transcriptional polyadenylation process of eukaryotic mRNAs in which the 3-end processing of the message either starts earlier in the transcripts or poly (A) tails of various lengths are added to the 3-terminus of heterogeneous nuclear RNA or mRNA transcribed from the same gene caused by usage of different poly (A) additional signals by poly polymerase. Thus the 3′end of a processed gene transcript is characterized by a poly (A) tail and a poly (A) signal. Besides the two classical poly (A) signal (AATAA and ATTAAA) which were defined as a stretch of 10 consecutive As at the end of the sequence of 10 consecutive Ts at the start of the sequence, other poly (A) signals have been reported such as the nine possible alternative poly (A) signals (AATTAA, AATAAC, AATAAT, AATACA, ACTAAA, AGTAAA, CATAAA, GATAAA) and six random signal. Polyadenylation signals are thought to occur within 50 to 100bp from the poly (A) addition site (Salamov and Solovyev, 1997). Therefore the 150 nucleotides adjacent to the poly (A) or poly (T) stretch should be analyzed for the two classical polyadenylation signals and nine possible alternative poly (A) signals. Additional information to identify 3′ end sequences is obtained from depositors of the cDNA sequences which have assigned a label (3′ or 5′) to the GenBank sequence in the cloning and sequencing procedures.

3. Splicing—Splicing (removal) of introns is done. Splicing can either occur before or after the cleavage and polyadenylation. Nuclear mRNA precursors have a third class of introns that are spliced with the aid of RNA-protein complexes called SnRNPs assemble into complex structures called **spliceosomes**. Neither all mRNAs are capped at 5′ ends nor all are polyadenylated at 3′ends. The mRNAs from prokaryotes do not undergo such extensive processing simply because they do not contain introns. The mRNA processing takes place within the nucleus. In eukaryotes it is easy to study mRNA as it is more stable and half –life ranges from 1 to 24 hours. It is easy to purify because of 3′ polyadenylated tails. In prokaryotes transcription and translation are tightly coupled and ribosome will attach to the 5′end of mRNA and commence translation before synthesis its 3′ synthesis is complete. The half life is about 2 minutes and they lack 3′-poly (A) tail. Like eukaryotes they contain additional non-coding sequences at their 5′- and 3′ ends. 5′ or leader sequence in case of eukaryotes does not include the **S.D. (Shine and Dalgarno)** sequence which is involved in the binding of mRNA to the ribosome and which is present in prokaryotes but is absent in eukaryotes. Additional intercistronic sequences of variable length (from 1 to 100bp in some bacteriophage) also occur between the coding regions of polycistronic mRNA.

Beyond splicing and promoter-based regulation mRNAs are controlled by regulatory elements in their 5′ and 3′UTRs. Transport signals direct the mRNA to specific subcellular location whereas RNA instability signals mark RNAs for rapid destruction unless specifically protected. Other signals affect translational efficiency. The amount of information in mRNAs used to distinguish ribosome binding sites from other sequences is large and is spread over nearly 35 nucleotides. There are at least six elements within RNAs in E. coli that facilitate translation initiation and these include the initiation codon (AUG is the best initiation codon), the S-D region, spacing between initiation codon and S-D region (spacing >5 nts and < 13 has little effect on translation and spacing of ~9nts may be optimal), other primary sequences (non-random distribution of nucleotides at several positions other than the initiation and S-D region, on the 3′ side of the initiation codon, the information spreads out to +13 while the information extends 5′ to position -20), second codons (second most abundant codons in E.coli mRNAs are AAA and GCU) and mRNA secondary structures. Consensus motifs have been identified for many of these factors and usually correspond to short oligonucleotide tracts which generally fold in specific secondary structures which are binding sites for various regulatory proteins. Some of these regulatory signals tend to be protein family specific while others have a more general effect on diverse mRNAs. Sequences affecting the translation and stability of mRNAs are found in the 5′ and 3′ UTR.

1.23   SENSE/ANTISENSE TRANSCRIPTION

The sense strand of DNA generally provides the template for the production of mRNA which in turn encodes proteins. Transcription from the opposite (antisense) strand can produce transcripts that can hybridize with the coding DNA strand or with the antisense transcript to interfere with transcription or mRNA stability. Antisense transcription (transcription from the opposite strand to a protein coding or sense strand) has been ascribed roles in gene regulation involving degradation of the corresponding sense transcripts (RNA interference) as well as gene silencing at the chromatin level (see chapter 23 for detail). Global transcriptome analysis provides evidence that a large proportion of the genome can produce transcripts from both strands and that antisense transcripts commonly link neighboring 'genes' in complex loci into chains of linked transcriptional units. Perturbation of an antisense RNA can alter expression of sense mRNAs suggesting that antisense transcription contributes to control of transcriptional output in mammals (Katayama et al., 2005).

Transcripts may be derived from either or both strands and they may be overlapping and interlaced (Cheng et al., 2005). Many transcripts (including some noncoding transcripts) are alternatively spliced. Both exons and introns may transmit information. Many miRNAs and all small nucleolar RNAs in animals are sourced from introns (Mattick and Makunin, 2005). The range of types and functions of noncoding RNAs is unknown. Further, genome is transcribe at many more sites than is needed to provide these different RNA species. There is production of large quantity of non-coding and non-functional RNAs.

In case of T7 or SP8 only one strand is transcribed whereas both strands of the T4 and λ chromosomes are transcribed, one strand serving as template for some genes (usually the contiguous groups) while the remaining genes are transcribed along the other strand. The picture holds for the E.coli chromosome which can be transcribed both clockwise and anticlockwise. Many more genes such as lactose, tryptophan and galactose groups are read counterclockwise than clockwise.

1.24   ABERRANT mRNA

Aberrant mRNA contains a premature translational stop signal (nonsense mRNA) or lacks a translational signal altogether (nonstop mRNA). Such defective mRNAs can arise through a variety of mechanisms including genetic mutation, mis-splicing and premature plyadenylation. These mRNAs, if translated, can result in potential consequences of inappropriately terminated proteins.

1.25   mRNA DEGRADATION

mRNAs have limited lifetimes. mRNAs during their lifetime are escorted by a number of associated factors (CBC20/80, e IF4E, e IF4G, PABPN1, PABPCs, Hn RNP proteins, EJC, Y-box proteins, TIA-1/TIAR, mi RNA), some of which remain stably bound while others are subject to dynamic change. Together with mRNA this complement of proteins and small noncoding RNAs (mi RNAs) constitute the m RNP. How long mRNA lives depends on how efficiently the mRNA degradation machinery is recruited to that mRNP. The core degradation machinery attacks mRNA from its ends. The 3' poly (A) tail is removed by a host of deadenylases while 5' cap is removed by specific decapping enzymes. The body of the message is then degraded by 5'-3' and 3'-5' exonucleases. Whether a particular mRNA is destroyed in one direction or the other depends on what set of enzymes is most active in that particular cell type and which set is recruited most efficiently to that mRNP, the messenger ribonucleoprotein particle.

Endonucelolytic degradation mechanisms also exist, most notable sequence-specific mRNA cleavage by RNA-induced silencing complex (RISC) in association with endogenous small interfering RNA.

A general mRNA decay machinery is needed for elimination of aberrant mRNAs. Recognition of nonsense and nonstop mRNAs as abnormals requires their functional engagement to ribosomes which fail to terminate properly on both types of abnormal mRNA. The improper termination leads to recruitment of decay machinery presumably through interactions with ribosome release factors and/or the empty A site tRNA binding pocket on the ribosome.

1.26 P-BODIES

P-bodies are also referred to as GW or Dcp-bodies. In both yeast and mammals much of the decay machinery is concentrated in discrete cytoplasmic foci. These so-called cytoplasmic processing bodies or 'P-bodies' (PBs) appear to form around aggregates of mRNPs not actively involved in translation. (Teixeira et al., 2005). Translationally repressed messenger ribonucleoprotein (mRNP) is lacking translational initiation factors and containing the decapping enzyme and several accessory proteins (Collar and Parker, 2003). Targetting of mRNPs to these structures requires their removal from their translationally active pool and one mechanism for which appears to be interaction with miRNAs and the RISC. PBs also serve as routing stations that can temporarily store mRNAs before sending them out for translation into proteins. In *S. cerevisiae* PBs serve dual roles as way stations for transcriptionally inactive mRNAs and sites of mRNA degradation. P-bodies are linked with RNA interference (RNAi). In this phenomenon, short segments of double stranded RNA shutt off gene expression by directing the destruction of the corresponding mRNAs. The RNA breakdown which helps cell fight off viruses and genetic damage may also take place in P-bodies. The proteins Argonaut1 and -2 which are key components of RNAi machinery (known as RISC) are concentrated in P-bodies implicating the particles as the site of degradation.

1.27 STRESS GRANULES (SGS)

Another structures called stress granules (SGs) serve as temporary retirement homes. Prions like domains in TA1/T1AR (the mRNA binding proteins) are thought to self-oligomerize and promote SG assembly. When mammalian cells are exposed to environmental stresses, the global translational arrest of 'house keeping genes' transcript is accompanied by the formation of distinct cytoplasmic structures containing translationally inactive mRNAs. But there is selective translation of heat shock proteins as well as some transcription factors which allow the cells to repair the stress-induced damage. When stress is relieved SGs deassemble and sequestered mRNAs either return to the transcriptionally active pool or are targeted for degradation in BPs. So far SGs have not been observed in yeast.

1.28 HIERARCHICAL STRUCTURE OF RNA

Similar to those used in describing protein structure RNA structure can be described in hierarchical terms such as primary, secondary, tertiary and quarternary structures. The primary structure refers to the sequence of an RNA molecule. RNA secondary structure is dominated by the formation of double helices stabilized by Watson-Crick base pairs between complementary stretches. Unlike DNA these helices are relatively small, seldom more than 8 to 10 bp in length and are interrupted by single stranded nucleotides forming loop elements such as **hairpins, bulges** and **internal loops**. These together with the helical junction which is formed when more than two double helices come together are the secondary structure motives and building blocks upon which tertiary structure is built. Helical regions generally define the RNA secondary structure. RNA double helices adopt the A-form structure which differs significantly from the B-form adopted by DNA double helices. These differences lead to very diverse shapes with consequences for recognition by proteins and other ligands. The most common element of RNA secondary structure is the hairpin (or stem-loop). A hair-pin is formed when the phosphodiester backbone folds back on itself to form a double –helical tract (called the **stem**), leaving unpaired nucleotides to form a single stranded region, called the **loop**. Hair-pin loops represent the most extensively studied RNA structure motif (apart from double helices). Bulges and internal lops are formed when two double-helical tracts are separated on either one (bulge) or both strands (internal loops) by one or more unpaired nucleotides. Internal loops can be either symmetrical or asymmetrical depending upon the presence of equal or different number of bases on each strand, respectively. **Non-Watson-Crick base pairs** readily form within internal loops whereas unpaired nucleotides within a bulge may stack within the helix or be bulged outside. The formation of secondary structure dominates the process of RNA folding. RNA secondary structure is often stable on its own and because of this property RNA secondary structure can often be predicted successfully from **thermodynamics** or **phylogeny**.

The tertiary structure refers to the three-dimensional structure. Interactions between two or more RNA secondary structure elements give rise to RNA three-dimensional structure. Tertiary interactions (coaxial stacking of double helices) lead to the final three-dimensional structure. The noncanonical base pairs, unpaired bases and the backbone functional groups (the negatively charged phosphate groups and the unique 2-hydroxyl groups of RNA) are very important for tertiary interactions. Divalent metal ions, especially hydrated magnesium ions, are often used to screen the negatively charged phosphate groups along the helical backbone. Tertiary interactions often consist of base stacking and hydrogen bond interactions. Helical stacking between the terminal base pairs of two helices enables the building of the molecule into an extended helix. Hydrogen bonds between bases and backbone are observed when double helices are packed together in compact structure. RNA tertiary structure forms through relatively weak interactions between pre-formed secondary structure motives. The geometry of the four way junction in tRNA, combined with a conserved length of the double-helical regions help in positioning T-loop and D-loop nucleotides in close proximity to facilitate the tertiary loop-loop base pairs which define the L-shape of all tRNAs.

1.28.1 Pseudoknots

Pseudoknots are formed when complementary primary sequences of a hairpin or internal loop and a single stranded region interact with each other by Watson-Crick base pairing (Drapper, 1990). There is formation of two alternative hairpin structures when interaction occurs between a hairpin and a single stranded region. In other words, the pseudoknot is an RNA tertiary interaction formed when nucleotides form a hairpin loop and an adjacent single stranded region pair to form a second stem and loop region (Figure 1.6). The formation of pseudoknots leads to creation of an extended helical region. Pseudoknot is only marginally more stable than the two hair pins. Addition of Mg^{2+} stabilizes folded tertiary structure relative to more extended secondary structures. Secondary structure prediction appears to be a tractable problem but nothing is known about the sequence requirements for tertiary interactions. An oligonucleotide of 19 nucleotides can form a pseudoknot and it was shown to form using nuclease mapping and absorbance melting studies. The RNA from various plant viruses have structures at the 3′ end that are recognized by tRNA processing enzymes even though the secondary structures of these viral RNAs do not resemble the typical tRNA cloverleaf. A pseudoknot allows turnip yellow mosaic virus RNA to adopt a 3-dimensional structure similar to that of tRNA. The tRNA like structure and thus the pseudoknot is required for viral replication in brome mosaic viruses and may be a relic of a primitive telomere. Pseudoknots have been proposed to play a functional role in a number of RNAs. There is existence of regions of universally conserved nucleotide sequence in rRNA in a framework of highly conserved secondary structural elements. These sequences are some how crucial for translation process and most probably they help to form RNA 'active' sites.

1.28.2 Folding of RNA

All possible local RNA folding motifs (Moore and Steitz, 2003) such as U turns, T loops, S turns, Kink turns, hook turns, A-minor interactions, A platforms and tetraloops are all recurring features of the structures of globular RNAs. Together with A-form double helix and more than 20 types of noncanonical base pairs, they comprise the building blocks of RNA architecture. Functional capabilities of a number of cellular RNAs including the hammer-head ribozyme, group I intron and spliceosomal RNAs also appear to depend on their structural dynamics. The dynamic structure of ribosome is facilitated by the flexibility of its RNA. The folding of RNA differs from protein in many ways. There are only four types of nucleotide monomers, there are six backbone torsion angles instead of two; and RNA structure is not nucleated by hydrophobic core as are most proteins. Instead RNA folding uses two principle mechanisms- hydrogen bonding and base stacking. Many of these structural features of RNA can be predicted with good accuracy with the use of only sequence information.

tRNA contains many noncanonical base pairs and even base triplets which allow it to fold into its unique three-dimensional structure. Its strands follow an A-helical path even in

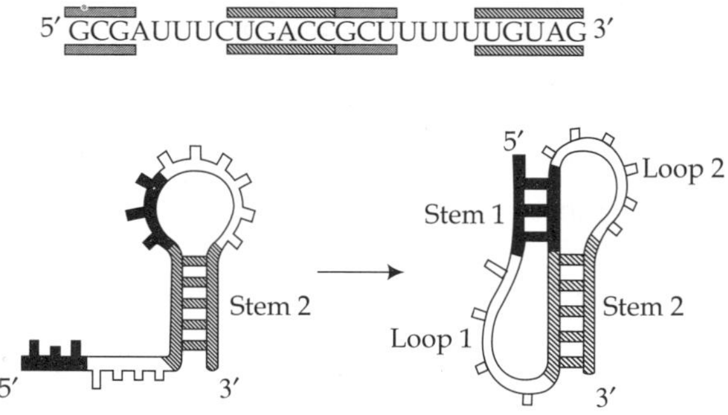

FIGURE 1.6 Showing stem 1 (black) and stem 2 (hatched) and loop regions in the pseudoknot structure.

nonbase paired region. Hairpins are accomplished not by incremental bends in the RNA chain but by abrupt local changes in direction usually centered around one or two nucleotides. A commonly observed motif is the U turn seen in anticodon loop of tRNA which involves hydrogen bonding of the N_3 position of a Uridine with the phosphate group of a nucleotide 3 positions downstream, causing an abrupt reversal in the direction of the RNA chain. It also revealed the coaxial stacking of RNA helices. A seven base pair acceptor stem stacks on the five base T stem to form one continuous A form helical arm of 12 bp. The other two helices, the D-stem and anticodon stem also stack although imperfectly to form a second helical arm. The two coaxially stacked arms form a familiar L form of tRNA. Coaxial stacking is a common feature of RNA and is more wide spread in rRNA where continuous coaxial stacking of as many as 70 bp is found.

In the A-minor motifs single stranded adenosines reach into the minor groove of a helix, making both hydrogen bonding and van der Waals contact. They are not simply base-base interactions but nucleoside-nucleoside interactions as crucial contacts are also made with ribose as well as bases. A-minor interaction, an abundant tertiary structural motif, has been shown to participate in both aminoacyl transfer RNA selection and in peptidyl transferase (Noller, 2005).

Many of the essential features of rRNA secondary structure have been predicted by using comparative sequence analyses. With the availability of many thousands of rRNA sequences plus high resolution crystal structures, prediction of RNA structure can be made by bioinformatics approaches (see Roy, 2009).

1.28.3 Quarternary structure of RNA

RNA molecules are associated with proteins to form supramolecular quarternary structure. For example, mRNAs are associated with five major ribonucleoprotein (RNP) particles called **small nuclear proteins, RNPs**. Such RNPs interact with one another and with mRNAs. These interactions and their dynamic disruption and formation by means of RNA-RNA quarternary interactions are essential for RNA splicing. The quarternary association of RNA molecules occurs by conventional base pairing. For example, small regulatory RNAs with longer complementary sequences within RNA molecules (antisense RNAs) form intermolecular duplexes during the control of gene expression in both prokaryotes and eukaryotes. Similarly, guide RNA recognizes complementary sequences to identify sites where mRNAs are edited post-transcriptionally. Supramolecular association of RNA molecules also occurs by means of non-conventional base pairing and is shown by so-called 'G-quartet structures'. RNA and DNA sequences containing stretches of guanidines or uracils form this structure *in vitro* (Cheong and Moore, 1992) but it is not clear whether these structures occur at all *in vivo*.

Unlike proteins which in most cases function only when properly folded, many RNAs function as unpaired structured single stranded species. For example, mRNA must be unfolded for the genetic message to be translated and stable RNA secondary structure inhibit protein biosynthesis. Further, protein secondary structure is generally only marginally stable whereas RNA secondary structure is often stable on its own. **RNA topology-** It refers to the three dimensional arrangement of a single stranded RNA chain by the formation of internal fold-backs (hairpin loops) and stem-and-loop structures, its folding into tertiary structures and the charges of these structures in response to physical (e.g. temperature) or chemical parameters (e.g. intercalating agents).

1.29 RNA LOCALIZATION

It is a process by which mRNA, mostly in a complex with protein as hn RNP, is concentrated in specific compartment of a cell, tissue, organ or organism. It has more effect than protein trafficking as a single mRNA can give rise to many proteins. Localized RNAs contain *cis*-acting signals ' (RNA zipcodes) ' targeting them to specific domains within cells. These localization signals usually lie in the 3'UTR of mRNAs that are rich in secondary structures and function as address sites for RNA-binding proteins. The actual transport of the RNA from its place of synthesis to site of localization is mediated by both microtubular and actin cytoskeleton elements.

Spatial segregation of individual species of RNA to their correct distributions in the cell constitutes a key element in eukaryotic gene expression. Certain mRNAs have been shown to have non-uniform distributions within cytoplasm linked to sequence elements in the mRNA's 3' –UTR. In case of snRNA species RNase MRP RNA- a discrete element near the 5' end is necessary and sufficient for localization in the nucleolus.

1.30 BASE PAIRING BETWEEN CODON OF AN mRNA AND ANTI-CODON OF tRNA

1. The first two bases of an mRNA (5'-3') codon always base pair with corresponding bases of tRNA anticodon. 2. The third base of an mRNA could pair with different bases of the anti-codon. In other words, the first base of the antocodon (reading in 5'-3') which pairs with 3rd base of the mRNA, determines the number of codons recognized by the tRNA. When the third base of the anticodon is C or A, the base pairing is specific and C pairs with G (C≡G) and A pairs with U (A= U) but if it is U or G, binding is less specific and two different codons may be read, U can pair with either A or G and G may pair with either C or U. When the first base is inosine (I), i.e, when I is the wobble nucleotide three different codons can be recognized, A or U or C.3. When an amino acid is specified by several different codons, codons that differ in either of the first two tRNA bases require different tRNAs.4. A minimum of 32 tRNAs are required to translate cell 61 codons.

1.31 SMALL NUCLEAR RNA (snRNA)

The small nuclear RNAs (snRNAs) are metabolically stable class of RNA molecules which reside in eukaryotic nuclei. The best known snRNAs (U1, U1, U4, U5 and U6) are involved in pre-mRNA splicing. In most eukaryotes the U1, U2, U4 and U5 snRNAs are synthesized by RNA polymerase (RNAP) II but U5 is synthesized by RNAP III. U6 snRNA genes have promoters that are closely related in structure to those described by RNAPII. In all organisms (vertebrates, sea urchins, insects and plants) transcription of snRNA genes is by RNAP II or RNAP III and requires a unique and essential proximal sequence element (PSE) or USE in plants) which is located upstream of -40 relative to the transcription start site. In vertebrates, difference between RNAPII transcribed and RNAPIII transcribe snRNA gene is that U6 gene promoter contains a TATA box whereas RNAP II snRNA promoter lacks a TATA box. In plants transcription of snRNA genes by either RNAPII or RNAP III requires two basal promoter elements the USE and a TATA box and a RNAP specificity is determined by differences in the spacing between the core promoter elements (32-26bp spacing between USE and TATA box for RNAPII and 23-26bp for RNAP III) (Jensen et al., 1998).

U3 and U8 snRNAs are members of a family of RNAs which are defined by their nucleolar localization and association with nucleolar protein, fibrillarin. Both are essential for pre-rRNA processing.

1.32   VARIATION IN DNA/RNA QUANTITY IN PLANTS

The RNA of cells show striking quantitative changes during growth and these changes more or less parallel the rate of growth of cells. DNA is much more stable quantity than the RNA and thus does not exhibit such a correlation with growth of cells. Increase in nuclear size is due mainly to the increase in protein and not exclusively to the chromosomal mass. Changes in the quantity of DNA occur regularly during the division of cells- stage of mitotic cycle at which DNA is synthesized varies for different tissues. There are, however, instances of variations of DNA content correlated with differentiation, morphogenesis and metabolic activity of cells (Leuchtenberger, 1954). It is based on the reason that varying numbers of chromosomes of cells of the same tissue. Great variation in mean chromosome size and DNA content exists among plant species with the same haploid number of chromosome belonging to the same family (up to 36 fold difference) (Stebbins, 1971). The total quantity of DNA of a haploid genome is often called the c-value of the organism. A comparison of c-values of different organism (see Roy, 2009) showed three things.1. The difference between prokaryote and eukaryote is over 1000-fold. 2. Within eukaryotes there is a difference of ~300 fold between the genome of smallest (yeast) and largest eukaryotic genome (maize). 3. In general, there is correlation between the c-value and the stage of evolutionary development. 4. Within amphibian the newt with n=12 has much higher c-value than the mammals with higher n values. Similarly, within flowering plants the lilies have much higher c-value than other plants with higher n values. This phenomenon is referred to as **C-paradox**. The higher content of DNA could be due to either a large number of structural gene or a larger amount of non-sense or repetitive sequence. Renaturation kinetics support the second view. Further DNAs from various organisms have shown that while the mammalian DNAs show G+C contents varying between 40 to 45% whereas the bacterial DNAs show variation between 30 to 75% with *Plasmodium falciparum*, the malarial parasite showing 19% G+C content and *Mycobacterium phlei* showing 73% G+C content.

Sequencing of human genomes has shown that there are places on the genome where a certain base varies among individuals. In some regions the single nucleotide polymorphism (SNP) density is higher than expected and elsewhere it is lower than expected. Regulatory regions called CpG islands which shut down nearby genes are denser in gene-rich regions than in the stretches of the gene-less DNA. Rate of recombination differs among different chromosomes and also between two arms of the same chromosome. Finally, 1.1% of the genome codes for proteins.

1.33   ANALYSIS OF NUCLEIC ACID

1.33.1   Techniques for Separation of Cell Organelles

Cell fractionation—The isolation of nucleic acid or the chromosomes requires separation of subcellular fractions (e.g. nucleus, plastids, mitochondria, dicotyosomes, lysosomes or vacuoles, membrane and cytosol) by differential centrifugation, gradient centrifugation or gel filtration. That part of a centrifuged solution which can not be sedimented by centrifugal force employed but remains above the precipitate (sediment, pellet) is called **supernatant**.

Ultracentrifugation—Ultracentrifugation is a process of sedimenting cells, subcellular particles or molecules in extreme gravity fields (>500,000xg) through density gradients using the instrument called ultracentrifuge. The ultracentrifuge drives the rotors of various designs (fixed angle rotor, swing out rotor, vertical rotor) up to 100,000 revolution per minute (rpm) corresponding to the gravity fields of >500,000xg. These gravitational forces serve to separate cells, subcellular particles or molecules on the basis of either their density or their sizes.

Sedimentation—Sedimentation refers to the migration of molecules in natural gravitational fields or artificial gravitational fields (generated by centrifugal force) that is driven by mass attraction.

Differential centrifugation—It is a technique for the separation of cell organelles on the basis of sedimentation coefficients. The large and small particles are separated by centrifugation at different speeds. In this technique the cell or tissues in solution are mechanically homogenized which leads to breakage of cells and cell contents come out in the aquous buffer. The homogenate is then centrifuged at progressively higher centrifugation speeds which results in larger particles such as nuclei, plastids, mitochondria being precipitated at lower centrifugal forces and smaller particles such as ribosomes being precipitated at higher sedimentation rates. Organelles such as nuclei, mitochondria and lysosomes differ in size and so sediment at different rates. These organelles also differ in specific gravity and so they will float at different levels in a density gradient. When a liver tissue is homogenized and subjected to a low speed centrifugation (1000g for 10 minutes) the **pellets** contain whole cells, nuclei, cytoskeleton (actin filaments, microtubules, intermediate filaments) and plasma membrane. Pellet refers to any packed material sedimented by centrifugation. When the supernatant from this centrifugation is subjected to a medium speed centrifugation (20,000g for 20 minutes) the pellets obtained contains mitochondria, lysosomes and peroxisomes. The supernatant from this centrifugation upon again high speed centrifugation (80,000g, 1hr) results in pellets containing microsomes, fragments of ER and small vesicles. A very high speed centrifugation (150,000g for 3hrs) of the supernatant obtained from previous centrifugation results in pellets containing ribosomes and large macromolecules. The supernatant from this stage of centrifugation contains soluble proteins.

Aliquot—It refers to any part or fraction of a whole.

Differential gradient centrifugation— It is a technique used for the separation of macromolecules or cellular organelles on the basis of their differential sedimentation through a gradient either preformed (sucrose gradient centrifugation) or formed during ultracentrifugation.

Sucrose gradient centrifugation—Density gradients are either formed by progressive mixing of solutions with different density (e.g. linear sucrose gradients), overlayering of solutions with decreasing density (step gradient) or by redistributing solutes during ultrcentrifugation. In this technique a linear or exponential or step gradients are formed with a gradient former which mixes two sucrose solutions of differing concentration (density). Step gradients are formed after layering solutions of decreasing concentration on top of each other during ultracentrifugation. The particles are separated by their sedimentation through the gradient. The sedimentation velocity in sucrose gradient is largely determined by molecular size and shape of the molecule. In isopycnic centrifugation a centrifuge tube is filled with a solution whose density increases from top to bottom. The solute such as sucrose is dissolved at different concentrations to produce the density gradient. When a mixture of organelles is layered on the top of the density gradient and the tube is centrifuged then the individual organelles sediment until their density exactly matches with that of the gradient suspension and different layers are formed in the tube. Finally each layer consisting of particular type of organelle is collected separately.

Phase lock procedure — It is a technique used for separating the organic phase and interphase material (containing the denatured proteins) from the aqueous phase (containing nucleic acids) during phenol-chloroform extractions. In this technique phase lock gel- a chemically inert and hydrophobic gel is used to trap the organic and interphase material during centrifugation of the phenol-chloroform mixture and thus leading to complete separation of the phases and easy recovery of nucleic acids from aqueous phase.

DNA extractor — It is an equipment used for the rapid, automated and simultaneous extraction of ungraded DNA from several samples. The DNA is purified by a cycle of proteolytic cell lysis, extraction of non-nucleic acid material with phenol/chloroform, ethanol precipitation of the DNA from the aqueous phase and filtration to collect the purified nucleic acid.

Isopycnic centrifugation — Isopycnic centrifugation is a technique used for separation of molecules (especially nucleic acids) according to their buoyant densities in salt (Cesium chloride, Cesium sulfate) gradients. In the **isopycnic** (same density) centrifugation a centrifuge tube is filled with a solution of heavy salt, cesium chloride, CsCl the density of which increases from top to bottom and thus a density gradient develops. When a mixture of DNAs or DNAs and RNAs is layered on the top of the density gradient and the tube is centrifuged at high speed the molecules assume a position on the gradient corresponding to their own intrinsic buoyant density. At this position the nucleic acid molecules accumulate and form a band which can be collected. Separation of DNA by isopycnic ultracentrifugation is largely confined to the different chromosomal species in a given cell. DNA fragments generated by restriction endonuclease or mechanical shearing can not be separated by isopycnic ultracentrifugation as the fragmentation is random and base compositions of fragments are generally similar.

Ethanol precipitation — It is a technique for purifying nucleic acids from oligonucleotides, nucleotides, salts and other impurities using alcohol (ethanol, isopropanol, butanol) and high salt concentrations (0.2-0.6M NaCl). Under these conditions nucleic acid polymers aggregate and precipitate whereas the low molecular weight components remain solubilized. The precipitate nucleic acid can then be collected through centrifugation.

Phenol extraction — It is a technique for the denaturation and removal of proteins from solutions containing nucleic acids and proteins using buffer saturated phenol.

Separation of bacterial plasmid — Isopycnic ultracentrifugation can also be used to separate bacterial plasmids from chromosomal DNA. Separation is not on the basis of differing GC contents but on the fact that the plasmid DNA is circular whereas the chromosomal DNA is a linear molecule. This difference is exploited by ethidium bromide which stains DNA. The staining DNA is through intercalation which requires that the DNA strands be forced apart thereby resulting in decrease in buoyant density and partial, unwinding of the DNA double helix. This later process (partial unwinding) is hindered in case of closed circular plasmid DNA with the result that they bind less EtBr and thus have a higher buoyant density. Thus in case of separation of plasmid DNA through isopycnic ultracentrifugation the RNA will take the lowest position, plasmid will take the middle position and the chromosomal DNA will take the upper most position.

Separation of telomeres — Telomeres are the DNA sequences found at the ends of linear chromosomes. They define the boundaries of the genetical and physical map of such chromosomes and so are particularly important for the complete mapping of large genomes. Human telomeres contain tandem arrays of the sequence $(TTAGGG)_n$. These satellite DNA sequences can be purified from bulk of the genomic DNA by isopycnic centrifugation on a Cs_2SO_4 gradient in the presence of silver ions.

Qiagen column — It is a small disposable anion exchange column which is used for the fast separation and simple isolation and purification of nucleotides, oligonucleotides and polynucleotides especially plasmids.

Separation of RNA molecules — One of the most widely used methods of separating RNA molecules is through rate zonal ultracentrifugation employing sucrose density gradients. In this method the separation of RNA is based primarily on size, although in principle, shape of RNA can also have an effect but in practice differences in shape makes no significant contribution to the separation. But this rate zonal ultrcentrifugation is not appropriate for the separation of different species of tRNA as they are broadly similar in size. The major problem in isolation of mRNA is purification from other species of RNA which are approximately 20 times more abundant.

Affinity chromatography — It is a chromatographic method for separation of mRNAs from other species of RNA on the basis of their specific affinity to ligands (oligo (dT) or poly (U)) bound on the insoluble matrix. For example, oligo (dT) stretches of about 10-20 nucleotides can be bound as ligands to the inert matrix cellulose to yield olgo (dT) cellulose which is then used to fill a column. Now if a mixture of cellular RNAs (rRNA, tRNA and mRNA) is applied then only those RNAs which contain poly (A) tail at their 3′ end (mRNAs) will be bound to the column by affinity which in this case is A=T base pairing. After extensive washing to remove non-bound-RNAs the poly (A) $^+$-RNAs are eluted. The matrix could be a sepharose matrix to which poly (U) residues are covalently bound and is used for binding, isolation and purification of Poly (A) -mRNA in affinity chromatography. The minimum size required for binding to poly (U) -Sepharose is smaller than that for oligo (dT) -cellulose (~6-10 residues as compared to15). Although oligo (dT) -cellulose in more widely used but poly (U) -sepharose for separation of mRNA having very small poly (A) tails. Mixture of different mRNAs can be separated to a certain extent on the basis of size, either by sucrose density gradient centrifugation or on a smaller scale by agarose gel electrophoresis under denaturing conditions. Both agarose gel electrophoresis and PAGE can be used to separate other types of RNA. The isolation of individual mRNA species is usually accomplished by recombinant DNA techniques.

Hydroxyapatite chromatography—It is a technique for separating double stranded DNA or DNA-RNA hybrids from single stranded DNA. In this procedure the column is filled with calcium phosphate mineral-hydroxyapatite which binds to nucleic acids by electrostatic interaction between the calcium residues of the mineral and phosphate backbone of the polynucleotide. Single stranded nucleic acids bind comparatively weakly and are eluted from the column at lower phosphate concentrations than do double stranded. Single stranded cDNA will pass through the hydroxylapatite whereas cDNA-mRNA hybrids will bind.

Magnetic bead—It consists of paramagnetic material, for example, iron oxide, coated with polyacrylamide and agarose and is packed into submicron sized particles. These particles have no magnetic field but form a magnetic dipole when exposed to a magnetic field. Magnetic beads serve as solid phase support for the separation of DNA or RNA molecules from complex mixtures of biomolecules. Specific binding is usually achieved through specifically designed DNA fragments coupled to the magnetic beads.

1.34 DETERMINATION OF G+C CONTENT

From the analysis of genome we would like to have the following information. 1. What is the C-value of an individual? or What is the genome size? 2. What is the percentage of (G+C) content in the genome? 3. Whether the genome is complex? The C-value is a measure of genome size and is generally expressed in base pair (bp) of the DNA per haploid genome or in picograms (pg) of DNA per haploid cell. Each species has a characteristic C-value (see Roy, 2009). As the physical properties of DNA are influenced by the G+C content it would be desirable to estimate the G+C content first. The average genomic GC content for prokaryote and eukaryote varies widely. It ranges from less than 22% in *Plasmodium falciparum* to greater than 68% in the large amplicon of *Halobacterium* sp NRCl. (G+C) % is 41.4 (in tomato), 37.4 (in potato) and between 42.6-45.6 in barley. Local heterogeneity within GC content can be enormous ranging from 26 to 65% in the human genome. In contrast the AG content (purine) is homogeneous (40-43%) fluctuating by just a few per cent about a mean of 50%. The relative (G+C) content can be determined from the,

   I. thermal denaturation temperature

  II. ultraviolet spectrum of DNA

 III. buoyant density of DNA

 IV. paper chromatography.

Thermal denaturation of DNA—The DNA is highly viscous at pH7.0 and room temperature of 25^0C. When the double helix DNA is subjected to high temperature ($>80^0$C) or extremes of pH the hydrogen bonds between bases break and the two strands get separated. The DNA is said to be denatured. When the temperature is used as denaturant the DNA is said to be melted and the temperature at which this separation of strand occurs is called **melting or transition temperature, Tm**. Tm is the temperature at the mid-point of transition. The determination of Tm under fixed conditions of pH and ionic strength can yield an estimate of base composition. Tm depends on pH, ionic strength, size and base composition of DNA. There is a relation between the melting temperature and the G+C content of the DNA. % GC= (Tm- X) x 2.44 where X-depends on the nature (concentration) of solvent and nature of DNA. In lower concentration the denaturation will occur at lower temperature (i.e., Tm is low) but over a broad range of temperature whereas in higher concentration the Tm will be higher and the range is very narrow and thus transition is sharp. Although higher the G+C content of DNA the more stable will be the DNA molecule and higher will be the Tm but as most DNA molecules have different regions rich in either A+T or G+C, A+T regions will melt (separate) first (such regions are called **bubbles** and can be visualized with electron microscopy) and G+C regions will start melting only after all the A-T regions have melted. Strand separation is essential for replication and transcription and DNA sites where these processes are initiated are often rich in A-T sequences. Duplexes of two RNA strands are more stable than DNA duplexes and at pH= 7.0 it often requires temperature 20^0 or more higher than DNA duplex with comparable sequence.

Ultra-violet absorption spectrum — The pattern of U.V. light absorption is different in double helix, single stranded DNA, DNA-DNA hybrid and nucleotides. The study of U.V. light absorption at 260nm shows that absorption increases by 20-30% in single stranded DNA and thus it shows hyperchromic effect whereas the double helix DNA shows reduced absorption of U.V.light and thus it shows hypochromic effect. The DNA-DNA hybrids have reduced absorption in comparison to the double helix DNA but free nucleotides have higher absorption of U.V.light than the double helix DNA.

Buoyant density — Buoyant density refers to the intrinsic density of a molecule in salt (CsCl), or sugar solution and it depends on the number of G=C pairs. Higher the number of G=C pairs, higher the buoyant density. The G+C rich DNA has higher buoyant density than the A-T rich sequences. There is a linear relationship between the buoyant density (p) of different DNAs and their G+C contents. P= 1.660 + 0.98 (GC) where GC is the mole fraction of G+C. The buoyant density of dsDNA varies with the mole fraction of G+C. In general, RNA has a much higher buoyant density (~1.98g/ml) than dsDNA (~1.7g/ml). Single stranded DNA has a slightly higher buoyant density (1.73g/ml) than a double stranded DNA (1.703g/ml).

1.35 RENATURATION KINETICS

The application of DNA renaturation is used to determine the copy number of different portions of eukaryotic genomes. So long as all hydrogen bonds between bases are not broken (partial denaturation), the separated strands upon cooling start reannealing (or rewinding) and finally it reverts to the original (native) double helix form. Renaturation is very rapid if a segment of 12 or more residues is still paired and renaturation is spontaneous if pH or temperature changes back to normal. Only when all the bonds are broken (i.e., complete separation), melting become irreversible on rapid cooling and the two separated strands form random coils. However, when the temperature or pH is changed to the normal range then the renaturaton occurs slowly. Renaturation occurs in two steps. I. At first the two strands come together by random collision (chance collision between two single strands leads to surface contact and effective pairing) and form short segment of complex double helix and II. then the remaining unpaired bases of one strand start pairing with the complementary bases of the other strand and thus zipper themselves to form a double helix. The minimum recognition length (the least number of successive bases. for reannealing) is ~ 12 nucleotides long. Two DNA samples with identical concentration will take different times to reanneal (reassociate) depending upon the their complexicity. Eukaryotic DNA is more complex than viral DNA in that the former contains more different sequences and each sequence is present in much more lower concentration and thus will take longer time to find its complementary strand. The complexicity of DNA is determined by the time it takes in a solution of DNA of a given concentration to reanneal and is measured as **Cot value** of DNA where Co refers to the initial concentration in moles nucleotide per litre and t is the time in seconds for DNA to reanneal. The renaturation rate will be slow in the unique sequence whereas it will be fast in highly repeated sequence and renaturation rate will fall in between for the moderately repetitive sequences. Single stranded molecule with many complementary partners will reassociate more quickly than single copy sequences. Fraction of genomic DNA that reassociate more quickly is the highly repetitive sequences. Thus renaturation kinetics of DNA fragmented into small pieces will indicate the percentages of genome consisting of unique sequences and repeated nucleotide sequence, respectively. Britten and Kohne (1968) analysed the genome in terms of unique versus repetitive sequences. They assessed the complexity of genome in terms of the degree of repetitiveness of different fragments of genome which in turn were assessed by different rates (Cot values) at which they re-associated to form double stranded structure. The degree of double strandedness is determined by passing the partially reannealed sample through a column of hydroxy apatite which binds strongly with duplex DNA but lets single stranded DNA through. The complexity of a DNA sample is thus determined by the time it takes a solution of DNA of a given concentration to reanneal, i.e., by the Cot value of DNA.

Theory: If c is the concentration of single stranded DNA in terms of number of copies (moles per liter), the rate of re-association will be directly proportional to c^2.

$$dc/dt = - kc^2$$

where k is a constant.

After integration

$$\int 1/c^2 \, dc = - k \, t \, dt, \text{ and } (1/c_o - 1/c) = -kt$$

where c_o is the initial concentration .

When $c = c_o/2$ (i.e., DNA is 50% reassociated),

$$1/c_o = kt_{1/2} \text{ and } c_o t_{1/2} = 1/k$$

where $t_{1/2}$ is the time taken for 50% reannealing and $cot_{1/2} = 1/k$.

If C_o is the initial DNA concentration expressed in terms of mass of DNA per unit volume and m is the molecular weight of the sequence reannealing then

$$C_o = mc_o \text{ and } C_o t_{1/2} = mc_o t_{1/2} = mx \, 1/k.$$

Hence $C_o t_{1/2}$ (usually written as Cot) is directly proportional to m and also to sequence complexity in terms of base pairs. Thus theoretically the number of copies of a given DNA in the genome should be inversely proportional to the Cot value. The Cot value is used for definition of genomic organization. In other words, higher the Cot value, higher the number. of copies and the more complex is the DNA. Thus the complexicity of a DNA sample, defined as the length of the repeating unit, is directly proportional of its Cot value. A comparison of Cot values for DNA samples obtained from different organisms shows that the rate of association decreases with the complexicity of the organism and its genome (Figure 1.7). DNA renaturing at low Cot value (10^{-4} to 10^{-1}) is composed of highly repetitive sequences, DNA renaturing at intermediate Cot values (10^{0} to 10^{2}) is moderately repetitive and DNA renaturing at high Cot values is minimally or nonrepetitive. The reassociation in case of eukaryotic bovine DNA (Figure 1.7e) shows that several per cent of the DNA reassociates almost instantaneously which indicates that it is very highly repetitive DNA or simple sequence DNA. Another 40% or so appears to be moderarively repetitive, reassociating over a broad range of Cot values and only a little more than a half has a Cot value of order (~1000 moles of nucleotide seconds litre^{-1}) indicative of unique sequence DNA. The Cot values (Figure 1.7 b,c) in case of viral and bacterial genomes show that they consist predominantly of unique copy (single copy) DNA with little repetitive sequence.

This technique will provide only rough estimates as the proportion depends on the stringency of renaturation conditions, also because individual members of families of repeated sequences are frequently not identical and lastly it is not easy to distinguish single copy nucleotide sequences from small families of sequences consisting of 3 to 4 members (Ananiev, 1992).

If the DNA is sheared to fragments of different average sizes and if Cot analysis carried out of these DNA fragments then it is possible to know the interspersion of repetitive sequences with single copy sequences. If both single copy sequence and repetitive sequence are present in the same fragment then it will be counted as repetitive as partially reannealed fragment is retained on the separating column along with fully reannealed fragments. Now if the amount of a repetitive fraction is less at smaller fragment sizes it can be concluded that at least some of it is closely associated with single copy sequences. This analysis indicated that most highly repetitive sequences occur in big blocks in the genome. However some of it appeared to be more

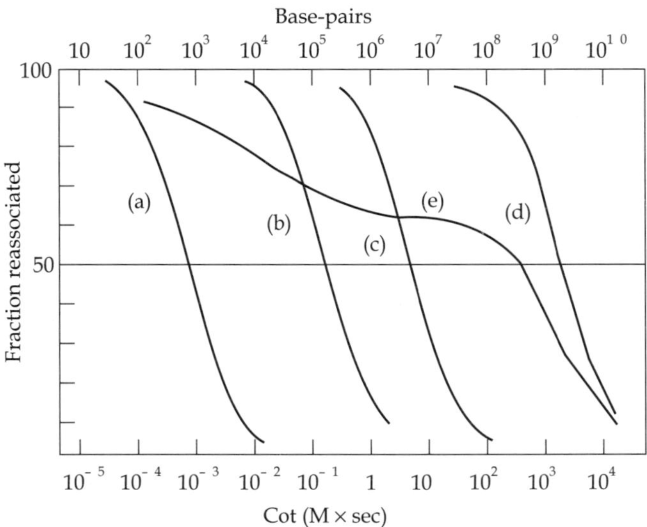

FIGURE 1.7 **Showing kinetics of reannealing of different DNA Samples: a. mouse satellite; b. bacteriophage T4; c. E.coli and d. 'single copy' component of (e) total bovine DNA. The upper scale shows the complexity corresponding to the Cot values at 50% reannealing.**

widely dispersed over the genome. Britten-Davidson (1969) proposed a model for the integrated control of gene batteries in eukaryotic organisms. The model tries to explain the presence of abundant repetitive DNA in eukaryotes. Repeats associated with co-ordinately regulated but otherwise unlinked genes were assumed to carry sites for *trans-acting* regulatory factors.

Test for the presence of repetitive sequences — For testing the presence of highly repetitive sequences, labeled total genomic DNA is used as probe for Southern transfers of restricted λ or cosmid clones (a technique called **'reverse' genomic Southern blots**). Restriction fragments of unique DNA sequences generally fail to give detectable signals whereas fragments containing repeated sequences give detectable hybridization signals, corresponding in intensity to the size and frequency of repeat.

Moderately repeated sequences can be identified by direct Southern blots. When a fragment containing such a repeat is hybridized to a Southern transfer of restricted genomic fragments, a multitude of bands are detected. Some bands are very intense, produced by restriction fragments internal to the repetitive element whereas majority of the bands are much weaker and correspond to junction fragments extending into adjacent chromosomal DNA.

1.36 QUANTIFICATION OF CYTOSINE METHYLATION

Combined Bisulfite Restriction Analysis (**COBRA**) is a technique used for detection and quantification of cytosine methylation at a specific loci (Xiong and Laird, 1997) in small amounts of genomic DNA. This technique combines the sodium bisulfite treatment of genomic DNA (conversion of unmethylated cytosine residues to cytosine) with restriction of bisulfite modified target DNA and its subsequent amplification. Silencing of tumor-suppressing gene expression by DNA methylation has emphasized the need for accurate, sensitive, reliable and quantitative methods to measure levels of DNA methylation at specific loci.

Transcriptional silencing of genes mediated by the epigenetic effects of DNA methylation at CpG-island containing promoters is well documented. Changes in methylation patterns at specific CpG sites are routinely monitored by digestion of genomic DNA with methylation - sensitive restriction enzymes (see chapter 6) followed by Southern Analysis of the region (s) of interest. This technique has disadvantages including the requirement for large amounts of DNA (≥5μg) and limited scope for analysis of CpG sites as determined by the presence of recognition sites for methylation-sensitive restriction enzymes.

A PCR based methylation assay which utilizes digestion of genomic DNA with methylation-sensitive restriction enzymes prior to PCR amplification (Sanger-Sam et al., 1990) has also been described but this technique has the potential to generate false positive signals (methylation present) because of inefficient enzyme digestion or overamplification in the subsequent PCR reaction.

Bisulfite treatment of DNA distinguishes methylated from unmethylated cytosines but the original bisulfite genomic sequencing technique described by Frommer et al (1992) requires large scale sequencing of multiple plasmid clones to determine overall methylation pattern. The Methylation Specific PCR (MSP) and restriction enzyme digestion of PCR products amplified from bisulfite converted DNA utilize bisulfite treatment of DNA as a starting point for methylation analyses. A bisulfite-PCR method called MSP (Herman et al., 1996) is an excellent alternative but tends to be more qualitative rather than quantitatively accurate method.

In higher order eukaryotes DNA is methylated only at cytosines located 5' to guanosine in the CpG dinucleotide (Holliday and Grigg, 1993). This modification has important regulatory effects on gene expression especially when involving CpG-rich areas known as CpG islands, located in the promoter region of many genes (Bird, 1992). While almost all gene associated islands are protected from methylation on autosomal chromosomes, extensive methylation of CpG islands has been associated with transcriptional inactivation of selected imprinted genes (Li et al., 1993) and genes inactive X-chromosomes of females (Pfeifer et al., 1989). Aberrant methylation of normally unmethylated CpG islands has been reported as a relatively frequent event in immortalized and transformed cells and has been associated with transcription inactivation of defined tumor suppressor genes in human cancers. In this last situation, promoter region hypermethylation stands as an alternative to coding region mutations in eliminating tumor suppressor gene function. Therefore, mapping of methylation patterns in CpG island has become an important tool for understanding of both normal and pathogenic gene expression events. Mapping of methylated regions in DNA has relied primarily on Southern Hybridization Approaches based on the inability of methylation-sensitive restriction enzymes to cleave sequences that contain one or more methylated CpG islands including some quantitative analysis but requires large amounts of DNA (=5μg), can detect methylation only if present in greater than a few percent of alleles and can only provide information about CpG sites found within sequences recognized by methylation-sensitive restriction enzymes.

A more sensitive method of methylation detection combines the use of methylation-sensitive enzymes and PCR (Singer-Sam et al., 1990). After digestion of DNA with the enzyme PCR will amplify from primers flanking the restriction site only if DNA cleavage has been prevented by methylation. Like Southern based approaches this method can only monitor CpG methylation in methylation-sensitive restriction sites. Moreover, the restriction of unmethylated DNA must be complete since any uncleaved DNA will be amplified by PCR yielding a false positive result for methylation. This approach has been useful in studying samples where a high percentage of alleles of interest are methylated such as the study of imprinted genes and X-chromosome inactivated genes. However, difficulties in distinguishing between incomplete restriction and low numbers of methylated alleles make this approach unreliable for detection of tumor suppressor gene hypermethylation in small samples or in samples where methylated alleles represent a small fraction of the population.

Bisulfite treatment—The chemical modification of cytosine to uracil by bisulfite treatment has provided another method for the study of DNA methylation which avoids use of restriction enzymes (Frommer et al., 1992). In this reaction all cytosines are converted to uracil but those that are methylated (5-methylcytosine) are resistant to this modification and remain as cytosine. This altered DNA can then be amplified and sequenced, providing detailed information within the amplified region of the methylation status of all CpG sites. However, this method is technologically rather difficult and labor intensive and without cloning of the amplified products, the technique is less sensitive than Southern analysis, requiring about 25% of the alleles to be methylated for detection.

Methyl specific PCR (MSP)—For understanding the regulation of imprinted genes, X-chromosome inactivation and tumor suppressor gene –silencing in human cancer, precise mapping of DNA methylation in CpG island is required and this is where methyl specific PCR (MSP) is important. It can rapidly assess the methylation status of virtually any group of CpG sites within a CpG island, independent of the use of methylation-sensitive restriction endonucleases. This assay entails initial modification of DNA by sodium bisulfite, converting all unmethylated but not methylated cytosines to uracil and subsequent amplification with primers specific for methylated vs unmethylated DNA (primers to distinguish methylated from unmethylated DNA in bisulfite-modified DNA take advantage of sequence differences resulting from bisulfite modification). Unmodified or DNA incompletely reacted with bisulfite can also be distinguished since marked sequence differences exist between these two DNAs. The frequency of CpG sites in CpG islands renders this technique uniquely useful and extremely sensitive for such regions. It requires only a small quantity of DNA, is sensitive to 0.1% methylated alleles of a given CpG island locus and can be performed on DNA extracted from paraffin-embedded samples. MSP eliminates the false positive results inherent to previous PCR-based approaches which relied on differential restriction enzyme cleavage to distinguish methylated from unmethylated DNA. It has identified promoter region hypermethylation changes associated with transcription inactivation in four tumor suppressor genes.

MsSNuPE—Methylation sensitive single nucleotide primer extension (Ms SNuPE) assesses methylation differences at specific CpG sites based on bisulfite treatment of DNA followed by single nucleotide primer extension. In this technique genomic DNA is first reacted with sodium bisulfite to convert unmethylated cytosine to uracil while leaving 5-methylcytosine unchanged. Amplification of the desired target sequence is then performed using PCR primers specific for bisulfite-converted DNA and the resulting product is isolated and used as template for methylation analysis at the CpG site (s) of interest. This technique is quantitative, does not use restriction enzymes and many CpG sites can be analyzed in each primer extension reaction using a multiplex primer strategy.

High throughput analysis of DNA methylation—A number of medium to high throughput assays have been described to analyze DNA methylation (Schmacher et al., 2006). Hellman and Chess (2007) performed allele-specific analysis of more than 1000 informative loci along the human X-chromosome. The Affymetrix 500, 000 SNP mapping array was modified to allow allele-specific analysis of DNA methylation. The genomic DNA was digested with a mixture of five methyl-sensitive restriction enzymes (MSRE) (AciI, BsaHI, Hha, HpII and HpyCH4IV). Together these frequently recognize 40% of all the CG dinucleotides in the genome thus allowing an efficient analysis of the regions of both high and low GC content. After this pretreatment fragments of 200 to 1100bps containing the polymorphic sites were PCR amplified and the resulting amplicons were then labeled and hybridized to the array. Thus an unmethylaed MSRE site present on a given amplicon will lead to allelespecific reduced intensity corresponding to the resident SNP. This permitted them to use either genotype calling or copy number algorithms to identify transitions from heterozygous state to a homozygous state. Single locus validation included DNA sequencing of specific PCR products before and after MSRE treatment and also bisulfite analysis.

The Xa (active X) showed more than two times as much allele-specific methylation as Xi (inactive X). This methylation is concentrated at gene bodies (transcribable regions), affecting multiple neighboring CpGs. Before X inactivation all of these Xa gene body-methylated sites are biallelically methylated. Thus a bipartite methylation-demethylation program results in Xa-specific hypomethylation at gene promoters and hypermethylation at gene bodies. These results suggest a relationship between global methylation and expression ability.

Constantly inactive regions such as gene –poor regions and the entire Xi may be more prone to loss of methylation maintenance (even if originally highly methylated). The resulting methylation decrease for the entire Xi and for Xa intergenic regions, would thus highlight Xa gene body-specific methylation. At the same time promoter CpG islands which are protected from methylation on Xa would remain more methylated on Xi. In contrast to the widely held view that X chromosome allele-specific methylation is restricted to CpG islands on the inactive X chromosome, the global allele-specific analysis uncovered extensive methylation specifically affecting transcribable regions (gene bodies) on the inactive X whether it is in the male or female. One aspect of sex chromosome dosage compensation is the requirement for a chromosome-wide, likely epigenetic mechanism with the ability to double X-linked expression when necessary (i.e., when somatic cell but not in haploid germ line cell). Gene body methylation in plants has also been reported (Zilberman et al., 2007).

When examining the perfect match (PM) probes one can observe both alleles in genomic DNA but only one in each clone after MSRE treatment. This is due to methylation differences at an MSRE site present within the amplicon. Mismatch (MM) probes contain a single MM at a site distinct from the polymorphic site and serves as control. Potential monoallelic methylation at potentially imprinted regions as well as technical artifacts resulting from MSRE site polymorphism or allelic bias from genotyping algorithm were filtered out by requiring at least one clone switch from AB to AO and one clone switch from AB to BO per SNP.

Denaturation mapping—It is a technique which is used for localization of A-T regions in the double stranded DNA. The DNA is first heated until melting begins. The resulting single stranded regions are then stabilized by formaldehyde and visualized in the electron microscope.

Acetylation mapping—It refers to determination of the number and precise location of acetyl groups in various histones of chromatin of a nucleus at a given time probed with chromatin immunoprecipitation.

1.37   NEAREST NEIGHBOR SEQUENCE OR FREQUENCY ANALYSIS

It is a method for the characterization of DNA molecule which is based on the estimation of the relative frequencies with which pairs of each of the four bases lie next to one another. Any deoxyribonucleotide can be covalently bound to any of the three others or to a nucleotide of the same type by its 3' or 5' hydroxyl group to form a dinucleotide molecule. Since there are four different deoxyribonucleotides (dATP, dCTP, dGTP, dTTP), there is possibility of formation of sixteen dinucleotides. The frequency with which each of these combinations occurs is characteristic of a particular DNA.

Nearest neighbor sequence can be determined by incubating a DNA template with E. coli polymerase and the four deoxyribonucleotides, one of which is labeled with 32p at the α (innermost) phosphate position. The labeled α-32p then links the labeled nucleotide with its nearest neighbor nucleotide. After synthesis the isolated DNA is digested with micrococcal nuclease and spleen phosphodiesterase which yields deoxyribonucleoside 3' monophosphates. The 32p is now attached to the 3' carbon atom of the neighboring nucleoside. The four deoxyribonucleoside 3' monophosphates are separated by paper electrophoresis and their radioactivity is measured. This measure gives the frequency with which the originally labeled nucleotide has been bound to the other nucleotides. By using all four α-32p labeled deoxyribonucletides in repeated nearest neighbor analyses the frequency of all sixteen dinucleotides can be determined. The utility of this type of information is discussed in the book, Bioinformatics (Roy, 2009).

2

RNA Editing

RNA editing is the modification of RNA sequence that is distinct from other RNA processing events such as splicing, capping or 3′ end processing. The modifications include both the insertion and deletion of nucleotides and conversion of one base to another. RNA editing is a site-specific alteration in an RNA leading to sequence revision. Editing is known to occur at multiple sites in the same gene (Luo et al., 1990). RNA editing will lead to changes in gene expression and can change the function or coding potential of the modified transcripts. RNA editing can affect mRNA, tRNA and rRNAs. It is a wide spread phenomenon and has been identified in numerous species including a range of metazoans, plants, protozoa and bacteria. RNA editing has shown to increase protein diversity by introducing altered splicing patterns, frame shifts, alternative start site or stop codons or directly affecting the protein sequence (Schaub and Keller, 2002) (Figure 2.1). However, recent evidences suggest that large majority of human A-I editing sites are situated in introns and UTRs which could either have regulatory functions or could be non functional. In mammals, A-I is the most prevalent type of editing. I is read by the translational machinery and reverse transcriptase as a guanosine. So, A-I edits appear as A-G mismatches between genomic and expressed sequences.

2.1 EDITING OF tRNAS

tRNA editing allows to synthesize two different tRNAs from the a single tRNA as a result of post-transcriptional exchange of distinct bases in tRNA. In Marsupialia C is exchanged by U. The unedited tRNA contains a GCC codon (specifying the amino acid glycine) which is replaced by GUC (specifying aspartate). Although tRNAs are subject to a large number of base modifications (79 different bases) changes in tRNA primary sequence that involve the four canonical nucleotides have only recently emerged. Editing creates essential structural elements at the primary, secondary and tertiary levels and involves both loop nucleotide and base-paired stems. Some editing events change the tRNA identity, i.e. alter recognition by aminoacyl synthetases whereas others affect 5′ and/or 3′ processing or subsequent base modifications. Mitochondrial tRNAs encoded by overlapping genes are often substrates for three nucleotide addition which is distinct from CAA addition (Gott and Emerson, 2000).

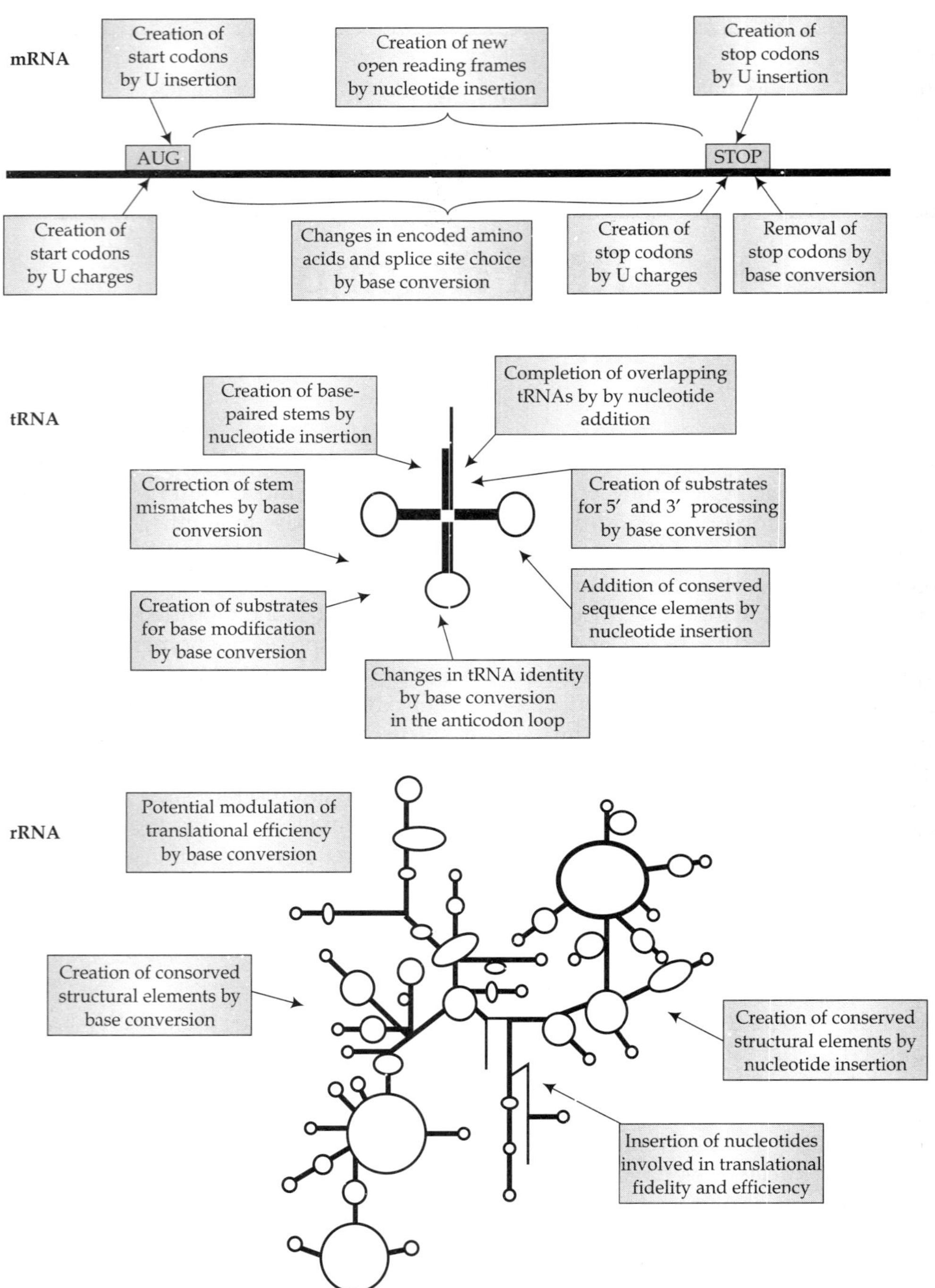

FIGURE 2.1 Showing functions of editing of mRNA, tRNA and rRNA.

2.2 EDITING OF rRNA

rRNA editing appears to be less frequent than mRNA and tRNA editing but where it occurs, it seems to be functionally important as we know that rRNA is a component of translational machinery. The single C to U change in the small subunit rRNA in Dictyostelium mitochondria falls within the highly conserved 530 loop which is believed to play a critical role in tRNA selection and proof reading. Further, both large and small rRNAs in Physarum mitochondria contain nucleotide insertions along the entire length. These nucleotides take part in the formation of conserved helices or are found at critical sites including the 530 loop and A dinucleotide which is involved in tRNA binding in the A site of the ribosome in E. coli. Thus in either case editing would have a major impact on translational fidelity.

2.3 EDITING OF mRNA

Majority of the RNA editing events involve changes in mRNA sequences. Eukaryotic mRNA undergoes various co- and post transcriptional modifications. The 5′ end of mRNA is generally capped and the 3′-end polyadenylated and up to 98% of the internal sequences can be eliminated by the splicing. Although capping and polyadenylation generation do not affect the protein coding capacity of mRNA, splicing generally determines or alters its coding potential. Various types of mRNA editing are given in the Table 2.1 (Cattaneo, 1991).

The table 2.1 shows that RNA editing can be **mononucleotide editing**-a variant of RNA editing in the mitochondria of *Physarum polycephalum* characterized by insertion of mononucleotides in mRNAs relative to their mtDNA template. The **dinucleotide editing** is characterized by the insertion of dinucleotides in mRNA of P. polycephalum. The different common insertions are GC, GU, CU, AU and AA. It creates ORFs in mRNA and contributes to highly conserved structural features of ribosomal and tRNA. Partial editing (5′ editing) is a special type of editing in which only 5′ termini of editing domains are edited in contrast to pan editing in which entire genes are edited.

2.3.1 Mechanisms of Editing of mRNAs

The mechanisms involved are : **A**. Post-transcriptional editing which involves insertion of four uridine that shifts the translational reading frame of the transcript. **B**. There can be synthesis of two proteins from the same gene in a tissue specific manner. For example, mRNA of the same gene apoB-100 produces two different proteins in liver and intestine. In the intestine cytosine changes to uridine through deaminase enzyme and thus stops at that point in the reading frame. When the mRNA editing process results in different translated polypeptides the resulting proteins are known as **splice variants or alternatively spliced forms**.

RNA editing systems programmatically alter mRNA sequences after transcription from genomic template and are found enigmatically scattered among phyla. It refers to the post-transcriptional modification of mitochondrial and chloroplast and nuclear mRNA. Like RNA splicing RNA editing alters the sequence of an RNA from that encoded in the DNA. But RNA splicing differs from RNA editing in that typically a single RNA splicing reaction removes a large block of contiguous sequence whereas each RNA editing reaction changes only one or two nucleotides. In other words, splicing is a cut and paste mechanism and editing is one of fine tuning. RNA editing refers to any site specific alteration in an RNA.

Table 2.1 Showing various types of editing, their mechanisms and consequences.

	Types of editing	Mechanism	Result	Occurrence
Co- and post transcriptional	Insertion or deletion of U residues	Guide RNA (transesterifications)	Allows generation of sensible RNAs from transcripts lacking features essential for correct translation such as initiation signal or an appropriate reading frame	Mitochondria of Trypanosomes
	Insertion of C, A or U	?	Single C residues are add edat multiple of several transcripts, allowing there constitution of reading frames	Mtochondria of slime mold, Physarum polycephalum
Co-transcriptional or immediately following transcription	Insertion of G residues	Polymerase stuttering	One or more guanylate (G) residues are inserted at a precise location of a transcript, allowing the production of at least two and sometimes three proteins with a common amino terminus and different carboxyl terminus.	Paramyxoviruses (P genes)
	3' terminal A residues addition polyadenylation	Polyadenylation	UAA stops codons are produced by	Mitochondria of vertebrates
	Conversion of C to U	Deamination	Results in tissue specific generation of a stop codon (the only example of change of expression of a nuclear gene)	Apolipoprotein B genes of mammals
	Conversion of Cs to Us or vice versa	Deamination-amination?		Mitochondria of higher plants

Types of editing—There are mechanically diverse types of RNA editing. RNA editing can create, delete or alter the meaning of a codon, create a splice site or alter RNA structure. Some types of editing can repair or correct the information encoded by genome whereas others act to diversify this information offering an organism the potential for greater complexity. It leads to creation of new codons or alteration of reading frames. Editing follows a 3′ to 5′ polarity and either inserts or deletes uracil residues (kinetoplast, mitochondria of trypanosomes), replaces cytosine by uracil (in mitochondria of mosses and higher plants) or transform an adenine to inosine which is read out as guanosine, guided by base pairing between a guide RNA and the edited RNA. Editing takes place predominantly within coding sequence and may create start codon, AUG from ACG as in maize chloroplast and mitochondrial mRNA and new stop codon. RNA editing is thus involved in regulation of gene expression by way of improving base pairing in secondary structures of the mRNA and may be necessary for processing of introns. RNA editing was discovered in mRNA encoded by the kinetoplastid mitochondria of trypanosomes (Benne et al., 1986) and in mammals in nuclear encoded mRNA (Powell et al., 1987). In the mitochondria of trypanosomatid protozoa the precursors of mRNAs (pre-mRNAs) have their coding information remodeled by the site-specific insertion and deletion of uridylate (U) residues. Small *trans*-acting guide RNAs (gRNAs) supply the genetic information for this RNA editing. A U-insertion editing occurs through a series of enzymatic steps which begin with gRNA-directed pre-mRNA cleavage. Inserted U's are derived from free uridine triphosphate and are added to the 3' terminus of a 5' pre-mRNA cleavage product. gRNA specifies edited RNA sequence at the subsequent ligation step by base pairing-mediated juxtaposition of the 3' cleavage product and the 5' cleavage product (Kable et al., 1996).

Mechanism of editing—There are two types of RNA editing in nuclear encoded mRNA and both involve deamination of encoded nucleotides (Gerber and Keller, 2001). The first type involves deamination of cytidine (C) to create uridine (U) and the second type involves deamination of adenosine (A) to create inosine (I). Adenosine deamination is catalyzed by members of an enzyme family called adenosine deaminases which act on RNA (ADARs). ADARs act on RNA which is completely or largely double stranded and catalyze the deamination of adenosine to produce inosine through hydrolytic deamination. Inosine is translated as a guanosine and most enzymes recognize inosine as a guanosine and thus deamination can result in informational recoding. In other words, the ADARs change the primary sequence information. Because inosine base pairs with cytidine, ADARs can change the structure of an RNA by changing an AU base pair to an IU mismatch. Thus ADARs can affect any biological processes which involve sequence- or structure-specific interactions with RNA. ADARs have been definitely shown to alter the meaning of codons, create splice sites and sequester an RNA to the nucleus. The biological consequences of ADAR action at a particular site can vary markedly between genes and between species. A-to-I RNA editing varies between anthropod species. Although the mechanism of gene recoding frequently involves imperfect base pairing of exonic and intronic sequences, the molecular basis of species-specific editing is unknown. Most RNA editing systems are mechanistically diverse, informationally restorative and scatter slots in eukaryotic lineages. In contrast, genetic recoding by A to I RNA editing seems common in animals altering highly conserved or invariant coding positions in proteins. Fruitflies, butterflies and mosquito possess shared and species specific styt I editing sites, all within a single exon. The editing machinery is usually directed to modify particular As by information stored in intron-mediated RNA structures. Complex, multidomain pre-mRNA structures solely determine species-appropriate RNA editing. One of these is a previously unreported long range pseudoknot. Small changes in intron sequences, far removed from an editing site can transfer the species-specificity of editing between RNA substrate. RNA editing sites are thought to arise through step wise addition of structural domains or by short walks through sequence space from ancestral structures (Reenam, 2005).

Creation of splice site—Editing alters an AA dinucleotide to generate AI dinucleotide which is recognized as the canonical AG typically found at 3' splice site. Mammalian ADAR2 acts on its own pre-mRNA to create a new splice sile (Rueter et al., 1999). In theory ADARs could create both 5′ and 3′ splice sites as well as destroy a 3′ splice site.

Function of ADARs — ADARs, the double stranded RNA binding proteins (dsRBPs) target double stranded structures which encompass coding sequences (exons), introns and 5′ and 3′ UTRs.

1. Many of the editing sites within coding sequences alter meaning so that more than one protein isoform can be synthesized from a single gene.
2. The most common site of editing is within non-coding sequences (Morse and Bass, 1999). In one case editing in an intron is known to create a splice site. A hallmark of an ADAR editing site is a genomically encoded A that appears as a G in a cDNA. An exception is the substrate with inosines in noncoding regions.
3. Editing results in fine tuning the myriad of protein-protein interactions. ADARs can create amino acid changes which alter the affinity of interacting proteins.
4. ADARs also act on UTRs of mRNAs. Editing may sometimes regulate the actual levels of an RNA or its translatability. In this way ADAR could alter the amount of a complex by changing the concentration of one of its protein partners.

2.4   EDITING IN NON-CODING RNAs, 5′ UTRS AND 3′ UTRS

Inosines were identified in non-coding sequences rather than in coding sequence. Inosines were found in sequences that were completely non-coding as well as 5 and 3UTRs and introns. Further, inosines were found in stable structures which in many cases contain hundreds of nearly contiguous base pairs and possibly these structures have roles in RNA translation, localization or stability and ADARs serve to modulate these functions. Why a method did not identify inosines in coding sequences is that these sequences are much more rare than those in non-coding sequences. RNAs with inosines clearly exist in human RNAs.

Preferences — The term preferences and selectivity have been used to describe the specificity that allows an ADAR to target a specific adenosine over others. Although ADARs will bind to any double stranded RNA they show slightly sequence preferences for deaminating certain adenosines over others. *In vitro* studies suggest that the enzyme ADAR1 in Xenopus targets adenosines with a 5 neighbor preferences of U=A>C>G and rarely targets adenosines less than 3 nucleotides from the 5 terminus or eight nucleotides from the 3 terminus (Polson and Bass, 1994). Human ADAR1 shows a distinct 5 neighbor preference (U~A>C=G). ADAR2 can deaminate adenosines as close as 3 nucleotides from either termini, if not closer, since other studies show deamination directly at the 5 terminus. Unlike human ADAR1 human ADAR2 also has a 3 neighbor preference (U=G>C=A). These mammalian ADAR1 and ADAR2 have overlapping but distinct preferences. In most cases an AC mismatch was observed to be optimal with an AU base pair.

Preferences for the identity of neighboring nucleotides can not explain the exquisite specificity observed in some ADAR substrate. Additional specificity comes, derives from the structure and thermodynamic stability of the substrate. Thus preferences are dictated by the active site, the discrimination between various pairing partners is determined by the catalytic domain rather than double stranded RBMs. The amino acid sequence similarity between

ADAR (found only in metazoan) catalytic domain and the ADATs, adenosine deaminases which act on tRNAs, suggests that the ADARs evolved from the tRNA-modifying enzymes.

2.5   miRNA EDITING

Many developmental and cellular processes are regulated by miRNA mediated RNAi. After incorporation into the RNAi induced silencing complex, miRNAs guide the RNAi machinery to target genes by forming RNA duplexes resulting in sequence-specific degradation or translational repression. The generation of mature miRNAs requires the processing of primary transcript (pre-miRNAs) and A → I RNA editing occurs to certain pri-miRNAs (Kawahara et al., 2007). Because of the double stranded nature of the pre-miRNA, miRNAs are potential targets for RNA editing. Pre-miRNAs are edited in their seed regions, the regions which determine their target specificity. The edited site is positioned in the middle of the 5′-proximal half 'seed' region of the miRNA. The editing is tissue-specific and changes the potential target range of the miRNAs. The edited miRNAs but not the unedited version, act to repress an enzyme involved in uric-acid synthesis pathway.

2.6   TRANSLATIONAL EDITING

Once an amino acid is attached to a tRNA, the position of the amino acid in a growing polypeptide chain is determined by the interaction between the anticodon of the tRNA and the complementary codon of the mRNA. Errors in protein synthesis is relatively rare (10^{-4}) but errors occur at higher frequency because of limitations to molecular recognition at the two critical points in translation. (i) Certain amino acids are difficult for synthetases to discriminate from each other and for this reason can be misactivated and attached to the tRNA cognate to the synthetase but noncognate to the amino acid (ii). Anticodon-codon interaction on the ribosome is determined by base pairing between the complementary triplets and this interaction is prone to error because, for example, pairing of two rather then three bases although less possible, is still possible.

Translational editing specifically corrects particular errors that occur at a much higher frequency. Most tRNA synthetases can activate amino acids in the absence or presence of tRNA. The activated amino acid then reacts with the 3′end of the tRNA cognate to the amino acid. Thus two steps are involved. The fist step is amino acid activation (E + AA + ATP ↔ E (AA-AMP) PP_i) and the second step is transfer reaction (E (AA-AMP) + tRNA ↔ AA-tRNA + AMP + E) where the aminoacyl moiety is transferred from the adenylate to the 3′end of the tRNA. This transfer reaction is the determining step in amino acylation. Errors in aminoacylation occur because of inherent difficulties in molecular recognition associated with step 1. For example, valine and isoleucine both are branched at the β-carbon and differ by one methylene group and are difficult for isoleucyl-tRNA synthetase to discriminate on the basis of binding interactions alone.

Translational editing by tRNA synthetases requires the presence of the cognate tRNA. This system of proof reading is an example of RNA-dependent amino acid recognition. In case of misactivation of amino acids by synthetases there is extra hydrolysis of ATP. For ribosome dependent system the cost of proof reading is an extra hydrolysis of GTP whereas in case of errors in DNA replication there is hydrolysis of nucleoside triphosphate. Translational editing can be measured as a tRNA^{Ile}-dependent ATPase activity of IleRS where the enzyme discriminates between isoleucine and valine in a reaction that depends on tRNA^{Ile}. Translation editing reactions share in common the use of nucleoside triphosphate hydrolysis to correct and prevent errors in protein synthesis. In both the RNA-dependent amino acid discrimination by tRNA synthetases and the mechanism for codon-anticodon fidelity on the ribosome, RNA plays a critical role as an effector in the editing response. The mechanism for the effector system is not known but the tRNA synthetase system needs specific nucleotides. But how these nucleotides (at the corner of the L-shaped structure) actually affect the editing response is not clear. In case of ribosomal system the slight difference, for example, in a perfect three out-of-three match vs an imperfect two out-of-three is sensed. Whether specific nucleotides (other than anticodon) in a particular part of the tRNA structure have a role in the effector response is also a question needing further investigation (Schimmel, 1999).

2.7 DNA EDITING

It refers to recombination process of the immune system genes which involves precise recognition of recombination signal sequences on the DNA at splicing sites, production of double strand cuts and ligation of previously non-adjacent sequences in new combinations. The intervening DNA is cutout as a ring. DNA splicing thus leads to gene arrangements. The recombinants generated encode receptor sequences having potential to recognize millions of different antigens.

3

Genetic Code

3.1 DEFINITION AND PROPERTIES OF GENETIC CODE

The answer for the question "How the genetic information encoded in the 4 letter (A,T (or U), C and G) language could be translated into 20-letter language of protein?" became clear by 1960s with the discovery of co-linearity of protein and gene structure (i.e., the linear sequence of the amino acid (AA) in protein is specified by the linear sequence of nucleotides in the nucleic acid) (experimental evidence for the co-linearity was provided by Yanofsky and co-workers in 1964) and that there are 20 amino acids led to further study of determining how the four nucleotides of DNA could be translated into 20 different amino acids of protein. If one nucleotide determines one AA then there would be formation of only four AAs. If a sequence of two nucleotides encodes one AA then it will form only 16 AAs as there will be only $4^2=16$ combinations (sequences) of two nucleotides and if a sequence of three nucleotides encode an AA then it could specify 64 AAs as there will be $4^3=64$ ways in which a combination of three of the four nucleotides can be arranged and thus in the simplest case a group of three nucleotides can specify one AA. This group of three nucleotides (or bases) is called a **code** or **codon** (or word) and thus the code is triplet (word of three letters or nucleotides) and there are 64 codons (or words). The genetic information in the DNA or RNA is in the form of code words or codon. F.G. C. Crick and co-workers in 1961 provided the evidence for the triplet code. The standard genetic code showing which codons specify which amino acids is given in the Table 3.1. The codon AUG specifies methionone and it also serves as the **start codon** for polypeptide synthesis in eukaryotes but in prokaryotes AUG, GUG function as start codon and any of the three codons, UAA,UAG, or UGA specify the end of polepeptide synthesis. The table shows that amino acids, methionine and tryptophan are specified by single codons, nine amino acids have two codons, amino acid, Ile has three codons, five amino acids (Ala, Gly, Pro, Thr, Val) have four codons and three amino acids, Leu, Ser and Arg have six codons and thus these 20 AAs use 61 codons. Thus it can be said that all amino acids except methionine and tryptophan have more than one codon. The third base of each codon plays a lesser role in specifying an amino acid than the first two (5′ and middle base) and in most cases codons that code for the same AA differ only at the third position. Besides being triplet the other properties of the codon are:

1. **degeneracy**
2. **non-overlapping**
3. **universality**

As there are 64 codons and there are only 20AAs to be formed, it was hypothesized that more than one codon could specify a specific AA and this is called degeneracy (redundancy) of genetic code and is shown in Table 3.1. For example, both GAU and GAC specify aspartic acid. A specific first codon (initiator codon, which initiates protein synthesis) in the sequence of nucleotides determining a polypeptide establishes the reading frame in which new codon occurs after every three nucleotide residues. There is no comma between codons for successive amino acid residues. The last codon of the reading frame is a termination codon. The **initiator codon**, AUG starts the polypeptide synthesis whereas the termination codons (also called stop or non-sense codons) terminates the peptide synthesis. The three **termination codons** are UAA, UAG and UGA. These terminator codons were first called 'nonsense' triplets as they did not code for an amino acid. In a random sequence of nucleotides termination codon would be expected to occur every 50 to 100 bases but the distribution of these termination codons is not random. UAA is the most frequently found termination codon and UAG is highly disfavored in prokaryotes. Non-overlapping means codons do not share nucleotides. Had there been overlapping of codons then number of codons would have been reduced to about twenty. Further, considering lysine and phenylalanine the overlapping could have produced protein in which lysine could follow phenylalanine or vice- versa and further a change in single nucleotide of a triplet could have produced changes in three consecutive amino acids of a protein and all these have not been observed and thus point to the non-overlapping nature of the codons. With a reading frame made up of six nucleotides would specify 4 AAs if codons are overlapping but with non-overlapping only two amino acids will be specified. The coding region (made up of exons) provides information for appropriate amino acid sequence and is also referred to as open reading frame (ORF).Reading frame refers to the system of reading in threes that defines where codons begin and end. A reading frame without a termination codon among 50 or more codons is called an ORF. The reading frame thus consists of contiguous and non-overlapping codons. Beyond AUG, the initiating codon the coding sequence or ORF continues until it ends with one of the three terminator codons. The number of AA coded is equal to the number of codons whereas in case of overlapping number of amino acids specified is more than the number of codons. There is no intervening nucleotide (spacer) between the codons. The codon is universal in nature in the sense that a code specifying a particular AA in bacterial protein can also specify the same amino acid in human protein. The gene code thus appears to be essentially the same in all organisms. For example, UUU codes for phenylalanine in bacteria, mouse, man and plant. The standard genetic code words are universal in all species with minor deviations in mitochondria and a few single-celled organisms. The codon is ambiguous in that there is one codon which can specify more than one amino acid, e.g. GGA specifies glycine and glutamic acid. The genetic code is defined by two elements:

1. Antocodons on tRNA which determines where an amino acid is placed in a group of polypeptide.

2. The specificity of the enzymes-amino acyl tRNA synthetases (that charge the tRNAs) which determine the identity of the AA attached to a given tRNA. Codon on mRNA and anticodon of tRNA pair in antiparallel orientation (see chapter 4). The third position in each codon is much less specific than the first and second and is said to be **wobble**. The termination codons do not generally occur within the genes. Li et al. (1985) defined different types of degenerate sites according to the synonymous codons in the standard genetic code shown in Table 3.1. The codons for eight amino acids can have any nucleotide (U,C,A, or G) in their third position which is called a **fourfold degenerate site**. The codons for seven amino acids can terminate in either U or C and those for five amino acids can terminate in either A or G; and these third position sites are called **twofold degenerate sites**. Three of amino acids such as Leu, Ser and Arg are encoded by two sets of triplets each, with two different doublets in the first and second positions. Only two residues, Met and Trp are each encoded by a single, nondegenerate triplet. A site is nondegenerate if any nucleotide substitution changes the amino acid encoded. Because of codon degeneracy nucleotides in a coding sequence can change without affecting the amino acid sequence. These changes are called **synonymous substitutions**. On the other hand, nucleotide substitutions that do change amino acids are called **nonsynonymous substitutions or amino acid replacements** (Hartl, 2000). Among the nonsynonymous substitutions there is great variation in the rate of nonsynonymous

Table 3.1   Showing the standard genetic code.

First nucleotide codon (at 5' end)	Second nucleotide in codon				Third nucleotide incodon (at5' end)
	U	C	A	G	
U	UUU Phe/F (GAA)	UCU Ser/S (IGA)	UAU Tyr/Y (GUA)	UGU Cys/C (GCA)	U
	UUC Phe/F (GAA)	UCC Ser/S (IGA)	UAC Tyr/Y)GUA)	UGC Cys/C (GCA)	C
	UUA Leu/L (UAG)	UCA ser/S (UGA)	UAA Stop	UGA Stop	A
	UUG Leu/L (CAA)	UCG ser/S (CGA)	UAG Stop	UGG Trip/W (CCA)	G
C	CUU Leu/L (IAG)	CCU Pro/P (ICG)	CAU His/H (GUG)	CGU Arg/R (ICG)	U
	CUC Leu/L (IAG)	CCC Pro/P (ICG)	CAC His/H (GUG)	CGC Arg/R (ICG)	C
	CUA Leu/L (UAG)	CCA Pro/P (UGG)	CAA Gln/Q (UUG)	CGA Arg/R ((UCG)	A
	CUG Leu/L (CAG)	CCG Pro/P (CGG)	CAG Gln/Q (CUG)	CGC Arg/R (CCG)	G
A	AUU Ile/I (IAU)	ACU Thr/T (IGU)	AAU Asn/N (GUU)	AGU Ser/S (GCU)	U
	AUC Ile/I (IAU)	ACC Thr/T (IGU)	ACC Asn/N (GUU)	AGC Ser/S (GCU)	A
	AUA Ile/I (UAU)	ACC Thr/T	AAA Lys/K (UUU)	AGA Arg/R (UCU)	C
	AUG Met/M (CAU)	ACG Thr/T	AAG Lys/K (CUU)	AGG Arg/R (CCU)	G
G	GUU Val/V (IAC)	GCU Ala/A (IGC)	GAU Asp/D (GUC)	GGU Gly/G (GGC)	U
	GUC Val/V (IAC)	GCC Ala/A (IGC)	GACAsp/D (GUC)	GGC Gly/G (GGC)	C
	GUA Val/V (UAC)	GCA Ala/A (UGC)	GAA Glu/E (UUC)	GGA Gly/G (UCC)	A
	GUG Val/V (CAC)	GCG Ala/A	GAG Glu/E (CUC)	GGG Gly/G (CCC)	G

*Anticodons are given in the brackets. This table shows that many tRNAs can serve two different codons, usually because G or I in the first position of their anti-codons will pair with U as well as C in the third codon position.

substitution from one gene to the next. The large variation is in rates is due to selective constraints on amino acid replacement that do not operate as strongly on synonymous nucleotide substitutions. There is a tendency for certain amino acids to be conserved at particular sites because of natural selection for optimal function (selection constraint). Not just any amino acid will serve at a particular position in a protein molecule because each amino acid must participate in the chemical interactions that fold the peptide chain into its three-dimensional shape and give the molecule its specificity and ability to function. Thus the need for proper interactions and folding constrains the acceptable amino acids that can occupy each position. The selection of synonymous codons is distinctly non-random.

3.2 CODON USAGE BIAS

Codon bias refers to the preference with which a specific organism uses a particular codon for a particular amino acid. Since there are only 20 amino acids but there are as many as 64 different codons, a given amino acid can be specified by more than one codon. The preference to use a specific codon is different from organism to organism and might be an obstacle towards efficient expression of foreign genes. Furthermore, within one species codon usage in strongly expressed genes is different from that in weakly expressed genes. In highly expressed genes of corn (*Zea mays*) codon usage is biased towards codons ending in C or G. The synonymous substitutions occupy a relatively narrow range of rates but some synonymous substitutions are also constrained. One type of constraints occurs through secondary structures that form in some RNAs due to internal base pairing (Parsch et al. 1998). A second type of constraint occurs through synonymous codon preferences which are correlated with the relative abundance of tRNA molecules used to translate the codons. Thus bias in codon usage has two components: correlation with tRNA availability in the cell and non-random choices between pyrimidine-ending codons. In bacteria and yeast, highly expressed proteins tend to use synonymous codons for tRNA molecules that are abundant in the cell whereas proteins produced in small amounts do not show such codon usage bias (Ikemura, 1985; Eyre-Walker, 1996). Natural selection also acts to optimize codon usage in such organisms as Drosophila (Akashi,1993,1995) but in warm-blooded vertebrates codon usage bias is confounded with large scale differences in base composition (G+C content) and patterns of nucleotide substitution across different regions of the genome (Bernardi, 1995). The bias in codon usage even extends to the stop codons. UAA is favored in genes expressed at high levels whereas UAG, and UGA are used more frequently in genes expressed at a lower level. Codon bias is thought to enhance the efficiency and/or accuracy of translation.

3.3 CODON USAGE FREQUENCY

Organisms are highly selective in the particular codons they use. In other words there is considerable variability in selection of codons that different organisms employ for a particular amino acid. For example, there are six possible codons for amino acid, serine and theoretically all should be used with equal frequency whenever a serine is specified in a CDS (also known as exons) but in practice there are characteristic differences in the use of particular codons as shown in Table 3.2 given below (Attwood and Parry Smith, 1999).

Table 3.2 Percentage use of codons for serine in a variety of organisms.

Codon	E. coli	D. melanogaster	H. sapiens	Z. mays	S. cerevisiae
AGT	3	1	10	3	5
AGC	20	23	34	30	4
TGC	4	17	9	22	1
TCA	2	2	5	4	6
TCT	34	9	13	4	52
TCC	37	48	28	37	33

Further, codon frequencies vary widely from organism to organism.Codon frequency is expressed as codons used per 1000 codons encountered. Amongst the three codons for arginine (AGG, AGA,CGA) the frequency of AGG is highest in *Closeridium pasteurianum* (32.8), *Homo sapiens* (21.3), *S. cerevisiae* (21.3), *Arabidposis thaliana* (18.4) and lowest in E.coli (1.4) whereas *Thermus aquaticus* has the highest frequency og AGA (13.7) followed by *A. thaliana* (10.9). *Xenopus laevis* has highest CGA (7.6).

Codon adaptation index—It provides a relative measure for the frequency of use of all codons in a gene. Generally alternative synonymous codons for any amino acid are not used randomly but rather in a species-specific preference bias (codon preference bias). The CAI permits to calculate the frequency of use of all codons in a coding sequence and to predict the efficiency of the expression of this sequence. Genes with a high CAI are highly expressed whereas genes with low CAI are either expressed at a low rate or are simply pseudogenes.

Codon usage efficiency—Whether mRNA requires tRNAs with prevalent or rare anticodons affect the translation. The levels of tRNA species within cells can be different and results in different translation rate. Further CCU, CCC and CCA are the preferred codons in humans for proline whereas CCG is preferred for proline in E. coli.

3.4 VARIATION IN GENETIC CODE

One would expect no variation in the genetic code as even a single AA substitution can have profoundly deleterious effects on the structure of protein but variations in code do occur particularly in small genomes encoding few proteins. The genetic code is nearly universal with the exception of a few minor variations in mitochondria, some bacteria and some single celled eukaryotes (Table 3.3). Most of the variation in the genetic code occurs in the mitochondria (which encode only 10 to 20 proteins) but the changes are restricted to termination codons, UGA specifies not termination but tryptophan. ATA is a methionine codon and AGA and AGG are not used as codons at all. Further, in ciliated protozoa UAA and UAG code for glutamine rather than terminating signal whereas in case of prokaryotes, Mycoplasma capricolum these codons are chain termination but UGA codes for tryptophan. The usage of 'synonymous codon' is definitely non-random and differs between prokaryotes and eukaryotes, viruses and their hosts. Further, codon usage for different proteins of the organism may be different. Codons need not always encode the same AA. GUG which encodes valine is used occasionally as initiating codon by *E.coli* and UGA is a termination codon but it is sometimes used for seleno cystein in *E. coli*. Overlapping genes in different reading frames are found in some viral DNAs.

Table 3.3 Deviations of some mitochondrial genetic codes from the standard code.

Codon	Standard code	Vertebrate	Plant	Yeast	Drosophila
CUU	Leu	n.c	n.c	Thr	n.c
CUC	Leu	n.c	n.c	Thr	n.c
CUA	Leu	n.c	n.c	Thr	n.c
CUG	Leu	n.c	n.c	Thr	n.c
AUU	Ile	n.c	n.c	Ile	Ile
AUC	Ile	n.c	n.c	Ile	Ile
AUA	Ile	Met	n.c	Met	Met
AUG	Met	Met	n.c	Met	Met
UGA	Stop	Trp	n.c	Trp	Trp
UGG	Trp	Trp	n.c	Trp	Trp
CGU	Arg	n.c	Arg	n.c	n.c
CGC	Arg	n.c	Arg	n.c	n.c
CGA	Arg	n.c	Arg	n.c	n.c
CGG	Arg	n.c	Trp	n.c	n.c
AGU	Ser	Ser	n.c	n.c	Ser
AGC	Ser	Ser	n.c	n.c	Ser
AGA	Arg	Stop	n.c	n.c	Ser
AGG	Arg	Stop	n.c	n.c	Ser

In Rous sarcoma virus, the genes 'gag and 'pol' are overlapping. The reading frame for pol gene is shifted to the left by one base pair relative to the reading frame for gag. The wobble rules of prokaryotes and eukaryotes are not followed by mt tRNAs and the groups of four synonymous codons are recognized by a single tRNA (Table 3.4).

3.5 ASSIGNMENT OF AMINO ACID CODONS

The problems of assigning triplets of bases to each of the 20 AAs were solved by,

    I. use of biosynthetic messengers

   II. use of ribosome-binding technique and

  III. use of mutations.

Since RNA molecules of any base composition can be made, mixtures of nucleotides were used to synthesize messeges. Synthetic polymer was added to cell-free extract of *E.coli* with a mixture of 20 AAs and the proportions of different AAs in the polypeptide were recorded (Nirenberg and Matthaei, 1961). With a mixture containing U and A in the ratio of 5U:1A and assuming random distribution of bases in the polynucleotide, the statistical probability of different codons could be calculated as UUU (5x5x5=125), UUA=UAU=AUU (5x5x1=25) AAU=AUA=UAA (5x1x1=5) and AAA (1x1x1=1) and correlated with the relative incorporation of different amino acids in the polypeptide synthesized. With this mixture of 5U:1A the proportions of phenylalanine: tyrosine: asparagine: lysine were 125:25:5:1 in the polypeptide which indicates that the code for phenylalanine could be UUU, for tyrosine either UUA, UAU, or AUU while for asparagines, either AAU, AUA, or UAA and for lysine, AAA but

Table 3.4 Showing observed pattern of base pairing at wobble position in mitochondria.

Original wobble position			Observed base pairing pattern at wobble position in mitochondria		
5′-antocodon base	3′-codon bases read	Observed pattern of base pairing in the wobble position	5′-antocodon base	3′-codon bases read	Remarks
A	U	A never found as 5′-antocodon base	C	G	Normal
C	G	Occurred as predicted	C	A,G	Unique to mt tRNA$_m^{Met}$
G	U, G	Wobbles occurred as predicted. Also when G is replaced by G$_m$ or Q	G	U,C	Normal
I	U,C,A	Wobble occurred as predicted	U	U,C,A,G	Unique
U	A,G	U never found as 5′ anticodon base	U*	A,G	Unique

U* indicates an unknown modification of U; A, as in other rRNA, is never found in the wobble position;

then this approach could not specify the order of nucleotides in the triplet. This technique established the AA assignments to codons. Khorana and his colleagues synthesized RNA of defined sequences. Nirenberg and Leder used specific codon- antocodon interaction to identify the AA corresponding to a triplet. They allowed the mixture of ribosome, a acyl tRNA and a triplet to react under suitable condition and was then poured on to a nitrocellulose filter which retained the ribosome and any bound tRNA. By using a series of 20 different AAs, each containing one radioactively labeled amino acid they could identify the amino acid by identifying amino acyltRNA-triplet retained on the filter. Confirmatory evidence came from the study of genetic mutations. When TMV made up of RNA was treated with HNO2, C changed to U and A changed to hypoxanthin (equivalent to G) and thus the mutant TMV produced protein in which the single AAs were replaced by different amino acids at certain positions which were correlated with the changes in C and A.

3.6 READING FRAME

It refers to coding sequence determined by reading nucleotides in groups of three starting at a specific start codon (AUG).

5′-AUGACGACAAAAGGGCUGCACUGA-3′

5′-AUG ACG ACA AAA GGG CUC CAC UGA-3′

Reading frame shift refers to an alteration of the reading frame caused by the insertion or deletion of nucleotides in a number not divisible by 3. Reading frame shift leads to the synthesis of a missense mRNA which usually results in non-functional protein. **'in frame'** means the proper reading frame. For example, if two genes are ligated 'in frame' then they can be translated into a single mRNA and further translated into a single fusion protein.

4

Function of DNA

4.1 FUNCTIONS OF DNA

DNA has got two very important functions to perform: 1. Replication 2. Protein synthesis.

4.2 REPLICATION

Replication occurs within a designated time period in the cell cycle (see Roy, 2009). The classic experiment of Meselson and Stahl (1958) using the density gradient technique showed that the mode of replication is **semi-conservative** (Figure 4.1 A). Each strand acts as a **template** for a new daughter strand. **Template** refers to a pattern or mold. DNA carries the genetic information and serves as a model or template from which information is taken by messenger RNA. Either strand can serve as the template. Which strand becomes the template depends on a combination of transcription initiation and termination signals such as promoters and enhancers sequences that are present in the DNA. Replication starts at a DNA sequence called the origin and usually proceeds in both directions. The DNA is synthesized in the 5′-3′ direction by an enzyme called DNA polymerases. There are three types of DNA polymerases.

1. DNA polymerase I
2. DNA polymerase II
3. DNA polymerase III.

DNA polymerase I has special function during replication, recombination and repair whereas **DNA polymerase III** is the primary enzyme involved in the polymerization of nucleotides and it also has 3′-5′ property and thus is engaged in proof reading (error correction) activity and removes mismatches. The mismatch refers to incorrect positioning of bases (e.g., a T opposite C or A = C, C is a rare tautomeric form of cytosine that pairs with A) and 3′-5′ exonuclease activity of DNA polymerase I removes this nucleotide C. Poly III is the primary enzyme involved in replication in bacteria. DNA polymerase I is involved in removing the RNA primers (and thus has got 5′-3′ exonuclease property) and replace them with new DNA strands (an thus also has got 5′-3′ polymerase property). It is also involved in DNA repair. **DNA polymerase II** is also involved in DNA repair. It has got 3′-5′ exonuclease property. 3′-5′ exonuclease activity is essential for proof reading and editing (where it permits the enzyme to remove a newly added nucleotide) whereas 5′-3′ exonuclease activity helps in removal of RNA primers from the DNA as well as removal of DNA segments damaged by U.V. light and other mutagenic agents in a process called 'nick translation'. U.V. light causes the formation of pyrimidine dimmers which block the replication. There are three types of pyrimidine dimmers of which thymine-thymine dimmer is the most important. The mismatch 3′ OH end of the growing strand blocks further elongation Thus all DNA polymerases have 3′-5′ exonuclease (proof reading) activity but only Poly I has 3′-5′ exonuclease activity. First of all, the DNA double helix unwinds and the separated strands are quickly replicated and thus there appears a loop like structure. One or both ends of the loops are dynamic points called replication forks. Replication loop always initiates at a unique point called **origin** and the origins are found in A-T rich regions. The regions that denature are highly reproducible and are rich in A = T base pairs. Initiation of replication requires binding of DNA A protein to the origin. Initiation is the only phase of replication which is being regulated. The enzyme, DNA helicase and DNA gyrase catalyses and facilitates, respectively the unwinding of the DNA. The DNA strands are relieved by topoisomerases and single stranded DNA binding proteins stabilize the separated strands. The Dna B and Dna C proteins are required to set up two replication forks. Then there is synthesis of RNA primer. The enzyme primase catalyzes the synthesis of RNA primer. A **primer** is a strand segment (complementary to the template) with a free 3′-OH group to which a nucleotide can be added. Primers are often oligonucleotides of RNA. Priming is the process of initiation of the synthesis of DNA strand by the formation of an RNA primer or by self priming (discussed in chapter). **Primosome** is an enzyme complex that synthesizes the primers required for lagging DNA strand synthesis. DnaB helicase(which unwinds the DNA) and DnaG primase (that synthesizes short RNA primer) constitute a functional unit within primosome. The leading strand (5′-3′) using the template or + strand (3′-5′ strand of DNA) is synthesized continuously whereas the lagging strand (3′-5′) using the template or –strand (5′-3′ strand of DNA) is synthesized discontinuously as Okazaki fragments which subsequently join by the enzyme DNA ligase (Figure 4.1B). Initiation of Okazaki fragments may occur at particular sites on the lagging strand. Okazaki fragments are 150 to 200 nucleotides in length in eukaryotes whereas it is ~1000 to 2000 nucleotides long in prokaryotes. The fidelity of DNA is maintained by (i). base selection by the polymerase (ii). a 3′-5′ proof reading exonucleaese activity that is a part of most of DNA polymerase and (iii). specific repair systems for the mismatches occurred during replication. In eukaryotes replication is initiated at many points together (**multiple replication origins**). In other words, DNA replication starts at specific origins to which a pre-replicative complex binds (Bell and Datta, 2002). Chromosomal DNA replication is of the order of 100-1000 per chromosome with separate replication 'bubbles' eventually merging. Nevertheless, a consensus sequence for origins has yet to be identified suggesting that there are epigenetic influences.

In case of prokaryote, plasmids and eukaryotic mitochondria and chloroplasts organelles the DNA is generally in the form of a closed loop and replication takes place in both directions from a single origin with divergent replication forks. The initial step in duplication involves unwinding of a limited region of the double stranded DNA to form a small single stranded DNA bubble. In bacteria, archea and lower eukaryotes the unwinding process begins at a sequence called the **replication origin** which contains several conserved binding

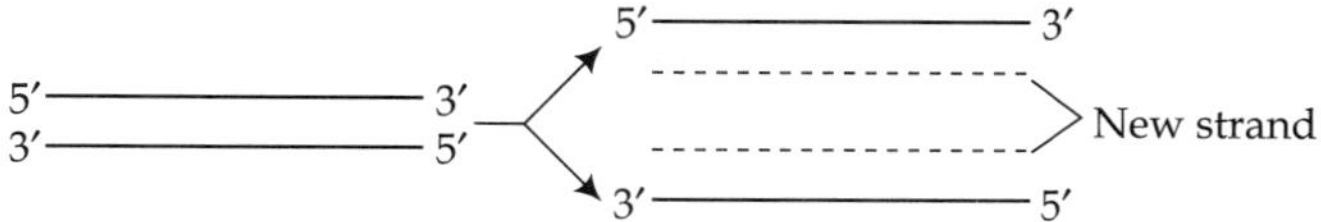

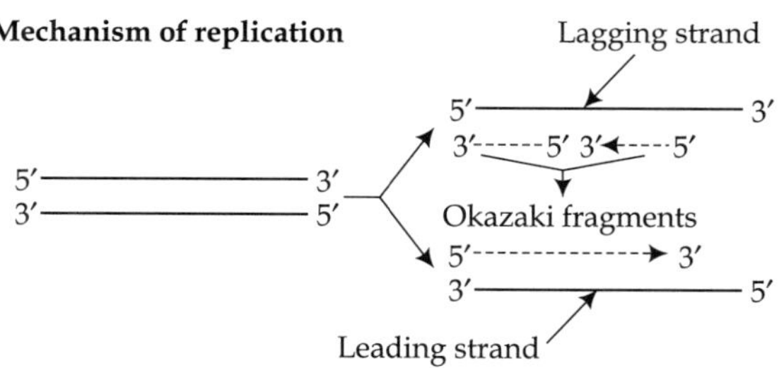

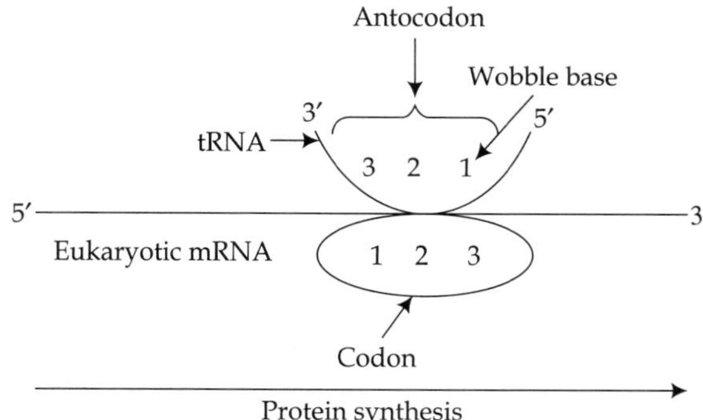

FIGURE 4.1   Showing (A) mode of replication; (B) mechanism of replication and (C) wobble base and the direction of protein synthesis.

sites for a protein called initiator and an A and T-rich unwinding elements. Binding of the initiator results in a nucleoprotein complex which melts DNA forming the DNA bubble onto which the replication machinery assembles. In other words, once bound to sites within a replication origin, the initiator proteins oligomerize and use ATP to separate DNA strands in a nearby region which is enriched with A and T nucleotides (Georgescue and O'Donnell, 2007). Like eukaryotes there are several types of DNA polymerases in prokaryotes. DNA polymerases, α, β, γ, δ and ε. DNA polymerase α appears to be involved in the synthesis of RNA primer for Okazaki fragments. DNA polymerase δ is involved in the polymerase activity. Further it has got 3'-5' exonuclease activity and thus it is involved in proof reading. DNA polymerase ε is involved in DNA repair. It has got 5'-3' exonuclease activity and thus involved in removing primers of Okazaki fragments. At intervals primase synthesizes an RNA primer for a new Okazaki fragment which is extended by DNA polymerase III. Synthesis of Okazaki fragment continues until it meets the primer of the previously added

O.fragment. RNA primers in the lagging strand is removed by 5'-3' exonuclease activity of DNA polymerase I and replaced with DNA by the same enzyme and the nick is sealed by DNA ligase.

Rolling circle mechanism — It is a mechanism for replication of circular, double stranded DNA molecules (e.g. of bacterial DNA, of λ phage) which generates concatamers of duplex molecules (Gilbert and Dressler, 1968). In this mechanism of replication, the 'negative' strand remains a closed circle. The 'positive' strand is nicked which exposes a free 3'OH end for strand extension by DNA polymerase. The original positive strand is peeled off from the rolling circle as a single strand tail. The newly generated strand displaces the original parental strand and polymerized on the positive single strand tail. After several cycles, a long DNA strand containing several units of genome is synthesized. The displaced strand then serves as template for the synthesis of a complementary strand.

Termination of replication — The termination of replication involves the synthesis of a structure called telomere at the end of chromosome. This is achieved by enzyme called telomerase which has got reverse transcriptase property and is ~ 150bp in length. Somatic cells lack telomerase whereas the germ cells have telomerase. Telomerase is composed of RNA and protein and contains about 1.5 copies of the appropriate C_yA_x repeats which act as template for the synthesis of T_xG_y strand of the telomere. The TG strand is longer than its complement leaving a region of single stranded DNA of up to a few hundred nucleotides at the 3' end. Beyond the end of a linear DNA molecule no template is available for the pairing of an RNA primer but the DNA replication requires a template and primer. Telomerase acts like reverse transcriptase that contains internal RNA template and thus provides the active site for RNA dependent DNA synthesis. The 3' end of the chromosome is used as primer which provides free hydroxyl group at the 3' end to which a new nucleotide unit is added. The complementary strand C_YA_X appears to be synthesized by the DNA polymerase starting with an RNA primer. The single stranded tail synthesized by the telomerase is folded back and paired with its component in the duplex portion of the telomere. In prokaryotes in which the telomere is of less than a few hundred base pairs the single stranded region is protected by specific binding protein whereas in case of eukaryotes where the telomere may be thousands of base pairs long the single stranded end is sequestrated in a specialized structure called a **T loop**. The formation of a T loop involves invasion of the 3' end of the telomere's single strand into the duplex DNA (Figure 4.2). In other words, the single stranded end is folded back and pair with its complement in the double stranded portion of the telomere. Thus there are four steps involved in the synthesis of telomere:

1. Binding of the internal template RNA of telomerase to and base pairing with the DNA's TG primer (TxGy)
2. Adding of more T and G residues to the TG primer by telomerase
3. Repositioning of internal template RNA to allow addition of more T and G residues
4. Folding back of the single stranded tail and pairing with its complement in the duplex of the telomere.

4.2.1 Mismatch Repair

Mismatch refers to the occurrence of incorrectly paired (mismatched) bases in DNA duplex molecules. Such mismatches arise by errors of replication and/or repair systems and are

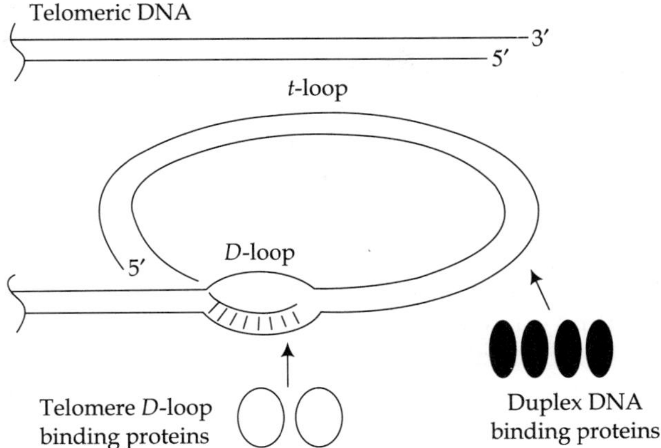

FIGURE 4.2 Showing of a telomere. The DNA with the 3′ overhang of the G strand is shown at the top. The double-stranded part of telomere DNA loops back on itself, forming a lariat structure. The 3′ G strand extension invades the duplex telomeric repeats and forms a D loop (displacement loop). Duplex telomere-binding proteins bind along the length of the telomere repeats, and a specialized telomere-binding protein binds the D loop at the junction of the lariat. Modified from Greider (1999).

sources of mutations. Mismatch repair refers to the detection and replacement of incorrectly paired(mismatched) bases in newly synthesized DNA. Replication itself can very occasionally damage the information content in the DNA when error introduces mismatched base pairs such as G paired with T. *Copy error* refers to an error in the DNA replication process giving rise to a gene mutation. In E. coli, a mismatch repair system encoded by the genes mut H, mut L, mut S, uvr D and uvrE, screens the newly synthesized DNA strand for the mismatched bases. These mispaired bases and a short region surrounding them are excised and then replaced by the DNA polymerase. This repair mechanism acts before the newly replicated DNA is methylated. Only after its completion the *de novo* synthesized strand is modified, e.g. by dam methylase according to the methylation pattern of the complementary strand (-maintenance methylation). Mismatch repair in *E.coli* is directed by transient nonmethylation of 5′ GATC sequences on the newly synthesized strand.

Besides replication error the DNA can be damaged by a variety of processes, some spontaneous whereas others are catalyzed by environmental agents. These processes will lead to the changes in DNA sequence and thus there is a need to repair the DNA in order to maintain the integrity of the information in DNA.

4.2.2 Base-Excision Repair Systems

They recognize and repair damage caused by environmental agents such as radiation, alkylating agents and spontaneous reactions of the nucleotides. The DNA repair enzymes find aberrant nucleotides among the myriad of normal ones. Information in double helix is preserved by DNA repair enzyme. A specialized pathway called base excision repair plays a role in rectifying base damage (Barnes and Lindahl, 2004). The damaged base is removed by BER glycosylases, followed by excision of the remaining sugar fragment and installation of an undamaged nucleotide by a DNA polymerase. The type of damage includes abnormal bases (uracil, hypoxanthine, xanthine), alkylated bases and in some organisms pyrimidine dimmers. Some repair systems recognize and excise only damaged or incorrect bases leaving an AP (abasic) site in the DNA. This is repaired by excision and replacement of the DNA segment containing the AP site. The enzymes involved in this repair system are DNA glycosylases, AP endonucleases besides DNA polymerase and ligase.

4.2.3 Nucelotide-Excision Repair Systems

They recognize and remove a variety of bulky lesions and pyrimidine dimmers. They excise a segment of the DNA strand including the lesion, leaving a gap which is filled by DNA polymerase I and DNA ligase reactions.

4.2.4 Direct Repair

Some DNA damage is repaired by direct reversal of the reaction causing the damage. For example, pyrimidine dimmers are directly converted into monomeric pyrimidines by a DNA photolyase and the methyl group of O^6-methylguanine is removed by a O^6-methylguanine-DNA methyltransferase. In bacteria, error-prone translesion DNA synthesis (TLS) involving TLS DNA polymerase occurs in response to very heavy DNA damage.

In eukaryotes the initial recognition of mismatches is accomplished by a complex of two proteins MSH2 and MSH6 which binds to mismatched bases. A second complex consisting of MLH1 and PMS1 proteins then joins the mismatch-MSH2/6 complex and catalyzes excision and repair of the mismatch.

Tsukuda et al. (2005) found that the Mre 11-Rad50-Xrs 2DNA-repair complex and INO80 chromatin remodeling complex are rapidly recruited to breaks and evict nucleosomes in the immediate vicinity of the break. Meanwhile, a histone H2AX in a large region around the break is phosphorylated forming γH2AX. Exonucleases chew back one strand of the DNA, forming a single stranded overhang. Before completion of repair by recombination, an unknown remodeling complex mediates the removal of γH2AX from 50kbs of chromatin. Koeghl et al.(2006) showed that subsequently γH2AX is dephosphorylated by a protein complex containing the enzyme Pph3, an event that is crucial for release from damage check point. It is unclear whether entire nucleosomes or just certain histones (γH2AX/H2B dimmers) are displaced (Nussenweig and Paull, 2006).

4.2.5 The Central Dogma

The central dogma describes the fundamental principles of molecular biology (Francis Crick, 1958). It states that the transfer of information is from nucleic acid to nucleic acid or from nucleic acid to protein. The central dogma states that DNA is transcribed into RNA which is then translated into protein. Figure 4.3 shows the central principle describing the functional relationship among DNA, RNA and protein. It shows that DNA is not the direct template for protein synthesis. In other words, the genetic information present in the DNA is transcribed (or copied) into mRNA which acts as template for protein synthesis, i.e., the genetic message encoded in mRNA is translated into polypeptide on the ribosome. That strand of DNA which is used for genetic transcription is called sense strand. The transcription process is catalyzed by DNA dependent RNA polymerase. The arrow encircling DNA shows that it also acts as

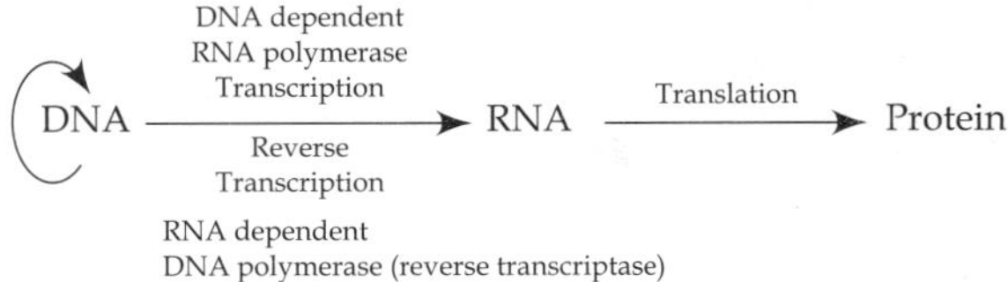

FIGURE 4.3 Central dogma.

template for its self-replication, i.e., the parental DNA is also copied to form daughter DNA with identical nucleotide sequences. Similarly, RNA can also act as template for its own replication. The back arrows (←) indicates that the genetic information in RNA is also transcribed in the form of DNA through a process called reverse transcription and it is catalysed by RNA dependent DNA polymerase (reverse transcriptase). The arrow(→) shows the direction of transfer of genetic information.

4.2.6   Synthesis of RNA

RNA is produced from DNA by a process called **transcription** which is catalyzed by DNA dependent RNA polymerase. Prokaryote possesses a single RNA polymerase to synthesize all RNAs (tRNA, mRNA and rRNA). The several steps involved in transcription are: I. Binding of RNA polymerase to promoter II. Initiation of transcript synthesis III. elongation and IV. Termination. Eukaryotes have three types of RNA polymerase (see chapter 1 for detail). Binding of RNA polymerase II requires activity of a large number of proteins called transcriptional factors, TFs. Transcriptional factors stops at sequences called terminator. Different parts of chromosomes are transcribed to form rRNAs, tRNAs and mRNAs (see chapter 1 for detail). rRNAs and tRNAs are derived from longer precursor RNAs that are trimmed by nucleases. Some bases are also modified enzymatically during maturation. The eukaryotic mRNAs are also made from longer precursors. The primary transcripts often contain introns which are removed by splicing. During transcription, RNA polymerase catalyzes the addition of nucleotides to the growing RNA chain. High-resolution structural snapshots show that the polymerase first identifies the substrate and then incorporates it (Cramer, 2007; Vassylycv, 2007).

4.2.7   Prokaryotic RNA Polymerase

RNA polymerase consists of five subunits, 2α, β_1, β and σ. 2α, β_1, β constitute the core enzyme, 2α subunits maintain the structural conformation and β_1 and β' form catalytic centre. And sigma associates with the core to form the holoenzyme and confers promoter specificity and allows accurate initiation of transcription. Accurate initiation from a prokaryotic promoter requires RNA polymerase holoenzyme. Core polymerase forms non-productive complexes with either non-specific or promoter DNA sequences. The sigma factor is a subunit of DNA-dependent RNA polymerase of *E.coli* which recognizes the Pribnow box in the promoter regions of genes and thereby directs the core enzyme (consisting of 2α, β_1, β' and ω subunits) to initiate transcription correctly. Once transcription has started the sigma subunit leaves the holoenzyme complex and may serve another core enzyme to initiate transcription of the same (or any other) gene provided it contains a recognition sequence. In other words, when the RNA product reaches a length of 10 to 15 nucleotides, sigma factor is recycled in other initiation events. A series of σ factors are known for example, in *B. subtilis* where different factors recognize different sequences, for example, σ^{28} CCGATAT; σ^{29}: CATATT, σ^{32}: TANTGTTNTA; σ^{37}: GGNATTGNT; σ^{43}: TATAAT.

4.2.8 Eukaryotic Polymerases

Each enzyme RNA pol I, II, III is composed of at least 12(8-14) subunits, three of which are common to all enzymes.

4.3   PROTEIN SYNTHESIS

The different steps involved in protein synthesis are as follows:
1. Activation of amino acids
2. Initiation of protein synthesis
3. Elongation of polypeptide chain
4. Termination of protein synthesis and release of polypeptide chain
5. Folding and post-transcriptional modification of the protein.

Amino acids are present in the cytosol and protein synthesis takes place on the ribosome. As we know there are 20 types of amino acids and so each of these AAs attaches to specific tRNA. There is at least one type of tRNA for each amino acid and at least 32 tRNAs are required to recognize all the amino acid codons. Some tRNAs can recognize more than one codon and some cells use more than 32 tRNAs. The protein synthesis starts with attachment of the 20 different AAs with their corresponding tRNAs by amino acyl tRNA synthetase in the presence of ATP and Mg^{2+} which activates the enzyme and in this process there is cleavage of ATP to AMP and PPi. There are 20 different types of amino acyl tRNA synthetase. Each enzyme is specific for one AA and one or more corresponding tRNAs. Translation is carried out by an RNA protein complex called a ribosome. Ribosome is just a large ribozyme/enzyme with active sites for its substrate, mRNA and fMet-tRNA. Ribosomes have three sites, A(amino acyl), P(peptidyl) and E(exit site). Both 30S and 50S ribosomes contribute to the characteristics of A and P sites. These two units are held together by a strand of mRNA. A-site (amonoacyl-tRNA binding site) refers to the site on the ribosome that first binds the amainoacyl tRNA whereas P-site (peptidyl-tRNA binding site) refers to the site on the ribosome to which the growing peptide chain is attached during protein synthesis. In the two site model for elongation, the peptidyl-tRNA is held on the P site, base paired through the anticodon to its cognate codon and in the ribosomal A site lies the next codon, ready to be inspected by charged tRNA. E site is largely confined to 50S sub unit of ribosome. The first amino acid attaches to P- site directly but the subsequent incoming AA first come at A-site and then at P and E. 5′ AUG is positioned at the P-site and fMet tRNA first binds to P-site. As AUG is the initiation codon it specifies methionine so there are two types of tRNA, tRNA^{Met} and tRNA^{fMet}. Methionin is always the first amino acid in the polypeptide chain but then it also occurs as internal addition. The amino acid incorporated in response to 5′ AUG is fMet tRNA^{fMet} in prokaryotes as well as in mitochondria and chloroplast but in eukaryotes polypeptide begins with Met residue rather than fMet residue. The initiating 5′ (AUG) is guided to its correct position by Shine-Dalgarno sequence in the mRNA. The anticodon of tRNA$^{\text{f Met}}$

now pairs with mRNA's initiating codon near the 3′ end of 16 S rRNA. The mRNA binds to the smaller of the two ribosomal subunits and to the initiating amino acyl tRNA. The large subunit (50S) of the ribosome then binds to smaller subunit (30S) bound to mRNA and tRNA then forms an initiation complex (70S) and thus the process of polypeptide synthesis begins. This initiation of polypeptide synthesis is promoted by a number of initiation factors such as IF-1, IF-2, IF-3, GTP and Mg^{2+}. Thus initiation of protein synthesis involves formation of a complex among the 30S ribosomal unit, mRNA, GPT, fMet tRNA^{fMet}, three initiation factors and 50S ribosomal unit. A major difference between prokaryotic and eukaryotic cells is observed at the level of the initiation of protein synthesis. In eukaryotes, the 40S ribosomal subunit, in association with initiator tRNA^{Met} and initiation factors (IFs) scans from the capped 5′ end of the mRNA to the first AUG codon (Cigan et al., 1988). In prokaryotes, the 30S ribosomal subunit directly binds a start site of translation on the mRNA, together with the IFs and initiator tRNA^{Met} (Gualerzi et al., 1990). Another difference appears at the level of initiator tRNA^{Met} which, in P-sites, displays a formyl group in addition to the esterified methionine. Formaylation was also shown to occur in the organelles (chloroplast, mitochondria) of many organisms but not in the cytoplasm of eukaryotes. The functional relevance of such a formylation of methionyl-tRNA^{fMet} for the initiation of translation is not yet precisely understood.

After initiation the next amino acyl tRNA in sequence binds to A site. The ribosome moves along the mRNA (to the next codon) and there binds to the next amino acyl tRNA in sequence. The two amino acids join through the formation of a peptide bond (between carboxyl terminal of first AA and amino terminal of second AA) and there is hydrolysis of GTP to GDP and PPi. The growth of polypeptides on ribosomes begins with the amino- terminal AA and proceeds by successive addition of new residues to the carboxyl-terminal end. N-formyl methionyl group is transferred from its tRNA to amino group of the second AA. The peptidyl bond is formed through peptidyl transferase and in the process tRNA (first tRNA) is displaced but still in P site. Ribosomes move one codon towards the 3′ end of mRNA which shifts the anticodon of dipeptidyl tRNA from A to P site and shifts the deacylated tRNA from P site to E site. Again is binding of the incoming amino acyl tRNA to mRNA and there is movement of the ribosome along the mRNA. There are a number of factors called elongation factors which promote the elongation of polypeptide chain, such as EF-TU, EF-Ts, EF-G. After many such elongation cycles the synthesis of polypeptide is terminated with the help of release factors. Completion of polepeptide chain is signalled by any of the termination codons (UAA, UAG and UGA) in the mRNA. After the termination of polypeptide chain the new polypeptide chain is released from the ribosome which is again promoted by a number of polypeptide release factors such as RF1, RF2, RF3. RF1 recognises UAG, UAA whereas RF2 recognises UGA and UAA. RF3 is thought to release the ribosome. In eukaryotes, termination of mRNA translation is mediated by the release of release factors eRF1 and eRF3. Gross et al. (2007) identified in *S. cerevisiae*, a member of DEAD-box protein, the DEAD-BOX RNA helicase and mRNA export factor Dbp5 as player in translation termination.

4.4 FOLDING OF PROTEINS AND DEGRADATION

The newly synthesized polypeptide folds into the active three-dimensional forms. Many proteins are further processed by post-transcriptional modification reactions. Protein folding involves multiple pathways. Initially regions of secondary structure may form followed by folding into super secondary structures (see chapter 14 for detail). Finally, large *ensambles* of folding intermediates are rapidly brought to a single native conformation. For many proteins folding is facilitated by specialized proteins called **molecular chaperones**. The two classes of chaperones are Hsp 70 and chaperorons. Disulphide bond formation and the *cis-trans* isomerization of praline peptide bonds are catalysed by specific enzyme, protein disulfide isomerase (PDI) and peptidyl prolyl *cis-trans* isomerase (PPI). The protein synthesis in prokaryotes is similar to that of eukaryotes.

The eukaryotic cell consists of many structures, components and organelles which require different types of proteins and enzymes for their functioning. So many proteins are directed to particular locations. Certain amino acid sequences serve as signals which determine the cellular location, chemical modification and half-life of a protein. Some proteins are targeted for export from the cell, some other proteins are targeted for distribution to the nucleus, the cell surface, the cytosol and other cellular locations

4.4.1 Targeting Mechanisms

There are two targeting mechanisms:

1. The targeting mechanism involves a **peptide signal sequence(short sequence of AA)** generally found at the amino terminus of a newly synthesized protein which directs protein to its appropriate locations in the cell. Signal peptide is essential for efficient and selective targeting of nascent protein chains to cytoplasmic memebrane in prokaryotes or the endoplasmic reticulum in eukaryotes. Signal peptide (leader peptide) consists of about 15-30 mainly hydrophobic amino acid residues.

2. In eukaryotes one class of signal sequences is recognized and bound by a large protein-RNA complex called **signal recognition particle** (SRP).SRP is a universally conserved cytoplasmic 11S ribonucleoprotein complex that mediates co-translational targeting of secretory and membrane proteins to cellular proteins. A small RNA called 7SL is a component of SRP. SRP binds the signal sequence as soon as it appears on the ribosome and transfers the entire ribosome and incomplete polypeptide to the endoplasmin reticulum (ER) and there they may be modified and move to Golgi complex and then sorted and sent to lysosomes, the plasma membrane or transport vesicle. Post transcriptional modification of many eukaryotic proteins begins in the ER. Proteins targeted to mitochondria and chloroplast and for export in bacteria also make use of amino-terminal signal sequence. Proteins targeted to the nucleus have an internal signal sequence which is not cleaved once the protein is successfully targeted.

Other known targeting signals include carbohydrates and signal patches. Other sequences act as attachment sites for prosthetic groups such as sugar groups in glycoproteins and lipids in lipoproteins. Some proteins are imported into the cell by receptors-mediated endocytosis.

Protein degradation—Protein degradation is mediated by specialized proteolytic system in eukaryotes. The proteins are first tagged by linking them to a highly conserved protein called ubiquitin. The ubiquitin dependent proteolysis is critical to the regulation of many cellular processes.

4.5 DIFFERENCES IN MECHANISM OF PROTEIN SYNTHESIS BETWEEN EUKARYOTES AND PROKARYOTES

Ribosomes in eukaryotes are bound with over 80 different types of proteins and are larger 80S. The various sources of ribosomes along with their sedimentation coefficients are given in the table 4.1 given below.

Table 4.1 Showing ribosomal components of ribosomes from different sources.

| Ribosome source | Ribosomal components | | |
	Average sedimentation coefficient (range)	rRNA chains(large sub/small sub)	Approximate number of r-proteins
Bacterial	70S	5S,23S/16S	50-60
Mitochondria (mammals)	55S	16S/12S	80-90
Chloroplast	70S	23S,5S,4.5 S/16S	50-60
Mitochondria (fungi, protozoans)	70S(55S-80S)	21-24S/15-17S	65-90
Arachae bacteria	70S	5S,23S/16S	65-75
Plant mitochondria	75S	5S,26S/16S	70
Eukaryotes (cytoplasm)	80S	5S,5.8S/17S-18S	70-90

The significant differences between the two appear in the mechanism of initiation. There are 9 initiation factors in eukaryotes such as elF2, 2B, 3,4A, 4B, 4E, 4G, 5 and 6. Eukaryotic ribosomes do not have E site and tRNA is released from P site itself. The three elongation factors of eukaryotes eEF1α, Eef1β and Eef2 are similar to EF-Tu, EF-Ts and EF-G1, respectively. Finally, a single release factor, eRF recognizes all three termination codons.

4.6 PLACE OF TRANSCRIPTION/TRANSLATION

The nuclear envelop was originally thought to functionally demarcate the nuclear compartment where transcription and RNA processing occurs from the cytoplasmic compartment where translation occurs. But low level of translation occurs within the nucleus (Shelby et al., 1996). The large subunit of RNA pol II as well as a number of general transcription factors (GTFs) and promoter specific transcription factors localize at transcription sites which are distributed throughout the nucleoplasm. A significant number of TFs are localized in the cytoplasm and only upon activation by a signaling molecule are they transported into the nucleus where they bind DNA to activate transcription. RNA pol II transcription has been reported to occur at nearly 8000 sites scattered throughout the nucleoplasm (Pombo et al., 1999). Transcription related proteins also localize to other nucleardomains such as nuclear speckles, OPT domain, cajal bodies, promyelocytic leukemia (PML) nuclear bodies and heat shock transcription factor 1 (HSF1) granules (Spector, 2003). Eukaryotic transcription complex from yeast that included RNA polymerase II-the primary transcription enzyme and five associated protein called **general transcription factors** have been purified. Further, there is a mediator- a complex of some 20 proteins that relays signals from proteins that turn on specific genes to pol II. The images of how pol II selects the right RNA bases to how it recognizes proteins that turn on expression of specific genes have now been obtained. Thus biology's central dogma, DNA to RNA to proteins has now been observed.

4.7 MOLECULAR MECHANICS UNDERLYING KEY STEPS IN EUKARYOTIC TRANSLATION INITIATION

1. Ternary complex formation and dissolution
2. Loading the ternary complex onto the 40S subunit
3. Getting mRNA on the ribosome
4. Scanning and AUG recognition
5. Subunit joining

Ternary complex formation: It involves the assembly of e $IF2$-GTP-Met-$tRNA_i$ ternary complex but first there is formation of e $IF2.GDP$ complex which is recycled to e $IF2.GTP$ and this is facilitated by e $IF2$. The ternary complex then binds to small 40S ribosomal subunit resulting in 43S complex. Binding is facilitated by at least e IF_s1, 1A and 3. e IF4F complex assembles on the 5' cap of the mRNA and unwinds structures found in the 5'-UTR. This is achieved by7 ATP-dependent action of e IF_4A assisted by the RNA-binding proteins e IF_4B. In mammals, e IF_4H, e IF_4F in conjunction with e IF_3 and poly (A) binding protein bound to the 3'-poly (A) tail loads the mRNA onto the 43S complex. The 43S complex then starts scanning down the message in the 5'-3' direction, looking for the initiation codon. The scanning requires ATP hydrolysis although ATPase involved has not been identified. e IF_1 and e IF_1A may play a role in the scanning process as well. When 43S complex encounters an AUG codon embedded in a favorable sequence context(e.g. the kozak sequence, usually the first AUG, codon-anti-codon base pairing takes place between the start codon and the initiation tRNA in the ternary complex. This then triggers GTP hydrolysis by e IF_2- a reaction facilitated by the GTPase-activating protein eIF_5. e IF_1 in detecting the correct start codon. After GTP hydrolysis by eIF_2 $eIF_2.GDP$ releases the Met-RNA_i into the P site of the 40S subunit and then dissociates from the complex. In the current model e $IF_s1,1A,3$ and 5 also dissociate at this stage. At some point either before or after GTP hydrolysis by e IF_2, e $IF_5B.GTP$ binds to the complex. Once other factors are gone it facilitates the joining of the large ribosomal subunit (60S) to the 40S-Met-$tRNA_i.mRNA$ complex. This event triggers GTP hydrolysis by e IF_5B and because the e $IF_5B.GDP$ complex has the affinity for the ribosome it dissociates from the complex.

Bacterial initiation involves three initiation factors IF1, IF2 and IF3. IF1 binds over the small ribosomal subunit and promoting the initiator tRNA binding to P site. IF2 is a GTPase which enhances binding of the initiator tRNA to the small ribosomal subunit. IF3 ensures i. fidelity of the intiation site selection, ii. selecting the formylated methionyl initiator (fMet-tRNA$_i$) along with IF2 for use in initiator rather than the elongator methionyl tRNA used to insert methionine residues into polypeptides during elongation, iii. dissociating 70S ribosomes into 30S and 50S subunits and iv. acting as ribosome recycling to remove deacetylated tRNA from P site of the 30S subunit. The purine-rich S.D. sequence is located upstream of the initiation codon in the mRNA. S.D. sequence is usually about 10 bases upstream of the initiation codon. In bacteria recruiting of the mRNA to the small ribosomal subunit takes place through a base pairing interaction between the 3′ end of the 16S rRNA (the anti-S.D.) and the purine-rich S.D. sequence.

Translation initiation: It involves getting the initiator tRNA and mRNA onto the small ribosomal subunit, finding the AUG codon and joining the ribosomal subunits to form an initiation complex.

4.8   COMPARISON BETWEEN EUKARYOTE AND PROKARYOTE

Eukaryotic mRNAs do not have a S.D. sequence which allows easy identification of the initiation codon. Further translation of degraded mRNA is less of a problem in prokaryotes because of coupling between transcription and translation. As the 3′ end of the mesasage is being made, ribosomes are rapidly moving down from the 5′ end to translate it. In bacteria transcription and translation are coupled processes and thus as soon as S.D. sequence emerges from the transcriptional apparatus it can be bound by the small ribosomal subunit whereas in eukaryotes the mRNA substrate for translation initiation is thought to be a completely, fully processed mRNA, potentially rife with secondary and tertiary structures and coated with RNA binding proteins such as hnRNP$_S$. It is apparently sufficient for bacterial to deal with the problem after the fact(i.e. release the stalled ribosome and tag the incomplete protein for degradation) whereas eukaryotes have had to develop system to prevent the translation of truncated mRNAs from happening in the first place.

Bacteria have three translation initiation factors whereas eukaryotes have at least 12 translation initiation factors which composed of at least 23 different polypeptides. Two of the bacterial initiation factors have clear orthologs in eukaryotes, IF1 = e IF$_1$A, IF2 = e IF$_5$B (nonessential in yeast). E IF$_1$ = IF1 based on functional and structural similarities.

The 3′-poly (A) tail stimulates translation initiation both *in vivo* and *in vitro* just as the 5′cap. This effect is mediated by poly (A) binding protein. Stability of mRNA is not affected by adding either a cap or poly (A) tail. Their effects result in differences in translation rather than mRNA stability.

Translation in prokaryotes is fast, of the order of 10 amino acids per second (Soresen et al., 1989) whereas splicing by spliceosome is slow in the range of 0.005-0.01 intron per second in globins (Audibert et al., 2002). Although translation in modern eukaryotes is slower about one amino acid per second (Palmiter, 1975) it is still much faster than splicing. The origin of the eukaryotic nucleus marked a seminal evolutionary transition. Nuclear envelope's incipient function was to allow mRNA splicing which is slow, to go to completion so that translation which is fast, would occur only on mRNA with intact reading frame. Prokaryotic model of co-transcriptional translation and forcing nucleus-cytosol compartmentalization which allowed translation to occur on properly matured mRNA only. The rapid fortituous spread of introns following the origin of mitochondria is cited as the selective pressure which forced nucleus-cytosol compartmentalization (Martin and Koonin, 2006).

Elongation—Mechanisms underlying elongation are the same in both eukaryotes and prokaryotes. Elongation factors share high functional and structural similarities in prokaryotes and eukaryotes. Peptide elongation starts with a peptidyl tRNA with ribosomal P site next to a vacant A site. An aminoacyl tRNA is carried to the A site as a part of a ternary complex with GTP and the elongation factor 1A (e FF1A;EE-Tu in bacteria). E FF1A.GTP.aa.tRNA ternary complexes with either the correct or incorrect aminoacyl tRNA can bind to the ribosomal A site. Codon-anticodon base pairing between mRNA and the tRNA, conformational changes in the decoding centre of the small ribosomal subunit and GTP hydrolysis by e FF1A/EF-Tu ensure selection of only correct tRNA. e FF1A.GDP releases the aminoacyl tRNA into the A site in a form then can continue with peptide bond formation. The ribosomal peptidyl transferase centre then catalyzes the formation of a peptide bond between the incoming amino acid and the peptidyl t RNA. The result is a deacetylated tRNA in a hybrid state with its acceptor end in the exist (E) site of the large ribosomal subunit and its anticodon end in the P site of the small subunit. The peptidyl tRNA is in a similar hybrid situation with its acceptor end in the P site of the large subunit and its anticodon end in the A site of the small subunit. This complex must be translocated such that the deacetylated tRNA is completely in the E site, the peptidyl tRNA completely in the P site and the mRNA moved into the A site. This task is accomplished by elongation factor 2 (EE-G in bacteria) which hydrolyses GTP as it facilitates translocation. This cycle is repeated until a stop codon is encountered and the process of termination is started. There is existence of elongation factor 3 (eFF3) exclusively in fungi. Table 4.2 shows characteristics of prokaryotic elongation factors.

Termination—Termination occurs in response to presence of a stop codon in the ribosomal A site. The result of termination is the release of the finished polypeptide following the hydrolysis of the ester bond linking the polypeptide chain to the P site tRNA. The peptidyl transferase

Table 4.2   Showing characteristics of prokaryotic elongation factors.

Prokaryotic Elongation factors					
Factor	Molecular mass	Number of amno acids	Gene	Function	
EF-Tu	43,195	393	tufA	Bind amino acyl tRNA to A site, GTPases	
	43,225	393	tufB	-do-	
EF-Ts	30,257	282	tsf	GDP/GTP exchange on EF-Tu	
EF-G	77,444	703	fus	Translocation, GTPase	

Prokaryotes	Eukaryotes	
EF-Tu =	EF-1α	
EF-Ts =	EF-1β	
EF-Ts =	EF-1γ	
EF-G =	EF-2	

centre of the ribosome is thought to catalyze the hydrolysis reaction in response to the activity of class I release factors which decode stop codons present in the A site. Class 2 release factors are GTPase which stimulate the activity of class 1 release factors (RF_s) regardless of which stop codon the class 1 factor has engaged. Bacteria possess two class 1 release factors-RF1 and RF2 with overlapping codon specificity such that each responds to UAA whereas UAG is decoded only by RF1 and UGA is decoded only by RF2. RF3 is a class 2 release factor which not only stimulates the activities of RF1 and RF2 but is also required to eject them from the ribosome following peptidyl tRNA hydrolysis. In eukaryotes only class 1 release factor eRF1 is present. It has an omnipotent decoding capacity and can promote the hydrolysis of peptidyl-tRNA in response to any of the three stop codons. In ciliates in which either UGA or both UAA and UAG have been reassigned as sense codons, e RF1 decoding potential is appropriately restricted by mechanism not currently known. There is also single class 2 release factor e RF3 present in eukaryotes (Table 4.3). It is essential protein in eukaryotes. e RF1 and e RF3 bind to each other in the absence of the ribosome and the interaction is required for optimum efficiency of termination in yeast *S. cerevisiae* but no such interaction between RFs 1 or 2 and RF3 has been observed.

Table 4.3 Showing release factors of prokaryotes and eukaryotes.

	Release factors	
Class	Prokaryotes	Eukaryotes
I	*RF-1 (UAG/UAA specific)*	*e RF-1 (reads all three stop codons, omnipotent)*
	RF-2 (UGA/UAA specific)	
II	*RF-3 (Stimulates RF-1/2 activity, recycle factor, GTPase)*	*e RF-3 (stimulates e RF-1 activity of yeast e RF-3)*

4.9 MECHANISMS OF TRANSCRIPTION TERMINATION IN PROKARYOTE

There are two mechanisms of termination of transcription.

1. Rho-independent termination — The signal of arrest at the end of the gene is provided by a G-C-rich palindromic sequence followed by a poly (A) sequence. Transcription of these sequences leads to formation of a hair-pin structure (secondary structure) and to a poly (U) sequence, respectively. Hair-pin induces pausing of the polymerase together with the adjacent poly (A) hybrid. This promotes transcription arrest and release of RNA transcript and RNA polymerase. In other words, formation of secondary structure within the nascent RNA leads to the termination of transcription in the factor independent termination.

2. Rho-dependent termination — Termination may require the presence of a hexameric releasing factor rho which shows an RNA-dependent ATPase and an RNA-DNA helicase activity. Termination factor rho binds the RNA transcript upstream of the rho-dependent terminator and subsequently translocates along the transcript (i.e., migrates towards the elongating polymerase). Termination occurs when rho catches up the polymerase pausing at the terminator. The factor NusA facilitates termination by increasing the frequency of pausing.

4.10 REGULATION OF TRANSCRIPTION IN PROKARYOTE

Regulation of transcription occurs at two stages.

1. At the initiation level, variation in the pattern of transcribed operons can result from the use of different σ factors, each specific to a type of promoter. Heat shock promoter involved in sporulation or response to heat shock, is driven by a cascade of σ factors. An alternative pathway is the *trans*-acting factors that bind to the operator, a specific sequence generally located nearby the promoter to enhance (activators) or repress (repressors) of the initiation of transcription.

2. During the process of transcription, when the elongating RNA polymerase encounters a first termination sequence it may either stop or continue to transcribe adjacent genes. This will depend on factors which fasten on the elongating polymerase after having bound to a specific site on the DNA or on the transcript. In case of antitermination factors (or **antiterminators**), RNA polymerase will read through the stop signal but some other factors exist that will increase its propensity to stop, probably by inducing frequent pausing. The phenomenon of attenuation can also occur when transcription is intimately coupled with translation. Progression of the ribosome along the nascent RNA modulates the formation of secondary structures enabling RNA polymerase to read through intrinsic terminators (**attenuators**).

Transcription efficiency — In prokaryotes the efficiency of transcription is further modulated by *trans*-acting factors known as transcription factors. Because of the relatively high efficiency of RNA synthesis by prokaryotic RNAPs, the main theme for regulation in prokaryotes is repression. Gene expression is a tightly controlled phenomenon in eukaryotes with respect to both time and space. Although there are many control points during gene expression the control of initiation of transcription is for most gene the most prominent among them. For initiation of RNA synthesis this is achieved through the action of TFs. These proteins bind with high affinity to their target DNA sequences and regulate the rate of transcription through interaction with basal machinery. In other words, the specificity of transcriptional initiation is controlled through the interaction of TFs with their DNA-binding sites in the promoter and enhancer regions of the genes. Within TFs there exists two independent domains-DNA-binding domain and transcriptional activation domain. Because of low efficiency of the basal transcriptional machinery , most eukaryotic TFs are activators.

Prokaryotic RNAPs are capable of recognizing gene promoters and initiating transcription from the proper start site whereas in eukaryotes the purified RNAPs can catalyze low-level template-dependent transcription of RNA but despite the structural complexity these enzymes are unable to recognize the promoter of genes and depend fully on a set of auxillary proteins called GTFs to initiate the transcription from the corresponding class I, II, III gene promoter.

4.11 ANTISENSE TRANSCRIPTION

The sense strand of DNA generally provides the template for the production of mRNA which in turn encodes proteins. Transcription from opposite strand (antisense) can produce transcripts that hybridize with the coding DNA strand or with the sense transcripts to interfere with the transcription or mRNA stability. Thus role of antisense transcription in gene regulation involves degradation of the corresponding sense transcripts (RNAi) as well as gene silencing at the chromatin level. It controls transcriptional outputs in mammals (RIKEN and FANTOM Consortium, 2005). Global transcriptome analysis has shown that a large proportion of the genome can produce transcripts from both strands and that antisense transcripts commonly link neighboring genes in complex loci into chains of linked transcriptional units. Expression profiling reveals frequent concordant regulation of sense/antisense pairs.

4.12 TRANSLATIONAL ERROR

Once an amino acid is attached to a tRNA, the position of the amino acid in a growing polypeptide chain is determined by the interaction between the anticodon of the tRNA and the matching complementary codon of mRNA. Error in protein synthesis is relatively rare (10^{-4}) *in vivo*. Where errors occur at high frequency, translational editing occurs to overcome this problem. Errors occur because of limitations to molecular recognition at two critical points in translation: 1. Certain amino acids are difficult for synthetases to discriminate from each other and for that reason can be misactivated and potentially attached to the tRNA cognate to the synthetase but noncognate to the amino acid. 2. Anticodon-codon interaction on the ribosome is determined by base pairing between the complementary triplets and this interaction is prone to error as, for example, pairing of two rather than three bases, although less possible, is still possible.

4.13 ROLE OF tRNA IN PROTEIN SYNTHESIS

The specificity of tRNA in protein synthesis depends not only on its recognition of the codon in the mRNA but also on its recognition of the correct aminoacyl-t-RNA synthetase enzyme (Mc Clain, 1993). The specificity of tRNA in aminoacylation (tRNA identity) depends on the tRNAs productive interaction with the correct enzyme and non-productive interaction with other enzymes. Although extensive regions of the tRNA interact with the enzyme, only a small number of nucleotides comprise the major determinants of tRNA identity. They often end in the same positions (acceptor and anticodon and variable pocket less often) in different tRNAs. Specificity of the acceptor end of the tRNA is achieved in part by the specific amino acid sequence with protein binding pocket domains which are part of all aminoacyl-tRNAsynthetase. These domains also bind the other two substrates of the enzyme, amino acid and ATP. Specificity for the anticodon and variable pocket of the tRNA is more idiosyncratic. Irrespective of their location in the tRNA, the determinants either interact directly with the enzyme or give the tRNA a conformation for a complementary fit to the enzyme.

The specificity of tRNA in translation depends on anticodon—codon interactions while that in aminoacyltation depends on its interaction with the cognate aminoacyl tRNA synthetase. There are twenty types of aminoacyl tRNA synthetase (aaRS), one for each amino acid and tRNA type. The specificity of aaRS for amino acid is determined by protein-amino acid interaction. However, specificity of aaRS for the tRNA is determined by protein-RNA interactions. The aminoacylation specificity, termed tRNA identity is determined by tRNA structure. Several isoacceptor forms of tRNA may exist for a particular amino acid but there is only one aaRS type for each tRNA acceptor type. Discrimination between tRNAs is determined by a set of rules as all tRNAs are structurally quite similar. tRNAs are about 76 nucleotide long and form a typical clover leaf secondary structure (Holley et al., 1965) (see chapter 1). The number of residues in stem and loop regions are so conserved that they can be referenced by a standard numbering system (Schimmel et al., 1979). Length variation occurs in the variable loop and the D-loop. The α- and β segments of the D-loop contain one to three residues each in different tRNAs. The first nucleotide at the 5′ terminus of tRNA is residue 1 and the last residue at the 3′ terminus is residue 76. The site of amino acid attachment is adenosine 76 (A76) which is part of the common 3′ CCA hydroxyl terminus; the anticodon residue nucleotides 34, 35 and 36. tRNA contains a number of constant nucleotides and post-transcriptionally modified nucleotides at the same relative positions in all molecules. Nucleotide pairing yield the same L-shaped tertiary structure of tRNA (see chapter 1).

Circles represent nucleotides which are always present; ovals, nucleotides which are not present in each structure. These are nucleotides before position 1 on the 5′ end, before and after the two invariant GMP residues 18 and 19 in the D-loop and the nucleotides in the variable loop. The nucleotides to be added at a given site are indicated by the number of the preceding nucleotide followed by a colon and a letter in alphabetical order. The nucleotides in the variable stem have the prefix 'e' and are located between positions 45 and 46 obeying the base-pair rules. The nucleotides in the 5′ strand and the 3′ strand are numbered by e11, e12, e134 and e21, e22, e23…, respectively. The second digit identifies the base pair. In case of a long variable region the loop can be formed by up to five nucleotides; e1, e2, e3, e4 and e5. Positions in which invariant nucleotides usually occur are indicated by a thick line.

In general, modified nucleotides arising post—transcriptionally contribute to a compact, native tRNA conformation and are the not major determinants of aminoacyltation specificity. However, unusual substitution of the amino acid lysine for the 2-oxygen group of the pyrimidine ring of C34 in *E.coli* tRNA^{Ile} and the 2-thiouridine at position 34 in E.coli tRNA^{Glu} contributes to the productive interaction of tRNA. In contrast with tRNAs the aaRSs show considerable variability (several hundred to more than one thousand amino acids) in their primary sequences and one to four subunits in their oligomeric states.

4.14 DISSECTING THE MECHANISM OF TRANSLATION

Antibiotics is a powerful tool in studies aimed at dissecting the mechanism of translation. Many antibiotics exert their effect by interfering with protein biosynthesis and most act directly on the ribosome. Many antibiotics interact directly with or otherwise perturb the structure of RNA. rRNA is the essential element in ribosome. Although studies have correlated the effects of certain antibiotics with perturbation of specific ribosomal events, direct correlation of structure with function has not yet been drawn. Moazed and Noller(1988) showed connections between certain structural features of 16S rRNA and ribosome function. Sites protected by antibiotics are closely correlated with those protected by tRNA. Nucleotides protected by tRNA can be classified operationally according to whether their protection is strictly mRNA-dependent (class I) or mRNA-independent at high Mg^{2+} concentrations (class II) or whether they are also protected by 50S subunits in the absence of tRNA (class III). Classes I and II are likely to correspond to A-site and P-site binding of tRNA, respectively. Several classes of antibiotics protect concise sets of highly conserved nucleotides in 16S rRNA when bound to ribosomes. And thus have strong implications for the mechanism of action of these antibiotics and for the assignment of function to specific structural features. Apart from the S.D. mechanism for mRNA selection, no defined model is known for how rRNA participates in translation.

5

Gene, Gene Concept and Gene Organization

5.1 GENE DEFINITION

In 1866 Mendel proposed the basic rules of inheritance and hypothesized that traits are determined by discrete units that are passed from one generation to the next. The term 'gene' was coined by Johannsen (1909) for the unit associated with an inherited trait. The gene thus appeared as an abstract to explain the hereditary basis of traits. Phenotypic traits were associated to hereditary factors. The gene is the basic unit of heredity. In 1910, Thomas Hunt Morgan showed that genes reside on chromosomes leading to the ideas of genes as beads on a string. In other words, he and others associated heritable traits with specific chromosomal regions (i.e. demonstration of chromosomal inheritance). The position of a gene along a chromosome is called a **locus**. Thus a gene can be defined as a unit that (i) resides at a unique locus and (ii) is responsible for a specific function. Further, genes are arranged in the linear order along the chromosome. In 1944 Avery, Mc Cleod and Mc Carty showed that genes are made of DNA. However, nucleic acid was discovered in 1871. Gene is defined as the stretch of DNA(or in a few cases RNA) which produces a functional protein. Beadle and Tatum (1941) proposed **one gene–one enzyme** hypothesis and defined gene as the segment of DNA that codes for one enzyme which was later modified to **one gene –one protein hypothesis.** The original hypothesis was incomplete in many ways. Firstly, nor all genes produce protein. Secondly, many genes code for proteins that are not enzymes. Thirdly, it was found that enzyme (tryptophan synthetase in *E.coli*) consists of two different polypeptide chains, each encoded by a distinct gene. First protein sequence was obtained in 1951 and the first protein to be sequenced was insulin. Later on it was found that haemoglobin A and many proteins are made up of multiple polypeptide chains. Haemoglobin A is particularly made up of two types of polypeptide chains- α and β which differ in the amino acid sequence and are encoded by the two different genes and thus one gene- one protein hypothesis is more accurately described as **one gene-one polypeptide** concept. The 'colinearity' of protein structure and gene structure was demonstrated by showing that changes in primary structure of a protein always corresponded to changes in the gene. In other words, the linear sequence of mutable and recombinable sites within the gene corresponds to the linear amino acid sequence of the polypeptide chain of the protein. Further evidence of colinearity came from the molecular identification of the gene as DNA, the demonstration of transcription of genic DNA into RNA and working out of the genetic code through which mRNA is translated into polypeptide sequence. In 1953 Watson and Crick gave the chemical structure of DNA and the central dogma of molecular biology in which information flows from DNA to RNA to protein. Thus a gene is defined as a specific segment of DNA, the end product of which is a protein but then gene can also be defined as a specific segment of DNA which produces nucleic acid as the end product.

The classical definition of the gene defines it as a unit which (i) resides at a unique chromosome locus and (ii) is responsible for a specific function. Another classical view of the gene was that it was an indivisible unit of hereditary transmission, i.e. recombination could occur only between genes and never within them. And if recombination could occur only between genes then crossing the two allelic mutants (mutants with defects in the same gene) could only produce mutant meiotic products. On the other hand, if mutants fall at different sites within a gene and recombination occurs between them then the heterozygote will produce wild-type recombinant although at a low frequency. Benzer's complementation and recombination tests in 1955 demonstrated the difference between what he proposed '**cistron**', the genetic unit of function as defined by Lewis's (1952) *cis/trans* comparison and **recon**, the minimal unit of genetic recombination. The word 'gene' was used to refer to both kinds of unit (cistron and recon), i.e., gene- a functional and indivisible unit (a unit of recombination) of genetic transmission before the distinction between them was recognized but after Benzer's(1959) complementation test on bacteriophage T4 rII mutants the word '**gene**' came to be used as synonymous with cistron- a functional unit subdivisible by recombination. Thus gene refers to *cistron*, the functional unit of DNA or RNA and one cistron in the DNA specifies one polypeptide. The smallest unit of DNA capable of mutation is called a ***muton*** and is a single base contained in a nucleotide. The smallest indivisible unit of DNA, also a single base, which gives rise to a new form by recombination is called ***recon.*** Thus the classical gene was replaced by the cistron, a unit of function consisting of a linear array of individually mutable and recombinable sites. During late 1940s and early 1950s the dogma was that crossing over does not occur within genes. 1960s saw the elucidation of the genetic code.

The *cis*-principle — It provides criteria for deciding whether two linked mutations a and b fall into the same or different genes. The two assumptions are that different genes function independently of their spatial relationship to each other and that a single gene is functionally indivisible. Now, if a and b are in different genes then considering the first assumption the *cis* and *trans* double heterozygotes, ab/++ and a+/+b must be phenotypically identical and approximating to wild type, if a and b are recessive. But if a and b are in the same gene, the *trans* combination a+/+b will be no more like wild type than the extreme of the individual mutants. This is called the **complementation test**. The *cis*- principle applies at several different levels. This complementation principle is of great value as it allows assigning mutations to the same gene or to different genes without a detailed analysis of gene functions. Although this concept is universally accepted but the term cistron has been dropped out of use and 'gene' is now generally used in the same sense (Fincham, 1999).

5.2 GENE AS A UNIT OF TRANSCRIPTION

The protein-encoding genes are *cis*-units of translation and all active genes are *cis*-units of transcription though these turn out to be far more complex and versatile than was initially thought (see chapter 24). The structural gene requires often very extensive flanking sequence for its expression and regulation. Beyond the transcription unit (sometimes far beyond), there are *cis*-units of control of transcription. But at each level there are some exceptions as we will see. The same enhancer may service two or more different transcriptional units. Thus the boundaries of functional genomic units tend to be become fuzzy. The gene thus can be defined as the minimum chromosomal unit which can be removed from its usual position and inserted elsewhere without loss of its function. But there are many examples of transcription units which although function to some extent with very limited specific flanking sequence but require distant enhancers for full and properly regulated levels of expression. Further, normal gene expression as we will see later, is dependent on chromosome architecture- euchromatin and heterochromatin.

The **gene** is thus a unique sequence DNA (nucleotide residues are linked in a specific sequence) as described above. In species for which the genetic material is double stranded DNA genes may appear on either strand. Different sequences of nucleotides that may occur in a gene are called **alleles**. The functional unit of the DNA contains one or more intervening sequences of DNA that do not code for amino acid sequence of the polypeptide and thus these *intervening sequences* are not translated. Such transcribed but non-translated non-coding sequences are called **introns** and the coding sequences are called **exons**. Introns occur in many genes encoding for proteins, tRNAs and ribosomal RNAs in nuclear, mitochondrial and chloroplast DNA. A gene thus can be defined now as a locus of cotranscribed exons. The coding sequence (CDS) is also referred to as the **open reading frame**. In other words, an ORF is the part of a gene encoding the amino acid sequence of its protein product. Genes without introns have unbroken ORFs from an initiation codon ATG to a termination codon TAA, TAG or TGA, with codons in between corresponding precisely to the known amino acid sequence. Thus the gene can consist of a number of exons interspersed with a number of introns. An exon is a stretch of DNA retained in the mature mRNA and translated into protein. **Number and size of introns**- The different genes can vary with respect to the number of introns, their positions and the fraction of the total gene length they occupy. The exon, the protein coding portion may be interrupted from one to a dozen times. Introns are highly variable in size and can range from a few nucleotides pairs to thousands of nucleotides pairs in length(from 60 to 20,000 nucleotides). In other words, introns are of variable length from 50 to 100 bases in fungi to tens of kilobases for many mammalian and Drosophila genes and can appear either between or within codons. Thus many introns are very long and in some cases they are substantially longer then the exons. A mammalian gene encoding a polypeptide chain of just a few hundred amino acid residues may be spread over a tract of some 100, 000 base pairs. In mammals, exons tend to be small (<300nt) and introns very large. The genes are thus split into separated segments in the genomic DNA and the genes containing introns are called **split genes**. In contrast bacterial genes are contiguous regions of DNA. The genes account for 80-85% of the DNA (in bacteria, *E.coli*), 70% (in yeast, *S. cerevisiae*), 25% (in fruit fly and nematodes), and 3-5% in human. The table 5.1 given below shows the percentage of genome coding for proteins in various organisms (Cavallier and Smith, 1985).

Table 5.1   Showing percentage of genome coding for protein in different organisms.

Organisms	Percentage of genome coding for protein
Bacteria (*E.coli*)	~100
Yeast (*S. cerevisiae*)	69
Nematode (*Ceanorhabditis elegans*)	25
(Newt) *Tritunus cristatus*	1.5-4.5
(Lung fish) *Protopterus aethiopicus*	0.4- 1.2
Drosophila	33
Homosapiens	9-27
Arabidopsis thaliana (an annual dicot weed)	31
Fritillaria assyriaca (a monocot)	0.02

The structure of the gene refers to exons (that parts which make their way into a functional mRNA) and introns (that parts which do not). Walter Gilbert (1978) coined the term introns and exons. Introns are characterized by dinucleotides, GT at 5′ and AG at 3′ end when applied to the genomic DNA and there are 20-30 nucleotides around the actual splicing sites. Cellular machinery splices together proper segments in RNA transcripts based on signal sequences flanking the exons in the sequence itself. There are genes such as genes for histones which contain no introns. The gene in eukaryotes thus contains one or more introns whereas only a few prokaryotic genes contain introns. There are only a few eukaryotic genes, α -and β-interferon genes which do not contain introns.

In Arabidopsis the average size of introns is about 200bp and on average, 6 or 7 introns are found per gene. Introns vary in (i). number per gene, from one in some tRNA genes to more than 30 in Xenopus yolk egg protein genes (ii). size and ranges from less than 50 (50 to 100 bases in fungi) to more than 12000bp (e.g. in many mammals and Drosophila) and (iii). in sequence. Intron can be present either between or within codons. Only the borders between exon and intron are identical in most introns and these boundries direct the correct excision of the introns and splicing of exons.

5.3   EFFECT OF EXON/INTRON

One consequence of the presence of exons and introns in eukaryotic genes is that the potential gene products can be of different lengths because not all exons may be represented in the final transcribed mRNA (although the order of exons that are included is preserved). When mRNA editing results in different polypeptides the resulting proteins are known as **spliced variants** or **alternatively sliced forms.** A single gene can encode multiple proteins by alternative splicing by varying translation or stop sites, or by frameshift during which a different set of triplet codons in the mRNA is translated

5.4   FUNCTIONS OF INTRON

Introns are also transcribed when RNA is first made in the cell nucleus but later on they are removed during splicing when the initial product of transcription (primary transcript) is processed to mature mRNA. The interruptions in the coding sequence are characteristic and different for every gene. In a number of genes introns have been found to separate regions of DNA sequences that encode distinct functional and/or structural domains (a region of a complex proteins sequence that can be associated with a particular structure or function). Two introns and one intron rat insulin genes are equally well expressed indicating no ill effect of introns. With the exception of the splicing regions the base sequences of introns exhibit a much greater divergence then that of exons. In some but not all genes the splicing out of intron transcripts may be a pre-requisite for stable mRNA formation. Intron can be functional in cases of differential processing. The product of one gene may be differentially spliced to yield more than one mRNA and the intron of message can become exon of another. Thus intron provides flexibility to gene function. Splicing allows a gene to encode two or more proteins to perform different functions at different times. Some sequences in the proteins are common and some are different. Introns can be involved in controlling transcription. There are many examples of presence of downstream enhancers located in the introns of genes. One such example is the bithorax enhancer within Drosophila Ubx gene.They have implications for the timing of transcription and they some times contain enhancer sequences necessary for normal gene expression and finally, they may allow the same gene to perform more than one function. In case of gene, Ubx (ultrabithorax) of Drosophila there seem to be two regulatory regions, each 20kb or more in length, one upstream of the transcription start site and another downstream in the long intron. Within these extensive regulatory regions there are shorter sequences apparently associated with expression in specific para segments. One aspect of gene expression is the lapse of time between the initiation of transcription and the appearance of the protein product. Considering the estimated rate of RNA synthesis in Drosophila as 1.1kb per minute and presence of about 50 kb of introns (or gene expanded to 60kb) in a transcription unit in higher organisms, it will take more than half an hour to express itself in terms of protein from the time of initiation of transcript. Thus introns are sometimes used as timing devices for gene expression. Now a positive transcriptional signal acting on along transcription unit is a sort of time- switch set to produce an effect at some fairly precisely defined future. There is one example where intron length is crucial (Sellem et al., 1996). Two genes, knirps (knr) and knirps-like (knr-1) encode very similar proteins but differ greatly in intron length, ~1kb in knr and 19kb in knr-1. Thus changes in intron length could be a way of fine tuning the system and combination of genes with different lead times for their expression could provide precise temporal patterns of gene activity (cycles of activity on a time scale of minutes or hours) (Fincham, 1999). Intron contains splice enhancers, splice donor and acceptor sites, branch point sequences as well as less well defined accessory sequences which are described in this section. There are examples of ORFs nested within the introns of other genes. One example is of the Drosophila Gart gene which functions in purine biosynthesis and has an ORF in its largest intron which encodes cuticle protein for use at pupation stage. Here the mRNAs of Gart and the nested gene are transcribed from opposite strands of the DNA and therefore, presumably can not both be produced simultaneously. There are also few examples of genes nested within mitochondrial introns. The sea anemone Metridium senile contains a group I intron within its mitochondrial gene for subunit 5 of NADH dehydrogenase and this intron includes the genes for subunits 1 and 3 of the same enzyme. Here all three genes use the same DNA strand for transcription and so the excised intron gives rise to a bicistronic mRNA for the nested genes.

5.5 TYPES OF INTRONS

There are four classes of introns: Groups I and II and Class III and IV. **Group I introns** are found in some nuclear, mitochondrial and chloroplast genes coding for rRNA, mRNA and tRNA whereas GroupII introns are found in primary transcript of mitochondrial or chloroplast mRNA of fungi, algae and plants. Neither requires ATP for splicing. In both groups splicing involves two *trans* esterification reactions steps.

1. Identification of nucleophile is the first step.

2. In case of GI splicing requires a guanine nucleoside or nucleotide co-factor (GMP, GDP or GTP).The guanosine 3′- hydroxyl group forms a normal 3′-5′ phosphodiester bond with 5′ strand of the intron. In case of **GII introns**, 2′-hydroxyl group of an adenylate residue within the intron is the nucleophile and there is formation of a lariate like intermediate (a branched lariate structure is formed as intermediate) in which the 5′ terminal guanosine of the intron transcript is joined via a 2′-5′ phosphodiester bond to an internal adenosine residue of the intron thus detaching the 5′ end of the intron from exon 1 and leaving the intron attached to exon 2 as a loop. The free 3′-OH of exon 1 attacks the 5′ C at the terminus of exon2 leading to the ligation of exons 1 and 2 and the exclusion of the intron as a free lariat which is subsequently degraded(Figure 5.1). The splicing of group II introns is similar to that of nuclear pre-mRNA introns. Groups I and II introns are distinguished from one another in their secondary structures, in nature of the proteins encoded by their ORFs and in the chemistry of their splicing reactions. Group I and group II introns differ from the common introns of nuclear pre-mRNAs in their requirements for splicing. Many GI and GII introns are also **mobile genetic elements**. In addition to their self-splicing activities they encode DNA endonucleases that promote their movement. These endonucleases promote insertion of the introns into

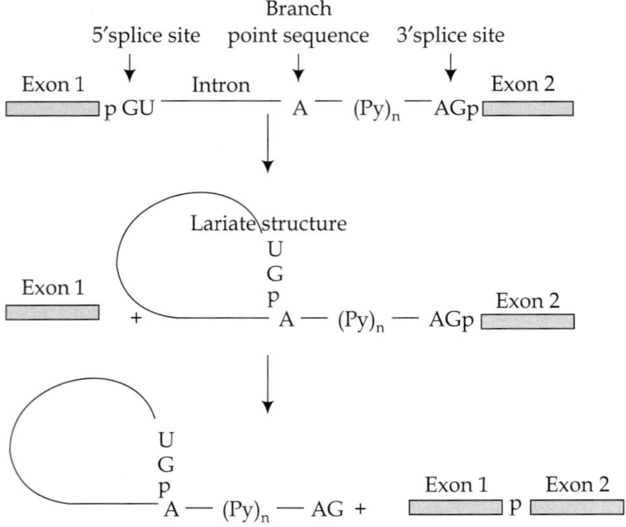

FIGURE 5.1 Showing two transesterification steps in splicing.

an identical site in another DNA copy of a homologous gene that does not contain the intron. In case of GI homing is DNA based whereas in case of GII homing occurs via RNA intermediate and have associated reverse transcriptase activity. Thus certain introns include a gene for enzymes that promote homing (type I introns) or retrohoming (type II introns). The gene within the spliced intron is bound by a ribosome and thus translation occurs. Type I homing introns specify a site specific endonuclease called a homing endonuclease whereas type II retrohoming introns specify a protein with both endonuclease and reverse transcriptase activities. **Homoing –** Intron homing is the process of insertion of an intron at a particular homologous site of an intronless gene and is catalyzed by by intron encoded homing endonuclease. Homing endonuclease refers to any one of a series of extremely rare cutting endonuclease recognizing large asymmetric recognition sites of 12-40bp. It causes a double strand break in the target double stranded DNA at or near the intron insertion site and generate staggered cuts with 4 nucleotide extension at the 3′ end of the cuts. The double strand break repairs and completes the integration of the intron. When allele **a** of a gene X containing homing intron is present in a cell containing **b** of the same gene which lacks the intron the homing endonuclease produced by **a** cleaves **b** at the position corresponding to the intron in **a** and double strand break repair(recombination with allele **a**) occurs leading to creation of a new copy of the intron in **b. Retrohoming-** When allele **a** of gene Y contains a retrohoming type II introns, and allele **b** lacks the intron, the spliced intron inserts itself into the coding strand of **b** in a reaction that is the reverse of the splicing that excised the intron from the primary transcript, except that here the insertion is into DNA rather than in mRNA. The noncoding DNA strand of **b** is then cleaved by the intron encoded endonuclease/reverse transcriptase. This same enzyme uses the inserted RNA as a template to synthesize a complementary DNA strand. The RNA is then degraded by cellular ribonucleases and replaced with DNA.

Class III introns are found in nuclear mRNA primary transcripts, are not self-splicing and follow lariate forming mechanism of splicing as in case of group II introns but requires action of specialized RNA-protein complexes called **small nuclear ribonucleoprotein** (SnRNP, snurps), each SnRNP contains one of a class of eukaryotic RNAs, 100 to 200 bp long called small nuclear RNAs (SnRNAs). Splicing occurs within and is catalyzed by a large protein complex called 'spliceosome'. Splicing of nuclear pre-mRNA involves sequential trans-etherification reactions which take place in the spliceosome. The spliceosome consists of many proteins and there are at least six snRNAs of which the five snRNAs (U1, U2,U4,U5 and U6) are found in abundance in eukaryotic nuclei. Some of these proteins are tightly associated with the snRNAs, forming small nuclear ribonucleoproteins (snRNPs) which assemble in a stepwise manner onto the pre-mRNA to form the spliceosome. The U1 snRNA has a sequence complementary to sequence near the 5′ site of mRNA introns and the U1snRNP binds to this region in the primary transcript. Further, addition of the U2, U4, U5 and U6 snRNPs leads to the formation of spliceosome. The U2snRNA interacts with branch site. U5snRNA interacts with exonic sequences immediately 5′and 3′ to the splice junctions whereas the U6snRNA interacts with the 5′splice site (Sontheimer and Steitz, 1993). Besides the snRNPs subunits, a large number of proteins perform various functions during the splicing reaction.

Class IV introns are found in certain tRNAs and differs from I and II in that the splicing requires ATP and an endonuclease which cleaves the phosphodiester bonds at both ends of the introns and the two exons are joined by DNA ligase. Class III and IV introns are not self-splicing.

Regulatory elements in introns—Recently many regulatory elements have been discovered within introns which indicate their role in regulating alternative splicing. Repeat sequences especially complementary repeats (simple sequence repeats(mono-through hexanucleotide repeats)) possess sequence motifs and symmetry elements that potentially allow the formation of secondary structure on single strand pre-mRNA such as hair-pins and triplexes (Mitas, 1997). These repetitive elements are frequently polymorphic. It has been suggested that the formation of secondary structure can participate in the regulation of splice site selection. A very high local repeat density certainly suggests where exons are unlikely but the reverse is not true, i.e. the absence of repeats does not prove the presence of the genes.

5.6 SPLICING OF INTRONS

Although capping (5′ capping) and polyadenylation (3′ polyadenylation) generally do not affect the protein coding capacity of mRNA but splicing generally determines or alters its coding potential. The pre-mRNA (primary transcript) contains introns and the mature mRNA is formed through removal of introns from the primary transcripts (pre-mRNA) by the process of splicing (Burge, et al., 1999). Splicing is done in a preferred but not in obligatory manner. There are three critical nucleotide sequences (the extreme5′ and3′ ends, the branch point or 'lariat' sequence, UACUAACA and spliceosome) within the introns which are critical in the processing of the POlII transcript.

Steps in splicing—The first step in splicing is the assembly of a large ribonucleoprotein complex called a spliceosome on the pre-mRNA. Before an intron is released it forms an intermediate structure called a lariat. In the final step of splicing the two exons are joined and the intron is released as lariat RNA.

Loss of activity of specific genes is not due to mutations affecting the spliceosome as these would affect so many genes that they would be certainly lethal. But a single gene mutation can be effectively silenced by a mutation within one of its own introns which prevents that intron from getting spliced out.

Types of splicing—There are different types of splicing. **Constitutive splicing** refers to the conventional splicing that follows a fixed rule and splices always the same exons in the same order. But in case of less than 1% of all the genes of an organism **alternative splicing** (Lopez, 1998) occurs and non-adjacent exons are spliced, exons are skipped and introns are retained. Alternative splicing produces either longer or shorter mRNA variants in comparison with the wild type mRNA and consequently the encoded protein contains more or less functional domains. A single gene may give rise to multiple transcripts and thus multiple distinct proteins with multiple functions by means of alternative splicing and alternative initiation and termination sites. Alternate splicing is thus used to enhance the information contained within a gene and to control its expression. The consequences range from switching expression of a protein on and off (e.g. by including or excluding stop codons as in Sxl (Sex-lethal proteins in Drosophila melanogaster) (Bomze and Lopez, 1991) to structural and functional diversification of protein products (by including or excluding elements as small as a single amino acid), for example in Pax-3 and Pax-7 (Vogan et al., 1996). An alternate splicing event in Pax-3 paired domain identifies the linker region as the key determinant of paired domain DNA-binding activity. It is frequent in metazoans as diverse as the nematode, *C.elegans*, the fruit fly and humans. Some alternate splicing appear to be constitutive with mRNA-variants co-existing at constant ratios in the same cells whereas other are regulated in response to developmental or physiological cues. Splicing is known to be location and time-specific.The splicing of a specific set of exons of a particular gene in one cell type and the splicing of another set of exons of the same gene in another cell type is referred to as **cell specific splicing**. The Drosophila transposable P-element is a clear example of tissue specific splicing. This element has a single ORF with three introns. The third intron is spliced out only in the germ line and consequently P-element movement is detected only in germ cells and not in somatic tissues. Sexual development in Drosophila is also controlled by different intron splicing. Alternative splicing and post-transcriptional modification of RNA transcripts (RNA-editing) can generate many more proteins than encoded by genes. In Drosophila, both these could theoretically generate 1, 032,192 mRNA transcripts (each encoding a slightly different protein) from a single para gene which encodes a sodium channel. In yeast only three genes are known to be alternatively spliced whereas in human at least 35% of the gene transcripts undergo alternative splicing.

Alternate exons—It refers to any of a series of nearly identical , tandemly arranged exons of mosaic genes which remains in the final mRNA whereas all its other alternatives also appear in the primary transcript but eliminated during the splicing. For example, the 61.2kb DSCAM gene of Drosophila melanogaster encodes a 7.8 kb mRNA after splicing which contains 24 exons. However, the itself contains an array of potential alternative (mutually exclusive) exons. Exon4 has 12, exon 6 has 48, exon 9 has 33 and exon 17 has 2 alternatives. This shows that each mRNA contains one of the 12 possible alternatives from exon4, of the 48 alternatives from exon6 and so on. If all possible combinations of a single exon from cluster 4,6,9 and 17 are used by alternate splicing the DSCAM gene would produce 38,016 different mRNAs and thus so many proteins.

Trans-splicing—It refers to the ligation of exons from two or more different mRNAs to form one mature message with a new combination of coding sequences. There are two categories of *trans*-splicing. The **spliced leader type** of splicing is a characteristic of protozoa, for example, trypanosomes and lower invertebrates such as nematodes where it results in addition of short capped 5′ non-coding sequences to the mRNA. The **discontinuous group II intron types** observed in chroloplasts of algae and higher plants (and plant mitochondria) involves joining of the independently transcribed coding sequences through unusual interactions between intronic RNA stretches. Splice sites are the nucleotide sequences of intron-exon boundaries and these are highly conserved and thus consensus sequences are found and any known sequence deviates marginally from consensus sequence. Splice sites are at the 5′ and 3′ ends of intron. The most invariant aspect of consensus sequence is the GU at the beginning of intron transcript and the AG at its end. The sequences immediately before the intron-exon splice site consensus sequence is always rich in pyrimidine and free from the dinucleotide AG. The spliced intron is eventually degraded. In genes (*C. elegans*) undergoing *trans*-splicing the 5′exon begins with a splice acceptor sequence, making this 5′exon more difficult to distinguish from internal exons. This combination of factors may result in two genes being merged into one (Blumenthal, 1995). *Cis*–splicing- It refers to recognition of exon-intron boundaries at the 5′ and 3′ splice sites, removal of the corresponding intron and ligation of the adjacent exons within the same pre-mRNA.

Splicing control element—It refers to any sequence or genomic region which is essential for correct splicing. Correct splicing requires exon recognition with accurate cleavage and rejoining at exon boundaries designated by the invariant intronic GT and AG dinucleotides, respectively known as the splice donor and splice acceptor sites. Other more variable consensus motifs have been identified in adjacent locations to the donor and acceptor sites. These elements can be found within an intron or also exon and at their 5′ or 3′ ends. They are frequently arranged in repeats.

For example, a weak exonic 'CACCAG' consensus flanking the SD site, an intronic GGG triplet, purine-rich elements or polypyrimidine (Y:C or T)-rich tract in the 3′ intronic regions (flanking the SA site) function such as splicing control elements which are usually located in the vicinity of a splice site but can be away as far as 150kb. A weakly conserved intronic 'YNYURAY' consensus 18-40bp from A site acts as a branch site for lariat formation.

Splicing enhancer—It is a purine-rich sequence element in exon (also intron) downstream of a 3′ splice site in pre-mRNA which promotes splicing. Splicing enhancers are address sites for SR proteins and help to assemble the spliceosome complex. There are exonic splicing enhancers (ESE) and intronic splicing enhancers (ISE), both of which promote exon recognition. **Splicing silencers** refer to any sequence in genomic DNA which suppresses splicing of a particular mRNA. They reside in exons (exonic splicing silencers, ESS) and comprise from 4 to 74 nucleotides in length and reduce the assembly of early spliceosomal complex and thereby preventing conventional splicing. Splicing silencers function in alternative slicing. Intronic splicing silencers (ISS) also exist. These silencers inhibit recognition of exons. Silencer sequences are binding sites for nuclear proteins (e.g. hnRNPA1 or hnRNPI). DNA recognition motifs for splicing enhancers and silencers are generally quite degenerate. The degeneracy of these consensus recognition motifs points to fairly promiscuous binding by SR proteins. These interactions can also be explained by the use of alternative and inefficient splicing sites which may be influenced by competitive binding of SR proteins and hn RNP determined by the relative ratio of hnRNP to SR proteins in the nucleus.

Splice junction—The junction between an exon and an intron at the 3′ end of the intron in eukaryotic split gene has the following consensus sequence. $^{c}_{A}$AG:G where semicolon indicates the splice point. Consensus sequences at the ends of introns are involved in excision and splicing reactions during post-transcriptional modification of primary transcripts from eukaryotic split genes. The junction signal at the 5′end of an intron transcript is called the **donor splice junction (or site)** and the signal at the 3′ end of the intron is called the **acceptor splice junction.** In other words, the exact phosphodiester bonds to be cleaved and ligated, i.e., the socalled 5′ and 3′ splice sites are determined primarily by short conserved sequences at the intron exon junctions. By comparing the sequences of nuclear pre-mRNA introns consensus sequences surrounding the 5′ and 3′ splice sites have been identified.

The internal sequences of all introns of eukaryotes start with 5′-GT…. and end with …AG-3′ with the G nearly always preceded by a pyrimidine. Following the GT at the 5′ end there is a degree of consensus over the next six or so bases, which varies somewhat from one group of organisms to another. At the 3′ end there is very little consensus upstream of the AG except for a tendency to short runs of pyrimidine residues. In the interior of the intron, usually within a few tens of bases of the 3′ end there is a sequence playing an important role in the splicing out

$$5' \underline{\quad\quad\quad} \overset{C}{_A}AGGT\overset{A}{_G}AGT \underline{\quad\quad} N\overset{C}{_T}AGG \underline{\quad\quad\quad} 3'$$

INTRON ‖ exon   Donor splice junction   Acceptor splice junction

FIGURE 5.2 Showing donor splice junction and acceptor splice junction.

process by providing the **branch point or 'lariat'** junction. This sequence in *S. cerevisiae* is always 5'-TACTAACA-3'. In most other eukaryotes a shorter and less constant versions of the sequence, 5'-CT ^A_GAC-3' (in filamentous fungus and to some extent in animals) are found (Fincham, 1994). Thus with these landmarks it is possible to identify with fair certainty the reasonably short introns interrupting an otherwise satisfactory ORF and so deduce the amino acid sequence of the product encoded after intron excision even where this was not previously known.

Nuclear pre-mRNA introns have been divided into two classes; (i)U2-dependent introns are found in all eukaryotes and (ii) U-12 dependent introns are less abundant and absent from yeast *S. cerevisiae*. U2-mammalian 5' splice sites are marked by the consensus sequence AG\GURAU whereas yeast by AG\GUAUGU where \ indicates the splice site, R= purine, Y= pyrimidine and N= any nucleotide. 3' splice sites can be divided into three distinct sequence elements which are typically found within the first 40 nucleotides upstream of the actual exon-intron junction. These include the so-called branch site which is characterized by the consensus sequence YNYUAC (in mammals) and UACUAAC (in yeast) where **A** represents the site of branch formation during pre-mRNA splicing and the actual 3' splice junction which conforms to the consensus sequences YAG/G and CAG/G in animals and yeast, respectively (Table 5.2). In addition, many vertebrate U2-dependent introns contain a 10-to-20-nucleotide stretch of pyrimidines (designated the polypyrimidine tract) that is found just upstream of the 3' splice junction. U-12 –dependent introns contain distinct 5' and 3' splice site consensus sequences. It consists of two subtypes, AU-AC and GU-AG which differ from one another solely in their terminal dinucleotides. It comprises, 0.2% of all nuclear pre-mRNA introns (Will and Luhrmann, 1999). Shorter introns are identified, at least tentatively as segments interrupting ORF with intron consensus sequences at their ends.

GT-AG rule – Splice site consensus sequences are highly conserved in yeast whereas in mammals the consensus represents the most prevalent sequence but often fluctuates considerable from one intron to the next. However, the terminal intron dinucleotides (GU and AG) are almost absolutely conserved in both higher and lower eukaryotes. In other words, there is obligate appearance of the dinucleotide consensus GT at the 5' donor splice junction and the dinucleotide AG at the 3' acceptor splicing site of introns in eukaryotic split genes. Greater than

Table 5.2 Intron consensus sequences in different organisms (Filipowicz).

Organisms	5' splice	Branch point	3' splice
Vertebrates	AG/GUAAGU	UNCURAC	(Y-rich)NCAG/G
Saccharomyces	AG/GUAUGU	UACUAAC	(Y-rich)YAG/G
Plants	AG/GUAAGU	UA-RICH	(U-rich)UGCAG/GU

Slash(/) divides exon from intron, Y= pyrimidine, R= purine and N= any nucleotide.

99% of U2-dependent introns follow this so-called GU-AG rule (Breatnach et al., 1978). The only exceptions are the few genes of *S. prombe* which possesses GC donor sites. Also, the unusual 5′ splicing border GC is used in myrosinase genes of the *Brassicaceae* and involved in hydrolysis of glucosinolates. *Genes encoding myrosinase* contains 12 exons and 11 introns. There is preferential use of a GC dinucleotide as the 5′ splicing border in intron 1 instead of an adjacent GT dinucleotide four base pair further 3′.

Cryptic splice site — It refers to a splice junction located within an intron and may be used for splicing of mRNA, if the normal slice junctions get mutated, deleted or otherwise become nonfunctional. In other words, they are only used in the absence of the authentic splice site. Cryptic splice sites may be also be used for the generation of different mRNAs from the same DNA sequence in alternate splicing. Splice selection in vertebrate pre-mRNA involves small nuclear ribonucleoproteins (snRNPs) recognizing sequences around the exon-intron borders; U1 snRNPs recognize 5′ (donor) sites and U2 and U5 snRNPs interact with the 3′ end of the intron where a 3′ consensus signal (acceptor site) and a branch signal are located.

Weak splice site/junction — It refers to any splice junction which is only used in the presence of splicing enhancer. Weak splice sites are underlying several alternative splicing events.

5.7 DIFFERENT PRODUCTS FROM THE SAME TRANSCRIPT

A gene subjected to different modes of splicing can generate a family of different but overlapping mRNAs. In other words a single gene can encode a whole family of proteins depending on the mode of intron splicing. This type of genes has large number of exons, some of them extremely short and separated by much longer introns. Such genes have been identified in mouse, rat and human genomes. Different mRNAs from the same transcript can also be generated as in case of RNA splicing in sexual development in Drosophila. If the ribonucleoprotein spliceosome complex fails to recognize one of these rather minimal landmarks (discussed above) of the introns and fastens instead on the acceptor sequence of the next intron downstream and thus splicing out two introns along with the intervening exon. Errors of this kind could result in two or more alternative RNAs encoding different polypeptide chains with perhaps different functions.

Overlapping genes — Genes with overlapping nucleotide sequences, for example, gene E of phage FX174 which overlaps with gene D. Overlapping genes produce two different polypeptides as the corresponding mRNAs are transcribed in two different reading frames as shown below.

Gene-1	AUG	GUG	CAU	UAU	AAU	GCA	AUU	GUC	ACA	GGG	UUU
	Met	gly	gln	tyr	asn	ala	ile	val	thr	gly	phe
Gene-2						**AUG**	**CAU**	**UUG**	**UCA**	**CAG**	**GGU**
						Met	**gln**	**leu**	**ser**	**gln**	**gly**

In spite of casual opinion that gene overlaps are likely to happen only in phase and virus genome where requirements for tight gene packing are vitally important, the complete bacterial genomes show quite a few gene overlaps. The gene overlaps cause several difficulties for a high accuracy prediction. Some overlapping genes could be missed. It might be hard to exactly predict the 5′ end of the gene whose translation initiation codon and ribosome binding site fall into the overlap region where oligonucleotide statistics may not fit to regularly used models (see Roy, 2009).

Differential analysis of transcripts with alternate splicing—It is a technique which is used for the isolation and characterization of multiple splice variants. It identifies alternatively used exons and introns in two or more different populations of mRNAs. In this technique mRNAs from two samples, say A and B are first isolated and divided into two aliquots each. One aliquot of each population is reverse transcribed using biotinylated oligo (dT) primer which allows subsequent isolation of the double stranded cDNA by streptavidin capture on magnetic beads. The residual mRNA aliquot from sample A is then hybridized to the cDNA of population B and vice-versa. Splice variants will then form loops, for example, where an exon from mRNA of population A has no counterpart in an mRNA from population B. The hybrids are then treated with RNase H to release the looped RNA which is then isolated and reverse transcribed into cDNA. This cDNA is then cloned and the clones thus collected form a so-called DATAS library. Sequencing of differentially regulated clones identifies the genes which are transcribed into alternatively spliced messages and further provides information on the location of the affected domains. This technique thus allows to define the splicing status of mRNAs in the material under investigation, to monitor the changes of the status during development and also to identify the specific splicing strategy between two different cells, tissues, organs or organisms. Further, if the identified splicing variants are known, a cDNA chip can be designed onto which the corresponding cDNAs are spotted. This chip can then be hybridized to cDNAs or mRNAs from various samples to identify the splicing status.

Formation of introns—There are two types of introns: associative type intron (type A) and divisive type intron (type B). Associative type introns are created during the creation of new gene (s) through genetic recombination between different exons. Divisive type introns arise by aberrant recombination through the insertion of a retrovirus or transposon or by retropositional insertion of DNA copies of cellular mRNA. Almost all prokaryotic genes are precisely co-linear with amino acid sequences.

5.8   CHANGING CONCEPT OF GENE

The concept of the gene as a unit of function becomes more complex considering *cis*-acting regions which control transcription and transcribed sequence itself (Fincham, 1994, 1999). In 1977, Robert and Sharp discovered alternate splicing in viruses. This showed that gene can be split into segments leading to the idea that one gene can make several proteins- a deviation from the one gene-one polypeptide hypothesis. In other words, there is a DNA sequence which uses some of the same protein coding sequences to produce two entirely different proteins with distinct functions. The genome is thus full of overlapping transcripts. 1977 also saw the advent of DNA sequencing and sequencing was automated in 1986. The first human gene was isolated during 1975-1979. It was then observed that instead of discrete genes producing identical RNA transcripts, a mass of transcription converts many segments of the genome into multiple ribbons of differing lengths. These ribbons can be generated from both strands of DNA rather than just one strand. Further, many transcripts come from the regions of DNA no encoding proteins. In 1993 the first microRNA was identified in worm *C. elegans*. In 1995, the first whole genome of bacterium *Haemophilius influenzae* was sequenced. The first human chromosome was sequenced in 1999 and the human genome was sequenced in 2002.

There is also example of fused transcripts in which transcription can start at a DNA sequence associated with one protein and run straight into the gene for another completely different protein. In other words, each gene sequence reaches into the next and beyond, i.e. descriptions of proteins encoded in DNA know no borders. There are also many examples of transcripts in which protein coding exons from one part of the genome combine with exons from another part of the genome that can be thousand of bases away with several genes in between. A protein is assembled when four different RNA molecules made from DNA scattered over 40, 000 base pairs are assembled into one transcript. There is an example in which one gene is nested within the noncoding intron of another gene. This continuum of genes might even spill over boundaries of chromosomes. There are examples of immune systems in which genes seem to be under control of regulatory regions from another chromosomes. All these examples show that the concept of discrete genes is starting to vanish.

Studies (see chapter 24) have shown that a vast amount of RNA do not code for proteins. Although 63% of mouse genome have been shown to be transcribed, only 1-2% of the genome carries the unique sequence DNA. Further, miRNAs and other RNA molecules are now known to be vital in controlling many cellular processes in plants and animals(see chapter 23 for detail). All these again indicate that RNA actively processes and carries out the instructions in genome. Thus non-coding RNAs should also be given the status of gene.

Besides genetic inheritance there is presence of epigenetic inheritance in which information is passed from parent to offspring independent of DNA sequence (see chapter 6 for detail). There is also example of RNA as carrier of information across generations. The progeny receives regular DNA plus RNA as back up copy of the grand parent's genetic information. RNA is here used as template to correct certain genes in the offspring.

Many transcripts generated by a sequence at one location spot landmark sequences in DNA that signal where one gene ends and the next begins. This type of gene is predicted in bioinformatics. Thus definition of a gene should be a locatable region of genomic sequence, corresponding to a unit of inheritance which is associated with regulatory regions, transcribed regions and/or other functional regions. In 2003 there came the gene sweep and human geneticists came up with a definition of protein coding genes. In 2006 there came the idea that genes are a **long continuum**. Gene in molecular terms thus can be defined as a complete chromosomal segment responsible for making a functional product. This definition has several components such as expression of a gene product, requirement that it be functional and inclusion of both coding and regulatory regions.

5.9 CHANGE OF LOCUS/GENE RESTRUCTURING

There are many examples of genomic DNA rearrangements. Many cells have the potential to engage in the rearrangement of segments of the genome (Sakano et al., 1979; Roeder et al., 1980). The rearrangements may involve segments coding for entire polypeptide chains or alternatively, segments that code only a part of a polypeptide chain. Rearrangements of the latter type could result in different part of various enzymes and other proteins becoming genetically fused. The various combinations of binding sites and catalytic units might be put to use in the generation of new enzymes or other proteins. Thus genes can be restructured. In eukaryotes, gene rearrangements and splicing seem to be intimately associated with the existence of untranslated intervening nucleotide sequences, the introns. In case of vertebrate immune system, the genes are put together from a repertoire of modules and a large variety of cell types responding to different antigens is generated by a large number of alternative modes of DNA splicing in the differentiating cells. Regulated somatic DNA rearrangements which give rise to functional immunoglobulin genes and T cell receptor genes involve deletions of whole segments of chromosomes. Immunoglobulin molecules are generated by a series of gene rearrangements and RNA splicing events which result in polypeptide chains consisting of variable and constant regions (Tonegawa, 1983). Both light and heavy chains of immunoglobulin molecules consist of two regions-variable region and constant region. Since variable and constant gene sequences exist in a single mRNA molecule as a contiguous stretch, such integration must take place at either the DNA or the RNA level. Integration at RNA level could result from 'copy choice' event during transcription or from joining of two RNA molecules after transcription (Hozumi and Tonegawa, 1976). In case of *S. cerevisiae*, the process of mating type switching involves the transposition of two alternative blocks of sequence, held at loci where they are not transcribed. Controlled DNA restructuring is fairly common in prokaryotes as a means of switching gene function into one path or another but unlike eukaryotic rearrangements, it is reversible. Then there is editing of transcript. The primary transcripts of some genes have their sequences changed by a form of selective copying from short 'editing' RNA sequences transcribed from elsewhere (see chapter 2 for detail). Thus the definition of gene as presented by classical genetics has changed with the advancement in molecular biology. Genes can be transposed to other regions, that they have been duplicated or deleted or that they have diverged significantly. In maize centromeric sites have been subject to breakage and fusion events. Finally, extrachromosomal elements can transpose segments between the genomes of different organisms. Extrachromosomal elements have played a role in gene transfer in prokaryotes (Jones and Sneath, 1970). For more on DNA rearrangements see Trends in Genetics(1992) 8, 403-61, a special issue on programmed DNA Rearrangement.

Classes of genes — There are three classes of genes.

1. **Class I gene** refers to eukaryotic genes which encode rRNAs and are transcribed by DNA-dependent RNA polymerase I. Nucleolar organizers (NORs) contain array of unit repeats encoding the 18S, 5.8S and 25S ribosomal RNA genes and are transcribed by RNA polymerase I.

2. **Class II gene** refers to eukaryotic genes encoding proteins and are transcribed by DNA dependent RNA polymerase II. Any gene that encodes the amino acid (primary structure of a protein) is called **structural gene**.

3. **Class III genes**- It includes genes encoding 5S and tRNAs and are transcribed by DNA dependent RNA polymerase III.

Thus in addition to the protein-coding genes the genome contains genes for non-coding RNAs. There are widely dispersed tRNA genes (cytoplasmic tRNA and organelle-derived tRNA) and tRNA-derived pseudogenes. Several other non-coding RNA genes occur in dispersed multigene families. The U1, U2, U4, U5 and U6 spliceosomal RNA genes are in 14, 21, 5, 12 and 20 dispersed copies in *C.elegans*. There are five dispersed copies of signal recognition particles RNA genes and at least four dispersed copies of splice leader2 (SL2). A striking feature of these dispersed gene families is their high sequence homogeneity. For example, of the 20 U6 RNA genes 17 are 100% identical to each other. Either gene conversion or gene duplication may account for this homogeneity. Several of these RNA genes occur in the introns of protein-coding genes which may indicate RNA transposition. In general, RNA genes in introns do not appear to occur preferentially in the coding orientation of the encompassing transcript which indicates that these RNA genes are probably expressed independently. Other non-coding RNA genes occur in long tandem arrays. The ribosomal RNA genes occur solely in such an array. The 5S RNA genes occur in a tandem array with array members separated by SL1 splice leader in *C.elegans*. A few other known RNA genes such as the small cytoplasmic Ro-associated Y RNA and the lin-4 regulatory RNA are found only once in the genome . Some of the RNA genes expected to be present in the genome are yet to be identified, probably because they are poorly conserved at both the sequence and secondary structure level. These include ribonuclease P RNA, telomerase RNA and small nucleolar RNA genes. The small nucleolar RNAs (snoRNAs) consists of two subfamilies, the C/D box snoRNAs which includes 36 genes in Arabidopsis and H/ACA box snoRNAs for which no members have been identified. U3 is most numerous of the C/D box snoRNAs with 8 copies found in the genome. Small nuclear RNAs (snRNAs) were found dispersed as singletons or in small groups in Arabidopsis. The snRNA genes are transcribed by either RNA polymeraseII or III. Although transcribed by two polymerases their promoters are structurally related. These genes differ from other classes of genes in having unique transcriptional factors. Majority of the splicing related genes are known in human, yeast, Drosophila and Arabidopsis but unlike other organisms, the knowledge of splicing is scanty in plants because of non-availablity of *in vitro* splicing system.

Housekeeping vs tissue-specific gene—Genes are often characterized dichotomously as either house keeping or a single tissue-specific. House keeping genes are expressed in all cell types whereas other genes are expressed in a more restricted selection of tissues.

Repetitive sequences—Some of the sequence that does not code for protein or RNA is undoubtedly involved in gene regulation or in the maintenance and movement of chromosomes. A significant fraction of the sequence is repetitive in *C.elegans* as in other multicellular organisms. The repeat sequences are either local (that is tandem, inverted or simple sequence repeats) or dispersed. Genes need not be more than a few thousand bps long, however, human have about 30kb of DNA per gene if the current estimate of ~100,000 genes is correct and it is because of repetitive and functionless sequences such as SINE, LINE and satellite DNA.

5.10 REGULATORY ELEMENTS

When a gene is to be activated, to which stimuli (hormonal or environmental) the gene should respond, in which cell it should be active, how much mRNA is to be made and when the gene is to be turned off is determined by DNA sequences on each side of the protein coding region and are called regulatory regions. In essence, the regulatory sequences are responsible for control and the rate of production of mRNA. The regulatory sequences include **attenuator**, **enhancer** and **promoter**. The gene has been defined as a segment of DNA involved in the production of a polypeptide. So first of all it (segment of DNA or gene) must have a starting point (initiation) and an end (termination) point. The gene thus is composed of a coding region (may also contain intervening sequences, non-coding regions within the coding region) and the regions preceding (upstream regulatory sequences) and following it containing the downstream regulatory sequences. Most of the regulatory elements are thought to be embedded in the down stream non-coding part of the genome. Among non-coding regions the 5'- and 3'-UTRs of eukaryotic mRNAs have often been experimentally demonstrated to contain sequence elements crucial for many aspects of gene regulation and expression. Regulatory sequences provide signal that may denote the beginning (AUG) or end of the gene (UAA, UGA or UAG) or influence the transcription of the gene or function as an initiation point for replication or recombination. These specific DNA sequences act as recognition sites for various proteins (transcription factors).

5.11 GENE STRUCTURE- GENE COMPONENTS AND THEIR OLIGONUCLEOTIDE COMPOSITIONS

The key components of a gene include 5'-region, exon, intron, 3'-region and non-coding regions. In other words, the non-coding regions of a gene include the 5' flanking (untranscribed) region (5' –FLR), 5' untranslated region (5' UTR), 3' untranstaed region (3'- UTR), 3' flanking region (3' –FLR) and introns and exons. The gene related features include coding regions, introns, UTRs, upstream region and CpG islands. Figures 5.3 and 5.4 show structure of a typical enkaryotic gene, its primary transcript and matured mRNA.

Thus what we see is that **structural elements** such as coding, noncoding and **regulatory elements** are organized within genes. The essential components of a protein-encoding gene include promoter, a transcription terminator and a polyadenylation signal. Within the transcribed region, there are distinct sequences corresponding to the different sections of the mRNA which include untranslated leader of various lengths in different genes and sometimes essential for controlling translation, a ribosome binding site and initiation codon, an open reading frame and a termination codon.

5' flanking region and **3' flanking region**

The eukaryotic gene is flanked by regions, **5' flanking region** and **3' flanking region**. The 5' flanking region consists of sequences upstream of the coding part of a eukaryotic gene. This region is not transcribed but it contains sequence elements essential for the control of the gene expression (promoters, TATA or CAAT boxes, enhancers and specific binding sites for transcription factors). The 5' **untranslated region (5' UTR)** is the transcribed part of the eukaryotic gene, follows the cap site and precedes the start codon. In other words it is the untranslated part of an mRNA extending from its 5' terminus to the translational start codon, ATG and called the **leader sequence**. The leader may contain an attenuator sequence or SD sequences. Leader sequence length is remarkably consistent from fungi and plants, to invertebrates and vertebrates (including humans) as shown in table 5.3 given below.

3' flanking region comprises sequences downstream of the so called **trailer** of eukaryotic genes. This region contains signals for the precise transcription and processing of the 3' end of the transcript. ATG triplet marks the 5' end of the coding region of a gene. Its corresponding codon in mRNA is AUG which functions as start codon at which polypeptide synthesis is initiated.

Thus 5'UTR sequences are defined as the mRNA region spanning from the cap site to the starting codon (excluded) whereas 3'UTR sequences are defined as the mRNA region spanning

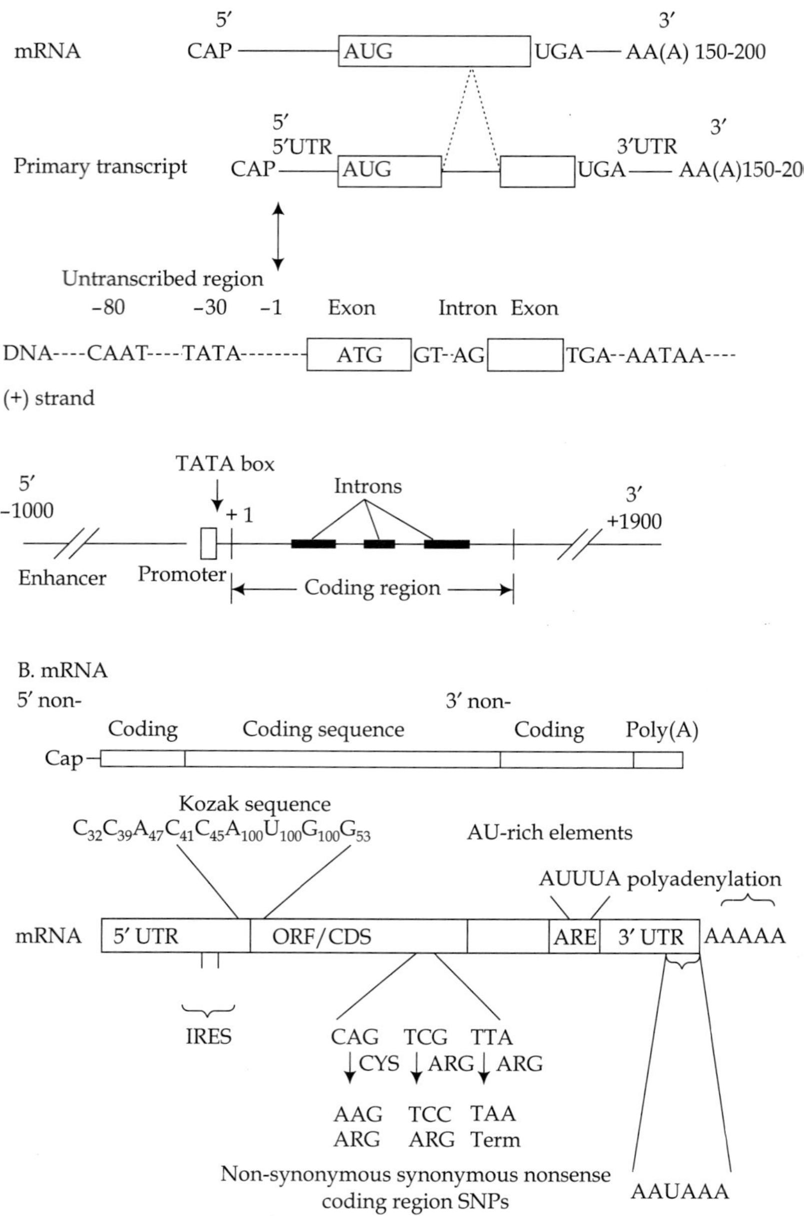

FIGURE 5.3   Structure of a eukaryotic gene and its transcript. The figure B shows some of the key regulatory and structural elements which control the translation, stability and post-transcriptional processing of mRNA transcripts. Further there exists polymorphisms in these regions.

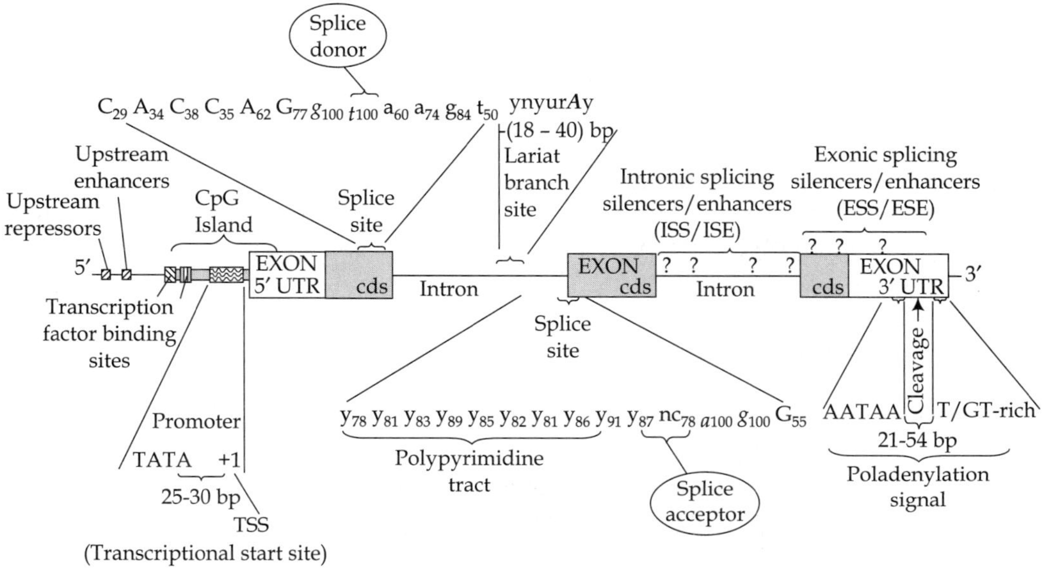

FIGURE 5.4 **Showing anatomy of a gene (adapted from Barnes, (2003).**

from the stop codon (excluded) to the poly (A) addition site. Several regulatory elements have already been identified in 5′ or 3′ UTR sequence, usually corresponding to short oligonucleotide tracts, able to fold in specific secondary structures which are protein binding sites for various regulatory proteins.

The main roles so far demonstrated for 5′- and 3′-UTR sequences are: 1. control of mRNA cellular and subcellular localization 2. control of mRNA stability and 3. control of mRNA translation efficiency (Pesole et al., 1998).

The **transcription initiation site** of a eukaryotic gene (its location on linear gene map is indicated by + 1) has the following consensus sequence.

5′-Py-Py-C-A(Py) 5-3′ (animals)

5′-Py-Py-C-A(Py) n-3′ (plants)

The sequence element also contains the start codon ATG. The term 'cap site' refers to the addition of a 7-methyl-guanosine –cap to the first nucleotide (mostly an A) of the primary transcript during its processing. The transcription initiation site is located downstream of the TATAbox and upstream of the translation initiation site. The initiator represents the simplest

Table 5.3 Showing leader sequence length in some organisms.

Leader sequence length	Organism
Plants	80bp
Cold blood vertebrates	100bp
Fungi, rodents	120bp
Invertebrates, humans	150bp

functional promoter and is present in many TATA-box containing promoter. It may also substitute the TATA sequence in TATA- less promoter of house keeping genes. The Inr elements are address sites for a series of proteins (CISF, E2F, USF, transcription factor IIB and RNA polymerase II).

Scanning model — It provides a hypothetical description of transcription initiation in which enhancers and/or upstream regulatory sequences are recognized by RNA polymerase II (or one of its subunits or a transcription factor). After binding the protein slides along DNA until it reaches proximal promoter element where it stimulates the formation of a transcription initiation complex.

5.12 MECHANISMS OF TRANSLATION INITIATION

The accepted convention is that the initiation codon will be the first inframe AUG encoding the largest ORF in the transcript. There is evidence for scanning mechanism for initiation of translation (Kozak, 1983). According to the scanning hypothesis for initiation of translation of mRNA, eukaryotic ribosomes bind to the 5' end of the mRNA and subsequently migrate along the mRNA until they meet an initiation codon AUG embedded in an appropriate context whereupon translation initiation occurs. The initiator codon generally conforms to a 'CCACCaugG' consensus motif known as the Kozak sequence (Kozak, 1996). However, Peri and Pandey (2001) found more than 40% of known transcripts containing inframe AUG codons upstream of the actual initiator codon, some of which conformed more closely to the Kozak motif than the authentic initiator codon. Their revised Kozak consensus C32C39147C41C45A100U1OOG100G53 was much weaker (Figure 5.3) and cast doubt on the validity of (Figure 5.3) scanning mechanism for the initiation of translation. Mechanism for translation initiation is still not known. Steps involved in the initiation of translation of most mRNAs in mammalian cells include recognition of the 5' cap structure and its interaction with it of e IF-4F. During this process the small ribosomal subunit with associated factors bind to the mRNA and scans to the initiation codon. Scanning model postulates that the 40S ribosomal subunit stops at the first AUG when that codon occurs in a favorable context. But if the first AUG codon occurs in a suboptimal context, for example, in the absence of the critical purine at position -3 or a G at position < +4 then some 40S subunits will bypass the first AUG and initiate instead at a downstream site. Thus two independently initiated proteins can be produced from one mRNA by context- dependent leaky scanning. The translation of the hepatitis B virus pol gene from the viral pre-genome RNA is a result of this ribosomal leaky scanning mechanism (Lin and Lo, 1992) (Nakamura and Karamysheva, 1999). In addition, internal start codons in eukaryotes can be reached by a reinitiation mechanism which was thought to be characteristic of eubacterial ribosome. Reinitiation at a downstream codon may be possible if the 5'-proximal AUG triplet is followed shortly by a terminator codon and thus fails to form the proper initiation complex (Hinnebusch, 1996).

Alternative mechanism which does not obey the '**first AUG rule**' involves cap-independent internal ribosome binding mediated by a Y-shaped secondary structure, denoted the internal ribosome entry site (IRES) located at the 5' UTR of 5-10% of human mRNA molecules (Le and Maizel, 1997). Capped eukaryotic mRNA can be translated without significant amounts of the intact protein complex e IF-4F. In other words, initiation and translation by an internal ribosome binding mechanism is independent of an intact E IF-4F holoenzyme complex. In yet another words, the internal ribosome binding mechanism is independent of 5′ cap structure and of the capping binding protein complex e IF-4F. Internal ribosome binding is dependent on the interactions between a *cis-* acting IRES and a *trans*-acting, cell type specific factor such as p 57/PTB(polyprimidine tract binding protein) required in the internal initiation of entero viruses.

Kozak consensus sequence — The consensue sequence 5′-ANNATGG-3′ encodes the Kozak consensus translation initiation sequence. The A at position -3 and G at position +4 are most critical for function. For example, if a purine base at -3 is replaced by a pyrimidine base, translation of the mRNA becomes more sensitive to any change of nucleotides in positions, -1, -2 and +4 and translation levels may be reduced by 95%. Thus for optimization of eukaryotic *in vivo* and *in vitro* protein expression the inclusion of Kozak's consensus sequence is important.

Kozak's rule- It refers to the prediction that eukaryotic ribosomes screen each mRNA starting from its 5′ terminus for the first translation start codon 5′ –AUG-3' in a defined sequence context to start translation. All other AUGs located further 3′ downstream are not used for translation initiation.

Sequences around initiation site

Sequences around initiation site in different organisms are as follows.

AAAAAAUGUCU(Yeast)

AA(A/C)AAUGCG(Plant)

GCCACCAUGG(Mammals)

The only common nucleotide being A at position -3′ with A of the AUG as 0. The first AUG in mRNA can be bypassed if it is in unfavorable sequence context and a downstream AUG imbedded in a favorable sequence context will be used instead. In mammals the sequences surrounding AUG codons play a role in specifying which one is used as the initiation site.

The **trailer** refers to 3′ noncoding region and 3′ untranslated reion.

The 3′ noncoding region encodes the mRNA 3′ **untranslated region (UTR)**. The 3′ noncoding refers to the sequences at the 3′ end of eukaryotic genes that do not code for proteins but are also transcribed and contain important signals. For examples, the 3′ UTR contains

1. the sequence AATAAA coding for mRNA-poly (A) addition signal which may occur in more than one copy per message and

2. a consensus sequence, YGTGTTYY ("**GT- box**") which is located about 30 bases downstream of the poly (A) signal and possibly involved in transcription termination and polyadenylation of the message.

3. UTR comprises sequences of eukaryotic mRNAs flanked by the coding part to the 5′ and the poly (A) tail to the 3′ side. This region contains Poly (A) addition signal which is located about 5 to 30 nucleotides upstream of the poly (A) sequence and serves as signal sequence both for an endonuclease to cleave the mRNA at a site 14-20 bases down stream of it and for a poly (A) polymerase to add a poly (A) tail to the 3′ terminus of the molecule. The length of **trailer sequence** increases with the evolutionary complexity of the organism as can be seen from the table 5.4 given on next page.

Thus we have seen that untranslated regions (UTRs), the portions of the sequence flanking the coding sequence which are not translated into protein, occur in both DNA and RNA and untranslated regions, particularly at the 3′ end is highly specific to both the gene and to the species from which the sequence is derived. 5′ UTR sequences are important as they are known to accommodate the translational machinery while 3′ UTR is involved in the regulation of gene expression. Finally, most organisms prefer a specific sequence adjoining the termination codon.

Table 5.4   Showing length of trailer in some organisms.

Organism	Length of trailer
Fungi, plants	200
Invertebrates	300
Cold blood vertebrate	400
Human	500

The base immediately down of this codon (UAA in case of *E.coli*, yeast and insects and UGA in mammals and monocotyledonous plants) exerts a strong influence on the translation efficiency. The reduction in efficiency is in the order for UAA, G> U, A>C and for UAG, U,A >C>G. In *E.coli* the translation efficiency varies from 80% (for UAAU) to only 7% in case of UGAC.

5.13   TRANSCRIPTION TERMINATION SEQUENCE

The end point of transcription contains a characteristic type of sequences which determines the end point of transcription. It is composed of an inverted repeat followed by an oligo (T) sequences such as TATATTCTT in DNA sequence. In mRNA the termination sequence appears as a GC-rich hair-pin and a run of several uridine residues, $5'\text{-}(GC)_{10}\text{-}(N)_5\text{-}(GC)_{10}\text{-}(A)_{4\text{-}8}\text{-}U\text{-}3'$ which is able to form stem and loop structure and thus serves as signal for RNA polymerase to terminate transcription. In bacteria, the termination sequence 5′-TTTTTATA-3′ is effective, 5′-TATATA-3′ sequence is very effective, 5′-TACATA-3′ is less effective and 5′-TAGTAGTA-3′ is also less effective. The transcription termination process in eukaryotes appears to involve different signals for different DNA-dependent RNA polymerase. RNA polymerase I recognizes a termination sequence of 18bp to which an auxillary protein is bound RNA polymerase II possibly leaves the template strand after contact with a specific secondary structure at the termination site which is a specific sequence, 5′-TTTTTATA-3′ and RNA polymerase III terminates transcription at a U_4 sequence embedded in a GC-rich region. In yeast the transcription termination of polyII gene occurs at AT-rich sequences, a 38 bp region including the sequence5′-TTTTTATA-3′.

Near-up elements—It comprises a DNA sequence that is located 4-40 nucleotides upstream of poly (A) signal in many plants and some plant viruses and functions as a part of termination signal complex in transcription. The NEUs of different genes contain a different core sequence as shown below.

5′-AAUAAA-3′ (CaMV35S gene complex)

5′-AAUGAA-3′(zein of maize)

5′-AAUGGAAUGGA-3′ (ribulose5phosphate carboxylase/oxygenase gene)

Far –upstream element (FUE)—It comprises a DNA sequence which is located from 40 to 160 nucleotides upstream of the poly (A) signal site in many plants and some plant virus genes and functions as a part of termination signal complex. EUFs of different genes contain a UG-rich core, for example, 5′-UAUUUGUA-3′ in the CaMV35S.

Poly (A) addition signal (poly (A) signal sequence)

It refers to a hexanucleotide consensus sequence, 5′-AATAAA-3′ in animal and 5′-AATAAN-3′ in plant and generally 5′-AATAA-3′, close to the 3′-end of most eukaryotic genes transcribed by RNA polymerase II. It also includes the consensus sequence 5′-AAUAAA-3′ in an mRNA molecule which directs the cleavage of message 10-30 bases 3′ of the element. The cleaved mRNA then serves as a substrate for processive polyadenylation.

Chi-like sequence—Recombination is not random within the human β-globin gene cluster. There are specific sequences promoting recombination. Chi-like sequences are promoters of recombination. The 5′ flanking region of the gene is known to have a chi-like sequence, 5′-GTCGCTGG-3′. There is a '**hot spot**' of recombination between the α and β genes of human globin gene cluster (Kazazian et al., 1982). Chi-like sequences are also involved in recombination in bacteriophage λ and murine immunoglobulin genes. Further, the sequence GGXGGX has been implicated to function as a hot spot for gene conversion events in an early chorion multigene family (Hibner et al.; Kravariti et al., 1995). Specific recognition sequences which promote the conversion were suggested to explain the conversion pattern observed (Rudikoff et al., 1992).

Conserved sequences—Various steps of gene expression involve the binding of factors on short signals (TATA-box, polyadenylation signal, etc.) or the formation of a particular RNA spatial structure such the stem-loop structure at the end of replication dependent histones mRNA. Sequence comparison of evolutionary related genes may reveal different patterns of biologically significant conservation. A region of strong similarity over a short sequence (<15 nt) may correspond to a target of regulatory DNA or RNA-binding proteins such as signals involved in polyadenylation, splicing, transcription or translation. Conversely, weak similarity over long regions (>10 nt) may reveal sequences coding distantly related proteins. Profiles of similarity in the 5′ noncoding regions show two areas preferentially conserved, sequences surrounding the transcription start site and sequences of the 5′ UTR just upstream of the initiation codon. There is presence of highly conserved regions (HCRs) in noncoding parts of genes. HCRs are relatively frequent in particular genes essential to cell life. In mutigene families, conserved regions are specific of each isotype and are probably involved in the control of their specific pattern of expression. Functional constraints are generally much stronger in the 3′ noncoding regions than in promoters or introns. The 3′ HCRs are particularly A-T-rich and are always located in the transcribed untranslated regions of genes which suggest that they are involved in post-transcriptional processes such as mRNA nucleocytoplasmic export, mRNA transport and localization, mRNA translation and mRNA degradation and thus noncoding regions are of functional importance. 3′ HCRs are clearly associated with widely expressed genes (Duret et al., 1993).

5.14 TYPES OF REGULATORY ELEMENTS

There are two types of regulatory elements.

1. *Cis*-regulatory elements(found on the same chromosome)
2. *Trans*-regulatory elements.

The *cis*-elements refer to any DNA sequence which exerts an effect on other DNA sequences located on the same molecule. In other words, *cis* acting DNA sequences are physically linked to the genes in question to influence their activity. Such *cis*-elements do not encode proteins but are binding sites for transcription factors. The *Cis* –regulatory elements include **promoters** and closely associated DNA such as **operators**. These two short adjacent segments are immediately upstream of the structure-determining part of the gene. These two affect the quantity of the enzyme produced. Other examples include enhancers and origin of replications. The promoter is a DNA sequence at which RNA polymerase may bind leading to initiation of transcription and determines the rate at which mRNA is synthesized. The operator is a region of DNA that interacts with repressor or protein (the product of **regulator gene** which controls the production of another structural gene or genes) to control the expression of a gene or group of genes. The *trans*-regulatory elements need not be in close proximity with the target gene. In eukaryotes *trans*-regulatory elements are often located on a different chromosome. *Trans*-acting sequences exert regulatory effect regardless of whether or not they are present on the same chromosome as the regulated gene. *Trans*-acting sequences generally encode proteins which bind to the *cis*-acting DNA sequences, for example, transcription factors.

***Cis*-regulatory sequences** — The eukaryotic promoters include

   I. Transcriptional start(AUG)

  II. A sequence 20-30 bps upstream of this start point (TATA box at -25)

 III. A region found further upstream at about -75.

The transcriptional start site is designated base position +1. The positions with minus signs indicate their occurrence at base positions before or upstream of 5′ end of the 5′- 3′ strand of either DNA or mRNA and the regulatory sequences are called '**upstream sequences'**. The regulatory sequences occurring at positions after the termination codon (UGA) at 3′ end either in DNA or mRNA are called 'downstream regulatory sequences'. The AUG triplet, the initiating codon is closest to the 5′ end of the gene. The Goldberg –**Hogness box** or **TATA box** has the consensus sequence TATA A_T A A_T, an AT-rich region (TATAAATA in plants) and is surrounded by GC-rich sequences. The TATA box is most frequently located at about 15-32bp (in yeast 60-120bp) upstream of the transcription initiation site of eukaryotic structural genes. In the gene for zein, the maize storage protein, two TATA boxes have been found as close as 10bp apart. The function of the TATA box is to direct RNA polymerase to the correct transcriptional start site. The TATA sequence is analogous to prokaryotic **Pribnow box** and represents the binding site for transcription factors. Although it is essential for accurate initiation of transcription but it is not necessary for quantitative expression.

Structure of core promoter — Promoter was thought to be the transcription start point, presumably the binding site for RNA polymerase. But it is now used to refer to the whole upstream segment related with assembly of a transcription initiation complex, including an RNA polymerase and various protein factors that assist (or restrict) its function. Although core promoter structure was initially thought to be invariant but a remarkable degree of structural and functional diversity has been apparent. The core promoter which extends about 35bp upstream and/or downstream of this site (the correct initiation site where the RNA polymerase II positions itself) plays a central role in regulating initiation. Specific DNA elements within the core promoter bind the factors which nucleate the assembly of a functional pre-initiation complex and integrate stimulatory and repressive signals from factors bound are distal sites

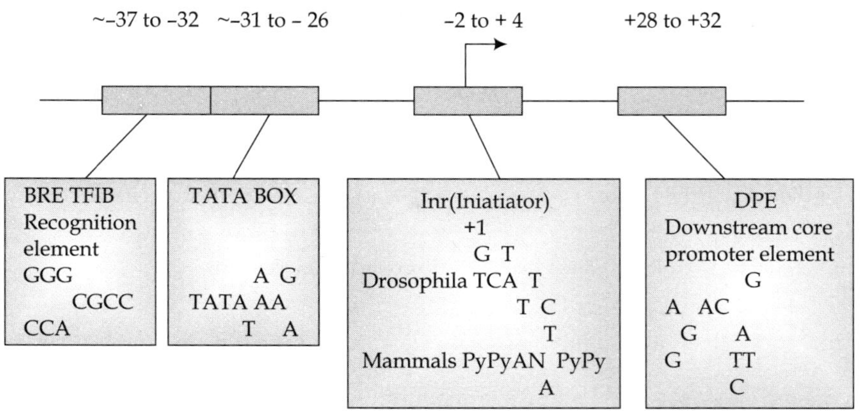

FIGURE 5.5 Showing some of the sequence elements which can contribute to basal transcription from a core promoter.

(Smale and Kadonaga, 2003) (Figure 5.5). The different core promoter elements include the following:

1. BRE
2. TATA box
3. Inr
4. DPE

Figure 5.6 shows some of the sequence elements which contribute to basal transcription from the core promoter. The core promoters comprise DNA sequence motifs -40 to + 40 nucleotides relative to the RNA start site (such as the TATA box, BRE, Inr and DPE) which in appropriate combinations are sufficient to direct transcription initiation by the basal RNA polymerase II transcriptional machinery. Immediately upstream of the core promoter (from about -50 to -200bp relative to transcriptional start site), there are typically multiple recognition sites for a subgroup of sequence-specific transcription factor which include Sp1, CTF (CCAAT-binding TF, also called nuclear factor-1(NF-1)) and CBF(CCAAT-box binding factor, also called nuclear factor-Y or NF-Y). The core promoter together with the promoter-proximal region is referred at as the promoter.

BRE is located at about -37 to -32, TAT box at about -31 to -26, Inr at -2 to + 4 and DPE at +28 to +32. All of which can contribute to basal transcription from a core promoter. Each of these sequence motifs is found in only a subset of core promoters. A particular core promoter may contain some, all or none of these elements. The TATA box (TATAAAAG) can function in the absence of BRE (TF II recognition element, GGA CGCC), Inr (Initiator, consensus sequence in Drosophila TCAGTTC, Mammals $(P_YP_YANTP_YP_Y)$) and DPE (Downstream core promoter element) motifs (AGACG). In contrast, the DPE motif requires the presence of an Inr. BRE is located immediately upstream of a subset of TATA box motif. The Inr element is functionally similar to TATA box and can function independently of a TATA box.

Function and occurrence of TATA box—TATA boxes or AT-rich sequences are located at a fixed distance upstream of the transcription start site and have been found in all animals ,

General structure of a Pol II core promoter module

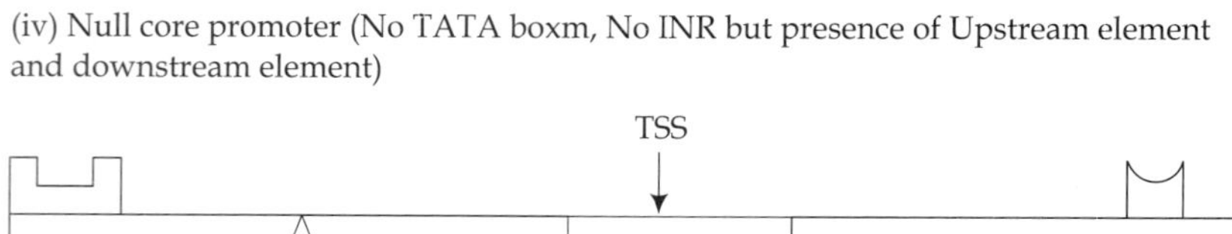

(i) Distinct TATAbox promoter (No INR, No downstream element)

(ii) Distinct INR promoter or TATA-less promoter (No TATA box, No downstream element)

(iii) Composite core promoter (N Upstream element, No downstream element but presence of TATA box plus INR)

(iv) Null core promoter (No TATA boxm, No INR but presence of Upstream element and downstream element)

FIGURE 5.6   Showing general structure of a Pol II promoter containing four elements (UPE, TATA box, INR and DPE) along with four different setups (i – iv). The presence of four groups of promoters indicate that the simultaneous presence of all four elements for functioning of promoter is not essential. The shapes above bar show additional binding sites for protein.

plant and fungi. Mutation in TATA box usually results in reduction or abolition of promoter's activity. Where reduction or abolution does not result initiation site can be displaced from the correct location. Although it was speculated that TATA box might be strictly conserved and essential for transcription initiation from all protein coding genes but the database analysis of human genes has revealed TATA boxes to be present in only about 32% cases and database analysis of Drosophila genes has revealed TATA boxes to be present in 43% of 205 core promoters and 33% of 1941 potential promoters. The TATA consensus sequence 5′-TATATAAG-3′ is the optimal TBP (TATA-binding protein) recognition sequence.

Initiation element—The Inr is defined as a discrete core promoter element which is functionally similar to the TATA box and can function independently of a TATA box. The two elements (TATA box and Inr element) act synergistically with one another when separated by 25-30bp but act independently when separated by more than 30bp. When they are separated by 15-20bp synergy is retained but the location of start site is dictated by the location of TATA box rather than the location of Inr(i.e. initiation occurs 25bp downstream of TATA box). Inr element is recognized by TFIID.

Downstream Promoter Element(DPE)—A downstream core promoter motif is required for the binding of TFIID to a subset of TATA-less promoters. The DPE is conserved from Drosophila to humans and is typically but not exclusively found in TATA-less promoters. The DPE acts in conjunction with Inr and the core sequence is located at precisely -28 to 32 relative to A_{+1} nucleotide in the Inr motif (Kutach and Kodonagar, 2000). Both DPE and TATA box are recognition sites for the binding of TFIID. On the other hand the TATA box but not DPE can function independently of an Inr. NC2/Dr1-Drap 1, a multifunctional factor activates DPE – dependent transcription and suppresses TATA-dependent transcription and thus can discriminate between DPE and TATA-dependent core promoters. RNA polymerase III works with downstream promoters.

TFIIB recognition element(BRE)—It is the only well characterized element in the core promoter of protein coding genes and is recognized by a factor other than TFIID (or the TRF1 and TRF2). Study with human TFIID established the existence of eukaryotic BRE which prefers a 7-bp sequence (Lagrange et al., 1998). Recognition of BRE is mediated by a helix-turn-helix motif at the C-terminus of TFIIB and as this motif is absent in yeast and plants BRE may not contribute to gene regulation in these organisms.

Most promoter elements appear to interact directly with the components of basal transcription machinery. The basal machinery is defined as the factors including RNA polymerase II itself and that are minimally essential for transcription *in vitro* from an isolated core promoter. Studies have been conducted with promoters containing TATA box as core element which show that a stable pre-initiation complex can form *in vitro* on TATA-dependent core promoters by association of the basal factors in the following order: TFIID/TFIIA, TFIIB, RNA polymerase II/TFIIF, TFIIE and then TFIIH.

G-box—It refers to a sequence motif in promoters of eukaryotic genes that functions as address site for transcription factors binding to the conserved core sequence 5′ $^{C}_{G}$ ACGTG-3′. G-box (CCACGTGG) is found in promoters of Arabidopsis RbcS-1 A and alcohol dehydrogenase and of parsley chalcone synthase. Interaction of such binding proteins with the G-box and other transcription factors is a pre-requisite for transcriptional activation of the linked genes.

GC-box—The upstream promoter elements include CCAAT and GC boxes. GC-box refers to the hexanucleotide consensus sequence element, 5′ –gggcgg-3′ in promoters of eukaryotic class II genes which is recognized by the transcription factor SP1and typically occurs in house keeping genes. GC-boxes may occur single or in tandem arrangement and function in either orientation with a positive effect extended by flanking bases. The most effective SP1 binding motif is therefore 5' –GGGGCGGGC-3′ or 5′ –TGGGCGGGGC-3′. Together with related GT/CACC boxes GC –boxes are ubiquitous sequence elements of many promoters and enhancers.

Organ-specific element (OSEs)—It is a *cis-* acting DNA sequence motif of 20-100bp in promoters of eukaryotic genes which is responsible for their organ-specific expression. When deleted, transcription from the resulting mutant promoter is no longer organ-specifically regulated. The OSEs are target sites for the binding of specific transcription factors.

Response element—It refers to any one of a series of short consensus sequences in DNA occurring in the promoters or enhancers of a number of genes that are controlled by the same external stimulus (e.g. temperature: heat shock element, hormones: glucocorticoid response element (GRE), auxin response element (AuxRE), interferon stimulated response element (ISRE), heavy metals (e.g., cadmium, copper, lead, mercury, tellurium or zinc)-metal regulatory element). Response elements are address sites for binding of transcription factors. AP-1 responsive element (ARE) is activated by the presence of oxygen radicals such as O^{2-} and OH^- (gene encoding superoxide dismutase, thioredoxin reductase, thioredoxin peroxidase). Heat-shock element consisting of a conserved palindromic recognition sequence 5′-CNNGAANTTCNNG-3′ is found in promoter of all heat shock genes from yeast to man.

5.15　TWO CLASSES OF TRANSCRIPTIONAL CONTROL SEQUENCES

One class of control sequences specify the position where the RNA synthesis is to start. It probably also determines which strand of the DNA duplex is to be transcribed. These sequences have qualitative effect and are evolutionary conserved. For example, TATA box which designates the transcription start site. The second class of transcriptional control sequences determine the efficiency of transcription initiation but signals do not show evolutionary conservation. The basic mechanism of transcriptional control operates through sequence-specific interactions of a special class of proteins, the TFs, with relatively short DNA elements of ~5-25 bp (Wingender, 1994). There is evidence showing that DNA sequence signals which ultimately provide for the regulation of structural gene expression can lie far removed from the site where transcription starts.

Enhancer—The two important points in the regulation of gene transcription are the point of transcription initiation and the rate of transcription. In prokaryotes, both these are determined by promoters which are in close proximity to the initiation site. In mammals and other eukaryotes, the point of transcription initiation of protein coding genes is determined by proximal promoter sequences whereas the frequency of transcription is influenced by DNA sequences called enhancers which can be located thousands of base pair away from the promoter (Figure 5.7). Although enhancer may occur within introns, they are probably more often outside the transcribed gene sequence, often upstream and sometimes downstream. Enhancers are typically 100 to 300bp long and represent an array of binding sites for DNA-binding transcription factors. Because of its specific DNA sequence any given enhancer binds a subset of the many hundreds of TFs in a eukaryotic organism. Some of these proteins are

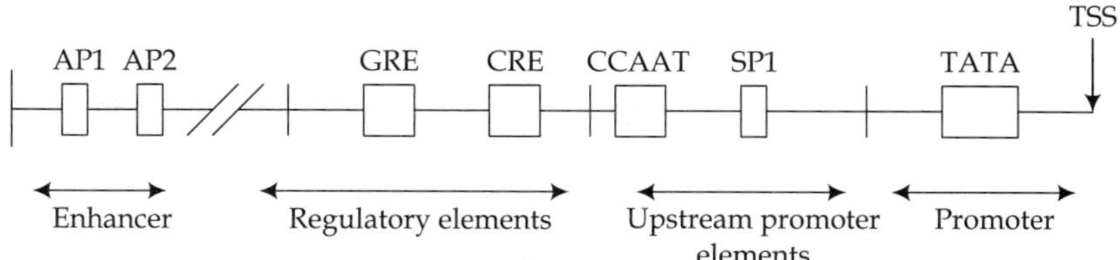

FIGURE 5.7   **Showing structure of a typical eukaryotic gene. AP (Activating protein) 1 and AP2 are binding sites for factors involved in gene expression in response to specific stimuli. GRE (glucocorticoid response element), CRE (cyclic AMP) along with MRE(metal response element) and HSE(heat shock element) are binding sites for different types of response elements. TATA and upstream promoter elements (SP1 and CCAAT)are binding sites for factors involved in constitutive transcription. The TATA box is about 20-30 nucleotide upstream of TSS. The proximal promoter elements is 250-500 nucleotides upstream of TSS and AP1- a complex of two transcription factors, one from the *Fos* and one from the *Jun* family, is located about 900 nucleotides upstream of TSS.**

present only in a given cell type whereas others become active only upon an external stimulus such as by a hormone. Thus an enhancer can activate a linked gene only in the appropriate cell type or in response to a specific stimulus. Many genes are controlled by several distinct enhancers which ensure gene activation in response to different cues (Schaffer, 1999). Enhancers were originally identified as *cis*-acting regulatory elements which increase transcription in a way that is independent of their orientation and distance relative to the RNA start site. For example, the wing margin enhancer of the cut locus in Drosophila lies 85kb upstream of the promoter whereas the murine immunoglobulin Hμ core enhancer lies in the second intron of the transcription unit. The enhancer of the T-cell receptor α-chain gene found up to 69kb downstream of the promoter. In case of transvection, chromosome pairing allows an enhancer on one chromosome to activate transcription from an allelic promoter on the other chromosome (*trans*-acting).

The position of the promoter elements relative to the transcription start site of gene is inflexible whereas position and orientation of enhancers appears to be flexible. Enhancer can increase transcription of a linked gene from a position either upstream or downstream of a transcription unit, in an orientation-independent manner (Figure 5.8). Enhancer have been identified in the gene introns (as described above) or in regions several thousands of bases away from a transcription unit. Further, unlike promoter elements, enhancers do not share extensive sequence homology and therefore can not be easily identified on the basis of sequence data alone. Enhancers fall into two categories: those that are generally active in a wide variety of differentiated and undifferentiated cells and those whose optimal activity is confined to a particular cell type (Rosenthal, 1987).

Transcriptional control regions often contain multiple, autonomous enhancer modules which vary from about 50bp to 1.5kbp in size (Blackwood and Kadonaga, 1998). Each of these enhancer modules seems to be designed to perform a particular function such as activation of its cognate gene in a specific cell type or at a particular stage in development. For example, one enhancer might activate its cognate gene in the liver whereas another might activate the same gene in brain. Thus one gene might contain many enhancer modules and each of which contributes in a more or less cumulative manner, to the overall spatial and temporal regulation of the gene. Enhancers are also distinct from other transcriptional regulatory elements such as

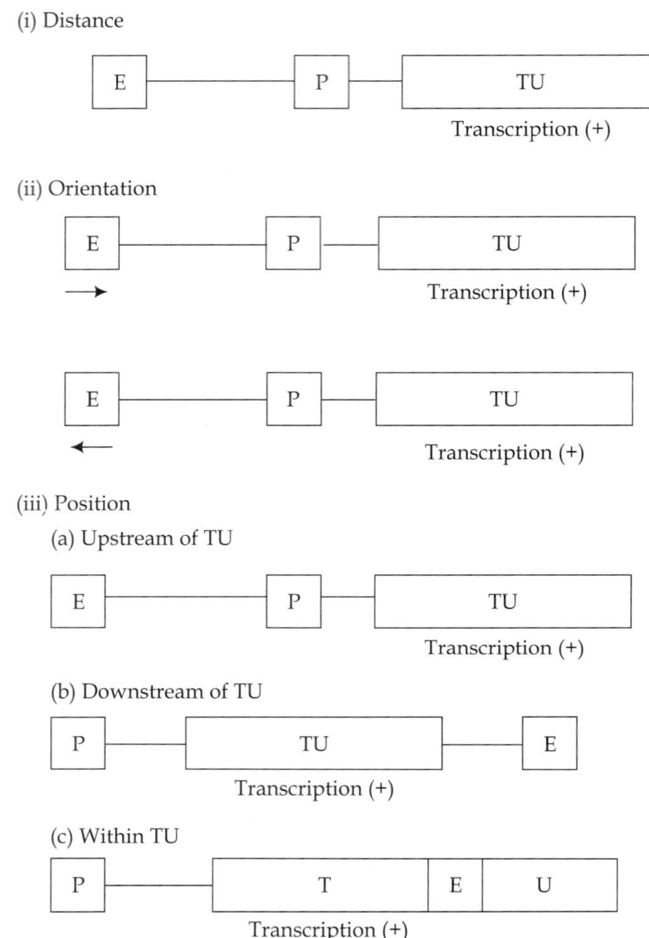

FIGURE 5.8 Showing the effect of distance, orientation and position of enhancer on transcription. E= enhancer; P= promoter and TU= transcription unit.

core promoters (discussed above) and boundary elements (also called **insulator**) discussed below.

Mechanisms of action of enhancer—An enhancer has to do two things. It must select the correct promoter over a large distance and also it must activate only one of the multiple promoters present in its vicinity. There are two possible mechanisms through which enhancer-promoter selectivity could be achieved and either or both could be used. First, there could be specific interactions between enhancer-binding proteins and factors interacting with the promoter. Second, the transcriptional boundary elements could assist in blocking undesirable enhancer-promoter interactions. The autoregulatory element 1 (AE1) enhancer in Drosophila provides an example of preferential interaction between an enhancer and a core promoter. In natural state, this enhancer although lies at equidistance from promoters, Scr (sex comb reduced) and ftz (fushi tarazu), but activates only the ftz expression. The two genes Scr and ftz differ in their core promoter elements. The Scr promoter contains a TATA box whereas the ftz promoter lacks a TATA box but contains Inr and DPE. *In vitro* AE1 can activate transcription from a TATA-less promoter in the absence of a competing TATA-containing promoter but in the presence of both TATA-containing and TATA-less promoters, AE1 activates only the TATA-containing promoter. This shows the importance of components of promoter for productive and specific enhancer-promoter interactions.

Models for enhancer function — Two models have been proposed for enhancer function. In case of 'on or off' model, enhancers can increase the probability that a gene will be transcribed in any particular cell while not affecting the level of transcription in the cells in which the gene is active. In other words, enhancers increase the probability of transcription but not the amount of transcription in a cell. The proportion of cells in which the gene is activated may reflect enhancer strength which is a function of the type and number of its associated TFs. The other model called 'progressive' model is based on observation that there is a uniform and progressive response of a gene to enhancer activity. In other words, enhancers uniformly increase the amount of transcription in a cell in a continuous manner. In this model, genes are uniformly activated and the amount of transcription is proportional to the strength of the enhancer.

Factors involved in activation of enhancer — Various studies suggest a number of factors to be involved in the activation of transcription by enhancer. These include sequence-specific DNA-binding proteins, numerous coactivators, chromatin template as well as chromatin remodeling factors which appear to influence the transcription process. Sequence-specific DNA-binding proteins (activators) interact first with sequences in the enhancer. There appear direct protein-protein contacts between enhancer associated factors and components of the basal transcription machinery formed by DNA looping, leading to the formation of a large transcription complex. Then, there are numerous coactivators which interact with the DNA-bound factors. Coactivators such as those that catalyze protein phosphorylation or acetylation and the enzymatic nature of the TFs can modify properties of the protein, the RNA polymerase II which shows extensive phosphorylation in the early stages of transcription. Also, acetylation of histone seems to reduce the repressive nature of chromatin. The packaging of DNA into chromatin structure appears to promote long distance interaction as a result of compaction of DNA. The ATP-dependent chromatin remodeling factors such as SWI-SNF, NURF, RSC, ACF and CHRAC which can change chromatin structure and increase the mobility of nucleosomes are likely to facilitate the function of transcription factors. Finally, there are a number of observations which suggest the involvement of nuclear architecture in transcription initiation. First, the nuclear matrix appears to be the principal site of active transcription. Secondly, euchromatic genes reside in nonrandom positions within the interphase nucleus whereas both telomeres and centromeres lie at the nuclear pheriphery. Thirdly, there is a relation between nuclear location and transcriptional competence. This has been supported by the finding that insertion of satellite DNA at a euchromatic locus not only directs the localization of the normally euchromatic gene to the 'heterochromatic compartment' but also results in position-effect variegation- a form of heterochromatic transcriptional repression.

Models for establishing enhancer-promoter contact — Three models have been proposed for enhancer function. Formation of a loop between the enhancer and promoter has been the hypothesis behind establishing a productive interaction between enhancer binding proteins, their associated coactivators and the cognate promoter (Wang and Giaever,1988). The problem with this hypothesis is that if the distance between the enhancer and promoter is large than a single large loop has to be formed in order to bring these two close. In other words, considering large distance between enhancer and promoter, the probability of forming small loops is greater than that of a single large loop. Another model of establishing enhancer-promoter contact is DNA scanning. In this model enhancer-binding factors bind to their recognition sequences and then move along the DNA until they meet their cognate promoter. The limitation with this model is that it can not explain how an enhancer could activate transcription from a tailed hairpin that extends outwardly from a double stranded circle or how it could explain the phenomenon of transvection. The third model proposed for enhancer function is called 'facilitated tracking'model. In this model an enhancer-bound complex containing DNA-binding factors and coactivators tracks through small steps (and perhaps scanning) along the chromatin until it encounters the cognate promoter, at which there is formation of a stable looped structure. Chromatin structure is another important component of the tracking mechanism. Coactivators facilitate enhancer-promoter communication and alter a repressive chromatin structure by recognizing and modifying chromatin structure. Furthermore, ATP-dependent remodeling factors might facilitate the DNA-binding factors-enhancer interaction and tracking of factor-coactivator complex along the chromatin template. This model explains most of the phenomena associated with enhancer function such as long-distance and orientation-independent transcriptional activation, the action of boundary elements and transvection.

5.16 OTHER CIS-REGULATORY ELEMENTS

These include **insulator** and **local control region**. Insulator (or boundary element) refers to any DNA sequence element which represents a binding site for specific protein (s), marks the boundary between active and inactive chromatin domains and shields the inactive region from being activated (or *vice versa*). Boundary elements are DNA sequences, vary from about 0.5 to 3kbp in size and are believed to function as transcriptionally neutral DNA elements which block or insulate, the spreading of the influence of either positive regulatory elements such as enhancer or negative regulatory elements (such as silencers or heterochromatin-like repressive effects). When located between an enhancer and a promoter, insulator elements impair the ability of the enhancer to activate transcription from the promoter. It is believed that boundary elements might function to demarcate regulatory domains. For example, the CCCTC- binding factor (CTCF) of vertebrates binds to its cognate sequence in promoters of various genes through its DNA-binding domain containing zinc finger motifs and thereby prevents the stimulating action of enhancer elements in the transcription of genes. The exact molecular mechanism of insulator is not known but it may result from an entrapment of the enhancer by the insulator so that there is no interaction with promoters. Also, specific proteins bind to enhancer (enhancer factors, EFs) or so-called local control regions recruit other proteins and among them is histone acetyl transferase which acetylates histone and thereby induces a local relaxation (opening) of the otherwise tight chromatin configuration. This in turn leads to the transmission of the open configuration along the chromatin fiber which is stopped at insulator elements. Here CTCF binds and recruits histone acetylases which revert histone hyperacetylation and restore the open chromatin configuration. Insulators have been characterized in a variety of organisms ranging from yeast to human. Insulators or chromatin boundaries are thus DNA sequences defined operationally by two characteristics (Gerasimova and Corces, 2001). 1. They interfere with enhancer-promoter interactions when present between them and thus this suggests that insulator might be one more regulatory sequence similar to enhancers and promoters. 2. They buffer transgenes from chromosomal position effects and thus this suggests that insulator might play a role in the organization of chromatin fiber into functional domains. Insulator elements in Drosophila include Mcp, Fab-6, 7 and 8, present in bithorax complex and the scs and scs' elements flanking the 87A7 heat shock gene locus. The gypsy boundary element comprises multiple binding sites for the Suppressor of Hairy-wing (Su(Hw)) protein.

Local control region — The analysis of the transcriptional elements that regulate the β-globulin locus, has led to the identification of a locus control region (LCR). LCR refers to any DNA sequence in a large chromatin domain (10-100kb) which exerts a dominant activating effect on the transcription of genes. LCR prevents the influence of heterochromatic silencing on neighboring sequences. It is used in transgenic experiments as insulator that protects itself and linked genes against the repressive action of heterochromatin. Specialized chromosome structure (SCS) refers to DNA sequences located at the junctions between euchromatin and heterochromatin. SCS elements have neither an inhibitory or activating effect on the genes within the chromosomal domain they flank but they prevent the spread of the repressive influence of heterochromatin or activating influence of an enhancer on these genes. SCSs are local control regions.

LCRs are similar to enhancers in that they consist of multiple activator binding sites but are often complex arrangements of multiple regulatory elements. The two elements differ in that the classical enhancers are orientation- and distance- independent, yet their effect can depend on the site of integration into native chromatin apparently because the effects of chromatin structure can dominate the function of enhancer. In contrast, LCRs stimulate transcription independent of their site of integration into native chromatin although their effects are limited by their orientation and distance. Tissue specific local control element refers to any LCR which directs the tissue specific expression of a family of genes nearby. The human β-globulin LCR corresponds to several (4 or 5) DNaseI-hypersensitive sites which are distributed throughout a 15kbp region that is located upstream of the genes (Figure 5.9). Thus like enhancers, the LCR contains multiple binding sites for sequence-specific transcriptional activators but unlike typical enhancers, LCR first acts at a stage in development before the genes are transcribed and renders all five genes in the locus competent but transcriptionally inactive. Once in this competent state, each of the genes is further regulated by stage-specific enhancers or repressors. Thus we see that the pathway leading to transcriptional activation involves the conversion of an inactive locus to a 'preactivated'(competent) state. This state was detected by an increase in sensitivity to DNaseI digestion from which the genes are subsequently activated. This multistep model in which transcriptional competence precedes transcriptional activation is the currently accepted model. Studies of the chiken β-globulin locus have revealed that the establishment of the competent state correlates with an increase in histone acetylation and general sensitivity to DNaseI digestion. These effects are correlated as hyperacetylation of histones leads to unfolding of the chromatin which should facilitate the accessibility of factors (e.g., DNase I or TFs) to DNA.

Enhancers, LCRs and silencers can increase or decrease expression of multiple genes within regions of genome. Additional DNA elements can limit the effects of these regulatory elements thus subdividing chromosomes into active and inactive regions.

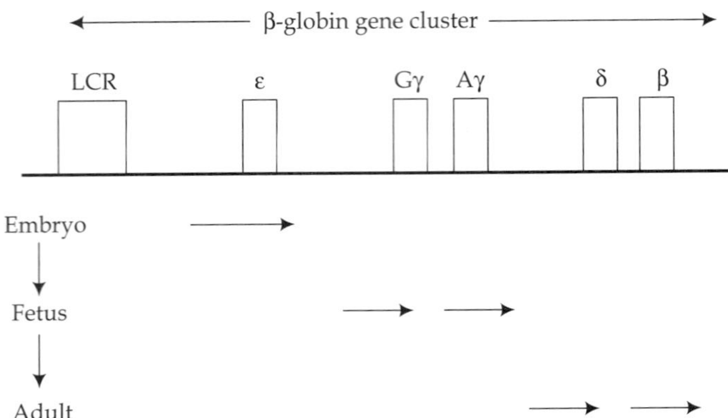

FIGURE 5.9   Showing β-globin gene cluster comprising of five genes and LCR in the gene cluster directs the correct pattern of chromatin opening in erythroid cells. LCR opens the entire region of the β- gene cluster but ε gene is expressed in embryo, γ in fetus and δ and β genes in adult.

Upstream repressing sequences(URSs) — The URSs refer to DNA elements bound by sequence specific gene repressors. URSs bound factors inhibit transcription through various mechanisms including interfering with activator binding, preventing recruitment of the transcription apparatus by the activator and modifying chromatin structure.

5.17   TRANSCRIPTION FACTORS

Transcription factors are proteins which are synthesized in the cytoplasm but are located in the nucleus. Thus they carry nuclear localization signals – a sequence of amino acid that target them to the nucleus after post translational modifications in the Endoplasmic reticulum-Golgi complex. (TFs) show several conserved domains. Because they bind DNA they carry a sequence of amino acid usually basic amino acids that bind to the major or minor groove in the DNA double helix, called DNA-binding domain. Many of TFs are active in DNA binding as homo or heterodimers. Thus besides DNA-binding domain they carry specific sequences that are involved in binding to other proteins and thus involved in protein-protein interactions. Dimerization of these proteins increases the tightness of binding to DNA and further it provides a greater specificity of binding as two heterodimers recognize different sequences in DNA. Many TFs serve to activate transcription . They carry yet another domain, the activator domain which interacts with basal initiation complex or other TFs to activate transcription.

Grouping of transcription factors — More than 2000 TFs are encoded in the human genome. Such proteins have often been classified on the basis of common structural elements. Grouping of eukaryotic TFs can be done on the basis of characteristics that describe their functional roles within cellular regulatory circuits (Brivanlou and Darnell, 2002). A host of proteins crucial to transcription initiation are assembled into the RNA polymerase, the general transcription factors, co-activators, co-repressors, chromatin remodelers, histone acetylases, deacetylases, kinases and methylases. These main proteins mentioned above are present in all eukaryotic cells and contribute to the initiation of RNA polymerase II primary transcript that eventually generates mRNA. About 200 to 300 proteins that constitute the co-activators and transcriptional machinery are also required for survival of cells or organisms and regulation of choice of specific initiation site for transcription is not vested in these proteins.

Transcriptional regulation depends on members of an even larger number of proteins, in mammals perhaps 2000 to 3000 with two characteristic domains. 1. A DNA binding domain (BD) that binds to gene specific regulatory sites directly and 2. a second domain, activation domain (AD) that shows transcriptional activation potential. In some cases the dual requirement is shared between partner proteins so that the site specific binding domain and transcription activation domain occur on separate proteins. These- site specific transcription factors recruit co-activators and transcription machinery to start gene –specific transcription.

Selection among 2000 plus transcription factors for the regulation of cell-specific gene expression involves (i) a cascade of transcriptional control of transcription factor genes and (ii) signals from outside the cell that activate, post-transcriptionally, already formed TFs.

In the regulatory regions in the DNA of a well-characterized vertebrate gene, as many as six to eight different protein chains, acting on one enhancer (together forming an 'enhanceosome') are required for gene-specific regulation (and thus likely to be true for many other genes (Grosschedl, 1995; Thanos and Maniatis, 1995). The combinatorial use of subsets of the 2000 plus proteins could easily mean that the complete set of regulation for each gene is unique, ensuring the right amount of the right protein at the right time during developmental processes. There are also proteins (a small number) that interact with co-activators or positive-acting TFs and/or a member of the transcription machinery to interdict their activity. But in general the negative acting proteins in animals do not bind DNA in a gene-specific manner to sterically inhibit polymerase binding. This is quite different from the classic Jacob-Monod (1961) model of bacterial repressor. The functional classification of positive acting transctriptional factors is given Figure 5.10.

Initial trigger for cell-specific transcription in complex animals often comes from the two groups of regulatory TFs: (i) The factors synthesized in sequence as dependent proceeds and called developmental factors and (ii) the factors activated by extracellular signals, chiefly the

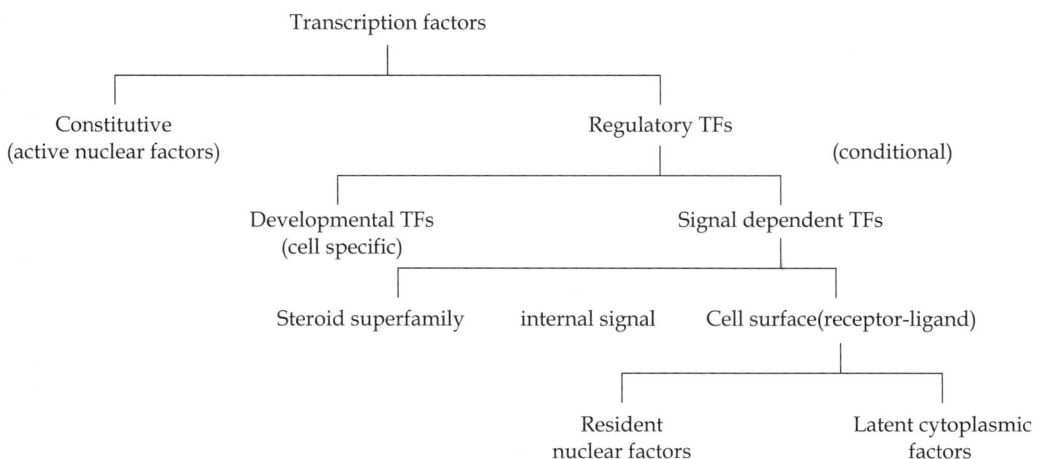

FIGURE 5.10 Showing classification of positive acting TFs.

steroid receptor superfamily and the latent cytoplasmic factors activated by extracellular polypeptodes. Individual genes require multiple pathways and developmental events are affected by mutations in multiple pathways. The evolutionarily earliest major regulatory circuits involved serine phosphorylation of resident nuclears factors and that the duplication of serine kinases provided increasingly complex effects on nuclear function. Constitutive nuclear factors are activated by serine phosphorylation. The proper regulation of genes in fungi and plants is dependent on these pathways. In animals the number of serine kinases, crucial for proper balanced transcription has been amplified.

Plants have a large number of TFs, more than 1500 (5% of the genome) encode TFs and half of these are plant-specific (Riechmann et al., 2000). Worm genome has 500 TF genes, fly has about 700 and the human genome contains more than 2000. The wide variety of TF in plants is because of their complex secondary metabolism. 25% of all plant genes are associated with a unique array of secondary metabolites. The total number of secondary metabolites is close to 50,000.

5.18 PROKARYOTIC PROMOTER

Prokaryotic promoters are characterized by four conserved features.
1. Start point , also called initiator or start site(+1) and it is usually a purine.
2. Pribnow box- a short A-T-rich sequence located at -10 position.
3. An element located around -35 position
4. A fixed distance that separates both of these elements(16 to 18bp).

-35 sequence is recognized by RNA polymerase whereas subsequent opening of the promoter from nucleotides -10 to +3 depends on the Pribnow box.

5.19 EUKARYOTIC PROMOTERS

As the three classes of eukaryotic genes are transcribed by three different RNA polymerases, RNA pol I, II and II, respectively there are three types of promoters.
1. RNA pol I promoters include a core promoter (or selector) and a regulatory upstream control element (UCE). Both elements contain a G-C-rich sequence and are recognized by the ancillary factor UBF1.
2. RNA pol III promoters show more diversity in structure. Downstream (or internal) promoters contain two conserved sequences (the box A sequence associated with either box B or box C) whereas upstream promoters possess a TATA-box in addition to Oct (octamer) and proximal sequence element (PSE) regulatory elements. Positioning of RNA pol III in these promoters is mediated by the transcription factor TFIIIB.
3. RNA pol II (or core promoter) is generally composed of two conserved elements either alone or functioning in synergy. A TATA-box element is usually located 30 nucleotides upstream of the start site in higher eukaryotes and an initiator element (Inr) emcompassing the start site.

Thus we see that are different classes of eukaryotic promoters. TATA-box (TATA) and initiator (Inr) constitute the RNA pol II core promoter and then there is UCE in RNA pol I promoters and PSE and Oct elements in RNA pol III promoters. Eukaryotic promoters show a more complex architecture , usually consisting of a multitude of binding sites for distinct initiation factors.

5.20 MAMMALIAN RNA POLYMERASE II PROMOTERS

They are known to occur in al least two major forms: (i) CpG-rich promoters which are associated with both ubiquitously expressed 'house keeping' genes and genes with more complex expression patterns particularly those expressed during embryonic development. They are the targets of trithorax-group (trxG) activity. (ii) CpG-poor promoters which are associated with highly tissue specific genes. They seem to be inactive by default, independent of repression by polycomb-group (PcG) proteins and may instead be selectively activated by cell type or tissue-specific factors. H3K4me3 is catalyzed by trxG proteins and is associated with activation whereas H3K27 me3 is catalyzed by PcG proteins and is associated with silencing of genes. **Bivalent promoters** are associated with genes with more complex expression patterns including key developmental TFs, morphogens and cell surface molecules. Bivalent promoters are a mixture of the two major classes of promoters and are intermediate CpG content promoters. They show low activity despite the presence of H3K4me3 suggesting that the repressive effect of the PcG activity is generally dominant over the ubiquitous trx G activity. **Alternative promoter usage-** Genes with alternative promoters have multiple, distinct chromatin states. An active site at any of these promoters may be sufficient to drive gene expression. A common situation involves genes with one major CpG promoter and one or more alternative low CpG promoters. Capping studies suggest that alternative promoters use may be substantially more common (Carninci et al., 2005).

5.21 TRANSCRIPTION IN EUKARYOTES

Transcription in eukaryotes requires two sets of factors.

1. The basal or general transcription factors (GTFs) are essential for transcription from a minimal promoter. They assemble onto the promoter to form together with RNA pol II the pre-initiation complex.

2. The sequence specific factors bind their cognate upstream elements or enhancers and then interact with the GTFs in order to regulate transcription. These interactions can be mediated by certain intermediary factors or co-regulators.

   In case of transcription by RNAP-II the GTFs necessary for low-level transcriptional initiation from the proper start point have been separated into TF IIA, -B, D, -E, -F, H and –J (11-14). Many of these factors consist of multiproteins. TFIID consists of the TATA-BOX binding protein (TBP) and at least 8 TAFs (TBP-associated factors).

3. Transcription activators- Some transcription factors do not have a DNA-binding domain but even then they are able to modulate activities of other transcription factors through protein-protein interactions. These transcription factors are sometimes called **transcription activators** although transcription activator is a term more commonly used

independently of a hormonal signal to activate a gene in a tissue- and/or organ-specific manner or may act in concert with a hormonal signal to activate a specific gene.

Transcription is a multistep, multilevel process involving many TFs. A fundamental step of transcription initiation is an interaction of basal transcription machinery(also called pre-initiation complex (PIC)) with a core promoter area of DNA spanning ± 40bp around the TSS. So far a few core promoter elements have been found to be a target for the basal machinery. The most common elements are TAT-box, Initiator (Inr), downstream promoter element(DPE) and TF IIB recognition elements (BRE). There are a few general TFs (TF IIA, B, D,E,F and H) necessary for successful initiation of transcription. In other words, these components of the polymerase RNA polymerase II initiation complex are common to many, if not all genes. TFIID plays the central role in the process acting in co-operation with the core promoter elements and/or specific TFs (Nikolov et al., 1997). TFIID is thus a general and unspecific factor; specificity is often provided by successive binding of other proteins which are specific for particular genes or set of genes. TFIID consists of TAT binding protein (TBP) and at least 12 transcription associated factors (TAFs). RNA polymerase II and several auxillary (or general) factors which are referred to as the RNA polymerase II transcriptional machinery are required for basal transcription. The activity of the general transcriptional machinery is controlled by sequence-specific DNA binding factors that interact with promoters and enhancers elements. In the TAT-box containing TAT+ promoters, TBP binding starts with the process of PIC formation. In the absence of the TAT-box (TAT less promoters) TAFs bind to DNA and/or other TFs in order to involve TFIID (and TBP) in PIC. Several combinations of the core promoter elements were found to be synergistically advantageous for transcription initiation such as TATA-box and Inr, DPE and Inr and BRE and TATA-box.

5.22 TRANSCRIPTION FACTOR BINDING SITES

A crucial component of gene regulation is the binding of transcription factor proteins to specific locations on the genomic sequence in close proximity (100-1000bps) to a target gene which leads to changes in transcriptional activity for that gene. These specific locations termed **binding sites** (BSs) are short (less than equal to 30bp) and share a common sequence of nucleotides though there is usually some variability in sequence between binding sites. These binding sites must have a specific enough shared pattern so that the transcription factor protein does not bind to many random locations throughout the genome but specificity can not be absolute in that varying binding affinities between transcription factors and its target sites are required for different genes (Wei and Jensen, 2006). The collection of sequences that can act as binding sites for a particular transcription factor is called its **binding motif**.

5.23 DIFFERENT TYPES OF GENES WITH DIFFERENT CORE ELEMENTS

The existence of promoter diversity shows that specific classes of genes will contain specific core elements.

1. C_PG-rich promoters are frequently associated with ubiquitously expressed house keeping genes.

2. In retrotransposons termed long interspersed nuclear elements (LINEs) the entire promoter region is located downstream of the transcription start site because these elements are propagated via an RNA intermediate in the absence of LTR The retrotransposons of Drosophila include the jockey, Doc, G,I and F elements, all of which contain a DPE. Thus Drosophila LINE is an example of a class of genes which could not function with an upstream TATA box and are entirely dependent on downstream promoter elements.

3. A third class of genes that may be associated with a specific core promoter structure are genes expressed during the earliest stages of mammalian embryogenesis. TATA boxes may be nonfunctional during early development which suggests that if the expressed genes contain core promoters which function in the absence of a TATA box.

CAAT box—There is one or more regulatory elements centered round -75 which are involved in regulating the frequency of initiation. Two of the most common of these regulatory elements are GC and CAAT boxes which have the following consensus sequences (in animal). GGGCGGG and $GGC_TCAATCT$. These upstream regulatory elements often show homologies between genes. CAAT box is a part of a conserved DNA sequence at about 80bp upstream of transcription start site of many but not all eukaryotic genes. It is a address site for transcription proteins and mediates binding of RNA polymerase II to its template. CAAT boxes are often absent in most yeast and plant's genes. The cereal genes CATC box is at position -90 and has the consensus sequence, CCATCTCNACC which may serve as substitute for CAAT (Kreis et al., 1986). In zein gene AGGA box replaces CAAT box. Eukaryotic promoters differ for different DNA dependent RNA polymerases. RNA polymerase I recognizes one single promoter for rDNA transcription. RNA polymerase II transcribes a multiple of genes from very different promoters which have specific sequences in common (e.g. TATA box at about position -25 and –CAAT box at about position 90). House keeping genes contain promoters (constitutive) with multiple GC-rich stretches with a consensus core sequence, 5'-GGGCGG-3'. RNA polymerase III recognizes either single elements (e.g. in 5SRNA genes) or two blocks of elements (e.g. in all tRNA genes) within the gene (internal control region). All these consensus sequences function as address sites for DNA affine proteins (transcription factors) that promote or reduce transcription. The strength of a promoter refers to the frequency with which an RNA polymerase molecule can bind to specific consensus sequences within a promoter and expresses the linked gene and it depends on the specific sequences (TATA box and –CAAT box) and their exact spacing within the promoter region.

Proximal sequence elements (PSE)—It is a highly conserved element present within the core promoters of small nuclear RNA genes of eukaryotes. It is located between -45 and -60 relative to the transcription start site of the snRNA genes. The PSE is essential for basal transcription and dictates the location of transcription start site. The PSE consensus sequence has been found to vary among organisms. In humans the consensus sequence is TCACCNTNAC/GTNAAAAGT/G. Within a single organism the PSE supports transcription initiation by RNA polymerase II at some snRNA genes and by RNA polymerase III at a distinct subset of snRNA genes.

In human the presence of a TATA box, 15-20nucleotides downstream of the PSE leads to the recruitment of RNA polymerase III to the promoter whereas RNA polymerase II transcribes snRNA promoters containing a PSE in the absence of a TATA box. The PSE is recognized by a unique multiprotein complex called SNAPC, PBP or PTF.

C$_p$G islands—CpG islands are stretches of unmethylated DNA with a higher frequency of CpG dinucleotides. In other words, CpG islands are regions within DNA that often occur near the start of genes where the frequency of the dinucleotide CG is more than in the rest of the genome. C$_p$G dinucleotide is a DNA methyl transferase substrate. 5 methylcytosine can under go deamination to form thymine which is not repaired by DNA repair enzyme. There exists 0.5-2kb stretches of DNA which possess a relatively high density of C$_p$G islands. Human genome contains about 29000 of these islands. CpG islands are believed to preferentially occur at the transcriptional start of genes and it has been observed that most house keeping genes have CpG islands at the 5' end of the transcript. Evidence shows that CpG island methylation is correlated with gene inactivation and has been shown to be important during gene imprinting and tissue-specific gene expression. In mammals the islands are associated with almost half of the promoters for protein coding genes. The islands usually lack consensus sequence or near consensus TAT boxes, DPE elements or In elements. In addition, they are characterized by the presence of multiple transcription start sites which span a region of 100bp or more. In other words, common feature of C$_p$G islands is the presence of multiple binding sites for transcription factor Sp1 (200-202). Transcription start sites are often located 40-80bp downstream of the Sp1 sites. In general there is difficulty in identifying core promoter elements within C$_p$G island which are essential for promoter function.

5.24   GENE TRANSCRIPTION REQUIRING PRIOR PROTEIN SYNTHESIS

It has been found that the transcription of some genes is inhibited if protein synthesis inhibitors such as cycloheximide are included in the medium. This indicates that the transcription of that gene requires the prior synthesis of a protein. When the expression of a gene is not inhibited by cycloheximide, that gene is considered a **primary response gene**, i.e., its induction does not require the prior synthesis of any other protein. As expected many of the primary response genes encode proteins which have turned out to be transcription factors.

5.25   REGULATION OF TRANSCRIPTION INITIATION VIA CHANGES IN DNA TOPOLOGY

The template for transcription is a negatively supercoiled DNA. The efficiency of some promoters is influenced by the degree of supercoiling. Most of these promoters are stimulated by negative supercoiling although few are inhibited. Some promoters are sensitive to the degree of supercoiling whereas others are not. Various regions on the bacterial chromosome are believed to have different degrees of supercoiling, the location of the promoter might determine whether it is sensitive to changes in superhelicity.

The molecular events that change a quiescent gene within compacted chromatin into actively transcribing RNAs are not yet clear (Haince et al., 2006) and unexpectedly enzymes that create and repair DNA breaks appear to be necessary.

5.26 GENE WITH TWO PROMOTERS

In case of human RCC1 gene transcription is initiated at two different promoters about 9kb apart. One promoter is located 5′ upstream of the gene whereas the other lies at the 3′ end of the gene. Initiation at the downstream promoter located at 3′end produces a pre-mRNA in which a 5′ terminal signal noncoding exon is spliced to downstream exons encoding the RCC1 protein. Initiation at the upstream promoter located at 5′ end leads to synthesis of a transcript containing four short noncoding exons spliced to the coding part of the mRNA.

Weak positive element (WPE)—It refers to a *cis* acting DNA sequence motif of 20-100bp in promoters of eukaryotic genes which enhances their expressions. If this motif is deleted then transcription is slightly reduced. WPEs are probably the target sites for binding of specific transcription factors.

In prokaryote, *E. coli*, the promoters consist of two sites of sequences which are at -35 and -10 regions and which have the consensus sequences as 5′-TTGACA-3′ and 5′- TATAAT-3′ (**Pribnow box,**) respectively. Pribnow box a six bp DNA has consensus sequence, 5′-TA-TAATG-3′, functions as binding site of the sigma factor of *E. coli* RNA polymerase. The Pribnow box facilitates correct initiation and is equivalent to the eukaryotic TATA box. **Consensus sequence** is a nucleotide sequence used to describe a large number of related though non-identical sequence. When each position of the consensus sequence represents the nucleotide most often found at that position in the real sequence it is called a **conserved sequence**. In case of prokaryotic promoter at -10 the consensus sequence 5′-TATAAT-3′ comprises the most frequent occurring nucleotides. In this consensus sequence the first two nucleotides (TA) are highly conserved in more than 90% of promoters. The three middle nucleotides, TAA show considerable variation and every promoter contains the last nucleotide T. Similarly, in the consensus sequence found at -35 the first three nucleotides, TTG are highly conserved sequence. The former is a recognition sequence for RNA polymerase recognition and binding site whereas the later is thought to be involved in opening of the double stranded DNA for transcription of the mRNA from the complementary (–) strand. The distance between the two sequences varies between 15-20bps. The promoter of a gene thus contains the switch that determines in which tissue the gene should be off and on and how much of mRNA will be produced. If more mRNA is produced and if more stable then more protein will be produced. Strong promoter means the synthesis of large amount of mRNA. The promoters are position and orientation dependent. It is present at upstream of the gene at 5′ end and it is ineffective if its orientation is changed (in reverted orientation). Jacob and Monod's original distinction between operator and promoter has not been found useful and the word promoter now refers to the whole upstream element concerned with the assembly of a transcription initiation complex including an RNA polymerase and various other protein factors that assists (or restricts) its function. Some of the components of the polymerase II initiation complex are common to many, if not all genes. One of these is transcription factor, TFIID. It is itself a complex proteins and one of which binds to a sequence immediately upstream of the translation start point that usually is close to TATAA box.

Types of promoters—Split promoter   In case of tRNA genes the promoter sequence elements are not contiguous and are arranged as two or more sequence blocks separated by spacer DNA. PolIII promoters are located downstream of the transcriprional startpoint. The promoter is bipartites, consisting of two approximately 20bp sequences with a space of about 20bp in between. **Tandem promoter-** When a promoter is duplicated then two promoters are placed one after another. Such tandem promoters are characteristic of rDNA genes. Tandem promoters result in efficient transcription of the linked genes and thus ensure high expression of linked genes. Promoters in tandem can also be found in histone genes. rrnB promoters are tandem promoters(P1 and P2), are ˜118 nucleotides apart. Both P1 and P2 are active *in vivo*(Brosius and Holy, 1987). **One promoter-two genes-** In *C. elegans* two or more genes can be transcribed from the same promoter with one gene separated form another by no more than a few hundred nucleotides. **Regulated promoter-** It is a promoter which is active as along as an inducer is present. **Constitutive promoter-** It is a promoter that permanently drives the expression of linked genes. The **house keeping genes** are under control of constitutive promoters. The house keeping genes are constitutively active in all cells of an organism and encode enzymes of general anabolic or catabolic pathways. Their expression is not influenced by the environmental conditions and is kept at a constant level. **Synthetic promoter-** The synthetic promoter is synthesized *in vitro* and may include TATA boxes, enhancer cores, CAAT boxes, negative elements. **Hybrid promoter-**This type of promoter is constituted using a consensus sequence(Pribnow box or TATA box) from one promoter and a second consensus sequence(-35 region TTGACA in bacteria or CAAT box in eukaryotes) from another promoter and is designed to direct maximal expression of linked genes. The tac promoter is a synthetic promoter. **Tissue specific promoter-** It refers to any promoter which is activated in specific tissues only and drives the expression of a luxury gene. The barnase gene(killer gene whose gene product leads to the death of the cell where it is expressed) encoding the extracellular ribonuclease barnase when fused to a tapetum- specific promoter is transferred to plants where it is expressed in the tapetum only, produces barnase protein which prevents the formation of microspores causing male sterility(see Roy, 2000). Luxury gene refers to a tissue specific gene which is expressed in one or a few cell types of an organism only. For example, the gene for hemoglobin which is expressed only in erythrocytes, Ornithine transcarbamylase(OTC) encoded by a single X-linked gene is expressed in only liver and intestine.

Strength of promoters—The strength of a promoter refers to the frequency with which an RNA polymerase can bind to specific consensus sequences within that promoter and express the linked gene. It depends on the specific sequences (e.g. TATA box, CAAT-box) and the exact spacing within the promoter region. Thus promoters can be weak or strong. A **weak promoter** is one which does not allow the frequent attachment of DNA-dependent RNA polymerase and thus the adjacent gene can only be transcribed at a low frequency. **Minimal promoter** refers to any promoter which consists only of the essential sequences for correct transcription initiation of the adjacent genes, for example the TATA box and cap site.

Enhancers—Enhancer (strong positive element) increases transcription at the nearby promoters. Enhancer elements are also involved in the regulation of transcription. They are 100-200bp in length and can be 5′ (upstream) or 3′ (downstream) to the transcriptional start and they stimulate the nearest promoters. Enhancers generally contain repeated sequences and no DNA consensus sequences are common to all enhancers. In dicot plants many rbc S genes contain the following sequence. $GTGTGGTT_{CC}AA_TTAT_{AG}G$ at -140 from the transcriptional start (Kuhlemeier et al., 1987) which is similar to enhancer, $NTGTGGAA_{TT}$ of animal gene. A number of other enhancers regions have been identified in specific plant genes. In yeast these additional sequences which are much farther away from the transcriptional site are called Upstream Activation Site (or UAS). UAS must be positioned upstream and within a few hundred base pairs of transcriptional start site. The term UAS is thus typically used to describe elements bound by transcriptional activators that influence transcription from nearby start site. Enhancer is position (orientation) dependent but orientation independent. It can be either at 5′ (upstream enhancer) or 3′ (downstream enhancer) end of the expressed gene, also within introns (internal enhancer) and the coding region, exons, itself (downstream enhancer). It is effective even when far away from the gene (greater than 1000 nucleotide base pairs) and thus function over large distances. Enhancers are thus clusters of DNA-binding sites for transcriptional regulators that influence transcription independent of their orientation and at distance as great as 85kb from the start site (Lee and Young, 2000).

How Enhancers Affect the Efficiency of Promoters?

The possible explanation is that they make contact with promoters through looping of the intervening DNA. In metaphase chromosomes, chromatin fibres are attached in loops to a chromosome scaffold. Some of the scaffold proteins notably the topoisomerase II, are also present in the interphase nuclei and they appear to form a more loosely organized nuclear matrix to which chromatin loops are still attached. Such attachment is in itself thought to be conducive to transcription. Promoters and enhancers could both be within scaffold/matrix attachment region.

Silencers—They are *cis*-elements located upstream in gene. They inhibit transcription and thus repress gene activity. Silencers appear to work against enhancers. Classical silencers are defined as sequence elements that can repress promoter activity in an orientation- and position (or distance) dependent fashion. In higher eukaryotes a CpG motif has been implicated in silencing mediated through methylation. The best example of silencers are the two DNA elements, E and I which contain binding sites for Rap1, Abf1 and the origin recognition complex and are necessary for silencing that requires Sir proteins and histones H3 and H4. These two elements are involved in the silencing of mating type genes at the HMR and HML loci in *S.cerevisiae*. Leader is a short sequence (an RNA sequence) of 162 bps near the translation initiation codon (AUG) or the amino terminus of a protein or the 5′ end of an RNA that has a specialized function. Leader sequence is thus a transcribed part of a eukaryotic gene that follows the cap site and precedes the start codon. The consensus sequence is 5′ A$_G$ NN-3′ (in animals) and 5′C$_G$AANN-3′ (in higher plants). The untranslated part of an mRNA (5′ UTR) that extends from its 5′-terminus to the translational start codon, ATG. Within leader lies a region consisting of G-C rich 3 -4 sequences, called **attenuator** that acts like transcription terminator or **S-D sequences**.

Attenuation—Transcription of prokaryotic (e.g. *E.coli*) gene can be regulated by mechanisms operating at two levels. The first level regulation occurs at the initiation step of RNA in which specific proteins interact with regulatory DNA sequences located close to the site where the transcription starts. Such regulatory sequences can be acted on either to repress or to activate gene expression. The second level regulation relates to premature termination of an elongating RNA chain. This regulatory mechanism referred to as **attenuation** is mediated by a process involving intra strand ribonucleotide sequence complementarity (Yanofsky, 1981). This mechanism for the control of gene expression in eubacteria (transcription attenuation) ensures that a specific amino acid is synthesized only when it is required that is when its concentration in growth medium is low. Attenuation thus can be defined as the regulation of gene expression by selective reduction of the transcription of distal portions of an operon. Attenuation region refers to the region of the leader sequence of a gene which prevents most DNA dependent RNA polymerase molecules from elongating the nascent transcript. The leader region refers to the segment of an operon between the transcription start site and the structural gene (s). Attenuator refers to a transcription termination site within an operon. There are specific proteins such as anti-termination factors which enable the polymerase to transcribe the message beyond the attenuator sequence. It involves the cessation of transcription after the translation of only a small leader region of 14 to 32 residues in length of the total length of less than 300bp of DNA. In prokaryotes it is found at a site between operator and first structural gene and thus lying downstream of promoter. Leader sequence is the transcript that forms the 5' end of the polycistronic mRNA and is believed to be translated but does not form a functional protein. In case of eukaryotes, between the cap and AUG there is a length of non-translated RNA known as the **leader sequence**. The sequence varies from 3 to 256 bp. The translation product of the initial segment of the transcript of each operon is a peptide rich in amino acid that the particular operon controls. If the amino acids is in short supply translation is stalled at the relevant codons of the transcript long enough for the succeeding segment of the transcript to form secondary structures that allow the transcribing RNA polymerase to proceed through a site that otherwise dictates termination of transcription. The genes for pea chlorophyll a/b binding protein and rbc-3A have upstream sequences which inhibit transcription but both these silencers appear to have enhancer properties in the light. The best characterized example of silencer comes from SV40 which is called 'T antigen'.

S-D sequence—There are sequences within genes that do not function to initiate translation and that contain SD regions followed by appropriately spaced initiation codons. The S-D sequence is a ribosome binding sequence located 3-11 nucleotides upstream of the AUG initiation codon in the leader sequence on mRNAs. It consists of seven (or eight) or fewer nucleotides. Its S-D core sequence is complementary to a highly conserved sequence at the 3' end of 16S ribosomal RNA. The 3'end of the 16S rRNA is used to form binary complex required for initiation of translation.

5'-AAGGAGGU-3' (*E. coli* mRNA)

3' UUCCUCCA-5' (*E. coli* 16S rRNA)

The number of complementary bases in the S-D region varies from 3 – to -9 bases. Distance separating the S-D region from the initiation codon also varies from 5 to 9 bases. These variations account in part for differences in mRNA translatability, i.e. control the frequency of translation initiation. Oligonucleotides complementary to the 3' terminus of 16S rRNA inhibit translation. Base substitutions in the S-D sequence which reduce complementarity to 16SrRNA severely restrict translation while complementary changes in the 16S rRNA sequence restore translatability. Gene expression is regulated at the translational level by sequestering or exposing the S-D sequence with a regulatory protein or mRNA structure. Together with start codon AUG the SD box forms the ribosome binding site of prokaryotic mRNA. Most of the bacterial genes use S-D sequence to mark initiation site, however, there are many others which do not use the S-D sequence. In other cases alternate sequences which are complementary to other sites of the 16S-rRNA are located immediately upstream or downstream of the initiation codon. The later signal is referred to as the downstream box (Ito et al., 1993).

Polycistronic operons in bacteria often contain genes that are expressed at different levels. This differential expression of genes has been attributed to RNA processing and/or segmental decay in some cases whereas in others, to differences in the efficiency of mRNA translation. In the later case, particularly the Shine-Dalgarno sequence and its flanking regions determine the level of expression.

5.27 *TRANS*-ACTING FACTORS

Trans-activity of genes through their gene products is the general rule. *Trans* acting refers to the effect of one gene on the activity of another regardless of whether or not the two genes are physically linked whereas in case of *cis* acting physical linkage (usually close) of genetic elements is essential for interaction. Among the *trans*-acting genes are all those that encode transcription factors and proteins whose binding is essential for the function of enhancer sequences, proteins involved in intron splicing and proteins of chromatin complexes. In general, proteins that function by binding to DNA or RNA (for splicing control) are all *trans*-acting and nucleic acid sequences to which they bind are *cis*-acting. However, in Drosophila melanogaster, there are some examples of DNA sequences which act in *trans*, enhancing the activity of cis-positioned gene and also, to some extent of the *trans* homologue if the later lacks an enhancer of its own. This unusual phenomenon called **transvection**, depends on the close pairing of homologous chromosomes in somatic cells of Drosophila. The *trans*-factor proteins such as steroid hormone-receptor protein complexes act by binding to *cis*-regulatory sequences. Steroid hormones react directly with intracellular receptors which are regulatory protein. They can have either positive or negative effect on transcription. Non-steroid hormones bind to cell surface receptor thereby triggering a signaling pathway. No *trans*-acting factors affecting the upstream elements of plant's gene have been identified so far but two *trans*-acting factors affecting enhancers have been identified. They consists of short DNA sequences which may either form binding sites for *trans*-acting substances such as hormone-receptor protein complexes or when linked to a gene can bring that gene under the control of a hormone, heat shock or some other stimulus as in case of control of gene expression by glucocorticoid hormone. The first factor has the sequence, CACATGTGTAAAGGT (which is similar to the enhancer consensus sequence of animal gene) which binds with 15bp conserved region in zein storage protein genes (Maier et al., 1987). The another factor is four GT-rich DNA sequences which share homology with animal enhancers.

Thus the general characteristics of a plant's genes are that they are flanked by 5′ and 3′-non-coding sequences interrupted by small introns and have defined promoters elements such as TATA and CAAT or AGGA and sequences that direct polyadenylation of mRNA precursors or more simply a gene consists of a 5′ untranslated region, a coding sequences which may be interrupted by introns and a 3′ untranslated region. In eukaryotes each gene is attached with regulatory elements and thus the mRNA is monocistronic. The organelle genes (mitochondrial, plastid, etc.) are essentially prokaryotic in nature and they all have S.D. promoter sequences in their mRNA. The mRNAs produced are polycistronic but they do occasionally contain introns. The S.D. region (Shine-Dalgarno region) consists of 6 to 8 bps rich in purine- nucleotides and found at just 3 to 11 bases upstream from the AUG start codon and it is required for transcription initiation. The translation is efficient if this sequence is exactly 8 bases upstream of the start codon. S.D sequences together with initiation codons AUG or GUG form the ribosome binding site (RBS).It pairs with pyrimidine rich sequence near the 3′ end of the 16Sr RNA of the 30S ribosome. S.D region also acts as 'silencer' and prevents the initiation of translation. It is one of the post-transcriptional mechanisms for controlling gene expression.

REP (repetitive extragenic palindromic sequence) — It refers to the highly conserved 38bp consensus sequence, 5-GCCGGATGNCGGCGCNNNNNGCGCCTTATCCGGCCTAC-3 of *E.coli, Salmonella typhimurium* and related bacteria that is located within untranslated regions of operons and is palindromic (i.e., forms a stable stem and loop structure). REP elements are dispersed throughout the bacterial genome, bind DNA topoisomerase II and DNA polymerase I and are potential sites for transcription termination, mRNA stability or chromosomal domain organization.

AU-rich element (ARE) — It refers to a 50-100 nucleotides long conserved sequence elements located in the 3′ UTR in many of the mammalian mRNAs which is rich in adenine and uridine (5′ –AUUUA-3′) and serves as address site for binding of proteins such as human protein AUF1 and also hnRNPD that contains two RNA recognition motifs responsible for an interaction with the usually repeated AREs. The ARE-protein complex functions to shorten the poly (A) tail of the mRNA and thereby destabilizes it. In other words, AREs are the largest class of *cis*-acting 3′UTR-located regulatory molecules that control the cytoplasmic half-life of a variety of mRNA molecules. ARE containing mRNAs are short lived and encode oncoproteins, cytokinins or inflammation mediators.

Nucleotide substitution — Study of the rates of nucleotide substitution (Li, 1997) has shown that the most rapidly evolving DNA sequence is the pseudogenes. Table 5.5 shows the average rates for different types of DNA sequences.

5.28 SYSTEMS OF GENE ORGANIZATION

In the great majority of eukaryotic organisms, a protein encoding gene is defined as a segment of DNA that is transcribed into a single RNA molecule (moncistonic) but then there are many systems of gene organization. In case of prokaryotes (e.g. bacteria, *E.coli*) and in nematode *Caenorhabditis elegans*, many genes are organized in operons. We have also examples of overlapping genes.

Cluster gene — Among the gene products-enzymes (polypeptides), many are multifunctional (see chapter 25) and one function can often be lost by mutation without eliminating other functions of the same polypeptide chain. As a consequence, analysis by mutation and complementation and recombination may subdivide a gene into several functionally distinct domains which may at first be confused with separate genes. In this condition, different mutant alleles can in principle complement one another without mutual conformational correction or even the formation of mixed oligomers though these interactions may also occur. Although most of the known examples of cluster genes are in filamentous fungi, *Neurospora crassa* and budding yeast *Saccharomyces cerevisiae*, they also occur in Drosophila and mammals. The multifunctional, multidomain gene is sometimes called a cluster gene and thus it can be distinguished from a gene cluster.

Pleiotropy — Pleiotropy refers to multiple(or manifold) effects of a gene at the phenotypic level. If a pleiotropic gene mutates so as to loose one of its function but not others then there are

Tabel 5.5 Showing rate of nucleotide substitution in different types, of sequences.

DNA sequence	Rate of nucleotide substitution (Number of nucleotide substitutions per site per 10^9 years.
Pseudogenes	4
Fourfold degenerate sites	>3
Introns	>3
3′ flanking region	>3
Synonymous sites	>3
3′ UTR	>2
Twofold degenerate sites	>2
5′ flanking region	>2
5′ UTR	>2
Nondegenerate sites	0.7
Nonsynonymous sites	0.4-0.5

obvious chances for allelic complementation. The result will be similar to what is observed in case of the genes that encode multifunctional enzymes. There are examples of pleiotropic genes (e.g. dp(dumpy), Ubx (ultrabithorax)) having effects at the morphological level in Drosophila.

Complex loci — In classical genetics, the locus refers to the position of a gene on a chromosome, identified as the site of an allelic difference arising by gene mutation. The locus and gene were treated almost as synonymous as neither the locus nor the gene were thought to be subdivisible by recombination. At molecular level, the mapped site of a mutation is quite likely to be no more than a single base pair change within the long DNA sequence of the gene. The finding of different mutations apparently at the same locus that give a range of different phenotypic effects suggested the presence of complex loci. Examples of complex loci includes R locus from maize, mating type loci in fungi, the scute (sc) and acaete (ac) families of mutations in Drosophila. The R locus in maize which encodes transcriptional activators for genes that encode pigment synthesizing enzymes has been shown to contain three genes. One gene P is specific for pigment and the other two (S1 and S2) apparently duplicate the function in seeds. Thus complex loci can mean a relatively short chromosomal segment (not frequently split up by genetic recombination) that contains a cluster of genes among which there is some functional connection. To qualify as a complex locus, a gene cluster must have some kind of similarity in structure, in function or in both. Then there is presence of groups of functional QTL (see chapter 20).

Operons — It is a system of gene organization in which genes are clustered in operons. The genes (structural genes) performing the sequential functions in a biochemical pathway (e.g., for protein catalyzing related activities) are often arranged in tandem so that they are transcribed on a single **polycistronic mRNA(a single transcriptional unit)** in which each cistron is a messenger template for a single protein. Such a cluster of structural genes is attached with a **promoter** plus other regulatory sequences and function together in regulation. This unit of closely linked structural genes and regulatory elements (regulator, an **operator**) constitutes the 'operon' (Jacob, Monod and Wollman, 1960). In other words, an operon is controlled by a single promoter and contains several genes devoted to the same metabolic pathway such as lactose catabolism in case of the lac operon. Operators are always located close to and in fact, may be part of one of the structural genes they control. Regulators may or may not be located close to the genes they regulate. Lactose or lac operon encodes 3 enzymes of lactose metabolism and trp operon contains genes for 5 polypeptides which make up three enzymes that catalyses tryptophan synthesis. Transcription of the genes is blocked by binding of a specific repressor protein at a DNA site called an operator. Dissociation of the repressor from the operator is mediated by a specific small molecule called an **inducer**. Regulatory regions of bacterial genes do not work in plants because the transcription factors of bacteria are different from those of plants. Further, regulatory system of the gene from a monocot may not work in dicot. **Regulon-** It is a set of structural genes or operons located in different regions within the same genome but having a common mechanism(a common regulator) for the regulation of their expression (e.g. the 8 genes for arginine synthesis in *E. coli*).

trp **operon** — The amino acid tryptophan is synthesized by plants and many microorganisms but not by animals. The trp operon of both *E.coli* and *Salmonella enterica* consists of five major structural genes trpE, trpD, trpC, trpB, trpA (Figure 5.11) which encode five polypeptides that catalyze tryptophan biosynthesis and a transcription regulatory region. The structural genes trp D and trp C have two genetic segments and encode bifunctional proteins (TrpG.trpD and trpC.trpF), respectively. The structural genes are preceeded by a transcribed leader region which contains trpL, region coding for 14-residue leader peptide. Transcription of operon is initiated at the principal promoter trp P1 and at an internal promoter, trp P2 (an efficient promoter) which is located at the distal end of trp D. trp P2 directs the synthesis of a minor transcript containing trp C, trp B and trp A. The operon terminus is defined by two adjacent sites of transcription termination, t and t'. Initiation of transcription at the principal promoter, trp P1 is regulated by tryptophan-activated trp repressor. Continuation of transcription beyond regulatory region into the structural genes of the operon is regulated by transcription attenuation. Transcription of the operon can be terminated at the attenuator in the regulatory region or it can continue to the end of the operon and stop at either of the tandemly termination sites. The t is a relatively inefficient factor-dependent termination site whereas t' is a site of efficient *Rho*-dependent termination. The use of this mechanism permits these bacteria to regulate operon expression in response to changes in the intracellular level of Trp-tRNA^{Trp}. Tryptophan biosynthesis from Chorismate requires catalysis of seven reactions by enzyme domains designated TrpA through TrpG. The first two reactions are performed by a complex containing the TrpE and TrpG domains. The next three reactions are catalyzed by independent functional domains TrpD, Trpf and TrpG domains although TrpG and TrpD are covalently joined as are TrpC and TrpF. The last two reactions occur at the separate active sites of TrpA and TrpB polypeptides of the tryptophan synthase complex with the product of the TrpA reaction, indole channeled to the active site of TrpB.

The attenuator refers to a DNA sequence upstream of a eubacterial operon encoding enzymes for amino acid biosynthesis. The attenuation sequence determines whether the mRNA of the adjacent operon will be truncated or complete depending upon the supply of the amino acid in the growth medium. If the concentration of amino acid encoded by the operon is limiting then the translation of the nascent message will not proceed beyond the attenuator sequence where a number of codons for this specific amino acids are accumulated. The presence of a ribosome at the attenuator region induces the formation of one of the two possible

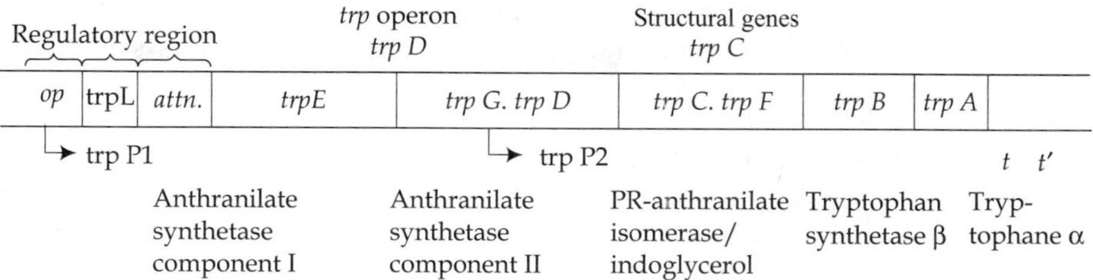

FIGURE 5.11 Showing *trp* operon.

secondary structures in the attenuator transcript. This specific secondary structure in turn permits RNA polymerase to read through the operon. On the contrary, if the amino acid is present in high amount then the alternative secondary structure is formed by the attenuator transcript. This structure signals termination for RNA polymerase at the *att* sequences. This phenomenon thus regulates the expression of operon coding for enzymes essential for the synthesis of a certain amino acid and this mechanism has been extensively studied in trp-operon of *E. coli*.

lac **operon** — The lac operon is polycistronic. It consists of three genes, the Z gene coding for β-galactosidase, Y-gene for lac permease and A gene for transacetylase plus the promoters P1 and Plac and operators, O1, O2 and O3 (Figure 5.12). Two RNA products are formed, one containing Z,Y and A ORFs and another RNA containg Pi, I and Plac. The Plac operator region, the promoter region of the lac operon contains the promoter site(P), the site for binding catalytic activator protein (CAP) and three operator regions (O1,O2 and O3). Two operators like sequences called **pseudooperators** are O2 and O3. The lac operon looked so perfect that a function for its two additional operator-like sequences was not considered for almost 30 years when in 1961 Jacob and Monod proposed this operon. The lac promoter itself needs the presence of CAP and Cyclic AMP to be fully active. The CAP binding site of the lac operon is located 615bp upstream of the transcription start site. Transcription of the lac operon drops down to 2% in the absence of CAP protein bound there. In the absence of both O2 and O3 repression drops 70-fold but if only one is destroyed then repression drops by 2-to-3- fold. The unrepressed transcription of the operon shows RNA polymerase binding to the promoter site

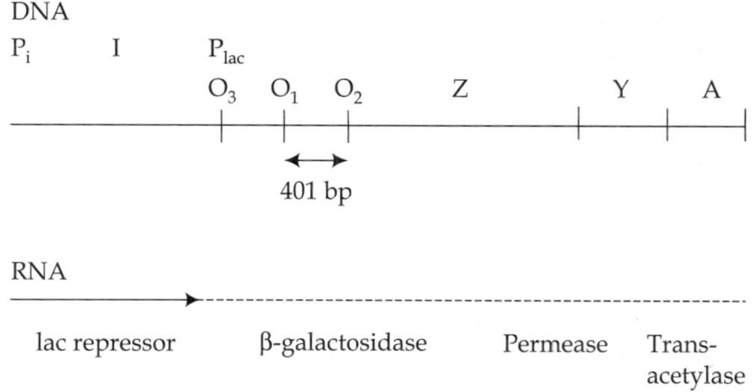

FIGURE 5.12 Showing lac operon.

and dimeric CAP protein with bound cyclic AMP binding to the site. Repression by the tetrameric Lac repressor binding simultaneously to O1 and O3; the CAP and Psites are shown unoccupied although there is no evidence for or against it. Repression is also by the Lac repressor which binds simultaneously to O1 and O2. It is likely that both the CAP protein and RNA polymerase are bound to their sites as shown although this has not been demonstrated. A specific repressor, the product of the lacI gene interacts with the lac operator and thereby inhibits the production of lac mRNA. Mutations in a linked but functionally independent gene, lacI give the same constitutive phenotype as the operator mutations. The critical difference between the operator and the lacI gene is that the former only influence lacZ when it is actually joined to the coding sequence, i.e., in *cis* whereas lacI$^+$ restores normal control to a lacI$^-$ mutant even when it is present on a separate DNA fragment in the bacterial cell. The lacI$^+$ encodes a DNA-binding protein that specifically binds to the lacZ operator, inhibiting transcription and it does through a phenomenon called **allosteric transition**- a change in the 3D structure of a protein in response to binding of a smaller molecule. Allolactose, a derivative of lactose, binds to the lacI$^+$ protein , inducing a conformational change in protein which prevents it from binding to the operator. The lac operon is an example of negative regulation. The inducer, lactose, specifically inactivates the Lac repressor. Initially positive control of gene expression was ruled out but it has been shown that it was wrong. Lac repressor forms tetramers and that two subunits bind to one operator. Dimeric Lac repressor represses only weakly as compared to tetrameric wild type. Tetrameric Lac repressor forms stable loop between either O1 and O2 or between O1 and O3. and thus O2 and O3 are termed **auxillary operators**.

Redundant gene—They are the genes with duplicative or back up functions or functions needed only under special conditions. These genes would not easily be found in the conventional mutant analysis.

5.29   SPACING/SPACERS

Spacing refers to the separation of two specific sequence elements in double stranded DNA molecule by a spacer sequence. Spacing is one of the structural pre-requisites for the correct functioning of protein target sites. For example, if the two target sites for different regulatory proteins are very close to one another then the binding of one protein and interaction between two proteins may be impaired. Thus a specific distance between start codon and TATAbox or between the stop codon and termination sequence is required for the correct functioning. Correct spacing is also important in the construction of expression vector. Spacer is an transcribed repetitive sequences (repetitive DNA) flanking functional genes of eukaryotic and some viral genomes. It also refers to DNA sequences found in between functional address sites (boxes) for DNA –binding protein in eukaryotic promoters. DNA sequences separating specific repeated gene units, for example in histone genes or rDNA are also spacers. Finally, spacers also include DNA region which separates the two parts of a fusion protein in frame and provides a cleavage point for the post-translational separation of these two parts.

Orientation of genes—Two or more genes in the genome be tandemly oriented(can have identical orientation), convergently orientated or divergently oriented as shown on next page.

5' start codon- stop codon- intergenic space-start codon-stop codon-3'

5' stop codon- start codon- intergenic space-start codon-stop codon-3'

5' start codon- stop codon- intergenic space-stop codon-stop codon-3'

5.30 GENE CONSTRUCT

A construct refers to a recombinant DNA molecule. Recombinant DNA technology (see chapter 8) can be used to construct the so called **switched gene** and **fused gene** (or **chimeric gene**). In case of 'Sandwitched gene' the coding sequence is cloned between two regulatory sequences, for example between a start codon or promoter and stop codon or trailer sequence from different sources. The reporter genes for plant transformation are frequently driven by a promoter from cauliflower mosaic virus (CaMV 35S promoter) and terminated by a sequence from nopaline synthase gene of *Agrobacteriun tumefaciens* Ti plasmid. The fused or hybrid or chimeric gene can be constructed in either of the two ways. 1. The coding sequence from gene say B is fused with promoter from say gene A. 2. The gene construct consists of coding sequences from two different genes, fused to each other and transcribed from the same promoter. In other words, coding sequences from genes say A and B are fused and both are further fused to promoter from gene say A. The protein product encoded by fused genes is called '**fusion protein'**. A fusion protein , therefore, consists of an N-terminal part encoded by the 5' end of gene A and a C-terminal part encoded by gene B. Fusion proteins are constructed to combine two different sequence with different catalytic properties, to include an affinity tag in a target protein, or to add a peptide sequence to a target protein that can be cleaved from the latter.

Fusion tag— The fusion tag facilitates detection and purification of recombinant proteins. It is a small stretch of amino acids joined to the N- or C-terminus of a recombinant protein (encoded by a gene fusion). Some fusion tags encode epitope recognized by a specific antibody which in turn is recognized by a secondary antibody labeled with a fluorochrome. Other fusion tags may adopt a specific physical conformation that binds to an immobilized ligands, for example, 5-10 histidine residues(see chapter 26 for more).

5.31 REGULATION OF GENE EXPRESSION

Gene regulation refers to controlling of activity of a particular gene. The expression of gene is regulated by processes which affect the rates at which gene products are synthesized and degraded. Much of the regulation occurs at the level of transcription initiation. Gene regulation can be either **down regulation** (attenuation of expression of a particular gene by environmental factors or cellular factors, e.g. decrease in the number of *trans* acting proteins) or **upregulation** (increase in the activity of a particular gene). **Negative gene control**- This type of gene control refers to termination of gene expression (repression of transcription) by binding of a specific repressor or protein to operator site which prevents the simultaneous binding of RNA polymerase. In case of **positive control** the regulator protein activates the transcription of a gene. In eukaryotes positive regulation is more common then negative regulation. A regulator gene codes for a protein called **repressor** that blocks the activity of an operator gene and thereby preventing the expression of the adjacent gene (s). Repressor binds with high affinity to an associated operator gene. The affinity of repressors for the operators is modulated by small molecules called effectors. In other words, the induction of transcription can be effected by chemical compound called **effector** which is recognized by the repressor protein. The binding of effector with the repressor results in alteration of three dimensional structure (conformation) of the repressor and thus prevents operator-repressor interaction, i.e. repressor protein no longer binds to operator and thus there is transcription of the adjacent operon. The effector is called an **inducer** when it induces the transcription of a specific gene whereas it is called a **co-repressor** when the effector enhances the activity of a repressor protein. The regulatory proteins are DNA-binding proteins which recognize specific DNA sequences and most of these have distinct DNA-binding domains. Within these domains common structural motifs that bind DNA are the helix-turn-helix, zinc-finger and homeodomain. Regulatory proteins also contain domains for protein-protein interactions including the leucine zipper and helix-loop-helix which are involved in dimerization and other motifs involved in activation of transcription (see Roy, 2009).

The **inducible gene** is thus a gene which is generally repressed (under negative gene control) but it can be activated either by a physical factor such as light or UV radiation or chemical stimulus such as hormone, ion or heavy metal. Hormone affects the regulation of gene expression in one of the two ways. Steroid hormones directly interact with the intracellular receptors which are DNA-binding proteins and binding of the hormone can have either positive or negative effects on the transcription of genes targeted by the hormone. Nonsteroid hormones bind to the cell-surface receptors, triggering a signaling pathway which can lead to phosphorylation of the regulatory proteins and thereby affecting its activity. Once induced this inducible gene remains active as along as the inducer is present.

The *lac* operon is an example of inducible operon.

Lactose operon—The lac operon stretches over 6kb DNA segment of *E.coli*. It consists of an operator and three structural genes, Z (coding for β-galactosidase), Y (encoding β-galactoside permease) and A (encoding β-galactoside transacetylase). These genes are transcribed into a single polycistronic mRNA. If a repressor protein encoded by lac I (a gene located 5′ upstream of the promoter is bound to its recognition sequence in the lac operator it effectively blocks lac transcription. This block is released by allolactose (a side product of the β-galactosidase enzyme reaction which converts the disaccharide lactose into the monosaccharides galactose and glucose) which functions as an inducer. It binds to the repressor and induces conformational changes in it. The altered repressor has reduced affinity towards the operator and thus gets dissociated from it and thereby induces lac expression.

House keeping gene—It is also called constitutive gene. It refers to any one of a set of genes constitutively active in all cells of an organism and encodes enzymes of general anabolic or catabolic pathways. The expression of the genes is independent of environmental conditions and is kept at a constant rate.

Regulated gene—It refers to any gene whose expression is conditional, i.e. both the transcription and translation of the corresponding mRNA are controlled by intra- and inter-cellular parameters.

MADS box gene—MADS box refers to a 56bp conserved motif of transcription factors which functions in the regulation of various genes and MADS box gene refers to any one of a series of genes encoding transcription factors containing a highly conserved domain of 56 amino acids and functions as DNA-binding protein.

Homeobox gene—Homeobox refers to a highly conserved short homologous sequence of about 180bp with the core consensus sequence, 5′-TACATTAAT-3′, initially found in 3′exon of homeotic selector genes in Drosophila. They control the pattern of morphogenesis in the developing adult organism by specifying the determination of cells in imaginal discs which give rise to specific body structure of the fly. Homeobox genes occur in a wide variety of eukaryotes from yeast to mammals. The homeobox encodes a protein domain (homeodomain) of about 60 amino acids which are involved in DNA-binding.

5.32 METHODS FOR IDENTIFICATION OF GENES

Methods for identification of genes become important as we have seen that coding sequences comprise only 3% of the human genome. The various methods for gene identification including positional cloning, candidate gene approach and expression screening are discussed in chapter 16 . Computational methods are better in prediction of protein coding genes (see Roy, 2009). There the aim is to combine numerous, often weak, signals such as promoters, RNA splicing junction and branch sites, polyadenylation signals, etc. in order to recognize exons and introns and to assemble the putative exons into complete genes (Guigo, 1997). Gene prediction methods currently have an accuracy of 60-70%. There are five criteria to identify genes in DNA sequence of a genome (Snyder and Gerstein, 2003) which are as follows:

1. ORFs
2. Sequence features
3. Sequence conservation
4. Evidence for transcription
5. Gene inactivation

Open reading frames—An ORF is defined as a string of codons bounded by start and stop signals where codons are nucleotide triplets encoding amino acids. Thus identification of protein-coding genes can be made through identification of large ORFs in the genome. This is particularly applicable to prokaryotes and other organisms with few introns in their genes. Because many genes are short, so difficult to identify in this way. In eukaryotes, organisms with genes undergoing an appreciable amount of RNA splicing often have small exons sandwitched between large introns making ORFs difficult to find.

Sequence features—Even when an ORF identified, codon bias often is used to determine whether the ORF is a gene. The value of codon bias arises from the fact that genes, particularly highly expressed genes show biased non-random use of codons. However, for many genes the bias is weak and small ORFs (or exons) contain too few codons to show statistically significant bias. Besides codon bias splice sites also help to locate genes. Although ORF and favorable sequence features may imply the presence of gene product they say nothing about the function of product.

Sequence conservation—Gene identification can be made through comparing multiple sequences among organisms. DNA sequence conservation among species is an excellent method to guage the importance of the gene products. This method requires sequences of related organisms that are separated by appropriate evolutionary distance. The estimate of gene number by this method can never be absolute (unchanging) as it depends on the specific related organism used for comparison.

Evidence for transcription—To identify genes is to search for RNA or protein expression- the hallmark of gene product. This is commonly accomplished using microarray hybridization, SAGE, cDNA mapping or a sequencing of ESTs (see chapter 24 for details). Large-scale tagging with transposons reveals many regions capable of producing proteins in yeast genome. Further, hybridization of cDNAs to microarray containing sequences of entire chromosomes shows sizable fraction of the chromosomes being expressed. There appears to be conserved ORFs which are not transcribed. Finally, function of many transcribed regions is not known.

Gene inactivation—One method for ascertaining a gene function is to mutate or inactivate its product. This can be achieved by direct gene disruption or RNA interference. However, many coding sequences make products whose inactivation does not result in any obvious phenotype. For instance, only $1/6^{th}$ of the yeast genes are essential and mutations in the remainder usually do not cause any obvious phenotype as long as yeast are grown in rich medium. This presumably reflects functional redundancy among gene products, assay sensitivity or the failure to find conditions under which the product is important. Thus many genes, if not most, genes are difficult to identify solely by this method.

5.33   ADDITIONAL ISSUES IN GENE IDENTIFICATION

These include overlap, alternative splicing and pseudogenes.

Overlap—There are many examples of overlapping reading frames of protein coding genes, overlapping transcriptional units (e.g. where the exon of one gene is encoded within the intron of another) and even overlapping protein-coding and RNA coding genes (see chapter 25 for details).

Alternative splicing—Each gene has a unique functional sequence and is thus distinct. But study of alternative splicing has shown more than half of human genes have sliced isoforms. Gene products from alternatively spliced mRNAs have functionally unique and distinct sequences. So ultimately it may be better to define a molecular part list based on functional protein domains (the protein'domainone') rather than whole gene.

Pseudogenes—Pseudogenes are similar in sequence to normal genes but they usually contain obvious disablements such as frameshifts or stop codons in the middle of coding domains. Thus no functional product is produced or having detectable effect on organism's phenotype. Another problem is their number, for example, there are 80 ribosomal protein genes in human genome but there are >2000 associated pseudogenes. Thirdly, the boundary between living and dead genes is not often sharp and so a pseudogene in one individual can be functional in a different isolate of the same species (for example, FLO8 is active in one strain of yeast but inactive in another). Fourthly, pseudogenes can be transcribed but there are pseudogenes that have entire coding regions without obvious disablements but which do not appear to be expressed. Human ribosomal pseudogenes lack the regulatory elements required for transcription. Further, each transcription experiment does not provide information about every possible gene in a genome and so the likelihood that a gene encodes a functional product is best weighted using multiple criteria. In summary, use of functional genomics along with conservation information obtained from cross-genome comparisons is to be made to identify genes. Pseudogenes in eukaryotes can be classified into those that arose by gene duplication and divergence and those reinserted into the genome from mRNA by a retrovirus, called processed pseudogenes. Processed pseudogenes can be identified by the absence of introns.

5.34   INEFFICIENT TRANSLATION

An inefficient translation is essential for mRNAs encoding transcription factors, growth regulators and other key proteins. The low translation rate of mRNs is achieved by a G-C rich leader sequence present in it.

Gene regulation can occur at any of the following stages.

1. Transcription
2. Post-transcriptional processing of mRNA
3. Degradation of mRNA
4. Translation
5. Post-transcriptional modification of proteins
6. Protein degradation
7. Protein targeting and transport.

Out of these transcription and translation are the two major genetic events where control of gene expression can be asserted.

Transcriptional control — It refers to the regulation of the expression of a particular gene through controlling the number of transcript per unit time. It occurs through transcription factors, TF (*Trans* acting protein, transacting factor) such as TFIIA, TF IIB, TFIIB-RE (regulation element), TFIID (TATA binding protein),TFIIE, TFIIF and TFIIH. These transcription factors are DNA binding proteins that interact with its recognition sequence and facilitate the initiation of transcription by eukaryotic DNA dependent RNA polymerase. TF may bind to URS (upstream repressing sequence), to the TATA box or also sequences within the coding region. Further, *trans*-proteins encoded by sequences on one chromosome but acting on positively(*trans*-activating proteins , activator, *trans* activating factor (TAF) or negatively (*trans*–silencing proteins, repressor) on the expression of genes on another chromosome. Transcriptional silencers (or negative elements) are *cis*- acting DNA sequence motif of 20-100bp, found in promoter of eukaryotic genes that reduce or abolish their expression and are probably the target sites for binding of specific *trans*-acting factors. In case of development of a multicellular organism the fate of cells in the early embryos is determined by establishment of anterior-posterior and dorsal-ventral gradients of proteins which act as transcriptional transactivators or translational repressors and thereby regulating the genes required for the development of structures in the particular part of the organism. The different sets of regulatory genes operate in temporal and spatial succession. In other words, developmental changes are effected through the regulation of transcription and to a large extent, differentiation is determined by the initiation of transcriptions of certain genes at particular times in development, with corresponding termination of transcriptions of other genes.

Transcription initiation is the first and the most highly regulated process in gene expression. In the first step of transcription initiation, RNAP binds to promoter DNA and unwinds ~14bp surrounding the transcription start site to yield a catalytically comptetent RNAP-promoter open complex (RP_0). In the subsequent step RNAP enters into initial synthesis of RNA as an RNAP-promoter initial transcribing complex (RP_{itc}), typically engaging in abortive cycles of synthesis and release of short RNA products and on synthesis of an RNA product of ~9 to 11 nts, breaks its interactions with promoter DNA, breaks or weakens its interactions with initiation factors, leaves the promoter and enters into processive synthesis of RNA as RNAP-DNA elongation complex (RP_e). The mechanism by which the RNAP active center translocates relative to DNA in initial transcription has remained controversial. Kapanides et al. (2006) showed that initial transcription proceeds through a 'scrunching' mechanism in which RNA polymerase (RNAP) remains fixed on promoter DNA and pulls downstream DNA into itself and past its active center. They also showed that putative alternative mechanisms for RNAP active-center translocation in initial transcription, involving 'transient excursion' of RNAP relative to DNA or 'inch worming' of RNAP relative to DNA do not occur. A stressed intermediate with DNA unwinding stress and DNA-compaction stress is formed during initial transcription and in which accumulated stress is used to drive breakage of interaction (s) between RNAP and promoter DNA and between RNAP and initiation factors during promoter escape.

In case of *E.coli* transcription termination can be **Rho factor (p) dependent** or **Rho factor-independent**. Rho factor (p) refers to an oligomeric protein which binds to specific sequences at the 3′ end of a gene and signals stop of transcription. In case of Rho factor-dependent transcription termination the RNA polymerase molecule recognizes the complex between the rho factor and specific termination sequences at the 3′ end of the gene whereas in case of Rho factor-independent termination the RNA polymerase recognizes specific termination sequences as the termination signal.

Alternate splicing—It refers to the unconventional ligation of exons of a particular part of mRNA to form a functional mRNA which differs in information content from normal messege. It produces either longer or shorter mRNA variants when compared to the wild type mRNA and as a consequence the encoded protein contains more or less functional domains. Splicing reaction is a versatile means of genetic regulation. It is a mechanism to produce protein diversity in eukaryotes. Alteration in splice site choice can have many effects on the mRNA and protein products of a gene. Alternative splicing pattern determines the inclusion of a portion of coding sequence in the mRNA giving rise to protein isoforms which differ in peptide sequence and hence chemical and biological activity. Alternate splicing is a major contributor to protein diversity. 60% of human gene products are undergoing alternate splicing. Assembly of eukaryotic mRNA is from their much longer precursors. Moreover, many gene transcripts have multiple splicing patterns and some have thousands.

How changes in splice site choice come about? Splicing pattern can be changed in many ways in a typical multi exon mRNA:

1. Most exons are constitutive, i.e., they are always spliced at or included in the final mRNA.

2. A regulated exon that is sometimes included and sometimes excluded from the mRNA is called a '**cassette exon**'.

3. In certain case multiple cassette exons are mutually exclusive, producing mRNAs that always include one of the several possible exon choices but no more. So special mechanisms must enforce the exclusive choice. Exons can also be lengthened or shortened by altering the position of one of their splice sites. One sees both alternative 5′ and alternative 3′ splice sites. The 5′terminal exons of an mRNA can be switched through the use of alternative promoters and alternative splicing. Primarily, the 3′ terminal exons can be switched by combining alternative splicing with alternative

5.35 ROLE OF INTERGENIC RNA/PROCESS OF TRANSCRIPTION OF DNA IN REGULATION OF GENE EXPRESSION

The process of transcribing DNA can itself regulate gene expression. RNA polymerase II transcribes DNA sequences upstream of a gene and so controls the production of mRNA from that gene. There lies an intergenic promoter, a DNA sequence which prompts the initiation of transcription that produces an RNA without protein coding capacity. The processive transcription starts at this promoter and reads through the downstream promoter of the gene concerned thereby shutting it off. This has been found in case of most of the known genes. These noncoding sequences can influence gene expression (Martens et al., 2004). Gene regulatory processes controlled by transcription of intergenic sequences relied on structural functions of the noncoding RNAs. These RNAs form part of complexes that modify or remodel chromatin by changing the degree of packaging or the epigenetic code of chromatin which is heritable information carried in forms other than DNA sequence. Thus intergenic transcription plays an important role in regulation of gene activity (Schmitt and Paro, 2004).

There is another gene regulatory system which is not intergenic RNA but rather the process of transcription itself triggers changes in the packaging of chromatin. Transcription through regulatory sequences activates certain chromatin domains. Possibly an increase in DNA accessibility is achieved by chromatin-modification and chromatin re-modelling factors that ride piggy-back on RNA polymerase complex.

Translational control — The regulation of gene expression is by modulation of the rate of translation of a specific mRNA into a protein. It results in either qualitative or quantitative translational control. Translational control is through different mechanism such as mRNA degradation, mRNA modification, subcellular localization of mRNA, availability of mRNA for competent ribosome and translational control of RNA and RNA binding proteins. There are three mechanisms of translational regulation in eukaryotes.

1. The initiation factors are subject to phosphorylation by a number of protein kinases and the phosphorylated forms are often less active and thus results in depression of translation.

2. Some proteins called '**translational repressors**' bind directly to mRNA whereas many others bind only at specific sites in 3' UTR.

3. Binding proteins disrupt the interaction between elF4E (binds to the 5' cap of mRNA) and elF4G (binds to elF4E and to poly (A) binding protein) which is essential for the translation.

There are two classes of eukaryotic RNAs which modulate the translation of mRNA. tcRNA isolated from mRNA ribonucleoprotein (mRNP) particles inhibits the translation of mRNA but has no effect on polysomal mRNA. Polysome-tcRNA stimulates the translation of polysomal RNA (a linear array of ribosomes attached to a molecule of mRNA) but has no effect on mRNAp-mRNA. The interaction may be mediated through poly (U) tracts on the tcRNA that forms hybrids with the poly (A) tail of mRNA.

In prokaryotes transcription is controlled through transcriptional attenuator (the transcription termination site) in the mRNA. Formation of the attenuator is modulated by a mechanism that couples transcription and translation while responding to small changes in amino acid concentration. In the SOS system, multiple unlinked genes repressed by a single repressor are induced simultaneously when DNA damage triggers Rec A protein-facilitated autocatalytic proteolysis of the repressor. An SOS response refers to the induction of many distantly located genes in response to extensive damage of bacterial chromosome and error-prone DNA synthesis is a part of the SOS response.

The translational control can be through enhancer of translation, silencers, RNA secondary structure known as hairpins (which can modify translation), amount of tRNAs present and regulation by antisense RNA. In case of synthesis of ribosomal proteins one protein in each r-protein operon acts as a translational repressor. The mRNA is bound by the repressor and translation is blocked only when the r-protein is present in excess of the available rRNA. Some genes are regulated by genetic recombination processes which move promoters relative to the genes being regulated.

Role of conserved noncoding elements—'Extreme' conserved non-coding elements tend to be enriched and clustered near genes expressed during embryonic development. Conserved non-coding regions significantly outnumber coding regions and that these elements are mostly involved in gene regulation. The origins of these elements are largely unknown. One group of these conserved genetic elements has now been identified as originating from a novel SINE family of retrotransposons. Some have acquired function and acting as an enhancer for the expression of a neuro developmental genes ISL1 and another as exon in the mRNA processing gene, PCBP2.

Messenger RNA circularization—It refers to the interaction between the 3' end (trailer) and 5' end (leader sequence) of a eukaryotic mRNA which is mediated by the protein-protein interactions leading to the formation of a loop structure. Poly (A) binding protein once bound to the poly (A) tail of a particular mRNA contacts translation initiation factor e1F4G which in turn interacts with the cap-binding protein e1F4E thereby resulting in circularizing the mRNA in a head-to-tail loop. Transcript circularization increases translation efficiency which can be compromised by the intervention of a protein or proteins bound to the trailer.

5.36 REGULATION AT TRANSLATION LEVEL CAN OCCUR THROUGH TWO MECHANISMS

1. **Translational frameshift**
2. **Editing of mRNA**

Translational shift—There are mechanisms which either allows a single mRNA transcript to produce two or more related but distinct proteins or to regulate the synthesis of a protein. In case of production of *gag-pol* protein which upon maturation becomes reverse transcriptase the reading frame for '*pol*' is changed to the left by one base (-1 reading frame). There is also example of reading frame changing to the right by one base pair. In case of E.coli the release factor (RF2) regulates its own synthesis by changing to the right by one base pair.

Factors Affecting the Rate of Translation

1. Interaction between the ribosome and the bases immediately upstream of the initiation codon of the gene and in prokaryote the key sequence is called the ribosome binding site

or Shine-Dalgarno (S-D) sequence. The degree of complementarity of S-D sequence with the 16S rRNA can affect the rate of translation and the maximum complementarity is with the sequence, 5'-UAAGGAGG-3'.

2. The spacing between the S-D sequence and the initiation codon also affect the rate of translation. The spacing usually ranges between 5 to 50 bases with 8 being optimal. Decrease in spacing below 4bp or increase above 14 bp result in reduction of translation by several orders of magnitude.

3. Sequence of bases following the S-D sequence affects the translation. The presence of four A residues or four T residues gives the highest translation efficiency

4. The codon composition immediately preceding or following the AUG start codon affects the rate of translation

5. Sequences upstream of the S-D site can also affect the rate of translation of certain genes

6. An element downstream of the initiation codon and involved in the formation of translation initiation complex also affects the rate of translation

7. The 3' UTR of the mRNA can also affect the translation. If this region is complementary to sequences within the genes then hairpin-loop can form and hinder ribosome movement along the mRNA.

Editing of mRNA—The factors affecting translation can be genetic and/or environmental. Theoretically factors affecting translation could include amino acid themselves, tRNAs and the ribosomes. Experiment has shown that an amino acid does not determine its position in the polypeptide chain. Further, alteration in either of the two recognition sites of a tRNA molecule could result in faulty translation. Similarly, any change in ribosome will result in misreading of an mRNA codon. Editing of mRNA will affect translation. For detail see chapter 2.

5.37 GENE FAMILIES

Genes may occur as members of structurally related gene families. In other words, many of the structural genes in eukaryotes are members of gene families. The evolutionary importance of gene families depends heavily on their specific function. Gene families are groups of genes with similar sequences that have been derived from a common evolutionary ancestor. The number of members in a gene family can range from 2 to hundreds of thousands per genome. Further, a genome can contain hundreds of gene families and members of a family can either be scattered over several chromosomes or clustered together on a single chromosome. Genes for proteins such as histones (100-1000), ribosomal protein, collagen, actin (5-30) and tubulin (3-15), globin gene (5), myosine heavy chains (5-10), ovalbumins (3), Alu sequence family (>500,000 copies displaying minor variation in the nucleotide sequences) constitute gene families and out of which ribosomal RNA genes in the NOR constitute clusters and also histone genes **tandem gene families** (group of genes whose members are arranged in tandem) and the rest **dispersed gene families** (spread all over the genome). In all these cases although there is only one functional gene but 7 to >20 gene copies per protein are found. Further different isozymes may represent the product of gene families that have evolved from a common ancestor. Isozymes may be also be generated by differential processing of one gene product. The members of the gene family may occur either as closely coupled like operon at one place or spread as individual gene or as small clusters. In case of histones the highly repetitive genes are clumped in caesium chloride- actinomycine density gradients and migrate as satellite separated from the main band of DNA. Using restriction endonucleases such as EcoRI or Hind III DNA fragment containing genes for all five histones H_4 – H_{2B}- H_3- H_{2A}-H_1 were found on the same DNA strand. Each gene was separated by non-transcribed, A-T rich spacer and each gene has its own TATA box promoter in its 5'-flanking sequences. Histones genes are intron-less genes and are repetitive (from about 50 to over 800 copies per haploid genome depending on the species) and clustered. The succession of genes coding for the different histones is species specific and they are separted by spacer DNA. One complete set of histone genes form a transcription unit and transcribed into a polycistronic mRNA which is post-transcriptionally processed. In case of ribosomal RNA (rRNA) gene families gene members are highly homogeneous (translation of genes require a functional ribosome). The **homeotic genes** also form a gene family but are relatively heterogeneous. The homeotic genes are directly involved in the control of genes related to the development of organism. In case of mammalian homeotic gene families, a single Hox gene cluster generally contains 13 genes and this gene cluster is repeated four times. The nucleotide sequences of all gene members are not homogeneous except a small portion (about 60 codons) of the gene called a **homeobox**. The homeobox genes occur in a wide variety of eukaryotes from yeast to mammals. It encodes a protein domain (homeodomain) of 60amino acids which is involved in DNA-binding. In case of homeotic selector gene in Drosophila the homeobox is a highly conserved short homologous sequence of about 180bp with the core sequence 5'-TCAATTAAT-3', found in the 3' exon. The heterogeneity among the gene members in this family is due to each gene having a specialized role in their cascade to control the development of organism. Mechanisms of maintaining among gene family members are transposition, unequal crossing over and gene conversion.

Conserved DNA sequence motifs—Conserved DNA sequence motifs are important for gene function and regulation (parts of promoters, consensus sequence for DNA-binding proteins or regulatory domains of gene families).

Gene battery/gene cluster—It refers to a cluster of closely linked but functionally related genes that are coordinately regulated. Histone genes are organized in such a battery where the sequential arrangement of different genes is a characteristic for the species. For example, histone genes are arranged in the fashion, H2A, H2B, H3, H4 and H1 in sea urchins. Human haemoglobin genes and pseudogenes appear in clusters on chromosomes 11 and 16. All haemoglobin and myoglobin genes have the same intron/exon structure. They contain three exons separated by two introns.

Multi gene family—It refers to a set of closely related genes originating from the same ancestral gene by duplication and mutation (gene amplification). They may either be clustered on the same chromosome, for example, genes encoding ribosomal RNAs) or be dispersed through out the genome (e.g. heat shock protein gene). Most of the members of such families retain a far reaching homology in the coding region but are divergent in the intron and promoter regions.

Super gene family—It refers to a category of genes with different chromosomal locations and restricted (limited) nucleotide sequence homology and which encodes protein with structurally (and functionally) similar domains (e.g. immunoglobulin genes).

Characteristics of a gene family—I. All members of each gene family have two introns inserted at the same position in the coding sequence. II. Genes are arranged in the order in which they are expressed. III. Gene families include pseudogenes i.e., some members of the family are not expressed or are only expressed in specific tissues. Multiple copies of a gene are required for a product to be produced in large quantity. A family is indicated when cDNA hybridizes to more than one area of the genome. It can also be detected through the use of Cot analysis and use of cloned cDNA copy of an mRNA as a probe for the gene from which the original mRNA was transcribed.

5.38 PSEUDOGENES

A region of DNA which shows extensive similarity to a known gene but which can not itself function, either because it has lost the signal (the promoter sequence) required for transcription or because it carries mutations that prevent it from being translated into protein. Pseudogenes are homologous to known genes but which have undergone one or more mutations or frameshifts or deletions and thus eliminating their ability to be expressed. They are closely related to fuctional genes but are inactivated. They show degenerative features such as premature stop codons. Pseudogenes are thought to arise by tandem duplication of genes with following loss of function because of gradual accumulation of distablizing mutations.

Types of pseudogenes—There are two types of pseudogenes. The first types appear as a redundant and nonfunctional copy in a series of tandemly arranged functional repeats and are found in the α- and β-globin gene cluster of mammals. Pseudogenes are found as counterparts of genes that encode protein product and these are thought to be completely non-functional relics of mutational inactivation and could have arisen as a result of unequal exchange during crossing over between different members of the tandem array. Non-reciprocal recombinants arise as a result of nonreciprocal recombination following crossing over between near by sites in the same gene (recombination within a gene). One product of this unequal exchange would have lost a gene and the other product would have an extra and redundant gene that can suffer mutational inactivation without detrimental to the organism. Pseudogenes of the second type are not members of the tandem arrays and they appear at positions quite different then their functional relatives have. Their structures indicate that they have lost all introns and typically have 3′ poly-A sequences at their downstream and because of this (**truncation**) they are also called **processed pseudogenes**. Processed pseudogenes presumably arise by retro transposition (reverse transcription) of processed mRNA followed by integration into genome. Pseudogenes are thus DNA copies of the mRNA and are found in multigene family. They might have arisen by insertion of reverse transcripts of mRNAs. As they do not have polymerase II (Pol II) promoter sequences they are not transcribed. Processed pseudogenes in mammals exceed the corresponding active genes in copy number. New genes may evolve by either grouping of previously unrelated exons into a new transcriptional unit or through gene duplication. Gene duplication is a common method which may give rise to multiple copy and/ or allow evolution of new genes. Partially processed pseudogenes generated by an RNA – mediated mechanism containing or not the complete coding region have been described in tomato and human. Chimeric non-functional genes combining parts of a functional and its pseudogene have been found in lemurs and humans. Pseudogenes are rare in Drosophila relative to some other animals especially vertebrates. Pseudogenes are considered potogenes, i.e., DNA sequences with a potentiality for becoming new genes (Balakirer and Ayala, 2003). Pseudogenes often show functional roles such as gene expression, gene regulation, generation of genetic (antibody, antigen and other) diversity. Pseudogenes are involved in gene conversion or recombination with functional genes. They show evolutionary conservation of gene sequence, reduced nucleotide variability, excess synonymous over non-synonymous nucleotide polymorphism and other features that are expected in genes or DNA sequences that have functional roles.

Eukaryotic genomes are not as tidy as the prokaryotic genomes. Introns, huge intergenic regions, satellite sequences, pseudogenes and many families of transposable elements present in eukaryotic genomes suggest that excess DNA is often not subject to strong negative selection. Pseudogenes are defined as non-functional sequences of genomic DNA derived from functional genes. Pseudogenes genes that have been investigated often show functional roles such as gene expression, gene regulation, generation of genetic (antibody, antigenic and other) diversity. Functional genes are based on several lines of evidence which includes transcriptional activity, intact splicing sites, no premature termination codons and presence of initiation and termination codons. Eukaryotic genomes contain many pseudogenes which have avoided strong degeneration as they are important components of the genome. Pseudogenes may loose specific functions but retain others and even acquire new ones which may not simply be recognizable. Pseudogenes that have been characterized only by their sequence include tRNA, U4 RNA, glutathione S-transferase and histone H2B and pseudogenes for which there is some functional or evolutionary information include ψ Lcp (Larval cuticle protein in *Drosophila melanogaster,*).

Most pseudogenes retain or acquire some functionality and thus it may not be appropriate to define pseudogenes as nonfunctional sequences of genomic DNA originally derived from function agenes or as genes that are no longer expressed but bear sequence similarity to active genes. Rather pseudogenes might be defined as DNA sequences derived by duplication or retroposition from functional genes which are often subject to natural selection and therefore retain much of the original sequence and structure as they have acquired new regulatory or other function or may serve as reservoirs of genetic variability. Pseudogenes are thought to arise by tandem duplication of genes with following loss of function as a result of gradual accumulation of destabilizing mutations. Studies of various pseudogenes have shown the following alterations.

1. Elimination of TATA box due to deletion of bases
2. Occurrence of two premature stop codons
3. Presence of a mutated splicing acceptor sequence
4. Frameshift mutations
5. Several substitutions and indels
6. Frame shift deletion in the signal peptide coding region
7. No consensus polyadenylation sites
8. Deletions in coding region
9. Intronless pseudogenes, Pglym87(phosphoglycuromutase pseudogene)

A processed Adh pseudogene ψAdh, found in *D. yakuba* and *D. teissieri* shows i. retention of reading frame ii. coding bias and iii. a higher rate of substitution at synonymous than at nonsynonymous nucleotide positions. Further, a ratio of 27:4 (silent: replacement polymorphism) was found which is a pattern atypical of active genes. Molecular evolution of ψAdh shows characteristics which are atypical of pseudogenes: i. rate of evolution is substantially lower in exons of ψAdh than in the intergenic region and only slightly faster than the rate of exons of the functional Adh genes ii. Codon bias is retained in most species studied iii. silent substitution in ψAdh significantly exceeds replacement substitution. Ks/Ka ratio (silent: replacement substitution) ranges from 10 to 14 in pairwise interspecific comparisons involving sequences from seven species. These ratios greatly depart from unity which is the value expected for a pseudogene(assuming no selective constraints) and are only slightly lower than those obtained from equivalent comparisons of the functional Adh genes. The replete group ψAdh thus may be a chimeric functional gene. If pseudogenes are devoid of functions then their pattern of nucleotide substitution is expected to reflect the pattern of spontaneous point mutations and 50% of the residues remain identical, have no consensus splice signals and contain multiple premature stop codons and deletions

Pseudogenes may recover the full original function of the genes from which they derive. Pseudogene function has been restored *in vitro* by mutagenesis, transfection or *in vivo* site specific (intermolecular) recombination. In cholella virus PBCV-1, a nonfunctional, nontranscribed cytosine DNA methyl transferase pseudogene differs by eight amino acids from the functional counterpart but a single amino acid change or deletion by site directed mutagenesis restores the pseudogene's activity. Further, a strain of *E.coli* deleted for the lacZ gene has been shown to give rise to spontaneous lactose utilizing mutants as a result of the activation of the ebg pseudogenes.

Sequence Comparison Between Pseudogene and Functional Genes

The pseudogenes have generally relatively minor defects, most frequently single nucleotide substitutions. Indels of single nucleotides which lead to frameshifts or stop codons or loss of an initiation codon are not common. In case of the 'preserved' pseudogene ψP_giC in wild flower of genus Clarkia the exons have not diverged faster in the pseudogene than in the homologous functional gene. The ratio of synonymous to replacement substitutions in the pseudogene is low as expected if there is selection. The conservation of pseudogene sequences retaining 90% and more homology with their functional counterparts has been reported in many organisms. The extent of similarity between the pseudogenes and their homologous functional gene is often not uniform along the sequences. In the human ferrochelatase gene and pseudogene the nucleotide identity is 82-93% in exons 2-11 but only 63% in the 5'-flanking region (Whitcombe et al., 1994). Between a human lactate dehydrogenase gene and its pseudogene the 5' flanking region has twice as many differences as the coding and 3' flanking region. In salmon there is high conservation of up to 80% between intron sequences of the growth hormone pseudogene and the functional gene but much less homology in the coding region. There is also high conservation of intronic regions but much less in the protein-coding regions in the Mojave rattlesnake (John et al., 1996). On the contrary, in the green algae *C. reinhardii*, a cystine-rich protein pseudogene and the functional gene are similar within the coding regions but the similarity drops dramatically in the intron and in the 5' and 3'-non coding regions. The first and second exon sequences show a 91% and 74% similarity, respectively whereas the 5' and 3' flanking regions are only 51% and 41% similar. Similarity between the intron sequences is even lower and is clustered toward the exon-intron boundaries (Matters and Goodenough, 1992).

Exon shuffling—It refers to the production of new genes through intron-mediated recombination of coding sequences(exons) which were previously specifying different proteins or different parts of one and the same protein.

5.39 GENE AMPLIFICATION/GENE DUPLICATION

It refers to the disproportionate replication of specific DNA sequences which leads to their overrepresentation in genome. For example, ribosomal genes are amplified during oocyte formation in amphibians (rDNA amplification in Xenopus the basic set of 500 rRNA). Gene is replicated over 400 times to an amount of about 2000000 copies during oogenesis. Amplification also occurs in insects. There is increase in 18S and 28S rRNA genes which occurs in germ line cells of rDNA-deficient Drosophila flies. The amplification process probably occurs via unequal sister chromatid exchange (**unequal crossingover**). The unequal crossingover refers to the non-reciprocal exchange of segments of homologous chromosomes which are not precisely paired. It leads to the loss of sequences on one chromosome so that chromosomes of unequal lengths are produced. Such an event is enhanced in chromosomal regions where clusters of tandemly repeated sequences are located. The amplified DNA can either be part of chromosome (s) or extrachromosomal (double minute chromosome). The amplified DNA of double minute chromosomes is unstable and only transiently expressed. Gene amplification also involves **gene duplication** which is a process by which an ancestral gene is copied so that the corresponding genome contains two identical gene sequences. One of these genes subsequently undergoes mutation (s) which may convert it into a pseudogene or retains its function inspite of changed sequence composition. Gene duplications occur in all kinds of organism, prokaryotic and eukaryotic alike. The duplicated segment stops signals for protein synthesis. The duplication may lead either to (i) an elongated polypeptide chain or (ii) two separate copies of the protein Thus there are two kinds of duplication- contiguous and discrete. The contiguous duplication results in a larger protein fashioned at the expense of the pre-existing gene product and this process has been the major route to development of larger proteins. In case of discrete duplication, two independent gene products result. One the two is free to mutate, most often to random oblivion but occasionally to adapted to some new roles. The mutations that lead to divergence include single base substitutions besides deletions or insertions. Although they retain many of their pre-existing characteristics, these small differences in their structure affect their interactions and their oxygen-binding properties.

5.40 ORDER OF GENES ON CHROMOSOMES

The precise order of genes on the chromosome is not required to enable organisms to function precisely during development. Drosophila and many other organisms have shown little or no effect of extensive chromosome inversions on the phenotype. Chromosome was earlier believed to be like a series of genes strung together like beads on a string and recombination was thought to occur by breaking and rejoining the string between beads. This distinction between beads and string disappeared with the discovery of DNA and genes are arranged end to end and recombination occurs within genes. Genes can occur as cluster as in case of haemoglobin. Many proteins are present in several different forms coded for by distinct but similar genes. Then genes can occur as tandemly repeated genes with identical function, for example, genes coding for ribosomal RNA. We have also seen in chapter 1 that repeated DNA can also occur as middle repetitive, dispersed and highly repetitive forms.

6

Imprinting

6.1 DEFINTION

In diploid eukaryotic organisms, it is generally assumed that the maternally and paternally derived copies of each gene are simultaneously expressed at comparative levels. However, there are some exceptions where only one of the alleles is expressed. Monoallelic expression with random choice between the maternal and paternal alleles defines an unusual class of genes which comprises X-inactivated genes and a few autosomal gene families. Xist noncoding RNA is the key initiator of the process of X-chromosome inactivation. In eutherians, this silencing involves the Xist gene which is located in the X-inactivation center (Xic) and encodes a long untranslated RNA. Xic is located on the long arm of the human X-chromosome. Monoallelically expressed genes fall into the following three classes (Gimelbrant, 2007).

1. One class is the autosomal imprinted genes such as IGF2 and H19 whose mono allelic expression is regulated in a parent-of-origin-specific manner (Reik,2001).

2. The second class is the X-inactivated genes regulated by random process. Early in development around the time of implantation, half of the cells inactivate the maternal X chromosome and half inactivate the paternal chromosome (Lyon, 1986).

3. The third class consists of autosomal genes subject to random monoallelic expression, including the odorant receptor genes as well as genes encoding the immunoglobulins, T cell receptors, interleukins and natural killer receptors. Thus the third class comprises genes involved in the immune or nervous system. For some genes in this class, some cells express the maternal allele and other cells express the paternal allele whereas for some genes, cells express both alleles.

The definition of **'imprinting'** as the transcription of only one parent's allele is both arbitrary and restrictive. The true purpose of imprinting and indeed the prevalent definition of imprinting is to control transcription of a small number of genes in certain somatic cells(Baroux et al., 2002). Sexual reproduction results from the fusion of gametes in which the chromatin configuration of maternal and paternal chromosomes is distinct at fertilization. In other words, zygote contains two pronuclei with different chromosome features. The chromatin structure refers to a complex system of DNA and protein interactions and modifications which result in a particular three-dimensional architecture that is being envisioned. Although many of the differences are erased during successive cell divisions and chromatin modifications as a consequence of extreme chromatin changes which might be required for genome activation and embryonic development, some are retained in both somatic and germ line cells. These epigenetic modifications can confer different characteristics (**epigenetic marks**) on the maternal and paternal chromosomes and such differences can be selected during any process that has the ability to distinguish between homologues. The end results of these selective forces are parental origin effects. In other words, any epigenetic difference that is maintained between maternal and paternal genomes will lead to the possibility of a parental origin effect. There are three classes of 'genomic' parental origin effects as follows.

1. Those resulting in transcriptional silencing of one parent's allele at some loci.

2. Those resulting in non-random segregation, heterochromatization or destruction of chromosomes from one parent.

   In fact, parental origin dependent chromosome elimination or heterochromatization is used by many anthropods as a mechanism of sex determination

3. Those in which alleles of each parent are clearly distinguished on the basis of DNA methylation or other aspects of chromatin structure but for which 'functional' significance may or may not have been described. Association between differential DNA methylation itself and transcriptional imprinting is complex. Although transcriptionally imprinted genes generally show differential DNA methylarion, differential DNA methylation of imprinted genes is retained even in tissues in which no allele is expressed. On the other hand, differential methylation of genes such as insulin-like growth factor 2 receptor occurs in both the mouse and human even though transcriptional imprinting (of only maternal alleles) occurs in mice whereas both alleles are expressed in the vast majority of humans. In some cases the ability to identify the parental origin of chromosomes is retained even after differential DNA methylation is erased or has yet to be established. Therefore, whatever imprint is involved in parental origin memory must involve additional chromatin marks such as histone modifications (acetylation, phosphorylation and methylation) that have been correlated with DNA methylation, gene silencing and transcriptional imprinting. With the advancement in molecular genetics it is hoped that additional epigenetic marks may be found to contribute to the differential chromatin configuration between paternal and maternal genomes.

In fact, parental origin dependent chromosome elimination or heterochromatization is used by many anthropods as a mechanism of sex determination. The range of effects observed include transcriptional imprinting and effects on chromosome segregation and heterochromatization, reflects the diversity of selective forces in operation. A closer look at these effects suggests that parental origin dependent differences in chromatin structure might be subject to some common forces and that these forces may explain many of the 'nontranscriptional' parental origin effects observed in mammals.

Differential methylation, histone modification, chromosome condensation or heterochromatization are observed between homologues in both somatic and germ line cells of many organisms.

6.2   IMPRINTING AS MECHANISM FOR HOMOLOGUE RECOGNITION

Parental origin-dependent chromatin differences or parent–dependent marking of chromosomes may be utilized during pairing, recombination and segregation. During meiosis, maternal and paternal homologues of each chromosome must find each other (pair), synapse and recombine, i.e., DNA double-strand break repair between maternal and paternal chromosomes. Although there are four members of a homologous pairing complex at meiosis, two maternal chromatids in one dyad and two paternal chromatids in the other, double strand break repair must take place between maternal and paternal chromatids and not between paternal and paternal or maternal and maternal (sister) chromatids. Failure of pairing and recombination gives rise to high rates of non-disjunction and aneuploidy leading to an obvious mechanism by which natural selection could distinguish between homologues was degraded. Clusters of imprinted genes provide complementary 'imprinting/pairing code' which allows specific recognition to occur between homologues and to initiate pairing during meiosis. In somatic cells these marks may also provide the mechanism to distinguish homologues from sister chromatids during DNA repairing processes.

6.3   THEORY OF IMPRINTING

The kinship theory of genomic imprinting (Trivers and Burt, 1999) proposes that parent specific gene expression evolves at a locus when changes in the level of expression in one individual influences a trade-off between the individual's own fitness and the fitness of one or more of individual's non-descendent kin. This fitness trade-off will be expressed within the individual in which the gene is expressed as a physiological trade-off (e.g. between allocation of lipid to BAT or WAT) or a developmental trade-off (e.g., between the allocation of cells to spongiotrophoblast or the gian cell layer of the mouse placenta). Thus what is required is to identify physiological and developmental trade-off which are influenced by expression of an imprinted gene and relate these to fitness trade-off among kin (Haig, 2004). Imprinted genes in four 'clusters' are associated with the imprinted loci Igf2, Igf2r, callipyge and Gnas, respectively. Igf2 and Igf2r contain paternally expressed transcripts which act as enhancers of prenatal growth and maternally expressed transcripts which act as inhibitors of prenatal growth respectively. For expression at these loci a fitness trade-off exists between a fetus' own fitness and its mother's fitness via other offspring. This fitness trade-off is expressed within the mother as a physiological trade-off between resources transferred to the fetus vs resources retained for other uses. It may be expressed within the fetus as a trade-off between allocating resources to the development of placental tissues that directly increase maternal investment vs allocating resources to other tissues. However, clusters also contain imprinted genes whose phenotypes as yet remain unexplained by the theory. The principal effects of imprinted genes in the callipyge and Gnas clusters appear to involve lipid and energy metabolism. The kinship theory predicts that maternally expressed transcripts will favor higher levels of noshivering thermogenesis (NST) in BAT of animals that huddle for warmth as offspring. The phenotypes of reciprocal heterozygotes for Gnas knock outs provide provisional support for the hypothesis as does some evidence from other imprinted genes. The theory thus has been able to provide an evolutionary explanation for the imprinting of some loci but is yet to provide such an explanation for the majority of the imprinted loci.

6.4   TERMS RELATED TO EPIGENETICS

Epinucleic is a type of heredity based upon the restrictive information transmitted in tissue lines of multicellular organisms and it can not sensibly be attributed to DNA–sequence codes.

Epigenesis — Alterations in genome heritable states occur through a change in the primary sequence of DNA and thus a genetic alteration or mutation can be said to have occurred. There is another type of change (epigenetic change) which is an alteration in the heritable states of DNA function produced by altering the chromatin which provides the packaging for the read out of the genome. Epigenetic changes generates different types of tissue under different conditions and in cellular responses of the organism to environmental factors. Epigenetic variation refers to the occurrence of differences between two genomes of two related organisms that are based on non-heredity and also reversible modifications as, for example, methylation of cytidyl residues in genomic DNA. Molecular mechanisms underlying epigenetic phenomena include modifications of DNA and changes in chromatin structure which are thought to be propagated through templating (Russo et al., 1996). **Epigenetic modification** refers to any reversible but inherited change of DNA beyond the level of the base sequences, for example, methylation of cytosyl residues in C_pG dinucleotides of the target DNA is reversible by demethylase but methylation pattern is heritable. Epigenetic mechanisms include all processes such as changes in chromatin structure or DNA methylation relating to the gene expression (transcription, translation) and interaction of the genetic material. It may act at three levels: 1. It may directly regulate gene function by switching on or off or may modulate the synthesis of specific type of proteins 2. They may regulate cell differentiation by modifying the translation of RNA into proteins 3. They may regulate the topographic distribution and function of proteins. Epigenetic modification of the genome ensures proper gene activation during development and involves, i.e. genome methylation changes ii. the assembly of histones and histone variants into nucleosomes and iii. remodeling of other chromatin associated proteins such as linker histones, polycomb group, nuclear scaffolding proteins and transcription factors. Epigenetic changes to chromatin thus include DNA methylation and histone-post-translational modifications. These marks were initially classified as either activators or repressors of gene transcription. In this way simple combinations of marks would have predictable outcomes. But evidence now shows that some marks have opposite roles depending on their context so chromatin marks represent a more complex 'language' than has been appreciated previously. Trimethylation of H3K4 enriched near active genes has improved the ability to accurately predict gene activity based on this and other histone modifications.

The two parental genomes are formatted during gametogenesis to respond to the oocyte environment and proceed through development. The zygote biochemically remodels the parental genome shortly after fertilization and before embryonic genome activation (Ride out III et al., 2003).

Epiallele can be defined as any one of the two or more different forms of a gene which are epigenetically regulated. Epigenetic regulation may involve methylation of cytosol residues in the DNA underlying the allele or chromatin remodeling of the affected allele. Remodelling involves the repositioning of nucleosomes through modifications of the core histones by acetylation, methylation, ubiquitimylation or phosphorylation and a subsequent change in chromatin compaction. Alternately a change in DNA topology may change its accessibility to regulatory proteins. In other words, epigenetic regulation involves reversible changes in DNA methylation and/ or histone modification patterns. Short interfering RNAs (siRNAs) can detect DNA methylation and heterochromatic histone modifications causing sequence-specific transcriptional gene silencing. In animal and yeast, histone H2B is known to be monoubiquitinylated and this regulates the methylation of H3. However, the relationship between histone ubiquitination and DNA methylation has not been investigated. H2B deubiquitination by SUP 32/UBP26 is required for heterochromatic histone H3 methylation and DNA methylation.

Epigenotype—It refers to a normally stable and heritable(during at least many cell generations) genotype upon which the socalled epigenetic code is superimposed, i.e. the specific pattern of methylated or otherwise modified bases are present in the genome. For example, the genotype in all cells of a multicellular organism is identical but the different types of cells have different distribution patterns of 5-methyl cytosine and thus these cells have different epigenotypes. In other words, the mode of impression is over and above or in addition to the classical genotype, i.e. the base sequence in DNA.

Epiphenotype—It comprises the totality of interactions among genes constituting the genotype and between genes and non-genetic environment resulting in the phenotype (epiphenotype).

Epigenome—It refers to a complete set of genes involved in the genetic imprinting.

Epimutant—It refers to phenotypic variants which are epigenetically rather than genetically different from their parents.

Epigenetics—Epigenetics is derived from the Greek word 'epigenesis'. In the epigenetic mode of development an organism develops progressively through a series of causal interactions of various parts. The epigenetic theory differs from the preformation theory that the development occurred by a simple enlargement of a preformed homunculus in the germ cell. Epigenetics has been defined in many ways. Waddington (1942) defined it as the study of processes by which genotype gives rise to phenotype. Holliday (1994) defines epigenetics to be the study of changes in gene expression which will occur in organisms with differentiated cells and the mitotic inheritance of given patterns of gene expression. He gave the second definition of epigenetics as the transmission of information from generation to generation other than the DNA sequence itself. In comparison to 'genetics' epigenetics studies those form of inheritance which do not rest on the chemical DNA. In other words, epigenetics deals with nuclear inheritance not based on difference in DNA sequence. But if there are elements aside from DNA which are responsible for the inheritance of traits then it should be considered central to the particulate theory of inheritance. The properties of genes can be studied at two levels. 1. Mechanism of transmission of genes from generation to generation- a science of genetics and 2. mode of action of genes during development of organism from the fertilized egg to adult- a field of epigenetics. Thus in the theory of gene(Morgan,1912) which deals with the study of heredity a geneticist must separate issues of inheritance from issues of development. 'Epi' means besides, around, upon or over and thus **Epigenetics** implies substances other than DNA that impinge on heredity. In other words, epigenetics refers to this control of gene expression based on chromatin organization rather than on primary (genetic) DNA sequence information (Bender, 2004). The current definition is that it deals with the study of changes in gene function that are heritable (mitotically and meiotically) but that do not entail a change in DNA sequence. It is defined as the science of study of mechanisms generating the phenotypic complexity of an organism. In other words, it deals with the causal analysis of development. The changes in gene activity during development are generally referred to as **epigenetic**. DNA replication and repair provide high fidelity inheritance over evolutionary period whereas epigenetic inheritance is less rigid. DNA methylation can mediate epigenetic inheritance during development and even between generations. Stable protein based inheritance is well established for prions and chromatin-based mechanisms are thought to maintain developmental states. Epigenetic effects provide a mechanism by which the environment can very stably change things. Epigenetic changes enable cells to respond to environmental signals conveyed by hormones, growth factors and other regulatory molecules without having to alter the DNA itself. Epigenetic changes are crucial for the development and differentiation of the various cell types in an organism as well as the normal processes such as X-chromosome inactivation in female mammals and silencing of the mating type loci in yeast(Loo and Rine, 1995; Holmes et al., 1996). However, epigenetic states can be disrupted by environmental conditions or aging. Epigenetics encompasses the effects on gene regulation of chromatin marks (modifications to DNA or its associated proteins, RNAi and the higher order structure of chromosome effects that are not necessarily heritable).

Epigenomics — It refers to a collection of techniques which allow to analyze epigenetic parameters such as methylation pattern of cytosine residues in genes or promoters during the activation or silencing of the genes in different developmental stages or after specific treatments of cells (e.g. abiotic stress) and identification of genes involved in genetic imprinting. High resolution mapping of cytosine methylation is done by using whole-genome microarrays. This modification co-localizes with repeat sequences and with centromeric regions.

Epigenome — It refers to the complete set of genes involved in genetic imprinting. The **DNA topology** refers to the three dimensional arrangement of the double helix and its changes in response to physical (temperature, irradiation) or chemical (e.g. EtBr, restriction endonuclease, DNA modifying proteins or regulatory proteins).

6.5   GENETIC/GENOME IMPRINTING

Genomic imprinting is a phenomenon which causes some genes to be expressed according to their parental origin and results in a developmental asymmetry in the function of parental genomes (Ferguson-Smith and Surani, 2001). Genetic imprinting found in both plants and mammals uses DNA methylation to yield gene expression that is dependent on the parent of origin. Imprinted genes play roles in prenatal growth, development of particular lineages and in behaviour as well as being implicated in human diseases. Imprints are initiated during gametogenesis and are inherited by mature gametes and then transmitted to embryos. Thus there could be parent-specific effects in the offspring. Some researchers have suggested that genes passed on by each parent had some how been **permanently marked** or **imprinted** as it is eventually to be known so that the expression patterns of the maternal and paternal genes differ in their progeny. These so called **imprints** have been found in angiosperm, mammals and some protozoa. It refers to an epigenetic process through which the male and female germ line of viviparous taxa confer a specific mark on certain chromosomal regions so that only the paternal or maternal allele of a gene is active in somatic cell. The imprinted regions are characterized by parent-of-origin DNA methylation at distinct C_pG dinucleotides and asynchronous replication, i.e. the paternal copies replicate earlier than maternal copies and are controlled by imprinting centers (ICs). Imprinting process has three steps. It starts with an imprint switch of the parental chromosomes in the germ line such that the previous imprint is replaced by a new one (germ cells of male or female individual possess chromosomes with paternal or maternal imprinting).This imprinting is replicated at each mitosis (**imprinting replication**). In interphase nuclei the imprinting is recognized by the transcription machinery such that either the paternal or maternal allele of an imprinted gene is transcribed (**imprinting recognition**). In mammals some 100-200 genes are imprinted. Imprinted genes are a small set of genes whose expression depends on the parent from which they are inherited. So, for some of these genes only maternally inherited copy is expressed and for others only the copy inherited from the father is expressed. Such selective gene expression is regulated by selective addition of methyl groups to the two equivalent parental chromosomes during the development of gametes (eggs and sperms). These chemical marks are then propagated in the resulting offspring.

Genetic imprinting is thus a process in which epigenetic information (i.e. information other than that encoded in the nucleotide sequence of genes) is introduced into chromosomes and is stably replicated together with chromosomes during cell division. Imprinting is carried over to progeny, can affect gene expression (often many cell generations later) and is reversible. Unlike heritable changes due to mutation or directed gene rearrangement (as in case of immunoglobulin genes), imprints can normally be removed from the chromosomes without leaving behind any permanent alteration of the genetic material. Thus imprinting can be seen as a remote control switch which once turned on, may have no effects for many cell generations but eventually gene expression and thus the phenotype will depend on this initial switch. There is a particular class of imprints which mark the parental origin of genomes, chromosomes and genes in mammals. Parentally imprinted genes are those genes whose expression depends on whether they are carried on a maternally or paternally derived chromosome. In other words, some are expressed only from maternal chromosomes and others only from paternal chromosomes. Androgenetic embryos-embryos possessing two paternal sets of chromosomes are useful for the studies of imprinting of maternal and paternal genomes.

Imprinting mutation—It refers to any mutation in one or (more) of the imprinted genes. Such mutations usually occur in so called imprinting centers and either lead to complete silencing of genes in adjacent gene cluster or the expression of both parental alleles and thus increase in transcript doses. Either of these events triggers physiological, frequently pathological changes in the carriers. For example, aberrant biallelic expression of human growth factor, IFG2 on the chromosome 11 leads to a series of phenotypes of the so called Beckwith- Weidemann syndrome (the significantly bigger size of the newlyborn children).Epigenetic changes, so-called **epimutations** occur because nucleotide sequence is not the only form of genetic information in cells. Chromosomal proteins and DNA methylation can also be inherited with important phenotypic consequences. In plants and filamentous fungi, genomic methylation is restricted mostly to transposons and other repeats. In mammals by contrast coding sequences are methylated as well except for the so-called C_pG islands which often encompass the first exons of genes. This difference likely reflects the colonization of mammalian introns by transposable elements and the possibility of methylation spreading into flanking exons. Human genome contains far less exonix DNA (<2%) than transposons (>45%) which thus contribute more to the level of cytosine methylation overall.

6.6 CHROMATIN CHEMISTRY(HISTONE MODIFICATION) AND CHROMATIN REMODELLING

Cells transmit information to the next generation via two distinct routes, genetic and epigenetic. Genetic inheritance is based on DNA code whereas epigenetic information comprises modifications occurring directly on DNA or on chromatin- a proteinaceous complex associated with DNA. The major type of DNA modification is the methylation at cytosines but there are multiple modifications associated with chromatin and their inheritability has been demonstrated only in rare cases. **Chemical modifications—** acetylation or methylation of histone proteins determines whether genes on the surrounding DNA are active. In other words, histone modification discriminates heterochromatin from euchromatin in eukaryotic genomes. Methylation, acetylation, phosphorylation and other changes are likely to occur in combinations that define a **'histone code'** which fine-tunes gene expression. Chemical modifications change the chromatin structure, sometimes clearing the way for transcription and other times blocking it. In other words, a 'histone code' directs the recruitment of chromatin remodeling complexes appropriate for either euchromatin or heterochromatin formation. Thus chromatin proteins are much more than static scaffolding and they form an interface between DNA and the rest of the organism. The distinct levels of chromatin organization are dependent on the dynamic high order structuring of nucleosomes, the basic or fundamental unit of chromatin. In each nucleosome roughly two superhelical turns of DNA (~150 bp) wrap around an octamer of core histone proteins formed by four histone partners: an H3-H4 tetramer and two H2A-H2B and this DNA/ protein complex is further compacted by high-order folding. Adjacent nucleosomes are connected by histones of H1 linker class. Histones are small basic proteins consisting of a globular domain and a more flexible and charged NH_2-terminus (**histone tail**) which protrudes from the nucleosome. Chromatin is subject to a diverse array of posttranslational modifications that largely impinges on histone amino termini thereby regulating access to the underlying DNA. In other words, the N-terminals tails of different histone subunits which extend outward from the nucleosome core can be post-translationally modified at several different lysine, serine and arginine residues. Distinct histone amino terminal modifications can generate synergistic or antagonistic interaction affinities for chromatin associated proteins which in turn dictate dynamic transitions between transcriptionally active or transcriptionally silent chromatin states. This combinatorial nature of histone amino terminal modifications thus shows a **'histone code'** which considerably extends the information potential of the genetic code (Jenuwein and Allis, 2002). Chromatin structure plays an important regulatory role and multiple signaling pathways converge on histones. All histone proteins themselves come in generic or specialized forms and exquisite variation is provided by covalent modifications(acetylation, phosphorylation, methylation and ubiquitination) of the histone tail domains which allows regulatable contacts underlying DNA. The enzymes transducing these histone tail modifications are highly specific for particular amino acid positions thereby extending the information content of the genome. Combinations of these are thought to contribute in various ways to chromatin organization and gene expression. The effect of histone methylation is context dependent and is shown in Table 6.1. Similarly, chromatin modifications play different roles as can be seen in Table 6.2.

Table 6.1 Context dependence of histone methylation.

Methylated histone	Location	Transcriptional activity	Trancriptional status
H3K4me3 H3K9me3	ORFs	Inducible	Dynamic
H3K9me3 H3K20me3	Pericentric heterochromatin	Constitutive	Silenced
H3K4me3 H3K27me3	Embryonic stem cell bivalent domains	Constitutive	Poised for transcription

Table 6.2 Chromatin modifications.

Marks	Transcriptionally relevant sites	Transcriptional role
DNA methylation		
Methylated cytosine (meC)	CpG islands	Repression
Histone post-translational modifications		
Acetylated lysine(Kac)		Activation
Phosphorylated serine/threonine(S/Tph)		Activation
Methylated arginine(Rme)		Activation
Methylated lysine(Kme)		Activation
Ubiquitylated lysine(Kub)		Repression
Sumoylated lysine(Ksu)		Activation/ Repression
Isomerized praline(Pisom)		Activation/Repression

6.7 HISTONE CODE HYPOTHESIS

Histone code hypothesis predicts that the modification marks on the histone tails should provide binding sites for effector proteins. In other words, it predicts that distinct modifications of histone tails would induce interactions affinities for chromatin associated protein. Bromodomain the first protein selectively interacts with a covalent mark (acetylated lysine) in the histone NH2-terminal tail. Chromodomains of HP1 is highly selective for methylated H3 at lysine 9 and thus is a targeting module for methylation mark. Histone code hypothesis further predicts that modifications on the same or different histone tails may be interdependent and generate various combinations on any one nucleosome. All four NH2-termini of the core histones contain short 'basic patches' which often comprise acetylation, phosphorylation and mathylation marks in close proximity in one individual tail. The hypothesis further predicts that distinct qualities of high order chromatin such as euchromatin or heterochromatin domains are largely dependent on the local concentration and combination of differentially modified nucleosomes. Thus **nucleosome code** then allows assembly of different epigenetic states leading the distinct readouts of the genetic information such as gene activation vs gene silencing.

Methyl groups in particular methyl lysine have a lower turn over than do acetyl or phosphoryl groups. Acetyl or phosphoryl groups can be removed from histone tails by the activity of HDACs or phosphotases whereas histone demethylases (HDMases) are yet to be characterized. If it does not exist than histone lysine methylation would be very nearly perfect long term epigenetic mark for maintaining chromatin states. In contrast to DNA methylation where the methylated imprint can be removed by nucleotide excision followed by repair-DNA replication and semiconservative nucleosome distribution appears as the role means to dilute histone lysine methylation below a critical threshold value. Another potential mechanism for removing methylation marks from histone tails is **proteolytic processing**.

In plants, animals and most species of fungi euchromatin is associated with hyperacetylated histones H3 an H4 and methylation of lysine at position 4 of histone H3 whereas heterochromatin is associated with underacetylated histones H3 and H4 and methylation of lysine at position 9 of H3.

6.8   GENE EXPRESSION

Genes constitute only a small proportion of the total genome. Precise control of their expression in the presence of overwhelming background of non-coding DNA presents a problem for their regulation. Non-coding DNA containing introns, repetitive elements and potentially active transposable elements require effective mechanism for its longer term silencing.

Higher order chromatin organization plays a key role in determining patterns of gene expression. The less compacted euchromatin regions are the most accessible for transcription whereas highly compacted heterochromatin regions are refractory to transcription. Thus same gene can be either well expressed or transcriptionally silent depending on whether it lies in euchromatin or heterochromatin. Plants can inherit epigenetic changes through meiosis but mammals erase and reset genomic methylation pattern early in embryogenesis (Yonder et al., 1997). In mammals, the two periods of epigenetic programming are gametogenesis and early embryonic development. Sex specific changes to epigenome occur as a wave of DNA demethylation followed by DNA methylation and chromatin modifications. DNA-methylation plays a prominent role in allele-specific gene expression that occurs during parental genomic imprinting and X-chromosome inactivation. It is a unique feature of germ line and there is control of gene function from one gene to the next. In germ cells there is diverse set of histone variants and highly abundant proteins are thought to be important for chromatin organization. In plants there is absence of cytosine methylation in euchromatin but heterochromatin domains are determined in part by methylation of cytosines. The methylated 5′position in cytosine is exposed in the major groove of B-form DNA where it can serve as a recognition determinant for DNA-bind protein. The methylation of C5′ position of cytosine in DNA in higher plants and animals is associated with transcriptional gene silencing or so-called epigenetic gene inactivation.

Cytosine methylation modification is inherited on one of the two daughter strands after DNA replication, providing an epigenetic memory of the original methylation imprint on the parental DNA duplex. In such hemi-methylated daughter strand the methylation imprint can be re-established on the newly synthesized strand using the pre-existing methyl groups on the complementary strand as a guide. Thus once methylation is established it can be perpetuated indefinitely by 'maintenance' methyltransferases which use hemi-methylated DNA as a substrate even if the original stimulus for the methylation is no longer present. In most cases where methylation imprints are lost form plant genomes, it is not due to direct excision of the methylated cytosines but instead by failure to maintain. Cytosine methylation provides a heritable mechanism for altering DNA protein interactions to help in such silencing. Genes can be transcribed from methylation free promoters even though adjacent transcribed and non-transcribed regions are extensively methylated. In Arabidopsis fewer than 5% of expressed genes have methylated promoters although about 1/3 of genes were methylated in their ORFs. Such methylation in the body of genes was found to correlate with genes that are both highly transcribed and constitutively expressed. By contrast genes with methylated promoters had lower expression levels and frequently had tissue-specific expression patterns. The distribution of cytosine methylation is in contrast to that observed in mammalian genomes which are often densely methylated but have hypomethylated CG islands in gene promoters (Henderson and Jacobsen, 2007). Gene promoters can be used and regulated while keeping non-coding DNA including transposable elements suppressed. Methylation can be used for long term eipigenetic silencing of X-linked and imprinted genes and can either increase or decrease the level of transcription depending on whether the methylation inactivates a positive or negatively regulatory elements(Jones and Takai, 2001). Methylation blocks gene expression, is an oversimplification. Methylation changes the interactions between proteins and DNA which leads to changes in chromatin structure and either a decrease or an increase in the rate of transcription. Methylation of a promoter C_pG island leads to binding of methylated C_pG binding protein (MBDs) and transcription repressors including HADcs and to a block of transcription initiation. DNA methylation is a major epigenetic modification of the genome. Methylation occurs predominantly at the symmetrical dinucleotide C_pG. Methylation of silencer or insulator elements blocks the binding of the cognate binding proteins potentially abolishing their repressive activities on gene expression. Methylation changes the appearance of the major groove of DNA which can result in alternative effects as transcription through modulation of the interaction with DNA binding proteins. There are several examples of epigenetic silencing in plant kingdom. 1. Paramutation in maize, PAI and SUPERMAN gene silencing in Arabidopsis and transgene silencing in many plant species. Further, proper DNA methylation lelevs are required for normal development in Arabidopsis.

Gene expression is not determined solely by the DNA code itself. Gene activity also depends on a host of epigenetic phenomenon which is defined as any gene regulating activity that does not involve changes in DNA code and that can persist through one or more generations. Gene activity is influenced by

(i) proteins that package the DNA into chromatin, the protein-DNA complex which helps the genome fit nicely into the nucleus,

(ii) enzymes that modify both those proteins and the DNA itself and

(iii) even by RNAs.

In other words, proteins and RNAs can alter gene activity and tell the genes when and where to turn on or off. A variety of RNAs can interfere with gene expression at multipoints along the road from DNA to protein. 1. PTGS 2. RNA can also act directly on chromatin, binding to specific regions to shut down gene expression (see chapter 23). Sometimes RNA can shut down an entire chromosome. Females have XX chromosome but only one X is active. If both would have been active then it would have been producing twice as much of the X-encoded protein as male's cells. An RNA called XIST, translated from an x-chromosome gene, binds to one copy of the X-chromosome brings about a series of modifications of its chromatin that shuts the chromosome down. Male fruit flies also use an RNA to solve the dosage compensation problem but now by turning up the gene activity of male's single X-chromosome to match that of female's two X. So the unit of inheritance, i.e., a gene now extends beyond the sequence to epigenetic modifications of that sequence. The role of proteins and RNAs in gene expression is based on the observations that some genes from one parent are 'silenced' in embryo so that certain traits are determined only by the other parent's genes and further some tumor suppressor genes are inactivated without any mutations thus increasing the propensity for cancer. Chromatin modifying enzymes are now considered the 'master puppeteers' of gene expression.

6.9 DNA METHYLATION- THE MOLECULAR BASIS OF IMPRINTING

DNA methylation is a key epigenetic determinant that controls parent of origin-specific gene expression (imprinting) in mammals where methylation is erased and re-established in each generation. In contrast, epigenetic states of gene expression in flowering plants are often inherited unchanged over many generations. Methylation in mammalian genome is generally restricted to C_pG sequences. In plants, in general, methylation is preferentially targeted to repeated sequences including centromere associated repeats, ribosomal RNA encoding (rRNA) repeats and transposable element sequences. Plants and fungal genomes have methylated cytosines in the symmetric dinucleotide context 5′ CG 3′ are the most abundant species but there can also be substantial methylation of cytosines in the symmetric context 5′ CNG 3′ and even methylation of cytosines in other asymmetric contexts depending on the locus being examined. Thus the density of methylation varies from locus to locus. The SUP and PAI have extensive asymmetric and C_pX_pG methylation in Arabidopsis. The post-synthetic addition of methyl groups to the 5′ positions of cytosines alters the appearance of the major groove of DNA to which the DNA-binding proteins bind. These epigenetic 'markers' on DNA can be copied after DNA synthesis, resulting in heritable changes in chromatin structure. Methylation of C_pG –rich promoters is used by mammals to prevent transcriptional initiation and to ensure silencing of genes on the inactive X-chromosome, imprinted genes and parasitic DNAs. Cytosine methyl transferase enzyme catalyzes the transfer of an activated methyl group from S-adenosyl methionine to the 5′-position of the cytosine ring (5′-Me-C). In mammals, there are two distinct types of methyl transferases.

Dnmt1 and Dnmt3 have been fully characterized. Dnmt1 is primarily involved in maintenance whereas Dnmt3 is involved in establishment of genomic methylation patterns. In plants, Arabidopsis, MET class encodes genes related to the mammalian Dnmt1. The CMT-chromomethylase is also related to Dnmt1 except that a novel chromo-domain amino acid motif is inserted between 2 of the canonical methyltransferase motifs. The domain rearranged methyl transferase or DRM class is mostly related to Dnmt3. RNA-dependent DNA methyltransferases may serve to monitor transcribed transposable elements. DNA methylation by the enzyme Dnmt1 plays a central role in genomic imprinting. Genomic imprinting may be due to other reason. There is a potential connection between imprinting, chromatin structure and DNA methylation. The links between general epigenetic mechanisms and chromatin structure and function also include those between histone methylation and heterochromatin protein, HP1. Thus chromatin structure modifications in conjunction with DNA methylation seem a plausible mechanism.

Foreign sequences such as transgenes, parasitic repetitive elements including retrotransposons, retroviruses and some repetitive elements are shut down by epigenetic modifications. One epigenetic modification is DNA-methylation is important in the regulation of genomic parasites such as TEs, RTs and RVs. The majority of 5′-methyl cytosine lies within these elements. This methylation facilitates a host defense mechanism to counter the deleterious effects of such elements. Promoter methylation of the elements inhibits their transcription and methylation directed heterochromatization of repetitive elements is proposed to prevent chromosomal rearrangement. Some transgenes become parentally imprinted. Imprinting of transgenes in some cases seems to depend on where the transgene has become integrated into the chromosome and in other cases, DNA sequence of the transgene seems to be important. For some tarnsgenes imprinting is reversible in germ line, in others epigenetic modifications can persist and in fact accumulate over several generations. In most cases parental transmission results in undermethylation (hypomethylation) whereas maternal transmission leads to hypermethylation and hypermethylated transgenes are repressed. Differences in DNA methylation and as well as chromatin structure have also been detected in some endogenous imprinted genes. A number of genes have been cloned, some of which encode proteins that may be involved in the formation of heterochromatin chromosome domains (which can form stable epigenetic switches). A family of genes with homology to the Drosophila heterochromatin protein gene HP1 has been identified in mouse.

The state of activity is inherited clonally through subsequent cell divisions by templating of the methylated state. Particular CpG sequences in imprinted genes are methylated in gametes of one sex but no the other. The zygote formed is heterozygous for the imprinted genes- carrying a methylated allele and a non-methylated allele. During development methylated alleles are not expressed while the homologous copies that are non-methylated are expressed and provide the necessary functions. Further, in primordial germ cells (the precursors of eggs and sperms), imprinting is erased by demethylation of appropriate residues in the imprinted genes. These residues are remethylated *de novo* in gametes in either the maternal or paternal mode.

Types of methyl transferase activities—From the above it can be said that there are two types of DNA-methyl transferase activities. 1. Maintenance and 2. *de novo* methylation. The methylation of hemimethylated symmetrical sequences (CpG and CpXpG) after DNA replication is **maintenance methylation** that are maintained through out development or in many cases between generations. Methylation that occurs at previously unmethylated cytosines is known as *de novo* methylation. For symmetric sites, *de novo* methylation occurs only once after which methylation can be preserved by maintenance activity. However, for maintenance methylation at asymmetric sites (cytosines in context other than CpG and CpXpG) *de novo* methylation must occur continually. All known methyl transferases are characterized by the presence of several conserved motifs in the region of protein involved in catalysis.

Classes of DNA methyl transferases — Again from what has been discussed above it can be said that DNA methyl transferases can be grouped into four distinct classes based on sequence homology and function. 1. Dnmt1/MET1class 2. Dnmt2 3. CMT class and 4. Dnmt3 class. Dnmt1/MET1 class of methyl transferases are predominant maintenance methyl transferases in mouse and Arabidopsis. In other words, they act primarily as maintenance methyl transferase. Dnmt2 class of methyl transferases are found in mammals, fission yeast and Drosophila. They do no appear to play a significant role in establishing or maintaining DNA methylation. Further, neither Drosophila nor fission yeast contain a detectable amounts of cytosine methylation in the genomes. The CMT class of methyl transferases are found only in plants. Plants contain a high level of CpXpG methylation in their genomes relative to animals and it suggests that CMT may act as a specialized type of plant specific maintenance methyl transferase.

Varying patterns of methylation — There are two models which could account for the varying patterns of methylation.

1. An exclusion of access to methylation sites by proteins bound to specific DNA regions. This is based on the observation that removal of Sp1 binding sites flanking a $C_P G$ island led to its *de novo* methylation during development.

2. A methylation–targetting mechanism streered by sequence specific binding proteins (targeting model). This model is based on the finding that there is an association of the DNMT1 and DNMT3a enzymes with protein including Rb,E2F1, histone deacetylases (HDACs) and transcriptional repressors RP58.

Sequence-specificity of DNA methyltransferase — In vertebrates methylation of DNA is associated with shut-down of local gene expression and is mediated by DNA methyltransferase enzymes. DNA methylation has a central role in imprinting but in neither case (plants or mammals) it is clear how imprinted genes are targeted for methylation. In other words, how these enzymes discriminate between different sequences between the genome. DNA methyl transferases have shown sequence–specificity (Jia et al., 2007). In higher organisms DNA methyl transferases mainly methylate C-G dinucleotides but then the CpG islands are generally unmethylated in normal cells. Further, transposable elements which are not particularly rich in CG dinucleotides are usually methylated. It has been shown that DNA methyltransferase 3a (Dnmt3a) and its regulatory factor, DNA methyltransferase 3-like protein (Dnmt3L) are both required for the *de novo* DNA methylation of imprinted genes in mammalian germ cells. Dnmt3L interacts specifically with unmethylated lysine 4 of histone H3 via its amino terminal plant homeodomain like domain (PHD). Imprinted genes in mammals are often associated with differentially methylated regions (DMRs) which show DNA methylation patterns which depend on the parent of origin. How imprinting machinery recognizes DMRs is unknown. Structural analysis (Jia et al., 2007) has revealed that the enzyme complex mediating this process shows sequence-specificity. Dnmt3a and Dnmt3L form a tetramer in the order, Dnmt3L- Dnmt3a- Dnmt3a- Dnmt3L. The central Dnmt3a dimmer in the tetramer can methylate two CG dinucleotides in one binding event and that it preferentially methylates pairs of CGs that are 8-10 base pairs apart with other CGs in between remaining unmethylated. In imprinting control regions that contain differentially methylated sequences, CG dinucleotides do occur with a periodicity of 8-10 base pairs. Thus there is a remarkable specificity of the Dnmt3a-Dnmt3L complex. In human genome there is striking overrepresentation of CG dinucleotides with an 8bp periodicity. Thus potential targets for Dnmt3a-Dnmt3L occur in the genome and not seem to be limited to differentially methylated regions containing imprinted genes. Mutation in the gene encoding Dnmt3L affects DNA methylation and further Dnmt3-L does not interact with DNA when it is associated with a methylated form of the core histone, H3. This protection from DNA methylation on the basis of a particular histone modification suggests another layer of genomic recognition, linking the DNA methylation machinery to the modification state of histones. Thus they showed an underlying functional periodicity to the seemingly random distribution of CG dinucleotides within crucial regulatory sequences and thus different patterns of CG periodicity might reveal different functional specificities within the genome.

Concept of spread of silencing—There is *de novo* methylation of an integrated retroviral sequence spread into host sequences flanking the integration site. Further, it has been proposed that repetitive elements such as short and long interspersed nuclear elements (SINEs and LINEs) might serve as foci for *de novo* methylation and that methylation may spread from such attractors of modification. The mechanisms of spreading remain unknown but it may be related to the fact that partially methylated DNA is a better substrate than unmethylated DNA for DNMT1 *in vitro*. Here the methylated DNA allosterically activates the catalytic center of DNMT1. Thus it can be said that methylation attracts more methylation and thus ensuring self-perpetuating methylate state.

Methods of studying gene regulation and epigenetic control—One method of investigating gene regulation and epigenetic control is to produce a tiling path microarray, using PCR products or oligonucleotide probes and hybridize labeled samples enriched by CHIP or other experimental strategies to look for possible binding sites, modified chromatin or other useful features (Mockler and Ecker, 2005)

Evolution of epigenetics—RNA-mediated silencing and DNA methylation evolved as part of host defense mechanism against viruses and parasitic DNAs (Matzke et al., 1999). The double stranded RNA for both post transcriptional gene silencing(or RNA interference, RNai) and transcriptional gene silencing (TGS) observed in plant is a common intermediate in the life cycle of many viruses and transposons (see chapter 23 for detail). Plant viruses are known to be targets and elicitors of PTGS and suppressors of PTGS have been identified in the genomes of many of these viruses. These suppressors were previously identified as pathogenicity determinants. In TGS in plants cytoplasmic dsRNA containing promoter sequences is able to direct the silencing and *de novo* methylation of homologous DNA. In plants and filamentous fungi DNA methylation is mainly restricted to transposons and other repeat units but in mammals, coding sequences are also methylated, possibly reflecting the presence of transposons in introns. Drosophila does not methylate its genome but suffers a very high rate of spontaneous mutations from 50 to 85% through the action of transposable elements. Further support came from DDM1 gene in Arabidopsis which results in the reactivation of silent repeat sequences and the activation of a family of transposons. DDM1(decrease in DNA methylation) gene encodes a protein similar to the chromatin remodeling factor, SW12/SNF2 which suggests link between DNA methylation and chromatin structure. The processes of repeat-induced point mutation (RIP) and methylation induced premeiotically (MIP), both found in fungi and of repeat induced gene silencing, found in flowering plants, all silence and methylate repeat sequences, a common characteristic of parasitic DNA elements. They protect their genomes from TEs through mechanisms known as RIP and MIP, respectively (Goyon et al., 1994). REs often present in multiple copies trigger their mutual silencing in a process called quelling by methylation or C to T transition mutations of the repeated sequences. RIP is a mechanism to inactivate repetitive sequences, to reduce their copy numbers or both by introducing transition mutation (s), affecting predominantly G = C(and some G = C pairs are more than others) which are replaced by A = T pairs in Neurospora crassa and some other fungi. RIP involves deamination of cytosines or 5′-methylcytosines to uracil or thymine, respectively and occurs after fertilization specially in the subsequent mitosis prior to nuclear fusion. In MIP there is extensive methylation of cytosyl residues in duplicated DNA segments in *Ascobolus immerses*, filamentous fungus. MIP inactivates the genes located in the duplicated segments. Once established, both the C-methylation and gene silencing are stably maintained via vegetative and sexual reproduction even after segregation of the duplicated segments. The minimum duplicated size for MIP-induced methylation is ~300-400bp and so smaller duplications are not methylated. MIP is a special type of epimutation and functions to protect genome against TEs and recombination of ectopic repeats which could be lethal. Genomic imprinting in placental mammals and X-chromosome dosage compensation may themselves have evolved from such host defense mechanism directed against parasitic DNA.

6.10 PRION HYPOTHESIS

This hypothesis proposes that proteins can act as infectious agents. It is based on the idea that a protein conformation can replicate itself and therefore serves as a genetic element. TSEs (transmissible spongiform encephalopathies) were originally formulated to explain prion hypothesis. There are several non-Mendelain traits in fungi are due to heritable changes in protein conformation. Mechanistic parallels between the mammalian and yeast prion phenomena point to universal features of conformation based infection and inheritance involving propagation of ordered β-sheet-rich protein aggregates commonly referred to as amyloid (Chien et al., 2000).

7

The Other Types of Code

7.1 PROPERTIES OF NUCLEOTIDES AND NUCLEOTIDE SEQUENCES

Besides being the subunits of nucleic acids nucleotides perform a variety of functions such as energy carriers, components of enzyme cofactors and chemical messengers. ATP is the central carrier of chemical energy in cells. DNA is a template/ substrate for enzyme system which performs replication, transcription, repair and recombination. The presence of an adenosine moiety in a variety of enzyme cofactors may be related to binding-energy requirements. Cyclic AMP (cAMP) formed in a reaction catalyzed by adenylyl cyclase is a common **second messenger** in response to hormones and other chemical signals. Cyclic cGMP occurs in many cells and also has regulatory functions. Another regulatory nucleotide, ppGpp(guanosine tetraphosphate) is present in bacteria which prevents the unnecessary synthesis of nucleic acids by inhibiting the synthesis of rRNA and tRNA during amino acid starvation. The extracellular signalling molecules such as hormones and neurotransmitters are called 'first messengers'

7.2 THE OTHER TYPES OF CODES

Nucleotide sequence is a sequence of four letters. Sequences of As, Gs, Ts and Cs make up the genome. Code (genetic code) was cracked in 1950s and 1960s. The genetic code parses the passage of DNA into three letter combinations that correspond to particular amino acid (see chapter 3). The nucleotide sequences thus specify codons. Further, it also specifies proteins. This is the first layer of biological information in the DNA. As the cell is performing a number of functions so there are a number of other layers of information in DNA interspersed between or superimposed on the passage written in triplet code. A unique property of the nucleotide sequences is the superimposition of the codes they carry, i.e. a given base in a given position along the sequence may be involved simultaneously in several messages of the same or a different nature as was theoretically predicted by Schaap (1971). For example, the coding sequences of the genome of hepatitis B virus massively overlap, so that the same regions of the sequence simultaneously code for two different proteins with different amino acids. Another example is location of promoters of 5S rRNA genes within the genes themselves. Further, sequences at the ends of the exons are involved not only in the proteins coding but also in gene splicing sequence patterns. Thus the aim is to extract codes that regulate, control or describe all sorts of cellular processes. Other types of codes include regulatory code, structural code, code allowing interactions of DNA with protein and RNA.

Coding sequences can be further grouped into those encoding amino acid (exon) and those which do not code (intron). Besides these types of sequences there are those which are clubbed as regulatory sequences such as promoter, operator, etc. involved in the regulation of expression of genes. Nucleotide sequence in the DNA molecule can be said to be a sequence of triplet codons encoding amino acids. Specific-nucleotide sequence can also affect the shape of DNA molecule and is involved in positioning of nucleosomes. Thus what we see is that there is superposition of all kinds of codes on the nucleotide sequence. Studies of structure and function of nucleic acids, genes and proteins (chapters 1, 4, 5, 16, 24) indicate that there are numerous sequence-directed processes and sequence-dependent structures and there exists interactions (Trifonov, 1999). The corresponding sequence interactions are reads from the DNA or RNA molecule in its characteristics way through one or another specific molecular interactions or a whole network of such interactions. In case of coding sequences encoding amino acids the triplet genetic code is read by the ribosome in combination with tRNA and various other protein factors(see chapter 4). The sequence signals in case of gene splicing is read by the spliceosome, nuclear RNA transcript and snRNA and proteins (see chapter 5). Besides these there are numerous relatively simple specific- DNA or RNA sequences which are recognized by simple protein, DNA-binding protein and RNA-binding protein via interaction with these specific-sequences.

7.3 REGULATORY CODE

DNA contains tissue specific information which instructs a particular cell type to produce a particular type of proteins. In other words, regulatory code directs the production of a type of proteins tailored to specific cell types and used at specific times and thereby switching on in of many genes. DNA contains signature sequences in promoter regions nearby and enhancer regions which may be millions of bases away (see chapter 5). Thus regulatory codes are involved in gene regulation. DNA-binding transcriptional regulators interpret the genome's regulatory code by binding to specific sequences to induce or repress gene expression (Jacob and Monod, 1961). Construction of yeast's transcriptional regulatory code has been made by identifying the sequence elements that are bound by regulators under various conditions and that are conserved among Saccharomyces species. In human there are more than 20, 000 proteins and about 1500-2,000 transcription factors (TFs). Regulatory proteins and RNAs direct the production of TFs. Thus the objective would be to find all positions on DNA where each of the regulatory protein binds. TFs show liking for specific short sequence motif in DNA such as the six letter sequence with which they bind. A given six base pair sequence can be sometimes a binding site and sometimes not depending on whether the DNA is folded up in a way that prevents transcription factors from gaining access. The way these sites are recognized is not as specific as binding between the bases that translate the triplet code into protein. TFs recognize DNA sequence from the effects of the sequence on the outside of the helix and although this recognition is still sequence dependent it is not quite so precise (see chapter 24). Some of these proteins will bind to a range of related sequences, sometimes more tightly, sometimes less or so. Thus the objective would be to identify regulatory code for all TFs.

Presence of sequence-specific patterns — Each sequence-specific interaction is represented by the presence of a specific sequence pattern in a sequence. Identification of specific sequence pattern permits localization of the corresponding functional site (or region) in the DNA or RNA sequence. A sequence pattern thus corresponds to one or another function. In other words, sequence pattern serves as a code for the recognition of function sites in the sequence. The sequence pattern can be general, even universal, of broad use in various species, or specific, e.g. species or gene specific. The general sequence codes include transcription signals in promoters such as TATAAA-box in eukaryotes and TATAAT- and TTGACA-boxes in bacteria. Another example of sequence code is the gene-splicing code, the GT-AG rule and some sequence preferences around the intron-exon boundaries. Eukaryotic sequences around these junctions resemble to varying degrees the consensus sequences AGGUAAG and y_{10-15}ncAG where n denotes any nucleotide, y denotes a pyrimidine and the capitalized nucleotides are most conserved. As the recognition algorithms based on the above sequence patterns are not as precise as the natural gene –splicing process the present sequence features are likely to change in future. Further, there are patterns in DNA spanning hundreds and thousands of bases, for example, between AG regions there are hundreds or thousands of bases.

A section of DNA can contain two or more layers of information which are used at different times or in different ways depending on cell's requirement. So a given sequence, read as binding site for TF to some extent depends on how DNA involved is packaged at that point in chromosome and that packaging depends on a different code stored in the DNA.

7.4 STRUCTURAL CODE

Other signals in the sequence decide at what point DNA should coil around its scaffold of structural proteins. The human genome is about two metres of DNA which is packed up in the nucleus of a few micrometres in diameter. About 90% of the DNA is bundled up into nucleosomes and their position influences the DNA's activity. Nucelosome consists of 147bp of DNA wrapped around a globule of eight proteins called histone (see Roy, 2009). The chromatin fiber consists of long arrays of such nucleosomes-coiled into higher order structure. The folding of chromatin fibers can be tuned to either permit or prevent the use of genetic information. The sequences wrapped up in nucleosomes are often less accessible to TFs. This is largely determined by the histones which in addition to the globular parts that comprise the DNA spool, having long 'tails' domains which reach out to contact other proteins contributing to structure or function. These interactions may be reduced or enhanced by chemical modifications of the histone tails-the most common of which is the addition or removal of the phosphate, methyl or acetyl groups. Status of the cell may be signaled to chromatin through the placement of such chemical tags. The marked chromatin is then reorganized by effector proteins that may regulate how underlying genetic information is used. The regulators of chromatin structure recognize histone marks through protein domains. Bromo- and chromodomains are prominent domains which are able to bind to histone tails that carry acetyl or methyl marks at certain amino acid residues, respectively(Becker, 2006). Structural conversion of the nucleosome has critical effects on DNA-mediated reactions. The mechanism of nucleosome assembly and disassembly has been elusive but now the crystal structure of a histone chaperone in complex with histones H3 and H4 has been determined. Human and Xenopus histones H3 and H4 differ by only one amino acid in total (Ryo Natsume et al., 2007). A 'hallmark' property of all HP1 proteins is the combination of a chromodomain with a chromoshadow domain that are separated by a short but variable hinge region.

7.4.1 Sequence Conservation in Non-Coding Sequences

There are numerous examples of sequence conservation in non-coding sequences. A comparison of human and mouse T-cell receptor alpha locus shows greater than 70% similarity in non-coding sequences. High conservation has also been observed in the 5′ and 3′ untranslated regions as well as in introns (Duret et al., 1993). These conservations in non-coding sequences suggest encoding of some message(s) by it.

7.4.2 Sequence-Dependent Shape of DNA

Eukaryotic genomic DNA exists as highly compacted nucleosome arrays called chromatin. Each nucleosome contains a 147 bp of DNA which is simply bent and tightly wrapped around histone protein octamer. This sharp bending occurs at every DNA helical repeat (~10bp) when the major groove of the DNA faces inwards the histone octamer and again ~5bp away with opposite direction when the major groove faces outward. Bends of each direction are facilitated by specific dinucleotides (Satchwell et al., 1986). Neighboring nucleosomes are separated from each other by 10-50bp long stretches of unwrapped linker DNA (van Holde, 1989) and thus 75-90% of genomic DNA is wrapped in nucleosomes. The shape of DNA plays an important role in DNA-protein interaction and DNA folding in the cell. The shape of DNA is sequence dependent. The sequence dependent preferences that correlate most closely with rotational orientation of the DNA relative to the surface of the protein are of two kinds: ApApA/TpTpT and ApApT/ApTpT, the minor grooves of which face predominantly towards the protein and also GpGpC/CpCpC and ApGpC/CpCpT whose minor grooves face outwards. Long runs of homopolymer (dA):d(T) prefer to occupy the ends of the core DNA, five to six turns away from the dyad(the mid point of the bound DNA). These same sequences are apparently excluded from the near centre of core DNA, two to three turns from the dyad. Hence, the translational positioning of any single histone octamer along a DNA molecule of defined sequence may be strongly influenced by the placement of (dA):d(T) sequences. It may also be influenced by any version of the protein for sequences in the 'linker' region, the sequence content of which remains to be determined. The DNA structure is not monotonously uniform. It is modulated by sequence –dependent local deviations from the standard geometry which, for example, may accumulate to make it curved. DNA curvature results from local deflections of the DNA axis at certain dinucleotide steps, primarily at AA(TT) dinucleotides repeated at the DNA helical pitch distance. Intrinsic DNA curvature was observed in the predicted 10.5 base periodicity. By deflecting the DNA axis at every step according to the wedge and twist angles from the table of the dinucleotides, one can calculate the predicted path of DNA axis for any given sequence.

Homopolymeric (dA:dT) stretches have unique structural and functional properties. They adopt a rigid structure which is characterized by a high level of propeller twist and an increased base stacking. This allows the formation of additional non-Watson-Crick bifurcated hydrogen bonds. Phasing of short (dA:dT) tracts within the helical repeat of normal-B-DNA results in a curvature of the DNA. It has been proposed that the general sequence B-DNA gently writhes with a net effect of all local bends being a straight helix. Introduction of a straight (d A:dT) tracts distorts the array of compensating writhes and results in a curvature of DNA. This curvature can play a role in the modulation of the transcriptional activity of the genes and enhance the affinity of the DNA for transcription factors such the TATA-binding protein. Alternatively, (dA:dT) can modulate the access of TFs to the DNA via local distortion of a nucleosome. In yeast dA:d T tracts are functional promoter elements. Homopolymeric(dA:dT) stretches of about 200-500bp in length are part of scaffold associated regions(SARs) which are supposed to anchor the chromatin loops in the nucleus. These DNA segments closely flank active or potentially active genes or their 5′ regulatory and also 3′ UTRs. The SARs are also the place of residence of topoisomearse II which control the topology of DNA during replication, recombination and transcription. It has been proposed that the curvature induced by homopolymeric tracts in the SAR defines the sequence characteristics preferred by topoisomerase. The consensus sequence consists of a 10bp oligo(dA) stretch[A-box; 5′ –AATAAATCAA-3′] and a 10bp(dT) stretch[T-box; 5′ –TTATAATTTATTT-3′]. In yeast SARs co-map with autonomously replicating sequences and centromere domain elements.

Homopolymeric DNA tracts or more generally simple repetitive sequences can give rise to slippage of the polymerase during replication. The internally repetitive DNA sequences allow the nascent strand to slip back or forward on the parental strand with one or more repeat units, resulting in an expansion or contraction of the new DNA strand. Slipped strand replication is a major force in the evolution of genes and genomes and it is supposed to be implicated in a large number of human genetic diseases. In addition to replication slippage, unequal crossingover, mutation and selection affect the persistence of simple sequences. Enrichment for dA:dT tracts is restricted to noncoding DNA. In majority of genomes d A: d T tract overrepresentation proved to be an exponential function of the tract length (Dechering et al., 1998).

7.4.3   TA Under Representation

It refers to the occurrence of the dinucleotide, 5′-TpA-3′ in a genome at a significantly lower level than inferred from statistical distribution. TA dinucleotide is under represented in both pro and eukaryotes with the exceptions such as metazoan, fungal and plant mitochondrial DNA and chloroplast genomes. TA under representation probably protects the DNA from unwinding and bending of the double helix at too many sites and instead reduces the occurrences of TA to locations where it is functionally essential, for example, at the transcription initiation sites, TATA boxes, polyadenylation signals.

7.4.4   Sequence –Specific Positioning of Nucleosomes

The structural and mechanical properties of DNA change according to its base sequence and therefore the ability of a DNA molecule to bend around a histone octamer is thought to be a major determinant of nucleosome positioning. Access to DNA wrapped in a nucleosome is occluded for polymerase, regulatory, repair and recombination complexes, yet nucleosomes also recruit other proteins through interactions with their histone tail domains (Jenuwein and Allis, 2001).Thus detailed locations of nucleosomes may have important inhibitory or facilitatory roles in regulating gene expression. DNA sequences differ greatly in their ability to bend sharply. Consequently, the ability of the histone octamer to wrap differing DNA sequences into nucleosomes is highly dependent on the specific DNA sequences. Thus nucleosomes have substantial DNA sequence preferences (Segal et al., 2006).

Sequence plays a substantial role in nucleosome positioning. Sequence-dependent deformational anisotropy (bendability) of DNA appears to be an underlying principle of the nucleosome sequence specificity. The anisotropy can be attributed to certain sequence elements, particularly dinucleotides AA and TT periodically distributed along the nucleosome DNA following the DNA helical repeat. Other dinucleotides such as CC(GG), TA and perhaps some additional sequence elements also contribute to the nucleosome sequence pattern. In other words, distinct sequence motifs recur periodically at the DNA helical repeat and are known to facilitate the sharp bending of DNA around the nucleosome and these include ~10bp periodic AA/TT/TA dinucleotides that oscillate in phase with each other and out of phase with ~10bp GC dinucleotides. Study of nucleosome sequence pattern is important for understanding of the role of nucleosome in gene expression.

Thus we see that **nucelosome code** and **regulatory code** are both interdependent. DNA's overlapping codes means that an individual nucleosome might be usurped if regulatory proteins are already tightly bound there. Then objective would be to find nucleosome code involved. How nucleosomes are twisted up into a cable of chromatin and eventually coiled into the tightly interwoven ropes of chromosomes (see Roy, 2009)? In human 1-2% of the genome are protein coding sequences and thus other sequences hold other information. Many stretches of DNA do multitask and thus hold multiple information. A sequence can code for a protein and still it manages to guide the position of nucleosome. This multitask is possible as the genetic code is degenerate. Slightly different triplets code of same amino acid and many positions in a protein can be filled by different amino acids. So different sequences can effectively mean the same thing. This allows other signals to be imprinted on top of the first especially when these signals are themselves encoded with some slack.

Higher level structural organization of genome is important for compaction of chromosomes in the nucleus and for signaling genome functions. The genome is folded into structural domains (loops) bases of which are attached to a protenacious nuclear skeleton (matrix). Such loops are believed to provide an additional 1000-fold compaction of the genome necessary for its accommodation into the interphase nucleus. The DNA sequences (MARs) anchoring loops of heterogeneous size to this matrix are considered to be structural elements of the genome. Their ability to affect gene expression has been shown. In plants MARs have been reported to play a role in reducing both position effects and homology-dependent gene silencing. Somatic cell chromatin is organized into loops that span ~50-100kb. The points of attachment of these chromatin loops serve as specific sequence landmarks as they anchor the DNA sequence of the fibers of chromosomal scaffold. These sites of DNA attachment to nuclear scaffold are termed scaffold (metaphase) or matrix (interphase) attachment region (SAR or MAR) (Figure 7.1). They are known to facilitate the

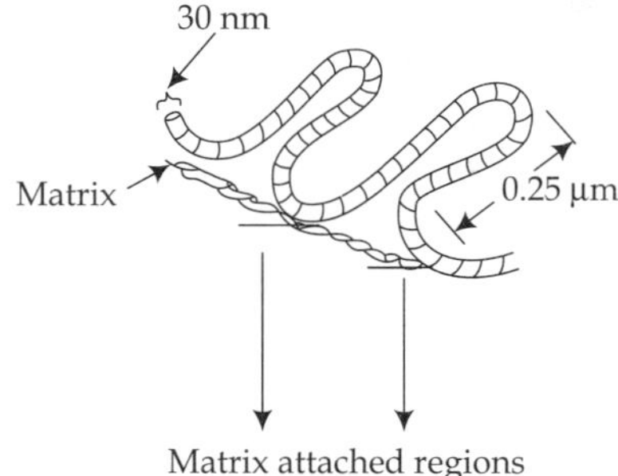

FIGURE 7.1 Showing matrix attached regions (MARs).

expression of genes and may function as the origin of replication (Singh et al., 1997). The MAR sequences are 100-1000bp long and anchor chromatin loops to the nuclear matrix. Adjacent MARs often delineate areas where transcription factor binding sites are concentrated and thus can be used to identify areas within a sequence to search for coding potential.

In search for a characteristic MAR sequence several motifs have been reported as elements clustered in MAR regions: the A-box (AATAAAYAA), The T-box (TTWTWTTWTT), BURs (base unpairing sequences) (AATATATT/AATATT) and topoisomerase II consensus binding sites from Drosophila(GTNWAYATTNANNG) or mouse (RNYNNCNNGYNGGKTNYNY). MARs are identified on the basis of the probability that the MAR motifs can occur at random in a given window of the sequence being analyzed. Detection of MARs can be used to identify regions of high coding potential.

Genome-wide positioning of nucleosomes—The DNA binding proteins have high binding specificity but are present at lower concentrations whereas the nucleosomes have lower binding specificity but are present at high concentrations covering 75-90% of the DNA. Thus both are expected to make important contribution. Nucleosome –DNA interaction model predicts the genome wide organization of nucleosomes and that this intrinsic organization can explain ~50% of the *in vivo* nucleosome position, ie. ~50% of the *in vivo* nucleosome organization can be explained by sequence preferences of nucleosomes. This nucleosome positioning code may facilitate specific functions including transcription factor binding, transcription initiation and even remodeling of the nucleosomes themselves. Nucleosome organization is encoded in eukaryotic genomes, i.e. genome encode the positioning and stability of nucleosomes in the regions that are critical for gene regulation and for other specific chromosome functions and this nucleosome positioning code can be successfully decoded. Transcription factors bind preferentially to appropriate sites in promoters rather than to the excess of irrelevant sites in the genome. Nucleosome organization varies by type of genomic regions. Nucleosome occupancy varies across different types of chromosomal regions including centromeres, telomeres, intergenic and coding and specific gene classes. Highest predicted occupancy was found over centromeres. rRNA and tRNA genes have markedly low predicted nucleosome occupancy. Low nucleosome occupancy is encoded at functional binding sites and at transcriptional start sites. Nucleosome depletions observed at coding and intergenic regions are attributable in part to unstable nucleosomes (i.e. positions on the DNA sequence that nucleosomes have a low probability of occupying) encoded in these regions. Genomes facilitate their own chromatin remodeling by encoding intrinsically low occupancy at sites destined for remodeling. SWI-SNF complex is a multi subunit protein assembly which remodels chromatin at enhancer (or silencer) regions thus opening it for the binding of transcriptional activator or repressor proteins. Other large subunit protein complexes which carry out chromatin remodeling are RSC, INO 80 and SWR. SWI/SNF is a member of ATP-dependent chromatin remodeling activities and is implicated in DNA damage response. RSC remodels the structure of chromatin and is encoded by gene STH1 in *S. cerevisiae*. RSC activity is important for transcriptional regulation of genes that are involved in responses to stress and cell cycle progression. INO80 has been shown to reposition nucleosomes *in vitro* and is important in regulation of certain genes including involvement in inositol metabolism. SWR removes H1A-H2B dimmers from the nucleosome and replaces them with dimmers containing H2B and H2AZ type variant in Htz1 in *S. cerevisiae*.

The genome wide prediction of nucleosome occupancy and stability should facilitate the understanding of specific natural gene regulatory phenomena such as the mechanism by which TFs bind preferentially to appropriate sites in genome. It will be helpful for improving the performance of engineered transgenes. It will help in quantitatively integrating chromatin structure into models of gene regulation, i.e. quantitative prediction of understanding of transcriptional regulation in all eukaryotes.

7.5 HISTONE CODE

Histone code refers to the pattern of different modifications of the histones determining the chromatin structure of a particular gene.

Chromatin, the protein wrapping of the genome, harbours information about how the genes it contains are to be regulated. For example, a triplet methylation mark on lysine 4 of histone H3 (designated as H3K4me 3) is a hallmark of all active genes (Figure 7.2). The set rules governing the recognition and interpretation of methylation marks constitute a 'histone code'. Histone methylation alone can not specify the selectivity of nucleosome interactions observed *in vivo*. Additional specificity could be provided by combinations of histone modifications that may be recognized by corresponding combinations of interacting domains. For example, the PHD finger of BPTF may cooperate with a neighboring bromodomain in docking with a combined methyl-acetyl mark. In *S. cerevisiae*, methyl H3K36 nucleosome by Rpd3S also requires the plant homeodomain (PHD) of its Rco1 subunit. Thus the coupled chromo and PHD domains of Rpd3S specify recognition of the methyl H3K36 mark, demonstrating the first combinatorial domain requirement within a protein complex to read a specific hostone code. Nucelosome remodeling factor called NURF can be recruited to H3K4 trimethylated chromatin through a selective interaction with NURF's largest subunit, BPTF. BPTF recognizes the methylation mark through a structure called the PHD finger, a well known protein structural fold co-ordinated by two zinc atoms. How this triple methyl mark can contribute to gene activation remains unknown. On the other hand, alternative recruitment strategies involving sequence-specific DNA binding proteins are known for all regulators.

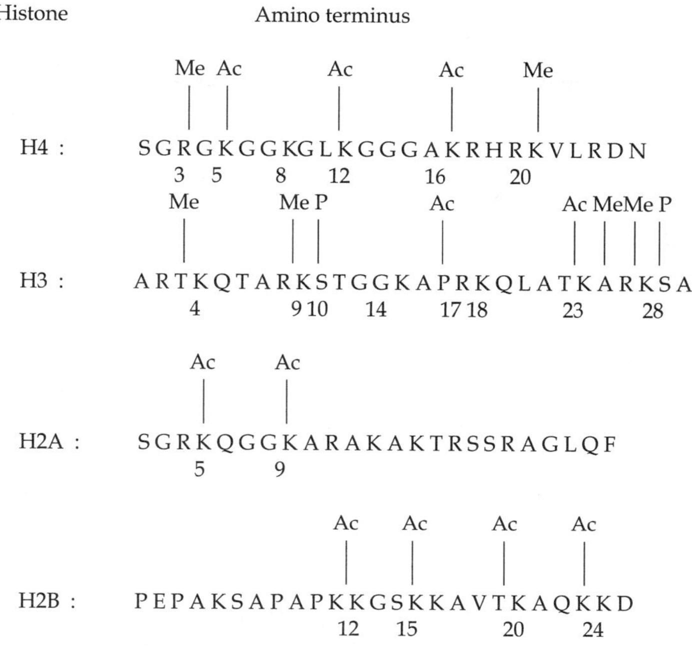

Figure 7.2   Showing histone methylation, acetylation and phosphorylation.

Histone modifications are important for almost all DNA-related processes. However, role of histone methylation in transcription regulation is a major focus. Methylated lysine K4me, K3me and K39me are enriched around regions of active transcription. The methylation marks do not simply facilitate transcription because K4me does not affect transcription *per se in vitro*. Moreover, for majority of the yeast genome transcription occurs normally in the absence of K4me, K36me or K79me whereas methylation appears to be dependent on active transcription. This hypothesis implies that methylation acts in maintaining the architecture of transcribed chromatin templates rather than directly facilitating transcription. Consistent with this hypothesis recent studies have demonstrated that K36me is recognized by the chromodomain of its Eaf3 subunit (CHD Eaf3) within Rpd3S thereby tethering Rpd3S to the coding region of the actively transcribed genes. Once targeted Rpd3S creates a hypoacetylated state which in turn suppresses transcription initiated within the body of the gene. Nucleosome must be deacetylated behind alongating RNA polII to prevent cryptic initiation transcription within the coding regions. RNA pol II signals for deactylation through methylation of histone H3 lysine 36(H3K36) which provides the recruitment signal for Rpd3S hostone deacetylated complex (Li et al., 2007). Post-translational histone modifications play important regulatory roles. One example of such modification is histone ubiquitination which occurs predominantly on histone H2A and H2B.Recent identification of the ubiquitin ligase for histone H2A has revealed important roles for H2A ubiquitination in Hox gene silencing as well as X-chromosome inactivation.

However, the enzymes (s) involved in H2A ubiquitination and the function of H2A deubiquitination are not known. To identify the deubiquitinase for histone H2A, Joo et al.(2007) developed an *in vitro* assay using ubiquitinated H2A (ubH2A)-containing mononucleosomes as substrate. When these nucleosomes are incubated with HeLa nuclear protein fraction, a decrease of ubH2A levels and an increase of the released intact ubiquitin are detected suggesting that an H2A-specific deubiquitinase is present in the corresponding fractions.

ARS-consensus sequence — It refers to a degenerate 11bp long sequence element found in most, if not all eukaryotic autonomously replicating sequences. ACS elements are recognized by a multiprotein complex(origin replication complex) which presumably acts as a replication initiation protein *in vivo*.

DNase I hypersensitive site (DHS) — It refers to a region of the genome showing a sharply different sensitivity to DNase I compared with its immediate local. More exactly it refers to a region of chromatin spanning from 50-200bp that is more sensitive to DNase I digestion by a factor of 100 as compared to neighboring regions. Hypersensitive sites are free of nucleosomes and frequently map within 5′ but also 3′ conding and noncoding regions of genes. They are necessary but not sufficient prerequisites for transcription by RNA polymerase II. DHSs have characteristic histone modification patterns which reliably distinguish them from promoters. Some of these distal sites show marks consistent with insulator function. It is possible to distinguish between constitutive DHSs (permanently present in chromatin) and inducible DHSs(appearing before or during the activation of a gene). Most probably DHSs are address sites for transcription factors.

Within a given cell type DNase I hypersensitive sites fall into two major categories: constitutive and inducible. Constitutive sites are often present in promoter regions of genes, poised for transcriptional induction and their presence precedes transcriptional activation and is independent of gene expression. Inducible sites can also appear prior to transcription and often persist long after removal of the inducing agent and/or continued transcription. Sites which differ between cell types are tissue specific whereas tissue-specific sites that appear transiently during embryogenesis are called developmental. A wide variety of functional genomic sequences are associated with hypersensitive sites. These DHSs are nearly always associated with *cis*-acting DNA sequences. In *S. cerevisiae*, DHSs occur around centromere, silencers, recombination sequences, replication origins, UASs, promoter elements and presumptive transcription terminators (Gross and Garrard, 1988). A number of proteins such as topoisomerase I and II, RNA pol II and TFs are associated with a subset of DHSs. Often, nucleosomes are positioned specifically along the DNA sequences residing adjacent to or between DHSs.

7.6 SEQUENCE-DEPENDENT DNA STRUCTURE

The idea that sequence defines DNA structure has gained acceptance and thus the root of sequence dependent conformational variations has become an important problem. Results from crystallographic screens to address this problem show that variations from mean structural features may provide proteins with information required for indirect readout and for specifying altered structures (Sarai and Kono, 2005). Sequence-dependence of DNA conformation plays an important role in the protein-DNA recognition process during the regulation of gene expression. Proteins recognize specific DNA sequences not only directly through contact between bases and amino acids but also indirectly through sequence-dependent conformation of DNA.

Sequence defines structure but does it in a different complex way since the same neighbor perturbs the conformational state of each central dinucleotide in a different manner. The conformational multimodality plays an important role in DNA recognition since the different conformational modes induced by the neighbors of a central base-pair step can work as a signal for the binding of protein or other ligand. Coarse predictions of the DNA structure from nucleic acid sequence using knowledge-based techniques are possible (Arauzo-Bravo and Sarai, 2005) but such an approach requires data of high quantity and quanlity. Arauzo-Bravo et al. (2006) proposed a simple model expressed in a linear way how different bases that embrace a central dinucleotide perturb its conformational state emphasizing how the conformational role of each base depends on its relative position(left, central, right) in the final tetranucleotide and how the same peripheral base plays a different role depending on which is the central dinucleotide. To what extent the DNA sequence defines the DNA structure they analyzed the conformational space of all unique tetranucleotides. Large quantity of data for such study was generated by Molecular Dynamics (MD) Simulations which permitted to explain multimodal conformational states of some dinucleotides as aggregations of tetranucleotide conformational states that have such a dinucleotide inside their centre.

Constrained sequence—It refers to a genomic region associated with evidence of negative selection (i.e., rejection of mutation relative to neutral regions).

8

Recombinant DNA Technology and its Applications

8.1 BIOTECHNOLOGY

Biotechnology refers to the development and improvement of biological systems (usually microorganisms and plants or the enzymes) for the production of industrially useful compounds and thus includes also the development and design of instruments, incubators, specific culture vessels etc. **Genetic engineering** (GE) is considered as a discipline of biotechnology. The conventional plant breeding programmes deal with **genomic** and **subgenomic** manipulation (see Roy, 2009). The former deals with the manipulation of whole genetic material while the later deals with either the whole chromosomes (single chromosome) or the chromosome segments. In the sub-genomic manipulation, the alteration of specific part of the genetic material of an organism can be used for characterization of gene (gene structure, function), increasing copy number, studying the potential for directed mutagenesis and for expression of gene (s) in widely different species (expression of eukaryotic gene in prokaryotic species). The manipulation of genetic material at the level of DNA molecule is the subject of genetic engineering. The *in vitro* DNA recombination technology deals with the use of cutting and joining DNA molecules. The *in vitro* methodology is used to i. isolate gene (s) ii. change the structure of gene iii. design new genes or to construct chimeric genes. Genetic engineering also includes the technology to transfer these genes into any organism of choice and to express them in foreign environments. In basic research GE is used to study i. gene structure and regulation ii. as a means to provide organism with new trait (s) to produce more and better chemicals or drugs or additional function (s).

 Cloning is the process of producing identical clones and **clone** is a group of genetically identical individuals, the genetic similarity applies to organisms, cells and molecules (DNA). In other words, cloning is the separation and individual propagation of a single element that is capable of reproduction from a mixture of similar elements. Thus the word 'cloning' has several different meanings. For example, it is possible to clone cells, i.e. to cause cells to reproduce themselves so as to make a population of identical cells- a group of cells with common ancestor. The cloning of cellular organisms requires only nutrient medium and cloning of intracellular parasites or symbionts such as viruses and autonomous genetic elements requires a specific host cell system plus a nutrient medium whereas cloning of genes requires a vector replicon, a specific host and a nutrient medium. In molecular biology, any piece of DNA can be cloned- inserted into a vector which replicates in a host to produce many copies of the same recombinant vector. Cloning is the process of generating sufficient copies of a particular piece of DNA to allow to be sequenced or studied in some other way. Gene cloning is the process of identifying and isolating specific gene of interest. **Gene cloning** involves the separation and individual propagation of a single DNA segment from a single DNA molecule in a suitable vector-host system. In other words, DNA cloning involves isolation of a specific gene or DNA segment from a larger chromosome or genome, its attachment to a self replicating small DNA molecule (called **cloning vector**), introduction of this **recombinant DNA molecule** (cloning vector linked covalently to the DNA fragment) in the host and replication of the recombinant DNA in the host through increase in the cell number and creation of multiple copies of the cloned DNA and finally, selecting the host cell containing the recombinant DNA. In other words, cloning is used to describe the process of isolation and replication of individual genes. **Gene cloning** refers to the process of isolating a particular gene, usually from a DNA library. The appropriate cloned DNA is detected and isolated in various ways including hybridization with a complementary nucleic acid, *in vitro* translation and detection of gene product by a labeled antibody, complementation by the cloned DNA of a genetic defect in a host cell or organism and expression in Xenopus oocyte expression system. The **recombinant DNA** thus refers to a novel DNA sequence formed *in vitro* through ligation of two or more nonhomologous DNA molecules. An **insert** is any foreign DNA that has been integrated into a cloning vector molecule and can be propagated in a host cell and a **construct** is a laboratory slang term used for any recombinant DNA molecule. A **hybrid plasmid (recombinant plasmid)** refers to any plasmid which contains sequences derived from any other plasmid, a bacterial genome, a virus or a higher organism. Gene cloning, therefore, requires breakage of the DNA molecules to generate gene sized segments, followed by the covalent linkage of such segments to vector molecules in which they can be propagated and introduction of the hybrid DNA molecules into suitable host cells. The different methods used to accomplish these (different methods of cloning, methods of cutting and joining of DNA molecules and finally diverse techniques of direct and indirect gene transfer) and other related tasks are collectively referred to as **recombinant DNA technology** or **genetic engineering.** Recombinant DNA technology grew out of classical molecular genetics, a field that involved studies of bacteria (especially *E. coli*) and bacteriophages. The bacterial system provided not only the materials for recombinant technology (such as plasmids and phage vectors, suitable hosts and expression systems) but also the very idea of plasmid and replication origin, the concept of promoter and the notion as well as the identity of the signals for start and end of transcription and translation. Although bacteria share fundamental properties with other organisms (including DNA genes, mRNA and ribosome-based protein synthesis and metabolic economy based on ATP and nicotine-adenine dinucleotide) there are limitations to the applications of paradigms of bacterial genetics to higher eukaryotes directly because the biology of bacteria differs from eukaryotes in the rules for transmission of genetic material, the type and the function of subcellular organelles such as the mitochondria and basic aspects of metabolism and regulation. It is where fungal genetics has played a role. Fungi being eukaryotic, share many fundamental characteristics of cell biology with higher eukaryotes (such as cytoskeletal organization, subcellular organelles, secretary systems, receptors and second messenger arrangements, metabolic regulation and chromosome mechanics).

8.1.1 Applications of Cloning

Once a gene is identified, cloned into an appropriate vector, it can be manipulated in different ways.

1. **Generation of large amounts of DNA:** A cloned gene can be used to generate large amounts of DNA for carrying out further analyses such as sequencing and mutational analysis.

2. **Sequencing:** Cloning amplifies the DNA so that it can be sequenced.

3. **As a probe:** A cloned gene is also useful as a probe in gene expression studies

4. **Production of protein:** A cloned gene inserted into an expression vector system produces up to 25% of its total protein or more as the gene product thus allowing large scale production of the protein.

5. **Structure-function analysis of gene product:** A cloned gene can be mutated and mutated form inserted back into an organism that lacks a functional copy of gene and this will allow structure-function analysis of the gene product. Cloning of gene is one of the first steps in the study of its function. Besides mutational analysis in some cases a cloned gene is used to disrupt the endogenous gene by homologous recombination.

6. **Creating new or modifying metabolic pathway:** A cloned gene or set of genes can be introduced into a new host to create a new metabolic pathway or to modify the existing pathway.

8.1.2 Practical Applications of Biotechnology

The practical applications of genetic engineering in plants will be as follows:

(i) To maximize the efficiency of photosynthesis and growth.

(ii) To increase nitrogen fixation either by improving the ability of the plant or by increasing the nitrogen fixation capacity of the associated organism and to broaden their host specificities.

(iii) Introduction of specific genes such as gene for resistance to diseases, insects pest or any other trait from other sources particularly other species and genera of plants to cultivated form (s).

Besides the *in vitro* DNA technology has the potential for increasing the yield of an enzyme and thus making it easy to purify.

8.2 GENETIC ENGINEERING METHODS OF CLONING GENES

Gene cloning is the process of isolating a particular gene. The recombinant DNA technology deals with two processes, transformation and transduction through which the DNA can be transferred to the host. The major advantage of gene transfer is that potentially useful genes from any organism (virus, bacteria, insects, etc.) can be transferred to a plant species. Thus genotypes can be improved by the introduction of a gene that may code for one trait like disease or insect resistance. The application of plant transformation techniques is dependent on plant tissue culture protocols to regenerate transformed plants. Since the first reports of gene transfer with species that are relatively easy to tissue culture, petunia, tomato, procedures have been developed for a wide range of species including trees crops.

Transformation

In the transformation, the DNA from a particular species is isolated and added to the protoplast (the plant cell culture) of another species to which the foreign DNA is intended to be transferred. During this process of treating the plant tissues with foreign DNA, there are some chances that the foreign DNA is introduced into the genome of other species. As a result of transcription and translation of the introduced DNA, the phenotypic changes will occur in the host plant which is generated from the host cell. This genetic transformation or modification of plant cells through uptake of DNA, cell organelles like nuclei, chloroplasts, plastids, nitrogen fixing bacteria, etc. has been suggested as a possible means of crop improvement. The problem with this method of transfer of DNA to host is that whether these transferred DNA, chloroplast or bacteria remain biological functional or are suitably stabilized (or integrated) in the host cell or protoplast and transmitted regularly in sexual progeny is not clear.

Transduction

In this process the transfer of DNA or gene is through vector which is introduced to plant cells. The procedure of transfer of foreign DNA which codes for specific information from a donor species into a recipient plant species by means of vectors such as bacterial plasmids, viruses or other-vectors is also called transformation. **Trans gene** is defined as a segment of DNA of defined function from a source that has been introduced into crop germplasm using transgenic technology (e.g., biolistics, *A. tumefaciens*) and the resulting plant is called **transgenic plant**.

8.3   INTRODUCTION OF SPECIFIC GENES

The various steps involved in the process of cloning genome are as follows:
1. Generation and isolation of DNA fragments of a genome.
2. Joining of the fragments to a vector.
3. Introduction of hybrid (fragment + vector) into host.
4. Identification of clones containing the desirable hybrid molecule.
5. Amplification, isolation and characterization of the hybrid molecule.

The cloning strategy thus involves detailed protocol from preparation of a high molecular weight DNA, its restriction into fragments, the treatments of fragments (e.g. ligation of adaptors, filling in, polishing) and their insertion into the appropriately cut vector.

Filling in reaction—The complementation of single stranded 5′ –overhanging ends (5′ extensions) of the double stranded DNA molecule to form a completely base-paired duplex using DNA polymerase I. Filling in is used to produce blunt ends in DNA fragments which can then be ligated by T_4 DNA-ligase or to mutagenize DNA molecules *in vitro* or to generate new ends at a duplex molecule by an incomplete fill in reaction. Thus besides often being used to join fragments with different cohesive ends by converting the cohesive ends into blunt ends, fill-in reaction is used to introduce frame shift mutation in a gene. The **polishing** refers to synthesis of short sequences complementary to single stranded protrusions (tails), sticky ends generated in DNA molecules by either mechanical shearing or digestion with restriction endonucleases II. Polishing is accomplished by Klenow polymerase or T_4 DNA polymerase and leads to complete duplexes (double stranded termini of duplex DNA).

8.3.1   Generation of Fragment Genome

The DNA chain of an individual can be cut into pieces either through use of restriction enzymes or by mechanical shearing. Restriction enzymes are endonucleases. They cut the DNA into pieces. There are two classes of endonuclease. Class I restriction endonucleases recognize specific sequence in the DNA chain but make non-specific cleavage where- as class II restriction endonucleases are sequence specific and cut the DNA within their recognition sequence. It cuts both strands of DNA. The most commonly used restriction endonucleases are EcoRI, Barn HI, HpaI, HaeIII and Hae II. Many endonucleases generate single stranded cohesive 5′ or 3′ ends for complementation. There are certain enzymes which will not cut at restriction sites where certain bases (particularly cytosine in aCpG nucleotide sequence) have been methylated. Although recognition sequences vary from one restriction enzyme to another but the common to different restriction enzymes is the axis of diasymmetry. The size of fragment genome desired is 10-20 kb. Average restriction enzymes have 4 base pair recognition sequence through which the size of the fragment genome generated will be of 44 (= 256 bp) in length. There are restriction enzymes with 5, 6 and 8 base pair recognition sequence. The DNA polymer after partial digestion with 4 base pair recognition sequence restriction enzymes yields a series of overlapping DNA fragment. For more on use of restriction enzymes see chapter 9. The populations of DNA fragments are then joined to a vector.

8.3.2   Isolation of Gene (s)

The isolation of individual genes is most conveniently accomplished by cloning of DNA fragments generated by a single specific restriction endonuclease but the isolation of a group of genes or all genes present on a chromosome or genome can only be achieved by the cloning of randomly or near randomly cleaved DNA fragment because of the presence of restriction endonuclease cleavage sites within some of the genes. Nearly randomly cleaved DNA fragments may be conveniently generated by partial digestion with a 'frequent cutter' which recognizes a tetranucleotide sequence whereas randomly cleaved DNA fragments are generated by controlled mechanical shearing. The ease of cloning a DNA fragment that contains specific gene depends upon its relative concentration in the total population of DNA fragments. Purification or selective enrichment of the required fragment will greatly simplify its cloning and may be achieved if the fragment exhibits a physical property such as size or buoyant density that distinguishes it from other fragments, or contains a sequence that has affinity for an available DNA,RNA or protein probe (Timmis, Cabello and Cohen, 1975;

Gautier, Mayer and Goebel, 1976 Weideli et al., 1977). A high resolution enrichment procedure involving RPC5 column chromatography and preparative electrophoresis was developed that enabled 100-1000-fold enrichment of any DNA segment in a complex mixture to be attained. Recently, polymerase chain reaction (PCR) technique is being used for such enrichment of a DNA fragment (see chapter 31 for detail on PCR).

Cloning of synthetic DNA — In addition to DNA from natural sources, synthetic DNA can also be cloned. There are two types of synthetic DNA.

1. cDNA that has been ezymatically synthesized by reverse transcriptase and DNA polymerase I from purified or partially purified mRNA (see chapter 31 for detail)

2. DNA that has been chemically synthesized in the absence of a template (see chapter 31 for detail).

Methods of isolating genes — The gene (s) determining a particular trait can be isolated by different strategies. The strategies for identification and isolation of gene for which no knowledge about the gene function is available are different than when the gene product is known. Gene (s) can also be identified and isolated using nucleotide sequence characteristics. In case where the gene product is known, the isolation and cloning of gene is relatively straightforward. An array of techniques is now available for the cloning of genes. The cloned gene may be isolated on the basis of its pattern of expression or by its ability to hybridize to heterologous probes or to code for a particular antigen. Thus there are strategies for direct isolation of genes, strategies based on gene product and strategies based on nucleotide sequence. The four different strategies that allow direct isolation of plant gene (s) without considering the nature of gene product are as follows:

(A) Strategies for direct isolation

1. Map-based cloning

2. Molecular tagging

3. Genomic subtraction

4. Anti-sense mRNA technology

1. Map-based cloning

Map based cloning strategies rely on the knowledge of the phenotype conferred by the gene of interest and its genetic map position. Genes are identified as they are an unique phenotype when mutated. Thus gene (s) determining a particular trait is genetically mapped using the morphological and isozyme markers 'first' and then with molecular markers such as RFLP or RAPD. Thus location of gene (s) on a particular chromosome is established by physical mapping and molecular mapping. The first step in this strategy is to identify the tightly linked DNA sequences located at both sides of the target gene. These linked sequences are used as starting points for cloning of the chromosomal region where the target gene is located, RFLP or RAPD markers (described in Chapter 15) that flank a locus (DNA region determining a trait) provides starting point for chromosomal walking strategies (Steinmetz et al., 1982; Tanksley, 1995). Thus the second step is to isolate DNA clones covering the entire region between the two markers by a process called **chromosomal walking.** Chromosomal walking involves use of previously isolated DNA fragment to isolate the next overlapping fragment in a directional manner. In other words, DNA sequences progressively further away from the RFLP or RAPD markers are identified by a series of overlapping cloning steps. Thus walking from one of the flanking sequences to the one on the other side will result in cloning all the DNA between these sequences including the target gene. This step is facilitated by the use of genomic libraries produced using yeast artificial chromosome vector (YAC) which can carry inserts up to 700 kb (Burke et al. 1987. Anand et al.1990). One restriction fragment length polymorphism (RFLP) clone can be used to identify a YAC which includes the clone by screening a large number of candidate YACs. The inverse PCR (Ochman et al. 1988) can be employed to obtain a small bit DNA lying close at the end of a YAC. This terminal DNA is used as a probe to identify a second YAC which overlaps with but also extends beyond the first and the cycle is repeated until a YAC is identified which includes the second RFLP or RAPD. In practice, one does not know the direction of walk and thus this walking experiment is rather difficult than it looks. In principle, however, all the genomic DNA lying between the two starting DNA clones can be identified (Paterson et al. 1991). Such DNA segment can then be further dissected into smaller pieces in PI phase (Sternberg. 1990) and smaller cloning vectors such as cosmids, lambda phage, plasmids and M 13 phage (Sambrook et al. 1989) and finally pieces measuring 300 nucleotides can be sequenced to determine the identity of each nucleotide in the region: The last step in map based cloning is to isolate the clone harboring the target gene from among all the overlapping clones and this is achieved by carrying out another experiment. All of this genomic DNA can then be introduced, a small bit at a time into plants lacking the trait of interest (Schell, 1987). Those bits which cause the plant to exhibit the traits of interest are inferred to contain the target gene.

Map-based cloning of Pto gene in tomato—Map based cloning of protein kinase gene conferring disease resistance in tomato is shown in the following example. The Pto gene was genetically mapped by scoring of RFLPs (Figure 8.1 shows gene mapping with RELP marker) and resistance reactions to bacterial inoculation of 251 F2 progeny were derived from a cross between resistant (Pto/Pto) and a near isogenic susceptible line (pto/pto). One RFLP marker, TG538 cosegregated with the Pto locus and was used to screen a tomato YAC library. A 400-kb clone, PTY538-1 was identified which hybridized to this marker. End-specific probes corresponding to the right (PTY538-1R) and left (PTY538-1L) arms of PTY538-1 were isolated by inverse-PCR (IPCR) and placed on the high resolution linkage map of the region. The PTY-538-1L arm mapped 1.8 centimorgans from Pto whereas PTY538-1R cosegregated with Pto (Figure 8.2). The bold line shows the genetic map of the Pto region on chromosome 5. The locations are indicated of genomic RFLP markers (TG), the YAC end clones PTY538-1L andPTY538-1R, cDNA clones CD127 and CD186 and Pto. The YAC, PTY538-1 which spans this region is shown below. Genetic distance in cMs is based on linkage analysis. The size of the YAC in kbs was determined by PFGE with yeast strain AB972 as standard.

To confirm that PTY538-1 encompassed Pto, it was necessary to identify a plant with a recombination event between PTY538-1R and Pto. For this Martin et al. (1993) used markers TG538 and PTY538-1R to analyze a total of 1300 plants from various F2 populations, F3 families and over 50 cultivars. One plant was identified that was homozygous for the TG538 allele associated with Pto (resistant allele) but was also homozygous for the PTY538-1R allele linked with pto (susceptible allele). All the progeny from this plant were resistant to *Pseudomonas syringae* pv. tomato showing that the plant was homozygous Pto/Pto. This result shows that PTY538-1 spans the Pto locus.

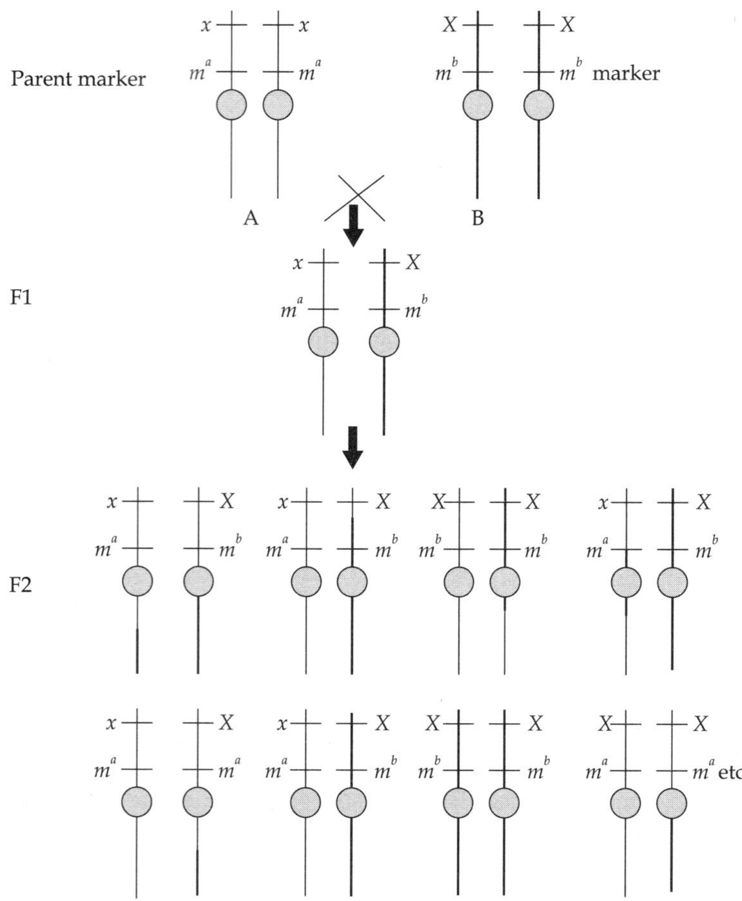

FIGURE 8.1   **Showing map based cloning. The distance between the gene X and the marker, m is determined by meiotic mapping. X is wild type allele whereas x is the mutant allele. Meiosis in the F1 will produce gametes which will either contain one of the parental marker combination (eg., x, m^a or X, m^b) or through crossing over new recombinant marker combinations (eg., x, m^b; X, m^a). In case the molecular marker (m) is very closely linked to gene X, hardly any recombinations will be produced. But if the marker is away from the gene X or on another chromosome, parental and recombinant marker combinations will be prodcued at random frequencies.**

A sample of DNA from PTY538-1 was isolated from agarose after separation on a clamped homogeneous electric field gel and used to probe approximately plaques forming units of a leaf cDNA library. Of about 200 hybridizing plaques 30 were investigated further. The cDNA inserts were amplified by PCR and used to probe a tomato mapping population comprising 85 plants with recombination events in the Pto region. Two of the clones CD127 and CD146 (both 1.2kb in length) contained sequences which cross hybridized with each other. When CD127 was mapped it co-segregated with Pto. The genetic co-segregation of CD127 with Pto and the fact that the clone was isolated from a leaf tissue library made the cDNA a strong candidate for the Pto gene.

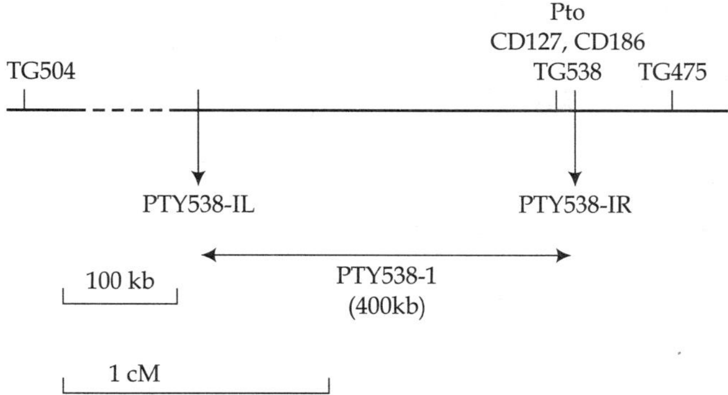

FIGURE 8.2   Showing genetic map- of Pto region on chromosome 5. The locations are shown of RFPL markers (TG), the YAC end clones. PTY538-1L and PTY 538-1R, cDNA clones CD127 and CD186 and Pto. The YAC, PTY538-1 which spans this region is shown below.

The CD127 clone detected numerous polymorphic fragments when hybridized with blots of genomic DNA from resistant (Pto/Pto) and susceptible (pto/pto) plants. This detection showed that clone might contain exons spanning a large region or that it represents a family of repeated genes. To distinguish between these possibilities, leaf cDNAlibrary was probed with CD127 insert and isolated an additional 14 cross-hybridizing clones ranging from 0.6 to 2.4kb. Analysis of these clones indicated that at least six different expressed genes with homology to CD127 exist in Pto/Pto. To investigate the genome location of the family members, total DNA from the YAC transformant YAC538-1 was digested with BstNI and analyzed by DNA blot hybridization. The YAC contained all of the CD127-hybridizing fragments with the exception of a 5kb band that is common to both resistant and susceptible line. Therefore, CD127 represents a gene family which is clustered primarily at the Pto locus. The experiment does not end over there. One can go ahead to determine the following.

1. To determine if homologs of CD127 gene family are present in other plant species. This is achieved using Southern analysis on genomic DNA.

2. To determine if there are differences in transcript size or abundance produced by the CD127 family members among resistant or susceptible tomato lines. This is done using RNA blots.

3. To determine if a member of the CD127 gene family confers resistance to *P.syringae* pv. tomato by conducting genetic complementation tests.

4. Sequencing the entire 2.4kb DNA insert of CD186 insert and a 963 bp ORF was found in the region nearest the 35S CaMV promoter in vector p^{PTC8} encoding 321 amino acid hydrophobic protein.

5. Comparison of deduced amino acid sequence of Pto with the protein sequences in GenBank which in this case uncovered similarity with the catalytic domain of many plant, mammalian and low eukaryotic serine-threonine protein kinase.

Positional candidate method—Positional candidate gene refers to any gene linked to a DNA marker co-segregating with a phenotype of interest and meeting the criteria for a gene which could be responsible for the trait. It is usually isolated by positional cloning. Thus gene cloning can be done by either positional cloning or positional candidate methods.

Gene golfing — This technique is used for isolation of a gene (genes) from a genome for which both dense genetic and physical maps (based on, for example, BACs) are available. The gene isolation procedure starts with the ordering of clones from a BAC library by BAC end sequencing and identification of BAC clones with overlapping sequences. These clones are usually fingerprinted (BAC fingerprinting) such that each clone by its specific restriction fragment pattern and overlapping clones by partial identical fingerprint images. A genetic marker in the genomic region of interest (e.g. a sequence tagged microsatellite marker) is then localized on a BAC clone by molecular hybridization techniques. After that the adjacent marker on the genetic map, say at a distance of 2cM from the first one, is localized on another BAC clone and the physical distance between the both markers is calculated by summing up the lengths of the spanning BAC clones and thus genetic distance can be converted into physical distance. Any subsequent golfing step then exploits the deduced relationship between the genetic and physical distance, i.e. the genetic distance between markers on the genetic map and the gene (s) of interest can be converted into physical distance. This technique permits golfing over large genetic and physical distances on a chromosome to identify BAC clones at or around the target gene without the tedious subcloning procedures.

This approach of cloning of genes for which the products (m RNA, proteins) are not known as in case of disease resistance, is known as **reverse genetics**. The reverse genetics approach is used for the isolation of a gene without reference to a specific protein or without any functional assays which could be exploited for its detection. This approach has been successfully employed in search of hereditary disorders. It involves the following steps.

1. Establishment of the map positions of the gene using cytogenetic method, supported by molecular marker, RFLP with a resolution of roughly several million base pair.
2. Identification of a specific gene within this region of interest in which mutations are strictly correlated with specific disease using cloning and chromosomal walking procedures
3. Search for an mRNA transcript and its disruption or altered expression in a disorder.

Reverse genetics starts with a cloned gene and through *in vitro* mutagenesis and transformation techniques generates mutant strains in an organism, for example, yeast. The result is a set of strains that differ only at the loci selected for the mutagenesis. With this method one can create an unlimited number of mutant alleles from a single cloned gene. A collection of alleles allows one not only to probe the full spectrum of mutant phenotypes which can be produced from a cloned gene (some of which may not be anticipated) but allows one to screen for those alleles which possess the most 'desirable ' phenotypes. For example, an experimenter studying kinetic aspects of DNA synthesis may need specifically those DNA polymerase mutants which most rapidly cease DNA synthesis.

Forward genetics — It is a strategy to isolate gene after its knockout (loss-of-function) by the insertion of say transposon with concomitant change of phenotype. The transposon sequence is then used as a tag which permits the isolation of the corresponding genes. Typically, in standard forward genetics, one mutagenizes the entire organism and isolates mutants affected in the process of interest. These mutants are usually examined for interesting phenotypes (dominance or recessiveness, conditional lethality, etc) and the corresponding wild type genes are cloned. The deduced amino acid sequence of these cloned genes often yields clues as to the function of the mutated loci.

2. Molecular tagging

The process of locating genes through linkage to moleculer markers is referred to as gene tagging. Gene tagging, here, involves the transfer of mobile genetic elements to the location of the gene determining a particular trait which then will lead to the change in phenotypic expression of the gene and will result in the transformation of phenotype. Transposons and retrotransposons are used as inertional mutagens to tag genes. Two types of mobile genetic elements have been used to isolate plant gene (s), namely, transposon, retrotransposons and the T- DNA of *Agrobacterium tumefaciens.*

(a) Transposon tagging

The insertional mutagenesis tags a gene by inserting a known sequence which can be used as a probe to screen a library, into the gene. Thus known sequence can be a transposon which is naturally present in maize or retrotransposons. Transposon is a movable genetic element which is capable of jumping from one chromosome to another chromosome. Structurally transposons are genes which are flanked by identical nucleotide sequence in opposite orientation. The insertion of a transposible element (e.g. Ds/Ac and others) may inactive a gene by producing a mutation and thus mutant plants observed are indicative of the insertion having gone into the target gene. The genomic library produced from the mutated plant is screened using transposon as a probe and the affected gene is isolated. The loss of resistant phenotype (in case of gene determining resistance) of the primary transformants and its subsequent progeny confirms the insertion near the locus determining resistance. The element thus can be used as a probe to isolate the flanking DNA from the library of the mutant. The flanking DNA is in turn used as a probe to isolate the wild type gene. Transposon tagging has been demonstrated as effective means of isolating targeted gene in maize and other crops. The transposable elements of maize have been shown to be mobile in *Nicotiana* and thus can be used to locate genes for disease resistance and other genes. Techniques such as IPCR, plasmid rescue and library screening (using the element as a probe) are all successful in identifying flanking DNA from a 'tagged' individual. Figure 8.3 shows a general gene tagging procedure.

(b) T-DNA tagging

The known sequence in insertional mutagenesis can also be a sequence which is introduced into the plant by transformation such as T-DNA from *Agrobacterium tumefaciens.* Insertion of T-DNA in the genome of plant cells causes mutation and gene mutated by the T-DNA is identified and isolated using T-DNA as a probe. A population of transgenic plants produced is screened for the desired phenotype. The affected gene is isolated from the genomic library produced from the mutated plant using T-DNA as a probe.

Although the gene tagging by T-DNA or transposon has worked well in some cases but in cases where transformation is not efficient, naturally occurring transposon is not available and for genes exhibiting phenotype which cannot be easily scored, this method cannot be employed.

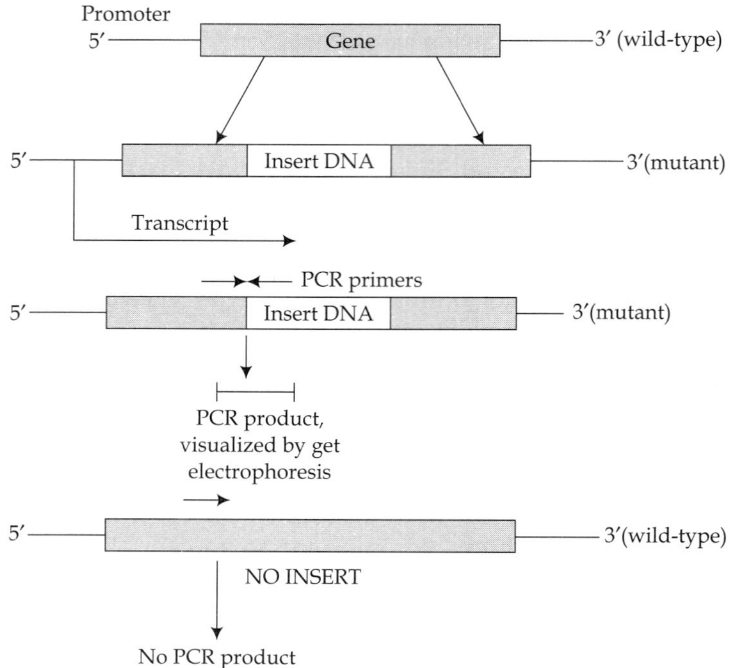

FIGURE 8.3   Showing gene tagging procedure.

3. Genome subtraction

Genomic subtraction method is used to clone genes corresponding to deletion mutations. In this gene isolation strategy, two lines that are isogenic except for a particular trait, say gene for resistance, are developed. The DNA sequences of the two isogenic lines will be same except for the resistance gene. The common DNA sequences can be eliminated after several rounds of denaturation and reassociation and the remaining DNA representing the resistance gene is multiplied by polymerase chain reaction.

4. Antisense mRNA technology

The principle underlying antisense mRNA technology is the down regulation of gene expression by interrupting the flow of information from DNA to protein through the biological inactivation of mRNA. This is achieved by introducing a gene that when expressed, encodes an antisense mRNA molecule that is complementary to the target RNA sequence. Subsequent base pairing results in a non-functional RNA duplex (Van der Krol et al.,1988). This allows identification of gene and is achieved by analysis of the changed phenotype produced by the presence of an antisense mRNA of the unknown gene in transgenic plants. A cDNA encoding an unknown function is isolated and then a chimeric gene (antisense gene) is constructed which directs the production of antisense strand of mRNA. This chimeric gene is then introduced in plant cells by gene transformation procedure. The limitation with antisense technology is that the trait to be altered must be well characterized genetically, biochemically and physiologically in order to allow for selection of specific targets (Fisk and Dandekar, 1993).

B. Strategies based on gene product

The strategies for identification and isolation of gene based on the knowledge of gene product are as follows.

1. Use of antibody.

2. Use of amino acid sequence of the gene product — Use of oligonucleotide probes predicted from the protein sequence of the gene can be used for gene cloning (see chapter 14 for gene designing). Further, short amino acid sequences which are highly conserved among members of a family of proteins (or TF) can be selected for cloning gene. In other words, cloning of novel transcription factors can be done through homology to known proteins (or TFs).

3. Use of oligonucleotide probes derived from the DNA binding site of the protein that the gene produces- There are limitations with the second method. Purification in case of transcription factor requires large quantity of cells and is technically difficult. Further, this method is costly in the sense that determination of the partial amino acid sequence of the protein requires expensive protein sequencing apparatus. This method is based on the fact that information is usually available about the specific DNA sequence to which a particular TF binds (Singh et al., 1988). So, a cDNA clone expressing the TF can be identified in a library by its ability to bind the appropriate DNA sequence. Thus this method relies on DNA-protein binding rather than DNA-DNA binding. A cDNA library is constructed in such a way that the cloned inserts are translated by bacteria into their corresponding proteins. This is done by inserting the cDNA into the coding region of the bacteriophage λ β-galactosidase gene which results in fusion protein (TF protein as part of the bacteriophage protein).

1. Use of antibody

We know that the gene of interest will produce polypeptide (enzyme or protein) which is purified and against this protein. specific antibody is raised. This antibody is used to select specific clone from cDNA library. The cDNA library is constructed with plasmid or phage expression vector so that the target gene upon transfer synthesizes protein in *E. coli* which binds with specific antibody. The plaque or colonies with which the antibody is bound are identified by means of an anti-antibody that is linked to an enzyme that produces a color reaction such as alkaline phosphatase or peroxidase (Huynh et al. 1985; Maniatis et al. 1989). With this technique 33 kDa protein of oxygen encoding complex (GEE1) from tomato and Arabidopsis have been identified (Ko et al. 1990).

2. Use of amino acid sequence

In this method of gene isolation, the protein produced by the target gene is purified and its amino acid sequence is determined and from this gene is designed. The synthetic oligonucleotides are labelled and used as probe to detect the corresponding gene from a cDNA library. This strategy has been used to identify cDNA clone for UDP-glucose pyrophosphorylase from potato (Katsube et al. 1990), betaine-aldehyde dehydrogenase from salt stressed spinach leaf (Wesetilnyk and Hanson, 1990), glutamine synthetase from barley (Freeman et al., 1990) and cM 16 protein in wheat (Gautier et al., 1990).

C. Strategies based on nucleotide sequence

There are two methods of isolation of gene based on nucleotide sequence: (i) Use of heterologous probe (ii) Differential screening (iii) PCR strategy.

1. Use of heterologous probe

Target gene in a particular organism can be identified and isolated through the use of heterologous probes, isolated and characterized in other organism. The cDNA or genomic library of the desired organism is screened using the heterologous probes. This strategy is based on the principle that proteins with simple function from different organisms often have similar primary structure and thus similar nucleotide sequence. Using this strategy nucleotide sequences isolated from animal cells, drosophila and yeast have been used to isolate plant related genes. It is thus a **homology-based cloning** technique. Several proteins involved ion metabolic pathways or signal transduction serve common functions in bacteria, yeast, animal and plants. Thus conserved sequences of genes encoding these proteins can be taken and used as probes to isolate clones from a plant cDNA library. This technique has been successful in isolation of many plant cDNAs and genes encoding components of signaling systems such as protein kinases, protein phosphatases, small GTP-binding proteins, heterotrimeric G protein, calmodulin and Ca^{2+} Calmodulin-dependent protein kinase. But this technique can not be employed to isolate genes occurring uniquely in plants. Alternatively, the conserved sequences may be amplified by PCR to obtain related plant sequences directly.

Functional complementation — The technique of functional complementation has been employed to isolate D-type cyclins $\delta 1$, $\delta 2$ and $\delta 3$ (also referred to as D1, D2, D3) from Arabidopsis (Sony et al., 1995). Cyclines for the G/S transition show much more sequence heterogeneity among species and taxa than mitotic cyclins or CDKs. Thus routine methods of fishing for them with a cDNA or oligonucleotide probe based on conserved sequences are often unsuccessful. This technique consists of expressing heterologous proteins in a suitable vector that is defective in some particular function. If that function is restored by one of the heterologous proteins then not only is the function of that heterologous protein is determined but the cDNA clone that expressed it can be used to isolate the gene in the original organism. The first CDK gene isolated in plants was also isolated by functional complementation of a yeast mutant defective in Cdc2 protein (Hirt et al., 1991). Functional complementation has been used for isolation of mitotic cyclins and CDKs.

2. Differential screening

In this strategy two cDNA libraries are prepared from different organs of the same plant or two similar plant raised in different environments or different developmental stages. From these populations of cDNA mRNAs are produced which are then allowed to hybridize. mRNAs from the two populations having homologous sequences will form double stranded molecule which can be separated from the RNA molecule that are present in only one of the two populations. The specific mRNAs thus isolated can be used as a probe to screen the cDNA libraries. The aim in this strategy is to isolate mRNAs (or cDNAs) which are expressed only in particular environment or at particular stage of development.

3. PCR strategy

When sequence information is available for a gene it is possible to design PCR primers which can amplify all or part of a gene directly from DNA isolated from desired organism. The PCR product is used as a probe to screen for the clone or the PCR product is cloned. PCR strategy was used to generate the hybridization probe for isolation of phosphatidylinositol 4 kinase (PIK1) gene in *S. cerevisiae* (Flanagan et al., 1993). Sequence of a 21-residue peptide was derived from purified p125 and four sets of corresponding degenerate oligonucleotides (primers 1 to 4). Peptides from p125 were generated in sufficient quantity for microsequencing. One 21-residue sequence obtained was used to design degenerate oligonucleotides for cloning by PCR. Reactions containing yeast genomic DNA and primers 2 and 3 generated a single product which was authentic on the basis of several criteria. First, it was obtained only when primers 2 (and not primer 1) was used in combination with primer 3 and only when template DNA was also provided, indicating a specific requirement for the correct Arg codon to amplify a product from genomic DNA. Second, its length (50bps) was precisely that predicted from the peptide sequence. After labeling with $[\gamma\text{-}^{32}P]$ ATP and polynucleotide kinase, a fourth set of oligonucleotides (primer 4) which corresponded to the least ambiguous region of the peptide sequence (residues 9 to 15), hybridized specifically to the 50bp PCR product blotted on a nitrocellulose filter.

The 50bp product was labeled by the random primer method and used to screen by hybridization bacterial colonies containing a yeast genomic DNA library in a plasmid vector. A single clone with an insert of about 10kb was isolated. Restriction fragments from the insert hybridized to oligonucleotides that corresponded to portions of two other peptide sequences (residues 31 to 39 and residues 297 to 304) derived from p125. The nucleotide sequence spanning this entire region was determined and comprised a continuous ORF that lacked any obvious introns. The predicted sequence contained perfect matches to 13 separate peptides obtained by proteolytic cleavage of purified p125 (with endoproteinase Lys-C or trypsin) from two independently derived preparations. Thus the gene isolated corresponded to the enzyme that was purified and was designated as PIK1.

The problem with these conventional cloning techniques is that in most cases nothing is known about the product of the gene and also there is no efficient method available for assaying the product and no homologous cloned genes are available from other organisms to serve as heterologous probes.

Cloning tissue or treatment specific genes — It involves identification of a certain class of gene such as liver specific genes or genes induced by a hormone rather than a specific gene. The numerous methods of identifying such specific types of genes include **subtractive hybridization, differential display, database analysis, analysis of microarrays** (PCR and expressed sequence tag) and are discussed in chapter 24.

Selection of a full length clone in a phage vector when a partial clone is available — It is done by inserting the sequence next to supF gene in a plasmid, infect recA$^+$ cells carrying the plasmid with the phage library and plate the resulting phage on an *E.coli* strain where only phage containing supF grows. Recombination between the complementary sequences introduces the supF only into the desired phage.

8.3.3 Selection of Vector and Joining of Fragment to Vector

After selecting a suitable vector depending upon the size of the insert to be cloned (Table 8.1) and on the host organism to be used, the next step is to join the fragment genome to a vector. The joining of DNA fragments to cloning vector molecules is achieved by treatment of the DNA mixture with *E. coli* or bacteriophage T_4 DNA ligase which forms phosphodiester linkages between the fragment termini. The circular plasmid vector molecules are first converted to linear plasmid in order to introduce the fragment of genome and this is obtained by digesting the plasmids with restriction enzymes which results in single stranded cohesive ends. DNA fragments and vector molecules that have identical cohesive termini (i.e. that were generated by the same restriction endonuclease) are readily ligated. DNA fragments with blunt ends (lacking single stranded termini) may also be joined by T_4 (but not by E.coli) ligase but the reaction is inefficient because joining depends upon the random collision of fragment ends that are unable to form stable associations. For more on cloning vehicles see chapter 10.

Table 8.1 Showing insert size and the type of vector used.

Vector	Insert size
Plasmid	Up to 8kb
Lambda λ	Up to 20kb
Phage T4	Up to 120kb
Phage P1	Up to 80kb
Cosmid	Up to 40kb
BAC	>300kb
YAC	=500kb

Gene cloning strategy can be said to be appropriate only if it allows subsequent recovery of the cloned fragment from a hybrid molecule. The precise excision of the cloned fragment is possible only when both the linear vector molecule and cloned fragments, are generated by the same precision. This is because of the reason that restriction endonuclease recognition sequence is reconstituted at the two vector: fragment junction and hence will again be susceptible to cleavage by that restriction enzyme.

Thus same restriction enzyme generating single stranded cohesive ends is used for generation of chromosomal DNA fragments and cleavage of the vector DNA molecule as it allows annealing of the generated DNA fragments to vector molecules and facilitates the linkage by DNA ligase as both DNA fragments and vector molecule have identical cohesive termini. The advantage of the flush-end ligation procedure is that the DNA fragments generated by different restriction endonucleases can be linked and so the experimentor has much wider flexibility in his choice of enzyme for cleavage of DNA. But if different endonucleases are used for generation of DNA fragment and vector molecule, the two DNA fragments will form hybrid recognition sequence at the junction of the two fragments which cannot be cleaved by either of the two endonucleases used in the construction of hybrid molecule. This problem can be overcome by the use of

  (i) **linker and polylinker**

 (ii) **adaptors and**

(iii) **homopolymer tailing**.

In other words, insertion of cDNA into vector DNA to construct recombinant plasmids can be accomplished by either of the two methods: ligation of 'bridges' molecules such as linkers or adaptors and polymerization of homopolymeric tails.

Linker tailing—Linkers are short synthetic double stranded oligodeoxyribonucleotide molecules which are blunt at both ends and contain one or more restriction endonuclease recognition sequences **(polylinker)** (Bahl et al., 1981). The linkers are ligated to termini of DNA duplex molecule having blunt ends using DNA ligase. Subsequent restriction enzyme cleavage generates cohesive ends suitable for cloning into an appropriate restriction site of a vector. For example, the ligation of EcoRI linkers to a DNA fragment having flush ends provides it with EcoRI termini (5′ –AATT) and thus enables it to be cloned into an EcoRI site of a vector plasmid and to be precisely excised subsequently by cleavage of the hybrid molecule with EcoRI. Thus there are two steps involved in use of linkers. 1. Blunt end ligation of the linker to the DNA fragment and 2. cleavage of the linker with the appropriate endonuclease to generate the specific termini. The later step will result in the internal cleavage of the DNA fragment if it contains one or more of the restriction sites specified by the linker employed. In other words, restriction site (s) on the linkers will permit the internal cleavage of the DNA fragments to be cloned. As some of the randomly or near randomly generated fragments of genomic DNA must contain internal cleavage sites of the type present in the linker, these have to be protected from cleavage by treatment of the fragment mixture with a specific modification methylase (e.g. the EcoRI methylase) prior to addition of linkers. The masking of specific restriction endonuclease recognition sites (e.g. EcoRI sites) within a clonable genomic DNA fragment or cDNA by specific methylation of C or A residues using site specific methyl transferase (e.g. EcoRI methylase). The protected DNA can then be modified by linker tailing and restricted without internal cuts. The synthesis of asymmetric linkers that contain single, preformed, specific cohesive termini and that do not require endonuclease cleavage subsequent to their ligation to DNA fragment have been developed which will solve the problem of internal cleavage of the DNA fragment. Thus linker tailing can also be performed by ligating one strand of an unphosphorylated linker duplex to a normal 5′ phosphorylated terminus of a target DNA duplex molecule. This results in termini carrying covalently linked single stranded self complementary tails that can be annealed to produce a hybrid molecule. Thus DNA fragment termini need to be modified prior to cloning. In a symmetrical arrangement, identical polylinkers flank a duplex DNA fragment whereas in case of non-symmetrical arrangement the polylinkers at both ends of a duplex DNA fragment are not identical. Polylinkers allow the cloning of foreign gene into any of the restriction sites. Table 8.2 shows some linker molecules with recognition sequence.

Adaptors—Like linkers, adaptors are also short synthetic oligonucleotides but differs from linker in that the later have a preformed cohesive terminus. Such adaptor molecules are used to join one duplex DNA fragment with blunt ends to another DNA duplex with sticky ends. The adaptors possess one blunt end with a 5′ phosphate group and a cohesive end which is not phosphorylated in order to prevent self-ligation (Bahl et al., 1978). The adaptor is ligated to the blunt ended DNA target fragment and the construct is phosphorylated at the 5′ termini with

Table 8.2 Some linkers molecules with recognition sequences.

Linker	Recognition sequence
5'-GGAATTCC-3'	EcoRI
3'-CCTTAAGG-5'	
5'-GGTCGACC-3'	Sall
3'-CCAGCTGG-5'	
5'-CCGGATCCGG-3'	BamHI; Hpall
3'-GGCCTAGGCC-5'	
5'-GCTGCAGC-3'	Pstl
3'-CGACGTCG-5'	
5'-CCAAGCTTGG-3'	HindIII
3'-GGTTCGAACC-5'	
5'-AATTCCTGCAGAAGCTTCCGCATCCCCGGG-3'	Pstl, HindIII, BamHI, Saml

polynucleotide kinase. After that the hybrid molecule is ligated into a corresponding restriction site of the second DNA molecule.

Cloning blunt ended DNA fragments involve two ligation reactions: (i) ligation of linker or adaptor to the fragments to be cloned and (ii) ligation of these tailored ends to suitable vectors. In case of linker ligation, additional steps are required which are not needed in adaptor ligation. With linkers cloning efficiency can be as much as 5-fold higher than with adaptors, however, adaptors are safer and require fewer manipulations. Unlike adaptors linkers can be ligated to one another to form blunt ended concatamers comprising linkers alone or linkers attached to cDNA. Thus there is a need to generate cohesive ends flanking each DNA fragment by cleaving with restriction enzyme but during this process DNA molecule can also be cleaved at their internal restriction sites. This problem can be solved in some cases by enzymatically methylating the internal restriction sites before hand so that they become resistant to subsequent cleavage. For example, protection of internal restriction enzyme sites (e.g., EcoRI sites) can be accomplished by treating the DNA with EcoRI methylase (Wu et al., 1987). To verify that methylation has in fact occurred, control DNA such as p BR322 linearized with an irrelevant enzyme (nor EcoRI) is methylated with EcoRI methylase and restricted with EcoRI. When the methylated sample is compared by electrophoretic analysis in a 1% agarose gel with a similar restricted unmethylated sample, the latter should yield two fragments whereas the former should yield the intact linearized plasmid used in the starting material. The last step before ligation to the vector molecule is the removal of excess linkers and adaptors by gel chromatography. Thus cloning using linkers require sequential methylation of the insert, ligation of linkers, cleavage at an internal site of the linker, purification of the cDNA in order to remove excess linker molecules and ligation of the tailored insert to a suitably cleaved vector and cloning using adaptors need only the initial ligation of the adaptor, purification of the cDNA and final ligation to cleaved vector DNA.

Homopolymer tailing—Blunt end ligation (blun-end ligation directly into a blunt –ended site) is much less efficient *per se* than ligation of fragments with cohesive ends. However, the high molar concentration of the small linkers can drive this reaction in a manner impossible with large molecules. An alternate more widely used strategy has been to add homopolymeric tails of complementary oligonucleotides to the 3'OH in both DNA fragment and linearized plasmid. The terminal ends of DNA fragments can also be modified by the addition of homopolymer tails of a nucleotide (dA or dC) to their 3' ends by means of terminal deoxynucleotidyl transferase. The tailed DNA fragment joins with the linear vector molecule tailed with the complementary nucleotides dT and dG, respectively. Judicious choice of the restriction endonuclease used to cleave the vector molecule and the nucleotides used in tailing results in reconstruction of the original cleavage sites in the vector molecule thereby allowing subsequent excision of the cloned fragment. The advantage of the homoplymer tailing is that the circularization of individual fragments,i.e. intramolecular joining, is completely precluded as the tails of both ends of any particular fragments are identical.

Tailing is catalyzed by terminal deoxynucleotidyl transferase in which nucleoside monophosphate residues donated by the dNTPs are polymerized onto 3'-OH termini of either single or double stranded DNA. The length and size distribution of homopolymeric tails are functions of the concentration of enzyme, the concentration of selected dNTPs and the concentration of termini to be tailed and a high concentration is essential to avoid synthesizing long tails on selected molecules while others acquire no tail. Tailing cDNA with dG is recommended as this favors narrow distribution of sizes regardless of the type of termini to be tailed (Eschenfeldt et al., 1987).

Advantages of tailing over linkers and adaptors — The advantages of tailing in cloning are as follows.

1. Multiple ligation steps are unnecessary
2. Methylated restriction of DNA accompanying the use of linkers is avoided
3. Requires fewer cloning steps
4. Takes less time

However, the disadvantages of tailing are: (i) GC stretches are detrimental when screening libraries with GC-rich oligomers (ii) Special sequencing methods are required to 'read through' the tails flanking the insert (iii) Gaps and nicks of poorly constructed cDNA may be tailed with the resultant loss of cloning efficiency (iv) Tailing is not recommended when the identification of clones of interest depends on expression of fusion proteins able to bind to antibodies.

Method of cloning cDNA

Cloning of cDNA from poly (A)mRNA — First of all mRNA is isolated using affinity chromatography as discussed in chapter 30 and reverse transcribed using retroviral reverse transcriptase with oligo (dT) hybridized to the poly (A) tail of the mRNA as a primer. This results in the generation of a single stranded cDNA.

Classical method of synthesis of second strand of the cDNA — The usual procedure for synthesizing the second strand of cDNA depended on the **self-priming** which occurs after the mRNA is removed either by alkali or heat denaturation from the mRNA-cDNA hybrid. The liberated single stranded cDNA folds back on itself and forms a hair pin loop at the 3'end corresponding to the 5' end of the mRNA template which is used as primer for the synthesis of a second strand of cDNA. DNA polymerase I from E.coli is used to generate the second strand. The resulting products are double stranded cDNA molecules which are covalently closed at the end corresponding to the 5' end of the original mRNA. The loop is subsequently removed with single strand-specific S1 nuclease which leads to loss of additional sequences at the 3′ end of the mRNA and thereby resulting in loss of some information of the original mRNA and hence to the synthesis of truncated cDNA. Thus the use of S1 nuclease resulted in destroying the extreme 5′ end of the cDNA (Efstratiadis et al., 1976; Rougeon and Mach, 1976). Figures 8.4 and 8.5 show generation of double stranded cDNA.

Okayama and Berg (1982) devised a method of producing the full length cDNA. Their technique avoids the use of S1 nuclease commonly used in conventional cDNA cloning. The full length cDNA refers to any cDNA that contains a complete reading frame from start codon to stop codon or more precisely the 5′ UTR as well.

For synthesizing the full length cDNA an oligo (dC) tail is annealed to the first strand which allows the priming of the second strand synthesis by oligo (dG). This process does not lead to hair pin formation and so that the S1 nuclease treatment is superfluous and consequently full length cDNAs clones are generated. This method uses the standard conditions of reverse transcription for the first strand synthesis but upon completion of this reaction the RNA-DNA hybrid is, in contrast to classical second strand synthesis method discussed above, left intact. E.coli RNaseH is used to nick the RNA in the hybrids, leaving only small RNA primers with a free 3′-OH group attached to the cDNA. These 3′-OH groups can subsequently be used by DNA polymerase I for synthesizing a second strand all along the first strand (original cDNA). Figure 8.6 shows generation of full length cDNA. The resulting double stranded cDNA molecules are ready for cloning. The cloning plasmid contains an ampicillin-resistance gene (amp^r) and three restriction sites for Kpn, HpaI and HindIII designated as K, Hp and H, respectively.

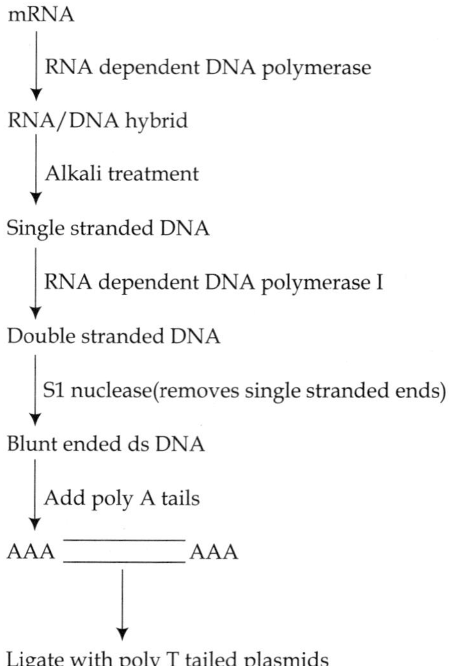

FIGURE 8.4 Shows generation of double stranded cDNA.

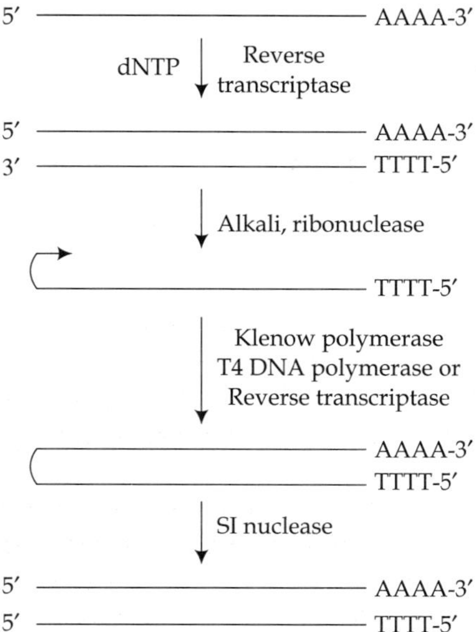

FIGURE 8.5   Showing classical approach for construction of double stranded cDNA.

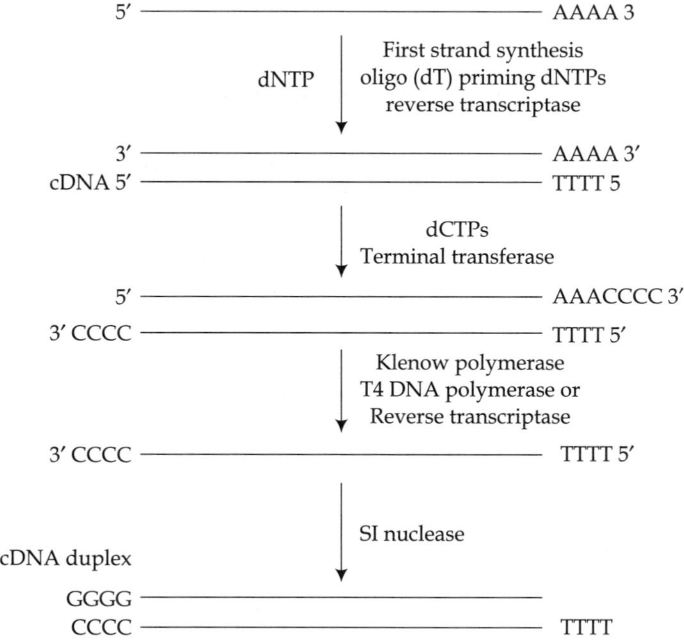

FIGURE 8.6   Showing generation of full length cDNA.

Steps involved in cDNA cloning—The different steps involved in cloning cDNA are as follows.

1. A tail of oligo (dT) is joined at each and then one end is removed by restriction enzyme, HpaI.
2. The messenger RNA molecules are annealed through poly (A) tails to the oligo (dT) tailed plasmid. cDNA complementary to the mRNA is synthesized by reverse transcriptase, primed by the 3'-end of the oligo (dT).
3. Oligo (dC) tails are added to both ends and then a segment at the plasmid end of the construct is removed by the restriction endonuclease HindIII.
4. A plasmid fragment tailed with oligo (dG) at one end and with a cut at HindIII site at the other end is annealed and ligated to the oligo (dC) tailed end of the cDNA.
5. The construct is circularized through its mutually cohesive HindIII cut ends with DNA ligase. The mRNA is nicked and digested with RNaseH which acts on mRNA hybridized and replaced by the second cDNA, synthesized by DNA polymerase, primed by the 3'-OH ends of transient RNA fragments. At the end of the procedure the double stranded cDNA is already built in to the cloning vector.

cDNA from poly (A)⁻ mRNA—cDNA from the poly (A)⁻ mRNA can be obtained by polyadenylating with *E.coli* poly (A) polymerase or more usually, by priming with a random mixture of oligonucleotides although in this case DNA corresponding to the 3' end of the mRNA may be lacking. In the technique of primer extension a single stranded oligonucleotide fragment derived from the 5' end of a cDNA clone that is less than full length may be used as a primer in order to obtain a cDNA clone corresponding to the 3' that is lacking.

Random/forced cloning strategy—It is a technique to clone cDNA into a specific vector (λ gt 10 or gt11) where the orientation of the cDNA insert is totally at random.

Forced/directional cloning—It is a strategy especially used in cloning cDNA to prevent self-ligation of the termini of a recombinant DNA molecule. A directional vector and a directional cDNA insert, both are required in forced cloning. Forced cloning refers to the orientation specific insertion of the cDNA into the appropriate cut polylinker of a directional vector so that the expression is effected. **Directional vector** refers to any plasmid, phagemid or phage vector that allows the cloning of cDNA in a predictable orientation. In addition to sequences characteristics for a vector such as origin of replication, selectable marker gene, the intergenic region of phage, f1 for the production of single stranded DNA, opposing SP6 and T_7 RNA polymerase promoters for the synthesis of RNA from cDNA inserts such directional vectors carry polylinker, usually within a lacZ/sequence which permits the generation of linearlized vector DNA with non-compatible termini, for example, NotI at 3' end and EcoRI at the 5' end. These asymmetric ends inhibit vector recircularization and force an appropriately cut cDNA (directional cDNA) to be cloned in a defined orientation. The **directional cDNA** is a cDNA whose 5' terminus differs in sequence from its 3' terminus, for example, XhoI recognition site at 5' end and EcoRI recognition site at the 3' end. For cloning in expression vector such non-complementary sites can be introduced by the ligation of linker or adaptor oligonucleotide. In one strategy the first strand synthesis is started with oligo (dT)XbaI primer adaptor. After second strand synthesis an EcoRI adaptor is ligated to both ends of the cDNA duplex molecule. Subsequent digestion with XbaI yields a molecule with asymmetric termini, EcoRI and XbaI created cohesive ends. A corresponding directional vector is also digested with this pair of restriction enzymes. Thus ligation occurs in the intended orientation. The vectors for forced cloning are lambda vectors-lambda ORF and lambdaZAP. **Non-palindromic cloning**—It is a technique in which DNA fragment is inserted into non-complementary cloning sites of a cloning vector called non-palindromic vector. Non-palindromic sites on the vector can be created by the ligation of non-palindromic linkers to the termini of a linearized vector molecule. Non-palindromic events prevent the self-ligation of the vector molecules and the concatemerization of linkers.

Plant cloning vector—It refers to any cloning vector that is designed to introduce foreign DNA into a plant's genome. Such vector may be based on Ti plasmid of *Agrobacterium tumefaciens* or DNA plant viruses (see chapter 10 for detail).

Plant expression vector—It refers to any plasmid cloning vehicle specially constructed to achieve efficient transcription of the cloned DNA fragment (s) and translation of the corresponding transcript (s) within a target plant cell. Such cloning vectors contain either a constitutive and strong (highly active) promoter sequence (for example, CaMV35S promoter, nopaline synthase promoter) or inducible (for example, hormone or light inducible) or regulated promoter and immediately downstream of the promoter appropriate cloning site (s) and a plant transcriptional termination sequence are inserted. Any promoter less foreign gene cloned into such vector will be expressed at a high level in the transgenic plant.

The steps involved in introduction of gene in the host are annealing, filling the gap and joining the ends with DNA ligase and then *in vitro* packaging the fragment genome into the host cell. Once fragment genome (foreign DNA) has been inserted into the cloning vehicle, one will have to follow the presence of hybrid molecule or vector through the use of selective markers. The foreign DNA can be inserted into Ti plasmid using binary vectors. The binary vectors involve foreign genes being located on an additional plasmid capable of autonomous replication in Agrobacterium cells. It has high frequency of introduction into Agrobacterium in comparison to the co-integrated vector which is produced as a result of co-integration between the two plamsids. Here the foreign DNA is inserted into a vector that can not replicate in Agrobacterium cells but can recombine' with Ti plasmid through a single or double crossover event at a homologous site to produce recombinant plasmid.

8.3.4   Introduction to Host

The *in vitro* construction mixture used for transformation in bacteria usually consists of a heterogeneous mixture of DNA molecules formed by the random joining of the DNA fragments and vector molecules and only a small proportion of the total molecules will be of the required type, i.e. those consisting of a single vector molecule linked to a single DNA fragment. Therefore the problem in gene cloning experiment will be to select and individually propagate the required hybrid molecule. This is usually achieved in two stages. Firstly, the crude preparation is used to infect suitable host bacteria. Any molecule not containing the cloning vector and hence not possessing the ability to replicate will be lost at this stage. Secondly, the transformant bacteria that contain the desired hybrid molecule are identified by a specific procedure, purified, and their hybrid molecule isolated and characterized. The recombinant DNA molecule can be introduced into the host bacteria through the processes of **transformation** and **transfection**. The transformation involves the introduction of DNA fragment into bacteria involving plasmid DNA whereas the transfection involves bacteriophage genome. During usual transformation procedure little or no multiplication occurs until transformant bacteria are plated out on a selective medium for single colony isolation. Rarely two or more primary transformant clones contain identical recombinant DNA molecules that are progeny of the single DNA molecule from the original cloning mixture. With some procedures only hybrid DNA molecules transform, transfect, infect, or propagate in host cells whereas in others reconstituted cloning vector molecules are the majority of those that transform host cells. Thus specific procedure followed to isolate clones carrying hybrid DNA molecules depends on the properties of the cloning vector employed. In both processes, the DNA fragments are mixed with bacterial cells and incubated for some time to allow the expression of the introduced DNA. Prior to incubation the bacterial cells are made permeable (competent) by treatment with a cold solution of calcium chloride. The cells subsequently receive a heat shock during which they take up exogenous DNA. In transformation experiments the bacteria are plated on a medium (containing an antibiotic) that selects the transformants. The transferred cells do not usually survive and must be detected by plating with indicator bacteria.

8.3.5 Identification of Transformant Clones

There are different approaches available for screening a gene:

1. **Complementation** — Complementation of a mutant gene in the host organism permits direct selection of the desired clone.

2. **Screening transformant colonies for specific DNA sequence or gene product** — The identification of transformant clones carrying hybrid DNA molecules depends on the availability of an assay for the specific DNA sequence or its gene products (Table 8.3). Thus methods of screening hybrid clones include screening for DNA sequence through *in situ* hybridization using probes, unique gene sequences. repetitive DNA or total genomic DNA and autoradiography and for gene products the functional protein or immunologically detectable protein. Several colony and plaque hybridization tests are available for the quick detection of clones or plaques containing specific DNA sequence for which a homologous radioactive DNA or RNA probe is available. In colony hybridization plasmid DNA in transformed bacteria is immobilized on nitrocellulose and hybridization of this to a suitable 32P-labelled probe permits identification of

Table 8.3 Detection of transformant clones carrying required hybrid molecules (Adapted from Timmis, 1981).

Technique	Detection	Requirement
Colony hybridization	DNA fragment	Radioactive DNA or RNA probe homologous to required sequence
Immunodetection	Protein (polypeptide)	Specific antibody probe against required polypeptide
Phenotypic change	Protein (gene product)	Specific assay or host cell system for a change in phenotype
Microscale hybrid DNA analysis	Specific size insert	–

colonies carrying the inserts of interest. Antigen-antibody formation in the immunological detection method is used for detecting the clones products of specific genes against which an antibody is available. The probes are labelled with radio isotopes such as 32p, 3H or 35S. Non-isotopic methods of labelling of DNA includes sulphonation and the use of biotin or digoxigenin. The vector may encode a readily assayed property such as resistance to an antibiotic which may be inactivated by the insertion of DNA fragment into the vector. These tests are highly sensitive and very effective in cloning of double stranded cDNAs, genes from genomic DNA whose cDNAs have already been cloned and genes whose products have been immunologically characterized. Screening transformant colonies for the gene product either with specific antibodies or by an enzymatic assay requires i. a very sensitive, specific assay as foreign genes are often expressed at low level, ii. enzymes from the host organism must not interfere with the assay and iii. the gene to be cloned need not be expressed in the host organism.

3. **Insertional inactivation** — It refers to the transposon tagged gene where the selection is for antibiotic resistance gene encoded in the transposon. Identification of transformant is done by **insertional inactivation** in which insertion of Pst I fragment in the Pst I cleavage site of the A_p^r makes the pBR 322 vector sensitive to amplicillin. The p BR322 vector contains genes for resistance to ampicillin (theA_p^r) and tetracycline (Tc^r). The transformant carrying hybrid pBR322 molecules containing PstI fragments inserted into the unique PstI cleavage site of the A_pr gene are selected on Tc-containing medium and identified by their sensititvity to Ap. Similarly, clones carrying hybrid plasmids containing HindIII, BamHI or SalI fragments inserted in the unique HindIII, BamHI or SalI cleavage sites of the Tc^r gene are ampicillin resitant but tetracycline susceptible (Ap^r Tc^s). Thus a single replica plating of transformant clones from a medium containing one antibiotic onto plates containing another rapidly identifies clones carrying hybrid molecules. The technique of insertional inactivation can also be employed in the direct selection of hybrid molecules in case the vector specifies a function that is conditionally lethal for the host. Genes for antibiotic resistance frequently reside in or are derived from transposon (Kleckner, 1981) and are flanked by 0.8 to 1.4 kb repetitions. These flanking regions are called insertion sequences (IS1, 1S2, 1S4, etc.) which can integrate into various sites of the host or plasmid chromosome. Thus the spread of resistance is not only due to mobilization of the plasmid but also due to the insertion sequences flanking resistance gene which render such genes particularly mobile and confer on them the ability to jump to other chromosome locations (Jinda et al.1983). Finally, the cloning procedure may include a biological size selection of DNA molecules. The selection of DNA molecules that are larger in size than the vector molecules confirms the hybrid clone. In many instances no DNA/RNA or antibody detection probes are available and in those circumstances the hybrid identification must rely on the known size or restriction endonuclease cleavage pattern of the desired DNA fragment. Digestion of clones with a set of restriction endonucleases followed by separation of the resulting fragments on agarose gels and staining with EtBr, different patterns of different clones represent different numbers of restriction sites or different distribution of these sites along the clones and thus allows to discriminate between different clones of a cDNA library or BAC library. To facilitate the characterization of hybrid plasmids from many

clones, microscale procedures for isolation of plasmid DNA are available. In one procedure called 'toothpick assay' the single bacterial colonies from plates are transferred to the wells of an agarose gel, lysed and their plasmid DNAs are sized by gel electrophoresis.

8.3.6 Amplification, Isolation and Characterization of the Hybrid Molecule

Once a transformant clone containing the desired hybrid molecule has been identified. A small amount of plasmid DNA is isolated and used for a preliminary analysis and if this analysis confirms the identification made by the screening procedure then a small amount of the DNA is used to re-transform a plasmid free host cell to obtain a clone that is certain to carry only one DNA species. This step is of paramount importance as sometimes the initial transformation involving saturating amount of ligation mixture DNA, produces primary transformant clones which contain two or three distinct hybrid molecules and so plasmid DNA is prepared from secondary transformant clone and used for a detailed characterization of the hybrid DNA molecule.

The cloned DNA fragment can be isolated through excision by cleavage with restriction endonuclease used in the construction of the recombinant DNA molecule and separation of the cloned fragment from its vector molecule by agarose gel electrophoresis. The UV irradiation of the ethidium bromide-stained gel shows the DNA bands and enables the band containing segments of the gel to be cut out. The gel pieces may then be dissolved in the a saturated solution of KI, followed by absorption of the DNA to and subsequent elution from glass beads (Vogelstein and Gillespie, 1979). In case of agarose gel electrophoresis for analyzing preparations of plasmid DNA, the closed circular supercoiled form of the plasmid migrates more rapidly than the less compact form produced by single stranded nicks whereas the linearized plasmid molecules generally migrate at an intermediate position between these two. Figure 8.7 shows cloning of a gene in *E.Coli*.

8.4 POTENTIAL PROBLEMS WITH LIBRARY-BASED CLONING

1. There is contamination of library with host-DNA leading to cloning of a host gene rather than the gene from the original organism. So to see that genes isolated by complementation are from the desired organism, are checked by Southern blotting.

2. Some DNA fragments can not be cloned as they contain **poison sequence**. The DNA fragments containing repeated DNA sequence are often unstable as they recombine easily. So recA mutant strains of E.coli are often used as a host for cloning to minimize recombination.

3. Problem with DNA fragment having identical ends (cohesive or blunt) is circularization of vector molecules which do not contain insert as two ends of the cut vector are also identical. This gives a high frequency of transformants or plaques lacking inserts. This problem can be minimized by two methods.

   A. Treat the cut vector with calf alkaline phosphatase to remove its 5'-phosphate

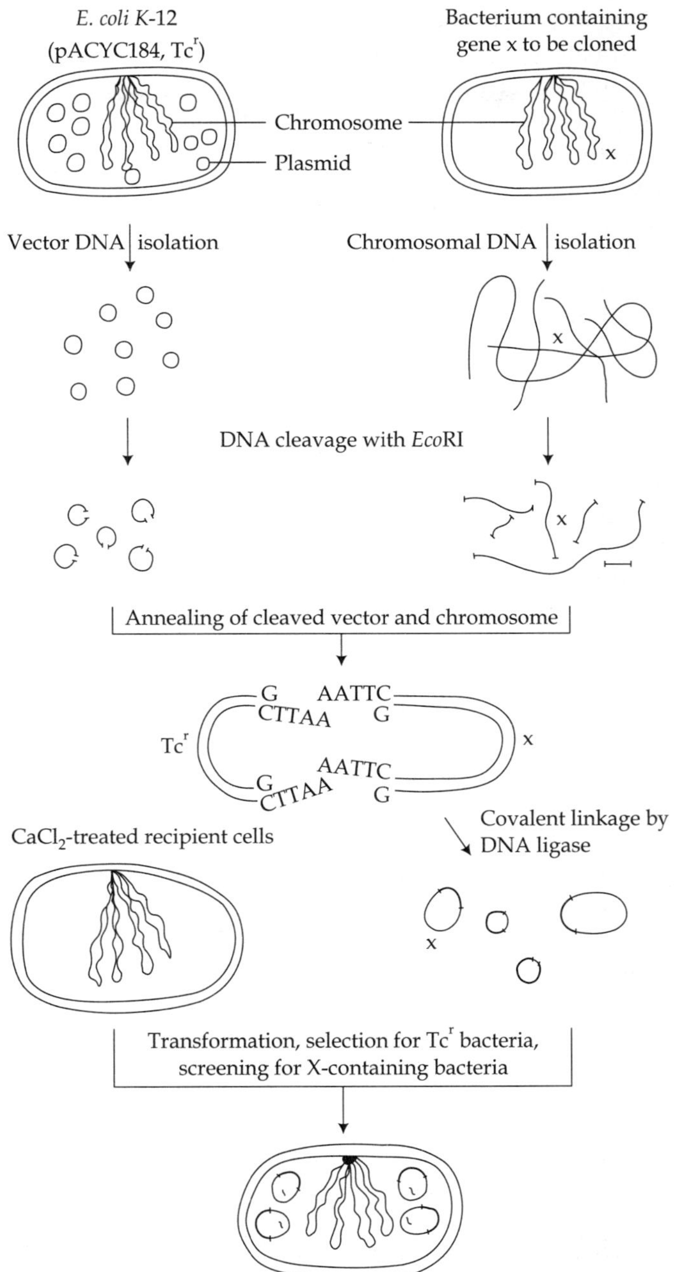

FIGURE 8.7 **Showing gene cloning procedure (adapted from Timmis, 1981).**

groups. Dephosphorylated vector molecules can not recircularize but they ligate to the 5′-phosphate group with DNA fragment to be cloned. The resulting recombinant molecule circularizes producing a circular molecule with single stranded nick in each strand. These molecules give transformants.

B. Another method of dealing with problem of vector molecules lacking insert is **blue-white screening**. Many *E.coli* vectors contain an N-terminal portion of β-galactose gene encoding for the α-fragment which complements the inactive α-fragment. A polylinker is introduced into the β-galactose gene so that the α-fragment is inactivated when a DNA molecule is ligated to polylinker. The ligation mixture is transformed into an *E.coli* host which produces the β-fragment and transformants are plated on selective plates. The colonies which contain an insert are white whereas colonies without an insert are blue.

The DNA is first cloned into appropriate vectors (a cointegrate or binary vector system, see chapter 10 for detail on vectors) whose cloning site is flanked by Ti plasmid sequences (-TDNA borders). An appropriate *E.coli* strain is then transformed with this construct. By conjugal transfer the plasmid carrying the foreign DNA can now be introduced into *Agrobacterium tumefaciens* cells. And upon incubation of these transformed cells with plant protoplast (coculture) the vir region of *Agrobacterium tumefaciens* Ti plasmid is activated by plant wound substances such as acetosyringone. As a result the T-region of the Ti plasmid together with foreign DNA flanked by T-DNA borders is precisely excised. A single stranded copy of the T-DNA is then packaged with vir gene encoded enzyme and transferred into the recipient plant cells and integrate into nuclear genome. Infection of plants with this recombinant Ti plasmid will lead to transfer of foreign DNA along with the rest of T-DNA to plant chromosome.

The explants such as cotyledons, leaves, thin cell layers, cotyledonary petioles, peduncles, hypocotyl, stems, microspores or proembryo or root section is dipped into a culture of modified Agrobacterium, the Agrobacterium with Ti plasmids having the desired gene to be transferred along with marker and reporter genes. After 24 -72 hours incubation, the explants are transferred to medium containing an antibiotic usually carbenicillin to suppress the growth of Agrobacterium. Further transgressive plants are selected by using selective agent, kanamycin. Callus and/or shoots growing on the selection medium are transferred to fresh medium (root inducing medium) as required for the development of complete plant. The selective agent is used in all the media so that it ensures selection and development of only transgenic plants. The concentration of selective agent varies from species to species and from genotype to genotype within a species even from one explant source to the other within a genotype. The frequency of transformation varies from species to species and within a species from genotype to genotype so that the frequency of transgenic plants varies considerably.

8.5 EXPRESSION OF CLONED GENE

The expression of cloned gene in host can be known through the phenotypic changes (trans-formation) occurred in the plant tissue as a result of translation of the introduced DNA. The expression of the introduced gene (s) can be known by various analyses such as Northern analysis which confirms the presence of the expected RNA transcript. Western blotting which allows detection of the expected product and enzyme assays which when possible confirms the presence of active gene product. Independently selected transgenic plants often show varying degrees of gene expression. The magnitude of gene expression depends on several factors such as copy number, position effects resulting from the site of insertion and methylation of the transformed genes besides the regulatory sequences. As the number of integrated copies of genes is not controlled and rearrangements during integration are relatively frequent, the degree of expression of introduced gene (s) varies among different individuals within the population of transformants. Generally the conventional transgenes comprise complementary DNA (cDNA) or intron-less gene, expressed under the control of viral promoter or cell specific regulatory elements and such transgenes are highly sensitive to position effect. So for correct high level expression of transgenes it would be better to use genomic constructs which include introns besides flanking sequences containing enhancer and regulatory elements such as locus control regions (LCRs) and boundary elements which all act together to eliminate the position effects on transgenes. Transgenes vary in their sensitivity to position effects and in some cases two transgenes should be introduced simultaneously in order to overcome the problem of position effects. Finally, site specific recombination systems can be used to introduce transgenes into a locus showing no position effects.

The evidence for the integration of transferred gene (s) into nuclear genome which is a random process can be obtained through different analysis such as Southern analyses which can determine the number of copies and nature of the integration of specific gene (s) or DNA segment and the use of polymerase chain reaction (PCR) for determining the position of the transgene.

Once the expression and integration of the transferred gene (s) is confirmed, one should carry out the segregation study in order to examine the inheritance and the genetic stability of the transferred gene (s). The primary transgenic plants (plants containing DNA from another species or organism) are expected to be heterozygous for the T-DNA insertion and so it -is expected to segregate in the subsequent generations and so the selection is practiced in the segregation generations until transgenic plants are obtained which are homozygous for T-DNA insertion and thus until true breeding lines are obtained.

Specific genes such as gene for resistance to disease and insect can be transferred from wild species to the cultivated one by recombinant DNA technology. Besides above such genes there are bacteria and viruses which are pathogenic to insects and also there are strains of fungus whose product is toxic to insects such as brown plant hopper in rice, such genes from bacteria, viruses or fungi can be transferred through genetic engineering into plants. There are crops where weeds are a problem and in such crops, plant gene for herbicide resistance can be introduced from different species or genera. It has been found in case of frost resistance that the bacteria *Pseudomonas syrenge,* found on the surface catalyses ice nucleating bacterial contaminants on most of the crop plants so gene from the bacteria can be transferred to crop plants where frost resistance is required. In spite of natural ability to transfer its T-DNA into plant genome, Agrobacterium mediated transfer has been restricted to dicot. The reason for this is problem of combining selection of transformed cells and their subsequent regeneration. Agrobacterium mediated gene transfer, has been routinely used in Solanaceous crops like potato, tomato and tobacco.

What is required for successful application of in *vitro* DNA recombinant technology is that we should have efficient methods available for identification of genes or packet of genes for isolating specific genes and for achieving frequency of transformation high enough to be detected and large scale selection of transformed cells and rapid regeneration of the transgenic plant. The problems complicating this method of plant breeding is that the T-DNA insertion which is a random event causes mutation by position effects, interferes with functioning of existing genes in plants and the independently transformed cells and in turn the plants differ in the degree of gene expression and this causes the problem of discarding the undesirable ones and selecting the original elite type with the desirable transformed traits. Secondly, the occurrence of genetic changes (somaclonal variation) during cell culture and plant regeneration phase in association with the Agrobacterium mediated transformed cells complicates the selection of truly transformed cell and in turn the selection of the transgenic plants. Thirdly, the number of transgenic plants produced is also important. In seed propagated crops, fewer number of primary transgenic plants can be expected to meet our expectation of ultimately isolating a true breeding transgenic line. However, in clonally propagated crops, the genetic purity of the clones will be lost by inbreeding and so it is essential to have large number of primary transgenic plants in the hope of isolating a single transgenic plant having the original features. The limitation with this technique is the problem of transfer of polygenes determing a desirable trait (quantitative trait) from a different species as it requires arrangement of their coordinate regulation to give a functional pathway in order to produce a multi-steps product. The other application of gene cloning is isolation and selective amplification of a DNA segment for detailed physical structure and their functional analysis for manipulation of expression of a particular gene. Besides elucidation of gene control mechanism in eukaryotic cells and characterization of eukaryotic chromosomes, the recombinant DNA technology can be used for genetic mapping. The potential risks of conducting the recombinant DNA researches are creation of unique pathogen due to change in virulence, antibiotic resistance, resistance to surface antigen due to production of new foreign protein, due to production of toxins and due to presence of tumor genes and change of the environment due to presence of tumor genes and change of the environment due to creation of organism with growth or evolutionary advantages.

Applications of gene cloning—The gene cloning is used for isolation and selective amplification of the DNA segment for a detailed structural and/or functional analysis or for manipulation of the expression of a particular gene. In other words it provides information regarding gene structure and function and/or provides products of direct use. It will answer whether a gene producing protein in a biochemical pathway has its own promoter or it is expressed from a promoter that serves genes of other enzymes in the pathway as well. If a gene assayed is transcribed from its own promoter or at least from a promoter contained within cloned fragment then it will be expressed irrespective of the orientation of the fragment within the vector molecule but then if it is not then it will be expressed only when the DNA fragment is cloned in the orientation that brings the assayed gene under control of promoter of the vector. Cloning can also be used to study the cellular location of the gene products. It has been particularly useful in the analysis of replication functions of determinants and its organization. Gene cloning is being applied in the dissection of complex, multifactorial traits (quantitative traits) such as yields in crop plants, bacterial pathogenicity, symbiosis, etc. The cloning of individual pathogenicity factor will allow manipulation of the cellular levels of individual pathogenicity factor and construction of pathogenic derivatives from non-pathogenic bacteria for evaluation of the roles and relative importance of each factor. DNA cloning can be used for the production of DNA hybridization probes for identification, mapping and characterization of homologous or partly homologous sequences in or from biological material. Gene separation in the *in vitro* gene cloning procedure is achieved at the following stages:

  (i) During DNA fragmentation through both physical and spatial separation of genes from one another

 (ii) During linkage of DNA fragment to vector molecules through enzymatic separation as DNA fragments which do not join the vector molecules are not subsequently propagated

(iii) During transformation biological separation of the different hybrid molecules occurs as the experimental conditions are usually so arranged that only one or at most two or three hybrid DNA molecules are taken up by each bacterial cell that is transformed

(iv) Physical (dilution) and biological (selection) separation of transformed bacteria through plating of bacteria on solid medium that is selective for expression of a vector function (e.g. antibiotic resistance).

8.5.1 Use of Other Species of Bacteria for Gene Transfer in Plants

Agrobacterium is the most effective vector for gene transfer in plant biotechnology. However, complexity of patent landscape has created obstacles to the effective use of this technique. The plant associated symbiotic bacteria were made competent for gene transfer by acquisition of both a disarmed Ti plasmid and a suitable binary vector. Several species of bacteria outside the Agrobacterium genus can be modified to mediate gene transfer to a number of diverse plants (Broothaert et al., 2005).

Horizontal/vertical gene transfer—In case of vertical gene transfer the DNA (or gene) is transferred from parent to progeny through meiosis or mitosis. Horizontal/lateral gene transfer refers to the transfer of DNA from one cell or organism to another cell or organism of different species through methods other then meiosis or mitosis. For example, transfer of DNA from virus to cell or organism (virus-mediated gene transfer) or from bacteria to plant (bacterium-mediated gene transfer) or from bacterium to yeast. In prokaryotes (bacteria and archaea), some genes are there through the acquisition of sequences from distantly related species via horizontal/lateral gene transfer whereas eukaryotes evolve mainly through lineal descent and mutations.

8.6 METHODS OF DIRECT GENE TRANSFER

The major hurdle in the way of transfer of foreign genes into plant cells is the cell wall which does not allow the DNA from the surrounding medium to enter into the cell. Agrobacterium species and some plant viruses can transfer DNA across the cell wall and then into the genome. However, the transfer of foreign gene (s) using Agrobacterium as vector is limited primarily to dicots. Most monocots including cereals, grasses and some dicots (many legumes) are not susceptible to Agrobacterium mediated gene transfer. Now different techniques have been developed which do not require biological vector and gene (s) can be directly transferred to the plant genome. There are four methods of direct gene transfer:

  (a) Micro-injection
  (b) Microprojectile techniques,
  (c) Electroporation
  (d) Silcon carbide fibre

(a) Microinjection — In this method DNA is directly injected into the nucleus through the use of a very fine needle. Although this method has high rate of success if done by skilled technician, the total number of cells that can be treated is small as each cell must be injected individually. Microinjection is a technique for mechanical transfer of genes into living cells. It uses a micromanipulator comprising a glass capillary with a blunt end (holding capillary) to position the recipient cell and to hold it under slight suction and a finely drawn glass capillary into which the injection fluid (a DNA solution) is sucked under slight vacuum (injection capillary). The injection capillary is then moved with mechanical or hydraulic devices (chopsticks) towards the fixed cells until it penetrates the cell membrane. The injection capillary can also be directed to enter the nucleus. The injection capillary is then emptied and releases the DNA solution which is frequently mixed with an indicator dye such as red oil or fluorescence Lucifer-Yellow. The recipient cells may also be glued to a glass surface with poly-L-Lysine or partly embedded in a thin layer of agarose. This technique of direct gene transfer is comparably accurate and highly efficient.

(b) Microprojectile technique — Techniques of direct gene transfer which do not require removal of the cell wall are known as micro projectile techniques. In this technique (Sanford, 1998, 1990) the DNA coated microprojectile travels at higher velocity to the target tissues and delivers nucleic acid (the gene) into plant cells. The tungsten and gold projectiles have been widely used in gene transfer studies. The DNA coated heavy metal projectiles are delivered either by gun powder charge, a gas pulse or electrical discharge' energy. This technique of micro-projectile bombardment is also known as **particle gun** or **gene gun** system or **biolistics**. In this technique of gene transfer, selection of cells for expression of foreign genes can be made within days without the need to use cells capable of further division. By this technique, transgenic plants in virtually any species can be produced. Transgenic plants generated by microprojectile techniques have already been reported in crops such as cotton, maize, soybean.

Both organized meristematic tissue such as embryo from seed or proembryoids or shoot produced through tissue culture and de-differentiated totipotent cells such as cell suspension, callus or tissue explants which regenerate plants via callus stages can be used. Although we have seen that the problem with meristematic tissue culture is of somaclonal variation and production of chimeric plants but there is likelihood of uniformly transformed plants with dedifferentiated cells. Even anther and pollen culture can be used for gene transfer by this technique.

The limitation with micro-projectile technique (particularly gun delivery method) has been the production of inconsistent results which can be attributed to various factors such as irregularities in microprojectile shape and size, DNA coating methodology to generate uniformly covered particle which do not agglutinate rapidly and engineering of devices which can provide more uniform acceleration and distribution of bombarding particles (Herman, 1990; Sanford, 1990).

(c) Electroporation — Another method of direct gene transfer which requires the removal of the cell wall from the cell is electroporation. This technique thus involves protoplast isolation and culture system. In this technique high voltage electric pulses are applied to cell to induce transient membrane pores through which the foreign DNA in the surrounding solution can enter into the cell. Different electrical conditions (the applied voltage) are required for efficient permeabilization of different cell types and sizes. The equipment called gene pulsar is used in electroporation for transfer of macromolecules (e.g. DNA) into bacterial, animal and plant cells. In this technique the degree of expression of gene depends on the source of the protoplast, electrical voltage and the buffer. Protoplast isolated from suspension culture is perferred as it ensures consistent expression of genes because of greater uniformity of suspension cells. This technique of gene transfer can be applied only to species in which regeneration of plants from protoplast is possible. Production of transgenic plants is further plagued by the culture induced or somaclonal variation which comes in the way of improving the elite cultivars.

Electroporation technique differs from Agrobacterium mediated gene transfer in that it does not require the use of antibiotics to eliminate bacteria from cell culture. High copy number plasmids can be used and selectable and non-selectable genes can be incorporated on different chromosomes which facilitates elimination of selectable genes in the F_1 generation.

Recently this technique was modified and electroporation was used to deliver DNA into intact organised embryo and callus tissue from two different inbred lines of maize after either enzymatic or mechanical wounding (D' Halluin et al., 1992). This approach to transformation may find more utility as it has been used to express transgenes in intact, organised tissue of rice, barely and wheat (Dekeyser et al., 1990). Several additional methods have been used to produce the transgenic plants. Stable transformations of a number of species have been achieved after permeabilizing cells with polyethylene glycol (PEG) (Shillito et al., 1985). Microinjection of DNA directly into cell nucleus has been used (Neuhaus and Spangenberg, 1990).

Lipid-mediated gene transfer — There is a report of lipid mediated transformation in which fusion of negatively charged DNA-containing liposomes with plant protoplasts have been used as a method of transformation (Riggs and Bates, 1986) which although offers no advantage over other methods of direct DNA uptake utilizing protoplast (Potrykus, 1991). Lipofection introduces up to 120kb of DNA into eukaryotic cells. Liposomes are small spherical vesicles consisting of a lipid (phospholipid) bilayer that encloses a recombinant DNA molecule. The unilamellar lysosome consists of synthetic cationic lipids, for example, N, [1- (2,3-dioleyloxy)-propyl]-N,N,N-trimethyl-ammonium chloride (DOTMA). Liposomes are formed spontaneously when phospholipids are suspended in an aquous buffer. DNA can be entrapped in such lysosomes by simple mixing phospholipids (e.g. phosphatidylseine) and buffer containing DNA followed by brief sonication. The loaded vesicles are then fused with membrane of the target cells. In case of plants the cellulose wall is removed by enzyme. The sonication step is very critical in this technique as it might break high molecular weight DNA molecule. The efficiency of this method is poor as only one in 100-10000 lyposomes conatin DNA and further most of the lyposomes do not fuse with the cell membrane.

Sonication/ultrasonication — A method of direct DNA transfer based on ultrasound has been developed (Joersbo and Brunsteid. 1992).The foreign genes (DNA sequences) can be introduced into the target cells, if they are exposed to ultrasound waves which facilitates the uptake of DNA by the cells and thereby its integration into the target genome. Ultrasonication has been used to successfully transform protoplasts and more importantly, intact, organised plant tissue (Zhang et al., 1991). In sonication audible sound waves are produced by sonifer (sonicator) to disintegrate cells in liquid medium and is a rapid method for the introduction of DNA into cells in suspension (e.g. the amoebae of Dicotyostelium discoideum). This method of sonication loading is less costly and simpler than particle bombardment. Electrophoretic transfer of DNA has also been used to introduce gene into plant cells (Ahokas, 1989). Unlike electroporation in which diffusion is thought to be responsible for DNA uptake following a short electrical discharge, electrophoresis involves lower voltage applied for a relatively much longer period of time and moves the negatively charged DNA into plant cells. Through this technique DNA has been delivered to barley seed meristem and orchid protocorms.

Hypo-osmatic shock loading—It is a method of introduction of DNA or protein into culture cells which exhibit high rates of fluid phage pinocytes (fluid phage endocytosis). When the cultured target cells when exposed to 0.5 sucrose and 1% PEG6000, the DNA or protein is taken up by via pinocytes.The loaded pinocytotic vesicles (pinosomes) are then lysed *in vivo* with a hypotonic medium resulting in release of the DNA/protein into the target cells.

(d) Silicon fiber mediated gene transfer—Fiber-mediated DNA transfer is a method of dicer gene transfer into the cells in which silicon fibers are used for the perforation of cellular membranes. These fibers are softly ground with the cells (or tissues) and the DNA enters the cells through fibers induced pores in the membranes. Since the silicon fibers are potentially hazardous the survival of the target cells is not satisfactory and the transformation frequency is not better than with other techniques of direct gene transfer and thus fiber mediated gene transfer would remain exotic.

(e) Irradiation and fusion gene transfer—In this technique chromosomes are broken into large fragments by controlled exposure to ionizing radiation. The irradiated cells are then fused with non-irradiated cells of the same or different organism.

(f) Chromosome-mediated gene transfer—In this method of direct gene transfer large fragment of a DNA (usually in megabase range) is inserted into an artificial chromosome which is then transferred into mammalian cells either through microinjection or lipid vesicles.

(g) Calcium phosphate/DEAE dextran precipitation—This technique of direct gene transfer is based on precipitation of DNA in insoluble calcium phosphate (Graham and van der Eb, 1973) or diethylaminoethyl (DEAE) dextran complexes directly onto the membrane of the target cells. Thus calcium phosphate precipitates are produced by adding a DNA $CaCl_2$ solution into an isotonic phosphate buffer. The precipitates which are formed after about an hour are used to transform bacteria, plant or animal cells. The calcium phosphate co-precipitation technique also results in the stable integration of DNA into the chromosomal DNA of a small proportion of cells. Dextran is a high molecular weight polysaccharide produced by specific microsorganism and consists of D-glocose linked by α (1,6) and more rarely by α (1,2) or α (1,3) and α (1,4) bonds. It accelerates the hybridization rate in nucleic acid hybridization.

Microcell mediated gene transfer—In this technique single chromosome from one mammalian somatic cell is transferred to another cell using so-called mini cells. In this method the donor cells are treated with colcemid which blocks mitosis, leading to reorganization of the nuclear membrane which engulfs single or small group of chromosomes (micronuclei). Addition of cytochalasin B and centrifugation of these multinucleated cells generate minicells (micronuclei surrounded by the plasma membrane) which is then fused to the target normal sized cells with the help of PEG. The table 8.4 given below summarizes the methods of direct gene transfer in different organisms.

Table 8.4 Methods of direct gene transfer.

Species	Procedure
Bacteria	
Diplococcus pneumoniae	Will take up DNA from culture medium during certain phases of growth
Hemophilus influenzae	-do-
Bacillus subtilis	-do-
E. coli	Treatment with calcium chloride prior to exposure to DNA
Fungi	
Saccharomycesd cerevisiae	1. Enzymatic removal of cell walls, Treatment of resulting protoplasts with DNA, calcium chloride and PEG
	2. Treatment of intact cells with lithium acetate- then DNA followed by PEG
Neurospora crassa	Same as method 1 of Saccharomyces
Drosophila	Insert gene into transposable element P and inject the construct into early embryos for integration into the genome
Mammalian cells	1. Direct injection of DNA into egg pronuclei
	2. DNA supplied to cultured cells as a calcium precipitate
	3. Cells infected with a retroviral vector
	4. Electroporation- fluctuating electrical potential applied to cells with DNA
Plants	1. Electroporation of protoplasts
	2. Use of Agrobacterium T-DNA as vector
	3. Penetration of cell or tissues by explosively projected tiny gold or tungsten beads coated with DNA (also may be useful for animal cell transformation)

8.7 TRANSFORMATION OF ANIMAL CELLS

Many types of animal cells can be isolated and manipulated *in vitro*. Cells derived from particular tissue such as liver can be grown under appropriate tissue culture conditions where it maintains its differentiated properties for weeks or even months. There is no suitable plasmid like vector available for introducing DNA into an animal cell and so transformation usually require the integration of the foreign DNA into the host chromosome. Although some success has been achieved with spontaneous uptake of DNA or electroporation techniques which are common methods used to transform bacteria (Table 8.4) but these techniques are inefficient (transform only one in hundred to ten thousand cells).

Transformation in animal cells—The efficacy, specificity and safety of the vector systems used for gene transfer are important factors in view of their application in human including gene therapy and DNA immunization. The most commonly used techniques for transformation in animal cells are:

1. microinjection
2. Liposome mediated transfer
3. Viral vectors.

Of the three methods the two, liposome and viral vectors are the most efficient and widely used techniques for transforming animal cells. The use of DNA-coated microprojectiles or the use of a non-viral gene delivery system such as the high mobility group I (HMG1) chromosomal proteins, the α-2-macroglobulin and peptides have also been investigated. Alternative means of for gene targeting within cells have been developed based on the concept of receptor targeting (Wagner et al., 1990; Perales et al., 1994). The efficiency of such a delivery system has been enhanced by the addition of or coupling to viruses or virus proteins.

Liposome mediated transfer — Liposomes are small vesicles consisting of a lipid bilayer that encloses an aqueous compartment. Liposomes containing a recombinant DNA are fused with the membrane of the target cells to deliver the foreign DNA into the cell (Felgner et al., 1987; Felgner and Ringold, 1989). The DNA sometimes reaches the nucleus where it is integrated into chromosome at random locations. The liposomes could contain viral fusion peptides or whole viruses in view to facilitate the release of DNA into cytoplasm (Kamata et al., 1994; Wagner et al., 1994).

Viral vectors — Viruses are infact a natural and most effective vector system for gene transfer. Viral vectors are even more efficient than liposome at delivering DNA into the animal cells. Animal viruses have very effective mechanisms for introducing DNA into the host cell and several types of animal viruses have also mechanisms to integrate their nucleic acid into cells. Viruses like adenoviruses, adeno-associated viruses, Epstein-Barr viruses and retroviruses (Danos and Mulligan, 1988; Naldini et al., 1996) have been engineered to carry therapeutic genes and to deliver their own genomes into target cells. Moreover a tissue specific gene transfer could be obtained through the cell specificity of the viral cell receptors. The problems associated with the use of viral vectors are as follows (Touze and Coursaget, 1998).

1. Potential production of replication-competent viruses if a recombination event occurs in the target cells.
2. Inactivation of the vectors by pre-existing neutralizing antibodies to these viruses in the host.
3. Immunological response induced by the viruses.

Mammalian cells and cells of other higher eukaryotes are able to take up and express exogenously added DNA. Only a small portion of the cells exposed to the transfection solution are 'competent' to take up DNA. These transgenomes are expressed by the transfected cells and but as they lack centromeres, are lost from the cell population. The expression of transfected DNA for a brief period is termed transient expression. When transgenome becomes randomly linked to the cellular genome, it is then stably maintained. When an excess of a non-selected gene of interest is mixed with a dominant selectable gene marker before transfection, these genes normally become associated during transfection process and the non-selected gene is present and expressed in the majority of the selected colonies. This process of **co-transfection** (Cullen, 1987) results in cells which demonstrate stable or constitutive expression of the gene of interest.

Production of transgenic in animals — As shown in the table the transgenic animals can be generated through i. microinjection of DNA into male pronucleus of one-celled embryos ii. infection of pre- and post-implantation embryos by retroviruses and iii. transfer of desired genetic material using embryonic stem cells. There is more controllable introduction of DNA in stem cells over the first method. ESCs (embryonic stem cells) are the ideal vehicles for the introduction of foreign DNA into mouse genome. This technique is used for the establishment of loss-of-function mutations through site-specific integration of foreign DNA sequence by homologous recombination. Embryonic stem cells are derived from the inner cellular mass of the mouse blastocyst which is cultured on a feeder layer of primary embryonic fibroblasts. The ESCs when cultured on embryonic fibroblasts and/or in the presence of leukemia inhibitory factor can be propagated *in vitro* and retain their undifferentiated state and pluripotency. ESCs can be genetically manipulated *in vitro*. ESCs with stably integrated exogenous DNA are introduced into the embryo by blastocyte injection or by aggregation with eight-cell embryos and manipulated embryos are then transferred to foster mothers (Mansouri, 1999).

8.8 APPLICATIONS OF RECOMBINANT DNA TECHNOLOGY

The recombinant DNA technology is being used in the production of transgenic plants showing resistance to insects and diseases, resistance to herbicides and production of male sterility and fertility restorer system for producing hybrids crop plants and can be used for molecular farming (Roy, 2000).

8.8.1 Manipulation of Plant Metabolism

Manipulation of plant metabolism involves changing metabolic activities such as increasing photosynthetic capacity or the ability to fix atmospheric nitrogen in order to enhance productivity. Chloroplasts resemble photosynthetic prokaryotes both genetically and cytologically and thus can be manipulated using techniques of recombinant DNA technology. Since chloroplasts replicate semi-autonomously, chloroplast genome alteration are heritable by daughter cells and progeny plants (For more on chloroplast see chapter 1). Steps involved in manipulation of chloroplast are as follows (Miller, 1978).

1. Isolation and *in vitro* maintenance of chloroplasts.
2. Manipulation of the chloroplast through uptake of specific isolated prokaryotic DNA's to generate either a recombinant chloroplast genome or a plasmid carrying chloroplast. By analogy with blue-green algae, chloroplast may be capable of under-going transformation and of maintaining and replicating plasmid DNA. The DNA of interest could be carried on a plasmid containing a prokaryote-specific and therefore chloroplast specific drug resistance.
3. Introduction of recombinant chloroplast into cultured plant cells by uptake or fusion followed by selection and cloning of plant cells carrying the trait of interest. Cells carrying recombinant chloroplasts could be selected for drug resistance and screened for the trait of interest.
4. Regeneration of whole plant from a cultured plant cell clone.

This suggested genetic manipulation is based on the assumption that if chloroplast evolved from a blue-green algal ancestor then there could exist homology of recombination mechanisms and of DNA sequence between prokaryotic and chloroplast genome and further, if the blue-green algal ancestor of chloroplasts fixed nitrogen then the function of nitrogen fixation may be compatible with chloroplast metabolism.

8.8.2 Molecular Farming

By genetic engineering techniques specific polypeptides or proteins that can be produced in large quantity are as follows.

1. Enzymes (microbial, eukaryotic)
2. Hormones
3. Interferon
4. Antigens

Enzymes — The enzymes could be enzyme of microbial origin such as DNA ligase, enzymes for industrial purposes such as penicillin acylase which is used in the production of antibiotics. Enzymes produced by eukaryotic genes could be industrial enzymes and enzymes having use in medicine in order to correct the defects in specific gene. Only low level of specific enzyme is required-HPRT. The problems associated with the production of these specific polypeptides or proteins are :

1. How to get sequence transcription in an organism different then in which it is found?
2. Translation of the information contained in the gene
3. Protein stability.

Production of human insulin — Insulin is a small protein used to treat diabetes. Diabetic patients have excess of sugar in their blood and urine. It is synthesized in the pancreas as an inactive single chain precursor, pre-proinsulin with an amino terminal sequence that directs its passage into secretory vesicles. Proteolytic removal of the signal sequence (23 amino acids long at the amino terminus of preproinsulin) and the formation of three disulphide bonds produces pro-insulin which is stored in pancreas β cell. When raised blood sugar triggers insulin secretion proinsulin is converted to insulin by specific protease which cleaves two peptide bonds to form mature insulin which convert glucose into storage compounds such as glycogen and triacylglycerol and thus insulin lowers blood glucose. Insulin thus keeps blood sugar concentration constant. Insulin regulates both metabolism and gene expression. The gene sequences of insulin genes are known. It consists of 2 polypeptide chains, A and B. A polypeptide is 20 AA long whereas B polypeptide chain is 30AA long and two polypeptide chains are joined through two disulfide bonds. The steps involved in the production of insulin by *E.coli* are as follows (Goeddel et al, 1979).

1. Synthesize the genes for insulin A and B chains
2. Splice them into plasmids p BR322
3. Put the gene under β- galactosidase to provide efficient transcription and translation and a stable precursor protein
4. Cleaving of insulin peptides from β- galactosidase.

Insulin peptide chains *in vivo* are obtained as short tails joined by a methionine to the end of β- galactosidase. Insulin chains which contain no methionine can be cleaved off efficiently by cyanogens bromide.

So put one gene, say gene for A chain in one plasmid and gene for B chain in another plasmid. Thus two separate bacterial strains, one for each of the two peptide chains of insulin are constructed. The insulin peptides are cleaved from β- galactosidase. Finally, allow these two isolated polypeptide chains to interact and get the functional insulin. Joining of the chains *in vitro* is by air oxidation. The presence of insulin is detected by radioimmunoassay.

Itakura et al (1977) designed and synthesized a gene that produced a small polypeptide hormone, somatostatin. Somatostatin, a mammalian peptide (14amino acid) hormone inhibits the secretion of a number of hormones including growth hormone, insulin and glucagons. Glucagon signals that the blood sugar is too low. The synthetic somatostatin gene was fused to the *E.coli* β- galactosidase gene (*lac* operon) on the plasmid p BR322. After transformation the chimeric plasmid directs the synthesis of a chimeric protein which can be cleaved *in vitro* at methionine residues by cyanogens bromide to yield active somatostatin. Translation was predicted to result in a somatostatin polypeptide preceded either by 10 amino acids (p SOM1) or by virtually the whole β- galactosidase subunit structure (p SOM11-3). The peptide hormones insulin, glucagon and somatostatin are produced by specialized pancreatic cells β, α and δ, respectively.

Interferons (INFs) — It refers to a group of proteins that form a closely related group of non-viral proteins of molecular weight in range of 15, 000 to 30, 000 which are produced and released by animal cells following exposure to a variety of microorganisms and the viruses are the most potent inducing agents. Interferon exerts non-specific and anti-viral activity at least in homologous cells through cellular metabolic processes involving the synthesis of RNA and protein. Human interferons are a class of glycoprotein with antiviral properties. It interferes with viral production and thus used for treating some cancer. Thus human interferons have antiviral, anticellular and immunomodulating activities. In other words, IFNs have the ability to confer virus resistance, inhibit cell proliferation and modulate immune response. Interferons can also be made by *E. coli*. Here carbohydrate does not seem to be a problem. The problem is to get the proteins against viral infection which can stimulate immune responses of the cellular tissues against viruses. IF Le is produced in WBC (Leucocytes) whereas IF (F) is produced in fibroblasts. Interferon IF N-β-1 is produced in ovary. Same proteins produced in the two cells have different glycosylation pattern. There are two methods of selecting the clone. So to start with clone cDNA and screen the cDNAs to pull out the cDNA that carries the interferon genes. Protein is made up of AA sequences. From the knowledge of AA sequence predict a relatively small number of basic sequence of bases in DNA. Construct a four base pair synthetic oligonucleotides and the radioactively labeled (32 p) probe could be used in hybridization to isolate interferon genes. Purify that mRNA which is complementary to the labeled probe. In the second method called mRNA hybridization assay first make DNA and hybridize to mRNA and isolate DNA/RNA hybrid. Finally, dissociate RNA and transfer to frog oocyte and check for interferon activity. Of pulled RNA at least one out of 512 carries cDNA carrying interferon activity. DNA sequence codes for 23 AA signal peptide followed by an interferon polypeptide of 165 AA. The interferon gene in *E.coli* was under *trp* promotor control and the synthesis in *E. coli* was of the order of 2.5×10^{8} units/meter per litre of culture. Goeddel et al. (1980)

synthesized human leukocyte interferon cDNA enzymatically, inserted this into the vector p BR322 and cloned in *E.coli*. Derynck et al. (1980) inserted human fibroblast interferon (HF-IF) coding sequence into site on a thermoinducible expression plasmid under the control of leftward promoter (P_L) of bacteriophage λ. The primary translation products predicted on the basis of the plasmid constructions were hybrid proteins starting with β-lactamase or phage MS2 polymerase information followed by the total pre-interferon.

Antigens — It is a molecule capable of eliciting the synthesis of a specific antibody in vertebrates. It can be produced to use for diagnostic purposes or controlling pathogens such as hepatitis B virus, influenza virus, foot and mouth disease virus, etc. The immunity against influenza lies in HA surface glycoprotein. Emtage et al. (1980) constructed bacterial strains that had potential of synthesizing large quantities of viral proteins (antigens) which may ultimately be useful as vaccines. In other words, they obtained haemagglutinin antigen from an HA gene cloned in *E.coli*.

Expression of an eukaryotic gene in *E.coli* can be obtained by fusion between the foreign gene and a bacterial gene carried by a plasmid or a phage. Thus lacZ gene has been used for hormone biosynthesis. In case of Hepatitis B virus, gene S codes for polypeptide carrying the surface antigen (HBsAg).Expression of S gene in *E.coli* was obtained by gene fusion between gene S and gene lacZ. Charnay et al. (1980) constructed a derivative of bacteriophage λ (λ lacHBs-1)carrying a fusion between the β- galactosidase (lacZ) and the HBsAg coding sequence. Infection of *E.coli* with λ lacHBs-1 led to the biosynthesis of a polypeptide of molecular weight 138,000 carrying antigenic determinants of HBV surface antigen. In case of foot and mouth disease virus (FMDV) the coding sequence for structural protein VP1, the major antigen of virus was inserted into a plasmid vector, p BR322 where the expression of this sequence was under control of λ P_L promoter (Kupper et al. 1981)

Antibody is a defense protein (immunoglobulin) synthesized by the immune system of vertebrates.

Monoclonal antibodies — Antibodies produced by a cloned hybridoma cells (antibody-producing cell can be fused to cancer cells to found immortal hybrid cell clones) which therefore are identical and directed against the same epitope of the antigen.

Single cell protein (SCP) — It refers to a protein-rich material produced by cultures of bacteria, yeast or other fungi or algal and it is extracted for use as a protein supplement in animal or human food. The suitability of microorganisms for using as source of protein depends on the efficiency with which they convert carbon into cellular carbon. Single cell proteins include food yeasts such as species of Candida, Saccharomyces, Torula using sugar, starch, molasses, etc. substrate as a growth medium and the moulds, Fusarium and Penicillium notatum grown on starch. Other sources of SCP include species of chlorophyta chlorella and Scenedesmus and Cyanobacteria, Arthrospira and Spirulina (Wickens, 2001).

8.9   RECOMBINEERING

Genetic recombination involves the exchange of DNA sequences between two chromosomes or DNA molecules. It creates variability and repair damaged DNA. There are two major classes of recombination: Site-specific recombination and general homologous recombination (see chapter 21). In homologous recombination the joining of the DNA duplexes shows a similar degree of precision or fidelity but does not take place at specific sites. As exchange can take place any where along the length of the two homologous chromosomes so the proteins catalyzing homologous recombination are not sequence or site specific binding proteins. Recombineering is a term used to describe homology-dependent, recombination-mediated genetic engineering (Court et al., 2002). Recombinant DNA constructs are made *in vivo* via homologous recombination. Homologous recombination (HR) thus provides new ways to modify genes and segments of chromosome. In this technique constructs can be made on plasmids or directly on the *E.coli* chromosome from PCR products or oligonucleotides by homologous recombination. It is possible as bacteripphage encoded recombination function efficiently recombine sequences with homologies as short as 35 to 50 base pairs.Homologous recombination is defined as the process of exchanging DNA between two molecules through regions of identical sequence. It ensures precise exchange and joining of two DNA molecules within the limits of exchange events defined by homologies. Homologous DNA recombination systems are useful for moving mutations into and out of bacterial chromosome. There are two pathways available for homologous recombination: Standard *E.coli* Rec A-dependent recombination pathway and phage system. Phage encoded homologous recombination systems (such as bacteriophage λ Red system and other technological advances have eliminated the need for restriction enzymes and DNA ligase for modifying or subcloning DNA and thereby eliminating many of the time consuming steps *in vitro* steps of genetic engineering. The advantages with phage system are as follows.

1. Phage system can catalyze efficient recombination with very short regions of sequence homology (<50bp).
2. This system functions even in the absence of RecA, a protein essential for *E.coli* homologous recombination. Rec-A action can lead to unwanted recombination and rearrangement of large genomic clones on BACs.

Precision in generating recombinant DNA molecules by standard genetic engineering techniques with restriction enzymes and DNA ligase is lost when working with large DNA molecules. Most cloning techniques depend on unique restriction sites but with large DNA molecules even rare cutter such as NotI enzyme may have many sites of action. Further, large DNA can be difficult to work with *in vitro* as it is prone to breakage. Thus homologous recombination *in vivo* is a precise way to engineer large DNA molecules.

In contrast to genetic engineering the recombineering does not require construction of plasmid or phage DNA intermediates containing approximately pre-engineered homology segment. All that is required *in vitro* is the synthesis of standard oligonucleotides (oligos) which provide the homology. These oligos can be used directly for recombineering or for construction of PCR products that are used for recombineering. For effective gene replacements the PCR products are generated with about 50bp ends which are homologous to sequence targets in the genome. Figure 8.8 shows the difference in steps involved in genetic engineering and recombineering. recA protein is a multifunctional 40kDa protein of *E.coli* whose expression is induced by the damage of DNA by, for example, UV light. recA catalyzes the HR of a single stranded DNA with a superhelical DNA duplex, during which a loop of single stranded DNA (displacement loop) is formed. Displacement loop is a single stranded looped structure formed when a short sequence of a supercoiled DNA molecule is displaced either by a protein (as, for example, in the replication of cDNA) or by a single stranded DNA fragment with homology to the displaced region (as, for example, in site specific mutagenesis). The *E.coli* mutants with a defective recA (recA⁻) are frequently used as hosts in recombinant DNA experiments as they do not permit any recombination of introduced foreign DNA with host's chromosome. For *in vitro* experiments recA protein is used to catalyze the HR. The protein first binds preferentially to single stranded DNA. The resulting nucleoprotein filament complex then in turn binds to duplex DNA and scans the DNA for regions of homology. If such homologous sequences are found, strand displacement and exchange starts. Thus recA protein is used for site-directed mutagenesis through displacement loops, targeted site-specific cleavage of DNA and visualization of specific DNA sequences is done by electron microscopy. The designation 'R-loop' refers to the form of this structure observed in electron microscopy.

8.10　GENE THERAPY

Gene therapy involves replacement of a mutated, non-functional gene by a functional gene using genetic engineering techniques in mammals especially human. In other words, gene therapy attempts to fix various problems in the DNA script by artificially providing a properly edited version to compensate in those cases where both copies are undesirable or it can attempt to provide an added piece of script to counter the actions of the undesirable gene. Thus the

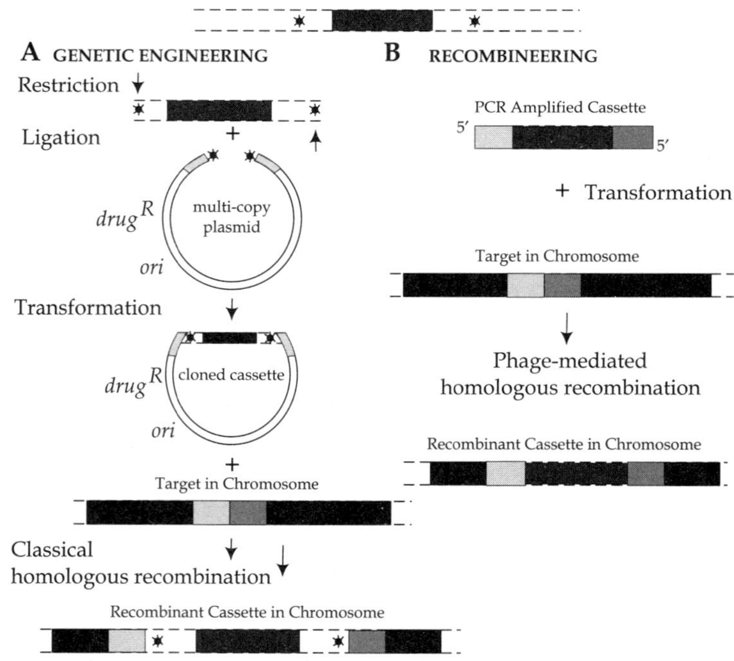

FIGURE 8.8 Showing comparison of steps required to create a recombinant DNA in the chromosome using standard genetic engineering (A) or recombineering (B). The cassette is indicated in its native site at the top flanked by restriction sites (stars). The horizontal arrows in part B indicate the primers used to PCR amplify the cassette with flanking homologies (Adapted from Court et al., 2002).

objective in gene therapy is to make the undesirable copy unreadable so that the undesirable enzyme is not coded or to produce another gene product which interferes with its action. Gene therapy is thought of a possibility to correct monogenic hereditary diseases or disorders where the diseased condition is as a result of production of defective protein by the mutant allele. There are two approaches of gene therapy.

1. Germ cell therapy
2. Somatic gene therapy.

Germ cell gene therapy refers to the replacement or repair of a defective gene within the reproductive cells (eggs, sperm) such that the introduced correct gene is transmitted to the progeny. As germ cell therapy involves manipulation of germ cells, especially zygotes, the application is technically difficult and has roused ethical controversies. However, somatic gene therapy appears feasible. Somatic gene therapy involves introduction of new, foreign or modified genes into somatic cells in order to correct the defects in somatic cells. There are two approaches of somatic cell therapy.

1. *in vivo* gene therapy
2. *ex vivo* gene therapy.

***In vivo* gene therapy** — It involves highly selective non-viral vectors which when injected, find their target organs or cells within the organism and regulate the synthesis of correct protein. Thus in this gene therapy the correct gene (s) are directly delivered into the cells or tissues of the individual containing the mutated gene. The electro-gene therapy involves injection of a therapeutic DNA (cloned into a plasmid or similar vector) into an organism or one of its organ and subsequent electroporation of the DNA into the target cells.

Ex vivo gene therapy — In this technique the target tissues are first isolated from the organism, cultured and maintained in cell culture. The cells are thus mitotically active. The cell culture is then transformed by viral vectors with high infection rate (but low selectivity), containing the correct gene (s). The transformed cells are subsequently cultured and transfected, infused or injected into the patient. The introduced genes are expected to integrate into host cell's genome and assumed to produce the correct protein and thereby curing the genetic disease.

Molecular scissors — Editing of human genome *in vivo* is hindered by the low homologous recombination. This can be circumvented using engineered 'zinc finger' nucleases as molecular scissors to cut DNA inside cells at a specific sequence. The DNA break is then patched using new genetic information (High, 2005).

9

Restriction Endonuclease

9.1 TYPES OF ENZYMES USED IN RECOMBINANT DNA TECHNOLOGY

The construction of recombinant DNA molecules, vector preparation and cloning involves various manipulations such as cutting of DNA, modification (lengthening or shortening of fragments, generation of cDNA or RNA) and joining of DNA fragments. All these DNA manipulations require different enzymes which can be grouped into the following five classes (Table 9.1).

Table 9.1 Showing different types of enzymes, their functions and source.

Enzymes	Function	Source
A. Nucleases		
1. Endonucleases(Type II restriction endonuclease)	Cutting of DNA at specific base sequence	Given in Table 9.2
2. Exonucleases (Exonuclease III, bacteriophage lambda exonuclease	i. Removes nucleotides from the 3′ end of a DNA strand ii. Removes nucleotides from the 5′ends of a duplex	
B. Ligases	Joining of two DNA fragments	E.coli and T4 phage
C. Polymerases		
1. DNA polymerase I	Filling of gaps in duplex DNA molecules	E.coli
2. Klenow fragment	Filling in nicks	E.coli
3. Reverse transcriptase	Makes a DNA copy of an RNA molecule	Avian myeloblastosis virus(AMV), T4, T2 and T7 phages
D. Modifying enzymes		
1. Alkaline phosphatase	Remove terminal phosphates from either 5′ or 3′ end or both	E. coli or calf intestinal tissue
2. Polynucleotide kinase	Adds phosphate groups on the 5′-OH end for labeling or ligation	E.coli infected with T4 phage
3. Terminal transferase	Add homopolymer tails to the 3′ end	Calf, rat, mouse thymus tissue

The other DNA modifying enzymes are Bal 31, DNA methylase, S1 nuclease which are being used for different purposes discussed in chapter 16.

Nucleases or DNases are the enzymes which cut, shorten or degrade nucleic acid molecules. There are two types of nucleases: 1. Exonucleases which remove nucleotides one at a time from the end of a DNA molecule. Exonuclease III removes nucleotide residues from 3′ end of one strand of the double stranded DNA whereas bacteriophage λ exonuclease removes nucleotides from 5′ end of a duplex to expose single stranded 3′ end. 2. Endonucleases which break internal phosphodiester bonds within a DNA molecule and thus the DNA fragments containing specific gene (s) are generated by cleaving of genome (the chromosomes) or other long DNA molecules. Endonucleases degrade at any site in a nucleic acid strand or molecule. A few exonucleases and endonucleases can degrade only single stranded DNA. Restriction endonucleases cleave only at specific nucleotide sequences. The cleavage is done using enzymes called restriction endonucleases and the process is called **digestion**. There are three classes of restriction endonucleases, class I, II and III but only class II restriction endonuclease are used in genetic engineering and which have got the following important properties. More than 100 types of restriction enzymes are found naturally in bacteria, each with its own restriction site. Exonuclease III removes nucleotide residue from the 3′ ends of a DNA strand and λ exonuclease removes nucleotides from 5′ ends of a duplex to expose single stranded 3′ ends.

1. They recognize and cleave the DNA at specific base sequences, called **recognition sequences** (Table 9.2). For class II restriction enzymes this recognition site is identical to the target site (i.e. where the enzyme cleaves the DNA duplex). In other words, recognition site refers to the DNA sequence to which a certain restriction endonuclease binds and target site refers to the specific sequence which is cleaved by a restriction endonuclease. The **arrow** ↑ or ↓ indicates the cleavage site within the recognition sequence. Cleaving occurs when a base sequence in DNA matches a particular short nucleotide sequence called the **restriction site** of the enzyme. The characteristics of recognition sequences are as follows.

   (a) The recognition sequences are usually four to six base pairs in length and all most all recognition sequences are palindromic (a segment of duplex DNA in which base sequences of the two strands exhibit two fold rotational symmetry about an axis).

   (b) Most recognition sequences have an axis of rotational symmetry, i.e., they read the same in either direction (5′ – 3′ or 3′ – 5′) in opposite strands.

   (c) The G-C rich sequences predominate the recognition sequences

   (d) Cleavage occurs within the recognition sequence.

2. There are restriction enzymes which generate **blunt or flush ends** (i.e., leave no unpaired bases on either end). In other words they cut both strands of DNA at the opposing phosphodiester bonds at the same location. For example, Bal I, Hpa I, Hae III, Alu1 and Pvu II (Table 9.2). There is another group of restriction enzymes which make staggered cuts in the two DNA strands (Table 9.2) and thereby generating **cohesive or sticky** ends with protruding or overhanging short 5′ – or 3′ – single strands consisting of 2 to 4 nucleotides termini. For example, Bam H1, Bgl II, Pvu I, Hind III, Hinf I, Sau 3A, Taq I, Cla I, Eco R I, Eco RV, Not 1, Pst I and Tth 1Hl produce

Table 9.2 Showing different restriction enzymes, their recognition sites and types of end gene-rated.

Enzyme	Source Recognition	Recognition Sequence	Types of ends	
EcoRI	Escherichia coli	* 5'- GAATTC-3' 3'-CTTAAG-5' *↑	5' extension	Hexameric
HindIII	Haemophilus influenzae Rd	↓ 5'-AAGCTT-3' 3'-TTCGAA-5' *↑	5' extension	Hexameric
BamHI	Bacillus amyloliquefaciens	↓ 5'- GGATCC-3' 3'-CCTAGG-5'	5' extension	Hexameric
PstI	Providencia stuatrii 164	* ↓ 5'-CTGCAG-3' 3'-GACGTC-5' *↑	3' extension	Hexameric
SalI	Streptomyces albus	↓ 5'-GTCGAC-3' 3'-CAGCTG-5'	5' extension	Hexameric
HaeIII		* ↓ 5'-GGCC-3' 3'-CCGG-5'	Blunt	Tetrameric
AluI		↓ 5'-AGCT-3' 3'-TCGA-5'	Blunt	Tetrameric
Sau3A		5'-GATC-3' 3'-CTAG-5'	5' extension	Tetrameric
NotI		↓ 5'-GCGGCCGC –3' 3'-CGCCGGCG-5'	5' extension	Octameric
PuvII		↓ 5'-CAGCTG-3' 3'-GTCGAC-5'	5' extension	Hexameric
EcoRV		5'-GATATC-3' CTATAG	Blunt	Hexameric
TthIIII		5'-GACNNNGTC-3' CTGNNNCAG	5' extension (endonuclease cuts interrupted palindromes)	

(Contd.)

ClaI	5′-ATCGAT-3′ TAGCTA	5′ extension	Hexameric
Hinf I	G ↓ AN TC	5′ extension (endonuclease cuts interrupted palindromes)	
Hpa II	C ↓ C G G	5′extension	Tetrameric
Kpn I	G G T A C ↓ C	5′ extension	Hexameric
Aat II	G A C G T ↓ C	3′ extension	Hexameric
Sca I	A G T A C ↓	Blunt	Hexameric
Ssp I	A A T ↓ A T T	Blunt	Hexameric
Xmn	GA ANN ↓ TTC	Blunt	

Name of each enzyme is denoted by a three letter abbreviation of the bacterial species from it was isolated(for example, Bam stands for *Bacillus amyloliquefaciens*, Eco for *E. coli* etc) ; The Roman numerals differentiates different restriction endonucleases isolated from the same species; Arrows indicate the phosphodiester bonds cleaved by each restriction endonuclease; Asterisks indicate the bases that are methylated.

staggered cuts (Table 9.2) (Brooks, 1987) of which Bam HI, EcoRI and Hind III produce sticky ends with 5′ overhangs. First three letters abbreviate the organism from which they are extracted such as Eco for **E. coli**, Taq for *Thermus aquaticus,* etc. These restriction enzymes recognize symmetrical DNA sequences (palindromic) and cleave the DNA somewhere within the palindromic sequence.

3. Some restriction endonucleases have multiple recognition sequences.

4. There are several enzymes (known as isoschizomers) possess identical recognition sequence. Hind III and Hsu I (A/AGCTT) share the same cleavage and recognition sequence.

5. Some restriction endonucleases possess same recognition sequence but have difference in their cleavage sites. Sma I and Xma havecleavage sites, CCC/GGG and C/CCGGG, respectively.

6. Some restriction enzymes recognize only unmethylated target sequence but their isoschizomers cleave both methylated and unmethylated sites. Hpa II (C/CGG) cleaves the unmethylated tetranucleotide whereas its isoschizomer Msp I cleaves both the methylated and unmethylated tetranucleotide (CG* GG). Similarly FnuD111 is capable of cleaving the c-methylated recognition sequence (GC* G/C) and FnuD111 and its isoschizomer Hha I cleave both methylated and unmethylated recognition sequence. There are other restriction endonucleases that can be used for testing methylation of adenine residues. These are 1. Sau 3A which cleaves both (/GATC) and 5′ – GA* TC-3′ 2. Mbo 1 which cleaves only the unmethylated recognition sequence and 3. Dpn I which cleaves only the methylated GA*TC. A hemimethylated sequence, i.e., a sequence methylated in only one strand is resistant to Dpn I.

7. Some restriction enzymes cleave not only double stranded DNA but also DNA-RNA hybrid. Further some restriction enzymes are capable of cleaving or hydrolyzing single stranded DNA.

9.2 GENERATION OF FRAGMENT OF SUITABLE SIZE

The average size of DNA fragments generated depends on the frequency with which a particular restriction site occurs in the DNA molecule which in turn depends on the size of the recognition sequence. The average size of DNA fragments suitable for cloning is ~ 10-20kb in length. Assuming a DNA molecule of average 50% (G + C) content (i.e. all four nucleotides are equally present) and a random sequence of nucleotide bases (i.e., nucleotides are ordered in a random fashion) a four base pair recognition sequence will occur after every 4^4 = 256 bp whereas a 6 bp recognition sequence will occur after every 4^6 = 4096 bp. Similarly a 8bp recognition sequence will occur after every 65, 556bp. Thus the DNA fragments will be much smaller in case of tetrameric restriction endonucleases (**frequent cutter**) in comparison to hexa or octameric restriction endonuclease (**infrequent cutter**). In other words the tetrameric enzymes cleave 16 times more frequently then hexameric restriction enzyme and the probability of generating DNA fragments of ~ 20kb increases with the increase of the number of cleavage sites on the DNA. But then restriction sites are not evenly spaced out along a DNA and there is considerable spread of fragment sizes. Now as the base sequence in the DNA is not random and all the four nucleotides are not equally abundant the particular recognition sequence will occur less frequently than this. Hexanucleotide sequence containing only two of the four bases (A, T, C, G) will occur less frequently, one after every 65, 536bp $(1/(4^6 \times 4^2))$ and with only 25% (G + C) content the hexaploid recognition sequence will occur every 262, 144bp $(1/(4^6 \times 8^2))$. The octomeric restriction endonuclease generates fragments of tens or hundreds of kilobase. When the objective is to construct the restriction site maps (physical map) of genome, the four cutters are not used for this purpose as they are too small-scale for most purposes. Six-cutting enzymes which on an average cleave once every few kilobases, are suitable for mapping tracts of DNA in the range of 10-15kb. The eight-cutting enzymes are extremely useful for coarser scale mapping. The 'rare cutter' restriction endonuclease is one which recognizes and cleaves sequences in DNA duplexes occurring only rarely. FseI (GGCCGGCC), NotI (GC/GGCCGC), SwaI (ATTT/AAAT), IScI (TAGGGATAA/CAGGGTAAAT), AscI, Asal I, Sfi I and Srf I are such enzymes. Thus the scale of restriction site mapping depends on the type of restriction enzyme used. Restriction endonucleases with the same sequence specificity and cut site are called **isoschizomers**. For example, restriction endonucleases Msp I from *Moraxella* species and HpaII from *Haemophilus parainfluenzae* are such isoschizomers.

5' – CCGG – 3'
3' – GGCC – 5'

9.3 DIGESTION WITH RESTRICTION ENDONUCLEASES

The digestion with restriction endonucleases enzyme could be total, partial or double digestion. In the **total digestion** the incubation time is long and the fragments generated differ greatly in size. Some fragments are very long whereas others are very small. Thus some fragments are too large for cloning and some are too small to be detected. In complete digestion all the potential cleavage sites are restricted. In case of **partial digestion** either the incubation period is short, so the enzyme does not have time to cut all the restriction sites or the incubation temperature is kept low (e.g. 4^0C rather than 37^0C) and so the enzyme activity is low and thus partial digestion results in generation of fragments of more or less uniform. In other words, partial digestion means that the reaction is not carried out for long enough to allow every possible cleavage to be made. The average size of DNA fragments generated by restriction endonucleases can thus be increased by partial digestion, i.e., terminating the reaction before completion. In partial digestion not all the potential cleavage sites are restricted and it results in generation of a collection of overlapping DNA fragments for the establishment of an ordered gene library. In **double digestions** either the two restriction endonucleases are used simultaneously provided they have similar requirements for pH, Mg^{2+} concentration, etc. or two digestions are carried out one after another and thus a series of double digestions involving different restriction endonucleases can be carried out. The double digestion will produce a random set of overlapping fragments.

9.4 PURPOSE OF GENERATION OF FRAGMENTS

There are two purposes of generation of DNA fragments. First the DNA fragments generated can be used for construction of restriction map (DNA mapping) and secondly the generated fragments from a genome can be cloned which can be further studied (sequence determination, construction of probes, genomic library, transformation, *in vitro* mutagenesis, study of gene function, etc.). DNA digested with a mixture of two restriction enzymes in a complete digestion would produce fragments averaging <1kb. In case of partial digestion majority of the fragments are in the range of 10-30kb.

Separation of fragments—The generated fragments are separated (size fractionated) by gel electrophoresis and a population of fragments of ~20kb are isolated which are suitable for insertion into a λ-replacement vector.

9.5 CONSTRUCTION OF A RESTRICTION MAP

A restriction map indicates the positions of a particular type II restriction endonuclease sites along a fragment of DNA. Restriction mapping method mapping can be done under the following two conditions.

1. When restriction enzymes have not too many recognition sites
2. When restriction enzymes have many recognition sites

In this method the cloned DNA fragment is isolated, their ends radioactively labeled and digested separately and in combination with different restriction endonucleases.. The resulting fragments are separated by agarose (or sometimes polyacrylamide) gel electrophoresis. These fragments are stained with ethidium bromide and bands are visualized under U.V. light. Construction of a restriction map is shown below with the help of hypothetical example. Suppose that the cloned DNA fragment measures 10kb

(Figure 9.1(i)) and further suppose that it contains restriction sites for two restriction endonucleases, E1 and E2. The DNA fragment upon digestion with restriction endonuclease, E1 yielded two fragments of sizes 4 and 6kb, respectively. The ends of which are shown as A and A' (Figure 9.1(ii)). The same DNA fragment upon digestion with another restriction endonuclease, E2 produced fragments of sizes, 1 and 9kb, respectively and there is thus two possible positions of cleavage (I and II) of a second enzyme, E2 (Figure 9.1(iii)). The two alternatives can be distinguished by the different sized fragments (fragments of sizes, 6, 3 and 1kb) obtained from the double digestion with E1 and E2. The result of sizes of fragments generated by double digestion is consistent with the alternative I and thus represents a restriction map of the 10kb fragment.

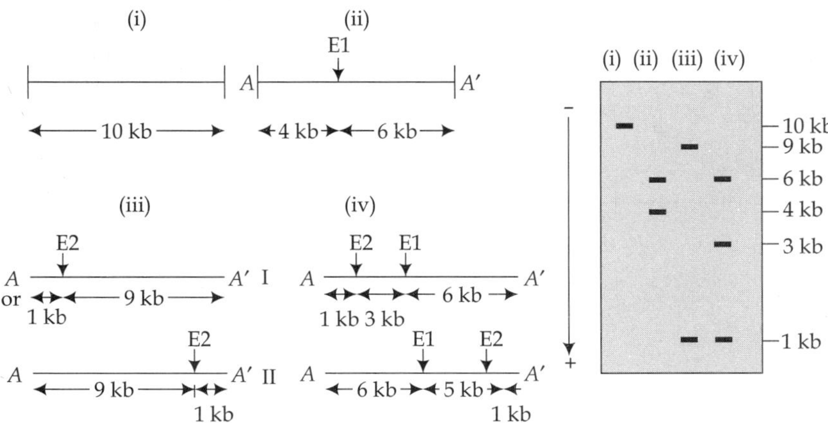

FIGURE 9.1   Showing complete digestion restriction mapping of a hypothetical 10 kb DNA fragment using two restriction endonucleases E1 and E2.

9.6   MAPPING WITH RESTRICTION ENZYMES WITH MANY RECOGNITION SITES

The above method of constructing a restriction map will be difficult if there exists many cleavage sites for restriction enzymes. In case of large pieces of DNA cloned on vectors like bacteriophage lambda or cosmid, most restriction endonuclease will have many recognition sites and in this situation it will be very difficult to overcome this problem by choice of enzyme. In such cases Smith and Birnsteil (1976) restriction mapping will be employed. In this technique the long cloned fragment is first isolated and labeled at one end. Then one aliquot is completely digested and another aliquot is partially digested with an appropriate restriction enzyme. The fragments of both digests are separated by agarose gel electrophoresis and the labeled fragments are detected by autoradiography of the gel (or a blot) and their sizes are determined in relation to the standards. Now by comparison of both autoradiograms and an analysis of the length differences the correct order of the various restriction fragments can be arranged in their original order and a restriction map of the DNA can be established. Supposing the same 10kb fragment considered in the first example, has four restriction sites for restriction endonuclease, E3 and one restriction site for restriction enzyme, E2. The end lebeled fragment is first digested with E2 which generated fragments of sizes, 9 and 1kb, respectively. The long fragment of 9kb is isolated and subjected to partial digestion with E3 which generated a mixture of end-labeled fragments with non-end lebeled fragments of sizes, 9, 7, 4.5, 3 and 2kb, respectively. The distances of the E3 recognition sites from A′ end can be read off from the autoradiography ladder using molecular size marker which allows the construction of the restriction map (Figure 9.2).

9.7 MAPPING OF CIRCULAR DNA MOLECULE

Bacteriophages, viruses, motochondria, plastids or bacteria all have circular genomes. The circular DNA molecule containing multiple restriction sites for a restriction endonuclease is digested in the presence of ethidium bromide (Paker et al., 1977). There is thus generation of linear molecules. Suppose there is a circular DNA molecule containing four restriction sites

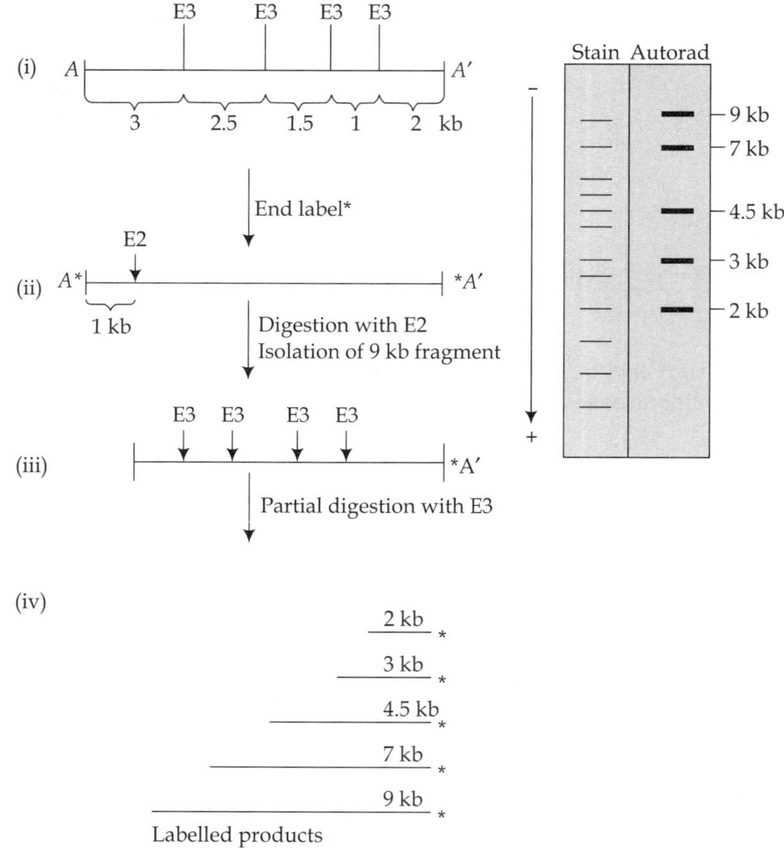

FIGURE 9.2(i-iv) Showing partial digestion restriction mapping using E2 and E3 restriction endonucleases (after Adams, Knowler and Leader, 1986).

for a restriction endonuclease enzyme (Figure 9.3), the digestion will produce four different linear fragments, the total length equaling to that of the circular DNA molecule and all linear fragments will be of equal length. These linear fragments are again digested with a restriction enzyme which cuts each of the fragments only once at a position between A′ and A″ and thus in all eight fragments are generated. When the mixture of eight fragments is subjected to gel electrophoresis, eight bands will appear and they can be arranged in such a way that the total of the largest fragment (upper most band) and the smallest (lower most band) will be equal to the second largest and the second smallest (middle ones) and so on and thus the different pairs of fragments are, 1 8 3 6, 2 7, 4 5.

9.7.1   Preparation of Map

The smallest fragment no. 8 is taken and placed at the LHS whereas the largest fragment no. 1 is put at the RHS. Those fragments which differ in their molecular weights by a value corresponding to their molecular weight of a fragment occurring in the total digest will come to lie on top of each other. In our case 8 and 7 differ in their molecular weight by a value, 4.5 which does not occur among the molecular weights of the fragments observed in the complete digest and so 8 and 7 will be positioned opposite to each other, i.e. to the left and right, respectively. The next pair is of 6 and 3. The fragment no. 6 differs from 8 by a value corresponding to the fragment (C) observed in the total digest and so 6 will occupy the left position whereas no. 3 to the right and C is left to A, i.e. A′C. Similarly nos. 5 and 4 are taken up and eventually the sequence A′CBDA″ is determined. Thus in this method of mapping it is essential to have precise estimates of length and molecular weight of different restriction fragments.

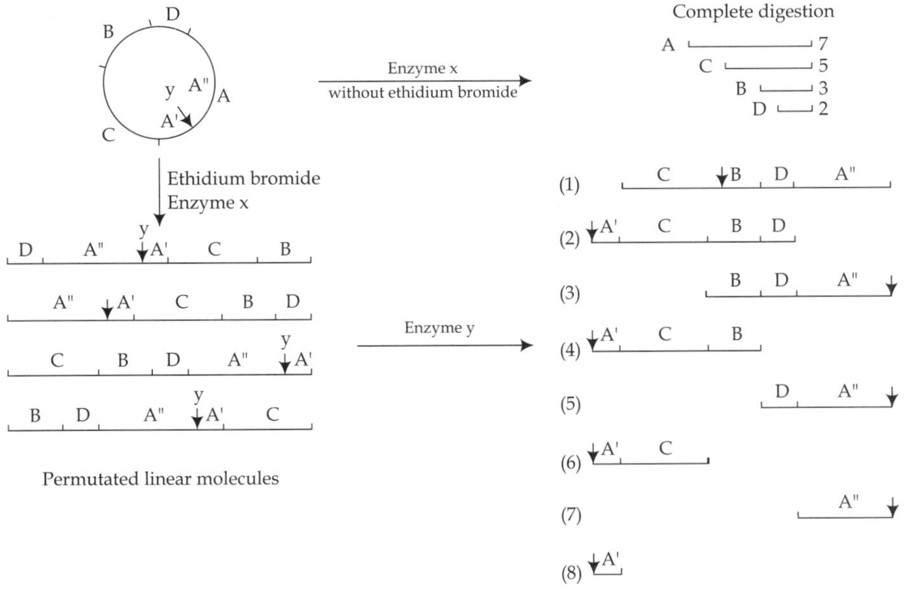

FIGURE 9.3   Showing mapping of restriction endonuclease cleavage sites on circular molecules.

9.7.2   Mapping of Individual Restriction Fragment

In this method (Baralle et al., 1980a) the restriction fragments are generated with a particular restriction endonuclease, say Hinf I and thus the objective is to map Hinf I fragment. Digestion with Hinf I which cleaves the dsDNA at G\ANTC sequence results in generation of protruding 5′ ends which are subsequently filled in using Klenow fragment of DNA polymeraseI and α-^{32} P d ATP (radioactively labeled deoxyribonucleoside triphosphate). These blunt ends fragments are then separated by gel electrophoresis and each of the fragments is hybridized with a single stranded DNA (original) cloned in M13 vector. Each fragment now acts as a primer in the presence of Klenow fragment and unlabelled deoxyribonucleoside triphosphate synthesizes the complementary strand. This double stranded DNA is now again digested with HinfI restriction endonuclease which again produces a number of fragments with 5′ protruding ends and a fragment is again filled in with labeled deoxyribonucleoside triphosphate. But this time only those fragments which were near to the DNA fragment initially labeled with α– $+32^{\,P\,d\,ATP}$ will be labeled by dA. If this experiment is repeated with all the generated Hinf I fragments then it will yield a physical map for Hinf I on the original cloned fragment.

When the objective is to map the DNA fragment cloned in bacteriophage or cosmid vector the end labeling is replaced by hybridization of a specific^{32}P labeled oligonucleotide to either the right or the left cohesive end of the molecule and thus there is no need for secondary cleavage and gel fractionation to produce a fragment of DNA labeled at one end.

9.8   METHYLATION ASSAY

This is a technique used for the detection of methylated nucleotides within recognition sequences of restriction endonucleases in genomic DNA using methyl sensitive endonucleases or pairs of endonucleases recognizing the same sequence but differing in methylation sensitivity. Foe example, the endonucleases, MboI and HpaII recognize and cut the same cleavage site, 5′-CCGG-3′. Msp I also recognizes this site is the internal cytosine is methylated, i.e. 5′-CC^{m}GG-3′ whereas HpaII does not. Now by comparing the cleavage pattern obtained from the same genomic DNA with either MspI or HpaII differences in the methylation of CCGG sequences can be detected.

9.9   MECHANICAL SHEARING

The DNA molecule can be sheared by high speed stirring in a blender or through the use of ultrasound (ultrasonication). It can also be sheared by forcing it through a French press or a hypodermic needle. High molecular weight DNA can be cut into small fragments of approximately 250-450 nucleotides. Mechanical shearing is the most random and thus the most unbiased method of fragmenting DNA molecule as the DNA sequence has no effect on breakage sites. However, sheared DNA can not be cloned directly. The sheared ends must be filled in and linker must be added before DNA is cloned into a vector.

10

Vectors and Expression Vectors

10.1 DEFINITION AND TYPES OF VECTOR

Vectors are the vehicles on which the fragment genome (gene) is cloned and through which it is introduced in prokaryotes (bacteria) or eukaryotes (plants and animals).Vector is a DNA molecule known to replicate autonomously in a host cell to which a segment of foreign DNA may be spliced to allow its replication. The various cloning vehicles used are as follows.

1. Plasmids
2. Plant viruses/Animal viruses
3. Phages
4. Cosmids
5. Yeast/bacterial artificial chromosome (YAC/BAC).
6. Plant artificial chromosome (PAC)
7. Human artificial chromosome
8. Mammalian artificial chromosome.

The bacterial cloning vectors are plasmids, bacteriophage and BAC.

10.1.1 Plasmids

Plasmids are circular double stranded DNA molecules that are found in a variety of microorganisms including prokaryotes and eukaryotes. There are two types of bacterial plasmids. One class of plasmids is able to replicate (multiply) itself within the cell quite independently of the host bacterial chromosome. They all have one gene that acts as origin of replication. Such plasmids lead an independent existence in the cell. Within this class the smaller plasmids make use of replicating enzymes produced by host chromosome for its replication whereas the bigger plasmids code for their own replicating enzymes for making copies. There is another class of plasmids which not only exist as independent element in the cell but also able to insert itself into the bacterial chromosome and they replicate with the replication of host chromosome and these plasmids are called **episomes**. Two features of plasmids that are particularly important as far as cloning is concerned are size and copy number. The **copy number** refers to the number of molecules of an individual plasmid that are normally found in a single bacterial cell. The copy number ranges from 1 to 100. Plasmids range in size from ~1.0 kb to 500kb. The cloning vehicle should be relatively small, ideally less than 10kb and should have high copy number. Low copy number plasmid is a plasmid with one or a few copies per bacterial cell, for example pSC101. It is useful for cloning of genes (genes encoding surface membrane proteins) which disturb the host cell's normal metabolism if present in high copy number. With more number of genes copy per cell (in case of multicopy plasmid) there is reasonable expression of cloned gene. Copy number can be increased spontaneously to10-100 or artificially to 20-40, 000. All plasmids possess one gene which acts as an origin of replication. Further, they carry one or more genes which are responsible for the unique characteristics associated with the host bacterium and on the basis of genes plasmids have been classified into five types (see Roy. 2009). The most commonly used vector is pBR322 (Figure 10.1) (Bolivar et al., 1977) for *E. coli* and *Serratia marcescens*. The notations 'BR' is derived from Bolivar and Rodriguez, the biologists who synthesized the plasmid. The features of the p BR322 are as follows.

1. It has an origin of replication (derived from Co1E1) which maintains it at a level of 10 to 20 copies per cell.

2. It is a small plasmid with 4.4kb in length

3. It has got two genes, Tc^R (derived from plasmid, R1 (Tn3) and Ap^R (derived from plasmid, Psc101 which had Tc^R gene derived from R6-5 plasmid) that confer resistance to ampicillin and tetracycline, respectively. These genes are used as selectable markers.

4. It has got unique recognition sequences for different restriction endonucleases such as Pst, (Figure 10.1).

5. It is amplifiable to ~ 1000 copies per cell when chloramphenicol is added.

6. It can clone up to about 8kb of foreign DNA.

Plasmid vectors are especially suitable for studying the expression and regulation of gene in host cell system used for cloning. To obtain active transcription of the cloned DNA

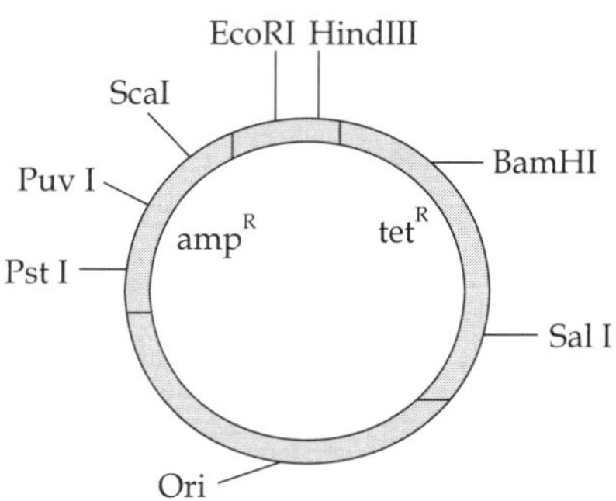

FIGURE 10.1 Map of a plasmid vector, p BR322 (4363bp in length) showing the positions of ampicillin resistence (amp^R) and tetracycline resistance (tet^R) genes, origin of replication (ori) and a few most important restriction sites.

the vector should have one or more strong promoters located near to and oriented towards the cloning sites. The unique cloning sites are located within one or the other resistance gene so that the insertion of foreign DNA can be detected by insertional inactivation of the antibiotic resistance function. The important variants of pBR322 are pBR327, pUC8 and GEM3Z.

pBR327 — pBR327 is a high copy number plasmid (~30-45 copies per cell) than pBR322 (~10-20 copies per cell) and is more suitable for studying the function of cloned gene as the effect of the cloned gene is more likely to be detected because of higher gene dosage. pBR327 was constructed by deleting 1089bp segment from pBR322 which resulted in change of replicative and conjugative abilities but it has Ap^R and Tc^R genes. As pBR327 is a non-conjugative plasmid it can not direct its own transfer to other *E. coli* cells which is an ideal property from safety view point as the recombinant will not be able to escape and thus this vector is suitable for cloning harmful gene.

pUC8 — This is one of the most popular *E. coli* cloning vectors. pUC8 is also a derivative of pBR322 containing only the replication origin *(ori)* and the Ap^R gene of the later. Further, the nucleotide sequence of Ap^R has been changed so that it no longer contains the unique restriction sites and all these cloning sites of pBR 322 are now clustered into short segment of lacZ′ gene. The lacZ gene which resides on the *E. coli* chromosome and codes for the α-peptide portion of β-galactosidase produces β-galactosidase which splits lactose into glucose and galactose. There are strains of *E. coli* which have a modified lac Z gene that lacks the segment called lacZ′ which codes for the α- peptide portion of β-galactosidase. Thus a mutant strain will produce β-galactosidase only when the vector plasmid pUC8 contains lacZ′ gene. Cloning of gene in the lacZ′ region will lead to its inactivation and the recombinant pUC8 molecule will be identified by its inability to synthesize β-galactosidase. Consequently the host cell produces colorless colonies if grown on media with ampicillin and X-gal whereas strains that are transformed with non-recombinant vectors, develop blue colonies on the same. Chance mutation within the origin of replication occurred during development of pUC8 resulted in plasmid having a natural copy number of 500-700 per cell and this will increase the yield of the cloned gene. Further, identification of recombinant pUC8 is easy in comparison to either recombinant pBR322 or pBR327 which involves two steps requiring replica plating from one antibiotics medium to another. Finally, the multiple restriction sites in pUC8 allows a DNA fragment with two different sticky ends (say EcoRI at one end and BamHI at the other) to be cloned and thus it provides greater flexibility in choice of restriction endonucleases for generating DNA fragments. The advantage with plasmid vector is that DNA fragment with two different sticky ends can be ligated.

10.1.2 Plasmids from Eukaryotes

Yeast vector systems are of three generic types (Botstein and Fink, 1988). When introduced into *S. cerevisiae*, they allow the propagation of the cloned DNA in three different forms: as low copy number, autonomously replicating, stable and properly segregated plasmids and such vectors carry a yeast centromere and are called yeast centromere plasmids (YCPs) ; as high copy number, autonomously replicating, unstable and irregularly segregated plasmids and such vectors carry a replication origin from the yeast 2μm plasmid and called yeast episomal plasmids (YEPs) and as independent segment of DNA integrated into yeast genome by homologous recombination and such vectors are called yeast integrating plasmids (YIPs). All the yeast vectors are also shuttle vectors as they allow propagation and large scale preparation of their DNA in *E. coli*.

The best characterized eukaryotic plasmid is the yeast, *Saccharomyces cerevisiae*, plasmid which is 2μm circle. It is 6kb in size and thus ideal as cloning vector with a copy number ~ 70-200. It has an origin of replication and two genes REP1 and REP2 involved in replication. The plasmid has no appropriate selectable marker genes and most of the populations of yeast vectors make use of yeast gene, Leu2 which encodes β- isopropyl malate dehydrogenase which is one of the many enzymes involved in conversion of pyruvic acid to leucine. Yeast plasmid vectors are unsuitable for expression of genes of higher eukaryotes because of differences in the signals for transcription and processing. It is because of this reason that only autotrophic mutants which lack leu2 gene are used as host and the transformed cells are able to grow on minimal medium which lacks A.A leucine (which contains no added AAs). One or two yeast vectors carry gene conferring resistance to methotrexate and copper.

YEP13—It has been derived from yeast episomal plasmid (YEP) and it contains the entire pBR322 sequence and thus it can replicate and be selected in both yeast and *E. coli* and is thus a shuttle vector. It has highest transformation frequency and highest copy number 20-50.The cloned DNA can be either present in host yeast cell as plasmid or integrate into one of the yeast chromosome as YEP can replicate as an independent plasmid. The plasmid may remain integrated or excised again later as a result of recombination. Recombination occurs as a result of homology between leu2 gene on the plasmid and the corresponding mutant locus on the yeast chromosome which results in insertion of whole of the YEP 13 sequence into one of the yeast chromosome. In case of recombination it is difficult to recover the recombinant DNA and thus purification is impossible and this is the disadvantage as purification of recombinant DNA is essential for the correct construct to be identified by, for example, DNA sequencing. To overcome this problem the initial cloning experiment is done with *E. coli* and the recombinant plasmid is selected in this organism. The recombinant plasmid is then purified, characterized and the correct molecules then introduced into the yeast. YEPs produce unstable recombinants.

YIPs—Yeast integrated plasmids or YIPs are basically bacterial plasmids carrying a yeast gene. For example, YIP5 is pBR322 containing a yeast UAR3 gene which codes for enzyme ortidine-5′-phosphate decarboxylase which is involved in the biosynthesis pathway of pyrimidine nucleotides which act like a selectable marker as Leu2. As the vector must be integrated into the host (yeast) genome for survival as it does not contain 2μ origin of replication and so it can replicate as plasmid. YIPs produce very stable recombinants and thus these are the vectors of choice even though it has lower transformation frequency and have low copy number (1-10).

YRPs—As the name suggests it can replicate as independent plasmid. YRP or yeast replicative plasmids are also made up of pBR322 but it contains the yeast gene TRP1 which carries with itself the chromosomal origin of replication and thus the vector can replicate as plasmid and also it can be integrated into host chromosome. YRPs recombinants are extremely unstable. YRP has highest copy number (5-100). It has transformation frequency lower than YEPs but higher than YIPs.

10.2 SHUTTLE VECTOR

A shuttle vector (or **bifunctional** vector) is one that will replicate in either of the two different organisms and can therefore be shuttled from one to another. In other words, a shuttle vector contains two different origins of replication which allows its selection and autonomous replication in two different organisms. The two organisms can be a eukaryote, yeast and prokaryote, bacteria (*S.cerevisiae* and *E.coli*) or two different prokaryotes (*A. tumefaciens* and *E. coli* or *E. coli* and *Bacillus subtilis*). The Yeast- *E. coli* shuttle vectors include the replication origins of both Co1E1 and the 2μm plasmid. Through the use of such shuttle vectors it is possible to use the *E.coli* as the host for cloning and amplification of a fragment of eukaryotic DNA (yeast) and then without change of vector, to transfer it back to yeast to find out more about its function. The shuttle vector also has a gene for tetracycline resistance for selection in *E. coli*. Fragment of yeast chromosomal DNA (such as gene for leucine, LEU2) can be inserted at unique sites and the constructs are introduced into the yeast mutant cells (having leu2 mutant gene) that require a certain gene to enable them to grow under restrictive condition. These shuttle vectors can then be extracted and propagated in *E.coli*. Yeast episomal plasmid variant pJDM219 replicates in both *E. coli* (via PMB9 origin) and *S. cerevisiae* (via the two micron circle origin). Similarly, this principle can be used to shuttling between *E. coli* or *Saccharomyces* and mammalian cells using the replication origin of a mammalian virus such as simian virus 40 (SV40).

10.3 ARTIFICIAL CHROMOSOME (MINICHROMOSOME)

Artificial chromosome refers to a plasmid shuttle vector which is engineered to replicate autonomously in both organisms, prokaryote (*E.coli*) and eukaryote (*S.cerevisiae*) and functions in a eukaryotic host as a chromosome and thus replicates autonomously, segregates in mitosis and meiosis. Gene transformation technologies have been successful in introducing and fixing genes determining monogenic traits as the process is relatively simple but is has not been successful in case of polygenic traits such as yield or drought resistance. In case of these quantitative traits an entire suite of genes that function together in one or several pathways need to be transferred. Carlson et al. have constructed a minichromosome which can be introduced into maize. The autonomous minichromosome which is built around an array of maize centromeric sequences, remains stable through meiosis and mitosis (Figure 10.2). Thus minichromosomes might offer a vehicle with which groups of functionally related genes can be introduced rapidly while minimizing disruption of the existing genome.

10.4 Ti-PLASMID

The Ti (tumor inducing) plasmids from *Agrobacterium tumefaciens* and Ri (root inducing) plasmids from *Agrobacterium rhizogenes* are the vectors for gene transfer in higher plants. The Ti plasmids cause crown gall disease in higher plants. The genes inducing crown gall disease are present on the plasmid called Ti-plasmid (Tumor inducing plasmid) which is about 200kb in length and the segment of Ti plasmid which is transferred to plant is called T-DNA

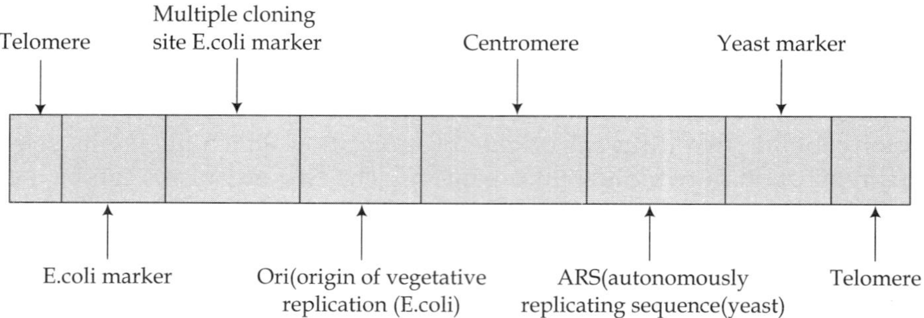

FIGURE 10.2 Artifical chromosome (mini chromosome).

(for transferred DNA). Ti plasmid has wide host range and is effective not only in dicotyledons but also in monocots. At wound site in plants infection by *Agrobacterium tumefaciens* occurs and as a result of which crown gall tumors are formed. Because of the natural ability of *A. tumefaciens* to transfer a segment of DNA into plant the research workers are now manipulating the T-DNA fo *A. tumefaciens* for the development of cloning vehicle to produce transgenic plants. The various steps involved in the crown gall formation are: 1. Attachment of *A. tumefaciens* to cells of a wounded plant 2. Anchorage of bacterial cells by bacterial cellulose 3. Induction of vir genes by plant exudates, acetosyringone 4. Formation of a single stranded linear T-DNA which is transferred to the plant cell 5. Integration of T-DNA into the plant genome 6. Synthesis of plant hormones (the auxin, indoleacetic acid and the cytokinin, isopentenyl adenosine) responsible for tumorous growth 7. Synthesis of special amino acids, opines (octopine, nopaline). The synthesis of nopaline or octopine depends on *A. tumefaciens* strain involved. The strain of *A. tumefaciens* which induces the synthesis of nopaline can grow only on the nopaline but not on octopine and vice versa. 8. Preferential growth of *A. tumefaciesns* because of opine nutrition. 9. Transfer of Ti plasmids from one strain of a. tumefaciens to another through conjugation induced by opines. The octopine Ti plasmids producing tumors are rough whereas the nopaline Ti plasmids producing tumors are smooth. The Ti plasmids can also be transferred to Rhizobium through conjugal transfer. Two components, the T-DNA and the *vir* region of the Ti-plasmid are essential for transformation. In fact, infectivity is mainly controlled by the *vir* region of the Ti plasmid. It encodes the genes responsible for excision, transfer and integration of T-DNA from *A. tumefaciens* into the genome of plants. During transformation the T-DNA is excised from the Ti plasmid, transferred to the plant cell and integrate (covalently inserted) into the DNA of the plant. Integration of T-DNA occurs at a random chromosomal site. In some cases multiple T-DNA integration events occur in the same cell. The Ti plasmid is a large plasmid which ranges in size from 140 to 235kb and the T-DNA which is transferred from Ti plasmid to plant cell ranges from 14 to 42kb. In the nopalin Ti plasmid the T-DNA is 23kb segment which carries 13 known genes that are expressed in the plant and are responsible for the rapid cell division, synthesis of hormones, opines, etc leading to the production of crown gall tumors. In octopine Ti plasmids there are two separate T-DNA segments, a 12kb- T_L-DNA (left) and a 7kb-T_R DNA (right) with an intervening pant sequence. The T_L-DNA codes for 8 transcripts whereas T_R DNA codes for 5 transcripts and upon infection of plant cells T_R is transferred independently from the left part, the T_L- DNA and integrated into the nuclear genome (Figure 10.3). The T-DNA is bordered by 25-nucleotide pairs imperfect repeats which are required in *cis-* orientation to T-DNA excision and transfer. The *vir* genes can supply the functions needed for T-DNA transfer when located either *cis* or *trans* to the T-DNA. Acetosyringone acts as inducer of the *vir* operons.

10.4.1   Disarmed T$_i$ Vectors

As the genes in the T-DNA were responsible for tumor formation (oncogenicity of T-DNA) and as the foreign DNA inserted anywhere between the T-DNA border sequences is transferred to the plant, the received T-DNA would form crown gall tumor, so disarming of Ti vector is essential so that the transformed cells do not show cancerous properties. Disarming refers to the deletion of one gene or more of these genes from the T-DNA resulting in disarmed or non-oncogenic Ti plasmid. In fact only parts of the T-DNA that are involved in the infectivity are the two 25bp repeat sequences found at the left and right borders of the region, integrated into plant's chromosome and thus any DNA placed between the two repeat sequences will be transferred to plant. To identify the cell which has received T-DNA segment of a disarmed Ti plasmid selectable marker is put within the T-DNA region of the disarmed Ti plasmid.

10.5   SELECTABLE MARKER GENE

A critical step in the development and evaluation of transformation strategies in plants is the construction of vectors with genes that could act as dominant selectable marker. A good selectable marker gene is one that will provide resistance to a drug, antibiotic or other agents that arrest the growth of normal plant cells (plant cells not harbouring the marker gene) and

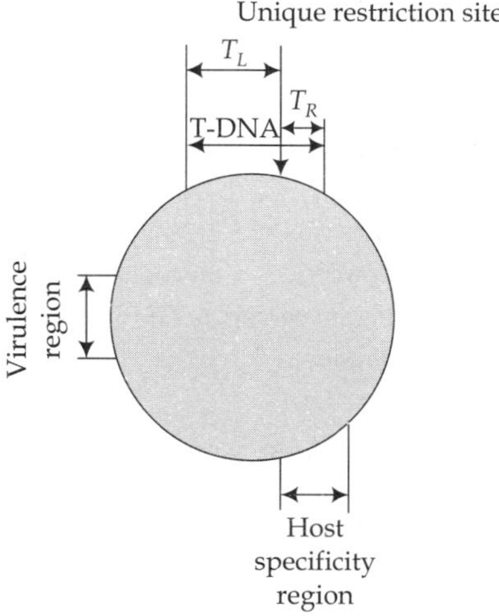

FIGURE 10.3   Showing a typical Ti plasmid, ~200kb in length, carrying T-DNA (oncogenes), virulence region and host specificity region.

thus the cells that have integrated the vector and express the marker gene will be selected. In other words, selectable marker gene allows to select for its presence in an organism. Three selectable marker genes that are being extensively used in plant systems are: 1. kanamycin and glycoside G418 2. hygromycin-HPH is the coding sequence of enzyme hygromycin phosphotransferase and 3. methotrexate of which kanamycin is most widely used in plants. Kan^r gene is obtained from *E.coli* transposon Tn^5 which encodes an enzyme called neomycin phosphotransferase type II (NPT II). NPT II coding sequence is provided with a plant promoter (5′ to the coding sequence) and plant termination and polyadenylation signals (3′ to the coding sequence). The most frequently used termination and polyadenylation sequences have been obtained from the nos gene of Ti plasmid. Different plant species and particular tissues of a plant may differ widely in their sensitivities to different selective agents. Other selectable markers are genes conferring resistance against ampicillin, chloramphenicol and tetracycline (Table 10.1). Selectable marker genes are commonly localized on or cloned into plasmids which are then transferred into bacterial cells. As transformation frequency is rather low only few transformants will arise and that can be identified by their new resistant phenotype.

10.5.1   Yeast Chromosomal Marker

It refers to any yeast chromosome gene which serves as a selectable marker gene in yeast transformation experiments. Yeast markers are central genes in, for example, amino acid (e.g., arg8, his3, leu2) or uracil biosynthesis pathway (ura3). If it is transformed into a specific auxotrophic yeast cell as a part of yeast cloning vector, the gene complements the deficient genome and consequently the cell becomes prototrophic and is easily selected.

Table 10.1   Showing various drug resistance and herbicide resistance markers.

Marker	Selective agent	Source
Drug resistance markers		
Aad (preferred for chloroplast transformation)	Trimethoprim, streptomycin, spectinomycin, sulphonamides	
Ble	Bleomycin	
Dhfr	Methotrexate	
Hpt	Hygromycin	
nptII and aphII	Kanamycin, neomycin, G418	
gat	Gentamycin	
Herbicide resistance markers		
Bar and pat	Phosphinothricin (bialaphos, glufosinate ammonium, Basta)	
Csr1-1	Chlorsulphuron	
Dhps (sul)	Sulphonamides (Asualam)	
epsp	Glyphosate	

10.6   PROMOTERS

Two most frequently used promoter sequences are: 1. nopaline synthase (*nos*) gene of Ti plasmid and 2. the 35 S transcript of the CaMV. The octopine synthase (*ocs*) genes of the Ti plasmids and 19S transcript of CaMV can also be used as promoter. These are constitutive and are also moderately induced by wounding. These promoters are widely used in dicots but have a much lower activity in monocots. In monocots rice actin-1 and maize ubiquitin-1 promoters have been widely used. The one widely used chimeric selectable marker gene has the structure CaMV 35S promoter/NPT II coding sequence/Ti nos termination sequence (35S/NPT II/nos). Thus the tumor producing gene of T-DNA is replaced by a chimeric selectable marker gene. The **chimeric** (or fused gene) refers to a construct consisting of coding sequence from one gene (reporter gene) that are transcribed and/or translated under the control of sequences (promoter) of another gene (controlling gene). It also includes gene construct comprising coding sequences from two different genes, fused to each other and transcribed from the same promoter.

10.7   OTHER COMPONENTS OF TRANSGENES

In plants all transgenes must include a polyadenylation site which in most cases is derived from Agrobacterium nos gene or the CaMV 35S RNA whereas in case of animals it is conventional to delete the UTRs from the expression construct. Further as in animals the translational start site should conform to Kozak's consensus sequence and the transgene should be codon optimized for the expression host. Finally, targeting information should be there on the transgenes for accumulation of the recombinant protein which is achieved through using an N-terminal signal sequence to direct the ribosome to the endoplasmic reticulum for folding of the protein and a C-terminal tetrapeptide retrieval signal, KDEL for accumulation in the endoplasmic reticulum.

10.8   REPORTER GENE

Reporter genes are genes that can not be used for selection as in case of selectable marker genes but that code for enzymes which can be easily detected with great sensitivity and thus they are used to monitor the transformation or for detection of promoter activity. In other words, reporter gene is any gene which may be easily fused to regulatory regions of the other genes and whose activity is normally not detectable in the target organism into which it is transferred. The most versatile reporter genes being used are the firefly luciferase gene (Ow et al., 1986) and the *E. coli* β-glucuronidase gene (uid A locus, Jefferson et al., 1987). The luciferase catalyses the oxidative carboxylation of luciferin which results in the emission of light which can be detected either by luminometry or autoradiography. β-glucuronidase cleaves a variety of β-glucuronides including fluorogenic sunstances that can be detected by fluorometry. β-glucuronidase activity can also be detected histochemically using substrates such as X-glu (5- bromo-4-chloro-3-indolyl glucuronide) and thus offers greater precision in identifying the specific cells and tissues in which the promoters are active. The other examples of reporter genes are opine synthase genes of Ti plasmids, *nos* and *ocs* and chloramphenicol acetyltransferase (CAT) gene from bacteria (Table 10.2). *Ocs* and *nos*

Table 10.2 Showing reporter genes, their products and source.

Gene		Origin	
NOS	Nopaline synthase	T-NDA	Expresses various bacterial resistance
OCS	Octopine synthase	T-DNA	-do-
NPTII	Neomycin phosphotransferase II	Tn^5	
CAT	Chloramphenicol	Tn^9	
GUS	β-glucoronidase		Q
LUC	Luciferase	Fire fly luciferase (Photinus pyralis)	Q
LUX		Vibrio harveyi or V. fischeri or Photobacterium phosphoreum	
LacZ	b-galactosidase	E. coli	
Lc	Anthocyanin biosynthesis		Q
Neo	Genes for neomycin		Q
Acetyl transferase			

catalyze the reductive condensation of arginine with α-ketoglutarate and pyruvate, respectively. These reactions can be performed with cell- free extracts of the transformed tissues. After paper electrophoresis which separate nopaline from substrate arginine, nopalin is detected by staining of guanidines with a fluorescent dye. In the early transformation experiment this method provided clear evidence of transformation with Ti plasmid. The acetylation of chloramphenicol catalyzed by the CAT is detected by subjecting the cell-free extracts to thin layer chromatography which separates them into three forms, 1-AcCM, 3-AcCM and 1, 3-AcCM and detected by autoradiography. This system is suitable for monitoring transient expression and the regulation of transcription by promoters.

Also, a series of reporters are autofluorescent proteins, e.g. green fluorescent protein (GFP) and the various analogues.

10.9 Ri-PLASMID

Ri- and Ti-plasmids are very similar but the main difference is that the transfer of T-DNA from an Ri-plasmid to plants does not result in crown gall but in a hairy root disease in which there is a massive proliferation of a highly branched root system. In both cases T-DNA is a part of Ti plasmid or Ri plasmid that is transferred from bacterium to nuclear genome of host plant cells.

10.10 PROBLEMS WITH Ti PLASMIDS

Converting Ti plasmid into a useful vector is difficult. The Ti-plasmid can not easily be manipulated like the small cloning vector (3 to 10kb) due to 1. its large size of the genome which is ~ 200kb in length and 2. the large 200+ kb Ti plasmid can be cut at several sites by almost all restriction enzymes and thus one can not open up and insert the foreign DNA or chimeric gene. Thus the genetic manipulation of the molecule is difficult and the main problem is to find a unique restriction site within a plasmid of 200kb in size. So the following two strategies were developed to insert a foreign DNA into the plasmid.

1. The binary vector strategy—This strategy is based on the observation that T-DNA does not need to be physically attached to Ti-plasmid for its transfer as vir genes can supply the functions needed transfer even if they are on separate plasmid. This led to the development of two-plasmid system in which the T—DNA is placed on a very smaller plasmid and the other plasmid is the normal Ti-plasmid excluding the T-DNA but containing the vir genes. The foreign DNA is placed between the two 25bp repeat sequences found at the left and right borders of the T-DNA region and transferred and integrated into the plant DNA using unique restriction site in the smaller molecule. The two plasmids (also called a binary Ti vector), one with T-DNA and one without it complement each other when present together in the same *A. tumefaciens* cell and the foreign DNA inserted in between the 25bp repeat sequences flanking T-DNA is transferred to the plant chromosome by proteins coded by genes carried by the normal plasmid without T-DNA but with vir genes which is inserted into a unique restriction site in the plasmid.

2. The co-integration strategy—In this strategy a new smaller plasmid based on pBR322 is constructed which besides carrying a small portion of T-DNA carries a foreign DNA fragment and selectable marker to be transferred and there is another normal functional Ti-plasmid. When these two plasmids are present in the same *A. tumefaciens* cell, recombination takes place between the Ti segment (crossingovers within regions of homology) and as a result of which this smaller new pBR plasmid is integrated into the T-DNA region of normal Ti- plasmid and thus there is generation of a larger recombinant Ti-plasmid. Thus a compound plasmid is formed by fusion of two smaller plasmids through a single crossover. Co-integrate plasmids are generally unstable but they can be stably maintained if each of the original plasmids contains a selectable marker gene and if one of the plasmids contains an origin of replication that is functional in the host cell. Infection of plants with this recombinant Ti-plasmid leads to transfer of foreign DNA along with the rest of T-DNA to plant chromosome. This has been accomplished by at least four slightly different procedures and all these procedures involve putting either T-DNA sequences into smaller plasmid or plasmid sequences into the T-DNA segment of Ti-plasmid and the crossover results in insertion of genes. In another approach (Zambrysky et al., 1983) the tumor inducing genes of a nopalin Ti-plasmid is replaced by DNA from the *E. coli* plasmid, pBR322, the resulting plasmid is called pGV3850. This disarmed Ti plasmid is used as an acceptor plasmid. Another smaller plasmid called pBR325 is a derivative constructed from pBR322 which carries origin of replication, ori (which will not replicate in *A. tumefaciens*) and foreign gene to be cloned along with selectable marker gene (35S/NPT II/NOS) and it provides a region of homology to the pBR322 sequence, the acceptor Ti-plasmid. And a single crossover between the pBR322 sequence and a homologous sequence on the donor plasmid (pBR325) will result in a co-integrate plasmid—in which the genes to be transferred to plant are now contained within the right and left borders. As pBR325 does not contain an origin of replication which is functional in Agribacterium this plasmid can persist in *A. tumefaciens* cells only by forming a co-integrate. In other words, the DNA sequences on the pBR325 will be only maintained in *A. tumefaciens* if they are incorporated into the T-DNA on pGV3850.

The disadvantage with co-integrate plasmid is that a single crossover is a must to generate it and as the frequency of occurrence of such event is 10^{-4} to 10^{-5} it would be worthwhile to produce a vector that does not require cointegrate formation prior to T-DNA transfer to plant cells. The binary vectors differ from the co-integrate vector in that 1.it contains T-DNA border sequences (sometimes only one) and 2. it has a broad host range and has origin of replication that allows them to replicate autonomously in the *A. tumefaciens* cells. The binary Ti vectors have the advantages that as they are of smaller sizes (usually ~8 to 12kb), the manipulation (i.e., insertion of selectable marker and foreign gene) *in vitro* is easy and they do not require formation of co-integrate. Further although binary vectors contain both T-DNA border sequences it has been demonstrated that a single border sequence is sufficient and thus the single border sequence can be used twice to construct binary Ti vector. The use of single border sequence at the two ends functions like the direct tandem repeats at the ends of the T-DNA of a wild type Ti-plasmid. The typical binary vector, pMON410 contains two selectable marker genes, nos/NPT II/, nos providing resistance to kanamycin and 35S/HPH/nos, conferring resistance to hygromycin.

10.11 PLANT VIRUSES

Only two classes of DNA viruses, Caulimo viruses and Gemini viruses, are known to infect higher plants but neither is ideally suited for cloning. The problem with plant viruses in using them as cloning vehicle is that the majority of plant viruses are RNA viruses rather than DNA viruses and it is more difficult to carry out genetic manipulation with RNA. Thus it would be advantageous to use DNA viruses.

10.11.1 Caulimo Viruses

The Caulimo viruses have limited host range and is restricted to *Cruciferae* and *Solanaceae*. The Caulimo viruses mainly infect Brassicas such as turnip, cabbages and cauliflower. Caulimo viruses have a smaller genome (8016bp in length) and have a double stranded circular DNA and like λ is constrained by the need to package it into its protein coat. Only very short pieces of DNA can be cloned.

10.11.2 Gemini Viruses

The Gemini viruses infect important crops like maize and wheat. They have got extremely broad host range. They can infect both mono- and dicotyledonous plant. The genome of Gemini virus is ~2kb in length and has a single stranded circular DNA. Each virus particle contains two DNA molecules which are identical in length but differ in their sequences (Haber et al., 1981). The problem with the use of Gemini viruses is that it undergoes rearrangements and deletions during infection which could scramble up any additional DNA that has been inserted and further the Gemini viruses cause damaging infection to crops and thus not ideal as cloning vector. The viral vectors have the following limitations. 1. They are generally poorly transmitted through pollen or seed to progeny and thus viral vector can not be used for the development of a transgenic variety. 2. The distribution of viruses throughout the host may not be uniform. 3.It may not be possible to separate viral disease symptom from that of the effect of vectors. 4. Relatively high rates of recombination and error accumulation may create instability. Further they are largely confined to vascular tissues of infected plants. Some are not mechanically transmissible but are instead transmitted by insects and finally the amount of packageable foreign DNA is rather limited. But unlike plasmid vectors, viral vectors have the advantage that the expression of specific gene products can be studied quickly after transmission after transduction without the need of generating plants in tissue culture. The use of plant DNA viruses as vectors for transfer of foreign genes into plants offers two advantages over that existing methods of producing transgenic plants (Hayes et al., 1988). First, as mentioned above these viruses systemically infect whole plants thus obviating the need for the difficult and time consuming step of regeneration from transformed single cells or protoplasts. Secondly, the viruses replicate as separate, autonomous entities within the plant's cell so that any gene cloned in a plant DNA virus vector would be amplified to high copy number, a feature that differs from methods producing transgenic plants by chromosomal integration of foreign DNA. Although vector based on CaMV has been developed but its use is hampered by the narrow range of plants infected by the CaMV and by practical limitations on inserting foreign DNA imposed by the biology of CaMV. But then the gemini virus-derived vector is not only useful for amplification of gene expression by the systemic infection of plants but also for heritable gene amplification by the integration of stable master copies of the vector into the plant chromosomal DNA.

Plasmid vectors are especially suitable for the cloning of genes whose expression and regulation is to be studied in the host cell system used for cloning whereas phage vectors are particularly useful for the cloning of poorly expressed, autoregulated genes or genes that are lethal for the host cell.

10.12 BACTERIOPHAGE λ AND M13

Bacteriophages are viruses that specially infect bacteria. The bacteriophage like viruses consists of mostly DNA but occasionally RNA molecule. The two main types of phage structure are 1. Head and Tail (e.g. bacteriophage λ) and 2. Filamentous (e.g. M13)

10.12.1 Bacteriophage λ

The DNA molecule is ~49kb (48502bp) in length and it delivers all its 48502bp of DNA into bacterium. The λ genome can be cleaved into three pieces (A, B and C). The two pieces (A and B), the left and the right contain the essential genes and are together ~ 30, 000bp in length (constitute ~2/3 of genome). The left piece contains all 10 genes (A to J) coding for capsid components and assembly whereas the right piece contains ten genes, six regulatory genes (cI, cII, cIII, cro, N&Q), two genes (O&P) for DNA replication and two genes (S&R) for lysis of the cell. A and B regions are required for lytic cycle. The central region, the third piece, the middle one, C (~1/3 of genome) contains all the non-essential genes and can be replaced by foreign genes to be cloned. The central region contains genes such as int, xis, exo, etc. determining lysogenation, i.e. the process leading to the integration of viral DNA and excision of lambda prophage from *E. coli* chromosome. Much of this central region is not essential as they are not required for lytic growth and infection and this can be deleted during construction of suitable vector. The deleted λ genome is thus non-lysogenic and can infect *E. coli* and follow the lytic cycle. The linear DNA molecule is present in the phage head structure. The ends of the double stranded linear molecule consist of 12 nucleotide long single strand. The two single strands are complementary and thus they can base pair with one another to form a circular completely double stranded molecule. These two complementary single strands are often referred to as sticky ends or cohesive ends or *cos* sites. The *cos* sites perform two important functions. First, it allows the formation of a circular double stranded DNA molecule. The λ genome replicates by '**rolling circle**' mechanism of replication in which a continuous DNA strand is 'rolled off' of the template strand and thus there is production of a chain *(concatamer)* of linear λ genomes joined together at *cos* sites. This *cos* sites act as recognition sequences for the endonuclease produced by A gene of λ DNA molecule which cleaves the catenane at the cos sites producing individual genomes with single stranded sticky ends (Sambrook et al., 1989).

The recombinant phage vector is introduced into a bacterial cell through 1. transfection and 2. *in vitro* packaging.

Transfection—Transfection involves mixing of recombinant phage molecule with the bacterial cell and DNA uptake is induced by a short heat sock. When recombinant plasmid DNA is used in place of recombinant phage DNA then this is called **transformation**.

10.12.2 *In vitro* Packaging

Before infecting the host cell it would be desirable to package the recombinant λ DNA molecule into λ- head and tail structure in the test tube. Thus *in vitro* packaging is done by adding the recombinant λ- DNA molecule to crude bacterial cell extracts that contain all the proteins needed to assemble a complete phage and there is thus generation of mature phage particle which can be introduced into the host cell by just adding the assembled phage to bacterial cell and the normal λ infection process will lead to delivery of its genome to host cell. The crude bacterial extracts containing all the proteins can be obtained through either of the two methods. In one method called single strain system the defective λ phage carrying a mutation in the cos site is used. The mutant will produce λ concatamers but as the cos sites will be recognized by the endonuclease so it will not produce individual λ genomes and thus not replicate but it will produce all the proteins needed for the packaging. The proteins isolated from *E. coli* infected with defective λ phage will be used for *in vitro* packaging. In the second method called the two strains system the two strains carry a mutation in two different genes involved in the synthesis of components of protein coat (for example, one strain carries a mutation in gene D whereas the other strain carries a mutation in gene E). Neither strain will be able to complete an infection cycle as there will not be capsid formation because of absence of a particular component of protein coat. However, a mixture of extracts from these two cultures of cells infected with two different strain will have all the required components of protein coat and thus can be used for *in vitro* packaging.

10.12.3 Packaging Constraints

There are problems associated with the development of λ vector. 1. There is a size limitation of the foreign DNA to be cloned into an unmodified lambda vector. Packaging requires a small amount of DNA than the normal lambda. DNA will be packaged into infectious lambda particle only if it is between 40, 000 and 53, 000bp long. A genome must be of ~ 47kb in length in order to be packaged and it must not be greater than 52kb otherwise it will not package into head and tail structure. A+C regions with the deletion of central region (B) can not be packed up into a phage particle. Thus as much as 10-23kb of foreign DNA can be inserted before the cut off points for packing are reached. 2. As the lambda genome has multiple recognition sites for almost all restriction endonucleases restriction can not be used to cleave the normal lambda genome for inserting the foreign DNA. So before development of lambda vector what is first required is to remove the middle region from the lambda genome but even this deleted λ genome has multiple recognition sites for most restriction endonucleases. Thus what is required is to develop a vector with a few recognition sites for a few restriction endonucleases. One way of achieving this will be to go for *in vitro* mutagenesis and change the recognition sites so that they are not recognized by a specific enzyme. For example, Eco RI site, GAATTC can be changed to GGATTC which will not be recognized by EcoRI. This technique is suitable for removing just one or two recognition sites. The another way is to go for selection of strain of *E. coli* lambda which lacks the undesirable restriction sites. This could be achieved by infecting E. coli strain which produces EcoRI with different lambda phage molecules. Although most of the λ DNA molecules will be destroyed by restriction endonuclease of the host cell but a few will survive and produce plaque and these will be the mutants having lost one or more EcoRI. Several cycles of infection will lead to the development of λ molecules lacking all or most of the EcoRI sites. Thus mutants can be obtained lacking all or most of the sites for other restriction endonucleases.

10.12.4. Types of λ vectors

Two types of λ vectors have been developed.

1. Insertion vectors

2. Replacement vectors.

The two types of vector are called gene replacement vectors. **Gene replacement vector** refers to any vector into which a gene can be cloned and transferred to a target cell nucleus where it replaces a specific resident but a mutated gene through homologous recombination.

Insertion vectors — It is a derivative of a wild-type cloning vector (a **λ vector**) with a single restriction site or a polylinker with several restriction sites into which the foreign DNA can be inserted. Frequently the restriction site is located within the functional gene so that any insertion will lead to its inactivation. In the insertion vectors most part of the middle region, the non-essential region is deleted and the two arms are ligated. The '**stuffer fragment**', an internal region of λ phages code for recombination (red) and integration (att, int), is not essential for phage growth and thus this can be replaced by foreign gene. The insertion vectors possess at least one unique restriction site in which the foreign DNA is inserted. The two popular insertion vectors are 1. λgt10 and 2. λZAPII. The λgt10 has an EcoRI site located in the cI gene and thus when foreign DNA is inserted in the EcoRI site it leads to the insertional inactivation of this gene. Insertional inactivation of the cI gene causes a change in the plaque morphology. And recombinants are distinguished as clear rather than turbid plaques. The vector λZAPII has a polylinker with multiple (6) restriction sites. Insertion of foreign DNA into any of the restriction site will lead to the inactivation of the lacZ′ carried by the vector. Insertion of foreign DNA in lacZ$^+$ inactivates β-galactosidase synthesis. A clear plaque on X-gal agar confirms the presence of recombinant λ molecule whereas blue plaque indicates the non-recombinant. With insertional vectors 8-10kb of foreign DNA can be cloned (Figure 10.4).

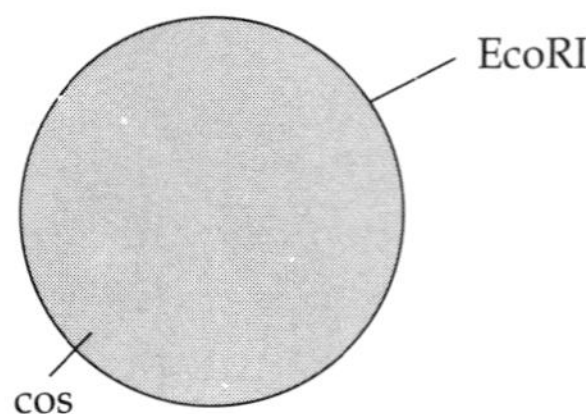

FIGURE 10.4 Showing circular form of λ insertion vector.

Replacement vectors—The replacement vector has two recognition sites for a restriction endonuclease used for cloning. These restriction sites flanking a segment of DNA can be replaced by the foreign DNA to be cloned. Replacement vector can carry up to 20kb of foreign DNA. The recombinant λ molecules are selected on the basis of size. The non-recombinant λ molecule (λ molecule without foreign DNA) will be smaller in size and thus will not be packaged into λ phage head. The two popular replacement vectors are λWES.λB′ and λEMBL4. In the vector λWES.λB′ there are two EcoRI sites. These EcoRI sites flank the middle or central or non-essential region of λ genome which can be replaced by the foreign DNA. The vector with two EcoRI sites (the non-recombinant) is about 35kb in length and thus the recombinant λ molecule can be selected on the basis of size determination.

In case of λEMBL4 a region is flanked by pairs of EcoRI, BamHI and SalI which can be cleaved by either EcoRI, BamHI, SalI or a combination of these and replaced with foreign DNA. Using this vector DNA fragments obtained with restriction endonucleases generating sticky ends can be cloned. The recombinant λEMBL4 can be identified using the Spi (Sensitive to P^2 prophage inhibition) phenotype. When *E.coli* cells containing the P2 prophages are infected with recombinant λEMBL4, recombinant (Spi⁻) will form plaques whereas the non-recombinant (Spi⁺) will not. The insertion of foreign DNA into vector causes a change from Spi⁺ to Spi⁻.

10.12.5 M13

M13 is a filamentous phage and has a much smaller genome (6407bp in length) than λ genome and contains 10 closely packed genes which are all essential for replication. M13 is a circular single stranded DNA molecule (Figure 10.5). During infection M13 DNA molecule is injected into the bacterial cell through the pilus, the structure that connects the two bacterial cells during the conjugation. After introduction into the host cell there is synthesis of the second strand and thus the circular single stranded molecule is converted into a double stranded replicative form DNA. This circular double stranded replicative form (RF) DNA replicates inside the cell through the rolling circle mechanism as in case of λ phage and maintains ~100 copies per cell. When the bacterium divides the daughter bacterium cell also receive copies of the phage and there it again replicates, produces multiple copies of itself and maintain its overall numbers per cell. The single stranded molecules are synthesized through the rolling circle mechanism from the double stranded DNA replicative form and used in the assembly of new mature M13 phage which is produced continuously. As the genome is less than 10kb in size it can be used as a vector for transferring genes. The double stranded replicative form of M13 is very much like a plasmid and can be reintroduced by

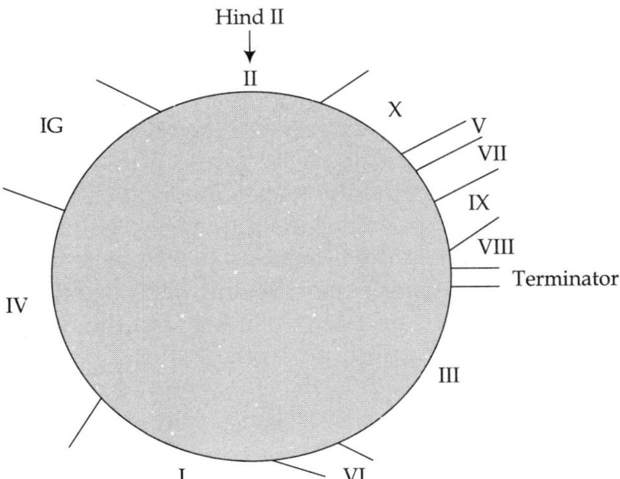

FIGURE 10.5 Showing map of bacteriophase M13 with 10 genes essential for replication. IG signifies the intergenic region into which foregin DNA can be inserted. IG is involved in initiation of replicatio. The Hind II site is used as reference point.

transfection. The advantage of using M13 as a vector is that as the M13 vector can be obtained in the form of single stranded DNA, the inserted gene in the vector can be sequenced by the Sanger's dideoxy method and *in vitro* mutagenesis can be easily be practiced as the single stranded recombinant can easily be generated. Further as cloning with different restriction endonucleases is into the same ' region' containing multiple cloning sites the same primers can be used with different recombinants.

The problem with development of M13 vector is that unlike plasmids where short DNA sequences serve as the origins of replications, several genes are responsible for replication in M13 and alteration or deletion of any of these genes will drastically reduce the replicative ability of M13. There is only one stretch of 507 nucleotide intergenic sequence into which foreign DNA can be inserted and thus there is limited scope for manipulation of M13.

Cloning vector M13mp[1&2] — The cloning vector M13mp[1] contains a lacZ$'$ gene which does not contain any unique restriction sites and it has a hexanucleotide GGATTC site near the start of the gene. The M13mp[2] vector on the other hand contains EcoRI site, GAATTC near the lacZ$'$ gene which is slightly altered in that its 6th codon now specifies asparagines rather than aspartic acid but the gene product is functional. The DNA fragment thus can be generated by EcoRI and recombinant M13mp[2] can be distinguished as a clear plaque on X-gal agar.

M13mp[7] — The vector M13mp[7] was developed following insertion of a polylinker with multiple restriction sites (BamHI, SalI and PstI) into the EcoRI site of M13mp[2] and gave a greater flexibility in choice of restriction endonucleases for generating DNA fragments. Insertion of foreign DNA in any of the four possible sites did not lead to the inactivation of the lacZ$'$ and so the recombinant M13mp[7] will produce clear plaques on X-gal agar. The advantage with M13mp[7] with its symmetrical cloning sites is that the DNA inserted into either BamHI, SalI or PstI sites can be excised from the recombinant molecule using EcoRI.

Vector pairs — The other more complex vector pairs are M13mp$^{8/9}$, M13mp$^{10/11}$ and M13mp$^{18/19}$.

M13mp$^{8/9}$ — The M13mp^8 has a more complex polylinker inserted into the lacZ$'$ whereas M13mp^9 has the same poly linker inserted in reverse orientation. Further M13mp^8 can ligate with DNA fragment with two different sticky ends. These two vectors will help in sequencing of a foreign DNA fragment > 400 nucleotide in length as only ~ 400 nucleotides can be read from the sequencing experiment. As sequencing is done from one end of poly linker into the inserted DNA fragment it will read up to 400 nucleotide. After that this insert is excised and reinserted into the sister vector M13mp^9 and sequencing is done. Now the sequencing will start again from one end of polylinker into the other end of the fragment which could not be sequenced using M13mp^8.

10.13   COSMIDS

As enzymes that package the λ- DNA molecule into the phage protein coat need only the *cos* sites for functioning, packaging can be done not only with λ genome but with any DNA molecule that carry the *cos* sites separated by 37-52kb of DNA. Using this information a type of vector (cosmid) was designed which is a hybrid between plasmid and *cos* site of λ genome. In other words, cosmid is a plasmid that carries *cos* site. It also carries a selectable marker such as the ampicillin resistance gene and a plasmid origin of replication (Figure 10.6) (Fairweather, 1997). As cosmid lacks all the λ genes so they do not produce plaque and instead the colonies are formed on selective media (containing ampicillin). **Cloning with cosmid-** The cosmid is opened at unique restriction sites, BamHI and the DNA fragments are also generated through partial digestion of genomic DNA using restriction endonuclease, BamHI. The ligation through the use of DNA ligase results in formation of concatamers. If the foreign DNA fragment inserted is of right size then *in vitro* packaging will cleave the *cos* sites and place the recombinant cosmids in mature phage particle which is used to infect an *E. coli* culture. When infected cells are placed on to a selection medium containing antibiotics, ampicillin, only the recombinants form the resistant colonies.

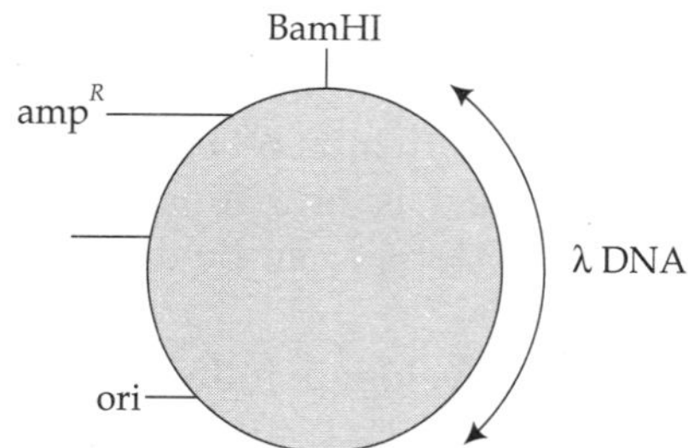

FIGURE 10.6   Showing a typical cosmid consisting of plasmid and cos site derived from λ.

10.14 FOSMID

It is an F-based cosmid. It is a single copy F-factor based cosmid cloniong vector which allows the packaging of cosmid sized DNA fragments. These fragments are comparatively stable, i.e., they do not undergo deletions or rearrangements. A collection of DNA fragments cloned in a fosmid constitutes a fosmid library.

10.15 BACTERIAL ARTIFICIAL CHROMOSOME

The problem with many bacterial vectors based on high-to-medium copy number replicons is that they often show structural instability of inserts, deleting or rearranging portions of cloned DNA particularly DNA inserts of eukaryotic organism which may contain families of repeated sequences. Thus it is difficult to clone and maintain large intact DNA in bacteria. With bacterial artificial chromosome (BAC) DNA fragment up to 300kb in size can be cloned. The BAC is based on the F (fertility factor) plasmid of *E. coli* (pMBO131) (Shizuya et al., 1992). It is a plasmid with a replication origin, ori which maintains 1-2 copies per cell (low copy number) plus *par* genes derived from the F plasmid which facilitates the even distribution of plasmids (even when few copies are present) to daughter cells during cell division. The low copy number plasmid vector is useful in cloning large segment of DNA. It reduces the opportunity for unwanted recombination between DNA fragments carried by the plasmid which one could expect when the copy number is higher. It also carries chloramphenicol resistance gene and a LacZ gene (Figure 10.7). The LacZ gene contains restriction site (s) for cloning foreign DNA. Insertion of foreign DNA in LacZ gene inactivates this gene. The recombinant BAC is introduced into cell through electroporation. Further the recombinant BAC is screened for chloramphenicol resistance (Cm). The plate also contains X-gal (5-bromo-4chloro-3lactose) analogue the substrate for β-galactosidase that yields a colored product. The recombinant will show Cm resistance and form white colors whereas the non-recombinant will form colored colonies.

10.16 YEAST ARTIFICIAL CHROMOSOME

Yeast artificial chromosomes (YACs), a plasmid cloning vector, can be used to clone very large foreign (non-yeast) DNA segment (up to 2×10^6 bp) in yeast, *S.cerevisiae* (Burke et al., 1987). Prior to use in cloning the vector is propagated as a circular bacterial plasmid. The bacterial plasmid, pBR322 acts like vector, YAC (pYAC3) when it contains all the key components of an yeast (eukaryotic) chromosome (a centromere, telomere and a sequence for initiation of DNA replication) that are required to maintain a stable eukaryotic chromosome in the yeast nucleus. When these components are combined on artificial chromosomes that are approximately 50kb or larger, these chromosomes show enough mitotic and meiotic stabilities. Thus a YAC vector has the following elements (Figure 10.8). 1. An ARS (autonomous replication sequence)- an yeast origin of replication, *ori* 2. A centromere (CEN) 3. two telomeres (TEL) and 4. two selectable markers (X and Y) on both sides of the centromere. Besides it contains restriction sites for EcoRI and BamHI. In case of each component, DNA segments that display full functional activity *in vivo*, are confined at most

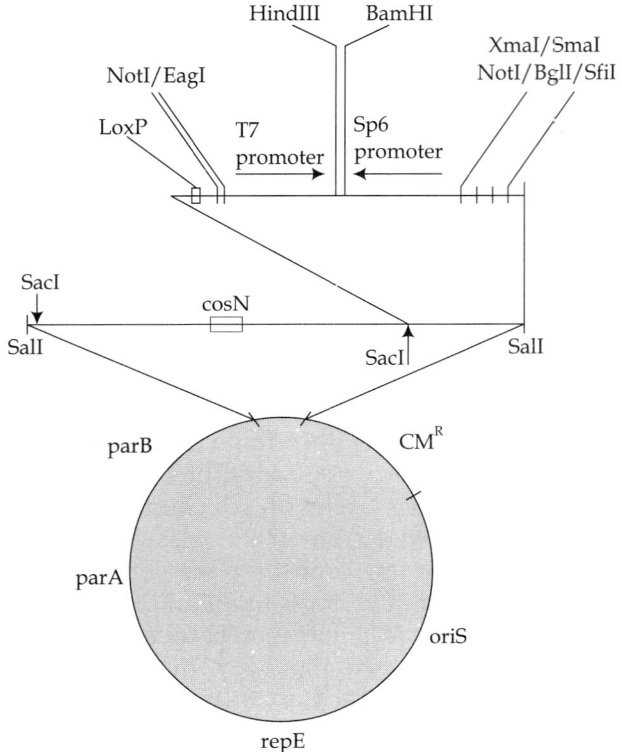

FIGURE 10.7 **Showing vector-bacterial artificial chromosome.**

to a few hundred base pairs. The telomeres provided by the two TEL sequences are not themselves complete telomeric sequences but once inside the yeast they act as seeding sequence on which yeast telomeres will be built. The individual components of yeast chromosomes are isolated by recombinant DNA techniques. The YAC is digested with BamHI and EcoRI which generate two separate DNA arms, each are containing one TEL sequence and one selectable marker (Figure 10.8). The genomic DNA is partially digested with EcoRI and the genomic fragments are then separated by PFGE. The DNA fragments of appropriate size are then mixed with the prepared vector arms. The DNA fragment is ligated between the two arms and thus yeast artificial chromosome is created. For introducing the YAC into the yeast cell, the method of protoplast transformation is used. The cell wall of yeast is enzymatically digested and which is then mixed with YAC and thus yeast with YAC is selected for X and Y by plating on minimal medium on which only cells containing the YAC will be able to grow. Stability of YAC clones increases with size up to a point. DNA inserts of size >150, 000bps are stable whereas those with inserts of <100, 000bps are gradually lost during mitosis. Also YAC lacking a telomere at either end is rapidly degraded. YAC vectors are routinely used for cloning 600kb fragments and special types of YAC can handle up to 1400kb fragments. YACs are used in genome sequencing and positional cloning of genes. The problem is with YAC is that a fraction of the clones result from co-cloning

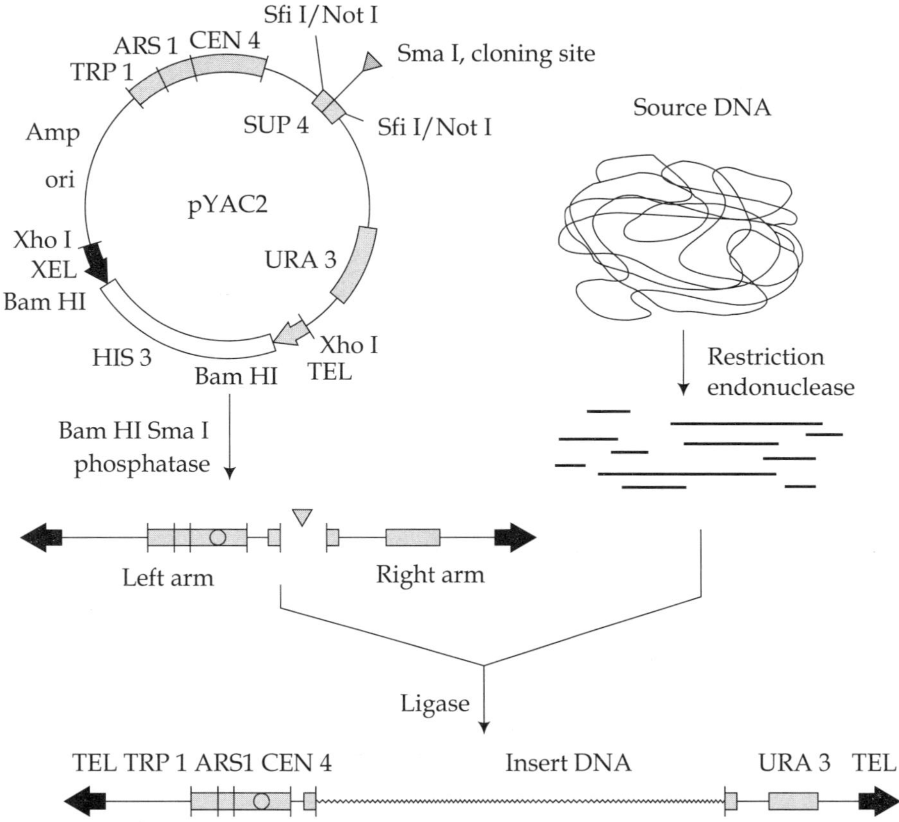

FIGURE 10.8   **Showing use of YAC vector (adapted from Burke et al., 1987).**

events, i.e., they include in a single clone noncontiguous DNA fragments (Schizuya et al., 1992).

10.17   PLANT ARTIFICIAL CHROMOSOME (PLAC)

It is a cloning vector containing plant centromere DNA and telomere repeat. It can be introduced and maintained by both yeast and a target plant as a stable autonomous minichromosome. It also contains selectable marker genes and genomic DNA in the megabase range can be inserted.

10.18   HUMAN ARTIFICIAL CHROMOSOMES (HACs)

HAC is a high capacity vector and can accommodate genomic DNA in the magabase range. It contains human centromere DNA and telomere repeats. The centromeric DNA consists of alpha-satellite DNA isolated from any human chromosome (preferably Y chromosome) which confers mitotic stability onto HACs. It is maintained in human cells as a mitotically stable autonomous minichromosome. It also contains selectable marker genes and functions in mammalian cells although designed to optimally function in human cells. HACs are being developed for treatment of genetic disease by somatic gene therapy.

10.19 MAMMALIAN ARTIFICIAL CHROMOSOME (MAC)

It is also a high capacity cloning vector for mammalian cells. It contains a mammalian origin of replication (*ori*), telomere, centromere and other sequences necessary for its function in mammalian cells. MACs are not integrated into host genome but are stably maintained at one copy per cell in host so they can be used in gene therapy.

10.20 RETROVIRUSES

Retroviruses and adenoviruses are used as vectors in the transformation of mammalian cells. The retrovirus has RNA genome. When it enters a cell its RNA genome is transcribed to DNA by reverse transcriptase and the DNA is then integrated into the host chromosomes at random locations. The integration requires two regions of DNA: LTR sequences for integration of retroviral DNA into the host chromosome and ψ (psi) sequence for packaging the viral RNA into viral particles. The regions containing genes such as *gag*, *pol* and *env* which are required for retroviral replication and assembly of viral particles are replaced with foreign DNA. For assembling viruses containing recombinant DNA the recombinant DNA is introduced into cells in tissue culture that are simultaneously infected with a 'helper virus'. The helper virus has the genes to produce viral particles but lacks the ψ (psi) sequence required for packaging. The recombinant DNA is transcribed and thus RNA copies of the recombinant viruses are produced in cells containing helper virus and packaged into viral particles. Along with RNA the enzymes, viral reverse transcriptase and integrase are also packaged into the viral particle. These viral particles now act as vectors to introduce the recombinant RNA into host chromosome. Once the virus is inside the cell, these above mentioned enzymes create a DNA copy of the recombinant viral RNA genome and integrate it into the host chromosome. The integrated recombinant DNA then is replicated with the host chromosomes at every cell division. The use of retrovirus is the best method for introducing DNA into a large number of mammalian cells. The problem with adenoviruses is that they lack the mechanism for integrating DNA into a chromosome and thus recombinant DNA introduced using adenoviral vector is expressed only for a short time and then destroyed. So the adenoviral vector can be used for studying transient expression of a gene.

There are problems associated with transformation of animal cells. First, the introduced DNA is generally integrated into chromosomes at random locations and so even if foreign DNA contains a homologous sequence, nonhomologous integrants still outnumber the targeted ones. If integration disrupts essential genes then the cellular functions will be changed. Sometimes integration might activate a gene leading to increase in cell division and thus producing a cancer cell. As the site of integration determines the level of expression so integrants are not transcribed equally well everywhere in the genome.

10.21 RETROVIRAL VECTOR

It refers to plasmid cloning vector containing elements of a retrovirus. Viral genes for gag, pol and env are deleted and replaced by a selectable marker such as neomycin phosphotransferase gene. Additionally a polylinker is inserted which permits cloning of foreign genes. The plasmid background (pBR322) allows the replication of this vector in *E. coli*. Since the vector still contains the viral sequences essential for integration (e.g. the LTR sequences) the cloned foreign gene can be introduced into the host cell's genome and transcribed from the promoter of the 5′-LTR region.

10.21.1 Adenoviruses

Adenovirus has a double stranded genome of about 36kb and its termini carry inverted terminal repeats of variable length (from 60 to 160bp). Both strands of the double stranded DNA molecule encode proteins. Adenoviruses infect a number of animal and human cells and viral DNA is inserted into the recipient genome. The adenoviruses cause upper respiratory tract infection in some invertebrates. The advantages with adenoviruses are as follows.

1. A considerable part of the viral genome can be deleted without interfering with viral function and thus large segment of foreign DNA can be inserted.
2. Necessary functions of the virus can be deleted provided helper virus complements them.
3. It has a broad host range.
4. They possess several strong promoters (e.g. major late promoter, MLP that drives the late transcription of genes encoding the capsid protein) which permit the expression of foreign genes.

10.22 P1 CLONING VECTOR

P1 cloning vector is derived from the phage P1 of *E. coli*. It allows packaging of up to 100kb of foreign DNA without interfering with the phage function (Sternberg, 1990) and thus has much wider cloning capacity then λ-derived or cosmid vectors. This vector contains a P1 packaging site (*pac* site) for initiation of packaging of vector and cloned DNA into phage P1 particles., two directly repeated P1 recombination sites (lox P), flanking the cloned insert and essential for circularization of the packaged DNA after its entry into the host cell and selectable markers as such gene for ampicillin and kenamycin (Figure 10.9). P1 artificial chromosomes, **PACs** are single copy cloning vectors derived from P1 plasmid (Pierce and Sternberg, 1992; Ioannou et al., 1994). They contain P1 replication and partitioning systems which ensure low copy number and faithful segregation. PACs have combined the features of the bacteriophage P1 and the F- factor based BAC cloning system.

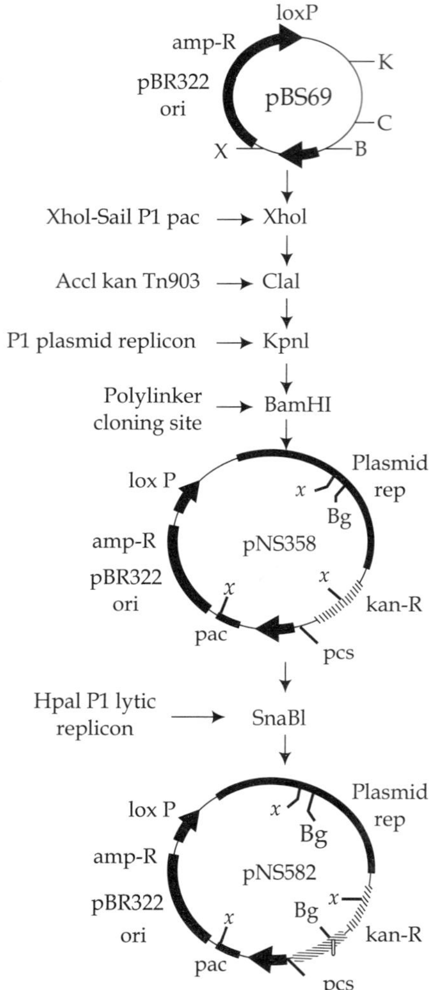

FIGURE 10.9   Showing construction of P1 cloning plasmid Pns358 and Pns583 (adapted from Sternberg, 1990).

10.23   BACULOVIRUS

It refers to any group of viruses with a closed circular double strands DNA genome of about 129kb. This virus affects only anthropods. They are classified according to whether the **virions** (inert complete virus particle) become embedded (occluded) in so called inclusion bodies and according to the shape and size of the inclusion bodies. Baculovirus is a nuclear polyhedrosis virus characterized by polyhedral inclusion bodies. Since the baculoviruses do not infect vertebrates they are of potential value in biological control pest control.

10.24 SIMIAN VIRUS 40

It is small icosahedral papova virus. SV40 contains a small circular double strand DNA molecule and the genome is 5243bp in length. It was originally detected in African green monkey, *Cercopithecus aethiops* and infect cultured cells of non-human primates. The genome of SV40 after its transfer into the host cell nucleus becomes complexed with host cell hostones, H2A, H2B, H3 and H4 and resembles a small chromsome **'minichromsome'**. There are three different forms of SV40 and all three forms can be distinguished by their sedimentation behaviour as shown in the table 10.3 given below. A series of cloning vectors have been constructed from SV40 for use in animal cells and these vectors contain the foreign DNA either instead of deleted early region or early gene, gene expressed early in the life cycle (early region replacement vector) or the deleted late region or late gene, genes expressed late in the life cycle (late region replacement vector). In each case the missing functions are complemented by propagation in the Cos cell line or by a helper virus. Cos cell line is a derivative of the permissive (allowing replication of virus) CV-1 monkey cell line. Cos cells are transformed by a segment of SV40 early region DNA in which the SV40 –origin of replication has been inactivated. Cos (conserved ortholog set) refers to a series of nucleic acid fragments whose sequence and copy numbers are highly conserved between two or more species.

Table 10.3 Showing forms of SV40.

DNA	Sedimentation rate	
	Neutral pH	Alkaline pH
Form I	20S (closed circular DNA)	53S
Form II	16S (relaxed circle)	18S (circular single strand), 16S (linear single strand)
Form III	14S (linear DNA)	16S circular single strand

10.25 TRANSFORMATION COMPETENT ARTIFICIAL CHROMOSOME VECTOR (TAC VECTOR)

It is a plant transformation vector used for cloning large genomic DNA fragments which are both stable in *E.coli* and *Agrobacterium tumefaciens* and can be effectively mobilized from *A. tumefaciens* into plant target cells to produce transgenic plants. The TAC vector consists of the pRiA4 replicon of Ri plasmid, the P1 bacteriophage replicon (both of which are responsible for a single plasmid copy in the host cells that stably maintains foreign DNA), a plant selectable marker gene (e.g a hygromycin phosphotransferase gene driven by the nopalin synthase promoter and terminated by the nopalin synthase 3-terminator sequence, located at the right rather than left-TDNA border so that hygromycing resistant plant can be expected to contain the complete TDNA), a multiple cloning site with various rare cutter recognition sequences, a kanamycin resistance gene, an overdrive sequence and a left border sequence. Genomic DNA is ligated into the Hind III site of the vector, transformed into *E.coli*, mobilized into *A. tumefaciens* by **triparental mating** and then used to transfer the passenger

DNA into the recipient cell. The triparental mating is a method of gene transfer from *E.coli* to an unrelated host, for example, *Agrobacterium tumefaciens* which is mediated by *E.coli* strain which provides a wide host range origin of transfer. The process starts with mixing of *E.coli* cells containing a plasmid vector carrying the DNA to be transferred and *E.coli* cells containing a broad host range plasmid. This mixture is the added to suspension of freshly grown *A. tumefaciens* cells containing a helper Ti plasmid which provides the *vir* functions required for gene transfer into plants. All three kinds of cells are filtered out and the filter is placed on agar-rich medium to allow three –way conjugation to occur. The wide host range plasmid is transferred to *E.coli* containing the vector plasmid where it provides transfer proteins (*tra* genes) and other factors which act on the origin of transfer of the vector plasmid and mobilizes it into Agrobacterium cells. These cells can then be used to transform susceptible plants.

10.26 PHAGEMIDS (PHAGE-PLASMID)

It is a chimeric plasmid vector (hybrid vector) which contains an origin of single stranded DNA replication such as the f1 (a coli phage with 6.408kb ss circular DNA genome which is about 97% homologous to the M13 genome) or M13 intergenic region. Phagemids replicate as normal plasmids in *E. coli*. If the host cells are infected with helper virus (e.g. M13KO7) which supplies the function necessary for single stranded DNA replication and packaging then phage like particles are synthesized and released through the bacterial cells in a non-lytic process.

10.27 EXPRESSION VECTORS

There are different ways of increasing the level of gene expression. A significant increase in the level of gene expression is frequently achieved through gene dosage and which is obtained by cloning the gene into a high copy number vector plasmid. A further increase may be obtained by insertion of the gene in a cloning vector carrying a strong promoter located close to and oriented towards the cloning vector.

The cloning of gene (s) means that a foreign gene from one prokaryote is introduced into another prokaryote or foreign gene from one eukaryote is transferred to either a prokaryote or another eukaryote. Prokaryotic genes are usually cloned from genomic DNA whereas eukaryotic genes are often cloned from cDNA. Most eukaryotic genes contain introns which prevent them from being expressed in prokaryotes but introns are not present in the cDNA. Thus we have homologous systems (*E.coli* genes cloned in an *E.coli* host/vector system) and heterologous system (plant or animal genes cloned in *E.coli*).

10.27.1 Homologous Systems

In case of homologous systems the constitutive β-lactamase promoter of pBR322 and p ACYC177 is reasonably strong and promotes the expression of many genes on DNA fragments cloned in the PstI cleavage site of these vectors. This PstI cleavage site is located 181-182 codons downstream of the N-terminal end of the β-lactamase gene. Fusion proteins consisting of the N-terminal end of β-lactamase and the C-terminal end of a protein partly coded by the cloned fragment may therefore be created by this type of construction, if the cloned gene is present in the same orientation and reading frame as the β-lactamase gene. Although the β-lactamase promoter may be active for some purposes, it is inadequate for others and inappropriate where conditional (inducible) expression of a cloned gene is required.

10.27.2 Heterologous Systems

Here a foreign gene for an important plant or animal protein can be taken from its normal host, inserted into a suitable cloning vector and introduced into a bacterium. The foreign gene is expressed (transcribed and translated) and the recombinant protein is synthesized in the host cell and thus it may be possible to obtain large amounts of the protein. The heterologous gene expression in bacterial hosts, notably *E.coli* is exploited for the production of pharmacologically interesting proteins such as insulin, human growth hormone, blood coagulation proteins, etc. To obtain a large amount of a recombinant protein the foreign gene must be highly expressed and finely regulated. Expression vectors have been successfully employed to obtain transcription rates of cloned genes in heterologous systems. The expression of eukaryotic genes in prokaryotes which do not specify mRNA splicing systems requires that the gene does not contain intervening sequences (introns), In other words, a DNA copy of the matured (spliced) mRNA (cDNA) rather than genomic DNA must be cloned in order to express split genes. The eukaryotic genes (non bacteria) are also surrounded by the regulatory sequences in their hosts but these sequences are not the same as in Prokaryotes (*E. coli*) and so the bacterium is unable to recognize its expression signals. Promoter sequences and transcription factors including transcription termination signals of bacteria are different than plants or animals. Thus translational start and stop signals that are recognized by the translation apparatus of the host cell must bracket the cloned structural gene. This may be accomplished by the use of vectors containing such signals or by *in vitro* fusion of synthetic oligonucleotides containing these signals to the fragments before cloning. The eukaryotic genes (non bacteria) are also surrounded by the regulatory sequences in their hosts but these sequences are not the same as in *E. coli*, the prokaryote and the bacterium is unable to recognize its expression signals. Thus it is unlikely that a significant amount of recombinant protein will be synthesized in the host cell because the expression of a non bacterial gene depends on a number of important elements involved in the regulation of expression of cloned gene and the foreign gene must be surrounded by these bacterial regulatory sequences are as follows:

1. **Promotor**—It is the sequence of DNA at which the transcription starts. It is the point at which the sigma unit of RNA polymerase II binds in *E. coli*. The promoter thus allows efficient transcription of the inserted gene.

2. **Terminator**—It indicates the end point of the gene. In other words it is the point at which the transcription should stop. Termination codons generally do not occur within genes. It is usually a nucleotide sequence that can base pair with itself to form a hair pin structure (stem-loop). Termination signal is present on the other side of gene insertion site. The transcription terminator sometimes improves the amount and stability of the mRNA produced.

3. **The ribosome binding site** — It is a short nucleotide sequence (Shine-Dalgarno like sequence in prokaryotes) recognized by the ribosome as a point at which it should attach to mRNA. The ribosome binding site provides sequence signal needed for efficient translation of the mRNA produced from the gene.

4. **Signal peptide sequences** — The signal peptide sequence is required so that protein may be secreted. Some proteins particularly animal proteins are synthesized as precursors which are subject to post-translational modifications, for example, by specific cleavage to form an active gene product. Human hormone like many secreted polypeptides is synthesized as a precursor molecule containing an N-terminal signal polypeptide which is cleaved off during secretion from the producing cells. Thus if a protein or hormone requires post-translational modification then a signal peptide (or synthetic oligonucletide) coding from the N-terminal end amino acids must be linked to a cDNA segment coding for the major C-terminal portion of the hormone or any protein. The signal peptide directs the transport of translated protein into cell organelles such as mitochondria and plastids, or intraluminal cisternae of the endoplasmic reticulum or extracellular space.

   Further, foreign mRNA and proteins are often rapidly degraded in *E. coli* but protein degradation has been shown to be greatly reduced if the foreign protein is fused to an *E.coli* protein. In other words, gene fusion may overcome problems of gene product degradation and a lack of correct translation start and stop signals.

5. **Initiation codon** — The initiation codon of the gene (5′-ATG-3′) is always a few nucleotides downstream of ribosome binding site.

6. **Operator** — It is a sequence of DNA that interacts with repressor protein to control the expression of gene through induction and repression and thus it regulates the promoter. The operator permits regulation by means of a repressor that binds to it. Although gene regulation only indirectly involves the promoter itself many genes important in induction and repression lies in the region surrounding the promotor (see chapter 5) and therefore should be present in the expression vector. Insertion of regulatory sequences with a foreign gene is essential for two reasons. Suppose the foreign gene is producing toxin which is harmful to host then the regulatory gene will not allow to increase the concentration of the toxin beyond a point when it is toxic. Also when a foreign gene is allowed to produce a product continuously then this higher level of transcription may affect the ability of recombinant plasmid to replicate which will result in eventual loss of recombinant protein from the culture. Thus what is required is to replace the gene's own promoter and other regulatory sequences with more efficient and convenient version. Different promoters differ in their transcription efficiency and thus there are strong and weak promoters. **Strong promoters** with higher transcription frequency are required when large quantity of products are required but when small quantity of products are required then **weak promoters** with lower transcription efficiency should be used. The foreign gene to be cloned is inserted in one of the restriction sites in the polylinker with unique sites for several restriction endonucleases near the promoter with the end encoding the amino acid terminus proximal to the promoter. Finally, the selectable markers allow selection of cells containing the recombinant DNA. Cloning vectors which provide the transcription,

translation and regulatory signals needed for regulated expression of a cloned gene and thereby producing recombinant proteins are called **expression vectors**. Thus expression vector differs from the cloning vector in having the polylinker sandwiched between a strong promoter (lacUV5, trp, lambda P_L or CaMV35S) and a terminator sequence. Expression vector may also contain ribosome binding site (SD sequence) to ensure efficient protein synthesis and sequence peptide sequence so that the protein may be excreted. If the cloned fragment includes complete polypeptide-coding sequence and inserted in right orientation it will be both transcribed and translated. The reading frame of the inserted gene must correctly match the reading frame of the initiation codon. Thus expression vectors are frequently available in three versions, each differing by 1bp which shifts the reading frame.

Since many gene products are detrimental to the cell if over produced, expression system demands an inducible strong promoter. Such a promoter can be generated by two approaches: (i) an inducible but weak promoter is modified to strengthen the promoter or (ii) a strong constitutive promoter is made inducible by addition of a regulatory element. A double mutant in the Pribnow box of lac promoter results in the stronger lacUV5 promoter. Strength of this promoter was further increased by at least a factor of 5 when the -35 region of the lacUV5 promoter was replaced with the corresponding sequences from the trp promoter, yielding the tac promoter. Promoter for *E.coli* outermembrane lipoprotein (lpp) gene was made inducible by a downstream placement of the lacUV5 promoter and operator. However, this promoter renders the lpp promoter of *E.coli* less efficient by a factor of at least 3.

10.28 EXPRESSION SHUTTLE VECTOR

It refers to any plasmid vector which permits the expression of foreign genes into two or more different organisms (eukaryote and prokaryote). **Expression secretion vector**- It allows transcription of its cloned gene, translation of its mRNA and excretion of the synthesized protein and this type of vector requires insertion of signal peptide sequence close to restriction endonuclease recognition site into which the foreign gene is inserted. The **hybrid promoter** can be a strong promoter and will be more effective than either. In case of vector designed for expression of a cloned gene in *E. coli* a hybrid promoter is constructed from the *E. coli* lac and trp promoters. Hybrid promoter is an artificial promoter which is engineered to contain a consensus sequence (Pribnow box or generally TATA box) from one promoter and a second sequence, for example, the -35 region TTGACA in bacteria or the CAAT box in eukaryote from another promoter. It is designed to maximize the expression of linked genes. Besides production of recombinant proteins on a commercial scale cloned gene can be finely altered by site directed mutagenesis and thus structure and function of protein can be studied.

The commonly used highly active (strong) promoters in expression vectors are as follows:

1. The Lac promoter (Plac)
2. The trp promoter (Ptrp)
3. The tac promoter (Ptac)
4. The λPL promoter (PL) or CaMV35S.

The lac promoter controls the expression of the lac Z gene coding for β-galactosidase. The promoter is induced by IPTG (isopropyl-thiogalactoside). The trp promoter is involved in the biosynthesis of amino acid tryptophan. This promoter is repressed by tryptophan. It is a very efficient promoter and is frequently used for higher level of expression of foreign gene in *E. coli*. The tac promoter is a hybrid between lac and trp promoters and it is induced by IPTG. This promoter is stronger than either of the two promotors. The λPL- The leftward promoter of bacteriophage λ is one of the promoters involved in the transcription of λ DNA molecule. This promoter is repressed by the product of the gene, λcI. The expression vector using λPL is used with a mutant *E. coli* host that synthesizes a temperature sensitive form of the cI protein. At the temperature <30°C the mutant protein represses the λPL promoter and the cloned gene is not transcribed whereas at higher temperature the protein is denatured and thus inactivated resulting in transcription of the cloned gene. Thus the foreign gene along with strong regulatable promotor, ribosome binding site and a terminator is inserted into the unique restriction site in the middle of the signal cluster (Figure). The expression vector also contains **heat-shock promoter**. Such promoters drive the heat-induced expression of the linked genes in their natural and in transgenic environment and are therefore, used for construction of expression vectors.

As the recombinant protein has to be ultimately purified it must not be degraded by the host cell and it must be transported either to the culture medium or to the periplasmic region between the inner and outer cell membranes. If recombinant protein is retained in the cell then it will be difficult to purify. To overcome this problem associated with the recombinant protein the foreign gene (DNA segment) must be inserted in the expression vector along with a segment of DNA from the beginning of an *E. coli* gene which forms the short peptide fused to amino terminus of the foreign protein. In other words, what is required is a *gene fusion*. The short segment of DNA from the beginning of an *E. coli* gene attached to the foreign gene will overcome the problem of secondary structure resulting from the intra strand base pairs of foreign gene which could interfere with the attachment of ribosome to its binding site and thus this short segment from the gene will improve the translation efficiency. The two DNA segments are inserted in such a way that the two reading frames fuse and read like one reading frame. The presence of a bacterial peptide will provide stability to the fused protein and if the bacterial segment is from a gene whose protein is exported by the cell then this fused protein may itself be transported either into the culture medium or into the space between membranes. The fused protein can be purified and the fusion protein can be recovered by affinity chromatography. Gene fusion requires that the vector and cloned genes are spliced in the same orientation and reading frame. The disadvantage with this fusion system is that the presence of bacterial peptide may change the properties of the recombinant protein. The two series of cloning vectors which allow fusion of genes to a bacterial gene in all three reading frames are available. The pPCf1, 2 and 3 vectors permit the cloning of EcoRI fragments and the fusion of genes on these fragments to the *E. coli* β-galactosidase gene and the expression of fusion protein is under control of promoter, P_{lac}. Another series of vectors, Pwt111, 121 and 131 permit the cloning of HindIII fragments and the fusion of genes to the trpE gene and expression is under contro of promoter, Ptrp A. Fusion of the foreign gene to a secreted protein such as β-galactosidase or β-lactamase may result in secretion of the fusion protein from the producing cells and thus its purification is simplified.

10.29 PROBLEMS ASSOCIATED WITH PRODUCTION OF RECOMBINANT PROTEIN ARE DUE TO

1. foreign gene and
2. host *E. coli.*

10.29.1 Problems with Foreign Gene

The foreign (non bacterial) gene contains introns whereas the bacterial gene do not and bacteria do not have the machinery to splice these introns from the mRNA transcript and so the foreign gene is poorly expressed. 2. The foreign gene must contain sequences which act as terminator signals in *E. coli.* This again affects the transcription of the gene (to be added more). 3. The codon usage of the gene may not be ideal for translation in *E. coli.* Although genetic code is universal each organism has a bias towards preferred codons. This bias reflects the efficiency with which the tRNA molecules in the organism are able to recognize the different codons. If a cloned gene contains a high proportion of unfavourable codons then the host's tRNAs may encounter difficulties in translating the gene and thereby reducing the amount of protein that is synthesized. These problems can be solved by manipulations shown below which is time consuming and costly. 1. Use of cDNA instead of gene will solve the problem of introns 2. Through use of oligonucleotide directed mutagenesis the sequence of possible terminators can be changed and also the unfavoured codons can be replaced with those preferred by *E. coli.* In case the gene is <2kb in size it is better to go for artificial gene through chemical synthesis using the AA sequence of the protein.

10.29.2 Problems due to Host *E. coli*

Chemical modification of AAs within the polypeptide- After translation there is processing of the proteins and during which there is chemical modification of AAs within the polypeptide. The modification is essential for the correct biological activity of the protein. Proteins of prokaryotes (bateria) and eukaryotes are not processed indentically. For example, in animal (eukaryote) the protein is attached to a sugar group and thus called glycosylated which is not the case in bacteria. Secondly, the peptide chains in eukaryotes ultimately acquire tertiary structure whereas in bacteria protein might not fold into tertiary structure. Now if the tertiary structure is not formed then the protein will be present as inclusion body which is insoluble in bacteria. Further the polypeptide can not be converted into the correctly folded form in the test tube and thus the recombinant protein formed in bacteria is inactive. Thirdly, there is degradation of recombinant protein by *E. coli* but fusion protein is less degraded than the unaltered recombinant protein. How *E. coli* recognizes the foreign protein and thereby subjecting it to preferential turnover, is not known so far. There are two ways to overcome this problem. 1. Use special strain of *E. coli* which is deficient in one or more of the protease responsible for protein degradation. 2. Use of strain which oversynthesizes chaperone proteins thought to be responsible for folding of proteins (see chapter 14).

10.30 OVER EXPRESSION

Over expression refers to the transcription of a gene at an extremely high rate so that its mRNA is more abundant than under normal conditions. Over expression can be obtained by using a cloning vector containing gene driven by a very strong promoter. Over expression can also be obtained by using a **runway plasmid**- a plasmid whose replication is tightly controlled below but uncontrolled above a certain temperature threshold.

10.31 EXPRESSION VECTORS FOR EUKARYOTES

The problems associated with expression of eukaryotic gene in prokaryotes led to the development of suitable expression vectors for eukaryotes. It is assumed that gene from eukaryote will be expressed and the recombinant protein will be synthesized more efficiently when it is transferred to yeast or filamentous fungus, the eukaryotes than in *E. coli*, the prokaryote and now microbial eukaryotes are routinely used for production of several animal proteins. The two most important species of yeast and filamentous fungus used for producing recombinant proteins are *S. cerevisiae* and *P. pastoris* and *Aspergillus nidulans* and *Trichoderma reesei*, respectively. Particularly the two filamentous fungi have the good glycosylation abilities. The yeast S. cerevisiae is currently the most popular eukaryotic host for the production of recombinant protein. Further as the higher eukaryotic gene's expression signals will not work efficiently in lower eukaryotes, expression vectors for lower eukaryotes are required. Some of the promoters used in the expression vectors are as follows: 1. The **GAL promoter**- It is involved in the control of gene encoding galactose epimerase which is involved in the metabolism of galactose which is converted to galactose-6-phosphate by a pathway involving 6 genes (GAL1, GAL2, PGM2, GAL7, GAL10 and MEL1). All the GAL genes have similar promoter. Each of the GAL gene is transcribed separately. The GAL regulatory genes are GAL3, GAL4 and GAL80 of which GAL is playing the neutral role. Binding of galactose to GAL3p and its interaction with Gal80p produces a conformational change in GAL80p that allows GAL4p to function in transcription activation. Gal promoter is induced by galactose. But this yeast, *S. cerevisiae* is unable to glycosylate animal proteins. The problem of glycosylation is overcome by use of another species of yeast, *Pichia pastoris* with AOX promoter. 2. **AOX promoter**- Expression vector of *P. pastoris* makes use of alcohol oxidase (AOX) promoter which controls the gene encoding alcohol oxidase involved in the metabolism of methanol. 3. **The glucoamylase promoter**- The expression vector for Aspergillus nidulans usually carry the glucoamylase promoter and it is induced by starch and repressed by xylose. 4. **The cellobiohydrolase promoter**- The expression vectors for *T. reesei* carry the cellobiohydrolase promoter which is induced by cellulose.

The features of eukaryotic expression vector include (i) a prokaryotic origin of replication and a selectable marker functional in *E. coli* (ii) an SV40 origin of replication and a plasmid backbone deleted of 'poison' sequences (iii) a powerful transcriptional control region active a wide range of eukaryotic cells (iv) a genomic (viral or cellular) polyadenylation signal and sites and (v) intronic sequence. Such a dual-host replicon (SV40-p BR322 plasmid) also offers the possibility of rescuing and subsequently cloning genes which are properly expressed and stably integrated into the cell genome. There were problems with this SV40-p BR322 plasmids. Such plasmids propagated in *E. coli* replicate poorly if at all after transfection of simian cells. Furthermore, recombinant plasmids isolated from the transfected simian cells subsequently show reduced potential to retransform *E. coli*. Both these phenomena were associated with the presence of a *cis*-acting DNA sequence (called 'poison' sequence) in the plasmid, p BR322. The 'poison' sequence in the plasmid lies between the Pvu II site and the p BR322 origin of replication (Lusky and Botchan, 1981).

Transgene expression in eukaryotic host cells encounters a number of problems. First, the very AT-rich gene sequences are hardy expressed or not at all expressed. However, many signal sequences such as poly (A) addition signal, transcription termination sequences, mRNA destabilizing sequences and also some introns are AT-rich. Further the transgene may not be expressed (or at a very low level only) if it lacks introns. The underlying mechanism may involve post-transcriptional processes. Some introns, for example, immunoglobulin gene introns also contain transcriptional enhancers. Finally even if the transgene is fully expressed the protein may be toxic to the host cell so that the transgene expression can not be exploited.

10.32   STUDY OF GENE REGULATION

One purpose of cloning vector is to study the regulation of gene expression. The regulatory region of a gene includes promoters, enhancers, etc. These regulatory sequences are found upstream of the gene (see chapter 5). The objective here would be to know what regulatory sequences within the upstream regulatory region are responsible for regulating gene expression. To achieve this objective one will have to 1. change the DNA sequences of the regulatory region through deletion or base pair substitution and 2. assay the gene product. In other words, one will have to study the effect of change in regulatory sequence on the expression of gene. As many gene products are difficult to assay it would be essential to fuse the regulatory region of a gene to the coding region of another gene whose gene product can easily be assayed. This type of gene is called a reporter gene. Thus what is done is to first identify the controlling region of a gene in the cloned segment through gene sequencing and then mutations are created by deletions or base pair substitution and fused to reporter gene. In practice, the regulatory region (altered) is then subcloned into the cloning site next to a reporter gene that lacks its own regulatory region of an expression vector. This combination of altered regulatory region and a reporter gene is then inserted into a bacterial cell (or yeast or eukaryotic cell) and the level of reporter gene product is assayed. It is essential that the gene/regulatory region (altered) construct must be put back into the same cell from which the original segment was cloned so that the specific transcription factors are there to interact with the controlling regions (promotor, enhancer). The reporter gene can be used for study of regulatory regions of a gene family. Besides studying the regulatory sequences to determine which base pairs control gene expression it can be used to determine where a gene is expressed in an organism. It can also be used to determine the timing of gene expression during the development of the organism. As some genes are expressed only in specific tissues or sites and only at specific stages of development so the reporter gene can be used. LacZ structural gene is an excellent reporter gene whose product, β-galactosidase can be easily assayed (color metric assay). Chloramphenicol acetyltransferase (CAT) is another excellent reporter gene. The study of the level of rRNA promoter activity can be shown with the help of reporter gene, β-galactosidase. Each eukaryotic rRNA gene with its regulatory gene is inserted into a cloning vector. The regulatory region of this gene is then identified by gene sequencing which is then subcloned into the site next to the β-galactosidase structural gene (reporter gene) in a plasmid vehicle (expression vector) which contains origin of replication (ori) and selectable marker gene amp^R. The regulatory region fused to regulatory gene is reintroduced into the organism's genome and the level of β-galactosidase expression is measured and thus level of rRNA promoter activity is determined.

10.33 STUDY OF GENE FAMILY

All the genes in the gene family produce products that have similar or identical functions. For example, α -and β-globin genes making the protein components of haemoglobin and genes for rRNA form gene family. Multiple copies of genes are present in most animals, plants and bacteria. Using a transcriptional fusion vector with a reporter gene each gene can be separated and studied independently and its effects can be characterized.

10.34 ENHANCER TRAP/GENE TRAP VECTOR

Enhancer probe vector is a cloning vector which contains appropriate cloning site (s) located at 5′ (upstream) or at 3′ (down stream) of a function reporter gene, gusA that is driven by a minimal promoter. In other words, the construct comprises a visible marker gene downstream of a minimal promoter. Any foreign gene cloned into such vector and possessing enhancer elements will increase the expression of the reporter gene over that of an enhancer-less control gene, if transformed into target organism. In other words, under normal conditions the promoter is too weak to activate the marker gene and is not expressed. However, if the construct integrates in the vicinity of an endogenous enhancer, the marker is activated and reports the expression profile driven by the enhancer. Such enhancer trap vector will thus permit isolation of enhancer sequences, their characterization and is used in gene expression studies. As enhancer is far away from the gene so enhancer trapping is not a convenient method for cloning novel genes (see chapter 5). However, it can be used to drive the expression of a toxin and thus ablate a specific group of cells. Enhancer trapping has been widely used in Drosophila (O'Kane and Moffat, 1992).

Gene trap vector — It is usually a promoter-less cloning vector carrying a reporter gene (e.g. a green fluorescent protein gene), fused to a splice acceptor cassette that is used to detect a target gene. After its integration into a genome the gene trap vector inserts into a gene and inactivates it. The genomic location of this gene is thus easily identified by the reporter gene sequence ('gene trap') which also permits to be amplified by conventional PCR using a primer directed towards the reporter gene and another primer (a primer of arbitrary sequence) towards the flanking region.

Cassette — Cassette refers to a potentially protein coding DNA sequence which requires transposition for its expression. It is an *in vitro* construct of two or more genes and their regulatory sequences that is used as one unit in vector constructions. Such gene cassettes can easily be inserted into and removed from any vector. The cassette usually contains a reporter gene, a selectable marker gene and one or more genes (or generally sequences) of interest for the particular experiment and cloned into a unique restriction site of an expression vector. The cloned gene is transcribed from a promoter 5′ upstream of the cloned gene. The cassette may additionally contain a ribosome binding site and termination sequences (Figure 10.10). In the gene trap vector the insertion element contains a visible marker gene such as lacZ or gusA downstream of a splice acceptor site. The marker gene is therefore activated only when the element inserts within the transcription unit of a gene and generates a transcriptional fusion. This strategy selects for insertions into genes and is useful in animal and plants with large amount of non-geneic DNA (Evans et al., 1997; Springer, 2000). Early gene trap vectors depended on in-frame insertion and so up to two-thirds of all hits on genes were not recognized. Furthermore, expression of the marker relied on the transcriptional activity of the surrounding genes and so inserts into non-expressed genes were not detected. The use of internal ribosome entry sites has obviated the need for in-frame insertion and has greatly increased the hit rate of gene traps. The incorporation of a second marker which is driven by its own promoter but carries a downstream splice donor making it dependent on the surrounding genes for polyadenylation, has facilitated the detection of non-expressed genes (Zambrpwicz et al., 1998).

Gene catridges—Dual promoter vector—It is a coupled promoter that are a part of dual promoter vector and are separated from each other by a polylinker and driving the transcription of the inserted DNA in opposite directions. Such systems frequently consist of T_7 and an SP_6 promoter and allow the synthesis of sense and an anti-sense RNA.

Exon trap vector—Exon trap vector is used in exon trapping procedure. It is a shuttle vector capable of replicating in *E.coli* as a plasmid and in mammalian cells as a defective retrovirus. The vector carries a donor splice junction and a polylinker for cloning of the DNA to be tested for the presence of acceptor splicing site. The α- complementing fragment of the β-galactosidase gene is inserted into an intron between polylinker and donor splice site. When a functional acceptor splice sequence is cloned into the polylinker the β-galactosidase sequence will be absent from the spliced RNA transcript.

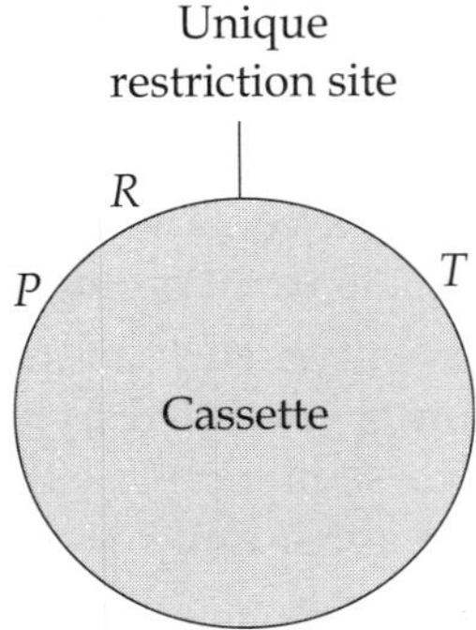

FIGURE 10.10   Showing a cassette containing a promoter (P), ribosome binding site (R) and a terminator (T). It also contains a unique restriction site where the foreign gene is inserted.

10.35 INDUCIBLE EXPRESSION VECTOR

A cloning vehicle (plasmid or phage) which is especially constructed to achieve efficient transcription of the cloned DNA fragment and translation of its mRNA if the coupled promoter is activated. This type of vector usually contains a repressed promoter (the synthetic trp/lac (trc)-hybrid promoter which can be depressed by the inducer (IPTG), polylinker, transcription termination and an appropriate selectable marker gene (Table 10.4).

Table 10.4 Showing various expression vectors for use in human and memmals.

Expression vector	Host
Plasmids with viral replicons	
SV40 replicons	COS cells
BPV replicons	Various humans
EBV replicons	Various humans
Viral transduction vectors	
Adenovirus E1 replacement	293 cells
Adenovirus amplicons	Various mammalian
Adeno-associated virus	Various mammalian
Baculovirus	Insects
	Various mammalian
Oncoretrovirus	Various mammalian and avian ES cells
Lentivirus	Non-dividing cells, mammalian
Sindbis, Semliki Forest virus	Various mammals
Vaccinia virus	Various mammals

For maximum expression of transgene in animal cells the construct should have the following elements.

1. The construct should have strong, very active promoters. In viral vectors the transgenes are expressed under the influence of endogenous promoters such as the baculovirus polyhedron promoter, adenoviral E1 promoter and the vaccinia virus p7.5 promoter. In case of plasmid vectors the commonly used promoters are the SV40 early promoter and enhancer, the Rosa sarcoma virus LTR promoter and enhancer and the human cytomegalovirus immediate early promoter.

2. The presence of a heterologous intron such as the SV40 small t-antigen intron or the human growth hormone intron or modified hybrid introns matching the consensus splice donor and acceptor site sequences as the presence of an intron has been observed to enhance transgene expression.

3. Polyadenylation signal- This polyadenylated tail, poly (A) is required for the export of mRNA into the cytoplasm and also increases its stability besides it generates a defined 3′ end to the mRNA.

4. Removal of UTR sequences has been generally found to maximize expression in animal systems. The 5′ –UTR may contain one or more start codon upstream of the

authentic translation start site which is detrimental to translation initiation. Further, 3'-UTR may contain AU-rich sequences which reduce the stability of mRNA. Besides these 5' and 3' may be rich in secondary structure which prevents efficient translation.

5. Selection of right codon in the transgene for an amino acid is a must because if the transgene contains a codon which is commonly used source organism but not in host then expression of transgene will be affected because of scarcity of the corresponding tRNA in the host.

6. The construct should have the signal peptide sequence then only the transgene protein product will be exported to the right location and will get modified.

10.36 OTHER TYPES OF VECTORS

Integration vector — It is a cloning vector which allows the covalent integration of any cloned DNA fragment into the genome of the host. In other words, only the cloned DNA and short flanking sequences are inserted into the host genome. Integration vectors carry integrative DNA sequences such as long terminal repeat sequences of retroviruses, other repetitive elements such as Alu I sequences, Kpn I sequences or more generally sequences with homology to genomic sequences of the host. These sequences facilitate the integration of the foreign DNA through homologous recombination, i.e. single cross over event. Ti-plasmid derived plant transformation vectors are the integration vectors.

Integrative vector — It refers to the whole vector molecule plus the insert stably integrates into the host genome.

Sequence insertion vector — It is a cloning vector (e.g.) which is able to insert *in toto* into an into an endogenous locus thereby disrupting original sequences and thus leads to mutations as a result of inactivation of target genes.

Sequence replacement vector — It refers to a cloning vector (*e.g*) which replaces the endogenous DNA sequences with exogenous sequences thereby disrupting original sequences and thus leads to mutations following inactivation of the target genes.

Reporter vector — It refers to any cloning or expression vector which contains one or more reporter gene (s). For example, genes encoding β-galactosidase, β-gluccuronidase, chloramphenicol acetyltransferase, green fluorescent protein and its derivatives, human growth hormone, luciferase, secreted alkaline phosphatase. Reporter plasmid is any plasmid vector that contains a reporter gene driven by a minimal promoter, fused to multiple enhancer elements (e.g intereferon stimulated response elements, gamma activating sequence or others). If a plasmid expressing a desired gene is cotransfected into host cell together with a reporter plasmid, an increased reporter gene expression demonstrates presence of transcription factors binding to and activating both promoters.

Transient expression assays — Transient expression of a gene is observed when the cloning vector containing it is introduced into a non-permissive host system. Such assays are preferentially used to test the functionality of gene constructs in host cells, especially their promoter strength and their compatibility with TFs. A transient expression vector allows the transient over expression of the cloned foreign genes in suitable host cells. The expression is transient only because the host cell's transcription-translation machinery is soon inactivated and the host cell dies after a few days. Such vectors are employed to monitor cloned genes for their function (s) and to detect effect (s) of promoter or gene mutations on gene expression. They also serve to optimize DNA delivery into the host cell. For example, when the foreign genes are introduced into cells such as spheroplasts or protoplasts by direct gene transfer, they are not covalently integrated into host DNA. These genes are transcribed but their expression is transient as they are degraded by cytoplasmic and/or nuclear nucleases.

11

DNA Library and its Construction

11.1 DNA LIBRARY

The genome of an organism consists of genes. In the genetic engineering the objective is also to study the structure and function of individual genes and thus it is required that each gene is isolated, put (cloned) on vectors and maintained for years, propagated when required then transferred to another organism and shared with other researchers in this field of study. Further these isolated genes from one species can be used as probes for detection of a similar gene (s) in other species. So what is required is the construction of library of DNA fragments or genes. A clone library has entire collection of DNA fragments covering a total genome one or several times and each DNA fragment inserted into a cloning vector. There are two types (or forms) of libraries.

1. **DNA library.**
2. **cDNA library.**

11.1.1 Genomic Library

The fragments of DNA comprising the genome are introduced into vectors (plasmids or phage) and the genomic library is thus a collection of plasmids or phage vectors containing the recombinant DNA molecules. The number of DNA fragments (clones) that are required for a genome library can be calculated using the formula as

$$P = \ln (1 - P)/\ln (1 - a/b)$$

where N is the number of clones. P is the probability, a is the average size of the DNA insert and b is the total size of the genome. Thus with an average fragment length of 17 kb the number of clones required in case of tomato of genome size 7.0×10^9 bp will be 123, 500. A fragment size range is chosen that is both compatible with the cloning vector and ensures that virtually all the sequences will be represented among clones within the library. With constant genome size the number of clones in the genomic library depends on the type of vector used. With vectors able to handle larger DNA inserts, the number of clones will be reduced drastically. With constant length of DNA inserts the number of clones in the genomic library will increase with the increase in complexity of genomes, the larger the genome the more complex it is. The probability of finding a gene in the genome library depends on the length of DNA insert and the complexity of the genome. DNA fragments of appropriate length and suitable vectors are required for construction of genomic library and the number of the hybrid molecules in the library.

11.1.2 Generation of Fragments

Complete genome of a particular organism is cleaved into thousands of fragments. The fragments can be obtained by either mechanical shearing or digestion with restriction endonucleases. If the genomic fragments are obtained by complete digestion then the resulting fragments will be very heterogenous in size, some of them will be very large whereas some others will be very small and might be deleted and thus lost. Considering these problems the DNA fragments must be obtained by either shearing or partial digestion with restriction enzymes. In case of restriction endonuclease enzyme with tetrameric recognition sequence can be used for generating fragments. If one wants to use λ vectors then restriction enzyme with hexameric or octameric recognition sequences will lead to generation of fragments too large to be accommodated in λ vectors. The tetrameric enzymes cleaves 16 times more frequently than hexameric restriction enzyme and the probability of generating DNA fragments of ~20 kb increases with increase of the number of cleavage sites on the DNA. The tetrameric restriction enzymes such as AluI (AG/CT), Hae III (GG/CC), or Sau 3A (/GATC) can be used either simply or in mixtures. Constant amount of DNA samples are either digested with various amounts of enzyme for a fixed incubation period or the different samples are incubated for different periods with constant amount of enzyme. The products of digestion are subjected to gel electrophoresis or centrifugation and digested samples having more fragments of uniform length are pooled. The DNA fragments of the desired class are then separated from such pools by either agarose gel electrophoresis or separation in sucrose gradient. The genomic library prepared with the type of fragments generated above indicates that certain DNA fragments will be expected to be over or under represented or even lost. Further it has been observed that different tetrameric restriction enzymes differ in their efficiency in recognizing the endonuclease cleavage sites. The loss of certain fragments in the genomic library can be due to loss during amplification which can result because it might be producing a toxic product, replicating slowly, or has changed due to recombination. The chance of recovering a complete functional gene rather than just a fragment of a genome, depends on the method used to generate the large fragments. The three general methods are: i. moderate mechanical shearing ii. digestion with a rarely cutting enzyme or iii. incomplete digestion with one of the 'six-cutting' restriction enzymes. The use of randomly sheared fragments or incomplete restriction digests may thus be necessary for cloning of some genes.

11.1.3 Suitable vector

Cloning vector is also cleaved with the same restriction endonuclease enzyme and ligated to the genomic DNA fragment. The ligated mixture is used to transform bacterial cells or is packed into a bacteriophage particle to generate bacteria or bacteriophages each harbouring a different recombinant DNA molecule. Different vectors differ in their capacity of carrying the inserts. Some vectors carry smaller inserts whereas others can carry larger inserts. A replacement vector such as λ EMBL4 can carry up to 20kb of DNA insert whereas some cosmids can manage segments up to 40kb. The maximum insert size of ~ 8kb is for most plasmids and less than 3kb of M13 vectors (Table 11.1). The bacteriophage can squeeze up to 11kb of DNA whereas cosmids type vector based on PI can be used to clone DNA fragments ranging in size from 75 to 100kb. The BAC based on F plasmids can handle DNA inserts up to 300kb in size. So in principle all the vectors (plasmids, cosmids, BAC, YAC) employed in genetic engineering can be used but in practice λ vectors and cosmids are the most suitable ones considering the size of the fragments to be inserted in these vehicles and the efficiency of transformation. In case of plasmids the transmission efficiency decreases with increasing length of the inserts and inserts of > 15kb can hardly ever be cloned. The situation is just the reverse in phage and cosmids. Foreign DNA up to 23-25 kb can be inserted in suitably adapted λ genomes in particular replacement vectors. Such DNA molecules can be incorporated into infectitious phage particles *in vitro* or used directly. DNA is highly infectious and may yield up to $10^5 - 10^6$ pfu per mg of recombinant phage DNA. Phage DNAs can be stored and screened more than plasmids DNA easily. Cosmids in comparison to λ replacement vectors can accommodate even larger DNA. In cosmid MUA- made of plasmid pBR322 plus a 403bp *cos* fragments are inserted at the PstI site. In case of cosmidMUA-3 the yield of non-recombinant clones is very low as cosmid dimers are too small for packaging (see chapter 10).

Suppose with human genome of length of 3.0×10^9 bp one wants to obtain 1000 fragments then those fragments will be of approximately 3000kb in length. The fragments of this size of course, can be cloned but these fragments are still extremely long for a molecular characterization, e.g. by DNA sequencing. In such situation analyses like subcloning, chromosome walking and chromosome hopping will lead to the molecular characterization of the fragments. The genomic clones can be sequenced to identify the nature and location of the introns and upstream and downstream sequences of the genes.

11.1.4 Difficulties in Cloning Certain DNA Sequences

Recombinant DNA libraries in bacteriophage λ vectors are generally found to contain most of the sequences of the genome from which they are derived but reports of sequences that can not be found in libraries of eukaryotic genomes are also fairly common. The hypervariable region 3' to the α-globin locus resists efforts cloning. The junction fragment from a translocation

Table 11.1 Showing different vectors, their hosts and the insert size they can carry.

Vector	Host	Insert size
λ phage	E.coli	5-25kb
λ cosmids	E.coli	35-45
PI phage	-do-	70-100
PACs	-do-	100-300
BACs	E.coli	-<300kb
YACs	S. cerevisiae	200-2000kb
M13		10-20kb
T4 lambda phage vector		>150kb
Plasmid		8-16kb

between chromosomes 6 and 10 in murine plasmacytoma is difficult to clone. The discrepancies between cloned and genomic sequences can also be due to deletions occurring during cloning of mammalian DNA. There is also problem in cloning of sequences surrounding and including the highly polymorphic site. Also, inverted repetitions can not be cloned into phage vectors unless mutant hosts are used. Inverted repetitions cloned in phage λ can be propagated in rec B, recC, sbcB host but not in rec+ host. Recent results suggest that the genotype of the vector as well as of the host can influence the viability of the subset of human DNA clone. Growth of clones of human genomic DNA in a bacteriophage λ vector was examined in a number of *E.coli* hosts such as rec+, recB, recC, sbcB genes carrying mutants. A large proportion (8.9%) of human genome fail to grow on a standard rec+ host but will grow on hosts carrying mutations in recB, recC, sbcB genes. Heteroduplex analysis in electron microscope of DNA from four of these phages revealed substantial secondary structure including snap-back regions 200-500bps in length. Such structures were not found in phages from the same DNA library that grow in rec+ hosts.

Sequences prone to under representation in genomic libraries — Two types of sequences are under represented in genomic libraries (Wyman and Wertman, 1987). 1. Those which contain inverted repeats 2.Those which contain direct repetitions. Palindromes or inverted repetitions are common features of eukaryotic DNA. The so called Alu family of repeated sequences represents these. Most inverted repetitions consist of elements of about 300bp in length. These sequences are able to form extensive secondary structure and can not be cloned in standard rec$^+$ hosts. These sequences have been seen to be unstable or completely inviable in bacteripophage λ as well as plasmids, in wild type *E.coli* host.

Direct repetitions are sequences which are present in multiple copies within the same λ clone may gradually be lost when the clones are grown on a wild type host. λ clones of α-globin gene in human DNA readily give rise to deletions which is brought about by unequal crossingover between homologous regions. In other words, these sequences are prone to rearrangements such as deletion. Thus these sequences which are inviable as λ clones on wild type hosts can be recovered using mutant hosts which maximize stability. Similar deletions are observed in immunoglobulin α light chain genes which are tandemly duplicated in the genome. Such deletion can be prevented by subcloning and growing on recA⁻ host. Another example of instability of a tandem array is found in the 5′-upstream region of the human insulin gene. Repetitive sequences include highly repetitive sequences which can be short (1kb) and highly abundant (even greater than 10^4-fold) and moderately repetitive elements such as multigene families or transposons. Inserted transposons or the subsequent deletion variants are thus the cause of heterogeneity in the genomic library.

11.2 cDNA LIBRARY

In comparison to genomic library cDNA or gene library is a collection of plasmid or phage vectors with inserted cDNA and maintained as population of bacterial transformants. The cDNA is synthesized from mRNA through reverse transcriptase. cDNA library is thus a collection of mRNAs expressed in a cell at a specific time. In comparison to the genomic library the DNA fragment is the gene consisting of exons. The DNA fragments in genomic library consist of exons as well as introns. As only 0.2 to 0.3% of the genome is involved in protein synthesis so the number of clones constituting a cDNA or gene library is comparatively very small. As the housekeeping genes are actively expressed during all stages of plant growth but then there are also genes which are expressed at only specific stages of plant's growth so the cDNA libraries prepared at different stages of plant growth will contain different sets of genes in the libraries. Further as the plant genome requires a higher amount of a particular product by a particular gene or group of genes at a particular stage so more copies of a particular type of mRNA will be found at particular plant growth stage and so the cDNA library constructed from mRNAs at that stage will have more the number of cDNA clones for that particular gene or group of genes in the library. In the genomic library, however, the number of clones for the same particular gene or group of genes will be relatively very small. cDNA library can be said to be complete only and only when if the number of clones carrying a particular DNA insert reflects the relative fraction of the corresponding mRNAs in the cell. The genomic library whereas contains all the genes of a given species.

11.2.1 Equalized/normalized cDNA Library

It refers to any cDNA library which contains less clones derived from a redundant mRNAs then conventional cDNA library. In other words, it refers to cDNA library generated such that all the genes in the library are represented at the same frequency. It starts with the ligation of a DNA adaptor to both ends of the double stranded DNA. This adaptor is then amplified through PCR using a primer complementary to the template primer. After this it is denatured and allowed to reanneal under specific conditions (described below) which preferably allows the reannealing of abundant cDNAs while less abundant cDNAs remain single stranded. The double stranded cDNA is then separated from single stranded cDNA by hydroxyapatite chromatography and the single stranded cDNA is reamplified byPCR and repeated cycles yield double stranded cDNA originating from rare mRNAs.

11.2.2 Application of cDNA Library

The uses of physical cDNA resources are as follows:

1. Over expression studies using cell lines or transgenic (yeast, plants or mice) in which the product of gene concerned is made in abundance to allow its function to be investigated.
2. Production of proteins for structural studies or antigens to obtain antibodies for investigating gene expression.
3. Studies of interactions between different proteins.
4. Design of primers. Knowledge of the end points of mRNA is sufficient to design primers to rapidly retrieve a cDNA by reverse transcriptase.

Besides the cDNA library can be used for sequencing. cDNA clone is often used as a probe to isolate gene from a genomic library. The number of clones in the library reflects the efficiency of mRNA extraction from the source of cells. Good libraries contain at least 1 million clones and probably substantially more. Some tissues and cell types are difficult to deal with and the resulting libraries will tend to be less representative. The actual number of genes expressed in a cell at a given time may be a few thousands and this number varies with the cell type. The human brain cell expresses up to 15,000 different genes whereas the gut expresses about 2000. Thus there is a small number of different genes (expressed genome) represented in the library of 1 million clones. A relatively small single random sample of clones rather than multiple samples is taken for sequencing, for example, 10,000 from a library of 2 million clones. Sequencing of 10,000 clones will result in 10,000 sequences, each between 200 and 400 bases in length and representing part of the sequence of each clone. In practice some sequencing runs will fail altogether, some will fail to generate sufficient sequence data and some will fail to produce good quality data. The sequences generated from the sequencing are called ESTs. Further sequencing vectors are microbial in origin and there is a possibility that some of the vector sequence could contaminate the results unless they are identified and removed. So while preparing EST data the ESTs are subjected to quality control screening in order to remove vector contamination, along with poly-A, poly-T and poly-CT sequences, if any. The minimum length accepted is 100bp with less than 3% N base calls. Sequencing of different types of libraries provides different types of information.

By choosing normalized libraries the quantitative information on the levels of gene expression in the source tissue is sacrificed in an attempt to increase the sampling of different genes. The normalized libraries represent a broad cross-section of the tissue types of interest to a wide variety of researchers. In case of use of standard cDNA libraries for sequencing the emphasis is on the quantitative information derived by sequencing. The aim here is to obtain information on the relative copy number of transcribed genes in healthy and diseased tissues to facilitate the elucidation of potential therapeutic targets. Here the library sample sizes tend to be small and the objective is to find the difference in gene expression between samples rather than searching for every last gene by EST analysis in case of normalized libraries. By creating normalized library the redundant entries for the same gene are removed.

11.2.3   Full Length cDNA Cloning

Full length cDNA–It refers to any cDNA which contains a complete reading frame (from the start codon, ATG to the stop codon) or more precisely the 5′ UTR as well.

It is a cloning procedure which allows the synthesis of a complete (full length copy, cDNA) of an mRNA. This procedure avoids the used of S1 nuclease commonly used in conventional cDNA cloning where the second strand synthesis is self-primed by the formation of a linear loop. This loop is subsequently removed with S1 nuclease which leads to the loss at the 5′-end of the mRNA and thus results in the synthesis of truncated cDNA. For synthesis of full length cDNA an oligo (dC) tail is annealed to the first strand which allows the priming of second strand synthesis by oligo (dG). This process does not lead to hair-pin formation and so S1 treatment is superfluous and consequently full length cDNA clones are produced. A library of non-redundant full length sequences cDNAs is essential for the study of gene structure and function. They allow the accurate prediction of gene structures, particularly of 5′ and 3′ UTRs which are refractory to computational prediction based on genomic DNA sequence alone. The different steps in the construction of full length cDNA library are as follows (Rubin et al., 2000).

I. The oligo (dT)-primed cDNA libraries derived from the RNA isolated from a variety of developmental stages and tissues using established procedure is constructed but normalization was not attempted to decrease the contribution of abundant mRNAs to these libraries. **Normalization**-It refers to equalization of the concentrations of various transcripts (mRNAs) in a cell at different levels (e.g. single copy or rare or least abundant versus abundant or highly abundant RNAs). As the difference between single copy and highly abundant messages is more than 10^5 in most cells, any cloning of cDNAs will inevitable lead to an over representation of clones from strongly expressed genes whereas least abundant messages will probably escape cloning. Normalization thus balances the otherwise unequal representation of the various messages in a cDNA library through reducing the proportion of highly expressed mRNAs with simultaneous enrichment of rarely expressed messages. An efficient technique for normalization is called '**phenol reassociation technique**'. This technique involves the amplification of cDNAs, its precipitation with ethanol and resuspension in hybridization solution containing 8% phenol which reduces the aqueous phase and increases the hybridization rate. Vigorous shaking leads to a mixing of the phases. The resulting emulsion then permits hybridization of abundant cDNAs. Subsequently chloroform-isoamyl alcohol extraction and desalting is performed and the single stranded cDNAs, representing single copy mRNA is enriched by restriction of double stranded cDNAs representing abundant mRNAs. The efficiency of normalization can be monitored by the loss of distinct bands (over represented cDNAs) and an increase of the background smear in EtBr-stained agarose gels (normalization of the previously under represented messages). Normalization protocols are difficult to perform without compromising cDNA length and so it was not attempted.

II. Next the ESTs from the 5′ ends of the cDNAs were generated. The use of 5′ ESTs helps to evaluate the quality of libraries and to identify the clone that extends farthest toward the 5′ end of each gene.

III. The 5′ ESTs are clustered by sequence and a clone representing each gene that extends farthest 5′ is selected.

IV. The sequence of 3′ends of the selected clone (3′ESTs) is then obtained next. At this point two quality control tests are performed.

    1. The clones for which a poly (A) tail is not apparent are discarded.

    2. The 5′ and 3′ sequences of each clone are aligned to the genomic DNA sequence and clones for which the two sequences were not in proximity were discarded as well. This test eliminated the clones that contain two unrelated cDNAs colligated into the same cloning vector. Also discarded were the clones for which a data tracking error occurred such that the 5′ and 3′ reads of that clone were not appropriately associated in the database.

V. The remaining clones are clustered on the basis of their 3′ end- sequences. This allows eliminating remaining duplicate clones. Such clones might escape detection in the 5′ end clustering if they differ in length that their 5′ESTs do not overlap. In these cases the longer clone is retained.

These steps resulted in generation of non-redundant, full length cDNA library representing 42% of all predicted Drosophila genes. Selection of replacements, if they exist in EST collection, for clones that failed to pass quality control tests will further increase representation from 42% to over 50% of all genes. For isolating cDNA representing the remaining genes, additional 5′ ESTs will have to be generated and there is need of 50 to 100 bp sequence in order to obtain an unambiguous aligment with the genome rather than 500-bp ESTs which were used in the initial work. The promising clones would then be sequenced from the 3' end. The additional ESTs will increase the representation to 80% of all genes. The remaining 20% of clones can be isolated by library screening with PCR based techniques (Monroe et al., 1995).

11.2.4 Rapid Amplification of cDNA Ends (RACE)

cDNA cloing strategy based on the DNA PCR technique was developed by Saiki et al. (1985). PCR employs two oligonucleotide primers, one complementary to a sequence on the (+) strand and the other to a downstream sequence on the (-) strand. Repeated cycles of denaturation, annealing and extension are used to generate multiple copies of DNA that lies between the two primers. Wherever PCR is used, it requires the use of primers designed to match two known or presumed genomic or cDNA sequences. Amplification and cloning of the region between a single short sequence in a cDNA molecule and its unknown 3' or 5' end can be made by a technique called RACE. Separation of amplified cDNAs by gel electrophoresis allows precise selection by size prior to cloning and thus facilitates the isolation of cDNAs representing variant mRNAs such as those produced by alternative splicing or by the use of alternative promoters.

RACE is a technique for obtaining full length cDNAs from a known partial cDNA sequence and it is used to obtain either the 3'- or 5'- ends of cDNA. For isolation of 3'–ends mRNA is reverse transcribed into cDNA using a primer (the RT primer) consisting of oligo (dT) followed by a unique sequence. Then PCR amplification is carried out using a primer specific to the known sequence and a primer (PRIMER U) specific to the unique sequence of the RT primer. The nested PCR is performed to improve specificity.

To clone 5'-ends reverse transcription is carried out using a primer (PRIMER K) specific to unknown DNA sequence, reading forward the 5'-end. A poly (A) tail is joined to the first strand cDNA and the RT primer utilized for 3'RACE is used to synthesize the second strand of cDNA. Then PCR amplification is carried out using a gene specific primer upstream of Primer K in conjunction with Primer U. Again this amplification is followed by nested PCR.5'- and 3'-RACE are used to isolate full length cDNs in a few days in comparison to the weeks required to screen cDNA libraries and analyze library clones.

11.2.5 Directional cDNA

It refers to cDNA whose 5'-terminus differs in sequence from its 3'-terminus, for example, 5'-CTCGAG-3' at the 5'-end represents an Xho I recognition site and 5'-AATTC-3' at the 3'-end represents an EcoRI recognition site. Such directional cDNAs are employed for forced cloning, i.e. orientation specific insertion of the cDNA into an appropriately cut polylinker of a directional vector.

11.2.6 Specialized cDNA Libraries

cDNA libraries can be more or less specialized depending upon the degree of differentiation of the cells from which mRNA was isolated. Highly specialized cDNA containing only one or a few gene transcripts can provide ideal probes for cloning of the corresponding chromosomal genes. Cloning a cDNA into a vector that fuses the cDNA with the sequence for a marker or reporter gene, the fused genes forming a reporter construct. The two useful markers are the genes for green fluorescent protein and epitope tags. A target gene fused with a gene for GFP generates a fusion protein that is highly fluorescent and thus its location and movement in a cell can be monitored. An epitope tag is a short protein sequence bound tightly by a well characterized monoclonal antibody. Any protein bound to the tagged protein can be precipitated providing information about the protein-protein interaction in the cell. The tagged protein can be specifically precipitated from a crude protein extract by interaction with the antibody. cDNA can be constructed using expression vectors (λgt11) which allows efficient detection of individual cDNA and also clones can be screened directly for synthesis of a specific protein using immunological method (specific antibody). In principle both λgt10 and λ gt11 can be used for cDNA cloning. Both are insertional vectors which accommodate DNA of similar size. The former, λgt10 not only allows easy identification of recombinant phage but also selection for such recombinants. λgt11 allows the expression of cDNA insertion as a fusion protein with N-terminal part of bacterial-β-galactosidase but inspite of presence of lac 1 repressor a small amount of toxic protein is encoded by a cDNA. If no other structural information on gene in question is available then go for λgt11 expression vector. Vectors used for cDNA libraries are lambda ORF8 and lambdaZAP, cosmids, bacteriophage and occasionally plasmids.

11.2.7 Specialized DNA Libraries-Subgenomic Library

It refers to gene library which contains parts of a genome of an organism, for example, one chromosome or part of a chromosome.

Chromosomal gene libraries—As intact chromosomal DNA molecules of many eukaryotes including yeast, filamentous fungi and protozoans can be separated by PFGE and of higher eukaryotes by chromosome sorting (Davies et al., 1981; Labo et al., 1984) it would be better to construct chromosomal gene libraries rather than complete genomic libraries. As chromosome specific libraries will contain genes from individual chromosomes, each library will be substantially smaller and easier to handle or maintain than a complete genomic library. Further chromosomal DNA molecules can be immobilized on a nitrocellulose or nylon membrane by Southern Transfer and studied by hybridization analysis. Thus chromosome carrying a cloned gene can be identified and thus will help in physical localization of single copy or low copy DNA sequences.

11.3 STS (SEQUENCE TAGGED SITE) LIBRARY

STS is any known sequence that has been mapped within a chromosome and/or clones derived from it. A short segment of ~200-500bps that is unique to a genomic DNA fragment and that serves to identify the fragment among the thousands of other fragments used to construct a genetic and physical mapping of eukaryotic genome. There can be sequence tagged restriction site (STAR) or sequence tagged microsatellite (STMS). Thus in STAR the STSs are derived from the RFLP markers i.e. STAR is a short DNA sequence that identifies the sequences flanking the recognition site of a restriction enzyme. STMS is a genomic sequence that flanks the microsatellite clusters or preferentially at specific microsatellite sequence in eukaryotic genome. An STS mapped to the region of interest will allow design of primers for PCR amplification of the STS. The amplification product is then labeled and used as a probe to identify the corresponding DNA fragment from a gene library. STS database refers to a database containing sequences, mapping data and other relevant information on STSs. STSs if known for all genomic DNA fragments can be used for mapping procedure and will eliminate the need to store and exchange clones.

11.4 EST (EXPRESSED SEQUENCE TAG) LIBRARY

It refers to any collection of expressed sequence tags cloned into a cloning vector (plasmid). ETS works as a marker for expressed gene. It is a specific type of sequence tagged site representing a gene that is expressed. EST represents tags for the state of expression of genes at a given time and cell or tissue type. EST is a short synthetic oligonucleotide complementary to 5′ or 3′ end of a specific mRNA and usually derived from a cDNA library. Thus we can have 3′ ESTs and 5′ESTs. It ranges from a few dozens to several hundred bps in length. cDNAs in a library are partially sequenced at random to produce a useful type of STS called ESTs which can unambiguously identify the corresponding genes in most cases. Single pass-sequences of 300-500bps are determined from one or both ends of the randomly chosen cDNA clones and are used as ESTs. Locating cDNAs on the YAC physical map using a three dimensional PCR screening method generates the EST markers (Wu et al., 1999). A large number of ESTs from various cDNA libraries have been produced from microarray expression experiment.

11.5 LIBRARY AMPLIFICATION

It refers to the growth of host cells containing the recombinant plasmids belonging to a gene library or the development of the corresponding recombinant phages within the host cells with the simultaneous multiplication of foreign DNA inserted into these plasmids or phages, respectively.

12

DNA Synthesis and DNA Sequencing Technology

12.1 CHEMICAL SYNTHESIS OF NUCLEIC ACID AND NUCLEOTIDE SEQUENCING

Chemically synthesized oligonucleotides are required in the following conditions:
   I.   Cloning DNA
   II. Labeling DNA
   III. As a primer for sequencing DNA or RNA in M13 bacteriophage
   IV. Producing adaptors and mutagenesis *in vitro* site directed mutagenesis
   V.   Studying specific interaction between nucleic acid and protein.
        An oligonucleotide of up to 250 bases in length can be synthesized *in vitro*.

12.2 METHODS OF NUCLEIC ACID SYNTHESIS

There are three pre-requisites for chemical DNA synthesis:
   1. The nucleotide derivatives should be soluble in non-polar organic solution in which synthetic reaction proceeds optimally.
   2. The reactive amino groups of bases, namely, adenine, guanine and cytosine and 5′-OH group of deoxyribose should be blocked.
   3. The free amino groups of bases are protected either by benzoylation in case of adenine and cytosine or by isobutryrylation in case of guanine. The base, thymine does not need protection.

Finally, 5′ and 3′-OH groups of deoxyribose are blocked by monoethoxytrityl-, 4, 4-dimethoxytrityl- or trityl residues.

Approaches to the synthesis of DNA

There are two approaches to the chemical synthesis of DNA:
   1. Phosphodiester approach
   2. Phosphotriester approach.

In both approaches 3′ and 5′- OH of the deoxyrobose moiety are protected, however, the phosphate group between the two nucleosides is unprotected in diester approach when a third protecting group is used for the hydroxyl group at the internucleotide bond. The phosphate triester approach differs from the phosphodiester approach in that the former employs compounds with trivalent phosphorus and thus can be regarded as a triester method. However, the term phosphotriester method is reserved exclusively for method using pentavalent phosphorus. There is a variety of solubility problems associated with phosphodiester method and this method has now been completely replaced by triester method.

Phosphotriester method — Phosphotriester method (Figure 12.1) is useful for synthesizing nucleotides containing up to 50 bases whereas oligonucleotides containing more than 150 bases can be synthesized by phosphitetriester method. The phosphitetriester method of DNA synthesis allows to react a protected nucleoside phosphoroamidite with a nucleoside to form a phosphate triester which is then oxidized to the corresponding triester. Further phosphitetriester method uses a more reactive form of phosphorus which allows the condensation to be completed in about 2 minutes in comparison to about one hour in phosphotriester method and thus is well suited for mechanization. The building blocks in chemical synthesis of DNA are protected nucleotides derivatives. Diester bonds are formed between the 5′-OH group of one nucleotide and 3′-OH group of a second nucleotide. The first nucleoside is directly linked to a solid support (e.g., silica gel) packed in a column. It is attached at the 3′-OH (through a linking group, R) and its 5′-OH position is protected by dimethyl trityl group (DMT). The reactive groups in all the bases are also protected. The protecting DMT is now removed by washing the column with acid. The next nucleoside is activated with diisopropylamino group to form a 5′, 3′ linkage (phosphate triester) which is oxidized with iodine to produce a stable phosphotriester linkage. One of the phosphate oxygen carries a cyanoethyl protecting group. Again DMT is removed and next activated nucleotide is reacted and finally oxidized to produce a phosphotriester linkage between the second and the third nucleotide. This cycle(Figure 12.1) is repeated till all nucleotides are added. After each step (cycle) the excess nucleotide is washed off the column before addition of next nucleotide. The protecting groups from the bases and cyanoethyl groups from phosphates are removed only after the end of overall synthesis and finally the newly synthesized nucleotide is released from the support and purified. This so-called **Phosphoroamidite method** of chemical synthesis of DNA is fully automated and most '**gene machines**' use this process for oligonucleotide synthesis.

Khorana's method — The technique of chemical synthesis of longer DNA sequences starts with the synthesis of single stranded oligodeoxynucleotides, the pairs of which are complementary to each other and anneal to form double stranded DNA molecules. These double stranded molecules carry single stranded protrusions of 4-6 bases at their termini. In each case the two duplex molecules have complementary overhangs and can be linked to each other (starter molecule). The next two oligonucleotides with termini complementary to the ends of the starter molecule will anneal and so on. Finally, the gaps are closed using DNA ligase and thus a fully base paired DNA duplex is generated. In 1972 Khorana first synthesized a gene, the structural gene for an alanyl transfer RNA from yeast, *in vitro*. Khorana's method has now been superseded by phosphotriester and phosphitetriester methods.

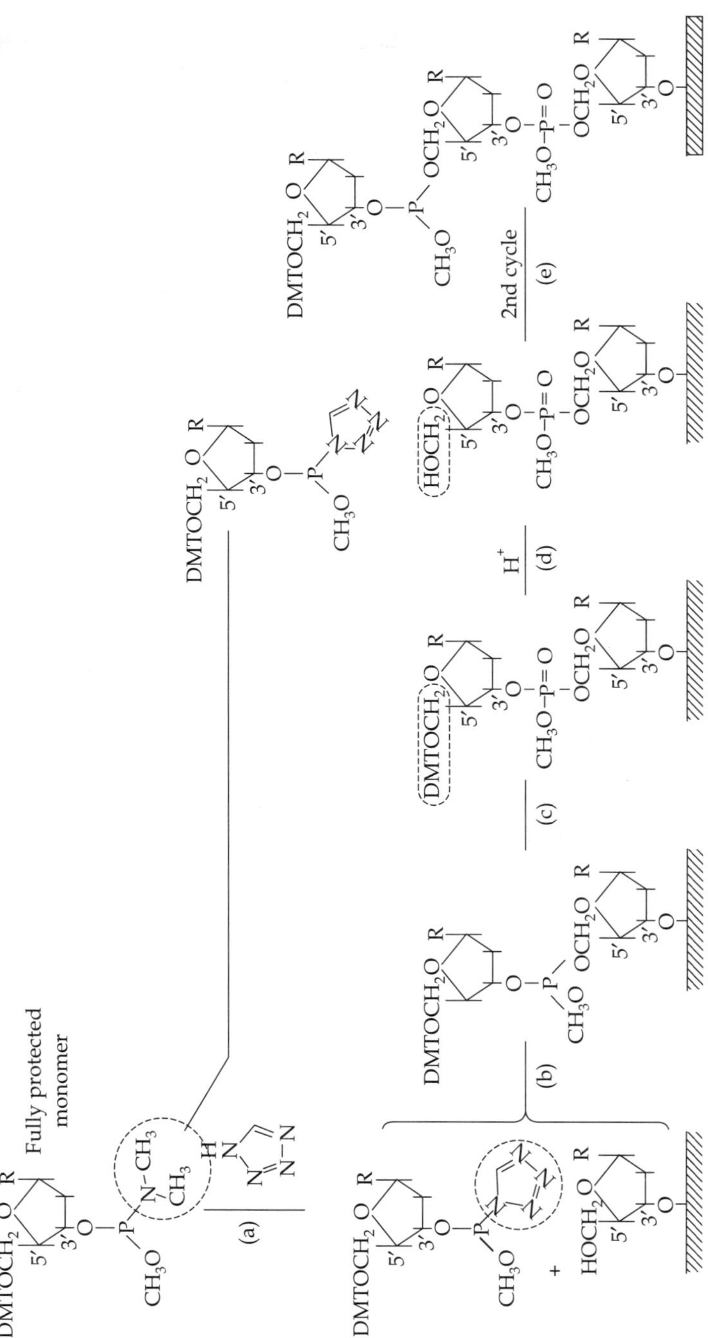

FIGURE 12.1 Showing phosphotriester method of DNA synthesis. (a) The fully protected monomer is unblocked at the 3′-phosphate position by treating with triethylamine permitting (b) condensation with a second molecule of monomer attached through its 3′-O to a solid support and having its 5′-OH unprotected. The dinucleotide produced is (c) treated with acid to remove the dimethoxytrityl group (DMT) protecting the 5′-OH before (d) condensation with another molecule of 3′-activated monomer can occur R = benzoyl adenine, benzoyl cytosine, isobutyryl guanine, on thiamine.

Fill-in synthesis—It is a variant of chemical synthesis of DNA sequences in which long single stranded oligodeoxynucleotides are synthesized with each carrying specific short sequences at its 3′ end. The termini of these oilgonucleotides match perfectly in each case. After annealing these cohesive ends DNA polymerase is added to synthesize the complementary strands using the short double strand of regions as primers.

Synthesis of RNA—In the chemical synthesis of RNA the 2′ hydroxyl of ribose is protected without affecting the reactivity of the 3′-hydroxyl.

Synthetic gene—Gene can be synthesized *in vitro* which encodes a peptide, protein or RNA. The synthetic gene than can be ligated to promoter and terminator and finally it can be transferred to a target organism through various gene transfer techniques. Here the objective is to synthesize a particular coding sequence which does not occur in nature. An artificial gene which codes for protein with high content of essential amino acids such as lysine and methionine has been constructed and transferred to potato through Agrobacterium mediated transfer and the transformed plants have shown accumulation of a relatively high lysine and methionine content.

Sequence analysis—Sequence analysis involves the determination of the sequences of bases in DNA or RNA or of amino acids in protein. Sequence analyses also involve the study of secondary and tertiary structures and occurrence of consensus sequences and other molecular characterization using computer analyses (see Roy, 2009).

12.3 DNA SEQUENCING

The ordering of bases along each chromosome will allow identification of genes and which in turn provide a framework for studying how certain DNA variations among humans predispose towards various diseases in human. DNA sequencing technology decodes the genome of an organism whereas DNA synthesis and genome construction technologies enable the opposite process. The nucleotide sequence of long DNA fragments can be determined by the following two methods:

1. Maxam and Gilbert's (1977) method
2. Sanger and Coulson's (1975) method and Sanger, Nicklen and Coulson's (1977).

Of these two methods Sanger and Coulson's method also called **dideoxy** method of DNA sequencing is easier and more widely used. This method makes use of mechanism of DNA synthesis by enzyme DNA polymerase complementary to the DNA strand under investigation. DNA strand synthesis by DNA polymerase requires a template strand (single stranded) and a short oligonucleotide primer which has a free hydroxyl (OH) group at the 3′ end to which a new nucleotide unit is added. Synthesis starts at a fixed point defined by a primer (**sequencing primer**), often a sequence complementary to the cloning vector into which the unknown sequence has been inserted. The commonly used cloning vectors are derived from the single stranded bacteriophage M13. The new nucleotide unit in the cell is deoxyribonucleoside triphosphate (dNTP) which reacts with 3′-OH to form a phosphodiester bond.

$$(dNMP)_n + dNTP \rightarrow (dNMP)_{n+1} + PPi$$

In Sanger's method 2′ 3′ dideoxyribonucleoside triphosphate (ddNTP) analogues are used to disrupt the DNA synthesis. The ddNTP is incorporated into the growing polynucleotide and blocks further strand synthesis. The ddNTP lacks a 3′-OH and thus chain termination occurs. This **chain termination method** requires single stranded DNA. The general principle underlying both methods is to cleave the DNA fragment to be sequenced at Cs, Gs, As and Ts which lead to generation of four sets of labeled fragments. Suppose there is a DNA fragment with the sequences 5′pATCGACT-3′ which is labeled at 5′ end. A reaction that breaks this fragment after each G will produce only one fragment containing four nucleotides, 5′pATGCG, will be seen on the gel. Similarly if a reaction breaks this fragment after each C will produce two fragments, one containing three nucleotides and another containing six nucleotides. A reaction breaking the fragment after each A will produce two fragments, one containing one nucleotide and the other containing 5 nucleotides and a reaction breaking the fragment after each T will produce two fragments, one having two nucleotides and one having 7 nucleotides. When the sets of fragments corresponding to each of the four bases are electrophoretically separated side by side they produce ladder of bands from which the sequences of nucleotides can be directly read (Figure 12.2). The fragment sizes correspond to the relative positions of G, C, A and T residues obtained by changing the gel properties (length and composition) and electrophoretic conditions. Higher resolution PAGE can be obtained by very thin PAG (less than 0.5mm thick) is used. The gel contains urea which denatures the DNA so that the newly synthesized strands dissociate from the templates. Further, electophoresis is carried out at a high voltage in order to heat up the gel up to 60° C and above to make sure that the strands do not reassociate. PAGE works with short DNA molecules(up to a few hundred nucleotides) whereas agarose gel is generally used for longer pieces of DNA. Higher resolution PAGE can be used for separating DNA fragments differing in size by only one nucleotide. A conventional gel is composed of 8% PA/8M urea which is 0.3 mm thick and 400 mm long. A 12% PA/7M urea (0.3 mm thick) was run at 1350V and 35Ma. Resolution can be further improved by the use of gradient gels and [35]S- labeled dNTPs. In practice the DNA fragment to be sequenced is used as a template strand and a short primer strand which is either radioactively or fluorescently labeled is annealed to it. To these Klenow fragment of *E. coli* DNA polymerase I, dCTP, d GTP, d ATP and d TTP along with a small amount of dd NTP are added. There are four reaction mixtures. In one reaction mixture dd ATP is added, in the second dd CTP, in the third dd GTP and in the fourth reaction mixture dd TTP is added. In the normal reaction mixture when there is no dd NTP there will be synthesis of complementary strand but when different dd NTPs are added in different reactions, the synthesized strands will be prematurely terminated at some locations depending upon what kind of dd NTP was used. For example, in the reaction mixture containing dd CTP the synthesized strand will terminate at position (s) opposite guanine in the template strand or in other words where d C normally occurs. The dd NTP thus acts like chain terminator and all fragments end in dd XTP. The ratio of d XTP: dd XTP used is ~ 100:1. Thus in the above reaction excess of d XTP (d CTP) ensures that the probability that the analog dd XTP (dd ctp) will be incorporated when ever d XTP (d CTP) is added is small. Considering this ratio of 100:1 the probability of termination at a given X is 1/100 and thus dd XTP will add only after more than 100 nucleotides from the original

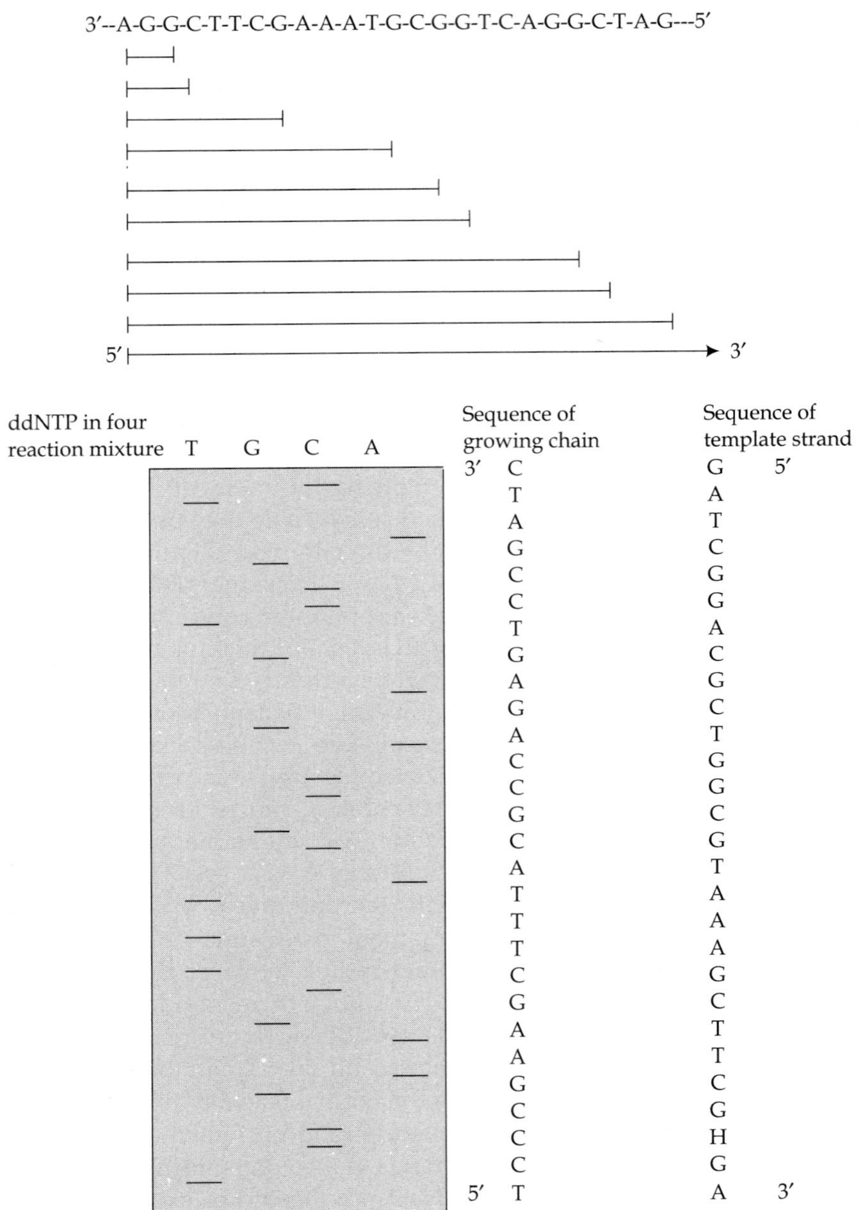

FIGURE 12.2 Showing dideoxy method of DNA sequence determination (chains terminated by ddCTP).

primer terminus. This solution will thus contain a mixture of labeled fragments of different length each ending with a C residue. Similarly, three other solutions will generate mixture of labeled fragments each ending with G, A and T, respectively. There is thus generation of four populations of fragments that terminates at As, Gs, Cs and Ts, respectively. These solutions are then subjected to gel electrophoresis and autoradiography of the gel is done. The different sized fragments ending with a C are separated by electrophoresis and which will reveal the locations of C residues. Similarly the other lanes (G, A and T) or tracks on the gel will reveal the locations of G, A and T, respectively and thus the sequence can be read directly from the autoradiogram of the gel. The fragment (or band) that moved the fastest represents the smallest piece of DNA terminated by dd XTP at the first position in the template. Thus see the track (A, T, C or G) in which it appeared. Suppose it appeared in the G then the first nucleotide in the sequence is G. The next fragment (band) is one that corresponds to a DNA molecule one nucleotide longer than the first and now again we see the track in which it appeared. Now suppose it appeared in T so this second nucleotide in the sequence is T. This process is continued till all the sequence of all the nucleotide is determined. A sequence of ~ 400 nucleotides can generally be read from one autoradiograph. The generated fragments can also be run on fluorescent sequencing machines besides on standard radioactive sequencing gels to determine the order of bases in the sequence.

Sanger et al. (1977) determined the order of nucleotides of DNA using chain- terminating nucleotide analogs. DNA sequencing by Sanger's dideoxynucleotide method has undergone significant refinement including development of additional vectors (Yanish-Perron et al., 1985), base analogs (Mills and Kramer, 1979; Barre et al., 1986), enzymes (Tabor and Richardson, 1987) and instruments for partial automation of DNA sequence analysis (Smith et al., 1986;Prober et al., 1987; Angorge, 1987). The basic procedure involves

1. Hybridizing an oligonucleotide primer to a suitable single- or denatured double stranded DNA template

2. Extending the primer with DNA polymerase in four separate reaction mixtures, each containing one α-labeled dNTP, a mixture of unlabeled dNTPs and one chain-terminating ddNTP.

3. Resolving the four sets of reaction products on a high resolution polyacrylamide/urea gel and

4. Producing an autoradiographic image of the gel which can be examined to infer the DNA sequence.

Success of large scale sequencing depends on the speed and automation of technology- automatic preparation of DNA samples and performing the sequencing reaction. Automatic DNA template preparation is the selective amplification of DNA by PCR. With this method, segments of single-copy genomic DNA can be amplified > 10 million-fold with very high specificity and fidelity. The PCR product can then either be subcloned into a suitable vector(M13, pUC-based vectors, λ phage and other cloning vectors) for sequence analysis or alternative PCR products can be sequenced. Direct sequencing of PCR products by any methods produces a 'consensus' sequence, those bases that occur at a given position in the majority of the molecule. The misincorporation rate of PCR DNA synthesis is low (1 in 600bp or less) but in a cloning experiment only 1-20% of the amplified DNA corresponded to the desired sequence with the remainder arising from a background of priming from other genomic sequences. Engelke et al., (1988) used a 'nested set of PCR amplification primers and dideoynucleotide chain-termination sequencing primers that hybridize internally on the amplified DNA. Analysis of unknown sequences can be carried out by amplifying a cloned insert with vector specific primers that flank the insertion site (Wong et al., 1987).

PCR procedure (Gullensten and Erlich, 1988) — The PCR procedure involves repeated cycles of denaturation of DNA, annealing of oligonucleotides homologous to sequences flanking the segment of interest and primer extension by a DNA polymerase, resulting in a doubling of the amount of the specific DNA fragment with each cycle. This procedure results in a double stranded DNA fragment whose sequence can be identified indirectly by hybridization to allele-specific oligonucleotide probes representing the various alleles studied or whose sequence can be determined. The sequence of DNA fragments generated by PCR has been determined either by cloning them into M13 (Scharf et al., 1986; Horn et al., 1988) or by direct sequencing of the double stranded template using a third 'internal primer (Wong et al., 1987; Wrischnik et al., 1987). Acquisition of large amounts of DNA sequence data involve the random molecular cloning of small kilobases fragments of target DNA sequence into M13 phage, adjacent to a primer homology that can be repeatedly used to drive 300-500 nucleotides of sequence data per sequencing experiment (Sanger et al., 1980). The method requires single stranded DNA to act as a template for the DNA polymerase. Further, it requires a complete random collection of clones in which the different regions of the DNA are equally represented. Random fragments from restriction enzyme digestion of the DNA are inserted into the EcoRI site of the modified bacteriophage M13 using a linker oligonucleotide. The nucleotide sequence in DNA is determined using DNA polymerase and specific chain terminating inhibitors. However, the M13 cloning method is time consuming and requires that several sequences be determined to distinguish between mutations occurring in the original sequence from:

1. random point mutations introduced by lack of fidelity of the DNA polymerase and
2. PCR artifacts such as the formation of mosaic alleles by *in vitro* recombination. Direct sequencing of double stranded templates often presents difficulties due to rapid annealing of strands and the presence of sequences on both strands homologous to that of the sequencing primers resulting in compound sequence ladders. By modifying the PCR reaction there is generation of single stranded DNA by PCR. Two amplification primers are present in different molar amounts. During the first 10-15 cycles exclusively double stranded DNA will be produced. However, when the primer added in limiting amounts has been used up, an excess of single stranded DNA will be produced in each cycle. Theoretically, the amount of double stranded DNA should increase exponentially whereas the production of single stranded DNA should only follow a linear growth. Direct sequencing procedure is capable of identifying both alleles in a heterozygous condition. Direct sequencing method facilitates the analysis of alleleic variants at a known locus as well as the determination of unknown sequences. It provides a rapid and simple approach to the analysis of nucleotide sequence. The rapid identification of mutant or allelic variants can be accomplished by amplifying DNA segments by using locus-specific primers.

Production of single stranded DNA templates of PCR products for direct sequencing (Stoflet et al., 1988) — This method involves attaching a phage promoter to one of the PCR primers, transcribing the PCR product to obtain an RNA copy and sequencing this with reverse transcriptase. This procedure has more limited applicability since additional enzymatic steps after the amplification reaction and is restricted to using reverse transcriptase as the sequencing enzymes.

Sequencing primer is a short synthetic oligonucleotide that is complementary to a sequence at one end of a single stranded DNA target, hybridizes to it and permits the Klenow fragment of the DNA polymerase I to synthesize a complemantary copy of the target DNA fragment. Usually the plus strand is sequenced first starting at its 3′ end (3′-5′)using a normal sequencing primer. The **reverse sequencing primer** which is complementary to the minus strand of the sequencing vector, M13, permits to sequence this minus strand from its 5′ terminus and minus –strand sequencing serves to confirm the base sequence determined by the plus strand sequencing. **Sequencing primer linker (splinker)**- It refers to a synthetic oligodeoxynucleotide which contains an inverted repeat sequence which forms a double stranded stem and loop structure with a restriction endonuclease recognition site located in the double stranded region. Splinker serves both as linker and sequencing primer in direct sequencing of DNA restriction fragments.

Primer directed sequencing- It is a technique to sequence DNA fragments up to 1kb in length. In this technique the target fragment is first cloned into an appropriate cloning vector (M13) and a forward and reverse sequencing primer complementary to flanking vector sequences are used to sequence the inserts from both ends by Sanger sequencing technique. This technique provides information of about 600-800bp on both ends of the insert. Now primers are designed from the outermost 100bp at both ends (walking primers) and used for second sequencing step and so on. **Primer walking** thus permits sequencing of long stretches of DNA which can not be sequenced by classical sequencing techniques. Walking primers refer to any synthetic oligodeoxynucleotide of a defined sequence, for example, TmGn; as 5′-T11GT3GT2GTG5TGT-3′ which can be used as primer in unpredictably primed PCR. The walking primer is employed in a low stringency amplification of unknown regions of a genome. The generated products are subsequently amplified with sequence specific primers, for example, gene specific primer at high stringency. The products of this second amplification are amplified again using the **nested primer** to reamplify the target sequence at sites different from the original primer sites and thus it increases the specificity of the amplification. The nested primer thus has the sequence complementary to an internal site of DNA fragment that has been amplified with other primers in a conventional PCR. The products of the second amplification can also be reamplified with nested (inner) specific primer and a short walking primer (sWP) complementary to the part of the original walking primer. The final product then contains part of the known sequence complementary to the iSP and the unknown flanking regions extending to the sequence complementary to the short walking primer. **Degenerate primer**- It refers to one of a mixture of oligonucleotides which possess the same number of bases but vary in base sequence. Such mixtures are used to amplify a specific genomic region whose sequence has been deduced from the amino acid sequence of a protein. Since the genetic code is degenerate and most amino acids are encoded by more than one codon, therefore, a set of oligonucleotides with the same number of bases but varying sequences has to be used to amplify the coding sequence for that protein.

SP6/T7 sequencing primer—The synthetic oligodeoxyribonucleotide, 5′-CATACGAT TTAGG-3′ which hybridizes to a conserved 20bp sequence of bacteriophage-SP6 RNA polymerase promoter and permits sequencing of double stranded DNA inserted into vectors containing this promoter without the need of subcloning into an M13 vector. If the bacteriophage contains T7 RNA polymerase promoter then in that case T7 sequencing primer is used.

Plasmid sequencing—It is a technique of double strand sequencing, the sequencing of linearized plasmid DNA which has been denatured, i.e., made single stranded. The single strands can either be separated and each annealed to a synthetic, strand specific oligonucleotide primer or both strands can remain in the same mixture, if only one strand specific primer is used. The primer –annealed plasmid strands can then be sequenced following Sanger's sequencing procedure.

Preparation of single stranded DNA for sequencing—Single stranded DNA template for sequencing can also be generated by strand-specific elution from a solid support. In this technique one of the PCR primer is biotinylated at the 5'-end and following symmetric amplification the PCR product is immobilized by specific binding of the biotin group to streptavidin-coated magnetic beads. The non-biotinylated strand is then removed by alkali treatment and the immobilized single strand DNA can then be sequenced directly.

Nucleotide degradation—In Maxam-Gilbert's method the sequencing is not by way of synthesizing new strands as in case of Sanger and Coulson's method but by way of cleaving the existing DNA (either single or double stranded) molecules by specific chemical reagents that act specifically at a particular nucleotide. There are chemical reagents which are specific for guanine (G) residues and cytosine (C) residues, respectively and also there are those which are specific for only purines (A + G) or pyrimidines(T + C), respectively. There are several variants of Maxam-Gilbert method based on the way in which the DNA fragment is labeled and chemical reagents used. In one version of M-G method the double or single stranded DNA fragment to be sequenced is labeled using radioactive phosphorus at 5' end of each strand. To this labeled DNA molecule DMSO (dimethylsulphoxide) is added and whole thing is heated to 90°C. The two strands thus get dissociated which are then separated by gel electrophoresis. The separation of two strands is based on weight (mass) for which the strands differ. One strand probably contains more purine bases and thus is slightly heavier than other strand. One strand is separated from the gel and purified and divided into four samples. To each sample is added one of the chemical reagents (one specific to cytosine, another specific to guanine, third specific to purine and fourth specific to pyrimidine). The first set of reagents cause modification of base in the specific nucleotide which makes it susceptible to cleavage. When the second chemical reagents are added to all samples which results in the removal of the modified base from the deoxyribose moiety and a strand break occurs at the position of the modified base. The modification and cleavage reactions are controlled in such a way that it results in only one breakage per strand and it should occur at one (G or C) simultaneously at two bases (G and A or T and C). The chemical reagents for the first and second reactions are dimethylsulfate and paperidine in case of purines and hydrazine and piperidin for pyrimidines. The DNA fragment is thus partially cleaved at each of the four bases in four different reactions. These four samples containing population of fragments are then subjected to denaturing polyacrlamide gel electrophoresis and autoradiography. The nucleotide sequence can be read from the autoradiograph by determining which of the base specific reagents cleaved adjacent nucleotides. Both methods of sequencing described above normally use cloned DNA but Maxam and Gilbert method can be applied directly to genomic DNA (Church and Gilbert, 1984), i.e. sequencing *in vivo*. This technique is very suitable for study of modified bases and interactions of proteins with DNA and for sequencing of oligonucleotides.

Wandering-spot method (Mobility shift) — Although the Maxam-Gilbert and Sanger methods are extremely powerful for sequencing large restriction fragment of DNA these are not well suited to sequencing of smaller oligonucleotides. A partial pancreatic DNase digest of GGTGAATTCTTTCTT labelled with 32p at its 5′ end is subjected to two-dimensional thin layer homochromatography on DEAE-cellulose (Gait, 1984). The shortening of the oligodeoxynucleotide by a single nucleotide causes a change in mobility which is a characteristic of the nucleotide removed and which permits the sequence to be read. The identity of the 5′ nucleotide is determined by comparison with standards.

Sources of sequencing error — The main source of sequencing error is due to '**compressions**' on the sequencing gels. In stead of being a regular distance between each band, some bands run closer together or occupy the same position. Compressions appear to be caused by the secondary structure effects which are not broken by the denaturing conditions in the gel. A small region at the 3′ of the newly synthesized DNA becomes double stranded and the corresponding bands run abnormally fast in the electrophoresis. Compression can be overcome by determining the sequences in both directions, i.e. on 2 clones with the sequence in opposite orientations. Two other techniques for overcoming this problem are as follows.

1. Replace dGTP in the mixtures with dITP. The secondary structures are less likely to form because I-C pairing is weaker than G-C. This method will not work with clones constructed with the vector M13mp7 as secondary structure which can be formed due to complementary sequences in the vector inhibits the progress of the DNA polymerase in the presence of dITP.

2. Run the gels in 25%(v/v) formamide

Another occasional source of error occurs mainly at G-T sequences. At position of G-bands there appear in both G and T channels with T band sometimes stronger than the G. This error is usually associated with the use of unsatisfactory polymerases and this can be eliminated by using a higher concentration of enzyme(Sanger et al., 1982).

12.4   AUTOMATIC SEQUENCING

The four color fluorescence-based sequence detection, improved fluorescent dyes, dye-labelled terminators, polymerase specifically designed for sequencing, cycle sequencing and capillary gel electrophoresis have brought about improvements in automation, quality and throughput of collected raw DNA sequence. Automatic sequencing with the development of automated fluorescent sequencer in the mid 1980s (Smith et al., 1986) allows DNA sequences containing thousands of nucleotides to be determined in a few hours. It is possible to sequence up to 750, 000 nucleotides per day. In the automatic sequencing procedure the four different dd NTPs are labeled with differently colored fluorescent tags. The reaction mixture containing template of unknown sequence primers, DNA polymerase, four d NTPs and four dd NTPs added to a single test tube. The resulting labeled DNA fragments are then separated by single electrophoretic gel contained in a capillary tube (**capillary sequencing**).

This is a refinement of gel electrophoresis which allows faster separation of DNA fragments. The slab gel is replaced with 48 or 96 capillaries filled with the gel matrix. Fragments of equal length produce a single peak associated of a color and each peak is detected by the Laser beam. When the bands (the capillary tube) migrate past the detector which is connected to a computer which is used to interpret the laser-activated fluorescence and convert into a digital form suitable for further analysis. The sequence of colors in the peak determines the nucleotide sequence and the computer comes up with the information on the DNA sequence. In other word, the output consists of a series of (colored-coded) peaks, below which is a string of base symbols–the particular base shown is determined by the highest peak at that position of the trace. Sometimes the software which interprets the chromatogram (base calling software) is unable to determine which base should be called at a specific position and so a '-' appears. Such ambiguous positions are replaced by 'N' in the resulting sequence file. The result is a sequence in which a portion of the symbols will be Ns. A list of the base-ambiguity symbols as defined by the IUB-IUPAC is given in Bioinformatics (Roy, 2009). The quality control criteria keep the number of Ns in a production sequence to less than 5% of the total length. The start and the end of the sequences are trimmed in order to reduce the ambiguity further. Although the base calling software assigns an N it is sometimes possible to call a base by eye. When the sequence is being used in the assembly process the assembly editor normally provides the facility to view the chromatogram to increase confidence in base calling in particularly difficult areas of sequence (especially GC-rich regions which are areas of high secondary structure content and difficult to sequence).

Church-Gilbert sequencing—It is genomic DNA sequencing technique that permits the determination of both the sequences of bases in an uncloned genomic DNA and the position of 5′- methylcytosines which escape detection by normal chemical sequencing procedures.

Bisulphite genomic sequencing—It is a technique for the detection of methylated cytosines in genomic DNA based on the reaction of DNA with bisulphate under conditions such that cytosine is deaminated to uracil but 5′-methylcytosine remains unreacted.

Mutiplex sequencing(Church sequencing)—It is a DNA sequencing technique which permits the determination of base sequences in 10-15 different fragments simultaneously.

Thermal cycle sequencing—It is a technique for the sequencing of an amplification product generated in the PCR reaction. Either single stranded or double stranded template DNA is prepared and primers are then annealed to the template in four separate reactions which contain all the ingredients (described above) for amplification. Additionally each reaction is supported with one of the dideoxynucleotides and Sanger sequencing is started. After this procedure denaturing PAGE and autoradiography is used to detect the sequence of the amplified product. It requires only small amounts of template and a high temperature profile during cycling ensures high specificity of primer annealing and good resolution of intramolecular secondary structures which otherwise interfere with sequencing.

Transposon walking—It is a technique for sequencing of unknown DNA sequences that contains long stretches of either adenine or thymine which are difficult to sequence. In this technique transposons are inserted into the target DNA at random and used as starting points for sequencing using transposon- specific primers.

Genome priming—It is an *in vitro* technique for the sequencing of long stretches of target DNA. In this technique the so-called transprimers (transposons) are inserted into the target molecule at random which then serves as binding sites for sequencing primers and thus allow the determination of the base sequence. First, the target DNA is cloned into a plasmid vector(able to replicate in *E. coli*, ie., ori$^+$) and then a transposase (e.g. the TnsABC transposase) is used to mobilize the transprimer *in vitro* from a donor plasmid (unable to replicate in *E. coli*, i.e., ori$^-$) into the target DNA at random. The transprimer consists of a selectable marker gene such as kanamycin or chrloramphenicol resistance gene allowing selection of insertion mutants, two rare cutter recognition sites and flanking unique priming sites. Generally two transprimers with different antibiotic resistance genes are used. Statistically, only one transprimer insertion per target DNA occurs. Finally, a population of target molecule is generated. Each of which contains the transprimer at a different position. After their transformation into competent *E.coli* cells, only target DNA molecules containing the transprimer insertion survive selection. After plasmid isolation, a specific primer complementary to the priming site of the transprimer is used to sequence the insert. Since unique priming sites are positioned on both ends of the transprimer both strands of the inserts can be sequenced.

Pyrosequencing—This technique of sequencing is only suitable for the diagonistic sequencing of relatively short DNA fragment (up to 200 bases). It is a technique to determine the sequence of bases in DNA without using sequencing gel electrophoresis and radioactivity or fluorescence needed in other sequencing procedures. Pyrosequencing quantitatively measures the pyrophosphate (PPi) released during DNA polymerase reaction by coupling it to the generation of light by firefly-luciferase. In this technique single stranded DNA template is first annealed to a short sequencing primer. The DNA polymerase together with an apyrase ATP sulfurylase and fire fly luciferase and only one DNTP (e.g dGTP) are added. If dG does not form a base pair with the first free base on the template then it is rapidly removed by a apyrase (a mixture of nucleoside 5′-triphosphate and nucleoside 5′-diphosphate). The next base is added (e.g dTTP). If a base pair T = A can be formed then the DNA polymerase extends the primer and releases PPi which is quantitatively converted to ATP by ATP sulfurylase. Now luciferase utilizes the ATP to oxidatively decarboxylate luciferin and the light is generated which is detected by a sensitive luminometer or a charge-coupled device (CCD) camera. The light pulse signaling incorporation of a base is shown in real-rime on a PC.

12.5   MASSIVELY PARALLEL SYSTEM

Sanger sequencing (Sanger et al., 1977) and fluorescence-based electrophoresis technologies (Prober et al., 1987) have been primarily employed in DNA sequencing. But then there was a search for alternative methods which could reduce time and cost. Massively parallel picolitre-scale sequencing was developed by Rothberg et al. (2005). This apparatus uses a novel fibre-optic slide of individual wells and is able to sequence 25 million bases at 99% or better accuracy in one four-hour run. To achieve an approximately 100-fold increase in the throughput over current Sanger sequencing technology they developed an emulsion method for DNA amplification and an instrument for sequencing by synthesis using pyrosequencing (pyrophosphate-based sequencing, Ronaghi et al., 1996; Ronaghi, Uhlen and Nyren, 1998) protocol optimized for solid support and pico litre-scale volumes (Margulies et al., 2005).

The steps involved in the traditional micro-litre-scale Sanger DNA sequencing and electrophoresis involve preparation of purified DNA, construction of a library of bacterial clones-each containing a DNA fragment (two weeks), plating and growth of bacteria (one day), bacterial colony picking and growth (one day), template isolation (half-day), Sanger sequencing reaction (half-day), purification of reaction products (half-day) and electrophoresis of product to detect sequence (1 million bases per 24 hours per machine). In case of massively parallel sequencing, the steps involved are isolation of purified DNA, sample/library preparation (one day), emulsion-based amplification, bead recovery and enrichment (one day) and sequencing reaction and detection (25 million bases per 4 hours per machine (Rogers and Venter, 2005). Thus the traditional microlitre-scale approach requires a longer time per production cycle, more equipment, large facility and more labour than picolitre-scale approach.

The complexity of this system lies primarily in the sample preparation and in the microfabricated, massively parallel platform which contains 1.6 milliom picolitre sized reactors in a 6.4 cm^2 slide. The sample preparation starts with isolation of genomic DNA, then fragmentation of the genomic DNA, followed by attachment of adapter sequences to the ends of the DNA fragments and finally, separation into single strands. The adapters permit the DNA fragments to bind to tiny beads which are around 28 µm in diameter. This is done under conditions to allow only one fragment of DNA to bind to each bead. The beads are then encashed in droplets (100 µm) of oil which contain all the reactants required to amplify the DNA using PCR. The oil droplets form part of an emulsion so that each bead is kept apart from its neighbor thereby ensuring that the amplification is not contaminated. After PCR amplification each bead carries up about 10 million copies of its initial DNA fragment.

To perform the sequencing reaction the emulsion is broken, the DNA strands are denatured and beads carrying single stranded DNA clones are loaded into the pico-litre reactor wells, each well having space just for one bead. The technology uses a sequencing-by synthesis (SBS) developed by Uhlen and colleagues (Nyren et al., 1993) in which DNA complementary to each template strand is synthesized. Smaller beads carrying immobilized enzymes required for pyrophosphate sequencing are deposited into each well. The nucleotide bases used for sequencing release a chemical group as the base forms a bond with the growing DNA chain and this group drives a light emitting reaction in the presence of specific enzymes and luciferin. In other words, nucleotide incorporation is detected by the associated release of inorganic pyrophosphate and the generation of photons. Wells containing template-carrying beads are identified by detecting a known four-nucleotide 'key' sequence at the beginning of the read. Finally, sequential washes of each of the four possible nucleotides are run over the plate and a detector senses which of the wells emit light with each wash to determine the sequence of the growing strand. Poor quality reads have a high proportion of signals that do not allow clear distinction between a flow during which no nucleotide is incorporated and a few during which one or more nucleotide is incorporated. In sequencing-by–synthesis a very small number of templates on each bead loose synchronism(i.e. either get ahead or fall behind all other templates in sequence). This effect is primarily due to leftover nucleotide in a well creating 'carry forward'. This sequencing technology has several applications including resequencing, *de novo* sequencing of smaller bacterial and viral genomes. This technique generates over 25 million bases with a Phred quality score of 20 or better (predicted to have an accuracy of 99% of higher).

Sequencing instrument—It consists of four major subsystems: A fluid assembly, a flow chamber that includes the well-containing fibre-optic slide, a CCD cameras-based imaging assembly and a computer which provides the necessary user interface and instrument control.

Limitations—There are limitations with MPSS technology such as shorter reads and lower average individual read accuracy. First, this technique can read comparatively short lengths of DNA, averaging 80-120 bases per read which is approximately 1/10th of the read lengths possible with Sanger sequencing. Thus more reads will need to be done to cover the same sequence and piecing the sequences together into larger genomic sequence is lot more complicated and so is a problem in sequencing particularly genomes containing long repetitive sequences. Second, the accuracy of each individual read is not as good as with Sanger sequencing particularly in genomic regions in which single bases are constantly repeated (homopolymer). Phred quality throughput is significantly higher than that of Sanger sequencing by capillary electrophoresis. Sanger-based capillary electrophoresis sequencing systems produce up to 700 bases of sequence information from each of 96 DNA templates at an average read accuracy of 99.4% in one hour or 67, 000 bases per hour with substantially all of the bases having Phred 20 or better quality. Third, as the DNA library is currently prepared in single stranded format unlike the double stranded inserts of DNA libraries used for Sanger sequencing, this technique can not produce paired-end reads for each DNA fragment. The paired-end information is crucial for assembling and orientating the individual sequence reads into a complete genomic map and for *de novo* sequencing applications. Finally, the sample preparation and amplification are still quite complex and require automation and/or simplification (Rogers and Venter, 2005). Considering these limitations this technology can not yet replace Sanger sequencing approach for some of the more demanding applications such as sequencing of mammalian genome.

12.6 MALDI-TOFMS(MATRIX-ASSISTED LASER DESORPTION/ IONIZATION TIME-OFF-FLIGHT MASS SPECTROSCOPY

It is used to analyze large macromolecules of biological significance. It has shown potential as an alternative to gel-based DNA sequencing providing very high speed analyses with no requirement for fluorescence or radioactive labeling.

Analysis of SNPs by MALDI-TOFMS—In this method a single dideoxynucleotide complementary to the base at the polymorphic position in the DNA template is added to the 3′ end of an oligonucleotide primer by polymerase extension (Haff and Smirnov, 1997). Determination of the mass of the extended primer identifies the added nucleotide and thus reveals the nature of base at the polymorphic position in the template. The limitation of this approach is that the small mass differences between certain nucleotides render it difficult to unambiguously identify the added nucleotide due to a lack of resolution in the mass spectrum. For example, in case of A/T transversion because of small mass difference (9Da) between ddA and ddT, the polymerase extension MALDI-TOFMS procedure described above was unable to distinguish the extension products of an A/T heterozygote. Oligo (dT) sequences varying in length have been employed as mass-tags in primers to permit multiplexing of the procedure (Rose and Belgrader, 1997). A mass-tagged dideoxynucleoside triphosphate is employed in the strand extension reaction in place of the unmodified dideoxynucleoside triphosphate (ddNTPs). The increased mass difference due to the presence of dyes greatly facilitates the accurate identification of the added nucleotide and is particularly useful for typing heterozygous samples (Fei et al., 1998).

Resequencing—In this technique sequencing of a genomic, subgenomic or organeller DNA region from an individual, say A is done and its comparison is made to the same, already known region from individual, say B (the reference). Thus it is a technique of detecting sequence variations which is a pre-requisite for SNP profiling, i.e. detection of single nucleotide polymorphisms in many individuals of a population.

Sequencing array—It is a high density DNA chip onto which multiple short oligonucleotides are fixed that permits to determine the sequence of a DNA fragment which is hybridized to it. As the sequence of each oligonucleotide is known, hybridization patterns can be directly converted into base sequence information.

12.7   SEQUENCING BY HYBRIDIZATION (SBH)

Conventional DNA sequencing technology is a laborious procedure requiring electrophoretic size separation of labeled DNA fragments. An alternative approach to *de novo* DNA sequencing is termed sequencing by hybridization (Pease et al., 1994). It is a technique used for quick sequence determination of DNA fragments. This method uses a set of short oligonucleotide probes of defined sequence to search for complementary sequences on a larger target strand of DNA. The hybridization pattern is then used to reconstruct the target DNA sequence. It is based on the hybridization of short oligodeoxynucleotides (6-10bases) to a target sequence under extremely stringent conditions and the detection of hybrids. The techniques can be classified into two categories. Technique of category I works with an array of synthetic hexa-to-decameric oligodeoxynucleotides fixed onto a solid support (a DNA chip). A labeled DNA fragment with unknown sequence of some hundred bases long is then exposed to such a chip where all base sequences complementary to the set of oligonucleotides (which may be 65, 536 in number) will hybridize. The oligodeoxynucleotides produces a nested array. For example, the octamer produces a nested array, i.e. the first complementary to bases 1-8 of the unknown DNA, the second to bases 2-9, third to bases 3-10 and so on. The overlap then can be used to put the octamers in order so that a computer can determine the full sequence of the unknown DNA. Suppose a 12-mer target DNA sequence, AGCCTAGCTGAA, is mixed with a complete set of octanucleotide probes. If only perfect hybridization complementarity is considered, five of the 65, 536 octamer probes TCGGATCG, CGGATCGA, GGATCGAC, GATCGACT and ATCGACTT will hybridize to the target. Alignment of overlapping sequences from hybridization probes reconstructs the complement of the original 12-mer target.

TCGGATCG

CGGATCGA

GGATCGAC

GATCGACT

ATCGACTT

TCGGATCGACTT

Category II is bases on hybridization of the labeled oligodeoxynucleotides to the immobilized target sequence. The drawbacks of SBH involve mismatched oligonucleotides and repetitive sequences.

A variant of SBH called SBH to oligonuceoltide microchip (**SHOM**) works with oligodeoxynucleotides immobilized on a gel-based microchip element. This technique allows to detect mismatch mutations and gene polymorphisms and is used for the diagnosis of genetic diseases.

Applications of SBH—Hybridization analysis of a large number of probes can be used to sequence long stretches of DNA. The more immediate application of hybridization methodology is that a small number of probes can be used to interrogate local DNA structure. Hybridization methodology can be carried out by attaching target DNA to a surface. The target is then interrogated with a set of oligonucleotide probes, one at a time. This approach can be implemented with well-established methods of immobilization and hybridization detection but involves a large number of manipulations. Alternatively, SBH can be carried out by attaching probes to a surface in an array format where the identity of the probe at each site is known. The target DNA is then added to the array of probes. The hybridization pattern determined in a single experiment directly reveals the identity of all complementary probes.

12.8   PROBES AND PRIMERS AND LABELING OF NUCLEIC ACIDS

12.8.1   Probe

Probe can be defined as the labeled (radioactively or non-radioactively) fragment of nucleic acid containing a nucleotide sequence complementary to a gene or genomic sequence that one wishes to detect in hybridization experiment. In other words, probe is used in molecular cloning to identify specific DNA molecules with complementary sequence (s) by autoradiography or with non-radioactive DNA detection systems. The term, probe is also used for proteins, for example, a monoclonal antibody reacting with its target protein. Probe also refers to any defined nucleic acid sequence (e.g. an oligonucleotide, a cDNA) which is covalently bound to a support (chip) and hybridized to a target nucleic acid (mostly cDNAs). Thousands of such probes can be assembled on socalled cDNA expression arrays, DNA chips, microarrays, sequencing arrays and serve for the simultaneous detection of multiple hybridization events ('massively parallel). In chip technology the probes are termed reporters.

Sources of probe—There are various methods of designing probe that specifically hybridizes to the desired gene. In other words, probes can be obtained from various sources as follows.

1. **For cloning a gene for a known protein**—A major procedure in a cDNA cloning involves the synthesis of oligonucleotide probes to a known peptide sequence. If the protein product of gene is known and purified then probe can be designed and synthesized using the AA sequence and the knowledge of the genetic code (Wallace et al., 1979, 1981). As one AA can be coded for by a number of codons (called the degeneracy of genetic code) the correct DNA sequence can not be known. The probe is designed to be complementary to a region of the gene with minimal degeneracy (fewest possible codons). The first step would be to work out

the AA sequence along with the number of codons for each AA and then find out the region of AA sequence for which there is minimal degeneracy of codons. Probe is designed with DNA sequences making up these regions of AA sequence. The synthesized nucleotide can thus be a mixture of a number of different sequences one of which at least will complement the gene perfectly and all will match at least most of the positions. The oligonucleotides making the probe are synthesized with selectively randomized sequences so that they contain either of the two possible nucleotides at each position of potential degeneracy. For example, cytochrome c from from yeast is made up of a polypeptide containing 103 AAs. The region of minimal degeneracy is shown in the (Figure 12.3), contains 6 AAs with 18 codons. The synthetic nucleotide will be actually a mixture of 16 different sequences and 14 of 18 nucleotides can be predicted with certainty and all 16 will match at least 14 of 18 positions and one of 16 will complement the gene perfectly. Thus the problem of wobble bases in the genetic code, i.e., ambiguity observed in the third base of many codons can be overcome by using mixtures of derivatives nucleotide bases rather than individual bases at the position in question during the chemical synthesis of primer. The resulting oligonucleotide mixtures will always contain the desired primer species in sufficient concentration.

The protein is sequenced usually at the N-terminus (Edman degradation) and **reverse translation** is used to determine the probe sequence. However, sequences from internal peptides produced by specific cleavage of the polypeptide chain (e.g. by an arginine specific proteinase or by cyanogens bromide cleavage at methionine residues) are also used if the N-terminus is blocked or if the N-terminal sequence is unsuitable for reverse translation. Reverse translation refers to inferring DNA sequence from the amino acid sequence of protein. It is used to make hybridization probe or a PCR primer and thus used to clone a gene. In most cases amino acid sequence of a protein is directly inferred from the DNA or mRNA as each codon specifies either a single amino acid or a termination signal. But determining the DNA or RNA sequence coding for a specific amino acid is more complex because of degeneracy of genetic code. 61 codons specify 20 amino acids and so many amino acids are coded by more than one codon. The degeneracy of the genetic code for all amino acids except methionine and tryptophan requires synthesis of oligonucleotide mixture for use as hybridization probes. Thus the reverse translation of protein will produce a population of different sequences which if translated would all code for the same amino acid sequence. Synthesize a mixture of oligonucleotides which corresponds to all of the potential coding sequences determined by reverse translation. This pool of olgonucleotides (oligos) is used as a degenerate (mixed) hybridization probe to isolate the corresponding DNA or

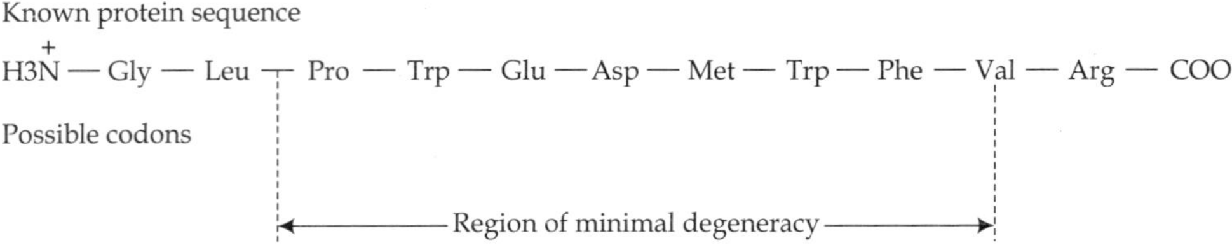

FIGURE 12.3   Showing synthesis of a probe.

cDNA clone from a library. Alternatively reverse translation is used to design two sets of degenerate PCR primers to amplify the gene from the genomic DNA. A procedure based on PCR was developed for cDNA probe generation from amino acid sequences with high degeneracy by Lee et al (1988). Specific cDNA probes can be rapidly generated by the PCR when mixed oligonucleotides derived from amino acid sequences are used as primers. The cDNA probe generated by mixed oligonucleotide primed amplification of cDNA can be used for hybridization studies or for screening a cDNA library for a full length clone.

Factors considered during designing of oligos for gene cloning—The factors to be considered while designing probes are oligonucleotide length, G + C content, self complementarity and complexity (in case of mixed oligonucleotide probes) (Nei and Li, 1979; Suggs et al., 1981). Length of the probe determines hybridization specificity and duplex stability. When used as probes for cDNA clones oligonucleotide probes should be made as long as possible, preferably 20 or longer. Probes shorter than 14 bases will tend to bind non-specifically to non complementary DNA sequence. G content of oligonucleotides has an effect on DNA synthesis itself particularly purification of oligonucleotide. G-rich oligonucleotides are difficult to purify. In this situation synthesize complementary sequence which would serve the same purpose for many applications. Self complementarity of oligonucleotides creates like G content problems for oligonucleotide purification. Complexity of a probe refers to the number of different oligonucleotides in the mixture. With the increase of mixture complexity hybridization specificity decreases and there is a decrease abundance of the single correct probe in the mixtures.

The following must be kept in mind while designing probes:

1. A 14-base oligo is sufficiently long to identify a gene.
2. The 5- residues (amino acids) stretch of protein which is reverse translated to produce this oligo must be chosen carefully.
3. Serine, Leucine and Arginine are each coded by six different codons and so these residues should be avoided.
4. Protein sequences containing tryptophan and methionine residues are preferred as they are each coded by only one codon.
5. Fewer different oligonucleotide sequences are required to cover all reverse translation possibilities if less 'degenerate' amino acids are selected and a less complex of oligos makes a more efficient probe or PCR primer.
6. Different organisms preferentially use particular codons to specify amino acids (see chapter 3) and thus codon usage bias should be taken into account when designing oligos by reverse translation.

Computer programs are available to design synthetic genes and degenerate probes and primers by using reverse translation (Tamura et al., 1991). With genome sequencing projects making rapid progress and with increasing number of EST available, reverse translation will be used to isolate genes from organism where little sequence information is available. Database searches where a computer program compares known protein sequences and translated nucleotide sequences to look for similarities will replace need to reverse translation and clone to determine the nucleotide sequence.

2. **The homologous gene cloned from another species can be used as a probe- called heterologous probe**—Two genes for the same protein from different organisms will have

substantial amount of nucleotide homology and probe prepared from one gene will produce enough pairing to produce a stable structure. This method thus uses a segment of DNA from a gene of a closely related organism as a hybridization probe. A modification of this approach is to look at homologous sequences from a set of organisms to identify the highly conserved regions that are used to design an oligonucleotide probe specific for the desired gene. It is important to select organisms with DNA base compositions close to that of the organism from which the gene is to be cloned.

3. **Use of genomic and cDNA sequence information (databases)** — The DNA sequence information can be obtained from the sequence database searches which detail the structure of millions of genes from different organisms. Hybridization probes are designed from sequence information and used to screen a library to to amplify the gene directly using EST (expressed sequence tag).

4. Probe can be obtained from the cDNA library. The cDNA libraries are prepared from mRNA transcripts at different stages of growth and development of plants starting from the germinating seeds to developing seeds. The individual cDNAs from cDNA library developed from a particular stage of growth and development can be selected, labeled and used as probes to detect a possible clone expected to be expressed at that stage of growth. This type of probe can be used for detecting gene (s) which produce abundant amount of gene product at a particular stage of development. For example, cDNA library prepared from developing wheat seed will contain a large proportion of the clones which are copies of mRNA transcripts of the gliadin gene.

Preparation of cDNA — There are different methods of cDNA preparation:

1. The cDNA is the DNA copy of mRNA. The mRNA is isolated from the mixture by chromatography using affinity column of oligo (dT) cellulose and it is reverse transcribed using retroviral reverse transcriptase and dd NTPs with primer oligo (dT) which hybridizes with poly (A) tail of mRNA. This results in a single stranded cDNA.

2. This method utilizes self priming mechanism. The double stranded cDNA can be prepared by first dissociating the mRNA from mRNA-cDNA hybrid using alkali or heat denaturationor using ribonuclease. The separated single stranded cDNA by the 3′ end of the single stranded DNA folding back on itself and the second strand is obtained using Klenow polymerase, T4 DNA polymerase or reverse transcriptase and dNTPs. Finally the loop is digested using S1 nuclease.

3. In another method RNA is first removed from mRNA-DNA hybrid by alkaline hydrolysis or digestion with ribonuclease. This method relies on the addition of a short homopolymeric tail to the single stranded DNA using dCTP and terminal transferase and then copying this strand by using a complementary homopolymeric primer (oligo (dG) 12-18, Klenow polymerase or reverse transcriptase). Finally, S1 nuclease is used to digest any protruding ends on the double stranded DNA. In yet another method the mRNA-DNA hybrid is subjected to RNaseH (H stands for hybrid) which creates nicks in the mRNA strand and now when it is subjected to *E.coli* polymerase I in the presence of dNTPs, the double stranded cDNA is formed.

Probe set — It refers to any collection of probes which altogether represent a target sequence, for example, gene. A set of about 20 oligonucleotides of about 50 nucleotides each derived from different regions of a known gene and thus representing the gene can be spotted onto a microarray. Then labeled cDNAs can be hybridized to the microarray and the expression of the gene of interest can be detected. Thus the presence of probes(the probe set) complementary to various regions of one distinct gene permits to discriminate between various gene homologues or splicing variants.

Universal probe—It refers to any nucleic acid sequence which detects homologous sequences in a whole variety of different organisms, for example, ribosomal RNA or rDNA which has been conserved during evolution.

RNA probes—RNA probe refers to any RNA which is used as probe in hybridization experiments. Usually such probes are generated by *in vitro* transcription of cloned genes by T7 RNA polymerase and labeled by either radioisotopes, for example, 32P or 125I, fluorochromes or biotin. The RNA probes exist in three versions, plus-sense, minus-sense and anti-sense.

12.8.2 Primers

Primer refers to a short DNA or RNA oligonucleotide which is complementary to a stretch of a larger DNA or RNA molecule and provides the 3′-OH end of a substrate to which any DNA polymerase can add the nucleotides of a growing DNA chain in the 5′-3′ direction. In prokaryotes a specific RNA polymerase (RNA primase) catalyzes the synthesis of such primer RNAs used by DNA polymerase for DNA replication (especially the Okazaki fragments of lagging strand). Primers are also required by RNA-dependent polymerases (reverse transcriptase). In prokaryotes primer synthesis starts at preferred sequences on the template, 3′-GTC (*E.coli* primase, DNAG), 3′-CTG(G/T) (T7 phage primase) whereas in case of eukaryotes sequence preference is uncertain. Primase extends the primer chain to a particular length, 8-12 nucleotides by eukaryotic primase and 10-12 by DNAG protein. The primer RNA which is made with a much higher error rate than DNA is removed upon completion of Okazaki fragments. DNA polymerase can start DNA synthesis from RNA primer without an abortive proofreading process occurring. In other words, 3′-end of RNA annealed to template DNA is not subjected to proofreading by DNA polymerase.

In vitro synthetic primers, usually about 10bp in length are needed for any DNA polymerization reaction using DNA polymerases or reverse transcriptase. Thus they are essential for cDNA synthesis, Sanger sequencing, the PCR, primer extension and similar techniques. The various types of primer employed in PCR are discussed in chapter 31.

Universal primer—It refers to a synthetic oligonucleotide complementary to sequences which are conserved among widely divergent species. It may be used in PCR as primer for amplification of a particular nuclear or organeller gene fragment from nearly all members of a major taxonomic group (e.g. fungi, plants).

A common primer for DNA replication of chromosomal DNA is a short RNA transcript synthesized *de novo* by a primase in both eukaryotic and prokaryotic cells. But there are several ways to make primer other than priming process by primase.

1. RNA polymerase can act as a priming enzyme producing a special kind of RNA primer for initiation of replication of some bacteriophage and plasmid DNA.

2. In bacteriophage *B. subtilis* Φ 29 and animal viruses such as Adenovirus, protein priming promotes the initiation step in which the 3′-OH group of an amino acid

residue within the termination protein is recognized as primer end by DNA polymerase (Hermose et al., 1985).

3. The end of a pre-existing DNA can also serve as a primer.

4. The 3'-OH terminus generated in nicks or gaps of DNA is a common primer for DNA synthesis to complete DNA repair or the recombination process.

Self priming/auto priming—The conventional method of second strand synthesis in cDNA is the self priming. The presence of hair-pin structures (fold-back DNA) at the 3' terminus of the first strand of cDNA is used for DNA polymerase I catalyzed synthesis of second strand. After the synthesis of the first cDNA strand the corresponding mRNA is removed from the cDNA-mRNA hybrid by treating with either alkali or heat. The liberated single stranded cDNA forms a hair-pin loop at its 3' end which is used as primer for the synthesis of double stranded cDNA molecule. The remaining hair-pin can be removed with S1 nuclease. The S1 nuclease frequently removes additional sequences from the 3' end and so some information of the original mRNA may be lost.

In cDNA synthesis the first strand (the DNA strand complementary to the mRNA) is synthesized by reverse transcriptase. Synthesis of first strand can only proceed if a free 3'OH group of primer based to the template mRNA is available. Usually the eukaryotic mRNAs contain up to 200 adenylate residues at their 3' termini to which an oligo (dT) primer can be annealed. After completion of synthesis of first strand, the mRNA is removed by enzymatic or alkaline hydrolysis. It then serves as a template for the second strand synthesis to form a duplex cDNA in a manner described above(through self priming).

RNA priming (Gubler- Hoffman procedure)—It is a modification of the conventional second strand synthesis in cDNA cloning procedure. For RNA priming the mRNA template is partially removed from the first strand-mRNA hybrid by treating with RNase H or alkali which leaves short RNA stretches with free OH groups. These RNAs then serve as primers for second strand synthesis catalyzed by reverse transcriptase.

An accurate knowledge of oligo thermodynamic qualities is of great importance. Knowledge of melting temperature (Tm), free energy of hybridization (GoT) and fraction of oligos hybridized (Fb) greatly facilitates optimization of oligo sensitivity and specificity for a variety of applications including PCR primers and probe design.

12.8.3 Labeling of Nucleic Acid

Labeling refers to introduction of radioactive or non-radioactive markers into DNA, RNA or protein. The labeled nucleic acid is used as probe in DNA-DNA or DNA-RNA hybridization experiments, sequence analysis, mapping of mRNA transcript and foot printing. In case of hybridization analysis **probe** or **cloned gene** is labeled and applied to the chromosome preparations (single stranded DNA molecules). Hybridization occurs between the probe/cloned gene and its chromosomal copy which when autoradiographed results in a dark spot. The position of spot indicates the location of the cloned gene on the chromosome. In case of labeling by fluorescent marker the hybridization between fluorescent labeled probe and its chromosomal copy is observed directly under a special type of light microscope. The technique of fluorescene *in situ* hybridization (FISH) is generally used for detecting chromosomal rearrangements. Rearrangements such as chromosomal duplications or translocations of a segment of one chromosome to another can be detected more quickly than by conventional techniques. The nucleic acid could be DNA (double stranded DNA or single stranded DNA) or RNA (mRNA or tRNA or sRNA). The nucleic acid can be labeled either throughout the sequence or specifically at ends. It can be labeled with either **radioactive** or **non-radioactive material**. The radioactive labeling consists of the introduction of radioactive nucleotides into the probe. The radioactive material commonly used are usually 32p and less commonly 35s or 3H or 14C. Nucleic acid probes when labeled isotopically are detected post-hybridization by autoradiography. As the radioactive materials are hazardous to the researcher the probe may be labeled in a non-radioactive manner. The non-radioactive labeling introduces non-radioactive chemical compounds. The non-radioactive material involves use of biotin-labelled derivatives of dNTPs or digoxigenin-labeled nucleotides or fluorescent dye, etc.

12.8.3.1 Methods of Labeling

Although several methods of labeling are available the most popular are 1. Nick translation 2. End-labelling 3. Random priming.

Nick translation—When the objective is incorporation of radioactivity through out the molecule the method of nick translation can be used for labeling of double stranded DNA using α(33p) dNTPs. In this method the double stranded DNA is treated with a low concentrate of DNase and thus nicks are produced in the DNA molecule. This nicking produces 5′ phosphate ends which serve as substrate for exonuclease activity of DNA polymerase I which initially fills in nicks by addition of (33p) dNTPs to the 3′OH end of nick but then continue to synthesize a new strand, degrading the existing one as it proceeds.

Nick translation is preferred to random primer extension labeling procedure as the size of the generated probe molecules is easier to control and labeling plan be carried out at higher DNA concentrations. The critical size range of probe molecules (smaller than 500bp, preferably 150-to-250bp long) is achieved by empirically varying the amount of the amount of deoxyribonuclease in the nick translation. The size distribution of the probe DNA is verified by analyzing a portion of DNA (denatured) on an agarose gel run under denaturing condition. The intensity and specificity of the hybridization signal is influenced by several factors including the purity of the DNA probe and the size of the probe fragments used in the hybridization reaction.

Copying method—The single stranded labeled probe may be prepared by copying the DNA after cloning into the single stranded phage vector M13.

End-labelling—The labeling generally involves either the 5′-phosphate or 3′-hydroxyl ends of the DNA fragments. End labeling can only be used to label DNA molecules that have sticky ends. End labeling is essential for sequence analysis and restriction mapping. The method of labeling depends on whether blunt or sticky ends are available for labeling and the label is co-valently incorporated. The DNA fragment to be labeled is generated by cleavage with a restriction endonuclease and the label is covalently incorporated so that it produces homogeneous products.

3′ -end labeling—In case of 3′-OH overhang Klenow fragment of *E.coli* DNA polymerase I is used to fill the gap using α (32p) dNTPs. The blunt end of DNA can also be labeled by Klenow fragment of *E. coli* DNA polymerase I. Currently the most effective method of end-labelling 3′-OH overhangs for sequencing involve the use of terminal transferase but with α

(32p) 2′ 3′- ddATP to ensure addition of a single nucleotide. The terminal deoxynucleotidyl transferase adds four ribonucleoside triphosphates to the 3′- OH termini of either double stranded or single stranded DNAs. The product obtained is treated with alkali which removes all but one added ribonucleotide but two labeled phosphate residues at the 5′ ends of the terminal ribonucleotide are retained. If α-cordycepine triphosphate (2′ 3′-dideoxyadenosine triphosphate, ddATP) is used in the reaction then only a single nucleotide is added and in this case there is no need for alkaline hydrolysis.

5′-end labeling—Restriction endonucleases always generate 5′phosphote ends and these ends are dephosphorylated (i.e., 5′p is removed) by alkaline phosphatase and after that ends are labeled. End labeling at the 5′ end is usually achieved by transferring the Y-phosphate residue of Y-32p-ATP using the enzyme T4 polynucleotide kinase. In the second method of labeling 5′ end overhangs the Klenow fragment of *E. coli* DNA polymerase I or T4 bacteriophage polymerase is used to fill the gap in the strand the recessed 3′-OH end using an appropriate α (32p) dNTP.

Random priming—It is a technique to label single stranded DNA molecules to high specific activity(i.e. greater than10^8cpm/ìg DNA) using a mixture of short hexameric primers (hexameric oligonucleotides) which anneal to specific complementary sequences of the denatured target DNA (serving as template) and serve as a primer for the synthesis of the second strand by the Klenow fragment of DNA polymerase I. Radiolabeled nucleotides are added to the reaction mixture and that are incorporated into the growing second strand and thus gets labeled.

Labelling of RNA—There are different ways of labeling RNAs. The DNA put into a special vector can be placed under control of suitable promoter which can then transcribe to produce radioactively labeled RNA in the presence of radioactively labelled33p NTPs. The labeled mRNA can also be produced by producing single stranded cDNA using the α (33p) dNTPs and reverse transcriptase. In case of mRNA the end can be labeled by extension of the 3′ end with E. coli Poly (A) polymerase with α (32p) ATP or α (32p) cordecypin (3′-deoxy ATP) if addition of a single nucleotide is required. mRNA can also be labeled by hybridization of the labeled fragment (3H poly (U)) to suitable region (e.g. poly (A) tails).

Labeling of proteins—**Gold labeling**-It is a method to detect protein *in situ* and uses a colloidal suspension of gold chloride particles of 5-20 nm in diameter. These particles interact electrostatically with proteins and can be detected by light microscopic or electron microscopic techniques. **Immunogold technique**-This technique detects specific protein using gold particles coated with biotin binding protein and can be detected with biotin conjugated antibodies.

Non-isotopic labeling—A variety of methods for labeling and detecting DNA probes by non-radioactive in situ hybridization are available (see chapter 31). This enables one to visualize more than one target sequence in a single hybridization experiment. Probes can be modified with reporter molecule (e.g. biotin) both enzymatically and chemically and detection of bound probes is generally mediated by reporter binding protein (e.g. avidin). Direct coupling of fluorophores to probe molecules has also been accomplished. Haptenated probes, e.g. probes labeled with biotin, digozigenin or dintrophenol have been most frequently detected by direct or indirect immunofluroscence leading to the acronym FISH. Three sets of distinguishable fluorophores emitting the green (fluorescein), red (rhodamine or Texas red) and blue (AMCA or Cascade blue) have been used for FISH to date. Three separate chromosomal DNA sequences have been delineated simultaneously by combining appropriate flurophores with three differentially labeled probes. Combinatorial labeling of probes (i.e. with two or more different reporters) increases the number of targets which can be detected simultaneously by FISH (Reid et al., 1992). Nederlof et al., (1990) extended the number of simultaneously detectable targets to four by double haptenization of one probe molecule by using a combination of chemical and enzymatic labeling procedures. In principle, combinatorial labeling of probes with two or more different reporters can markedly increase the number of targets relative to the number of available fluorophore detectors. For example, with three heptans and three fluorophores a total of seven probes should be resolvable.With four labeling and fluorescent detection system the number of different targets could be increased to 15. But then the practical problem is that the multiplex exposure of the color film can not adequately display and resolve images from combinatorial labeled probes. This problem can be overcome with digital imaging camera system such as silicon intensified tube or charged coupled device (CCD) cameras. Multiparameter hybridization analysis should facilitate molecular cytogenetics probe –based pathogen diagnosis and gene mapping studies. Different reporter/detection systems can be used to facilitate the analysis of several probes simultaneously. To date reporter group/detection systems have been described for probes labeled with biotin, digoxigenin, DNP, bromodeoxyuridine, acetylaminofluororene, mercury, sulphone groups and antibodies for the specific visualization of DNA/RNA hybrids have been prepared.

SP6 *in vitro* transcription system — It is an *in vitro* system for generating large quantities of specific, homogeneous, active and radiolabeled RNA. In this technique, the DNA template is first cloned into the polylinker of a plasmid expression vector which contains a phage promoter located 5-8bp upstream of the polylinker and thus allows the transcription of the cloned insert. After linearization of the vector the transcription is mediated by adding SP6 RNA polymerase (or coliphage, T3, T5 or T7 TNA polymerase) which possess a high degree of specificity for their own promoter *in vitro*. SP6 vector is a cloning vector which contains a promoter from the bacteriophage SP6.

Generally two phage promoters in opposite orientation and separated by the multiple cloning site (dual promoter vectors) are used which allow transcription from either strand of the insert and thus sense or antisense RNAs may be produced. In the SP6 system, SP6 and T7 RNA polymerase promoters on either side of the polylinker are combined. The SP6 *in vitro* transcription system is used to produce single stranded RNA for the analyses of post-transcriptional modifications, *in vitro* translation and RNA sequencing and radioactive probes for nucleic acid hybridization (Kahl, 2004). SP6 sequencing primer refers to the synthetic oligodeoxyribonucleotide 5′-CACATACGATTTAGG-3′ which hybridizes to a conserved 20bp sequence of the SP6 RNA polymerase promoter and permits Sanger sequencing of the double stranded DNA inserted into vectors containing this promoter without the need for subcloning (into, e.g. an M13 vector).

13

Genome Sequencing and Genome Assembly

13.1 SEQUENCING OF GENOME

Sanger and co-workers developed strategy of random shot gun sequencing in the early 1980s. It involves producing random sequence reads, generating a preliminary assembly on the basis of sequence overlaps and then performing directed sequencing to obtain 'finished sequence' with gaps closed and ambiguities resolved. Ansorage and colleagues (1990) extended the technique by the use of 'paired-end sequencing' in which sequencing is performed on both ends of a cloned insert to obtain linking information. Myers and co-workers developed algorithms for using such linking information (Figure 13.1). The steps used by most large-scale sequencing centers in sequence determination include randomly fragmenting the source clone into small (~1500 base pairs) pieces, subcloning the small pieces into a sequence ready vector, sequencing 10 to 30 subclones per kilobase of the source clone and assembling the overlapping sequence reads into a contiguous multiple sequence alignment from which a consensus sequence can be inferred from the highest quality reads. Improvements in sequencing chemistries (better polymerases and high sensitivity dyes) resulted in high quality, more accurate sequence data. Finally, more powerful computer combined with more sophisticated assembly programs have facilitated the delineation of a consensus sequence from a given set of sequence reads.

13.1. 1 Approaches to Sequencing

Shot gun experiments are based on the theory of 'randomly' divide up into manageable fragments and conquer (Wilson, 1999). By sequencing large number of random library inserts, the original shot gun cloned sequence is pieced together by using the overlaps between different clones to order them. Thus shot gun sequencing refers to the determination of the sequence of bases in a complete genome which involves the fragmentation either by physical or enzymatic means, of the target genome, cloning and sequencing of the resulting fragments and the reconstruction of the whole sequence from individual sequences. Besides

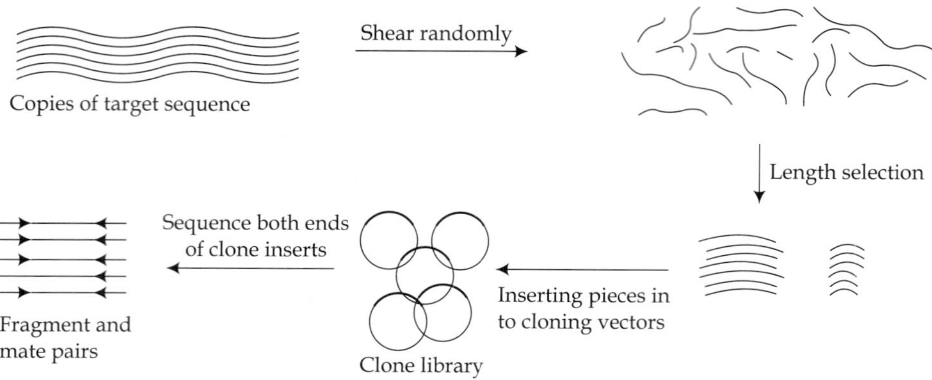

FIGURE 13.1 **Experimental protocal of paired-end shotgun sequencing.**

being used for sequencing entire genome, it can be used for sequencing smaller fragments of DNA such as YAC inserts or cosmid inserts. The practice of sequencing from both ends of double stranded clones ('double-barrelled' shotgun sequencing) introduced by Ansorage and others in 1990 allowed the use of 'linking information' between sequence fragments. Shotgun strategy has remained the fundamental method for large scale genome sequencing for the past 25 years. The application of shotgun sequencing was extended by applying it to larger and larger DNA molecules- from plasmids (~4kb) to cosmid clones (40kb), to artificial chromosomes cloned in bacteria and yeast (100-500kb) and bacterial genomes (1-2Megabases). In principle, a genome of arbitrary size may be directly sequenced by the shotgun method provided it contains no repeated sequence and can be uniformly sampled at random.

Practical difficulties arise because of repeated sequences and cloning bias. Small amounts of repeated sequence pose little problem for shotgun sequencing and one can readily assemble typical bacterial genomes (~1.5% repeat) or euchromatic portion of the fly genome (~3% repeat). By contrast human genome is filled with >50% repeated sequences including interspersed repeats derived from transposable elements and long genomic regions that have been duplicated in tandem, palindromic or dispersed fashion. These include large duplicated segments (50-500kb) with high sequence similarity (98-99.9%) at which mispairing during recombination creates deletions responsible for genetic syndrome. In other words, there are regions, often highly repetitive, that are difficult or impossible to clone (one of the initial steps in a sequencing project) or sequence with current technology. Such regions are expected to contain relatively few protein coding genes. The extent of these regions varies widely in different organisms and so a universal gold standard can not be applied to what constitute a sufficient level of coverage for a particular genome. 1/3 of the sequence of fruit fly was not stable in the cloning system used. Such features complicate the assembly of a correct and finished genome sequence.

There are two approaches of genome sequencing depending upon whether to perform shot gun sequencing on the entire genome at once (whole genome sequencing, WGS) or to first break the genome into overlapping large insert clones and then to perform the shot gun sequencing on these intermediates (hierarchical shot gun). The two approaches are for sequencing large repeat-rich genomes.

1. Whole genome sequencing (Weber and Myers, 1997)
2. Clone-by –clone or BAC-by-BAC approach (Hierarchical shotgun sequencing)(Venter and Smith, 1996)
3. Sequencing of specific chromosomes or chromosome segments

Clone-by-clone approach was adopted by public sequencing project supported by the U.S. National Institutes of Health and Department of Energy and U.K. Wellcome Trust whereas Celera Genomics from U.S., a private sequencing project adopted the sequencing of whole human genome directly by the 'random-shotgun' method. Clone-by-clone sequencing is sequencing applied to 150, 000bp fragments of human DNA (the clones), one at a time. The entire sequence is then reassembled. The 'random-shotgun' approach simple goes on directly to the totality of human DNA by sequencing at random. Why two strategies? It is because that the human genome is full of duplications of DNA sequences and contains about a million interspersed repeats. These repeats are short, near identical copies of the same sequence and they make assembly of DNA sequences difficult as the wrong repeats can get matched together if sequences are being assembled by computer. In the clone-by–clone technique working on one DNA clone at a time simplifies the repeat problem by concentrating on just those which are found in a single clone. The advantage with this method is that useful information can be generated well before completion of the entire sequences. This goal will be achieved by continuing a clone-by –clone approach as before but not completing each 150, 000bp fragment. This will result in a 'draft' sequence that will not be continuous in that each 150, 000bps will be contained in 10, 000-15, 000bp pieces. But then this 'draft' sequence will be less useful than the fully 'finished' sequence. Celera sequenced fruit fly genome using random shot gun approach and was successful because fruit fly genome is smaller, much less rich in interspersed repeats and duplications in comparison to human genome (Little, 1999).

The traditional approach of genome sequencing involves step wise sequence analysis of a minimal tiling path of overlapping clones containing large inserts of DNA and contiguous maps (contigs) are produced from the genomic DNA in larger-insert clone libraries. In other words, it involves a low redundancy sequencing of a contig of BAC clones that cover about 260Mb of the genome. The sequencing of specific chromosomes or chromosome segments is slow and extensive but provides the most precise and complete sequence with a goal of 99.99 % accuracy. The shot gun sequence analysis of small-insert clones is the fastest and least expensive method and it involves 6-fold redundancy (6X) coverage) for the entire genome. Shot gun sequencing alone does not provide locations of the sequenced segments on the genetic or physical genome maps. Both WGS and clone-by-clone approach involve shot gun sequencing.

Whole genome sequencing—It is used for the repeat-poor genomes of viruses, bacteria and flies using linking information and computational analysis to avoid misassemblies.

Hierarchical shotgun sequencing—It is a map-based strategy to sequence genome. It involves generating and organizing a set of large insert clones (typically 100-200kb each), covering the genome and separately performing shotgun sequencing on appropriately chosen clones. Because the sequence information is local the issue of long-range misassembly is eliminated and the risk of short-range misassembly is reduced. One problem is that some large-insert clones may suffer rearrangement although this risk may be reduced by appropriate quality control measures involving clone finger prints. Both these methods must also deal with cloning biases resulting in under representation of some regions in either large-insert or small insert clone libraries. Clone-by-clone approach is modular, allows efficient gap filling, avoids problem arising from the distant repetitive sequences and results in early completion of larger contiguous segments of a genome (Feng et al., 2002). The major rationale for the BAC-by –BAC approach is to make easier the finishing phase of human genome project. A two step process has been used to sequence genome in this approach.

1. Shot- gun sequencing and assembly of random fragments from each clone (Shot gun phase)—In this step first cloning of DNA fragments randomly generated by restriction enzyme from a genome by either physical or enzymatic means is done and then each is cloned onto BAC (insert size of 130-160kb) (and thus a BAC library is constructed) whose relative position on a physical map of genome is known by **genome mapping**. The key to human genome project's strategy (clone-by-clone approach) is the subsequent 'mapping' step in which the BACs are each positioned on the genome's chromosomes by looking for distinctive marker sequences, called STSs whose location has already been pinpointed. In this way the BACs provide a high resolution map of the entire genome. Then the BAC clones in turn are fragmented into pieces of 500-800 bases (in a process called **shot gunning**) which are then subcloned and finally used for sequencing. In other words, the genome is divided into appropriately sized segments and each segment is covered to a high degree of redundancy (typically 8 to 10 fold) through the sequencing of randomly selected subfragnents. Thus the clone based shot gun sequencing refers to the determination of the sequence of bases in a genome which involves the establishment of a collection of large insert clones covering the genome and the shot gun sequencing of each individual clone. Shot gunning the human genome in large clones such as BACs which have a capacity of 80 to 350kbs will do away with the extra overlap that is needed to line up smaller clones, would increase efficiency by about 25%. For even greater efficiency, Venter suggests that the ends of each large clone be sequenced before the whole clone is tackled to ensure no more than a 2kb overlap with any clone already sequenced. It is not the size of the clone but the speed of sequencing which is a major issue. Whether one uses BACs or smaller cones is not going to fundamentally impact the change of scale.

BAC end sequencing (BES)—It refers to the estimation of nucleotide sequence of about 400 nucleotides of one or both ends of a BAC. This technique is used as a first step to estimate the sequence of whole genomes, starting from a few completely sequences BAC clones that are extended into overlapping BAC clones selected from a set of end-sequenced BACs. The BES therefore serves to find overlapping clones in BAC library. Genome mapping is a procedure required for the generation of an ordered clone library which completely represents a genome (or a defined part of it) together with sufficient genetic marker positions (RFPLs) to allow an accurate alignment of both the physical and the genetic map. BAC end sequence tag (BEST) refers to any specific sequence derived from the 5' or 3' terminus of a BAC clone which is unique to this clone and serves as a tag to identify it among thousands of other clones in BAC library.

Tiling path—It refers to the ordered arrangement of BAC or PAC or YAC clones using sequence overlaps of neighboring clones such that they completely cover the corresponding region of the genomic DNA. In case of *S. pombe*, end sequencing and restriction digestion of cosmids were used to construct a **minimal tile path**. In case of nematode both bacterial-based and yeast-based cloning systems were required to achieve contiguity in the clone map. 80% of genome is represented in bacterially-based cosmid clones and the average size of cosmid tiling path (contigs) is 150 to 200kb. YACs (yeast-based clones) bridge the cosmid contigs with only three gaps remaining in the 100-Mb genome (see section 13.5).

BAC shot gun sequencing—In this technique the sequence of bases in a BAC clone is determined by fragmentation of the clone by either mechanically or using restriction endonuclease and subcloning of the fragments. Each sub-clone is then individually but out of context sequenced by Sanger sequencing and the resulting sequences are ordered by superimposing overlaps. The sequence of an individual BAC clone is assembled from the sequences of an 'over sampled' set of subclones (in other words, enough subclones are sequenced to ensure that each part of the original clone is analyzed several times). Finally, the whole genome is assembled by melding together the sequences of a set of BACs that span the genome.

2. Finishing process—It involves closing of gaps and resolution of ambiguities and thus finally construction of the whole sequence from individual sequences. In other words, it is the phase in which the gaps are closed and the remaining ambiguities are resolved through directed analysis. Finishers examine the data, resolve discrepancies between conflicting sequence reads, direct any required and additional sequencing and evaluate the precision of a consensus sequences by comparing its consistency with finished products and sequences of overlapping clones. Finishing is easier when the quality of sequence reads is high (long reads, few errors or ambiguities). However, even with good data there are usually gaps which must be filled and often there are difficulties in the assembly of sequence reads into correct alignment owing to the presence of repeats.

This approach has been used to sequence genomes of *S.cerevisiae, Caenorhabditis elegans* and *Homo sapiens*.

The sequence of insert of a clone, for example, BAC or YAC clone, which is not yet assembled, i.e, not linked to other clones in a tiling path is referred to as **'raw sequence'**. In other words, raw sequence refers to individual unassembled sequence reads produced by sequencing of clones containing DNA inserts.

Paired-end sequence—It refers to raw sequence obtained from both ends of a cloned insert in any vector such as a plasmid or bacterial artificial chromosome. A **draft clone** refers to a large insert clone for which roughly half shot gun sequence has been produced. Operationally, the collection of draft clone produced by each centre was required to have an average coverage of 4-fold for the entire set and a minimum coverage of 3-fold for each clone.

Finished clone—It refers to any large insert BAC or YAC that has been completely sequenced with a 99.99 % fidelity. Clone-by-clone approach starts with identification of tiling path, sequencing of each clone by random shot gun approach, sequence assembly and its verification and collinearity of assembled sequence using ESTs.

Shotgun approach—In this approach the entire genome is broken into random fragments or 'shotgunned', sequenced and then reassembled in one go. This approach was developed by

TIGR (The institute for Genomic Research) -Hopkins group (Venter and Smith). They shot gunned the entire 1800-kb genome of *H. influenzae*. This approach needed more computational power to reassemble the 24,000 fragments once they were sequenced. But the team's own software program, the TIGR assembler reassembled the whole genome. This approach differed from the method developed until then. In the conventional sequencing the genome is laboriously broken down to ordered, overlapping segments, each containing up to 40kilobases of DNA, which are then shattered or shotgunned into smaller pieces. After these smaller fragments are sequenced, they are ordered according to how their sequences overlap and the original segments are used to reconstruct the genome (Nowak, 1995).

13.2   HUMAN GENOME SEQUENCING

Human chromosome can not be sequenced directly. Rather DNA must be isolated, randomly fragmented and cloned into vectors capable of stable propagation in a suitable host such as *E.coli* or yeast. Existing approaches to sequencing the human genome is based on the assumption that each region to be sequenced must first be mapped (**map-based strategies).** The most common approach, **random or shot gun approach** to sequencing involves a three-stage strategy (Figure 13.2) (Venter, Smith and Hood, 1996).

1. **Generation of fragments, clones and clone library** — It involves the construction of three different clone libraries from chromosomal DNA. This is achieved by randomly cutting the DNA into fragments, separating these into differing size classes and then inserting the fragments into distinct cloning vectors capable of propagating them in appropriate host such as bacteria or yeast (see table 13.1). A **clone** here refers to a vector with a single inserted fragment of DNA whereas a **clone library** comprises the entire collection of DNA fragments each inserted into a vector molecule. The **ideal clone library** for genome sequencing should have the following features. I. The clones are highly redundant, covering the entire genome many times. II. The clone coverage is random and not biased toward or against any specific regions of the genome. III. The clones are stable and not subject to deletion or rearrangement during the propagation process. Before sequencing clone must be selected from libraries with chromosomal markers as probes, verified for their fidelity to the genome and ordered in a minimal-overlapping tiling path spanning a portion of a chromosome. Genome should be sequenced from multiple libraries so that no individual's chromosomes are dominantly representing the final sequence. **Clone validation** is judged by internal consistency among overlapping clones assayed by restriction enzyme fingerprinting. The rate of polymorphism in human population is ~one in 500bp with ~15% of the variations being insertions/deletions. Sequence polymorphisms lead to differences in the fingerprints among overlapping clones that can not be distinguished easily from the differences in fingerprints that arise from deletions or rearrangements of artifactual clones and thus one should go for using minimum number of highly redundant clones rather than multiple libraries. A rate limiting step in large scale sequencing is the identification of contiguous array of sequence ready-clones across each chromosome from which a set of minimally overlapping clones (minimal tiling path) can be sequenced. So the challenges in large scale sequencing are as follows.

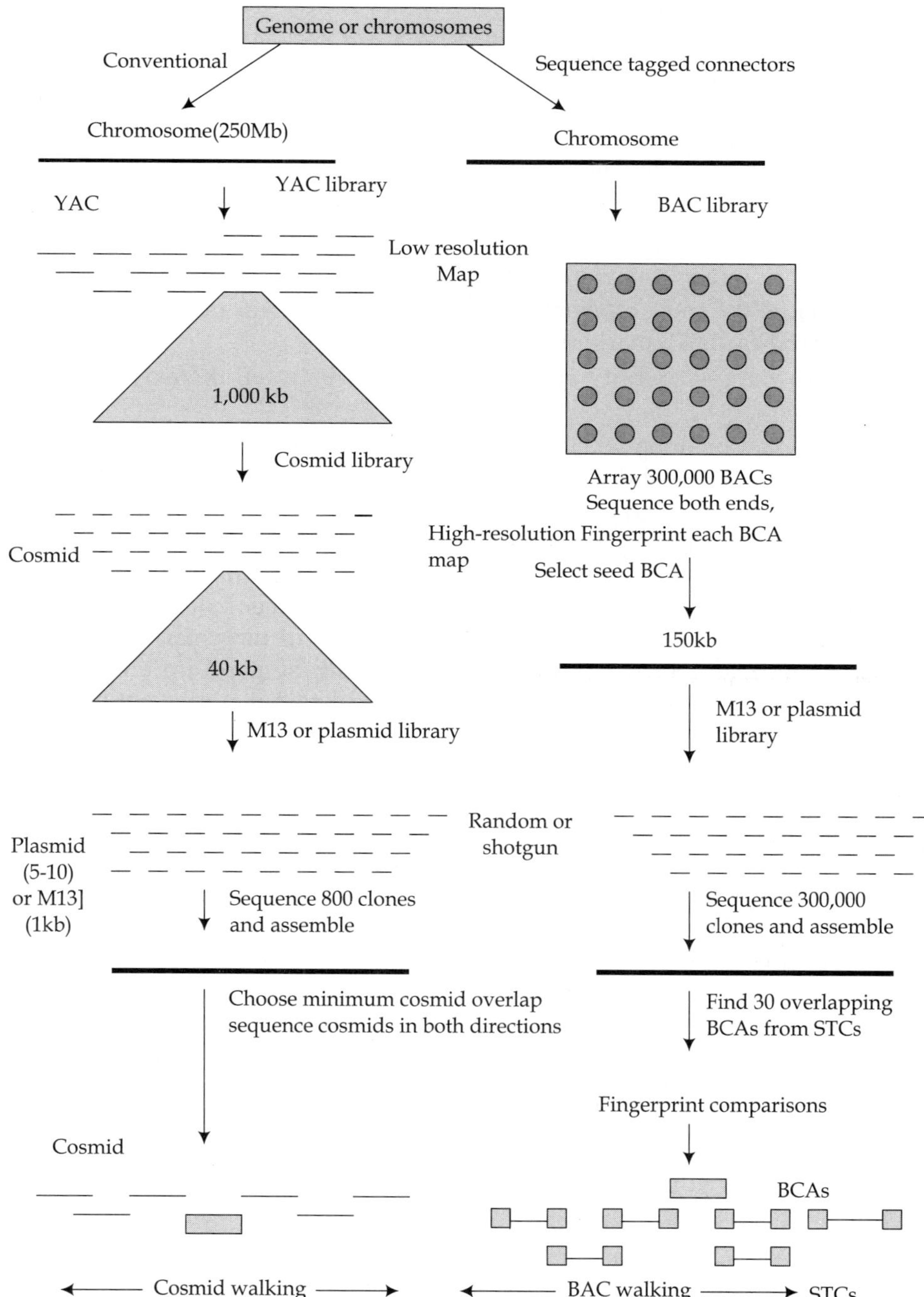

FIGURE 13.2 **Showing conventional sequencing approach and sequence-tagged connector approach (After Venter, Smith and Hood, 1996).**

Table 13.1 Showing vectors, the insert size they can carry and the number of clone required.

Vector	Human DNA insert size	No of clones required to cover the human genome
YAC	100-2000kb	3,000(1000kb)
BAC	80-350kb	20,000(150kb)
Cosmid	30-45kb	75,000(40kb)
Plasmid	3-10kb	600,000(5kb_
M13 phage	1kb	3,000,000(1kb)

I. What is the most efficient means of obtaining local minimum tiling path (e.g. around each STS marker)?

II. How one can identify and cover the gaps between the existing chromosomal markers?

III. What can be done when a clones are missing from a highly redundant library because of non-random coverage or sequence specific-instability?

Minimum tiling paths of contiguous clones—The minimum tiling paths can be built using STSs or genetic markers already mapped to a 200kb to 2Mb region of a chromosome. With these markers as probes clones are selected from a library and ordered by restriction enzyme fingerprints. To obtain a minimum tiling path, gaps between contigs must be filled by additional round of library screening after identification of new markers from the contig ends. As the rounds of library screening required to fill the gaps are time consuming the construction of physical maps which contain minimal tiling paths may not be able to match the required throughput of sequencing. The up-front characterization of a highly redundant BAC clone library would be an automable approach to the construction of minimum tiling paths (Venter and Smith, 1996).

2. **Construction of physical maps**—Construction of low-resolution physical maps of each chromosome by identifying shared landmarks such as unique sites which can be amplified by PCR (sequence tagged sites, STSs or restriction enzyme digestion sites) on overlapping YAC clones. The high-resolution or sequence ready maps are then constructed by randomly cutting and subcloning YAC inserts into cosmid vector and then a map is constructed by identifying their landmark overlaps. After that a minimally overlapping path of cosmid clones is chosen and the DNA from each clone is randomly fragmented into small pieces and subcloned into M13 phage vectors. Celera Genomics's approach was to prepare small insert clones directly from genomic DNA rather than from mapped BACs. Celera Genomics used a mixed strategy involving combining some coverage with whole genome shot gun data with hierarchical shot gun data. If the raw sequence reads from whole genome shot gun components are made available then it may be possible to evaluate the extent to which the sequence of human genome can be assembled without the need for clone-based information. Mapping step was unnecessary (Myers and Weber, 1997) and they said that algorithms used to reassemble shot gunned DNA fragments could be applied to cloned random fragments taken from the genome as a whole.

3. **Sequencing of each clone**—Sequencing of insert of M13 phage vector is done by Sanger sequencing techniques. Thus for each cosmid clone, about 800 M13 phage

clones are sequenced (roughly 400 base pairs per clone) and assembled into the sequence of the 40kb cosmid insert. This approach ensures a high degree of accuracy as every nucleotide is sequenced about 8 times (400 bases per clone x 800 = 320, 000 bases of sequence). Before sequencing selection of clones from the library using chromosomal marker as probes is made, verified for their fidelity to the genome and ordered in a minimal-overlapping tiling path spanning a portion of a chromosome. **Standards for sequencing** — The current standards for sequencing set are: I. Error rate of not more than 1 in 10,000. (The current error rate is not more than 1 in 100,000). II. Sequence contiguity, i.e. sequence without gaps III. Clone validation is a demonstration that clones faithfully represent the genome. **Precision of sequencing** — I. Probable error is estimated by comparing the sequences of overlapping clones II. Assign a quality measure to each base in a consensus sequence III. Contiguity is judged by sequence assembly program and by the successful overlapping of sequences from adjacent clones IV. Finishing product comparisons of the clones against other overlapping clones and in some cases directly against genomic DNA verifies clone integrity and validity.

The limitations with this approach are as follows.

1. It is difficult to obtain complete maps without any gaps.
2. Some 50% of YAC clones show structural instability of inserts resulting in deletions or rearrangements of portions of cloned DNA or are chimeras in which two or more DNA fragments have become incorporated into a clone and thus unsuitable for mapping and sequencing. Cosmid inserts sometimes contain the similar aberrations and are similarly difficult to detect.
3. Human genome contains tandem (adjacent) arrays of DNA units with high sequence similarity and tandem arrayed dispersed repeats and poses problems for high resolution mapping when the size of the clone insert is less than that of the tandem array because the landmarks are similar such as a 40kb cosmid insert against a 105kb DNA array.
4. Conventional sequencing procedure is complex and difficult to automate fully.
5. Requires extensive infrastructure for high resolution physical mapping and collaboration among large and small groups.

Clone libraries used for genome mapping and sequencing — Considering the average size of the cloned insert as 20kb, the number of clones required in a genomic library of human genome will be

$$n = 2.8 \times 10^6 / 20 = 1.4 \times 10^5$$

The number of independent recombinants required in the library must be greater than n. This is because the sampling variation will lead to the inclusion of some sequences several times and exclusion of other sequences in a library of just n recombinants. Clark and Carbon (1976) derived a formula which relates the probability (P) of including any DNA sequence in a random library of N independent recombinants.

$$N = \ln (1\text{-}P) / \ln (1\text{-}1/n)$$

Thus in order to achieve a 95% probability (P = 0.95) of including any particular sequence in a random human genomic library of 20kb fragment size

$$N = \ln (1\text{-}0.95)/ \ln (1\text{-}1/1.4 \times 10^5) = 4.2 \times 10^5$$

If the probability is to be increased to 99%, then the number of recombinants (N) required becomes 6.5×10^5. Thus a three-fold coverage gives a 95% probability of including any sequence whereas a five-fold coverage gives a 99% probability. These calculations are based on assumption of equal representation of sequences which is not true in practice.

The new approach is called '**sequence tagged connectors (STC) approach**- It is an approach to sequencing whole genome. Two advances such as the development of BAC libraries that can accept inserts of up to 350kb and carry large ~150kbp inserts stably and BAC clones proving excellent substrate for shot gun sequencing and thus BAC clones appearing ideal for producing an accurate contiguous sequence and advances made in sequencing and assembling of prokaryotic genomes led to development of this new approach. Cosmid clones which served the basis for sequencing of genomes of yeast and nematode, *C. elegans* are less stable and carry much shorter ~35kbp inserts. This new approach eliminates the need for any prior physical mapping and uses BAC clones as the basic sequencing reagents (Figure 13.2). The procedure starts with construction of a BAC library with an average insert size of 150kb and about 15-fold coverage of the human genome contains 300,000 clones. These clones are arrayed into microtitre wells. Both ends (starting at the vector-insert points) of each BAC clone are then sequenced to generate 500 bases from each end. The generated 600,000 sequences are scattered roughly every 5kb across the genome and make up 10% of the genome sequence. These end sequences of BAC are termed '**sequence tagged connectors**' or STCs as they allows any one BAC clone to be connected to about 30 others (for example, a 150kb insert divided by 5kb will be represented in 30BACs). STCs are ideal potential chromosomal markers for creating a more dense physical map. STCs could be localized to human chromosomes by radiation hybrid mapping at an average spacing of 100kb. This would facilitate the identification of gaps in the tiling path so additional markers in the gaps regions could be identified.

Each BAC clone is then fingerprinted using one restriction enzyme to provide the insert size and detect artifactual clones by comparing the fingerprints with those of overlapping clones. A seed BAC of interest is sequenced and checked against the database of STCs to identify the 30 or so overlapping clones. Two BAC clones showing internal consistency among the fingerprints and minimal overlap at either end are then sequenced. In this way the entire genome could be sequenced with just over 20,000 BAC clones.

Assembling of sequences—It refers to reconstruction of assembled DNA sequence with the proper order and orientation.

Genome assembly strategy—There are two approaches to genome assembly. The first approach involves computational combination of all sequence reads with shredded data from GenBank to generate an independent and unbiased view of genome. The second approach involves clustering of all the fragments to a region or chromosome on the basis of mapping information. This approach provided fewer gaps and slightly greater sequence coverage. And finally, comparison of completeness and correctness with public genome sequence reconstructed by BAC-BAC approach.

Whole genome assembly (WGA)—WGA uses Celera data and PFP data in the form of additional synthetic shot gun data. Given a set of reads randomly sampled from a target sequence, the order and the position of these reads in the target is reconstructed. The Cerela assembly consists of a set of contigs that are ordered and oriented into scaffolds which are then mapped to chromosomal locations by using known markers. In other words, in this 'whole genome shot gun' strategy fragments are first assembled by algorithms into larger scaffolds. The correct position of these scaffolds on the genome is then worked out using STS. The **contigs** consist of a collection of overlapping sequence reads which provides a consensus reconstruction for a contiguous interval of the genome. In other words, contig is the result of joining an overlapping collection of sequences or clones. **Mate pairs** are a central component of the assembly strategy. Mayers described the analysis of *'mate pairs'* which correspond to sequences from either end of an individual cloned DNA segment. The lengths of these segments are known (Figure 13.3). So if mate pairs subsequently end up the wrong distance apart within the final sequence, it suggests a problem with the assembly. They are used to produce scaffolds in which the size of the gaps between consecutive contigs is known with reasonable precision. This is accomplished by observing that a pair of reads, one of which is in one contig and the other of which is in another contig, implies an orientation and distance between the two contigs (Figure 13.4).

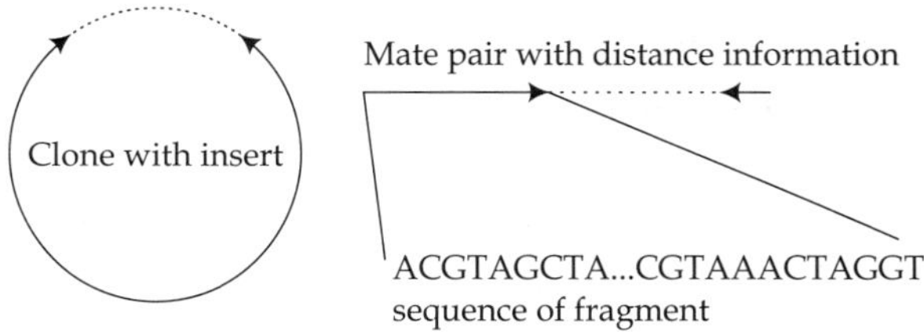

FIGURE 13.3   Fragments and mate-pairs.

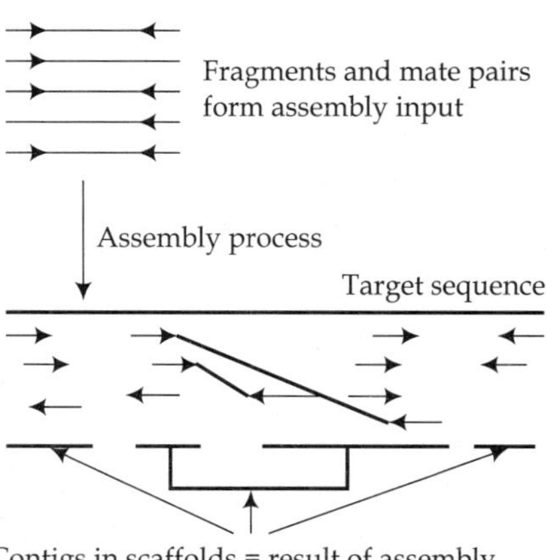

FIGURE 13.4   Pair-end reads are assembled into contigs based on how the reads overlap with each other. The contigs are then organized into scaffolds using the mate-pair information.

Compartmentalized shot gun assembly (CSA) — CSA is a localized assembly approach. It subdivides the genome into segments, each of which could be shot gun assembled individually. The CSA process first partitions the Celera (random shot gun data) and PFP (publicly funded human genome project data primarily based on BAC clones) data into sets localized to large chromosomal segments and then performed *ab initio* shot gun assembly on each set. In other words, the data are partitioned into the largest possible chromosomal segments or components which could be determined with confidence and then shot gun assembly is applied to each partitioned subset wherein the bactig data are again shredded into faux reads to ensure an independent *ab initio* assembly of the component. Thus, construction of components from the largest scaffolds of the sequence from each BAC and assembled scaffolds of the data unique to Celera's dataset was carried out. In other words, one put the genome together region by region making some use of mapping information. In comparison to WGA the CSA assembly was found a few percentage points better in terms of coverage and slightly more consistent. Further, CSA improves the statistic for calculating U-unitigs.

Anatomy of WGA is shown in (Figure 13.5). Overlapping shredded bactig fragments and internally derived reads from five different individuals are combined to produce a contig and a consensus sequence. Contigs are connected into scaffold. A **scaffold** is a set of contigs that are ordered, oriented and positioned with respect to each other by mate pairs whose reads are in adjacent contigs (Figure 13.5). In other words, scaffold is the result of connecting contigs by linking information from paired-end reads from plasmids, paired-end reads from

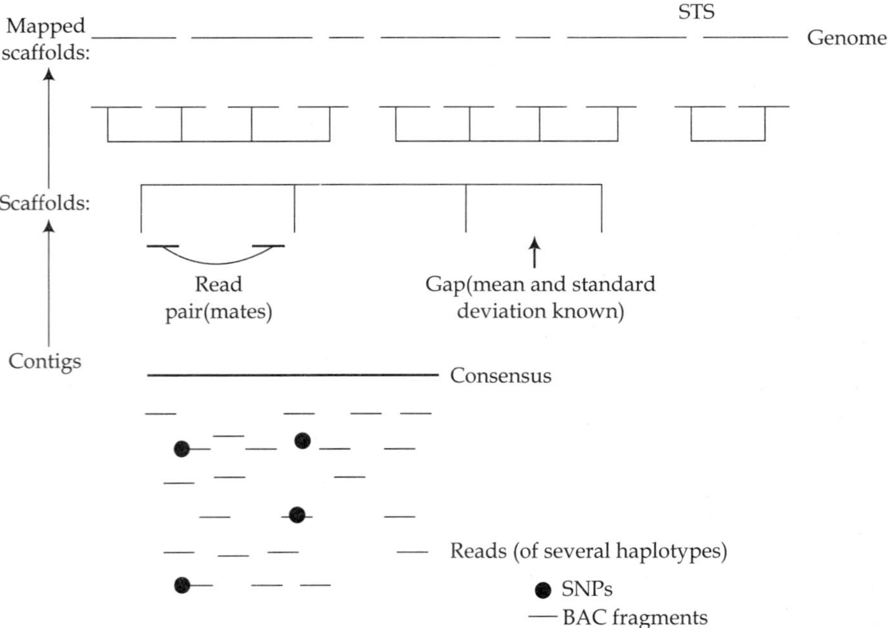

FIGURE 13.5   **Anatomy of whole-genome assembly (adapted from Venter et al., 2001).**

BACs, known mRNAs or other sources. Contigs separated by gap of known length (i.e. with approximately known distances between them). In other words, a scaffold is a collection of ordered contigs with approximately known distances between them. A scaffold is produced by joining sequenced clone contigs on the basis of overlapping sequence information. Scaffolds are then mapped to the genome with STS physical map information.

WGA assembler—WGA assembler consists of a pipeline composed of five principal stages.

1. Screener
2. Overlapper
3. Unitigger
4. Scaffolder and
5. Repeat resolver.

Screener—It finds and marks all microsatellite repeats with less than 6bp element and screenouts all interspersed repeat elements including Alu, *LINE* and ribosomal DNA.

Overlapper—It compares every read against other read in search of a complete end-to-end overlaps of at least 40bp and with no more than 60% differences in the match. The primary difficulty in building an assembler for a whole genome shot-gun data set is to develop an algorithmic process which detects and is not confused by stretches of repeats. The key to not being confused by repeats is the exploitation of mate pair information to circumnavigate and to fill them.

True vs repeat overlap—When fragments are sampled from overlapping segments of the genome they belong together in an assembly, a true overlap. When overlapping portion is a part of a repeated sequence which occurs multiple times in the genome and then the two reads do not belong together, it is a repeat overlap. Because of this repeat the result of Cerela assembly is a set of scaffolds of contigs versus a set of contigs as customarily produced by other assembler. The clone-end pairing information is used to construct scaffolds-sets of nonoverlapping contigs linked together in the correct ordered and orientation. The assembler thus must avoid choosing repeat induced overlap especially early in the process.

Unitigger—It generates a set of correctly assembled subcontigs. Contigs are formed from these subassemblies unitigs(uniquely assembled contigs). Contigs are built from **U-unitigs** which form a scaffold via **bundles** and then have a series of **rocks, stones** and **pebbles** filled into the gaps between them where possible.(Myers et at., 2000) (Figure 13.6). In other words, contig is a series of overlapping **U-unitgs**. Collections of fragments whose arrangement is uncontested by overlaps from other fragments are assembled into what is called **unitigs**. In other words, unitig refers to a set of genomic sequence reads assembled into a single contiguous sequence such that no fragment in this unitig overlaps any fragment not present in the unitig. Each unitig is assessed as to whether it represents unique or repetitive sequence. Unitig representing unique DNA(probably genes) is designated as U-unitig. Potential boundaries of repeat sequences are sought at the tips of the U-unitigs and those found are used to extend U-unitig ends as far as possible into a repeat (Figure 13.7). Logarithm of the odds ratio that a unitig is composed of unique DNA or of a repeat consisting of two or more copies is the statistical discriminator. This identifies a subset of the unitigs and further identifies a subset of remaining unitigs very likely to be correctly assembled of which one selects those that will consistently scaffold. The union of these sets is called U-unitigs.

Scaffolder—It uses mate-pair information to connect these contigs together into scaffold. A scaffold is a collection of ordered contigs with approximately known distances between them. All possible U-unitigs with mutually confirming pairs of mates or BAC ends are linked into scaffolds, consisting of a set of ordered, oriented contigs for which the size of the intervening gaps is approximately known. So, link together all U-unitigs which are linked by at least two or 10 kbp mate pairs producing intermediate sized scaffolds that are then recursively linked together by confirming 50kbp mate pairs and BAC end sequences. When the left and right reads of a mate are in different unitigs, their distance relation orients the

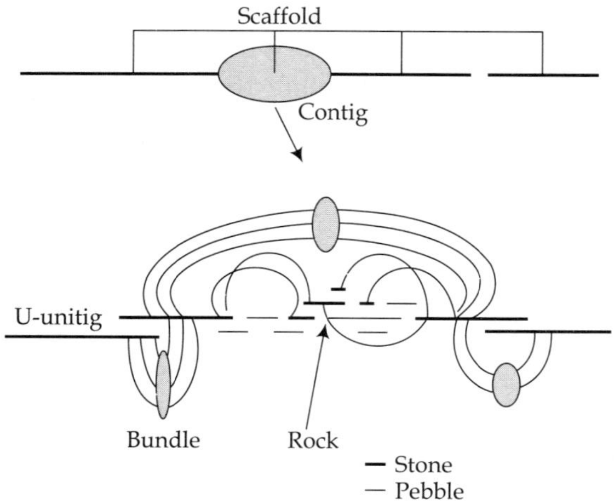

FIGURE 13.6   Showing anatomy of a scaffold (After Myers et al., 2000).

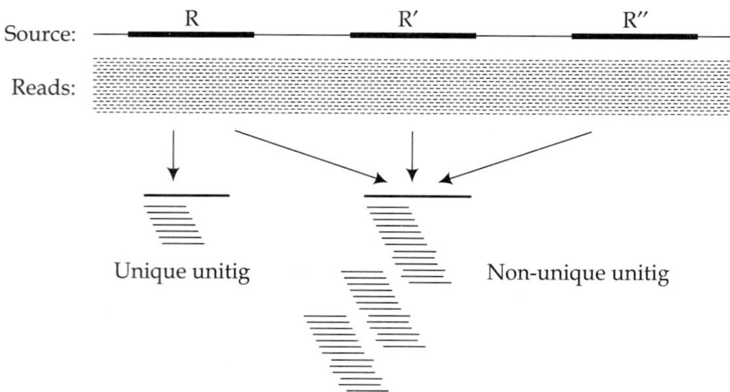

FIGURE 13.7   A unitig represents a chain of consistently overlapping reads. However, a unitig does not necessarily represent a segment of unique source sequence. For example, its fragments may come from the interior of the different instances of a long repeat, as shown here. R, R′ and R″ represent three instances of the the same repeat.

two unitigs and provides an estimate of the distance between them. This process yields scaffolds which are on the correct order of megabase pair in size with gaps between their contigs that generally correspond to repetitive elements and occasionally to small sequencing gaps. These scaffolds reconstruct the majority of the unique sequence within a genome.

Firm scaffolds — A scaffold can be described as firm if it contains at least one U-unitig. By definition all scaffolds which are not firm are unitigs with A statistic less than 10 and without exceptions these unitigs are i. unrelated to the firm scaffold by either link or overlap relations ii. localized to repeat-induced gaps in the firm scaffolds or iii. pebbles that are relevant but not used in late stage repeat solution.

Bundles — All sets of U-unitigs which are consistently ordered and placed by confirmed bundles, that is, containing 2 or more 2kbp or 10kbp links are assembled into a scaffold of contigs where a contig at this stage is a series of overlapping U-unitigs. Thus bundles of mate pairs and overlaps consistently place contigs relative to each other.

A hypothetical genome consisting of three unique stretches A, B and C with two nearly identical interspersed repeats, X′ and X″ will result in the four unitigs (A, B, C and X′+X″) and overlaps as shown in the Figure. The unitig (X′ + X″) is overcollapsed and the U-unitigs for regions A, B and C have repeat boundaries indicating the tail portions which extend into X. Unitigs with an A-statistic greater than 10 are termed 'U-unitigs' as they almost certainly represent unique DNA sequence in the genome that has been assembled. A gap within a scaffold is called a **sequence gap** and a gap between scaffolds is called a **physical gap** (no clones available that span the gap). In other words, **sequence gap** refers to any gap between adjacent sequences clone contigs in a draft genome sequence which does not represent a fingerprint clone contig gap. The fingerprint clone contig gap refers to any gap between adjacent **fingerprint clone contigs** (FCCs) in a genome. The FCCs refer to any two contigs with matching(overlapping) sequences that are inferred from the restriction digest fingerprints. In other words, it refers to contigs produced by joining clones inferred to overlap on the basis of their restriction digest fingerprint. The **fingerprinted contig** (FPC) refers to one of a multitude of BAC clones in a BAC library that is characterized by BAC fingerprinting, i.e the unequivocal assignment of a specific restriction pattern to a particular clone employing 2 or better 3 double digest, for example, HindII/HaeIII, HindIII/Dpn I, HindIII/RsaI. The number of gaps increases linearly with the target size and so is the number of interspersed repetitive sequences which tend to found the assembly. The bundle implies a small amount of overlap between two contigs because it is actually very short whereas the reality is that there is a small gap at that location. Both intra- and interscaffold gaps are filled by a series of three levels of repeat resolution. The rock phase places unitigs that are consistently positioned by at least two mate pairs. Rocks are unitigs which have a positive a positive A-statistics and have either two mate links that consistently link it to contigs on one or both sides or four or more links The stone phase places unitigs that are positioned by a single mate pair and confirmable by an overlap tiling across the gap containing it and the pebble phase attempts to find the best tiling across gaps using a quality-value based measure of significance.

Sequenced clone layout — It refers to the assignment of sequenced clones to the physical map of fingerprint clone contigs.

Initial sequence contigs—It refers to contigs produced by merging overlapping sequence reads obtained from a single clone, in a process called sequence assembly.

Merged sequence contigs—It refers to contigs produced by taking the initial sequence contigs contained in the overlapping clones and merging those found to overlap. These are also referred to simply as '**sequence contig**' where no confusion will result.

Sequence contig scaffold—It refers to scaffolds produced by connecting sequence contigs on the basis of linking information.

Sequenced clone contig—It refers to contigs produced by merging overlapping sequenced clones. **Sequenced clone contig scaffolds**- It refers to scaffolds produced by joining sequenced clone contigs on the basis of linking information.

Estimation of A-statistic—Assuming F fragments in a database and estimated genome size as G, for a unitig with k fragments and distance p between the start of its first fragment and the start of its last fragment, the probability of finding the k-1 start points in the interval of length, p, given unitig is not oversampled is $[(Pf/G)^K/K\exp(-p\ F/G)$. If the unitig is the result of collapsing two repeats then the probability is $[(2pF/G)^K/K]\exp(-2p\ F/G)$. The log of ratio of these two probabilities $((\log e)p\ F/G-(\log 2)k$ is the A-statistic. Unitigs with an A statistic greater than 10 are termed 'U-unitigs' as they almost certainly represent unique DNA in the genome that has been correctly assembled.

After computer assembly the finishing phase follows.

Repeat resolver—Here both intra- and interscaffold gaps are filled. The WGA assembler is engaged in following three stage repeat resolution strategy.

1. 'Rocks' substage- Where all unitigs with a good but not definitive discriminator score are placed in a scaffold. In other words, the rock-phase places unitigs which are consistently positioned by at least two mate pairs.

2. 'Stone' substage- For each gap every read R that is placed in the gap by virtue of its mated pair M being in a contig of the scaffold implying R's placement is collected. Almost every but all of the reads in the set belong in the gap and when a read does belong it rarely agrees with the remainder of the reads. In other words, the stone phase places unitigs which are positioned by a single mate pair and confirmable by an overlap tiling across the gap containing it. Final method to resolve gap is to fill then with assembled BAC data which cover the gap and this method is called external gap 'walking'.

3. 'Pebble substage- This phase attempts to find the best-tiling quality across gaps using a quality-value based measure of significance. It was not included in Human genome analysis as was done in Drosophila because it produced repeat reconstructions for long interspersed element whose quality was only 99.62%.

   At the final stage of assembly process and also at several intermediate points a consensus sequence of every contig is produced. The algorithm used is driven by the principle of maximum parsimony with quality value-weighted measures for evaluating each base. The net effect is a Bayesian estimate of the correct base to report at each base position.

In case of human genome two unfinished 'draft' sequences have been produced by different methods, one by International Human Genome Sequencing Consortiun (IHGSC) and one by Cerela Genomics (CG).

IHGSC procedure involved the following steps.

1. It started with BAC clone-based physical map of the genome.

2. A map was constructed by digesting each clone with restriction enzymes and deriving a characteristics pattern or fingerprints.

3. All fingerprints were then processed by FPC which generated BAC clone contigs on the basis of shared fragments in their fingerprints (Soderlund et al., 2000).

4. A selection of clones from this map covering the vast majority of the genome were than shot gun sequenced (Sanger et al., 1982).

5. The fragments of each clone were than assembled into initial sequence contigs based upon overlaps between shot gun sequencing reads.

The collected initial sequence contigs from a single clone made up the sequence data for a BAC clone in GenBank. As more shot gun sequencing of the clone has been carried out the initial sequence contigs are reassembled with new sequences and the database entries for the clone is updated accordingly until the sequence of the clone is finished and is represented by a single contig of 100-200kb. The program, Phrap was used to assemble the initial sequence contigs and it takes into account the sequencing quality estimate for each base.

CG procedure used the whole genome shot gun method in which the entire genome is randomly fragmented and each of the cloned fragment is sequenced (Venter et al., 2001). Sequences from these cloned fragments are produced as mate pairs: 15–800bp sequencing reads from either end of the clone with relative orientation and approximate spacing. A mixture of clones of different sizes was used (2, 10, 15 and 100kb).

In both cases the strategy was to merge overlapping sequences into contigs and then to order contigs relative to one another using various types of mapping data. Both groups had to address the problem of assembling incomplete data.

Gig assembler—IHGSC used a program called 'Gig–Assembler' produced by the university of California, Santa Cruz (UCSC). Gig assembler starts with initial sequence contigs from GenBank at a given point (called a 'freeze' dataset). All sequences were repeatmasked using repeatMasker. With each IHGSC physical map contig the initial sequence contigs from BAC clones belonging to it were assembled into '**raft**' sequences using the sequence overlaps between fragments. The first joins were made between the best matching fragments. These rafts were ordered and oriented relative to one another using bridge sequences from other sources (mRNA, EST, plasmid and BAC end pairs) and FPC contig data.

CG assembly—The draft genome assembly was carried out by a compartmentalized shot gun assembler (CSA) (Huson et al., 2001) using both CG data and IHGSC initial sequence contigs from GenBank fragmented into smaller sequences (a few hundred base pair long). CSA involved the following steps.

1. It started by comparing all CG mate-pair fragments with all the initial sequence contig fragments and avoiding matches based upon repetitive sequence. Repetitive sequence was identified using comparisons to a library of unknown repeat and also by additional procedures.

2. The mate-pair fragment pairs matching more than one initial sequence contigs were then used as bridging sequences to order and orient the initial sequence contig fragments within and between BAC clones. The paired CG fragments are used as high

resolution mapping data to reassemble both IHGSC and BAC sequences and the broader genomic regions they originate from. The result is a set of 'scaffolds' consisting of ordered, oriented sequence contigs separated by gaps of estimated sizes. CG fragments not matching IHGSC initial sequence contigs were also assembled using a different algorithm (Myers et al., 2000) to provide additional scaffolds containing sequence not represented in IHGSC data.

3. Scaffolds are then positioned relative to one another based upon sequence overlaps and bridging mate-pair fragments.

4. The derived order of scaffolds was then manually curated to identify mistakes by combining sequence alignments by eye and confirming or rejecting orders based on physical mapping data such as those from IHGSC.

Third assembly method — It used repeat masked data from IHGSC produced by NCBI using a computational tool based on the BLAST algorithm. This method starts by finding an order for adjacent BACs but in this case it was derived from BAC sequence overlaps (detected using a variant of BLAST), FISH chromosome assignment and STS content. The sequence fragments from these overlapping BACs were then merged into consensus '**meld**' sequences. Melds were then ordered and oriented based on ESTs, mRNA and paired plasmid reads before being combined into a single genomic sequence contigs with melds separated by runs of letter N. NCBI contigs were ordered and oriented relative to one another according to matches to mapped STS markers and paired BAC end sequences.

Differences between assemblies — From the above it is clear that the protocols used by the different assembly methods differ in terms of amount and variety of inputs and so differences between assemblies will result in differences in finished sequences. Semple et al.(2002) found variable amounts of tandemly duplicated and interspersed repeat sequence between UCSG, NCBI and CG derived assembly. Further, large difference between genes found in CG and UCGS assemblies such that one-third of the genes identified in one assembly were not found in the other (Hogenesch et al., 2001)

Completeness of assembly — It is defined as the percentage of euchromatic sequences represented in the assembly and is measured on the basis of i. the estimated sizes of the intrascaffolds ii. coverage of the two published chromosomes ii. analysis of the percentage of an independent set of random sequences (STS) markers contained in the assembly. **Correctness of assembly** — It is defined as the structural and sequence accuracy of the assembly. Structural consistency is measured by mate-pair analyses. In a correct assembly every mated pair of the sequencing reads should be located on the consensus sequence with correct separation and orientation between the pairs. A pair is valid if the reads are in the correct orientation and the distance between them is with ± 3 SD of the distribution of the insert sizes of the library from which the pair was sampled. The pair is otherwise ' misoriented' if the reads are not correctly oriented and 'misseparated' when the distance between the reads is not in the correct range. The consensus sequence of the assembly is then compared against other finished sequence for determining sequence accuracy.

The publicly available human genome sequence is still categorized as draft or unfinished. Relatively small but problematic regions of gapped draft may persist. Gaps arise because of a number of reasons. Certain regions of the genome are simply not present within existing clone libraries and are also recalcitrant to subcloning. It thus requires specialized techniques to close such gaps. Also, a fraction of the genome (perhaps 5%) consists of large (>10kb) duplicated segments with 90–98% sequence identity. Regions containing such duplicated segments are difficult to assemble and found not only in pericentric and subtelomeric region but also across the genome including the gene-rich regions.

The finished genomic DNA sequence refers to any contiguous DNA sequence generated by genomic sequencing which has an accuracy of at least 99%. **Finished sequence** refers to complete sequence of a DNA, for example, a BAC or YAC clone or a genome which has been determined with error less than 1/10,000 bases and does not contain any sequence gap. There are many definitions (Bork and Copley, 2001): one definition is that fewer than one base in 10,000 is incorrectly assigned, second definition is that more than 95% of the eukaryotic regions are sequenced and a third definition is that each gap is smaller than 150kb. In the finished genomic sequence the sequences are placed in the correct order and orientation along a chromosome. Finishing a sequence requires to sequence each and every base for 8–10 times. As the gaps are difficult to close a genome sequence is usually not finished in the above sense and because of this reason the term finished sequence is commonly replaced by the slang term **'done sequence'**. Finishing is the final phase of DNA sequencing in which the complete sequence is assembled by filing gaps and alignment of sequences.

Finishing rules — The following general rules must be kept in mind while sequence finishing.

1. All regions must be sequenced either on each strand or with dye primer and dye terminator technique with which extensive comparisons have shown to be at least as reliable as double stranding in revealing and correcting compressions and other base calling errors.

2. All regions must be represented by reads from two or more independent subclones or from PCR products across the region.

3. If subcloned PCR products are used for a region then three independent clones must be sequenced.

4. Rare exceptions to the general rules of double stranding or alternative sequencing are permitted on the basis of the following. For regions of <50 bases where despite efforts, a finisher is unable to achieve double stranding or double chemistry, the sequence may be submitted provided the sequence is of high quality. When editing in XGAR all sequence data must be resolved at the 75% consensus level either by collection of additional data or by the editing of **poorly** called traces. In CONSED any consensus base with a quality <25% must be manually reviewed to determine are the available data sufficient to unambiguously support the derived sequence.

Each finished sequence is submitted to a series of quality tests including verification of all the finishing rules and verification that the assembly is consistent with all restriction digest. In addition, every finished sequence undergoes an automatic process of base calling and reassembly with different algorithms than those that were used for the initial assembly and comparison of the resultant consensus by a banded Smith-Waterman analysis against the sequence that was obtained by the finisher. Finally, any discrepancies in assembly or sequence along with any regions failing to meet the finishing criteria, are manually reviewed. The sequence is passed on for annotation only when all discrepancies are accounted for. The term **annotation** means obtaining biological information from unprocessed sequence data.

There are two kinds of annotation- **structural** and **functional annotation**. Structural annotation means the identification of genes and other important sequence elements and functional annotation means the determination of their functional roles in the organism. If annotation indicates any suspicious regions then these are again passed back to the finisher for resolution either through additional data collection or editing.

Finished high quality sequence is required for the following three reasons.

1. The ability to determine gene function is highly dependent on having accurate sequences as most of the essential regulatory sequences fall outside the transcribed regions

2. Comparative genomics — In case of a model plant for cereal grasses the complete rice sequence will directly affect what can be accomplished with other cereal grasses

3. Identification of genes responsible for agronomic traits of economic importance requires precise map based genomic sequences

Draft genome sequence — It refers to the preliminary sequence information of individual clones, by creating merged sequence contigs and exploiting overlap sequence information to establish scaffolds and positioning the resulting sequence along the physical map of the individual chromosomes (called **'golden path'**). In other words, draft genome sequence refers to the sequence produced by combining information from the individual sequenced clones (by creating merged sequence contigs and then employing the linking information to create scaffolds) and positioning the sequence along the physical map of the chromosome. Golden path is a term applied to the first and subsequent assemblies of human genome.

Working draft genomic sequence — It refers to a set of DNA sequences representing about 90% of the genome in question. The mapped clones are first shot gun sequenced and the data are assembled in a **'draft sequence'** (covering most of the regions of interest but still containing ambiguities and gaps). The gaps are then filled and a finished genomic is sequence produced. The working draft of the human genome combines both finished genomic DNA sequence and draft sequence where the later has an average length of about 15kb with about one sequencing error per 5000bases.

Rough draft — It refers to an incomplete sequence of any genome, interrupted by gaps, most of them containing repetitive DNA. The presence of too many gaps makes it impossible to order and orient the relatively small runs of bases which are the raw products of genome sequencing. Thus a genome sequence can not be said to be finished completely if even one single gap exists and it is rather called a **rough draft**. In case of human it referred to a condition when around 90% of the gene-rich euchromatic portion of the genome had been sequenced and assembled and each base pair of this 90% was sequenced 4 times on average ensuring reasonable precision. This term is used to describe the process of using a computer to join up bits of sequence into a larger whole. In other words, rough draft refers to the fact that the sequences are not continuous and there are gaps. In case of finished draft each base pair is sequenced 8 to 10 times on average with gaps in the sequence.

Alignment — It refers to the ordered linear arrangement of two or more DNA, RNA or protein sequences such that matches or mis-matches between neighbouring sequences can be detected. Simple sequence repeats and G-C rich regions and especially the G-C rich regions are the most difficult to sequence. Genomes with high GC(~65%) and high AT(~82%) composition present special problems for sequencing and assembly.

Whole genome shot gun (WGS) sequencing vs clone by clone (hierarchical shot gun)— WGS delivers excellent quality reconstructions of the unique regions of the genome but as the genome size increases and more importantly the repetitive content increases the WGS approach delivers less of the repetitive sequence. In other words, WGS works well for simple genomes with few repeats but is a problem in case of genomes containing highly repetitive sequences such as human genome. The centromere which is important in cell division but is repetitive, is nearly impossible to sequence. Further, gaps may contain sequences which are rich in G and C bases and these G+C regions are often difficult to sequences but are of interest as they tend to contain more genes. Hierarchical shot gun sequencing overcomes these problems by suing local assembly thus decreasing the number of repeat copies in each assembly and allowing comparison of larger regions of overlap between clones. The cost and efficiency of clone-by-clone approach is difficult to justify it as a stand-alone strategy for future large scale genome sequencing projects. But application of BAC-based or other clone mapping and sequencing strategies to resolve ambiguities in sequence assembly which can not be resolved with computational approaches is worth exploring. Production of finished sequences has relied on either HSGS or a combination of WGS and HSGS. Hybrid approaches to whole genome sequencing will only work if there is sufficient coverage in both Whole Genome shotgun phase and BAC clone sequencing phase (Venter et al., 2001). A hybrid strategy for sequencing involves 1. production of a BAC-based physical map of the genome by fingerprinting and sequencing the ends of clones of a BAC library 2. whole genome shot gun sequencing to approximately 7X coverage and assembly to generate an initial draft sequence 3. hierarchical shot gun sequencing of BAC clones covering the genome combined with the WGS data to create a hybrid WGS-BAC assembly and 4. production of a finished sequence using the BAC clones as a template for directed sequencing. Megabase-sized genomes can be sequenced efficiently without any input other than the *de novo* mate-paired sequences. In case of Drosophila and human genomes map information in the form of well ordered markers is critical for long range ordering of scaffolds. For joining scaffolds into chromosomes the quality of markers in terms of order of markers is more important then the number of markers *per se*. In case of Arabidopsis BAC clones allowed extension of sequence well into centromeric region and permitted high quality resolution of complex repeat regions. Similarly, BAC physical map is most useful in regions near the highly repetitive centromeres and telomeres.

13.3 WHOLE CHROMOSOME SEQUENCING

It is a technique for the sequencing of isolated DNA from a single chromosome and so it requires separation of all chromosomes from a target organism with PFGE or flow cytometry (FACS, chromosome sorting), isolation of single chromosomal bands from PFG, shearing of the chosen chromosome into fragments of defined length, cloning of the fragments into a suitable vector(e.g plasmid, PAC, BAC or YAC) and subsequent sequencing of the inserts. Usually as many clones are required as necessary for a 5-10 times coverage of the chromosome. Plasmids or bacteriophage vectors are especially adapted for sequencing through having a sequencing primer site just upstream (in terms of the direction of primed replication) of the fragment insertion site. The sequences are ordered via overlapping fragments and the complete chromosome DNA sequence is reconstructed. The pachytene spread fluorescence in situ hybridization (FISH) and fiber FISH analyses will indicate the size of gaps in the sequence contigs. The sequencing of very long DNA molecule or a whole chromosome can be done by obtaining a number of restriction fragments (containing overlapping fragments) using restriction endonucleases. These overlapping restriction fragments can be subjected to either Maxam-Gilbert or Sanger-Coulson's method or automatic sequencing to produce a master sequence. Often 4-5 different restriction endonucleases are used to generate restriction fragments in order to overcome the problem of overlapping sequences. Similarly all the chromosomes constituting the genome are sequenced and genetic and physical maps are prepared.

Whole genome shot gun (random sequencing) sequencing (WGS) — The problem with determining the sequence of a very large genome is that the with current sequencing technology one can directly determine the sequence of at most thousand consecutive pairs at a time. Given the limitation on read length, shot gun sequencing approach is employed in which random sampling of sequence reads is collected from a larger target DNA sequence and with sufficient oversampling the sequence of the target can be inferred by piecing the sequence reads together into an assembly. It is a technique for direct genome sequencing. WGS obviates the need for a sequence reading map before sequencing. It is an effective and efficient method to sequence the genomes of prokaryotes which are generally between 0.5 and 6Mb in size and it could be effectively applied to large eukaryotic genomes. This technique is employed for determining the sequence of large genomes such as Drosophila melanogaster without the traditional BAC, PAC OR YAC cloning, subcloning, sequencing and overlap screening. The principle underlying the whole genome sequencing is to subdivide the genome (random breaking) into small fragments of a suitable size for sequencing and other analyses. In this strategy the genome of an organism is sheared into segments of a few thousand base pairs in length and cloned directly into a plasmid vector suitable for sequencing. The presence of repeat elements, regions which are unclonable in a particular vector and the advantage of having more DNA available in clones than is actually sequenced requires the use of **multiple vector libraries**. A library of p UC18-based plasmid (high copy plasmid) containing about 2kb insert will provide most of the sequencing templates and sequencing of clones from both ends to produce pairs of linked sequences representing ~ 500bp at the ends of each insert. End sequences from a library of low copy plasmid clones containing ~ 10kbp inserts will provide medium range linking including spanning the common repetitive elements, **LINE-1** and **THE repeat** (Venter et al., 1998). Sufficient DNA sequencing is performed so that each base pair is covered numerous times in fragments of about 500bp. After sequencing the fragments are assembled in overlapping segments to reconstruct the complete genome sequence(Adams et al., 2000). Besides much longer size of their genomes the eukaryotic genomes often contain substantial amounts of repetitive sequences which interfere with correct sequence assembly. Weber and Myers(1997) examined the impact of repetitive sequences and suggested strategies to mitigate their effect on sequence assembly so that whole genome sequencing approach could be effectively applied to larger eukaryotic genomes and one key strategy is to obtain sequence data from each end of the cloned DNA inserts, the juxtaposition of these end sequences(**'mate pairs'**) is a critical element in producing a correct assembly. The heterochromatic DNA is not stably cloned in the small insert vectors used for the WGS libraries. The various steps involved in this technique are as follows.

1. Mechanical shearing of the whole genome with overlapping fragments of about 500bp each

2. Cloning of fragments and sequencing of the ends of these fragments

3. Assembling into a complete genome using powerful computer programs such as.

A. **Sequence base calling software, PHRED (Ewing and Green, 1998; Ewing et al., 1998)** which assists in unequivocal conversion of peaks into a sequence fluorogram to correct bases. The PHRED introduced the concept of assigning a 'base quality score' to each base on the basis of probability of an errorneous call. The quality scores make it possible to monitor raw data quality and also assist in determining whether two similar sequences truly overlaps. The PHRED- a base caller is computer program that analyzes raw sequence to produce a 'base call' with an associated 'quality score' for each position in the sequence. A PHRED quality score of X corresponds to an error probability of approximately $10^{-X/10}$. Thus a Phred quality score of 30 corresponds to 99.9% accuracy for the base call in the raw read. Phred provides improved base calling and a measure of data quality (range 0-50) for each base in a trace. Sequence quality is linked to an error probability, i.e. accuracy of each base call. Three criteria are used to generate quality measures in Phred including peak spacing, the relative size of the uncalled and called peaks and the change in signal between called peaks.

B. the **software, PHRAP** or Gap4 for assembly of many short fragments into a longer sequence exploiting overlaps (i.e., for contig assembly). The computer package PHRAP- a sequence assembler then systematically assembles the sequence data using the base quality scores. In other words, assembly of the sequencing reads and generation of the consensus genome sequence for each individual is performed using the Phrap program. Phrap uses the base calls and quality information obtained from the Phred and aligns each of the overlapping reads. The program assigns 'assembly quality scores' to each base in the assembled sequence, providing an objective criterion to guide sequence finishing. In other words, it assembles raw sequence into sequence contigs and assigns to each position in the sequence an associated 'quality score' on the basis of the Phred scores of the raw sequence reads. Phrap uses sequence quality to generate a consensus sequence from the highest quality base calls in the final assembly and generates an adjusted quality (range 0-90) at each position by taking sequence context quality and opposite strand confirmation into account. A Phrap quality score of X corresponds to an error probability of about $10^{-X/10}$. Thus a Phrap quality score of 30 corresponds to 99.9% accuracy for a base in the assembled sequence.

The average number of times with which a base in a defined DNA sequence is independently read with a base quality score of at least 99% accuracy is referred to as **sequence coverage**. **Base quality score** is a confirmative function of the PHRED software package which determines the probability of an erroneous call in a raw sequence data and help to determine whether the two similar sequences truly overlap. The base quality scores are then used by PHRAP software package to systematically assemble the sequence data. These two programs are widely used in generation of reference genome sequences. They are applied to assess the quality of each base call across the sequence.

C. the programme, PRIMO for detection of gaps in the sequences and the design of primers for closing the gaps and

D. FINISH- a programme for filling-in the gaps and calling for more data if required.

The efficiency of this direct genome sequencing technique increases if multicapillary gene sequencers are used which employ capillary electrophoresis and fluorescence detection by an argon laser beam. WGS libraries are produced with three different insert sizes of cloned DNA (e.g. 2kb, 10kb and 130kb). End sequence from the BAC provided long range linking information which can be used to confirm the overall structure of the assembly. The BAC-end sequencing was originally proposed to accelerate the genome sequencing by providing markers every 5kb throughout the genome. Although BAC-end sequencing will be the primary scaffold onto which the end sequences from the smaller clones will be assembled. One of the sources which will be used to verify the alignments and to place contigs on individual chromosomes is the STSs (PCR specific sites) markers which constitute the physical map. A BAC–based physical map spanning>95% of the euchromatic portions of the genome was constructed by screening a BAC library with STSs markers. 30,000 STSs in case of human was well ordered along the chromosome ad provided a defined marker approximately every 100kb. The clone-based draft sequence served two purposes. First it improved the likelihood of accurate assembly and secondly it allowed the identification of templates and primers for filling gaps which remained after assembly. The other sources such as ESTs which tag 50 to 80% of the human genome and full length cDNA sequences spanning up to 5Mb of genomic sequence will be used to verify the final assemblies. Search for STSs, full length cDNA and ESTs in assembled contigs is done through BLAST. Complete contiguity of the clone map should theoretically be achieved by about 9X coverage and so 46X coverage (clone coverage) allows for substantial deviation from the statistical model. There are likely to be contigs that are misassembled or incorrectly linked together because of the presence of long duplicated segment of the genome. Through use of a manual inspection and directed experimental effort these will be recognized and ambiguous or conflicting assembly structure corrected. The mitotic chromosomes of Drosophila show euchromatic regions, heterochromatic regions and centromeres. The euchromatic length is derived from the sequence analysis. The heterochromatic lengths are estimated from direct measurements of mitotic chromosomes lengths (Yamamoto et al., 1990).

Repetitive sequences are usually removed from other sequences before whole genome shotgun assembly because they can cause global misassembly but we must know that functional genes and transposable elements are found in the repetitive sequences. Heterochromatic DNA is not stably cloned in small insert vectors used for the WGS library. The simple sequence repeats (heterochromatin) are not stable in YACs or other large insert cloning system. The euchromatic genome is that portion of the genome which can be cloned stably in BACs.

Genome assembly—Genome assembly was based on several types of data including clone-based sequence, whole genome sequence and a BAC-based STS content. For clone–base genomic sequencing, BAC, P1 and cosmid DNAs were prepared by alkaline lysis procedures and purified by CsCl gradient ultracentrifugation. DNA was randomly sheared and size selected on LMP agarose for fragments in the 3-kb range for plasmids and in the 2kb range for M13 clones. M13 clones were generated by double-adaptor protocol (Anderson et al., 1996). Plasmid sequence templates were prepared by alkaline lysis (Qiagen) or by PCR and M13 templates were prepared by the Sodium perchlorate fiber glass procedure (Anderson et al., 1996). Paired end-sequences of 3-kb plasmid sub-clones were generated with dye terminator chemistry or capillary sequencer. Additional M13 subclone sequence was generated using dye primer. There are two types of assembly: WGS assembly and joint assembly. An initial assembly is constructed using the WGS data and BAC end sequence. The joint assembly includes the clone-based draft sequence data.

Mapping of scaffolds to chromosomes—Two methods were used to map the scaffolds to chromosomes:

1. Cross referencing between STS markers present in the assembled sequence and the BAC-based STS content.

2. Cross referencing between assembled sequence and shot gun sequence data obtained from individual tiling path clones selected from the BAC physical map.

Methods for closing gaps—A number of methods were used to close gaps. Whenever possible gaps were localized to a chromosome region and a spanning genomic clone was identified. When a spanning clone could be identified, it was used as a template for sequencing. The sequencing approach was determined by the gap size. For gaps smaller than 1kb, BAC templates were sequenced directly with custom primers. For gaps larger than 1kb, 3kb plasmids or M13 clones from the clone-based draft sequencing were sequenced by the directed methods or 10kb plasmids from the WGS were sequenced by random transposon-based methods. When no 3kb or 10kb plasmid could be identified then PCR products were amplified from BAC clones or genomic DNA and end- sequenced directly with PCR primers. Closing gaps is a challenge for sequencing the human genome because of non-randomness of clone libraries and STS maps.

Validation of the Assembly—Validation can be carried out at three levels.

1. **STS-level validation**—STS maps for the chromosome arms were concatenated to give the whole genome map which orders STSs and thereby allowing comparison between this independent order and the WGS assembly.

2. **Clone-level validation and coverage**—The assembly of the WGS data set was compared with the finished and the draft sequence for the published clone-tiling path which covers most of the euchromatin of the genome. This comparison permitted to identify the appropriate clone reagents for gap closure and to verify the order and assembly of contigs in the scaffolds.

3. **Sequence –level validation**—This involves comparison of the published sequences of different genes against the contigs from WGS assembly that cover it.

Comparison with existing physical or genetic maps validates assembly accuracy on the Mb length scale but that is much larger than the size of most genes. Clone-end pairing information does validate a contig assembly on the Kb length of scale of genes. However, when clone ends are also used to assemble the sequence they do not qualify as an independent confirmation. This problem can be solved in the following ways.

1. Alignment of cDNA sequences with the genomic sequences

2. Remove the redundancies by eliminating any cDNA that is more than 90% contained another

3. Transposon sequences generally on the 3′ UTR regions are trimmed off to minimize the number of ambiguous hits. Alignments are allowed to span multiple contigs. Within any one contig a putative mis-assembly is flagged whenever an exon is

Table 13.2 Showing measures of eukaryotic genome assembly.

Measures of completion

Number of scaffolds mapped to chromosome arms

Number of scaffolds not mapped to chromosome arms

No. of base pairs in scaffolds mapped to chromosome arms

No. of base pairs in scaffolds not mapped to chromosome arms

Largest unmapped scaffold

% of total base pairs in mapped scaffolds>100kb

% of total base pairs in mapped scaffolds>1Mb

% of total base pairs in mapped scaffolds>10Mb

No. of gaps remaining among mapped scaffolds

Base pair accuracy against LBNL BAC (non repetitive sequence)

missing from the middle of the chain, in the wrong order or in the wrong orientation. Missing splice sites resulting from minor sequencing errors and partial alignments resulting from missing sequences at the end of a contig are not counted. All putative misassemblies are validated by visual inspection to ensure that no better alignment could be found. If in the end the best alignment remained problematic one conclude that there must have been a misassembly.

Criteria for describing the completion of eukaryotic genome—Because of the unclonable repetitive DNA surrounding the centromeres it is highly unlikely that the genomic sequence of chromosomes of eukaryotes such as Drosophila or human will ever be complete. The completion status of genome can be judged by different parameters shown in the Table 13.2 and by which improvements in future releases can be measured. In the future releases more and more gaps will be filled and overall sequence accuracy will be increased. One measure of completeness of the assembled sequence is the extent to which previously described genes can be found.

13.4   GENOME SEQUENCING OF NEMATODE *C. ELEGANS*

The sequencing process begins with the purification of DNA from selected clones of the tiling path. The DNA is sheared mechanically and after size selection the resulting fragments are subcloned into M13 or plasmid vectors. Random subclones are selected for sequence generation using the shot gun sequencing approach. The sequencing procedure involved two phases, the shot gun phase which is sequence acquisition from random subclones and the finishing phase which involves closing of gaps if, any, and to resolve ambiguities and low quality areas. Sequencing was started by the isolation and assembly of random cosmid clones with a 40kb insert which was the largest cloning system available at that time. Autosomes were divided into the genetically defined compartments of the left arm (L), the central cluster region (C) and the right arm (R). Restriction digests with several enzymes were performed on most of the cosmid and provided valuable checks on sequence assembly. When assembly was ambiguous because of repeats the digests were helpful in resolving the problem. PCR checks were conducted along the length of the sequence to confirm that the assembled sequence was an accurate representation of the genome. PCR checks were later abandoned when it became clear that there is the tendency of PCR to yield artifacts in repeat regions. At a six fold redundant coverage of the genome in cosmids non-random gaps persisted. When available cosmids exhausted fosmids which are similar to cosmids but are maintained at a single copy per cell and thus are potentially more stable, were screened. And with these a third of the gaps were bridged in the central region of the chromosomes but very few bridged in the outer regions. Long range PCR was used which recovered some of the central gaps. The remainder of the central gaps and all of the gaps in the outer regions were recovered by sequencing YACs. Thus as for the cosmids, a tiling path of YACs was chosen and DNA from selected clones was isolated by PFGE. Sequencing was performed as for cosmids. Restriction digests were carried out for assembly checks. The comparison of the assembled YAC sequences with the overlapping cosmid sequences showed few discrepancies which resulted from rearrangements in the cosmids. The success of YAC sequencing is because of the reason that nearly all the regions of the YACs can be cloned in bacteria as short fragments although cosmid and fosmid libraries failed to represent these regions. A clean separation of the YAC DNA from the host chromosomal DNA sometimes required the use of yeast strains in which specific yeast chromosomes are altered in size to provide a window around YAC that is free from native chromosomes.

Assembly of genome—The genome is assembled using the simple principle of computer science technique of 'hashing' (in which one detects overlaps by consulting an alphabetized look-up table of all k-letter words in the data). The mathematical analysis of expected number of gaps as a function of coverage is similarly straightforward (Lander and Waterman, 1988). **Sequence assembly** is a process of computerized linkage of small sequence stretches into larger regions. In other words, it is a process of aligning overlapping sequence fragments into a contig or a series of contigs. The assembler enables to assemble the genome. It simultaneously clusters and assembles fragments of the genome. A single fragment starts the initial contig. To extend the contig a candidate fragment is chosen with the best overlap based on oligonucleotide content. The current contig and the candidate fragment are aligned by a modified version of Smith-Watermann (1988) algorithm which provides the optimal gapped alignments. The contig is extended by the fragment only if the following strict criteria for the quality of match are met.

1. Minimum length of the overlap
2. The maximum length of an unmatched end
3. The minimum percentage match

The algorithm automatically lowers the criteria in the region of minimal coverage and raises them in regions with a possible repetitive element. The number of potential overlaps for each fragment determines which fragments are likely to fall into repetitive elements. Fragments representing the boundaries of repetitive elements and potentially chimeric fragments are often rejected on the basis of partial mismatches at the needs of alignments and excluded from the contig.

Closing sequence assemblies—The key steps are as follows.

Obtain subclones which bridge the gap remaining after the shot gun phase. Often gaps are spanned by the subclones used in the shot gun phase as the insert length is deliberately set at 2 to 4 times the typical sequence read length. The introduction of plasmid clones half way through the program greatly improved the coverage of inverted repeats and other unusual structures. In cases where shot gun phase failed to yield spanning subclone, plasmid clones that bridge gaps was obtained by isolating and subcloning restriction fragments from cosmids. Very short insert plasmid library was used to find gap bridging clones. The gap bridging clone was then either sequenced directly or a short insert library (SIL) was constructed by breaking the gap-bridging clone into smaller fragments (0.5kb or even smaller) with breakpoints interrupting the secondary structure. In some cases transposon insertion were used although SILs are generally preferred as first pass because of their ease of throughput. The 97-Mb sequence of *C.elegans* is a composite of 2527 cosmids, 257 YACs, 113 fosmids and 44 PCR products. Bases were determined with PHRED (Ewing et al, 1998). An assembly of these random sequences were generated with PHRAP (Green) which resulted in a number of contigs. Gap closure and resolution of sequence ambiguities were achieved during finishing phase using the editing package GAP (Bonfield et al., 1995) and CONSED (Gordon et al., 1998). Additional data were collected through longer reads and directed sequencing reactions. High quality finished sequence was analyzed through the use of a suite of programs including BLAST and GENEFINDER (see Roy, 2009).

Environmental genome shot gun sequencing—It refers to the application of whole genome shot gun sequencing to microbial populations collected *en masse*. WGSS has traditionally been applied to identify the genome sequence (s) from one particular organism whereas EGSS is intended to capture representative sequences from many diverse organisms simultaneously. Variation in genome size and relative abundance determines the depth of coverage of any particular organism in the sample at a given level of sequencing (Venter et al., 2004). The starting material consists of mixture of genomes of varying abundance. WGSS is thus applied to environmental-pooled DNA samples to test whether new genome approach can be effectively applied to gene and species discovery and to overall environmental characterization.

Direct sequencing (PCR sequencing)—In this technique of direct sequencing the determination of sequence of bases in DNA (or cDNA) is done without prior cloning of the DNA. Such direct sequencing is made possible by PCR techniques which permit amplification of a particular sequence. The amplified sequences are purified on an agarose gel, visualized by EtBr staining, isolated and used directly for sequencing.

Cycle sequencing—Cycle sequencing (Murray, 1989) generates a sequence reaction during repeated cycles of thermal denaturation and extension of an oligonucleotide using a thermostable DNA polymerase. It requires less template DNA than other DNA sequencing approaches. Since it does not exponentially amplify the target sequence as is achieved by PCR it has hitherto been impossible to use cycle sequencing procedure to directly determine nucleotide sequences from complex genomes. DEXAS (Direct exponential amplification and sequencing)- This technique is used to achieve the amplification of template DNA and sequencing reaction simultaneously in a single reaction using two primers, a mixture of dNTP and ddNTP and a thermostable DNA polymerase. It is possible to perform a combined amplification and sequencing reaction (DEXAS) directly from complex DNA mixtures by using two thermostable DNA polymerases, one (Taq polymerase) which favors the incorporation of dNTPs over ddNTPs and one which has a decreased ability to discriminate between these two nucleotide forms. During cycles of thermal denaturation, annealing and extension the former enzyme primarily amplifies the target sequence whereas the later enzyme primarily performs a sequencing reaction. This method permits the determination of single copy nuclear DNA sequences from amounts of human genomic DNA comparable to those used to amplify nucleotide sequences by PCR. Thus DNA sequences can be easily determined directly from total genomic DNA (Kilger and PAABO, 1997).

Genomic amplification with transcript sequencing—This is a rapid and sensitive method for direct sequencing of the genomic target DNA which combines the advantages of both the amplification of this DNA by PCR and of the phage T7 RNA polymerase promoter driven primer. This technique involves, annealing of primers to sites just out side the sequence to be amplified. One (or both) oligonucleotide primer contains a T7 promoter sequence. The subsequent repeated cycles of PCR amplify the target DNA to some 10^6 copies which are then transcribed with T7 RNA polymerase. This reaction then generates single stranded DNA which can be sequenced following Sanger sequencing. A specific oligonucleotide primer for reverse transcriptase is annealed and reverse transcriptase mediated sequencing is initiated.With GAWTS a specific genomic DNA segment can be amplified more than 10 million times.

RNA amplification with transcript sequencing—It is a method for amplification of a certain DNA by PCR and its amplification by phage promoter driven transcription. This method involves cDNA synthesis with oligo (dT) or an mRNA specific oligonucleotide primer, subsequent PCR with a primer or primer (s) containing a phage promoter, for example, promoter from phage T7, transcription from the phage promoter using T7 RNA polymerase and reverse transcriptase- mediated Sanger sequencing of the transcript which is primed with a nested (internal) oligonucleotide. This technique permits to amplify a specific RNA more than one billion times and thus can be used to detect very low abundance mRNAs.

Linear amplification of DNA sequencing—It is a technique for sequencing the native double stranded DNA after its amplification using PCR. The purified double stranded DNA and a 5′ end labeled sequencing primer is mixed in four Sanger sequencing reactions with Taq polymerase. It is then submitted to repeated programmed temperature cycles in a thermocycler which leads to repeated denaturing and reannealing of template DNA and primers. The Taq polymerase extends the primer until incorporation of ddNTP stops the reaction. Thus in this way a series of fragments are generated and amplified which are ectrophoretically separated in sequencing gels. This allows to determine the base sequences of the original DNA. Linear amplification of DNA sequencing permits to read more than 500 bases in one step and due to the signal amplification both background problems and the amount of DNA required for sequencing are reduced appreciably.

Single pass sequencing—It refers to the determination of sequence of bases in one strand of a double stranded DNA that is not confirmed by a simultaneous or subsequent sequencing of the complementary strand. Single pass sequences may contain '**sequencing error**' introduced by erroneous synthesis or reading.

DNA Sequence Assembly of a clone—Another aspect of analysis of DNA sequences is the process of determining the nucleotide sequence of a clone. If a specific gene whose sequence is already known then it is essential to check that the cloned sequence is identical to the published one. But if it is not identical then one must design the experiments to correct the sequence. Cloning errors can arise either as a result of use of incorrect primers or the use of low fidelity enzyme in a PCR.

A cDNA clone is generated using mRNA as a template. The cDNA clone is then sequenced by designing primers to known oligonucleotides present in the cloning vector flanking the inserted DNA as described above. It is unusual to be able to sequence a complete CDS in one run and so overlapping fragments are built up in a multiple alignment, a process called **sequence assembly**. The quality metrics that matter for gene identification are i. contiguity on the length of scale of a gene ii. single base-pair probability and iii. contig assembly accuracy on the length scale of a gene (Yu et al., 2002) Understanding the limitations of the sequencing protocol, effects of GC-rich regions (resulting in high secondary structure and consequently awkward reads) and repetitive sequences, etc. all make sequence assembly a highly skilled pursuit.

The assembler program builds a consensus sequence for the clone according to the weight given to each nucleotide position in the sequence. Parameters are set in the assembler for the number of mismatches allowed per position. Normally a degree of redundancy in sequence coverage is required, for example, at least two base reads per position on each strand(+ and -) give a high level of confidence in the resulting sequence. Three plus-strands and two minus strands reads helped to build a **consensus DNA sequence** as a result of sequence assembly. It showed two positions where mismatches have resulted in lower confidence in the consensus sequence and so further reads and / or visual inspection of the sequencing chromatogram are required to resolve the ambiguities. Normally for fully validated sequence confirmation complete reads on the plus and minus strands are required. Part of a consensus sequence derived from a single read on one strand (i.e., no overlapping fragments) would give only a low level of confidence in the assembled sequence.

13.5 SHOT GUN STRATEGY FOR WHOLE GENOME SEQUENCING OF BACTERIA, *HAEMOPHILIUS INFLUENZAE*

The different steps involved in the whole genome assembly are shown in the Table 13.3 given below (Fleischmann et al., 1995).

The shot gun sequencing strategy involved preparation of a single random DNA library fragments, sequencing of ends of a sufficient number of randomly selected fragments and assembling to produce the complete genome. So the genomic DNA was mechanically sheared, digested with BAL31 nuclease to produce blunt ends fragments and size fractionated by agarose gel electrophoresis. Mechanical shearing (e.g. sonication) maximizes the randomness of DNA fragments. It avoids potential bias which may be introduced by either non-random distribution of restriction sites or differential kinetics of cleavage, if the genomic DNA is enzymatically cleaved. The fragments between 1.6 and 2.0kb in size were excised and recovered. The maximum size of 2kb was selected in order to minimize the number of complete gene which might be present in a single fragment and thus might be lost as a result of expression of deleterious products. The fragments were ligated to SmI cut, phosphatase treated Puc18 plasmid vector and ligated products were fractionated on an agarose gel. The linear vector plus the insert band was excised and recovered. The ends of the linear recombinant molecules were repaired with T4 polymerase and molecules were

Table 13.3 Showing different steps involved in whole genome assembly.

Stage	Description
1. Construction of random small insert and large insert library	Shearing of genomic DNA randomly to generate about 2kb and 15-20kb fragments, respectively
2. Library plating	Verifying the random nature of library and maximizing random selection of small insert clones for template production
3. High-throughput DNA sequencing	Sequence sufficient number of sequence fragments from both ends for 6x coverage
4. Assembly	Assemble random sequence fragments and identify repeat regions
5. Gap closure	
(a) Physical gaps (no template DNA for the region)	Order all contigs(fingerprints, peptide links, lambda clones, PCR) and provide templates for closure
(b) Sequence gaps (template available for gap closure)	Complete the genome sequence by primer walking(discussed above)
6. Editing	Identify the sequence visually and resolve sequence ambiguities including frameshifts
7. Annotation	Identify and describe all predicted coding regions(putative identifications, starts and stops, role assignments, operons, regulatory regions)

then ligated into circles. Thus a plasmid library was created. An amplified library was constructed in vector λGEM-12 and also an unamplified library λDASH II was constructed. The use of λ clones has the following advantages.

1. Some fragments of the *H. influenzae* genome would be non-clonable in a high copy number plasmid because they would produce deleterious proteins in the *E. coli* cells. Lambda clones (lytic lambda clones) would provide DNA for these segments because such segments would not inhibit plaque production.
2. Sequence information from the ends of 15-20kb clones is particularly suitable for gap closure and providing general confirmation of the genome assembly.
3. Because of their size they would be likely to span any physical gap.
4. Finally, lambda clones are particularly useful for solving repeat structures.

Confirmation of global structure of the assembled circular genome was done by comparing the computer generated restriction map based on assembled sequence for the endonucleases, Apa I, Sma I, Rsr II with the predicted physical map.

Estimation of sequence coverage, number of gaps and gap size — The probability that a base is not sequenced is $P_0 = e^{-m}$, where m is the sequence coverage (Lander and Waterman, 1988). Thus after 1.83 Mb of the sequence has been randomly generated for the *H. influenzae* genome(m=1, 1X coverage), $P_0 = e^{-1} = 0.37$ and approximately 37% of the genome is unsequenced. Five fold coverage (approximately 9500 clones sequenced from both ends and an average sequence read length of 460bp) yields $P_0 = e^{-5} = 0.0067$ or 0.067% unsequenced. If

L is genome length and n is the number of random sequence segments done the total gap length is Le^{-m} and the average gap size is L/n. Five fold coverage would leave about 128 gaps averaging about 100bp in size. The 10X sequence coverage means that the accuracy of the sequence will be comparable to the standard now prevailing in genome sequencing (fewer than one error in 10,000bp). **Coverage(or depth)–** It refers to the average number of times a nucleotide is represented by a high quality base in a collection of random raw sequence. Operationally, a 'high quality base' is defined as one with an accuracy of at least 99% (corresponding to a Phred score of at least 20). **Half shot gun coverage-** It refers to 4-5 fold coverage required for the raw sequence of a large insert clone (e.g. a BAC) to be suitable for finishing. **Full shot gun coverage** refers to 8-10 fold coverage of the large insert clone to be suitable for finishing. In other words, it refers to the coverage in a random raw sequence needed from a large insert clone to ensure that it is ready for finishing. Clones with full shot gun average can usually be assembled with only a handful of gaps per 100kb. **Reading** refers to one way linear process by which the information encoded in nucleotide sequences is decoded(e.g. DNA-RNA, RNA-protein). **Reading length** refers to the number of nucleotides of a DNA or RNA sequence which can be read on a single sequencing gel. It is usually 500-600 bases and maximally 1-1.5kb.

Ordering of contigs — There are four strategies to order contigs separated by physical gaps. Oligonucleotide primers are designed and synthesized from the end of each contig group. These primers are then used in one or more of the following strategies.

1. **DNA hybridization analysis(Southern analysis)** — This procedure is based on the supposition that the labeled oligonucleotides homologous to the ends of adjacent contigs should hybridize to common DNA restriction fragments and thus share a similar or identical hybridization pattern or fingerprint. Adjacent contigs identified in this manner are targeted for specific PCR reactions.

2. **Identification of peptide links** — Peptide links were made by searching each contig end with BLASTX against a peptide database. If the ends of two contigs match then the two contigs can be tentatively considered to be adjacent.

3. **Forward and reverse sequence data from lambda clones** — Two lambda libraries constructed from the *H. influenzae* genomic DNA were probed with oligonucleotides designed from the ends of contigs groups. The positive plaques were then used to prepare templates and the sequence was determined from each end of lambda clone insert. These sequence fragments were searched with GRASTA against a database of all contigs. Two contigs that matched the sequence from the opposite ends of the same clone were ordered. The lambda clone then provided the template for closure of the sequence gap between the adjacent contigs.

4. **PCR data** — To confirm the order of contigs found by the other approaches and establish the order of the remaining contigs both standard and long range (XL) PCR amplifications are performed. Although a PCR reaction is done for essentially every combination of physical gap ends, techniques such as DNA fingerprinting, database matching and probing of large insert clones are particularly valuable in ordering contigs adjacent to each other and reduce the number of combinatorial PCRs necessary to achieve complete gap closure. In other words, Southern analysis data, forward and reverse sequence data from lambda clones, identification of peptide links and PCR data are used to establish the relative order of the contigs separated by physical gaps.

Closing sequence assemblies — The key steps involved in closing sequence assembly are the same as described in case of *C. eleganes*.

Editing — Editor is a prime tool for sequence viewing and editing for the purpose of genome assembly. Each contig is edited visually by reassembling overlapping 10kb sections of the contig by means of assembler. The assembler provides graphical interface to electropherogram data for editing. The electropherogram data is used for assigning the most likely base at each position.

13.6   GENOME SEQUENCING OF *S. POMBE*

In case of *S. pombe*, the problems with earlier maps included the existence of chimaeric clones, mismapped cosmids, bacterial insertion elements and unfilled gaps. Small gaps were covered using a long range PCR strategy, plasmid libraries and a BAC library provided clones for gap closure across regions were not represented in the cosmid library. Most sequencing was performed using random sequencing of subcloned DNA followed by directed sequencing. DNA from clones was fragmented usually by sonication and fragments of 1.4-2.0kb were cloned typically on M13 or pUC18. Random subclones were sequenced with dye-terminator chemistry. Gaps and low quality regions of sequences were resolved using primer walking, PCR and re-sequencing clones under conditions which gave increased read lengths. Some labs also used direct blotting procedures, classical radioactive sequencing and nested deletions. The depth of coverage was on average eight fold (8X). The sequencing error rate was less than 1 in 180,000bps which was calculated from the number of single base differences observed in overlapping sequences from different sources. The finished sequences had high degree of accuracy with at least two high quality reads on each strand or if this could not be accomplished, an additional read on the same strand was made using an alternative chemistry (Wood et al., 2002). Homopolymeric tracts located outside the coding regions, possibly generated by slippage during DNA replication.

Challenges for sequencing genome — The challenges for sequencing a genome are: (i) source material (ii) clone validation (iii) minimum tiling paths of contiguous clones (iv) sequence scale- up (v) standard of sequencing (Rowen et al., 1997). The source material involves construction of clone libraries with which physical maps covering significant portions of the genome can be constructed. Chromosomes can not be sequenced directly rather DNA is isolated, randomly fragmented and cloned into vectors capable of stable propagation in a suitable host such as *E. coli* or yeast and thus a clone library is constructed. Before sequencing the clones are selected from libraries with chromosomal markers as probe, verified for their fidelity to the genome and ordered in a minimal-overlapping tiling path spanning a portion of a chromosome. Use of multiple libraries is made for selecting source clone so that no individual's chromosomes are dominantly represented in the final sequence. But then the problem with use of multiple libraries is that it undermines the ability to validate the fidelity of the clone. Technical considerations suggest the use of a minimum number of highly redundant clone libraries whereas social considerations are for diversification of clone libraries. The ideal library for genome sequencing has the following features.

(i) The clones are highly redundant, covering the entire genome many times.

(ii) The clone coverage is random and not biased toward or against specific regions of the genome and

(iii) The clones are stable, not subject to deletion or rearrangement during the propagation process.

Clone validation refers to a demonstration that clones faithfully represent the genome. Clone validation is judged by internal consistency among overlapping clones assayed by restriction enzyme fingerprinting. Sequence polymorphisms will lead to differences in fingerprints among overlapping clones that can not be distinguished from differences in the fingerprints that arise from deletions or rearrangements of the artifactual clones. Construction of minimum tiling paths of contiguous clones involves identification of a contiguous array of sequence-ready clones across each chromosome for which a set of minimally overlapping clones (minimum tiling path) can be sequenced. It is a rate limiting step in large- scale sequencing. Obtaining minimum tiling path poses the following three challenges as described above.

In nematode, both bacterially based and yeast based cloning systems were required to achieve contiguity in the clone map. There is a need to scale-up sequencing which used to be a throughput of 2 to 30Mb/ year.

What after sequencing and assembling of genome? – After completion of genome assembly annotation of the assembled genome is done which includes prediction of the coding regions, identifying genes, tRNA and rRNA as well as other features of the DNA sequence such as repeats, regulatory sites, replication origin sites and nucleotide composition. The biological defined repeats (BDRs) include microsatellites, transposable elements, multigene families, duplicated chromosome segments or pseudogenes. Further prediction of transcript and protein sequence and prediction of function for each predicted protein is made. Each segment is subjected to a series of automatic analyses to reveal possible coding regions and transfer RNA genes, proteins, similarities to ESTs and other proteins (comparative genomics), repeat families and local repeats. BLAST is used to search for STSs, cDNA sequences and ESTs in assembled contigs and compared with physical map of STS markers. ESTs provides evidence for gene identification and for gene expression analysis. For detail see Roy, (2009).

Variation in genome sequencing techniques – From what is described above it can be said that the high throughput genome scale sequencing presents a variety of technical and social challenges. A variety of strategies from high redundancy shot gun to directed strategies such as ordered shot gun sequencing (Chen et al, 1993), sequence mapped gaps(Richards et al, 1994) or transposon- based walking (Martin et al., 1994) to sampling strategies that produce ordered sequence fragments separated by gaps of known length (Smith et al, 1994) are used by sequencing labs.

13.7   SEQUENCING STRATEGY FOR ARABIDOPSIS GENOME

The following strategy was adopted by 'The Arabidopsis Genome Initiative'. Large insert BAC, phage (P1) and transformation–competetent artificial chromosome(TAC) were used for sequencing. Early stages of genome sequencing used cosmid clones. The physical maps were assembled by restriction fragment 'fingerprint' analysis of BAC clones, by hybridization or polymerase chain reaction (PCR) of sequence-tagged sites and by hybridization and Southern blotting. The resulting maps were integrated with the genetic map and which in turn provided a foundation for assembling sets of contigs into sequence-ready tiling paths. End sequence of BAC clones was used to extend contigs from BACs anchored by marker content and to integrate contigs. Ten contigs representing the chromosome arms and centromeric heterochromatin were assembled from 1,569BAC, TAC, cosmid and P1 clones with average insert size of 100kb. 22 PCR products were amplified directly from the genomic DNA and sequenced to link regions not covered by cloned DNA or to optimize the minimal tiling path. Telomeric clones were obtained from specific YAC and phage clones and from inverse polymerase chain reaction (IPCR) products derived from genomic DNA. Clone fingerprints together with BAC-end sequences were adequate for selection of clones for sequencing over most of the genome. In centromeric region, these physical mapping methods were supplemented with genetic mapping to identify contig positions and orientation.

Selected clone were sequenced on both strands and assembled using the standard techniques described above. Comparison of independently derived sequence of overlapping regions and independent reassembly sequenced clones showed accuracy rates between 99.99 and 99.999%. Over half of the sequence differences were between genomic and BAC clone sequence. All available genetic markers were integrated into sequence assemblies to verify sequence contigs.

13.8 RICE GENOME ANALYSIS

Rice genome with 430Mb in size is the smallest of the cereal crops. It is 3.7 times larger than that of Arabidopsis and 6.7 times smaller than human. The analysis involves isolation of individual chromosomes. A 4', 6-diamidino-2-phenylindole dihydriochloride (DAP) stained pachytene spread of rice chromosome showed heterochromatic(bright blue) and euchromatic (light blue) regions. The euchromatic genome comprises that part of a genome which is mostly composed of euchromatin, i.e. actively transcribed genes and can be cloned into BAC. The centromere (green) is detected by FISH with a centromere-specific probe (p RCS2 clone). A tiling path consisting of BACs and phage (P1)-derived artificial chromosome was identified. Each clone was sequenced by a random shot gun approach on both strands of small insert subclones to achieve a 10-fold coverage. Shot gun sequence analysis of small insert clones is the fastest and least expensive method for generating draft genome in rice.Sequence assemblies were verified by comparing Not I and Hind III restriction profiles predicted from the sequence against experimentally determined restriction profile. The collinearity of the assembled sequences with the chromosome was confirmed by integrating sequenced genetic and ESTs markers. Sequence overlaps of adjacent clones were used to assess potential sequence differences between them. Sequence mismatches were corrected further by manual checking or by additional sequencing reactions. The pachytene spread fluorescence in situ hybridization (FISH) and fiber FISH analyses showed the sizes of gaps. Low-voltage electrophoresis was used for resequencing which provided longer sequences with better quality and in many cases resulted in closing gaps between contigs (Goff et al., 2002).

13.9 MAIZE GENOME ANALYSIS

In maize approximately 80% of the genome comprises highly repetitive sequences interspersed with single copy, gene rich sequences and standard genome sequencing strategies are not readily adaptable to this type of genome. A technique called 'methylation filtering' in which hypermethylated sequences are excluded with the use of bacterial restriction systems that cleave methylated sequences, has been used to produce libraries which are gene-enriched. Another technique- high cot selection allows separation of DNA fractions into low copy (high cot) or high copy (low cot) sequences where concentration (co) and annealing time (t) determine the composition of the fractions. The most repetitive DNA renatures first and the double stranded DNA can be separated from lower copy number, unnatured DNA. The low copy number fraction from maize can be enriched four fold in genes in comparison to random shot gun libraries (Giot et al., 2003).

13.10 GENOMIC ANALYSIS OF HETEROCHROMATIN

Genomic sequences for most metazoans and plant genomes are incomplete because of presence of repeated DNA in the heterochromatin which is a major component of metazoan and plant chromosomes. The functions of heterochromatin include regulation of chromosome segregation, nuclear organization and gene expression. So the description of sequence and organization of heterochromatin is essential in order to understand the function. The genomic analysis of heterochromatin has been checked by the difficulties in cloning, mapping and assembling regions rich in repetitive elements. Heterochromatin contains tandemly repeated simple sequences (including satellite DNAs), middle repetitive elements (such as TEs, ribosomal RNA) and some single copy DNA. Hoskin et al. (2007) produced 15Mb of finished or improved heterochromatic sequences of Drosophila melanogaster with the use of available clone resources and assembly methods. WGS3 is an excellent assembly of Drosophila euchromatic sequence but has lower contiguity and quality in repeat-rich heterochromatin. Moderately repetitive sequences are well represented in WGS clones and sequence reads but they tend to be assembled into shorter scaffolds with many gaps and low quality regions because of difficulty of accurately assigning data to a specific copy of a repeat. The sequence consists largely of ends of fragmented TEs and most remaining gaps are bound by TEs or simple sequence repeat. WGS heterochromatic scaffolds have 5.8 times as many sequence gaps per MB as well as lower sequence quality. They constructed a bacterial artificial chromosome-based physical map that spans 13Mb of the pericentromeric heterochromatin and a cytogenetics map which positioned 11Mb in specific chromosomal locations. This map was necessary for ordering, orienting and linking WGS sequence scaffolds into larger BAC contigs and Release-5-scaffold. They used a set of 10-Kb genomic clones from a library to fill small gaps and improve low quality regions. Higher – level sequence assembly into Mb-sized linked scaffolds (longer scaffold with high quality sequence) constructed from smaller scaffolds used relationships determined from BAC-based STS physical mapping and BAC end sequences. The STS content mapping experiments benefited greatly from the availability of large-inserts BACs libraries generated by fragmenting genomic DNA with three restriction enzymes and with physical shearing. FISH was used to map BAC contigs and sequence scaffolds to specific cytogenetics locations in mitotic chromosomes. The high repeat content of heterochromatin needed the use of single copy probes (P-element insertions and cDNA clones) which could be assigned to specific sequence scaffolds. Use of BAC probes that had sufficient single copy sequences provided unambiguous localization. Restriction fingerprints of tiling path BACs will provide an independent benchmark for evaluation of the accuracy of the finished sequence assemblies. Thus with current technologies production of complete map and sequence assemblies for the single copy and middle repetitive components of the heterochromatin can be obtained. In other words, heterochromatic regions containing single copy genes and a high density of TEs can be assembled into high quality, contiguous sequence but the remaining gaps, if exist, would be due to extensive simple sequence arrays. Thus, in order to substantially improve the sequence and maps, three components are required are: (i) High quality WGS sequence assembly (ii) a high-depth collection of precisely sized and aligned genomic clones for sequence finishing and gap closure and (iii) physical and cytogenetics mapping to deduce relationships between WGS scaffolds.

13.11   GENOMIC SEQUENCE ANNOTATION

Once the raw sequence data of an organism is obtained from sequencing (see chapter 12), there is a need to relate it with what we already know about the genetics of that organism and biology and this is the process of genome annotation. Stein(2001) has defined three hierarchical levels of annotation.

1. Nucleotide level
2. Protein level
3. Process level.

Nucleotide level annotation—Preliminary annotation consists of determining the position of known markers, known genes and repetitive sequences in combination with efforts to delineate the structure of novel gene. Further, we would like to know much more including the multiferous interaction of genome contents with one another and the environment, their expression in the biology of the cell and role in physiology. In the first step, identify as many known genomic landmarks as possible. These are generally markers from previous studies, repeats and known genes already stored in the public databases. Markers from cytological and physical maps are placed upon the genomic sequence through the use of algorithms designed to find short, almost exact sequence matches such as ePCR, BLAST, SSHA and BLAT . Further, information on gene structure are used to identify the positions of known mRNAs using the algorithms such as Spidey (Wheelam et al., 2001), SIM4 (Florea et al., 1998) and est2 genome of EMBOSS package (Rice et al., 2001). Identification of interspersed and simple repeats is an important part of annotation. This is done by a widely used program, RepeatMasker. Use of comparative genomics is made to predict gene structures using Twinscan program (Korfe et al., 2001). The usual categories of nucleotide level annotation include repetitive sequences, CpG islands, Poly (A) sites and marker positions but the cutting edge of annotation is defining regulatory regions, transcription start site, TF binding sites and promoter modules.

At a higher level gene expression is also regulated by the large scale topology of chromosomes and annotations may eventually indicate features such as chromosome domains (genomic regions that bind histone-modifying proteins) and matrix attachment sites (regions that facilitate the organization of DNA within a chromosome into loops). However, defining the genes whose transcription is regulated from such features may be an insoluble problem computationally as they may regulate transcription from a given TSS, from several different TSS of the same gene or multiple genes in a region.

Problems with nucleotide annotation—The nucleotide level annotation refers to prediction of gene structure Correctly delineate every exon of every gene is a problem in the prediction of gene structure. This task is made more difficult in repeat-rich eukaryotic genomes. There is also problem with *ab initio* gene prediction. In yeast, only 202 genes have been discovered and most genes appear to have been missed because they are relatively short or overlap a previously annotated gene on the opposite strand. Use of sequence similarity is now being used in gene prediction and in practice, genome annotators use a combination of information to make predictions of gene stricture. But even after using a variety of statistical analyses and comparative genomics several hundred of originally annotated genes may be spurious. New methods aiming at the prediction of non-coding features of genome such as regulatory regions and noncoding RNAs are being advanced.

Protein level annotation—Once a gene has been predicted the protein level annotation assigns possible function to encoded protein. Most computationally assigned functions are derived from sequence similarity. A pair of proteins that align along 60% or more of their lengths with significant similarity (e.g. E<0.01 in BLAST search) are likely to be homologous derived from common ancestor. Such a pair of sister protein may be paralogous derived from a duplication event or orthologous that exist as a result of a speciation event. Additional searches are needed to verify that each member of the pair identifies the other member as the best match within the organism of interest. Pairs identified are likely to be orthologues (Huynen and Bork, 1998) which is desirable as orthologues are likely to share the same function (Jordan et al., 2001) whereas functional diversification between paralogues is thought to be common (Li, 1997). In most cases this strategy of reciprocal sequence similarity searches to identify orthologues is successful and is the principle underlying the construction of the clusters of orthologous Groups of Proteins (COGs) database (Tatusov et al., 2000). One should be cautious while dealing with the results of such analysis. For example, a novel human gene may be directly descended from a common ancestor of yeast gene (in which case the two genes are orthologues and likely to share the same function) or it may be descended from a duplicated sister yeast gene (and the two sister genes are really paralogues with a different function). Similarity may sometimes imply functional similarity. For example, two proteins with only 30% identity may share much of their biochemistry but have different substrates requirements (Todd et al., 2001). In spite of their divergence they may share a common functional domain. A protein domain database, Pfam is widely used in genome annotation and HMMER software package is used for this purpose (Eddy, 1988).

Process level annotation—The main source of process level annotation is the literature. As mentioned above the process level annotation relates to understanding of how a given protein influences phenotypes. How protein interacts with other proteins? Where it is localized? Which cellular process and organelle is it involved with and in which tissue and at which developmental stage does it act? The information about all these will come from the high throughput large scale technologies to generate data on many genes simultaneously. The large scale parallel measurement of gene expression for entire genomes will generate data on the developmental timing and tissue-specificity of many genes from which it is possible to infer process level annotation (see chapter 24 for detail). An important step on the way to designating the process a protein is involved in is to define the proteins with which it interacts and works in a well. For more on annotation see Bioinformatics (Roy, 2009).

Chromosome organization—The GC content is unchanged across all the chromosomes of *C. elegans*, unlike the GC content in vertebrate genomes such as human or yeast.

Counting the gene—Complete sequencing will reveal all possible protein-encoding genes in the form of ORFs. An estimate of gene count can be obtained in the following ways.

1. A preliminary gene count can be obtained from the mean gene size, for comparison against the number of genes identified by gene prediction programs discussed in Bioinformatics (Roy, 2009). The estimated gene size of 4.5kb for rice is similar to maximal gene density of 1 gene per 4 to 5kb which was obtained from the analyses of syntenic loci across many plant species provided information on gene density.

2. Knowing the proportion of intergenic fraction (42% in rice) and genome size, one can obtain the gene count

3. One could include information from CpG island counts in the gene count although not every CpG island is associated with a gene. Identification of CpG island is done by standard algorithm (Larsen et al., 1992)

After a new ORF is identified, one will have to find out

(i) Whether it represents a functional gene producing protein

(ii) Whether the DNA sequence is actually transcribed.

(iii) What happens to the organism if the ORF is disrupted by mutation either induced or natural?

For answering the first question one will have to find out whether the predicted amino acid sequence resembles anything in the database of sequenced genes of known function. The degree of resemblance will provide some clue about the function of the new ORF but further evidence of functionality will still be required. Finding out the answer of the second question is difficult as it is not easy to determine because genes may be only weakly transcribed but still play essential roles and in complex organisms, many genes are transcribed only in certain specialized tissues.

Annotated genome—It refers to the sequence of itself besides a description of the likely function of each gene product.

Gene prediction and annotation—There are two general approaches to gene finding.

1. The homology-based methods include the use of known mRNA sequences as well as gene families and interspecific sequence comparisons.

2. The *ab initio* methods include detection of exons and other sequence signals like splice sites by various computational methods within the sequence being analyzed.

Finding and dissecting these important sequence regions can be done by means of motifs which are known to be conserved in transcription factor-binding regions. Interspecific genome comparisons are one of the ways of getting at these regions (those regions as being conserved). Array-based technique can be used to locate genome-wide sites of action of transcription factors (Tatusov, et al., 1997) which will sort out *cis*-regulatory signals in the genome.

Tools for automated annotation — There are a number of approaches which use a combination of statistical and heuristic methods to recognize genes and gene features. These are HMM, neural nets and Bayesian networks. They are most effective in finding genes and are used in concert with homology-based methods. Efficiency of these methods are strongly affected by errors in sequencing and statistical biases in base composition. Thus draft sequence in which error rate is higher can be markedly inferior to finished sequence for *ab initio* prediction. Genscan is a widely used software for gene finding and prediction. Another software Genie which is a HMM system and allows for the integration of information from different sources such as signal sensors (splice sites, start codon, etc), sensors of introns, exons and alignments of mRNA, EST and peptide sequence, gives promising result. Ensemble follows *ab initio* predictions by GENSCAN with mRNA, EST and protein motif information comparisons for initial predictions. It then uses a program called GeneWise to extend protein matches.

Rule-based expert system for annotation — Venter et al., (2001) reported the development of a rule-based expert system for annotation. They called 'otto' which attempts to embed some human curational function in software. The evidence-based approach- otto is used to increase the probability of identifying genes and it uses regions conserved between different genomes (e.g. mouse and human), similarity to ESTs or other mRNAs-derived data or similarity to other proteins.

De novo gene prediction although less accurate is the only way to find genes that are nor represented by homologous proteins or ESTs. Problems with computational identification of transcriptional units in genomic DNA sequence can be divided into two phases.

1. Partitioning the sequence into segments that are likely to correspond to individual genes (a problem with *de novo* gene finding algorithm).
2. Construction of a gene model reflecting the probable structure of the transcript (s) encoded in the region. This can be done with higher accuracy when a full length cDNA has been sequenced or a highly homologous protein sequence is known.

Mining information — The genome sequence data from draft genome can be mined to help understand the following.

1. How gene families are created, amplified and diverged to create new biological activities and specificities?
2. How different classes of repetitive DNAs become dominant in higher eukaryotes?
3. Comparison of the gene complements in rice and related species to see which pathways they share, which are unique and how these pathways may have been modified.
4. Use of full set of genes to permit comprehensive characterization of gene expression by any of the several high throughput approaches.
5. Comparison of the complete set of predicted and known peptides to identify those that are of most interest for 3D characterization.

6. Identification of genes through '**candidate gene approach**'. A completed genome sequence is the first step toward a 'candidate gene approach' of biology. Identification of genes and characterization of gene variation by genomic sequencing only provides a correlation with a particular process. Use of identified genes can be made for studying allelic variability and distribution in an organism. Relate specific changes in gene structure and content to differences in the evolved biology of different species. Genes identified will provide the raw material to determine why particular characteristics are shared or not shared by specific lineages of organisms. Thus identify both the shared and diverged genic complements to understand the difference between species.

14

Protein Synthesis and Sequencing

14.1 DEFINITION

Proteins are long polymers of amino acids (AAs). The sequencing of insulin in 1955 by F. Sanger first proved that proteins have definite amino acid sequence. Some proteins have catalytic functions and serve as enzymes and others have structural function and serve as structural elements, signal receptors or transporters carrying specific substances into and out of the cells. Proteins are more complicated than DNA. It gets phosphorylated, glycosylated, acetylated, ubiquitinated, farnesylated, sulphated, or linked to glycophosphatidylinositol anchors. Proteins response to altered conditions by changing the location within the cell, getting cleaved into pieces and adjusting their stability as well as changing what they bind to (other proteins, nucleic acids, lipids, small molecules or other ligands). Protein levels often do not reflect mRNA levels and even the presence of an ORF does not guarantee the existence of a protein. Finally, single protein may be involved in more than one process and conversely, similar functions may be performed by different proteins (Fields, 2001). The following are the facts about proteins.

1. Proteins with different functions always have different AA sequence.
2. Changes in primary structure lead to change in function of proteins. Proteins polymorphism results from change in the primary structure, i.e., change in the AA sequence. About 20-30% of proteins in human are polymorphic and variation in AA sequence have little or no effect on the function of proteins
3. Proteins with similar function from different species often have similar AA sequence.
4. Proteins that carry out a broadly similar function in distantly related species can differ greatly in over all size and AA sequence.

Like other macromolecules such as polysaccharides and nucleic acids, proteins are also macromolecules. Polysaccharides are polymers of simple sugars such as glucose, nucleic acids are polymers of nucleotides and similarly proteins are polymers of AAs. All these macromolecules are formed as a result of polymerization of compounds with molecular weights of 500 or less. Thus amino acids are the basic unit of protein. There are twenty different AAs. Each AA is composed of C, H, O, N and sometimes S. The different amino acids can be classified as either **polar**, charged and water soluble or **non-polar**, non-charged and water insoluble. Polar and positively charged AAs are **basic (lysine, arginine, histidine)** whereas the polar and negatively charged AAs are **acidic** in nature. Non-polar AAs (lacking electronegative atoms such as N, O ands S) can have either aliphatic (having aliphatic R groups, H, CH2, etc.) or aromatic side chain (aromatic R groups, -< >). Aromatic (tryptophan, tyrosine, phenylalanine) and aliphatic amino acids are classified as hydrophobic (valine, leucine, isoleucine, methionine). Acidic and basic AAs confer a charge to protein (depending upon the pH of the medium) when present in high concentration and thus can be considered polar. The acid-acid amide group includes glutamine, glutamic acid, asparagines, aspartic acid. Groups of amino acids which are small but not strongly hydrophilic or hydrophobic are glycine, alanine, proline, threonine and serine and cysteine.

Thus amino acids can be classified into above mentioned broad groups (Table 14.1). But in fact their properties are far more complex and there is considerable overlap between them as can be seen in venn diagram.

Table 14.1   Showing different groups of aminoacid.

Groups	Name of AA	Polar/Non-polar	Three letter code	Single letter code
Aliphatic	Glycine	P	Gly	G
	Alanine	NP	Ala	A
	Valine	NP	Val	V
	Leucine	NP	Leu	L
	Isoleucine	NP	Ile	I
Hydroxylic	Serine	P	Ser	S
	Threonine	P	Thr	T
Sulphur	Cysteine	P	Cys	C
	Cystine			
	Methionine	NP	Met	M
Acidic	Aspartic acid	P	Asp	D
	Glutamic acid	P	Glu	E
Basic	Lysine		Lys	K
	Arginine		Arg	R
	Histidine		His	H
Aromatic and heterocyclic	Phenylalanine	NP	Phe	F
	Tyrosine	P	Tyr	Y
	Tryptophan	P	Trp	W
Imino acid	Proline	NP	Pro	P
	Hydroxyproline			
	Glutamine		Gln	Q
	Asparagine		Asn	N
	Asparagine/Aspartate		Asx	B
	Glutamine/Glutamate		Glz	Z

The individual AAs are linked together through a linkage called **peptide bond** (-CO = NH-) which is formed by condensation of á-COOH(carboxyl group-negatively charged) of one AA with α- NH2(amino group- positively charged) group of another. The α carbons of adjacent amino acid residues are separated by three covalent bonds, C_α-C-N-C_α. Out of the three bonds, C-N bond is unable to rotate whereas the N- C_α and the C_α –C bonds can rotate and these bonds rotate with angles φ and ψ, respectively. The rigid peptide bond limits the range of conformations that can be assumed by a polypeptide chain. The length of peptide ranges from 2 AAs. to many thousands of AAs(polypeptide). For example, a number of hormones in vertebrate are small peptides active at very low concentration. Similarly toxic mushroom poisons and many antibiotics are small peptides but majority of the naturally occurring polypeptides are made up of ~ 2,000AA residues. The peptide has an amino terminal containing free α- amino group and a carboxyl terminal containing free carboxyl group. The amino acid unit in a peptide is often called a AA residue (after loosing a hydrogen atom from its amino group and a hydrogen moiety from its carboxyl from its carboxyl group. *In vivo* the protein synthesis proceeds from amino terminus to carboxyl terminus, the reverse of the direction of chemical synthesis of protein. First a chain of polypeptide is synthesized. Disulphide bond formation and *cis-trans* isomerization of proline peptide bonds are catalysed by specific enzymes. Most proteins contain about 100 to 10,000 amino acid residues. Typically proteins are 200-400 amino acids long. The order and the number of AA residues determine the special property of each protein. The –C-N- atoms of the peptide bond forms the back bone and the side chains of the invidual AAs are arranged *trans* to each other across the back bone. The formula for all the AAs are the same except the nature of the side chain R(Figure 14.1).

14.2   VENN DIAGRAM

One approach of studying protein structure is to consider amino acid residues as members of groups as defined by shared biochemical properties. Venn diagram refers to a graphical description of the chemical similarities between the 20 or more amino acids based on their physico-chemical properties which determine the protein structure. Venn diagram shows (Figure 14.2) overlapping groups of amino acids, for example, polar, basic, hydrophobic, small and so on. The basic and acid-acid amide groups tend to replace one another to some extent and phenylalanine interchanges with the hydrophobic amino acids more often than chance expectation predicts. These patterns are imposed principally by natural selection and only secondarily by constraints of the genetic code. They reflect the similarity of functions of the amino acids in their interactions in the 3-dimensional conformation of proteins. Some of these properties of an amino acid residue which determine these interactions are size, shape and local concentrations of electric charge, conformation of van der Waals surface, ability to form salt bonds, hydrophobic bonds and hydrogen bonds.

14.3   PRIMARY STRUCTURE

The primary structure of protein is determined by the sequence of AAs. The long chain of AA can not remain as such in the cell and so after synthesis of polypeptide there is shortening of the length of AA chain by way of adopting secondary structure.

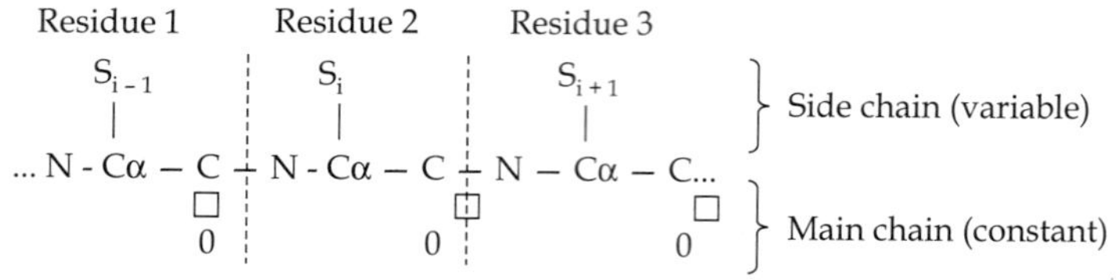

$$\text{Residue 1} \qquad \text{Residue 2} \qquad \text{Residue 3}$$

FIGURE 14.1   **Showing the polypeptide chain of a protein with amino terminal (N) and carboxyl terminal (CO).**

14.4   SECONDARY STRUCTURE

It refers to the spatial inter- relationship of AAs in segments of polypeptide. A variety of forces are possible between AAs and these lead the AAs along many portions of the polypeptide backbone to adopt relatively simple specific orientations with respect to one another. These are known as secondary structures. The string of AAs can wind into a either α-helix (or coil) or take the shape of a pleated sheet (β-sheets) (Pauling and Corey) (Figure 14.3) and beta band (turn). The secondary structure thus refers to recurring structural patterns. The local spatial structure of small numbers of AAs, independent of the orientations of their side groups generates secondary structure. Protein has a regular structure that repeats every 5.15 to 5.2 Å (5.4 Å). The helical form (right handed) has 3.6 AA residues per complete turn and rise along the central axis of 1.5Å per residue. The polypeptide back bone

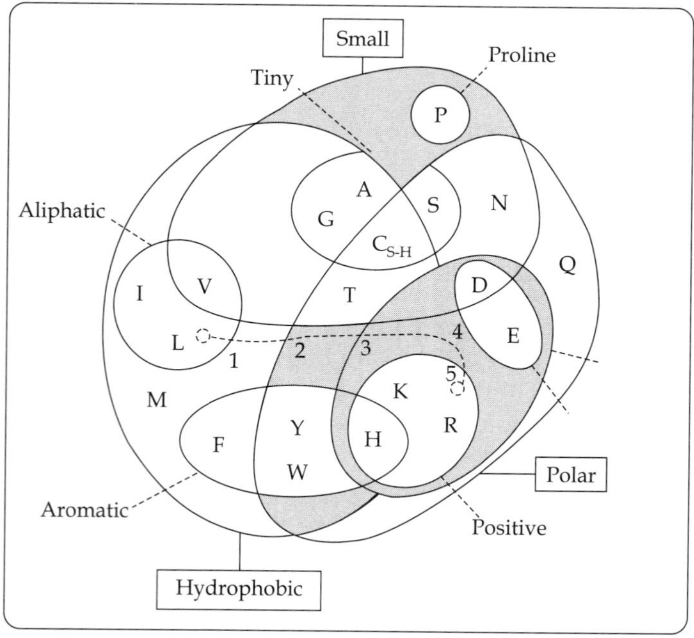

FIGURE 14.2   **Showing Venn diagram showing the properties of amino acids.**

is tightly wound around an imaginary axis (central axis) drawn longitudinally through the middle of the helix and the R groups of AA residues protrude outward from the helical backbone. The helical structure is maintained by hydrogen between carboxy oxygen (-CO) and the amide nitrogen residue (-NH2) apart in the peptide backbone. The α- helix has been found in both fibrous and globular proteins. The zig-zag polypeptide chain can be arranged side by side to form a structure resembling a series of plates called β-sheets. Here the hydrogen bond is between the two polypeptide chains. The two peptide chains may run parallel in that their N -atoms point in the same direction or anti parallel and in this case alternate chains are oriented in the same way. β-sheets have been found in the fibrous proteins. Protein structures are divided into four classes, all α, all β, α/β (alternate) and α + β and α - β regions are somewhat separated. Every type of secondary structure can be completely described if the bond angles Φ and ψ at each amino acid residue in a polypeptide segment are known.

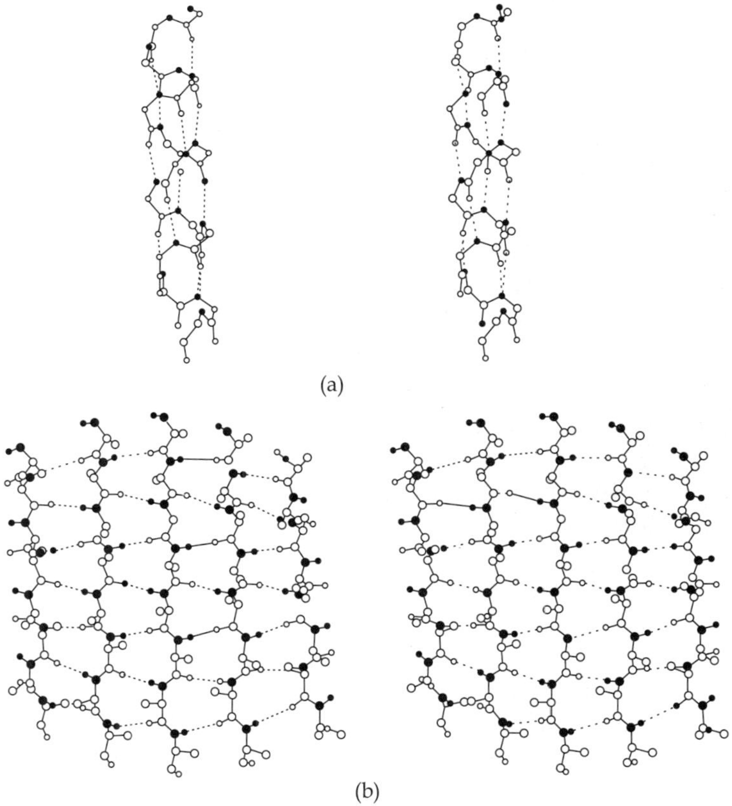

(a)

(b)

FIGURE 14.3   Showing secondary structures of proteins. (a) α-helix (b) β-sheet. Broken lines show H-bonds. (b) shows a parallel β-sheet in which all strands point in the same direction. Antiparallel β-sheets are also common. β-sheet can be formed by any combination of parallel and antiparallel strands.

14.5 TERTIARY STRUCTURE

It refers to the over all shape of the protein molecule which in turn shows the arrangement in space of polypeptide. In case of a typical globular protein tertiary structure refers to the assemblage of polypeptide segments in the alpha-helix and beta-sheet conformations linked by connecting segments. The tertiary structure thus can be described by defining how these segments stack on one another and how the connecting segments are arranged. Both the arrangement of the secondary elements and the spatial arrangement of all the atoms of the molecule are referred to as the tertiary structure. It also refers to the folding of polypeptide in three dimensional space. Each protein has unique three dimensional structure that reflects its function. The three dimensional structure of a protein is determined by its AA sequence. Specificity of enzymes depends on their unique, complex three dimensional structures which in turn are determined by primary structures- the AA sequences which are determined by base pair sequences in the structural genes. The spatial arrangement of atoms in a protein is called its **conformation.** Each of the AA could take up on average 10 different conformations. So a polypeptide containing 100 AA residues could take up 10^{100} different conformations. Polypeptide chain is folded into characteristic and specific conformation depending upon the AA sequence. Protein molecules may take the form of a very compact molecule as in case of globular protein or may form long thin threads as in case of fibrous proteins. The tertiary structure is maintained by a number of bonds such as hydrogen bond(-C= O-H-N-) or ^\N-H-O-CH2- between side chains of amino acids, ionic bond between acidic and basic AAs(-COOO-NH3-) or between basic proteins with other acidic macromolecules such as nucleic acid etc., disulphide bond(-S-S-), hydrophobic bonds(-CH3= CH3-), —<><>- between non-polar side chains of AA residues, hydrophilic bond between proteins and burried water molecules on the surface of protein molecule, dipole-dipole interactions and phosphodiester bonds.

14.6 QUARTERNARY STRUCTURE

It refers to association of two or more polypeptides in multimeric forms. Thus it refers to the sub unit structure of proteins. Multi-subunit proteins have two or more proteins. Further they may be made up of two or more identical or dissimilar polypeptide chains. For example, Haemoglobin is made up of 4 polypeptide chains($2\alpha + 2\beta$ chains). The peptides are associated non-covalently.

Considering these higher levels of structure, protein can be classified into two major groups; fibrous proteins (structural proteins) having polypeptide chains arranged in long strands or sheets and globular proteins (regulatory proteins) having polypeptide chains folded into a spherical or globular shape. Fibrous proteins usually consist largely of a single type of secondary structure whereas globular proteins often consist of several types of secondary structure. There can be aggregation of protein based on the repetition of identical or near identical amino acid sequences and this kind of specific intermolecular association is known as **amyloid fibrils** or plaques.

14.7 FUNCTION OF PROTEINS

Proteins can be classified as **simple proteins** containing only AA residues or **conjugated proteins** which contain chemical components (non-amino acid part) called **prosthetic groups**. So proteins might be associated with lipids called lipoproteins, with sugar-called glycoproteins, phosphate, flavin nucleotides or metal ions(Fe, Zn, Ca, Mol, Cu). Proteins often contain regions within their AA sequence that are essential for their biological function. AA sequence in other regions might vary considerably without affecting these functions. Thus a polypeptide is made up of invariant AA residues and regions of variable AA residues. In the region (s) of variable AA residues changes could be either conserved substitution (positively charged AA is replaced by positively charged or the negative charged AA replaced by negatively charged AA) or non-conserved substitution. Also there could be variation in the number of differences observed in the variable positions. Homologous proteins with similar function from different species will have the same AA residues at many positions (invariant AA residue) but show considerable variation at other positions in the AA residue (called variable AA residue). Invariant residues are most critical to the structure and function of a protein than the variable ones and the number of AA residues that differ in homologous proteins from two species is in proportion to the phylogenetic difference between these species. For example, Cytochrome c, an iron containing protein of duck differs from that of chicken in only 2 AA residues whereas Cytochrome c of horse differs from yeast in 46 AA residues. Pigs, cow and sheep have indentical AA sequence in cytochrome c. The proportion of sequence that is critical varies from protein to protein. Those regions of polypeptide chain that can fold into one, two or more stable and independent globular units are called **domains.** Domain refers to any particular or specific two or three dimensional structure of a macromolecule usually a protein which forms a structural or functional niche within the remainder of the protein molecule. In other words, domain is a structural unit of a protein consisting of compact folded region and usually encoded by a single exon. For example, DNA binding proteins possess specific feature (DNA binding domain, e.g., helix-turn-helix, helix-loop-helix configuration or Zn^{2+} fingers which enable them to recognize and bind to specific structures or sequences in the target DNA with high specificity (see chapter 25). It is a local group of AAs that have many few interactions with other portions of the protein than they have among themselves. Small proteins generally have only a single domain whereas large proteins may have several domains. Polypeptides with more than a few hundred AA residues often fold into two or more stable globular units (domains), hence the majority of proteins contain more than one domain. Domains within the protein may have separate functions such as binding of small molecules or interaction with other proteins. In many cases a domain from a very large protein retains its correct three dimensional structure even when it is separated from the remainder of the polypeptide chain through the use of proteolytic cleavage. Thus the study of a protein's structure can be done on a domain-by-domain basis. In summary, domain is a compact, local, semi-independent folding unit, supposed to have arisen through gene fusion and gene duplication. Domains need not be formed from contiguous regions of an amino acid sequence. They may be discrete entities, joined only by a flexible linking region of the chain. They may have extensive interfaces, sharing many close contacts and they may exchange chains with domain neighbours. The combination of domains within a protein determines its overall structure and function. A domain is best characterized in structural terms because physical parameters can be applied (i.e., ratio of diameter versus surface area) which can provide clear thresholds for domain definition. If proteins are known to harbor a certain domain or such a domain has been identified in a database search then one should perform separate searches with the remaining part of the proteins. Homologous domains can be found in a diverse set of protein with unrelated overall functions. Genetic mechanisms exist which allow their horizontal spread within genomes. In case of animals, exon shuffling (see chapter 5) appears to be one such mechanism especially in case of extracellular proteins but because it requires intron so this mechanism is not operating in bacterial proteins. Both plasmid transfer and recombination mechanisms are likely to contribute to the horizontal transfer of domains between bacterial genomes.

14.8   HOMOEOBOX PROTEIN

It refers to any one of the series of proteins which share a conserved homeodomain. These proteins are selectively expressed in both fetal and adult cells and tissues and play fundamental roles in the development of an organism. For example, the HOX proteins (HOXA, B, Hex and Phox) are involved in cell differentiation, Pitx2 establishes left-right asymmetry and six proteins regulate muscle formation.

Homeodomain—It refers to a stretch of about 60 amino acids in a specific class of DNA-binding proteins which contact its target sequence, usually an AT-rich sequence. It is folded into an alpha helical structure. After binding to the target DNA, homeodomain protein activates transcription of neighboring genes. Homeodomain motifs are found in such diverse factors as OCT-1, OCT-2.

14.9   PROTEIN FOLDING

After the synthesis of primary structure regions of secondary structure(α sheets or β sheets) are formed which is followed by folding into **super secondary structures or motifs or folds**. Supersecondary structures (Figure 14.4) also called motifs or simply folds are particularly stable arrangement of secondary structure and the connections between them. In other words it refers to the arrangement of α-helices and/or β strands into discrete folding units, for example, α-barrels, β α β–units, etc. These three terms are often used interchangeably. Motifs sometimes appear in repeating units or combinations. A single large motif may comprise the entire protein as in case of the coiled coil α-keratin. Thus a protein complete three-dimensional structure is built upon protein substructures, the supersecondary structures and domains. Finally, large collections of folding intermediates are rapidly brought to a single conformation.

Folding rules

1. Hydrophobic interactions contribute more to the stability of protein structure. The buried hydrophobic amino acid R groups require at least two layers of secondary structures and the motifs creating these layers are the β-α-β loop and α-α corner.

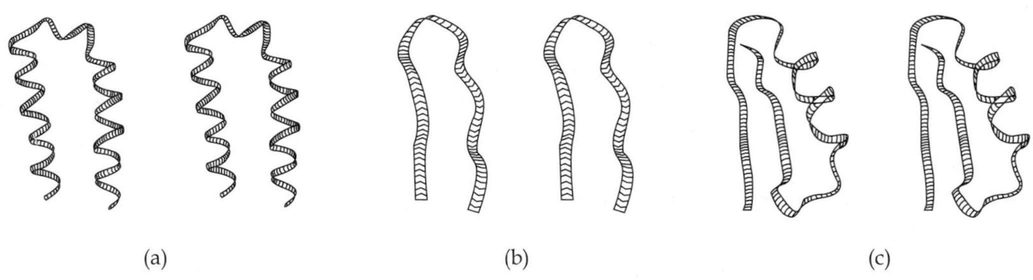

FIGURE 14.4   **Showing super secoondary structures. (a) α-helix hairpin (b) β-hairpin (c) β-α-β unit. The chervons show the direction of the chain.**

2. α helices and β sheets are found in different structural layers when they occur together in a protein. Backbone of polypeptide segment of β conformation can not readily bond with an α helix.

3. Polypeptide segments adjacent to each other in the primary sequence are usually stacked adjacent to each other in the tertiary structure. Distant polypeptide segments may come together but this is not the norm.

4. The connecting elements of secondary structure do not cross or form knots.

5. The β conformation is most stable if the individual segments are slightly twisted right handed. Two parallel β sheets are connected by a crossover strand and the crossover is almost always right handed. The twisting of β sheets leads to the characteristic twisting of the structure.

The folding in many proteins is facilitated by Hsp 70 chaperones, the molecular proteins helping in providing micro-environment for folding, and chaperonins.

Protein function often brings about interactions with other molecules. A molecule bound by a protein is called a **ligand** and the site on the protein is **called binding** *site*. Ligand binding can be regulated. Proteins are not rigid and may under go conformational changes when a ligand binds. Proteins thus owe their specificity and catalytic activity to the correct arrangement of AAs on the space for the substrate to bind.

14.10   PREDICTION OF STRUCTURE OF A PROTEIN

Knowledge of proteins' structure is essential for understanding their functions and interactions with other proteins. It will tell us how the repertoire of protein folds is used and how it is distributed among different functional categories in different species. Further, for interspecies comparisons, protein structure can reveal relationships not visible in highly diverged sequences. The objective of protein structure prediction is to infer the unique three-dimensional protein structure of any protein simply from its amino acid sequence or primary structure. The primary structure of protein of protein is determined by its linear sequence of amino acids. The secondary structure of many proteins can be rather accurately predicted by empirical rules.

There are three main approaches to secondary structure prediction.

1. Empirical statistical methods which use parameters derived from known 3D structures

2. Methods based on physiochemical criteria such as fold compactness, hydrophobicity, charge, hydrogen bonding potential, etc.

3. Prediction algorithms that use known structures of homologous proteins to assign secondary structure (Kyngas and Valjakka, 1998).

One of the standard empirical statistical methods is that of Chou and Fasman (1978), which is based on observed amino acid conformational preferences in non-homologous proteins. They computed structure parameters (considering all the atomic interactions) whose structure had been determined by X-ray diffraction. These parameters give a probability that an amino acid will be found in alpha helix, beta sheet or beta turn. For example, the probability of leucine is P alpha(Leu) =Frequency that leucine is in alpha helix/ Frequency of all AAs being in alpha helix. The probabilities were calculated for each of the AA in each of the structures alpha helix and beta sheet and in each of four positions in a reverse or beta turn. Using a table of these probabilities and a few simple rules for combining the probabilities over neighboring AA residues, the secondary structures of regions of a protein may be determined with about 75% accuracy. In case of prediction algorithms the use of sequence alignment information can improve the accuracy percentage (about 70%) but is of little practical value because in a blind prediction we do not know which 70% is correct. The tertiary structure of a protein can not be predicted and what is required is to formulate a code by comparing the tertiary structures of many proteins to the amino acids involved. The direct prediction of structure from sequence so still remains very far.

The secondary structure of a polypeptide segment can be completely defined if the φ and ψ angles are known for all amino acid residues of that segment. Structure prediction of protein uses conserved hydrophobic residues whereas for RNA, conserved base-pairing residues are used.

14.11 PREDICTION OF THREE-DIMENSIONAL PROTEIN STRUCTURE

There are several approaches to protein structural prediction.

1. Folding simulation approach

2. Use of known protein structures

3. Threading (aligning) protein sequences.

The folding simulation approach suffers from the problem of size effects. There is extremely large number of conformations that a polypeptide chain can adopt. Although the entire protein folding process can be simulated on the computer but this type of computer simulation is effective only for small-sized polypeptide chain. This is because of the reason that there is extremely large number of conformations that a polypeptide chain can adopt. The second problem arises from the intrinsic nature of protein folding. Folding is a co-operative transition between denatured and folded states. During forming folded structure each part of the protein molecule is influenced by the entire structure and vice-versa and because of this reason there is no **one-to-one correspondence** between the sequence and the structure. For example, a segment of polypeptide with a certain sequence may adopt an alpha helical conformation in one protein whereas a beta-strand in another. If this method of prediction succeeds then it would also explain the mechanisms of protein folding. Prediction of novel folds can be made either by a priori or knowledge-based methods.

Another type of approach uses rules relating sequence and structure which are derived from the known protein structures. This approach is successful with secondary structure prediction but has not been able to predict tertiary structures. The **threading (aligning) protein sequences** approach is based on the assumption that the folded conformation is known and present in the structure database and one can simply examine whether the new sequence is likely to fit any of these known structures. Threading is a method for fold recognition. This method is an extension of the methods which search for homology between protein sequences (sequence analysis) (see Roy, 2009). Homology modelling is a method of prediction the 3-dimensional structure of a protein from the known structures of one or more proteins. Sequence comparison methods such as PSI-BLAST or HMMs can identify relationships between proteins both within an organism and between species. Although several proteins of unknown structures have been successfully predicted by this method, there is an inherent limitation to this method. It can not predict any truly novel structures as it assumes that the structure is already known. Further, it does reveal the mechanism of protein folding.

Prediction of secondary structure is an intermediate step in the prediction of its tertiary structure. Most secondary structure prediction methods predict only three states, α-helix, β-sheet and coil (Chou, 2000). However, in addition to these three repetitive structural states tight turn is a significant element frequently occurring in protein structure. Based on the number of their constituent amino acid residues tight turns are categorized as δ, γ, β-, α- and π-turns. Of these five turns, the occurrence of β-turn is the most frequent, constituting approximately 25-30% of the residues in globular proteins (Kaur and Raghava, 2002). In contrast the second most frequently occurring tight turn, γ-turn takes up only 3.4% of the total residues. β-turn is also an important stage in protein folding and because they usually occur on solvent-exposed surfaces, they often participate in molecular recognition processes in interactions between peptide substrates and receptors (Rose et al., 1985). Accurate β-turn prediction would increase the accuracy and reliability of secondary structure prediction. β-turn prediction algorithms are very few. Most of the beta-turn prediction methods use statistical approaches.

14.12   MACHINE LEARNING APPROACHES

Most of the recent protein structure prediction algorithms which perform better than earlier statistical approaches have been developed via machine learning and amongst them neural network and support vector machine (SVM) are most notable (see Roy, 2009). Neural network algorithms usually use a segment of peptide sequence as the basis for prediction where it looks automatically for subtle correlations between input amino acids and their structural preference via back-propagation training process. In these approaches each of the segment resides is transformed into 20 (or 21) nodes of numerical data which are then used as 20 (or 21) numerical values for the input nodes (or neurons) of the neural network. During the training process, correlations between each set of the input nodes and output data are automatically adjusted to be in line with the relationship between the structure and the preference of amino acid (Xie and Hwang, 2006).

14.13   SEQUENCING OF POLYPEPTIDE

Sequencing is done to determine the primary structure of a protein. The different steps involved in the sequencing of a very large polypeptide are as follows.

1. **Determination of AA composition and amino terminal of an intact polypeptide. AA composition**—The sample protein is hydrolysed. The hydrolysis of protein results in production of free AAs and the AA composition (types and amounts of AAs) can be determined using HPCL or ion exchange chromatography.

$$\text{Polypeptide} \xrightarrow{\text{6M HCL}} \text{Free AAs} \xrightarrow[\substack{\text{Ion exchange} \\ \text{chromatography}}]{\text{HPLC or}} \text{AA composition}$$

This method is useful in interpreting the results obtained from other procedures.
Identification of amino terminal residue—The amino terminal residue of an intact sample(polypeptide, $NH3^+$- R_1-R_2-R_3——R_n-$COOH^-$) can be determined by Sanger's method. In this method the polypeptide is labeled with a reagent, FDNB (1-fluro, 2-4, dintrogenbenzene) which is then hydrolysed through the use of HCL to its constituents AA and the labelled AA is identified. It will also help in determining the number of chemically distinct polypeptides.

$$\substack{\text{Polypeptide} \\ \text{terminal residue}} + \text{FNDB} \xrightarrow[\text{+ Free AA}]{} \xrightarrow{\text{6M HCL}} \substack{\text{2, 4 Dinitrophenyl} \\ \text{derivative of amino}}$$

Besides FNDB the other reagents used are dansylchloride and dabsylchloride. These two yield derivatives which are more easily detected than Sanger's reagent, FNDB.

2. **Breaking of disulfide bonds**—There are two methods of breaking of disulphide bonds, if any.

   A. Oxidation of a cystine residue with performic acid will yield two cysteic acid residues.
   B. Reduction by dithiothreoitol to form cystine residues followed by acetylation by iodoacetate to modify the relative group-SH to prevent reformation of disulfide bond. This results in formation of acetylated cystein residues.

3. **Fragmentation of larger polypeptide**—The larger polypeptide chain must be broken into a number of smaller fragments and after that each fragment will be separated and sequenced. There are different methods for obtaining smaller fragments. These methods use either enzyme or chemical agents to cleave the peptide bond adjacent to specific AA residue. **Enzymatic procedure-** In the enzymatic procedure such proteases are used which catalyse the hydrolytic cleavage of peptide bonds. Some proteases cleave only the peptide bonds adjacent to particular AA residue. Table 14.2 – shows the different kinds of proteases along

with the cleavage points. Cleavage of peptide bond occurs either on carbonyl (C) or amino (N) side of the initial AA.

4. **Separation of fragments** — The smaller fragments generated above is separated by chromatography or electrophoretic methods (see chapter 30).

5. **Sequencing by Edman procedure** — Edman procedure reveals the entire sequence of a peptide. For shorter peptides it alone yields the entire sequence and there is no need to go for hydrolysis and apply Sanger's method for detecting amino terminal residue. Edman degradation procedure labels and removes only the amino terminal residue from a peptide leaving all other peptide bonds intact and one AA is detected a time. In this procedure the polypeptide is reacted with phenyl isothiocyanate which is then hydrolysed by HCL producing phenyl thiohydantoin derivative of AA (Amino terminal residue) which is identified and purified.

$$\text{Polypeptide} \quad + \quad \text{Phenylisothiocynate} \quad \xrightarrow{\text{6M HCL}} \quad \text{Phenylisothiocynate} + \text{Phenylthiohydanton}$$

The left peptide fragment is recycled through Edman process and after every cycle one AA is identified. The Edman process has now been automated and modern machine used for this purpose is called '**Sequennator**' which works with 99% efficiency and it can sequence a polypeptide of more than 50 AA residues. **Microsequencing**- It is a technique to increase the sensitivity of the conventional protein sequencing by 2 to 3 order of magnitude into the picmol range. In gas phase microsequencing the reagents of the Edman degradation are carried out by a steam of argon to the protein which is bound to a polybren film. This reduces the amounts of solvents, reagents and byproducts which leads to increased sensitivity of the technique.

6. **Ordering the fragment** — While ordering the fragments there comes the problem of overlapping and in view of this one sample should be cleaved with one enzyme or reagent and the another sample should be cleaved with a different enzyme or reagent that cleaves the peptide bond at points other than that cleaved by first enzyme. Thus the overlapping between fragments produced by 1st cleavage procedure will be corrected by the overlapping peptide obtained from the second fragmentation. In case overlapping problem still remains one should go for third or even a fourth cleavage method. Suppose there is a peptide containing 38 AA residues and Glutamate (E) is the amino terminal residue. Cleavage with Trypsin produces four fragments (T1, T2, T3 and T4) whereas with cyanogens bromide three

Table 14.2   Showing various enzymes and their cleavage points.

Amino acids	Cleavage points
Trypsin	Lys, Arg (C)
Submaxillaris protease	Arg (C)
Chymotrypsin	Phe, Trp, Tyr (C)
Staphylococcus aureus V8 protease	Asp, Glu (C)
Asp-N-protease	Asp, Glu (N)
Pepsin	Phe, Trp, Tyr (N)
Endoproteinase	Lys (C)
Cyanogen compound	Met (C)

fragments (C1, C2 and C3) are generated.Trypsin will cleave at Lysine and Arginine. As amino terminal residue is Glu (E) so T2 is the first fragment of the peptide and as T3 fragment does not contain either R or K so it is the last fragment and problem is now to distinguish between T1 and T4 and which one is the second fragment is difficult to say at this point. Now as the cleavage with cyanogens bromide produced three fragments, C1 is the fragment having starting amino terminal residue, Glu and ending with M, so it is the first fragment. T3 was the fourth fragment so C2 which includes T3 is the last fragment and C3 becomes the second fragment which overlaps with T1 and T4 and thus allowed it to be ordered (Figure 14.5).

Location of disulfide bond — Once sequencing is completed then one can go for locating the disulfide bond. One protein sample is treated with trypsin to generate fragments which are then separated by electrophoresis. After that in another sample of protein disulfide bond is broken first and then trypsin is used to generate the sets of peptide which are again separated by electrophoresis. Now the two sets of fragments are compared. In the later set two of the original peptides will be missing and a new larger peptide will appear. The two missing peptides represent the regions of the intact polypeptide liked by disulfide bond.

Use of polypeptide sequencing — Once the sequence of a polypeptide is known the sequence of the nucleotide in the gene can be determined. In case a gene is available it is better to do sequencing of the gene rather than the protein as sequencing of gene is faster and more accurate than sequencing of the protein. When a gene has not been isolated then it is better to go for direct sequencing of polypeptide as it provides information on the location of disulfide bonds, the information for which is not available in the DNA sequence.

14.14   DESIGNING GENE SPECIFIC PROBES

Amino acid sequence can provide information about the structure of gene and thus help in getting them cloned. Information about the AA sequence of even a part of a polypeptide can greatly facilitate the isolation of the corresponding gene. Knowledge of the sequence of a segment of 6 to 8 AA residues is enough to pinpoint the gene encoding the entire protein. Although because of degeneracy of codons the mRNA sequence can not be deduced exactly from the amino acid sequence, however, by choosing a part of the amino acid sequence with minimum coding degeneracy and considering the strong bias in favour of certain codons by different organisms it is possible to select a panel of candidate coding sequence, say the most probable sequences of 25 bases. A sequence of 25 bases is long enough to have a good chance

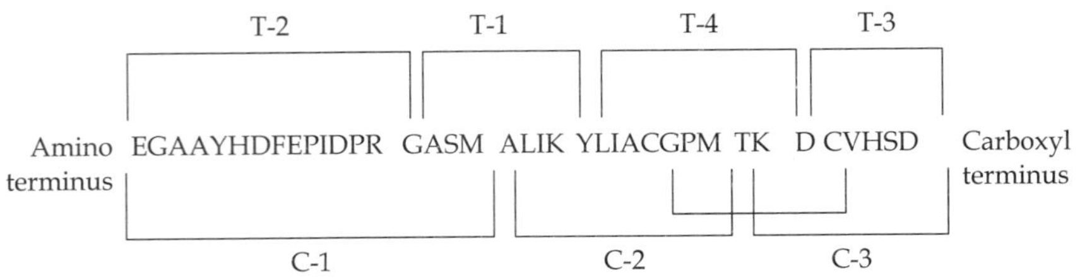

FIGURE 14.5   Showing ordering of peptides.

of identifying a unique gene. These sequences can now easily be synthesized by automatic DNA-synthesis machine and thus **gene-specific probes** can be designed. This method can be used for isolation of genes for which no viable mutants are known. It is just like use of EST (expressed sequence tag) for isolation of a gene. This method is extremely powerful and has been increasingly employed as protein micro-sequencing technology has improved.

14.15 DOUBLE FRAMESHIFT ANALYSIS

A DNA sequence of a gene can be deduced by double frameshift analysis (Montogomery et al., 1978) who used the deduced sequence as a probe for cloning of yeast gene, CYC1 which encodes cytochrome c. Some of the mutations consist of single base pair deletions or insertions and thus the reading frame will be shifted on which decoding of the message depends. At several places along the gene a frameshift mutation was identified that could be reverted to apparent wild type by a compensating insertion or deletion occurring at some distance away from the primary one. The effect of the second frameshift is to restore the correct reading frame but in between the two compensating changes the messenger is read out of frame so that a series of unusual amino acids interrupts the other wise normal amino acid sequence. Knowing the possible codons for each amino acid one can see that only one particular codon sequence is compatible with the replacement of one amino acid sequence by the other as a result of frame shift. The altered amino acid sequence in the wild type derived from a null cyc1 mutant (Figure 14.6) can be explained by two successive frame shift mutations, one is a single base deletion in codon 3 and the other is the insertion of a codon G in codon 11 or 12. Of the several mRNA sequences between codons 4 and 11 consistent with the wild type amino acid sequence only one is consistent with the amino acid sequence in the double frameshift strain. Thus the predicted codon sequence was synthesized and the CYC1 gene in yeast was cloned.

14.16 PROTEOME/PROTEOMICS

The entire protein complement produced (or encoded) by an organism's DNA is called **proteome** and the corresponding complete sequences of an organism's DNA is called the **genome**. It includes not only identification and quantification of proteins but also the determination of their localization, modifications, interactions, activities and ultimately their function. Then there is called **structural proteomics** studying protein structure. 2D gel electrophoresis for protein separation and identification, mass spectrometry, global 2-hybrid techniques and spin offs from DNA arrays and computational tools and methods to process, analyze and interpret the data.

14.16.1 Isolation of a Polypeptide

The cell's proteins can be separated by two dimensional gel electrophoresis and individual protein spots can be extracted from the gel. Small peptides derived from such process can be identified by mass spectrometry (MS) (see chapter 26 for detail) and these sequences can be compared with the genomic sequences to identify the protein. Thus changes in cell's protein complement brought about by environmental, nutritional, biotic and abiotic stress can be examined and genes can be identified and further isolated.

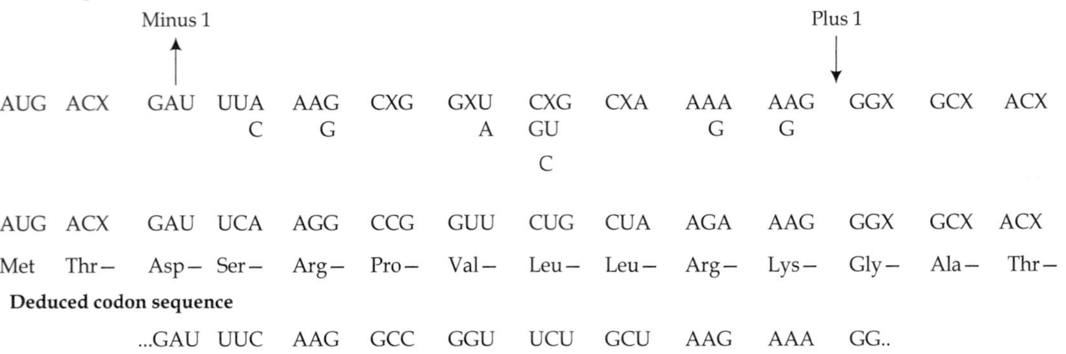

FIGURE 14.6 Showing synthesis of a proble through double frameshift analysis (After Montogomery et al., 1978).

14.16.2 Chemical Synthesis of Protein

Chemical synthesis of protein proceeds from the carboxyl terminus to the amino terminus the reverse of direction of protein synthesis. The different steps involved in the synthesis of protein are as follows.

1. The first amino acid with α-amino group is protected by Fmoc group (9-fluroenyl methoxy carbonyl) and this amino acid is attached to the insoluble polystyrene bead, the insoluble polymer resin through the carboxy terminal of amino acid.

2. The protecting group is removed by flushing with solution containing a mild organic acid.

3. The amino acid-2 with protected α-amino group is activated at carboxyl group by DCC.

4. The α-amino group of amino acid-1 now attacks the activated carboxyl group of AA-2 to form a peptide bond.

5. The steps 2 and 4 are repeated depending upon the length and constitution of amino acids of polypeptide chain to be desired by the experimentor.

6. The α-amino group protected by Fmoc in the completed polypeptide is now removed by treatment with mild organic acid as was done is step 2 and finally, the completed polypeptide chain is cleaved from the resin by HF which cleaves the ester linkage between peptide and resin.

14.17   GENE DESIGN

The computer program is used to design proteins with special structures and/or functions and to translate the amino acid sequence information of such proteins into the corresponding DNA sequences. The DNA sequences are then optimized to function as genes in the target organism (insertion of the correct regulatory sequences and accounting for the codon bias of the organism) and finally, the sequences are synthesized *in vitro* (synthetic gene or chemical DNA synthesis).

Gene isolation—Reverse translation is a technique for the isolation of genes or their mRNAs which uses short oligodeoxynucleotides which are synthesized according to known protein sequences. The oligonucleotides are employed as probes to fish the corresponding genes (in genomic libraries) or mRNA (in cDNA libraries) through hybridization.

The degeneracy of the code allows for a large number of possible sequences that could code for the same sequence of amino acids. While gene designing one must keep in mind the following (Itakura et al., 1977).

1. Where this designed gene is to be expressed? Amino acid codons known to be favored in *E. coli* (if the gene is to be expressed in this organism) should be used.

2. Since the complete sequence of the gene is constructed from a number of overlapping fragments, the fragments are designed to eliminate undesirable inter- and intramolecular pairing.

3. G-C rich followed by A-T rich sequences are avoided as they might terminate transcription.

Further, to facilitate insertion into plasmid the 5'ends have single stranded cohesive termini for the EcoRI and Bam HI restriction endonucleases. A methionine codon precedes the normal NH_2-terminal amino acid of somatostatin and the COOH is followed by the two nonsense codons.

14.18   IDENTIFICATION OF PROTEINS

Proteins can be identified immunologically, by the biological properties is possesses, by its amino acid sequence or merely by its electrophoretic behaviour in polyacrylamide gels (see chapter 30 for detail).

Protein splicing—The precursor protein consists of exteins and inteins. Extein is analogous to exon whereas intein is analogous to intron but it can also be considered as a selfish genetic element. The different exteins are designated as **extein-1, 2, 3** starting from the N-terminus of the precursor protein and the different inteins are designated as **intein-1, 2, 3, 4** starting from the N-terminal of the precursor protein. Any amino acid sequence that flanks an intein in a precursor protein is called extein and any amino acid sequence of a precursor protein that is removed during protein splicing is called intein. In other words, protein splicing is defined as the excision of an insert or intervening protein sequence, the intein from a protein precursor. In other words in the process of protein splicing the exteins of precursor protein are precisely cut out and joined to produce mature protein. Studies of sequence comparison and structural analysis have indicated that about 100 amino acid residues at the N terminus and about 50 amino acid residues at the C terminus of the intein are responsible for the splicing. These two splicing regions are separated by a gene encoding a homing endonuclease. Most (or all) inteins have homology to homing endonucleases which are here called protein insert or protein intervening sequence endonuclease.

14.19 PROTEIN SIGNATURE

It refers to the complex pattern of all proteins and peptides of a cell at a given time as revealed by two dimensional or three dimensional PAGE and subsequent characterization of the individual spots by electrospray ionization or time-of-flight MS. **ESI-MS** is a technique used for the production and mass analysis of peptides, proteins, oligonucleotides and metal complexes.

TOF-MS (Karas and Hillenkamp, 1988)—It is a special type of mass spectrometer which permits the separation of fragment ions based on their mass-to-charge ration (m/z). The traveling time of an ion in the flight tube is correlated to its m/z, the lighter ion arriving sooner than heavier ions.

ESI-TOF—It is a technique that combines electrospary ionization of target molecules (peptides, proteins) with the precision of TOF detection for accurate determination of fragment ion spectra and masses of the analytes.

TOF analysis—Each particle which intersects the focus of the laser beam stays within the beam for a time period that is affected by the stream velocity, the laser beam width and size of the particle. This time period can be directly measured by the analysis, in real time, of the shape of the pulse of fluorescence or light scatter that is provided by the particle. The time interval described by the pulse can be then used as a measure of the particle size (Leary et al., 1979). This process is commonly termed Time-of-Flight analysis.

There are two 'soft ionization' (non-destructive methods to convert proteins into volatile ions) methods for analyzing large biomolecules such as peptides and proteins.

1. **MALDI**
2. **ESI**

Thus protein mixtures are first separated using 2-D gel electrophoresis followed by the excision of protein spots from the gel. The next step is digestion using a sequence-specific protease such as trypsin and then the resulting peptides are analyzed by MS.

In fact, the resulting tryptic peptides are analyzed by ESI or MALADI. These mass spectrometry techniques allow peptide and proteins molecular ions to be put into the gas phase without fragmentation. In the widely used **bottom-up approach**, proteins of interest are digested in solution with an enzyme such as trypsin and the resulting tryptic peptides are analyzed in the gas phase by mass spectrometry (MS) in two stages. In the MS the masses of the intact tryptic peptides are determined. In the MS/MS these peptides ions are fragmented to produce information on the identity and sequence of the protein as well as its post-translational modifications. In the **top-down approach** intact protein ions are introduced in the gas phase and are fragmented and analyzed in MS yielding the molecular mass of the protein as well as protein ion fragment ladders. This information can be used to deduce the complete primary structure of proteins. Both methods make extensive use of correlations of the mass spectrometric data with protein and whole genome sequence database (Chait, 2006). The 'top down' methodology analyzes intact proteins by combining electrospray ionization with high performance mass spectrometry such as **Fourier transform mass spectrometry (FTMS)**.

MALDI ionizes molecules with molecular masses of 100-1,000,000 Da for analysis by MS and provides high sensitivity, high throughput and simplicity in operation. In case of MALDI, the desired samples are solidified within an acidified matrix which absorbs energy in a specific UV range and dissipates the energy thermally. This quickly transferred energy generates a vaporized plume of matrix and thereby simultaneously ejects the analytes into gas phase where they acquire charge. **MALDI-MS** (Matrix supported laser description ionization MS) (Karas and Hillenkamp, 1988) is a technique for the production and mass analysis of intact gas phase ions from a wide variety of biomolecules such as peptides, proteins, oligonucleotides, carbohydrates, glycolipids. It provides a rapid means to identify proteins when a fully decoded genome is available because the deduced masses of the resolved analytes can be compared with those calculated from the predicted products of all the genes in the genomes of an organism. It appears to work well for RNA analysis. **MALDI-TOFMS** can quickly and accurately determine unfractionated mixtures at concentrations below 100fmol per liter. MALDI combined with enzymatic reactions and protein chemistry can provide very useful information on molecular masses, peptide maps and primary structure.

ESI — This method is also widely used to introduce mixtures of biomolecules in the MS instrument. It allows the rapid transfer of analytes from the liquid phase to the gas phase at atmospheric pressure. The spray device creates droplets which once in the MS go through a repetitive process of solvent evaporation until the solvent has disappeared and charged analytes are left in the gas phase. Normally, the ESI is coupled with either a triple quadrupole, ion trap or hybrid TOF MS. The advantage of ESI in comparison with MALDI lies in the ease of coupling to separation techniques such as liquid chromatography (LC) and high-pressure LC (HPLC) allowing high throughput and on-line analysis of peptide or protein mixtures. LC-MS has been applied to large scale protein characterization and identification and has proven to be an important alternative method to 2-D gels. In addition to its role in protein profiling LC-MS is the most powerful technique for the monitoring, characterization and identification of impurities in pharmaceuticals.

Three dimensional proteomics — It refers to combination of subcellular fractionation, protein chromatography and/or affinity capture of more abundant proteins (first dimension) with 2-dimensional PAGE (second and third dimension).

Fourier transform ion cyclotron resonance MS(FITCR-MS) — This instrument combines the benefits of high mass accuracy and highly sensitive detection. It is used to identify low abundance compounds or proteins in complex mixtures and to resolve species of closely related m/z ratios. FITCR-MS when coupled with HPLC and ESI is able to characterize single compounds (up to 500 Da) from a large combinatorial chemistry libraries and to accurately detect the mass of peptides in a complex protein sample in a high through-put mode. The combination of electrospray ionization (ESI) and Fourier transform MS (FTMS) is becoming a powerful method for structural analysis of large molecules. ESI-FTMS and FTMS performances increase with the magnetic filed strength and newer methods are being developed. A high throughput and LC coupled FITCR-MS can be used to resolve peptides (termed accurate mass tag, AMT) with high confidence. Here ATM corresponds to predicted ORFs and thus with this technique the genes 'products, peptides can be characterized.

Isotope-coded affinity tag technology (ICAT)—Because of the complexity of proteome and the separation limits of both 2-D gel electrophoresis and liquid chromatography, only a fraction of the proteome can be analyzed. In order to reduce the complexity prior to protein separation and characterization, the Aebersold group (Gygi et al., 1999) designed a pair of ICAT reagents which differentially label protein samples on their cysteine residues. The ICAT reagent contains a biotin moiety and a linker chain with either eight deuterium or eight hydrogen atoms. When two samples, each labeled with the ICAT reagent carrying one of the two different isotopes, are mixed and subjected to site-specific protease digestion, the labeled peptides containing cysteine are enriched by binding the biotin tags to streptavidin. Thus there is generation of simplified peptide mixture which can be characterized by LC-MS approach. Quantification of differential protein expression level can be obtained by comparing the areas under the doublet peaks which are separated by eight mass units. This method works well for the differential analysis of many proteins in a complex mixture but the limitation with this approach is that a protein has to contain at least one cysteine residue to get detected.

14.20 PROTEIN POSTTRANSLATIONAL MODIFICATION

Modifications to protein structures play a pivotal role in regulating protein activity. Identification of the type of modification and its location often provide crucial information for the understanding the function or regulation of a given protein in pathways. Most of the strategies developed for analyzing protein modifications focus on only protein phosphorylation modification. To identify phosphoproteins, one strategy is to enrich the phosphorylated peptides using either immunoprecipitation with phosphopeptide-specific antibodies or by meta-chelate affinity chromatography. The latter uses resins with chelated trivalent metal ions such as Fe (III) and Ge (III) to bind the phosphopeptides or phosphoproteins. The enriched proteins are then subjected to trypsin digestion and the resulting fragments are identified using MS techniques. This technique will provide information on the sites of phosphorylation.

In order to identify multiple types of modification in a single experiment a shot gun MS approach can be used (MacCoss et al.,). This approach uses a multidimensional liquid chromatography (LC/LC), tandem mass spectrometry (MS/MS) and database searching algorithms. In this technique, the protein mixture is first digested with one specific and two non-specific proteases. The multiple proteases generate overlapping peptides of a given protein thereby providing thorough coverage of the protein and increasing the probability of pinpointing a modification on a specific amino acid residue. The resulting peptides are then separated by LC/LC and finally characterized by MS/MS.

14.21 STRUCTURAL CLASSIFICATION OF PROTEINS

Chromosomes evolve by transposition of mobile elements; by gross rearrangements such as inversions, translocations, deletions and duplications; by homologous recombination; by slippage of DNA polymerase during replication. All these factors contribute to the proliferation and dispersal of protein building blocks. Ancient duplications and rearrangements of protein coding segments have resulted in complex gene family-relationships. Duplications can be tandem or dispersed and can evolve entire coding regions or modules that correspond to folded protein domains. As a consequence of which gene product may acquire new specificities, altered recognition properties or modified functions. The complexity of protein tertiary structure can be reduced by considering its substructures. The hierarchical classification of proteins reflects their structural and evolutionary relatedness. The presence of recurring domains or motif structures indicate that protein tertiary structure is more reliably conserved than primary sequence. Thus comparison of protein structures can provide information about evolution. The complete contents of databases have been organized according to hierarchical levels of structure. In other words, protein structure information is known to be more conserved than amino acid sequences and serves as ideal references to study protein **structure-function relationships**. Similar protein folds may suggest similar biochemical functions (Zarembinski et al., 1998). Most reliable structural comparison method is to manually inspect similar protein structures such as SCOP. The development of database, **Structural Classification of Proteins (SCOP) database** describes the structural and evolutionary relationships between proteins of known structures. The SCOP is a powerful complement to sequence analyses in tracing many evolutionary relationships (Murzin et al., 1995). Proteins with high structural similarity will be classified into the same hierarchical SCP domain. Even though manual inspection provides more accurate structural classification, it is laborious for a large number of protein tertiary structures. Annotated structural comparison methods such as **Distance Alignment** (DALI) based on extraction of similar structures from distance matrices and combinatorial extension (CE- a database of structural alignments) algorithms globally find a structural alignment between two peptide chains such that superimposed segments of amino acids can have structural matching within a small root mean square deviation thresholds. Due to large combination of possible local alignments exhaustively searching a local optimal solution is known to be computationally expensive, proving a complexity of NP-Hard. So there is a need to develop an efficient domain classification method with sufficiently high accuracy to streamline the labour-intensive classification process. Even though structural alignment methods present satisfactory classification accuracies, the process of performing multiple pair wise alignments between an unknown protein and known protein databases is still incapable of providing fast predictions.

The SCOP database is curated manually and place proteins in the correct evolutionary framework based on conserved structural features. Two similar enterprises, the **CATH** (class, architecture, topology and homologous superfamilies) (Orengo et al., 1997) and **FSSP** (fold classification based on structure-structure alignments of proteins) databases make use of automated methods and provide additional information. SCOP and CATH provide tools for clustering protein structures that have evolved from a common progenitor with that evolutionary process retaining discernible structure similarities even if sequence identity/similarity is insignificant. Identification of consensus residues from sequence alignments as well as structural properties help explain the functional properties of the super family. But functional relationships based on consensus sequences or even motifs are also unreliable in the absence of structure-functional correlations that describe their functional roles. There exist no structurally contextual definitions of enzyme functions.

Structural motifs—Protein local structure refers to a set of protein peptides that share common physicochemical and structural properties. Usually cluster protein fragments by different criteria such as solvent accessibility, residual burial and backbone geometry and represent these fragment clusters by alphabet called a local structure alphabet (structural alphabet or structural motifs). Local structure prediction predicts the local structure of a protein fragment expressed by a letter of the structural alphabet from its amino acid sequence. Number of letters in each structural alphabet is large. It is 100 (Unger et al.,), 40 and 100 (Micheletti et al.,), 100(Schuchhardt et al.,) and 20-300 with fragment lengths from 5 to 7 (Kolodry et al.,). Though large alphabet sets can better approximate protein tertiary structures, predicting protein local structure from amino acid sequence is much more challenging. Various local structure libraries have been constructed.

Local structure can facilitate *ab initio* structure prediction, protein threading and remote homology detection. It also helps to improve the performance of both profile and threading/fold-recognition models for tertiary structure prediction.

Structural motifs become important in defining protein families and superfamilies. The three main levels within the hierarchy are: the **family**, **superfamily** and **fold**. Higher levels of gene or protein classification are families, subfamilies and superfamilies. The term 'family' generically is used to describe among collection of genes or proteins that are presumed to share common ancestry. Proteins are grouped into families if they have sequence identities >_ 30%. But in case of globins some members of the family share only 15% identity. In other words, proteins with significant primary sequence similarity and/or with similar structure and function are said to be in the same protein family. The members of protein families are called homologous proteins or **homologs**. When the 3-dimensional structure of one of the proteins encoded by one gene family is known it can be assumed that all of the other homologous protein members of the family adopt essentially the same fold. When two homologs are present in the same species they are referred to as **paralogs**. When several homologues are found within one species then the query sequence belongs to a multiple gene family. Paralogous relationships have been known for α-globin, β-globin and myoglobin and the genes arose from duplications of ancestral globin genes in the vertebrate lineage. Homologs from the different species are called **orthologs**. Orthologs can only be determined definitely with a complete inventory of genes in an organism. A hierarchical classification of proteins with definitions, criteria or main features is given in the Table 14.3 (Koonin, et al., 2002).

Homologous enzymes are derived from a common ancestor and are therefore structurally related. Frequently such enzymes show a high degree of sequence similarity, easily identifiable from simple database searches and sufficient for inference of function by analogy. Homologous enzymes can also be highly divergent and require more sophisticated searches for reliable identification of highly diverged sequences. Motifs or profile search methods frequently much more sensitive than pair-wise comparison methods at detecting distant relationships (Altschule et al., 1997). In other cases homologous enzymes can not be identified from sequence information alone but require comparisons of three-dimensional structures. At higher levels of divergence homologous enzymes often do not catalyze the same biochemical reactions. Furthermore, conserved sequence elements need not be located in the active site but may be responsible for maintenance or stabilization of common folds which support different reaction mechanisms, for example, triose phosphate isomerase (TIM)-barrels.

Table 14.3 Hierarchical classification of proteins.

Category	example	Definition, criteria or main features
Structural class	a/β	Overall composition of structural elements. No evolutionary relationship.
Fold	TIM barrel	Topology of the folded protein backbone. Monophyletic origin?
Superfamily	Aldolase	Recognizable sequence similarity (at least a conserved motif), conservation of basic biochemical properties. Monophyletic origin.
Family	Class I aldolase	Significant sequence similarity; conservation of biochemical activity.
Group of orthologues (COG)	2-keto-3-deoxy-6-phosphogluconate aldolase	Orthologous relationships within the given set of species; conservation of the biochemical activity and, most often, also the biological function.
Lineage- specific expansion	PA3131 and PA3181 in *Pseudomonas aeruginosa*	Paralogues originating from a lineage-specific duplication; possible functional specialization.

Orthologs are homologous enzymes found in different species and which catalyze the same reaction. **Paralogs** refer to homologous enzymes in the same species and are likely to have diverged from the one another by gene duplication after speciation. **Analogous enzymes** catalyze the same reaction but are not structurally related.

Proteins can be placed in superfamilies even if the sequence identity is low but their structural and functional characteristics indicate a common evolutionary origin. The term 'super family' is commonly used in both structural and functional contexts, i.e., it has correlated structure-function based definition. In other words two or more families with little primary sequence similarity sometimes make use of the same major structural motif and have functional similarities and these families are grouped as superfamilies. Superfamilies have a higher level of classification. Typically they include conserved motifs that are determinants of a distinct biochemical activity which, however, may be required for a variety of cellular functions. Superfamiles and their signature motifs may be useful in classifying proteins that have evolved to an extent that they can not be assayed to any clusters of orthologous groups (COGs) (Tatusov et al., 1997) but still retain a conserved motif. Proteins can be said to be having a common fold if they have the same major secondary structures in the same arrangement and with the same topology. Such proteins may or may not have common evolutionary origin. In these proteins structural similarities could have arisen because of physical principles which favour particular packing arrangements and fold topologies. **Homeomorphic superfamily-** It refers to a class of proteins which are homologous over their whole or an appreciable part of their sequence (lower threshold: 30%) and possess identical domains of homology arranged in the same order.

The definition of super family used in SCOP is proteins (that) have (the same fold and) low sequence identities whose structures and in many cases features suggest a common evolutionary origin is possible. With this definition enzymes are often grouped according to substrate specificity but without regard to conservation of active site functional groups that mediate the chemical transformations. As a result this definition of superfamily may partition groups of divergently related enzymes into separate superfamilies.

Families refer to group of homologous (frequently orthologous) enzymes that catalyze the same reaction (mechanism and substrate specificity). Often the members of families share greater than 30% sequence identity, however, orthologs may show sequence identities well below 30% and therefore can be difficult to discern in the absence of structural information for at least one member of the family.

Superfamily refers to group of homologous enzymes that catalyze either (i). the same chemical reaction with differing substrate specificity or (ii). different overall reactions that share a common mechanistic attribute (partial reaction, intermediate or transition state) enabled by conserved active site residues that perform the same function in all members of the superfamily. Typically the members of superfamilies share less than 50% sequence identity and often share less than 20% sequence identity. The randomized sequences of a typical protein share 6-10% sequence identity. Dayhoff et al (1978) group proteins belonging to the same family if they differ at fewer than half of their amino acid positions whereas superfamilies comprise those sequences that are demonstrably homologous but differ at more than half of their positions. Assigning protein sequences to family and superfamily is arbitrary (Doolittle, 1981) and let any sequences that are demonstrably related to each other belong to the same family regardless of per cent identity that they show. Superfamilies would then be composed of two or more families, not all of whose different proteins are demonstrably homologous with all the members of the other families in the set1, i.e Superfamily 1= Family A U Family B.

Suprafamily refers to groups of homologous enzymes that catalyze different overall reactions but whose reactions do not share any common mechanistic attribute. Although active site residues may be conserved these perform different functions in the members of the family.

Membership in families, super families or suprafamilies may not be assigned from sequence data alone but often requires correlated functional and structural characterization. A consequence is that annotations of function for members of mechanically diverse superfamilies and functionally distinct suprafamilies identified in genome sequencing projects are not possible from sequence data alone. In case of mechanically diverse superfamilies the structure elements which are common to all members of the superfamilies allow identification of the underlying structural strategy used to catalyze reactions sharing a common mechanical attribute.

Functional diversity is more complex than sequence diversity. So functional annotations of protein identified in genome sequencing projects based on sequence similarity without regard to the existence of mechanistically diverse superfamilies or functionally distinct suprafamilies will lead to many assignments which are either misleading or incorrect. As such knowledge of the 3-D structure of the complete repertoire of protein folds will be insufficient to assign specific function to homologs that possess a given fold. This conclusion supports that the ongoing efforts in structural genomics may have a more limited impact on determination and prediction of specific function than expected. Genomic enzymology requires the interplay of both chemical (functional) and structural (and/or sequence) analyses as neither structure-only based nor function-only-based investigation can provide an accurate picture of how enzyme structures evolved and deliver new functions.

14.22 HIERARCHICAL DOMAIN CLASSIFICATION OF PROTEIN STRUCTURES

Protein's tertiary structure is more reliably conserved than primary sequence and this it can provide more information about the evolution. The three-dimensional structure is built upon the analysis of its part. The five levels within the hierarchy are: **class, architecture, topology, homology and sequence**. At the highest level of classification protein structure has been divided into four classes. Class is derived from gross secondary structure content and packing. The four classes of recognized domains are i. all-α ii. all –β iii. α-β which includes both the α and β segments are interspersed or alternate α/β and α+ β in which the α and β regions are somewhat segregated iv. domain with low secondary structure content. Within each of the first three classes there are tens to hundreds of different folding arrangements built up from identifiable substructures. The architecture describes the gross arrangement of secondary structures without considering their connectivities. The various secondary structure arrangements are **barrel, roll, sandwich**, etc. The top two levels of organization, class and fold are purely structural and below the fold level categorization is based on evolutionary relationships. **Topology** refers to the overall shape and the connectivity of secondary structures. Topology refers to the study of properties of an object that do not change under continuous deformations such as twisting or bending. It is obtained by using structure comparison algorithms which can cluster the domains. Structures in which at least 60% of the larger protein matches the smaller are grouped to the same topology level. Homology groups domains which share -> 35% sequence identity and are thought to share a common ancestor, i.e. are homologous. Similarities are first detected by sequence comparison and subsequently through a structure comparison algorithm. Sequence is the final level within the hierarchy in which the structures within the homology groups are further grouped on the basis of sequence identity. Domains with sequence identities >35% (with at least 60% of the larger domains equivalent to the smaller) show highly similar structures and functions.

Modules—Modules are contiguous in sequence whereas structural domains are independent folding units which need not to be contiguous (Henikoff et al., 1997). Modules may be thought of as a subset of protein domains. Protein building blocks or modules have duplicated and evolved in complex ways through a variety of gene–rearrangement mechanisms. As a result, composite proteins consisting of multiple modules (**chimeras or mosaics**) constitute a large proportion of the protein complement of an organism. Typically modules consist of multiple motifs which form the structural core of proteins. Modules are composed of a single or multiple motifs. Modules are most useful for protein classification. Modules present in larger protein might have dispersed by transpositions. Tandemly repeated modules including the C_2H_2 Zinc fingers and many examples of extracellular modules most likely arose by recombinational mechanisms such as unequal crossingover and gene conversion.

Motif — The term 'motif' has different interpretations (see Bioinformatics Roy, 2009) . The smallest sequence units of protein families are termed 'motifs' which are identified as highly similar regions in alignments of protein sequences. Motifs can be as simple as the hexamer repeat unit that forms a left-handed parallel β-helix formed in uridine5-diphosphate (UDP)-N-acetyl glucosamine acetyl transferase. Motifs are used to identify functional regions of proteins and where they share common ancestry, are useful for family classification. A highly conserved motif often hints at a functional site usually overlayed on a conserved structural unit. Enzyme active site residues which are usually highly conserved are often found within motifs. C_2H_2 Zinc finger DNA binding motif forms a contiguous independently folded structure, the finger is itself a module. The larger homeobox module consists of ~ 60amino acid motif and is also involved in binding DNA. Motif can reflect either common ancestry or convergence from independent origins. Identification of motifs can be important for drawing structural and functional inferences. For example, a 'P-loop' motif is present in nucleotide binding domains from families as diverse as kinesin moter proteins and ATP-binding cassette (ABC) transporters. Motifs contributing to a structural core can be widely separated within the primary sequence(e.g.'HIGH and KMSKS' motifs of class I amino acyl tRNA synthetases are hundreds of amino acids apart). Motifs are often sufficient to predict the presence of a functional activity , so sequences can also be scanned against motif databases which contain signatures of well characterized protein families.

14.23 EVOLUTION OF PROTEINS

It is easier to duplicate and modify proteins genetically than to assemble appropriate amino acid combinations *de novo* from random beginning. In other words, evolution of new protein depends on gene duplications which is the result of the various breakage and reunion events which occur more or less randomly in the genome (see Roy, 2009). Breaks occur randomly but misalignment most frequently involves base-pairing of similar as opposed to random base sequences. As a consequence gene duplication begets more gene duplication. The duplicated segments stop signals for protein synthesis. Duplication leads to either an elongated polypeptide chain or two separate copies of the same protein. Duplication can be contiguous or discrete. Contiguous duplication is the major route to larger proteins. The discrete type of duplication will lead to two independent gene products, one of which is free to mutate leading to formation of protein with a new function. Mutations (single base substitutions that engender individual amino acid replacements, deletions or insertions) will lead to divergence but the new protein retains many of its pre-existing features. The general shapes and folding patterns of these proteins are quite similar but there is small difference in their structures which affect their interactions and their oxygen-binding properties. The fundamental catalytic machinery is virtually identical for all enzymes but differences in substrate binding regions allow for an elegant selectivity of action for the diverse gene products. There is radical change in function of new proteins but key structural features are retained.

Gene duplication leading to amino acid sequence redundancies, is a force in both elongation of small primitive polypeptides and in expanding the available repertoire of gene products. All proteins and enzymes have evolved from a small number of prototypes.

14.24 REARRANGEMENT OF SEGMENTS OF GENOME

Rearrangement involves segments coding for the entire polypeptide chains or alternatively segments that code for only a particular part of the polypeptide chain. The former results in elongation of some polypeptide chains whereas the latter results in different parts of various enzymes and other proteins become genetically fused and ultimately result in construction of new enzymes or proteins because rearrangement results in various combinations of binding sites and catalytic units. In eukaryotes gene rearrangements and splicing appear to be intimately associated with the existence of untranslated intervening sequences (introns). Assignment of boundaries in a given sequence, i.e. where do different sequences start and stop? Although it may be easier to find a match for a short sequence, it will be proportionately more difficult to prove its validity. Existence of introns in eukaryotes may be the mechanistic basis for internal gaps in proteins and small deletions resulting from base substitutions changing the splice points. If this proves to be so then the junction between exon and intron will be blurred and the comparison will become more difficult. Finally, extrachromosomal elements can transpose segments between genomes of different organisms which further complicate the situation.

15

Molecular Marker Technology and its Applications

15.1 DEFINITION- MARKERS

Genes of known functions and known locations are called genetic markers. Markers are useful for performing detailed genetic mapping analysis. There are three types of marker: **Phenotypic or morphological markers, isozyme markers, cytological markers and DNA markers**. A genetic marker is a phenotypic variant due to a variation in the nucleotide sequence of a genome or in the phenotypic manifestation of an allele. Genetic markers may affect morphology, behavior, physiology or chemistry of an organism and they are defined in relation to the phenotype of the standard wild-type. A marker gene is a gene whose phenotypic variants are used as markers.

15.2 TYPES OF GENETIC MARKERS

There are four types of genetic markers considering the practical view point (Cerda-Olemedo and Avalos, 1999).

1. **Selectable markers** are markers which enable the individuals possessing them to grow under experimental conditions that hinder similar individuals lacking the marker. These markers are useful in the genetic analysis of very large populations of viruses, bacteria, unicellular eukaryotes and cultured cells. Selectable markers are particularly useful when they are also counter selectable, i.e., certain experimental conditions allow the growth of individuals that lack them and hinder individuals that carry them.

2. **Genetic markers** which are easily recognized by visual inspection of the shape, color and sometimes the movement of cells and organisms are called **Morphological markers** (any easily identifiable trait (eye color in Drosophila or flower color in plant) that is characteristic for in an individual is called morphological marker). They have been useful in the analysis of small populations of plants and animals. The limitations

with the phenotypic markers are that their number is low in most of the species and there is problem of constructing multiple-marker line and thus resulting in the low resolution of map produced.

3. The third class of genetic markers are those whose detection require specific tests. Conditional lethal markers impede growth under normal conditions but are rescued by special circumstances of the environment or genetic background (conditional lethal mutations, suppressor mutation, temperature-sensitive mutation, auxotroph).

Chemical analyses are often indispensable, for example, to detect the presence or absence of a metabolite or to distinguish the electrophoretic variants of a protein. Behavioral markers require their own specific tests. Some of the adenine auxotrophs of *S. cerevisiae* develop pink color under certain growth conditions. The presence of certain chemical or certain enzymes in a cell may be revealed when they produce a colored product with certain reagents (reporter gene).

Although **isozyme markers** have been used in the study of genetic variation between organisms, identification of hybrids, paternity testing, genetic diagnostic and genetic mapping but problem with isozyme marker is that they are not so numerous to allow high resolution mapping.

4. **Cytogenetic markers**–The cytological markers refer to the unique characteristic of a chromosome, for example, knob, satellite, translocation that can be visualized by the techniques of cytogentics.

5. **Molecular genetic markers**–Molecular markers refer to variations in the DNA sequence that have no other phenotypic expression than the results of direct tests on DNA (eg. Restriction maps, PCR). Their main advantage is that they occur all over the genome even in regions in which no other genetic markers have been located. A marker could be any genomic element with a uniquely identifiable sequence or property. **Molecular markers** are specific nucleotide sequences found on the chromosomes. They are not traits. Genes are mixed with these sequences. The markers are small sequence differences in the long strand of DNA. Any specific DNA segment whose base sequence is different (polymorphic) in different organism is diagnostic for each of them. Molecular markers can also be defined as small heritable sequence of DNA. These can be recognition sites for restriction enzymes. The various molecular markers or the different types of DNA polymorphisms are AFLP, RFLP, RAPD, SCAR, SNP, SSCP and VTNR among others. There is abundance of molecular polymorphisms within species and differences among species. Variations in genomes can be measured at different scales by different molecular markers. In comparison with morphological markers molecular markers have the advantages that they are neither affected by environment nor show any epistatic interaction. Molecular markers have the ability to detect a large number of polymorphisms with a single primer pair without any prior knowledge of target genome. Finally, the results are highly repeatable. Information about various molecular markers at DNA level can be obtained by studying the DNA polymorphism. Marker can exist in different forms such as polymorphic or nonpolymorphic. A **monomorphic** gene is one that is not polymorphic whereas a **polymorphic gene** is one for which there exists several alternatives, all of them fairly common. Polymorphic markers are used to construct

genetic map whereas polymorphic and nonpolymorphic markers both can be used to construct physical maps. A great deal of variation is because of the presence of this type of polymorphism (DNA polymorphism) at many of the loci. Genetic differences that are common among individuals of the same species are called genetic polymorphism whereas genetic differences that accumulate between species constitute genetic divergence. Polymorphism can be observed at higher resolution (at nucleotide and amino acid sequence level) as well as at lower resolution (at DNA fragment level). Polymorphism shown at the level of individual base pair is called higher resolution polymorphism and polymorphism shown at the level of restriction fragments or amplified fragments constitute lower resolution polymorphism. DNA polymorphism refers to the difference in the base sequence of a distinct region between two or more different genomes. Such polymorphisms are generated by deletions, insertions, inversions or generally sequence rearrangements. These mutations lead to, for example, the existence of different alleles for a specific locus. In case of repetitive DNA, variations in the number of repeats may lead to a RFLP polymorphism (VTNR). DNA polymorphisms may be detected by various DNA fingerprinting techniques or by DNA sequencing. The different types of DNA polymorphism are RFLP, AFLP, RAPD, SCAR, SNP, SSCP and VTNR among others which are discussed below. The type of polymorphism depends on the method of DNA manipulation (electrophoresis, southern blot, DNA hybridization, DNA amplification, PCR, etc). Ideal molecular markers are those which are highly polymorphic, co-dominant, distributed evenly throughout the genome and easily visualized and should be stable over generations. The molecular markers can be visualized by hybridization based techniques (e.g. DNA fingerprinting, RFLP) or PCR based methods (e.g. DNA amplification fingerprinting, RAPD, sequence tagged microsatellite). **Transcriptome- derived marker (transcriptome marker)**- It refers to any transcript band detected by cDNA-AFLP or other transcript profiling techniques which allow to identify polymorphisms between parents (and the progeny from a cross between these parents and can then be mapped). Transcriptome markers differ from conventional molecular markers in that they are not anonymous but derived from active genes.

1. **SNPs (Single nucleotide polymorphisms, pronounced 'snips')**–The different individuals may differ in the identity of a nucleotide pair at a particular defined site in the DNA. In the following example SNP is in the 5′- flanking region of the gene and is a T=C polymorphism. Some individuals can have A=T base pair at this position whereas others can have G=C base pair. The SNP locus has thus two alleles and there will thus be three types of individuals in the population considering this locus, homozygous for A=T/A=T, homozygous for C=G/C=G and heterozygous A=T/C=G. The SNP can be found either in the coding or non-coding region. Studies suggest that one SNP site is likely to be found in every 1000-3000 bp in coding sequence and about one SNP site in every 500-1000bp in non-coding DNA (Chakravarty, 1999). The relatively good coverage and the distribution of SNPs throughput the genome in both coding and non-coding regions make SNPs highly informative markers for mapping studies. Since specific SNPs correlate with increased risk for a particular disease these polymorphisms are diagnostic (disease associated SNPs). In some cases, these

polymorphisms result in creation or abolition of a restriction enzyme site and thus can be used diagnostically. In case of the genetic disorder, sickle-cell anemia the mutation from GAG to GTG eliminates restriction sites for the enzymes Ddel (CTANG) and MstII (CCTNAGG). The basic difference between point mutation and SNPs lies in their different frequencies (point mutation ≤ 1%; SNP ≥ 1%). SNPs can be detected directly using microarrays ('DNA chips'). There has been discovery of over two million SNPs in human genome (Venter et al., 2001).Statistically an SNP will occur every kilobase in human genome and thus 3-30 millons SNPs are expected to occur. SNP frequency varies between genomic regions in the same individual and between related individuals. In humans, the SNP frequency is about 1/200-300−bp in coding and 1/100-200bp in the regulatory and intronic sequences. SNP frequency is about 1/65bp in some region of maize but in other region it is 1/85bp. In soybean the frequency is 3-4 SNP per kilobase. SNP –MS (snp-spectrometry) is a technique for the detection of SNPs between two or more genomes. A **Consensus patterns**-which refer to the similar or identical distribution of specific sequences (SNPs) or mutations along a specific piece of DNA, e.g., a chromosome of two or more individuals will allow to group individuals into specified haplotype classes (which may share common properties as, for example, the same sensitivity towards drugs). By simply comparing the sequences in, for example, a distinct gene from two or more individuals identical bases can be assigned a1 and non-identical bases at a specific base a2. A cluster analysis will then find consensus patterns which permit to group individuals into da specified class. SNP can occur either in coding or noncoding DNA. Noncoding SNP may have effects on various mechanisms such as transcription, translation and splicing. Coding SNP can be divided into two main categories, **synonymous** where there is no change in the amino acid coded for. **NSP (Non-synonymous polymorphism)**−It is a SNP present in the coding region. This type of polymorphism will result in an AA substitution and thus results in AA polymorphism. For example, if the polymorphism occurs for A/U in a gene then there could be either A or U at this position in the mRNA and thus there will be two types of codons GAG and GUG which specify glutamic acid and valine ,respectively. In human a Glu/Val polymorphism at amino acid position 6 in the beta-globin gene results in sickle-cell anemia. Non-synonymous polymorphisms occur much less frequently than silent substitution an SNPs in non-coding regions. Nonsynonymous SNP tend to occur at lower frequencies than synonymous SNPs.

2. **SNP**−It is also a SNP occurring in the coding region. As the polymorphism leads to formation of synonymous codon and this type of polymorphism does not result in AA replacement. The polymorphism occurs at the third base of the codon. For example, both codons GAU and GAC which can occur as a result of polymorphism at a particular base code for the same AA, aspartic acid. Such type of polymorphism does not result in change of AA in the polypeptide chain is called **silent polymorphism**. Which one of the two codons is widely used depends on the preference by the organism. Some organisms prefer certain codons, particularly when they encode proteins produced in large quantity. The main reason for preference towards certain codons is that the translational accuracy and speed and precision with which a certain codon is to be produced (Akashi, 1993; 1995; Hartl et al., 1994; Eyre-Walker, 1996).

3. **Indel (Insertion/deletion)** — It is an insertion/deletion polymorphism. The indel's size can range from <10bp to 1-5kbp. Most indels are less than 10bp in length and the largest insertions are of transposable elements (TEs) (1-5kb, see chapter 22). Retrotransposon (RT) sequences can be used as molecular markers as they are ubiquitous, present in high copy numbers, are widely dispersed on chromosomes and show insertional polymorphism both between and within species in plants. Active RTs will produce new insertions in the genome leading to polymorphism. Many RTs are widely distributed within the euchromatic domains of chromosomes and so it should be possible to generate markers linked to agronomically important traits. RT Tnd-1 has provided marker that is linked to black root resistance in tobacco (Kenward et al., 1999). Sequence analysis of clones enriched barley microsatellite library has revealed RT sequences to be intimately associated with microsatellite (Ramsay et al., 1999) and thus REMAP (RT Microsatellite Amplified Polymorphism) can be studied. NonLTR–RTs have also been used as molecular markers. Unlike SNP it is difficult to determine the ancestral sequence of insertion or deletion in most cases of indel polymorphism. It is a co-dominant polymorphism.

4. **Sequence tagged site (STS)** — STS is a region of known sequence that is present once per haploid genome found in a reproductive cell (Olson et al., 1989). STSs are specific chromosome sites which can be recognized by probes. It is a type of DNA polymorphism that affects DNA fragments ranging from 300 to 3000bp in length. The STS marker is used in identifying the cloned DNA fragments that contain particular STS markers known to be in or close to a gene of interest. The advantage of STS marker is that the ends of known sequence can be used to design primers oligonucleotides to specifically amplify the sequence and PCR using these primers (18-24 bases) enables one to detect the presence of a particular STS in a DNA fragment. **Sequence tagged connector (STC)-** It refers to any short DNA sequence of about 500 bases generated by sequencing the end of a BCA (BES). It can be used for the design of STS which represents DNA stretches for the PCR based amplification of specific loci without amplifying unwanted genomic regions. STC can also serves as molecular marker. **Sequence–tagged microsatellite site (STMS)-** It refers to any genomic DNA sequence that flanks microsatellite clusters or preferentially a specific microsatellite sequence in eukaryotic genome. **STAR (sequence–tagged restriction site)-** It is a short DNA sequence that identifies the sequences flanking the recognition site of a restriction endonuclease.

5. **RFLP(Restriction fragment length polymorphism)** — It is also called restriction site polymorphism, i.e. some DNA molecules in the population contain a particular restriction site whereas others lack it. The most easily identified type of restriction site polymorphism is one that results in change in size of a restriction fragment and this is called RFLP (Botstein et al., 1980). Supposing a DNA molecule with three restriction sites digestion with a restriction endonuclease will yield two fragments. Cleavage at the middle site yields a short fragment whereas the absence of the middle restriction site will result in a single long fragment which can be identified by Southern blot using a probe that hybridizes with the restriction fragment. Changes in restriction fragment length may be due to 1. single base pair changes that either creats or

eliminates restriction sites (mutation) 2. rearrangements such as insertions or deletions or inversions 3. the presence of repetitive DNA in different copy numbers on a specific chromosomal region. Thus RFLP is the variation (s) in the length of DNA fragments produced by a specific restriction endonuclease from genomic DNAs of two or more individuals of a species. The differing sizes of RFs define two alleles, short and long and there could be three possible genotypes for this RFLP, short/short (homozygote), long/long (homozygote) and short/long (heterozygote). RFLP is a co-dominant marker and thus homozygotes as well as heterozygotes can be identified. RFLPs are especially good for detecting changes in numbers of repetitive sequences and they reveal only a small fraction of base pair substitutions. For every sequence difference that affects a restriction site there must be many more that do not invite attention to themselves in this way. The limitations with RFLP are that it requires large amount of DNA and needs a probe, particularly radioactive probe to achieve the most sensitive detection. Restricted fragment length polymorphisms are identified by the use of restriction endonuclease. Restriction endo-nucleases are enzymes that recognise specific nucleic acid sequence in DNA and cleave the DNA at these sites or at adjacent sites. The more prevalent the recognition sequence in the DNA, the more frequently the DNA will be cleaved by the enzyme which recognises the sequence. When the DNA from a strain is digested by restriction enzymes, many different size fragments are produced. The DNA fragments so formed are identified by Southern blotting (Southern, 1975), a procedure whereby DNA fragments can be separated by gel electrophoresis, since smaller fragments migrate more rapidly through the pores of gel than larger fragments. Specific fragments are detected as bands following hybridization and autoradiography. The DNA is then transferred from the gel to a permanent filter in a single stranded state. The radioactively labelled probe consisting of a cloned DNA sequence (single stranded) hybridizes with wholly or partially homologous DNA fragments and this DNA-DNA hybrid is visualized as bands on the film (X-ray). Thus, the specific homologous DNA fragment now called the RFLP is detected.

6. **RAPD(Random amplified polymorphism DNA)** — It is a PCR based marker (Welsh and McClelland, 1990; Williams et al., 1990). It requires small amount of genomic DNA, requires no probe and also no advance information about the genome of the organism. It uses random primers of 8 to 10bp in length and the primers are used singly or in pairs. The primers are so short that they often anneal to the template DNA at multiple site. Some primers anneal in the proper orientation and at a suitable distance from each other to support amplification of the unknown sequence between them. Among the set of fragments will be ones that can be amplified from some genomic DNA samples (individuals) but not from others and thus show polymorphism for the DNA fragment (absence or presence of fragment). The presence of fragment is dominant over absence of the fragment. In other words, if one allele (+) supports amplification but the alternative allele (-) does not and + allele is dominant over – allele, then DNA from genotypes +/+ and +/- will support amplification whereas DNA from genotype -/- will not support amplification. Most RAPD fragments are dominant markers and thus only presence or absence of a band can be

scored and one does not know whether a band is present in a homozygous or heterozygous condition. Therefore allele frequencies can not be estimated accurately. RAPD markers are well suited for genetic mapping and for DNA fingerprinting with particularly utility of population genetics study. It can also provide an assay for polymorphisms which will permit rapid identification and isolation of chromosome-specific DNA fragments. Hybrid cell lines or genetic stocks carrying deletions or additions of large chromosomal segments could be screened relative to appropriate controls to identify the region of the genome carrying addition or deletion. Many RAPD can be used to define (or characterize) a genome. Genetic mapping with RAPD markers has several advantages over other methods: (i) a universal set of primers can be used for analysis in a variety of species(ii) no isolation of cloned DNA probes, preparation of filters for hybridization or nucleotide sequencing is required (iii) each RAPD marker is the equivalent of Sequenced Tagged Site (STS) which can simplify information transfer amongst research programs. The determination of genotype can be automated and genetic maps using RAPD markers can be obtained with more efficiently and with greater marker density than by RFLP or targeted PCR based approaches.

Venugopal et al. (1993) provided the probable mechanism by which a single primer generates DNA polymorphism. They suggested that there are a number of sites in the genome flanked by perfect or imperfect invert repeats which permit multiple annealing of the primer to occur. Primer annealing sites are spread throughout the nuclear and cytoplasmic genomes, in all classes of DNA from single copy DNA to multiple copy DNA and in coding and non-coding regions. RAPD polymorphism between DNAs seems to be the result of a number of processes including nucleotide substitution that can create or abolish primer sites, formation of secondary structures between priming sites and insertion, deletion or inversion of either priming sites or segments between priming sites (Williams et al.,1993). In addition to polymorphism in product size distribution within a RAPD profile there is also polymorphism with respect to product intensity. This intensity polymorphism is the result of product copy number differences, competition between PCR products, heterozygosity, comigration or partial mismatching of primer sites.

Long-primer RAPD—It is a variant of RADP which uses comparatively long primers (18-25nucleotides) for amplification of anonymous genomic regions using PCR. These primers are designed from consensus sequences in several families of short interspersed repetitive elements of eubacteria and permit amplification to be performed at higher stringency in comparison to normal RAPD with primer of 10bp. This LP-RAPD detects sequence polymorphisms between genomes of different organisms and is used for identifying, testing, population studies and phylogenetic relationships.

7. **SPAR (single primer amplified region) and SCAR (sequence-characterized amplified region)**—The main problems associated with RAPD include reproducibility and product homology. The homology problem in RAPD can be solved in some cases by using improved systems of product resolution and gaining a detailed understanding of each of the products through genetic analysis and a combination of Southern blot and restriction enzyme studies. Such approaches have resulted in the generation of sequence characterized amplified regions (SCARs) (Paran and

Michelmore, 1993). When a fragment can be amplified using a single primer oligonucleotide then it is called a SPAR and when a particular amplified fragment is isolated and its nucleotide sequence determined it becomes a SCAR. The SCAR can be converted into a conventional sequence tagged site (STS) by the use of primers specific to the ends of the sequence. SCARs are superior to RAPD markers because they detect single loci only. It serves as genetic markers in physical mapping procedure and in MAS and plant cultivar identification.

8. **CAPS**—A particular amplified fragment may show restriction site polymorphism in the sense that it may show presence or absence of a particular restriction site and in which case the polymorphism is called a CAPS (cleaved amplified polymorphic site). A CAPS is the analogue of an RFLP except that the genotype is identified by amplification and enzyme digestion and not by a Southern blot with a radioactive probe. This is why a CAPS is sometimes called a PCR-RFLP (PCR product containing restriction site (S)). It is a **co-dominant marker** which means heterozygote can be distinguished from homozygotes. The amplified CAPS will be cleaved either into two fragment if it contains a restriction site or it will not be cleaved if the restriction site has mutated. If the allele with restriction site is designated as (+) and without restriction site as (-) then +/+ genotype will yield two bands upon amplification and cleavage; -/- will yield one band and +/- will yield three bands and thus all the genotypes (homozygotes as well as heterozygote) will be distinguished. Thus a polymorphism in the sequence of the same specific genomic locus of the two organisms can be detected by CAPS techniques. In this technique a set of oligonucleotide primers complementary to a known sequence within the locus of interest are synthesized. The primers are then used to amplify part of the locus with conventional PCR and the DNA of several different individual. The amplication products are then restricted with a number of restriction endonucleases to identify the RFLP among the individuals.

9. **AFLP (Amplified fragment length polymorphism)**—Any difference between corresponding DNA fragments from two organisms can be detected by this technique. Such DNA fragments (usually 50-1000bp in length) are obtained by restriction of the genome with two restriction endonucleases. The AFLP is detected with specific primers and not with random primers as in case of RAPD (Vos et al., 1995). The specific primers are homologous to the short double starnded DNA sequences(called adaptors) that are attached to the ends of the genomic restriction fragment through the enzyme DNA ligase and contain a specific restriction site and 1-3 selective nucleotides at their 3′ end. The genomic DNA is digested with two different restriction endonucleases and thus different genomic fragments are produced. These fragments are then joined to adaptor through ligation which results in generation of different genomic fragments flanked by adaptors which are in turn amplified simultaneously with specific primers. Some of the fragments will support amplification whereas others will not and such polymorphism is known as AFLP. Thus AFLPs show presence/absence type of polymorphism rather than a differing fragment length. It is a dominant marker and the allele supporting the amplification(+) is dominant over allele(-) not supporting the amplification and thus homozygote(+/+) can not be

distinguished from heterozygote(+/–). The experiment is carried out with primers that amplify 50-75 fragments. When genomic DNAs from two or more individuals are analysed, the variation in length of DNA fragments generated is generated by PCR using either one or several specific or arbitrary oligodeoxynucleotide primers. AFLPs arise from (i) restriction site polymorphisms where is a specific restriction endonuclease recognition sequence is present or absent (ii) sequence length polymorphisms where the number of tandemly arranged repetitive sequences at a given site varies and (iii) DNA base pair changes nor associated with restriction site.

AFLP is another approach based on endonuclease digestion of the total genomic DNA. Synthetic adaptors are ligated to the restriction fragments and selective PCR anchored on these adaptors is performed to amplify discrete DNA fragments. AFLP may be efficient in revealing polymorphism even between closely related individuals, each experiment implies three steps and four different primers for the analysis of the complex genomes. Furthermore, artifactual polymorphisms can be generated by incomplete cleavage of genomic DNA or imperfect ligation. Although AFLPs are much more reliable and reproducible than RAPDs, they suffer from the problems of data interpretation as in case of RAPD, for example, marker origin is unknown, locus/allele designations are unclear and homology statements regarding bands are uncertain and thus like RAPD data, analysis of AFLP data is a problem in population genetics study (Harris, 1999).

10. **Micro-and mini-satellite polymorphism**—The intergenic DNA consists of unique sequences and repetitive sequences. Many of the repetitive sequences are arranged in tandem and are called satellite DNA. There are three types of satellite DNA depending upon the level of repetition and the length of core repeat unit.

Mini- and micro- satellite polymorphisms are based on base sequences that are repeated in tandem at one or more places in the genome. Micro-satellite differs from mini-satellites in that the former has a very short core repeating unit of 2 to 9 base pairs whereas the later has a longer core unit of 10 to 60 bps (Table 15.1). An example of micro-satellite is $(5\text{-CAT-}3)_n$ where n is the number of repeats. Micro-satellite $(TG)_n$ is present in $5\text{-}10\times10^4$ copies per human genome spaced at 50-100kb intervals. Two micrsosatellites, one containing the AT dinucleotide motif and the other a TAT trinucleotide were found in soybean. Corn SSRs have been identified as containing $(AC)_n$ and $(AG)_n$ or $(TTG)_n$ and $(TTC)_n$ repeats. There is relative scarcity of $(AC)_n$ repeats in the plant genome, high degree of polymorphism within this class of genetic markers and the random distribution of SSRs cross the genome. Microsatellite in particular is present at many different places in the genome and each is flanked by restriction

Table 15.1 Showing types of repeat, their degree of repetition, no. of loci per genome and repeat unit length.

Types of repeat	Degree of repetition	No. of loci	Repeat unit length per locus
Satellite	$10^3\text{-}10^7$	one or two per chromosome	1000-3000bp
Minisatellite	$10\text{-}10^3$	Thousand per genome	9-100bp (14-500bp)
Microsatellite	$10\text{-}10^2$	Up to 10^5 per genome	1-6bp(1-13bp)

sites whose distance from the core repeat varies from one location to another and thus shows polymorphism. SSRs are highly polymorphic sequences and found throughout plant and animal genomes. SSR repeat lengths are easily detectable and have made SSRs a population type of **co-dominant** molecular marker. In rice majority of SSRs are mononucleotides, primarily $(A)_n$ or $(T)_n$ with n = 6 to 11. In contrast for human the greatest contributions come from dinucleotides and trinucleotides with n = 6 to 11 are a barometer of gene content. The genomic DNA is cleaved with restriction enzyme and the resulting fragments are separated by electrophoresis and hybridized in a Southern blot with a probe consisting of core repeats. Each location in the genome containing the core repeats will yield a separate band in the gel and probes for some core repeats may hybridize with 100 or more bands and thus yielding a bar-code sort of pattern often referred to as a **DNA fingerprint** (Geffreys et al., 1985). As each location in chromosome that contains core repeats may have different number (n) of copies of repeat so at any particular location a micro- or mini-satellite repeat has multiple alleles in the population. Each allele will yield a different fragment size and there is a distinct band in the gel. Alleles are co-dominant and +/– the heterozygous genotype will yield two bands; +/+ or –/– homozygous genotype will yield one band. Fragments derived from each genomic location can be assayed separately using a probe that hybridizes with the unique flanking sequence of the repeat or if PCR primers to the unique flanking sequences are used to identify the repeat. The micro-satellite polymorphism is also called a simple sequence length polymorphism (SSLP) and a mini-satellite polymorphism is called a **variable number of tandem repeats** (VTNR) and generically a polymorphism based on differences in the number of tandem repeats at a location on chromosome is called **simple tandem repeat polymorphism** (STRP). However, the alleles of different loci can not be distinguished by probing with the core repeat as each allele yields a band that is uninformative as to its origin in the genome. Minisatellites are dispersed throughout the human genome (but also occurs in animal and plant genomes) which share a common 10-35 bp consensus or core sequence (core repeat unit, tandem repeat unit). Minisatellites show substantial length polymorphisms which arise through 1. unequal crossingover and 2. mutation (gene conversion) (Goldstein and Schlotterer, 1999). Mutation process acts preferentially at the 3′ end of minisatellite so that most of the sequence variability originates from here. 5′ end belongs to a low mutable region which stabilizes the repeat. In microsatellites the predominant mutational mechanism is believed to be DNA slippage during replication. Simple tandem repeats (STRs) (also known as microsatellites) are not used as markers currently because they are less amenable to cheap, high throughput genotyping methods than SNPs. Further, STRs typically have a much higher mutation rate (10^{-3} per meiosis) than SNP (10^{-9}).

SSRs tend to be restricted to a few marker loci in a small number of species and although co-dominance is displayed, the homology of individual SSRs is based on product size and therefore suffers from the same difficulties as RAPD if used in phylogenetic studies (Jarne and Lagode, 1996). Furthermore, the repeated nature of SSR loci and their high mutation rates mean that it may not be possible to confirm that bands of identical size in two taxa or widely divergent populations are evolutionary homologous as different mutation events in the SSR may generate products of similar size (Jarne and Lagode, 1996; Provan et al., 1999).

11. **SSCP(Single stranded conformational polymorphism)** — An SSCP is one method for detecting single nucleotide polymorphism in any region of the genome without the

need for sequencing the homologous DNA fragment isolated from a large number of individuals (Hayashi, 1992). SSCP is a polymorphism within the fragment amplified from different organisms in a population. This type of polymorphism results from the tendency of the ssDNA to form a complex, three dimensional conformation. Single strands tend to forma certain amount of hydrogen-bonded secondary structure (stems and loops) by chance complementary matching between short segments. Even a single nucleotide difference in a short DNA molecule of order of 300bps may change the conformation of the single strands enough to change the electrophoretic mobility of strand in the gel. Change in conformation will make either easier or more difficult to pass through the pores of the gel and thus can be distinguished. The amplified fragments of ~ 300bps from the genomic DNA are denatured into single strands and subjected to electrophoresis. The different single strands differ in their electrophoretic mobility and thus different fragments will show bands at different places in the gel depending upon their electrophoretic mobility. SSCP alleles are co-dominant and so heterozygous genotypes can be distinguished from homozygous genotypes. SSCP is known to be technically unreliable and the interpretation of the data is problematic (Dowling et al.,1996) and thus it may not be an effective system to use for majority of systematic questions.

12. **Inter retrotransposon amplified polymorphism (IRAP)** — Any difference in DNA sequence between two genomes can be detected by PCR mediated amplification of the region between two neighbouring RTs using a left and a right facing primers directed to the conserved regions within the LTRs. The polymorphism observed here can be due to mutation (deletion/insertion) either primarily in the region between the two RTs or in one or both LTR (s). Since RTs are ubiquitous in eukaryotes the IRAP technique produces multilocus patterns and many bands in these patterns are polymorphic as the evolution of both RTs and inter-RT regions differ between organisms.

13. **DNA amplification fingerprinting (DAF)** — It is a variant of AP-PCR (Welsh and McClelland, 1990) and used for screening of the entire genome of an organism for the presence of highly polymorphic sequences (Caetano-Anolles et al., 1991). In this technique the genomic DNA is annealed to either a single or a group of short synthetic oligonucleotide of arbitrary (but usually GC rich) sequence. The polymorphism observed is usually fragment length polymorphism. The number of amplification products may be positively correlated with the evolutionary positions of the tested organisms (bacteria (0-20), human (0-60). Sequences flanked by the oligo-primer (s) are then amplified with PCR and the amplification products are separated by either agarose gel electrophoresis (and stained with EtBr, less sensitive) or polyacrylamide-urea gel electrophoresis (and stained with silver nitrate, AgNo3/HCOH). **Two dimensional DNA fingerprinting** — It is a technique for high resolution genotyping and generates two dimensional restriction fragments fingerprints. The genomic DNA is first restricted with, for example, HaeII or Hinf I and the restriction fragments generated are separated by electrophoresis in 6% neutral polyacrylamide gels. The relevant regions are cut out of the gel (region of fragment sizes from 0.3 to 3kb or from 1.0 to 10kb) and applied to polyacrylamide gel containing a 10-75% linear gradient of

denaturant (10% 0.7M urea, 4% formamide). This combination of neutral and denaturing gradient gel electrophoresis (DGGE) produces a resolution of up to 1000 or more spots per DNA sample on the stained denaturing gel. Specific sequences such as genes or repetitive sequences as in mini or microsatellites can then be detected by southern blotting hybridization of radiolabeled probes to blots and autoradiography.

14. **Retrotransposon-microsatellite amplified polymorphism (REMAP)**—Any difference in DNA sequence between two genomes is detected by PCR mediated amplification of the region between a long terminal repeat (LTR) of a retrotransposon and a nearby microsatellite. The polymorphism in the region between the retrotransposon and the microsatellite is primarily due to insertions or deletions.

15. **Selective amplification of microsatellite polymorphic loci (SAMPL)**—It is a PCR based technique used for detection of polymorphism that combines the advantages of AFLP and microsatellite marker technologies in one assay. It amplifies random genomic regions by using both an AFLP primer and a compound microsatellite primer and thus detects length polymorphisms in the microsatellite repeats. This technique does not require prior sequence information on the target DNA.

16. **EST polymorphism**—It refers to any difference in DNA sequence between two or more ESTs that can be detected by either restriction digestion of the ESTs or by separation of polymorphic sequences using DGGE. ESTPs can be used to screen for DNA polymorphism in populations or can serve as molecular markers in mapping and comparative mapping procedures. **ESTP mapping**—It is a technique for the conversion of ESTs into molecular markers that can be integrated into a genetic map based on SNPs in coding and non-coding regions adjacent to EST. In this technique the isolated genomic DNA is digested with either a four base cutter (AluI), six base cutter (DraI, SspI) or less frequent six base pair cutter (EcoRV). The vectorette like adaptors are then ligated to the termini of the restriction fragments using the DNA ligase. The resulting adaptored fragments are then amplified through conventional PCR using an EST specific and an adaptor specific primer. The EST primers are usually designed within the 5′ or 3′ UTRs (or near the start or stop codon within the coding region) such that amplification occurs towards the non-coding region, either 5′ or 3′ of the EST. The amplified fragments can be visualized by EtBr fluorescence. For fluorescent detection of the amplified fragments the adaptor-primer should be labeled by a fluorochrome (e.g. 6FAM or HEX). A second amplification is required if nested EST specific and nested adaptor specific primers are used. The resulting amplicons (amplified products) are separated on agarose or polyacrlamide gels. The presence of EST polymorphism between two parents can be used for the estimation of segregating pattern in the progeny of the cross and mapped. Thus the resulting EST map is based on genetic markers.

17. **EST-AFLP**—It is a variant of AFLP which permits scanning of the 3′ and/or 5′ flanking regions of a gene for sequence polymorphisms. The genomic DNA is first digested with a restriction endonuclease, say AluI. The generated fragments are then ligated to an adaptor and amplified with two primers, one complemantary to the adaptor and radioactively or non-radioactively labeled and the other complementary to particular part of a known gene, the EST using PCR. The amplified products

consisting of EST, non-coding region and the adaptor sequence are then separated in denaturation polyacrlamide gels and are detected by either autoradiography or fluorography. EST-AFLP of the two genomes may differ in the length due to SNPs, deletion, insertion and generally due to mutations in the non-coding region (s) adjacent to EST.

The various molecular techniques used for detection of the above mentioned molecular markers are described in chapter 30.

Direct amplification of length polymorphism (DALP) — The protocols of AP-PCR and RAPD are performed at a low annealing temperature (at least in the first cycles) and use either long oligonucletides for AP-PCR (>18mers) or a single short one for RAPD (10 mer). Both produce a fingerprint which can be useful in characterizing strains or species and in investigating the level of genetic polymorphisms of a new species. However, almost all the detected polymorphisms have a dominant inheritance (presence/absence of bands) and amplified regions can not be easily sequenced to design locus specific primers even with a cloning step.

In other words, detection of polymorphisms is possible but the sequencing of loci is difficult or even unrelialisable. A recent method called sequencing with arbitrary primer pairs (SWAPP, Burt et al., 1994) was developed to sequence polymorphic bands isolated from a multilocus pattern obtained by AP-PCR. It uses low resolution agarose gel electrophoresis for a first selection of potentially polymorphic bands and further SSCP analysis of each fragment. Although it is poor for characterization of polymorphism it can be simplified and improved for detection of more loci. Using DALP a much larger number of polymorphic loci can be detected and isolated for sequencing in one step (Desmarais et al., 1998). It is a technique used for the detection of sequence polymorphisms in genome that capitalizes on so-called 'selective' primers (sharing an identical 5′ core sequence: M13-40 universal sequencing primer) and reverse primers (DALPR: the conventional M13 reverse primers) used in combination to amplify genomic DNA in a conventional PCR (with high annealing temperature, lower Mg concentration and fewer cycles than in comparable arbitrary primed PCR). Amplification and subsequent separation of amplified fragments in denaturing polyacrylamide sequencing gels results in genome specific multiband patterns up to 30-40kb bands in the size range between 0.2 and 1.0kb. Obviously sequences complementary to the M13 primers are evenly distributed over genomes of a series of animals. Each amplified product can be directly sequenced. DALP is an arbitrarily primer PCR to produce genomic fingerprints and to enable sequencing of DNA polymorphisms in virtually any species. Oligonucleotide pairs are designed to each produce a specific multi-banded pattern and all the fragments thus generated can be directly sequenced with the same universal M13 sequencing primers. This strategy combines the advantage of a high resolution fingerprinting technique and the possibility of characterizing the polymorphism.

SNPs have long been known to be associated with phenotypic variation either through direct causal effects or by serving as proxies for other causal variants with which they are highly correlated (Linkage disequilibrium) and is quite common in human genome. One form of genetic variation namely, SNPs is associated with disease. Another type of common genetic variation, namely, structural variation including copy number variants (CNVs). Some genes have more than one copy and the copy number can differ among individuals and susceptibility to certain diseases can be altered. CNV is defined as DNA segments that are 1kb or larger in size present at variable copy number in comparison to a reference genome.

15.3 MOLECULAR MARKER TECHNOLOGY

The use of genetic markers for evaluation of polymorphic genetic loci affecting quantitative traits and subsequent manipulation of these loci in the genetic improvement is however limited by the presently available markers. Plant cultivars in particular inbreeders generally differ from one another with respect to only a small number of polymorphic or pigmentation markers and have been found to be remarkably uniform with respect to isozymes markers. Furthermore, many of the present markers have the disadvantage of being developmentally regulated i.e. they come to phenotypic expression only at a specific stage of development or only in some specific tissue or organ while morphological and pigmentation markers in particular, may have secondary pleiotropic effects on quantitative traits.

15.4 SELECTION OF MOLECULAR MARKER

Selection of a DNA marker system depends on the objectives (e.g., map construction, QTL mapping, marker assisted selection), population structure, the genomic diversity of the species under investigation, marker system availability and the cost per unit information (Staub et al., 1996).

RFLPs have advantages over PCR technology in systematic and evolutionary studies as it generates data based on homology among large fragments of DNA in contrast to PCR technology which detects variation in 20-40 bases but then the use of RFLP markers are time consuming, relatively more expensive and require technical expertise. Marker systems differ in their utility across population, species and genera and their efficacy in detection of polymorphism. For example, RFLPs mapped in one population can be used as probes for characterizing other population within the same species but polymorphic primers (SSRs) identified in one species are generally not useful in other species. RAPDs, SPARs and AFLPs can not be used across population because each marker is primarily defined by its length although they can be constructed in a relatively shorter period. Further, with inbreeders RAPDs are more useful in detecting the polymorphism within a gene pool than RFLPs. AFLPs have proved useful in studies where individual regions are sufficiently characterized to allow primer construction. The advantages with PCR based marker technology are that the small sample requirement, high throughput and early selection. STRs and SSRs marker systems are the choice where the level of polymorphism is low. The cost of molecular marker system depends on the time required for DNA extraction, amount of DNA required for analysis, necessity of cloning and sequencing, amount and type of genetic information required, types of marker (dominant vs co-dominant), automation of a marker system, usage of the resulting genetic map and proprietary status of the technique.

In trait mapping and marker assisted selection, application emphasis must be placed on attaining reduced time and cost of data acquisition. If a technology can not analyse more than 100-2000 individuals for each of the many populations that the plant breeders handle at one time, that technology will have limited applicability. Chromosomal segments associated with agronomic traits of economic importance involving fairly moderate genetic complexity can not be readily identified with sufficient precision or power by RFLPs. Further throughput using RFLPs is limiting especially during progeny selection with time constraints imposed prior to pollination or harvest (Smith and Beavis, 1996). AP-PCR based methods such as RAPDs or AFLPs are now increasingly used by plant breeders because these methods can provide speedier and cost effective marker assisted progeny selection (Ragot and Hoisington, 1993). No marker based selection scheme will be useful unless it advances the speed and efficiency of already rapid and significant progress made using conventional methods of breeding. In the marker assisted backcrossing programme the following two types of information are generally required.

1. Very detailed information on the introgressed region, usually markers that flank the gene or genes of interest as closely as possible.

2. More widely dispersed markers on the unlimited regions to ensure that the genetic material from the recurrent parent is carried forward. The recurrent parent analysis could be done using the SSRs for the genotyping offspring whereas analyses of introgression could use AFLP (Bates et al., 1996). AP-PCR methods can also be used to very quickly identify map and introgress chromosomal regions determining resistance to insects and pathogens. Profiling of near isogenic lines and bulk segregant analysis can be used to quickly map, characterize and introgress chromosomal regions (Michelmore et al., 1991; Paran et al., 1991). All gel based marker technologies have inherent problems of sample throughput efficiency and cost of data acquisition and thus only a limited number of marker assisted selection programme involving trait of even moderate genetic complexity can be run simultaneously and will not be effective in marker assisted selection programs involving traits of higher genetic complexity. The main problem with the mapping approach regardless of mapping technology is that the difficulty in obtaining reliable performance data for traits under control of more than 3 to 5 genes. The RFLP markers have been studied for their discriminational power and their utility to reveal association between genotypes that are reflective of pedigree of heterosis.

15.5 APPLICATIONS OF MOLECULAR MARKER TECHNOLOGY

The molecular markers can be used for fingerprinting of a variety, study of phylogeny, hybrid identification (intra-specific and interspecific), genome mapping, gene tagging, marker-assisted selection and gene cloning.

15.5.1 Pedigree Analysis

Molecular marker is used for carrying out linkage analysis. If a molecular marker is genetically linked to a trait, qualitative or quantitative (a gene determining disease in human or a gene determining disease or insect resistance or yield in plants) then the molecular markers scattered across the genome are tracked through pedigrees in the hope of identifying that linkage. Suppose there are two molecular marker genes, A and B with recessive alleles, a and b, respectively and the two families say 1 and 2 show segregation for these two markers. The molecular markers could be RAPD, AFLP or RFLP. As in human it is not always possible to work out the genotypic constitution of the parents because of impossibility of controlled crossing, the possibility of testing for linkage and estimating recombination frequency in the presence of linkage can best be done by asking what the probability would be of getting the observed distribution of phenotypes if the recombination frequency took some particular value and all known pedigrees that are informative with respect to the combination of markers of interest would be used. The calculation will be done by computer. The probabilities corresponding to the different pedigrees will be multiplied together to provide an overall probability. The computation is repeated for each of recombination values, spaced, for example, at intervals of 2% over some plausible range say 5 to 35%. Thus the overall probability for each recombination frequency which will usually be very small is divided by the probability of getting the data if there were no linkage at all which if linkage is real, should be even smaller. The resulting ratio which represents the relative odds against the data given the postulated recombination value is expressed as the logarithm to base 10 and called **LOD score** (Morton, 1955).

$$\text{LOD} = \log_{10} \frac{L\,(\theta = 0)}{L\,(\theta = 0.5)}$$

Where the recombination value is denoted by θ which is the probability of a recombination event between two loci of interest and as such it is a function of distance. $\theta = 0.5$ when the two loci are unlinked and closer a pair of loci, lower is their recombination value.

A plot of LOD against postulated recombination frequency may show a well defined maximum that can be taken as the best estimate of the recombination value. The significance of the evidence of linkage depends on the magnitude of the LOD. A value of 2 is about the minimum for acceptability. Let us consider an example of two families showing segregation of two dominant markers, A and B with recessive alleles a and b (Figure 15.1). In family 1 the father carries A and B and as it is known that he inherited A from his mother and B from his father, so if A and B are linked then he must be of genetic constitution Ab/aB. His wife's genotype is ab/ab and of the four children two are Ab/ab and two aB/ab. In family 2 the father carries both A and B but as it is not known whether he inherited them from the same parent or from different parents. Hence if there is linkage he may be of the genetic constitution AB/ab or Ab/aB. The mother's genotype is ab/ab and of the five children two are AB/ab, two are ab/ab and one Ab/ab. If the recombination frequency between two markers A and B is r then the probability of families with these children can be calculated. The doubly heterozygous parent AB/ab or Ab/aB will produce four types of gametes in the ratio $(1 - r)/2 : (1 - r)/2 : r/2 : r/2$ where r is the recombination frequency and $(1 - r)$ is the non-recombinant frequency. Thus by knowing the parental genotypes one can work out the probability of each observed pedigree in terms of the recombination frequency using binomial distribution. In family 1 the genotypic constitutions of the offspring show that they are all non-recombinant and the probability of observing offspring is $(1 - r)^4$. In case of family 2, there exists two possibilities. The first possibility is that the father is AB/ab and considering four non-recombinant and one recombinant children the probability of observing such offspring is $r(1 - r)^4$. When the father is of the genetic constitution Ab/aB then the probability of observing five children, four recombinant and one non-recombinant, will be

$r^4(1 - r)$. The average probability of the observed family 2 will be $= (r(1 - r)^4 + r^4(1 - r))/2$ as assuming A and B to be in linkage equilibrium both possibilities are equally likely to happen. Considering both families together the overall probability will be the product of the two single family probabilities as the pedigrees are independent and will be $= (r(1 - r)^8 + r^4 (1 - r)^5)/2$. There is another way of calculating the overall probability which is based on calculation of probability of the overall numbers of recombinants and non-recombinants regardless of their sequence in the family and although it would have values 5 times as great for all values of but it could have made no difference to the relative odds and the final LOD scores. One can now calculate the overall probability or odds against for different values of r.

The odds are strongly in favor of linkage and the recombination frequency is likely to be in the range of 5 to 20 %. As the estimate of recombination frequency in this example is based on nine gametes, its precision is very low. With the increase in the number of families in the analysis the LOD scores (log odds, relative to no linkage) would usually increase rapidly

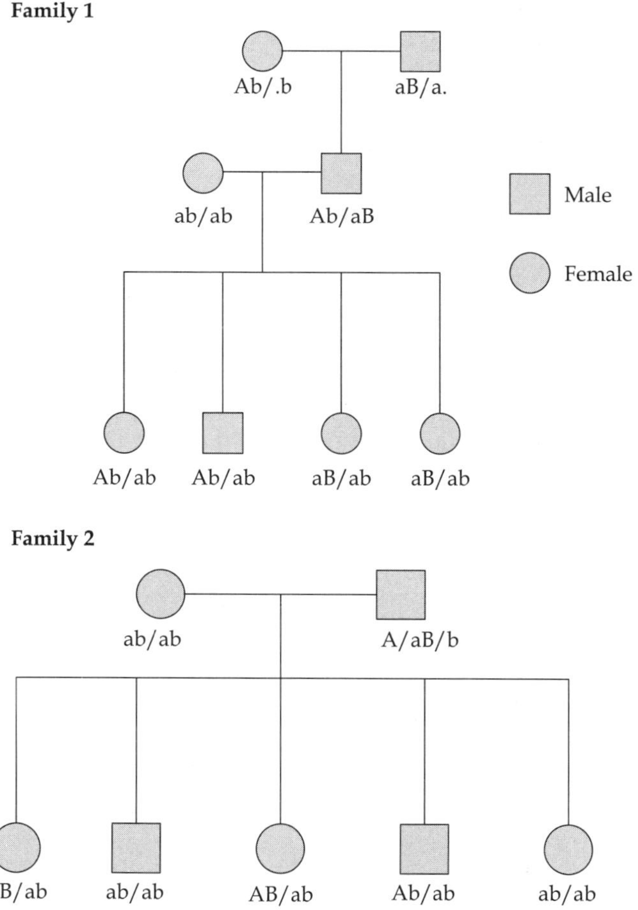

FIGURE 15.1   **Showing two hypothetical families used for the calculation of LOD scores.**

resulting into a more precise estimate of recombination frequency. The steps involved in the calculation of LOD scores are as follows:

1. Write down the probability of each observed pedigree by knowing the parental genotypes in terms of recombination frequency using the binomial distribution.

2. Again write down the probability of each pedigree assuming r = 0.5 (i.e. no linkage between gene and a molecular marker(or independent segregation, ie. either genes are on different chromosomes or in one chromosome but wide apart)).

3. For each pedigree calculate the likelihood ratio (the ratio of probability of the pedigree given an arbitrary value of r, Pr (Pedigree) r to that with r = 0.5, Pr (Pedigree) r = 0.5)

4. Calculate the LOD score which is the logarithm (base 10) of the likelihood ratio) for each pedigree.

5. As the pedigrees are independent the overall probability can be obtained by multiplying the probability of each pedigree. Similarly, the overall likelihood ratio can be obtained by multiplying the likelihood ratio of each pedigree but the overall LOD score is obtained by adding the lod score of each pedigree (lod scores are additive across pedigrees because $\log_{10} (xy) = \log_{10} (x) + \log_{10} (y)$).

6. The lod scores are calculated for different arbitrary values of r (say, for example, 0.3, 0.2, 0.1, 0.05, 0.02) to that with r = 0.5 (Table 15.2).

7. The value of r that maximizes the lod score for all the pedigrees is the estimated value of r, the recombination frequency.

8. A lod score greater than 3 is generally regarded as significant evidence for linkage; a lod score less than 2 is generally regarded as significant evidence against linkage and any value –2 lod <3 is considered uninformative in regard to linkage and requires additional data for analysis. Suppose that we have to find out linkage between a gene , D (D and d) determines a trait and a molecular marker, M with two alleles, M and m. Again suppose that the genotype of the doubly heterozygous parents is D M/d m and the recombinant individual will have, d M/d m constitution. The family 1 is having one recombinant and two parental types(d m/d m and D M/d m). In family 2 all three have d m/d m genotype (non-recombinant). Family 3 has parental types (d m/d m and D M/d m) individuals and family has two non-recombinant (D M/d m) type individuals. The Pr (pedigree) r, Pr (Pedigree) r = 0.5, Likelihood ratio and lod score for each family are given in the Table 15.3.

Table 15.2 Showing LOD scores for different arbitrary values of r.

r	Probabilities (each x 10^{-3})	Relative odds	Log odds(LOD)
0.5	1.95	-	-
0.3	9.33	4.78	0.68
0.2	17.03	8.72	0.94
0.1	21.55	11.04	1.04
0.05	16.58	8.49	0.93
0.02	8.51	4.36	0.64

Table 15.3   Showing likelihood ratio and lod score for four families.

	Family 1	*Family 2*	*Family 3*	*Family 4*
Pr(Pedigree) r	$3r(1-r)^2$	$(1-r)^2$	$(1-r)^2$	$(1-r)^2$
Pr(Pedigree) r = 0.5	$3(1/2)^3$	$(1/2)^3$	$(1/2)^2$	$(1/2)^2$
Likelihood ratio	$8r(1-r)^2$	$8(1-r)^3$	$4(1-r)^2$	$4(1-r)^2$

The LOD score for different families will be = $\log_{10}(8) + \log_{10}(r) + 2\log_{10}(1-r)$, = $\log_{10}(8) + 3\log_{10}(1-r)$, = $\log_{10}(4) + 2\log_{10}(1-r)$ and = $\log_{10}(4) + 2\log_{10}(1-r)$, respectively. The overall lod score then will be = $\log_{10}(8 \times 8 \times 4 \times 4) + \log_{10}(r) + 9\log_{10}(1-r)$. Thus for each position lod score is calculated and inference about the linkage is drawn using some arbitrary LOD score, say 2 has to be taken as cut off point below which the evidence for linkage is deemed to be too weak.

15.5.2   Applications of Isozyme/DNA Marker Analysis

1. **Cluster analysis-** One of the objectives of isozyme/DNA marker analysis is to classify or group together isolates on the basis of banding patterns. The different steps involved in cluster analysis are as follows.
    (a) Scoring of bands
    (b) Computation of similarity/distance matrix
    (c) Clustering.

Scoring the bands—The presence of a band should be scored as '1' and its absence as '0'. The precautions to be taken while scoring bands or comparing banding patterns are as follows.

1. As only a limited number of samples can be run on one gel so the conditions between runs should be made as uniform as possible and particularly the distance the running dye migrates should be consistent.
2. Load one or more identical samples in different gels in order to ascertain that the bands are identical between two samples on different gels.
3. To determine the identity of bands in different samples, gels or blots it is worthwhile to trace the banding pattern in a piece of transparent acetate. Multicolored pens can be used where one color designates bands are scored as identical. However, a ruler is indispensable.

Scoring of bands is more an art than a science. Although automation of the process has been done and computer programs are available that can take a scanned picture of an autoradiogram and produce a list of molecular weights of each band but if the background is too dark or the bands are too faint the autoradiogram can not be read. False positives are also a problem. The program refuses to read if the bands are quite distinct. In case the banding patterns are moderate or complex it is impossible to work with program and thus human eye is still the best judge.

Computation of similarity/distance (dissimilarity) matrix—The second step in cluster analysis is to compute the similarity or distance between all possible pairs of samples. Similarity and distance are measured in terms of similarity and distance coefficients. Similarity and distance coefficients provide estimates of association or resemblance between pairs of units. Similarities range from 0 to 1; 0 value indicates no likeness whereas 1 indicates highest level of likeness (self-similarity). Similarity increases with increasing likeness. The distances are just the opposite. Distance decreases with increasing likeness or similarity and they are usually negative. Thus the resulting array of coefficients is called a similarity or distance matrix.

Types of distance and similarity measures — In case of binary data, i.e. data encoded as presence or absence as in case of banding patterns, the data between pairs of genotypes can be arranged in a 2 x 2 contingency table as shown below where 1 indicates presence of band and 0 indicates absence of band.

Genotype i

	1	0
1	a	b
0	c	d

Genotype j

where m (= a + d) stands for the number of matches, u (= b + c) stands for the number of 'ummatches and n (= m + u) is the total number of bands.

Similarity measures — The simplest way to measure similarity (S_{SM}) is to simply divide the number of matches by the total number of bands.

$$S_{SM} = m/n$$

Here 'negative matches', i.e. both genotypes that have no bands are also counted. However, it is not done in case of banding pattern data where it is argued that a potentially infinite number of negative matches exists.

Jaccard's coefficient — Jaccard's coefficient (S_J) is calculated as

$$S_J = a/(a + u)$$

Here the similarity measure does not include negative matches.

Dice's coefficient — Dice's coefficient also does not include negative matches but it differs from Jaccard's coefficient in that the matched pairs carry twice the weight of unmatched pairs.

$$S_D = 2a/(2a + u)$$

In case of banding pattern data Dice's coefficient can be more easily computed using the following formula.

$$S_D = 2a/(n_1 + n_2)$$

Where n_1 is the number of bands in genotype i and n_2 is the number of bands in genotype j. Thus twice the number of positive matches is divided by the total number of bands in both genotypes. As Dice's coefficient is easy to compute, this coefficient is used in the subsequent analysis.

Distance Measures

Distance measures (ds) are derived from similarity coefficients (S) in the following way

$$d = 1 - S$$

Clustering by hand — The different steps involved in clustering by hand are as follows.

1. Count the number of bands for genotypes 1 and 3 and designate these as n_1 and n_2, respectively.
2. Count the number of shared bands between genotypes 1 and 2 and designate this as a_{12}.
3. Compute Dice's coefficient using the formula, $S_D = 2. \, a_{12}/(n_1 + n_2)$
4. Transform the Dice's coefficient into distance (d_D) as $d_D = 1 - S_D$
5. Repeat the process for each pair of genotypes, 1 and 2, 1 and 3, 1 and 4 and so on for a group of n genotypes where n equals 1 to n.

Clustering with UPGMA

Clustering analysis summarizes the similarity or distances among all pairs of genotypes in a set (i.e. a similarity or distance matrix) in terms of nested sets (clusters). The standard procedure for clustering is as follows.

1. Search the input matrix for the pair of genotypes (i, j) that have the smallest distance (or are most similar)
2. Merge these genotypes into a new cluster.
3. Update the matrix to reflect the deletion of the pair of genotypes i that were grouped and the addition of a new genotype corresponding to the new cluster. Now compute similarities or distances between the existing genotype and the new cluster.
4. Repeat the procedure until all genotypes have been clustered.

Different clustering methods have been developed simply by varying the computation of the similarity or distance of the new cluster described above in step 3 and one of the simplest method is the average distance method or **unweighted pair-group method with arithmetic mean (UPGMA)**. The various steps involved in UPGMA are given below.

1. Find the smallest distance value in the distance matrix.
2. Cluster together the 2 genotypes which give this distance. The branching point is at the midpoint of the 2 distances
3. Construct a new distance matrix combining the 2 nearest genotypes into a single cluster. The distance between this new cluster and another genotype is given by the average distance of the other genotype to each of the the genotypes in the cluster.
4. Repeat the procedure until all the genotypes have been clustered. When plotted the final dendogram should show branch lengths proportional to the distances among clusters.

Solved Example

Consider the following distance matrix of five genotypes:

Genotype	1	2	3	4
2	d_{12}			
3	d_{13}	d_{23}		
4	d_{14}	d_{24}	d_{34}	
5	d_{15}	d_{25}	d_{35}	d_{45}

Suppose that the distance between genotypes 3 and 4 (d_{34}) is the shortest. Thus the two genotypes (3 and 4) are grouped with a branching point located at distance $d_{34}/2$.

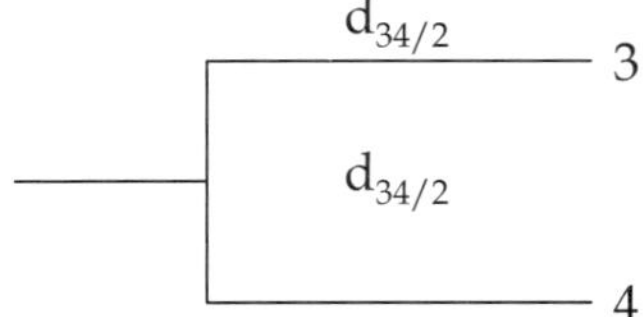

Now construct a new distance matrix by calculating the distances between the (3, 4) cluster and the other genotypes as shown below.

Genotypes	1	2	(3, 4)
2	d_{12}		
(3, 4)	$d_{1(34)}$	$d_{2(34)}$	
5	d_{15}	d_{25}	$d_{5(34)}$

The distances between 1 and (3, 4) ($d_{1(34)}$) , between 2 and (3, 4) ($d_{2(34)}$) and 5 and (3, 5) ($d_{5(34)}$) are calculated as follows.

$d_{1(34)} = (d_{13} + d_{34})/2$

$d_{2(34)} = (d_{23} + d_{34})/2$ and

$d_{5(34)} = (d_{35} + d_{45})/2$

In the above new matrix look for the smallest distance. Now if, for example, this were d_{12}, these are clustered together to form cluster (1, 2).

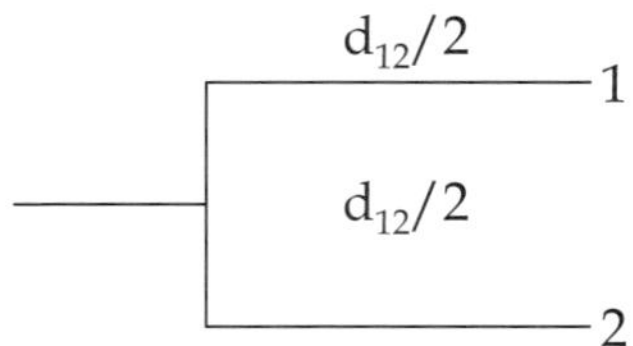

After formation of two clusters yet another distance matrix is constructed.

Genotype	1, 2	(3, 4)
(3, 4)	$d_{(12)(34)}$	
5	$d_{5(12)}$	$d5_{(34)}$

$$d_{5(12)} = (d_{51} + d_{52})/2 \text{ and } d_{(12)(34)} = (d_{1(34)} + d_{2(34)})/2$$

Assuming that the distance $d_{(12)\,(34)}$ is the shortest distance, the two clusters (12) and (34) are themselves grouped into cluster ((12), (34)) at distance $d_{12(34)}/2$.

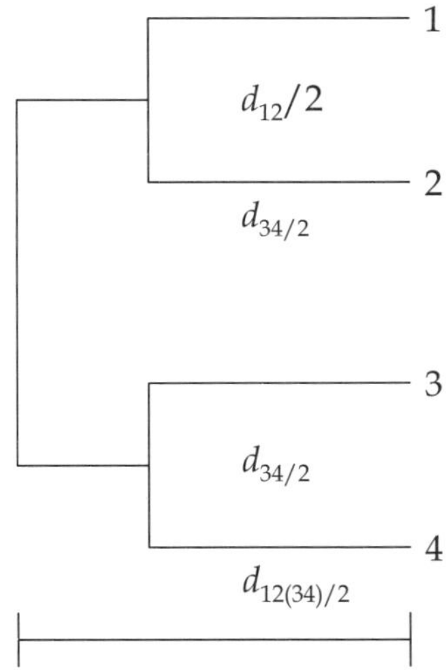

The distance $d_{12(34)}/2$ equation refers to the distance from the branching point of cluster ((1, 2), (3, 4)) to the ends of the branches and not to the next branching point. Now a new matrix is calculated by taking average of $d_{5(12)}$ and $d_{5(34)}$. This leaves only one genotype in the distance matrix as given below.

Genotype	((1, 2), (3, 4))
5	$d5((12)(34))$

The new distance $d_{5((12)(34))}$ is calculated as above by taking the average of $d_{5(12)}$ and d_{534}.
$d_{5((12)(34))} = (d_{5(12)} + d_{534})/2$.
Finally, the genotype 5 joins to cluster ((1, 2), (3, 4)) at distance $d_{5((12)(34))}/2$.

16

Structural Genomics and Structural Genomics Technology

16.1 DEFINITION

The study of genome involves (i). construction of genetic and physical maps of the genome (ii). determination of complete nucleotide sequence (iii). Determination of location of genes and (iv). estimation of effects of genes. The genome analysis thus aims at identifying the gene repertoire emphasizing similarity, differences and uniqueness among genes. All these require the development of new computational methods for analyzing the genetic map, sequence data and new techniques and instrumentations for detecting and analyzing the gene (DNA) and above all the development of means of information dissemination for the benefit of research workers through out the world.

Structural genomics refers to the study of systemically characterizing the architecture of genomes including sequence directed domain formations (fold back DNA-stem and loop structure), folding of DNA (in nucleosomes, solenoid, looped domains), condensation of DNA in chromosomes and fine and overall structure of chromosomes or chromatin. The total genome can be explored at the following four levels;

1. A heterogeneous mass of DNA
2. A set of visible chromosomes
3. Detailed linkage map correlated with chromosomes
4. As DNA sequence

The second objective in genome analysis is the assessments of genomic organization and sequence heterogeneity which include compositional biases of short oligonucleotides, dinucleotide relative abundances (the genome signature), codon and residue biases, rare and frequent words (oligonucleotides, peptides, codons), clustering, over dispersion or excessive evenness in the distribution of various markers, e.g. particular oligonucleotides, restriction sites, nucleosome placements, methylation targets, origin of replication, repair recognition site and a myriad of control sequences (Karlin et al., 1998) and repeat structure. The DNA heterogeneity include isochore compartments in vertebrate species and G + C – and A + T –rich halves of λ genome, TEs (such as Ty in yeast, IS in *E.coli* and Alu in human), centromeric satellite tandem repeats such as the 171-bp human alpha satellite DNA, characteristic telomeric sequences such as the hexanucleotide AGGGTT tandem repeat in human, repetitive extragenic palindromes (REPs) of *E.coli*, recombinant hot spots such the *chi* elements in *E.coli*, universal under representation of the dinucleotide TpA, under representation of the dinucleotide CpG in vertebrates and many thermophiles, HTF island-DNA sequences that generally occur upstream of vertebrate gene and are abundant with non-methylated CpG sequences, under representation of the tetranucleotide CTAG in proteobacterial genomes, GNN periodicity in coding sequences and methyl transferase modification. The under representation of CG has usually been ascribed due to the classical methylation/deamination/mutation scenario causing mutation of CG to TG/CA. The deficiency is because of structural constraints related to high dinucleotide stacking energy, super coiling and chromatin packing. The dinucleotide CC/ GG, TG/CA and AG/CT, all a single base mutation from CG are (except for dicots) overrepresented only in genomes with strong CG suppression. The under representation of TA may be due to its low thermodynamic stacking energy, high degree of degradation of UA dinucleotide by ribonucleases in mRNA tracts and the presence of TA as part of many regulatory signals(e.g. TATAbox, transcription terminators). TA suppression may help in avoiding inappropriate binding of regulatory factors. The dinucleotide TA is under represented in the bulk of prokaryotic and eukaryotic sequences. Eukaryotic chromosomes often contain tandem repeats formed by polymerase slippage or unequal crossingover and with direct and inverted repeats promoted in part by transposition, translocation, recombination, amplification and excision. Further, many genomic sequences show polymorphisms strain variation, DNA inversions and rearrangements thereby reflecting a state of influx. The study of genome organization thus involves determining genome size, base composition, gene distribution, dispersal of repetitive sequences, long tandem arrays and gene order. Large mammalian genomes have a CpG content of 1% which is lower than 4% expected for a genome that is 40% G + C rich and thus there is under representation of CpG. However, there are small regions, several hundred bases long (called CpG islands) which have a CpG/GpC dinucleotide ration of approximately one. CpG islands coincide with the promoter of genes. They are also marked by clusters of unmethylated CpG dinucleotides which therefore contain recognition sequences for methylation sensitive restriction enzymes. Thus methylation sensitive restriction enzymes define active genes. In mammalian genomes, methylation generally is restricted largely to the sequence m5CpG. 70 to 80% of CpG dinucleotides of mammalian DNA is methylated whereas >80% of CpG dinucleotides in wheat are methylated. In plants 5-methyl cytosine is not confined to CpG dinucleotides but also present in more than 80% of trinucleotides CpXpGs. In nuclear DNA of wheat and other higher plants, a higher proportion of cytosine residues is methylated, 30% in comparison with 1 to 8% in vertebrate. The G + C content in wheat genome is 45%

16.2   STRUCTURAL ANALYSIS OF CLONED DNA

The structural analysis of cloned DNA aims at obtaining the following information.

16.2.1 Genome Size

The haploid genome size for some different species are given in Table 16.1. There are various methods for estimating genome size. These include microdesitometry of Feulgen-stained nuclei(Bennett and Smith 1976, 1991; Bennett et al., 1982), DNA-reassociation kinetics (Goldberg et al., 1978; Kiper et al., 1979; Leutwiler et al., 1984), nuclear volume measurement, flow cytometry (Arumuganathan and Earle, 1991; Galbraith et al., 1983) and estimations from sampling genomic clone libraries (Hauge et al, 1991). These diverse methods discussed in chapter 31 give slightly different values for the same species as can be seen in the Table. The Table shows that there is a 300-fold difference between the smallest eukaryotic genome (yeast) and the largest (maize). Further, plant genomes vary in size between species by up to1500-fold and the fact that species with small genomes exist as fully functional flowering plants indicates the presence of the extra nongenic DNA (redundant or functionless) in species with larger genomes. *Arabidopsis thaliana* is considered as the model species for dicots and rice is the model species for monocots. In case of mammals, mouse and man have almost the same C-value. Certain amphibian such as Newt and flowering plant such as lilies have the highest C-values in their groups and have far more DNA than mammals and thus there is no correlation between the C-value and the apparent degree of evolutionary advancement. This phenomenon is referred to as the **C-value paradox**. The C-value paradox has been resolved to some extent in that there is a tendency for species with higher C-value to have higher proportions of repetitive DNA but then such species tend to have larger amounts of single copy DNA as well. The total quantity of DNA in the haploid genome is often called the C-value of the organism. The C-value can be estimated in picograms or number of base pairs. There is no relationship between the C-values of different species and their length of chromosomes measured in recombination map units (centimorgans). One centimorgan in human is equivalent to between one and two megabase pairs whereas it is about two kilobase pairs in yeast-nearly a thousand fold difference.

Estimation of genome size—The genome size can be estimated following Hulbert et al. (1988). Assuming a continuous genome the number of markers expected to fall within a given interval, say 10cM can be estimated as:

$$Yx = n(n-1)/2.\ 2x/G$$

Where Yx is the number of two point linkages at distance equal to or less than x, n is the number of markers mapped (at a given LOD score), x is the interval size and G is the estimated genome size.

Discarding the pairs that are not independent events(as the result of being located at homologous chromosomes) one can get the estimate of the genome size using the above formula. One estimate of G is generated with markers ordered at LOD > 3.0 and another estimate with markers ordered at LOD > 2.0 and the mean of the two estimates of G is an approximation of the final estimate of G.

A characteristic feature of eukaryotic genomes is the enormous variation in genome size which bears little relation to potential differences in organism complexity or to the number of genes that code for proteins (Orgel and Crick, 1980). The genomes of many important grasses are largely collinear yet vary extensively in size. For example, maize genome is 3.5 times as large as that of sorghum and more than 35 times as large as that of rice. More than 50% of the

Table 16.1   Haploid DNA content of a variety of organisms.

Plant species	Mb/1C	No of base pair
Thale cress (*Arabidopsis thaliana*)	70,100, 145,190	1.17×10^8
Oryza sativa (rice)	430, 580, 440	4.2×10^8
Lycopersicon esculentum (tomato)	950, 965	1.0×10^9
Zea mays (maize)	2300, 2500	2.5×10^9
Hordeum vulgare (Barley)	4900, 5300	4.8×10^9 (5×10^9)
Triticum estivum (wheat)	16000, 16700	1.6×10^{10} (1.7×10^9)
Fritillaria assyriaca (lily)	123000	
Nicotiana tabacum	3, 500, 000	4.8×10^9
Triturus cristatus (newt)		31.5×10^9
Sorghum (*Sorghum bicolor*)	750	$8x \ 10^8$
Bacteriophage T4 (Bacterial virus)	2×10^2 kb	
E. coli (Bacterium)	4.5×10^3 kb	4.7×10^6
S. cerevisiae (yeast)	1.8×10^4 kb	1.5×10^7
Caenorhabditis elegans (nematode worm)	c.1×10^5	8.0×10^7
Drosophila melanogaster	1.7×10^5	1.8×10^8
Mus musculis (mouse)	c.$1x \ 10^6$	2.7×10^9 =
Homo sapiens (human)	c.1×10^6	3.2×10^9
Lambda phage		48,502
Adenovirus 2		35,937
F x 174		5,386
SV40	0.000006 pg	5,243
P^{BR322} DNA		4363
Garden pea (*Pisum sativum*)		4.1×10^9
Soybean (*Glycin max*)		1.1×10^9
Potato (*Solanum tuberosum*)		1.8×10^9
Oilseed rape/canola (*B. napus*)		1.2×10^9
Lilium davidii	40,000	
Poplar (*Populus trichocarpa*)		485 Mbp
Grape vine (*Vitis vinifera*)		487, 475 Mbpf
Papaya (*Carica papaya*)		372 megabases

1 pg = 1000 Mbp = 1 Gbp (No. of bp = pg x 0.9869×10^9), 1 kp = 1000 bp (10^3), 1 Mbp = 10^6 bp and 1 Gbp = 10^9 bp. 1 pg = 10^{-12} g, 1 µg = 10^{-6} g, 1 ng = 10^{-9} g. M (Mega), G (giga), n (nano) and p (pico).

maize genome may consist of retrotransposon DNA. **Genome expansion**–It refers to the increase in size of a genome over evolutionary times, caused by either duplication, polyploidization or transposition or a combination of three processes whereas decrease in genome size is due to deletion (as seen in case of Ac and Spm transposons of maize). For example, comparison of the genomes of different members of the grass family such as rice, maize and sorghum shows that they differ greatly in their physical lengths (see Table 16.1) although they originate from a common ancestor. Since the gene content of the various grasses does not differ, the vast difference in the genome size is most probably due to expansion of non-genic sequences (e.g. by the insertion of retroelements).

16.2. 2 Analysis of Whole Genome

Genome by definition consists of the haploid set of chromosomes. The genome analysis aims at studying genomes or genomic DNA in general or the identification of genetic defects in particular using various DNA detection and analysis techniques, **genome mapping** and DNA sequencing. So there are two levels at which the genome analysis can be performed.

1. Whole genome level
2. Individual chromosome level

Euchromatin percentage in genome—Chromosome architecture comprises euchromatin and heterochromatin and so the genome can be partitioned into euchromatic and heterochromatin portion. In mouse the euchromatic portion consists of 2.5 Gb which is 14% smaller than that of human genome (2.9 Gb). In Arabidopsis genome there is about 105Mb of euchromatic DNA and around 15Mb of heterochromatic DNA.

Isochore organization—Isochore refers to any compositionally homogeneous DNA segment of >100-200 kb or more than 300 kb in the nuclear genome of vertebrate, di-and monocotyledonous plants that is distinct in its repeat content, gene density and G/C vs A/T ratios. For example, the A/T rich isochors generally harbour longer genes than G/C rich isochors. Further, different isochors differ in their pattern, for example, isochors of monocots are different from those of dicots. It includes the compositional patterns of DNA molecules and of coding sequences, compositional correlations between coding and non-coding sequences and the relationship between isochors and chromosomal bands. Isochore organization is very wide spread in eukaryotes in contrast to prokaryotes where as a rule genomes are compositionally very homogeneous. The different compositional heterogeneity and the accompanying abundance of repeated sequences are the major distinctive features of pro- and eukaryotic genomes.

16.3 GENE SPACE

It refers to that part of a genome where gene density is substantially higher than in the rest of the genome, usually representing gene clusters. In most genomes such gene spaces are interrupted by gene-empty, mostly repetitive sequences which frequently represent the majority of the sequences in the eukaryotic genomes (e.g. human genomes have about 95%). Gene-rich isochors are characterized by distinct structure and functional features like chromatin structure and levels of transcription and recombination. Gene distribution in genomes of vertebrates and of plants is highly conserved in evolution and so remarkable conservation in gene order is found at the chromosome level. Synteny (gene order) is extensive in mammals. Gene distribution along a cereal chromosome is not randm, with high density of genes concentrated in the distal regions. Long-range mapping with a 4-Mb length indicated five clusters of unmethylated CG-rich recognition sites indicative of active genes in wheat. Physical mapping of genes in wheat can be carried out with deletion stocks. A comparison of wheat chromosome 1 (group3) and rice chromosome 1 shows similar gene content and extensive conservation in gene order. Rice chromosome has more even distribution of clusters of gene-rich regions. Further, gene density in these gene-rich regions on rice chromosome is higher than in gene-rich region in wheat. The proximal regions of wheat chromosomes are composed largely of repetitive blocks whereas the distal composed of both cluster of gene-rich regions and repetitive blocks. There ia abundance of repetitive elements in wheat genome (~ 83%) (Chantret et al., 2005). As in maize, genes in wheat are embedded within long stretches of nested retroelements and other mobile sequences. Comparisons of species such as rice and wheat and rice and maize have shown a degree of conservation of gene order. Genes have been maintained in a similar order despite gross differences in the genome size and chromosome number of these species and thus genes on the rice genetic map could be grouped into sets and the genetic maps of wheat and maize could be described by the same sets of gene (rice linkage segments). This analysis could be extended to include sorghum, foxtail millet and sugarcane. Thus a series of sets of genes could be used to describe most of the major cereal genomes. Gene order in sorghum is similar to that of rice.

Correlations between coding and non-coding sequences are due to compositional constraints which are strong in the gene-rich, G-C isochors and weak in the gene poor, G-C poor isochores. Compositional correlations also exist between coding sequences (and their coding positions) and isochores or flanking sequences as well as between exons and introns from the same genes. These correlations concern coding and non-coding sequences as isochores are essentially made of noncoding sequences and introns obviously are also noncoding sequences.

Correlation holds among GC levels of different coding positions (GC_3 vs GC_{1+2}) again because of compositional constraints working in the same direction although not within the same amplitude. This correlation is a universal correlation which holds for all eukaryotic and prokaryotic genes (Bernardi, 1995). It is known that gene density in GC-rich regions is five times higher than in regions with moderate GC content and 10 times higher in AT-rich regions. Four categories of DNA were identified in humans based on their GC content as follows.

1. <43% GC
2. 43-51% GC
3. 51-57% GC
4. > 57% GC

These are known as **isochors**. GenScan besides utilizing the basic signals including transcriptional, translational and splicing signals (including elements present in most eukaryotic promoters such as the TATA box and cap site) as well as length distribution and compositional features of exons, introns and intergenic regions, makes use of many substantial differences in gene density and structure based on GC composition of human genome for building a complete gene structures (ie. exons + introns). If the input genomic sequence has a GC content of 45%, it can be said to have an isochore value of 2.

There is correlation between base composition and chromosomal bands (see Roy, 2009).

Grass genomes are characterized by a compositional compartmentalization with gene islands (also termed 'gene space') of 100-200 kb. In contrast, genes in the similarly complex human genome are scattered among G + C – rich and G + C – poor regions. Fluctuations in the G + C content of individual genes must somehow be compensated by intergenic sequences to explain the observed uniform base composition in 100-200 kb gene space regions.

16.4   GENERATING GENOME-WIDE CHROMATIN STATE MAP

Different types of cell in a multicellular organism differ markedly in their behavior. The cellular behavior (or state) may be closely related to '**chromatin state**', i.e., modifications to histones and other proteins that package the genome. Thus it would be worthwhile to construct 'chromatin state maps' for a variety of cell types. Chromatin state can be studied by CHIP (chromatin immunoprecipitation) in which an antibody is used to enrich DNA from genomic regions carrying a specific epitope. Enrichment at individual loci is commonly assayed by PCR but this method does not scale efficiently and thus the challenge in generating genome-wide chromatin maps is to characterize these enriched regions in a scalable manner. A more recent approach has been CHIP-chip in which enriched DNA is hybridized to a microarray (Buck and Lieb, 2004; Mockler, et al., 2005). Although CHIP-chip has been successfully used to study large genomic regions, this technique suffers from technical limitations. It requires large amounts of DNA, involves amplification which induces bias. It is subject to cross-hybridization which hinders the study of repeated sequences and allelic variants and is expensive.

In principle, chromatin could be readily mapped across the genome by sequencing CHIP-DNA and identifying regions that are over represented among these sequences. Sequence–based mapping could require relatively small amount of DNA and provide nucleotide level discrimination of similar sequences thereby maximizing genome coverage. But the limitation with this approach is that high resolution mapping requires millions of sequences. This approach is too costly with traditional technology, even with concatenation of multiple sequence tags (Roh et al., 2005). But the single molecule based sequencing technology promises to increase throughput and decrease cost. In this approach DNA molecules are assayed across a surface, locally amplified, subjected to successive cycles of primer-mediated single base extension, SBE (see chapter 12) using fluorescently labeled reversible terminators and imagined after each cycle to determine the inserted base. The read length is short (25-50bases) but tens of millions of DNA fragments may be read simultaneously. Kim et al. (2007) reported a method for mapping CHIP enrichment by sequencing (CHIP-seq) and described its application to create a chromatin-state map- a high resolution map of active promoters. ChIp-seq technique has the potential for using ChIp for genome-wide annotation of novel promoters and primary transcripts, active TEs, imprinting control regions and allele-specific transcription.

16.5   MAPPING HISTONE MARKS

Histone characteristics can define the immediate start sites of genes which are often regulatory in nature. It can also define discrete but distant regions that influence gene expression as well as regions that may encompass an entire gene to prompt its active or repressed transcription. The aim in the genome tiling approaches is to catalogue across the entire genome the locations not only of key histone modifications but also of proteins that respond to and mediate them. Solexa IG sequencing allows millions of short DNA 'sequence tags' to be assigned to individual histone marks, thus mapping the marks to their precise location in the genome. Genomic regions-within genes or between genes have been found to be unexpectedly marked for expression activity. This provides the evidence that besides the classical mRNA there is also a huge number of regulatory RNAs (which modulate genome read out by producing multiple forms of the same protein or without producing proteins at all).

Fluoresecent protein–based chromatin tagging—The chromatin can be labeled through fluorescent protein tagging of the nucleosome components. Using H2B-YFP or H2B-GFP one can mark and track specific regions of the genome. Because of slow exchange ratio of H2B-YFP or H2B-GFP from the nucleosome core, photobleaching can be used with H2B-YFP or H2B. This strategy to track specific chromosomal regions could revolutionize capability to observe chromatin based process in real time. The steps involved in this technique are as follows.

1. Construction of a fusion protein between GFP and the DNA binding domain (DBD) of a known transcription factor.

2. The binding sites for this particular DBD is multimerized to about 250 copies and is then inserted into the genome of animal and yeast cells.

3. Expression of the GFP-DBD fusion results in fluorescent tagging of the concatamer *in situ*.

4. This provides a 'beacon' which permits one to track the position of this region in the genome.

The chromatin needs to be denatured and hybridized with labeled probes in order to visualize the position of specific sequences. This H2B fluorescent protein fusion approach provides insights into behaviour of chromosomes during cell division. It can also track the movement of specific chromosomal locations within living cells.

Application of the GFP as an *in vivo* tag of specific chromosomal regions in living cells—In this technique concatamers of the lac operator are inserted in different chromosomal loci (Lac operator sites). GFP-LacI fusion proteins are then expressed in the nucleus using an inducible promoter system. The fusion proteins recognize the operator sequence and the bound arrays are visualized as bright fluorescent spots in the nuclei with the fluorescent microscope (Robenett et al., 1996; Straight et al., 1996).

16.6 MAPPING

Mapping refers to plotting of gene positions or other defined sites along a strand of DNA. **Genetic mapping** refers to a technique to determine the linear order of molecular markers or genes (generally loci) along a stretch of DNA (e.g. a BAC clone, a chromosome). In other words, it refers to any method used in the measurement of positions and relative distances between genes of a linkage group or sites within a gene (fine scale mapping). It requires the identification of unique genome markers, for example, ESTs or STSs and their localization to specific chromosomal sites. STSs are unique addresses generated by PCR primers which amplify just a single chromosome site. Thus a locus which is polymorphic in a population can be screened by PCR to determine its inheritance patterns in families (Murray et al., 1994). Adjacent loci showing similar inheritance pattern permits proximity to be inferred. Genetic linkage can be reliable detected over distances of about 30Mb given the recombination rate of human chromosome. A **high density map** or high resolution map refers to any genetic or physical map of a genome which contains a large number of mapped molecular markers such that markers are spaced at recombination frequencies of 2-5 cMs or ~ 100-200 kb physical distances. Some laboratories consider a map as a high density map if marker density is 0.5 cM (or less) (genetic map) and 5-10 kb or less (physical map) whereas an **ultra-high density map** refers to a genetic map which consists of a very dense placement of molecular markers, for example, 5-10 AFLPs or STMS per cM. The three techniques that have been used for marker localization in human chromosomes are:

1. **Genetic mapping** (generally 1-to-10 Mb resolution)
2. **FISH** (~ 1-Mb resolution). FISH using chromosomes in interphase when they are less compact increases map resolution further to around 100 kb.
3. **Radiation hybrid mapping** (down to 50-kb resolution).

Markers have been placed on average 200 kb across human genome (Rowe et al., 1997).

Genome mapping—Genome map refers to a collective set of markers with known positions. Genome mapping involves generation of an ordered clone library that completely represents a genome (or a defined part of it) together with sufficient genetic marker positions (e.g. RFLPs) to permit an accurate alignment of both the physical and genetic map. In other words, genome mapping consists of ordering genomic DNA fragments on their chromosomes using several methods such as FISH, somatic cell hybrid analysis or random clone fingerprinting. Chromosome mapping project requires a large number of clones to be rapidly analyzed for construction and ordering of contigs across a particular region. YAC clones identified by the PCR via STSs or isolated by colony hybridization are then ordered by determining common homologous sequences in each YAC by fingerprinting technique and STS content mapping. Thus what is required is subchromosomal localization and order of different YACs and to find which clones are non-chimeric. Figure 16.1 shows principle underlying RFLP analysis.

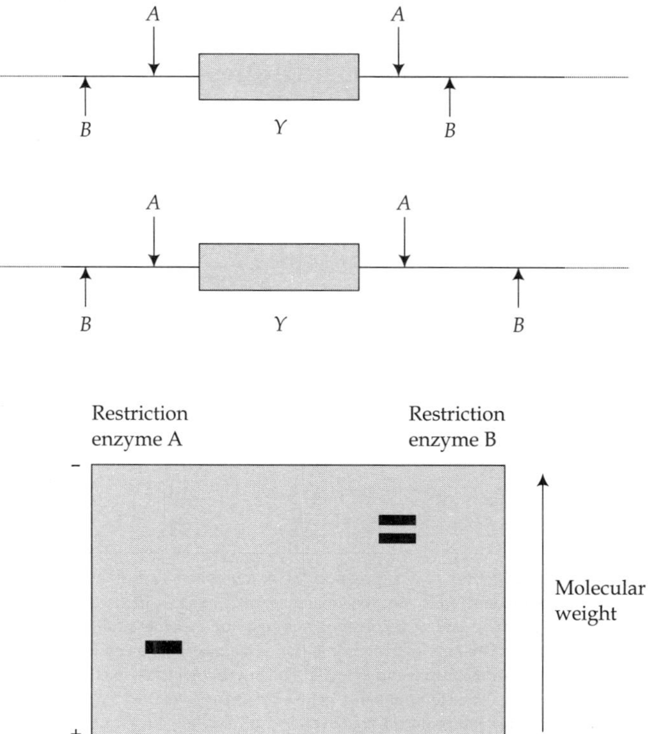

FIGURE 16.1   Shows principle underlying RFLP analysis.

Number of RFLPs to be required for a given degree of coverage — Assuming that every gene be within c Morgan of a marker and ignoring the effects of chromosome ends, the number of polymorphic markers required for a given degree of genome coverage is estimated in terms of the proportion, P, of a circular genome of total length k Morgans that will be covered to this degree by a set of n randomly occurring markers over the genome (Lange and Boehnke, 1982).

$$n = \log(1 - P)/\log(1 - 2c/k)$$

Thus the number of polymorphic markers required for a given proportion of the genome coverage depends on the total genome size (k) and maximum spacing (c) between marker and other loci. In case of genetic analysis, a QTL should not be more than 0.2 Morgan away from a marker locus while in case of introgression, a maker bracket of no more than 0.2 Morgan enclosing the QTL of interest would be desirable. Thus, for these purposes marker should be spaced every 0.2 to 0.4 Morgan, i.e., c = 0.1 to 0.2 Morgan. As a genome length of 10 Morgans would correspond to many cultivated plants and a genome length of 30 Morgans to many livestock or poultry species, the table 16.2 given (Beckmann and Soller, 1983) shows values of n for selected values of P (50%, 60%, 70%, 80%, 90% and 95%) considering spacing c = 0.1 and 0.2 and total genome length a genome length k = 10, 20 and 30 Morgan.

16.7 PHYSICAL MAPPING

The physical map refers to the linear arrangement of genes or other markers on a chromosome as determined by techniques (e.g. heteroduplex analysis, DNA sequencing) other than genetic recombination. Usually the map distances are expressed in numbers of nucleotide pairs between identifiable genomic sites (e.g. contigs, STSs, or restriction sites). Physical map results

Table 16.2 Showing the number of RFLP markers required considering different values of c and k.

Proportion of genome coverage	Total genome size and spacing (in Morgans)					
	k = 10		k = 20		k = 30	
	c = 0.1	c = 0.2	c = 0.1	c = 0.2	c = 0.1	c = 0.2
0.50	41	22	83	45	124	67
0.60	54	29	109	59	164	89
0.70	71	38	144	77	215	117
0.80	96	51	192	104	187	156
0.90	137	73	275	148	411	224
0.95	178	95	358	193	535	291

*Values given in the table are those obtained by the text expression for n, multiplied by 1.2 and 1.3 for c=0.1 and c= 0.2, respectively, taking into account the effects of chromosome ends. As the influence of a marker locus does not extend past the ends of the chromosome on which it falls, markers near the end of a chromosome will not provide a full 2c of the genome coverage and so the values of n given in Table are underestimates. The results from computer simulation suggested to add 20-30% top the required number of markers for the range of c values as chromosome end effects. The human genome has a length of approximately 33 Morgan units(Renwick, 1971). If it is assumed that the distance between a gene and a marker is 0.1 Morgan unit in order to guarantee that they segregate together, then the number of markers required to cover the entire human genome would be 0.1/33 = 165. This value represents the lower limit. For practical purposes, one could possibly use 1000 to 2000 markers as the above estimate is based on the assumption that markers are uniformly distributed along the genome which is not.

in genetic map of a genome. A map is a representation of the relationship among landmarks organized according to defined co-ordinate systems and mapping refers to description of order and spatial relationship among different genetic landmarks. The genetic landmarks include landmarks such as polymorphisms, genes and DNA sequences. It requires data relating to genome organization including comparisons between different organisms, functional groups of genes or dispersed gene families and chromosome regions associated with pathogens. Genetic linkage maps are based on co-inheritance of allele combinations across multiple polymorphic loci. A map unit of 1 cM corresponds to an observation of recombination in 1% of the gametes.

16.7.1 Physical Maps

Physical maps can be cytogenetically or molecularly based. Cytogenetically based physical maps order loci with respect to visible banding pattern or relative position along chromosomes, primarily by means of data from somatic cell hybrids and *in situ* hybridization. Thus there are many types of physical map. Quantitative DNA fiber mapping is another technique used for the construction of high resolution map at a resolution of few kilobases and relies on hybridization of specific fluorescently labeled probes to target DNA immobilized on a specially prepared glass surface and stretched by hydrodynamic forces to produce linear templates(fibers) of about 2-2.5kb per μm. The hybridization events and the distribution of different probes along stretched fibers can be detected by fluorescene and be imagined for further analysis.

Molecularly based physical maps directly characterize large tracts of DNA by establishing molecular landmarks such as restriction endonuclease sites and STSs. These maps are constructed from data generated by PFGE or by related techniques that size and order larger restriction fragments of genomic DNA. **Long-range map** refers to the linear array of rare cutter recognition sites on large DNA fragments isolated by PFGE. Another molecular strategy is to characterize cloned DNA (in the form of YACs), cosmids, or shorter phage vectors, sufficiently to establish overlapping assemblages of clones, known as contigs. Accuracy and efficiency of physical mapping increase progressively with the size of the clone fragments in these libraries. Physical map can be developed by hybridizing probes to YAC libraries using colony hybridization experiments. The probes can consist of markers genetically mapped to specific chromosome, previously unmapped genes, random genomic fragments, sequences flanking transposable elements and repetitive elements. Southern blot analysis of YAC clones can confirm the colony hybridization result and thus can reveal common restriction fragments in different YAC clones hybridizing to a given marker thereby suggesting overlaps between the inserts of the YAC clones. Generation of YAC end fragments can be done using either IPCR or plasmid rescue from YAC clones lying near the ends of each of the contigs. Conversion and comparison between physical maps with other different types of scales is a high priority but a common reference point will have to be mapped and conversion factors determined empirically. STSs have been proposed as the common reference points which could be used to co-ordinate information from different mapping strategies. Cytogenetic maps have a scale and co-ordinate system corresponding to the chromosome banding pattern. At best cytogenetic map could be used to locate a DNA fragment to a region of ~10Mb- the size of a typical chromosome band. It is thus the lowest resolution physical map. The large scale restriction maps have a scale on the order of kilobases of DNA and have not been related to specific chromosomal bands.

There exists no simple means to convert physical distances into recombination frequencies. Recombination frequencies per megabase of DNA vary considerably by sex and by chromosomal region in human. Patterns show that telomeric regions have proportionately more recombination in male and that centromeric regions have higher frequencies of recombination in female meiosis and overall, recombination frequency is higher in female than in male. The highest level of resolution for a molecularly based physical map is the DNA sequence which gives linear order of nucleotides for each of the 24 distinct human chromosomes.

Physical maps based on 'STS' landmarks are used to develop so-called 'sequence ready' clones consisting of overlapping cosmids or BACs. The purpose of creating high resolution physical maps is to create a scaffold for organizing large scale sequencing. In mammals STSs together with radiation hybrid panels and YAC libraries are used to construct dense landmark maps. Together the genetic and physical maps provide thousands of anchor points that can be used to tie clones or DNA sequence to specific locations in the genome.

16.7.2 Clone-Based Physical Mapping

The different steps involved in clone-based approach are as follows (Olson, 2001).

1. Partial digestion of genome with site-specific restriction endonuclease which results in generation of segments of about 150,000bps.
2. Large DNA segments are introduced into BACs and inserted into bacteria where they are copied exactly each time the bacteria divide. This process produces 'clones' of individual DNA molecules which can be purified for further analyses.
3. Each clone is completely digested with a restriction endonuclease chosen to produce a characteristic pattern of small segments or a 'fingerprint' for each clone.
4. Comparison of patterns reveals overlap between the clones permitting them to be lined up in order while the sites in the genome at which the restriction endonuclease cleaves are charted.

This results in a physical map. Thus what we see is that the clones are produced by first partially digesting many copies of the genome with different restriction endonucleases. The resulting large segments are then inserted into bacteria and replicated (cloned). Each clone is digested with a restriction endonuclease and the resulting fragments are separated, by size, on an electrophoretic gel (gel analysis of inserts). This process yields a distinctive patter (fingerprint) for each clone. The map assembly problem requires working backwards from the fingerprints to a clone-overlap map and restriction site map of the chromosome segment.

Limitation with clone-based mapping — Maps often have poor continuity. For example, there is not always a BAC clone to cover every part of genome and overlaps between clones can be obscured by data errors or the presence of large scale repeats in the genome.

16.8 LOW RESOLUTION MAPS

Two low resolution maps are the genetic and cytogenetic maps. The genetic map is based on the probability of the occurrence of recombination, the swapping of corresponding, nearly identical segments of DNA between maternally and paternally derived chromosomes as the genome is passed from one generation to the next. Cytogenetic map is based on subtle variations in the staining properties of different regions of the genome as viewed by light microscopy. One can construct a so-called cytological map for the same genes for which percentages of crossing-over have yielded a genetic map. Cytological map is constructed by physical or chemical mutagenic treatment which results in generation of a large number of deficiencies in a particular chromosome. The genes which on the basis of pseudodominance are known to be missing in the deficiency are located in the part of the chromosome that forms the buckle in the deficiency heterozygote.

Dynamic molecular combing—It is a technique for physical mapping of large fragments such as BAC , cosmid and YAC clones on a mechanically stretched chromosome fibers. The silanized coverslips are dipped into a buffered solution containing the genomic DNA and after a short incubation they are pulled out of the solution. During the immersion the DNA binds to the surface of the coverslip. The localized hydromechanical forces at the fixed horizontal meniscus that are directed downwards during removal, progressively unwind the DNA and align the unwound fiber parallel to each other. The silanized surface dries instantaneously as it emerges from the solution and thereby fixing the DNA fibers irreversibly. Most of the stretched fibers are longer than several hundred kb, the stretching factor being 2 kb/μm. Now different chromosomal probes such as BAC clones labeled with different fluorochromes are bound to the combed target DNA and visualized by fluorescence microscopy. DMC thus allows to map the positions of the clones along the fibers.

16.9 MAPPING STRATEGIES

There are different mapping strategies for physical mapping. Physical mapping techniques provide a complement to genetic linkage. Physical map can be constructed using restriction endonucleases (called restriction map) or overlapping clones (contig map).

Macro-restriction map—In this technique large DNA fragments are generated by rare cutting restriction endonucleases of the genomic DNA. Genome restriction maps show the order and distances between cleavage sites of site–specific restriction endonucleases. This type of mapping has been extended to much larger genomes such as that of *E. coli* by exploiting the ability to separate very large restriction fragments using PFGE. Standard agarose gel electrophoresis gives resolution in the range of 1-to-10-kb and pulsed field techniques extend this range to the megabase level. The physical map generated spans long DNA stretches and can be used to localize genes (sequences) by hybridization of fluorescent or radioactive labeled probes to the different cloned restriction fragments. Thus we can construct genome restriction map or restriction maps of individual chromosomes. Individual chromosome can be separated by FACS discussed in chapter 31 and can be fragmented using restriction endonucleases (see chapter 9). The restriction fragments generated of different lengths are sized through electrophoresis and then are cloned into vectors (see chapters 8 and 10 for information on different cloning vectors and cloning) for further studies. Pure restriction maps are difficult to construct primarily because the sites for the most suitable restriction endonucleases are distributed non-randomly and are sometimes blocked by the action of methylation system that covalently modify DNA *in vivo*. Furthermore, restriction maps fail to meet the need of most users for ready access to the cloned DNA.

Top-down mapping—It is a technique to construct long range genetic map. In this technique chromosomes are first separated with PFGE, transferred onto hybridization membranes, immobilized and hybridized to radioactively or non-radioactively labeled DNA probe. After autoradiography the probe can be assigned to specific chromosome. This mapping procedure can also be used for subchromosomal fragments.

Botton-up mapping—It is a technique to construct large range restriction map. A population of subchromosomal fragments is separated by PFGE, blotted onto hybridization membranes and finally hybridized as in case of top-down mapping. Thus fragments which are contiguous **in vivo** can be identified by their hybridization to their specific probe.

Microdissection—It is a technique to fragment a chromosome by physical microsurgery (e.g. by laser beam). Subchromosomal fragments thus generated can be used to establish subgenomic libraries.

Microcloning—Microcloning refers to the cloning of specific subchromosomal regions produced by microdissection. Microdissection and microcloning procedures are used to generate markers for specific chromosome regions that can serve as starting points to clone more extended region of the chromosome. In case of Drosophila the polytene chromosomes of the salivary galnds can be dissected physically with ultra-fine glass needles and a micromanipulator and sections containing just one or a few bands are transferred to small containers for extraction of DNA which can be used for constructing a library in a bacteriophage λ vector. This technique can be applied to human chromosome but here the degree of resolution is much less and the dissected piece contains a thousand genes rather than ten or fewer. The small amounts of DNA thus obtained can be selectively amplified by the polymerase chain reaction (PCR) technique.

Shot gun cloning—It is a technique to establish a cloned gene library containing the entire genomic DNA of an organism in the form of randomly generated fragments(restriction fragments) produced through restriction endonucleases. It is a technique for analyzing whole genomes or any DNA molecule too large to be analyzed directly.

Subcloning—It is a technique to subdivide a large cloned DNA fragment(e.g. 9 -25kb in genomic libraries) into smaller sized fragments which are easier to analyze. The original cloned fragment is first removed from the vector by restriction endonuclease which was used for insertion and then cleave with other restriction endonuclease. The resulting fragments of size up to 5kb are reinserted either in the same or another vector. Subcloning permits easy and fast characterization (sequencing, *in vitro* mutagenesis, *in vitro* transcription) of subclones. **Subclone** can thus be defined as a smaller DNA fragment derived from a larger cloned fragment by restriction endonuclease digest which is separately cloned into a vector.

16.10 CONTIG MAPPING

It is a technique for physical mapping of genome. In this technique the genomic DNA is divided into overlapping fragments which are then cloned and sequenced. The resulting sequences are then aligned using computer software package which searches for overlaps. **Contigs** refer to the genomic clones which contain mutually overlapping sequences and **contig map** refers to a library of overlapping clones (contigs) representing a complete stretch of DNA (e.g. a BAC clone, a chromosome). Contig maps represent the structure of contiguous regions of the genome by specifying the overlap relationship among a set of clones. Contig maps are dependent on the continuing existence of a particular clone collection. Their generation and uses of these maps depend on detailed analysis of individual clones. Pure contig maps are also difficult to construct because these maps loose continuity at any point where clones are not available or overlap relationships are nor clear. A map of contigs covering the chromosomes will speed up the identification of the causes of diseases. It provides an immediate access to the genomic segment including any locus as soon as it has been localized by genetic linkage or cytogenetic analysis.

Process of developing contigs map

I. Identification and localization of landmarks in the cloned genomic fragments
II. Where there are enough landmarks for the size of the cloned fragments, contigs are formed and landmarks are simultaneously ordered.

YACs provide the means to isolate large DNA fragments (~100 to 2, 000 kb). This fills the gap between genetically and physically measurable distances. Moderate size of YAC library makes them well adapted to screening by PCR. YAC allows to use STSs which are short stretches of DNA sequence that can be specifically detected by PCR. STS marker is operationally unique with the genome. The advantage with STSs is that it provides an ordered set of markers which can be stored as sequence information in a database. Other potentially informative physical landmarks can be readily converted into STS by DNA sequencing. Thus establish the common language for comparison with other physical mapping data. The ordering of markers on individual chromosomes can be derived by genetic linkage analysis, PFGE analysis of fragments generated by rare cutter enzyme, analysis of panels of somatic cell hybrids containing naturally occurring deletions or radiation induced deletions. There are two main sources of error in contig assembly. The trivial false positives can be excluded by checking candidates individually. False merging could be due to chimeric clones that contain more than one genomic region. In other words, redundancy of library is a source of false overlap. Another source of false overlap can be inclusion of data from PCR landmarks which are not STS in strict sense (not unique). This problem can be solved partially by assignment of STS to human chromosomes using a somatic cell hybrid panel.

Hybrid map — In a hybrid map, restriction map based on the direct analysis of uncloned DNA as well as data from other low resolution mapping sources such as linkage mapping, cytogenetics and somatic cell genetics are used to orient and align a series of contigs.

16.11 USE OF 'SEQUENCE-TAGGED SITES' FOR MAPPING

In the STS –content mapping YACs are screened by PCR to identify all clones containing a given locus. Nearby loci tend to be present in many of the same clones allows proximity to be inferred. STS-content linkage can be detected over distances of about 1Mb given the average insert size of the YAC library used there (Green and Olson, 1990). YACs are the best clones for covering large distances but it suffers from high rates of chimerism and rearrangement and thus are unstable for genomic sequencing. STS-based maps overcome this problem by having a sufficiently high density of landmarks that one can rapidly generate physical coverage of any region by PCR-based screening of clones appropriate for sequencing such as cosmids, BAC and P1-AC. The STSs fall into one of the following four categories depending on the sources of generation (Hudson et al., 1995).

1. Random loci- generated by random sequencing human genomic clones and discarding those appeared to contain repetitive sequences.
2. Expressed sequences- STSs generated from cDNA sequences in GenBank (EST database).
3. Genetic markers-polymorphic loci, primarily dinucleotide CA repeats or primarily tri and tetranucleotide repeats.
4. Other loci- includes STSs developed from other sources such as X-chromosome specific and Y-chromosome specific libraries.

STSs will enhance the hybrid mapping strategy. As virtually any mapping strategy uses cloned DNA fragments as landmarks, regardless of whether they are members of contigs, segments that contain an unusual restriction site, probes that detect genetically mapped DNA polymorphisms, or sequences that hybridize *in situ* to particular cytogenetic bands. Translation of any of these examples to produce an STS would require sequencing a short tract of DNA from the clones that defines the landmark. In most cases an STS varies from 200 to 500bp in length and it can be detected through PCR in the presence of all other genomic sequences. A PCR assay for an STS can be implemented simply by synthesizing two short (~20bp)oligonucleotides, chosen to be complementary to opposite strands and opposite ends of the sequence tract. A DNA sample is tested positive for the presence of an STS if it serves as template for the *in vitro* synthesis of the tract in the presence of these two oligonucleotide 'primers' and in case of positive such large amounts of product are made that it can be detected without radioactive labeling(Olson et al., 1989). If STSs are known for all genomic DNA fragments then it eliminates the need to store and exchange clones. If a clone from a specific part of the genome is required then a database search for an STS mapped to the region of interest will allow to design primers for the PCR amplification of the STS. The amplification product can then be labeled and used as probe for fish the corresponding DNA from a gene library.

Ordering of the clones in a DNA library—Ordering of the clones on the genetic map is a multistage process. First, the presence or absence of an STS on an individual clone is determined by hybridization, for example, by probing each clone with PCR amplified DNA from the STS. Once the STSs on each clone(e.g. a BAC clone) are identified, the clones can then be ordered on the map. In one hypothetical example, now suppose a segment of a chromosome from some organism has the STSs markers on it in the order A to Q. Again suppose after digestion with a restriction endonuclease nine BAC clones were constructed and STSs on each clone were identified. After that comparisons of clones are made for the position of STSs. In this example three clones, namely, 3,4 and 5 contain STSE ; STSF is found on clones 4 and 5 but not on 3 and STSG is positioned on clone 5. This shows that the order of STSs is E, F, G. The clones, 3, 4, 5 thus partially overlap and thus their order must be 3, 4, 5. So the order of the clones is 1, 2, 3, 4, 5, 6, 7, 8 and 9 and this resulting ordered series of clones is called a contig.

Ordering the STS landmarks—Construction of the YACs by STS content analysis simultaneously produces the relative order of these landmarks across the chromosome. The resolution of map depends not only on the number of STS tested but also on the number of YACs containing informative combinations of STS. The most useful ones are those which contain only pairs of adjacent landmarks. Contig assembly is more efficient with larger YACs and STS maps are more accurate with smaller clones. The landmark order is refined when the clone coverage increases. In certain regions there is high density of STS and thus no local order can be unequivocally deduced given the lack of informative clones. Another factor influencing the order of STS is the presence of clones not detected with a given STS but detected with its neighbor. Such false negative can create local errors when deducing order but this can be eliminated by checking clones for the presence of neighboring STS. The presence of rearranged, unstable clones creates errors which are difficult to eliminate when they occur at the same site. A few markers can be mapped at several places on contig which reflects the existence of homologous regions in this part of the genome.

Two STSs are said to be singly linked if they share at least one YAC in common and doubly linked if they share at least two YACs. Single linkage is an inadequate criterion for declaring adjacency of STSs because of high rate of YAC chimerism (~50%) and the possibility of laboratory error. Double linkage is a reliable indicator as two genomic regions are unlikely to be juxtaposed in multiple independent YACs and so a three step procedure is used in human genome mapping.

1. STSs are assembled into doubly linked contigs (groups of STSs connected by double linkage).
2. The doubly linked contigs are localized within the genome on the basis of RH and genetic map information about loci in the contig.
3. Single linkage is then used to join contigs localized to the same small genomic region.

Gap closure—Gap closure is done by suing non-STS based information. YAC overlaps can be inferred on the basis of fingerprint analysis and Alu-PCR hybridization. As Alu-PCR hybridization data have high false positive rate, gaps were closed only when there were at least seven hybridization links between adjacent contigs.

Database preparation—Spotting the PCR reactions on membranes, hybridizing them to a chemiluminescent probe specific for each STS, capturing the resulting signal directly by a Charge-coupled device (CCD) camera and uploading the result into database.

Cosmid insert restriction mapping—It is a technique for the construction of a restriction map of insert cloned into a cosmid. Cosmid vectors are used to clone genomic sequences up to 45kb in length and thus allow the isolation of overlapping clones which span large chromosomal intervals. This **cosmid walking** technique reduces the number of cloned necessary to characterize extremely long DNA fragments. In this technique the cloned DNA is first cleaved with lambda terminase and then partially digested with a restriction endonuclease. After that a labeled oligonucleotide, complementary to either *cos* site overhang, is hybridized to the digestion products. The hybrid formed are separated by agarose gel electrophoresis and visualized by autoradiography. Finally, a ladder of fragments allows to calculate the positions of restriction sites relative to the labeled *cos* site.

16.12   FINER DETAIL MAPPING

In search for a particular gene it may be essential to make a very fine map of a small region of a chromosome, for example, 1 or 2Mb. This requires a range of different techniques. These techniques include PFGE by which large fragments of DNA (up to several megabases in size) are separated after the genomic DNA has been treated with 'rare cutting' restriction enzymes. It is then possible to test for which genes are present on each fragment. Another approach is to isolate all the DNA from a particular region either in cosmids (fragment length about 40kb) or YAC which may include fragment length greater than a megabase in length. The idea is then order the overlapping clones into a physical map or contig so that all the genes in that region are included. A complete map of the region is assembled (see chapter 13) using sequence tagged sites (STSs), the short sequences of DNA which can be amplified by PCR. A final map may contain, for example, 810 YACs and spans region of about 50Mb. A YAC contig thus provides a valuable resource for identifying genes known to lie in a particular chromosomal region. It could also be utilized as a starting point for sequencing the whole genome (see chapter 13).

16.13   COMPARISON BETWEEN VECTORS FOR SUITABILITY FOR MAPPING AND FUNCTIONAL ANALYSIS

The ability to clone large (>100 kb) DNA fragments is crucial to physical mapping, positional cloning and molecular analysis of complex eukaryotic genomes. YACs developed were quickly adopted to research on genomes of human and other species and allowed cloning and maintenance of DNA fragments up to 1000 kb. However, the limitations with the YACs are as follows (Tao and Zhang, 1998).

1. High level of chimerism

2. Occasional insert instability

3. Difficulty in purifying insert DNA which tends to be contaminated with yeast host chromosomal DNA.

Both BAC and PAC are capable of cloning and stably maintaining DNA fragments >300 kb in *E. coli*. Although insert sizes of BACs and PCAs are somewhat smaller than YACs, they have low levels of chimerism, facilitate and speed the insert purification and have high stability in host cells. Because of these qualities they have assumed a central position in genome research. But the limitations with BACs and PACs are that there is limited variety and diversity. In case of BAC only pBe10BAC11 and its derivative pE CBAC1 have been developed so far. Only one binary vector BIBAC2 has been developed for large-insert DNA library development and plant transformation. Further, as just one or two copies of BAC are present per cell so it will not be suited for robot-assisted assembly of ordered library. The one-to-two copies of BACs and PCAs per cell also limit the yield of insert DNA and a large number and a large scale reproduction of cloned DNA fragments. Further, first p BAC108 L from which other BAC vectors were developed lacks selectable marker for recombinants and thus recombinant clones in p BAC108L must be identified by colony hybridization which is not well-suited for library development.

Many traits are controlled by clusters of genes (disease resistance traits). Transformation of such genes clusters will be most efficient using a binary vector which is capable of cloning and transferring very large DNA fragments. Such a large DNA fragment cloning and plant **binary transformation system** will also facilitate positional cloning of genes in plants if the major purpose of research is to isolate DNA fragments carrying genes of interest. Binary BAC vector BIBAC2 was developed based on p BAC108L for Agrobacterium mediated plant transformation. However, it has a single cloning site Bam HI and the SacB gene for recombinant selection. **Non-Artificial Chromosome Systems** such as conventional plasmids and conventional cosmids which have already been employed in molecular genetics and molecular biology, can be used for large DNA fragment cloning as the **Artificial Chromosome Systems**, i.e. BACs and PACs.

Clones in PACs and BACs — Many of these clones contain genes of biological interest. However, the large size of the insert makes further restriction endonuclease manipulation of these clones for functional analysis extremely difficult. In case of PAC, once the P1 insert is transferred into yeast- bacteria shuttle vector, p Clas per, researcher can use homologous recombination in yeast for reporter gene insertion or site directed mutagenesis (Bradshaw et al., 1996). Use of co-transformation (which refers to simultaneous transfer of two or more physically unlinked genes (or generally DNA fragments) into animal target cells) mediated recombination in yeast can be made to transfer the inserts from P1 clones into yeast-bacteria shuttle vector. Co-transformation is useful for the transfer of a non-selectable gene together with a selectable marker gene. If high DNA concentrations are transfected the procedure will lead to the transformation of the target cell with both genes so that transformants carrying non-selectable DNA fragment can be selected with co-transfer marker.

16.14 VISUAL MAPPING (OPTICAL MAPPING)

It is used to infer the genome map of the location of short sequence patterns called restriction sites. This technology developed by David Schwartz allows visualization of map of randomly located single molecules around a million bps in length. The genome map is constructed from overlapping these shorter maps. The visualization of genes generally DNA sequences along a chromosome or chromosome fiber or along a BAC or YAC clone that are extended by DNA combing is made by *in situ* hybridization of fluorochrome labeled probes (representing genes) and detection of fluorescence emission. The threshold of direct visual mapping is about 3kb so that single genes can be detected. Optical mapping can also be used for the construction of restriction maps from a series of single DNA clones. In this procedure large DNA molecules are first dropped into specially designed glass surfaces, linearized in parallel through a fluid flow across the surface and then fixed onto the glass. Subsequently restriction endonucleases are added to generate ordered patterns of restriction fragments which are then stained with fluorochrome and visualized with fluorescence microscope. The various microscopic images are captured one at a time, processed and images of the various restriction fragments are aligned to match the restriction sites. Finally, multiple maps are merged into large contigs using map assembly programmes.

16.15 HAPPY MAPPING

Various strategies have been applied to map complex genomes including those involving the use of radiation to break genomes but all suffer from a number of drawbacks. Random cloning and chromosome 'walking' are impeded by repeated sequences (Wyman et al., 1985) and inability to clone some loci or confused by rearranged or co-ligated inserts (Little, 1992). Difficulties in cutting DNA into large fragments (Drmanac et al., 1986) and resolving them prevent efficient restriction mapping. The infrequency of meiotic recombination also limits resolution of classical linkage mapping, at best, one marker per 10^6 bp. The resolution offered by *in situ* hybridization using metaphase is little better (Lawrence, 1990) though this can be improved when applied to interphase nuclei (Engh et al., 1992). Dear and Cook (1993) proposed a simple method for ordering markers on a chromosome and determining the distances between them. It uses haploid equivalents of DNA and the PCR and hence happy mapping. It should prove helpful in closing final gaps in maps, for mapping over distances that are inaccessible using other approaches and for generation of maps of diverse groups of organisms and individuals. This approach is analogous to classical linkage mapping where meiosis both breaks DNA by crossingover and segregates into aliquots containing haploid amounts of DNA (i.e. sperm or eggs). These processes of chromosome breakage and segregation are replaced by *in vitro* analogues. DNA from any source is broken by γ-irradiation or shearing (physically) and then diluted and divided it into aliquots which contain about 1 haploid genome equivalent. In other words, markers are then segregated by diluting the resulting fragments to give aliquots containing ~ 1 haploid genome equivalent. Linked markers tend to be found in an aliquot. After detecting markers using PCR, map order and distance can be deduced from the frequency with which markers 'co-segregate'. As the frequency of breakage can be controlled so the scale over which one maps can be controlled, from a few kilobases to a few megabases. Generally one introduces many more breaks into the chromosome then meiosis and so attain a higher resolution.

Happy mapping (Haploid equivalent of DNA and PCR) — It is an *in vitro* technique used for the determination of the order and spacing of DNA markers directly on genomic DNA and the construction of physical map of any density. This technique does not require the parental lines, polymorphic markers and segregating populations as *in vivo* process of crossingover and segregation is replaced by *in vitro* breakage of DNA and dilution into aliquots containing less than a haploid genome equivalent. In this technique genomic DNA (>2Mb) of a target organism is first isolated and then randomly sheared (mechanical shearing) or γ-irradiated. The resulting fragments are then size fractionated (450-500kb) using PAGE. The sized fractions are diluted and aliquoted into 96-well microtiter plates in such a way that each aliquot contains less than one haploid genome equivalent (mapping panel). The different panels may differ in the size of the genomic fragments they contain which in turn determine the resolution of the resulting map and the maximum distance over which linkage between consecutive markers can be detected. For example, a so called short panel with fragments less than ~100kb will detect linkage between markers up to 50kb apart (high resolution) whereas a panel with fragments of ~ 150kb and more permits to monitor linkage over distances up to ~ 100kb (low resolution). The mapping panels whatever its type is then amplified by PCR using 15-mer reverse primer. The resulting amplicons are then again amplified by PCR but this time with a forward internal primer and a corresponding reverse primer for each selected DNA marker. Since many markers are to be localized this second amplification works with a multitude of primer combinations. If two or more markers are physically linked in the genome they will be localized on a single fragment, i.e. they will remain linked after random breakage in one aliquot. The products are then analyzed by agarose gel electrophoresis. From the resulting data linkage between markers can be determined from the calculation of LOD scores and thus a map of the target genome can be constructed using Dg map or similar software.

16.16   GENE DENSITY

Gene density is defined as the number of genes per unit length of DNA. It varies from organism to organism and within genome it varies region to region. For example, in the gene space(part of genome where the gene density is substantially higher than in the rest of the genome usually representing gene clusters) the gene density is much higher than extrapolated from the a uniform distribution in the genome whereas in the intergenic space(gene space interrupted by gene empty, mostly repetitive sequences which constitute about 95% of the genome in human) the gene density is equal to or lower than predicted from randomness. In Arabidopsis thaliana chromosome 1 it is one gene / 4-5kb and varies from chromosome to chromosome in one organism(human chromosomes 4, 5,8,13,18 and X have considerably lower density than chromosomes 1, 11,17, 19 and 22). Gene densities in some of the organisms are given in Table 16.3.

Maize may have H"5-10 times lower gene density than in Arabidopsis. As the duplicated maize genome is 20 times the size of the Arabidopsis genome, it has been predicted that most genes in maize will be represented by an ortholog in Arabidopsis. Thus it is generally accepted that any gene in maize can be used to identify at least one homolog in Arabidopsis.

16.17   GENE NUMBER

It refers to the total number of genes in the genome of an organism. Gene numbers in some of the organisms are given below. The table 16.4 shows that the humans have only seven times as many genes as yeast, ~ 2.5 times as many as the *Drosophila melanogaster* and less than twice as many as the nematode worm *Caenorhabditid elegans*. This figure may increase as our understanding of the genome and gene prediction increases. Fewer genes in an organism does not necessarily equate to reduced complexity. Complexity can manifest at many levels including splicing, gene regulation, post-transcriptional editing and post-translation modifications. For example, the Drosophila DSCAM gene which has 115 exons but which are alternatively spliced to code for 38,016 related but distinct protein isoforms (Schmucker et al., 2000).

Traditionally a gene is defined as a protein-coding unit and a transcript which does not code for a protein is categorized as 'regulatory RNA'. Further, there is a large number of ESTs and cDNA showing no *in silico* evidence of splicing (i.e. by each end aligning either side of an intron in a genomic sequence). These could be derived from a real gene but simply do not span an intron and therefore show no evidence of splicing. Alternatively they could be *in vitro*

Table 16.3   Showing organisms and their gene density.

Organism	Gene density (number of genes/100 kb)
E.coli	87
S.cerevisiae	52
C.elegans	22
Homo sapiens	5
Arabidopsis thaliana	4-516.16 Gene density

Table 16.4 Showing gene number in different organisms.

Organism	Gene number
S. cerevisiae	~5000 (6278)
Drosophila	17,000 (13601)
Arabidopsis	25,000-28,000 (25498), 30 700 (Li et al., 2006)
Homo sapiens	30,000-40,000 (34, 000)
Mycoplasma genitalium	~ 500
Aquifex aeolicus	1500
Haemophilus influenzae	Over 1749
Rice	33,000-50,000 (37,544)
Wheat	30, 000
E.coli	4406
C. elegans	19099
Maize	50, 000
Poplar (*Populus trichocarpa*)	45, 555
Grape vine (*Vitis vinifera*)	30,434

artifacts generated during the construction of cDNA libraries or *in vivo* artifacts generated from cryptic promoters or pseudogenes. Finally, gene prediction and annotation tools generally disregard unspliced ESTs as supporting evidence for the existence of a gene. The use of unspliced ESTs as evidence for the transcribed gene is unreliable as they can arise from genomic contaimination.

16.18   GENE CONTENT

Gene content refers to the absolute number of genes per genome or chromosome. The content varies between genomes of different yet related organism and between different yet equally sized chromosomes within the same genome. Gene content of the various grasses (see table 6.14) do not differ but the vast difference in genome size is most probably due to an expansion of non-genic sequences (retroelements).

The physical arrangement and copy number of genes in DNA sequences along a chromosome can be measured by ISH of fluorochrome labeled probes (representing genes) to chromosome spreads and quantification of the emitted fluorescence. The higher the number of target sequences, the higher the number of bound probe molecules and higher is the fluorescence light emission which allows semi-quantitative analysis of gene content at the detected locus. Thus a quantitative chromosome map(idiogram) can be constructed.

16.19   CODING DENSITY

It refers to the number of coding sequences (genes) per length of a genome. Coding density in prokaryotes is relatively high as its genome does not contain any or only a very low number of non-coding sequences (i.e. DNA sequences that does not code an RNA or protein as opposed to coding sequence (a gene)). Major noncoding sequences in eukaryotic genomes are minisatellites, microsatellites, repetitive DNA, retrotransposons and satellites. Eukaryotic genomes harbor vast amounts of non-coding sequences (i.e. genes are like islands in an ocean of junk DNA) and possess low coding density. Coding sequences refers to that part of a gene which codes for the amino acid of a protein or for a functional RNA (tRNA, rRNA), CDs are determined by the start codon at the 5′ terminus and stop codon at the 3′ terminus.

Gene size: It is the length of a gene, measured in terms of number of base pairs starting from cap site to the poly (A) additional signal (eukaryotes) and varies from 21bp in case of enterobacteria gene, mccA encoding the antiobiotic heptpeptide microcin C7 to 2.34×10^6bp in length in case of dystrophin gene of man. An average gene is around 8.8kb in length in Drosophila.

16.20   ARCHITECTURE OF GENOMES OF SOME MODEL SPECIES

Arabidopsis thaliana

58% of the genome contains 24 duplicated segments of size = 100kb. The genes in the duplicated segments are usually not identifiable homologues. Further, 17% of all genes appear in 1528 tandem array of gene families containing up to 23 members. On average there is one gene per 4.6kb which is intermediate between prokaryotes and Drosophila and roughly similar to *C. elegans*. Exons are typically 250bp long and introns relatively small with mean length 170bp. The coding regions are rich in GC content which is typical of plant genes. 25% of the nuclear genes in plants have signal sequences determining their transport to organelles-mitochondria and chloroplasts in comparison to 5% of mitochondrial targeted nuclear genes in animals. The various features of the Arabidopsis genome are given in the tables (16.5 and 16.6) given (Lesk, 2002).

Grape vine (*Vitis vinifera*) is the fourth flowering plant, the genome of which have been sequenced (Consortium of grape vine characterization, 2007) after poplar (Tuskan et al., 2006), rice (International rice genome sequencing project, 2005) and Arabidopsis (Arabidopsis

Table 16.5   Showing characterishes of different chromosmes of Arabidopsis genome.

	Chromosome number (Arabidopsis thaliana genome)					
	1	*2*	*3*	*4*	*5*	*Total*
Length of chromosome	29105111	19649945	23172617	17549867	25353409	1154099409
No. of genes	6543	4036	5220	3825	5874	25498
Density (kb/gene)	4.0	4.9	4.5	4.6	4.4	
Mean gene length	2078	1949	1925	2138	1994	

Table 16.6   Gene distribution in Arabidopsis between nucleus and organelles.

	Nuclear	*chloroplast*	*Mitochondria*
Size	125100	154	367
No of protein genes	25498	79	58
Density (kb/protein gene)	4.5	1.2	6.25

genome initiative, 2000). *Vitis vinifera* genome (obtained from line derived by selfing of the heterozygous grape variety) contains 30,434 protein-coding genes with an average of 372 codons and five exons per gene. This value is considerably lower than 45, 555 protein coding genes reported for poplar genome which as a similar size at 485Mb and even lower than 37,544 protein coding genes identified in the 389Mb of the rice genome. In grape vine 41% of the genome is composed of repetitive/transposable elements, a slightly higher proportion than that found in rice genome which has a lower genome size. In eukaryotes with large genomes the coding and repeated elements are distributed over the chromosomes and may be more or less interlaced thereby forming gene-rich and gene-poor regions. However, the distribution of genes along the chromosomes of rice and Arabidopsis is fairly homogeneous. In contrast, in grape vine there are large regions that alternate between high and low gene density and density of TEs reflects a pattern substantially complementary to gene density. Similar characteristic is seen in the genome of poplar.

Rice (*Oryza sativa*) — The rice species *Oryza sativa* is also considered a model plant because of its small genome size, extensive genetic map, relative ease of transformation and synteny with other cereals. The analysis of 43.3 megabases of nonoverlapping sequence reveals 6,756 protein coding genes of which 3,161 show homology to proteins of another model flowering plant, *Arabidopsis thaliana*. Rice chromosome 1 is (G +C) –rich, especially in coding regions. Comparative genomic analysis shows little conservation in gene order between rice and Arabidopsis.

E.coli — About 89% of the genome codes for protein or structural RNAs. The average size of ORF is 317 amino acid (951bpb). The average distance between genes is 118bp but the longest intergenic region is 1730bp. The large intergenic regions contain regulatory and repeated sequences. Arabidopsis genome contains 630-700 operons. Operons vary in size and a few contain more than 5 genes. 3/4th of the transcribed units contain only one gene and the rest contain several consecutive genes or operons.

Drosophila melanogaster — One- third of genome is contained in heterochromatin and two-thirds is euchromatin. The heterochromatin is concentrated in pericentric and telomeric regions of chromosomes (X, 2, 3, 4 and Y). The heterochromatin contains tandemly repeated simple sequences (including satellite DNAs), middle repetitive elements (such as TEs and rDNA) and some single copy DNA. Drosophila contains 13601 genes which is approximately double the number in yeast but is very less than in *C. elegans* (Table 16.4). The average density of genes in eukaryotic regions is one gene per 9kb which is much lower than the typical one gene per kb densities of prokaryotic genes. Drosophila contains homologues of 289 human genes.

Yeast — Chromosome size ranges from 1352kb for chromosome IV to 230kb of chromosome 1. The number of predicted genes is 5885 (for protein coding genes), about 140 (for rRNA), 40 (for snRNA) and 275 for tRNA genes. Genome is denser in coding regions than genomes of more complex eukaryotes such as *C. elegans*, Drosophila and human. Introns are relatively rare and relatively small. Only 231 genes contain introns and there are fewer repeat sequences in comparison to more complex eukaryotes. One-third of proteins have identifiable homologues in human.

C. elegans — Genome is about eight times larger than that of yeast. Gene density is relatively low with one gene per five kb of DNA. Exons cover about 27% of the genome. Genes contains an average 5 introns each. Approximately 25% of the genes are in cluster. 2.6% of the genome

Table 16.7 Showing some human genes different in length.

Human Gene	Base pair	Exon size
B globin	1660	
g-interferon	4966	
Serum albumin	6682	
Tissue plasminogen activator	19534	Exons vary from 43 to 757bp long
HGPRT	42830	Exons vary from 18 to 640bp long

consists of tandem repeats and 2/6% contains inverted repeats which appear preferentially within introns rather than between genes.

Human—DNA content of autosomes ranges from 279Mb down to 48 Mbp. X-chromosome contains 163Mbp and the Y-chromosome 51Mbp.Heterochromatin, the repeat-rich regions concentrated in centric and telomeric regions of chromosomes makes up at least ~20% of the genome. Coding sequences form less than 5% of the genome whereas the repeat sequences over 50%. Thus a large part of human genome consists of repeated regions or low complexity DNA. The number of genes in human is only 2-3 fold over invertebrates. About 59% of the genes have alternative splicing patterns. Exons are relatively small compared to those in other known eukaryotic genomes. The introns are relatively long and as a result of which many protein coding genes span long stretches of DNA. For example, dystrophin gene, coding for a 3685 amino acid is >2.4Mbp long. Variation in length of some protein coding genes is given in the table 16.7. A typical exon is 150bp, a typical intron is several kbs and a complete gene can be hundreds of kilobases in length. Thus what is required to determine is given in the Table 16.8.

16.21 POSITIONAL CLONING

The only way to find and clone a gene of unknown coding function and unclear affinities is through the knowledge of its map position. Positional cloning (map-based cloning, map-assisted cloning) refers to the cloning of a specific gene in the absence of a transcript or a protein product using genetic markers tightly linked to the target gene and a directed or random chromosome walk by linking overlapping clones from a genomic library. If the mutation gene of interest can be mapped between two molecular markers separated by a megabase or less then it should be possible to clone the whole sequence in between. The region between two markers that appear to bracket the gene is explored by a procedure called **chromosome walking**. In other words, classical genetic mapping is used initially to determine the position of the gene relative to known genes and to the breakpoints of chromosomal deletions or other rearrangements. This approach thus relies on cytogenetic and molecular mapping and then the region of the genome containing the desired gene is isolated by chromosome walking (Bender et al., 1983) or chromosome microdissection (Scalenghe et al., 1982). In the chromosome walking overlapping segments of DNA are isolated, starting from the nearest known cloned DNA segment, by sequentially screening a genomic library so that one extends the cloned region in an ordered way toward the gene of interest. A cosmid vector is used to generate probes specific for the ends of cloned inserts. The vector contains bacteriophage T3 and T7 promoters, flanking a unique Bam HI cloning site. Not1 site facilitates

Table 16.8 Compositional analysis of sequence.

---

1. Overall physical length
     Short arm
     Long arm
2. Base composition (% GC)
     Overall
     Coding region
     Non-coding region
3. Gene number (predicted)
     Gene density
     Average gene size
     Exons
     Size of exons
     Number of exons per gene
     Introns
     Size of introns
     Number of introns/gene
     Repetitive sequences
     Number of retrotransposons (class I)
     Number of DNA transposons (class II)
     Autonomous type
     Non-autonomous type (MITEs)
     Total number of repeats

---

restriction mapping and excision of the insert DNA. Ori-origin of replication; Amp^R and Kan^R; *cos*–cohesive sites essential for *in vitro* packaging in λ particles. Chromosome walking thus yields DNA sequences from the region containing the gene of interest. The problem then reduces to identifying the DNA sequences comprising the gene within the larger cloned region. In some cases, the region defined by the positions of chromosomal rearrangements or restriction site polymorphisms may encompass only the gene of interest. In other cases, looking for DNA sequences that have a transcript accumulation pattern that matches the expected pattern of expression of the desired gene has allowed the localization of a gene in a given chromosome walk. The most rigorous test, however, is to find whether a cloned DNA segment can complement the mutant defect if introduced into the organism by DNA transformation methods.

Riordan et al., (1989) identified the gene involved in human genetic disease, cystic fibrosis using this procedure. The gene called CFTR (cystic fibrosis transmembrane conductance regulator) T codes for 1480 amino acid protein which normally forms a cyclic –AMP-regulated epithelial cl$^-$ channel. It comprises 24 exons and spans a 250kb region. The different steps involved in positional cloning or reverse genetics are as follows.

1. A search in family pedigree for a linked marker showed that the cystic fibrosis gene was close to a known variable number tandem repeat (VTNR), DOCR- 917. Somatic cell hybrid placed this gene on chromosome 7, band q3.

2. Other markers found were linked more tightly to the target genes. It was then bracketed by a VTNR in the MET oncogene and a second VTNR, D7S8. The target gene lies 1.3 cM from MET and 0.9 cM from D7S8-localizing it to a region of an approximately 1-2 million base pair. A region this long could contain 100-200 genes.

3. The inheritance patterns of additional markers from within this region localized the target more sharply to within 500kb.Chromosome jumping made the exploration of this region more efficient.

4. A 300kb region at the right distance from the markers was cloned. Probes were isolated from the region to look for active genes characterized by an upstream CCGG sequence. The restriction endonuclease Hpa II is useful for this step. It cleaves DNA at this sequence but only when the second C is not methylated, i.e. when the gene is active.

5. Identification of genes in this region is made by sequencing

6. Checking in animals for genes similar to the candidate genes turned up four likely possibilities. Checking these possibilities against a cDNA library from sweat glands of cystic fibrosis patients and normal (healthy) controls identified one probe with the right tissue distribution for the expected expression pattern of the gene responsible for cystic fibrosis. One long coding segment had the right properties and corresponded to an exon of the cystic fibrosis gene.

   The variants of cystic fibrosis gene had a common alteration in the sequence– a three-base pair deletion, deleting the residue 508Phe from the protein.

Proof that the gene was correctly identified

(a) 70% of the CF alleles had the deletion. It was not found in healthy or resistant individual.

(b) Expression of the wild-type gene in cells isolated from diseased individual restores normal Cl$^-$ transport in epithelial tissues.

(c) Knock out of homologous gene in mice produces the CF phenotype.

(d) The pattern of gene expression matches the organ in which it is expressed.

(e) The protein encoded by the gene would contain a transmembrane domain, consistent with involvement in transport.

Map-based cloning is often used for mutations caused by chemical mutagenesis which result in small changes (single nucleotide changes) in nucleotide sequences. To clone the gene mutation is genetically mapped relative to molecular markers (and to morphological markers if convenient). For mapping a cross is made with strain that carries nucleotide differences due to normal polymorphism between strains. The F1 is self-fertilized to produce F2 population which contains meiotic recombination between the chromosomes of the two parents. The position of the locus can be mapped on the basis of segregation with DNA markers. Given a genetic map of molecular markers and a large segregating population a mutation can be mapped to a chromosome interval encoding only one or a few genes. After that additional analysis such as chromosomal walking is carried out to verify the identity of gene and to ensure that the gene is the same as defined by mutation.

Construction of a detailed restriction endonuclease map of the cloned DNA fragment—The construction of a restriction endonucleas map of a cloned DNA fragment forms the basis of structural and genetic studies of a cloned DNA fragment and it is also necessary for designing of a DNA sequence strategy. The restriction map is constructed by determining the sizes of sub-fragments generated with single and pairs of infrequent cutter endonucleases. The cleavage sites of 'frequent cutters' may be mapped by measurement of the sizes of partial digestion products of a fragment that is radioactively labeled at one end (Smith and Birnstiel, 1976). For this purpose isolation of the cloned DNA fragment can be done by excision from the vector by cleavage with the restriction endonuclease used in the construction and separation of the DNA fragment from its vector molecules by gel (agarose) electrophoresis. UV irradiation of the ethidium bromide stained gel visualizes the DNA bands and enables band-containing segments of the gel to be cut out. The gel pieces are then dissolved in the saturated solution of KI, followed by absorption of the DNA to and subsequent elution from glass beads. The DNA fragments obtained in this way are of high purity and can be used in any biochemical analysis. Further, structural analysis of a cloned DNA fragment involves comparison with alternative DNA form (e.g. if it cDNA, it may be compared with genomic DNA and vice-versa) or with related fragments or purified RNA molecules, by hybridization methods such as heteroduplex formation or Southern blotting. Electron microscopic methods may also be employed to localize regions of sequence symmetry (inverted repeats) or RNA polymerase binding sites (potential transcription promoters). Alternatively filter binding techniques can be used to detect specific binding sites for RNA polymerase and other proteins. Finally, polynucleotide sequence of the cloned fragment or a part there of which has been identified during preliminary characterization of fragments may be determined by rapid sequencing methods (see chapter 13).

16.22 GENE SCREENING

It is a method to detect specific gene sequences in a genomic or cDNA libraries by using either a radioactively or non-radioactively labeled probe (colony hybridization, plaque hybridization) or an antibody in the procedure called **immunological screening**. Immunological screening is done for identification of a specific clone in a cDNA expression library by precipitating the corresponding protein with a specific radioactively labeled probe or non-radioactively labeled antibody. The resulting antigen-antibody complex can then be detected by autoradiography, different chromogenic methods, if enzyme conjugated antibodies are used or an ELISA or **RIA (radioimmunoassay,** a technique used for the quantitative determination of antigens in which the specific binding of radioactively(e.g. 125I) labeled antigen or antibody is measured.

There are three basic strategies for large scale identification of expressed sequences.

1. **Exon-amplification**—It employs *in vivo* splicing systems to identify and selection for functional splice sites in genomic DNAs, resulting in the generation of cloned exon libraries. Exon amplification is a technique for the rapid identification of exons and the corresponding genes from genomic DNA. In this procedure 1-4kb genomic restriction fragments are first cloned into an *in vivo* splicing plasmid vector, pSPL. The cloning site is localized within an intron of a gene flanked by splice sites from another gene. The transcription of the cloned gene is driven by an SV40 early promoter and terminated by polyadenylation signal from, for example, SV40. The construct is transfected into COS cells by electroporation, the expression plasmid vector amplified *in vivo* by its SV40

origin of replication and RNA transcripts produced at a high level. Then the splice sites generated by the fusion of the acceptor intron and the genomic exon are spliced to generate a polyadenylated cytoplasmic RNA. If the genomic insert does not contain an exon from the genomic DNA then it will not be spliced. After 2-3 days of transfection cytoplasmic RNA is isolated , converted to cDNA by reverse transcription which is then amplified with oligodeoxy nucleotide primers complementary to the flanking beta globin sequences and a standard PCR reaction. The amplification product contains the introduced exon sequence and can be analysed by , for example, cloning, sequencing or direct PCR sequencing. Additionally the exon sequences can be labeled and used as probe to fish out the corresponding full length gene from the genomic libraries. Generally genomic fragments of up to 20-40kb can be screened for exons in a single transfection experiment.

2. **Hybrid selection**—Hybridization of cDNA fragments to immobilized genomic DNAs to enrich for cDNAs encoded by specific genomic clones.

3. **Automated partial DNA sequencing of random cDNAs clones** providing catalogs of ESTs.

All these strategies result in identification of short (100-400bp) cDNA fragments which when localized relative to markers of known map positions, pinpoints the location of an expressed gene. Subsequent closure of regionalized transcription maps and/or in-depth analysis of genes identified as exons and or ESTs requires the recovery of a more full length clone or cDNA. PCR strategy in conjunction with techniques such as exon amplification which can be used to develop ESTs from defined physical locations without regard to the expression level of individual transcript.

Identification of exons—For identification of exons, cross-linking and immunoprecipitation (CLIP) (Ule et al, 2003) (see chapter 25 for detail) and splicing microarray (Ule et al., 2005) can be used (see chapter 24 for detail).

Comparison of the cloned fragment with an alternative DNA form–In this technique hybridization methods such as hetero duplex formation or Southern blotting assuming that the library is contained in plasmids, *E. coli* cells harbouring clones from a genomic library are grown on plates. A sample from each colony is transferred on to a nitrocellulose membrane. The cells on the DNA binding membrane are lysed and the released double stranded DNA bound to the membrane is made single stranded with alkali treatment which is then hybridized to radioactive probe, 32P and exposed to X-ray film. A radioactive spot observed on the film indicates the position of a colony harbouring a clone recognized by the probe. Thus genomic library can be screened for genes or classes of genes with DNA probes. Electro microscopic methods may also be used to localize regions of sequence symmetry (inverted repeats) or RNA polymerase binding sites (potential transcriptional promoters). Alternatively, filter binding can be used to detect specific binding sites for RNA polymerase and other proteins.

Identification of regions involved in the regulation of gene expression—The gene expression is regulated by interaction between proteins and DNA and so if a particular piece of genomic fragment has been cloned then it would be worthwhile to identify the regions of DNA fragment involved in such interaction and this can be achieved through a technique called MG sequencing (**foot printing** protein binding sites in DNA) strategy and finally polynucleotide sequencing of the cloned fragment or a part thereof which has been identified during preliminary characterization of the fragment. Specific promoter sequences in genomic DNA can be identified using promoter trap vectors in cloned DNA fragments using promoter probe vectors.

The development of technique of sequencing *in vivo* has permitted the 'foot printing technique' to be employed to the study of interactions between uncloned DNA and protein *in vivo*. In one approach the nuclei is irradiated with U.V. light, producing pyrimidine dimmers through formation of bonds between adjacent pyrimidine. Of the three possible types of pyrimidine dimmer, the thymine dimmer is most readily formed. The formation of such dimmers blocks the action of the DNA polymerase and so prevents replication. The resulting saturation of the 5, 6-double bond will allow ring opening through reduction with $NaBH_4$ (Becker and Wang, 1984). In another approach dimethyl sulphate can be used for methylation of purines to intact systems such as *E.coli* cell (Nick and Gilber, 1985) and mammalian nuclei (Church et al., 1985)

But the most precise identification of the features of DNA-protein interactions are done through X-ray crystallography of complexes formed between proteins and chemically synthesized oligonucleotides corresponding to their binding sites (Frederick et al., 1984).

The above mentioned interaction between DNA (single stranded or double stranded) and protein (DNA modifying enzymes, DNA dependent DNA– and RNA-polymerase and transcription factor) serves as **regulatory function** but then interactions may serve **structural function**, for example, nucleosomes.

Genomic foot printing—It is a technique to detect specific DNA –protein interactions at the base level *in vivo*. In this technique cells in suspension culture are treated with DMS, a chemical mutagen which reacts with unprotected guanosyl residues of DNA. Then the DNA is isolated and digested with a particular restriction endonuclease and subsequently with a peperidine. This treatment cleaves all methylated guanosine residues(β-elimination). Subsequent C-G sequencing identifies the modified bases. If proteins are bound at specific sequence of the target DNA these areas are protected from the modification by DMS thus leaving 'foot print' on the DNA sequencing gel. *In vivo* footprinting allows the detection of temporal and spatial interactions between DNA-binding proteins and their cognate elements. While the standard *in vivo* footprinting techniques are generally carried out using cell cultures, *in vivo* footprinting coupled with ligation-mediated PCR allows the analysis of TFs in the tissues from where they originate when an appropriate cell culture is not available. In this technique linkers are ligated to cleaved genomic DNA fragments which can then be used to amplify the resulting DNA fragments by PCR (Hammond-Kosak and Bevan, 1993) (Figure 16.2).

Psoralene foot printing—In this technique for detection of specific contacts between one or several proteins and a DNA duplex molecule psoralene and UV light are employed. In the presence of UV light psoralene reacts with DNA and forms a photo adducts and pyrimidine monoadducts. These complexes influences or prevent the binding of DNA and protein.

Photo-foot printing—This technique is employed to detect contacts between specific sequences of DNA and regulatory proteins *in vivo*. The DNA is isolated after irradiation of the intact cells but before cellular repairs of the DNA damage starts. After purification it is subjected to a series of chemical reactions that break its sugar-phosphate backbone only at the sites of UV damage. The DNA is then denatured and labeled. After electrophoresis on polyacrylamide sequencing gel the resulting fragment patterns is visualized by autoradiography. Since protein-DNA contact can inhibit or enhance UV photoproduct formation differences in the single breakage patterns of protein –free and protein bound DNA can be used to detect the protein-DNA contact at base pair level.

16.23 METHYLATION INTERFERENCE ASSAY

It is a method for testing the specificity of binding interaction (s) between a specific DNA sequence and a sequence specific binding protein. In this method the DNA sequence is partially methylated *in vitro* at purine residues by DMS, then mixed with a nuclear extract or a purified nuclear binding protein and tested for its binding properties in a mobility shift DNA binding assay described below. Methylation of purine residues within the target DNA interferes with the binding of specific protein which readily binds to the identical non-methylated sequence.

Mobility- shift -DNA binding assay (gel retardation assay) — In this method the principle of detection of specific DNA-protein interaction is based on an altered mobility of protein-DNA

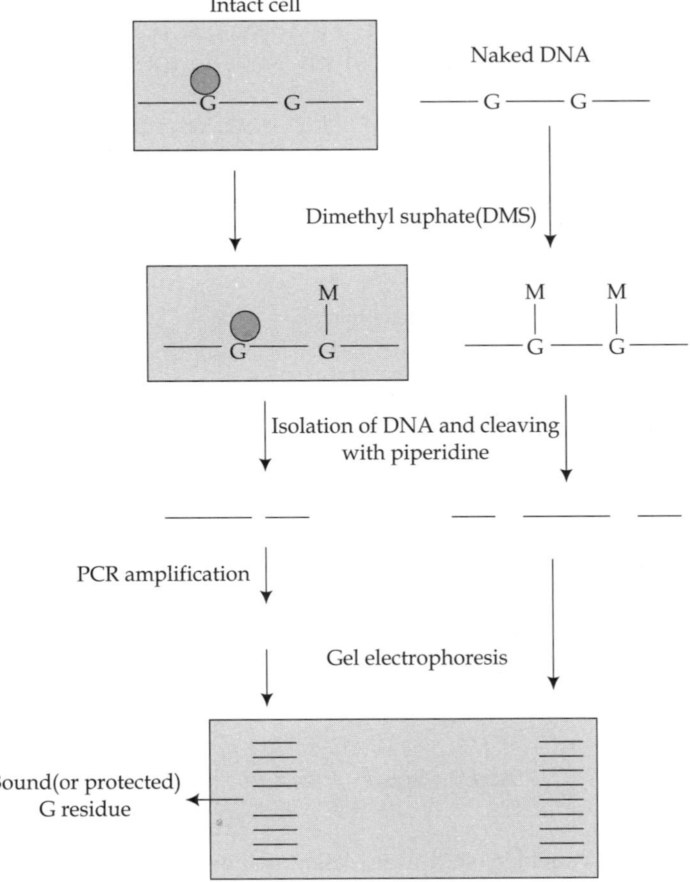

FIGURE 16.2 Showing *in vivo* footpriting using methylation interference assay.

complexes during non-denaturing gel electrophoresis as compared to free DNA. In this method the target DNA is first end labeled with y32P-dATP using deoxynucleotide transferase and then incubated with a nuclear extract. Specific DNA-protein complexes are detected by low ionic strength polyacrylamide gel electrophoresis and autoradiography (Figure 16.3). The free fragment moves faster than the protein-DNA complex which is retarded. Usually an excess of heterologous competitor DNA is added to saturate the more abundant, non-specific DNA binding proteins. **Analysis of methylation pattern-** Analysis of methylation pattern is done using bisulphate sequencing (see chapter 12).

Electrophoretic mobility shift assay (ESMA)—Most promoters contain an array of *cis*-elements which can be recognized by TFs. Promoters studies usually involve fusing promoter regions to a reporter gene and assaying reporter gene activity in transiently or stably transformed plants. Defined promoter regions can then be further characterized by EMSA assay (Carey, 1991). This assay is used for detection of sequence specific DNA binding proteins such as transcription factors. EMSA technique for studying DNA-protein interactions, is based on the ability of a DNA binding protein to alter the mobility of DNA in a non-denaturing acrylamide gel. In this method an end labeled DNA fragment containing binding site for the protein is electrophoresed through a non-denaturing polyacrylamide gel together with a nuclear protein extract. The labeled fragment of the promoter usually consists of a 200 to 300bp segment carrying a *cis-* sequence. Protein which binds to the DNA fragment decreases its

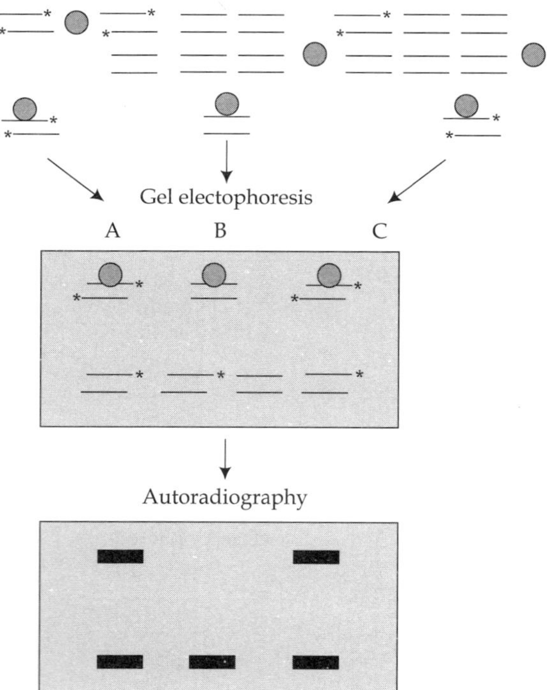

FIGURE 16.3 Showing DNA mobility shift assay using unlabelled competitor. Result of A is similar to C where as in case of b, the radioactive retarded band will not be observed.

electrophoretic mobility and thus allows its discrimination from the non-bound DNA fragment. Thus the difference in electrophoretic mobility of identical DNA fragments in two adjacent lanes of an agarose gel is due to the binding of specific proteins. The label is usually visualized by a phosphorimager or by an exposure on the X-ray film. In the absence of nuclear proteins all label ends up at the bottom of the gel but in the presence of proteins some proteins may bind to the labeled promoter fragments and retard their flow through the gel and thus they appear as 'retarded' bands. To minimize the non-specific binding of the proteins to the promoter fragment an excess of non-competing heterologous DNA may be added to the incubation mixture. The retarded bands are checked for specificity by adding an excess of unlabeled promoter segment (100-400bp) to the incubation mixture which reduces the intensity of labeling or eliminates the retarded band. EMSA can be used for identification of DNA-binding domains. Although EMSA demonstrates specific binding of a protein to a particular DNA sequence, it provides only limited information about the bases that are directly involved in protein binding. Using antibodies, it is possible to confirm the identity of protein present in a DNA-protein complex as the addition of the antibody to the binding reaction can supershift the complex or inhibit complex formation. More detailed analyses can be carried out using footprinting techniques or random binding site selection (RBSS) which allow identification of bases important for transcription factor binding (Schwechheimer et al., 1998). That binding occurs in a specific region of the promoter can be shown by **DNA footprinting** assays. DNA footpriting techniques have been used to determine TF binding sites on promoter fragments. Footprinting methods are based on either chemical modification and subsequent cleavage of the promoter or on DNase I treatment (Adhya et al. 1991; Chern, et al. 1996). When the resulting DNA fragments are separated on denaturing acrylamide gels, residues protected by the protein are identified by missing or underrepresented bands. The 'foot print(s)' assists in the identification and confirmation of a conserved sequence as response element. Thus nuclear proteins that bind to the promoter fragment in a gene- and tissue-specific manner can be isolated. *In vitro* footprinting with DNaseI is based on the ability of DNaseI to nick randomly either strand of DNA. Although treatment with DNaseI can take place before (interference) or after (protection) protein binding, but the latter technique is very widely used. **RBBS** is used to determine the range and degree of DNA-binding specificity of a TF. RBSS is based on the amplification of specific protein binding sites from a pool of randomized sequences. This technique is particularly useful for defining differential DNA-binding and the role of sequences flanking a core consensus site. The limitation with this technique is that it is difficult to interpret the results when nothing is known about the cognate binding site of the TF (Quellette and Wright, 1995).

16.24 IDENTIFICATION OF RNA-PROTEIN INTERACTION SITE

Here the objective is to identify minimal length of RNA sequences that bind specifically to a protein which is required to know in case of selection of rRNA binding site for the ribosomal protein. There are different approaches to accomplish this objective. The classical approach involving digestion of an RNA-protein complex with RNase usually does not give the minimal binding region as the protein(s) with the complex might hinder the access of an RNase. Alternatively, RNase can cut the RNA within the complex into short sequences that loose the binding capacity. Crosslinking approaches and protection experiments with the base modifying reagents show vicinity but not necessarily the binding sequence. *In vitro* selection method which can identify the best fitting RNA sequences from about 10^{15} variants (SELEX, Gold et al., 1995) but when randomized sequences are used as stating conditions, affinity selection methods usually do not lead to naturally occurring binding site for a target as such. A variant of SELEX called SERF(selection of random RNA fragments, 2000) in which a pool of random fragments from large RNAs directly reveals the native binding site when selected for affinity to a certain protein *in vitro*. In this technique a pool of random rRNA fragments is generated by random cutting. The rDNA fragments are transcribed into RNA in the second step. There are many advantages of SERF. First, selection does not work on just any RNA – proteins but rather on the native ones. Secondly, the size of selected RNA can be chosen and a series of overlapping fragments can reveal the minimal binding site and thirdly, the selection should be fast and efficient as the variability of the pool is low compared with the starting pool in classical affinity experiments.

16.25 DETERMINATION OF SPACING BETWEEN TWO REGULATORY SEQUENCES

The optimal spacing between the two adjacent regulatory sequences(boxes) of a promoter can be estimated using synthetic homopolymeric linkers of variable length such as oligo (dT), oligo (dG), oligo (dA) or oligo (dC). They are cloned between the two boxes and the effect of linker length on the expression of linked gene (s) is determined. This technique allows also the mapping of additional sites within spacer that function in the binding of transacting proteins.

16.26 ALLOCATION OF CLONED GENE (S) TO SPECIFIC CHROMOSOMES

Four methods are available for initial chromosome assignment and precise gene localization.
1. Somatic cell hybridization (somatic cell genetics)
2. ISH (Hybridization to fixed chromosomes *in situ*)
3. Hybridization to DNA from individual chromosomes separated by FACS
4. Use of RFLP.

The most direct method for identifying chromosomal locus of a segment of human DNA is by ISH. Although unique sequences less than 1kb long can be localized by isotopically labeled probes, autographic development times are long, extensive statistical analysis is required and mapping precision is limited by the necessity of having to capture the emitted isotopic signal by an emulsion overlay. In contrast, non-isotopically labeled probes offer markedly improved speed and spatial resolution but they lack sensitivity. However, there are reports of detection of unique sequence targets of 6kb or less by nonisotopic ISH. Hybridization signals from two probes can be spatially resolved on metaphase chromosomes when the probes are only several hundred kilobases apart but a minimum of 1-2Mbp separation is required to allow their physical order to be established. **Chromosomal in situ suppression (CISS) hybridization** (discussed in chapter 31) in conjunction with fluorescent detection of hybridized probes can be used for the rapid and precise localization of large number of cloned genomic DNA segments and development of a physical map of individual chromosome (Lichter et al.,1990). The use of deletions, duplications and translocations will be of value in mapping chromosomal regions of high interest with greater definition. Current strategies for using such reagents for mapping require techniques such as Southern blotting or segregation of aberrant chromosomes in somatic cell hybrids which are laborious and time consuming and this is where the CISS hybridization fits in. Gene order established by CISS hybridization can be confirmed by the use of two fluorophores simultaneously. The use of combinatorial analysis (see chapter 24 for detail) can be used to confirm map order for all of clones by two–fluorophores technique. The direct ordering of genomic clones by CISS hybridization shows the value of simultaneous multiparameter analytical approach. For determining the chromosome to which a probe has hybridized a technique for chromosome identification compatible with CISS hybridization is required. Conventioanl Giemsa bainding of chromosomes before hybridization lowers the hybridization efficiency and post-hybridization staining quenches fluorescence. Although it is possible to do Giemsa banding after hybridization but this process requires the relocalization of specific metaphase spread and two separate photographic steps. The alternative banding or labeling methods such as quinacrine banding, bromodeoxyuridine banding and diamidinophenylindole (DAPI) banding (see Roy, 2009) are compatible with fluorescent detection of probes by ISH and can be used for direct chromosome identification. Alternate strategy is to use cohybridization with a differently labeled probe (or a probe set) such as an Alu DNA BLUR clone which provides a banding pattern resembling R banding or with probe tagging a particular chromosome such a previously mapped cosmid clone or a chromosome-specific DNA repeat.

Although CISS hybridization has the potential to map a large number of cloned rapidly, efficiently and accurately, develop a detailed maps of specific chromosome regions and to order a large number of independently isolated clones or previously ordered overlapping clone sets that span a discrete region several megabases in size, it has got some limitations. The two parameters that affect the resolution of CISS hybridization are the position of the chromosome and the degree of chromosome condensation. Elongated prometaphase chromosomes give higher resolution than compact metaphase chromosomes. Further, regions of the chromosome that include the centromere and telomeres often show somewhat higher variability in map positions than probes in other chromosomal locations. The resolution of CISS hybridization is on the order of 1 Mbp and thus can be compared with pulsed-field gel analysis. Methods in which DNA can be analyzed in a more extended state such as premature chromosome condensation or chromosome shattering or by using DNA in interphase nuclei and refinements in optical imaging techniques provide means to improving spatial or lateral resolution.

Steps in mapping process—The different steps in mapping process are probe production, probe labeling, pre-annealing to suppress signals from interspersed repeat DNA and ISH

Somatic cell genetics

The somatic cells of two different animal species can be fused and there has been development of the human-mouse hybrids (see chapter 31) in which the human chromosomes are preferentially lost and the cell clones are produced which retain only one or a few human chromosomes. The clones can then be tested for specific enzymes and their presence can be related to those chromosomes present in the human-mouse hybrid. For this method to be successful one must be either able to select for a particular characteristic (for example, thymidine activity) or to be able to distinguish the human enzyme from the corresponding mouse enzymes which will be present in all clones. In other words, it is possible to correlate the loss of a particular phenotype with the loss of a particular chromosome. Somatic cell hybrids particularly radiation induced are useful in constructing a detailed hybrids map of a limited chromosomal region.

Radiation hybrid—It refers to any hybrid which contains small fragments of human chromosomes in a Chinese hamster ovary (CHO) genetic background. To produce radiation hybrid a CHO cell line which harbors a single but complete human chromosome with a wild type hypoxanthine ribosyl transferase (hprt) gene is selected. This line is exposed to very high dose of X-ray (8krads) which causes breaks in the chromosome (on average into 5 pieces). After healing the resulting cells are fused with a CHO cell line (hprt), the inactive hprt gene or allele is present. Cell clones showing hprt$^+$ phenotypes are selected and maintained on a selective HAT medium (growth on this medium requires an active hprt gene). The selected clones represent a randomly sampled library of fragments of particular human chromosome.

Radiation hybrid (RH) mapping—The use of RFLP in conjunction with genetic linkage analysis has allowed the construction of meiotic linkage maps for each of the 23 chromosomes with an average resolution of 10 to 15 cMs. The ability to separate human chromosomes from one another either in rodent-human somatic cell hybrids or by physical chromosome sorting has led to significant advances in defining a map of human genome. Hundreds of loci have been assigned to specific chromosomes with these techniques. Further, *in situ* hybridization now provides a means of locating molecular probes to specific positions on human chromosome. On average meiotic recombination between two markers on a human chromosome corresponds to 1 megabse of DNA. *In situ* hybridization can localize markers within 2% of the total chromosome length but in molecular terms this again represents several million base pairs. PFGE which can separate DNA fragments of several million base pairs in agarsoe gels provides a potentially powerful means for constructing long-range physical maps of chromosomes when used in conjunction with restriction enzymes which cut infrequently in human DNA. The problems with rare cutting enzymes are that there is paucity of rare cutter enzymes and the non-random distribution of rare cutter sites in human genomic DNA make it difficult to order DNA sequences more than a few hundred kilobase pairs apart with this technique. Thus long stretches of contiguous order information at the 100-500kilobase level of resolution remains a difficult task.

In the radiation hybrid mapping hybrid cell lines, each containing many large chromosomal fragments produced by radiation breakage are screened by PCR to identify those hybrids which have retained a given locus (Walter et al., 1994). Nearby loci which tend to show similar retention patterns allows proximity to be inferred. Radiation hybrid linkage can be detected for distances of about 10Mb, given the average fragment size of the RH panel used there. As RH mapping and genetic mapping can detect linkage over large regions (0.3 to 1% of the genome), comprehensive RH and genetic maps spanning all chromosomes can be assembled with a few thousand loci. It is a somatic cell hybrid technique for the construction of dense (high resolution) map of mammalian chromosomes and involves radiation hybrids and the statistical analyses of DNA markers to determine their relative order with respect to one another or the identification of chromosomal location of a specific gene. For gene localization two gene-specific primers are used to amplify a single PCR product (usually between 100 and 800bp). This product can be then labeled and hybridized to a panel of RH clones, consisting of donor DNA fused to hamster or mouse recipient DNA. Occurrence of hybridization indicates presence of the probe complementary to sequences in the RH clone. The advantages of the RH mapping are that the estimated distances are directly proportional to the physical distances and that non-polymorphic DNA markers can be used for mapping as opposed to genetic mapping (Figure 16.4).

Theory—Radiation hybrid mapping (Goss and Harris, 1975,1977; Cox et al., 1990) provides a general method for ordering DNA markers spanning millions of base pairs of DNA at the 500kb level of resolution. In this technique a high dose of X-rays (8000 rad of X-rays) is used to break the desired human chromosome into several fragments. These broken chromosome fragments are recovered in rodent cells and approximately a hundred such rodent-human hybrid clones are analyzed for the presence or absence of specific DNA markers. The farther apart two markers are on the chromosome the more likely a given dose of X-rays will break the

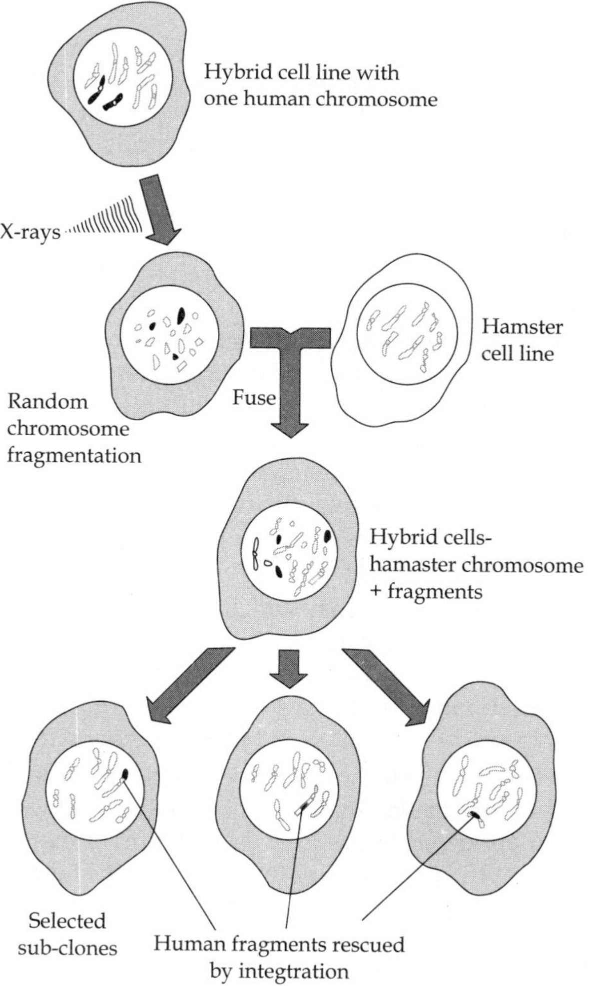

FIGURE 16.4 **Showing radiation hybrid mapping procedure.**

chromosome between them, placing the markers on two separate chromosomal fragments. There are three main radiation hybrid panels which have been used for STS mapping and constructing RH maps in human. The Stanford TNG panel is constructed using 50,000 rad of X-rays and it allows STS resolution down to 60-100kb with high confidence (Lunetta et al., 1996). These different panels offer different levels of resolution based on the dose of irradiation. GB4 RH panel is constructed by using 3000 rad of X-rays and it resolves marker at 1Mb intervals. The G3 RH panel is constructed by using 10, 000 rad of X-ray and it resolves markers at 260kb intervals. By estimating the frequency of the breakage and thus the distance, between markers, it is possible to determine their order in a manner similar to meiotic mapping.

Assuming breakage between two markers being independent of marker retention and that retention of one fragment being independent of the retention of any other one can calculate the frequency of breakage, θ, as

$$\theta = [(A^+B^-) + (A^-B^+)]/[T(R_A + R_B - 2 R_A R_B)]$$

where (A^+B^-) is the observed number of hybrid clones retaining marker A but not marker B, (A^-B^+) is the number of hybrid clones retaining marker B but not marker A, T is the total number of hybrids analyzed for both marker, A and B, R_A is the fraction of all hybrids analyzed for marker A which retain marker A and R_B is the fraction of all hybrids analyzed for marker B which retain marker B. Although θ is a good estimate of the distance between markers which are close together it can underestimate the distance between more distant markers. The mapping function, $D = -\ln(1 - \theta)$ which assumes no interference and is analogous to the Haldane mapping function in meiotic linkage analysis, can be used to make a more accurate estimate, D of distance between markers.

Radiation hybrid map — It is a genome map constructed with molecular markers, STSs that are positioned relative to one another on the basis of the frequency with which they are separated by radiation-induced breaks (expressed as centiRays (cR), defined as a 1% chance that break occurs between two loci. The frequency is determined with a panel of radiation hybrid. Because θ, the frequency of breakage between two markers and thus the distance (D) depends on the amount of irradiation so it is essential to include information about the X-rays dose such as a distance of 1 cR_{8000} between two markers corresponds to a 1 percent frequency of breakage between the markers after exposure to 8000 rad of X-rays.

Unlike a meiotic recombination frequency which can vary from 0 to 0.5 , θ varies from 0 to 1.0. A θ value of 0 shows that two markers are never broken apart whereas a θ value of 1.0 indicates that the two markers are always broken apart and thus are unlinked. A lod score (logarithm of the likelihood ratio for linkage) identifies those marker pairs which are significantly linked. A lod score of 3.0 or more is taken as evidence for significant linkage.

***In situ* hybridization (ISH)** — The ISH technique allows assignment of single copy genes as well as repetitive gene families to particular segments of specific chromosomes using radioactive gene probe which can be hybridized to the fixed preparation of metaphase or prometaphase chromosomes. Radioautography associated with chromosome staining methods will lead to illumination of gene on a particular chromosome. Further, the Southern transfers can be prepared of the DNA isolated from the various hybrid clones and which when hybridized with radioactive gene probe provides a physical rather than enzymic location of the gene on a particular chromosome. This technique in association with the use of deletion mutants and cotransfer of genes on chromosome fragments will result in more precise assignment of genes to chromosome regions. In other words *in situ* hybridization of cDNA or RNA probe to isolated chromosomes will lead to the identification of chromosomal loci. As now it is possible to isolate individual chromosome by fractionating a preparation of metaphase chromosomes using FACS, the individual chromosome fraction can be hybridized with the gene probe in order to assign genes to specific chromosomes. PFGE provides a means of assigning cloned genes to chromosomes that are too small for useful for microscopy. The different genomic bands can be blotted onto a nitrocellulose membrane but with treatment of the gel after electrophoresis with U.V. light to break the separated chromosome-sized DNA bands to smaller fragment that will blot more easily. Hybridization of the blot to gene-specific probes will then reveal which genes are located in which bands. Finally, analysis of restriction fragment length polymorphisms will provide information on the location of gene (s) on specific chromosomes.

RFLP — Analysis of RFLP in family pedigrees has allowed the construction of linkage maps as well as identification of the genetic loci for many diseases in human. Physical mapping provides a complement to genetic linkage. Standard agarose gel electrophoresis gives resolution in the 1-to-10kb range whereas the PFGE techniques extend this range to the megabase level. However, gel electrophoresis methods are not useful for producing an initial localization for a DNA sequence that has not been mapped previously. Furthermore, these techniques do not directly provide the ability to order DNA sequences more than a few hundred kilobases apart or the ability to assign DNA sequences to specific chromosomal regions. For initial chromosome assignment and precise gene localization, somatic cell genetics, FACS of metaphase chromosomes and ISH (discussed above) have been widely used. Figure-16.5 shows comparion of different types of maps.

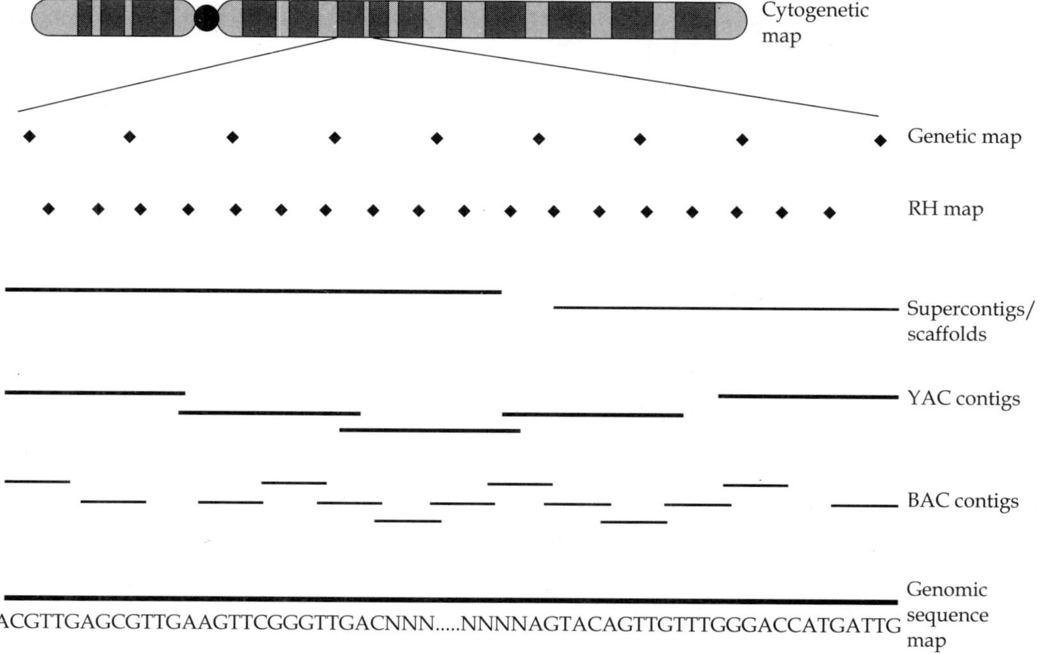

FIGURE 16.5   Comparison of different types of maps.

16.27 GENE IDENTIFICATION

Gene hunting refers to detection, isolation and characterization of a gene and involves whole genetic engineering techniques such as library construction, gene screening, vector design, gene cloning and sequence analysis. After cloning restriction fragment it is essential to identify whether or not the cloned sequence contains an open reading frame, ORF (or gene). In other words, whether or not the cloned sequence has contains coding sequence function. ORF is a sequence which starts with initiating codon, AUG and end some distance downstream with one of the three terminating codons. The first check on the identity of a putative gene clone is the determination of its DNA sequence. Sequencing will tell us whether it includes any **ORF.** The DNA sequences not having any protein coding function are unlikely to runs of codons extending for more than a few hundred nucleotides without a random encounter with a termination codon. Thus the presence of an ORF much longer than that is *prima facie* evidence of coding function. If the protein product of the putative gene is already known then the comparison of the amino acid sequence with the sequence of codons in the ORF will tell us whether or not the cloned sequence contains gene. Thus ORF can be detected in cDNA clones but the recognition of ORFs in the genomic clones may be complicated by the presence of introns. ORF is any nucleotide sequence in the DNA that potentially can be transcribed into a protein or be transcribed into an RNA and starts with an ATG (start codon) and terminates with any of the three termination codons. A good ORF candidate for coding a bonafide cellular protein has a set size requirement. It should have the potential to encode a protein of 100 amino acids or more. An ORF is not necessarily equivalent to a gene or locus unless a phenotype can be associated with a mutation in the ORF and/or a mRNA or generally a gene product produced by the ORF's DNA detected. In other words, one would like to know what happens to the organism if the ORF is disrupted by mutation either natural or engineered. There are different approaches to gene identification.

1. Gene identification by complementation
2. Screening of gene library with DNA probe
3. Gene identification through gene expression, positional cloning
4. Designing gene specific probe.
5. Detection of introns—It is done by comparing of genomic and cDNA restriction map.
6. Defining the end point of gene
7. Defining the starting point of the gene point
8. Use of genes from other species or group boundaries

Approaches to Gene Recognition

Task of identifying coding sequences is significantly difficult when the region of interest is 20-40kbs and become almost intractable when the region is several hundred kilobases long. There are three main approaches to gene recognition in cloned genomic DNA.

1. Exon trapping
2. cDNA selection
3. Identification of CpG islands.
4. Cross-species hybridization

Identification of CpG Islands

A CpG island is a relatively short stretch of a G+C-rich regions(up to 2kb) in which the frequency of nonmethylated CpG dinucleotides is substantially higher than elsewhere in genomic DNA. In vertebrates most constitutively expressed genes(house keeping genes) and some regulated coding sequences (genes expressed in specific tissues, tissue-specific genes) are marked at their 5′ ends by distinctive regions containing a high density of hypomethylated CpG residues (CpG island). Since CpG islands are closely associated with genes, their identification is a useful step toward isolation of genes. These 'CpG' islands are identified by clustering of regulation sites for rare cutting restriction enzymes and are confirmed by testing with methylated-sensitive and –insensitive isoschizomers (Bird, 1978). In other words, the criteria for inferring localization of CpG islands are the presence of high G+C density, correspondence with a region of rare cutter sites such as BssHII (GCGCGC), Eag(CGGCCG), and Sac II(CCGCGG), presence of Bst U1 sites, proximity to transcribed sequences and cross-species conservation. A method depending on the binding to a specific nuclear protein(Cross et al., 1994) has been reported for selective isolation of CpG islands from genomic DNA.

Another method is called segregation of partly melted molecules (**SPM**)) (Shiraishi et al., 1995, 1999) which allows preferential isolation of putative CpG islands. SPM is expected to provide a means for convenient and efficient isolation of genes from unsequenced DNA. It is a procedure for preferential isolation of DNA fragments with G+C-rich portions. Such fragments occur in known genes within or adjacent to CpG island. Isolation of these fragments permits detection and probing of many genes within much larger segments of DNA such as cosmids or YACs which have not been sequenced. SPM takes advantage of the reduced rate of strand dissociation in a denaturing gradient gel of DNA fragments derived from CpG islands. In this method the cloned DNA fragments digested with four restriction endonucleases are subjected to DGGE. Long G+C-rich sections in fragments inhibit strand dissociation after the fragments reach retardation level in the gradient. Such fragments are retained in the gel after most others disappear. This system depends on both extensive fragmentation at restriction sites that are infrequent in CpG islands and expected relative durability during DGGE of surviving fragments. Other fragments are expected to be dissipated from the sharp, retarded bands characteristic of DGGE through more or less rapid dissociation of their strands.

SPM is thus a variant of DGGE technique that allows separation and enrichment of DNA fragments with both G-C rich domains and a non G-C rich domain (as in the case with regions from the periphery of G-C rich islands). The strand association of such fragments is low in a denaturing gradient gel and such molecules are retained on the gel as a partly melted structure (helical at the G-C rich domain, dissociated at the non G-C rich domain) after long exposure to an electric field.

Cross-species hybridization—Another method for identifying coding sequences is interspecies cross-hybridization ('zooblots, Monaco et al., 1986). In other words, use of '**zooblots**' is made to detect cross-species conservation of genomic sequences and hybridization of radiolabelled cDNAs to assay genomic clones. The limitation with this method is that many but not all unique and low copy sequences conserved between species represent genes and can be detected by Southern hybridization techniques. It is used for screening of short genomic DNA segments for sequences that are evolutionary conserved.

The methods of CpG island and zooblots do not provide a direct means of purifying coding sequences from genomic DNA.

Direct screening of cDNA libraries — Large segments of DNA from complex genomes are becoming available as cosmids or phage contigs. The fragments are enriched by affinity capture or chromosomal segments are isolated in a foreign background in somatic cell hybrids. This method involves direct screening of cDNA libraries or Nothern blots with whole phage or cosmid clones.

Method of cDNA selection – It is a method of detection of coding regions within a large genomic DNA. In this method the YAC DNA is directly used to screen a cDNA library (Lovette et al., 1991). This method of cDNA selection is rapid and insensitive to the size of the DNA fragments to be screened. It is based on hybridization of cDNA fragments to immobilized DNA and recovery of the selected DNA by PCR.

PCR-based methods — PCR methods are targeted at the selective amplification of cDNAs that contain human sequences from somatic cell hybrids.

Other strategies — Other methods for gene recognition include DNA sequencing and genetic screens for ORFs (Grey et al., 1982), enhancers (Weber et al., 1984) or promoters (Allen et al., 1988; Gossler et al., 1989).

The value of methods of CpG island, zooblots, direct screening and other strategies appears limited for identifying genes in long segments of cloned genomic DNA.

16.28   DETECTION *IN SITU* OF GENOMIC REGULATORY ELEMENTS

Functional non-coding DNA sequences such as regulatory elements are difficult to identify as they are usually short, often degenerate and can reside on either strand of DNA at variable distances from the genes they control. As functional sequences tend to be conserved through evolution they appear as 'phylogenetic footprints; in alignment of genome sequences of different species. In the study of gene regulation it is frequently useful to join the promoter and the controlling elements of one gene to the structural part of another well-characterized gene whose product is easy to assay. This type of gene fusion is especially useful for studying the regulation of genes (or operons) whose products are difficult to assay or whose products are not even known. Thus there was a need to develop rapid and simple technique for joining the lactose (lac) structural genes to different promoters in *E.coli*. The lac genes are especially convenient for this because of the availability of well developed biochemical and genetic methods. These methods include sensitive enzyme assays and convenient mutant selection procedures. Previous methods for fusing the lac structural genes to other promoters have required several genetic steps. These steps have included translocation of the lac genes to be near the desired promoter and subsequent isolation of the precise fusion desired. Casadaben and Cohen (1979) described an *in vivo* procedure for translocating and fusing the lac genes to a random promoter and they used the integrative properties of the bacteriophage Mu genome.

The random generation of operon fusion in prokaryotes involves integrating a promoter less reporter gene whose expression can be easily detected and assayed at many different positions in a target genome so that it comes under the control of a random selection of chromosomal promoters. This approach enables the isolation and characterization of genes, simply by knowing or postulating their pattern of expression and it is not necessary initially to screen for mutant phenotypes. In eukaryotes, however, for efficient expression of the reporter gene, not only would an active promoter be required but the reporter gene would normally have to be the first gene in the fusion transcript. Position effects, i.e., dependence of expression levels on the genomic position of an integrate gene, have frequently been reported in eukaryotes. This sensitivity of promoter activity to adjacent genomic sequences suggested that a suitable detection system would allow to detect genomic elements that could regulate transcription at a distance. O' Kane and Gehring (1987) developed an approach for the *in situ* detection of genomic elements that regulate transcription in *Drosophila melanogaster*. This approach is similar to a powerful method of bacterial genetics, the random generation of operon fusions. They used the expression of the lac Z gene of *E.coli* from the P-element promoter in germ-line transformant flies to screen for chromosomal elements that can act at a distance to stimulate expression from this apparently weak promoter. The P-lacZ fusion gene is an efficient tool for the recovery of elements that may regulate gene expression in Drosophila and for the generation of a wide variety of cell type-specific markers.

Analysis of eukaryotic promoter regions—The classical approach to study the function of promoter is to change its components one at a time by mutation and see its effect. Now a days, mutations are generated mostly by *in vitro* mutagenesis and introduced artificially into cells or preferably whole organisms although some useful mutations have been obtained by *in vivo* mutagenesis. There are different experimental approaches to study the structure and function of promoter. The first approach is related to the study of regulation of acetamidase (amdS) gene in *Aspergillus nidulans* (Hayes and Davis, 1986). The amdS encodes acetamidase which hydrolyses actamide to acetic acid and ammonia. Transcription of amdS gene is regulated by at least four different *trans*-acting genes, areA, amdR, facB and amdA and thus is an example of multiple positive regulation. areA acts positively to activate transcription of amdS, amdR supplied a transcriptional activator, facB is essential for induction of transcription of a set of genes including amdS and amdA seems to encode another transcriptional activator and acts both on amdS and at least one another gene. Specific sites have been identified within the amdS promoter regions that affect the response of facB, amdR and amdA. All these genes have a specific binding site for areA protein. A deletion of 43bp about 160bp upstream of the amdS transcription start site eliminates the stimulation of amdS transcription by ω-amino acid. This mutation eliminates a binding site for the amdR protein was confirmed by a 'titration experiment in which multiple copies of a DNA fragment including the suspected sequences were integrated into the genome via transformation. These additional copies of the binding sequence titrated most of the amdR product and thereby leaving insufficient to support transcriptional activation of the genes involved in ω-amino acid utilization. A single base change about 200bp upstream of amdS TSS enhances the effect of acetate on amdS transcription and this effect is eliminated by certain mutations in facB. Titration experiment again showed that the -200 mutation was within a binding region for the facB protein. And a mutation of 17bp duplication about 100bp upstream of amdS TSS enhances the transcription. Thus transcriptional controls are concentrated within just a few hundred bases of the upstream sequence. This type of promoter functioning is becoming the rule for a wide range of eukaryotic genes.

Deletion analysis—Deletion analysis can be used for functional characterization of *cis* sequences. In this analysis fusion constructs of the promoter region of the gene of interest with the coding sequence of a reporter gene are made. Although a number of reporter genes such as UidA or GUS, LUC, CAT and GFP are available the GUS is a very common reporter gene used in plant work. Fusion constructs are made in a series with lesser and lesser parts of the original promoter. The series is referred to as a deletion series and the assay is called deletion analysis. These constructs can then be introduced into a suitable tissue or plant either transiently using different methods of transformation (for example, bombardment with particle gun, PEG-mediated transfer, electroporation) or stably using Agrobacterium-mediated transformation. For most routine work transient expression in protoplasts or single cells is used and the expression of the reporter gene is monitored after +/- hormone treatment. The deletion constructs identify the minimal length of the 5′upstream sequence that is necessary for the hormone-induced (or any other environmental signal such as light or temperature) expression of the gene. Figure 16.6 shows deletion mapping peocedure using GUS.

Modification of fusion constructs—Fusion constructs can be modified to determine which *cis* sequences are important for induction by a hormone. Usually a minimal promoter, a short sequence containing the TATA box and the initiation site, from an unrelated organism (a heterologous promoter) is substituted and fused at its 5′ end to the promoter fragment (s) of interest and at its 3′ end to the reporter gene. One of the commonly used minimal promoters is CaMV35S which is expressed constitutively without any need for any TF (s) other than those in the basal initiation complex. Other sources of constitutive promoters are the genes (e.g., ACTIN or UBIQUITIN genes) that are nearly universally expressed in eukaryotic cells. Besides the above mentioned information the following types of effects on a *cis* sequence can be obtained.

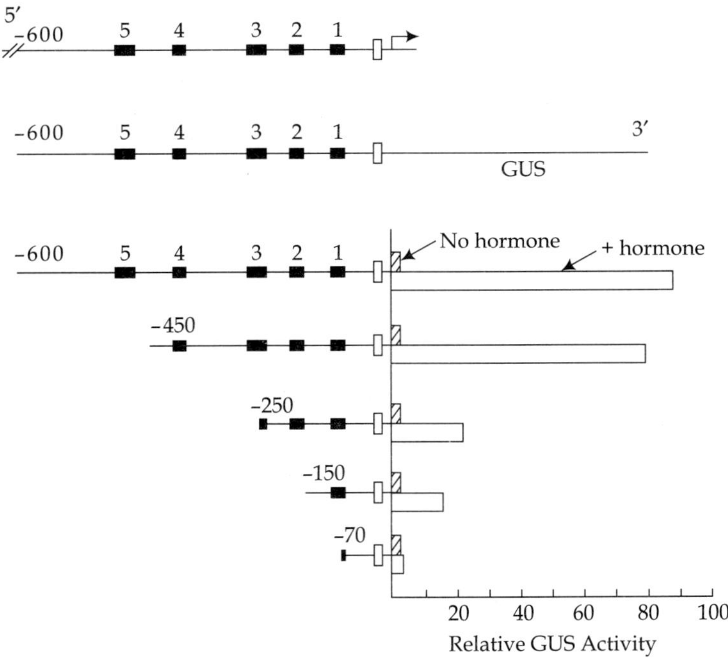

FIGURE 16.6   Deletion mapping of a promoter.

1. Effects of orientation and number of copies of a sequence- Constructs with reversed orientation or multiple copies of a *cis* sequence thought to be important can be fused to minimal promoter and to a reporter gene and the above effects can be studied.

2. Effect of unrelated sequences–Two unrelated *cis* sequences may be inserted along with the promoter in order to see whether the two sequences are complementary to each other and act in concert or are antagonistic to each other.

3. Effect of mutagenesis–A sequence can be mutagenized in a site specific manner and inserted along with the promoter and reporter gene construct to see if a sequence is important. If mutation causes deletion or loss in expression of the reporter gene then it means that the sequence is important.

Linker scan analysis—In this analysis a non coding sequence (an intron) is used to link the minimal promoter to the reporter gene and mutations are introduced in the promoter segment of interest every x number of bases. For example, in the Figure 16.7 mutations were introduced at intervals of 10bases in the promoter fragment and their effects on GUS expression were monitored. Mutations in segments 1 and 4 severely reduced the regulatory gene expression and thus these sequences are important. Linker scanning is a technique to estimate the optimal spacing between two adjacent regulatory sequences(boxes) of a promoter using synthetic homopolymeric linkers of variable lengths such as Oligo (dA), oligo (dT), oligo (dC) or oligo (dG). They are cloned between the two boxes and the effect of linker length on , for example, the expression of the linked gene (s) is determined. Linker scanning allows the determination of the distance between regulatory boxes which is optimal for transcription and the mapping of additional sites within the spacer that function in the binding of transacting proteins. In this way the sequence between the -35box and -10box in *E.coli* promoters has been found to be 16 to 17 base pairs in length and any variation, for example, 14, 15, 18 or 19 leads to a decrease in promoter efficiency. Linker scanning was developed by McKnight and Kingsbury (1982) to study the transcriptional control signals of the thymidine kinase gene of Herpes simplex virus. The study of regulation of gene expression involves analysis of nested sets of genetic deletions that enter the region of interest from both upstream and downstream directions. When the desired region is too large to be analyzed by substituting individual bases, there it would be better to use 'Linker scanning' mutagenesis to search within the regulatory region and locate sequences that are particularly important. In other words, instead of mutating separately each of the 50-100 nucleotides comprising the control sequences of interest, linker scanning method was developed to introduce clustered sets of point mutations at desired locations and test the retention of the transcriptional competence. This method of *in vitro* mutagenesis requires two opposing libraries of deletion mutants that terminate deletion with the same synthetic restriction endonuclease recognition sequence in the same general region of DNA. A conventional mutagenesis method consisting of sequential treatment of linearized DNA with exonuclease III and S1 nuclease was used to construct sets of 5′ and 3′ deletion mutants of the gene. The end points of different 5′ and 3′ deletion mutants terminating within a 140-nucleotide segment surrounding the 5′ terminus of the structural gene tk were identified by DNA sequencing. The end products of all 5′ and 3′ deletion mutants terminated with a synthetic BamHI restriction site. Through recombining matching 5′ and 3′ deletion mutants, clustered point mutations were introduced into the tk gene. Here matching refers to two opposing deletion mutants whose deletion termini are separated by 10 nucleotides. When two such

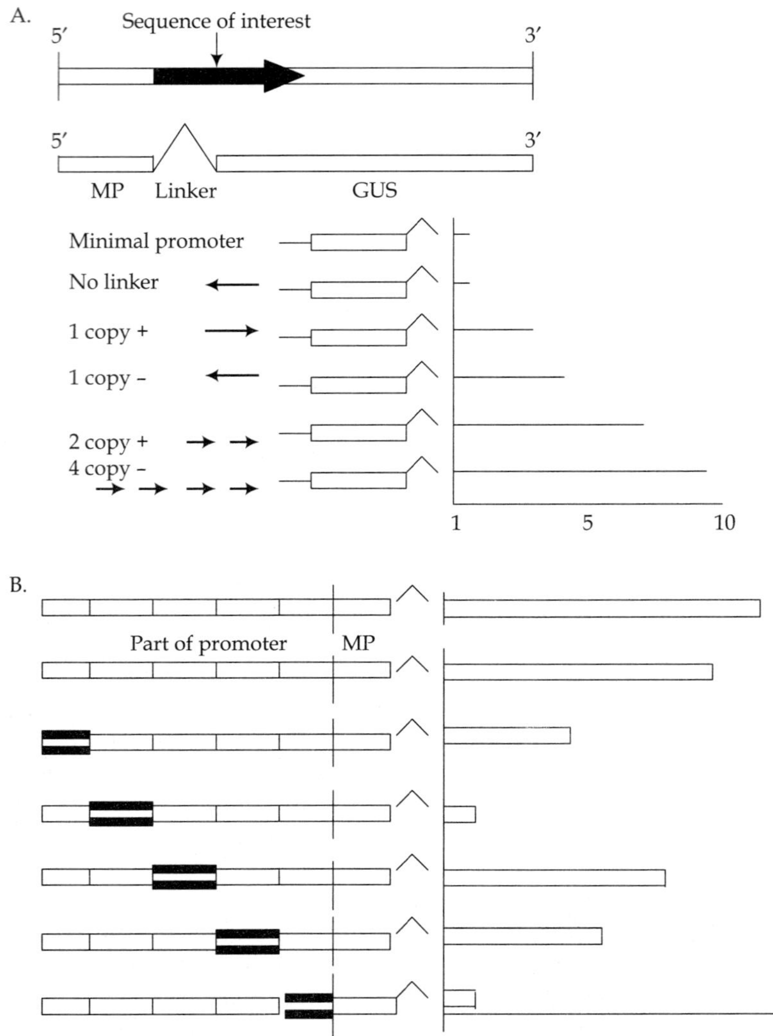

FIGURE 16.7 **A shows the effect of orientation, number of copies and mutagenesis in a promoter on the expression of a reporter gene. The minimal promoter (MP) is linked to the 4 coding sequence of β-glucuronidase gene for transient expression in an appropriate vector. Figure B shows linker scan analysis of a part of the promoter. Mutations were introduced at intervals of 10 bases in the promoter fragment and their effects were monitored by GUS expression.**

mutants are recombined at the synthetic BamHI restriction site, the ten-nucleotide residues of the linker replace the ten nucleotides that formerly separated the two deletion termini. Recombination of matching 5′ and 3′ deletion mutants does not result either a net increase or decrease in the number of nucleotide residues in the DNA sequence. Thus this method allows small clusters of nucleotide residues to be substituted in a site-specific manner without causing the addition or deletion of other sequences.

Subsequently, refinements of this technique have been made. The most commonly used method requires a unique restriction site adjacent to the region being mutagenized and it uses complementary oligonucleotides. The sequence of interest is initially cloned into a plasmid vector. The plasmid is linearized and a nested series of 5′ and 3′ deletion mutations is created using restriction enzymes. The fragments generated are ligated together with complementary oligonucleotides and filling in the gap between the sequence desired sequence and the nearby restriction site. Thus a series of mutants is created to scan the site of interest. Another method employs site-directed mutagenesis techniques to introduce smaller clusters of point mutations across the region of interest (Green, 1996).

In case snapdragon *Antirrhinum majus*, the regulation of purple flower color is by different segments of the promoter of the Pal gene (Almeida et al., 1989). The Del$^+$ gene is required for the transcription of Pal$^+$ in the flower tube. The Pal promoter region within 170 bases of the transcription start site was deleted to different extents by imprecise excision of the transposable element, Tam3 from its site of insertion in the pallida-recurrens allele. Deletion of the 10-bp region between -80 and -70bp gave a reduced level of lobe pigmentation which is further reduced in the absence of Del$^+$. A 20bp deletion between -9- and -70bp reduced lobe pigmentation to a very low level but completely dependent on Del$^+$. The longer deletion extending all the way back to -170bp eliminates all purple pigment, irrespective of presence or absence of Del$^+$. These results suggest existence of separate elements in the promoter region determining pigmentation in different parts of flower and the Del$^+$ product seems to be a DNA binding protein, binding upstream of -90bp to activate Pal transcription in the tube and to some extent in the lobes as well.

The function of the receptor binding sequence can be explored by testing the ability of a series of DNA constructs with different base pair replacements to promote transcription when introduced into cells. But in this type of studies the gene whose expression is monitored is not the one which is naturally attached to the promoter region under test but a rather reporter gene whose product can be measured easily and quickly and a favorite reporter is the bacterial gene encoding chloramphenicol transacylase (CAT). The promoter region of the ovalbumin gene producing ovalbumin, the egg-white protein in chiken, was studied using this approach. A sequence 40-50bases upstream of the TSS and just 16bases upstream of the TATA 'box' has been identified as the binding site for the oestrogen receptor, a protein which complexes with hormone and then activates transcription of a set of genes expressed in the oviduct. The entire coding sequence of the CAT gene was separated from its own promoter and fused in proper orientation to the ovalbumin gene (Tora et al., 1988). The variously modified sequences were introduced into cultured chicken fibroblast cells and the amounts of CAT produced were measured. This system is called **transient expression** as it does not rely on integration and replication of the introduced DNA but just on the transient transcription of those constructs in the cell nucleus.

16.29 ANALYSIS OF ENHANCER

The Adh gene encoding alcohol dehydrogenase in Drosophila is transcribed from two different start points depending on the stage of development. This two modes of transcription was shown to be dependent on much farther upstream sequences (enhancer). This was confirmed by inserting variously deleted derivatives of Adh gene bracketed by P-element (TE) termini into Drosophila genome. The transformation vector based on P-element contained plasmid sequence for replication in *E.coli*, the Drosophila rosy[+](eye color gene) flanked by P-element termini with characteristic inverted repeats and a polylinker site for inserting Adh gene (Corbin and Maniatis, 1990). The resulting plasmid was injected into homozygous rosy Drosophila plasmid containing a complete P-element to supply the transposase required for the excision and chromosomal integration of ry^+/Adh construct through cutting and splicing of its P-termini. Transformed flies were identified by their wild type (ry^+) eye color. The Adh construct was transferred into a genotype lacking active Adh gene through crossing. The mRNA from various larval tissues was analysed by 'Northern blotting(gel electrophoresis followed by specific probing for Adh transcripts). The results showed that two segments 660-2395 and 2830-5000bp upstream of the transcription start site enhanced larval transcripts and deletion of either upstream or downstream segments of this 4.34kb sequence resulted in reduction of an approximately eight fold reduction of larval Adh mRNA and thus these segments- one at least 600bp and other more than 3.5kb upstream of the transcription start site were identified as enhancers.

Local control region (LCR) — One enhancer can govern the activity of two or more genes. LCR is an example of enhancer and lies upstream of the human β-globin gene cluster. The β-globin gene cluster consists of genes encoding ε, γ G, γA, δ and β peptides in that order, upstream to downstream. The LCR rgion lies between 6 to 21kb upstream of the ε-globin gene and it contains several sites as shown by arrows that are hypersensitive to digestion by DNaseI. As DNaseI hypersensitivity in chromatin is an indicator of transcriptional activity, LCR can be said to be involved in controlling the activity of the β-globin gene cluster. The evidence for its role came from the analysis of γδβ - thalassaemia- a form of rare kind of anemia in which a long deletion including LCR but not the genes themselves or their promoters resulted in the under activity of whole β-globin gene cluster. The second piece of evidence came from the finding that the injection of DNA constructs including LCR greatly enhanced the expression of β-globin gene cluster in comparison to the trimmed-down version of the cluster. Further experiments showed LCR to be influencing the gene more closer to it. Such inference was made when the DNA construct with genes in the order LCR- γ- β was found to show strong expression of γ gene and almost zero expression for β gene. When the gene order was reversed, i.e. in case of DNA construct LCR- β- γ- β, β was expressed at a relatively higher level in comparison to γ-gene. These findings led to the formulation of two plausible hypotheses of LCR action. First, the LCR enhances the transcription of linked genes but then the second hypothesis could be that the distant DNA elements (genes) can be brought close to enhancer by looping out of the intervening DNA.

16.30   TRANSVECTION

Although it is generally assumed that the control regions act only in *cis* but it has been questioned by the observation of the phenomenon of transvection. The inactivation of a specific gene by the physical proximity of its homologous alleles is referred to as transvection. For example, in Drosophila, the aberrant phenotypes of certain bithorax gene mutants can be changed by chromosomal rearrangements involving breaks leading to the disruption of the alignment of the corresponding homologous chromosomes. The bithorax locus in Drosophila is involved in the regulation of the development of segmentation pattern and encompasses approximately 320kb. The term transvection was introduced by E.B.Lewis in 1954 to describe cases in which gene activity is influenced by homologous pairing. Homologs are intimately synapsed in somatic cells and involve action of enhancers in *trans*. The bithorax complex (BX-C) contains three genes, Ubx, abd-A and Abd-B. and the best studied of BX-C genes with respect to transvection is Ubx. E.B.Lewis observed that the *trans* heterozygote bx +/+ Ubx was much more normal than the bx +/bx + homozygote. He discovered that this enhancing effect in *trans* was dependent on the ability of the homologous loci to undergo somatic pairing as seen in polytene chromosome. Chromosomal rearrangements involving breaks close to the Ubx locus making close pairing of the Ubx alleles difficult, made the mutant phenotype of bx +/+ Ubx heterozygote much more extreme. Such transvection effects have also been observed in several other genes such as zesta, white, decapentaplegic, glue protein gene Sgs4, notch and cubitus interruptus. This phenomenon can be explained as due to an enhancer on one chromosome acting in *trans* to promote the transcription of a closely linked gene in the homologue. Sequence analysis has shown that the Ubx' mutation falls at the upstream end of the first exon and appears to nullify the gene function altogether. The bx mutations fall not within the Ubx coding region but rather within one of its long introns and they are believed to disrupt a region- a segment specific enhancer that is necessary for the expression of Ubx^+.

Two models have been proposed to explain transvection (Duncan, 2002). 1. Pairing-dependent interallelic complementation could be due to the production of joint RNAs either by *trans*-splicing or by template switching by RNA polymerase during transcription. Although results from experiments involving Ubx locus to test this model have evidence against this model but *trans*-splicing has been demonstrated at another locus (mdg4) in Drosophila. 2. Transvection results from the action in *trans* of what are usually considered to be *cis*-regulatory regions. This second model is supported by the finding that enhancers can activate promoters carried by different DNA molecules provided that they are brought together into close proximity either by an artificial protein bridge (Muller and Schaffner, 1990) or by catenation of plasmids (Rothberg et al., 1991). For genes showing transvection it has been suggested that regulatory regions could act in *trans* by recruiting a locally high concentration of regulatory RNAs from one homolog to other or by direct action of enhancer sequence in *trans*. When chromosome rearrangements were introduced to disrupt pairing this transvection was reduced dramatically. Consistent with analysis the Ubx protein expression is found in the wing discs of Cbx' Ubx/++ animals when pairing was allowed but not when pairing was disrupted. Thus transvection of Cbx and most or all other cases of transvection in the Ubx domain results from the action of regulatory regions in *trans*.

A striking feature of transvection is that different genes can have different pairing requirements. Transvection within Ubx is relatively easy to disrupt by chromosome rearrangement whereas iab-5,6,7 transvection is very difficult to disrupt. Thus different genes differ in their response to pairing disruption. Differences may simple reflect the length of the cell cycle in the tissue in which the gene is active- very long cell cycle times will facilitate pairing after each division even in the presence of rearrangement. Further, genes that interact by *trans*-splicing or by the diffusion of regulatory RNAs between homologs might appear resistant to pairing disruption as interactions could still occur when the genes are at some distance from one another. Another possibility is that the difference in response to rearrangement heterozygosity is an indirect consequence of the mechanisms underlying pairing. Pairing must be established within a restricted period of time. As pairing is established only early in salivary gland development it can not explain differing response of genes to pairing disruption. For genes requiring intimate synapsis for transvection, transvection will be completely blocked if a region that remains asynapsed after the pairing-competent stage. However, for genes that contain sites that can interact over long distances, transvection will still be possible. For example, sites within homologous genes could be brought together in the absence of pairing by the action of DNA-binding protein that interact with one another with high affinity.

Besides examples of transvection in which enhancers acting positively in *trans* there are also cases which involve repressive interactions in *trans*. Pairing disruption in diploid cells can be monitored by FISH or inferred from the effects of rearrangements on mitotic recombination.

Study of active chromosomal domain

The structures and functions of promoters which control specific patterns of gene expression can be investigated through the use of reporter constructs. Putative promoter sequences are joined to an ORF encoding an easily quantifiable enzyme such as LacZ or CAT and introduced into the organism.

Embryonic development entails a well defined temporal and spatial program of gene expression which may be influenced by active chromosomal domains. These chromosomal domains can be detected using transgenes which integrate randomly throughout the genome as their expression can be affected by chromosomal position. Position effects are probably exerted most strongly on transgenes that do not contain strong promoters, enhancers or other modulating sequences. Allen et al. (1988) systematically explored position effects using a transgene with the weak herpes-simplex-virus thymidine kinase promoter, linked to the readily visualized lacZ indicator gene (HSV-TK-lacZ). Each transgenic fetus with detectable expression construct displayed a unique lacZ staining pattern. Thus expression of this construct is apparently dictated entirely by its chromosomal position without any construct specificity. Furthermore, the transgene is faithfully transmitted to subsequent generations allowing for systematic mapping of the changes in expression during development and in adult life. These results demonstrate that transgenes can be indeed powerful tools to probe the genome for active chromosomal regions with the potential for identifying endogenous genes.

Integration of the lacZ gene linked to a weak promoter was shown to detect *cis*-acting elements in mouse genome that activated β-galactosidase expression in the developing spinal cord and caused neurological mutation (Kothary et al., 1988). Use of transgenic lines in animals to detect developmentally regulated genes, however, is limited by the number of integration events which can be readily analyzed. So, embryonic stem (ES) cells and two different types of the lacZ constructs were used to screen many integration events and to rapidly clone the associated genes. ES cells fully retain their pluripotent character after a variety of genetic manipulations in vitro and efficiently form chiaemeva after reinjection into blastocysts.

Introduction of reporter gene constructs into ES cells — Two types of reporter constructs were used to screen integration events and to rapidly clone the associated genes (Gossler et al., 1989).

1. **Enhancer trap construct** — The lacZ gene is fused in frame to a minimal promoter derived from mouse heat-shock protein (hsp68) gene that provides a TAT box and translation initiation codon but not sufficient on its own for expression of β-

galactosidase in the mouse ES cells. The construct also contained bacterial SUIII suppressor gene and bacterial neomycin resistance (*neo*) gene. In this construct the expression of the lacZ gene should depend on *cis*-acting regulatory elements close to the site of integration that activate the weak hsp68 promoter.

2. **Gene trap vector**—The vector contains the lacZ gene, lacking a promoter and translation initiation signal, inserted in frame into homeobox exon of the En-2 gene such that a splice acceptor is placed at the 5′ end of the lacZ gene. Integration of the construct into introns of the genes in the correct orientation should create a spliced lacZ fusion transcript and a functional fusion protein when the reading frame is maintained.

The enhancer trap and gene trap approaches are based on the assumption that the expression pattern of the reporter gene reflects the expression of endogenous host gene. Whether this assumption is correct will only become apparent after cloning of the host gene at the site of integration. However, the use of ES cells instead of transgenic mice should simplify cloning of the host genes. For the enhancer trap the simple structure of the integration site as well as the inclusion of the SU III gene should facilitate rapid cloning of the host flanking sequences. For the gene trap construct it should be possible to clone the expression sequences of the endogenous gene present in the lacZ fusion transcript directly from the cDNA and to purify the fusion proteins from cell extracts by means of antibody directed β-galactosidase.

Expression patterns provide clues as to the function of developmentally regulated genes but final proof will come from analyzing mutant phenotypes. Since enhancer trap vector must integrate near *cis*-acting regulatory sequences some integration events will cause a mutation in endogenous gene. Insertion of the gene trap vector should in all cases produce a mutation in the host gene. Thus one can study dominant and recessive mutations obtained from such integration events. Further, dominant mutation can be studied directly in chimaeric embryos and recessive can be studied after germ line transmission. Thus one can identify genes and mutations involved.

16.31   IDENTIFICATION OF RECEPTOR-BINDING SEQUENCE IN THE DNA

The identity of the receptor-binding sequence in the DNA was established by a technique called 'Footprinting'. In this assay the DNA fragment is generated through the use of restriction endonuclease. The restriction fragment containing the gene sequence of interest is radioactively labeled at one 5′ end with 32P. The protein is bound to this DNA fragment and the DNA-protein complex is partially digested with DNase1 which cleaves at random between any two successive nucleotide residues except where the DNA is bound to the protein and is thus protected from digestion. The DNA fragments thus generated are subjected to electrophoresis (denaturing polyacrylamide gel) where the DNA melts to single strands and the result is a ladder (series) of end labeled single stranded fragments extending from the labeled end to every position in the chain except within the DNA binding region. In other words, no end labelled single stranded fragment terminating in the DNA-protein bound region will be produced or the ladder is with a region of missing rungs. But when no protein is bound to the DNA (unprotected DNA) then in that condition a ladder of end labeled single stranded fragments emcompassing all possible sizes is produced on the autoradiography. Thus in comparison to unprotected DNA in case of protected DNA there appears a gap or hole on the electrophoretic gel, a '**foot print**'. The sizes of the missing digestion products which correspond to the distances from the labeled 5′ end of the original restriction fragment to the protected region are determined by reference to a ladder of fragments of known sizes and thus the part of the DNA sequence covered by the protein is defined (Figure 16.8).

Exonuclease III footprinting—It is a method for detection of specific contacts between one or more proteins and a duplex DNA molecule using exoIII. This enzyme catalyses the exonucleolytic removal of the 5′- phosphomononucleotide from each 3′-terminus of the duplex but stopped at sequences where a protein is bound.

16.32   IDENTIFICATION AND CHARACTERIZATION OF TRANSCRIPTS

Identification of the 5′ and 3′ boundaries of transcripts with extensive variation in transcripts arising from alternative promoter usage, splicing and polyadenylation (see chapter 5) is required. Fantom consortium and RIKEN genome exploitation research group and genome science group found 1.32 5′ start site for each 3′ end and 1.83 3′ ends for each 5′ end. They also found number of transcripts to be at least one order of magnitude larger than the estimated 22, 000 genes in the mouse genome. Production of mRNA from genomic DNA is directed by sequences which determine the start and stop of transcripts and splicing into mature mRNAs. Thus the transcriptional landscape includes pattern of transcriptional control signal plus transcripts which can be studied through the use of combined full-length cDNA isolation and 5′ and 3′ end sequencing of cloned cDNA with new cap analysis gene expression (CAGE) and gene identification signature (GIS) and gene signature cloning (GSC) ditag technologies for the identification of RNA and mRNA sequences corresponding to transcription initiation and termination sites (Shiraki et al., 2003). Gene identification signature refers to sequences characteristic of a gene and includes distinct promoter boxes, start codons or ORFs. There are significant differences in the lengths of the 5′ and 3′ends. Analysis of the 3′ ends by EST distribution profiles indicate three alternative polyadenylation positions (Southan, 2001).

Generation of cDNA set—cDNA set comprising full length transcripts obtained by merging the Fantom-3 cDNA set with mouse cDNAs from GenBank and clustering the cDNAs into transcriptional units in which members share sequence transcribed from the same strand.

Defining the end points of transcription—The positions of intron splice sites and the points at which transcription of mRNA starts and terminates are usually determined by the technique known as **nuclease S1 mapping**. This technique is based on nucleic acid hybridization (DNA/ RNA hybrid between the mRNA encoded by the cloned DNA and the complementary single strand of the latter). In this technique restriction fragments of the gene are cloned using a single-stranded phage vector such as one of those derived from phage M13. Thus the fragment can be amplified in radioactive single stranded form, ie., an appropriate (32P) end lebelled restriction fragment of the cloned DNA can be obtained and used in hybridization. Either strand of the original duplex can be amplified depending upon the orientation of the insertion into the vector. The appropriate DNA single strand can be annealed with a large amount of mRNA including the processed transcript of the gene under investigation. The annealed mixture is then digested with S1 nuclease. The endonuclease S1 digests the nonhybridized, single stranded regions of the DNA but not the double stranded DNA or DNA-RNA hybrid molecules and thus leaves the regions of the DNA complementary to the mRNA intact. In other words, the

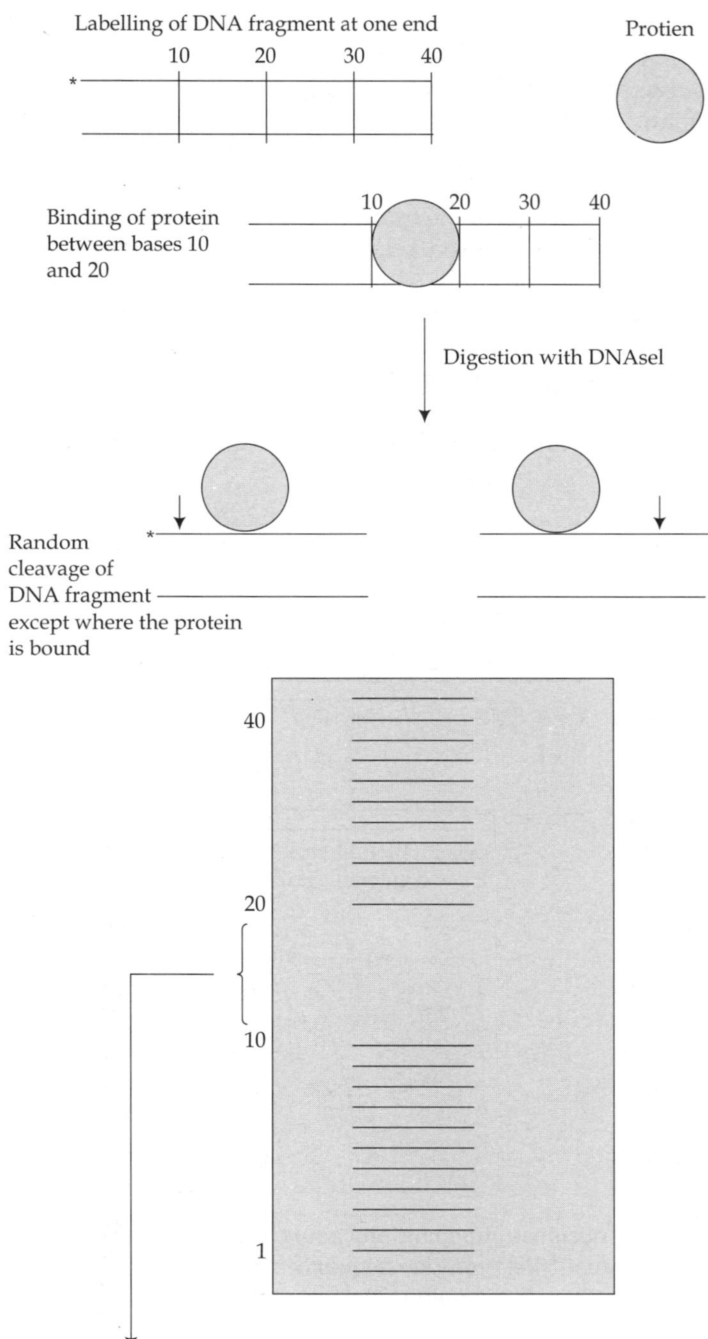

FIGURE 16.8 **Showing DNAseI footprinting assay for determining the region on DNA at which a protein binds.**

DNA sequence that has hybridized with mRNA will be protected and the rest will be digested. The protected DNA-RNA heteroduplexes are isolated. The protecting RNA is then enzymically removed through subsequent electrophoretic separation of the hybrid molecules in denaturing polyacrlamide gels and the length of the remaining single stranded DNA fragments are determined by comparison by with a 'ladder' sequences of known sizes (standard fragments) by gel electrophoresis. That length will be reduced from that of the original restriction fragment by the distance of the fragment end to the transcription start or stop point if either falls within the fragment. Thus the ends of the transcribed unit can be placed within the genomic restriction map (Figure 16.9). The length of the intron (s) can be determined after digestion of mRNA-DNA with **exonuclease VII** which removes specially single stranded termini but leaves the intron loop intact. The S1 mapping technique thus can be used to precisely determine the coding region of a gene and the number of its exons and introns, to map the transcriptional start (cap site) and transcriptional termination site of a gene and the direction of its transcription by forming mRNA-DNA hybrid and removing unpaired DNA with S1 nuclease.

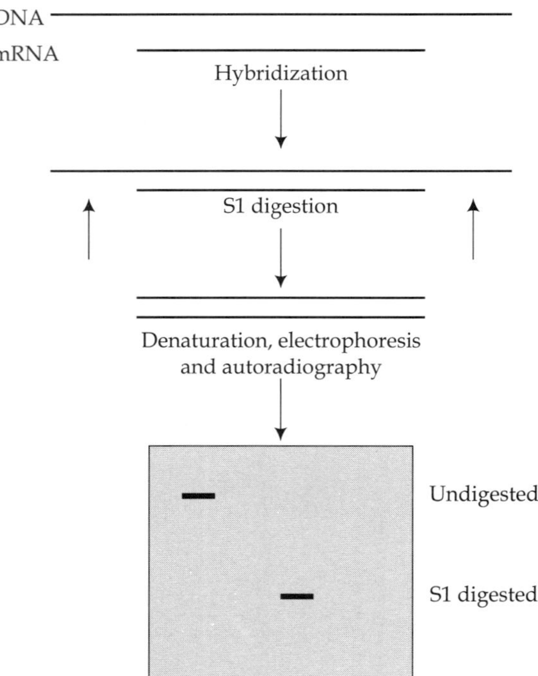

FIGURE 16.9 Showing S1 nuclease mapping procedure. This technique is used for mapping the 5′ end of mRNA (and thus identifying promoter sequences) and mapping of exons in spliced mRNA. Mapping with an end labelled probe will lead to determination of the polarity and the map position of the corresponding DNA sequences.

16.33 DETECTION OF INTRONS AND DETERMINATION OF POSITION OF INTRON

There are two ways of detecting introns. In the first method the introns are defined by comparisons between the sequences or restriction maps of chromosomal (genomic) DNA and cDNA(cloned cDNA) obtained by reverse transcription of the processed mRNA. In the second method the genomic DNA can be compared with cDNA and introns are detected by **heteroduplex formation by electron microscopy**. The cloned genomic DNA containing the gene and the corresponding cloned cDNA are made single stranded, mixed and annealed. The reconstituted DNA molecules, some of which will be hybrids are generally mixed with cytochrome c. The cytochrome c which is a basic protein sticks to the acidic DNA and covers it with a thick coat and thus the duplex structure can be seen by electron microscopy. The duplex structure in electron micrograph will look as relatively thick rigid filament and the thinner more flexible loops are single stranded introns not hybridizing to the cDNA.

Displacement loop mapping (R-loop mapping)—This technique can be used for precisely locating exons and introns in genomic DNA through hybridizing a genomic clone to its complementary DNA (cDNA) or mRNA. Introns can not base-pair with the cDNA and are displaced from the hybrid forming displacement loop, the length of which can be measured. The intron appears as unpaired single stranded loops (Dugaiczyk et al., 1979).

Heteroduplex (R-loop) mapping—It can be used for localization of regions of noncomplementarity in DNA or DNA-RNA heteroduplex with the help of electron microscopy. Such non-complementary regions give rise to single stranded loops that can be characterized by their contour length. Heteroduplex mapping can be used for the physical mapping of intron-regions in eukaryotic split genes (Figure 16.10).

Determination of position of intron–In determination of position of introns exonuclease VII in addition to the nuclease S1 is employed. Exonuclease VII will digest single stranded ends protruding from a mRNA-DNA hybrid molecule but unlike the S1 nuclease will not digest a single stranded intron looped out from such a hybrid molecule.

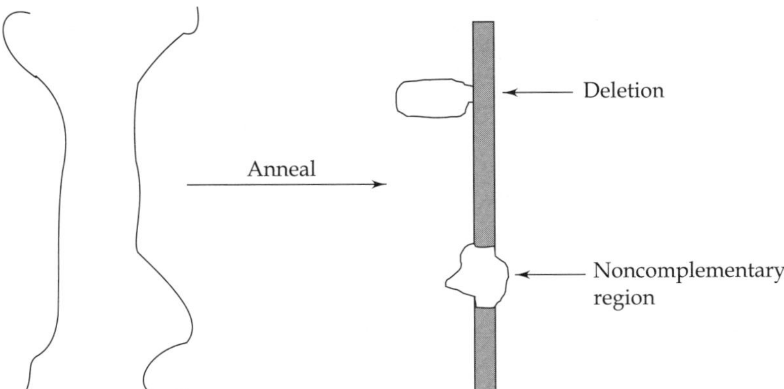

FIGURE 16.10 Loops in a heteroduplex indicate either a deletion or a region of noncomplementarity in the two strands.

Comparison of genomic and cDNA restriction map–Comparison of the structure of a gene with cDNA copy of its transcript is through their restriction sites map. This comparison will lead to demonstration of the presence on intron and will define the end points of transcription unit but without indicating which is the starting end and which is termination end.

Defining intron by sequence analysis–If one knows the amino acid sequence of the gene translation product protein then comparison of amino acid sequence with DNA nucleotide sequence will indicate whether or not the coding sequence in the DNA contains introns. Intronless genes will have unbroken ORFs from a start codon (ATG) to a termination codon (TAA, TAG or TGA) with codons in between corresponding precisely to the known amino acid sequence. ATG is a base triplet in DNA which marks the 5′-end of the coding region of a gene. Its corresponding codon in mRNA is AUG which functions as start codon at which polypeptide synthesis is started. Having known the common , though limited earmarks of introns it is often possible to identify short introns with fair certainty and so deduce the amino acid sequence of the product encoded after intron excision.

16.34 EXON TRAPPING

It isolates exon sequences from cloned genomic DNA by virtue of selection for functional 5′ and 3′ splice sites. Random segments of chromosomal DNA are inserted into an intron present within a mammalian expression vector and after transfection, cytoplasmic mRNA is screened by PCR amplification for the acquisition of an exon from the genomic fragment. The amplified exon is derived from the pairing of unrelated vector and genomic splice signals. Previous studies have shown that introns constructed with novel combinations of 5′ and 3′ splice sites from diverse genes are actively spliced (Chu and Sharp, 1981).

It is a method for the identification of **acceptor splice sites** found adjacent to exons. The exon trap vector carries a donor splice junction and a polylinker for the cloning of the DNA fragment to be tested for the presence of an acceptor splice site. Between the polylinker and donor splice junction the α-complementary fragment of *E.coli* β-galactosidase gene is inserted into an intron. When a functional acceptor splice sequence is cloned into the polylinker, β-galactosidase sequence will be absent from the spliced RNA transcript. To facilitate the detection of spliced RNAs the **exon trap vector** is based on a retrovirus genome. The exon trap vector is a shuttle vector capable of replicating in *E.coli* as a plasmid and in mammalian cells as a defective virus. The insert containing vector is transfected into a retroviral helper cell line (Psi2) in which clones with functional acceptor splice sites will produce progeny virus with spliced RNA genomes. After one round of retroviral growth in PA-317 cells the progeny virions are used to infect COS cells. Circular DNA genomes are then recovered from the COS cells and transformed into *E.coli*. Unspliced clones are then detected as blue colonies on chromogenic β-galactosidase substrate whereas the spliced clones will appear as white colonies as the β-galactosidase gene sequences are absent.

Limitations with exon trapping—Most genes with regulated splicing events have also introns that are constitutively removed and these would be identified by exon trapping. However, some exons will not be recovered with this method (Duyk et al, 1990; Buckler et al., 1991). For example, a small percentage of known genes do not contain introns and therefore will be missed by this screen. Some splicing events are temporally regulated or tissue specific and may not occur in the packaging cell lines used in exon trapping. Finally, all DNA sequences are propagated equally well in retrovirus vectors but it is possible that the library of recovered clones may not be fully representative of the exons in the starting cloned genomic DNA. Since most genes are composed of multiple exons and identification of a gene requires recovery of only a single exon, this consideration should not be a limiting factor.

Monitoring splicing events—Most of the genes are interrupted by noncoding regions can be variably spliced to yield diverse mRNA. Microarray or fiber-optic array probes can monitor splicing events. At least 74% of human multi-exon genes are alternatively spliced. Microarray containing ~125, 000 different 36 nucleotide junction probes can be used to monitor the exon-exon connections of 10, 000 multi-exon genes. Splicing predictions from the data are used to guide RT-PCR and sequencing validation effort to specific transcript locations.

Poly(A) tail signal selection—It refers to scanning of the mRNA for alternative poly(A) additional signals and the cleavage of one preferred site by an endonuclease recognizing the target sequence, 5'AAUAAA-3'. If several such signals are located on primary transcript molecule then poly (A) signal selection may result in multiple messenger RNAs derived from one transcript unit (if all sites are cut out) or the generation of one specific mRNA that may be different from cell to cell or tissue to tissue (if signal selection is cell or tissue specific).

Primer extension technique—This technique is used for mapping the 5' terminus of mRNA and to detect precursors and intermediates of processing of mRNA. In this technique the mRNA is hybridized to a synthetic 5' radiolabeled complementary oligodeoxynucleotide, 30-40 nucleotide in length which is then used as primer by retroviral reverse transcriptase. The enzyme completes the synthesis of the complementary strand (cDNA) at the 5' end terminus of the mRNA template. The length of the extended primer and consequently the 5' terminus of the transcript can be precisely determined by PAGE and autoradiography.

Differentially expressed coding sequence tag (dCST)—dCST refers to any sequence from the coding part of an mRNA (as opposed to 3'-noncoding part) which is expressed to different extents in different cells, tissues, organs or organisms and thus serves as molecular markers. dCSTs are generated by a technique called **preferential amplification of coding sequences, PACS**. This is a technique for detection of differentially expressed genes (or better cDNA) that specifically target at the coding regions of mRNA (rather than its 5' or 3'–noncoding parts). In this technique total RNA is isolated and the contaiminating DNA is removed by RNase-free DNaseI. The single stranded cDNA is generated by reverse transcriptase PCR with random primers. The ds CDA is synthesized with an ATG-containing forward primer and a double restriction site primer as a reverse primer in a PCR. The ATG-complementary primer specifically selects coding parts of mRNAs as ATG is the start codon of almost all organisms except some viral (human T-cell lymphotropic virus type I), chloroplast (mRNA encoded by infA gene), plant mitochondria (at p9-rp116 cotranscript) and bacterial mRNA encoding ribosomal proteins. Moreover ATG codons occur only rarely downstream of the proper initiation codon. Therefore, generation of multiple amplicons from a single mRNA by PACS is unlikely. ATG primer also contains a restriction site at its 5'-end for cloning of amplification product. The amplification products are then separated electrophoretically in 6% polyacrlamide/urea sequencing gels. The gels are then dried and autoradiographed. The resulting pattern represents a differential mRNA fingerprint of the cell, tissue, organ or organism.

Isolation of the 5′ termini of mRNAs- Ligation anchor—PCR is used for the isolation of the 5′ terminii of mRNAs. The mRNAs are first isolated and the first strands of cDNAs are synthesized with reverse transcriptase. Then T4 RNA ligase is employed to add an anchor oligonucleotide to the 3′ end of the first strand cDNA. Subsequently an anchor specific primer and a primer complementary to the known sequence within the cDNA are both used to amplify the 5′ end of the transcript via conventional PCR.

RACE-RNA ligase-rapid amplification of cDNAs ends—It is a technique used for the generation of cDNA libraries and for the analysis of the 5′ and 3′ ends of the generated cDNAs. RACE is a variant of the standard PCR technique that utilizes gene-specific oligonucleotide primers to amplify cDNAs, reverse transcribed from low abundance mRNA. In other words, RACE is used to obtain full length cDNA copies of low abundance mRNAs. Depending on whether the 3′end or 5′end of cDNA can be amplified, one can have 3′ RACE or 5′ RACE. The anchored PCR is used to synthesize cDNA corresponding to the 3′ and 5′ ends of the transcript (Frohman et al., 1988).

3′RACE starts with an oligo (dT) containing adaptor primer partly complementary to the poly (A) tail of mRNAs. This primer allows first strand synthesis with reverse transcriptase. After destruction of mRNA with RNase H, a gene specific primer complementary to a region at the 5′ end of the original mRNA and a universal adaptor-primer complementary to its 3′ end are used to amplify the cDNA with an intact 3′ end.

5′RACE—It a technique for amplifying cDNA sequences between a known internal position in a transcript and its 5′end. The different steps involved in this technique are i. the annealing of a gene–specific antisense primer complementary to the 3′ end of the mRNA ii. first strand synthesis with reverse transcriptase iii. degradation of the mRNA with RNase H iv. Purification of the cDNA v. homopolymer tailing of cDNA with dCTP vi. anchoring of an oligo (dG) sequence and the amplification of the cDNA using the anchored primer and a nested gene primer and PCR.

16.35   DETERMINATION OF THE LENGTH OF POLY(A) TAIL OF mRNA

Determination of the length of the poly (A) tail of specific mRNAs in a population of RNAs is done through a technique called **RNase H mapping**. In this technique RNA is first annealed to an oligonucelotide 20-25 nucleotides long and complementary to a region of about 300-400 nucleotides upstream of the 3′ end of the mRNA. The RNA-DNA hybrid is then cleaved with RNase H and the resulting fragments are electrophoretically separated on a denaturing agarose or polyacrylamide gel. After that they are transferred to membrane and hybridized to a radioactively labeled probe complementary to the 3′ fragment that contains the poly (A) tract. The mobility of the 3′ fragment provides a measure of the length of poly (A) tail.

Another technique for the determination of the length of poly (A) tail of specific mRNA in a population of mRNA is called **RNase T1 protection**. In this technique the RNA is annealed to a labeled RNA probe which spans the entire 3′ end of the mRNA. The RNA-DNA hybrids are cleaved with RNase T1 which specifically attacks non-hybridized RNA 3′ to guanine residues. The T1 resistant RNA including the poly (A) tract is elctrophoresed on a native gel that preserves the hybrid and detected by autoradiography.

16.36   GENE IDENTIFICATION BY COMPLEMENTATION

In microorganisms where it is easy to recover the whole organism from a single treated cell, virtually any functional deficiency resulting from mutation can be complemented (repaired) by treatment with wild type DNA provided there exists an efficient protocol for DNA uptake and a method of screening for cured cell. This method of gene identification has been used in *E.coli*, *S. cerevisiae* and *Neurospora crassa*. An obvious kind of defective mutant to use will be an auxotroph (one that has lost the ability to synthesize some essential metabolite but can grow if that compound is supplied in the medium). The other type of mutant can be temperature sensitive mutant growing at an either lower or higher temperature. So first of all a DNA library is screened for any gene for which one has a functionally defective gene. After that one takes a library of wild type DNA fragments cloned in the 2-μm plasmid. The DNA treated cells are then plated on a medium or at a temperature that permits growth of only transformed strains which have its functional defects cured. The transformants will harbour plasmids that carry the wild type allele of the gene that was defective in the original mutant. The plasmid is then separated from the total cell DNA on the basis of its distinctive size and the gene can be recovered.

16.37   GENE IDENTIFICATION BY GENE EXPRESSION

If the protein product of a gene is known or can be guessed at then this information can be used for identification of its gene. Gene product, the protein, is purified and is used to immunize mice. In this technique the antibody producing cells are fused to cancer cells to find hybrid cell clones (**hybridomas**), each of which produces a single **monoclonal antibody** (MAB). A suitably selected MAB is an extremely specific and powerful reagent for protein identification. With a cDNA library cloned in an expression vector (see chapter 10) it can be used to find the gene from which the protein was made. In this procedure colonies are transferred to the nitrocellulose membrane as in case of DNA probing of libraries but here one looks for the protein with the suitable lebelled immune reagent rather than DNA. Thus monoclonal antibodies are important tools for the characterization and identification of proteins. **Antibody** is a protein (immunoglobulin) usually found in serum whose presence can be shown by its specific reactivity with an antigen or hapten. The cell taking part in this immune responses are 1. lymphocytes and 2. various accessory cells (e.g. macrophages, cells of the reticuloendothelial system, various types of blood leucocytes).

Polyclonal antibody–See chapter 26.

16.38   GENE IDENTIFICATION USING GENE FROM
OTHER SPECIES OR GROUP BOUNDARIES

There is strong conservation of certain gene functions in evolution but even though amino acid sequences may be conserved over great evolutionary distances the degeneracy of the mRNA-amino acid codon permits mRNA and hence the gene sequences to vary more freely. So in this approach genes from eukaryotes (animals) can be identified through their complementation of yeast mutants. If gene structure and function is conserved then one can use a cloned gene or cDNA from one group of eukaryotes to identify the corresponding gene in another and one of the best examples is the identification of two Ras (rat sarcoma virus) analogues in Saccharomyces using cloned DNA from a human member of the Ras family of oncogenes.

16.39 GENE IDENTIFICATION THROUGH DISRUPTION

A direct way of obtaining information on the function of the gene identified by sequencing is to create a loss of function mutation and study the phenotype of the resulting mutant. One way of associating a cloned DNA sequence to a gene is to disrupt the corresponding genomic sequence and to see whether the engineered mutation is a member of the same gene through **complementation** and **recombination**. The disruption of the corresponding gene can be done through DNA-mediated transformation. The transforming DNA tends to integrate into the chromosomes **ectopically**, i.e. within **nonhomologous sequences** at many different loci. But occasionally integration occur by **homologous recombination** between the introduced DNA and the corresponding genomic sequence. In yeast, *Saccharomyces cerevisiae* there is no ecotypic integration and transformants have the transforming DNA integrated at a locus to which it shows homology. Homologous integration of transforming DNA can occur in two ways:

1. Through **meiotic crossingover** in which the crossingover between homologous sequences in plasmid (if the transforming DNA enters the nucleus as a closed circular plasmid) and the chromosome will result in the integration of the whole plasmid into the chromosome as a linear insert without any loss of chromosome (Tatchell et al., 1984). 2. Through meiotic gene conversion with **displacement** of a chromosome segment by a homologous sequence of transforming DNA. If a selectable marker is inserted into the cloned DNA sequence and thus has disrupted its ORF then integration by homology will result in the replacement of functional gene by the disrupted one (Mansour et al., 1988).

Repeat induced pont mutation (RIP) — This phenomenon of **repeat induced pont mutation** (RIP) can be applied to the fungus, *N. crassa*. This phenomenon provides a means of destroying the function of the genomic counterpart of any unknown segment of cloned DNA. Ectopic integration of cloned DNA is easily obtained in this species as homologous events rarely occur. Now if an ectopic integrated sequence duplicates a gene already present at its normal locus and the duplicated strain is crossed with any other strain then a high proportion of the resulting meiotic tetrads show extensive disruption of both of the duplicated sequences. If the duplicated sequences are closely linked then the frequency of these aberrant tetrads approaches 100% but only 30-50% if they are on different chromosomes. DNA sequencing has shown the disruption to be due to heavy methylation of cytosine residues as well as involvement of a large number of base pair substitutions and thus all transitions of G-C to A-T.

In many organisms (e.g. microbes and mice) homologous recombination can be used efficiently to target mutations into specific genes by replacing the wild–type with a mutated allele. In plants controlling homologous recombination has proven extremely difficult because of the prevalence of illegitimate recombination events (Puchta and Hohn, 1996). Further, although some success has been reported (Miao and Lam, 1995; Kempin et al., 1997), gene replacement by recombination is not considered feasible on a large scale and other strategies (targetted mutations through T-DNA or transposons) have to be employed to allow functional analysis of genes.

DNA sequencing—Two major methods for the identification of new genes which do not rely on cDNA cloning and which do not require prior knowledge of the site of expression of gene are: (i) direct sequencing of cloned genomic DNA and the interpretation of the raw sequence by 'neutral net' gene prediction program such as GRAIL (see Roy, 2009) and (ii) Exon amplification.

Structural analyses of genes—Gene Map' 99 gives the chromosomal positions of 45, 049 human ESTs and genes belonging to 24,106 UniGene clusters (Deloukas et al., 1998). By combining 'in silico' (through computer modeling) screen (BLAST search of GenBank against many databases) and direct sequencing of BAC clones specific to the chromosomal regions, the genomic structure of the genes is further compared. Structural analyses of these genes is further refined using rapid amplification of cDNA ends and PCR (RACE-PCR) and RT-PCR in conjugation with the genomic analysis.

16.40   ISOLATION OF CHROMOSOME ENDS

In this technique for isolation of telomeres and subtelomeric regions of eukaryotic chromosomes the high molecular weight DNA is treated with terminal deoxynucleotidyl transferase which adds adenosine nucleotides to its ends to form a poly (A) tail. Then this polyadenylated DNA is restricted with Sau3AI which neither cuts in the TTAGGG repeats of the telomere nor in the subtelomeric repeat so that fragments much longer than the bulk of genomic fragments are generated. These larger fragments contain the telomeres and are isolated by oligo (dT) affinity chromatography. After isolation the fragments are separated by agarose gel electrophoresis, blotted and hybridized to radioactive labeled $(TTAGGG)_n$ where n = 25-40, probe.

Telomere mapping—It refers to the localization of telomeres on the ends of the chromosome done by FISH and identification of other sequence elements with the telomeric (and also subtelomeric) regions. Telomere mapping generates a map called telomere map. **Telomeric map**-It refers to the graphical depiction of the order of sequences at the telomeres of a chromosome. Usually this region is missing in genetic and physical maps because a series of difficult-to-sequence elements are crowded at the termini of the chromosomes (e.g. minisatellite).

16.41   CENTROMERE MAPPING

The localization of centromere on individual chromosomes is also done through FISH.

16.42   ANALYSIS OF INDIVIDUAL GENES

In comparisons of genomes within species where one expects a relatively small number of sequence differences throughout the genome and so determining the entire sequence is unnecessary. Assess the extent and location of variation similar to comparative genomic hybridization which compares copy number changes between closely related genomes at genic resolution. An approach called DNA microarrays of short oligonucleotides designed to interrogate each base individually (i.e. resequencing arrays) is applied to analyses of individual genes (Pollack et al., 2002). But then this approach can not be extended to whole genomes as large number of probes are required. Another approach uses Microarrays that detect mismatches exploiting the fact that hybridization to a short oligonucleotide is quantitatively sensitive to the number and position of mismatches. Sequence level differences are detected without allele-specific probes by comparing hybridization intensities of individual features on the microarray(single feature polymorphisms, SFPs). High density Affymetrix yeast tiling microarrays (YTMS) with overlapping 25 nucleotide oligomers spaced an average of 5bp apart provides complete and a 5-fold redundant coverage of entire *S. cerevisiae* genome. The current high density tiling microarrays contain 6.5 million distinct features on a single chip and each feature measures 5 μm × 5μm. This array design was used to discover novel expressed sequences and to precisely map sites of transcription (Cheng et al., 2005). High density genomic tiling microarrays cover a complete genome or a large fraction of it with densely tiled oligonucleotide probes. The major applications of these arrays are for transcriptome analyses, DNA-protein binding and chromatin modification assays (CHIP-chip) and DNA sequence variation detection. In case of transcriptome analysis the different tasks to be performed are as follows.

1. Detection of transcript boundaries, i.e. transcript with start and stop sites. In addition, quantification of the relative level of transcript abundance.
2. Detection of the presence of architectural features within genes such as alternative transcription start and stop sites, alternative splicing and alternative partial degradation.

In case of first task, the challenge is to obtain optimal estimates of genomic co-ordinates of transcript boundaries from tiling array data. The hybridization signal corresponds to the sum of the target molecules at each probe position. The maximal precision is determined by the offset between the tiling features and can be as fine as a few bases. However, in practice, it is often limited by noise and the transcriptional activity of the genomic region surrounding the transcript.

In case of second task, one needs to compare the signal from probes that target different parts of the same gene. To achieve this one must address the problems of differential probe response. The different oligonucletide probes may report consistently different intensities even if the abundance of their target molecules is the same. As such sequence dependent variation can extend over several orders of magnitude, it can generate a great deal of noise in the data and thus must be taken into account otherwise it will obstruct the reliable detection of μtranscript boundaries, levels and architecture (Huber et al., 2006).

Genome-wide location analysis — Expression analysis with DNA microarrays allows to identify changes in mRNA levels in living cells but the inability to distinguish direct from indirect effects limits the interpretation of the data in terms of the genes that are controlled by specific regulatory factors. Genome wide binding analysis (also known as genome wide location analysis) is used to identify target genes bound *in vivo* by each of the transcriptional regulators encoded in the genome. In other words, it provides information on the binding sites at which proteins reside through the genome under various conditions *in vivo*. It is a powerful tool for the discovery of global regulatory network. This approach has been used to identify the genomic sites bound by nearly a dozen regulators of transcription. A combination of genome wide location analysis and expression analysis can identify the global set of genes whose expression is controlled directly by transcriptional activators.

16.43   CHROMATIN STRUCTURE

Besides regulatory proteins, chromatin structure is involved in transcriptional regulation. High- and low resolution characteristics of chromatin have been described through the use of three distinct assay types: micrococcal nuclease sensitivity, DNase I sensitivity and histone occupancy. Most commonly used method to map nuclease hypersensitive sites in chromatin is DNase I digestion of nuclei followed by indirect end-labeling of the resulting purified double stranded DNA (Wu, 1980). The resolution of this method has generally been overestimated and these regions are now being elucidated by a wide variety of enzymatic and chemical probes in conjunction with alternative mapping procedures which yield single nucleotide resolution. Micrococcal nuclease is an endonucleolytic enzyme from *Staphylococcus aureus* and catalyzes the Ca^{2+} dependent nucleolytic cleavage of linker DNA between adjacent nucleosomes in chromatin. This enzyme is frequently used to isolate nucleosome monomers (mononucleosomes) and linker free core particles. The complete digestion of DNA with this enzyme leads to 3′ mono nucleotides. Chromatin structure can be examined for DNaseI sensitivity by quantitative PCR (Dorschner et al., 2004) and tiling arrays, histone composition (Hogan et al., 2006), histone modifications (using ChIP-chip assays) and histone replacement using (FAIRE, formaldehybde-assisted isolation of regulatory elements-a method to assay open chromatin using formaldehyde cross-linking followed by detection of the products using a genomic tiling array). In case of human genome, distal DNaseI hypersensitive sites have characteristic histone modification patterns that reliably distinguish them from promoters. Some of these distal sites show marks consistent with insulator function. Chromatin accessibility and histone modification patterns are highly predictive of both the presence and activity of transcription start sites. DNA replication timing is correlated with chromatin structure.

Production of nucleosomes with specific modifications—The basic bulding block of chromatin is called nucleosome which contains two copies of histones H2A, H3B, H3 and H4. Nucleosomes are stacked on one another. 15 to 38 amino acids from each histone N terminus form the histone tails which provide a platform for post translational modifications that modulate the biological role played by the underlying DNA. One prevalent modification is H4-K16 acetylation and it has a role in transcriptional activation and maintenance of euchromatin. Acetylation renders the entire chromatin open for gene activity. Addition of a single acetyl group to a specific lysine located in that tail of histone H4 can prevent this folding, presumably by blocking the necessary N to N interactions. Random hyperacetylation of histone tails (>6 acetates per octamer) disrupts intramolecular folding of nucleosomal arrays into compact, 30nm thick fibers. Furthermore, the H4 tail and particularly residues 14 to 23 are uniquely important for the formation of these fibers. The acetylation of H4—K16 occurs within this region, thus providing a mechanism to regulate chromatin folding. Thus histone modifications seem to control higher order chromatin structure. Histone modification is one part of the histone code (see chapter 7 for detail). Various modifications of these proteins have been shown to influence gene activity. It involves chemically synthesizing the 22 amino acid H4 tail peptide with the desired modification. To this is added an acetyl group to the tails 16th amino acid, a lysine as it is among the amino acids commonly found acetylated in living organisms. (Shogren-Knaak et al., 2006).

DNA conformation in nucleosome core — The core comprises of 147bp of DNA and the histone octamer compared with nucleosome it lacks only 10-90bp of linker DNA envisaged to be naked or bound to histone H1 (see chapter 7). The histone-fold domains of the octamer organize the central 129 of 147bp in 1.59 left handed superhelical turns with a diameter of 4-fold that of the double helix. The relatively straight 9bp terminal segments contribute little to the curvature of the complete 1.67 turn superhelix. So far, the site specific regulatory factors that have been discovered, bind to the linker or terminal regions of the intact nucleosome (Richmiond and Davey, 2003). The lack of binding to the central region of the superhelix might simply be a result of bending the double helix or additionally of unusual DNA conformations introduced by histone binding.

Nucleosome positioning — Access to genetic information encoded in the chromosomes depends on the positions of nucleosomes along the DNA. Alternative locations just a few nucleotides apart can have profound effects on gene expression. Nucleosomes sometimes adopt precise, well defined locations with respect to the underlying DNA sequence. This 'positioning or phasing' is particularly evident in regions flanking hypersensitive sites (Gross and Garrard, 1988). Chromosomal elements ranging from telomeres to centromeres and transcriptional units are found to possess characteristic nucleosomal architecture which may be important for their function. Promoter regulatory elements including the TFBSs and TSSs (transcription start sites) show topological relationships with nucleosomes such that TFBSs tend to be rotationally exposed on the nucleosome surface near its border. TSSs tend to reside about one helical turn inside the nucleosome border. There is thus an intimate relationship between chromatin architecture and the underlying DNA sequence it regulates (Albert et al., 2007).

A nucleosome has two fundamental relationships with its DNA. A **translational setting** defines a nucleosomal mid point relative to a given DNA locus and a **rotational setting** defines the orientation of the DNA helix on the histone surface. Thus DNA regulatory elements may reside in linker regions between nucleosomes or along the nucleosome surface where they may face inward (potentially inaccessible) or outward (potentially accessible). Preferential occupancy of a specific DNA sites by the histone octamer is also known as **nucleosome positioning**. Translational positioning refers to the extent to which a histone octamer selects a particular contiguous stretch of 147bp of a DNA in preference to other stretches of the same length that are translated forwards or backwards along the DNA(Lowery and Widom, 1997). Rotational positioning is a degenerate form of translational positioning in which a set of discrete translational positions, differing by an integral multiples of the DNA helical repeats are all occupied in preference to the set of other possible locations. Nucleosome positioning arise from sequence dependences to histone –DNA interactions. In fact, a tendency of AA/TT dinucleotides to recur in 10-bp intervals and in counter phase with GC dinucleotides generates a curved DNA structure that favors nucleosome formation. If nucleosome positioning is not precise then essential DNA regulatory sequences will sometimes be buried when they need to be accessible or will sometimes be accessible when they need to be repressed (or buried). Nucleosome positioning is inherently statistical , not precise. The detection of low level periodic occurrences of different structural motifs in the genome suggests that the cell employs these structural properties of DNA to facilitate the incorporation of the eukaryotic genome in nucleosomes. Several examples of positioned nucleosomes in gene promoter regions have been described *in vivo* and *in vitro*. Preferential positioning could place factor binding sequences in nucleosome linker or terminal region DNA. Furthermore, nucleosomes are intrinsically mobile and yield access to their DNA *in vitro*, permitting RNA polymerase to transcribe nucleosomal DNA without causing dissociation of histone octamer. *In vivo* chromatin remodeling factors targetted by gene regulatory proteins and acting directly on the nucleosome core, augment nucleosome mobility. Their mechanism of action most probably derives from innate ability of the nucleosomes to slide along DNA without releasing it. Genome-wide maps of nucleosome locations have been generated but not at a resolution which would define translational and rotational settings.

Analysis of chromatin changes — Analysis of chromatin structure over a region of 12-20kb encompassing the double strand break by ChIP followed by real-time PCR which provides a sensitive measurement of the kinetics and spatial distribution of chromatin changes and the recruitment of repair proteins around the break. Nucleosome integrity is lost near the DSB. Nucleosome refers to 146bp of DNA wrapped twice around a histone octamer comprising a $(H3-H4)_2$ tetramer and two H2A-H2B dimmers. To determine whether nucleosome stability changes at the DSB, ChIP in strains expressing Flag-H2B or FlagH3 is done(Tsukuda et al., 2005). The quantities of both histone decreased 60-90 min after HO induction and were reduced 3-fold by 120min. The comparable loss of both histone suggests that whole nucleosomes were displaced from chromatin near the DSB.

Nucleosomes were remodeled near the DSB. This was shown by the analysis of the sensitivity of MATα chromatin to micrococcal nuclease. Before DSB induction, a strong MNase ladder reflects positioned nucleosomes. After DSB formation the nucleosome ladder became progressively less organized with time. The alteration in the nucleosome pattern closely paralleled histone depletion, indicating that nucleosome integrity is compromised around the DSB.

16.44   DNA FINGERPRINTING/PROFILING

DNA fingerprinting, DNA profiling or genetic fingerprinting refers to the construction of a DNA fingerprint. The genomic DNA is restricted with a 4- or 6-base cutter restriction endonuclease. The resulting fragments are separated by gel electrophoresis, transferred to membranes (nitrocellulose filters) and hybridized with a fingerprint probe (e.g. insert free wild type M13 DNA, various synthetic oligonucleotides, a variety of cDNAs or genomic DNA probe containing diverse sequences from genes, microsatellite and minisatellite DNA probes). Thus there are two types of probes: multilocus probe and single locus probe. **Multilocus probe** refers to any repetitive DNA sequence which allows to detect two or more, in extreme cases a multiple of loci in a genome whereas a **single locus probe** refers to any DNA sequence that allows to detect one single locus in a genome. In principle, a multilocus DNA fingerprint can be produced either by the simultaneous application of several probes, each one specific for a particular locus or by applying a single DNA probe which simultaneously detects several loci. In case of microsatellite fingerprinting the restriction fragments need not be transferred to a membrane but can remain in the gel which has to be dried. Before hybridization the gel has to be incubated in a small amount of hybridization buffer. The microsatellites are sufficiently small so that they can diffuse into the pore system of the gel. Since appropriate probes detect individual specific polymorphisms, this method can be used for the genetic identification of different individuals of one species with high certainty. It is, therefore, used for genome mapping and paternity testing. The advantage with use of single locus probes is that the generated DNA profile can be converted into a numerical format and thus a database can be established and new profiles can be matched with this database. Use of single locus probes(17 microsatellite loci) has shown that the different cultivars of grapes grown in different regions originated from the cross of two grapes cultivar (Bowers et al., 1999). As VTNRs are found in mtDNA as well as nuclear DNA, the mtDNA can be used to decide the parentage. The advantages with the use of mtDNA in historical genetics are three fold. First, mitochondrial sequences(genes) are passed from mother to child via the egg (maternal inheritance). The second advantage in comparison with nuclear DNA is the small size of DNA (16-20kb) which means there is less scope for variability but this is more than compensated by the high copy number (~ 10000). Thirdly, in old specimens even though the nuclear DNA is totally degraded, mtDNA from that specimen can still be recovered. This DNA fingerprint analysis yields genotypes showing specific banding patterns which can distinguish them from one another. In this analysis, average number of scorable bands per individual, number of specific bands and level of band sharing are calculated. Besides these, molecular sizes of bands are calculated which depend on the restriction enzyme used. The band sharing coefficient is calculated as

$$Sxy = (2nxy)/(nx + ny)$$

Where nx, ny are the number of bands in X and Y samples and nxy is the number of shared bands. The difference in electrophoretic migration of identical DNA fragments in two adjacent lanes of agarose gel is due for instance to the binding of specific proteins.

Problem with parental testing—Using a multilocus probe any band in the profile is found roughly 1 in 4 (25%) of unrelated individuals. It means that the same 2 bands will be found 1 in 4x4=16 individuals; the same 3 bands will be found in 1 in 4x4x4= 64 individuals and so on. An average DNA profile contains 11 bands and any given combination will be found in 1 in 2.5 million individuals.

Two-dimensional DNA fingerprinting—It is a technique for high resolution genotyping that produced two-dimensional restriction fragment fingerprints. In this technique, genomic DNA is first restricted with, for example, HaeIII or Hinf1 and the restriction fragments are separated by electrophoresis (6% neutral polyacrylamide gels). The relevant region (e.g. the region of fragment sizes from 0.3 to 3kb or from 1.0 to 10kb) is then cut of the gel and applied to a 6% polyacrylamide gel containing 10-75% linear gradient of denaturant (10%: 0.7M urea, 4% formamide). This combination of neutral and denaturing gradient gel electrophoresis leads to a resolution of up to 1000 (or more) spots per DNA sample on the stained denaturing gel. Specific sequences, for example, genes or repetitive sequences such as mini or microsatellites can then be detected by Southern blots and autoradiography. This technique allows to measure mutation frequencies in genes, to associate genetic variation with a trait (e.g. a disease), to detect genomic instability in cancer and aging, to establish linkages and to map target genes (e.g. disease genes).

16.45   WAYS OF FINDING GENES

There are different ways of finding genes. One of the ways is finding genes by way of **finding relatives**.

Search by finding relatives—Genes can be found through an implied relationship to something else. For example, being a putative orthologues (related to gene in another species). Thus search the genomic sequence or preferably its mRNA sequences and protein products using BLAST. Another way to finds a gene is by looking for paralogues (family members derived by gene duplication).

Search by position—One can find genes by their position in the genome rather than by sequence similarity. If genetic or cytogenetic analysis has implicated a particular region in the aetiology of a disease then see what genes lie in the region. A natural way to describe positions in a sequence would be by base co-ordinates but it is impractical for working with draft genome sequence as the sequence is still being revised.

Cytogenetic analysis—Cytogenetic band nomenclature is more commonly used to describe positions in the genome and many human diseases are linked to chromosomal deletions, amplifications and translocations. These designations must be related to the sequence. Here FISH can be used to localize BAC clones that also bear sequence tags which can be found in the draft sequence.

Use of STS markers—Another way to describe positions in the genome is relative to mapped sequence tagged site marker. This is particularly useful in positional cloning projects where candidate regions are usually defined by polymorphic STS used in genetic analysis. STS markers from several genetic and physical maps have been localized in the working draft sequence using a procedure called **electronic PCR (ePCR)** (Schuler, 1997) which is used for sequence mapping. Electronic PCR is an *in silico* method to test a DNA sequence (query sequence) for the presence of a STS. ePCR searches for subsequences that more or less perfectly match the PCR primers used for the amplification as well as spacing is determined and if they are meeting the expectations, the primers are expected to amplify a PCR product of the correct molecular weight in any other genome. In other words, e-PCR searches for subsequences within a query sequence that match known STS PCR primers and are in correct order, orientation and spacing to be consistent with the PCR product size. These criteria eliminate the possibility of false positives (e.g. hits to pseudogenes or repeat sequences) which occur with other similarity searching tools such as BLAST. e-PCR will help in the intergration of genomic sequence data with existing maps. It will also help in correlating genetic distances with physical distances (Barnes, 2005). ePCR is an *in silico* equivalent of laboratory based STS mapping process. The ePCR tool maps known STSs from the databases, GDB and RH databases to a user submitted sequence. Two primers are used to map sequence feature (e.g. a SNP). To validate the position both primers must map in the same vicinity spanning a defined distance, effectively producing an ePCR product.

17

Systems Biology

17.1 SYSTEMS APPROACH

It would be desirable to think not in terms of single molecules (or genes) but in terms of **'systems'**. Systems biology is sometimes loosely associated with the use of genomic technologies to understand specific biological processes when it should refer to the exercise of integrating the existing knowledge about biological components, constructing a model of the system as a whole and extracting the unifying organizational principles that explain the form and function of organisms as put forward by Bogdanov and Bertalanffy (Gutierrez et al., 2005) and developing biological hypotheses for further experimentation. Connecting different levels of analysis from molecules through modules to organism is essential for understanding of biology. The genome sequences and protein properties are not sufficient for interpreting biological systems. Rhythmic phenomena are manifestations of dynamic bahvior of biological systems. Rhythms are observed at all levels of biological organization with periods ranging from a fraction of a second to years. The rhythms find their roots in many regulatory mechanisms that control the dynamics of living systems. With the availability of finite codes, the genomes of different plant species, the task now is to decode this information into the components (genes) and their relationships (e.g. regulatory, physical interactions) and construct an *in silico* model of plants that encompasses all levels of organizations (Figure 17.1 and 7.2) (Gutierrez et al., 2005). In other words, a combination of experimental and computational approaches- a systems biology approach (Kitano, 2002) is expected to resolve this problem. Although presently the focus is on the simplest tractable system-individual cells but organization and regulation need to be defined at all levels of resolution if the source of stability is to be elucidated. Nesting systems which biologically corresponds to the association of cells into tissues, tissues into organs and organs into intact organism, could be spontaneous and would provide additional stability to the community of the self. Thus every system (irrespective of the level of resolution) be analyzed with respect to the system's structure, its dynamics, its method of control and its method of design. Cells, tissues, organs, organisms and ecological webs are systems of components whose specific interactions have been defined by evolution and thus a system –level understanding should be the prime goal in biology. Cells themselves provide the most obvious form of biological modularity by

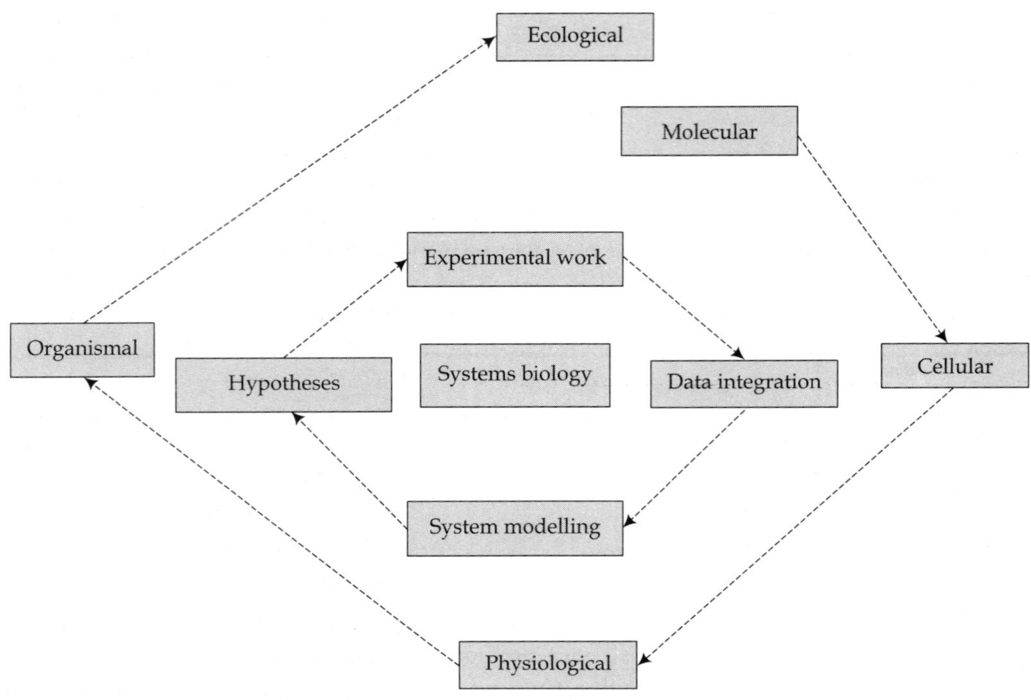

FIGURE 17.1 **Systems biology for the virtual plant. System biology aims at generating a model of the plant as a whole which describes processes across all layers of biological organizations such as molecular and ecological.**

physically partitioning off biochemical reactions. However, biochemical networks within cells also form modular compartments separated by spatial localization, anchoring of proteins to plasma membranes and by dynamics. Cells also provide redundancy with many autonomous units carrying out identical roles. Redundancy also appears at other levels by having multiple genes that encode similar proteins or multiple networks with complementary functions. Functions in biological systems rely on a combination of the network and specific elements involved. Mapping and understanding the molecular interaction networks represent the first step toward modeling how a cell actually functions in time and space. Although systems approach has been used to study plants for many years but with the advent of genomic technologies (microarray, yeast two-hybrid, Chip-chip) it has now been feasible for systems approach to reach to and includes molecular details. Further, it is now able to integrate knowledge (genomic sequences, expression data) across different levels of organization and to anchor this at molecular level. In other words, genomic-wide high throughput experiments are now generating a list of functional elements and a catalog of how these functional elements interact with and regulate each other. In biological systems large number of functional diverse and frequently multifunctional, sets of elements interact selectively and non-linearly to produce coherent rather than complex behavior. Having collected all these information the objective then would be to put all these back together to create predictive models of cellular behaviour which is the goal of systems biology. Systems

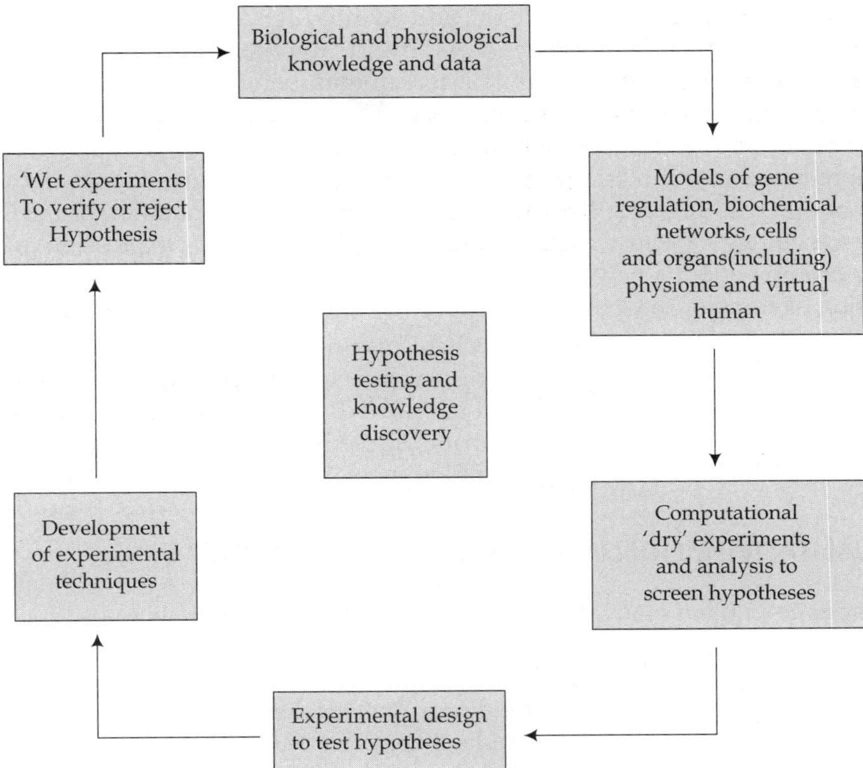

FIGURE 17.2 Showing basic systems biology in human.

biology is a field of biology that quantitatively measures and models the behavior of cell from a system perspective. Systems approach can be described as an iterative process that includes data collection and integration of all available information (all components and their relationships in the organisms), system modeling, experimentation at a global level and finally generation of new hypotheses. Thus systems biology will provide a holistic view of the form and function of biological systems which can be seen in the following example. Ideker et al. (2001b) applied the systems approach using genomic technologies to study the galactose utilization pathway in yeast (*S. cerevisiae*). This pathway was previously studied using one gene/protein at a time or reductionist approach. They constructed knock out strains for the nine genes involved in the GAL pathway. These genes code for four enzymes, one transporter and five transcription factors. Global mRNA expression levels were then examined in the mutant strains and compared with wild-type both in the presence and absence of the signal, galactose. By combining expression data in mutant and wild-type strains, protein levels measurements and protein-protein and protein-DNA interaction data they identified a number of regulated processes, their interconnections to each other and to potential regulatory proteins. In other words, their results provided insights into the regulation of metabolic pathways and the way distinct pathways are connected to each other and to other cellular processes in the yeast. Furthermore, through integrating the protein and

RNA measurements into a single model and identifying agreements and discrepancies between RNA and protein levels, they could hypothesize transcriptional versus posttranscriptional regulation. Finally, disagreements between the observed expression patterns and those predicted led to the generation of new hypotheses about regulation of the GAL pathway. In one striking example, the model generated with the knowledge prior to this work predicted that mutations in the genes encoding enzymes would not disturb the expression of other GAL genes. However, it was found that the mutation of the GAL10 and GAL7 genes encoding galactose metabolism enzymes did affect the expression of other GAL genes. This brought about a new hypothesis that a metabolite could act as a signal to control the regulation of some of the genes in the pathway which was later found correct.

For gaining insights into molecular mechanisms underlying the biological associations of genes in the gene catalogue generated using genomic techniques, one will have to analyze a gene's function, gene product associations and activities in the context of known biological pathways. As genomic data are coming from different sources, integrating heterogeneous databases has been acknowledged as one of the most important tasks in bioinformatics.

17.2   QUALITATIVE MODELING OF BIOLOGICAL SYSTEMS

Biological systems contain many non-linear subsystems and so there is a need for developing control theory for many complex biological systems as many results in control theory are based on linear algebra. Various mathematical models such as Boolean network have been proposed for modeling complex and non-linear biological systems. Further, identification of a set of perturbations that induce desirable changes in cellular behavior may be useful for systems-based drug discovery (Figure 17.3 shows the application of systems biology in drug discovery) and cancer treatment. For discussion on quantitative modeling see Bioinformatics(Roy, 2009).

17.3   APPLICATIONS OF SYSTEMS BIOLOGY

Systems biology approach attempts to understand biological systems as systems, especially aiming identification of their structures and dynamics and establishment of methods to control behaviours by external stimuli and to design genetic circuits with desirable properties (Kitano, 2002).

It will help to identify the gene networks that are important, form example, for plant development, metabolism or that are implicated in the plant response to biotic or abiotic stress. Furthermore, it will help in finding the connections between these different gene networks. As distinct processes like development have been mostly studied in isolation and segregated from other processes such as metabolism or responses to biotic or abiotic stress, it will help in bridging the boundaries. The models generated by the systems approach would help in making predictions, for example, about the effect of a perturbation (e.g. gene mutation) in economically important traits such as seed yield or plant growth, or for determining the conditions for optimizing the plant growth.

A systems approach integrating global datasets provide insights about biological networks. Before starting to construct a particular network, the following questions have to be answered.

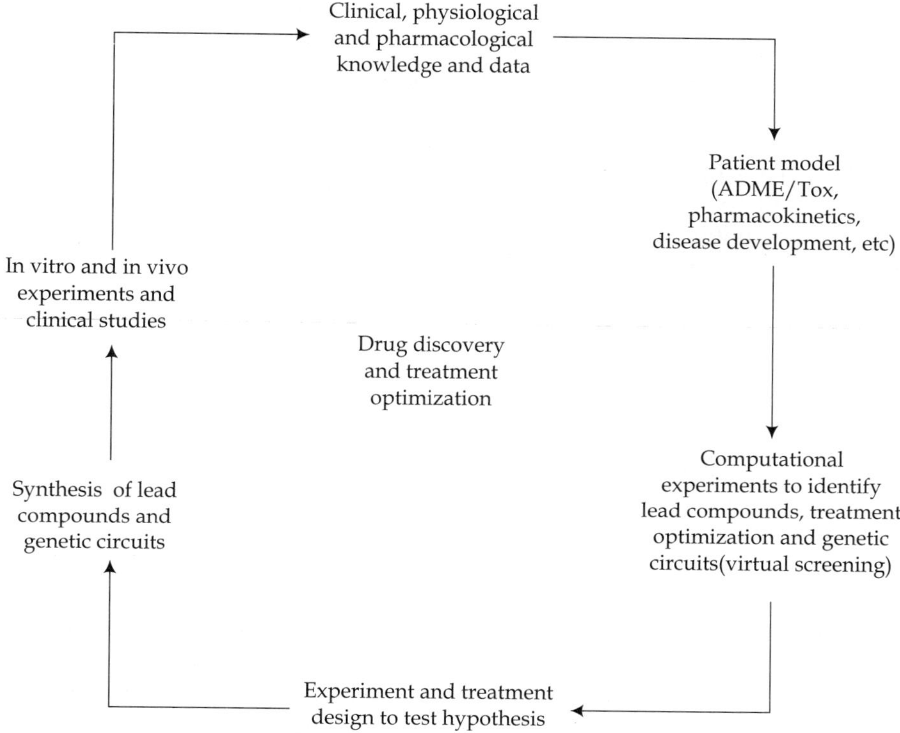

FIGURE 17.3   Showing linkage of systems biology with drug discovery.

1. What types of data are required for modeling networks?
2. What modeling approaches best represent networks?
3. What are the salient features of networks?
4. How robust are the networks, how internal and external perturbations affect processes, pathways and networks and how do the properties of network structure confer biological function?

The ultimate goal of systems biology in plants will be to generate a model of the plant as a whole that describes processes across all layers of biological organization- molecular, cellular, physiological, organismal and ecological. Systems biology embodies an iterative process of experimentation at global level, data integration, system modeling and generation of hypotheses. These hypotheses will lead to the design of new experiments that will start a new round of the cycle. Each iteration will refine the model and deepen our understanding of the system. Finally, such models may be employed in a predictive model and applied to improving economically important traits in plants.

Post-genomic technology provides the data to study regulatory processes at multiple levels of biological organizations (Spellman et al., 1998). One of the major challenges is to integrate and interpret these data to gain a understanding of the structure and function of the molecular processes that mediate adaptive and cell cycle driven changes in gene expression. The visualization and computational representation of these complex regulatory processes as networks structures facilitates the application of a range of analysis and simulation techniques which can, in turn, shed light on our understanding of their organization and behavior (Utez, et al., 2000). The knowledge gained from these analyses will then provide insights and allow the formulation of hypotheses which can then be tested in the laboratory. To achieve these objectives what is required is the techniques which will allow pathway data to be modeled and analyzed. The existing modeling approach for regulatory networks in which regulatory entities (i.e. genes, proteins and external signals) are viewed abstractly as binary switches is Boolean networks.

17.4 PROTEIN INTERACTION NETWORK

Protein interaction networks are principal components of a systems-level description of the cell. Most studies of protein network operate on a high level of abstraction, neglecting structural and chemical aspects of each interaction. Protein interaction networks do not differentiate between many types of relationships, for example, high affinity and direct vs loose and transient. Sometimes, in fact, interactions are reported that connect two proteins that never touch each other physically but are only linked via a third protein. Although global aspects of network topology, linking it to protein function, expression dynamics and other genomic features have been studied, there is a need for use of structural information for systems biology. In particular, a protein's degree (number of interacting partners) is an important factor and proteins with high degree (**hubs**) have been found to be essential. The relationships between network topology and genomic features are actually more reflective of a structural quantity, the number of distinct binding interfaces. Subdividing hubs with respect to this quantity provides insight into their evolutionary rate and shows that additional mechanisms of network growth are active in evolution (Kim et al., 2006).

Challenges in systems biology—The first challenge is to create a 3D view of molecular interaction networks as molecules are 3D objects and function and interact through spatially atomic precise interactions in the crowded micro-environments. The second challenge is to capture the dynamic and context dependent nature of interaction networks and requires mapping out all possible interactions in addition to knowing conditions (cellular states, environment type and protein modification type) under which each interaction works. The third challenge is to quantitatively measure interaction networks. Interaction networks tell whether or not two molecules interact but to model cellular behavior it is also important to know how strongly and quickly they interact. Researches in systems biology aim at understanding and modeling of cellular processes at molecular level.

17.5 APPROACHES TO STUDY SYSTEMS BIOLOGY

Traditional approaches to systems biology are based on a mathematical description of putative pathways in terms of coupled differential equations with the objective to obtain a deeper understanding of the exact nature of regulatory circuits and their regulation mechanisms. However, the availability of high throughput data has prompted interest in **reverse engineering** the networks (regulatory networks) and pathways (biochemical pathways) in an inferential way from the data themselves. A central question in reverse engineering of genetic networks consists in determining the dependencies and regulating relationships among genes. Although learning gene regulatory networks in *S. cerevisiae* from gene expression profiles has been made with the help of Bayesian networks, BNs there are various alternatives such as Relevance network, RN (Butte and Kohne, 2003) and Graphical Gaussian model, GGM (Schafer and Strimmer, 2005a) have been applied to the inference of gene regulatory networks from gene expression data. Bayesian networks and GGMs tend to outperform RNs but the difference is less pronounced for the non-linear simulated data and the measured protein concentrations than for Gaussian data. Also, there is insufficient evidence for any significant difference between BNs and GGMs on observational data (Werhil et al., 2006).

17.6 GENETIC NETWORK

Network refers to an interconnected group or system. In other words, it refers to a system of interconnected components or circuits. Genetic network refers to the network of genes producing mRNAs and thereby producing proteins. Studies of real genetic networks through transcriptome analyses of cells in diverse sets tend to support the prediction, i.e. large subsets of genes are often strongly induced or repressed in an all or none response as cell moves from one state to another. Recently a second genetic network has been discovered which consists of miRNAs. It refers to network of those genes which produce miRNAs/silencing RNAs. Microarray technology has accelerated the development of genetic networks.

A gene regulatory network is typically a complex biological system in which proteins and genes bind to each other and act in an input-output system for controlling various cellular processes. Living cells contain thousands of genes and each of which codes for one or more proteins. Many of these proteins in turn regulate the expression of some other genes through complex regulatory pathways to accommodate changes in different external environments or carry out the essential development programs. The key to understanding living processes is therefore to uncover the structures of these regulatory networks that underlie the regulations of cells (Chan and Ma, 2006). Thus large regulatory gene network (GRNs) determine the course of animal development.

17.6.1 GRN Structure

These networks consist largely of the functional linkages among regulatory genes which produce transcription factors and their target *cis*-regulatory modules in other regulatory genes, together with genes that express spatially important signaling components. They have a modular structure, consisting of assemblies of multigenic subcircuits of various forms. Each such subcircuit performs a distinct regulatory function in the process of development. GRN structure is inherently hierarchical as each phase of the development has beginnings, middle stages and progressively more fine scale terminal processes so that network linkages operating earlier have more pleiotropic effects than those controlling terminal events (Davidson and Ervin, 2006).

17.6.2 Classes of GRN Components

The different classes of GRN components are: (i) evolutionary inflexible subcircuits which perform upstream functions. They are called 'kernels' of the GRN. The kernels are network subcircuits that consist of regulatory genes, i.e. genes encoding transcription factors. (ii) certain small subcircuits called the 'plug-ins' (modular attachments) of the GRN, co-opted to diverse developmental purposes or pathways (iii) switches that allow or disallow developmental subcircuit to function in a given context and so act as input/output devices within the GRN and (iv) differentiation gene batteries.

17.7 APPROACHES TO MAPPING OF NETWORKS

There are different approaches to mapping networks (Xia et al., 2004; Briggs and Singer, 2005). Work has started to identify the DNA targets of transcription factors. Chromatin immunoprecipitation shows whether a particular protein is bound to a given gene in certain cells or tissues. Microarrays and chromatin immunoprecipitation can only predict changes in protein and metabolite levels. Whole-genome tiling arrays or promoter arrays may enable all of the genes that are bound by a particular protein to be identified although current methods are not robust. Promoter arrays are limited by the quality of genome annotation but they are smaller and less expensive than genome tiling arrays. A promoter array could facilitate the construction of genetic networks downstream of a DNA-binding protein. But then what about the network upstream of a gene? One can determine whether a gene that is immunoprecipitated by one protein is also bound by another, indicating that both bind upstream of the same gene. In this way only candidate proteins can be tested but unsuspected candidates can not be discovered. Search for mutants that change the expression of promoter reporter constructs can define an upstream network

Another approach to mapping networks is to repress the expression of one gene (**gene disruption**) and then look to see which other genes respond. Gene disruption (through eg. T-DNA insertion) is a powerful approach although it is very difficult to distinguish direct from indirect effect. Although a full set of gene disruption mutants is available in maize and T-DNA collections in Arabidopsis has been characterized, but without some way to multiplex the plant mutant collections, it will not be possible to build large transcription networks.

The use of **DNA barcodes (genes)** to mark deletion mutant strains has allowed construction of genetic networks for fitness in yeast (Giaever et al., 2002). In this approach each individual gene deletion mutant is combined in equal numbers and grown competitively in a population. The frequency of each strain remaining in the population is assayed using a microarray which detects barcodes. Thus the mutants which either disappear or become abundant define the fitness genes for growth under these conditions. This approach has been applied to the discovery of **synthetic lethals. Synthetic lethals** are strains that contain two defective genes that individually are viable but together are lethal. On average each gene had 30 lethal partners and typically their function was related to that of the query gene. Many genes formed synthetic lethals with the same set of 30 partners but not with each other and thus 30 genes form a pathway and their interactions are part of a protein complex. This method can be applied to plants and the first step is to create homozygous mutants for each member of the T-DNA collections. This will reveal the genes that are individually essential for survival. These individuals are then removed and the remaining mutants are intercrossed crossed to find synthetic lethals. Lethality will appear in the ratio of 1/16 in F1 progeny as aborted seed, dead seed or plants dying after germination. Pollen can be examined to detect gametophytic lethals. The limitation with this approach is the very high number of crosses to be made and their evaluation for fitness. So, an alternative approach which mimics the yeast system in plants, is transient expression of bar-coded, dominant gene suppressors and it could be applied to populations of dividing plant cells. A library containing suppressors for every gene could be mobilized into cells and then monitored by amplifying the barcodes and hybridizing to a microarray. Barcodes (genes) which disappear are linked to genes which are required for survival. Then a query gene could be added with the library and microarray results are compared. Barcodes which disappear when the query gene is added reveal synthetic lethals. **Virus induced gene silencing** or **artificial transcription factors** could be used as dominant gene suppressors. Thus the entire network could be revealed with only 27000 queries rather than almost impossible number of crosses to be made in the earlier approach of use of DNA barcodes.

17.8 NETWORK OF OVER EXPRESSING GENES

No information is available on networks revealed by over expressing genes. Artificial transcription factors can be used to activate genes. They can be changed from repressors to activators simply by swapping domains. It is generally much more difficult and expensive to activate genes than to disrupt. A library of artificial transcription factors is made in two versions, one to activate and one to repress the target genes. After that two sets of synthetic lethals are produced and compared. The advantage with artificial transcription factors is that when fused to a protein transduction domain, they can be easily purified from *E. coli* and added to cells as proteins and thus eliminating the cost and variability associated with transformation.

Allelic substitution—As an alternative to ectopic expression, wild-type but variant allelic substitutions can be made at every locus and their effects on mRNA levels of every gene can be measured. This approach has be applied by Singer and colleagues using recombinant inbred lines to multiplex the allelic substitutions and by using microarrays to measure mRNA levels. A remarkable degree of connectivity between genes was observed which shows that members of co-ordinately regulated protein complexes or biochemical pathways are themselves regulators. They concluded that **most regulators act in *trans* and they are not transcription factors**. Studies in yeast led to similar conclusions (Yvert et al., 2003).

17.9 LOW EXPRESSED GENES

A large fraction of regulation is posttranslational and this is undetected or counter indicated by changes in mRNA levels. For example, transfer of yeast from complete to minimal medium showed 8-fold increase of a transcriptional repressor without increasing mRNA level. Further, plants (rice) present no problem to this level of analysis. Like protein levels, the levels of metabolites in plants can be uncoupled from changes at the mRNA levels. These observations show a key difference between biochemical networks which are mechanistic but make unreliable prediction of relationships and genetic networks which reveals quantitative relationships but not mechanisms.

17.10 COMBINING NETWORKS

Combining multiple genetic networks in order to find genes at their intersection can be more effective than examining a single network. For example, in rice genome microarray detected the induction of suppressor of the G2 allele of gene SKp1 (SGT1) caused by rice blast fungus. The SGT1 is determining disease resistance in barley and Arabidopsis. Several other rice proteins interacted with this SGT1 in a yeast two-hybrid assay. The microarray experiment showed that one of these proteins, ERP (elicitor response protein) was induced by stress whereas others showed its induction by an elicitor from the rice blast fungus. ERP itself when used as bait, interacted with other proteins including one which was undefined due to lack of homology with defined protein. This shows that all five (undefined protein plus the four that interacted directly with SGT1) were predicted to be essential for disease resistance in rice. Orthologs of these rice genes were evaluated using Arabidopsis T-DNA mutants and each mutation was found disrupting disease resistance significantly (Cooper et al., 2003).

17.11 DEVELOPMENTAL NETWORKS

In most eukaryotes fertilized eggs undergo a series of cell divisions and differentiate to generate a multicellular organism. The developmental processes include differentiation (creation of distinction between different cells) and producing co-ordination among different cells so that they function as units. Cell differentiation starts with asymmetric cell divisions in which a cell divides to produce two daughter cells performing two different functions. One way to achieve asymmetric division is the unequal partitioning of cell contents at the time of cell division. For example, unequal partitioning of transcription factors can results in activation of different sets of genes in the two daughter cells. Alternatively, the two daughter cells may be identical in their cellular contents but because they are in different environments (the two daughter cells are surrounded by different cells) getting different types of signals from the surrounding cells and thus different sets of genes are activated. The two processes are not mutually exclusive and thus both processes can occur simultaneously resulting in different types of cells (Benfey, 2005).

In plants both processes (differentiation and co-ordination) rely heavily on cell-to-cell communication and activation and/or repression of subsets of genes. Signals that pass between cells frequently result in the activation of signaling pathways within cells. These signals can modify transcription factors which bind to specific sites on the DNA causing the activation or repression of genes. The products of activated genes are then synthesized. Among them are transcription factors which bind other DNA sequences activating or repressing the activity of a new set of genes. The term '**transcriptional network**' refers to the sequential activity of transcriptional factors that activate or repress the expression of other transcriptional factors responsible for developmental regulation. A high-throughput approach for Chip-chip will help provide the transcriptional wiring for the entire genome. A feature of developmental transcriptional networks is that at each stage some of the induced or repressed gene products are enzymes, receptors or other molecules whose activity ultimately results in differentiated functions that define specific cell types. Thus signals act to co-ordinate all of the major stages of development from asymmetric cell division to organ functioning necessary to maintain the integrity of the organism. There are three different types of signal: short range (for example, a ligand secreted from one cell binds a receptor on a neighbouring cell), mid range in which a secreted ligand diffuses through several layers of cells before activating a receptor on its target cell and long range in which, for example, a hormone is secreted from a cell within one tissue and passes from one organ to another where it activates a response. Frequently these different types of signal are interwined, forming a **signaling network**. It is important to understand the interplay and regulation of signaling and transcriptional networks and how they modulate cellular function. A system approach integrating global datasets can provide insights about biological networks. Although both signaling and transcription are equally important for development, high throughput techniques for identifying the nodes and links in the transcriptional networks have made rapid advances and critical datasets include global expression profiles at cell type-specific resolution and direct binding data for identification of targets of transcription factors. These two datasets have been combined to deduce transcriptional networks in yeast, *S. cerevisiae* (Lee et al., 2002; Harbison et al., 2004)

17.12   CONSTRUCTION OF GENETIC REGULATORY NETWORK

Gene regulatory networks consist of huge sets of regulatory genes that control one another's expression as well as the expression of the downstream effector genes via so-called *cis* regulatory elements to which transcription factors bind. Gene expression depends on recognition of specific promoter sequences by transcriptional regulatory proteins. These regulatory proteins recruit (modification-induced recruitment of chromatin associated proteins to acetylated and methylated histone NH2-termini) and regulate chromatin modifying complexes and components of the transcription apparatus. Knowledge of all transcriptional regulators encoded in the genome is required for the construction of transcriptional regulatory networks. There have been categorizations of the distinct components (subcircuits) of gene regulatory networks according to their function and the different degrees of evolutionary conservation. There are different functional categories of transcriptional regulators such as cell cycle regulators, regulators involved in metabolism and environmental response. Architecture and functioning of the *cis*-regulatory gene regions represents the basic nodes of gene regulatory network. The subcircuits consist of small subset of genes and *cis* regulatory elements.

DNA microarrays for large scale monitoring of gene expression have made the reconstruction of gene regulatory networks. Before one can infer the structures of these networks, it is important to identify, for example, for each gene in the network which genes can affect its expression and how they affect it. The identification of a signal transduction pathway could be traced back to the genetic regulatory level. The genome sequencing and DNA microarray technology make possible the quantitative analysis of signaling regulatory network besides the qualitative analysis (Hughes et al., 1999).

17.13 TRANCRIPTIONAL REGULATORY NETWORK ARCHITECTURE OR NETWORK MOTIFS

Lee et al. (2002) identified the following six regulatory network motifs within regulatory networks.

I. **Auto regulation** — Auto regulation motif consists of a regulator which binds to the promoter region of its own gene. It reduces response time to environmental stimuli, has decreased biosynthetic cost of regulation and increased stability of gene expression. Most of prokaryotic genes encoding transcriptional regulators are auto-regulated.

II. **Multiple component loops** — Multiple component loop motif consists of a regulatory circuit whose closure involves two or more factors. The closed-loop structure provides the capacity for feed back control and offers the potential to produce bistable systems that can switch between two alternative states. This system is yet to be identified in bacterial genetic network.

III. **Feed forward loop** — Feed forward loop motifs contain a regulator which controls a second regulator and have additional features that both regulators bind a common target gene. It may act as a switch which is designed to be sensitive to sustained rather than transient. FF loops have the potential to provide temporal control of a process as expression of the ultimate target gene may depend on the accumulation of adequate levels of their master and secondary regulators. FF loops may provide a form of multiple step ultra sensitivity as small changes in the level or activity of the master regulator at the top of loop might be amplified at the ultimate target gene because of the combined action of the master regulator and a second regulator which is under control of the master regulator.

IV. **Single input motif** — Single input motifs contain a single regulator that binds a set of genes under a specific condition. It is potentially useful for co-ordinating a discrete unit of biological function such as a set of genes that code for the subunits of a biosynthetic apparatus or enzymes that bind a metabolic pathway.

V. **Multiple input motifs** — Multiple input motifs consist of a set of regulators that bind together a set of genes. This motif offers the potential for co-ordinating gene expression across a wide variety of growth conditions.

VI. **Regulator chain motifs** — Regulator chain motifs consist of chains of three or more regulators in which one regulator binds the promoter for a second regulator, the second binds the promoter for a third regulator and so forth. This network motif is frequently observed in location data for yeast regulators. The chain represents the simplest circuit logic for ordering transcriptional events in a temporal sequence. The most straight forward form of this appears in the regulatory circuit of cell cycle where regulator functioning at one stage of the cell cycle regulates the expression of factors required for entry into the second stage of the cell cycle. Figure 17.4 (1-6) shows six basic regulatory network motifs.

17.14 GENE EXPRESSION REGULATION

Feed back occurs through autoregulation wherein a protein modifies directly or indirectly its own rate of production. Positive or negative feed back depends on the network dynamics. Modeling of gene networks is focused on stability characteristics of networks dominated by positive and negative feed back. Stability of networks refers to the tendency of a system to remain close to a steady state (a state in which production and decay rates are balanced). The genes regulated by negative feed back should be more stable then either unregulated genes or those regulated by positive feed back. A significant feature of the positive feed back is its role in the generation of bistability where two steady states of the system are stable. Positive feed back is important in generation of multiple stable states and has been implicated in the stability of differentiated and undifferentiated states in Xenopus oocytes.

Gene expression involves a series of single molecular events. As each of these molecular events is subject to significant thermal fluctuations, gene expression is best viewed as a stochastic process. Thus there is a proposal for use of stochastic model for gene expression and biochemistry in general. Now the question is how cells function and process information when the underlying molecular events are random. Single cell measurements often display significant heterogeneity. Biochemical rates of transcriptional and translation are proportional to the number of promoter sites and mRNA molecules. These rates are small and imply relative infrequent transcriptional and translational events compared with other interactions in the cells (e.g. protein-protein interactions). Such infrequent events lead to large fluctuations and these fluctuations are known as internal noise.

17.15 IDENTIFYING THE COMPONENTS OF REGULATORY GENE NETWORK

There are primarily two approaches to identify the components of these controls:

1. **Genetic approach**—Regulatory genes have been identified which influence the cell cycle timing of M-phase. Mutant in these genes lead to the advancement of cells into mitosis. Thus what is required in this approach is to generate mutants and study its influencing on a particular process.

2. **Biochemical approach**—In this approach, protein (s) affecting a particular process is identified. Proteins have been purified which enable Xenopus and star fish oocytes, naturally arrested just before meiotic M-phase, to enter meiosis. Thus mutating genes or adding proteins advances M—phase in cells proceeding through the cell cycle or enable M-phase to take place in cells naturally arrested in the cell cycle.

The underlying structure of gene regulatory networks can be inferred using biochemically driven approaches. The limitations with biochemically driven approach for inferring the underlying structures of gene regulatory networks is that most of the biochemical reactions under participation of proteins do not follow linear reaction kinetics and also gene expression data seem not sufficient to globally understand regulatory networks at this level.

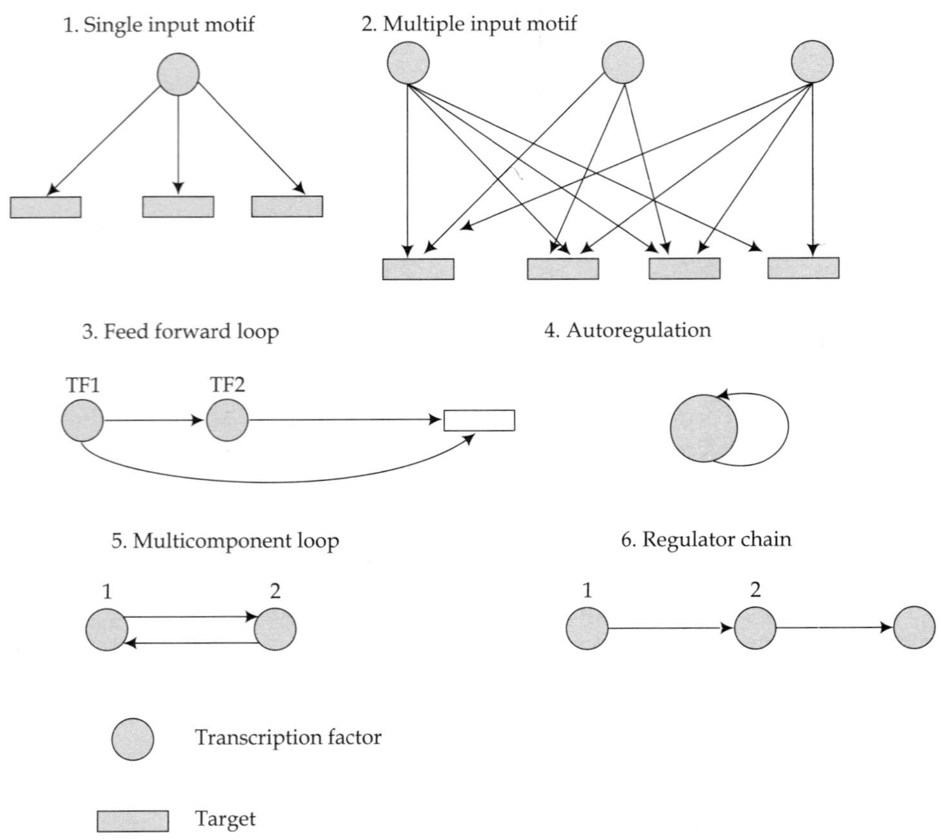

FIGURE 17.4 Showing six basic regulatory motifs.

17.16 DATA MINING APPROACH FOR INFERRING STRUCTURES OF GENE REGULATORY NETWORKS

The limitation with this approach is that the clustering of gene expression data only measures whether genes share a significant linear relationship with each other. The regulatory relationships such as which gene affects which other genes can not be discovered.

17.17 REDUNDANCY IN SIGNALING NETWORKS

Understanding of signal transduction networks requires the use of genetic analysis. Although knockout mutants are searched for generating the defined genotypes to dissect signal transduction pathway but the problem with this approach is that many mutants produce no obvious phenotype. Many signaling components are functionally redundant. Two reasons have been put forward for the functional redundancy of signaling networks. The trivial first reason is that the functional redundancy arises from the rich sequence redundancy found in the eukaryotic genomes. Deletion analysis in yeast *S cerevisiae* suggests that a quarter of functional redundancy can be explained by compensation by duplicate genes (Gu et al., 2003). The second but important reason is that it arises from the ability of the networks to buffer the effects of perturbations in neighboring nodes and related pathways. This property of network is called '**homeostasis**' which is difficult to predict from first principle (unlike the redundancy caused by sequence redundancy) (Cutler and McCourt, 2005).

The ultimate goal of genetics is to understand the relationship between genotype and phenotype. The problem with the traditional genetic analysis can be solved by squeezing more out of the classic equation,

Phenotype = Genotype + Environment.

Thus phenotype, genotype and environment all have to be precisely measured. Phenotype is being refined through improvements in molecular analysis and global transcript profiling of a mutant with 'no observable phenotype' can yield sufficient information to place a gene into a signaling pathway or reveal the compensatory changes in related pathways that enable buffering. The whole genome transcript profiling of deletion mutants in *S. cerevisiae* has shown effects on the transcription of at least one other gene besides the deleted gene (Hughes et al., 2000). Thus what is required is to develop a high throughput molecular fingerprinting as a phenotyping tool.

Genotype (s) will have to be precisely constructed as is now being done. In case of a systematic genetic analysis, called '**synthetic genetic array**', a mutant of interest is crossed to a large collection of deletion mutants in order to produce a synthetic genetic array in which the phenotypic consequences of double mutants are assayed (Tong et al., 2001). Construction of lines containing multiple mutations in related genes to probe the functional redundancy is being made by plant biologists but then problems come from the breeding systems and diploid genetics of plants. One productive strategy would be to use RNAi based system to create random assemblies of genes 'stitched' together to inhibit disparate pathways simultaneously—think of SAGE meets RNAi. Ten or more genes could be stitched together randomly, transformed *en masse* into Arabidopsis and the novel phenotypes generated could then dissected by deconvolution of the complex transgenes. Misexpressing multiple genes concurrently will similarly increase the phenotypic space available to probe signaling networks.

As the phenotype develops as a result of the interaction of between genotype and environment, environment can be changed to uncover previously hidden phenotypes. Systematic way of manipulating the environment is through the use of chemical genetics. This technology involves phenotype based screens of an organism against libraries of defined small molecules to identify compounds that perturb specific gene products (Stockwell, 2000; Blackwell and Zhao, 2003). Each compound in the library is thought of creating a unique environment and thus thousands of tractable environments can be created to probe a genotype for novel phenotypes. Thus an 'aphenotypic mutation' may reveal an obvious phenotype in the right chemical context which can be identified using a chemical genetic screen. Although 'aphenotypic mutant' can be isolated using the technique of 'reverse genetics', the application of chemical genetics leads to the identification of compounds which enable the Mendelization of loci that are otherwise quantitative in nature or more difficult to score due to weak penetrance. Thus the small molecule reagents (for more see chapter 27) are useful for both exposing phenotype and unraveling the pathways that normally buffer a gene of interest. This strategy, in essence, is a form of synthetic chemical genetics used analogously to SGA in *S. cerevisiae*. Since it is difficult to employ the SGA strategy in plants, the strategy of small molecules offer a powerful mechanism for overcoming the problems of strain construction and further it will expose phenotype and ultimately infer gene function using genetic analysis.

17.18   INTEGRATION OF PROTEIN-PROTEIN-INTERACTION DATASET

Data for the reconstruction of NW can come from different sources (experimental and computational) and so integration of data is required in order to complete understanding of the network. One can have multiple datasets of physical protein-protein interaction and genome-wide protein-protein interaction.

17.19   INTERGRATION OF MULTIPLE DATASETS OF PHYSICAL PROTEIN-PROTEIN INTERACTION

As the individual protein-protein interaction experiments tend to measure subsets of the potential interactions, so it is reasonable to consider their union. Further, as individual experiments may contain many false positives, it is necessary to combine the different datasets in order to reduce the error and to make reliable predictions of protein-protein interactions.

The simplest rules for integration of multiple datasets are the and- and or rules. The '**and rule**' predicts a positive interaction only when all datasets agree (intersection). It produces more accurate results but offers low coverage as few cases exist where all available datasets agree. The '**or rule**' predicts an interaction when at least one dataset gives a positive result. It thus yields maximum sensitivity, i.e. it discovers the highest number of true positives but it simultaneously produces the highest number of false positives

In the **majority voting procedure of combining the datasets,** each experimental result contributes an additive positive or negative vote toward the final result. If the majority of the datasets detect an interaction between a pair, that pair is predicted to interact whereas if the majority do not measure an interaction then that pair of proteins is considered noninteracting. Although voting method of prediction offers no improvements in accuracy compared with the results from individual experiments, it has higher coverage then the individual experiments. In this method each dataset carries the same weight despite the fact that some datasets contain more reliable results and other datasets may be redundant.

17.19.1   Machine–Learning Methods

Machine learning methods of data integration take into account data reliability and redundancy and often provide better results in both coverage and accuracy. An effective method is the Bayesian network, particularly the naïve Bayesian network. Bayesian networks have been applied in computational biology research ranging from the prediction of subcellular localization of proteins to the combination of different gene prediction algorithms. The Bayesian network combines different interaction datasets in a probabilistic manner and assigns a probability to the prediction result rather than just a binary classification. Each individual dataset is weighted by its accuracy and redundancy. In addition to integrating and correlating sets of data, i.e. besides being used as a tool for integration and classification Bayesian networks can be used to model the regulatory relationships between individual proteins where it aims at modeling the interdependency of gene and protein activities.

17.20 INTEGRATION OF GENOME-WIDE PROTEIN-PROTEIN INTERACTION DATA

The data integration methods described above can also be applied for combing dataset containing genome-wide protein-protein interaction data. It has been shown that a large number of false positives occur in individual genome-wide experiment. Further, as the number of false positives in high-throughput studies is on the same order of magnitude as the actual number of true positive interactions, so the number of interacting proteins in any cell is perhaps several orders of magnitude smaller than the number of all possible combinations between the proteins in the entire proteome.

17.21 *DE NOVO* PREDICTION OF PROTEIN COMPLEXES

Jansen et al. (2003) showed that protein complexes can be predicted *de novo* with high confidence by integrating multiple genomic datasets. The confidence with which proteins can be predicted to be in the same complex (which is measured in terms of the likelihood ratio) is low in the individual dataset but high in the combined data. Figure 17.5 (a-b) shows the definition and terminology that has been used to define protein complex architecture and the cross talk between different cellular processes.

17.22 RECONSTRUCTING BIOLOGICAL PATHWAYS AND REGULATORY NETWORKS FROM QUANTITATIVE MEASUREMENTS

Here the objective is construction of gene regulatory networks to model multivariate gene interactions. When the data is available in the form of mRNA expression levels (quantitative measurements), many computational methods including the following three have been developed to reconstruct the pathways and networks. In other words, genetic regulatory networks can be inferred from time series gene expression profiles through the following methods.

1. Correlation metric construction (Arkin et al., 1997)
2. Boolean networks (Liang et al., 1997; Akutsu et al., 2000, 2006; Shmulevich et al., 2002)
3. Bayesian networks (Friedman et al., 2000; Hartemink et al., 2002).

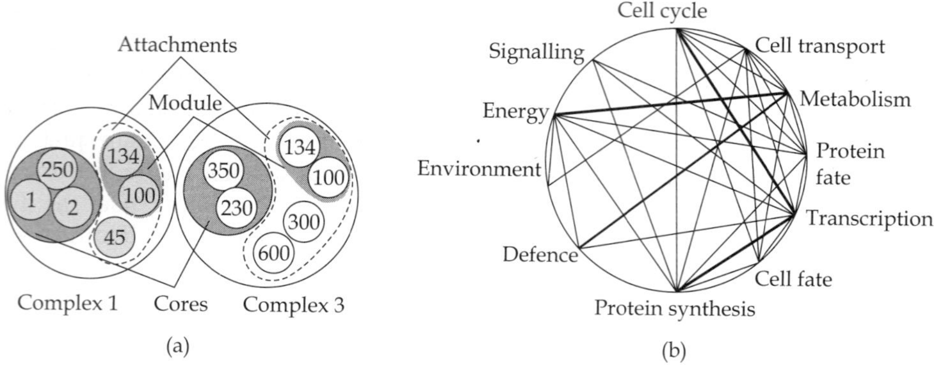

FIGURE 17.5   a. Definition and terminology used to define protein-complex architecture. (Adapted from Gavin et al., 2006). b. Frequency of cross-talk between different cellular processes. The thickness of the lines between the functional classes are proportional to frequency of coremodule interaction between them.

17.22.1   Boolean Networks

Boolean networks model regulatory relations in terms of Boolean relationships and combinatorial logic circuits. Boolean networks compose a class of discrete models where expression levels of each gene are assumed to have two possible values: ON or OFF (Kauffman, 1969). The Boolean network thus treats genes as binary switches. Such a model can not capture the underlying continuous and stochastic nature of protein production and gene regulation. However, one encounters genes that are essentially ON or OFF throughout a given biochemical pathway. Thus switch-like regulatory functions of these genes determine their role in regulation and this activity is well represented by a model such as a Boolean network. A Boolean NW is a system of interconnected binary elements defined by a set of nodes and a group of Boolean functions (a set of regulation rule for nodes). A Boolean network $B = (V, F)$ on genes is defined by a set of nodes/genes $V = \{x1, \ldots, xn\}$, xi $\{0, 1\}$, I $=1, \ldots, n$ and a vector of Boolean functions, $F = \{f1, \ldots, fn\}$, fi: $\{0, 1\}$, i= 1, $\ldots$, n. It is a simple model with each node (e.g. gene) existing in one of the two states off/on, active/inactive. In general these states are assigned numerical value of 1 (active) and 0 (inactive). In other words, each node takes either 0 or 1 at each discrete time t, a regulation rule for each model is given by a Boolean function and the states of nodes change synchronously. Boolean operation is a function taking input from a set of binary variables and producing output to a single binary variable. The function fi is the predictor function for the gene xi. A subset Wi V is called the predictor set for the gene xi if the restriction fi/wi of the predictor function fi equals fi. The cardinality of the set Wi is related to the number of edges incident with the vertex xi in the directed graph $\Gamma = (V, E)$ where an edge (xi, xj) E indicates that gene xi is one of the factors determining the value of the gene xj. $W = (W1, \ldots, Wn)$ is called the predictor set for the Boolean network. A state of Boolean is a vector (x1, $\ldots$xn) of gene values. The states of Boolean are interpreted as binary numbers and are ordered accordingly. Thus there are $N = 2^n$ states in a Boolean network and they are enumerated as 0,1, 2, , N-1. There is a Nx n truth table associated with and equivalent to Boolean where the rows correspond to the

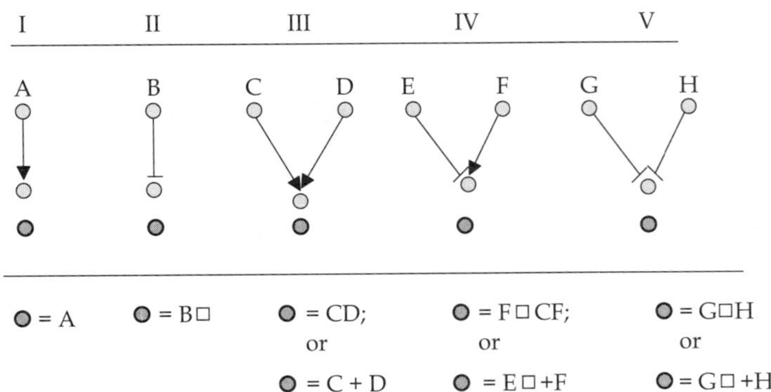

FIGURE 17.6 Shows Boolean representations of signalling interactions. Nodes (components) are represented by circles (○) and each is labelled with a letter: A–H are input and ○ is the output. Stimulatory links are designated by arrowheads and inhibitory connections by plungers. The two single input functions, I and II are IF (○ = A) and NOT (○ = B□), respectively. In Boolean notations, the prime symbol represents the complement, i.e. it flips a value between 0 and 1. The other three functions (III–V) are two input functions. In III, the two stimulatory inputs converge on an output and thus this interaction can be represented by an AND function (○ = CD), if both inputs are needed to stimulate the output and by an OR function (○ = C + D), if only one input is required to propagate the signal. In IV, dual regulation (positive and negative) of a node is indicated. The AND function (○ = E□F) describes this interaction if the inhibitor is off and the activator is one whereas the OR function (○ = E□ + F) describes stimulation either when the inhibitor is off OR the stimulator is on. In case of V the two inhibitors converge on the output. Here again AND (○ = G □ H) and OR (○ = G□ + H) function can be used to represent the interaction. In case of former both inputs are needed to be inactive whereas in case of latter both inputs required to be activated to inhibit the output (Adapted from Hasseldine et al., 2007).

states of Boolean and the columns correspond to the corresponding values for the predictor functions. The truth table of Boolean induces a directed graph $\Gamma = (V, E)$ with the states of Boolean as the set of V of its vertices and with edges (si, sj) E connecting the state si with the state sj if F (si) = sj. The truth table associated with Boolean determines Γ and vice versa. Γ is called the transition diagram of Boolean and Γ is called compatible with W if the truth table induced by Γ has W as the predictor set for the Boolean network associated with that truth table. The state transition diagram represents the dynamics of the network (Pal et al., 2005). Though Boolean networks can not model detailed behavior of biological systems, it may provide good approximations to non-linear functions appearing in biological systems. Boolean NWs are deterministic and do not reflect the inherent randomness which is an integral part of biology. The limitation with this approach is that the validity of the predefined assumptions and the values of the Boolean approach in general have been questioned. Figure 17.6 shows Boolean represontations of singnalling interactions.

17.22.2 Probabilistic Boolean networks (PBNs)

PBNs are composed of finite numbers of constituent Boolean networks, each of which corresponds to a contextual conditions determined by variables outside the model. PBNs and DBNs characterize the same probabilistic understanding with PBNs being more specific in that they specify functional relationship within their constituent Boolean networks.

While Boolean networks have proved to be successful in modeling real world genetic regulatory networks, they suffer from a number of shortcomings (Steggles, et al., 2007). Analysis can be problematic due to exponential growth in Boolean states and lack of tool support and they do not cope up well with the inconsistent and incomplete data that are often obtained in practice. To address these problems, Steggles et al (2007) proposed a new model for genetic regulatory networks based on Petri nets (Reisig, 1985; Murata, 1989) for modeling and analyzing complex concurrent systems (Reisig and Rozenberg, 1998).

17.22.3 Bayesian NWs

It is a suitable technique for studying gene expression data. It is graphical representation of a joint probability distribution consisting of two parts: Bs and Bp. Bs is a directed acylic graph where no path starts and ends at the same node and Bp is a set of local joint probability distributions describing associations. A Bayesian network is defined by a graphical structure M, a family of (conditional) probability distributions F and their parameters q which together specify a joint distribution over a set of random variables of interest. The graphical structure M of a BN comprises a set of nodes and a set of directed edges. The nodes represent random variables whereas the edges indicate conditional dependence relations. If we have a directed edge from node A to node B then A is called the parent of B and B is the child of A. The structure M of a BN has to be a directed acyclic graph (DAG), i.e. a network without any directed cycles. This structure defines a unique rule for expanding the joint probability in terms of simpler conditional probabilities. Suppose $X1$, $X2$, ..., Xn be a set of random variables represented by the nodes i ε { 1, ..., n} in the graph, pa[i] to be the parents of node Xi and let Xpa[i] represent the set of random variables associated with pa[i] then

$$P\ (X1,\ ....,\ Xn) = \prod P\ (Xi^2\ Xpa[i])$$

When adopting a score-based approach for inference, our objective is to sample model/structure M from the posterior distribution

$$P\ (M/D)\ P\ (D/M)\ P\ (M)$$

which requires a marginalization over the parameters q.

$$P\ (D/M) = \int P\ (D\ |q,\ M)\ P\ (q\ |M)\ dq$$

Causal inferences can be drawn from their associations by statistically testing the significance of the associations between variable or by using a certain measure to score all possible structures and searching for those with the high scores. The limitations with Bayesian approach are that all observations are assumed to arise from the same distribution and so it can not model the dynamics of biological system and further many distinct directed acylic graphs may result in the same joint probability distributions (identifiability problem). Another limitation with Bayesian approach is that the task of learning model parameters in NP-hard especially for high dimensional data. Moreover many parameters are required to be estimated accurately and thus requires a large amount of samples which may not be readily available.

Bayesian network, a probabilistic graphical model representation has been widely used for analyzing expression data. In comparison with cluster analysis, Bayesian network has the advantage of uncovering conditional independency among genes which provides a promising way to survey direct interaction of gene regulation. Moreover, by using statistical evaluation approaches, one can examine features of induced high score networks, i.e. the confidence of existence of an edge. Thus highly confident features provide a potential way to mine significant sub-networks from candidate Bayesian networks. By using Efron's non-parametric bootstrap approach with replacement, one can generate several best reasonable networks from microarray data. This is a computationally effective approach to estimate the confidence levels on features of the generated networks. Is the existence of an edge between two genes warranted? In other words, is the regulatory or binding relationship between two genes highly confident. By selecting edges whose confidence levels exceed the pre-determined threshold, obtain a set of highly confident edges whose encoding relationships are believable. Bayesian networks constrain the network model to be an acyclic graph which might not be always the case since feedback loops have been found to be motifs in gene regulation.

Computational problems can be in class P, i.e. problems for which algorithms of polynomial asymptotic order are known or they could be in NP-complete problems, i.e. a set of problems for which no algorithms of asymptotic polynomial-time order are known but which are reducible to one another in the sense that the discovery of an algorithm of asymptotic polynomial-time order for one of them(proving it to be class P) would show that all NP-complete problems are of class P or NP problems, for which all optimal algorithms are provably non-polynomial. If it is found that P=NP, then class P would be expanded to fill the entire class NP set.

17.22.4 Dynamic Bayesian Network (DBNs)

It is an important approach for predicting the gene regulatory networks from the time course expression data. In this network analysis regulator-target gene pairs are usually identified based on statistical analysis of their expression relationships across different time slices. For example, time slices T1 for the regulator and T2 for the target gene where T1 precedes T2. The time period between the time slices of the regulator and target (T2-T1) is considered as the transcriptional time lag. Specifically it is the time that it takes for the regulator gene to express its protein product and the transcription of the target gene to be affected (directly or indirectly) by this regulator protein. Consequently one is more likely to observe a significant correlation between the expression of a regulator and its target if biologically relevant time slices are used.

Problems with dynamic Bayesian network methods—Limitations include lack of systematically determining a biologically relevant transcriptional time lag which results in relatively low accuracy of predicting gene regulatory networks and excessive computational cost.

Thus better statistical methods are required to reconstruct biological pathways from quantitative measurements. Additional improvements in networks architecture obtained from quantitative measurements if

1. quantitative measurements are performed on a systematically perturbed NW as this will help define the NW architecture with increasing accuracy.
2. cross-species comparison is made which reveals the conserved core network
3. information from 1. and 2. is combined with shared functional classification, shared promoter motifs, protein-protein interaction data and protein-DNA binding data.

17.23 PREDICTION ON GENE EXPRESSION PATTERNS

Most of the existing approaches allow the user to generate biological hypotheses about transcriptional regulation of genes which can be tested in the laboratory. However, patterns discovered by these approaches are not adequate for making accurate prediction on gene expression patterns in new or held out experiments. Therefore, it is difficult to compare the performance of different approaches or decide which approach is likely to generate plausible hypotheses. So there is a need of an approach which can not only provide interpretable insight into the structures of gene regulatory networks but can also provide accurate prediction.

17.24 FUZZY LOGIC-BASED APPROACH

Fuzzy logic and fuzzy sets allow modeling of language-related uncertainties by providing a symbolic framework of knowledge comprehensibility. Fuzzy representation is becoming popular in dealing with problems of uncertainty, noise and inexact data. Fuzzy logic has been successfully used for clustering gene expression data. The fuzzy k-means algorithms have been applied to discover the clusters of co-expressed genes so that genes have similar function can be revealed. Fuzzy logic-based algorithm has been developed for the inference of gene regulatory networks.

17.25 LARGE NETWORK ANALYSIS

Once NW has been reconstructed, one can compare and contrast them in terms of global and local topology and to relate structural properties of NWs to protein properties such as function or essentiality.

17.25.1 Network Topology

Topological analysis of the networks provides quantitative insight into their basic organization. The network theory used to model complex network assumes that each node in a network is connected to another node randomly with probability p and the links (degrees) of nodes follow a Poisson distribution which has a strong peak at the average degree, K. Most random networks are highly homogeneous in that most nodes have the same number of degrees, $k_i \approx K$ where k_i is the degree of ith node. The probability of having nodes with k links falls off exponentially (i.e., $P(k) \approx e^{-k}$) for large k. For explaining the heterogeneous nature of complex networks, a 'scale-free' model was proposed by Barabasi and colleagues (1999) in which the degree distribution in many large networks follows a power-law distribution[$P(k) \approx k^{-1}$]. In this distribution most of the nodes within the networks have few links but a few (termed the **hubs**) are highly connected. Further, Watts and Strogatz ((1998) found that many networks are defined as being both highly clustered and containing small characteristic path lengths. The four topological parameters are as follows.

1. **Average degree (K)** — The degree of a node refers to the number of links that this node has with other nodes. The average degree of the whole network is the average of the degree of all of its individual nodes.

2. **Clustering coefficient** — It is defined as the ratio of the number of existing links between the neighbors of a node and the maximum possible number of links between them. The clustering coefficient of the network is the average of the individual coefficients. This statistic is used to determine the completeness of the network. The clustering coefficient can not be calculated for the directed networks.

3. **Characteristic path length (L)** — The graph theoretical distance between two nodes is the minimum number of edges that is necessary to traverse from one node to other. The characteristic path length of a network is the average of these minimum distances and thus gives a measure of how close nodes are connected within the network.

4. **Diameter (D)** — It refers to the longest graph theoretical distance between any two nodes in the graph.

Scale-free networks can arise when a network grows through new nodes being linked preferentially to the most highly connected existing nodes. There is a small number of highly connected nodes but the majority of the nodes have few connections.

17.26   SUBSTRUCTURES WITHIN NETWORKS

Protein-protein physical interaction networks and regulatory networks usually contain biologically meaningful substructures. In case of former, protein complexes will theoretically appear as a clique, a fully connected subgraph but because of the limitations with the identification techniques, some of the links within the same complex may be missing.

17.27   CROSS-REFERENCING DIFFERENT NETWORKS

Mapping the networks in model organism to other species by homology will provide insight into how to exploit the usefulness (and to prevent the potential pitfalls) of annotating unknown genes in other less characterized species. To this end, Walhout et al. (2000) introduced the concept of 'interlog' to transfer interaction network from one species to another. Interlogs are defined as orthologous pairs of interacting proteins in different organisms. For example, if proteins X and Y in one organism have interacting orthologs X' and Y' in another organism then the X-Y and X'-Y' interactions are called interlogs. Cross-species comparison of interaction networks will tell us how these networks evolve. Similarly, comparison of different networks within the same organism will throw light on the basic organization principles of the cell. Yu et al. (2003) showed that coregulated genes are generally coexpressed and the correlation in expression profiles is highest for genes targeted by multiple transcription factors. Furthermore, coregulated genes tend to share cellular functions and there are subdivisions within individual network motifs that separate the regulation of genes of distinct functions. The expression profiles of transcription factors and their target genes show more complex relationships than simple correlation with the regulatory response of target genes often being delayed.

17.28 NEURAL NETWORKS

There are other kinds of networks whose primary function is to transform a set of inputs into a second set as outputs and this is the case with neural network. Neural networks are set of interconnected nodes, each of which has a state that depends on the integrated inputs from other nodes. Neural networks have the ability to learn different patterns of inputs by changing the strengths of their connection and are used in a variety of tasks of machine recognition. The closets approximation to a neural network is probably found in the pathways of intracellular signals (Bray, 1995). As do protein signaling networks, neural networks function to process information between input and output nodes. Like biological networks, neural networks are optimized by an 'evolutionary' tinkering process of adding and removing arrows and changing their weight until the neural network performs a given computational goal (gives the correct output responses to input signals). Unlike biological networks, however, neural networks are non-modular. They typically have a highly interconnected architecture in which each node participates in many tasks. In case of a living cell, multiple receptors on the outside of a cell receive sets of stimuli from the environment and relay these through cascades of coupled molecular events to one or a number of target molecules (associated with DNA or the cytoskeleton). Because of the directed and highly interconnected nature of these reactions, the ensemble as a whole should perform many of the functions commonly seen in neural networks. The signaling pathways are capable of recognizing sets of inputs and responding appropriately with their connection 'strengths'. One combination of external conditions might trigger cell division whereas another might cause differentiation. Further, some signaling molecules in the pathway should perform the function of 'hidden units' which embody in their state of activity, an abstraction of the outside world. Finally, the networks of cell signaling reactions, like highly connected neural networks, be resistant to damage and continue to function even if some of their connections are severed. For more on neural networks, see Bioinformatics (Roy, 2009).

17.29 PATHWAYS AND NETWORKS

Each living cell can be treated as a network. Everything in a living cell is connected to everything else and interactions between macromolecules through multiple covalent binds are the very fabric of life. Thousands of components of a living cell are interconnected so that cell's functional properties are ultimately encoded into a complex intracellular web of molecular interactions. Cellular metabolism is a fully connected biochemical network in which hundreds of metabolic substrates are densly integrated through biochemical reactions. Within this network, however, modular organization (i.e. clear boundaries between subnetworks) is not immediately apparent. Revealing modular structures in biological networks will assist us in understanding how cells function. Thousands of enzyme catalyzed reactions in cells are functionally organized into many different sequences of consecutive

$$A \xrightarrow{\text{Enzyme 1}} B \xrightarrow{\text{Enzyme 2}} C \xrightarrow{\text{Enzyme 3}} D \xrightarrow{\text{Enzyme 4}} E \xrightarrow{\text{Enzyme 5}} F$$

FIGURE 17.7 **Pathway of biosynthesis of isoleucine.**

reactions called pathways in which the product of one reaction becomes the reactant of the next as shown in Figure 17.7.

This figure shows that between the input and the output there is an organized set of intermediates sub-reactions. *In E.coli* the amino acid threonine is converted to isoleucine in five steps where each stepis catalyzed by a separate enzyme as shown in the figure 17.7. The letters A through F represent the compounds or intermediates in this pathway of synthesis of isoleucine. This biosynthetic pathway is regulated via 'feddback inhibition'. If a cell produces more isoleucine than is required for protein synthesis then the higher concentrations of the accumulated unused isoleucine inhibit the catalytic activity of the first enzyme in the pathway and thus the production of the amino acid isoleucine slows down. Living cells also regulate the synthesis of their own catalysts (the enzymes). Pathway is thus referred to as macroscopic series of discrete events with a functional outcome. Two sets of pathway can be extensively interwined with several key intermediates in common. For example, there are many distinct signaling pathways which allow the cell to receive process and respond to information and often components of different pathways interact resulting in signaling network. In other words, signaling pathways interact with one another and their final biological response is shaped by interaction between pathways. These interactions results in networks. Similarly, higher order functions like cell division or vesicle trafficking which are composed of orchestrated interactions of one or more pathways. We need to know how proteins function together and organized into pathways and how these pathways are integrated into yet larger contexts. Major protein interactions can be organized as a pathway interactome. Networks and interactome pathways maps are useful tools to help explain a biological pathway but they need to be anchored in biological context and experimental facts. This can be done by embedding interactome within well characterized pathways which will further assist in other pathway investigations. Thus what is required to know is how proteins function together and are organized into pathways and how these pathways are integrated into yet larger context. Plotting the pathway protein interactome allows one to identify the key players (hubs) and approximately where they work in the pathway. A conceptual thinking coupled with advances in interactome mapping and in visual immunoprecipitation is required which will ultimately lead to a better understanding of complex biological processes.

In case of chemical reactions occurring in solution, a system is defined as all of the reactants and products, the solvent and the immediate environment, i.e. everything within a defined region of space. Chemical kinetics reveals the mechanism of a chemical reaction by establishing the sequence and rates of individual elementary steps.

In metabolic pathways (in multiple chemical reactions) complex networks of reactants and products may be set up that involve numerous elementary steps and multiple feedback whose significance varies at different overall concentrations. The reductionist approach of identifying and isolating individual reaction steps can be experimentally difficult in that it requires purifying the individual complements and finding all of the interactions that constitute the complete mechanism. These problems are especially prevalent in enzymatic and genetic reaction systems in which the catalyst for a single reaction can have tens of interacting effectors. Correlation matrix construction takes a different approach to the analysis of complex chemical kinetics. A chemical reaction system operating near a steady state (but far from equilibrium) is subjected to random perturbations in the concentrations of a set of input species. Given that we can measure chemical concentrations of species *in situ*, we can analyze the response of these concentrations to input variation. From these time series data, the method is used to deduce the reaction pathway underlying the response dynamics. Understanding the connections between the various metabolic and regulatory pathways that control growth and development is necessary. While progress depends on reductionist approaches toward analyzing individual pathways, much is to be learned from viewing the multitude of pathways as the interconnected regulatory networks.

Signal transduction—To unravel intracellular communication and elucidate the signaling pathways involved, it is essential to identify signaling mechanism (identification of components involved) evoked by environmental and biological changes. Factors such as hyperosmotic stress, wounding and pathogen attack trigger signaling which induces the appropriate response. Signal transduction refers to the transmission of molecular signal from the surface of the cell to the targets (cytoplasmic, nuclear or organelle targets) leading to a cellular response. Signal transduction is a process by which extracellular molecules influence intracellular events. Many pathways originate with the binding of extracellular ligands to cell surface receptors. In other words, transduction starts with receptors, often in the plasma membrane that perceives the change and relay the information into the cell. This response of the cell to extracellular stimuli is mediated by a wide variety of biochemical processes. Signal transduction from a plasma membrane can be organized into three components: **receptors, effectors** and **targets**. The membrane–bound receptor interacts with the extracellular molecule and transmits information to the cytoplasmic portion of the memebrane. This information is conveyed by an intermediary system referred to as an effector which generates different types of messengers. The second messenger system generates further specificity in response and amplifies the extracellular signal (chemical, mechanical or electrical). Second messenger refers to molecule which forms in the cytosol in response to an extracellular signal or released acting as second messenger and which helps to transfer the primary signal into the cell and amplify it (e.g. cAMP, IP3 and Ca^{2+}). Finally, the messages are received by targets which are often enzymes which may undergo co-valent modifications thereby altering the activity. In reality, the signal transduction system represents a complex network of pathways which allow hormone to elicit their characteristics effects on cells. In summary, the membrane –bound so-called receptor kinases interact with their ligand (e.g. a hormone) then they phosphorylate proteins which in turn are transferred to the nucleus where they bind to target DNA sequence (s) and turn gene (s) off or on (response).

In case of hormonal signal transduction the α-subunits of heterotrimeric G proteins interact directly with one or more effector proteins in the plasma membrane to induce release of soluble second messengers which in turn amplify hormonal signal into the cell. The four distinct types of effectors that have been studied are adenylate cyclases, phospholipases, ion channels and phosphodiesterases, all of which are transmembrane proteins or proteins closely associated with the plasma membrane and they interact with receptor system to generate second messengers especially, cyclic AMP, diacylglycerol, inositol phosphates and calcium. The four types of receptors that generate signals are G-protein-coupled receptors, ion channels, kinases and receptors binding to cytokinins. Signal transduction and control of gene expression have been merged and become an area of **developmental genetics**.

Types of receptors—Pathways are mediated by three types of receptors: brassinosteroid receptors (BR1) which control plant size, G-protein-coupled receptors that control mating responses in yeast and Notch receptors that control cell fate in animals. Brassinosteroids are plant hormones that contribute to cell growth and division, differentiation and reproductive development. BR1 is a plasma membrane-localized leucine rich-repeat receptor kinase which initiates a kinase cascade ultimately controlling gene expression (Belkhander and Chory, 2006.

In Gprotein-coupled receptor pathway the Gβγ subunits activate the mitogen-activated protein kinase cascade. Gα subunit participates in transmitting the mating signal by interacting with phosphoinositide 3-kinase (PI3K) at the endosome to stimulate the production of phosphoinositide 3-phosphate (Slessareva and Dohlman, 2006).

Notch signaling pathway is crucial to animal development and aberrant activity of this pathway is associated with certain type of leukemia. The transmembrane Notch receptor interacts with the transmembrane ligand on adjacent cells which leads to cleavage and release of the Notch intracellular domain (NICD) which translocates to the nucleus and regulates gene expression (Ehebauer et al., 2006).

In cells, signaling proteins which make up a pathway are often physically organized into complexes by scaffold proteins. Scaffolds direct information flow. They promoter signaling between proper protein partners and prevent improper cross talk. They may also play a role in shaping the quantitative response behavior of a pathway. The scaffold complex could serve as a central hub for feedback or activity of pathway members on the scaffold. Such feedback loops could tune pathway dose response and dynamics-the change in output over time. Quantitative response behavior is critical for signaling. The behavior of a pathway must match its specific physiological function (Bashor et al., 2008).

Apoptosis (programmed cell death) refers to an active process of cell death and is morphologically characterized by nuclear condensation and blebbing of the plasma membrane. Activation of endonucleases during the process of apoptosis leads to the fragmentation of chromatin into multiples of ~ 180bp. Chromatin condensation and membrane blebbing are hallmarks of apoptosis. Apoptosis is one of the central cellular processes in development, the stress response, aging and disease in multicellular eukaryotes. The signal transduction pathways involved in programmed cell death are mediated by interactions between well characterized domains that include extracellular domains, adaptor (protein-protein interaction) and those found in effector and regulatory enzymes. In case of mutants, basidia which experience problems at the beginning of meisosis (prophase I) undergoes mass apoptosis showing that the classical apoptotic hallmark of DNA fragmentation. Apoptosis is triggered at a single check point in the mushroom cell cycle. In case of mammals (mouse where the switch which activates apoptosis can be tripped at many steps throughout meiosis implying that there is an almost continuous molecular assessment of the viability of the gamete forming cells (Lu et al., 2003).

Whether the cell dies in response to diverse developmental clues or cellular stresses is determined largely by interactions between three factions of Bcl-2 protein family. Two factions promote apoptosis. The BH3-only proteins (including Bim, Bid, Puma, Bad and Noxa) sense cellular damage and Bax and Bak are critical downstream mediators of apoptosis as their combined absence abolishes most apoptotic responses (Willis et al, 2007).

Model for initiation of apoptosis proposes that a subset of BH3-only proteins termed 'activators' namely Bim and Bad and perhaps Puma directly engage Bax or Bak. The other BH3-only proteins termed 'sensitizers', for example, Bad and Noxa, purportedly act only by displacing the activators to bind Bax and Bak.

17.30   NETWORKS

Biological networks are abstract representations of biological systems which capture many of their essential characteristics (Alon, 2003). The function of a biological system relies on a combinatory effect of many semantic elements which interact non-linearly. There is a need to take global view of entire biological network at many levels of abstraction to manage complex biological states. In the network molecules are represented by nodes and their interactions are represented by edges (or arrows). The cell can be viewed as an overlay of at least three types of networks which describes protein-protein, protein-DNA and protein-metabolite interactions. The most challenging molecular network in a cell is that governs gene expression. Tens of thousands of genes direct the formation of proteins, many of which then collaborate to control, in reciprocal fashion, the expression of other genes. Many different mechanisms of transcription regulation may be described by a single type of arrow. Furthermore, interactions can be of different strengths so there should be numbers or weights on each arrow. Whenever two or more arrows converge on a node, an input function needs to be specified. Figure 17.8 shows some elementary interaction patterns. Biological systems viewed as networks can readily be compared with engineering systems and thus biological networks can be understood using concepts from engineering. It shares the principles of modularity, robustness to component tolerance and use of recurring circuit elements. Modularity is a property of biological network. Cellular functionality can be seamlessly

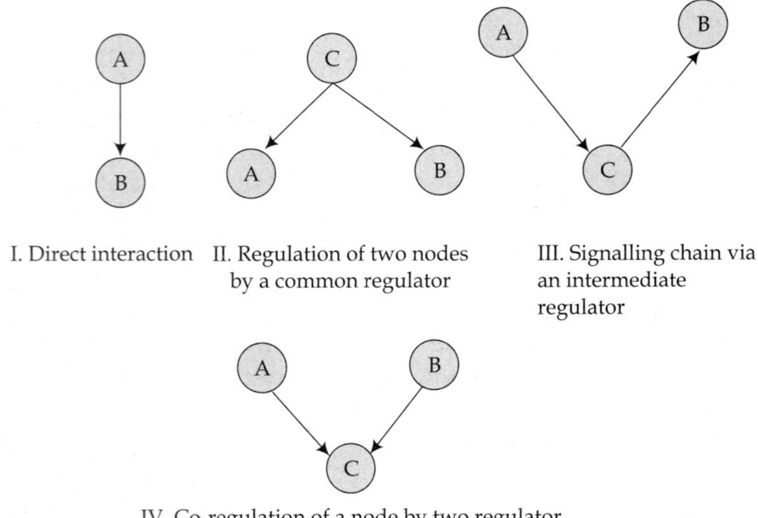

FIGURE 17.8   Showing elementary interaction patterns.

partitioned into a collection of **modules**. Modules are composed of many types of molecule. One discrete biological function can only rarely be attributable to an individual molecule and complex set of molecules interact to form functional modules. Each functional module is a discrete entity of several elementary components and performs an identifiable task, separated from the functions of other modules. Their discrete functions arise from interactions among their components (proteins, DNA, RNA and small molecules). Most functional properties of a module are collective properties arising from the properties of the underlying components and their interactions. Modules can be insulated from or connected to each other. Insulation permits cell to carry out many diverse reactions without crosstalk whereas connectivity allows one function to influence another. Spatially and chemically isolated molecular machines or protein complexes (such as ribosomes and flagella) are such functional units. A ribosome is a module which synthesizes proteins. A signal transduction system such as those that determine chemotaxis in bacteria or mating in yeast is an extended molecule (Hartwell et al., 1999). Modules for protein synthesis, DNA replication, glycolysis and even parts of the mitotic spindle have been reconstituted *in vitro*. Modularity is at the root of success of gene functional assignment by expression correlation. Proteins are known to work in slightly overlapping, co-regulated groups such as pathways and complexes. A module in a network is a set of nodes that have strong interactions and a common function. A module has defined input nodes and output nodes that control the interactions with the rest of the network. Further, a module also has internal nodes that do not significantly interact with nodes outside the module. Engineering system also use modules such as subroutines. Modular structures can be less optimal than neural net-style non-modular structures. Modules greatly limit the number of possible connections in the networks and usually a connection can be added which reduces modularity and increases the fitness of the network. Modular networks may have advantage over non-modular networks in real-life ecologies which changes over time. Modular networks can be readily configured to adapt to new conditions.

Modules are evolved to perform specific biological functions. Biochemical processes are taking place within the modules. The aim here is to determine the relationship of inputs to outputs of modules, their biochemical connectivity and the states of key intermediates within them. To achieve this task there are three complementary approaches. 1. Methods for perturbing and monitoring dynamic processes in cells and organism. 2. Reconstituting functional modules from constituent parts or designing and building new ones 3. Quantitative description and modelling of modules. Although monitoring of cellular processes can be made in genome-wide analyses of gene expression, there is a need of better methods of finding patterns which identify networks and their components, of finding possible connections among the components and of restructuring the evolution of modules by comparing information from various organisms. For understanding of cells , it is necessary to make quantitative predictions about behavior and their testing. This requires simulations of biochemical processes taking place inside the modules.

Engineers are also designing circuits to perform specific functions. The properties of module's components and molecular connections between them are similar to the circuit diagram. The circuitry of modules can be deduced by listing their component parts and determining how changing the input of the module affects its output. This is called reverse engineering. It is unlikely the one can deduce the circuity or higher level description of a module solely from genome-wide expression analysis and physical interactions between proteins.

Metabolic networks use regulatory circuits such as feedback inhibition in many different pathways. The transcriptional network of *E.coli* has been shown to display a small set of recurring circuit elements termed '**network motifs**'. Each network motif can perform a specific information processing task such as filtering out spurious input fluctuation and generating temporal programs of expression or accelerating the throughput of the network. Finding a network motif in a new network may help explain what systems-level function the network performs and how it performs. Analysis of binding sites of transcription factors not only documents potential pathways to regulate gene expression but also identifies network motifs, the simplest unit of network architecture.

17.31 HIERARCHICAL ORGANIZATION OF NETWORKS

Networks are invaluable tool for describing and quantifying complex systems. Networks often show hierarchical organization in which vertices divide into groups which further subdivide into groups of groups and so forth over multiple scales. In many cases groups are found to correspond to known functional units such as modules in biochemical networks (protein interaction networks, metabolic networks or genetic regulatory networks). Clauset et al (2008) presented a general technique for inferring hierarchical structure from network data and showed that existence of hierarchy can simultaneously explain and quantitatively reproduce many observed topological properties of networks such as right-skewed degree distribution, high clustering coefficients and short path lengths. Further, knowledge of hierarchical structure can be used to predict missing connections in partly known networks with high precison.

17.32 IDENTIFICATION OF NETWORK

Network identification by multiple regression (NIR) is derived from a branch of engineering called **system identification** in which a model of the connections and functional relationship between elements in a network is inferred from measurements of system dynamics (e.g., the response of genes and proteins to external perturbations). To apply a system identification method one must assume the behavior of a gene, protein and metabolite regulatory network can be modeled by a system of non-linear differential equations. Construction of a first order predictive model of gene and protein regulatory network can be done using only a steady state expression measurement. Use of multiple regression is made to determine the model from RNA expression changes resulting from a set of steady state transcriptional perturbations. The assumption is that the behavior of a gene, protein and metabolic regulatory network can be modeled by a system of non-linear differential equation.

$$dx/dt = Ax + u$$

where dx/dt denotes the rate of accumulation of the species in x; x is a vector denoting the concentrations of nuclear RNAs, proteins and metabolites in the network; u is the vector representing an external perturbation to the rate of accumulation of the species in x and A, the network model, is an $N \times N$ matrix of the coefficients describing the regulatory interactions.

Although NIR is highly effective in inferring small microbial gene networks, it requires prior knowledge of which genes are involved in the network of interest and the perturbations of all the genes in the network via the construction of the appropriate episomal plasmids. In addition, it requires the measurement of gene expressions at steady state (i.e. constant physiological conditions) after the perturbations. This step is challenging for large networks and so not easily applicable to higher organisms (Bansal et al., 2006).

17.33 FEATURE OF NETWORK

The most basic feature of any network is its architecture which places boundaries on how it acts and how it might have been formed. A large collection of nodes (representing molecules in biology) can be arranged is two-dimension and they can be connected in a variety of ways. Further, a few nodes can be given a very large number of connections and the rest can be allowed to have relatively few.

Types of architecture — A regular network with nearest neighbors connected tends to be 'cliquish', having local groups of highly interconnected nodes. A random network lacks cliquishness but is easily traversed because the number of steps between any two nodes is relatively small. Scale-free networks are distinguished by the presence of a few highly connected nodes and are both cliquish and easily traversed. The extent to which neighbors of a node are themselves connected (referred to as its' clustering coefficient' or cliquishness) is almost as large as in a regular network. The clustering coefficient, C_i, is defined as $C_i = 2n/K_i (K_i-1)$ (Watts and Strogatz, 1998) where n denotes the number of direct links connecting K_i nearest neighbors of node i. $C_i = 1$ for a node at the center of a fully interlinked cluster and is 0 for metabolite that is part of a loosely connected group. Therefore, C_i averaged over all nodes I of a metabolic network is a measure of network's potential modularity.

Complex network thus can be divided into two major classes based on connectivity distribution P (k), giving the probability that a node in the network is connected to k other nodes. 1. The **exponential network** is homogeneous: most nodes have approximately the same number of links. 2. The **scale-free network** is inhomogeneous: the majority of the nodes have only a few (one or two) links but a few nodes (called **hubs**) have a large number of links, guaranteeing that the system is fully connected (Figure 17.9). The interconnectedness of a network is described by its diameter d, defined as the average length of the shortest paths between any two nodes in the network. The diameter characterizes the ability of the two nodes to communicate with each other. The smaller the diameter d, the shorter is the expected path between them. Network with a very large number of nodes can have quite a small diameter. Thus identification of architecture of network involves determining whether it is best described by an inherently uniform exponential topology with proteins on average possessing the same number of links or by a highly heterogeneous scale-free topology in which proteins have widely different connectivities. An important consequence of the inhomogeneous structure is the network's simultaneous tolerance to random errors, coupled with fragility against the removal of the most connected nodes. Studies have shown the emergence of inhomogeneous structure in both metabolic and protein interaction networks.

The best characterized molecular network that shows scale-free characteristics is that of interlinked pathways of metabolism- pathways of enzymatically catalyzed reactions that

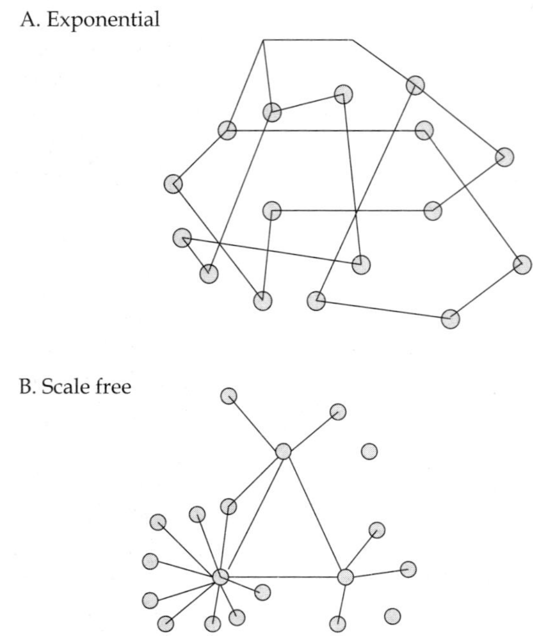

FIGURE 17.9   A and B showing exponential and scale free network, respectively.

interconvert the hundreds of small molecules of a cell. Small molecules such as pyruvate or coenzyme A, are large hubs whereas the average molecule undergoes just one or two reactions. The number of catalytic steps required to go from any one compound to any other is small and metabolic networks have a high clustering coefficient which suggests the presence of local cliques or clusters of connected molecules. The distinguishing feature of such a scale free metabolic network is the existence of a few highly connected nodes which participate in a very large number of metabolic reactions. With a large number of links, these hubs integrate all substrate into a single, integrated web in which the existence of fully separated module is prohibited by definition.

Hubs — The basic elements of the interactome (individual proteins and lipids and so on) are called nodes and their interactions are shown as links connecting the nodes. Stable protein complexes like the adaptor protein complex AP2 or a clathrin triskelion are considered as single nodes. Some protein complexes can have disproportionately more interactions than other proteins in a pathway and so fall into the definition of 'hubs'. There are two types of hubs: party and date. **Party hubs** are biomolecules which interact with most of their partners simultaneously (static hubs). Date hubs are biomolecules which have many partners but interact with them at different times or locations and so are dynamic hubs (Han et al., 2004). **Date hubs** are proposed to organize the proteome, connecting the biological processes to each other whereas party hubs are proposed to function with modules (which may be pathways) (Schmid and McMahon, 2007). Clustered hubs are a new subtype of hubs which refer to cases in which many pathway/party hub proteins oligomerize or cluster to function as pathway hubs. Hub centric organization of a pathway has at least four major consequences: Ease of module attachment, dynamic instability, pathway progression and fidelity and flexibility. Ease of module attachment is a feature of network. Hubs act as the control points where these modules can be plugged into the system to give extra capabilities. A model of organized modularity in which date hubs (represent global or higher level) organize the proteome, connecting biological processes or modules to each other whereas party hubs function inside modules (at lower level). A test for proteins that function as hubs only when clustered in space and time is that over expression of such a protein should not have any phenotype on the pathway. Conversely, over expression of nodes that bind directly to hubs would be predicted to have disastrous affects owing to titration of hub interaction points. The hubs in interactome network can be classified into date and party on the basis of their partner's mRNA expression profiles. Deletion of date hubs will lead to more severe effects on cell viability owing to the additive result of disrupting a frequently used module and thus affecting desparate pathways whereas deletion of a pathway hub will have more limited consequences. Hubs in proteome are generally not directly connected to other hubs and this means that the organization of each pathway is independent with the loss of one hub not affecting other pathways in the cell.

17.34 PATHWAY CONSTRUCTION

It is widely accepted that proteins rarely act in isolation while performing their functions. The analysis of proteins with known functions shows that proteins involved in the same pathway interact with each other and thus an approach for elucidating the function of an unknown protein is to identify other proteins (some of which with known functions) with which it interacts. Thus mapping protein-protein interactions on large –scale has not only provided insight into protein function but facilitated the modeling of functional pathways to explain the molecular mechanisms of cellular processes.

17.35 BIOCHEMICAL NETWORKS

In general, a biochemical network is a collection of enzymatic reactions that serve to process metabolites within the cell and to convert intercellular metabolites into intracellular metabolites and vice-versa. Biochemical networks include metabolism, cell cycle and intercellular signaling pathways. A metabolic pathway refers to a set of enzyme catalyzed biochemical reactions by which an organism transforms an initial (source) compound into a final (target) compound. A metabolic pathway converting one compound into another can be shown as a directed graph. A database of reactions/compounds can be utilized to enumerate possible paths fulfilling various constraints from the source compound to the target compound. One of these paths must then by definition correspond to the set of reactions/ compounds involved in the experimentally determined (observed) pathway. Any mathematical model for metabolic pathway recovery must accomplish two tasks, prediction and disruption. A model must be able to predict pathways, e.g. those resulting should a reaction not function due to a genetic defect meaning a catalyzing enzymes is not available. Secondly, it must assist in investigating how to disrupt pathways, e.g. by finding those reactions/enzymes that should be disabled so as to prevent efficient pathway functioning. Given sufficiently dense information on the concentrations of specific molecular species in a cell type under many conditions and multiple time points, it should be possible to investigate any biological pathways automatically. Quantitative data are available primarily for gene expression but not generally for translated proteins or metabolites. The proteomic methods are just beginning to offer a solution to quantifying protein levels. The problem with expression data is that they are noisy and the regulation of gene expression is complex, often controlled by combinatorial factors.

17.36 APPROACHES TO ESTIMATION OF PARAMETERS OF MODELS OF BIOCHEMICAL NETWORK

The estimations of parameters of mathematical models of biochemical networks have classically been done using reductionist approaches. Amongst such approaches are those that use isolation and those that use perturbations. In case of former, kinetic parameters are typically estimated one at a time, typically *in vitro*, by isolating a relevant network constituent. Substantial kinetic information typically in the form of traditional rate laws have been determined in this way, for example, from enzymatic assays. In case of the latter, kinetics are approximated based on system responses to single perturbations from a steady state.

Limitations with these approaches — Both these approaches are limited in their applications especially when taken in a systemic perspective and there are a number of reasons. One reason is that isolation is not always possible and even not desirable when feasible. Another reason is that there are constraints as to which singular perturbations are possible. Further, it is well known that component isolated and studied *in vitro* do not always behave in the manner when studied *in vivo*.

17.37 SYSTEM LEVEL METHODS FOR ESTIMATION OF PARAMETER ESTIMATION

Living organisms can be viewed as complex biochemical systems that require researches about their essential components and interactions. These can be studied through mathematical models and computer simulations. Molecular production, degradation and interactions are modeled with chemical kinetics. The availability of biochemical profiles has made possible the system-level methods for parameters estimation. Biochemical profiles are simultaneous measurements of biochemical which are currently taken as single snapshots or as a sequence of snapshots forming a time series. These profiles are valuable as they carry information regarding the network structure and dynamics of the underlying systems. Biochemical systems are conveniently modeled by non-linear dynamical systems called S-systems. S-systems are systems of coupled ordinary differential equations adhering to the power-law formalism. S-systems takes the following forms.

$$Xi = \alpha i \; II \; Xj - \beta \; II \; Xj \text{ , for } I = 1, 2, 3, \ldots, n.$$

Where n is the number of dependent variables, m is the number of independent variables, Xj's denote the current states (typically as concentrations) of the different metabolites, Xi's represent the rates of change in concs, rate parameters αi, βi and kinetic orders gij and hij are parameters measuring the effect of Xj on Xi.

18

Analysis of Single Nucleotide Polymorphism

18.1 STUDY OF SEQUENCE VARIATION

Sequence variants are responsible for the genetic component of individuality such as disease susceptibility and drug response. A sufficiently dense map of SNPs would allow the detection of sequence variants responsible for particular characteristics on the basis that they are associated with a specific SNP allele. SNPs can be defined as alleles which exist in normal individuals in a population with the least frequent allele having an abundance of at least 1%(Brookes, 1999). In principle, SNPs could be bi-, tri- or tetra-allelic variation but tri- and tetra-allelic SNPs are very rare in humans. In practice, the term SNP is often applied in a more generic context and may encompass disease-causing mutations which are recessive, or low-penetration dominant allele. SNPs are stable, bi-allelic sequence variants where two alternate bases occur at one position. Sequence variations which are due to substitutions that occur at individual base pair are called SNPs.´

Genes containing one or more SNPs can give rise to two or more allelic forms of mRNAs. mRNAs containing different bases at SNP sites may vary in their interactions with cellular components involved in mRNA synthesis, maturation, transport, translation and degradation. A number of single base pair substitutions can alter or create essential sequence elements for splicing, processing or translation of mRNA. In other words, these mRNA variants may possess different biological functions as a result of differences in primary or higher order structures that interact with cellular components. Evidences show that the folding of mRNA influences a diverse range of biological events such mRNA splicing and processing and translational control and regulation(Shen et al., 1999).

18.2 IDENTIFICATION OF SNPs

SNPs in transcribed regions were found by analyzing the clusters of ESTs(Buetow et al., 2001); Irizarry et al., 2000) or by aligning ESTs to human reference set. The genomic

SNP(single base pair variations found by analyzing genomic sequence clones without regard to whether they represent exonic DNA) were discovered in sequences from restricted genome representation libraries, random shot-gun reads aligned to genomic sequence and in the overlapping sections of large insert clones(mainly BAC) that make up to human reference genome.

18.2.1 Cloning Experiment

Fundamentally, one identifies a SNP by comparing two or more sequences from the same region on the chromosome. It can be done easily if the DNA sequence is of high quality and the sequence data are derived from cloned DNA because each clone comes from a single copy of one of the two chromosomes in the diploid cell.

18.2.2 Application of PCR for SNP Genotyping

Almost all SNP genotyping assay techniques use PCR to amplify DNA. In case of identifying SNPs in targeted regions in the genome one amplifies genomic DNA using PCR and sequences from the PCR products derived from different individuals. In this procedure the SNP discovery is complicated by the fact that the same region on both chromosomes in diploid cell are amplified by PCR and so some bases will be heterozygous in one or more individuals. A computer tool will be able to identify a SNP even when only heterozygotes and homozygotes of just one of the two alleles are present in the samples sequences. This is not a trivial problem as the commonly used dye-termination-based DNA sequencing methods yield peaks of uneven heights at the polymorphic sites and the base-calling algorithm will frequently miscall the base at these sites in the sequences of heterozygous individuals.

To discover SNPs, several copies of each locus must be sampled from a population and compared for sequence differences. The locus-specific PCR amplification requires the synthesis of oligonucleotide primers for each locus, limiting it to regions of known sequence and making it expensive for large-scale approaches. Moreover, it produces diploid genotypes. This requires identification of SNPs as heterozygotes which is technically challenging as described above.

Approaches exploring sequence variation of individual genes in depth involves analysis of sequence differences in clusters of ESTs or by re-sequencing DNA fragments after amplification from different individuals.

18.3 SINGLE NUCLEOTIDE POLYMORPHISM

Computational discovery of polymorphism in sequence data follows a four steps procedure.

1. Sequences of high similarity from multiple sequence individuals are identified using a BLAST program.
2. To avoid spurious similarity due to known repeats, sequences are masked for high copy number repetitive elements with REPEATMASKER. MASKERAID is a new resource which increases the speed of masking more than 30-fold(Bedell et al., 2000). Still there exists the possibility of sequences originating from regions of as yet uncharacterized chromosomal duplications.

3. Inclusion of a second, paralogue-filtering step into the procedure can reduce false positive SNP arising from comparing paralogous sequence copies. This will result in reduction of false prediction due to paralogy.

4. Construction of a base-wise multiple alignment of the sequences(discovering SNPs in clusters of cDNA sequences). Aligning expressed sequences is even more complicated because of exon-intron punctuation and possible alternative splice variants.

5. Sequences in the precise, base-to-base multiple alignment are scanned for nucleotide differences.

Discrimination between true polymorphism and sequencing error is through use of statistical tool based on measure of sequence accuracy or base quality values(Ewing and Green, 1998; Ewing et al., 1998). Prediction is accompanied by a measure of confidence. Accurate confidence values permit one to use the highest number of candidates with an acceptable false positive rate.

18.4   METHODS FOR IDENTIFYING SEQUENCE VARIATION

There are two methods for identifying sequence variants affecting phenotypes. Sequencing method involves direct sequencing of individual genomes and comparing the nucleotide sequences of genomes of two individuals. This method is not cost effective. The second method is called non-sequencing method.

18.5   NON SEQUENCING METHODS FOR SNP DISCOVERY

The most promising technique is the use of high density oligonucleotide arrays(DNA chips) for SNP discovery and validation(Dong et al., 2001). Patil et al. (2001) used high-density oligonucleotide arrays in combination with somatic cell genetics to identify a large fraction of all common human chromosome 21 SNPs and directly observe the haplotype structure defined by these SNPs.

18.6   TILING STRATEGY

As each DNA strand carries with it the capacity to recognize a uniquely complementary sequence through base pairing, the process of recognition is highly parallel as every nucleotide in a large sequence can in principle be queried simultaneously. Thus hybridization can be used to efficiently analyze large amounts of nucleotide sequence. There are two approaches in which sequence can be analyzed. In one approach, sequences are analyzed by hybridization to a set of oligonucleotides, representing all possible subsequences. Another approach involves hybridization to an array of oligonucleotide probes designed to match specific sequences. Thus in this approach most informative subset of probes is used. Implementation of these concepts relies on the combinatorial technologies to generate ordered array of a large number of oligonucleotide probes(Fodor et al., 1991). An array consisting of oligonucleotides complementary to subsequences of a target sequence can be used to determine the identity of a target sequence, measure its amount and detect differences between the target and a reference sequence. Many different arrays can be designed for this purpose. One such design, is called a **4L tiled array**. In each set of 4 probes, the perfect complement will hybridize more strongly than mismatched probes. By this approach a nucleic acid target of length L can be scanned for mutation with a tiled array containing 4L probes. For example, to query the 16, 569 bps of human mtDNA , only 66,276 probes of the possible ~10^9 15-nt oligomers need to be used(Chen, 1996). In other words, a 4L tiled array is one in which L corresponds to the length of the sequence to be analyzed. The sequence is probed with a series of oligomers of length P which exactly match the target sequence except for one position which is systematically substituted with each of the 4 bases, A, T, G or C. A tiled array of 15 mers varied at position 7 from the 3′ end is called a $p^{15.7}$ array. Another design can use eight unique oligonucleotides(forward and reverse strand probe quartets), each 25 bases long for interrogating each unique position of the genome sequence.

SNPchip refers to any microarray onto which specific PCR amplified regions of the genes known to contain one (or more) SNPs are spotted. Using primer extension or quantitative PCR techniques SNPs can be detected on the chip. First, the various SNPs are amplified with 16-25 bases long SBE hybrid primers, containing a generic sequence tag at the 5′ end and a locus specific sequence terminating at the base 5′ to the individual SNP. The extension reactions are catalyzed by DNA polymerase in the presence of fluorescently labeled ddNTPs. Subsequently the fluorescently labeled SBE products are hybridized to a microarray containing the full length reverse complements of the tags(generic tag array). The spotted oligonucleotides are synthesized with 15dT residues at their 5′ ends to facilitate their quadruplicate attachment to the poly-L-Lysine-located or selane-treated microscope slides. The hybridization is performed in a sealed hybridization chamber. The arrays are then washed at high stringency, dries by centrifugation and scanned with argon laser and appropriate filter sets to discriminate between different fluorochromes. The four replicate spots on the assay have to show identical signals to be accepted. The SBE tags with bi-functional primers (unique sequence tag function plus locus-specific sequence) allows single base extension at multiple loci. As each locus is identified by distinct tag sequence the SNP screening procedure is highly multiplexed. The resulting product mixture can then be demultiplexed by hybridization to the reverse complements of the sequence tags in the glass slide array. In other words, SBE tags is a technique for the highly parallel genotyping of hundreds or thousands of SNPs that combine SBE with the hybridization of the SBE products to generic microarray.

High density Affymetrix yeast tiling microarrays with overlapping 25-nt oligomers spaced an average of 5 base pairs apart to provide complete and ~five fold redundant coverage of the entire *Saccharomyces cerevisiae* genome was constructed. This array design was previously employed to discover novel expressed sequences and to precisely map sites of transcription in humans. This design provides five to seven measurements of a given nucleotide's effect on hybridization efficiency which were used to predict the presence and location of SNPs and deletion break points through the entire yeast genome.

SNPs can be identified by using a conventional gel-based DNA sequencing to examine STSs distributed across the human genome. STS are short genomic sequences which can be amplified from DNA samples through PCR. One-third of STSs came from random genomic sequences whereas two-thirds came from 3'ends of ESTs (3-ESTs) and primarily representing UTRs of genes.Large-scale SNP identification approach involves hybridization to high-density probe arrays (Chee et al., 1996). Such DNA chips can be produced with parallel light-directed chemistry to synthesize specified oligonucleotide probes covalently bound at defined locations on a glass surface or chip. A target DNA sequence of length L can be screened for a polymorphism by hybridizing a biotin-labeled sample to a variant detector array of size 8L. For each position on both strands, the array has four 25-nt oligomer probes complementary to the sequence centered at the position. The four differ only in that the central(13th position) is substituted by each of the four nucleotides. Homozygotes (AA) for the expected sequence should hybridize more strongly to the perfectly complementary probe than to the other three probes containing a central mismatch. The presence of an SNP would be expected to give rise to a different hybridization pattern, with homozygotes (BB) showing strong hybridization to an alternative base and heterozygotes (AB) showing strong hybridization to two probes. The variant detection array thus signals the presence of sequence variation (by a change in hybridization pattern) and in many cases can indicate the nature of change (by a gain of signal at a specific mismatch probe) (Wang et al., 1998).

18.7 SNP ASSAY

SNP assay has three major components.
1. Allele discrimination methods
2. Reaction formats
3. Detection methods

18.7.1 Allele Discrimination Methods

There are four methods of allele discrimination.
A. Allele-specific hybridization
B. Primer extension (includes single base extension)
C. ligation
D. Invasive cleavage

A. **Allele-specific hybridization**—In this technique a RFLP is first cloned into an appropriate vector, sequenced and locus-specific primers oligonucleotides for standard PCR is designed. These primers are then used to amplify the corresponding locus with genomic DNA from different closely related organisms as templates. The resulting amplicons, differing by a SNP are then sequenced and used as allele-specific probes to detect allelic differences by hybridization. This technique is used for the detection of small deletions or insertions in DNA sequences that allows to discriminate between wild-type and mutant.

B. **Primer extension**—This technique is used to detect SNPs in target DNA. The target DNA, for example a gene, is first amplified with specific primers (appropriate forward and reverse primers) in a standard PCR and subsequently denatured. A single stranded primer is annealed to the single stranded target DNA such that primer ends exactly at the SNP site. After annealing the duplex exposes a 3'OH group for an extension catalyzed by DNA polymerase in the presence of four ddNTPs labeled with a specific fluorochrome. The matching ddNTP will then be incorporated and stops extension. The incorporated ddNTP is then identified by the specific fluorescence emission. Finally, a comparison with wild-type sequence at the SNP site allows identification of the type of SNP.

Single base extension — This technique uses primers ending directly adjacent to the SNP mismatch and used to incorporate the complementary, fluorescently labeled ddNTPs in a standard PCR. After incorporation of the ddNTPs the reaction is stopped and the extension product is electrophoresed and the incorporated nucleotide can be detected by laser-induced fluorescence. If primers with different 5′ tails and different fluorochromes (green, blue, yellow, orange) are used then in a single reaction, several SNPs can be detected. Since each ddNTP differs in its mass and the extended primer varies correspondingly, incorporation of the various ddNTPs can also be detected by MS.

C. **Ligation detection reaction** — This technique relies on the potential of DNA ligase to ligate adjacent oligonucleotides hybridized to the target DNA only if a perfect complementarity exists at the junction site. In this technique the template is first amplified by PCR and the so-called common probe annealed to the template immediately downstream of the nucleotide in question. After that two allelic probes are added, one with a 3′ terminal nucleotide corresponding to the wild-type allele and another with a 3′ terminal nucleotide complementary to the wild-type allele. Both probes compete for the template, i.e. anneal to the template adjacent to the common probe. A double stranded region containing a nick(i.e. missing phosphodiester bond) is formed at the target nucleotide position. The DNA ligase can only ligate the allelic probe (with perfect complementarity) to the common probe but not the mismatched probe. The allele-specific probes are either designed such that each has a unique length so that the two ligation products can easily be separated by size or the allelic probes are labeled with different fluorochromes so that they can be discriminated on the basis of their emission wave lengths. This technique can also be multiplexed, ie. several SNPs can be typed simultaneously.

D. **Cleavage fragment length polymorphism** — In this technique genomic DNA is amplified by PCR using an end labeled primer. The amplified products are purified and incubated with cleavase. A cleavase is an endonuclease which recognizes, bind and cleaves hair pin structure in DNA 5′ of transition from the single to the double stranded molecule. If DNA from organism say A forms two and DNA from organism say B produces three hair-pin loops in the same target region, then cleavase will produce three fragments in A and four in B. The cleavage products are separated by denaturing PAGE. The number of fragments in a cleavage fingerprint will allow to calculate the number of hair-pin loops in the target DNA. In other words, by comparison of the two cleavase fingerprints from two organisms, sequence differences in the target region can be detected and localized.

18.7.2   Reaction Formats

The reaction formats are either homogeneous reactions or solid phase reaction and the detection methods use product light emissions, product-mass measurements and electrical property changes in product (Kwok, 2001).

18.8   DESIGNING SNP PRIMER

SNP primer is an oligonucletide that matches a SNP site at its 3' or 5' flank, hybridizes to this site and can be extended by one fluorescently labeled dideoxynucleotide. In order to design the correct primers, on must first determine the method of assay. However, there are some basic guidelines used while designing primers for genomic sequences. All designs require obtaining sequence, repeat masking, setting experimental and design primers, picking primers and formatting the information. The flanking sequences for each SNP can be obtained from a variety of resources such as dbSNP and SNPper, the public databases (Riva and Kohane, 2001).Masking of the repeat or making repeat sequence unavailable to automated primer picking program prevents most unwanted amplifications. RepeatMasker and new resource MaskerAid assist in this work. Computer programs available for use in SNP detection program are PolyBayes (Marth et al., 1999), PolyPhred and Sequencher. PolyBayes was developed for *de novo* SNP discovery in unambiguous clonal sequence data. As the name suggests this algorithm uses a Bayesian approach to combine prior knowledge such as average polymorphism rate or expected transition to transversion ratio, with the base calls and base quality values of the sequence in the multiple sequence alignment. Prediction of each SNP comes with a SNP score (or true positive rate). For further detail see chapter 24.

18.9   LARGE-SCALE STRATEGIES FOR MAPPING SNP

Two high throughput strategies are: 1. Reduced representation strategy 2. Genomic-alignment strategy.

18.9.1   Reduced Representation Strategy (RRS)

In this strategy specific subsets of restriction fragments made from equimolar mixture of DNA isolated from unrelated individuals are repeatedly sampled by sequencing. Reads derived from the same fragment in different genotypes are aligned into clusters or cliques and high confidence sequence differences between any two reads (that is candidate SNPs) recorded. Experimental verification of a subset of these candidate SNPs (by re-sequencing to identify the SNP in individual samples of the original DNA panel) tests the criteria used in the computational detection of candidate SNPs (Altshuler et al., 2000; Mullikin et al., 2000). The use of 'reduced representations' reproducibility prepared subsets of genome, each containing a manageable number of loci to facilitate resampling can be made to discover the SNPs. Many properties could be used for preparing reduced representations. One of the simplest is to purify restriction fragments in a given size range. Thus SNPs could be discovered by mixing DNA from many individuals, preparing a library of approximately sized restriction fragments and randomly sequencing clones.

18.9.2   Genomic-Alignment Strategy (GAS)

In this strategy single reads obtained by shot-gun sequencing of a library of DNA fragments are aligned directly to available genomic sequence to detect the candidate SNPs. The whole-genome shot gun sequences random clones from the genomes of many individuals. It provides haploid genotypes (does not require previous knowledge of genomic sequence or PCR). Whole-genome shot gun is inflexible, however, requiring several-fold coverage of the genome before SNPs are discovered.

Comparison between two strategies   The two strategies are complementary. RRS strategy allows detection of SNPs throughout the genome without genomic sequence but the SNPs are not automatically mapped whereas the GAS requires genomic sequence and provides a map for each SNP. Further, in contrast to the RRS in which the library must be sequenced to a sufficient depth to obtain clusters of multiple reads fro SNP detection, GAS analysis minimally requires alignment of just one read against finished genomic sequence. GAS therefore should result in a higher efficiency of SNP detection. Like RRS the GAS should also yield high confidence SNPs.

18.10   SCANNING MTDNA FOR SEQUENCE VARIATION IN HUMAN

The analysis of sequence variation in the D-loop has been applied in the study of evolutionary genetics and in forensic situations while identification of SNP mutations in the mitochondrial coding regions has been associated with various maternally inherited or sporadically occurring pathologies. A number of approaches have been used to scan mtDNA for sequence variants including DGGE, chemical or enzymatic treatment, single stranded DNA conformational analysis and hybridization with allele-specific oligonucleotides or on oligonucleotide arrays as well as direct DNA sequencing. Among these methods direct sequencing analysis has got many advantages.

19

Analysis of Haplotypes

19.1   ASSOCIATION OR LINKAGE DISEQUILIBRIUM (LD)

We will see in chapter 20 that for identifying the genetic basis of quantitative traits F_1-derived populations are used. The alternative is to use natural populations for mapping traits by **Association analysis** or **Linkage disequilibrium**. LD increases mapping resolution substantially over the current capabilities of standard mapping population. It has the potential to identify a single polymorphism within a gene that is responsible for the difference in phenotypes. Thus LD can be applied in the dissection of quantitative traits and it has been used extensively to dissect human disease. LD refers to the non-random association of alleles at different loci. Thus with the availability of genome-wide map of SNP, LD can be used to map genes that cause disease in human. LD refers to correlations among neighboring alleles, reflecting 'haplotypes' descended from single, ancestral chromosomes. LD plays a central role in association analysis. The distance over which LD persists will determine the number and density of markers and experimental designs required to perform an association analysis. LD differs from linkage in that the former refers to the correlation between alleles in a population whereas the later refers to the correlated inheritance of loci through the physical connection on a chromosome. In other words, LD is the correlation between polymorphisms (eg. SNPs) that is caused by their shared history of mutation and recombination. A SNP is a genetic variation when a single nucleotide(i.e. A, C, G or T) in the DNA sequence is altered and kept through heredity thereafter. A set of linked SNPs on one chromosome is called a '**haplotype**'. The patterns of LD reveal a block-like structure. The entire chromosome can be partitioned into high LD regions interspersed by low LD regions. The high LD regions are usually called '**haplotype blocks**' whereas the low LD regions are referred to as '**recombination hotspots**'. Within a haplotype block there is occurrence of little or no recombination and the SNPs are highly correlated. Consequently, a small subset of SNPs (called tag SNPs or haplotype tagging SNPs) is sufficient to capture the haplotype patterns of the block (Cheng et al., 2006). Tight linkage may result in higher levels of LD (Flint-Garcia et al., 2003). If two mutations occur within a few bases of one another, the presence of these SNPs is highly correlated because of rare occurrence of recombination between the two neighboring bases. On the other hand, the correlation between the SNPs occurring at long distance or on different chromosomes will have a much lower level of correlation or level of LD. The LD is proportional to recombination fraction and the LD decreases with genetic or physical distance between the genes.

For understanding genes and genome organization, experimental techniques can be focused on the physical and genetic composition of a region in terms of genetic markers, recombination frequency, and other characteristics rather than its functions. Phenotype-driven family based whole genome linkage scan used to identify genes determining the Mendelian traits is one such approach. Use of linkage disequilibrium to identify genomic regions of genetic association is another approach. A polymorphism associated with a disease state can either directly contribute to the disease state or may be a surrogate marker which is co-inherited with an adjacent functional variant (i.e., *a* which influences the phenotype) involved in the disease state. A SNP has usually two allelic states whereas a stretch of DNA can typically be represented by several different haplotypes and thus chances are more that one of the many haplotypes is associated with a functional variant than the odds of a strong correlation with one of the two alleles for a single SNP. In this sense a series of haplotypes is analogous to a multiallelic STR marker. But if the functional variant is itself under test or a polymorphism shows perfect co-segregation with the functional variant then haplotype analysis has no advantage. This co-inheritance of the surrogate marker with the disease allele can occur to a varying degree and is termed linkage disequilibrium. The LD can be said to be present if co-occurrence of the two polymorphisms is with a frequency greater than would be expected by chance. And greater the extent of LD, the larger is the chance of detecting the phenotypic influence of one by genotyping the other. The degree of LD is dependent on the history of two adjacent markers and is influenced by the relative frequencies of occurrence of the two polymorphisms in the population and the degree of recombination between them. Two most commonly used measures of LD are based on the difference between the observed and expected number of haplotypes bearing specified alleles of two markers. LD mapping is an important tool in the positional cloning of genes and its application will grow as complex phenotypes (quantitative traits) are dissected genetically. Linkage disequilibrium mapping will help gaining a view of the full depth of variation in a genome. Linkage disequilibrium can allow mapping of a genetic association to a very small region (typically 10-100kb) following the construction of a detailed population based LD map. LD map will ultimately make comprehensive SNP-based whole genome association scans a realistic possibility: selecting SNPs which tag all of the major haplotype block across the genome. The number of polymorphic sites and the linkage disequilibrium among them requires such discussions to focus on the DNA 'haplotypes' at a locus.

19.2 HAPLOTYPE

A haplotype is a string of co-inherited alleles of different markers which are arranged in a successive fashion along a given stretch of DNA (e.g. BAC clone, a restriction fragment, a chromosome). In other words, each haplotype represents a linear section of DNA rather than the single point corresponding to a single marker. Haplotype also refers to an individual with a specific arrangement of alleles in a given piece of its DNA (e.g. a gene, haplotype block) that is inherited as a block, probably because of lower recombination frequency than other parts of the genome. In other words, haplotype refers to set of allelic forms of closely linked genes usually inherited *en block*. In practice, a haplotype block is characterized by a series of SNPs in linkage disequilibrium. The extent of discernible haplotype length varies widely for different regions of the genome. Well defined haplotypes (characterized by moderate or high linkage disequilibrium) are punctuated by regions of extremely low LD suggesting that the recombination processes, selective pressures and other factors that dictate the linkage equilibrium vary widely in abrupt fashion across the genome (Goldstein, 2001). Although the length of preserved haplotype shows considerable variation the typical length of a discernible haplotypic block is 10-100kb in Caucasian population (Daly et al., 2001).

19.3   NUMBER OF HAPLOTYPES

The levels of polymorphism and consequent heterozygosity poses a problem- which haplotypes are present in a given diploid organism. This is because of the frequent occurrence of multiple heterozygous sites in a given individual. A specific DNA haplotype is the specific combination of variants at each of the polymorphic sites being studied. With two heterozygous sites there are four possible haplotypes and the number of possible haplotypes doubles for each additional heterozygous site. Ideally, an individual's genotype across multiple heterozygous loci should include resolution into the two constituent haplotypes.

19.4   SIZE OF LD BLOCKS

Computer simulations and empirical data suggest that LD extends only a few kilobases around common SNPs whereas other data have suggested that it can extend much further, in some cases greater than 100kb (Reich, 2001).

19.5   CONSTRUCTION OF HAPLOTYPE

Haplotypes are usually constructed by comparing the genotypes of closely related individuals at two or more linked markers and identifying groups of alleles which are co-inherited as a set from one generation to the next. Classically, haplotypes in multiple heterozygous individuals or organisms have been resolved by pedigree analysis, breeding program or such attributes as a haploid sex depending on the organism.

19.6   CONSENSUS PATTERN

It refers to the similar or identical distribution of specific sequences or mutations along a specific piece of DNA of two individuals. Consensus pattern can be established for SNP which occur at different frequencies in different regions of the genome. By comparing the sequences in, for example, a distinct gene from two or more individuals , identical bases can be assigned a1 and non-identical bases at a specific position a2. A cluster analysis will then find consensus patterns that allow to group these individuals into specified haplotype classes.

19.7   METHODOLOGY FOR MOLECULAR HAPLOTYPING

Haplotyping (or allelotyping) refers to the determination of the specific configuration of alleles in a genome. Highly polymorphic STR markers yield the most informative haplotypes for linkage disequilibrium but ambiguous linkage phase in a population sample compromises their value. Family materials may not be available or adequate to allow phase determination and so there is a need for haplotyping methodologies to establish linkage phase from an individual genomic DNA. Molecular cloning and restriction site mapping or DNA sequencing of relevant clones has been used to define haplotypes of multiallelic chromosomal regions. These strategies are too laborious to produce large sample sizes required in a population study and resolution of haplotypes would require constructing and screening a library for each individual so that clones containing either paternal or maternal alleles can be analyzed. Another alternative approach of resolution into the two constituent haplotypes takes advantage of any association either physical or otherwise among polymorphisms that might allow sequential resolution of alleles at each polymorphic site. For instance, sites of nucleotide substitution that are nearby sites of length variation (VTNR or dinucletide repeats) can be co-amplified with PCR. In this case amplification would resolve allelic size variants and alleles at the polymorphic nucleotide site could be resolved by dot blot or other means. One can also separate allelic bands on the basis of a RFLP or on a denaturing gradient gel, with secondary typing of the separated bands.

Molecular haplotyping accomplishes direct phase determination by generation of hemizygous templates from diploid genomic samples. Molecular haplotyping can be achieved by using either single molecule dilution of genotypic DNA to separate allele physically (Ruano et al., 1990) or allele discrimination by allele-specific PCR primers to amplify hemizygous DNA segments from a heterozygous template (Michalatos-Beloin et al., 1996).

19.7.1   Single Molecule Dilution

Ruano et al developed the SMD method to overcome the problem associated with allele-specific amplification. Alleles could be stochastically separated into maternal and paternal contributions by dilution of genomic DNA until only a single molecule of the desired region was expected in an aliquot. In case of humans, a million-fold dilution from 3μ per aliquot would bring DNA concentration down to one haploid equivalent per aliquot. Amplification from one molecule of template is possible with conventional PCR, would follow using 'booster' PCR, a more efficient biphasic approach that reduces artifactual formation of so-called primer dimmers. There are two stages I and II in booster PCR. In stage I an initial 10^7-fold molar excess of each primer to template is maintained. Further, it involves 20 cycles of amplification with longer annealing and polymerization. At the end of stage I additional dose of each amplimer is added followed by a further 50 cycles of amplification. In this method there is a need to optimize booster PCR conditions separately for each region to be studied.

19.7.2   Allele-Specific Amplification

It can be used to obtain haplotypes directly. In this case PCR primers are designed to anneal at their 3'ends to one of the polymorphic sites: one primer specific for each polymorphic base.

Amplification is allele specific. The other polymorphic sites nested between primers are typed and haplotypes are determined from allele specific PCR products. Such techniques are applicable to genomic DNA but do not seem to be applicable to regions much larger than a few kilobases. This method suffers from the need for strict empirical optimization of amplification conditions to ensure allele specificity.

Haplotypes can be derived by classical and molecular means. For example, the CD4 locus haplotype consists of two non-expressed markers 9.5kb apart- a biallelic polymorphism involving partial deletion of an Alu element and a multiple pentanucleotide STR. The classical Mendelian analysis involved genotyping of the Alu deletion polymorphism by PCR using markers (CD104 and CDR105) and amplification products were separated by size on a 1% agarose denaturing gel in order to prevent heteroduplex formation in heterozygous individuals and visualized under U.V. light. Genotyping of the STR was performed by PCR using markers (CD4A and CD4B) and amplification products were electrophoresed on a sequencing gel and autoradiographed.

In case of molecular haplotyping long range allele-specific amplification was performed utilizing either Int-ASO (designed to extend specifically the intact allele) or Del- ASO (designed to extend specifically deletion allele) and anchored primer CD4A. Allele specific amplification was performed using hot-start PCR. The 5'end of both ASOs (allele specific oligonucleotides) is complementary to the non-Alu genomic sequence downstream of the deletion site. Specificity of Int-ASO is achieved by having its 3' end complementary to Alu sequence only present in the 'intact' allele but absent in its 'deletion' counterpart. Specificity of Del-ASO is achieved by having its 3' contiguous to its 5' end only in the 'deletion' allele but interspersed by Alu sequence in its 'intact' counterpart.

These methods were developed for short (~ 500bp) segment only. It is necessary to physically separate the two copies of each stretch of DNA to allow unmixed analysis of haplotypes. In other words, for absolute definition of all haplotypes it is essential to physically separate the two copies of each stretch of DNA under analysis. For every short stretches of DNA (up to ~ 10kb) this can be achieved by use of allele-specific primers to a molecular haplotyping assay based on long range PCR (Barnes, 19..94). Long range PCR may not preserve intact the linkage phase of distant markers due to well known template jumping artifacts produced by Taq polymerase. A combination of rTth (from *Thermus thermophillus*) and Vent (from *Thermus litoralis*) DNA polymerases used for long-range PCR may minimize template switching as a result of the enzyme's joint proof-reading capability and processivity. Molecular haplotyping should prove useful in mapping disease genes and in establishing founder effects.

For large-scale haplotype construction it is necessary to separate entire chromosomes. In case of human a rodent-human somatic cell hybrid technique can be used to physically separate the two copies of chromosome (Douglas et al., 2001). High density oligonucleotide arrays in combination with somatic cell genetics is used to identify a large fraction of all common SNPs on a particular chromosome and to directly observe the haplotype structure defined by these SNPs. Global patterns of human DNA sequence variation (haplotypes) defined by the common SNPs have important implications for identifying disease associations and human traits.

Where family information is not available statistical methods can be used to infer haplotypes and haplotype frequencies. The most common method for estimation of haplotype is the expectation- maximization maximum likelihood estimate (Excoffier and Statkin, 1996). Although haplotype construction using family inheritance patterns methods is most robust than population-based MLH, it requires a degree of inference and the resulting haplotypes may be probable rather than actual.

19.8 MEASUREMENT OF LINKAGE DISEQUILIBRIUM (LD)

The two most common statistics for measuring LD are r^2 and D′. Considering a pair of loci with two alleles A and a at locus one and B and b at locus two with allele frequencies fA, fa, fB, fb, respectively, the resulting haplotype frequencies will be fAB, fAb, faB, fab. The difference between the observed and expected haplotype frequencies will be

$$\mathbf{D_{ab}} = (fAB\text{-} fA\, fB)$$

$\mathbf{D_{ab}}$ is the basic component of all LD statistics. Now, r^2 which is the square of the correlation coefficient between two loci, is calculated as

$$r^2 = (\mathbf{D_{ab)}}^2/ fA,\, fa,\, fB,\, fb$$

The value of r^2 will be 1 only when the two loci have identical allele frequencies. Another LD statistic D′ is calculated following Lewontin (1964) as

$$D' = (D_{ab})^2/ \min (fA,\, fb,\, fa,\, fB) \text{ for } D_{ab} < 0$$
$$D' = (D_{ab})^2/ \min (fA\, fB,\, fa,\, fb) \text{ for } D_{ab} > 0$$

D′ is scaled based on the observed allele frequencies and so it will range between 0 and 1 even if allele frequencies differ between the loci. D′ will be less than 1 when all possible haplotypes are observed and hence a presumed recombination event has occurred between loci.

The r^2 summarizes both recombinational frequency and mutational history whereas D′ measures only recombinational history. Although both statistics are affected by small sample size, D′ is strongly affected. Between the two statistics r^2 is favored for examining the resolution of association studies as it is indicative of how markers might correlate with the QTL of interest.

19.9 MEASUREMENT OF GENOME-WIDE LD

One of the goals to measure genome-wide LD is to estimate recombination in a population and its implications for gene mapping and association studies. Visual inspection of genome-wide LD requires graphical methods able to display large-scale patterns of LD. The commonly used graphical displays are (i) sliding windows (Dawson et al., 2002) and (ii) LD decay plots (Reich et al., 2001). The LD latter method displays average pair-wise LD inside a region. The first one displays average LD over regions determined by a window of constant size and the second one displays patterns of LD decays for increasing physical or genetic distances. These plots do not show pair-wise LD but compress information into one-dimensional plot and thus there is a need for a tool which could display a two-dimensional LD along a whole chromosome which would allow to examine recombination as well as distribution of the haplotype blocks (HapMap- Consortium, 2005).

Linakge analysis can be carried out using either parametric approach or non-parametric approach. In case of parametric approach, linkage analysis tests whether or not the inhetritance pattern fits a specific model of inheritance. In this approach, linkage is measured by the LOD score and here the recombination frequency is assessed. This approach is more informative for large, multiple affected predigrees.

The non-parametric (model-free) approach is more powerful when the mode of inheritance is unknown as in case of complex trait analysis for which small predigrees are often ascertained. This approachdoes not allow direct estimation of the recombination frequency. This approach is based on the principle that relatives who share similar trait values will show increased sharing of alleles at markers which are linked to a trait locus (Holmans, 2001). Allele sharing can be defined as identical by state or identical by descent. Two alleles can be said to be identical by state if they have the same DNA sequence. They are identical by descent besides being identical by state, if they have descended from (and are copies of) the same ancestral allele. A statistical test can be performed to compare the oberseved degree of sharing to that expected under the assumption that the marker and the trait are not linked. Although the test statistic can take the form of a chi-square, normal or F-statistic, it is often transformed to allow it to be expressed in LOD units.

19.10 TRANSMISSION/ DISEQUILIBRIUM TESTS

Association analysis may provide a test for the presence of a difference in allele frequency between cases and controls. However, a difference does not necessarily imply a causal relationship between traits, for example a disease and a trait as many other factors may yield this effect. But in a well designed study, association analysis does provide a flag for further study and in some cases the association can be due to the marker being physically close to the causal variant. Association testing is done using X^2 which is applied to a contingency table in which case/control state is tabulated by frequencies of either genotypes or alleles with the degrees of freedom being (r-1) (c-1) where r is the number of rows and c is the number columns in the table.

As ethnic mismatching of non-family and controls(population stratification) can sometimes yield false positive evidence of association, so interest is now on family-based testing for obtaining information about linkage. The transmission/disequilibrium test or TDT has gained importance as a test of linkage in the presence of association that does not give false evidence of linkage due to population stratification (Spielman et al., 1993). The TDT involves counting of alleles transmitted from heterozygous parents to one or more affected children in nuclear family. The alleles that are not transmitted to affected children may be regarded as control alleles. The test takes the form of a McNemar's test which under the null hypothesis of no linkage,follows a X^2 distribution with one degree of freedom. The TDT is also a valid test for association but only when applied to alleles transmitted from heterozygous parents to just one affected child per family.

Assuming a diallelic locus, let b be the counts of heterozygous parent-to-offspring transmission in which allele 1 passes on to an affected child while allele 2 is not transmitted. Further, let c be the counts of transmission the other way round in which allel 2 is not inherited in an affected child but allele 1 is not transmitted. The TDT test then takes the following form.

$$X_1^2 = \frac{(b-c)^2}{(b+c)}$$

19.11 GENOME-WIDE ASSOCIATION STUDY FOR IDENTIFICATION OF LOCI

A genome-wide association study identifies novel risk loci for type 2 diabetes in human. This has been made possible with the availability of high density genotyping arrays which combine the power of association studies with the systematic nature of a genome-wide search and this has led to undertake a two-stage genome-wide association study for identification of additional type 2 diabetes mellitus. The high density arrays allow genotyping of hundreds of thousands of polymorphisms. In this approach a number of SNPs are tested. Markers with the most significant difference in genotype frequencies between cases of type 2 diabetes and controls are fast-tracked for testing a second cohort. This led to identification of 4 loci containing variants which confer type 2 diabetes risk. This is in addition to the known association of type 2 diabetes with TCF7L2 gene. This study this constitutes a proof of principle for genome-wide approach for the elucidation of complex genetic traits.

Methods for detecting single base substitutions:

19.12 DETECTION OF STRUCTURAL VARIATION OF THE GENOME

In genomics, detecting all the differences in sequence among the genomes of individual members of a species is a challenge. Structural variation (SV) of the genome involves kilobase-to-megabase-sized deletions, duplications, insertions, inversions and complex combinations of rearrangements (see chapter 24). Structural variations in humans have been implicated in gene expression variation, female fertility, susceptibility to HIV infection, systemic autoimmunity and genomic disorders. Previous methods for detecting SVs used comparative genome hybridization-array CGH (Redon et al., 2006) which involved DNA and fosmid paired end sequencing (FPES) at relatively low resolution (Tuzun et al., 2003) (>50kb for CGH, >8kb for FPES). These methods map SVs below the resolution where breakpoints can be detected (for array-CGH) or are laborious (for FPES) and so breakpoint junction sequences of only a limited number of SVs and/or copy number variants have been reported.Thus how SVs affect genes and mechanisms by which SVs are formed, are not known. Korbel et al. (2007) used a high throughput and massive **paired- end mapping** (PEM) and a large scale genome sequencing method to identify SVs of about 3kb or larger. PEM for identifying SVs involves the preparation and isolation of paired ends of 3kb fragments and their massive sequencing with 454 technology. The different steps involved in this method are: (i) genomic DNA is sheared to yield DNA fragments ~ 3kb (ii) biotinylated hairpin adapters are ligated to the fragment ends (iii) fragments are circularized (iv) and randomly sheared (v) linker (plus) fragments are isolated (vi) the library is subjected to 454 sequencing

(Margulies et al., 2005) (vii) paired ends are analyzed computationally to determine (viii) the distribution of paired end spans. Structural rearrangements are identified as significant differences between the fragments identified by the PE reads and the corresponding regions of the reference sequence. Five signatures are used in the prediction of SVs. Deletion relative to the reference genome is identified by PEs spanning a genomic region in the reference genome longer than a specified cutoff. Simple insertions relative to the reference genome is predicted with PEs that span a region shorter than the cutoff. Mated insertions contain sequences connected to a distal locus on the basis of their paired ends. Inversions are detected via a relative orientation different from the reference genome.

19.13   COMPARATIVE GENOMIC HYBRIDIZATION (CGH)

It is a variant of chromosome painting and FISH techniques which allows the detection of major differences (gains or loss of whole chromosomal regions) between two or more genomes. In other words, CGH can be used to scan for deletion or amplification of chromosome fragments which often contain many genes. In this technique the genomic DNA from two organisms (a mutant is called a tester and a wild-type genome is called a driver) is labeled using two fluorochromes. Both differently labeled genomes are then simultaneously hybridized to a metaphase chromosome spread and the ISH is visualized by two-color detection using epifluorescence microscopy with selective filters. The ratio of fluorescence intensity of both fluorochromes reflects the relative amount of hybridized genome probes. If there exists regions deleted or amplified in one genome then the observable fluorescence ratio will change. CGH also allows mapping the amplifications or deletions in one single experiment.

Matrix CGH—It is a variant of CGH which works with a matrix of defined DNA fragments immobilized on a solid surface (e.g., a glass surface). Matrix CGH increases the resolution to 30-200kb from about 10Mbp in chromosome spreads. CGH can additionally be coupled to a further amplification of the target sequence using a degenerate oligonucleotide-primed PCR which increases the sensitivity of detection. CGH compares copy number changes between closely related genomes at genic resolution (Pollak et al., 2002).

DNA microarrays of short oligonucleotides designed to interrogate each base individually (i.e. **resequencing arrays**) have been applied to the analysis of individual human genes and small genomes such as ther human mitochondria and SARS coronavirus genomes (Gresham et al., 2006). However, extension of this approach to whole genomes is currently impractical as large number of probes is required for complete coverage.An alternative approach uses microarrays which detect mismatches, exploiting the facts that hybridization to a short oligonucleotide is quantitatively sensitive to the number and position of mismatches. Sequence-level differences are detected without allele-specific probes by comparing hybridization intensities of individual features on the microarray (referred to as single-feature polymorphisms, SFPs). This technique has been successfully applied in genetic diversity and gene mapping studies. Until recently, comprehensive detection of single base pair differences has been limited by probe density across the genome which is typically a few oligonucleotides per gene. Even complete single copy coverage of the genome is unlikely to be sufficient for finding all mutations as statistically detectable decreases in hybridization intensity usually requires that a variant nucleotide falls within the central fifteen bases of a 25-bp probe.

19.14   DETECTION OF SEQUENCE VARIANTS

An approach for automating the detection of sequence variants in mtDNA or nuclear DNA integrates the use of conventional fluorescence-based sequencing with DNA analysis software that measures sequence quality and improves the accuracy and automation of detecting DNA variantions (Reider et al., 1998).The two computer programs used are Phred and Phrap (see Roy, 2009).These two programs interfaced with the Consed program which provides a uniform environment for viewing sequence data and simplifies the localization of tagged DNA variants through its navigation system. Furthermore, as information is stored in Consed tags, annotated sequences containing the existing information on agene or sequence can be viewed along with the newly generated sequence. Finally, one can use a PolyPhred program which detects the presence of heterozygous bases in nuclear DNA sequences.

20

QTL Analysis

20.1 TYPES OF GENE INVOLVED IN QUANTITATIVE VARIATION

Thompson (1975) made the observation that a continuous variation does not necessarily mean the involvement of segregation of a large number of polygenes. Major effect genes (major genes) may be modified by the action of minor gene to produce a continuous distribution of resistance levels and as few as 3 or 4 genes may be present. Further, some of the variation found within the amino acid sequences of essential proteins which appears neutral but may well have small effects. Then there are variations in effects (strong or weak) of promoters and enhancers which could result in large as well as small quantitative differences. As the polygene concept is based on a different type of gene involved more in the 'fine tuning' of the phenotype than having essential and unique functions, highly repetitive genes such as genes for ribosomal RNA (rRNA) or histones could represent the polygenes. Finally, redundant genes could also be involved in continuous variation.

20.2 IDENTIFICATION OF MUTATION OF GENE WITH STRONG EFFECT

For identification of mutations in any organism it is essential to have knowledge of chromosomes and linkage groups and to determine the molecular nature of the mutations it is necessary to clone the genes in the organism in from which they have been isolated. The mutant gene can be identified by a method (**transformation and complementation**) described in chapter 16. It can also be identified by a technique called '**candidate gene approach**' described below in this chapter. But the only way to find and clone a gene of unknown function and unclear affinities is through knowledge of its map position and the method is called **positional cloning.** If a desired mutation can be mapped between two molecular markers separated by a megabase or less then it should be possible to clone the whole sequence in between. The region between two markers that appeared to bracket the gene is explored by a procedure called **chromosomal walking**.

Chromosomal walking—Chromosomal walking refers to the step-by-step establishment of successive overlaps between cloned chromosome fragments, starting at a fixed point. Starting with a cloned fragment say A one screens for a second fragment B that cross hybridizes and so overlaps with A and then for a third fragment C that overlaps with B but not with A. Chromosome walks can run into problem by the presence of repetitive DNA and cross hybridization between fragments will often be due not to a true overlap but rather to their both having the same repetitive element or motif. This problem can sometimes be overcome by a manoeuvre called 'jumping'. In this manoeuvre the difficult fragment is cut out of the vector and circularized by ligation of its ends. A small fragment bridging the two original ends is obtained by digestion with another enzyme and cloned for use as a probe for the next step of the walk. **Chromosme jumping** (or simply called jumping) (Poustka et al., 1987) is a technique to analyze DNA sequences that are separated from each other by more than 100kb. It is a procedure for isolating clones which contain regions of the same chromosome which are not contiguous as in case of chromosome walking (Collins and Weismann, 1984). In this technique the isolated genomic DNA (> 1000kb in length) is fragmented with an appropriate rare cutter restriction endonuclease such as Not 1 or partially digested with a frequent cutter enzyme (e.g. Mbo1). Fragments of ~ 100kb in lengths are isolated by PFGE, circularized at low concentrations that prevent multimers formation and again linearized by another restriction enzyme (e.g. EcoRI). This procedure leads to the following three types of populations of DNA fragments.

1. The first type of population will consist of sequences from within the original 100kb fragments.
2. The second type contains the termini of the original fragment ligated to one another (called **junction fragment**).
3. The third type contains sequences from elsewhere in the genome.

If instead of a rare cutter restriction enzyme, a frequent cutter enzyme is used, the genomic DNA will at first be only partially digested. The generated fragments can also be size selected using PFGE. In both cases the 'junction fragments' are then selectively cloned into a vector containing a selectable marker(e.g. λ phage derivatives with a suppressor tRNA-sup F gene). Alternatively, the marker gene sup F can also be integrated during circularization. After plating on a bacterial host with sup F⁻ genotype, only phage genotypes carrying their own sup F gene will replicate and form plaques and thus is easily identified ('**jumping clones**'). Junction fragments harbouring a known gene marker can be identified by nucleic acid hybridization using the appropriate probe. Once identified such a junction clone that also contains sequences some 100kb apart in the original context, can in turn be used to find another fragment that allows the repetition of the jumping process. The cystic fibrosis gene was cloned by a combination of chromosome jumping and chromosome walking from a known linked gene sequence.

This method (walking and jumping) of finding gene is becoming increasingly obsolete and more and more human genome is being physically mapped by overlapping yeast artificial chromosome clones (YAC). Positional cloning was used to identify the gene involved in common genetic disease, cystic fibrosis (CF) (Romens et al., 1989). Clones covering the region of interest will be available but even if the desired gene is present in the cloned segment there is still the problem of identifying it. In case of search for CF gene the responsible gene was found to start only 300kb away from one of the flanking molecular markers. The gene in this clone was thus identified by the following criteria.

1. It cross-hybridized to a sequence present in each of a number of different vertebrates. The positive result in such a laboratory test (also called 'zoo blot') indicated that the sequence is at least an important sequence. **'Zoo blot'** is a laboratory term for a Southern blot onto which several restricted genomic DNAs from different organisms have been transferred and which serves to detect sequences common to all species by hybridization to either oligonucleotide or DNA probes.

Reverse dot blot—It is technique to detect specific amplified DNA sequences using an oligonucleotide probe that is immobilized on a membrane and to which a labeled DNA fragment generated by the PCR is hybridized. In brief, the oligonucleotide probe is tailed with poly (dT) using terminal transferase and UVcross-linked to a nylon membrane. The amplified DNA fragment labeled during the amplification process by using biotinylated primers is then hybridized to the immobilized oligonucleotide probe. The presence of specifically bound PCR fragment is detected using streptavidin horseradish-peroxidase conjugate. Thus the reverse dot blot technique allows to screen different genomic fragments synchronously for mutations.

2. It hybridized to a human cDNA copy obtained from mRNA synthesized by salivary gland which again indicates of the role of this sequence in causing CF

3. Sequencing study showed that this mutation was due to a small(3-bp) deletion in one of the exons and chromosomes from normal individuals did not carry this mutation. Thus the mutation confirms the identification of the gene.

Chromosome walking—On a small scale this technique can be used to expand a clone containing a fragment of a gene to a region containing the entire gene whereas on large scale it is possible to clone arrays or clusters of genes of interest. For successful chromosome walking one should have some information regarding the starting clone to be used for walk and about the gene to be isolated. I. Is it a single copy gene or a member of a multiple gene family? II. Are mutants nearby genetic markers available? III. How large is the distance between starting clone and target? IV. Walking will extend in both directions from the starting point, unless a particular direction can be distinguished, for example, by using genetically marked chromosomal rearrangements. Walking is easier in Drosophila than in most other organisms because of existence of polytene chromosomes which allow ISH and conventional monitoring of the progress of walk. Secondly, a lot of genetic and cytogenetics information is available so even if no cloned probe is available near the target site, one can usually jump into the vicinity of it taking advantage of translocations, inversions or deficiencies with one end near the target site and the other near a cloned sequence. Thirdly, the Drosophila genome has few interspersed repetitive elements including repetitive TEs which can lead the walk astray. In organisms with extensively interspersed repetitive DNA the most useful genetic marker is defining RFLPs in the vicinity of the target gene. Successful walk also depends on the quality of the library used. The library should be complete, ie. it should represent the organism. It is essential otherwise there is danger of ending walk prematurely. Except human, use of homozygous line or inbreds of an organism is recommended for the construction of library in order to reduce the amount of heterogeneity encountered during the walk. Further, the library should not contain cloning artifacts such as rearrangements of the insert. Cloning artifacts can be prevented through the use of multiple isolates in conjunction with genomic Southern blots which establishes the correct organization of the DNA (Spoerel and Kafatos, 1987). Multiple libraries for distinct homozygous strains should be constructed in case of Drosophila which will solve the problem of repetitive TEs which tend to be in different locations in different strains. If more than one rearrangement is present in a certain region of the chromosome then it may be impossible to align individual clones unambiguously but one can jump over the scrambled region using a cosmid clone. Individual cosmid walking step can cover much longer distances along the chromosomes (up to ~ 40kb) than is possible with λ(~ 20kb). Most single genes ranging in size from a few kilobases to at most a few hundred kb can either be contained within a single λ or cosmid clones or can be covered by a limited amount of chromosome walking. For several gene families such as α- and β-globin gene clusters, the major histocompatibility complex and the component genes, it has been possible to generate a series of overlapping cosmid clones encompassing as much as 240kb of contiguous DNA. Neither of the complex loci such as the major histocompatibility complex which may occupy more than 1000kb of contiguous DNA and the heavy and light immunoglobin loci has yet been internally connected by standard cloning techniques. Thus what is required is a method which will take larger steps along the chromosome, preferably with the ability to specify in which direction the step will be taken. This is done with the help of a marker DNA fragment into the covalent circle.

Chromosome jumping or hopping—Chromosome jumping or hopping may improve the speed of chromosome walking sufficiently to tackle such a distance(even if useful clones to start a walk may be separated by more than 1000kb from the target gene). Jumps over hundreds of kb are possible as the length of jumps(in contrast to the steps in chromosome walking) is not limited by the capacity of the vector system used. Chromosome jumping libraries differ from libraries used for chromosome walking.

Use of chromosomal segmental interchange for precise location of gene—In some cases a mutant gene is marked more conspicuously by a microscopically visible chromosome rearrangement. Whenever a chromosome breaks there is a chance of a gene being damaged at the break point. The damage can be in the form of either an interruption in the coding sequence or separation of an essential regulatory element. The example of the use of a segmental interchange in locating a gene is from human. The disease called neurofibromatosis (*NF*) is as a result of a reciprocal exchange of long arms of chromosomes 17 and 22. The dominant condition neurofibromatosis (NF1) was known to map to chromosome 17. The break in 17 was presumed to disrupt the NF1 gene. One of the interchanged chromosome was isolated in human-mouse hybrid cell line and tested for hybridization to a series of chromosome 17 probes such as α, β, γ, δ that identify four closely linked sites on the chromosome 17. The probes γ and δ hybridized but probes α and β did not which indicated that the break was between α and β (Ledbetter et al., 1989). Thus the gene NF1 was located and found to be within a region bound by two very closely linked molecular markers.

20.3 SCREENING KNOWN GENES HAVING PHENOTYPIC EFFECTS FOR SEQUENCE VARIATION

If the wild-type gene has been cloned and its sequence determined then the primers can be constructed. The polymerase chain reaction (PCR) can be used with the constructed (appropriate) oligonucleotide primers to amplify each new allele in fragments from total genomic DNA and thus sequence comparison can be made segment-by- segment with the wild type gene. The work involved in searching for a gene for a mutation can be greatly reduced if one knows more precisely where to look. In case of organisms such as microorganisms or Drosophila where the fine structure genetic map is available hunting for mutants is easy but in case of higher organisms where such genetic map is not available the following three techniques can be employed to locate the sequence variants.

1. Denaturing gradient gel electrophoresis (DGGE)
2. Chemical cleavage of mismatches in normal/mutant heteroduplex
3. Single-strand conformational polymorphism (SSCP)

Denaturing gradient gel electrophoresis (DGGE) — In this method DDGE of normal/ mutant heteroduplex made by hybridizing polymerase chain reaction (PCR) fragments is carried out (Higuchi et al., 1991). Where the mutation is due to a base pair substitution the normal and mutant homoduplexes are often separable without hybridization. When the double stranded fragments (heteroduplexes) are run through an electrophoretic gel with an increasing gradient of denaturing reagents such as urea and/or formamide (which tends to melt the duplex to single strand) the fragments will run faster or slower depending upon their sizes until they reach the point in the gradient where their complementary strands start separate out. At this point their migration rate is slowed and fragments of even slightly different melting point will end up in distinctly different positions. As heteroduplexes are more sensitive to denaturation so their migration is slowed at a lower concentration of denaturant. The addition of A G + C rich terminal segment (G-C LAMP) to each PCR product helps to maintain at least some duplex structure and so extends the usable range of denaturant concentration. For detail on DGGE see chapter 30.

In addition, the use of heteroduplexes between normal and mutant DNA which contains a mismatch, increases the resolution by DGGE so that virtually all possible base changes can be detected even when a base change does not produce a shift in the mobility of the mutant homoduplex (Fisher et al., 1983; Myers et al., 1987; Sheffield et al., 1989).

Chemical cleavage of mismatches in normal/mutant heteroduplex — In this method a segment of the mutant allele is amplified by PCR with one radioactive labeled primer so as to end label the product in one strand. The labeled single strand is hybridized with the complementary normal sequence (wild-type) and thus mutant/wild- type heteroduplexes are made. These heteroduplexes are then subjected to a chemical reagent that cleaves at mismatched pyramidines- either at C or T depending upon the types of reagents used. The size of the radioactive cleaved product and hence the position of the cleavage is determined on an agarose gel electrophoretic gel in comparison with a 'ladder' of fragments of known sizes and the uncleaved strand as a control. Thus the point of mismatch in the DNA fragment and thereby the position of the mutational difference can be pinpointed (Roberts et al., 1992).

Single-strand conformational polymorphism (SSCP) — The principle underlying this technique is the distinguishable electrophoretic mobilities of DNA single strands with sequence differences- called single strand conformational polymorphism. The rate at which single stranded DNA fragments migrate through a non-denaturing electrophoretic gel depends not only on its size but also on its nucleotide sequence. In the absence of denaturing reagents single strands tend to form a certain amount of secondary structure (stems and loops) as a result of chance complementary matching between different short segments. Even single base changes will usually change the relative stabilities of alternative conformations and thus any mutation that increases or decreases the length of the sequences will cause a change in the average conformation and often thereby a change in mobility in the gel. In other words, it will make it either easier or more difficult for the partly folded strand to pass through the pores of the gel. Through this technique most variants in the single strand fragments of up to a few hundred base-pair in length can be distinguished. Thus in this procedure single strands of defined length isolated from mutant and wild-type genomes by PCR are run through the non-denaturing gel (Dryja et al., 1991).

Dynamic allele specific hybridization—It is a technique for detection of mutations (transition, transversion, especially SNPs). This method works without gel electrophoresis and it requires sequence information about wild-type and mutant allele. In this technique the target DNA containing the mutation is first amplified by PCR using two primers, one of which is biotinylated. The resulting amplified product is immobilized by binding to a streptavidin coated mictotiter plate. After that the biotinylated strand is removed by NaOH which leaves the single stranded DNA bound to the microtiter plate. Now an oligodeoxynucleotide probe complementary to one of the expected sequences (wild-type) is added to the well together with an intercalating dye (e.g. SYBR green). The sample is denaruted by heating and then gradually cooled to reanneal the complementary sequence. After removal of hybridization buffer, detection buffer is added and the fluorescence is continuously monitored while the microtiter plate is heated from the room temperature to beyond the denaturation temperature. The exact melting point of the hybrid(Tm) is shown by an abrupt decrease in fluorescence. The fluorescence values are then plotted against the temperature. By plotting the negative first derivative the denaturation point is represented as peak which is an indicator for the wild-type allele. The whole procedure is repeated with the probe containing the polymorphism in question and the presence of an SNP is derived from the position of the denaturation point of the hybrid.

Oligonucleotide ligation assay—It is a method to detect single substitution in a target DNA sequence. In this technique the target DNA is denatured and hybridized to two oligonucleotide probes in a way that the 3′-end of one of one oligonucleotide is immediately adjacent to the 5′-end of the other (head-to-tail juxtaposition). If DNA ligase is added it will covalently join the two oligonucleotides provided the nucleotides at the junction are correctly base-paired with the target DNA. This will not be the case if single base substitutions at the junction site have occurred. The ligation of the two oligonucleotides may be deduced more conveniently if one oligonucleotide is labeled with biotin and the other one with 32P. Ligation chain reaction is thus another PCR alternative. A thermostable ligase is used to specifically link two adjacent oligonucleotides which hybridize to a complementary target with perfect base pairing at the junction. The oligonucleotide dimmers can be exponentially amplified by repeated thermal cycling in the presence of a second set of adjacent primers complementary to the first.

   Restriction analysis of PCR products (Saiki et al., 1985), allele specific PCR (**AS-PCR**) and its derivatives, combined chain reaction (**CCR**) and proof-reading PCR (**PR-PCR**) have been developed to detect known mutations in genomic DNA. Oligomer restriction (OR) involves the stringent hybridization of a 32P end labeled oligonucleotide probe to the specific segment of the denatured genomic DNA which spans the target restriction site. The ability of a mismatch within a restriction site to prevent cleavage of the duplex formed between the probe and the target genomic sequence is the basis for detecting allelic variant. The presence of the restriction site in the target DNA is revealed by the appearance of a specific labeled fragment generated by the cleavage of the probe. AS-PCR (Wu et al., 1989) allows direct detection of normal or mutant allele in genomic DNA without additional steps of probe hybridization, ligation or restriction enzyme cleavage. Two allele specific oligoncleotide primers, one specific for the mutant allele and one specific for the wild/normal allele, together with another primer complementary to both alleles are used in PCR with genomic DNA templates. The allele specific primers differ from each other in their terminal 3′ nucleotide. Under proper annealing temperature and PCR conditions these primers only direct amplification on their corresponding complementary allele (Figure 20.1). P1 and P3 are two synthetic oligonucleotide primers that anneal to opposing strands of a single copy gene. P1 anneals to the region of a gene in the region of a DNA sequence variation such that its terminal 3′ nucleotide base pairs with the polymorphic nucleotide of the template. P1 is completely complementary to allele 1 (A) but forms a single-base pair mismatch with allele 2 at the 3′ terminal position due to one or more nucleotide differences relative to allele 1 (B).

Combined chain reaction PCR (Bi and Stambrook,1997)) is a DNA amplification strategy which combines the use of thermostable DNA polymerase and ligase chain reaction (LCR) which has been used for detecting point mutations (Landegreen et al., 1988; Barany, 1991) and for introducing mutations in DNA *in vitro*. Unlike most PCR based mutation detection systems it relies on mismatch between primer and template at the primer 5′ end. It requires neither the use of radioactivity nor PAGE nor autoradiography for mutation detection at the single base pair level.

PR-PCR differs from standard PCR in that one of the two primers has its 3′ end aligned with a putative mutation site and has its 3′ –OH replaced by a blocking group. Distinguishing a mutant gene from the wild type depends on preferential removal of the blocked 3′ terminal nucleotide by the polymerase proof reading activity when it is mismatched with the template. Preferential removal of the blocked nucleotide allows subsequent extension and selective amplification and provides a basis for distinguishing mutant from normal genes (Bi and Stambrook, 1998).

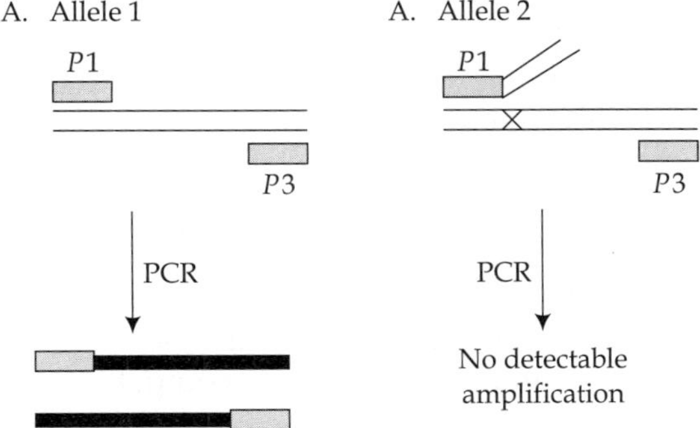

FIGURE 20.1   Showing allele-specific PCR procedure.

Diversity of alleles–In general, in human for each type of disease corresponding to a single gene there is not just one defective allele but many with some worst effects than others. Defective allele will produce disease condition as in case of CF whereas the normal individual will have the normal allele. When different types of mutations are there in the population then one can expect the existence of variety of defective alleles. Thus uncommon mutations in case of CF can be due either to a few more short deletions or single base pair changes. About 40% of the base pair changes result in amino acid substitutions, most of them to be very deleterious to protein function but some with relatively mild effects. Other base pair substitution either results in chain termination codons or changed sequences resulting in defective intron splicing. The kinds of mutations likely to be found in functionally defective alleles depend on the nature of gene product- proteins. Some proteins are sensitive to many single amino acid replacement whereas others may be robust to most changes short of losses of substantial lengths of polypeptide chain. But among all mutations both severe and mild single base pair substitutions are the most frequent. In case of base pair substitution in human there is a substantial bias favoring transition from G-C to A-T. Cytosines immediately 5′ to guanines seem to be particularly vulnerable to mutation. In mammals such Cs are prone to methylation to 5-methylcytosine which can be spontaneously deaminated to form uracil and an A-T base pair appears at G-C base pair in the DNA sequence after the next round of replication.

Screening for known mutants—If the mutant allele of a particular gene has been sequenced then a specific *allele-specific oligonucleotide* or ASO can be used to probe for its presence in individual genome. An ASO, a synthetic sequence of about 18 nucleotides with the site of mutation centrally placed within it is constructed to match one strand of the mutant allele. The radioactive labeled allele-specific oligonucleotides when used as probe on the membrane bound DNA samples then it will at right temperature and salt concentration hybridize much better to the mutant sequence than to the normal or wild-type sequence (Serre et al.,1991). Allele specific oligonucleotide (ASO) probe is a synthetic oligonucleotide of abour 20 bp in length designed to locate single base mismatches in complex genomes. Such probes are long enough to detect unique sequences in the genome but sufficiently short to be destabilized by a single internal mismatch in their hybridization to a target genome. This technique involves immobilization of the target DNA, hybridization with oligonucleotide probes and finally washing which allows to discriminate sequences with one single nucleotide mismatch from the wild-type on the basis of their different hybridization behaviour. Thus once a particular variation has been identified in an individual, an ASO probe can be used to test for the presence of that same sequence difference in any other individual and is relevant to people who from their family histories, seem likely to be carrying particular known deleterious recessive alleles and are at the risk of having defective children. In practical plant breeding programme once a QTL has been identified, it should be possible to use ASO probe to follow its transmission without ambiguity.

Sequence variants without phenotypic effects—A chain termination codon or one encoding an unacceptable amino acid can be converted by a single base mutation to nine other codons, several of which encode different amino acids may be compatible with normal function or at least sufficiently normal. This result can be explained as follows.

Conservative amino acid substitution, i.e. those that replace one residue by another of the similar properties. For example, isoleucine for valine or aspartate for glutamate at many positions within polypeptide chain will have little or no effect on growth. At some positions in the polypeptide chain even non-conservative changes have minimal effect on the function. But then laboratory findings might not be observed in the real world. Some of the base pair changes were changes in the third positions of codons that merely changed one codon for another encoding the same amino acid-a **silent** or **synonymous** change but some resulted in amino acid replacements. Then the question is whether the synonymous changes in codons are necessarily neutral. Because if the changes in the codons requires the use of a different and scarce tRNA (transfer RNA) then it might be difficult to maintain a translation rate sufficient for accomplishing the normal function. So far no observable effects of synonymous codon changes have been reported.

Allelic variation—For some genes show one gene two allele system, i.e., there is one wild-type (or common) allele and the other is defective (mutant) allele whereas in case of many others there does not seem to be a standard wild type so far as protein coding sequence is concerned and there exits a multiple of functionally normal alleles which have been revealed by electrophoretic differences between their enzymes products. The use of a number of conditions of electrophoresis is thought to reveal changes in protein shape as well as in electrical charge and which led to the identification of different alleles. In case of Drosophila pseudoobscura 37 different electrophoretic variants (alleles) have been recognized in case of enzyme, xanthine dehydrogenase and all these variants seemed quite adequately functional and thus was not possible to pin point to a single standard wild type.

20.4 VARIATION WITHIN THE GENE

Where the variation occurs in the gene sequence? The gene (functional) sequence includes 5′ flanking sequence (non-transcribed), 5′ (transcribed but not translated leader) exon, introns, exons (amino acid encoding), 3′ untranslated exon and 3′ flanking sequence. DNA sequencing of a sample of wild-type Adh (alcohol dehydrogenase) allele showed many cryptic differences (Kreitman,1983) as shown in Table 20.1 given below. There were numerous base substitutions in the third positions of codons as well as in introns and flanking sequences of the gene. This result support the general finding recorded in comparisons between related species in many groups of plants and animals.

1. Differences in codons that affect the nature of amino acid encoded are far less frequent than synonymous changes but they do occur.

Table 20.1 Showing variation within Adh gene of *Drosophila melanogaster.*

	5′flanking	5′ untranslated exon	introns	Amino acid encoding	3′ untranslated exon	3′flanking
Total base pairs	63	157	755	768	178	767
Insertions/ deletions	0	0	2	0	1	3
Variable bases	3	1	20	13	2	5
Amino acid replacement	0	0	1	1*	0	0

2. Differences in introns, falling outside the consensus sequences required for splicing, occur much more freely.

Thus we see that the gene sequence reveals a great deal of variation. Deletions/ frameshifts and chain termination codons within exons point clearly to functional deficiencies but many other variants such synonymous codon replacements and most changes within introns are almost certainly inconsequential. Amino acid replacements are difficult to assess. Many proteins, with some more than others will tolerate a considerable amount of variation at certain positions in the polypeptide chain without any observable effect on the function. It is very difficult to distinguish neutral changes from those of small effects and further whether or not there is an effect, depends on the environment. Silent variation in codons is so much prevalent in the population than amino acid replacement that one can conclude that most of the amino acid replacements reduce fitness.

Variation in non-coding sequences — In case of outbreeding populations there is an immense amount of cryptic variation in DNA sequences lacking specific function. For detail see the chapter 24.

20.5 QTL MAPPING

When trying to study quantitative variation in a particular trait, one will have to find the answers of the following questions.

1. Which are the genes affecting variation in a specific trait?
2. What is the nature of their allelic differences?

The answers are difficult to come by because of the reason that although major genetic differences that segregate as Mendelain factors are sometimes found, alleles with major effects are an exception. Even drastic genetic differences are often due to allelic variation at several loci (multiple factors or polygenes) and the contribution of each locus to the phenotypic value can be quire small. So while the most direct approach toward identifying the causal genes is genetic mapping, one will have to use statistical methods as discussed below to find regions of the genome linked with the trait of interest-an approach called **QTL mapping**. Initial QTL mapping in an F2, backcross or recombinant inbred lines (RILs) is usually followed by the generation of near-isogenic lines (NILs) in which only one QTL region segregates in an otherwise identical genetic background. Each of the NILs contains a single RFLP-defined chromosome segment and together these lines provide complete coverage of the genome. The development of NILs enables a more accurate estimate of the number of QTLs, their mode of inheritance and linkage relationship up to a YAC resolution. The conventional QTL mapping typically uses markers with an average spacing of several cMs and uses sample sizes insufficient to identify minor QTL. **Minor QTLs**- QTLs with small effects which may be responsible for a large proportion of trait variation. Many minor QTLs contribute to complex trait variation. As genes with large effects may not be representative for the majority of segregating polymorphisms so fine map phenotypic effects segregating within a one-centimorgan chromosome interval for which lines with mapped recombination break points are available and examine the sequence signature of historical polymorphism.

After that conventional fine-mapping methods as described below are applied to identify the gene (s) responsible for the QTL and ultimately the causal changes at the nucleotide level. In other words, once QTL regions have been identified the next step is to find the underlying genes and nucleotide polymorphisms causally associated with the trait variation. As the effects of individual genes (polygenes) are often small and because of presence of complex interactions between the genes it is not easy to obtain definitive proof for the identification of a QTL or quantitative trait nucleotide. In case of mammals the members of the complex trait consortium (2003) has proposed combining results from several types of experimentations to provide evidence for the identification of a QTL, once genetic linkage between a particular genomic region and a trait of interest is established.

1. DNA polymorphisms that distinguish alleles with phenotypic different effects.
2. A mechanistic link between function of the gene and the trait of interest.
3. Functional studies showing that one allele has, for example, different biochemical properties.
4. Transgenic complementation.
5. Allele replacement via homologous recombination.
6. Deficiency complementation test showing that one allele has a different phenotypic effect when in *trans* to a knockout of the QTL candidate.
7. Mutational analysis showing that a knockout of the gene affects the trait of interest.
8. Natural genetic variation at a homologous locus affecting the same trait in another species.

Some of the criteria will provide stronger evidence than others. Further, several of the criteria listed above need to be met in order to claim a causal link between allelic variation and variation at a trait. However, Fridman et al.(2000) used genetic mapping in combination with generation of NILs for the QTL, their hybrids(NIL hybrids) and fine mapping strategy and delimited the QTL affecting brix value in tomato to a single nucleotide polymorphism-defined recombination hotspot of 484bp spanning an exon and intron of a fruit-specific apoplastic invertase. Differences between the Brix 9-2-5 alleles of the two species were associated with a polymorphic intronic element which demonstrates a link between naturally occurring DNA variation and a Mendelian determinant of a complex phenotype for yield associated trait. Using this they mapped 23 QTLs which increased the brix value.

The QTL mapping started with crossing of cultivated species *L. esculentum* with the wild species, *L. pennellii* (having 15% TSS). From this cross 50 NILs were developed. Each of the NILs contained a single RFLP-defined *L. pennellii* chromosome segment and together the lines produced provided complete coverage of the genome. Using this resource it was possible to map 23QTLs which increased brix. One of these QTLs (Brix9-2-5) was mapped to a 9-cM segment on chromosome 9.

Map-based cloning of Brix 9-2-5 — To map the QTL, 7000 F_2 progenies of NIL hybrids produced and subjected to RFLP analysis with markers CP44 and TG225 which revealed 145 recombinants (1cM compared with 9cM between the same markers in the F_2 generation resulting from selfing of the interspecific hybrid). Such 10-fold reduction in recombination frequencies is often observed when exotic chromosome segments are introgressed into cultivated background. Of the 145 recombinants identified, 28 were further localized

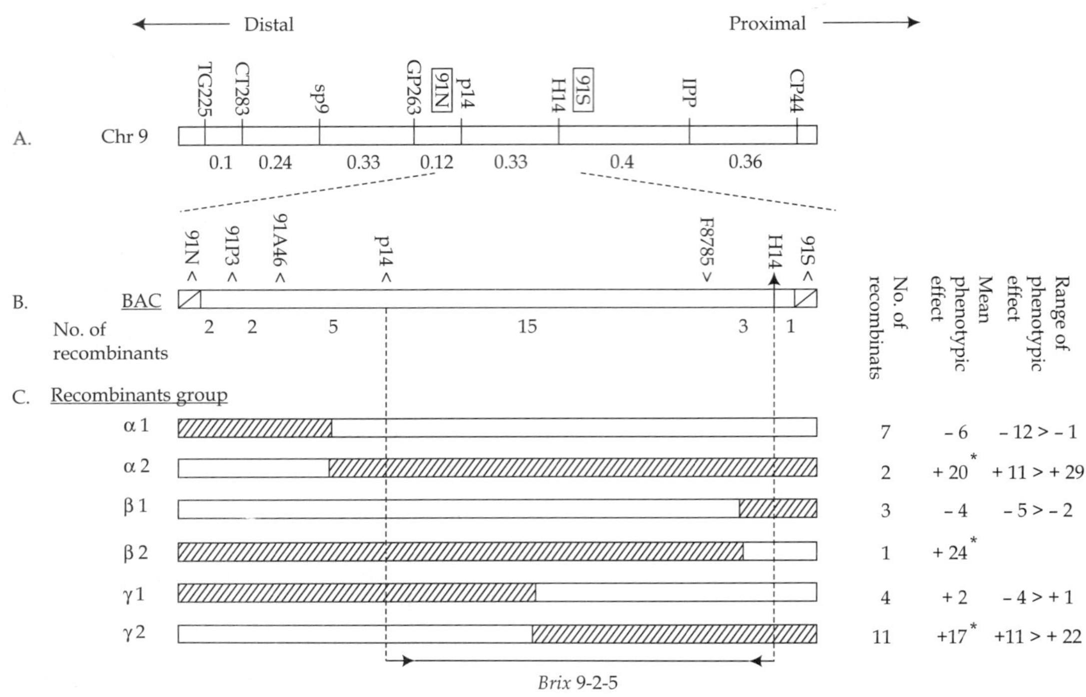

FIGURE 20.2 Showing fine mapping and physical positioning of Brix 9-2-5. (A) Genetic map (in CM) of the chromosomal region of Brix 9-2-5. The two end clones of BAC91A4, 91N and 91S are boxed. (B) Ordered markers on BAC91A4 and the number of recombinants between them (C) Phenotypic analysis of the recombination groups in the BAC.

between the two ends of BAC91A4 (Figure 20.2). For each of the 28 recombinant families, 48 selfed progenies were further genotypes with appropriate segregating markers and analyzed for brix. To simplify the data, the 28 recombinant families were divided into six recombination groups (Figure 20.2). Group α_1 included the recombinant between 91N and 91P3 (two families), 91P3 and 91A46 (two families) and between 91A46 and P4 (three families). The later three families which contain the largest introgressed in the α_1 were used to set the limits for the combined graphical representation of the group. None of the recombinants in group α_1 showed a significant effect on brix value. The reciprocal recombination group α_2 contained two families with a *L. pennellii* segment proximal to P14 and showed a significant increase in brix. The α groups placed the QTL proximal to 91A46. Using the same procedure groups β and γ located Brix9-2-5 between H14 and P14. To further narrow the position of Brix9-2-5 the 18kb spanning P14 and H14 were sequenced and used to design different primer pairs which amplified polymorphic products (in size or restriction pattern) between the parental lines. These products were genetically mapped by using the 28 recombinants and one of these primers (F 8785, Figure 20.2) which amplified a fragment of approximately 1kb, showed a complete co-segregation with the QTL. This interval was sequenced for the parental types and the recombinants and based on SNPs, 13 families were shown to be recombinants within this 1kb fragment. The phenotypic effects for each of the 13

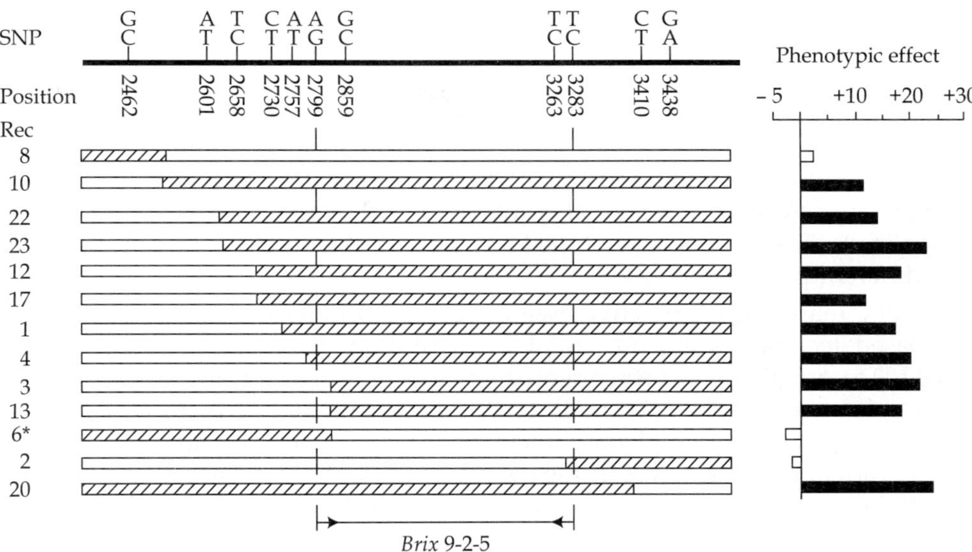

FIGURE 20.3   Showing nucleotide and phenotypic analysis of 13 recombinant families in Brix 9-2-5.

families were then used to determine the location of Brix9-2-5 on the SNP map (Figure 20.3). Recombinants 3, 13 and 6 delimited Brix9-2-5 to a region downstream of 2799 in a manner consistent with the mapping of the rest of the recombinant families. Recombinant2 delimited the QTL to the region upstream of 3283, a conclusion that is in agreement with the mapping of recombinants 5 (a member of group β_1, Figure 20.3) and 2. Thus SNP mapping allowed the unequivocal placement of Brix9-2-5 to a 484bp interval between positions 2799 and 3283.

Kroymann and Mitchell-Olds (2005) also applied fine mapping strategy to map QTL. One QTL was limited to a single gene. This approach can not be applied to mammals.

20.6   ASSOCIATION MAPPING/LINKAGE DISEQUILIBRIUM MAPPING

Like initial mapping and QTL identification which are the rate limiting step in QTL mapping, it is difficult to show a causal relation between a particular allelic variant and the phenotype. Several types of whole genome analyses across a large panel of wild strains have been proposed as shortcuts to show this correlation. One is the direct identification of genes whose expression is correlated with a trait. Another is the identification of sequence variants which correlate with a particular phenotype, so-called linkage disequilibrium or **association mapping**. Association mapping refers to the determination of the linkage disequilibrium of two or more linked loci(e.g. a gene and a marker locus) in a genome. Linkage disequilibrium is a measure of lack of independence between alleles at two loci. This phenomenon can provide information on locating disease variants for marker data as a marker in LD with casual variant provides a flag for its location. LD information also provides a means by which the efficiency of high density marker maps can be increased. If markers are in strong LD with each other, there is a need for genotyping only a subset of them. As LD typically decays rapidly in Arabidopsis (over 25-50kb; Nordborg et al., 2005) the sequence variants identified through this approach are expected to be very closely linked to the QTL. Panels for genome-wide association scans are becoming available through sequencing efforts targeting thousands of regions across the entire genome and through DNA-hybridization studies. Through the DNA hybridization technology one can genotype hundreds of strain for thousands to hundreds of thousands of markers which will greatly enhance the power of association studies. Arabidopsis and other inbreeding plant species are particularly suited to these kinds of studies and once genotyped a collection of strains can be repeatedly assayed for many different phenotypes.

LD genetic mapping is an active area of research in bioinformatics as more classical pedigree-based techniques are not suitable for finding the causal mutations. However, traditional LD techniques have been oriented towards fine mapping and this shows optimum properties only when applied to small genomic regions. But by focusing only on small regions LD methods can fail to identify interactions between loci or may result in inaccurate estimates as variation at other loci outside the region studied are not accounted for. Although LD-based techniques for fine scale gene mapping have a long history, their development and application to genome-wide studies are less well developed (Lin et al., 2004a; Marchini et al., 2005) and there is a need for developing methods that specifically account for this challenging increase in data complexity. By searching globally across the genome the probabilities of detecting loci with small effect and/or epistatic effects are increased (Perez-Enciso, 2006).

The limitation with LD-mapping is that it does not provide direct evidence of linkage. Marker trait associations may not be due to causal relationships but rather due to an unexpected statistical association between an unknown causal gene and other genes. As stains from the same geographical area are often more related than those from different areas as they are exposed to similar environments, strains that show a certain phenotype are more likely to be related to each other than expected by chance. Thus when many more markers are identified as associated than expected then one can safely assume that many of these associations are not due to causal marker-trait relationships. Statistical methods have been proposed which can correct this problem but it will require further experimental studies.

Association studies in combination with other approaches especially the analysis of experimental, will be a powerful approach. If association studies point to alleles with opposite effects on a trait of interest, then one can generate multiple F2 populations from parents containing contrasting alleles and determine whether differences in phenotypes co-segregate with the locus. Association studies in combination with the use of RILs will be a more powerful approach. Finally, expression studies can also provide information by further narrowing down the list of candidate genes likely to be causally linked to the trait.

Rigorous standards for proof for determination of the molecular bases of a QTL have been established (Weigel and Nordborg, 2005). Studies have used methods such as fine mapping and transgenic complementation to identify causal genes. However, until homologous recombination techniques for allelic replacement become available the variation introduced by position effect may limit the power to detect minor allelic effects. Molecular polymorphisms controlling quantitative trait variation are photoreceptor protein polymorphisms, transcription factors, *cis*-regulatory polymorphisms, insertion/deletion polymorphisms and copy number variation in tandem gene families. It is insufficient to draw generalization about the importance of regulatory vs coding polymorphisms especially as the QTLs that have been cloned so far have larger than average effects and may not be representative of small effect QTL (Mitchell-olds and Schmitt, 2006).

20.7 METHODS FOR REFINING THE IDENTIFICATION OF QTL

Linkage analysis is used for measuring genetic proximity of loci to each other, mapping qualitative traits and QTL. However, there are limitations with use of F2 populations for mapping QTL. The limited number of recombination events results in poor resolution for quantitative traits and further only two alleles at a given locus can be studied simultaneously. In order to increase the resolution of mapping populations-large recombinant inbred line populations which have undergone several rounds of recombination can be used. Increased rounds of recombination will increase the potential number of recombination events. Despite these the resolution of many QTL is still several centimorgans corresponding to hundreds of genes. Further, as low number of alleles are sampled per locus, so it is difficult to examine the full range of genetic diversity. A common method of refining the identification of QTL is the production of NILs and positional cloning. There are technical limitations with positional cloning. There is lack of contiguous coverage and the large amounts of repetitive DNA in genomes in many plant species prevents successful implementation of positional cloning by means of chromosomal walking. Positional cloning may not be efficient at identifying genes responsible for complex traits because of difficulty of developing NILs for loci that explain less than 20% of the variance and constraints created by only using two alleles. The majority of the genes cloned via positional cloning explain large portions of their phenotypic variance (fruit weight 2.2 in tomato, teosinte branched 19tb1 in maize, heading date in rice, FRIGIDA and CRYPTOCHROME 2 in Arabidopsis). Further, NIL is time consuming. But when nothing is known about the genes in a particular pathway, positional cloning may be the best option.

Working criteria for establishment of gene discovery in complex traits — The steps involved in the gene discovery in complex traits are as follows (Glazier et al., 2002).

1. Linkage and association
2. Fine mapping
3. Sequence analysis
4. Functional tests of candidate genes

Step I — Linkage analysis is applied in the early stages of gene localization and is one way by which a chromosomal interval of interest is defined. It is a process of tracking the inheritance pattern of genetic markers with the inheritance pattern of a trait. It involves mapping of the gene precisely and unambiguously to a small genetic interval. In case of simple traits because of strong correlation between genotype and phenotype, single recombinants are sufficient to define minimal intervals of less than 1 cM. As a result, discovery of coding sequence variants that are found only in one of a small number of candidate genes in affected individuals usually provides adequate evidence to establish gene identity. The same certainties do not apply to complex traits. In case of complex traits there is a need to establish statistically significant genome-wide evidence for linkage or association in a single study or consistent suggestive evidence in many independent studies. Although the Lander-Kruglyak guidelines for significant threshold address concerns about testing numerous genetic markers for linkage (multiple hypotheses) and about the correlated inheritance patterns among linked markers (autocorrelation), it is usual for the minimal interval of a QTL even in large human family collections or experimental crosses to be restricted to no less than 10 to 30 cM. in the primary genome screens for genetic linkage. The genetic interval of this size typically corresponds to 10 to 30 Mb of DNA or about 100 to 300 genes in human.

Step II—Fine mapping aims at reducing as much as possible the size of the interval. This can be achieved with the use of high resolution crosses, congenic strains, NILs and progeny testing or by **LD mapping**. The goal of the high resolution study is to reduce the size of the candidate interval sufficiently that the number of candidate genes is modest and functional studies can be undertaken. These approaches reduce the minimal interval to less than 1 cM. For conclusive evidence in LD studies, dense genetic markers covering the entire minimal interval should then be tested for disequilibrium with the trait phenotype in several populations. The density of markers needed depends on the extent of the local linkage disequilibrium. Use of haplotype blocks may simplify these studies.

Step III—DNA sequence analysis within the interval is needed to identify candidate nucleotide variants. Despite considerable efforts, minimal QTL intervals often include several genes and numerous DNA sequence variants. Some of these variants may reside in coding regions and other located in flanking genomic DNA. Some QTLs result from single nucleotide lesions, others result from several variant nucleotides either in the same gene or in closely linked and perhaps functionally unrelated genes. As a result, each candidate nucleotide variant as well as all combinations of candidate nucleotides in one or several genes must be identified, prioritized and functionally tested.

Step IV—The most conclusive evidence is a demonstration that replacement of the variant nucleotide results in changing one phenotypic variant for another. This test can be based on knock-in technology or a combination of gene targeting to create an engineered deficiency followed by transgenic complementation with the nucleotide or combination of nucleotides that is being tested. For cellular phenotypes, *in vitro* functional tests may be appropriate. The limitations with transgenic and gene targeting technologies are that they are not available for many species and further some variants may be specific to particular species or heavily dependent on genetic background in which case functional tests might not be informative. **Circumstantial evidence**—Like suggestive and significant linkage circumstantial evidence could include appropriate tissue expression pattern and cellular distribution, similar phenotypes associated with naturally occurring or engineered mutations in other species or a strong mechanistic support for the causal relationship between variant and nucleotide, altered protein expression or function and phenotype. In species where no *in vivo* functional tests can not be carries out, other lines of evidence, for example, *in vitro* complementation tests or reporter gene assays of gene expression combined with other formal evidence(Steps 1 to 3) may provide evidence for gene discovery.

In case of plant species it is possible to obtain strong evidence of gene discovery because of the following attributes. First, the crosses often involve large number of meiosis (up to 10, 000) which enable precise QTL localization. Secondly, the ratio of physical distance to genetic distance is generally smaller in plants than in mammals, for example, 250 kbp per cM in Arabidopsis vs 1970 kbp per cM in mice. Thirdly, genetic transformation techniques are available in several plant species which make it feasible to test whether candidate nucleotides are responsible for the phenotypic variants.

Mapping of genetic factors underlying polygenic traits in outbred populations has met with little success. This might be because of the size of individual locus effects to be modest owing to genetic interactions between loci and the prevalence of interactions has not been well characterized. Biometrical analyses have provided evidence for many interactions underlying transcript levels.

20.8 QTL MAPPING METHODS

There are two general approaches to QTL mapping.

1. Single marker
2. Flanking markers or interval mapping and
3. Multiple and marker regression approaches.

It two inbred parents differing in quantitative traits can be distinguished by a sufficiently large number of mapped molecular markers, then these two parents, their F_1, F_2 and back-cross progenies can be used to map some of the QTLs of more substantial effects.

Single marker approach—In this approach take each segregating marker in turn and see whether it correlates significantly with the QTL under investigation. Considering Q being the QTL affecting a particular quantitative trait where Q and q are alleles, linked to the molecular marker locus, M where M and m are the alleles the two parents are QQMM (QM/QM) and qqmm (qm/qm). The two parents upon hybridization produces F1, the heterozygote QqMm (QM/qm) which upon selfing produces F2 progenies. The molecular marker can be an RFLP which identifies the heterozygote. If the frequency of recombination between the QTL and the marker locus is r with $0 = r = 0.5$ then frequencies of the four types of gametes, QM, Qm, qM and qm will be 1-r/2, r/2, r/2 and 1-r/2, respectively and the progeny genotypes and their frequencies will be as shown in the Table 20.2.

Table 20.2 Showing the progeny genotypes and their frequencies.

	QM(1-r/2)	qm(1-r/2)	Qm(r/2)	qM(r/2)
QM(1-r/2)	QQMM $(1-r/2)^2$	QqMm $(1-r/2)^2$	QQMm $(1-r/2)(r/2)$	QqMM $(1-r/2)(r/2)$
qm(1-r/2)	QqMm $(1-r/2)^2$	qqmm $(1-r/2)^2$	Qqmm $(1-r/2)(r/2)$	qqMm $(1-r/2)(r/2)$
Qm(r/2)	QQMm $(1-r/2)(r/2)$	Qqmm $(1-r/2)(r/2)$	QQmm $(r/2)^2$	QqMm $(1-r/2)(r/2)$
qM(r/2)	QqMM $(1-r/2)(r/2)$	qqMm $(1-r/2)(r/2)$	QqMm $(r/2)^2$	qqMM $(r/2)^2$

From the above Table 20.2 the probabilities of individuals with particular QTL genotypes such as QQ, Qq and qq with respect to QTL locus and with particular marker genotypes such as MM, Mm and mm can be calculated as follows.

$Pr(QQ/MM) = (1-r)^2,$ $\quad$ $Pr(Qq/MM) = 2r(1-r),$ $\quad$ $Pr(qq/MM) = r^2$

$Pr(QQ/MM) = r(1-r),$ $\quad$ $Pr(Qq/Mm) = r^2 + (1-r)^2,$ $\quad$ $Pr(qq/Mm) = r(1-r)$

$Pr(QQ/mm) = r^2,$ $\quad$ $Pr(Qq/mm) = 2r(1-r),$ $\quad$ $Pr(qq/mm) = (1-r)^2$

Assuming the values of QTL genotypes, QQ, Qq and qq as a, d and –a, respectively the expected phenotypic value of each marker genotype can be estimated as follows.

$E(MM) = Pr(QQ/MM)) a + Pr(Qq/MM)) d + Pr(qq/MM)) –a = (1-2r)a +2r(1-r)d$

$E(Mm) = Pr(QQ/Mm)) a + Pr(Qq/Mm)) d + Pr(qq/Mm)) –a = [r^2 +(1-r)^2]d$

$E(mm) = Pr(QQ/mm))a + Pr(Qq/mm)) d + Pr(qq/mm)) –a = -(1-2r)a +2r(1-r)d$

From the phenotypic values of the marker classes of the F2 generation and their frequencies it can be shown that the phenotypic mean of the F2 population will be 1/4E(MM) + 1/2E(Mm) + 1/4E(mm) = d/2 where ¼, ½ and ¼ are the frequencies of marker classes, MM, Mm and mm, respectively.

Estimation of genetic effects — Estimates of a, additive effect and d, dominance effects can be obtained by the following comparisons.

Comparisions	*Estimates*
Additives effect = 1/2 (MM-mm)	= $(1-2r)a$
Dominance effect = Mm-1/2(MM-mm)	= $(1-2r)^2 d$
Dominance/Additive ratio = (Mm-(MM-mm)/2)/(MM-mm)/2	= $(1-2r)d/a$

It can be seen from the estimates that both additive additive and dominance effects are a function of r, the frequency of recombination. If r = 0 then the marker locus is tightly linked to the QTL locus and are inseparable by recombination and in this situation the mean values of the marker genotypes, MM, Mm and mm reduce to the assigned value of a, d and –a, respectively. Marker genotype will not be associated with the metric trait if it is not linked to the QTL and also if the QTL effect is too small. To see how linkage and QTL effects contribute jointly to any association one should study the regression coefficient of the phenotype on the marker genotype. There will be two types of regression to study. One type of regression will be the regression of the mean phenotype (P) against the number of marker alleles in the marker genotype. The genotypes, frequencies and deviations are set out in Table 20.3.

Table 20.3 Showing marker genotypes, their mean phenotype, the number of marker alleles and heterozygosity.

Genotype with frequency in bracket	Mean phenotype expressed as deviation from the population mean	Number of marker alleles (M)	Number of marker alleles expressed as deviation	Heterozygosity	Heterozygosity expressed as deviation
MM (1/4)	$(1-2r)a + 2r(1-r)d –d/2$	2	1	0	-1/2
Mm (1/2)	$(r^2 + (1-r)^2)d –d/2$	1	0	1	1/2
Mm (1/4)	$-(1-2r)a + 2r(1-r)d- d/2$	0	1	0	-1/2
Population mean	0	1	0	1/2	0

The regression coefficient of phenotype (P) on number of M alleles can be calculated as b_{PM} = Cov (P, M)/Var (M) = (Genotype frequency x Mean phenotype expressed as deviation x No. of alleles expressed as deviation) /(Genotype frequency x No. of alleles expressed as deviation)

$$= ((1/2)(1\text{-}2r)a) \,/\, (1/2) = (1\text{-}2r)a$$

This shows that the regression of phenotype on the number of M alleles depends on the frequency of recombination and the additive genetic effect parameter, a and if r is known then a can be estimated. The other type of regression will be the regression of mean phenotype on heterozygosity, i.e., whether or not the marker genotype is heterozygous and b_{PH} = Cov (P, H) / Var (H) = (Genotype frequency x Mean phenotype expressed as deviation x Heterozygosity expressed as deviation) / (Genotype frequency x Heterozygosity expressed as deviation) = $((1/4)\,(1\text{-}2r)^2\,d) \,/\, (1/4) = (1\text{-}2r)^2\,d$. The regression of phenotype on the heterozygosity also depends on the recombination frequency besides the dominance parameter, d and again if r is known then d can be determined. Thus for estimation of both a and d it is essential to know r, the recombination frequency and all these three parameters, a, d and r can be estimated from the mean phenotypic values of marker genotypes, MM, Mm and mm, respectively. As there are three statistics(means) and three parameters are to be estimated, this approach does not provide an independent test of goodness of fit of the model. The problems associated with the single marker approach can be overcome by another mapping strategy called **interval mapping** (Lander and Botstein, 1989).

Interval mapping — In this strategy each QTL is bracketed by marker genes and there are thus nine genotypes from which a, d and frequencies of recombination r1 and r2 between QTL and flanking markers are estimated and significant test and the estimation are done by maximizing the lod scores of likelihood ratio. In this method, the probability of a putative QTL being located in a given interval is estimated using the method of log-odds. In this approach a computer programme is set up to track along each inter-marker interval and to calculate for each position the ratio of the probability of getting the observed data with a QTL in that position to the probability if there were no QTL there. The logarithm of this ratio is called the LOD score. In other words, for any map position between the two markers the likelihood of the data is calculated assuming that a QTL is present. This likelihood is then compared with the likelihood of the data assuming no QTL. The log10 of this likelihood ratio(LOD) is then plotted along the genetic map and a support interval shown for the position of the assumed QTL, based on a given fall-off value from the maximum LOD Figure 20.4. Some arbitrary LOD score, say 2 is taken as the cut-off point below which the evidence for the presence of QTL is deemed to be too weak to be worth pursuing. This method allows efficient detection of QTLs whilst limiting the overall occurrence of false positives. There are limitations with this method (Stam, 1991). If two linked QTLS have opposite effects then there will be a downward bias of LOD values to the extent that both QTLs may go unnoticed. The presence of undirectional effects will result in upward bias in the map region between the actual QTL positions and may infer the presence of α single QTL.

For natural population it is essential to estimate the frequencies of QTL alleles. A more comprehensive approach is composite interval mapping in which multiple regression is carried out on all marker loci simultaneously (Zeng, 1994; Jansen and Stam, 1994). This

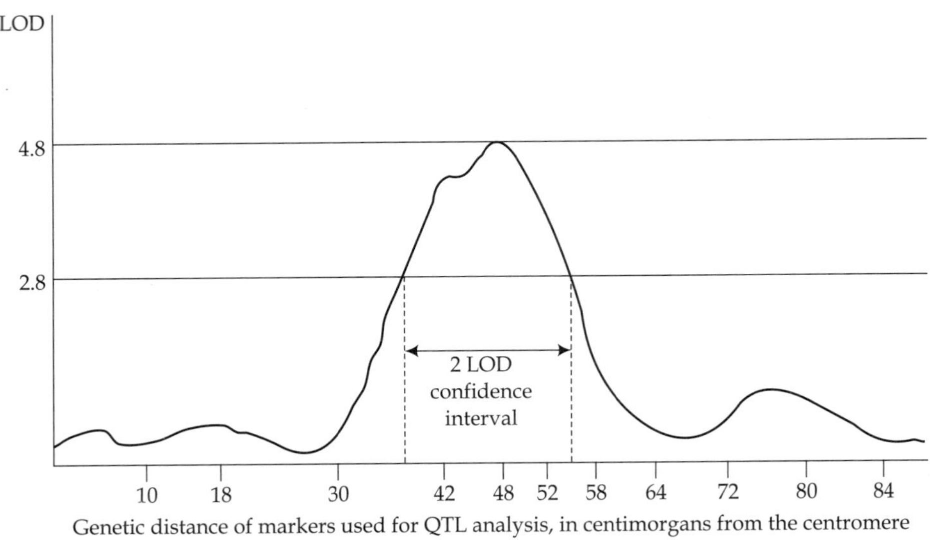

FIGURE 20.4   Showing two LOD score confidence for QTL.

method has advantages in that the effects of all QTLs are simultaneously absorbed in the regression coefficients and the method benefits from additional markers. This analysis is carried out in step –wise manner, identifying first the strongest effect and subtracting this out of the data, then identifying the next strongest effect and so forth. In this strategy choosing appropriate lod scores for statistical significance is difficult as hundreds or thousands of comparisons are made and so in practice it is often preferred to permute the phenotypes randomly among all the genotypes so as to obtain an empirical significance level (Nuzhdin et al., 1997; True et al., 1997).

20.9   USE OF CHROMOSOME SUBSTITUTION STRAIN FOR QTL ANALYSIS

Detection of QTLs and their molecular identification have proven to be serious bottlenecks in studies of complex traits. The traditional approach for QTL analysis involves two steps. The first step requires obtaining a large cross between at least two strains in which a large number of progeny are assayed for relevant phenotypes and genotyped for polymorphic markers spanning the genome. As such crosses involve the simultaneous segregation of multiple QTLs, the resulting 'phenotypic noise' limits both the power to detect individual QTLs to those with large effects and the precision to localize QTLs to large chromosomal regions. The second step involves molecular identification of the genetic variants which are responsible for each QTL. This step typically requires studying individual QTLs in isolation by performing 5 to 10 generations of backcrosses to construct strains with chromosomal segments carrying alternative alleles of the QTL on an otherwise isogenic background and then interbreeding the congenic strains to carry out fine-structure mapping and cloning. Nadeau et al., (2000) have proposed an approach for QTL analysis which involves prior construction of a panel of chromosome substitution strains (CSSs) between a donor strain (A) and a host strain (B). Strain CSS-I carries both copies of chromosome I from the donor strain but all other chromosomes from the host strain are intact and homozygous. A CSS panel partitions the variation between two strains and provides a permanent resource for studying the genetic control of phenotypic variation. Research workers can test individuals from each CSS for any phenotype of interest and immediately infer that a phenotypic difference between the CSS and the host strain implies that at least one QTL resides on the substituted chromosome. Fine structure mapping of the QTL can then be performed with crosses between the CSS and the host parental strain or with a panel of congenic strains derived directly from CSS.

Construction of CSS panels—Construction of CSS panels involves successive backcrossing to the host strain in which progeny carrying a non-recombinant copy of the desired chromosome are identified in each generation and used as parents for the next backcross to eventually progeny heterosomic (A/B) for the desired chromosome on an otherwise host (B/B) background. These progeny are then intercrossed to produce progeny which are homosomic (A/A) for the desired chromosome. Although the concept is straightforward the CSS construction has only become possible with the availability of complete genetic maps which can be used to trace inheritance throughout the genome. In case of *Drosophila melanogaster* in which special balancer chromosomes that suppress recombination can be used for chromosome substitution. Sanger et al., (2004) created a complete CSS panel in a vertebrate species using the inbred mouse strains, A.J and C57bl/6j as donor and host, respectively. CSS mapping is more efficient than traditional approaches in that it requires fewer animals to detect a given effect or allowing smaller effects to be detected with a given number of animals. Further, it is advantageous for detecting a given QTL in the presence of many other QTLs.

20.10 CANDIDATE GENE APPROACH

Another approach to QTL identification is to identify candidate genes in advance based on known functions. Candidate gene refers to any DNA sequence that by sequence homology to known genes can be expected to encode a protein with a specific function. If one or more proteins involved in the affected function are known then probes can be devised for the cloning of the corresponding genes and which can then be screened by different methods such as chromosome walking and jumping for mutations that accounts for the observed phenotypes. This is the '**candidate gene**' approach. Positional candidate gene refers to any gene linked to a DNA marker co-segregating with a phenotype of interest and meeting the criteria for a gene which could be responsible for the trait. Traditionally, the positional candidate gene strategy has relied on the ability to tailor a biological story to 'fit' the trait with a gene known to map into the identified QTL genomic region. This type of approach is possible in case of model organisms such as *Arabidposis thaliana*, *Drosophila melanogaster* and *Homo sapiens* which have been selected for genome sequencing. But as half of all the genes detected by genomic sequencing have no identified function and many genes have two or more than two functions of which only one is known so far and thus candidate gene approach is unlikely to replace QTL mapping but it will serve as support. Candidate gene approach has been used in the identification of a QTL in case of trait anxiety and depression in case of human. The serotonin uptake transporter gene is a candidate gene for traits related to anxiety or depression. The candidate gene is called SLC6A4 found on chromosome 17. Serotonin uptake helps to terminate neuronal stimulation caused by serotonin release.

QTL mapping has been widely used in plants and animals. However once a QTL has been located on a chromosomal subregion, identifying and isolating the QTL remains a difficult problem. The reason is that the LOD scores in QTL mapping typically have broad peaks as a function of chromosomal position, so additional data and often a greater density of molecular markers in the relevant regions are usually necessary to obtain greater precision in locating a QTL. Even then the isolation of the QTL may be problematic especially in a large genome. In human genome 1cM is approximately equal to 10^6 bp which is a lot of DNA to characterize given the rather wide confidence limits (5-10cM) typically accompanying a QTL localization.

Locus — a locus is identified as a part of published genome scan or a part of the researcher's own work. A locus could be syntenic with a mammalian disease model or it could contain a candidate gene with biological rationale. Locus needs to be defined as accurately as possible. Unlike monogenic disease(or qualitative trait) it is not possible to define a region by a clear recombination event between affected and unaffected family members. Analysis of complex traits generates an imprecise probabilistic signals based on increase observance of an allele in affected vs unaffected individuals. There is potential involvement of several genetic loci. The limits of each are defined by two genetic markers (usually STRs) spanning over several cMs. As 1cM equal 1Mb on average and so considering an estimate of 45, 000 genes to be present in the entire human genome (300Mb) each Mb contains an average of 15 genes.

The linkage across the locus may be defined by a broad flat peak or multiple peaks with no well defined apex. So the best approach is to define a core region with a maximal region based on LOD score thresholds which gives some margin for error. A core region is defined as any region with a LOD score of > 3 with a maximum region defined by markers with a LOD score >2 or perhaps >1 (respectively 10- and 100-fold drop in linkage probability). Where markers do not exactly define a locus it may be necessary to map markers on either side of the locus boundary. If the linkage peaks are very flat approximation to the nearest marker below the threshold might mean including a very large region. In such cases it would be worthwhile extrapolating between markers to identify the most probable region with an estimated LOD above the threshold.

High density map (high resolution map) — In general terms resolution can be defined as the extent to which closely juxtaposed objects can be distinguished as separate entities. The degree of resolution depends on the resolving power of the system. Resolution refers to the degree of density of molecular markers mapped on a stretch of DNA (e.g, a clone, BAC clone, a gene or a genome). The higher the marker density the higher is the resolution of the map. High resolution is a pre-requisite for map-based cloning of a distinct gene. It refers to any genetic or physical map of a genome which contains a large number of markers such that the markers are spaced at recombination frequency of 2-5cM or about 100-200kb physical distances. But some labs consider a map as a high density map if marker density is 0.5 (or less)cM (in case of genetic map) and 5010kb or less (physical map).

Saturation mapping is a technique for the enrichment of specific regions of an already constructed genetic map with molecular markers such that marker density is very high. This mapping strategy is essential as a saturated map is a pre-requisite for the map based cloning of gene of interest which is only possible when closely linked markers are available that were mapped in large segregating populations. Saturation mapping produces a so called high resolution map and usually involves markers generated with different marker systems (e.g. multilocus systems like AFLP, RAPD but also single locus systems like RFLP, STMS (see chapter 15).

Chromosome landing/ chromosome parachuting—It is a technique for the isolation of a gene (or genes) determining a specific trait, for example, pathogen resistance, that uses a high density of molecular markers. The multilocus marker systems such as AFLP, DAF, IARP, RAPD, REMAP, SAMPL serve to saturate the locus in question and one or more of these markers mapping at zero recombination (i.e. mapping within or very close to the responsible gene itself) are used to fish genomic clones carrying the genome (s).

21

In Vitro Mutagenesis and Directed Evolution

21.1 MUTATION-DEFINITION AND TYPES

Mutagenesis of DNA may involve minor or major polynucleotide sequence alterations at the site of mutation. Major sequence alterations do not provide useful information regarding the genetic structure of DNA whereas the minor sequence alterations such as deletion, insertion and point mutation do and amongst the three deletion represents the greatest sequence alteration (Table 21.1). Deletion mutants are particularly useful for genetic analysis where large DNA fragments are analyzed and approximate locations of specific genes and signals are to be determined. Deletion produces either an all or none response. Deletion of small DNA segments containing promoter sequence or positive regulatory sequences of polycistronic operons can completely inactivate the expression of large DNA fragments and may often result in production of fusion proteins, leading to phenotypic changes that are out of proportion to the size of the DNA segment involved in deletion event and thus not useful for analysis. **Types of mutation-** Forward mutation and reverse mutation. **Forward mutation** refers to any mutation that inactivates a gene (A $\rightarrow$ a). Such forward mutations occur at the rate of ~10^{-6} per locus per generation. **Reverse mutation** restores the original sequence of gene previously mutated by a forward mutation (A $\leftarrow$ a). **Recessive mutations** are loss-of-function mutations, the encoded product has lost its function or has reduced function. If a recessive mutation activates a response, its wild-type gene is thought to negatively regulate that phenotype. In contrast, **dominant mutations** may be gain-of-function or loss-of–function mutations. Gain-of-function mutations change the function of protein to give enhanced/ increased or new function (e.g., by over activity or ectopic expression). Dominant loss of function mutations can be caused by interference of the normal product by the mutant product and such mutations are referred to as dominant negative mutations. Mutation by activation tagging are used specifically to obtain gain of function mutations (see chapter 24 for detail).

Suppressor mutation — Suppressor mutation (second site mutation) refers to a secondary mutation which totally or partially restores function (s) lost by a primary mutation in a

Table 21.1 Showing different mutation types, procedures for generation and specificity.

Mutant type	Mutagenesis procedure	Site or regional specificity	Predictable change in change pattern
Deletion	Transformation of linearly cleaved monomers	None	−
	Transformation of specifically cleaved linear monomers	Regional	+
	S1 removal of cohesive ends of restriction endonuclease-generated linear monomers	Site	+
	Exonuclease treatment of specifically cleaved linear monomers	None	−
	Deletion of specific restriction endonuclease-generated fragments	Regional	(+)
Insertion	Random or specific linearlization; insertion of synthetic linkers	Site	+
	'Filling in; of cohesive termini	None or site	−
	Insertion of fragment encoding selectable function at specific cleavage site	Site	+
	Fusion of replicons at specific cleavage site	Site	+
Point	Hydroxyamine treatment	None	−
	(−) Sodium bi (−) sulphate treatment	Regional	(−)
	Incorporation of nucleotide analogues	Site	(−)
	Incorporation of synthetic oligonucleotide	Site	(−)

defined DNA sequence. The site of secondary mutation is distinctly different from the site of primary mutation and so a suppressor mutation does not eliminate the original mutation as is found in case of classical reverse mutations. Frequently suppressor mutation corrects a previous reading frame shift mutation. Any gene which reverses the effect of mutations in other genes is called suppressor mutation gene. A nonsense suppressor gene is a mutant gene coding for an abnormal tRNA which suppresses stop codons by reading them as encoding amino acids. For example, a supF gene of *E.coli* codes for a mutant tyrosil-tRNA (tRNA^{Try}) which reads the stop signal UAG as a tyrosine codon. Consequently polypeptide synthesis is not terminated at the UAG position but extends beyond it. Suppressor gene can be used as a selectable marker.

Forward mutations and metabolic suppressors obtained by reversion can provide insights into the function and relationships of normal gene products. Similarly, mutations and intragenic revertants provide the raw material for the analysis of gene product structure-function relationships. Reverse genetics and other methods based on recombinant DNA techniques are increasingly used for this purposes and they are the methods of choice where specific changes in specific genes or genetic sites are required.

Point mutations: Generating point mutations are essential for the analysis of individual genes and provide information on the functional parts of their products, i.e., the domains, regulatory elements and sequences involved in maturation and transport, etc. of proteins. Mutants that affect enzyme activity have in most cases been shown to change the amino acid sequence of the enzyme and exceptions to this generality are mutants that involve the ends of the gene and probably affect other functional characteristics of the enzymes. The **point**

mutation (single base change) refers to a mutation involving a chemical change only one single nucleotide (Transtition, Transversion). Point mutations are mostly single base-pair substitutions in the DNA and have the following characteristic features.

1. They are generally capable of mutating back to wild type
2. They can produce wild-type recombinants at some low frequency in the great majority of pair-wise combinations which indicates that they usually fall at different and non-overlapping sites.

21.2 CLASSES OF SINGLE BASE CHANGE MUTATIONS RESULTING FROM CHEMICAL MUTAGENESIS

I. **Nonsense mutations-** They arise from changes that convert an amino acid codon into a stop codon and thereby resulting in truncating the predicted protein

II. **Splice junction mutations-** They are also expected to truncate the protein. Both types (I and II) of lesions are subject to mRNA degradation and so no biologically active protein would be produced.

III. **Missense mutations-** They result when single base changes alter the amino acid encoded by a particular codon. These mutations can be further characterized as causing conservative or non-conservative substitutions. These classes of mutations have the potential to produce a phenotype. A missense mutation that causes a change in a conserved region of a protein will frequently destabilize it and in well-studied cases such lesions are conditional depending on temperature or other environmental factors. This feature of single nucleotide substitution mutation is exploited for reverse genetics (see chapter 24).

IV. **Silent mutations-**Mutations most likely to be phenotypically silent are called silent mutations and they include changes in codons that do not change the encoded amino acid, e.g., mutations in introns and in upstream and downstream regions.

21.3 SUPPRESSOR AND ENHANCER SCREENS

When a selected mutant is again mutagenized and the resulting population is screened for suppression of the mutant phenotype, i.e., looking for the original wild-type, it is termed suppressor screen. On the other hand, if a mutant line is again mutagenized and the resulting population of mutant line is screened for enhanced mutational phenotype, it is called enhancer screen. Suppressor mutations are similar in phenotype to the original but this similarity results from suppression of the first mutation , not from a restoration of the original allele. Such suppression may occur by a mutation in the same gene-an intragenic suppressor mutation or in a different gene-an extragenic suppressor mutation. Similarly, the enhancer mutant results from mutation in the same gene or at a new locus that enhances the effects of the earlier mutation. The products of extragenic suppressor or enhancer mutations act as *trans* factors to inhibit or enhance the expression of the affected genes. The cloning of their genes can be useful in elucidating the function of the original wild-type gene in a signal transduction pathway.

21.4 DELETION MAPPING

On the other hand, deletion mutants are rarer and are unable to revert to wild type because they have lost a more or less extensive tract of the unique genetic material and thus can not be recovered by random mutation. Null mutations of any cloned gene can be created by introducing *in vitro* generated deletion into the genome at the precise chromosomal position of the gene. Deletions are often combined within single genes but occasionally overlap adjacent genes.

Steps involved in deletion mapping:

1. A set of partly overlapping deletions can be mapped in a linear sequence on the principle that non-overlapping deletions can generate wild-type recombinants whereas overlapping deletions can not. The overlaps of the deletions can be used to define a series of segments in a linear map.

2. The second step is to cross each of the point mutants to the set of deletions. Each cross depending upon whether the site of the point mutation falls outside or inside the segment deleted in the other part, will or will not yield wild-type recombinants. On the basis of this each point mutation can be placed within one of the segments defined by the deletion overlaps (Fincham, 1994).

21.5 SATURATING A PARTICULAR LOCUS WITH MUTATIONS

The fist step in the genetic analysis of genes is to have a range of mutants available. If one wants to obtain many mutants of a particular kind or to saturate a particular locus with mutations, it is then essential to treat organisms with mutagenic agents. In **classical methods of mutation** the cells are treated with mutagens. The mutagenic agents can be chemical (EMS) or physical mutagens (X-rays, U.V light). The alkylating agents such as N-methyl-N'-nitro-N-nitrosoguanidine (MMNG) and EMS (ethyl methane sulfonate) induce high frequencies of base pair substitutions and little lethality. Transition is exclusively at G.C sites. EMS is a standard chemical mutagen used in plant and animal system because, it shows consistency, i.e. apparently similar levels of mutagenesis have been achieved in Arabidopsis and Drosophila. Further, base substitution rates are comparable for Arabidopsis seed soaked in EMS and Drosophila male fed with EMS and recessive lethals are estimated to occur at similar rates in both cases when the dose of EMS used is the maximum tolerated before levels of sterility and lethality become unacceptable. Ionizing radiations also induces chromosome breaks and segmental rearrangement. U.V light is most convenient mutagen for microorganisms and is effective on exposed cells. In case of U.V. light substitutions mostly occur in runs of pyramidines particularly T-T pairs. There is occurrence of transition and transversion. Frame shift mutations (exclusively nucleotide deletion) also occur. Mustard causes +1 frameshift mutations, i.e. G insertion in runs of two or more G's with preference for runs of three or more G's. Chemical mutagens induce less chromosome breaks comparatively. Whether it is physical or chemical mutagen, it needs to be sufficiently stringent to kill a substantial fraction of the treated cells, if there is to be a useful proportion of mutants among the survivors. The highest proportion of mutants per treated cell is usually found at doses giving 10 to 50% survival.

21.6 TILING MUTAGENESIS

In this strategy chemical mutagenesis is followed by screening for point mutations in pooled DNAs via the use of technologies that are being primarily developed for discovering SNPs. There are a number of advantages with chemical mutagenesis. It can provide allelic series of mutations including knockouts. High density of chemically induced point mutations makes TILING suitable for targeting small genes and it allows to focus on single protein domains when targeting larger genes. Saturation mutagenesis can be achieved with relatively few individuals. Finally, chemical mutagens produce a relatively high density of irreversible mutations. Steps involved in tiling mutagenesis in Arabidopsis are as follows (Figure 21.1).

1. Seeds are soaked in a 20-100mM solution of EMS for 10 to 20 hr. After that seeds are washed and sown.

2. Plants formed by this treated seed constitute the M1 generation. M1 plants are chimeric. They have mutated tissue sections that have descended from a single embryonic cell. Further, each mutated section is heterozygous for any mutation. M1 plants are selfed.

3. The M2 generation plants derived from selfed M1 plants are used to prepare DNA samples for screening while its seeds are inventoried. DNA samples are pooled and pools are arrayed on microtiter plates and subjected to gene specific PCR and mismatch detection in pools. Each mutation which is discovered in a pool is rescreened in individual plants that make up the pool. Its DNA is then sequenced to identify the particular mutation.

4. Phenotypic analysis is performed on plants raised using M3 seeds from M2 plants with crosses and genotyping as required.

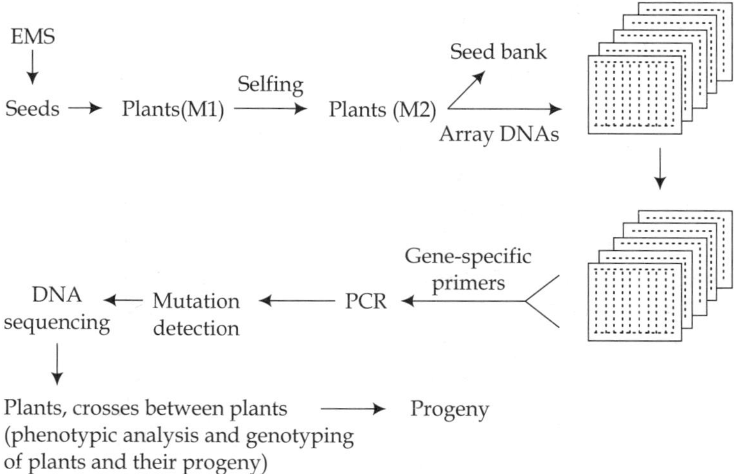

FIGURE 21.1   Showing of Tilling strategy – a reverse genetic strategy for plant functional genomics (After Henikoff and Comai, 2003).

21.7 ISOLATION OF MUTATIONS FROM PRECISELY DEFINED CHROMOSOME SEGMENTS

The principle is to place chromosomes from mutagenized flies (in Drosophila) in heterozygous combination with a chromosome with a deletion covering the segment of interest. Any recessive visible or lethal mutation within that segment will thereby be revealed. The method depends on the use of **balancer** chromosomes, of which a broad selection is available in Drosophila. Their main features are: (i) the presence of a recessive lethal mutation to prevent the balancer from becoming homozygous (ii) a dominant visible marker which may be the same as the recessive lethal to enable flies carrying the balancer to be identified and (iii) a long inversion or complex of inversions to prevent recombination between the balancer lethal and any mutations on the homologous chromosome with which the balancer is combined. Most balancer also carry one or more recessive viable markers. The advantage of balancer chromosomes is that they make it easy to maintain recessive lethal mutations. For example, any chromosome 3 carrying a recessive lethal can be kept in a permanently heterozygous stock in combination with a balancer chromosome 3 and such a stock will produce exclusively heterozygous progeny as both kinds of homozygous will be inviable. The inversion in the balancer prevents the generation of homozygous-viable chromosome by crossingover. Such technique is not feasible in mammals or flowering plants as it will prove to be very costly.

Mapping of the mutations:

21.8 GENOMIC SUBTRACTION

It is a technique used for the identification of DNA sequences (genes) missing in a deletion mutant but present in the wild-type. This technique is base on the removal of all sequences of the wild-type DNA (called **driver**) from the deletion mutant genome (called **tester**) such that only the deleted sequence is left over. In other words, genomic subtraction involves the isolation of the DNA of the deleted region. The enrichment for this region is achieved by mixing denatured wilt-type and denature biotinylated mutant DNA, reaasociation of both types of DNA and repeated removal of biotinylated sequences by binding to avidin coated magnetic beads. In each cycle the unbound wild-type DNA from the previous round is hybridized to new biotinylated deletion mutant DNA. The unbound DNA from the final cycle is then ligated to adaptors and amplified with adaptor specific primers using PCR. The amplified product can then either be cloned and sequences or used as a probe to find the corresponding deleted sequence in a genomic library of the wild-type genome.

Point mutations can provide a more detailed information regarding the structure and function of small DNA segments. The cloned gene can be sequenced and the regulatory region(genetic control elements, the regulatory gene) can be separated from the structural gene region.

Phenol emulsion reassociation technique (PERT)—It is a variant of the genomic subtraction technique in which phenol is used to increase the rate of hybridization. It is a form of competitive hybridization between two related but slightly different genomes, for example, genomes of male and female, which preserve only the unique sequences of one genome (e.g., the female one) in a clonable form.

21.9 INSERTION MUTAGENESIS

Insertion mutagenesis refers to the interruption of a DNA sequence by insertion of additional DNA. Additional DNA could be either single base pair or many base pairs. The single base pair insertions may be caused by certain chemicals, for example, acridine dyes while the integration of transposons or insertion sequence (IS) cause longer insertion mutation. Any insertion of bases in other numbers than 3 or multiples thereof may result in a shift in the reading, frameshift mutation. In any case insertion mutation may either lead to the loss of function of the original DNA (insertional inactivation) or the restoration of a previously defective DNA (insertional activation). *In vitro* insertion of DNA segments into target molecules may be site-specific or random.

In vitro mutagenesis is a technique for alteration of the base sequence of DNA in the test tube. It can be used to study the effect of changes in either regulatory region or structural region on the protein structure and function and regulation of the protein. Change in the primary structure of protein (AA sequence) will lead to change in folding pattern of protein, three dimensional structure and activity of the protein and change in regulatory region will lead to change in gene regulation. *In vitro* mutagenesis in combination with gene cloning is a powerful technique for the analysis and manipulation of gene structure and function and allows not only to carry out genetic , physiological and biochemical studies but also to carry out other very difficult experiments.

In *in vitro* mutagenesis the mutagenesis is directed to specific nucleotides of the coding regions of the gene. Although a variety of strategies such as Bal 31 mutagenesis, chemical mutagenesis, displacement loop mutagenesis, PCR mutagenesis and circular mutagenesis have been used to produce mutation in cloned gene, two of the most widely used techniques are:

1. Deletion mutagenesis and
2. Site directed mutagenesis.

21.9.1 Deletion mutagenesis

Deletion derivatives which are obtained primarily by *in vitro* manipulations are generated by nuclease digestion of genomes that have been linearized. At a unique site by digestion with a restriction endonuclease that cleaves the genome once ii. at one of several specific sites by partial digestion with a restriction enzyme that cuts the genome at multiple sites or iii. at random locations by cleavage with pancreatic DNaseI in the presence of $Mncl_2$ (Shenk et al., 1976). The deletion mutagenesis is carried out using Bal31 nuclease. This enzyme has exonuclease activity against double stranded DNA and it degrades both 3′ and 5′ ends. Bal 31 degrades both strands of the dsDNA having blunt ends and the resulting DNA fragments are shortened and possess blunt ends. Suppose the cloned DNA fragment in which the deletion is to be made is near a unique restriction enzyme (say E1) site in the plasmid. This circular plasmid is now linearized with this specific restriction enzyme. The monomeric molecules are usually separated from the uncleaved and more extensively degraded molecules by agarose gel electrophoresis. The purified full length linear molecules obtained are now digested with Bal31 nuclease for different times. Digestion with Bal 31 nuclease will produce deletion at both ends. The resulting fragments possess blunt ends. The size of deletion is determined by the time allowed for digestion and thus the digestion will generate a spectrum of deleted molecules. The other types of nucleases may be S1 if small deletions corresponding to loss of restriction endonuclease generated cohesive ends are required (Backman et al.,1979) or can be λ nucleases or exonucleaseIII if longer deletions are required. In case of later the degree of exonuclease digestion is monitored by following the appearance of acid soluble radioactivity from radioactive DNA to be derivative genomes showing deletions of the desired size. Digestion with λ exonuclease results in generation of 3′ single stranded termini whereas digestion with exonuclease III results in generation of 5′ single stranded termini and these single stranded termini can be removed by S1 nuclease. Thus these two nucleases (λ and exonuclease III) can convert blunt ends into protruding single strands. The S1 nuclease is single stranded specific, can cut either ss DNA or RNA. It can digest single stranded region in the deletion loop. It also removes up to 30bp of double strand. Thus S1 removal of cohesive ends does not result in generation of site specific deletions and the recircularization of the cleaved molecule is carried out *in vitro* by blunt end ligation, prior to transformation of the DNA into a host cell system. As for using S1 nuclease it is a must to have linear DNA molecules having single stranded termini and so the circular DNA molecule is linearized by digestion with a restriction endonuclease which cleaves at a unique site or partial digestion with a restriction endonuclease for which several sites are there. Circular DNA molecule can also be linearized through a cut at random location using the pancreatic DNaseI. The linear molecules are first treated with the Klenow fragment of *E. coli* DNA polymerase and d NTPs to fill in any overhangs and the blunt ends are then ligated to linkers which are cleaved and ligated together with DNA ligase prior to transformation of the DNA into the host cell system. As deletion of the terminal regions of linearized DNA fragment occurs readily *in vivo* after transformation, *in vitro* exonuclease digestion may dispensed with entirely provided a lower yield (1-20% in comparison with ~100% of all viable genome) of deletion derivatives is acceptable and this will be acceptable only if a rapid assay or screening procedure is available to detect the deletion derivatives. Where deletions originating from a unique restriction enzyme cleavage site are required the deletion derivatives can be detected by loss of that cleavage site and loss of susceptibility to the corresponding endonuclease. The location and extent of deletion of a DNA fragment is generally mapped by heteroduplex formation with the parental DNA fragment followed by visualization in the electron microscope or digestion of the single stranded region with nuclease (Bal31 or S1) and restriction endonuclease analysis of the remaining duplex DNA. Deletion analysis is used for the identification of control sequences for a gene by determining the effects on the gene expression of specific deletions in the upstream region. These *in vitro* deletion procedures have been widely used to examine functions of small genomes such as that of SV40 and have identified Tn3 encoded 'transposase' and the terminal inverted repeat sequences of the ampicillin resistance transposon, Tn3. Thus we see that a gene in plasmid is cut with restriction enzyme to produce ends with 5′ overhang. The overhang is filled in and the ends are relegated. This process changes the reading frame of the gene downstream of the restriction site.

Delete-a-gene method — Many knockout mutants can be discovered by screening for deletions using PCR amplification of pooled DNAs. Smaller fragments which result from kilobase-scale deletions within the amplicons are preferentially amplified and thus are enriched, permitting their electrophoretic detection even in pools of 1000 plants. Deletion mutagenesis may be the best way to knockout tandemly repeated genes which are common in plants (Achaz et al., 2001).

Displacement loop mutagenesis — It is a technique for the introduction of short deletions in a circular DNA duplex by annealing a synthetic oligonucleotide to the region which is likely to be deleted. Under appropriate conditions the oligonucleotide will induce a displacement loop (see chapter 16). It is then treated with single strand specific S1 nuclease which removes all non-paired region within the loop (and additionally the region where the oligonucleotide is hybridized). After S1 digestion the remainder is recircularized. The resulting duplex molecule carries deletion of about 10bp at the positions previously marked by termini of the oligonucleotide.

Base variation can be introduced with the use of **site-directed** or **random mutagenesis** (Green et al., 1990).

21.9.2   Random *in vitro* mutagenesis

Stemmer (1994) introduced the 'DNA shuffling' technique for random *in vitro* mutagenesis and recombination. DNA shuffling (or gene shuffling) is a technique for the *in vitro* recombination of homologous gene sequences and the generation of novel proteins with new properties. In this technique, a set of homologous genes are cut down to fragments of 50-100bp by DNase I first and then the fragments are recombined in a self priming process involving recombination at loci of high sequence homology. Subsequently, the reassembled gene fragments are amplified by PCR and the amplification products are cloned into vectors. Each clone is then subjected to a stringent screening process for a desirable trait. For example, multiple gene DNA shuffling with eight human interferon-α-genes generated a chimeric interferon-á with a 285, 000-fold high affinity than human interferon -2α and consisted of only five parental gene parts. This technique is more effective than site-directed mutagenesis since the later leads to single base exchanges with usually undesirable phenotypic consequences.

21.9.3   Random priming recombination (RPR)

RPR for *in vitro* mutagenesis and recombination of polynucleotide sequences is an effective alternative to DNA shuffling. This method involves priming template polynucleotide (s) with random sequence primers and extending to generate a pool of short DNA fragments which contain a controllable level of point mutations. These fragments are reassembled during cycles of denaturation, annealing and further enzyme-catalyzed DNA polymerization to produce a library of full length sequences. Screening or selecting the expressed gene products leads to new variants with improved functions. Random sequence primers are used to generate a large number of short DNA fragments complementary to different sections of the template sequence (s). Due to base misincorporation and mispriming these short DNA fragments also contain a low level of point mutations. The short DNA fragments can prime one another based on homology and be recombined and reassembled into full length genes by repeated thermocycling in the presence of thermostable DNA polymerase. These sequences can be further amplified by conventional PCR and cloned into a vector for expression followed by screening or selection. RPR and screening can be repeated over multiple cycles in order to evolve the desired properties.

Advantage over DNA shuffling

1. RPR can use single stranded polynucleotide templates without an intermediate step of synthesizing the whole second strand. Mutations and/or crossingovers can be introduced at the DNA level from ss or ds DNA template using DNA polymerases.

2. DNA shuffling requires fragmentation of the ds DNA template which is generally done with DNase I. The DNase I must be removed completely before the fragments can be reassembled. Gene reassembly is generally easier with RPR method which employs random priming synthesis to obtain the short DNA fragments. Furthermore, as DNase I hydrolyzes ds DNA preferentially at sites adjacent to pyrimidine nucleotides its use in template digestion may introduce a sequence bias into the recombination.

3. Synthetic random primers are uniform in their length and lack sequence bias. The sequence heterogeneity allows them to form hybrids with the template DNA strands at many positions so that at least in principle, every nucleotide of the template should be copied or mutated at a similar frequency during extension. The random distribution of the short, nascent DNA fragments along the template (s) and random distribution of point mutations within each nascent DNA fragment should guarantee the randomness of crossing overs and mutations in the full length progeny genes.

4. Random priming DNA synthesis is independent of the length of the DNA template and so DNA fragments as small as 200 bases can be primed equally well as large DNA molecules such as linearized plasmid or λ DNA. This has thus unique potential for engineering small peptides. As the average size of synthesized ssDNA is inversely related to the primer concentrations(Hodgson and Fisk, 1987), the proper conditions for random priming synthesis can be easily set for a given gene using this information.

5. As the parent polynucleotide serves solely as the template for the synthesis of nascent ss DNA, 10-20 times less parent DNA is needed in comparison to DNA shuffling.

21.9.4 Site-saturation mutagenesis

Site directed saturation mutagenesis is an efficient technique to study well conserved region of a protein which is assumed or known to have a functional role (Albert et al., 1988). In this SSM the resulting population of mutated gene is expected to consist of otherwise identical genes but a given codon should be random. In individual clones this codon thus codes for any amino acid and the position is said to be saturated. The site saturation principle gives a 2-fold advantage over the more commonly used simple substitution mutagenesis procedure. First, several amino acid substitutions are gained by the same effort and secondly, the desired additional information can be obtained concerning the nature of acceptable substitutive amino acids.

Maximal randomness for codons is particularly important in approaches where direct phenotypic selection (selection of viable or functional gene products) will be performed after transfecting the saturated DNA pool in appropriate cells. The actual codons in the saturated position of the products have to be sequenced which is more laboursome than sequencing before transfection. When substitutions are as random as possible, a minimal number of sequences have to be determined. True randomization may be even more important when studying nucleic acid sequences themselves, for example, sequence-specific binding or viral noncoding regions.

Techniques for SSM—Several techniques are available for SSM. In 'cassette mutagenesis' the entire target area is synthesized and then ligated between two restriction sites. Close to equal proportions of four nucleotides can be measured at a given position of oligonucleotide, simply by using 25% of each nucleoside phosphoramite when synthesizing that position. A pre-requisite for cassette mutagenesis is that there are two unique restriction sites very close to the target codon or that such sites can be generated by SDM.

All other techniques are based on either plasmid, bacteriophage, or phagemid vectors or PCR and involve a step where mutagenic synthetic oligonucleotide is annealed to the target area. In SSM studies the mutagenic oligonucleotide has been synthesized using equimolar concentrations of bases at the positions to be randomized.

Problems associated with these techniques

A problem arises here due to differential efficiency in annealing. Oligonucleotides containing up to three successive mismatching bases anneal less efficiently than those containing less or no mismatches and the mutated population of clones will be biased in favor of the original sequence. The difference in annealing efficiency is most pronounced when the mutagenic oligonucleotide is relatively short. Significantly longer oligonucleotides should be able to reduce the bias as there the mismatches have a relatively slighter effect on the melting temperature ™ and the association and dissociation constants (Kass, Kdiss). Mismatches in oligonucleotides seem to have a greater effect on Kdiss than on Kass.

Bias toward the original sequence

The bias towards the original sequence can be completely avoided by first adding the codon (loop out) and then re-introducing a random codon in the same place (loop- in). This strategy has the advantages of cassette mutagenesis without requiring the close by restriction sites but it requires two rounds of mutagenesis and two oligonucleotides for each codon to be saturated. Wild type background can also be avoided by introducing stop codons or restriction sites in the first round of mutagenesis and replacing those with saturated codons in the second round but these methods do not avoid the bias towards the original template sequence. One round mutagenesis strategy which avoids the bias is based on modified nucleoside phosphoramidite concentrations during the synthesis of the mutagenic oligonucleotide.

SSM using degenerate oligonucleotide primers is a frequently used technique for introducing various mutations in a selected target codon. Oligonucleotides which are synthesized using equimolar concentrations of nucleoside phosphoramidites (dA, dC, dG, dT) in the positions to be saturated result in a mutant population which is biased towards the original nucleotides. This bias could be eliminated by modifying the concentrations of nucleoside phosphoramidites during the oligonucleotide synthesis. The proportions of the 'conservative' phosphoramidites have to be reduced at the target positions. The proportions of conservative C and G in the oligonucleotide have to be reduced more than those of the conservative A and T to achieve maximal randomness in the resulting nucleotide distribution as there is tighter base pairing between C and G and so have greater advantage in annealing and so compensated by greater reduction in concentrations.

21.10 SITE DIRECTED MUTAGENESIS (SDM)

SDM is a powerful tool for exploring structure-function relationships of proteins and nucleic acids. It is used to produce mutation at **specific site** which is not possible with Bal13 nuclease. SDM can be used to i. uncover nucleotides in mRNA precursors which direct the RNA splicing of introns, ii. to characterize amino acid residues critical for protein-protein interactions and iii. to probe catalytic mechanism of enzymes. Site directed mutagenesis is carried out using synthetic oligonucleotides (Figure 21.2). In this technique it is essential that the DNA must be subcloned into a bacteriophage M13 or related vectors such as φX174 that allows the generation of single stranded circular recombinant DNA. φX174 has life cycle involving both single and double stranded phase. With this technique a specific DNA sequence change (desired base pair change) can be inserted. A simple method for the introduction of insertions at cleavage sites of the restriction endonucleases that generate fragments with 5′ cohesive single stranded termini involves linearization of the molecules with the appropriate endonuclease, 'filling in' of the single stranded termini with DNA polymerase I, followed by recircularization of the molecules by blunt end ligation and finally re-introduction of the molecules into the host cell system (Backman et al., 1979). If the fragment single stranded termini are four nucleotide long there will be a net insertion of 4bp at this cleavage site, if two nucleotides long, the insertion will be two bp, etc.

21.10.1 Approaches to SDM

There are two approaches to site directed mutagenesis. In both approaches it is required to have knowledge of the polynucleotide sequence of the DNA segment (cloned gene) containing the site to be mutated. In the first approach if an appropriate restriction site flanks the sequence (in which the mutation is to be created) of the cloned gene then this sequence can be removed by a restriction endonuclease and is replaced by a synthetic gene (DNA segment) identical with the original fragment except for the desired change which involves at least a change in one codon, i.e., a three base pair change.

Oligonucleotide directed mutagenesis — If there does not exist a suitable restriction site in the cloned gene then one can adopt this approach. Oligonucleotide-based site directed mutagenesis was developed by Smith and his colleagues (1978). In this approach an oligodeoxyribonucleotide (10-20 mer) is synthesized whose base sequence is similar to the region of cloned gene in which a desirable mutation is to be created except that it contains a change at one base position. DNA strand with a specific base change which acts as a primer is then annealed to a single stranded copy of the cloned gene and synthesizes a duplex DNA (with one mismatch) in the presence of DNA polymerase I(klenow fragment), d NTPs and DNA ligase. The mismatch of single base pair out of 15-20 does not prevent annealing if it is done at proper temperature. This duplex DNA is then used to transform cells of *E. coli*. When the transformed *E. coli* cells with plasmids are analysed, about half of the plasmids will have gene with desired base pair change (mutant) whereas the other half of the plasmids will have

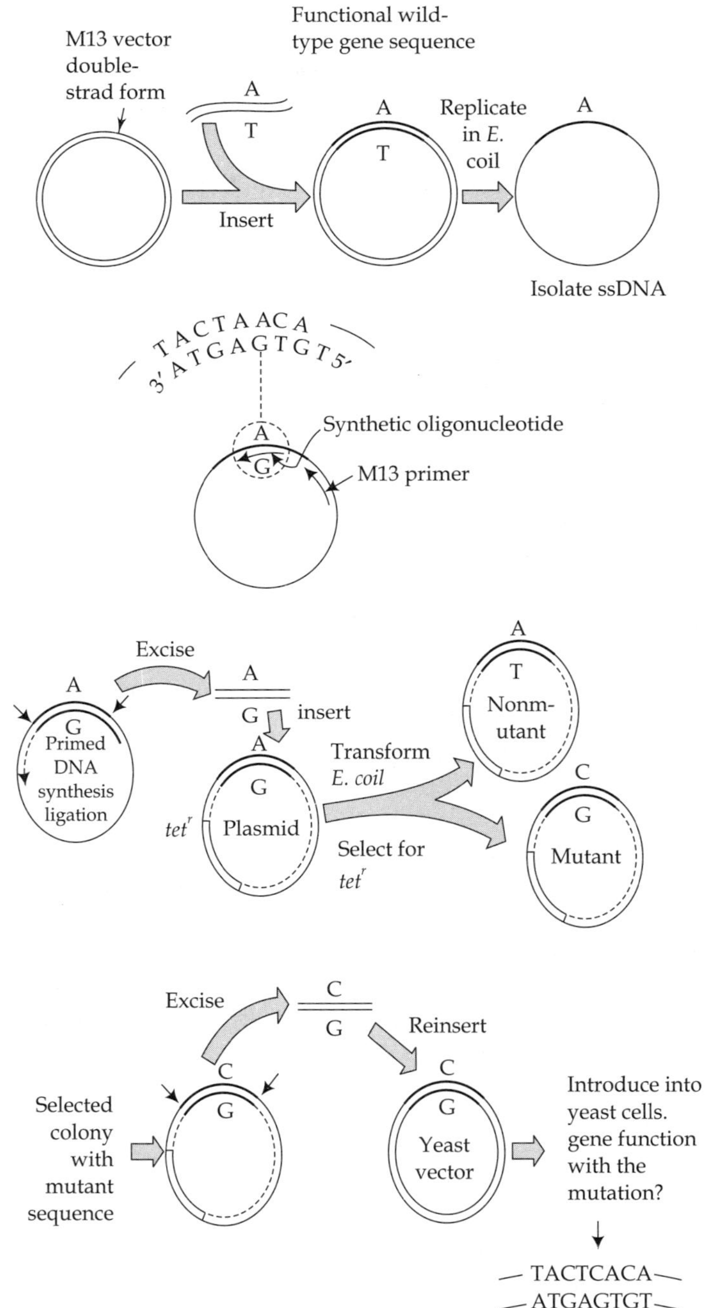

FIGURE 21.2 Showing *in vitro* mutagenesis procedure.

the original gene sequence (wild type). Although mismatch is repaired by cellular DNA repair enzymes it will generally convert about 50% of the mismatch to produce the desired sequence change (mutant type).

Oligonucleotide-direct site specific mutagenesis is used routinely to introduce desired mutations into target DNA sequences. Traditional methods of site-specific mutagenesis involve the hybridization of an oligonucleotide containing mismatched nucleotides to a region of ss target DNA (Smith, 1985; Zoller and Smith, 1987). DNA polymerization and ligation then complete the synthesis of the mutated strand. Following replication and selection of the double stranded mutant DNA, mutagenized vectors can then be propagated *in vivo*.

Although classical methods of SDM allow reliable production of mutated DNA necessary steps in the protocols are laboursome as selection and isolation of single stranded parent and mutated DNA constitute major rate limiting steps (Lawrence, 1991). There is a method of SDM which employs both T7 DNA polymerase for reliable polymerization of full length template DNA (Lundeberg et al., 1991) and a biotinylated primer for high affinity selections of full length double stranded mutated DNA (Uhlen, 1989). This scheme eliminates time consuming single strand selection procedure (Wang et al., 1997). A variety of protocols have been established to achieve efficient mutagenesis including several which use PCR (Madison et al., 1989).

There are two categories of SDM: 1. Cassette mutagenesis 2. Enzymatic extension of a mutagenic oligonucleotide annealed to a DNA template.

Cassette mutagenesis — It is a technique for site specific mutagenesis. First, a specific region flanked by two restriction sites is removed from the target gene and then a DNA fragment consisting of a chosen sequence, for example, a synthetic oligonucleotide, is inserted in its place which results in various types of mutations by way of amino acid replacement in the encoded protein.

Overlap extension-PCR — Mutagenesis by overlap extension has been previously described as a method for both SDM to create base substitution, insertions or deletions and production of chimeric genes by combining two DNA fragments without a need for restriction sites. In separate PCR reactions, two fragments of a target sequence are amplified by using, for each reaction, one universal and mutagenic primer. Two intermediate products with terminal complementarity form a new template DNA by duplexing in a second reaction. During this so-called overlap extension the fused product is amplified with the help of two universal primers. To obtain a high mutant yield, it is necessary to fractionate and purify the products of the 1st PCR reaction by gel electrophoresis.

OOE-PCR — A new method called one step overlap extension (OOE-PCR) has been developed to efficiently perform SDM based on OE-PCR. In this technique two template DNA molecules in different orientations relative to only one universal primer is amplified in parallel. By choosing a high dilution of mutagenic primers it is possible to run an overlap extension PCR in only one reaction without purification of intermediate products. In principle, OOE-PCR can be applied to every DNA fragment which can be cloned into a multiple cloning site of any common cloning vector (Urban et al., 1997)

Linker mutagenesis — Site-specific insertion procedures can be used to mutagenize a genome at random locations if the target DNA is linearized not with a restriction endonuclease but by

DNaseI in the presence of Mn^{2+} (Shenk et al., 1976). In this technique mutation is introduced into a DNA molecule by means of insertion of linkers. The blunt –end ligation of 'linkers' to randomly linearized hybrid molecules is a highly useful method for introducing short insertions at random locations in small genomes. To start with a circular DNA molecule is treated with DNase I under conditions which allows random cutting of the duplex molecule. There is thus generation of a set of linear molecules with different termini. The linkers are now ligated to the ends of linear molecules. The linker linked molecules are cut with restriction endonuclease whose recognition site is specified by the linker and which results in generation of single stranded overhangs which in turn are used to recircularize the molecules using DNA ligase. The newly created insertion can be easily localized by restriction mapping. Internal cleavage of the molecule is avoided by appropriate choice of linker, or by protection of internal cleavage sites by prior treatment with the specific methylase. Linker mutagenesis was employed to dissect the functions involved in the transposition of the ampicillin resistance transposon (Heffron et al., 1978). The use of linkers to generate readily mapped random insertions is a powerful method for the functional analysis of any cloned gene or operon.

SDM by unique restriction site elimination introduced by Deng and Nickoloff (1992) permits site-specific mutagenesis of a plasmid DNA without any subcloning step. This method uses two mutagenic primers. One carries the desired mutation and the second which acts as a selection primer, carries a mutation in unique, nonessential restriction site in the target plasmid. The mutagenesis method relies on the simultaneous annealing of two primers (mutagenic and selection primers) to one strand of the denatured double stranded plasmid. After DNA elongation and ligation the selection of plasmids lacking the selection restriction site will also encode the desired site mutation. However, in case of some plasmids such as p GEX-BTK, the recovery of the desired mutant is much lower than that of the selection restriction site selection mutants, yielding the desired mutant products at a frequency < 10%.This low frequency is due to the sequence of the target DNA which may acquire stable secondary structures such as stem-loops. These structures may interfere with annealing of the mutagenic primers and could result in low recovery of mutant plasmids. So the development of PCR fragment directed DNA synthesis relies on the use of selection and mutagenic primers to amplify a region of DNA lying between the annealed selection and mutagenic primers, generating the desired mutant DNA fragment. This mutant DNA fragment is then used as primer for the subsequent production of the heteroduplex DNA. As the amplified DNA fragment contains both the desired and the selection mutations, this technique results in a high efficiency of SDM (Chen et al., 1997). It is an efficient procedure for site specific mutagenesis of double stranded plasmids. It relies on a single PCR primer which incorporates both the mutations at the selection site and the desired single base substitution at the mutant site. This primer is annealed to the denatured plasmids and directs the synthesis of the mutant strand. After digestion with selection enzyme the plasmid DNA is amplified into E. coli strain BMH71-18 and subjected to a second digestion and amplification into bacterial strain DH5α.

21.11   INTRODUCTION OF POINT MUTATION

The introduction of point mutations at random locations in a DNA segment had been through the treatment of the DNA *in vitro* with hydroxylamine which introduces C $\rightarrow$ T transitions. Nathan (1978) introduced the bisulphate mutagenesis technique to produce localized point mutation. Sodium bisulphate is a single stranded specific mutagen which deaminates cytosine to uracil (U) but the double stranded DNAs are not easily deaminated to uracil. The gap filling *in vivo* or *in vitro* results in the substitution of a U/A pair for the original C/G pair and thus preventing the reversion of the mutation by cellular repair mechanism *in vivo*. The size of the gap and extent of deamination of exposed C residues within the gap can be precisely regulated by the investigator through appropriate choice of experimental conditions. Further, in this procedure only one strand in any given molecule is gapped and that different nucleotides are exposed in each of the two DNA strands. The steps involved in the *in vitro* local mutagenesis of DNA replicons are as follows (Shortle and Nathans, 1979a).

1. Introduction of a random or site specific nick in the circular genome by treatment of purified DNA with either DNase/or a cutter restriction endonuclease, both in the presence of ethidium bromide. Restriction endonuclease will break one phosphodiester bond in one strand of DNA in their cleavage site in the presence of ethidium bromide.

2. Extension of the nick into a small gap by limited exonucleolytic digestion with DNA polymerase I or exonuclease III. The DNA polymerase I from *Micrococcus luteus* is particularly suitable because it shows 5′-3′ exonuclease activity in the presence of only one deoxyribonucleotide triphosphate (d TTP).

3. Treatment of the short exposed single stranded region with sodium bisulphite (NaHSO$_3$).

4. The mutagenized DNA may be used directly for transformation or may be repaired with DNA polymerase plus all four normal triphosphates (4 dNTPs) and polynucleotide ligase and then used for infection of the host cells.

Two other *in vitro* techniques for the introduction of point mutations into small DNA segments are enzymatic incorporation of a nucleotide analogue at a predetermined location in a genome and enzymatic incorporation of a synthetic oligonucleotide containing a predetermined mutant sequence. These methods are time consumimg and require a prior knowledge of the sequence to be mutated. In principle, they are more powerful than simpler methods as they involve the introduction of a single pre-determined mutational change in a large proportion of the population of molecules treated and hence do not depend upon the availability of a specific phenotypic assay for the detection of rare mutant molecules. For this reason the sites of lethal mutations which can not be identified by other procedures are readily detected.

21.12   INSERTION OF NUCLEOTIDE ANALOGUE

This method of introducing site specific mutagenesis was developed by Weismann which involves insertion of a nucleotide analogue into RNA and DNA at a predetermined location.

The analogue used is N^4 hydroxy d CTP for DNA and N^4 hydroxy CTP for RNA synthesis. In case of DNA the synthesis is initiated either at a specific nick produced by a restriction endonuclease in the presence of ethidium bromide which leads to a strand break in either of the complementary strands or more generally at the 3′ terminus of a specific single stranded DNA fragment or oligonucleotide primer hybridized to an intact complementary DNA strand. The knowledge of nucleotide sequence of the DNA fragment containing the site to be mutated permits the conditions of DNA synthesis/nick translation to be chosen such that polymerization *in vitro* proceeds only as far as the nucleotide is to be mutated. The DNA synthesis will proceed along one of the two complementary strands in 5′-3′ direction. The incorporation of N-hydroxy-d CTP (OH-d CTP) is catalysed by DNA polymerase which will replace the TMP residues which will in turn direct the incorporation of complementary A or G residues and eventually AT residues will be replaced by TA or CG. This correction in this step is carried out in the presence of three out of four deoxyribonucleotide triphosphjate (d ATP, d CTP, d GTP). The triphosphates are then removed by chromatography and the nucleotide analogue is added to the replication complex. After the incorporation of the analogue all four normal triphosphates are added and the synthesis reaction is allowed to complete. Introduction of the mutated genome in the suitable host results in the production of 20-30% mutant progeny. After that identification and recloning of the pure mutant genomes is carried out.

21.12.1   PCR mutagenesis

Oligonucleotide-directed mutagenesis can be produced using PCR. There are dozens of variants for PCR-based SDM. It is a variant of conventional oligonucleotide directed mutagenesis (oligomismatch mutagenesis) which allows to introduce insertions, deletions or point mutations in a target DNA with its concomitant amplification using PCR. A few PCR-base methods are described below.

Overlap-extension method—This method requires four oligonucleotide primers and three separate reactions (Higuchi et al., 1988; Ho et al., 1989). Two complementary mutagenic primers induce the mutation into the desired sequence of DNA and two flanking primers amplify the mutant fragment and facilitate cloning of the PCR fragment into a suitable vector. A variety of mutations such as single base pair changes, deletions and insertions can be created by this technique. In this technique two separate PCR reactions are set up in parallel first. One reaction has the 'sense' mutant primer and an 'anti-sense' flanking primer 3′ to the mutation site. The other reaction contains the 'anti-sense' mutant primer and the sense flanking primer 5′ to the mutation site. The two amplified fragments contain mutations at the 5′ or 3′ terminus, respectively. In the second round of amplification the two fragments from the first round of PCR are purified and the used as templates for amplification using only the flanking primers. After amplification the mutation is contained within the target DNA segment which is cloned into appropriate vectors for sequencing and subsequent functional studies.

Patched circle mutagenesis (PC-PCR)—It is a variant of PCR and is used for SDM. In this technique the target DNA is cloned into a specific plasmid cloning vector between opposing T_3 and T_7 RNA polymerase promoters. Then one amplification primer (amplimer) is annealed to sequences within the T_7 promoter and a second amplimer, in opposite direction as compared to the first, is annealed to a sequence flanking the region to be deleted. The supercoiled plasmid is then serves as a template for PCR. During amplification the linear DNA molecules accumulate which lack the region to be deleted. A third oligodeoxynucleotide primer base pairing with the two ends of the linear molecules is then used to form patched circles for direct transformation of *E.coli*. Appropriate and rapid screening methods allow the isolation of clones that lack the deleted fragment.

Among PCR-based protocols the 'mega primer' method (Kammann et al., 1995) appears particularly simple and cost effective.

Megaprimer PCR mutagenesis—It is a technique used for the introduction of site specific mutations (single base exchanges) into a target DNA. This method uses only a single mutagenic primer to create mutations in the target template. In this first round of amplification the wild-type template is amplified using either a sense or anti-sense mutagenic primer and an appropriate flanking primer. The amplified product is then used in a second round of PCR with wild-type template and the other flanking primer to create a fragment of the same length as the original target DNA containing the desired mutation. The key feature of this technique is that the amplified product from the first round of PCR is used as primer in the second round of PCR. In comparison with the four-primer method this technique requires only a single mutagenic primer and yields more of the full length product. This is probably due to the instability of the 10-to 20-bp overlap between the two mutant templates during the second round of amplification when using the four-primer method. In mega primer method the overlap between the template and mutagenic strand is more extensive. In summary, this method involves two rounds of PCR which utilize two 'flanking' primers and one internal mutagenic primer containing the desired base substitution (s). The first PCR is carried out using the mutagenic internal primer and the first flanking primer. The product of this first PCR, the megaprimer, is purified and used along with the second flanking primer, a primer for the second PCR. The final PCR product contains the desired mutation in a particular DNA sequence.

Most mutagenesis strategies based on this two-step PCR scheme require an intermediate purification step of the first PCR reaction products to prevent left over primers from the first PCR from interfering with the second PCR step. This purification step is accomplished by agarose gel electrophoresis and subsequent elution of the DNA of interest and thus it is time consuming and laboursome. One approach to megaprimer PCR mutagenesis which eliminated the agarose gel electrophoresis step was reported by Landt et al. (1990). However this protocol employs several manipulations and enzyme treatments of PCR products. There is a rapid and efficient megaprimer PCR technique for SDM which does not require any intermediate purification of DNA between the two rounds of PCR. This protocol is based on the design of forward and reverse flanking primers with significantly different melting temperatures (Tm) (Ke and Madison, 1997). A megaprimer is synthesized in the first PCR reaction using a mutagenic primer, the low Tm flanking primer and a low annealing temperature. The second PCR reaction is performed in the same tube as the first PCR and utilizes the high Tm flanking primer, the megaprimer product of the first PCR and a high annealing temperature which prevents priming by the low Tm primer from the first PCR reaction.

Among many methods of SDM the PCR-based megaprimer strategy is widely used because of its flexibility and the yield of mutants approach 100%. This technique ir relatively cheap as only a single mutagenic primer and two flanking primers are used per mutant.

Limitations of megaprimer PCR are as follows.

1. Products of the first PCR need to be fractionated by gel electrophoresis and purified to remove leftovers before being used in the second round of PCR. This prevents the amplification of the wild type template during the second PCR reaction that would decrease the mutant yield drastically. This step is time consuming, exprensive, error prone, especially when multiple mutants are made in parallel. This step is also a potential barrier to automation.

2. The second problem is the non-templated addition of nucleotide, most frequently but not always an A at the 3′ end of the first PCR product. This can result in the presence of additional unwanted mutations.

Various strategies have been described to eliminate these unwanted mutations (Kupper et al., 1991) but they are time consuming and costly. Only one strategy is applicable under particular conditions or some time reduce but not eliminate these mutations.

An improved megaprimer mutagenesis strategy which overcomes the gel electrophoresis has been developed. This method is highly efficient and yield approaches 100%. All these steps are now performed in reaction tubes, making the method well suited to automation. This strategy relies on the observation that gel purification step is used in the megaprimer strategy is only required to remove the primer used during the first PCR reaction. Two modifications of the classical protocol will eliminate the amplification of the wild-type DNA template during the second PCR reaction rendering the gel purification step unnecessary. First, the templates used for PCR are cleaved with restriction enzymes to eliminate full length template. In this way wild-type DNA should not be amplified by the two external primers. Then comes the choice of restriction enzymes to be used. They should cut the fragment to be amplified once or multiple times on only one side of the mutagenic primer. The chosen enzymes should be the ones which will subsequently be used for the cloning of mutated PCR fragment. The second modification is the treatment of the product from first PCR reaction with Klenow enzyme. The proof-reading activity of the enzyme is used to reduce the concentration of the primers used during the first PCR reaction as well as reducing the level of non-templated nucleotide added by the Taq polymerase at the 3′ end of the PCR product. Using these modifications the product of the first PCR reaction could be used directly as a megaprimer in the second reaction. Both the cleavage of the template DNA and the reduction of the primer concentration after the treatment with Klenow enzyme are required to prevent amplification of the wild-type sequence. In fact, trace of full length DNA remaining after the digestion could still be amplified by standard amounts of primers (Seraphin and Kandels-Lewis, 1996).

Inverse-PCR mutagenesis — In this approach only two primers are used to create the desired mutations (Hemsley et al., 1989).The key feature of this technique is that in making mutation the entire vector is amplified. The two primers, one containing the desired mutation, are extended on the circular template DNA in opposite directions. Amplification ultimately yields a linear, double stranded DNA molecule containing the mutation at one end. Following amplification the ends are ligated and the resulting circular DNA molecule is transformed into *E. coli*. There are a number of variants of this technique which improve the efficiency of mutagenesis (Ling and Robinson, 1997).

PCR-ligation PCR mutagenesis — It is a technique for the production of fused genes, site directed mutagenesis or introduction of specific deletions, insertions or point mutations in the target DNA. The process of fusion of two or more genes starts with amplification of each gene in a separated PCR. The amplified products are then phosphyralated by T4 polynucleotide kinase and ligated with T4 ligase which lead to different combinations of the joined fragments. The fused gene is then specifically PCR amplified out of the heterogeneous mixture with a primer directed to 5′ end of the upstream gene and a primer complementary to the 3′ end of the downstream gene. The resulting amplification product is then subcloned into the original target sequence and thus thereby creates a type of insertion mutation. This technique relies on the DNA polymerase with 3′-5′ exonuclease (i.e. proof reading) activity so that the blunt ended fragments match exactly with the primer sequence.

SOE-PCR(Splice overlapping–PCR) — This technique differs from PCR — ligation mutagenesis in that the ends of the internal primers contain overlaps of complementary sequences. Then both PCR products are mixed, denatured and allowed to anneal. The complementary ends of the products hybridized serve as primers for the extension of the complementary strand and thus the two genes are spliced together.

Advantages/disadvantages of PCR-based methods — The low fidelity of Taq polymerase results in creation of undesired mutations. This problem can be solved by using fewer rounds of DNA amplification and conditions for the highest fidelity of DNA polymerization. Alternatively Pfu or Vent thermal stable polymerases which are less error prone than Taq polymerase can be used which result in fewer undesirable mutations. Regardless of which PCR-based method is used for mutagenesis, the DNA sequence of the entire amplified fragment should be determined to ensure that only the desired alterations have been made.

21.12.2   Distribution of mutations

The distribution of mutations following a mutagenesis procedure can be estimated on the basis of binomial distribution. For a sequence of length n that ε is mutagenized with an error rate ε, the probability of introducing k mutations is as follows.

$$P(k,n,\varepsilon) = (n! / [(n-k)!\, k!])\; \varepsilon^{k}(1-\varepsilon)^{n-k}$$

A target sequence of 100 nucleotides when subjected to mutagenic PCR ($\varepsilon = 0.0066$) would give rise to about 52% wild type, 34% one-error mutant, 11% two-error mutants, 2% three-error mutants and so on. The number of different sequences in each error class is given by $N_{k} = (n! / [(n-k)!\, k!]3^{k})$. Thus for the target sequence of 100 nucleotides there are 300 different one-error mutants, 44550 two-error mutants, 4365900 three-error mutants and so on. A mutagenized population of 10^{15} molecules will contain about 10^{12} copies of each of the possible one-error mutants, 10^{9} copies of each of the two-error mutants, 10^{7} copies of each of the three-error mutants and so on. All possible mutants containing five-error mutants or less would be represented in the population but only about 6% of the six-error mutants would be present as even a single copy with ever sparser representation of the higher error classes.

The most common method for introducing genetic diversity during *in vitro* evolution is error-prone PCR. There are several standardized protocols which utilize either altered reaction conditions or a mutant thermostable DNA polymerase to promote the incorporation and extension of mismatched nucleotides. One such procedure for mutagenic PCR achieves an error rate of 0.66% per nucleotide position with roughly equal probability of all possible transitions and transversions (Cadwell and joyce, 1992). Compared to standard PCR this procedure employs an increased concentration of Taq DNA polymerase, added $MnCl_2$, increased concentration of $MgCl_2$ and unbalanced concentrations of the four dNTPs. There is also a hyper mutagenic PCR procedure which achieves an error rate of about 10% per nucleotide position with only modest sequence bias but it has a greatly increased rate of introducing amplification artifact (Vartanian et al., 1996). This method also employs added $MnCl_2$ and increased concentration of $MgCl_2$ but relies on more heavily unbalanced concentrations of the four dNTPs and 50 rather than 30 temperature. Yet another procedure utilizes standard PCR conditions but a mutant thermostable DNA polymerase (Mutazyme ^{TM}) that produce an error rate of up to 0.7% per nucleotide position over the course of PCR (Cline and Hogrefe, 2000). This method generates a broad spectrum of mutations although with a substantial preference for transitions over transversions.

21.13 MUTATIONAL ANALYSIS

There are two techniques for rapid mutational analysis.
1. Ribonuclease A(RNaseA) cleavage and
2. PCR method for amplification of specific nucleotide sequences(Veres et al., 1987). These are simpler than the conventional cDNA library construction, screening and sequencing which often been used to find a new mutation.

21.13.1 RibonucleaseA cleavage

The conventional technique for finding a new mutation involves cDNA library construction, screening and sequencing. This technique involves the application of RNase A cleavage to localize the mutation followed by PCR amplification of the mutated site. RNase A cleavage was used to characterize the OTC (ornithine transcarbamylase) mRNA thus avoiding the inherent problems of analyzing the large gene (approximately 60kb). OTC catalyzes the conversion or ornithine and carbamyl phosphate to citrulline in a mammalian urea cycle. A radiolabelled, antisense RNA probe was synthesized *in vitro* from the normal mouse OTC cDNA and then annealed to total RNA isolated from wild-type and spf mice. The samples were treated with RNase A to digest the single stranded RNA and to cleave any mismatches between the wild-type probe and the spf OTC mRNA. Wild-type RNA protected an approximately 1270bp probe fragment which was the same length as the predicted length of homology with OTC mRNA.RNA from spf mice also protected a 1270 base fragment and in addition produced two distinct bands (920 and 350bases) from parental RNAase A cleavage of an internal mismatch site in the hybrids. Truncated RNA probes were used to unambiguously locate RNase A cleavage site relative to the ends of the OTC mRNA.

MAPREC (Mutant analysis by PCR and restriction enzyme cleavage)—It is a technique for the detection of point mutations in coding genomic DNA. In this technique RNAs of wild-type mutants are first isolated and reverse transcribed into cDNAs using a random hexanucleotide primer. The cDNA now serves as a template for asymmetric PCR with primers specific for the gene of interest (more precisely the gene region in which mutation has occurred). The primer is so designed that it creates a recognition site for restriction endonuclease Mbo1 (5′-GATC-3′) in the wild-type target DNA. The mutant sequence will give rise to, for example, a Hinf I restriction site. An excess of sense polarity primer ensures that the product is predominantly single stranded DNA. Now the second strand is synthesized using a labeled antisense primer (biotin labeling), the double stranded product is digested with Mbo1 and the resulting restriction fragemnts are separated using PAGE. After Southern blotting and fixation of fragments onto the blotting membrane, the fragments are visualized with streptavidin conjugated alkaline phosphatase. The wild-type sequence will be fragmented while the mutant sequence will remain uncut.

21.13.2 PCR-based method for mutation detection

DT-PCR (Directed termination PCR)

It integrates both selective PCR amplification and directed chain termination into a single PCR reaction. Because it exploits unbalanced nucleotide concentrations to induce the PCR reaction to terminate at specific nucleotide the method is called directed termination PCR. DT-PCR (Chen and Hebert, 1998) allows the generation of nested termination fragments by integrating both selective DNA amplification and directed chain termination into a single PCR reaction. These termination fragments can be examined for sequence variation in either denaturing or non-denatuing polyacrylamide gels. This technique provides a one step and highly effective approach for the detection of both insertions/deletions and single base pair substitutions in sequences up to 1kb in length.

Mutant allele specific amplification(MASA)—These PCR-based techniques allow specific amplification of an allele that has undergone a gene mutation such as deletion, inversion, insertion, transition or transversion.

PCR-clamping—This technique is used for the detection of deletion, insertion, mutant alleles or point mutation in the target DNA sequence. It is based on increased affinity and specificity of a peptide nucleic acids (PNAs) for their complementary target sequences and the inability of DNA polymerase to recognize and extend a PNA primer. PNA is a synthetic chimeric polymer with a neutral achiral polyamide (peptide like) backbone composed of N (2-aminoethyl) glycine units to which nucleic acid bases are covalently bound through carbonyl methylene (-CH2-CO-) linkers. It is used as substitutes of normal DNA.

MS-PCR—It is technique used for detection of point mutations in a known DNA sequence. This technique relies on conventional PCR. The wild-type and mutant alleles of a gene are simultaneously amplified in the same reaction using allele specific primers of different lengths. In addition, AS primers differing from each other at several nucleotide positions are used which introduce new and discriminating mutations into the allelic PCR products and this results in reducing cross reaction between amplification products during PCR. Since both products are of different lengths, MS-PCR separates both amplified alleles which can be identified by agarose gel electrophoresis and ethidium bromide staining.

21.14 METHODS FOR DETECTION OF SINGLE BASE SUBSTITUTIONS

The procedures for detecting base substitutions rely on differences in restriction endonuclease cleavage sites or on differences in the melting behaviour of wild-type and mutant DNA duplexes. These physical methods for detecting single base substitution (Orkin and Kazazian, 1984) are also helpful in detecting and analysis of single base mutations in regulatory or protein coding sequences.

Different methods have been developed to identify single base changes.

1. Sequencing complete genes to identify base changes is tedious.
2. Heteroduplexes formed between wild-type and variant DNAs have been treated with the single strand-specific S1 nuclease to cleave the DNA at the point of mismatched bases.
3. The differential mobility of native and denatured DNA-DNA heteroduplexes coupled with their differential melting temperatures has been exploited by Myers et al.(1985).
4. Since this was not generally applicable Myers et al.(1985) described a method in which mismatches in RNA-DNA heteroduplexes were cleaved by RNase A.
5. An alternative approach in which RNase A was used to cleave mismatches in RNA-RNA heteropduplexes has also been described.
6. Novack et al.(1986) reported that single base pair mismatches in DNA –DNA heteroduplexes react with chemical, carbodiimide.

1. **Application of restriction endonuclease** — Single base substitutions result in the loss or gain of a restriction endonucleavage site and can therefore be detected by Southern blotting. There are limitations with this method. First, it is usually necessary to use a large number of different restriction enzymes before a change is detected. Secondly, many substitutions can not be detected as they do not alter a restriction site.

2. **Heteroduplex analysis** — Another approach involves the use of synthetic oligodeoxyribonucleotides as differential hybridization probes. In this method a labeled synthetic oligonucleotide homologous to the mutant or wild type DNA is hybridized to blotted genomic DNA. Hybridization or washing conditions are then adjusted to allow the differential melting the mismatched and perfectly paired duplexes. This method is useful for scoring substitutions at specific locations but not practical for screening large regions of DNA for new mutations or polymorphisms.

Differential DNA melting is also the basis for detecting single base substitution by DGGE. In this method wild-type and mutant DNA molecules are separated by electrophoresis in polyacrylamide gels containing a gradient of formamide and urea. Duplex DNA fragments pass through these gels with a constant mobility determined by the molar weight until they migrate into a portion of the gel containing a denaturant concentration sufficient to melt the DNA. When the DNA duplex melts, its electrophoretic mobility abruptly decreases. Thus the final position of a DNA fragment in the gel is determined by its melting temperature. The difference in melting temperature between two fragments that differ by a single base change is sufficient to allow separation on the gel. Even greater separation is achieved with DNA duplexes containing a single base mismatch. With specially designed plasmid vectors virtually all possible base substitutions can be detected in cloned DNA fragments. This method suffers from technical limitations and only 25 to 40% of all possible substitutions can be detected directly in total genomic DNA.

3. **Application of ribonuclease (RNaseA)**—Alternative method for detecting base substitutions in cloned and genomic DNA involves the enzymatic cleavage of RNA at a single base mismatch in an RNA-DNA hybrid(Myers et al., 1985). This strategy is based on the development methods for synthesizing RNA probes(Melton et al., 1984) and on the observation that many ribonucleases are specific for single stranded RNA under appropriate reaction conditions. A similar strategy had been developed earlier to detect mutations in duplex DNA containing a single base mismatches(Shenk et al., 1975). In this case attempts were made to cleave DNA-DNA mismatches with single strand- specific nuclease S1. However, only a small amount of cleavage occurs at a few mismatches while most mismatches are not cleaved at all. Unlike this method, many single base mismatches in RNA-DNA hybrids are cleaved specifically by ribonuclease A(RNaseA).

In this method a 32P-labelled RNA probe is synthesized from a wild-type DNA template with SP6 transcription system. The RNA probe is hybridized to denatured test DNA in solution and the resulting RNA-DNA hybrid is treated with RNaseA. The RNA products are then analyzed by electrophoresis in a denaturing gel. It the test DNA is identical to the wild-type DNA, a single is observed in the autoradiogram as the RNA-DNA hybrid is not cleaved by RNase. However, if the test DNA contains a single base substitution, then it will result in a mismatch which will be recognized by RNaseA and two new RNA fragments will be detected. The total size of these fragments should be equal to the size of the single RNA fragment observed with the wild-type DNA. The mutant can be localized relative to the ends of the RNA probe by determining the sizes of the cleavage products. The end of the RNA probe mapping nearest to the substitution can be determined when the experiment is performed with DNA digested by additional restriction enzyme, thus localizing the substitution unambiguously.

Pre-gel hybridization—It is a technique for the detection of mutations, for example, deletions, insertions, SNPs which overcomes the various steps of Southern blotting(pre-hybridization, hybridization and stringency washes). In this technique the double stranded target DNA is denatured at low salt concentrations in the presence of a short 918-mer) labeled PNA(peptide nucleic acid). At these low concentrations of salt the DNA strands can not reanneal but PNA can bind to its complementary DNA target. The resulting PNA-DNA hybrids are then separated by agarose gel electrophoresis or capillary electrophoresis and detected after blotting by autoradiography, luminography or fluorography.

21.15 PROTEIN ENGINEERING

Protein engineering refers to the modification of the physico-chemical or biological characteristics of a naturally occurring protein with the ultimate objective of improving the quality of the protein for biotechnological process and one of the techniques used in protein engineering is *in vitro* mutagenesis which allows to alter the coding capacity of a gene at a defined location. Consequently the engineered protein adopts different properties that are useful for biotechnology.

The manipulation of gene expression and function is carried out to accomplish a variety of objectives. In homologous system, for example, *E.coli* genes cloned in an *E. coli* host/vector system, such manipulations may be carried out to provide information on the following.

1. Regulation of expression of a cloned gene
2. Influence of the cellular concentration of its product on the activity of related genes and products
3. Influence of cellular concentration of its product on cell physiology or
4. Cellular localization of the gene product.
5. Elevation of the level of gene expression or promotion of the excretion of an ordinarily intracellular gene product for the purpose of isolation of quantities of the product.

In case of heterologous systems, for example, mammalian genes cloned in *E.coli* the manipulation may be carried out to achieve the following.

1. Active transcription of the cloned gene
2. Correct translation and processing of the gene product
3. Increase in resistance of the gene product to endogenous proteolysis.

21.16   GENE TARGETING

Gene targeting refers to the introduction of genetic changes into pre-determined sites within the genome through **homologous recombination**. This technique allows the targeting of exogeneous DNA to specific chromosomal locations and is dependent on the cells's intrinsic ability to mediate homologous recombination process. This is exploited for example, in the method of transposon tagging. In this technique a DNA vector containing genomic sequences surrounding the required change is constructed by recombinant DNA techniques and introduced into the cell. Homologous recombination between the vector and the genome results in insertion of the genetic changes at the homologous endogenous location and production of a genetically modified cell. Gene targeting can be applied for mapping gene expression and sophisticated genetic alterations including single nucleotide changes. Integration of reporter gene into the genome in order to be expressed under the transcriptional control of the endogenous gene provides a means to visualize the normal cellular sites of gene transcription *in vivo*. Expression of reporter genes such as β-galactosidase (lacZ) can be resolved at the cellular level by histochemical techniques.

Gene targeting requires detailed knowledge of gene structure to ensure that the target locus has been effectively mutated. The availability of detailed gene structure information which enables mutations to be made with a full understanding of the likely function, functional consequences depends on the assembled genome sequence. Further, knowledge of genome sequence has made possible to index libraries of gene –targetting vectors thereby eliminating the need to screen a library to obtain genomic clone for targeting. Library indexing by end-sequencing significantly increases the rate at which knock out can be generated. Targeting is an inherently serial process. A single experiment typically produces one type of allele. **Gene trapping,** on other hand, in which genes are tagged for sequence retrieval by insertion mutagenesis, generates hundreds of different mutations from a single electroporation or viral infection. **Conditional gene tragetting**–See chapter 24 for detail.

21.16.1 Site directed gene targeting

It is a method of inserting a gene into a specific chromosomal sites using homologous recombination and success depends on the cells' ability to mediate homologous recombination. For example, a foreign gene (e.g. a reporter gene) can be cloned into another gene (e.g. an alcohol dehydrogenase gene) *in vitro* so that it is flanked by sequences derived from the alcohol dehydrogenase. This foreign gene will be transferred into the chromosomal region containing alcohol dehydrogenase gene containing the cellular counterpart of the targeting gene by homologous recombination (recombination between DNA sequences residing in the chromosome and newly introduced DNA sequences).

Targetting induced local lesions in genome(TILLING)—Conserved domains, expression patterns, biochemical activity, protein-protein interactions and molecular structure is inadequate to predict function. In many cases sequence-directed mutagenesis may not efficiently identify genes specific to certain functions or disease, for instance, the genes involved in diabetes by knocking out individual candidates. Mutational analyses can be focused to identify players in specific process by performing **genetic screens.** A reverse genetics technique which allows to introduce a high density point mutations into a genome by conventional chemical mutagens and a subsequent mutational screening to rapidly detect induced lesions. The targeted organism is first mutagenized by EMS which primarily induces GC-TA transitions (of which 50% are silent and most of the rest are missense mutations). Then the DNA is isolated and regions of specific interest is amplified by a PCR primer with region specific primers. The amplified products are denatured and allowed to reanneal to form heterodupleses which are analyzed by denaturing high pressure liquid chromatography (DHPLC). DHPLC detects mismatches in hetroduplexes molecules which appear as extra peaks in the chromatogram. The mutant allele then can be sequenced. As an alternative for mutant detection the plant endonuclease CELI from celery can be employed. This enzyme recognizes a mismatch and cleaves exactly at the 3 side of the mismatch. Therefore, cutting the mutated DNA by CELI and subsequent DHPLC or high resolution PAGE pinpoints the precise position of the mismatch. Tilling is used for functional genomics (e.g. the proof of the function of a gene of interest by introducing EMS mutations into it). For identifying gene function the underlying genetic lesions will have to be identified and the molecular mechanisms relating the lesions to the observed phenotypes must be understood. Colbert et al. (2001) developed a large-scale screening system , the so-called targeting induced local lesions in genomes.

In case of mouse ENU (N-ethyl-N-nitrosourea) mutagenesis is currently the method for random mutagenesis. Progeny of mice treated with ENU are screened for a mutation. Inheritance has to be worked out and mapping of the gene to a particular position of a chromosome is done. Finally, what is required is not only to identify the mutation (typically a nucleotide substitution) but also a detailed phenotypic understanding of each mutant (i.e. phenotyping mutants for a specific characteristic).

Chemical mutagenesis of ES (embryonic stem) cell *in vitro*—Isolation of embryonic stem cells and demonstration that these cultured cells can recolonize the mouse germ line has made the genetic tractability of ES cells uniquely accessible for genetic studies compared with every other multicellular organism.

Serial invasive signal amplification (SISAR)—It is an isothermal technique for the detection and high throughput analysis of single nucleotide polymorphisms and point mutations in a genome.

Hotspot recombination—It refers to any sequence within a gene or a chromosome at which mutations occur at a frequency significantly higher than usual. In case of Tn 10 mutagenesis insertion occurs at a hot spot with a symmetrical 6bp consensus sequence, 5-GCTNAGC-3 where the internal 5-methyl group at the third position of the pyrimidine is necessary for strong recombination. **Hypervariable regions**- It refers to highly polymorphic sequences scattered throughout the genome which consist of usually GC-rich, tandemly repeated units to which no special function can yet be attributed. HVRs are thought to be hot spots for recombination. Unequal exchanges at meiosis or mitosis, or slippage during DNA replication(**slipped strand mispairing**) may result in allelic differences in the number of repeated units present at an HVR site and consequently result in length polymorphism. HVR may be used for the establishment of a DNA fingerprint. The slipped-strand mispairing refers to an intrahelical process which leads to the mispairing of complementary bases at the site of short, tandemly repeated sequence motifs in viral, bacterial and eukaryotic DNA. In short, local denaturation(unwinding) of the DNA duplex, displacement of the two strands in the region of tandem repeats and a slip during renaturation may lead to non-paired single-stranded loops that are target sites for excision and repair. Both processes can lead to one or more duplications, deletions or insertions of tandem repeat units. Slipped strand mispairing is regarded as the driving force in the expansion of simple repetitive sequences in eukaryotic genomes.

21.17 DETECTION OF DNA SEQUENCE DIFFERENCES (DETECTION OF MUTATION)

DGGE and TGGE separate DNA molecules on the basis of differences in thermal stability caused by differences in base(Abrams and Stanton, 1992). The resolving power depends on the drop in electrophoretic mobility which occurs when part of a DNA molecule melts, resulting in the formation of a structure that is partly helical and partly random chain. PCR-mediated attachment of an artificial GC-rich sequence, a 'GC-clamp' provides a simple means for modulating the melting characteristics of most sequences into the two-domain profile that is considered optimal for detection of mutation(Sheffield et al., 1989). DGGE in combination with PCR and 'GC'-clamping has been highly efficient as a method of detection of DNA sequence difference.

22

Transposable Elements in Genome Organization

22.1 DEFINITION

Transposon (T) or transposable element (TE) is found in virtually all cells (prokaryotes as well as eukaryotes). In other words, TEs are found in genomes of numerous organisms from bacteria to human. Table 22.1 shows the percentage of TE found in genomes of different organisms. These genetic elements 'jump' from old to new DNA locations, a process that mutates genes, rearranges chromosomes and transmits genetic information between cells. TEs are specialized nucleotide sequences that are bounded by specific terminal sequences and that have the ability to replicate and move from one place to another. Transposable element is a segment of DNA that can move from one place on a chromosome (donor site) to another on the same chromosome or a different chromosome (target site) and from chromosome to extrachromosomal elements, i.e., plasmids or viruses and this process is called **transposition**. The new location is usually chosen more or less randomly but as the insertion of a transposon in an essential gene can disrupt its function and the cell could get killed, so the transposition is tightly regulated and usually less frequent. Transposon (or mobile genetic element) can increase in copy number because of its ability to replicate and transpose and thus considered a molecular parasite. Transposons can increase their copy number in more than one ways. Transposon can replicate passively within the chromosomes of host cells. In some cases they carry genes that are useful to host and thus there exists a 'symbiosis' with the host. Transposons can be classified as either class I or class II depending on whether they replicate through an RNA intermediate or directly through a DNA form. They are further classified by similarity either between their mobility genes or between their terminal and/or internal motifs as well as by the size and sequence of their target site. Both classes of transposons are used as a source of drug resistance genes and as mutagens.

Table 22.1   Showing different organisms with the percentage TE.

Organism	% TE
Frog	77
Maize	60-70
Human	45
Mouse	40
Drosophila	15-22
Nematode (*C.elegans*)	12
Yeast	3-5
Bacteria (*E.coli*)	0.3
Arabidopsis	14
Vicia faba	90

22.2   PROKARYOTIC TRANSPOSONS-BACTERIAL TRANSPOSONS

There are two classes of transposons.1 Insertion sequence (simple transposons) and 2. complex transposons.

Insertion sequence—They are simple transposons containing only the sequences required for transposition and the genes for proteins (transposes) that promote or regulate the process. They are typically 1-2kbp in length and contains at least one long open reading frame (ORF) coding the transposases. They mediate transfer of episomes into bacterial chromosome, i.e., recombination between non-homologous DNA molecular sequences and homologous recombination between IS elements in the episome and in the chromosome. IS elements have always short identical or nearly identical sequences (usually 9- 40bp long) at both ends in inverted orientation called **inverted terminal repeats** (ITR). IS elements cause a duplication of DNA at the site of insertion. One copy of duplication- the directed repeated sequence (3-12 np long) is located on each side of the IS elements. IS elements are non-replicating transposable DNA that can be inserted into the DNA of bacterial chromosome or bacteriophage or plasmids. They do not have independent existence. IS elements cause **polarity mutation** (abolishes the function of the mutated gene and impairs the function of other genes within the same operon) and thus mutant phenotype is reverted to wild type. The polarity mutation caused by in the insertion of an IS element differs from a point polarity mutation in that it easily reverts to wild –type condition. An IS element can be seen as single stranded loop in one of the two DNA strands under electron microscope. It can also be detected by using the heteroduplex technique, i.e., by examining the hybrid DNA double strands originating from the two different parent molecules. The different IS elements show different site specificity. The various IS elements found are IS1, 2, 4, 10, 50, etc. ISs can mobilize other sequences in the genome. Size and structure characterize a particular IS.

Complex transposons—When two copies of an IS (two homologous IS elements) insert near each other the unrelated sequence (also called '**passenger DNA'**) between them can be transposed by the joint action of the flanking inverted repeats at the extreme ends and thus they form a composite transposable element or transposon which transposes as a single unit. Both IS elements need not be functional and only the extreme terminal repeat of each member of the pair defines the limit of transposon. These transposable genetic elements or jumping genes transpose themselves from one genetic element or chromosome to another. In doing so they duplicate (replicate) themselves and leave one copy in the original condition and the second copy moves to the new location and thus there is rapid spread of transposons. Some TEs move via DNA intermediate while others move via an RNA intermediate. Transposons or composite transposable elements contain one or more genes in addition to those needed for transposition. These extra genes might, for example, confer resistance to antibiotics and thus enhance the survival of the host cell. The different transposons are Tn3, Tn5, Tn9 and Tn10. Tn3 does not contain IS elements but carries gene for ampicillin resistance.Tn3 has only one terminal repeat (~ 38bp) at its extremes. Besides gene for ampicillin it contains three genes, namely tnp A, tnp R and bla, encoding transposase, resolvase/repressor and beta lactamase. Tn5 consists of two IS50 elements flanking three genes conferring resistance to neomycin, streptomycin and bleomycin. Tn9 consists of two IS1 elements flanking a gene for chloramphenicol resistance. Tn10 consists of two IS10 elements flanking a gene for tetracycline resistance. Transposons also show different sites specificities. In the bacteriophage Mu transposon termini are not inverted repeat sequences. In addition to encoding a number of functions Mu possesses two genes, A and B of which A encodes transposase and B enhances the efficiency of transposition.

22.3 EUKARYOTIC TRANSPOSONS

Eukaryotic transposons are structurally similar to bacterial transposons and in some use similar mechanism of transposition. However in some other cases the mechanism of transposition appears to involve RNA intermediate. Transposons are also found in eukaryotes. Ac/Ds, Spm/sSpm (Mc Clintock 1951) and several other families of maize and P-elements of Drosophila are some examples.

22.3.1 Plant Transposons

The plant transposable elements cause unstable mutations in somatic cells, i.e., they cause loss and restoration of gene activity as the elements are inserted into and excised from gene loci and thus cause somatic variegation. There are somatically unstable lines in many species including horticultural variants with candy-striped or purple flecked petals, reflecting instability in anthocyanin pigmentation. Within each families of transposable elements some members are able to transpose autonomously and others transpose only in response to the activity of an autonomous element of the same family located elsewhere in the genome (Table 22.2). In case of Ac/Ds and Spm/dSpm systems Ac (activator) and Spm (suppressor or mutator) are autonomous elements whereas Ds (dissociation) and dSpm are non-autonomous elements. Autonomous transposition depends on the presence of an open reading frame (ORF) that encodes transposase required for excision and reinsertion. The ability to transpose whether independently or not depends on the special sequence- the cutting site, recognized by the transposase and present at the ends of the ITR. The non-autonomous elements such as Ds, dSpm are derived from their autonomous relatives (Ac, Spm) by deletion of essential part of the ORF (the sequence encoding transposase but they retain the inverted repeats that make them target for the transposase. Thus Ds or dSpm is

Table 22.2 Showing some plant transposons.

Family nomenclature		Host sequence duplication (bp)
Autonomous	Non-autonomous	
A. **maize**		
Ac	Ds (I-n)	+8
EN/Spm	dSpm	+3
Mu9 = MuRI	Mul-Mu8	+9
B. **Snapdragon**		
Tam I	Tam2	+3
Tam 3	?	+8
?	Tam4-10	?
		+8

incomplete in the sense that it is devoid of transposage gene and it results from deletion of internal sequence (containing gene for promoting transposition). Thus Ds only function in the presence of Ac. If both Ac-Ds are present then the chromosome breakage is increased in the organism which leads to chromosomal abnormalities like deficiency, duplication, translocation and ring chromosome. Ac is 4563bp long including 11bp inverted terminal direct repeats. Some of the maize elements increase their copy number by transposing from an already replicated locus to another locus just about to replicate thus loosing one copy but forming two. *Basho* and many Mutator like elements (MULs), first discovered in Arabidopsis, represent structurally unique transposons. *Basho* elements have a target site preference for mononucleotide'A' and wide distribution among plants.

Stowaway—It refers to a family of plant transposons like elements of about 100-300bp in length, occur in non-coding sequences of genes of both mono and dicots. It carries subterminal repeats with specificity for potential target sites (5'-TAA-3').

Tourist—It refers to a large family of plant transposon like insertion elements of about 130bp in length, occur in exons, introns or flanking sequences of maize, barley genes. The tourist elements have conserved 14bp terminal inverted repeats, a subterminal pentamer repeat (5'-GGATT-3') and are flanked by a 3bp target site direct repeat and possess insertion site specificity, i.e they prefer the target sequence, 5'-TAA-3'. Tourist sequences have the potential to form a hair-pin structure. The copy number of these elements in maize is very high (1 tourist per 30kb).

Mu (Mutator)—It refers to any one of a class of TEs in maize genome which increases the frequency of mutation of various loci by more than an order of magnitude. Mu elements are present in the genome in 10-100 copies and comprise maximally 2kb and are flanked by 200bp inverted repeats with adjacent 9bp direct repeats. Mu elements prevail in two size classes of which the shorter ones are derived from longer ones by internal deletions. Mu elements transpose by a replicative mechanism and can also occur in circular extracellular state Mu 1 (1.4kb) and Mu.7 (1.7kb). Methylation of the inserted Mu sequences prevents their transposition and stabilizes the mutation whereas incomplete methylation leads to transpositional activity.Few plant transposons are in introns. Plant transposons must be located in the intergenic regions between genes whereas most human transposons are in the introns.

RESites (Related to empty sites) — An *in silico* approach is used to search for a recent mobility of transposable elements. Sequences immediately flanking transposon insertions are used as querries to identify potential genomic target sites without inserts. These in turn are taken as evidence for recent transposon of the corresponding element.

22.3.2 Drosophila's Mobile Elements

P element is the best investigated transposon in Drosophila. The features of P-element are similar to that of plant transposon in that they contain a long ORF and have terminal inverted repeats and like the plant elements it gives rise to numerous non-autonomous derivatives. P- element of Drosophila differs from plant transposons in that the effects are seen only in the next generation (progeny generation) as the third intron interrupting its open reading frame is spliced out only in germ-line and not in somatic cells. The P-elements behaves like plant transposon and cause somatic variegation if the third intron is removed precisely by DNA manipulation *in vitro* P-element leads to genetic abnormality called P-M **hybrid dysgenesis** in Drosophila. It results from chromosome breakage and gene disruption due to hyperactive P-element transposition in the progeny of crosses between females lacking the element and males possessing it. P-element in Drosophila is 2907 bp long terminating in 31bp inverted repeats with 8bp target site duplication. In Drosophila as much as 15% of the DNA is mobile but the amount of P-element hardly exceeds 0.1%. P-elements tend to become quiescent in Drosophila strains that have harboured them over a number of generations and this acquired immunity is transmitted only through females. The increase in copy number of P-element is very fast and it is suspected that replication of P-element can occur outside the chromosome between excision and reinsertion. Transposons have not been reported in human or to other mammalian genomes.

Insertion of P-DNA into chromosomes — The two requirements for insertion of P-DNA into chromosomes often at multiple sites are, the specific sequences present at both termini of the element in reverse orientation and a source of the transposase for catalyzing the insertion. The transposase does not need to be encoded in the same piece of DNA as the one to be inserted. A transforming vector with P-element termini can be inserted into the genome by P-transposase supplied by a 'helper' plasmid injected together with the vector.

Hobo element — It refers to a class of transposable elements of *Drosophila melanogaster* which causes intra-chromosomal rearrangements and instability. A Hobo element is 3kb in length, has 12bp-inverted terminal repeats and is flanked by an 8bp duplication of the DNA at the site of integration.

22.4 RETROTRANSPOSONS

Retrotransposons, the other major type of mobile element resemble the retroviruses (the chromosomally integrated retroviral genomes). Retroviruses have RNA rather than DNA in their infective particles. They differ from transposons in that the former is flanked by the longer terminal direct repeats (LTR). Further, retrotransposons seem to move not by excision but by reverse transcription and chromosomal integration of reverse transcripts and thus they move through an RNA intermediate. Thus these elements can insert but do not excise and hence they create a permanently disrupted gene. Retroviral genome encodes reverse transcriptase and proteins of the infective virus particle. Although retrotransposons have been found in a wide range of organisms including Maize, Tobacco, Arabidopsis, Drosophila, Saccharomyces, mammals (mouse, human) they are major mutagens in mammalian genomes. Predominant type of transposable elements in maize is retroelement replicative through RNA intermediate and particularly prevalent are two classes of retrovirus-like retrotransposons- the socalled Ty1/copia and Ty3/gypsy. Ty1/copia group of retrotransposons is found in species throughout the plant kingdom. Mobile retrotransposons have been discovered as new insertion sequences in known genes of maize *(Bs1* in Alcohol dehydrogenase-1), *Nicotiana tabacum (Tnt1* in nitrate reductase) and as *Ta1* in *Arabidopsis thaliana.* Large blocks of retroelements (>50kb) have been found between single copy gene sequences in a 280kb region encompassing the maize adh1 gene. These blocks are made up of 10 diverse families with varying number of copies and the five most abundant families alone makes up about 25% of the maize genome. More than 20 characterized retroelement families have been found in Arabidopsis and most are present in one to five copies. Retroviral elements in plants are likely to cause null mutations and that is why these elements have not received more attention. In vertebrates the defective (or nonfunctional) proviruses are fairly common which arise as a result of mutation but generally they appear to be totally quiescent. In Drosophila and Saccharomyces the RV related elements are actively transcribed and transpose and unlike transposons they move to new location but leave one copy in the original condition.

22.5   YEAST TRANSPOSONS

There are several families of yeast RTs and together they account for ~ 0.1% of the genome. The most abundant RTs are the two closely related elements, Ty1 and Ty2, present in about 30 and 20 copies, respectively. In addition to complete Ty elements hundreds of incomplete Ty elements have been found and these incomplete Ty elements consist of only LTRs. These incomplete Ty elements are thought to have been derived through recombination (homologous recombination) between a site on one LTR and the corresponding site on the other LTR. Ty element of yeast resembles retroviruses which transposes by means of an RNA intermediate. RNA viruses that contain reverse transcriptase are called **retroviruses**. The RNA viruses through reverse transcriptase produce single stranded DNA and the product DNA-RNA hybrid is formed. Then the RNA is degraded and is replaced with DNA and thus double stranded DNA is produced which is integrated into the host genome as a **provirus**. The entire length of provirus sequence consisting of three functionally distinct regions namely *gag* (producing group-specific antigen), *pol* (polymerase, i.e., reverse transcriptase) and *env* (virus envelope proteins) is transcribed as a single unit. The processing of the single transcript can lead to production of two messengers, *env* messenger and *gag-pol* messenger. The env messenger is translated into a polypeptide chain which subsequently cleaved to two products, SU (surface protein) and TM (transmembrane protein required for penetrating the host cell membrane. The *gag-pol* messenger is translated into a gag-pol polypeptide. The boundary between the *gag* and *pol* parts of the messenger is marked by either a chain termination or a reading frame shift and so depending upon the requirements either *gag-*

polyprotein or pol-polyprotein or both are formed. The gag-polyprotein is cleaved into four inner components of virus particle and poly yields reverse transcriptase and a protease. Thus the retrovirus produces eight proteins. The LTR is made up of direct repeats which includes a short inverted repeats sequence (5-15 bp) at its end. It has the ability to acquire and transpose 'passenger DNA' of host. In case of Ty element of yeast the LTR is about 300bp terminal direct repeat (these do not contain terminal inverted repeats). In Drosophila, 10 types of RT have so far been characterized and this number is on increase. They are present in10-80 copies each and comprise at least 2% of the genome and constitute ~ 10% of the genome. In Copia like element of Drosophila the LTR is about 300 bp in length and each of which is flanked by a short (~ 17bp (inverted repeat. The size of the direct repeat is 5bp. Ty1from yeast and copia from Drosophila differ from a typical retroviral provirus in that they lack the *env* domain which encodes components necessary for cell to cell infection. Further, in comparison with some of the plant transposons or the P-element of Drosophila these yeast and Drosophila RTs move rather rarely but many spontaneous mutations in Drosophila have been found to be due to RT insertion. HIV (human immunodefieciency viruses) and AIDS (acquired immunodeficiency syndrome) are due to viruses –the retroviruses.

22.6 POSITION SPECIFIC PREFERENCES

There is preferential insertion of RTs into the vicinity of tRNA genes in the slime mold *Dictyostelium discoideum*. Similar integration preferences (targeted integration) are characteristic for the Ty elements of *S. cerevisiae*. The Ty element is preferentially inserted into the promoter region, e.g. 5′ sequences of tRNA genes. Ty 1 to Ty5 have strong bias for the sites into which they integrate. Ty1 to Ty4 families are located 750bp upstream of genes transcribed by RNA pol III, particularly tRNA genes and Ty5 elements are located at the telomeres or regions that have telomeric chromatin. Many plant gene sequences have adjacent retrotransposon insertion and in some cases insertions are located in promoters and contribute sequences important for promoter function. Regions between maize genes are packed with transposable elements inserted within transposable elements which together make up more than 50% of the maize nuclear DNA. Targeted integration by retroelements is suggested in maize and there is presence of intact target-site duplications flanking most insertions. These elements specifically avoid coding regions. Intergenic regions are hypermethylated relative to gene sequences and thus unique chromatin feature serves as homing device for maize retrotransposons during integration. Regions targeted by yeast retrotransposons are typically devoid of ORFs and reiterative integration can generate blocks of elements within elements.

Size of retrotransposon — Retrotransposons vary in sizes for plant genomes, 10^8bp for Arabidopsis (lack of interspersed repeat in this plant's nuclear DNA) to over 10^{11}bp for some species of lily.

Diagnostic features of transposition — The criteria for recognizing a DNA transposon are:

1. Recognition of a mutant phenotype from insertion event
2. Restoration of partial or full function sectors and gametes from excision events and

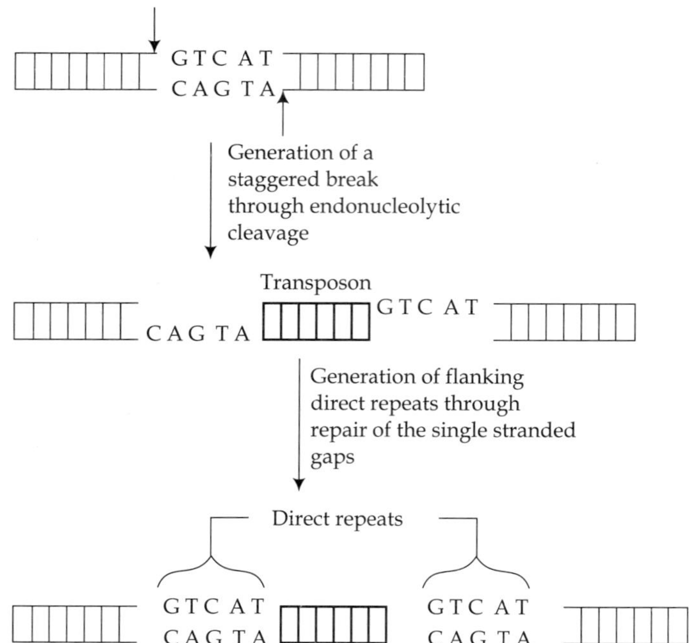

FIGURE 22.1 **Showing generation of direct repeats of the target site during transposon integration.**

3. There is presence of short direct repeat (usually 5bp in length) flanking the element at its site of integration (target site) (Figure 22.1). The size of the target direct repeat is characteristic of transposon rather than host. When LTR RTs are excised from the genome they leave behind an LTR sequence. Some genomes particularly those of plants are full of these lone LTRs and as they can affect gene regulation. These TE leftovers can contribute to genetic diversity. But not all TE excisions leave behind such a foot print.

Effect of transposons—Transposons are agents of mutation and cause chromosomal breakage and thereby chromosomal rearrangement. They cause gene mutation through integration of transposons into a gene and **polarity** mutation. They can also be used to tag genes with a readily identifiable DNA sequence. Antibiotic resistance transposons are important building blocks of infectitious bacterial plasmids called RTFs (resistance transfer factors). Inexact excision of transposons can lead to deletions next to the transposon site. Also duplications and inversions of regions of DNA have occurred next to tetracycline transposon, Tn10 (Botstein and Kecknery, 1977). Finally, transposons have the ability to acquire and transpose other genomic DNA (regions of DNA lying between them). Transposable elements are also useful in the genetic transformation of higher organisms. Frequency of transformation is significantly increased if DNA fragments are inserted on transposable elements used as vectors to transform the experimental organism. Many TEs are capable of horizontal transmission between reproductively isolated species (Robertson, 1993) (distantly related species) and show 95-99% nucleotide identity (Lohe et al., 1995)- have highly conserved sequence. Transmission of TEs in asexual organisms is through transmission of plasmids. Transmission is also through sexual means. Somatic mutations are typical of TEs but these elements also mutate genes in the germline leading to heritable genetic change. In some TEs an indefinite increase in TE copy number is checked in part by regulatory mechanism. The net rate of transposition is reduced as copy number increases (Engles, 1977; Hartl et al., 1997). RTs may be a driving force in genome evolution and gene duplication resulting in phenotypic changes in plants (Xiao, et al., 2008). They showed that the SUN locus in tomato arose as a result of an unusual 24.7-kb gene duplication event mediated by the LTR retrotransposon Rider.

Expression of TEs—The product of their encoded RNA is tissue specific. Some expressions are highly expressed during particular stage of development and some are even expressed differently in male and female reproductive cells. There is complex interactions between TE regulatory elements and the activity of host genes. The highly expressed host genes can have influence on the TEs adjacent to them through a '**read through phenomenon**'. TEs could have a role in genome reorganization and gene silencing. Further, it could have a marked impact on the relative contribution of the two sets of parental chromosomes in early development. DNA of SINES are methylated and thus inactivated in the oocyte but methylated in the male germ line. By contrast, intracisternal A particles (IAPs), RTs and a LINE called LINE1 show the opposite pattern. Silencing and activation of TEs in males and females can affect genes that are involved in RNAi pathway (see chapter 23 for detail) which is thought to have evolved as a form of nucleic acid based immunity to inactivate viruses and TEs. In addition to their effects throughout embryonic development TEs may act later in life. LINE-1 TRs seem to jump preferentially to regulatory region of certain neural genes and thus changing their expression. In other words, TEs act as promoter of genetic and phenotypic diversity. In amny organisms TEs are under 'epigenetic control' where gene regulatory instructions are laid out in the modification of DNA or its associated proteins (mainly histones) which do not change the DNA sequence itself. These changes includes methylation of certain DNA nucleotides and methylation or acetylation of histones. These epigenetic marks have a direct effect on the levels of expression of nearby genes (and TEs) and thus they are considered to be a second code in addition to genetic code. Although some epigenetic instructions as such as gene imprinting are long lasting and heritable but the epigenetic code is more sensitive to environmental stress than the DNA sequence itself. Environmental stress can modify the genomic methylation state and thus can change the structure of chromatin (the DNA and its associated proteins), inducing changes in gene expression. Different stresses can act differently on DNA methylation, histone methylation or acetylation, chromatin structure and production of small RNAs. Such processes which can affect genome function makes up the core of epigenetic 'memory'. The environmental effect on gene expression can also be felt through effects involving TEs. TEs can control genes epigenetically if inserted within or very close to them as TEs can act either directly by inducing methylation (and therefore silencing) of nearby DNA or indirectly by disrupting the normal epigenetic state of a nearby gene.

22.7 LONG INTERSPERSED ELEMENTS (LINES)

LINES occur widely in eukaryotes but are abundant in some mammals like human and mouse. LINES are scattered through out the genomes and in mouse it is likely to be present where an insertion is present without disruption of essential function. In case of human they appear to be present in abundance in chromosome segments showing 'G' bands. Mouse L1 element is 6kb long but occurs in a range of forms shortened from the 5' end. Human L1 is similar in size, structure, quantity and distribution to those found in mouse. Thus L1, LINES of human or mouse have no LTRs and two reading frames (ORF), almost contiguous in human and slightly overlapping (by 14 bases) in mouse. The upstream ORF is1137 bp long which does not code any recognizable product whereas the down stream ORF is 3900bp in length and encodes protein with strong similarity to reverse transcriptase and at the 3' the element terminates in a tail consisting of predominantly polyA. Thus this L1 resembles polymerase II (PoIII) and the large number of forms represent the incomplete reverse transcripts. In mouse there are up to 100000 copies of various lengths of L1 element (averaging about 2.5kbp/copy) which comprises ~ over 8% of the entire genome. Human L1s are sometimes called Kpn elements because they contain a Kpn1 restriction site. Lines have also been found in certain strains of Drosophila, maize and Neurospora. The best known LINE of Drosophila is I factor resembles P-element and causes hybrid dysgenesis. Human genome has about half a million copies of LINES of which 50-100 are still active.

22.8   SHORT INTERSPERSED ELEMENTS (SINES)

The Alu element which is abundantly found in human (> 1 million copies) is an example of SINE which contains a centrally located Alu restriction site. It is very similar in sequence to an abundant small RNA called 7SL which is a component of signal recognition particle, a ribonucleoprotein complex involved in the positioning of proteins in the cell membrane. The Alu sequence of ~300 bp corresponds to two somewhat diverged 7SL sequences arranged head to tail. The two parts thus show considerable similarity to each other except an insertion in one of them. Like LINE, Alu has an A-rich sequence at its tail. It is present in about 5000 000 copies, averaging about 6kbp/copy and accounts for about 5% of the genome. The various other SINE elements found in other mammalian species appear to be all related to small RNA molecules. Further, in all these RNA molecules the transcription is by RNA polymerase III (Pol III). PolIII transcription is initiated from promoters internal to the transcribed sequence. In the presence of cell's reverse transcriptase Pol III transcripts are reverse-transcribed and re-inserted into chromosome where it multiplies rapidly in the presence of its internal promoter and that is why it is present in abundant amount in the genome. The mechanism involved in the proliferation of SINES is that 1. the element is transcribed by its Pol III promoter 2. The transcript over-runs the A-rich 3' end of the SINE and terminates in a run of Us (as in template DNA) 3.The terminal U-rich sequence folds back and anneals to the A-rich sequence, priming reverse transcription 4. The reverse transcript is then ligated to chromosomal DNA at a staggered break and finally, 5. integrated in the double-stranded form by repair-filling of the single stranded gaps and ligation of loose ends. The repair of the staggered break generates the short flanking target site duplications characteristic of transposable elements (Deininger, 1989). Transposons identified by RepeatMasker can be assigned to three classes, class I (LINES, SINES, gypsy-like, copia-like are all retrotransposons), class II (Ac/Ds TEs, En/S$_p$m TEs, MULES are DNA transposons) and class III (stowaway-like, tourist-like are a previously unknown type of short DNA

Table 22.3 Showing the types of TE, their size, copy number and fraction of genome they constitute.

Types of transposable element	Size (bp)	Copy number	Fraction of genome%
SINES	100-300	1500 000	13
LINES	6000-8000	850 000	21
LTR	15 000-110 000	450 000	8
DNA transposon fossils	80-3000	300 000	3

transposons called MITES), each class has a number of families and each family has a number of subfamilies. Table 22.3 given shows different types of transposable elements found in the human genome (Lesk, 2002).

Due to insertional preferences to subtelomeric and telomeric regions the TEs have been detected in many ways. TEs are involved in the unusual organization of telomeres in *Drosophila melanogaster* and in the parasitic protozoan *Giardia camblia*. TEs contribute to chromosomes stability by expanding the buffer between the ends of the chromosomes and the genes. TEs–induced and other chromosome rearrangements (duplication, deletion, inversion or translocation) can lead to post-zygotic mechanisms resulting in almost total cross-fertilization barriers between different lines of the same species in experimental organisms in a relatively short time period as for instance, in *P. sativum*. So-called twin species, i.e., 'species' with no corresponding morphological differences to distinguish one systematic species from another have regularly been generated experimentally.

Because of the TE-induced changes in DNA the amounts of DNA in the haploid genomes of closely related species which are hardly distinguishable from each other morphologically can differ enormously (C-paradox). Species of the genus Vicia, for example, vary between 1.8 and 13.3 pg of DNA per haploid genome. Even within the same monoploid plant species the C-value can vary significantly.

TE activities can remain dormant over years of time, usually in a methylated state of inactivity and be awakened in new stressful environmental situations to produce the genetic flexibility essential for further adaptations and thus will lead to periodic induction of mutation. This periodic induction of mutation was formulated by Hugo de Vries.

22.9 CLASSIFICATION OF TRANSPOSABLE ELEMENTS

The term transposon is often used as a generic term instead of TE but it was originally coined to name the first characterized TE. Its classification is based on evolution (phylogeny) and the genetic modules that they contain (Biemont and Vieria, 2006). There are two main classes of TEs depending on the mode of transposition: 1. DNA transposons and 2. Retrotransposons (RTs). RTs are further subdivided into LTR retrotransposon (LTR RTs) and non LTR retrotransposon (non LTR RTs). The DNA transposable elements include simple insertion sequences, transposons and some bacteriophages in eubacteria and similar elements in archeobacteria and eukaryotes. They remodel genomes, facilitate lateral transmission of genetic information such as antibiotic resistance determinants. The DNA transposons act through a DNA intermediate using the transposase enzyme to splice itself in and out of the DNA (cut and paste) and constitute class II type elements.

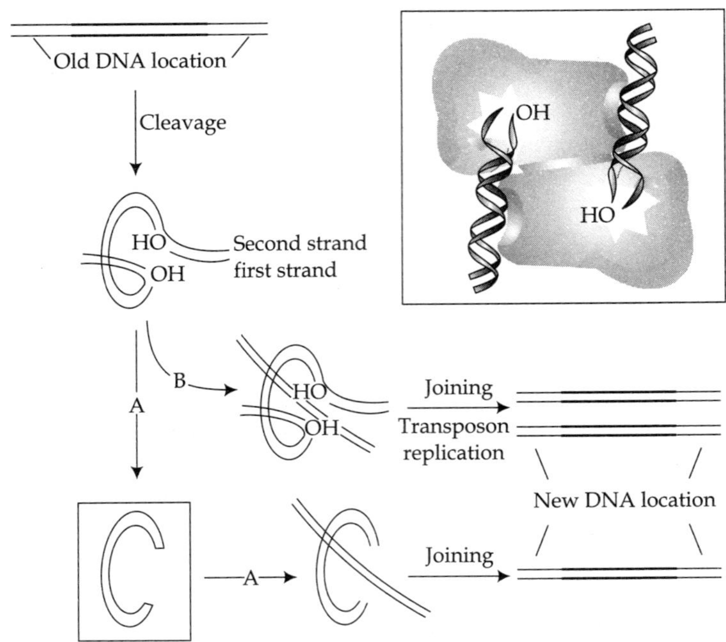

FIGURE 22.2   Showing two pathways (A and B) of transposon 'jumping'.

22.10   MECHANISM OF JUMPING

In transposon jumping, a transposon embedded with an old DNA location moves to a new DNA location via one of the two pathways. Both pathways involve two identical steps as follows (Williams and Baker, 2000) (Figure 22.2).

1. Cleavage of the first strand at each DNA end to yield a 3′-OH and

2. joining of this 3′-OH to a new DNA site.

The enzymes catalyzing these reactions (transposases and integrases) are encoded in the DNA of mobile element. Transposase protein is specific for its end sequences. Cut and paste transposons (pathway A) cleave both strands at each end of the element before joining step and include Tn5, Tn7 and Tn10. In other words, in case of cut and paste the 5′ ends are also cleaved within synaptic complex thereby releasing the TE from the donor. In pathway B after joining there is transposon replication. Transposition is a multi-step process. The first step involves transposase binding to specific terminal DNA sequence and the second step involves transposase-DNA oligomerization to form a synaptic complex. In case of Tn5 cut and paste transposition mechanism, transposase binds to specific 19bp recognition sequences at the ends of the TE. The 3′-OH generated from the initial strand cleavage step attacks the opposite strand of DNA, the 5′strand to form a 'hairpin' structure, followed by cleavage of the hairpin by attack from an activated water molecule and thereby excising the transposon from the donor DNA. The final steps involve target DNA capture in which the target DNA becomes bound to the synaptic complex, followed by strand transfer where the activated

3'OH groups at the ends of the TE perform nucleophilic attack on both strands of the target DNA which accomplishes strand transfer. The attack of 3'OH groups on the target DNA occurs with a staggered spacing between insertion sites which is specific for each TE (eg., 9bp in case of Tn5).

In case of replicative TEs such as Mu and Tn3 and RVs and LTR-RTs, the 5' strands are not cut before strand transfer of the 3' ends into the target DNA. Rather, the 5' ends are resolved by a replication/recombination process or through processing of the protruding ends after transposition. Identical steps are executed by transposition protein in retroviral integration and Mu replication. DNA cleavage reactions expose the 3'OH ends of the element followed by strand transfer reactions which covalently join these 3 elements. However, the products of retroviral integration and Mu transposition are different. Thus although all transposition reactions involve DNA breakage and joining, different types of recombination products can emerge depending on which DNA strands are broken and joined. In other words, difference arises not from the transposition reaction *per se* but rather from the state of the DNA substrate in particular at the 5'ends. Mu remains linked to the donor site while also inserting into the target site as no cleavage occurs at the 5' ends of the element. By contrast, in retroviruses the substrate DNA contains only the transposable DNA segment. Thus in cut and paste transposons (Tn7 and Tn10) and retroviruses, the transposition product is simple insertions whereas in phage Mu it is cointegration.

Retroviruses and retrotransposons containing long terminal repeats insert themselves into larger DNA through a mechanistically similar process. The RTs act through an RNA intermediate and use the reverse transcriptase enzyme and forms the class I elements (spread via replicative mechanism). Inverted terminal repeats (ITRs) are required for the movement of DNA transposons whereas *gag* gene specifies the components of molecular complex which is associated with the RNA transposition intermediate of RTs. RTs also encode reverse transcriptase enzyme which synthesizes a complementary single stranded RNA from the inserted DNA of the TE and converts it into a double stranded DNA which will be integrated into the genome elsewhere. This double stranded DNA is then cleaved by integrase to expose 3'OH ends at the embedded tips of the actual retroviral DNA. Strand transfer reactions then join these exposed 3' ends to staggered positions on the target DNA, one transposon end joining to one target strand and the other end joining to a displaced position on the other target strand. As a consequence the transposon is covalently joined to the target DNA but is flanked by short gaps which reflect the staggered positions of the target joining. The host DNA repair mechanisms then repair these flanking gaps.

In case of Mu replication the phage DNA is embedded in host chromosomal DNA. The DNA cleavage reaction is accomplished by MuA protein (Mu encoded transposase) which introduces single strand nicks at both tips of the element. These cleavages expose the 3'OH ends of transposon, separating them from flanking bacterial DNA but leaving the transposon covalently linked at its uncleaved 5' ends to the flanking DNA. The strand transfer reactions then join the exposed 3' ends of the transposon to staggered positions on the target DNA. This transposition product is then replicated by the host DNA replication mechanism to generate a product called a cointegrate in which the donor backbone, target and two transposon copies are linked. In some cases such as Tn3 another element encoded recombination system further processes the cointegrate to generate a target molecule containing a simple insertion and regenerate the donor.

RTs also encode ribonuclease H which degrades the DNA-RNA hybrids obtained during transposition. RTs have genes which encode integrase (INT) and protease (PR), respectively. The former (INT) splices the double stranded DNA into a new spot in the host genome whereas the later cuts up precursor protein and is involved in particle assembly. Some RTs have envelope gene (env) which encodes surface protein and confers infectious characteristics on the elements. A comparative structures of DNA and retrotransposons is shown in Figure 22.3. Transposable elements can be further classified into two groups on the basis of autonomy. 1. Autonomous TEs which encode reverse transcriptas and/or integrase activities. They are characterized by the presence of two ORFS in one strand, one of which encodes reverse transcriptase. Autonomous elements include DNA transposon (e.g. Tc1/

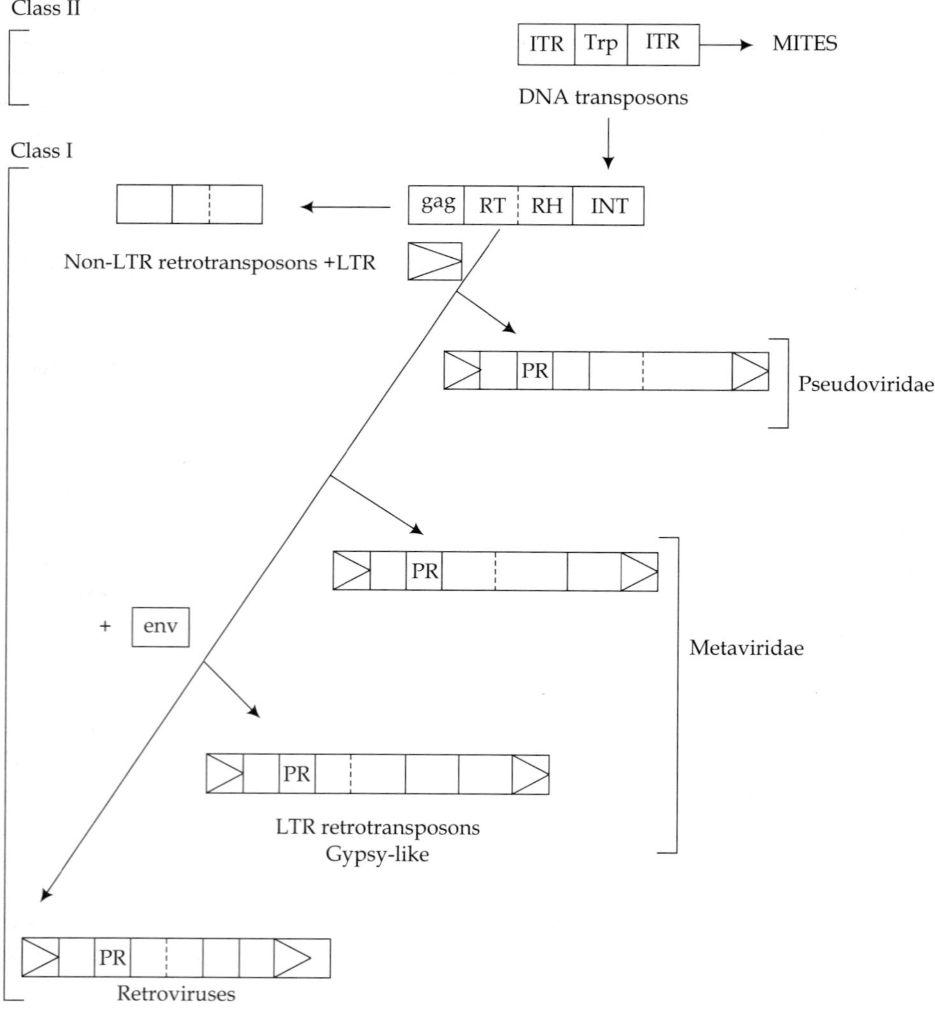

FIGURE 22.3 Showing classifications of transposons (After Biemont and Vieira, 2006).

mariner family), non-LTR retrotransposons (e.g. LINES) and LTR retrotransposons (e.g. Ty/copia/HERV) 2. Non-autonomous TEs are transcribed but not translated because they lack an ORF (Schmid, 1998) and they include SINES (e.g. Alu elements) and processed pseudogenes.

22.11 CHROMOSOME REARRANGEMENTS

There are evidences for the endless reshuffling of chromosome structures and gene sequences. There are models and mechanisms for gross and small chromosome rearrangements mediated by transposable elements. Gross chromosome rearrangements are induced specially in connection with the 'break-fusion- bridge cycle (in case of Ac/Ds system) as well as a wide array of small to large effects of the Ac/Ds, En/Spm and Dt systems on gene expression and the effects of transposons on the R and B loci and stress activations. Number of cases of transposable element target site selection (hot spot for TE integration) has been enlarged which implies pre-established rather than accidental rearrangements for nonhomologous recombination of host DNA. Although chromosomal rearrangements involve rearrangements of the linear sequence of chromosomes including transposition, duplication, deletion, inversion or translocation of nucleic acid segments, large chromosomal rearrangements are only tip of the icerberg (Lönning and Saedler, 2002). Homeobox, MADS-box and other regulatory gene families are involved in the development of animal and plant and these classes of 'controlling elements' do not generally move from one place to another within a set of chromosomes.

23

Gene Silencing

23.1 DEFINITION

Gene silencing refers to inactivation of a previously active gene. Gene or RNA silencing, a tool to reduce gene expression is triggered by dsRNA molecule. RNA silencing is a mechanism for cellular protection (of genome against molecular parasites such as viruses and transposons) and cleansing (removal of abundant and aberrant, nonfunctional mRNAs). Small noncoding RNAs trigger various forms of sequence-specific gene silencing including RNAi, translation repression and heterochromatin formation in a variety of eukaryotic organisms commonly referred to as **RNA silencing**. The mechanisms of RNA silencing are co-suppression in plants and RNAi (RNA interference) in animals. There are common RNA intermediates and similar genes are required in RNA silencing pathways in protozoa, plants, fungi and animals and thus indicating an ancient pathway. RNA silencing is directed by foreign dsRNA. All triggers which induce RNA silencing operate through a dsRNA intermediate giving rise to the formation of a siRNAs. The different direct sources of dsRNA are the inversely oriented transgenes, *in vitro* prepared dsRNA and viruses. The production of dsRNA is also from multiple arrays or highly transcribed single copy transgenes but it requires an additional step. This might occur via read-through transcription, either from an endogenous promoter at the site of integration or as a result of head-to-head organization of transcription units in a multiple array and thus giving rise to dsRNA directly or to antisense RNA which can pair with the sense to form dsRNA. Alternatively, sense (and probably antisense) aberrant RNAs are converted into dsRNA via *denovo* RNA synthesis, reminiscent of virus replication (Figure 23.1). Both sense and antisense RNA could silence gene expression. Double stranded RNA is the actual trigger of mRNA destruction with the sequence of the double stranded determining which mRNA is to be destroyed. The ds RNA is converted to siRNAs (fragments of ds RNAs) which guide protein complexes to complementary mRNA targets whose expression is then silenced. Hairpins are cut out of the pri-miRNA by double stranded RNA–specific endonuclease Drosha with its double stranded RNA binding protein partner DGCR8in humans or Pasha in flies to yield a pre-miRNA. Each mature miRNA resides in one of the two sides of the ~30bp stem of the pre-miRNA. The mature miRNA is excised from the pre-miRNA by another double stranded RNA-specific

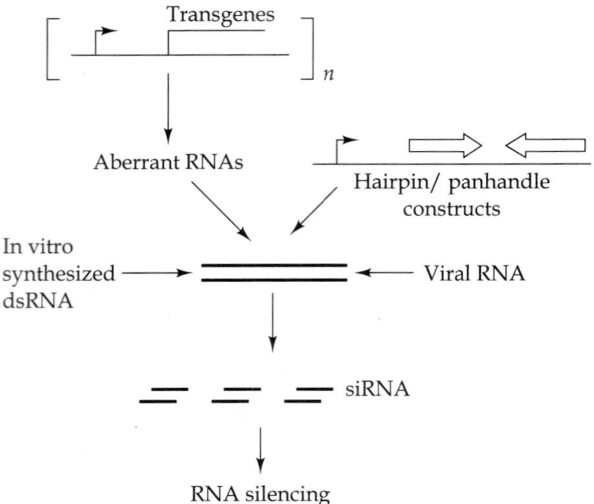

FIGURE 23.1 **Showing the process of RNA silencing.**

endonuclease, Dicer again with ds RNA binding protein partner, the tar binding protein (TRBP) in humans or Loquacious (Loqs) in flies.

Epigenetic silencing of genes (transgenes and endogenous genes) can occur at transcriptional (transcriptional gene silencing) or post-transcriptional level (post-transcriptional gene silencing). As they can be induced by transgenes and viruses, TGS and PTGS probably reflect alternative (although not exclusive) responses to two important stress factors that the plant's genome has to face: the stable integration of the added DNA into chromosomes and the extrachromosomal replication of a viral genome.

The fundamental strategies for silencing genes are similar between plants and other eukaryotes but the unique needs of plants have driven particular variations in epigenetic mechanisms (see chapter 6 for detail). For example, plants have evolved to depend on non-CG methylation as a reinforcement in the defense against invasive parasitic sequences in contrast to mammals. Furthermore, the majority of gene silencing events in plants are likely triggered by aberrant RNA whereas the role of RNA in mammalian gene silencing remains unclear (Bender , 2004)

23.2 TRANSCRIPTIONAL GENE SILENCING (TGS)

TGS refers to complete suppression of an endogenous gene or transgene (Mittelsten Scheid O and Paszkowski, 2000; Vaucheret and Fagard, 2001). It results from impairment of transcription initiation through methylation and/or chromatin condensation, could derive from the mechanisms by which transposed copies of mobile elements and T-DNA are tamed. TGS may be brought about by the following different mechanisms. TGS is mediated by

   i. surrounding heterochromatin

   ii. endogeneous repetitive sequences

iii. (*Trans*) gene repeats

iv. aberrant promoter transcript and

v. DNA viruses.

1. Chromatin structure may impose silencing (a transgene inserted into heterochromatin adopts this state and is silenced). Silencing can be position dependent, ie. there is influence of the chromosomal position on the activity of a gene. For example, if an active gene is introduced into a chromosome at a position adjacent to a heterochromatin domain, its transcription is repressed. In yeast, genes integrated close to the silent mating type loci or the telomeres are silenced. Gene inactivation spreads progressively from heterochromatin into the transposed euchromatic segment. The greater the distance of the gene from the euchromatin-heterochromatin junction, the lower the probability that it will be inactivated. This silencing influence of heterochromatin can spread from 5-10kb (Drosophila) to 20-30kb in yeast. Position effect variegation is a common observation in transgenic animals or plants. Reporter genes randomly introduced in the genome have highly variable transcription rate. Genes normally residing in euchromatic domains are silenced when packaged into heterochromatin. Silencing occurs when a transgene is integrated into a genomic region containing heterochromatin. The molecular features of heterochromatin including its characteristic nucleosome structure, deacetylated histones and in many cases hypermethylated DNA spreads into the transgene and caused its inactivation.

2. The presence of endogenous repetitive sequences may recruit chromatin components (e.g. protein) which induce silencing of neighbouring transgene.

3. A paramutation may be induced in one allele leading to TGS of the corresponding allele or integration of multiple copies of a transgene in a special spatial arrangement leads to cytosine methylation and TGS in both *cis* and *trans* (i.e. transgene loci silence-ectopic transgene driven by homologous promoter). *Trans*-TGS can be mediated by an aberrant RNA (truncated or nonpolyadenylated) or even by DNA viruses, for example, CaMV produces an aberrant RNA which impedes transcription through DNA-RNA interactions. TG-silenced transgenes are hypomethylated and hypermethylation attracts nuclear proteins (e.g. MeCP2) which assemble the local chromatin into a repressive heterochromatin complex. Many genes are involved in TGS. In *Arabidoposis thaliana* hog 1, sil 1, som/d dm 1 and mom 1 encode proteins which fix the suppressed state whereas mutations in these genes relax it by reverting cytosine methylation.

Telomeric silencing—When the genes are brought into close contact to telomeres by chromosomal rearrangements or transposition processes the telomeres impose an altered chromatin structure (heterochromatization) into the transposed genes so that they are no longer available for the regulatory proteins and thus gene silencing occurs. The change in chromatin conformation is mediated by telomere-binding protein RAP1 (repressor activator protein 1) and various silent information regulatory (SIR) proteins.

Methylation-mediated gene silencing—It refers to the down-regulation of the transcription of genes whose promoters carry cytosine with 5-methyl groups at strategic positions. These methylated cytosines may either sterically prevent the binding of transcription factors to their respective recognition sites or bind methyl-C_pG-binding proteins recruiting histonesdeacetylases. In both cases the adjacent genes are silenced. Cytosine methylation is an epigenetic mark used for silencing of TEs and for the regulation of development. Once established, DNA methylation is often stable through mitosis, in part, because CG methylation is faithfully maintained after DNA replication by DNMT1. The SRA domain present in Arabidopsis KRYPTONITE histone methyltarnsferase and also in UHRF1 family proteins found in plants and animals, was found to bind methylated DNA in either a CG, CHG (where H indicates A, T or C) or asymmetrical sequence context. KRYPTONITE is required for maintenance of CHG methylation and the Arabidopsis UHRF1 homolog VIMI/ORTH2 is needed for CG methylation at centromeric repeat sequences. Mammalian UHRF1 contains several domains including UBL, RING and PHD (plant homeo domain) and an SRA domain can bind methylated DNA. Bostick et al. (2007) showed that epigenetic DNA methylation patterns which persist through cell division depends on a protein that binds to hemimethylated DNA and a methyl transferase. They showed that UHRF1 may help recruit DNMT1 to hemimethylated DNA to facilitate faithful maintenance of DNA methylation.

Cytosine methylation is important in regulating gene expression and in silencing transposon and other repetitive sequences. Many endogenous genes are methylated either within promoters or within their transcribed regions and the gene methylation is highly correlated with transcription levels. However, plants have different types of methylation controlled by different genetic pathways and detailed information on the methylation status of each cytosine in any given genome is lacking. Cokus, S.J. et al. (2008) generated a map at single-base pair resolution of methylated cytosines for Arabidopsis by combining bisulphate treatment of genomic DNA with ultrahigh-throughput sequencing.

23.3 APPROACHES TO GENE SILENCING

Interaction between homologous DNA and or RNA sequences can silence genes and induce DNA methylation. There are three approaches to gene silencing.

1. Antisense RNA
2. Ribozyme gene binding
3. RNA interference (RNAi)

1. Antisense RNA—Possible mechanisms for gene regulation involves wide spread noncoding RNAs such as microRNA, Piwi-interacting RNA and antisense RNAs. Global transcriptome analysis shows that up to 70% of the transcripts have antisense partners and that perturbation of antisense RNA can change the expression of the sense gene (Yu et al., 2008). Antisense transcripts induced by genetic mutation leads to gene silencing and DNA methylation. DNA methylation is commonly associated with tumor suppressor gene silencing. Anti-sense gene is any gene which when introduced into a target organism, transcribed into an antisense RNA which forms a duplex with the target mRNA and thus the mRNA is inactivated and not translated. The antisense RNAs usually contain 1 to 3 stem-loop structures whereas the corresponding sense RNAs are frequently longer, possess the complementary stem-loops and additional structures. The loop determines the specificity of pairing between antisense RNA and sense RNA whereas the stems determine stability of antisense RNA. The different mechanisms involved in gene silencing are as follows.

1. The antisense RNA binds to sequences in the major groove of the duplex DNA and there is formation of triple helix structure which interferes with the binding of the DNA affinity protein, the transcription factors.

2. The antisense RNA forms double stranded RNA molecules with its sense mRNA and which is not processed and/or exported to cytoplasm as has been observed in al least mammals or is rapidly degraded as found in Drosophila or arrests translation by blocking the ribosome binding site as observed in bacteria. In other words, antisense RNA can anneal to a ribosome-binding site and block translation initiation. Antisense RNA can also anneal to an mRNA away from the ribosome-binding site so as to activate an RNaseIII attack and subsequent mRNA degradation.

3. The antisense RNAs can enter transcription bubble where the single stranded RNAs are available for RNA polymerase, bind to their own sequences and reduce the processivity of the transcription complex.

Control of gene expression by naturally occurring antisense transcripts was discovered long back in prokaryotes (Simons, 1988). In other words, regulation of expression of specific genes by antisense RNA is a naturally occurring mechanism in bacteria although gene regulation by this mechanism has not yet been observed in higher eukaryotes. However, antisense RNA has been shown to reduce expression of specific genes when injected into frog oocytes and Drosophila embryos. Inhibition of artificially introduced genes has been demonstrated by transient expression of antisense RNA constructs in mammalian cells and plant protoplasts and by stable expression in transgenic plants (Rothstein et al., 1987). Smith et al. (1988) developed homologous transformation system to introduce a construct designed to express PG antisense RNA constitutively in tomato plants. The PG (polygalacturonase) gene plays an important role in fruit softening and it does so by partially solublizing the pectin fraction of the cell wall. A 730bp HinfI fragment from the 5′ end of the PG cDNA including a 50-bp UTR and the translation start site, was fused in the inverted orientation to the CaMV 35S RNA promoter and the 3′ end of the nopaline synthase (nos) gene. This hybrid gene was introduced into the *Agrobacterium tumefaciens* binary transformation vector Bin 19. This plasmid, p JR16A was transferred from *E.coli* strain TG-2 to *A. tumefaciens* strain LBA 4404 by triparental mating with p RK2013 in *E.coli* HB101 and used to transform tomato segments. The steps involved in the construction of p JR16A are as follows.

I. A 730bp HinfI fragment from the 5′ end of the PG cDNA clone p TOM6 was isolated and the ends filled with Klenow polymerase. The fragment was cloned into the SmaI site of a p UC18 multiple cloning site.

II. The KpnI-PstI fragment from a clone (pC1) containing the cDNA fragment in the appropriate orientation was ligated into the same sites of the pUC based expression vector p JR1. This construct contained a 528bp CaMV 35S promoter fragment. The antisense gene was excised by a partial EcoRI-Hind III digest and inserted into EcoRI-Hind III digested Bin 19 to generate p JR16A. The construction of p JR16A was confirmed by DNA sequence analysis. Further, restriction analysis of plasmid DNA from *A.tumefaciens* confirmed that the construct was intact.

Global gene expression by microarray as well as computational analyses based on full length cDNA or EST sequences have suggested that genomes of higher vertebrates also containing a substantial number of genes that harbour oppositely oriented overlapping transcripts. However, the number of well characterized antisense genes is still small. Ontology of antisense genes, i.e., how many genes are associated with their antisense genes is largely unknown. Some of the identified sense-antisense gene pairs are artifacts owing to experimental error intrinsic to these approaches. Hybridization-based approaches are sensitive to cross-hybridization and the alignment of transcripts to sequences of the genome can be erroneous owing to the wrong annotation of the transcript orientation. Furthermore, about half of the antisense transcripts are single exon genes and therefore the correct orientation can not be confirmed by the analysis of canonical splice sites. SAGE is a promising method to evaluate antisense transcripts. The use of long SAGE is extremely powerful in the identification of thus-far unknown antisense transcripts (Wahl et al., 2005).

2. Ribozyme gene binding approach

Nucleic acid enzymes—They are made up of genetic material but possess enzyme like properties. These are also called **molecular scissors**. Ribozymes are short RNA sequences that catalyze their own cleavage (self-cleaving ribozymes). Thomas Cech (1982) discovered that RNA can act as an enzyme catalyzing specific biological reactions and shared Nobel prize with Altman in 1989. A small RNA segment can act as an enzyme clipping another piece of RNA. RNA scissors are remarkably precise, honing in on and snipping out specific sequences of foreign RNA. Only a few nucleic acid enzymes are known to occur in nature. All of which are RNA enzymes (ribozymes). These include the self-splicing group I and group II introns, the RNA component of RNase P which cleaves the precursor tRNAs to generate mature tRNAs and the various self-cleaving RNAs including the 'hammerhead', hairpin, hepatitis delta virus (HDV) and neurospora varkeed satellite (VS) motifs. In addition, the RNA component of the large ribosomal subunit is now recognized to be a ribozyme which catalyzes the peptidyl transferase step of translation. Ribosomal RNA is unique among known naturally occurring ribozymes as it contains several modified nucleotides. The U2/U6 complex within the eukaryotic spliceosome also may be a ribozyme, These two RNA molecules in isolation catalyze an unusual phosphoryl transfer reaction but occurs very slowly and may not reflect the natural catalytic activity of the spliceosome. Catalytic RNAs or ribozymes include naturally occurring phosphoryl transferases and the ribosomal peptidyl transferases as well as *in vitro* selected RNAs. Regulatory cellular ribozymes include bacterial cofactor-dependent ribozyme GlmS and the eukaryotic co-transcriptional cleavage (CoTC) ribozyme. Ribozyme is defined as an RNA molecule with the enzymatic properties of a sequence specific endonuclease (RNAase) which catalyses the cleavage of single stranded RNA (small nuclear RNAs which are associated with sn ribonucleoproteins and function in pre-mRNA splicing or the intron of rRNA precursor of Tetrahymena that excises itself from the large precursor in a process called '**self splicing**'). The ribozyme also catalyses the joining of the trimmed RNA fragments and peptide bond formation and additionally possess esterase activity. The Tetrahymena ribozyme can also ligate multiple oligonucleotides aligned on a template strand to generate a fully complementary daughter strand and thus is involved in RNA catalyzed RNA replication. Ribozymes promote cleavage and ligation at rate about 10^{10}-fold higher than the uncatalyzed reaction. RNA catalyzed reaction can be either intramolecular (for example, self-splicing or self-cleaving) or intermolecular using another RNA molecules as substrate and involving multiple turnovers of ribozymes. Ribozyme gene can be engineered to bind to target DNA sequences where it will produce ribozyme which in turn will bind to the target RNA sequences and thus gets cleaved and there by inactivated.

Ribozyme/catalytic RNAs are found in a broad variety of biological conditions. Most self-cleaving ribozymes are associated with exotic small replicating RNAs called replicons, many of which do not code for proteins. Recently, sequences for self-cleaving ribozymes have been found associated with protein-coding cellular genes and so it is thought to regulate or modify gene expression (Been, 2006).

DNA enzymes—Nucleic acid enzymes can be obtained starting from a population of random sequence RNA or DNA molecules. DNA enzymes catalyze the cleavage of DNA. The reaction involves deprotonation of the 2'-hydroxyl adjacent to the cleavage site resulting in cleaved products that bear a 2', 3'-cyclic phosphate and 5'-hydroxyl. The complexity of DNA enzymes depends on whether a catalytic cofactor such as a divalent metal cation or small molecule is present in the reaction mixture. DNA enzymes require larger catalytic motif containing more nucleotides than can be sampled exhaustively with a starting population of random sequence RNAs. RNA cleaving DNA enzymes cleave all RNA substrate under physiological conditions with high catalytic activity, high substrate sequence specificity and generality for almost any substrate sequence. DNA enzymes with these properties might be used to cleave and therefore inactivate, target RNAs in cells or whole organisms. 8-17 and 10-23 DNA enzymes were obtained by *in vitro* selection starting from a population of 10^{14} DNAs that contained 50 random sequence residues.

Applications—1. 10-23 DNA enzymes can be used to reduce expression of a target mRNA *in vivo* (Cairns et al., 2002; Khachigian, 2000). 2. 10-23 DNA enzyme has been applied in a diagonistic context as part of the DzyNA-PCR method for quantitative PCR. It has been applied to monitor the clinical course of acute promyelocyte leukemia by measuring genetic rearrangements associated with the disease. 3. RNA-cleaving DNA enzymes have been used to probe for oligonucleotide of a particular sequence. 4. 10-23 DNA enzymes have been used in the area of molecular computing (Joyce, 2004).

RNA enzymes—Ribozymes evolved from random sequence RNAs (Bartel and Szostak, 1993). They catalyze the ligation of an oligonucleotide substrate to the 5'end of the ribozyme, directed by an internal template region. The reaction involves attack of the terminal 2' or 3' hydroxyl of the substrate on the α-phosphate of the5'–triphosphate of the ribozyme forming a phosphodiester linkage and releasing inorganic pyrophosphate. This reaction is analogous to the reaction carried out by an RNA-dependent RNA polymerase.

23.4 CATEGORY OF RIBOZYMES

There are four major categories of ribozymes.

1. Self-cleaving RNAs from viroids and satellite virus RNAs—Four small catalytic RNAs were identified in satellite RNA from plants and animals. These include the hammerhead, hairpin, hepatitis delta virus and the Varkud satellite (VS) ribozymes. Hair-pin ribozyme is a naturally occurring or synthetic self-slicing ribozyme of about 50-70 nucleotide in length which binds and cleaves a 14-18 nt substrate RNA. The catalytic RNA is folded into a hair-pin like structure consisting of two helical domains, three loops. Two additional helices form between the ribozyme and its substrate. The substrate RNA must contain a sequence GUG in the loop 5 and cleavage occurs immediately 5 of the G in this motif. Hammerhead ribozyme is a small self-splicing RNA of various plant virusoids, virioids and viral satellite RNAs which functions in the self-splicing of multimeric RNA genomes. The endonucleolytic Mg^{2+}-dependent cleavage of the concatemers yields 5'OH termini and 2, 3 cyclosophosphates. The hammerhead ribozyme is folded into three stem-loop secondary structures (helix I, II and III) where the cleavage site is adjacent to helixI. The cleavage depends also on the sequence context around this cleavage site (e.g. on the flanking region, a moderately conserved trinucleotide sequence 5'-GUC-3', 5'-GUA-3', 5'-AUA-3'). Additional cleavage of helix I or III generates transribozymes. In genetic engineering the hammerhead ribozyme sequences are used to cleave RNAs at the specific sites. To achieve this the hammerehead sequences are put immediately up or downstream of the target sequences. The intramolecular splicing reaction either occurs during transcription or post-transcriptionally. A CuCu ribozyme is an RNA molecule which catalyzes the cleavage of single stranded RNA substrate at the nucleotide sequence 5'-CUCU-3'. All of these ribozymes differ substantially in their secondary and tertiary structures but thy all perform the same cleavage reaction. In each case the satellite RNA is proposed to replicate through a complementary RNA intermediate by a rolling circle mechanism. The ribozyme participate in the replication by self-cleaving the tandem satellite repeats into monomer units.

2. Self-splicing group I and group II introns — The second and more common category of ribozyme comprises the self-splicing introns. These RNAs provide the active site necessary to complete their own RNA splicing from an RNA transcript with simultaneous ligation of the flanking exons. Self-splicing introns are divided into two classes, group I and group II based on differences in their conserved structures and reaction mechanisms. Besides their ribozymic activity both classes of introns can catalyze reverse splicing reactions which permit them to act like mobile elements for horizontal transfer of genetic information.

3. The RNA moiety of Ribonuclease P (RNase) — It removes the nucleotides from the 5' – terminal ends of precursor or tRNA (pre-tRNA) which is one of the several steps involved in tRNA maturation or tRNA processing. The ribonuclease P in *E.coli* consists of an RNA, M1 RNA and a protein component, C5. Although both are biologically essential but the RNA subunit alone is sufficient to carry out the tRNA processing reaction albeit at a slower rate than holoenzyme.

Cis-**ribozyme** — It refers to a class of self-splicing ribozymes which autocatalytically cleave themselves out of precursor transcripts. This group of ribozymes includes the self-splicing introns, the hammerhead ribozymes and hairpin ribozymes which are involved in the autocatalytic cleavage of concatemeric RNA replication intermediates into monomeric units in different plant pathogens. The *cis*-ribozymes can be converted into *trans* ribozymes by the cleavage of the RNA strand into two separate single strands (a catalytic strand and a substrate strand). *Cis*-ribozymes are used to cut mRNA at specific sites and to inactivate them.

Trans-**ribozymes** — It refers to any ribozyme which catalyzes the cleavage of single stranded RNA substrates but does not cleave itself.

Ribozyme targeting — It refers to the development of specially designed synthetic ribozymes for the silencing of specific gene, their introduction into target cells and their action on the mRNA encoded by the target gene.

Ribozyme-PEI complex—It is a non-covalent and reversible association of a ribozyme molecule with low molecular weight polyethyleleimines (PEIs). Such complexes are effective transfection carriers for ribozyme (and DNA) as the ribozyme is compacted by PEI such that the resulting colloidal particles can easily be taken up by endocytosis and further the ribozyme is protected against degradation by intracellular ribonucleases since PEIs are acting as an efficient buffer system (proton sponge effect). The ribozyme-PEI complexes are therefore used to protect exogenous ribozymes from intracellular degradation and thus can be used to silence specific genes.

Ribozyme auto-cleavage vector—It is a plasmid cloning vector which produces large quantities of specific RNA molecules by ribozyme auto-cleavage. The vector contains a cloning site which is located in between two sequences encoding two ribozymes. *In vitro* transcription using the circular supercoiled plasmid vector results in RNA which initially contains both sequences from the cloned foreign DNA and the ribozymes. Large amounts of RNA containing exclusively the sequences transcribed from the insert DNA are subsequently generated by auto cleavage catalyzed by ribozymes.

23.5 DETECTION OF RIBOZYMES

All self cleaving ribozymes found to date were discovered by analysis of transcripts of single genes or RNA genomes of pathogens. Detection of ribozymes has also been made through computational method (see Bioinformatics(Roy, 2009)).

23.6 GENOME-WIDE SEARCH FOR RIBOZYMES

Salehi-Ashtiani et al. (2006) developed an *in vitro* selection scheme for the isolation of self-cleaving sequences without imposing fixed cleavage target sites. In this scheme, a genomic library is constructed in which library elements are uniform in size (~ 150 nucleotides) and are flanked by PCR primer sequences. The library is converted into a single stranded form, circularized by splint-ligation and converted back to double stranded DNA to form a relaxed double stranded circle that can be *in vitro* transcribed. Rolling circle transcription generates tandemly repeated RNA that preserved the co-valent linkage of the entire transcribed sequence even after self cleavage. If a sequence encoded a self cleaving motif, it could self cleave under appropriate conditions to produce unit-length copies, as well as kinetically trapped misfolded intermediates such as dimmers, trimers and so on. The size difference between the cleaved multimers and uncleaved sequences provide a positive selection criterion for the enrichment of active molecule. Dimers which contain one intact copy of the sequence, are isolated by PAGE, reverse transcribe and amplified to reinitiate the cycle. They used this technique to identify ribozymes in the human genome.

23.7 RNA INTEREFERENCE

It is a gene silencing process which is involved in diverse eukaryotic functions, viral defense, chromatin remodeling, genome rearrangement and developmental timing. In other words, it represents a protection mechanism against viruses, retrotransposons, transposons and also transgenes and aberrant single stranded RNAs. It is also involved in heterochromatin stability (e.g., regulation of histone H3 lysine-9-methylation) of fission yeast or genome rearrangement in Tetrahymena. It is a mechanism of sequence- specific **post-translational gene silencing** (PTGS) in eukaryotic organism and is triggered by dsRNA homologous to the silenced gene (Fire et al., 1998). RNAi thus refers to the silencing of gene expression by double stranded RNA molecules. Double stranded RNA is used as a reliable trigger for gene silencing. The mechanisms involved are: (i) silencing genes by marking out their mRNA intermediates for destruction (ii) blocking transcription (iii) inhibiting translation. Through RNA interference best known as RNA silencing pathway, it is possible to block expression of nearly any gene in a wide range of eukaryotes, knowing only part of gene's sequence. Small RNAs derived from over expressed gene might guide inactivation of co-suppressed genes. Over expression of a pigment synthesizing enzyme to produce deep purple petunia flowers resulted in white flowers because transgenic and endogeneous genes were co-ordinately repressed (co-suppression). In other words, introduction of numerous copies of a gene that codes for deep purple color in petunias led, not as expected to an even darker purple color but rather to plants with a while or patchy flowers. Transgenes, the introduced genes had silenced both themselves and the plant's own 'purple flower' genes. Further, when RNA virus engineered to contain fragments of a plant introduced into a plant, the plant's gene itself become silenced. It is because of the reason that RNA virus replicates through ds RNA intermediates and multiple copy transgenes can produce low levels of ds RNA as well. In plants and *C.elegans*, two distinct populations of small RNAs have been proposed to participate in RNAi (Pak and Fire, 2007; Sijen et al., 2007). Primary siRNAs-derived from DICER nuclease-mediated cleavage of the original trigger and secondary siRNAs (additional small RNAs whose synthesis requires an RNA-directed RNA polymerase, RdRP). They suggested that small antisense transcripts derived from cellular messenger RNAs by RdRP activity may have key roles in the formation of secondary siRNA during RNAi. Secondary siRNAs constitute the vast majority of siRNAs found during RNAi.

23.7.1   The RNAi Pathway

The RNAi pathway is divided into two phase steps. In initiation phase the dsRNA is processed by RNase III nuclease called Dicer and there is generation of 21-23 nt long double stranded small interfering RNAs (siRNAs) with symmetric 2 nt 3′ overhangs for local interaction and 24-26 nt long siRNAs for systemic interference. In the effector phase the siRNAs are incorporated into RNA induced silencing complex (RISC), a multiprotein complex. This siRNA-RISC complex then targets transcripts by base pairing between one of the siRNA strand and the endogenous RNA, the mRNA. In the second step in the process, the 'sense' strand of an siRNA-the strand that has exactly the same sequence as a target gene-is removed, leaving the 'antisense' strand which is complementary to the target gene to function in gene silencing. The nuclease associated with RISC complex then cleaves the mRNA-siRNA duplex and targets mRNA for destruction or translation inhibition. This RNAi process works at the level of post transcription (PTGS).

PTGS (Post transcriptional gene silencing)—Several defense mechanisms of eukaryotic cells against viroids, viruses and retrotransposons (generally RNAs) are grouped under a more general term called PTGS. In this protective mechanism the RNA is converted into ds RNA within the cell. These dsRNA is recognized by Dicer RNase III which cuts it into small RNA which in turn incites silencing of the genes encoding homologous RNA. PTGS was first described in *A. thaliana* and *Petunia hybrida* and the its variants are components of defense system in all eukaryotic cells. In fungi quelling suppresses transgenes, in invertebrates RNAi and co-suppressor are incited by ds RNA and in vertebrate ds RNA transgenes and short hair-pin RNA is the prime trigger for RNAi. PTGS can also be transmitted systemically from silenced to non-silenced plant tissues by degradation resistant signal RNA. The ds RNA is a small RNA with protective secondary structure packed with proteins to form nuclease resistant ribonucleoprotein complexes. These ribonucleoprotein complexes are released from the cells upon virus infection by RNAi interference and migrate to uninfected cells and prevent the spread of viral RNA by forming dsRNA. This dsRNA is in turn recognized by Dicer RNase III and cleaved it into siRNAs which then silence the viral gene in the way as described above.

Transcriptional gene silencing

RNAi process also works at the level of transcription (TGS, also called transgene silencing in plants), preventing a gene from even making its mRNA. Both strands of a transgene are probably transcribed rather than just one strand as in case for typical gene and thus there is generation of dsRNA rather than single stranded mRNA. This dsRNA is processed into siRNA which initiates transcriptional silencing. It involves reversible addition of chemical groups to nuclear DNA and the proteins that package it as directed by the sequence of dsRNA. Binding of a small inhibitory RNA(RNAi) to a gene results in the binding of proteins with DNA methyl transferase and histone methyl transferase activity. These enzymes methylate C residues in DNA and histone in proteins, respectively and thus produces a tightly packed chromatin structure and thereby no gene transcription.

The transcriptional silencing of transgenes by the sequence specific hypermethylation of cytosyl residues in promoter sequences of homologous genes (endogenous genes with higher or complete sequence similarity to the transgene) is induced by dsRNA. For example, viral or viroids RNAs (from e.g. potato spindle tuber viroids, poly- or potex viruses or viral satellite RNA) in the replicating double stranded form is processed to 21-25 nucleotide small dsRNAs by the invaded cell and these small RNA guide a methyl transferase complex to target gene promoters. The methyl transferase then methylates the cytosyl residues at strategic positions in the promoter and induces silencing of the adjacent gene. This phenomenon is exploited in plant gene silencing. If, for example, dsRNA with homology to promoter sequences is expressed in plants, it results in methylation of the corresponding promoter and the transcriptional silencing of the adjacent genes. RdDM has been evolved to protect plants from invading viroids or viruses and active in concert with RNA interference.

Double stranded RNAs (ds RNAs) that trigger PTGS/RNAi can be made in the nucleus or cytoplasm in a number of ways including transcription through inverted DNA repeats, simultaneous synthesis of sense and antisense RNAs, viral replication and activity of cellular or viral RNA-dependent RNA polymerases (RdRP) on single stranded RNA templates (Matzki et al., 2001).

RNAi activity limits transposons mobilization and provides an antiviral defense (Hannon, 2002). To produce enzymes that facilitate their jumping to a new location in the genome, the transposons must first be copied into mRNA. Small RNA guided silencing mechanisms which prevent copying of transposons into RNA sequences allows higher organisms to defend their genes against transposons. In some higher organism, the RNAi pathway provides a crucial defense mechanism against transposons. But for fruit flies the genome is protected by the Piwi-associated interfering RNA (piRNA) pathway (Vagin, et al. 2006). RNAi is required to establish silencing at heterochromatic domains in fission yeast (Volpe et al, 2002 and Hall et al., 2002). Loss of function mutations resulting in suppression of position effect variegation has identified critical components of heterochromatin including HP1, HP2 and Histone H3 lysine 9 methyltransferase. Loss of silencing is as a result of mutation in genes (piwi, aubergene or spindle-E (homless)) which encode RNAi components. The resulting mutations result in reduction of H3 lys 9 methylation and delocalization of HP1 and HP2 (Pal-Bhadra et al., 2004). Noncoding RNA and histone modifications are thus key players in the spreading of mammalian and fly dosage compensation complex as well as in the spreading of silenced centromeric chromatin in the fission yeast. RNAi-mediated heterochromatin assembly occurs by means of a self-enforcing loop mechanism and the central player of this loop is the RITS. RNAi effector complex termed RITS (RNA-induced initiation of transcriptional gene silencing) which is required for heterochromatin assembly in fission yeast. RITS complex contains Agro 1 (the fission yeast Agronaute homolog), chp1 (a heterochromatin- associated chromodomain protein) and Tas3 (a novel protein). Ago1 confers specificity by binding to siRNAs and recruits other chromatin protein to start heterochromatization. In addition, RITS complex contains small RNAs (siRNAs) which require the Dicer ribonuclease for their production. These small RNAs are homologous to centromeric repeats and are required for the localization of RITS to heterochromatic domains. These small interfering RNAs load into an effector complex called RISC (RNA induced silencing complex) which contains agronaute/PIWI family protein and targets cognate mRNAs for inactivation. The factors involved in RNAi pathway are:

    i. Dicer

   ii. Agronaute

  iii. RNA-dependent RNA polymerase (RdRP)

Members of Argonaute family of proteins are essential components of RNA silencing. To function miRNA associates with an Argonaute proteins. In Drosophila 5 genes encode distinct members of Argonaute family, AGO1, AGO2, Aubergene (Aub), Piwi and AG03. AGO1 and AG02 belong to AG0 family and bind miRNA and siRNA, respectively. Aub, Piwi and AG03 belong to PIWI family and they are involved in silencing retrotransposons and other repetitive elements and show target RNA cleavage (silencing) activity *in vitro*. Both Aub and Piwi associate with repeat associated siRNAs (rasiRNAs) and are derived from antisense strand of retrotransposons with little or no phasing and have a strong preference for uracil at the 5′ end. Small RNA processing factors such as Dicer and Drosha are known to cleave preferentially at the 5′ side of uracil. However, rasiRNAs are thought to be produced by Dicer-dependent pathway (Gunawardane et al., 2007). In mammals there are four members of Argonaute proteins (Argo1 to Argo4). The Argonaute super family thus segregates into two clades, the Ago clade and the Piwi clade. Single fission yeast Argonaute and all plant family members belong to the Ago clade whereas Ciliates and Slime molds contain members of the Piwi clade. Animals typically contain members of both clades (Aravin et al., 2007).

In some organisms, gene silencing by RNAi requires activity of an RdRP which uses the antisense strand of an siRNA as a primer with which to make more dsRNAs thereby amplifying the process. Amplification is not thought to occur in vertebrates as RdRP has not been found. There is thus a role of the RNAi machinery and small RNAs in targeting of heterochromatin complexes and epigenetic gene silencing at specific chromosomal loci (Vendel et al., 2004). The siRNAs thus silence transposable elements, repetitive genes (including transgenes) and viruses. Worms with mutations in genes of the RNAi pathway are unable to silence transposons in germline tissues. Further, plants with mutations in subsets of RNAi pathway genes show defects in the silencing of specific mobile elements. The same process could also limit the degree to which a gene can be expressed in certain tissues. For example, Drosophila shows 'dosage dependent' gene silencing. In all these cases the proteins that package DNA are modified to suppress transcription. This type of silencing can spread to nearby regions, repressing gene regions that are adjacent to the transposable elements.

RNAi has potential to engineer the specific control of gene expression and to serve as a potent tool for functional genetics (Study of gene function). For these purposes 21 nucleotide siRNAs with 2 nucleotides 3′ overhangs (19 nucleotides of dsRNA and two unpaired nucleotides at the ends) are designed for the inhibition of specific genes, i.e. for the degradation of the mRNA encoded by the genes.

Heterochromatin in many organisms is characterized by extensive DNA methylation and histone modifications. Plants show cytosine methylation in CG, CNG (N= any nucleotide) and CHH (H= A, C or T) sequence contexts. In Arabidopsis, small interfering RNAs (siRNAs) are involved in localizing and maintaining these chromatin modifications in processes requiring RNA-dependent RNA polymerase (RDR2), DICER-LIKE3 (DCL3), ARGONAUTE4 (AGO4) and two RNA polymerase IV isoforms, Pol IVa and b (Baurle et al., 2007). Heterochromatin, representing the silenced state of transcription consists largely of transposon-enriched and highly repetitive sequences. Piwi (P-element induced wimpy testis) and repeat-associated small interfering RNAs (rasiRNAs) are implicated in heterochromatin formation and transcriptional silencing in Drosophila (Yin and Lin, 2007).

Other functions of RNAi—There are other functions of RNAi. RNAi is involved in a protection mechanism against virus. Secondly, it causes a systemic response. RNAi spreads systemically in plants and nematodes to silence gene expression distant from the site of initiation. Systemic RNAi in *C. elegans* involves SID-1 mediated intercellular transport of dsRNA. SID is a multispan transmembrane protein which sensitizes Drosophila cells to soaking RNAi with a potency which depends on the length of dsRNA. SID-1 enables passive cellular uptake of dsRNA. RNAi is involved in heterochromatin stability of fission yeast or genome rearrangement in Tetrahymena. RNAi pathway silences specific genes and interferes with gene expression. siRNAs are stable over cell generation and further, they are effective at concentrations that are several orders of magnitude below those of the conventional antisense and ribozyme approaches. siRNA directs post-transcriptional mRNA destruction and also cause chromatin level gene silencing. RNA thus appears to be a general means of targeting *de novo* DNA methylation which may indicate how sequence-specific gene silencing is established in a variety of epigenetic phenomena.

The limitation with RNAi approach is that not all genes are sensitive to RNAi. An important factor that determines the therapeutic potential of RNAi is the longevity of its inhibitory effect. Previously the inheritance of RNAi induced transcriptional silencing was demonstrated for one generation in mice. Now a single RNAi in *C. elegans* is shown to induce TGS that is inhibited indefinitely over generations in the absence of the trigger and of RNAi machinery (Vastenhouw, 2006).

RNAi is an evolutionary conserved defense mechanism whereby genes are specifically silenced through degradation of mRNA. This process is mediated by homologous dsRNA molecules. In invertebrates long dsRNAs have been used for genome wide screens and have provided insights into gene functions. Because long dsRNAs trigger a non specific intereferon response in many vertebrates, short interefering RNA (siRNA) or short hairpin RNA (shRNA) must be used for these organisms to ensure specific gene silencing. (Zheng et al., 2004).

23.8 PREPARATION OF SiRNA

The double stranded RNA can be constructed. The siRNA can either be prepared by chemical synthesis, *in vitro* transcription by SP6 *in vitro* transcription system or digestion of long double stranded RNA by RNaseIII or Dicer. The use of construct containing adjacent sense and antisense transgenes producing hair pin RNA (Chuang and Meyerowitz, 2000 and Tavernarakis et al., 2000) or a single transgene with dual opposing promoters (Wang et al., 2000) provides a stable source of ds RNA.

23.9 DESIGNING SiRNA

The designing starts with selection of a region located 50-100 nucleotides downstream of the start codon, AUG of the corresponding mRNA. In this region the sequences AA (N19)TT or AA (N21) is searched and its G/C percentage is estimated. The G/C content should be around 50% (must be less than 70% but more than 30%). Then a BLAST for the nucleotide sequence fitting the above criteria is performed to ensure that only one single gene is silenced. But to be more effective it would be better to design more than one siRNA for a given target mRNA. Further, siRNAs consisting of negatively charged peptide nucleic acid can be employed for gene silencing as they are more resistant to nuclease and show better sequence specificity than conventional siRNAs.

The synthetic siRNA can be introduced into target cells by electroporation, lentiviral vectors, microinjection, retroviral vector, transfection or other techniques. Animals can be fed with bacteria that contain plasmid with siRNA expressing genes. siRNA producing cassettes can be steadily integrated into embryonic stem cells and transmitted in the germ line.

Dicer gene—It refers to any gene which encodes Dicer (Dicer nuclease). The gene is called CAF, SIN1 or SUS1 in *Arabidopsis thaliana*. The encoded protein is fundamental in RNA interference (RNAi). Dicer nuclease is a complex protein which cleaves double stranded RNA precursor or molecules into 21-22 bp miRNAs (micro or tiny RNA) or more specifically small interfering RNAs (siRNA) or short hair–pin RNA. The RNA III recognises the termini of dsRNA molecules, binds to them and cleaves the dsRNA successively into 21 nt long dsRNA fragments with 3′ overhangs of 2-3 nucleotides and 5′-phosphate and 3′-OH termini as it moves along RNA. This process occurs in nucleus or cytoplasm. The ds RNA bound to the dicer–ribonucleoprotein complex become denatured and guides the complex to target RNAs with complemenraty sequences in the cytoplasm. As a result of which mRNAs are endonucleolytically cleaved in the center of the recognized 21 nt sequence.

23.10   RNAS AS ACTIVATOR OF GENE EXPRESSION?

We have seen above the ability of short double strands of RNA to turn off specific genes, a process called RNAi. Recently, it has been shown that RNAs can have opposite effect and they can turn genes on and these RNAs are called **RNAa** (Li, 2006). RNAi typically silences genes for 5 to 7 days but RNAa boosted gene activity for up to 13 days. Now the questions can be asked whether RNAs are activating genes by silencing others which could just be RNAi by another name and how small RNAs could turn on genes especially for so long. The molecular mechanism underlying RNAi seems to be involved in RNAa, raising another question. What makes one siRNA a silencer and what makes the other one an activator. The difference could be due to targeting different parts of the gene's sequence and thus gene region may be the real key. If RNAa is a new phenomenon, researchers engaged in exploiting RNAi will be required to avoid activating other genes beyond the one they are trying to silence. Li used synthetic RNAs and if RNAa dies occur naturally then it could provide new insights into gene regulation. Steiz and her team (2007) have also found some miRNAs to stimulate protein production on cell cycle arrest and thus adds evidence linking miRNAs to various cancer.

23.11   ROLE OF POLYMERASE

Plants encode subunits for a fourth RNA polymerase (Pol IV) in addition to the well known DNA-dependent RNA polymerases I, II and III. By mutation of the two largest subunits (NRPD1a and NRPD2) polymerase IV silences certain transposons and repetitive DNA in a short interfering RNA pathway involving RNA-dependent RNA polymerase II and Dicer-like 3. The existence of this distinct silencing polymerase may explain the paradoxical involvement of an RNA silencing pathway in maintenance of transcriptional silencing.

Gene expression analysis has helped to find targets of an over-expressed plant miRNA but such an approach has not been suggested in animals where, in contrast to plants, miRNAs are believed to act mainly through translational repression rather than mRNA cleavage.

23.12   RNA WORLD

The central dogma has shown that the genetic information flows linearly from DNA to RNA to protein and never in reverse direction (see chapter 4). Further, the role of RNA in the cells has been limited to its function as mRNA, tRNA and rRNA. The discovery of a diverse array of transcripts which are not translated to proteins but rather function as RNA has changed this view. The existence of catalytic RNAs and pathways for the interconversion of RNA and DNA has led to speculation that an important stage in evolution was the appearance of an

RNA which could catalyze its own replication. In other words, RNA acts as a messenger and information transfer agent for DNA, the keeper of master genetic code. However, some types of RNA have ability to catalyze reactions as enzymes do. As a consequence, it has been thought that RNA could have been the original machinery of life. It refers to a pre-biotic era in which RNA was thought the **genetic template** (not DNA) and able to replicate itself (autocatalytically) and to modify other RNAs (hetericatylatically, analogous to ribozymes). In other words, the so called RNA world refers to the notion of RNA-based life. The discoveries of miRNA and siRNA biogenesis and effector pathways, the identification of hundreds of miRNAs (plants and animals) and diverse endogenous siRNA-generating loci (plants and fungi) and the finding of links between RNAi and heterochromatin formation (plants and fungi) has changed the view of RNA-based regulation in eukaryotes. Piwi (P-element induced wimpy testis) and rasiRNAs are implicated in heterochromatin formation and transcriptional silencing in Drosophila.

23.13 SMALL REGULATORY RNA

RNAs specifically tiny RNAs known as 'small RNAs' control plant and animal gene expression. Without these mRNAs transposons jump (wreaking havoc on genome), stem cells are lost, brain and muscle cells fail to develop, plant succumb to viral infection, flower shape changes, cell fail to divide for lack of functional centromere and insulin secretion is dysregulated. In worms, the two founding members of the miRNA class of genes-lin 4 and let-7 determine transitions between larval stages in development. In plants, miRNAs also control crucial developmental transitions and in flies they control cell division and cell death. The production and function of small RNAs require a common set of proteins: double stranded RNA-specific endonucleases such as Dicer (Bernstein et al., 2001), double strandedRNA-binding proteins and single stranded binding proteins called Argonaute proteins (Hammond et al., 2001). Together, the small RNAs and their associated proteins act in distinct but related 'RNA silencing' pathways which regulate transcription, chromatin structure, genomic integrity and most commonly mRNA stability (Zamore and Haley, 2005). Small regulatory RNA refers to any RNA of comparably small size (18-80nt) which regulates a nuclear process in eukaryotic cells.

23.13.1 Classes of Small RNA

There are three distinct classes of small RNAs: micro RNAs (miRNAs), small interfering RNA (siRNAs) and repeat associated small interfering RNAs (rasi RNAs). They are distinguished by their origins and by not their functions.

siRNA—David Baulcombe and colleagues discovered siRNAs (Hamilton and Baulcombe, 1999)approximating 21-24 nucleotides are associated with PTGS triggered by transgenes (inserted genes), mobile genetic elements and viruses in plants and thus has a significant function in the defense of an organism. siRNA targets complementary RNAs for degradation. siRNAs also guide the modification (by methylation) of DNA strands as well as the modification of the histone proteins around which DNA is wrapped, thus silencing gene expression (Matzke et al., 2004). siRNAs are derived from double stranded RNA hundreds or thousands of base pair long whereas miRNA are derived from long, largely unstructured transcripts (pri-miRNA), containing stem-loop or hairpin structures about 70nt in length, transcribed from one strand of distinct genomic loci. Both siRNAs and miRNAs are produced by RNase III like enzymes called Dicers. Both are single stranded molecules and both programme the activity of Agronaute proteins in RNA-induced silencing complex RISC. Unlike siRNA, miRNA is processed in a multistep pathway. Small RNA molecules have been found to play multiple roles in regulating gene expression. Small RNAs (21 to 30nt in length) provide specificity to range of biological pathways. These include targetted degradation of mRNA by small interfering RNAs (siRNAs) (post-transcriptional gene silencing, PGTS), developmentally regulated sequence-specific repression of mRNA by micro RNAs (miRNAs) and targetted transcriptional gene silencing (TGS).

Approaches for isolatioin of siRNAs—Lagos-Quintana et al. (2001) developed a directional cloning procedure for isolating siRNAs after processing of long double stranded RNAs in *Drosophila melanogaster* embryo lysate. In this technique, the 5′ and 3′ adaptor molecules were ligated to the ends of a size-fractionated RNA population, followed by reverse transcription PCR amplification, concatemerization, cloning and sequencing. This method which was originally aimed at isolation of siRNAs, led to the simultaneous identification of 16 novel 20-23 nt short RNAs which are encoded in Drosophila genome and are expressed in 0 to 2 hour embryos.

stRNA (small temporal RNA)—It refers to any one of a class of more than hundred noncoding single stranded highly conserved 20-25nt long small RNAs which is transcribed as about 75nt long precursor predicted to form a loop-stem structure which is processed into mature 22nt stRNA sequences by Dicer RNsaeIII and control major developmental transitions, developmental timing. The stRNAs are present transiently during development. One of the strands of the double stranded stems of the precursor then represents the stRNA. stRNA is encoded by genes. The antisense strand of stRNAs recognizes complementary sequences in the 3′UTR of the target mRNAs and bind there with bulges and mismatches such that it forms a characteristic interrupted hybrid which can not be translated into protein. In *C. elegans* the 22 nucleotide long stRNA encoded by genes, *lin*-4 and *let*-7 regulates developmental timing by blocking translation of the target mRNA. Further, the 21nucleotide long stRNA encoded by gene let-7 is involved in transition from the first to the second larval stage. Genes *lin*-4 and *let*-7 were identified by their mutant phenotypes.

23.14 IDENTIFICATION OF CANDIDATE SMALL REGULATORY RNA

Two approaches-Bioinformatics and cDNA cloning, were used to identify small RNA transcripts. In the bioinformatics approach the predicted *C. elegans* intergenic sequences that were also conserved in *C. briggsae* were analyzed using RNA folding program 'mfold'. 40 predicted sequences formed a stem loop similar in size and structure to *lin*-4 and *let*-7. Probes complementary to these sequences were tested against Nothern blots of total worm RNA and three of them detected small RNA transcripts.

In the second approach a cDNA library ($\sim$ 1.6 x 10^6 independent λ clones)was constructed from a size-selected ($\sim$ 22nt) fraction of *C.elegans* total RNA and the sequence was obtained for 5025 independent inserts, representing 3627 distinct sequences. Some of these sequences were represented by multiple (from 2 to 129) clones. Each of these multiple hit cDNA sequences was compared with the NCBI database using BLAST and this led to about 800,000 raw sequence traces of *C.briggsae* genomic sequence. Single copy cDNA sequences that corresponded to not previously known or previously detected transcripts and which were conserved in the *C.briggase* genome were analyzed using 'mfold' for a predicted stem loop structure. A total of 38 novel cDNA sequences fitted these criteria of which 13 were tested for expression by Nothern hybridization. In all 13 cases small transcripts (~ 22nt) and/or ~ 65nt) were detected.

These two approaches were used to select *C.elegans* genomic sequences that showed four characteristics of *lin*-4 and *let*-7: **(i)** expression of mature RNA of ~ 22nt in length **(ii)** location in intergenic (non-protein coding) sequences **(iii)** high DNA sequence similarity between orthologs in *C.elegans* and a related species *C.briggsae* and **(iv)** processing of the ~ 22 nt mature RNA from a stem loop precursor transcript of ~ 65nt.

23.14.1 Mi (micro)RNA

It refers to a class of non-protein coding RNAs which post-transcriptionally regulate gene expression in plants and animals. It guides development in an organism by regulating target genes (Bartel et al., 2004). It down-regulates the expression of mRNA by base pairing with partially complementary sequences in the mRNA and simply inhibiting their translation or by promoting the degradation of mRNA. It is thought to be formed from hairpin RNAs, the pre miRNAs (the precursor). Victor Ambrose and colleagues identified miRNA (Lee and Ambros, 2001), approximately 22 nucleotides in length, the small RNA product of heterochronic gene *lin*-4 in *C.elegans*. In most higher eukaryotes a class of small RNAs have been discovered which are involved in the post-transcriptional regulation of gene expression. They also mediate the silencing of particular genes. It refers to any one of a class of hundreds of ubiquitous, noncoding, usually single stranded, 16-24nt long regulatory eukaryotic RNAs which is processed from a longer transcript of 71-171 nt. It carries a stem-loop structure and is associated with proteins to form microribonucleoprotein complex. miRNA inhibits the translation of target mRNA containing 3'UTR sequences with partial complementarity. In other words, miRNA imperfectly base pair with the 3'UTR of target mRNA and directs either degradation of mRNA or inhibition of its translation. Such UTRs are important assembly sites for complexes which affect mRNA localization, translation and degradation. Although both miRNA and siRNA are generated by Dicer from longer precursor, however, siRNAs are not encoded by genes whereas miRNAs are encoded by genes. In other words, siRNAs generally target the genes or genetic elements from which they originated whereas miRNAs regulate separate genes, perhaps hundreds or more per miRNA. Further, the degree of translational inhibition by miRNA depends on how many of these molecules are bound to the target mRNA. miRNA contains many binding sites at one end (the 3'-UTR) and several different miRNAs can target the same 3' region. siRNAs usually silence a target mRNA efficiently (through degradation) if the mRNA contains a single site that is almost exactly complementary to the short RNA whereas mRNAi mode of action requires that the mRNA contains numerous partially complementary binding sites which function synergistically. miRNAs are almost never exact matches of their targets whereas siRNAs are perfectly complementary to theirs. But many plant miRNAs have shown perfect or near perfect complementarity to mRNA implying that these mRNAs might instead be targeted for degradation in a siRNA mode of action. In other words, not all miRNAs work by binding with imperfect complementarity to mRNAs and preventing their translation. Similarly, siRNAs could also work like miRNAs, i.e. inhibiting translation rather than targeting mRNAs for destruction. Use of chemically synthesized siRNAs partially complementary to numerous target sites in a 'reporter' gene showed that siRNA inhibited the expression of reporter gene at the stage of protein synthesis. Many of the plant mRNAs that are regulated by miRNA-directed cleavage, encode gene-transcription factors. Certain miRNA sequences are conserved throughput the animal kingdom or between distantly related plant species whereas siRNA varies widely. Bao et al. (2004) proposed miRNA-induced methylation to occur. Thus functional distribution between mRNAs and siRNAs is rather more blurred than was thought. In mammals, like plants, miRNA regulates the expression of a transcription factor. The demethylation shows various hallmarks of RNA-mediated gene silencing in plants (Chan et al., 2004).

The miRNAs are transcribed in the nucleus where they are processed into pre-miRNAs. Further processing occurs in the cytoplasm where the pre-miRNAs are cleaved into their final about 22 nucleotide single stranded form (Lund et al., 2004). Several hundred distinct miRNAs exist in animals and plants. miRNA can silence gene expression through inactivation or degradation of mRNA. The biogenesis of miRNA in mammal cells involves nuclear and cytoplasmic processing catalyzed by ribonuclease III (RNase III) like endonuclease which recognizes ds RNA. First nuclear Drosha cleaves long primary transcripts releasing 60 to 7- nt pre-miRNA which can be folded *in silico* into stem loop hair pins. Then cytoplasmic Dicer processes pre-miRNAs into ~ 20 to 22 nt duplexes bearing two nucleotides single stranded 3′ extensions. Generally, only one stranded of the duplex serves as the mature miRNA. The pre-miRNAs generated in the nucleus requires further processing in the cytoplasm. Exportin-5′ (Exp 5′) mediates efficient nuclear export of short miRNA precursors (pre-miRNAs) and its depletion by RNAi results in reduced miRNA levels. Exp 5′ binds correctly processed pre-miRNAd directly and specifically in a Ran guanosine triphosphate-dependent manner but interacts weakly with pre-miRNAs that yield incorrect miRNAs when processed by Dicer *in vitro*. Thus Exp 5′ is key to miRNA biogenesis and may help co-ordinate nuclear and cytoplasmic processing steps. miRNAs have been predicted to regulate at least one third of all human genes.

miRNA negatively regulate partially complementary target mRNAs. Target selection in animals is dictated primarily by sequences at the miRNA 5′ end. Specific miRNAs contain additional sequence elements that control their post-transcriptional behavior including their subcellular localization. In case of human miR-29b, hexanucleotide terminal motif acts as a transferable nuclear localization elements. miRNAs sharing common 5′ sequences are considered to be largely redundant and might have distinct functions because of the influence of *cis*-acting regulatory motif. Nucleotides 2 to 7 of miRNAs known as 'seed' sequences are considered most critical for selecting targets (Hwang et al., 2007). The exact position of 5′ cleavage of mature miRNAs is important as it dictates the core of the target recognition sequence. This leads to unique structural and evolutionary signatures including direct signals, present at the 5′ cleavage site and indirect signal stemming from the relationships of miRNAs with their target genes. Within a given species highly related miRNAs sharing a common seed sequence are grouped into miRNA families, are predicted to have overlapping targets and are considered largely redundant. Nevertheless, loss of function of miRNA family members with divergent 3′ end sequence results in overlapping but distinct phenotypes in *C. elegans* and in Drosophila. These distinct phenotypes often do not appear to be due to differences in miRNA expression patterns which raises the possibility that distinct sequences within miRNA family members confer upon them characteristics functional properties.

23.15 MiRNA ACTING AS ACTIVATOR OF TRANSLATION?

Although miRNAs are known to inhibit translation and promote mRNA degradation, miRNA can enhance or repress mRNA translation depending on whether cells are proliferating or arrested in the cell cycle (Vasudevon et al (2007). Small RNAs serving as activators and repressors of gene expression are perhaps not limited to miRNA. Specific Piwi-interacting RNAs believed to repress gene expression may enhance transcription in Drosophila. Moreover, when delivered into mammalian cells, some double stranded RNAs complementary to promoter sequences increase gene expression. If miRNA's role as activator of transcription is accepted then the many questions will have to be answered. What is the mechanism by which RNAi enhances translation? Does miRNA stimulation of translation raise a possible complication and opportunity in using miRNAs and small interfering RNAs as therapeutics. Finally, accepting miRNAs stimulating translation in cells exiting cell cycle, what is the role of miRNAs in development and terminal differentiation processes (Buchan and Parker, 2007).

Network of miRNAs—Besides genetic networks consisting of mRNAs, there is a second genetic network comprising miRNAs. In humans, 15% of the genes are miRNAs and they regulate protein levels of 10% of the genes. In Arabidopsis, up to 7% of the genes may be miRNA/silencing RNA. As target specificity is based upon nucleotide sequence homology it is possible to predict miRNA/target pairs (Adai et al., 2005). Microarrays that contain unique flanking sequence from both transcripts of all such pairs could be useful in sorting out the regulation of the miRNA network. Regulation of protein function by direct binding of short, sequence-specific double stranded RNA has been described in mammals but not yet in plants.

Complex networks of small RNAs act through RNAi pathways to regulate gene expression, to mediate antiviral responses, to organize chromosomal domains and to restrain the spread of selfish genetic element. Although RNAi has been defined as a response to the double stranded RNA, however, some small RNAs species may not stem from double stranded RNA precursors. Yet, like miRNAs and siRNAs such species guide Agronaute proteins to silencing target via complementary base pairing. Silencing is achieved by corecruitment of accessory factors or through the activity of Argonaute itself which often has endonucleolytic activity. The likely function of RNAi is to protect the genome from both pathogenic and parasitic invaders.

Hair-pin RNA—It refers to any RNA molecule with self-complementary sequences allowing intrastrand self annealing with hair-pin formation. Short hair pin RNAs (shRNAs) consist of short usually 20-30 bp stems and a loop of unpaired bases which suppress the expression of target genes through a RNA interference mechanism. They are endogenously transcribed from genes controlled by RNA polymerase III promoters.

Intron-containing hair-pin RNA—It refers to any hair-pin RNA (hpRNA) which contains intron sequence between its sense and antisense arms and is used for RNA interference (preferably in plants). For example, a specific ihpRNA encoding construct comprises an 800 nucleotides long intron flanked by sense and antisense sequences, complementary to the target sequence to be silenced, a 35S promoter and a nopaline synthase terminator. The intron is spliced out during pre-mRNA processing in the target cell, the ihpRNA is highly effective in post-transcriptional gene silencing in plants (**intron enhanced gene silencing**).

Adjacent hair-pin RNA—It is a variant of hair-pin RNA containing a single stranded sequences adjacent to a potential hair-pin forming structure. This configuration probably confers sequence–specificity to the gene silencing capacity of the hair-pin RNA.

Generation of shRNA library—Schlabach et al. (2007) generated barcoded miRNA-based shRNA library targeting the entire human genome which can be expressed from retrovial or lentiviral vectors in a variety of cell types for stable gene knockdown. They also developed a method of screening complete pools of shRNAs using barcodes coupled with microarray deconvolution. Barcodes are not essential for enrichment screens (positive selection) but they are critical for dropout screens (negative selection) such as those designed to identify cell lethal or drug sensitive shRNAs. Hairpins that are depleted over time can be identified through the complete hybridization of barcodes derived from the shRNA population before and after selection to microarray. They developed a methodology called **half-hairpin** (HH) barcoding for deconvoluting pooled shRNA. This alternative to the use of 60-nucleotide barcodes for pool deconvolution enables a more quick construction and screening of shRNA. They took advantage of 19 nucleotide hairpin loop of mir3—based platform and designed a PCR strategy which amplifies only the 3'half of the shRNA stem. In comparison to full-hairpin sequences for microarray hybridization, HH barcodes eliminates probe self-annealing during microarray hybridization thereby providing the critical dynamic range essential for pool-based dropout screens. HH barcode signals are highly reproducible in replicate PCRs.

rasiRNA (repeat associated small interfering RNA)—In plants and animals RNA silencing pathways defend against viruses, regulate endogenous gene expression and protect the genome against selfish retrotransposons and repetitive sequences. In Drosophila germ line, rasiRNAs ensure stability by silencing endogenous selfish genetic elements such retrotransposons and repetitive sequences. rasiRNAs protect the fly genome via a silencing mechanism distinct from both the miRNA and RNAi pathways. rasiRNAs arise mainly from the antisense strand whereas siRNAs are derived from both the sense and antisense strands from their double stranded RNA precursor. rasiRNA production does not appear to need Dicer-1 which makes miRNAs or Dicer-2 which makes siRNAs. rasiRNAs lack the 2', 3' hydroxyl termini characteristic of animal siRNA and miRNA. Finally, unlike siRNAs and miRNAs, rasiRNAs function through the Piwi rather than Argo, Argonaute proteinsubfamily (Vagin et al., 2006).

Small nucleolar RNAs (snoRNAs)—These are non-protein coding RNAs that are 60 to 300 nucleotides in length and function in guiding 2'-O-methylation and pseudouridylation in rRNAs, small nuclear RNAs (snRNAs) and tRNAs. These noncoding RNAs have been implicated in numerous biological processes including transcriptional regulation and the modulation of protein function.

Guide RNA (gRNA)—It is a small RNA of 50-57- nucleotides encoded by the minicircles of kinetoplasts from *Trypanosoma brucei* and is complementary to a segment of mt mRNA from trypanosomes, mosses and higher plants and base pair with target RNA (allowing for G:U base pairs). After pairing at the 'anchor sequence' RNA editing starts from a central core sequence and capitalizes on transesterification whereby an U residue from the 3′ terminal oligo (U) extension of gRNA ('U' TAIL) replaces cytosine (mt of mosses and higher plants) or simply inserted (mt of trypanosomes).

Guide sequence—It refers to a specific part of an RNA molecule which hybridizes to eukaryotic mRNA and facilitates splicing of the intron sequences. Guide sequences can be classified into external guide sequence (a small nuclear RNA, snRNA) and internal guide sequence depending upon whether RNA hybridizes to exon or intron sequences to form a splicing complex or whether the intron itself contains self-complementary regions which form the substrate for splicing.

23.16   ROLE OF SMALL RNA

A large proportion of plant miRNA families and some *trans*-acting siRNAs function as negative regulators of mRNAs coding for transcription factors with roles in development (Jones-Rhoades and Bartel, 2004) and disruption of negative regulation by miRNAs through mutation of target sites or miRNA sequences frequently results in developmental phenotypes (Baulcombe, 2004). Cell-specific imaging of expression patterns of specific miRNAs and siRNAs, target mRNAs and downstream genes over time is needed to place small RNAs into spatial and temporal context during development. This requires the deployment of sensor technology and microarray technology. In the sensor technology fluorescent proteins serve to reveal cell-specific patterns of miRNA gene expression and targeting activity. In the second technology of comparable methods, miRNAs, targets and downstream genes are analyzed in parallel. Virtual *in situ* expression analysis (Birnbaum et al., 2003) would be particularly suitable for understanding intergration of small RNAs in developmental pathways.

Small RNA–producing genes belong to gene families. Sequence duplication events resulting in fold-back or hair-pin transcripts seem to be one mechanism of generation of new small RNA regulators with novel specificities. Analysis of small RNA and the genes producing them in close relatives of Arabidopsis will provide information about the frequency of generation of new potential miRNA genes. Further, analysis of siRNAs in synthetic polypeptides may throw light on the mechanisms of heterochromatin formation through RNAi-dependent processes.

The eukaryotes possess genes encoding RNAi factors such as Dicer-like, Argonaute and RNA-dependent RNA polymerase proteins. There exist families for each of these three factors. The DCL protein contains two RNase III-like domains for dsRNA processing whereas the Argonaute protein possesses an RNaseH-like domain. There are specialized genes of miRNA, siRNA, heterochromatin-associated RNAi and antiviral RNAi pathways (Baulcombe, 2004). Functions in miRNA and siRNA-directed target cleavage have been assigned for only two family members AGO1 and AGO4 and ZIPPY–a factor out of 10-Argonaute members in Arabidopsis. Similarly, functions have been assigned to only some of the DCL factors.

How DCLs catalyze formation of different classes of small RNA is not understood. Also, not known is how RNAi complexes integrate with other cellular processes. For example, although it is known that a set of specialized RNAi components, DCL3, RDR2,AGO4 participate in formation of heterochromatin at many loci but then not clear is how these factors integrate with DNA and histone modification pathways.

Intercellular mobility of small RNA—Although it is known that sequence-specific information is transported over short (cell-to-cell) and long (phloem-phloem) distances during RNA silencing and further that the anti-viral response appears to depend, in part, on systemic transport of RNAi signals, what is not understood is the mechanistic basis of this pathway.

23.17 CO-SUPPRESSION

Co-suppression or HDGS (Homology dependent gene silencing) or **post-transcriptional gene silencing (PTGS)**refers to the co-ordinated and reciprocal post-translational inactivation of the resident gene and transgene or two or more transgenes, encoding the same sense RNA. PTGS results from degradation of mRNA, when aberrant sense, antisense, or double stranded forms of RNA are produced, could derive from the process of recovery by which cells eliminate pathogens (RNA viruses) or their undesirable products (RNA encoded by DNA viruses). Mechanisms involving DNA-DNA, DNA-RNA, or RNA-RNA interactions are described to explain the various pathways for triggering (*trans*) gene silencing in plants. TGS differs from PTGS in that the former results from a block of transcription, correlates with methylation in the promoter region, both mitotically and meiotically stable whereas the later results from degradation of mRNA, correlates with methylation in the coding sequence and is meiotically reversible. TGS events are not associated with methylation in yeast and Drosophila. The PTGS is mediated by

  i. sense Transgenes-strongly transcribed sense Transgenes, very weakly transcribed or untranscribed sense Transgenes

  ii. antisense Transgenes-Transcribed antisense transgenes, Untranscribed antisense trans genes

 iii. sense/antisense Transgenes and

  iv. DNA/RNA virus.

Both resident gene (s) and foreign gene (s) transcribe complementary mRNAs which anneal to each other and are subsequently cleaved by a nucleolytic enzyme recognizing ds RNA. When these genes, resident and foreign genes, are present alone in the recipient genome each gene is expressed. When transgenic Petunia plant is transformed with additional copies of flower pigmentation gene encoding chalcone synthase both the endogenous gene (s) and transgnes are transcribed at normal levels but the mRNAs are degraded. The same process is generally explained as RNA interference in animals and as 'quelling' specifically in *Caenorhabditis elegans*. Homology-dependent resistance refers to resistance of an organism towards a virus, based on the expression of a transgene encoding viral RNA sequences. The principle of homology-dependent gene silencing was developed by plant biologists in which various combinations of 'homologous' sequence interactions between DNA and/or RNA induce silencing at either the transcriptional or post-transcriptional level (Matzke and Matzke, 1995). Thus RNAi-mediated silencing pathways occur in both the cytoplasm and the nucleus. Epigenetic modifications induced by homologous sequence interactions including RNA-directed DNA methylation were identified in some of the earliest plant studies and paved the way for the discovery of RNAi-mediated heterochromatin formation in fission yeast. Connections between homology-dependent gene silencing and transposon control, virus resistance and development were made early on by plant scientist which is now considered at least in part to be RNA-mediated processes. The mechanism of resistance involves targeting the transgene mRNA to complex with the homologous invading viral RNA. The RNA-RNA complex would then become a substrate for degradation. Homologous co-suppression refers to attenuation or inhibition of the expression of members of a gene family by introduction of an additional identical gene using various gene transfer techniques. The mechanism of suppression of homologous resident gene by foreign counterparts is not clear but is probably caused by the transcriptional read-through of the incoming genes from promoters located on the opposite DNA strand which results in the accumulation of antisense RNA. Transcriptionally active genes are better inducers of PGTS than transcriptionally inactive genes and RNA viruses can induce silencing of homologous genes encoded by the host plant.

Systemic acquired silencing (SAS)—It refers to the systemic spread of gene silencing throughout the transgenic organism. When the gene, say A is transformed into a genome of a target gene (usually present in multiple copies) then it is observed that its own expression or that of the corresponding cellular homology is repressed frequently. This effect may be transmitted from cell to cell leading to SAS of all copies of gene A and its homologues in the organism. For example, if a plant leaf expressing GFP is infiltrated with *A. tumefaciens* carrying another GFP gene in the T-DNA of the Ti plasmid then the T-DNA together with inserted GFP integrates into the genome of the exposed leaf. As a consequence of which GFP expression is silenced throughout the whole plant although the agrobacterium is restricted to the infiltrated leaf only. Both PGTS and RNAi are systemic, i.e. silencing can spread from the site at which it was induced throughout the whole organism.

23.18 CLONING OF ANTISENSE GENES

Antisense oligonucleotides are currently widely used to interfere in a sequence-specific manner with gene expression (Sten and Cheng, 1993). An antisense oligonucleotide with the complementary sequence can be synthesized if the gene's nucleotide sequence is known. The different genes that have been cloned are polygalacturonase (delay of fruit spoilage in tomato), chalcone synthase (modification of flower color in petunia), polyphenol oxidase (prevention of discoloration in fruits and vegetables) and starch synthase (reduction of starch content). In case of cloning of antisense polygalacturonase gene , a 730bp restriction fragment was obtained from the 5′ region of the normal polygalacturonase gene representing just under half of the coding sequence. To this fragment a plant polyadenylation signal was attached to the beginning and a CaMV 35 promoter was ligated to the end, i.e., ligation to control sequences in reverse orientation and this construct was inserted into the Ti-plasmid vector, p BIN19 which carries a gene for kanamycin resistance. Transformation was carried out by introducing the recombinant p BIN19 molecules into *A. tumefaciens* and then the transformed bacteria was allowed to infect tomato stem. Kanamycin resistant plants were identified and allowed to develop into mature plants. When the tomato plant was transformed the transgene produced an antisense RNA complementary to the first half of the polygalacturonase mRNA.

The presence of antisense 'gene' was checked by Southern hybridization and its expression was measured by northern hybridization with single stranded probe that would hybridize only to antisense RNA. The effect of antisense RNA synthesis on the amount of polygalacturonase mRNA in the cells of ripening fruit was estimated by northern hybridization with a second single stranded probe which is specific for the sense mRNA. Finally, the amounts of polygalacturonase enzyme produced in the ripening fruits were estimated by studying the intensities of the relevant bands after separation of the fruit protein by PAGE and by directly measuring the enzyme activities in the fruits. This study of transfer of antisense gene showed that the antisense RNA had not completely inactivated the polygalacturonase gene but had nevertheless produced a sufficient reduction in gene expression to delay the ripening of fruits. Thus ripening of tomato can be delayed by the transfer of an antisense construct of polygalacturonidase encoding gene fused to a constitutively expressed or preferably an inducible promoter.

Antisense drugs—Such drugs are planned to have unique ability to bind to targeted mRNA while avoiding attachment onto other proteins. Antisense oligonucleotides are generally 20-30 bases in length.

dd RNAi (DNA directed RNAi)—It is a technique used for the knock out of the specific genes and requires introduction of a special DNA construct into the target cells. This special DNA construct is then transcribed in the target cell and then gets converted into a double stranded RNA. This double stranded RNA is in turn cleaved into small interfering RNAs (siRNAs) which destroys the target mRNA. This technique does not provoke an interferon response in the target cells as can happen in case of RNAi. This technique is used to reduce the mRNA level rather than to abolish it and produces only transient rather than permanent effect.

23.19   VIGS (VIRAL INDUCED GENE SILENCING)

Many plant viruses produce ds RNA intermediate during replication and thereby triggers VIGS (Angell and Baulcombe, 1997). Infection of a plant with a recombinant virus containing a sequence from host nuclear gene may lead to silencing of that gene or dsRNA as an indicator of a potential viral invasion in the cytoplasm of mammalian cells triggers as so called intereferon response. The dsRNA (>30bp) binds to and activates protein kinase PKR and a 2, 5 –oligoadenylates synthetase (2, 5-AS). The activated PKR phosphorylates the translation initiation factor elF2B and thereby stalls translation whereas the activated 2, 5-AS causes mRNA degradation by 2, 5-oligoadenylate-activated ribonuclease L. All these reactions are not dependent on any particular RNA sequence.

23.20 TARGETTED GENE SILENCING VIA RNA

There are two strategies. An RNA virus engineered to carry a target sequence in its genome can be used as the delivery vector for the double stranded RNA signal in a process termed VIGS. In the second strategy there is introduction of the double stranded RNA signal via a suitably integrated transgenes carrying a highly transcribed inverted repeat of the target sequence. In both strategies the target sequence is typically an exonic segment of the target coding region which directs the interference process to the appropriate transcripts. As a transcribed inverted repeat transgene can promote methylation of identical sequences, it led to a different strategy for targeted gene silencing. In this strategy a promoter sequence rather than a coding sequence is used in the trigger inverted repeats. A promoter sequence can be delivered by an RNA virus. Thus the goal is to direct methylation to target promoter region so that it blocks transcription initiation. A well characterized example of this strategy involves a target transgene and a RNA trigger transgene. The target transgene carried nopaline synthetase promoter (NOS pro) driving expression of neomycin phosphotransferase (NPTII) and a RNA trigger transgene 35S driving expression of an inverted repeat of NOS pro sequences. NOS pro-NPTII transgene showed expression of transgene and kanamycin resistance but when plants are supertransformed with NOS pro inverted repeat transgene, NOS pro sequence on both transgenes becomes methylated and transcription of NPS pro-NPTII transgne is suppressed via promoter methylation. NOS pro inverted repeat transgene produces both double stranded RNA and diced siRNAs. RNA virus and transgene RNai system also produce these types of RNA.

23.21 RIBOSWITCHES

Riboswtiches are involved in RNA-mediated gene regulation in bacteria. Riboswitches are structured RNAs typically located in the 5′ UTR region of bacterial mRNAs which bind to metabolites and control gene expression. In other words, riboswitches are sequences of nucleotides bases in mRNA that contain structural domains. Most riboswitches sense one metabolite and function as a simple genetic switch. Many riboswitches are composed of two functional domains- an **aptamer,** acting as sensor and an **expression platform** that work in concert to selectively bind to a target ligand (specific small molecule building block or metabolite) and modulate gene expression. In nearly all examples identified in bacteria, binding of a single metabolite by the aptamer controls transcription termination or translation initiation via structural changes in the expression platform. In other words, on binding to metabolite aptamers under go a conformational change that alters the mRNA access to the machinery required for either its transcription from a gene or its translation into a protein. Thus ribositch regulates the intracellular levels of bacterial metabolites (Blencowe and Khanna, 2007). Some riboswitches balance the kinetics of metabolite binding and RNA folding with the speed of RNA transcription to control gene expression appropriately. However, most metabolite sensing RNAs are generally believed to be unsophisticated in structure and action compared with the diversity of gene control factors made of proteins. Some riboswitches are known to control gene expression by using complex structures and mechanisms such as glm S genes in *B. subtilis* controlled by metabolite responsive self-cleaving ribozymes.

23.21.1 Tandem Arrangement of Riboswitches

Some mRNAs contain two riboswitches in tandem. There are examples of two riboswitches that respond to S-adenosylmethionine and coenzyme B_{12}. This tandem arrangement of riboswitches yields a composite gene control system that function as a two-input Boolean NOR logic gate (Sundarsan et al., 2006). Simple RNA elements can be assembled to make genetic decisions without involving protein factors (includes 5′ regions of *B. clausii* met E mRNA). Another complex riboswitches class uses two aptamers in tandem to bind two glycine molecules co-operatively. This tandem aptamer assembly functions with far greater efficiency to changing concentrations of ligand than do simple riboswitches.

23.21.2 Genome-Wide Search for Riboswitches

The technique for genome-wide search for reboswitches is similar to that for genome-wide search for rebozymes discussed in section 23.5.

Defense system in prokaryotes:

23.22 RESTRICTION-MODIFICATION SYSTEMS

R-M systems occur exclusively in unicellular organisms mainly bacteria. Certain strains of bacteria inhibit or restrict the propagation of viruses grown previously on different strains. This phenomenon has been attributed to the presence of R-M system in the organism (Wilson, 1991). This system is designed to restrict phage or plasmid DNA molecules. R-M systems comprise pairs of opposing intracellular enzyme activities, an endodeoxyribonuclease (ENase) and a DNA methyl transferase (MTase). These enzymes interact with specific sequences of nucleotides in DNA. The sequences usually comprise 4 to 8 defined nucleotide and they can be continuous or interrupted, symmetric or asymmetric, unique or degenerate. The enzymes recognize double stranded DNA and a few also recognize single stranded DNA. ENases and MTases from the same system ('cognate enzymes) recognize the same sequences. In some R-M systems the two activities are combined into a single, multisubunit enzyme but in most systems they are different.

ENases catalyze double stranded cleavage of DNA. Cleavage occurs for each occurrence of the recognition sequence and is accompanied by hydrolysis of one phosphate deoxyribose bond in the backbone of each DNA strand. Cleavage generally occurs on 5′ side of the phosphate leaving DNA fragments with 5′ phosphoryl and 3′ hydroxyl termini. In many systems cleavage occurs at a fixed position with respect to the recognition sequence, either within the sequence or a few bases to one side of it while in others hydrolysis takes place at an indefinite distance from the recognition sequence. The endonucleases of R-M systems are sensitive to methylation of the recognition sequence. In other words, the foreign DNA may become resistant towards the restriction enzymes by modifications, i.e., its sequence specific methylation as is also all the cellular DNA.

Modification methyltransferases catalyze the addition of a methyl group to one nucleotide in each strand of the recognition sequence. S-adenosyl methionine (AdoMet) always serves as a methyl donor and is thus an essential cofactor for methylation. The methylated nucleotides lie within the recognition sequence but their identities and positions vary from MTase to MTase. Usually the same base is methylated on the both strands and adenine and cytosine are only bases known to be methylated. The products of methylations are N6-methyladenine, 5-methylcytosine and N4-methylcytosine. Methyl groups lie in the major groove of the DNA helix. Although both strands of the recognition sequence usually become methylated, cognate methylation of just one strand (hemimethylation) is sufficient to prevent cleavage. Usually methylation prevents cleavage. Cognate methylation, the kind conferred by the natural partner of a restriction endonuclease, always prevents cleavage. This is an adaptation to protect the cell's own DNA from digestion. Noncognate methylation which occurs elsewhere in the recognition sequence sometimes prevents cleavage but not always.

R-M systems have been classified into Type I, Type II and Type III depending upon the enzyme composition and cofactor requirements, recognition sequence symmetry and cleavage positions (Yuan and Hamilton, 1984). R-M systems are genetically diverse. Type I systems are least diverse and occur as families whose members have similar enzymatic subunits but different specificity subunits. Type II systems do not occur as families, may have often similar organizations and the enzymes usually have dissimilar sequences. Type II endonucleases of different specificities are entirely different and Type II methyl transferases are somewhat similar but similarities are skeletal and may reflect mechanistic constraints rather than common ancestry.

Both activities of R-M systems are required to form a restriction barrier and thus only systems inherited in whole confer any benefit. Genes for R-M systems have been identified and cloned. Interdependent functions are efficiently co-regulated and more likely to be inherited as a unit if their genes are closely linked than separate. Close gene linkage is a characteristic of all R-M systems. The genes are usually adjacent and often overlap.

23.23 ROLE OF RNA IN DNA REARRANGEMENT

DNA rearrangement refers to any structural change in nucleotide sequence, a gene or a chromosome. RNA molecules can also organize DNA rearrangements expanding the epigenetic influence of RNA beyond gene expression, and priming or directing DNA and RNA synthesis, editing, modification or repair. Genome-wide DNA rearrangements occur in many eukaryotes but are most exaggerated in Ciliates. During development of the somatic macromolecules *Oxytricha trifallax* destroys 95% of its germ line, severely fragmenting its chromosomes and then unscrambles (re-ordering or inversion) hundreds of thousands of remaining fragments by permutation or inversion. In other words, the formation of somatic nucleus from the germ line nucleus involves extensive genome rearrangements including DNA deletion, fragmentation and amplification, to generate a greatly altered genome. DNA or RNA templates can orchestrate these genome rearrangements supporting an epigenetic model for sequence-dependent comparison between germ line and somatic genomes. A complete RNA cache of maternal somatic genome may be available at the specific stage during development to provide a template for correct and precise DNA rearrangement (Yao, 2008). Nowacki et al. (2008) showed existence of maternal RNA templates that could guide DNA assembly and that disruption of specific RNA molecules disables rearrangement of the corresponding gene. Further, injection of artificial templates reprogrammes the DNA rearrangement pathway, suggesting that RNA molecules guide genome rearrangement. Thus a new gene order induced by injection is passed on to subsequent generations as predicted. This is thus a new form of inheritance whereby the somatic genomic information is inherited independently of the germ line genome through a transient 'RNA genome'. This system of inheritance will assist to inherit adaptive immunity which requires rearrangement of DNA in somatic immune cells- lymphocytes.

24

Functional Genomics and Function Genomics Technology

24.1 DEFINITIONS

The term **genome** refers to an organism's complete set of genes and chromosomes. 'Omics' an abbreviation was coined by J. N. Weinstein, refers to 'in context'. The term 'genomics' refers to the scientific discipline of mapping, sequencing and analysis of genomes (Thomas Roderick, 1986). The genome analysis has now been divided into 'structural genomics' and 'functional genomics'. The structural genomics deals with the construction of high resolution genetic, physical, and transcript maps of an organism and the ultimate physical map of an organism is its complete DNA sequence. It also deals with determination of the three-dimensional structure of all macromolecules and the objective here is to reduce the cost of structure determination. The term functional genomics refers to the development and application of global (genome-wide or system-wide) experimental approaches to assess gene function by making use of the information and reagents provided by structural genomics (Hieter and Boguski, 1997). In other words, functional genomics refers to determining the role of DNA and RNA in the progression from information (DNA, the gene) to function (protein). The objective in functional genomics is to determine the function of all the transcribed sequences (all genes) in genome (the **transcriptome**) and all the proteins that are produced (the **proteome**).The transcriptome includes all RNA synthesized in an organism including protein-coding, non-protein coding, alternatively spliced, alternatively polyadenylated, alternatively initiated, sense, antisense and RNA-edited transcripts. Non-coding RNA constitute a major functional output of the genome. In other words, the collection of genes that are expressed or transcribed from genomic DNA, sometimes referred to as the expression profile or transcriptome. In addition to their roles in protein synthesis (rRNA and tRNAs) non-coding RNAs have been implicated in genomic imprinting and perhaps more globally in control of genetic networks. The term **'proteome'** was coined to refer to the total protein complement expressed by the genome. In other words, it refers to the complement of all the proteins produced by a particular cell. **Genomics** decode sequence information of an organism and provides the 'parts catalog' while **proteomics** attempts to elucidate the functions and relationships of the individual 'parts' and predict the outcomes of the molecules they form on a higher level. In other words, genomics deals with the study of the set of genes in genomes whereas proteomics deals with the analysis of proteins. Since protein is closer to biological function than DNA much emphasis has been devoted to the development of new tools for proteomics. It involves the use of high-throughput methods for the study of large number of genes (ideally the entire set) in parallel. Genomics, functional genomics and proteomics provide a way out to find what proteins are expressed by a microbe during infection and which of those are essential to microbes for survival so they can be targeted for developing new drugs. Selection of a few likely genes will be done using bioinformatics tools. Determine whether certain genes occur among a number of related pathogens. Data mining and new developments in bioinformatics software will be used for this purpose (see Bioinformatics by Roy, 2009). Thus we have three levels of genomic information:

i.   **Chromosomal genome**-the genetic information common to every cell in the organism

ii. The expressed genome (or **transcriptome**), that part of genome which is expressed in cell at a specific stage during its development thus transcript map of human or plant genome can be constructed. In case of human transcript map with more than 3235 expressed sequences have been localized.

iii. The **proteome**, the protein molecules that interact with internal and external environments to give the cell its individual character.

24.1.1   Gene Function

Gene function can mean the following types of functions:

1. Biochemical function (e.g., protein kinase) – It reflects the direct behavior of a molecule (s) whereas biological role is used to describe the consequence (s) of this function for the organism. Genome analysis techniques focus on biochemical functions but no necessarily on biological role. Further, biochemical function is amenable to large scale high throughput methods whereas the biological role is difficult to assay on large scale.

2. Cellular function (e.g., a role in signal transduction pathway)

3. Developmental function (e.g., a role in pattern formation) or adaptive function (the contribution of gene product to the fitness of the organism).

Thus assessment of function of gene product is a multidimensional endeavor and one may ascertain a number of properties including structure, the low level function of a protein (i.e. kinase, protease, etc) and a high level function describing the biological processes in which the protein participates.

Organisms store instructions for their own existence in DNA. Functional information is encoded in genome including gene, regulatory and structural elements. The functional sequences- genes and **regulatory** elements tend to be conserved across all species.

24.2 APPROACHES TO STUDY OF FUNCTIONAL GENOMICS

It refers to the study of component parts of genome- the individual genes and proteins. Thus the objective is to identify parts and understand their functions (Functional genomics). There are two systematic approaches to the study of functional genomics (Vukumirovic and Tilghman, 2002). The first is called the 'reductionist approach' and is the traditional approach. It is a powerful strategy for the study of functional genomics. It refers to the study of function of single gene at a time, i.e. assigning some element of function to each gene at a time (single gene approach). In other words, in this approach each part of genome is individually examined and thus is a small-scale study. Some parts (or genes) are easily recognizable as they have been encountered before and include the previously studied genes. Some can elicit an educated guess as their function because of their similarity to another gene. That is some genes showed resemblance to known genes, ie. genes had some match to proteins and sequences of random cDNAd (ESTs) from many other organisms and provided clues to the function of previously unstudied genes and the rest are **novel genes** with no match to any previously encountered genes in the DNA sequence. The second approach is called 'whole genome approach' and through this approach functions can be assigned to different genes simultaneously. This expansion from single gene to genome-wide analysis has been made possible because of the avaialbality of the whole genome sequence information and high throughput advances in experimental techniques. To understand the nature of complex biological processes such as development one must determine the specific gene expression patterns and biochemical interactions within an organism but equally important is to seek out the organizing principles that allow them to function in a coherent way. The difference between **pre-genomic** era and **post-genomic era** is that one can keep track of all components at once. Post-genomic refers to time after sequencing of human genome, ie. after 2001 when it should be the date of publication of sequence of genome of the first bacterium *Haemophilus influenzae* year 1995. It refers to be the time for genome-wide transcriptomics, proteomics and metabolomics. Metabolomics (or metabonomics) is employed towards the understanding of global systems biology (Nicholson and Wilson, 2003) and is defined as the quantitative measurement of the dynamic, multiparametric metabolite response of living systems to pathophysiological stimuli or genetic perturbation. Generally, functional genomics is regarded as the central topic of the post-genomic era and include bioinformatics. Further, as whole genome information (sequence) is available so one can identify the parts to understand their functions. Elucidation of the function of thousands of newly discovered genes is the prime goal of functional genomics. Whole genome approaches acquire functional information in the form of expression profiles (Lockhart and Winzeler, 2000), protein-protein interaction (Pandey and Mann, 2000),computational approaches (Eisenberg et al., 2000) and response to loss of function, sometimes called genetic fingerprinting (Lui et al., 1999; Smith et al., 1995).

Functions of genes can be studied at many different levels. One can identify an unknown gene as gene encoding enzyme, substrate specificity, intracellular localization and its target, pathways it affects and organs it affects. In case of organisms whose genome is known completely, one can monitor simultaneously potentially all events such as expression of genes at the RNA or protein level, all possible protein x protein interaction, all alleles of all genes that affect a particular trait or all protein binding sites in a genome.

While identifying the function of gene, one will have to see what its protein actually does in a living cell.The classical approach to study gene function is to inactivate the gene and see its effect. The function of gene can, in principle, be inactivated at any of several levels.Study of gene function by disruption of wild gene by targeted insertion has been successful in yeast and other lower eukaryotes but not able to achieve this in higher eukaryotes.Disruption of gene function at the level of RNA has met with mixed success.Use of antisense RNA can be made to block the expression of gene by preventing its translation of its sense transcripts. Finally, disruption of gene function at the level of protein, i.e., function of gene can be blocked at the protein level (Herskowitz, 1987).Use of antibodies against synthetic antigens based on predicted sequence of the gene product can be made to study gene function. The functional genomics thus involves the following types of analysis.

1. **Mutational analysis** — Mutational analysis can be carried out for examining gene function. Study of function of a gene through gene knock down (the reduction of gene's activity to a very low levels through various mechanisms (RNA interference) such that it can be conditionally expressed), **knock in** (disruption of gene by insertion of a sequence or mutation (s) that either activates the gene or restores its activity) and **knock out** (disruption of gene by insertion of DNA sequence or mutation (s) that abolish gene function). For judging the role and importance of a particular gene one will have to see what happens when sequence of that particular gene is altered by either random mutagenesis or controlled DNA manipulation (discussed in chapter 21). So one will be hunting for mutants in order to identify genes. Once genes identified and cloned further information can be obtained by controlled manipulation of the isolated DNA and its total knock out of gene function or **enhancement of gene expression** through increased copy number or attachment to a strong promoter. An imbalance of subunit concentration can have dire consequences for the proper formation of multiple protein structures such as the cytoskeleton and the histone scaffolding of eukaryotic chromosomes.

2. Determination of gene expression patterns of the expressed gene in a given cell, organ or organism at a time by microarrays (determining the expression profile of the gene, the sub-cellular location of the protein and the phenotype of a null strain lacking the protein)

3. Study of interactions between genes through transfer and integration of foreign genes which allows the study of the influence (s) on the activity of other gene (s).

4. Study of post-transcriptional events such as mRNA stability, frequency of translation of a specific mRNA, stability of protein product and protein-protein interaction (or **'interactome'**) of all cellular proteins (through two-hybrid analysis). There are proteins which interact with many other proteins such as histones, actin and tubulin.

5. **Localizing proteins** — A protein's location provides important clues about is function. Functional genomics technology involving microarrays and proteomics will provide added insights regarding gene function on the cellular level, thereby improving our ability to predict phenotypic effects of genes at the organismic level. The challenge of the so-called 'post-genomic' era is to extract biological information on a large scale from available genomic sequence data. The aim of the developmental geneticist would be to show exactly how information in gene can be used to construct the entire

organism. This endeavor includes annotation of genes to the genome (Stein, 2001) and large scale gene expression screens and might finally allow in conjugation with functional data to model and stimulate biological processes, i.e. **systems biology**. Systems biology may be broadly defined as the integration of diverse data into useful models that allow to easily observe complex cellular behaviors and to predict the outcomes of metabolic and genetic perturbations. Systems biology or more specifically network biology is driven by the gradual realization that a single gene is seldom accountable for a discrete biological function. In other words, the conditions, of cells and cellular processes are influenced by a large number of genes interwoven into networks rather than a few genes (Strohman, 2002). To understand individual biological components alone is not sufficient to discover the rules underlying complex biological systems. Systems biology focuses on understanding of the collective properties of biological networks and these networks in turn describe interactions between as many as thousands of unique elements. Dynamic systems are based on hierarchical control but there are also non-hierarchical systems such as embryos, brain, etc. A major goal of systems biology is to predict the function of biological networks. Although network topologies have been successfully determined in many cases, the quantitative parameters governing these networks generally have not. Systems biology approaches involve working out the underlying processes that can lead to changes in gene function and perhaps to discover processes and principles that will help us to understand the form and function of genes. The identification and characterization of systems-level features of biological organization is a key issue of post-genomic biology. In other words, a systems level study of genetic networks and identification of differentially expressed genetic networks holds the key to unraveling the relationship between genotype and phenotype (Xiong et al., 2004; Lu et al., 2005; Khalil and Hill, 2005). Reconstructing networks of biological entities such as genes, transcription factors, proteins, compounds and other regulatory metabolites is very important for understanding of the biological processes and the organizing principles of the biological systems (Barabasi and Oltvai, 2004). The tools of systems biology are well suited to investigate complex interactions. To understand complex biological systems what is requitred is the integration of experimental and computational research- in other words, a systems biology approach. Global transcription can be measured using cDNA microarrays and computational analyses of this information can lead to deeper understanding of the system as a whole.

24.3   BIOINFORMATICS APPROACHES

Classical bioinformatics approaches for assigning gene function include sequence homology, gene fusion events, gene order conservation and phylogenetic profiles (Huynen et al., 2000). Although these are extremely powerful approaches, one drawback is lack of experimental evidence. In addition, homology on its own may result in imprecise annotation or return weak similarity with well characterized genes. Thus complementary approaches are required to more precisely determine gene function.

DNA sequencing technology is a complementary approach for discovering genetic functions. By comparing DNA sequence information from different organisms one can identify sequences that have remained relatively constant thoughout millions of years of evolution. The conserved sequence likely encodes an important function. Besides this two additional approaches are required to confirm and exhaustively identify all functions encoded by a natural DNA sequence: (i) Specific DNA sequences thought to affect phenotypes must be purposefully changed and the expected affect confirmed and (ii) also, seemingly irrelevant DNA sequences must be removed, disrupted or otherwise modified and shown to be unnecessary.

24.4 HIGH THROUGHPUT APPROACHES FOR DETERMINING GENE FUNCTION

HTP approaches being developed for biological experimentation include the following.

1. mRNA expression profiling
2. Determination of gene-deletion phenotypes
3. Cellular localization of proteins
4. Protein-protein interactions
5. Assay for biological activity using protein arrays
6. Synthetic lethal screens and
7. RNAi screens

When these approaches are used on their own such HTP data sets have already proved their usefulness for inferring hypotheses about gene function. Combining HTP datasets is a powerful approach for functional genomic analyses. Combining several HTP data types will likely to aid gene function prediction.

Detection of location of mRNA and its product—Oligonucleotides or antibody probes can be obtained for locating the mRNAs and their products of their translation.

Study of structure and function of promoter—The structure and function of the promoter that control specific patterns of gene expression can be investigated through the use of **reporter constructs** and are discussed in chapters 16 and 24. Putative promoter sequences are joined to an ORF encoding an easily quantifiable enzyme such as E.coli Lac Z (β-galactosidase) or Chloramphenicol transcacetylase (CAT) and introduced into the organism by transformation procedure.

24.5 ARRAY TECHNOLOGY

An array refers to an ordered arrangement of the nucleic acids, proteins, small molecules or cells which allow parallel (or simultaneous) analysis of biological or chemical samples (usually on microscopic slides). Array technologies refer to designing arrays based on specific sequence information, a process sometimes called 'down loading the genome onto chip'. Traditionally the array consisted of fragments of DNA, often with unknown sequences spotted on a porous nylone membrane. The arrayed DNA fragments often came from cDNA, genomic DNA or plasmid libraries and hybridized material was often labeled with a radioactive group. DNA molecule was attached at specific locations on a surface. In other words, nucleic acid arrays work by hybridization of labeled RNA or DNA in a solution to DNA molecules attached at specific locations on a surface. The hybridization of a sample to an array is in effect, a highly parallel search by each molecule for a matching partner on an 'affinity matrix' with eventual pairing of molecules on the basis of recognition. Use of glass as substrate and fluorescence for detection, together with development of technologies for synthesizing or depositing nucleic acids on slides at very high densities have allowed the miniaturization of nucleic acid arrays with simultaneous increases in efficiency and information content. Several developments have taken place in the basic technical aspects of the array technology. The hybridization reaction may be driven by an electric field; other detection methods besides fluorescence can now be used; and surface may be made of materials other then glass such as plastic, silicon, gold, a gel or membrane or may even comprised of beads at the ends of optic fibre bundles. DNA microarrays, oligonucleotide arrays, Gene chip arrays or simple chips are high density arrays of nucleic acids on glass. An array of oligonucleotides complementary to subsequences of a target sequence can be used to identify a target sequence, measure its amount or relative expression level and detect differences between the target and a reference sequence. **High density oligonucleotide array** is a general term for any high density chip onto which hundreds of thousands or even millions of oligonucleotides are synthesized by a photolithographic method.

Ultra high-density microarray—It refers to any of a series of microarrays which contains millions of different immobilized probe molecules. For example, a so-called nanoarray produced by dip pen nanolithography (DNP) may harbour 50,000 dots (low resolution DPN) or even 100 million of dots (high resolution DPN) instead of a single dot of a conventional microarray. Arrays besides being used in gene expression analysis can be used to collect much of the data obtained presently by Southern or Northern blot hybridization.

Whole genome oligonucleotide array—It refers to any microarray that consists of thousands of oligonucleotide probes that are regularly spaced along a genomic DNA but not fully cover this genomic DNA. Such arrays serve to discover transcribed sequences that are not detectable by annotations of known exons.

Limitation with microarray

The limitation with microarray is that only genes/transcripts that can be studied are those that have already been identified by the sequencing

24.5.1 Principal Types of Array Used in Gene Expression Monitoring

Nucleic acid arrays are produced in one of the two ways, either by **robotic deposition** of nucleic acids (PCR products, plasmids or oligonucleotides) onto glass slide or *in situ* synthesis (using **photolithography**) (Fodor, 1997) of oligonucleotides. In the technique of photolithography, about 10^7 copies of each selected oligonucleotide (usually 20 to 25 nucleotides in length) are synthesized base by base in hundreds of thousands of different 24μm x 24μm areas on a 1.28cm x 1.28cm glass surface. Because the array is constructed on glass, it can be inverted and mounted in a temperature-controlled hybridization chamber. A target sequence is fluorescently tagged and then injected into the chamber where the target hybridizes to its complementary sequences on the array. In case of robotic deposition approximately one ng of the material is deposited at intervals of 100-300μm. Typically for oligonucleotide arrays, multiple probes per gene are placed on the array whereas in case of robotic deposition a single longer (up to 1,000bp) double stranded DNA probe is used for each gene or EST. In both cases probes are usually designed from sequence near the 3′ end of the gene (near the polyA tail in eukaryotic mRNA) and different probes can be used for different exons. After hybridization of the labeled sample (typically overnight) the arrays are scanned and the qualitative influorescence image along with the known identity in the probes is used to assess the 'presence' or 'absence' (more precisely the detectability above thresholds based on the background and noise levels) of a particular molecule (such as a transcript) and its relative abundance in one or more samples. The system thus consists of chips, a hybridization station to control hybridization and a reader and software to access the chip data.

Because the sequence of the oligonucleotide or cDNA at each physical location (or address) is generally known or can be determined and because the recognition rules that govern hybridization are well understood, the signal intensity at each position gives not only a measure of the number of molecules bound but also the likely identity of the molecules. Although oligonucleotide probes can vary systematically in the hybridization efficiency, qualitative estimates of the number of transcripts per cell can be obtained directly by averaging the signal from multiple probes.

The information obtained from spotted cDNA arrays gives the relative concentration (ratio) of a given transcript in the two different samples (derived from competitive, two-color hybridization). mRNA present at a few copies (relative abundance of ~ 1 : 100,000 or less) to thousands of copies per mammalian cell can be detected and changes as subtle as 1.3 to 2 can be reliably detected if replicated experiments are carried out. Thus DNA microarrays or DNA chips can be used to determine expression patterns of different proteins by detection of mRNAs or for genotyping by detection of different variant gene sequences including but not limited to SNPs. It provides measure of simple presence or absence (qualitative measurement) or quantitate relative abundance (quantitative measurement). Finally, the key to high-throughput analysis is to run many hybridization experiments in parallel and the DNA microarray achieves this. Gene expression is usually measured on a genome-wide scale using DNA microarrays which provides a tool for exploring the regulation of thousands of genes simultaneously. It allows the identification of co-regulated genes likely controlled by common regulatory mechanisms and thus the entire regulatory network of the genome of an organism can be dissected.

24.5.2   Combinatorial Technology

Solid phase chemistry, photolabile protecting groups and photolithography have been combined to achieve light-directed spatially and addressable paralle synthesis to yield a highly diverse set of chemical products. Foder et al. (1991) described a method to use light to direct the simultaneous synthesis of a many different chemical compounds. Synthesis occurs on a solid surface. The pattern of exposure to light or other form of energy through mask or by other spatially addressable means determines which regions of the support are activated for chemical coupling. Actvation by light results from the removal of photoliable protecting groups from the selected areas. After protection, the first of a set of 'building block' (eg. amino acid or nucleolic acids, each bearing a photolabile protecting group) is exposed to the entire surface but reaction occurs within regions that were addressed by the light on the preceding step. The substrate is then illuminated through a second mask which activates a different region for reaction with a second protected building block. The pattern of masks used in these illuminations and the sequence of reactants define the ultimate products and their locations. The number of compounds that can be synthesized is limited only by the number of synthesis sites that can be addressed with appropriate resolution. Combinatorial masking strategy can be used to synthesize a large number of compounds in a small number of chemical steps.

In the light directed chemical synthesis the products formed depend on the pattern and the order of masks and the order of reactants. For example, consider the four step synthesis with the ordered set of reactants as {A, B, C, D}. In the first cycle illumination through mask m1 activates the upper half of the synthesis area. Building block A is then added togive the product distribution A, A. Illumination through m2 which activates the lower half, followed by addition of B yields the next intermediate stage B, B. Building block C is added after illumination through m3 which activates the left half and D is added after illumination through m4 which activates the right half to yield the final product set {AC, AD, BC, BD}. Binary synthesis generates maximal number of products (2^n) for a given number of chemical steps (n). With four reactants, 16 compound are produced in the binary synthesis whereas only four are made when each round as two reactants. Further, in 10 steps an array of 1024 peptides can be synthesized and a 20steps binary synthesis yields 1,048,576. High density arrays formed by light directed synthesis are potentially rich sources of chemical diversity for discovering new ligands that bind to biological receptors and for elucidating principles governing molecular interactions.

24.5.3   Other Applications of Array Technology

Arrays can be used to collect much of the data that are obtained presently by Southern or northern blot hybridization techniques but in parallel fashion. The different applications are as follows.

1. Genomic DNA samples can be manipulated to select for particular regions before hybridization to obtain specific types of information. For example, location of hundreds of regions of origins of replication in yeast can be determined simultaneously by enriching for early-replicating regions using a variation of the Meselsohn and Stahl procedure and then hybridizing the resulting DNA to full genomic arrays.

2. Protein binding sites on DNA can be identified as probes for more intergenic regions are synthesized on arrays. Fragmented chromatin can be crosslinked to a protein and then immunoprecipitated with an antibody to that protein. The DNA fraction of the immunoprecipitated can be labeled and hybridized to identify the approximate location of the binding site (see chapter 25 for detail).

3. Full genomic arrays can also be used in the analyses of plasmid libraries in genetic selections such as a two-hybrid screen or in principle, for any type of expression in which the information is contained in the form of RNA or DNA.

4. The single stranded DNA arrays can be converted into double stranded DNA arrays to characterize interactions of protein and potentially other types of molecules with double stranded DNA.

Microarray analysis is a high-throughput method to measure abundance of multiple species of target DNA by simultaneous hybridization to an array of DNA probes. When the target DNA is cDNA corresponding to gene expression, it measures transcriptomic state of a cell under an experimental condition. When the target DNA is sampled from genomic DNA, it measures copy-number variations within a genome as polymorphisms or as chromosomal aberrations, for example, in case of tumor genome (Pollack et al., 1999). Finally, when the target is genomic DNA selected by immunoprecipitation with a protein, it identifies those regions of genome which interact with proteins such as TFs thereby elucidating regulatory genetic controls (Ren et al., 2000). All these applications use comparative methods. In a two-color scheme, simultaneous array hybridization detects target DNAs of two different experiments which are labeled with different fluorescent dyes. Target DNAs which have differential behavior from one experiment to the other are called 'enriched'. These are the objects sought after in these experiments. Enriched targets are found in two steps. The first step involves normalization of data which is obtained through transformation of measurements. This is done to assign similar local means (or medians) to 'presumed unenriched targets'. The purpose of transformation is to compensate for experimental sources of variation like dye-specific effects and hybridization unevenness in DNA arrays (Smyth et al., 2003; Buck and Leib, 2004). In the next step, the normalizaed data is subjected to statistical analysis aiming at identification of enriched targets which truely differ between the two analyzed DNA samples (Lerman, et al., 2007).

In practice, the targets are measured optically in terms of a raw intensity value and analyzed after logarithmic transformation. In case of just two samples, for every target the logarithms of intensities (according to two the samples) are averaged to generate an A value (log of their geometric mean) and subtracted to generate an M value (log of their ratio) and then plotted in an M vs A plot. The majority of the targets will belong to a stable unenriched set of targets and after suitable normalization their M values will be close to zero. The normalized M values of the enriched targets will lie either significantly above or below zero but not necessarily with any known distributions or even symmetry.

24.6 QUANTITATIVE MONITORING OF GENE EXPRESSION PATTERN WITH cDNA MICROARRAY

The temporal, developmental, topographical, histological and physiological patterns in which a gene is expressed provides clues to its biological role. cDNA sequences from many organisms present opportunity for defining these patterns at the level of whole genome. To fully understand gene expression, gene function and subtlelities of regulation, the quantitative levels of expressed genes under various conditions must be assayed. In addition, if a quantitative 'snapshots' of gene expression can be captured, the dynamics of cellular pathways can then be deciphered. Differential gene expression can be investigated with a simultaneous, two-color fluorescence hybridization scheme which serves to minimize the experimental variation inherent in the comparison of independent hybridizations. Microarrays prepared by high-speed robotic printing of cDNAs on glass are used for quantitative expression measurements of the corresponding genes. Fluorescent probes are prepared from two mRNA sources with the use of reverse transcriptase in the presence of fluorescein- and lissamine-labeled nucleotide analogs, respectively. The two probes are then mixed together in equal proportions, hybridized to a single array and scanned separately for fluorescein and lissamine emission after independent excitation of the two fluorophores. Quantitation of both scans revealed a range of expression levels spanning 3 orders of magnitude for 45 Arabidopsis genes tested (Schena et al., 1995). RNA blots for several genes supported the expression levels measured with the microarray to within a factor of five.

In this technique 45cDNAs of Arabidopsis averaging ~ 1.0kb were amplified with PCR and deposited into individual wells of a 96-well microtiter plate. Each sample was duplicated in two adjacent wells in order to allow the reproducibility of the arraying and hybridization process to be tested. Samples from the microtiter plate were then printed onto glass microscope slides in an area measuring 3.5 x 5.5 nm^2 with the use of high speed arraying tool. The arrays were processed by chemical and heat treatment in order to attach DNA sequences to the glass surface and denature them. Three arrays printed in a single lot were used for the experiments. Fluorescent probes were prepared from the total mRNA by a single round of reverse transcription. The mRNA was supplemented with human acetylcholin receptor (AChR) mRNA at a dilution of 1 : 10,000 (w/w) before cDNA synthesis to provide an internal standard for calibration. The resulting fluorescent labeled cDNA mixture was then hybridized to an array at high stringency and scanned with laser. Calibration relative to the AChR mRNA standard established a sensitivity limit of ~ 1:50,000. No detectable hybridization was observed to with the rat glucocorticoid receptor or the yeast TRP4 targets even at the highest scanning sensitivity. A high sensitivity scan gave signals that saturated the detector at nearly all of the target sites whereas a moderate sensitivity scan of the same array allowed linear detection of more abundant transcripts.

24.6.1   Uses of Microarray Data

Microarray data has important applications in biology and clinical studies. DNA microarray experiments are conducted to monitor the expression of a large number of genes (measuring the concentration of the corresponding mRNAs) under various conditions. Thus it provides the possibility and opportunity to understand gene's behaviors and functions indirectly and to find gene pairs with different kind of relationships and group genes together with similar biological function which usually show similar expression profiles against the time series and at least to construct a biological network. In other words, with increasing availability of RNA expression microarrays, the crurrent focus is now on elucidating networks of genomic regulation.

Identification of groups of co-expressed genes across diverse microarray data is although very promosing for assessing higher level function of gene products but such analysis is complicated by the fact that co-regulation is often condition-specific and may not extend across all conditions. The problem of context-specificity can be particularly pronounced when combining gene expression profiles across different experiments, tissue types or even different organisms to perform metacluster analysis. In these situations, measurements of gene expression under all conditions are not necessarily informative with regards to their co-regulation. Ignoring the local nature of co-regulation significantly reduces one's ability to detect co-regulated genes owing to the noise generated by non-informative measurements. The proposed solutions to this problem in terms of context-specific Bayesian networks and more general module networks rely on the specification or estimation of the 'correct' number of patterns (Liu et al., 2006).

The data generated from a set of microarray experiments is usually expressed as a large matrix with the expression levels of genes in rows and experimental conditions ordered in columns. Most of the microarray data analyses algorithms such as gene clustering, disease (experiment) classification and gene network design require the complete information. Thus microarray data can be used for quantitative as well as qualtitative analysis. It can also be used for classification, i.e. identifying types or subtypes of cancers. Machine learning techniques have been applied to microarray data for binary classification. SVMs and other supervised learning techniques use a training set to specify in advance which data should cluster together. SVM starts with a set of genes that have a common function, for example, genes coding for ribosomal proteins or genes coding for components of the proteasome. Further, a separate set of genes that are known not to be members of the functional class is specified. These two sets of genes are thus combined to form a set of training examples in which the genes are labeled positively if they are in the functional class and are labeled negatively if they are known not to be in the functional class. Sets of training examples can be assembled from literature and databases. Using this training set an SVM would learn to discriminate between the members and non-members of a given functional class based on expression data. Once the SVM has learnt the expression characteristics of the class, it would then recognize new genes as members or non-members. In other words, SVM uses the biological information contained in the training set to determine what expression features are characteristic of a given functional group and use this information to decide whether any given gene is likely to be a member of the group (Brown et al., 2000). SVM has been successfully applied to microarray data. SVMs have advantages over hierarchical clustering and SOM in that although all three methods employ distance (or similarity) functions to compare gene expression, SVs are capable of using a larger variety of interactions. SVM can take into account correlation between gene expression measurements. Further, SVM takes advantage of prior knowledge in making distinction between one type of gene and another. In case of an unsupervised method when related genes end up far apart according to distance function, the method has no way to know that the genes are related. Amongst machine learning tools neural network and SVM are most notable. SVMs are a class of supervised learning algorithms first introduced by Vapnik (1998). Given a set of labeled training vectors (positive and negative input examples), an SVM learns a linear decision boundary to discriminate between the two classes. The result is a linear classification rule that can be used to classify new test examples. The use of SVM is a prevalent technique for data classification based on linear decision rules (Vapnik, 1995, Burges, 1998;Boser et al., 1992). SVM takes as input i.i.d (independent and identically distributed) training samples $(x1, y1), \ldots\ldots, (xn, yn)$ where xi represents the sample attributes and yi {-1, +1) the class. SVMs will find a hyperplane separating the training instances by their classes and maximizing the distance from the closest examples to the hyperplane (maximum-margin hyperplane). The classification of a sample will be determined by the sign of the function.

$$F(x) = w^T x + b$$

where w and b are the parameters of the hyperplane. The examples closest to the hyperplane are called **support vectors** and are crucial for training. For many training sets it will not be possible to separate samples by a linear function in the original feature space so training instances are mapped to a higher dimensional pace by a function of Φ. SVM will then find a linear maximum-margin hyperplane in this higher dimensional space. For solving this problem it is not essential to directly define the mapping into higher dimensional space but it is sufficient to give the dot product of two instances in this space.

K (xi, xj) = Φ (xi)$^T\Phi$ (xj) is called a **kernel function**. Commonly used kernel functions comprise linear, polynomial, sigmoid and radial bases function. Molecular biology applications of the SVM have included gene expression classification, protein classification, protein-fold recognition and prediction of protein solvent accessibility, SNPs, protein secondary structure, protein quarternary structure and T-cell epitopes. SVM has recently been applied to protein-protein binding site prediction.

24.6.2   Identification of *Cis*-Regulatory Elements

For understanding biological systems at the molecular level the knowledge of gene sequences and function is important. Also important is the identification and characterization of the regulatory elements they govern the temporal and tissue-specific expression of any individual gene. In genomic projects, major challenges include deciphering gene function and regulation. Regulatory information is buried in the non-coding regions that constitue 95% of mammalian genomes. Testing the value of sequences by transgenesis is an important approach for finding regulatory information. Gene expression data can be used to identify new *cis*-regulatory elements (genomic sequence motifs which are overrepresented in the genomic DNA in the vicinity of similarly behaving genes) and regulons (sets of corregulated genes), basic units of the underlying cellular circuitry. The correlation between the presence of specific sequence motifs in promoter region and gene expression patterns may be stringer than the correlation between functional categories and gene expression patterns. For example, more than 50% of the genes in yeast which are transcribed in a cell cycle specific manner and whose transcript abundance peaks in the G1 phase of cell cycle have an Mlu cell cycle box within 500bps of their translational start site. Identification of regulatory elements is based on the principle of evolutionary conservation of functional binding sites in promoters. There are two approaches to assess such functional conservation. The first approach involves comparison of promoters from orthologous genes in several species (e.g. man, mouse, dog or mammals). Such analysis usually results in a conserved framework of about 3 to 8 binding sites which can be used for further evaluation. The second approach is the horizontal conservation. There are sets of genes within the same organism that are coupled functionally, i.e. by co-expression (Werner, 2001). The difference between co-regulation and mere co-expression is of great importance for this strategy. Co-regulated genes usually share partial promoter features, so-called modules responsible for the observed co-regulation. Co-expressed genes which just show up at the same time but are not co-regulated do not necessarily share such modules. Therefore, they may interfere with comparative sequence analysis and should be removed first. There are several possible ways to focus co-regulated rather than co-expressed genes.

Gene products that are involved in the same molecular process must be co-ordinately expressed. Transcriptional co-expression is achieved by regulatory proteins and their target *cis*-regulatory elements which promote gene transcription in overlapping spatiotemporal distributions. The function of a *cis*-element is encoded in its **molecular architecture**- a nucleotide sequence with instantiations (motifs) of the binding sites (motif types) for one or more transcription factors arranged with functionally significant motif combinations, orientations or spacing. As motifs are the functional units within a *cis*-element, analysis of a *cis*-element's molecular structure requires quantification of each motif's activity in a sufficiently large number of co-expressed genes whose functions have been maintained throughout evolution. Brown et al. (2007) studied the functional architecture of 19-*cis*-elements by mutagenesis coupled with a whole embryo expression assay.19 *cis*-elements came from the 19 genes which were found to be co-expressed in 36 muscle cells in Ciona. The three motifs that mediate muscle specific transcription in Ciona in general and of these loci in particular are the cyclic adenosine 5'-monophosphate response element (CRE), the MyoD motif and the Tbx6 motif. Each reporter construct harboring specifically mutagenized sequences was transfected into hundreds of developing embryos. Further, each *cis*-element was dissected using constructs with small deletions (5 to 10bp) or site directed mutants which removed putative motifs in isolation or different combinations. Spacing, order and relative orientation of motifs might explain the functions of individual motifs or of the element as a whole. They found these elements to be built from motifs of widely varying activity from different combinations of motif types and in divergence arrangements. Co-expression is driven by regulatory motifs of broadly varying activity assembled into a diverse array of *cis*-elements. In other words, individual motif activity ranged broadly within and among elements and among different instantiations of the same motif type. The activity of orthologous motif is strongly constrained although motif arrangement, type and activity varied greatly among the elements of different co-regulated genes. Polymorphisms in *cis*-elements will range in phenotype depending on the amount of activity that the affected motif contributes to the function of its element. They further conclude that a polymorphism in *trans*-acting factor would not affect expression of all targets equally but will instead have a target specific effect whose magnitude will be determined the architecture of the target's *cis*-element.

Comparative sequence analysis of such a set of co-expressed genes usually reveals the promoter module responsible for the observed co-expression and not a complete model. If the same promoter is analyzed in combination with different sets of expression related genes different modules may be found. New *cis*-regulatory elements may also be revealed by examining classes of co-regulated genes. Thus boundaries and all functioning sequence variants of *cis*-regulatory elements can be predicted using large gene expression data which are otherwise predicted using the more conventional approach–using site-directed mutagenesis (**promoter bashing**) (see chapter 21 for detail).

24.7　TRANSCRIPTIONAL NETWORK MODELING

Transcriptional regulatory networks specify the interactions among regulatory genes and between regulatory genes and their target genes. Its aim is to decipher the cellular regulation system. In other words, it helps to understand the underlying mechanism of complex cellular processes and responses. Two major tasks are involved in this type of modeling. 1. Finding the target genes for each transcription factor and 2. Correlating each transcriptional factor to its target transcripts as condition varies. The first task specifies the network configuration. Several methods are available for this. Computational approach includes the inference of transcription factor binding targets by drawing information from the TF binding motifs and gene expression dynamics. A more direct approach is experiment-based genome-wide location analysis or cross-linking chromatin immuno-precipitation (ChIP) which profiles each TF for its binding sites over the entire genome. This analysis is thus used to investigate how TFs bind to promoter sequences across genome and then construct TF-promoter binding networks to infer transcriptional networks. Location analysis provides *de novo* evidence of TF binding to genes. However, physical binding does not directly imply transcriptional regulatory activities. Further, location analysis experiments are often restricted to certain growth conditions and so TF-promoter binding network structures specific to other growth conditions may not be observed.

Combining ChIP data with microarray data can give more interpretable network connectivity estimates. It also elucidates the relationship between a TF's activity and the abundance of its target transcripts.

24.7.1. Combinatorial Control

Transcription is a complex process which involves in addition to basal transcription factors and RNA polymerase II, binding of many other proteins to specific sequences in its promoter and enhancer region of a genes. The transcription levels themselves are the result of a complex network of influences exerted by the genes on each other through gene products and signaling pathways. These sequences are conserved among different genes regulated by the same signal but variability in regulation is provided by combination of two, three or more sequences acting in concert in a modular fashion. In other words, genes are typically expressed in modular manners in biological processes. As a critical level of biology hierarchy, functional modules are cellular entities that perform certain biological functions which are relatively independent from each other (Barabasi and Oltvai, 2004). The modular nature of *cis*-elements is paralleled by a diversity of gene-specific transcription factors which may also form homo or heterodimers. The combinations of *cis*-elements and TFs provide diversity which is required for the transcription of different genes and of the same gene by a variety of signals in a tissue- and/organ-specific manner. There are many examples of the *trans*-acting protein product of one gene binding to the *cis*-acting segment of another in both prokaryotes and eukaryotes. However, in many of the prokaryotic and majority of the eukaryotic systems, the *trans*-acting protein acts as an activator of the transcription rather than as a repressor (or sometimes either as an activator or a repressor depending on the conditions). Frequently, more than one transcriptional activator is necessary for fully regulated gene transcription. There is a limited number of genome-wide regulatory motif but there is a large number of transcriptionally regulated process. This suggests regulation to be dependent on combinatorial control. Combinatorial control thus refers to a type of control in which combinations rather than an individual protein control a cellular process. It is accomplished to a large extent by combinations of *cis*-elements and heterodimeric regulatory proteins.

TFs can act alone or in combination to target promoters to control gene expression. The TF regulatory activity is controlled by higher-level cellular functions such as signaling pathways to reflect cellular physiology and environment, typically through post-transcriptional regulation such as phosphorylation, oligomerization, ligand-binding or subcellular translocation. Ativated TFs control transcriptional initiation through interaction with DNA or RNA polymerase. Thus TFAs in principle can be determined from the transcript levels of the genes they control. In the simplest case where one TF controls one gene, the activity of this TF is directly proportional to the gene it controls. In reality, however, TF-promoter connectivity is more complicated and requires advanced mathematics for deconvolution (Galbraith et al., 2006). Network component analyses, NCA (Liano et al., 2003; Tran et al., 2005) is a model-based decomposition method to deduce such information, namely, TFA and regulation control strength (CS) from transcriptome data and TF-promoter connectivity network. Connectivity networks are constructed by preprocessing CHIP-chip data, DamID methylation or extensive literature. NCA formulates gene expression as the product of the contribution of each regulating TFA using a combinatorial power-law model which can be viewed as a log-linear approximation of any non-linear kinetic system in multiple dimensions. Data decomposition can find independent features of related biological processes. The PCA (principle component analysis) and ICA (independent component analysis) prove useful for mapping highly co-ordinated transcriptional responses such as the cell cycle. ICAis an extension of PCA and like PCA, ICA can remove all linear correlation. By introducing a non-orthogonal basis, it also takes into account higher order dependencies in the data. Dimensionality reduction facilitates the classification, visualization, communication and storage of high dimesional data. The PCA finds the directions of greatest variance in the data set and represents each data point by its co-ordinate along each of these directions.

24.8 USE OF GENE SETS FOR RECONSTRUCTION OF PATHWAYS

Statistical methods have been developed for the analysis of microarray data based on single genes. Single gene–based methods do not take into account the interactions (dependencies) among genes and are often prone to multiple testing problems. On method would be to look at gene sets rather than all the genes at once or one at a time. Gene groupings can be based on GO. Pathways are sets of genes that serve a particular cellular or physiological function. Ranking pathways relevant to a particular phenotype, ie. cancer, will help focus on a few sets of genes.

24.9 STATISTICAL DATA MINING TOOLS FOR DISCOVERING GENE NETWORKS

These tools include reverse engineering approaches (Somogyi et al., 1997; Liang et al., 1998), differential equations and Bayesian networks. These approaches require large number of time-course data or rely on very greedy computational strategies. Some computation methods integrate gene expression data, DNA sequences and functional annotations into a comprehensive framework for discovering transcriptional regulatory networks. These methods allow one to infer motif-to-gene or to some extent gene-to-gene regulatory networks. Gene regulatory network states the interactions among the genes encoding transcription factors, as determined in an extensive perturbation analyses along with other data. Xiang and Lann (2005) described a statistical approach for constructing TRNWs using gene expression, promoter sequence and TF binding site data. In this approach the TFs that are significantly associated with changes in gene expression profiles under each experimental condition are identified first and then the strength with which a regulatory gene regulates a potential target gene is estimated and finally, average evidence across experiments are used to infer the TRNW structure.

The standard approach for modeling a known network is through a set of coupled differential equations (e.g. Voit, 2000). These equations describe how the concentrations of the various products evolve overtime. Although probably the most realistic model of the actual reactions taking place, such a model requires knowledge of the network structure and the various reaction rates. The simplest approach is to cluster genes by their expression profiles. Genes with similar expression profiles are likely to regulate each other or can be regulated by another shared parent gene. A major drawback of this approach is that it is impossible to infer any notion of causality–which gene is regulated and which gene is regulating. Classifying genes into different functional groups is a first step in order to gain more complex knowledge of different pathways and/or functions.

24.10 PROBLEM WITH EXPRESSION DATA

Expression data for genome-wide expression analysis is obtained using DNA microarray and SAGE. It is difficult to interpret the data, for example, in case of study of function of brain as they do not resolve cellular diversity within these structures. There is increasing evidence of complex, often cell type specific transcriptional splice variation that significantly increases the complexity and size of the transcriptome whereas a significant proportion of the transcriptome is also dynamically regulated, for example, by circadian rythms. Circadian rythms are intrinsic time-keeping mechanisms conserved throughout the animal kingdom. Many aspects of animal behaviour and physiology including sleep-wake cycles, blood pressure, body temperature and metabolic pathways are controlled by the circadian clock. At the molecular level cellular rythms are generated and maintained through interconnected transcriptional-translational feed back loops of the clock genes which are connected. Circadian rythms control gene expression, stomatal opening and the timing of component of photoperiodism which regulates seasonal reproduction. **Circadian clocks** produce an internal estimate of time that synchronizes biological events with external day-night cycle. Cicadian clocks are adaptations to the daily rotation of the planet. In other words, circadian rhythms allow organisms to adapt to periodic variations in the terrestrial environment. Regulation of gene expression underlies ciccadian rhythms. Circadian rhythms originate from negative feed back exerted by a protein on the expression of its gene. In plants and cyanobacteria benefits occurs when the clock is resonant with the environment. It requires the oscillator to be robust yet flexible which may explain the evolution of molecular clock with multiple feedback loops. Thus transcriptional feedback loops are a feature of circadian clocks in both animals and plants (Dodd et al., 2007).

24.11 TECHNIQUES FOR ASCERTAINING THE FUNCTION OF GENES FROM EXPRESSION DATA

The following four general techniques have been used to ascertain the function of genes from microarray data (Butte et al., 2000).

1. One method is to list genes by fold-increase or decrease after intervention. This method typically elucidates only the one regulatory network examined.

2. The second method involves assignment of each gene to a multidimensional point with co-ordinates equal to expression levels at various time points or experiments. Euclidean distances between points are calculated and then graphed using phylogenetic-type trees. Related genes are thought to be closer to each other in the multidimensional space.

3. The third method involves taking the same type of multidimensional space and constructing self organizing maps to find cluster of points. There are limitations with both methods 2 and 3. In case of method using Ecludian distances, there is difficulty in finding genes negatively associated with each other.

4. The fourth method involves phylogenetic-tree type clustering using branch lengths proportional to the correlation coefficient calculated between gene expression levels. This method also suffers from limitation in that it clusters genes into a single structure and pairs each gene with one another, when several regulatory pathways may be present in biological systems and expressed genes can participate in more than one pathways.

All these methods are described in detail below.

24.12 METHODS OF CLUSTERING GENES

Genes mostly interact with each other to form transcriptional modules for context-specific cellular activities or functions. Understanding such transcriptional modules is important for understanding biological networks, deciphering regulatory mechanisms and identifying biomarkers (Sun and Zhan, 2007). Genome-wide DNA microarray provides a glimpse of the signals and interactions within regulatory pathway of the cell. In a regulatory pathway of cells, genes or proteins interact with each other to control signal transduction and transcription. These regulatory interactions are multifaceted including DNA-protein or protein-protein interactions, one to one, one to many or many to one relationships, forward or feedback loops, etc. The dynamic activity of regulatory networks is constrained by the various forms of interactions and the network thus behaves only in certain ways and controlled manners in response to changing cellular conditions or external stimuli (Huang, 2001). There is simultaneous measurement of mRNA abundance of most, if not all, identified genes in a genome-wide under different physiological conditions. Because signaling pathways are dynamic events that take place over time, single time point expression profiles may not allow us to identify temporal events. Different pathways might resemble each other in expression level of genes for a limited period of time but separate at other time points. Clustering of genes based on time profiles of gene expression is leading more directly towards identification of the underlying mechanisms than clustering based on expression levels at a single time point. Groups of genes suitable for clustering can be derived from pathway information or directly from expression arrays. A natural basis for organizing gene expression data collected over multiple experiments is to first group together genes with similar expression patterns. To identify co-expressed genes transcripts with similar expression profiles from development time courses are clustered.

Various methods have been proposed for identifying gene transcriptional modules from microarray data. Clustering of genes into groups can be done either by manual examination of data (Cho et al., 1998) or by using statistical methods such as self organizing maps (Tamayo et al., 1999), k-tuple means clustering or hierarchical clustering (Eisen et al., 1998; Tavazoie et al., 1999, Wen et al. 1998) and model-based clustering (Yeung etal., 2001). These techniques are essentially different ways to cluster points in multidimensional space. They can be directly applied to gene expression data by regarding the quantitative expression levels of n genes in k samples as defining n points in k-dimensional space. The underlying assumptions under these methods are that genes with similar expression behavior (eg. increasing and decreasing together in the similar conditions) are likely to be related functionally. The clustering technique used by Cho et al uses direct visual inspection to group together genes with similar expression patterns whose expression correlated with particular phases of the cell cycle, i.e. clustering genes which are co-regulated and transcribed at a particular point in the cell cycle. This method is best suited for conditions in which the patterns of interest are clear in advance (such as a periodic fluctuation in phase with the cycle). The limitations with this technique are that first it does not scale well to larger data set and secondly it is less appropriate for discovering unexpected patterns. Clustering methods can be divided into two classes designated **supervised** and **unsupervised** clustering (Kohonen, 1997). In so called 'unsupervised' techniques such as hierarchical clustering one typically decides on a distance measure or perhaps a pair-wise correlation function as a yardstick but with no prior knowledge of the true functional classes of the genes. In supervised clustering vectors are classified with respect to known reference vectors whereas in case of unsupervised clustering no predefined reference vectors are used. As there is little a priori knowledge of complete repertoire of expected gene expression patterns for any condition, unsupervised methods or hybrid (unsupervised followed by supervised) approaches are favored. But what is desirable is the statistical organization coupled with graphical display which will allow biologists to assimilate and explore the data in a natural manner.

24.13 HIERARCHICAL CLUSTERING

Pair-wise average-linkage cluster analysis (Sokal and Michena, 1958) is a form of hierarchical clustering which has applications in sequence and phylogenetic analyses, can be used to gene expression data. Relationships among objects (or genes) are represented by a phylogenetic tree whose branch lengths reflect the degree of similarity between genes (or objects) as assessed by a pair-wise similarity function. Similarity is a measurement of resemblance or difference. Measures of similarity in the behavior of two genes that can be used are: Euclidean distance, angle or dot products of the two n-dimensional vectors representing a series of n measurements. Thus data points are forced into a strict hierarchy of nested subsets: the closest pair of points grouped and replaced by a single point representing their set average, the next closest pair of points is treated similarly and so on. In sequence comparisons these methods are used to infer the evolutionary history of sequences being compared whereas no such underlying tree exists for expression patterns of genes. Such methods are useful in that they represent varying degrees of similarity and more distant relationships among groups of closely related genes. Further, they require few assumptions about the nature of the data. The computed tree can be used to order genes in the original data table so that genes or groups of genes with similar expression patterns are adjacent. The ordered table can then be displayed graphically with a representation of the tree to show the relationships among genes.

Hierarchical clustering is a statistical method for finding relatively homogeneous clusters and determines clusters of similar data points in a multidimensional spaces based on empirical data and a certain kind of distance measures. It starts with each case in a separate cluster and then combines the clusters sequentially, reducing the number of clusters at each step until only one cluster is left. With N input data this involves N-1 clustering steps.

Limitations with hierarchical clustering—1.It is not designed to reflect the multiple distinct ways in which expression patterns can be similar. The problem gets aggravated as the size and complexity of data set grows. 2. It lacks robustness, nonuniqueness and inversion problem which complicate interpretation of the hierarchy. 3. The deterministic nature of clustering can cause points to be grouped based on local decisions with no opportunity to re-evaluate the clustering and thus the resulting tree can lock in accidental features reflecting idiosyncrasies of the agglomeration rules.

24.14 OTHER TECHNIQUES OF CLUSTERING

Other techniques applied to the analysis of gene expression include Bayesian clustering, k-means clustering and self-organization maps (SOMs). Bayesian clustering is a highly structured approach and is appropriate when a strong priori distribution on the data is available. K-means clustering is a completely unstructured approach which proceeds in an entirely local fashion and generates an unorganized collection of clusters which is difficult to interprete.

24.15 SOMs (SELF ORGANIZING MAPS)

It attempts to provide an 'executive summary' of a massive data set by extracting the n most prominent patterns (where n is the number of nodes in the geometry) and arranging them so that similar patterns occur as neighbors. SOM is ideally suited to exploratory data analyses allowing one to impose partial structure on the clusters (in contrast to the rigid structure of hierarchical clustering, the strong prior hypotheses used in Bayesian clustering and nonstructure of k-means clustering) and facilitates easy visualization and interpretations. SOM is fast and scalable to large data set. Computer package is available to generate and display SOMs of gene expression.

In this technique one chooses a geometry of 'nodes', for example, a 3x2 grid. The nodes are mapped into k-dimensional space, initially random and then iteratively adjusted. Each iteration involves randomly selecting a data point P and moving the nodes in the direction of P. The closest node N_P is moved to the most whereas other nodes are moved by a smaller amount depending on their distance from N_P in the initial geometry. In this fashion neighboring points in the initial geometry tend to be mapped to nearly points in the k-dimensional space. This process continues for 20,000-50,000 iterations.

SOMs impose structure on the data with neighboring nodes tending to define related clusters. An SOM based on rectangular grid is analogous to an entomologist's specimen drawer with adjacent compartments holding similar insects. Alternative structures which can be imposed on the data via different initial geometries include grids, rings and lines with different number of nodes.

24.16 STEPS INVOLVED IN GENE CLUSTERING

Data from microarray experiments are random snapshots with errors, inherently noisy and incomplete. The variation is caused by experimental design, experimental setup, image analysis and data analysis. Technical factors include quality of support (glass) coating, print, buffer, probe immobilization, arrayer (pins), blocking and probe concentration. Proper choice of probe sequences with optimal specificity and probe length are also cause variation. Experimental approach and image processing of any array, variability of signal intensities remains high and data interpretation may be misleading. Two preprocessing steps are involved which improves the ability to detect meaningful patterns.1. Data are passed through a variation filter to eliminate those genes with no significant change across the samples. This prevents nodes from being attracted to large sets of invariant genes. 2. The expression level of each gene across experiments is normalized in order to focus attention on the shape of the expression patterns rather than on absolute levels of expression. Data processing, i.e. data filtering, transformation and normalization may improve the situation by correcting for systematic variation. Data filtering is the assignment of pass/fail criteria to the quality control measures applied. Data processing takes care of overlapping spots, contamination, strong background and varying spot size, saturated measurements, scratches, donuts or spots that are not expected at the expected location. There is automatic detection of poor quality spots. Simple filtering method employ a fixed threshold which the background corrected signal intensity must exceed or use a cutt off value for signal-to-noise ratios (Jensen et al., 2002). Normalized hybridization signals are generated with Affimetrix Microarray Suite (MAS)5 software. It returns a single expression value per gene condensing the information from the hybridizing up to 18 pairs of perfect match (PM) and mismatch (MM) oligonucleotides (known collectively as a probe set) to complementary mRNA.

24.16.1 Steps Involves in Cluster Analysis

The three main steps involved in cluster analysis are as follows. First step is pre-processing of data which involves as discussed above a number of data transformations including feature selection, normalization and the choice of a distance function to ensure that the related data items cluster together in a data space. The second step consists of selection, parameterization and applications of one or several clustering methods. In the third step the resulting partitionings are evaluated and this requires cluster validation techniques.

Clustering genes with highly similar expression profiles locally or globally and time shifted is one of the most important steps to analyze microarray data which usually applies a kind of similarity measurement first such as Pearson correlation or Euclidean distance methods and then a clustering model follows which classifies genes on the basis of their pairwise similarities into different clusters. Time shift means a time lag between the local parts of the two gene profiles because the first gene–the regulator usually affects the downstream gene, the regulated with a time delay. The problem with Pearson correlation method is that it also calculates global similarity and ignores the time series characteristic. Euclidean method usually directly dismisses the time information and only focuses on the global profile distance calculation and rarely works well on microarray data. Besides these two methods other methods such as modified Pearson correlation, Fourier transformation (Spellman et al., 1998), Edge Detection method, Dominant Spectral component method and the Event method have been used as similarity measurement methods but all have limitations (Wang, 2006). Wang proposed a, method to measure similarity between microarray gene expression profiles which takes the local similarity and time shift characteristics into account yet simultaneously solved by the previous methods. This method starts with discretizing the data first and then applying a matrix to find all the local matching information including the time lag. Then an optimal combination of the local matches follows to attain a global similarity. A 3-value discretization method is used to preprocess the original gene microarray expression data matrix, Gnxm to get the discretized matrix Enx (m-1) where n represents gene number and m represents time condition points. The parameter t is a customized threshold which is empirically set to be 1.0. This value means that only an apparent change-increasing or decreasing in expression level can be assigned to the discretized value 1 or -1 and all others should be zero for the reason to maximally eliminate the potential noise in the original data.

The clustering structure will reveal the transcriptional information of the genes in the biological environment to help understand the biological control mechanism. Clustering results will provide usually with many other kinds of information and data such as TF and binding sites or the protein-protein interaction insight into gene function at the molecular level. Results from a microarray experiment yield a list of genes to be found to be differentially expressed. Such lists of differentially expressed genes are translated into a functional profile able to offer insight into the cellular mechanism relevant in the given condition. In other words, transcriptome analysis enabled by technology such as oligonucleotide microarray is a simultaneous interrogation of gene expression by measuring the transcriptional activity on a global scale and results in a set of differentially expressed genes. Then one will have to see whether some of these genes are functioning in a co-ordinated manner (a'**pathway**').

24.17 CLUSTERING RESULTS FROM *S. CEREVISIAE*

1. Redundant representations of genes cluster together—Genes represented by more than one array element or genes with high degree of sequence identity are clustered next to or in the immediate vicinity of each other. Thus the exact representation of a gene on the array makes little difference in the observed pattern of gene expression. They could be alternate cDNA

clones of differing length in case of human arrays or highly homologous genes in *S. cerevisiae*. Even though groups of genes may show very similar patterns of expression, individual genes can be distinguished from all other genes on the basis of subtle differences in their regulation. Genes of unrelated sequence but similar function cluster tightly together. There is strong tendency for these genes to share common roles in cellular function. The observed co-regulation occurs primarily at the level of cellular function and not only with the exact protein function (enzymatic reaction catalyzed) of the gene product. 2. Genes of similar function cluster together. A feature of gene expression is the tendency of expression data to organize genes into functional categories. 3. Patterns seen in genome-wide expression experiments can be interpreted as indications of the status of cellular processes. 4. Co-expression of genes of known functions with poorly characterized or novel genes may provide a simple means of gaining leads to the function of many genes for which no information is available (Eisen et al., 1998).

24.18 FEATURE SELECTION

Microarray experiments provide measurements of the expression levels of thousands of genes simultaneously. Now the question is how to relate the genes to the types of cancer. As many genes could correlate to a particular type of cancer so the focus would be on small subset of genes that demonstrates the outcome before conducting in-depth analyses and expensive experiments with a larger subset of genes. Thus automated discovery of this small subset (feature selection) is highly desirable.

24.19 METHODS FOR AUTOMATED FEATURE SELECTION

Methods for automated feature selection can be divided into two categories. 1. **Filtering approaches** 2. **Wrapper approaches**. In the filtering approaches the feature selection is carried out in a pre-processing step of classification, independent from the choice of the classification method. In the wrapper approaches a classifier is used to generate scores for features in the selection process and feature selection depends on the choice of classifier. For analysis of feature selection approaches, see the paper by Guyon et al. (2003). Both types of approaches have been applied to the extraction of gene subsets from DNA microarray data. Filtering methods such as correlation coefficient ranking (Golub et al., 1999) are obviously not the best choice because they score the importance of features independently, ignoring the correlation among them. More complex filtering methods such as Markov Blanket filtering (Xing et al., 2001) has been tried but it has not achieved the level of the best results of wrapper approaches. As a specific wrapper approach, recursive feature eliminating using SVM (SVM-REF) has been found to be successful (Li and Yang, 2005) in selection of important genes for cancer prediction.

24.20 INFERRING SIGNALING PATHWAYS FROM MICROARRAY DATA

The signaling pathway is a sequence of gene interactions leading to a specific biological end point function. Gene interactions are typically inferred through calculating the correlation between gene expression profiles over multiple physiological/genetical conditions. Gene pairs with high correlation (e.g. > 0.6) are hypothesized to be biologically relevant and to interact directly in the signaling pathways (Zhou et al., 2005; Lee et al., 2004). Signaling pathways, like metabolism are subdivided into smaller and more specialized subsystems with the added complications that these are frequently located into distinct regions of the cell. Analysis of signaling pathways is subject to all caveats noted for the measurement of protein-protein interactions and formation of protein complexes is an essential signaling mechanism.

It is typical that only a few genes are experimentally confirmed to be in a signaling pathway. Gene clustering is widely used approach that attempts to group all the genes in the pathway into a cluster such that functional prediction of unknown genes can be made based on functionally known genes. Some of the most popular clustering methods have been described above. These clustering methods have been successful in inferring many signaling pathways from gene microarray data. The ultimate goal of all gene clustering approaches is to group genes with similar function into a single cluster and in practice, most approaches simply group genes with similar expression profiles. However, many genes in the same functional pathway may not have similar expression profiles as measured by the correlation statistics or other pair-wise expression similarity measure. This is especially true for pairs of genes that are not in the same region of a signaling pathway. These genes will not be discernible using the traditional clustering methods. The limitation with the traditional clustering approaches is that it groups only functionally related genes with similar expression profiles but misses out on many others with dissimilar expression profiles.

In a gene co-expression (also called 'relevance') network, graph vertices (circles) represent genes and edges represent gene associations. The assumption underlying the traditional network is that the network is fully connected, i.e. biological function is executed through a direct interaction of a pair of genes. Direct pair-wise gene interactions represented by tightly connected subgraph (clique), only describes a small subset of gene interactions. In many cases an end point biological function is more commonly executed through a series of interconnected gene interactions. Consequently, for gene lying in a single pathway, traditional clustering approaches often group these into several different clusters, e.g. each cluster determined by a similarly co-expressed clique. This breaking of a pathway across several clusters makes it more difficult for biologists to identify groups of genes having common functions. Thus what is required is to group the whole pathway into a single tight cluster. Now then the question is how to extract the relevance network from microarray data and how to estimate the distance between two non-adjacent genes (genes that do not have similar expression profiles) in the network.

Genes co-expression networks typically use correlation statistics as pair-wise similarity measures (a decreasing function of the distance for clustering) between gene expression profiles, followed by either direct correlation thresholding or a combination of significance test levels with correlation thresholding. But then there is limitation with these methods. Direct thresholding only controls biological significance but not error rate whereas combining correlation thresholding allows one to control biological and statistical significance. Zhu et al. (2005) proposed a new network construction approach called False discovery Rate Confidence Interval (FDR-CI). It is able to identify both linearly and non-linearly co-expressed genes using the Kendall correlation coefficient combined with the Pearson correlation coefficient. Non-correlation measure is important when functionally related gene expression profiles are non-linearly correlated. Non-linear correlation can occur, for example, when a gene expression of different subunits of a whole enzyme are differentially regulated due to different enzyme efficiencies. As far as the estimation of distance between two non-adjacent genes in the relevance network is concerned the shortest path distance between then represents the most natural and parsimonious representation of biological interaction since genes along the shortest path are likely to have similar functions.

24.21 MATRIX DECOMPOSITION METHODS (MDMS) FOR DETECTING TRANSCRIPTIONAL MODULES

The clustering methods mostly measure correlation and linear relationships among genes and partitions genes into mutually exclusive clusters. The identification of transcriptional modules by various clustering methods described above is based on the assumption that genes with similar expression profiles share similar functions or the same pathway. However, genes involved in the same pathway or biological process can have different expression patterns (Zhau et al., 2005). Further, it is also important to capture contextual modularity which might result from highly non-linear interactions among genes. In fact, non-linear interactions among genes are often observed in gene regulatory networks such as negative feedback events, two consecutive biological events of threshold and saturation , etc. and so non-linear mixtures would provide a more realistic model for gene relationships and transcription modules. Furthermore, one gene may be part of several biological processes and thus should belong to different modules. Apparently cluster based methods are not suitable to address these problems and matrix decomposition methods (MDMs) have ben applied for uncovering transcriptional modules. MDMs treat microarray data as a mixture of unknown signals that may correspond to specific biological sources. As clustering of genes in MDMs is not based on pair-wise similarity measurement, functionally related genes showing different expression patterns can be clustered together. Further, these methods can also partition genes into non-mutually exclusive modules to reflect the fact that genes may have multiple functions or are active in multiple biological processes. MDMs can be grouped into two based on different approaches. Singular value decomposition, independent component analysis (Liebermeister, 2002; Lee and Batzoglou, 2003), non-negative matrix factorization and network component analysis are based on non-probabilistic approaches.The methods based on probabilistic approaches which take care of noise in the data include probabilistic matrix factorization. Most of these methods however, use linear models which describe gene expression data as linear combinations of latent biological sources.

24.22 SINGULAR VALUE DECOMPOSITION (SVD)

Gene clusters often contain genes that encode proteins required for a common function and hence co-clustering has been helpful in identifying the functions of unknown gene products. However, such cluster analyses provides little insights into the relationships among groups of co-regulated genes or the behaviour of biological networks as a whole (Hotler et al., 200).

They found that the highly complex set of gene expression profiles can be represented by a small number of "characteristic modes" which capture the temporal patterns of gene expression change. These patterns contribute unequally to the structure of expression profiles. This leads to the conclusion that transcriptional response of a genome is orchestrated in a few fundamental patterns of gene expression change. These patterns are simple and robust and dominate the alterations in expression of genes throughout the genome. They further observed that the characteristic modes of gene expression change in response to environmental perturbations are similar in yeast and human cells.

SVD is also known as Karhunen-Loeve expansion in pattern recognition and as principle component analysis in multivariate statistics. In multivariate statistics data are treated as vectors of discrete samples and permutation of components will not affect the analysis results. SVD is a linear transformation of the expression data from genes x arrays space to the reduced "eigengenes x eigenarrays" space where the eigengenes (or eigenarrays) are unique orthonormal superpositions of the genes (or arrays). Normalizing the data by filtering out the eigengenes (and eigenarrays) that are inferred to represent noise or experimental artifacts enables meaningful comparison of the expression of different genes across different arrays in different experiments. Sorting the data according to the eigengenes and eigenarrays provides a global picture of the dynamics of gene expression in which individual genes and arrays appear to be classified into groups of similar regulation and function or similar cellular state and sorting the significant eigengenes and eigenarrays can be associated with observed genome–wide effects of regulators or measured with samples in which these regulators are over active or under active, respectively (Alter et al., 2000). Thus SVD analysis provides insights into nature and behaviour of the underlying genetic networks.

24.23   FUNCTIONAL ANALYSES OF CLONED GENES

The gene identification through the sequence characteristics of a gene such as distinct promoter boxes, start codons and ORFs does not reveal the function of identified gene. The functional analyses of a cloned DNA fragment include

1. characterization of transcribed RNA and identification of operator-promoter sequences.
2. characterization of its gene product and regulation of their synthesis *in vivo* and *in vitro* systems.

There are three experimental approaches to detect gene expression based on detecting mRNA (Reymond et al., 2002).

1. *in situ* **hybridization**-In this approach a label is incorporated into DNA or RNA molecule that is complementary to a given mRNA. The complementary molecule is then incubated with a whole mouse embryo (for whole mount *in situ* hybridization) or with a tissue section and binds to the target mRNA. Developing the label produces a colored product which reveals where mRNA is present. In embryo the resolution is at the level of tissue or organ but tissue section (histological) provides higher resolution at the level of cell but with decreased sensitivity. ISH analyses contain more detailed information about each gene but the generation of these data is serial and significantly slower.

2. In this approach of **genome-wide analysis** thousands of genes can be analyzed in parallel. This approach involves isolation of mRNA and displaying the gene expression profile on a chip. The problem with this technique is that when it is applied to tissues then data is lost as some aspects of three-dimensional structure of multiple cell type is destroyed in biochemical extraction. Genomic analyses require miniaturization and integration of sample preparation. It requires 'chip array' and other technology which are useful for unraveling the complexities of gene expression

3. In the third approach mRNA is isolated from a tissue or organ. Then the mRNA is reverse transcribed to produce cDNA, followed by PCR amplification and thus allows detection of even a fewer copies of the mRNA. RT-PCR is more sensitive then ISH. ISH negative genes can be found expressed by this technique.

24.24   KINDS OF GENES

Number of different genes expressed in a given tissue varies greatly. There are kinds of genes that are expressed broadly or in restricted pattern. Genes with related sequences (paralogues) are found in yeast, nematodes and drosophila. Genes having paralogues are expressed broadly or ubiquitously in the mid-gestation embryo but a few are restricted to specific tissues. Genes might have general 'house keeping' functions that are required in all cells and the group of genes showing expression in tissue restricted patterns are involved in cell-to-cell communication and signal transduction. Genes whose expression is limited to spatially and/ or temporally are often linked with specific 'ontogenic' process and so analyze mRNA expression pattern at key stages of organism development.

24.25   ORFEOME

It refers to a complete set of ORFs in a particular genome. In other words, it refers to all possible (potentially overlapping) ORFs. It also refers to a slang term used for a set of full length cDNA clones transcribed from particular genome at a particular developmental stage of an organism, derived from cDNA chip analysis. Such ORFeomes circumvent cDNA library construction and ideally contain each transcribed gene sequence in equimolar concentrations.

ORF EST (OR EST+) — It refers to any EST which is derived from the central part of a cDNA rather than its 3′ or 5′ end. This is produced by first isolating total RNA which is then treated with DNase I to remove genomic DNA. After that poly $(A)^+$ is extracted and reverse transcribed with 18-25 nucleotide primer of random sequence with GC content of 50%. The resulting single stranded cDNA is again amplified with the same primer. The complexity of the preparation is checked by PAGE and the amplification pool with multibands cloned into a plasmid vector (e.g. p UC18). Finally, the inserts are sequenced. The ORESTES technique generates a better coverage of transcriptome of a cell and facilitates the construction of contigs of transcript sequences.

Amplification of an ORF — ORF mer refers to a set of two primers which allow the amplification of an ORF from genomic DNA using a standard PCR technique. One ORF mer (A-primer) contains a 13bp nonvariable sequence (adaptamer) including the start codon 5′-

ATG-3′ and a Sap I restriction site at its 5′ terminus followed by a 20-25bp gene-specific sequence (ORF sequence I). The adaptamer sequence is 5′TTGCTCTTCC**ATG**-3′. The other ORFmer (C-primer) also carries a 13bp adaptamer, contains a Sap I site and the stop codon 5′-TAA-3′ and thus the sequence 5′-TTGCTCTTCG**TAA**-3′ at the 5′ terminus adjacent to a 20-25bp gene-specific sequence (ORF sequence II). The start codon signals the amino terminus of the encoded protein whereas the stop codon signals the carboxy terminus of the encoded protein. The length of the gene-specific sequences in each primer is identical to achieve optimal melting temperature of template-primer duplexes. PCR products generated with both primers contain Sap I sites at both termini. Cleavage by SapI generates a product with an ATG start and TAA stop codons as 5′ overhangs which can be positionally cloned into vectors containg corresponding Sap sites. Thus using many different ORFmers, it is possible to perform expression profiling, expression vector cloning for characterization of the ORF and gene mutagenesis.

24.26 FUNCTIONAL CLUSTERING OF GENES

Interdependent functions are efficiently co-regulated and more likely to be inherited as a unit if their genes are closely than separate. There is occurrence of groups of genes that are co-ordinately expressed in the same tissue and such genes might be involved in common processes. Search for functional gene cluster can be made by examining gene expression databases called EST libraries where each EST (and each cDNA) library represents the mRNA expressed in a particular tissue at a specific point of time. Clusters of genes that are expressed in similar degree in different libraries suggest that the functional clustered genes are regulated together.

24.27 FINDING GENES

The large size of the genome makes finding the genes much more difficult. The protein coding parts of genes called exons are split into pieces in the genome and these pieces are separated by noncoding sequence called introns. Nearly all of the increase in gene size is human compared with fly or worm is due to the introns becoming much longer (about 50kb vs 5kb). The protein coding exons, on the other hand, are roughly the same size. This decrease in signal (exon) to noise (intron) ratio is human genome leads to misprediction by computational gene finding strategies. Many of methods of predicting genes are based on compositional signals that are found in the DNA sequence. These methods detect characteristics that are expected to be associated with genes such as splice sites and coding regions and then piece this information together to determine the complete or partial sequence of a gene (see Roy, 2009). *Ab initio* methods tend to produce false positives, leading to overestimates of gene numbers. These methods do not work with unfinished sequences that have gaps and errors which may give rise to frameshifts when the reading frame of the gene is disrupted by addition or removal of bases. Reliable gene predictions are based on information on EST and cDNAs and proteins from the same own and other organisms. The most effective algorithms integrate gene prediction methods with similarity comparisons. Such algorithms are integral to software programs such as GenWise, Genomescan and Genie whereas BLAST or FASTA programs typically require considerable manual efforts to determine the complete structure of a single gene.

24.27.1 Evidenced- Based Gene Prediction

The estimate of the number of genes is calculated as the ratio of estimated number of exons to the average number of exons per gene. The problem here is that the number of exons can be estimated only be comparing with exons in the new set of cDNAs. In other words, one can is unable to detect new exons without the support from known transcripts or homology to known cDNAs or ESTs in the same organism. In particular for genes that are expressed at very low levels or that are evolving rapidly are less likely to be present in the gene catalogue. Further, cDNAs can be probably less than full length and many tissues may not be sampled.

24.27.2 *De Novo* Gene Prediction

De novo gene prediction is made directly from genomic sequences by recognizing statistical properties of coding regions, splice sites, introns and other gene features. All this approach works well for small genomes with a higher proportion of coding sequence and has much lower specificity in higher eukaryotic genomes where coding sequences are sparser. Even the best *de novo* gene prediction program such as GENSCAN predict may apparently false positive exons. *De novo* gene prediction can be improved by analyzing aligned sequences (see Roy, 2009) from two related genomes to increase the signal-to-noise ration. Gene features such as splice sites which are conserved in different species can be given credence and partial gene models such as pairs of adjacent exons that fail to have counterparts in both species can be filtered out. Together these techniques can enhance **sensitivity** and **specificity**. Sensitivity is a measure of how well a classifier can allocate two classes correctly. Considering true positive (TP) or true negative (TN) and false positive (FP) or false negative (FN), sensitivity is defined as

Sensitivity = TP/(TP + FN), i.e. as the number of actual cases within a class

whereas specificity is defined as the number of actual positive cases of all these designated as being positive.

$$Specificity = TP/(TP + FP)$$

Validation can be made by performing PCR with RT between consecutive exons using RNA from organism and verifying resulting PCR products by direct sequencing. The second step of filtering *de novo* gene predictions (by requiring the presence of adjacent exons in both species) turn out to greatly enhance prediction specificity. Some true genes may fail to have been detected by RT-PCR owing to lack of sensitivity or tissue or developmental stage selection.

Assessment of likelihood of ORF in a stretch of genomic DNA

There are different lines of evidence for the detection of a gene in a stretch of genomic DNA sequence.

1. The most solid evidence of a gene is the experimental verification of the protein products by MS and or Edman sequencing. However, even this direct gene product verification is rarely sufficient to confirm the entire ORF. For example, secreted protein are characterized by the removal of signal peptides and frequent C-terminal

processing. This precludes the N and C translation termini by protein chemical means. Another piece of evidence is an extended mRNA. In case of human genome although there are as many as 48,681 mRNAs in GenBank, they share identity to a set only 13,429 transcripts. The transcript coverage is by no means complete and they are no full length transcripts. If the database does not contain an extended mRNA the assembly of overlapping and/or clone end clustered EST can be considered as a virtual mRNA (Schuler, 1997). 94% of known mRNAs are covered by at least one EST which provides a strong evidence for transcript especially if they include a plausible splice junction and are derived from multiple clones from different tissues cDNA libraries. Protein databases occupy the centre of evidence cascade for gene products. Those mRNAs which translate to an ORF are experimentally supported even if they are not full length and/or there can be ambiguity about the choice of potential initiating methionines. However, the fact that protein databases have now expanded to include human ORFs derived solely from the genomic prediction means that the evidence supporting them as gene products become circular.

2. Genomic prediction approach, i.e. where a cDNA, a translated ORF and a plausible gene splice patterns can be predicted from a stretch of genomic DNA (Burger and Karlin, 1997).

3. Similarity searches- This approach involves cross-species genome comparison. For example, genome of one vertebrate, say human can be compared with genomes of other vertebrates such as mouse or and fish. Cross-species data can be assembled at three levels. I. Similarity between genome at DNA level, ie. DNA-DNA comparison. This approach is termed phylogenetic footprinting (Susens and Borgmeyer, 2001) II. Translation similarity comparison III. Detecting conserved exons

4. *In silico* recognition of transcriptional control regions—This includes potential start sites in proximity to CpG islands, promoter elements, TF binding sites and potential polyadenylation acceptor sites in 3' UTR. When considered in isolation these signals have poor specificity but in combination with a consensus gene prediction programs and conservation of those putative control regions between closely related species, become useful part of the evidence chain.

24.28   STANDARD GENE MODEL AND ITS SHORTCOMINGS

The standard gene model of a defined gene locus - a single mRNA species and a single protein is no longer adequate to describe the complex relationship between the genome and its products. In other words, genome as having a defined set of isolated loci transcribed independently does not seem to be adequate. There is suggestion of the generation of numerous intercalated transcripts spanning the majority of the genome. Genome encodes a network of transcripts many of which are linked to protein-coding transcripts and to the majority of which we can not assign a biological role (ENCODE Project Consortium, 2007). But how splicing signals co-ordinated and used when there are so many overlapping primary transcripts is not understood. Fitting of transcription data into this standard gene model has highlighted lacking a number of clearly defined characteristics. In other words, there are a number of deviations from this simple gene model.

A. Delineating the extreme 5′ and 3′ ends of mRNA transcripts (Pesole et al., 2002; Suzuki et al., 2002).—Delineation of genes is the key step. At most basic level an mRNA molecule consists of a protein coding ORF flanked by 5′ and 3′ UTR. In case of polymorphism analysis these sequences should not be overlooked as the extreme 5′ and 3′ limits of UTR sequence delineate the true boundaries of genes. Once we known the location of a gene, all other functional elements fall in place based on their location in and around genes. Gene prediction and gene cloning generally have focuse on the ORF-the protein coding sequence (ORF/CDS) of the gene. Many mRNAs are labeled as partial is a testimony to the difficulty of finding library inserts that are complete at the 5′ end. Further, in many cases the mRNAs are considered finished when a plausible ORF has been delineated. However, very few cDNAs are full length in that they have been 'walked out' to determine the true 5′ most initiation of transcription in the 5′-UTR. The same problem applies to the 3′UTR. There may be substantial stretches of 3′UTR extending downstream of the first polyadenylation position at which further cloning attempts have ceased. This problem is compounded by the poor performance of gene prediction program for 5′ and 3′ ends. The uncertainties about transcript extremities can be resolved by surveying the coverage of all available cDNA sequences whether nominally full length or partial ESTs and patent sequences. These can often extend the UTR sections.

Translation of most eukaryotic mRNAs is likely to be initiated by a linear scanning although some other mechanisms are also possible (Kozak, 2000). According to scanning model 40S ribosomal subunits can either initiate translation at 5′-proximal AUG codon or miss it and initiate translation at downstream AUG (s). The initiation/scanthrough ratios depends on both the AUG nucleotide context and the features of downstream mRNA fragment (Kozak, 2002; Wang and Rothnagel, 2004). In case of mammalian and plant mRNAs the most crucial elements of AUG context are purine at position-3 and guanine at position +4 (Lukaszewick et al., 2000; Kozak, 2002). Although one might expect mRNA to be possessing the features providing efficient translation including the recognition of a genuine translation start site (TSS), however, the fraction of eukaryotic mRNAs with the start AUG codon in a suboptimal context is relatively large as well as the fraction of mRNAs with the AUG-containing 5′UTRs (Rogozin et al., 2001). Thus it is likely that at least some mRNAs with a suboptimal start codon context contain other signals providing additional information for efficient TSS recognition. Computational analysis of eukaryotic genes with either optimal or suboptimal start codon contexts showed that the eukaryotic mRNAs were characterized by a significantly higher occurrences of in-frame AUG codons (Kochetov, 2005). It is thus likely that the closely located additional start codons can often be used to increase the translation initiation efficiency and the corresponding proteins might show certain N-terminal heterogeneity.

B. Pseudogenes—In some cases genomic sequence is so severely degraded that transcription is unlikely. In a RefSeq of 1598 loci of pseudogenes at least 30 are recorded as having detectable transcript. Thus although pseudogenes are generally considered non-functional copies of genes but are sometimes transcribed and thus often complicate the analysis of transcription due to close sequence similarity to functional genes.

C. Gene product heterogeneity—In some cases there may be alternative upstream initiation methionines or alternatively spliced exons in the 5′UTR. The cause for 3′ heterogeneity includes variations in the pattern of intron splicing from a pre-mRNA as well as alternative polyadenylation positions inside the 3'UTR.

There is translational control of intron splicing in eukaryotes (Jaillon et al., 2008). Most eukaryotic genes are interrupted by noncoding introns. Normally the introns are spliced out of the pre-mRNA but in certain cases one or more than one intron can be left unspliced. The inclusion of an intron in final mRNA can have drastic consequences. For example, intron 3 retention in the P-element of mRNA of *Drosophila melanogaster* produces a repressor protein of transposition whereas splicing of this intron allows the expression of the transposase in the germ line. Splicing is guided locally by short conserved sequences but genes typically contain many potential splice sites and the mechanisms specifying the correct sites remain poorly understood. In most organisms, short introns are recognized by the intron definition mechanism and thus can not be efficiently predicted solely on the basis of sequence motifs. In multicellular eukaryotes, long introns are recognized by **exon definition** and most genes produce multiple RNA variants through alternate splicing. **Exon definition** refers to a process through which the initial recognition of splice sites is enhanced by interactions between the 3' and 5' splice sites across the exon. The nonsense-mediated mRNA decay pathway may further shape those observed set of variants by selectively degrading those containing premature termination codons which are frequently produced in mammals. Tiny introns of *Paramecium tetraurelia* cause premature termination of mRNA translation in the event of intron retention and the same bias is observed among the short introns of plants, fungi and animals. Intrinsic efficiency of splicing varies widely among introns and nonsense mediated decay activity can significantly reduce the fraction of unspliced mRNAs. Independently of alternate splicing, species with large intron numbers universally rely on nonsense- mediated decay to compensate for suboptimal splicing efficiency and accuracy.

D. Overlapping genes—There are many examples of both from gene products reading from opposite strands and same strand gene in close proximity.

Genes were thought to be compact, filled with information and hidden among billions of bases of junk DNA. Findings from the ENCODE project say that genes can be sprawling with farflung protein coding and regulatory regions that overlap with other genes. Although protein coding DNA makes up barely 2% of the genome, yet 80% of the bases show signs of being expressed. Analysis of transcriptome- the repertoire of RNA molecules showed half of the transcripts not to be translated into proteins. In other words, there are two types of genes: protein coding genes which produce mRNAs that are translated into chains of amino acids by ribosomes and other types of genes for which RNA is the end product. Results from this analysis showed a lot more of DNA turing as RNA and these RNA transcripts harbour short sequences, conserved across mice and humans, are likely to be important in gene regulation. Analysis of mRNAs of 400 protein coding genes contained in ENCODE's DNA showed presence of additional exons-the regions that encode for amino acid in more than 80%. Many of these found exons were located thousands of bases away from the gene's previously known exons, sometimes hidden in another gene. Furthermore, some mRNAs were derived from exons belonging to two genes.In addition, further extending and blurring gene boundaries, ENCODE discovered a slew of novel 'start sites' for genes with many located hundreds of thousands of bases away from the known start sites. Analysis of promoter showed higher number of promoters than known previously and one-quarter of the promoters discovered were at the ends of the genes instead of at the beginning. The distribution of exons, promoters, gene start site and other DNA features and the existence of widespread transcription suggest of a multidimensional network regulating gene expression. Considering these deviation it is essential to keep an open mind about extremities and plurality of gene products. Thus the original definition of gene seems to be changing and RNA transcripts and not genes can be considered as the fundamental functional units of genome (Pennisi, 2007). For more of deviation from this simple gene model see chapter 5.

24.29 ENSEMBLE GENE PREDICTION

Ensemble gene prediction pipeline (Hubbard et al., 2002) is augmented with Genie gene prediction pipeline (Kulp et al., 1997) and these are supported by experimental evidence such as ESTs. Ensemble system uses three tiers of input as follows.

1. Known protein coding cDNAs are mapped on to the genome
2. Additional protein coding genes are predicted on the basis of similarity to proteins in any organism using the GeneWise program (Birney and Durbin, 2000)
3. *de novo* gene predictions from GENSCAN program (Burge and Karlin, 1997)

These three pieces of evidence are reconciled into a single gene catalogue by using heuristic to merge overlapping predictions, detect pseudogenes and discard misassemblies. These results are then augmented by using conservative predictions from Genie system which predicts gene structures in the genomic regions determined by paired 5' and 3' ESTs on the basis of cDNA and EST information from the region.

Computational pipe line produces predicted transcripts which may represent fragmentary products or alternative products of a gene. It may also represent pseudogenes. Predicted transcripts are then be aggregated into predicted genes on the basis of sequence overlaps. Computational pipe line remains imperfect and the predictions are tentative. The incomplete nature of the gene catalogue is because of complexities arising from alternate splicing and the difficulty of interpreting evidence from fragmentary mRNAs such as ESTs and SAGE tags which may not represent protein coding genes (Kapranov et al., 2002).

The Ensemble analysis pipeline consists of a rule-based system designed to mimic decisions made by a human annotator. The idea is to identify 'confirmed' genes that are computationally predicted (by the GENSCAN gene prediction program) and also supported by a significant BLAST match to one or more expressed sequences or proteins. Ensemble also identifies the position of known genes from public database entries using GENEWISE to predict their exon structures. The total set of ensembled genes should therefore be a much more accurate reflection of reliability than *ab initio* prediction alone but many novels genes are missed (Hogensch et al., 2001). Further, many of the detected novel genes are expected to be incomplete for two reasons. Firstly, while GENSCAN can detect the presence of most genes in a genomic sequence it is substantially less successful in predicting their correct exonic structures (as with *ab initio* gene predictions). Secondly, any prediction is entirely dependent on the quality of the genomic sequence and where the sequence is gapped or wrongly assembled the missing exons may not be present for the software to find. In oder to improve preformance many other genomic features such as different repeat classes, cytogenetic bands, CpG islands prediction, tRNA gene prediction, expressed sequence clusters from UniGene database, SNPs from SNP database, disease gene in the draft genome from the On line Mendelain inheritance in man (OMIMdb) database and regions of homology to mouse draft genomic sequence have been included in the Ensemble. Once a gene is identified Ensemble can be linked to InterPro protein domains it encodes and its expression profile according to the SAGEmap repository (Lash et al., 2000).

24.30 ANALYZING A NOVEL GENE

Suppose there is a piece of GP (mouse) where there are no fully annotated known genes. The three viewers Ensemble, USCS and NCBI display it between the 3′side of the BACE gene and 5′ end of the next known gene PCSK7. Further, assume that the analysis of genomic region between these two known genes by genetic linkage study shows significant associations in this area either from the two STS markers or the 50 or more SNPs that lie in this interval but are outside the boundaries of the two neighboring genes. Now the question is what other gene product (s) might be located between the two genes. To answer this, the first step would be to check for the continuity of this section of GP and if there is complete clone overlap across this section then move forward and subject this segment to all three displays. Inspection of all three displays would indicate a possible novel gene product with a variety of supporting evidences using gene prediction and ETS coverage. GENSCAN discovers novel gene by finding predicted exons which can coincide with good matches from draft mouse genome or novel promoters by finding matches to the draft mouse genome that occur upstream of the 5′end of a gene.

Search with available mRNA and protein sequences is then made. An 81% protein identity to mouse orthologoue mRNA, BCO23073 was found. This level of similarity should result in this gene passing the GENWISE threshold for marking a novel gene product. At this level of similarity one can back-check this mouse sequence against human GP by BLAT search which supported both the orientation (5′ to 3′ relative to the BACE) and seven of the exons from C11002075. However, the mouse sequence was missing the 5′end. In the next step the entire genomic DNA section of 54kb was searched against human EST using MEGABLAST with 90% match stringency and masking the repeat sequences in the genomic query sequence using repeat masker. It resulted in identification of ETSs that bridge several exons. The same result will be obtained in principle to the UniGene clusters in the NCBI viewer and identify the exons, its number and its coverage. Performing similar search against mouse ESTs with an 80% identity cut-off yielded a long EST spanning over four central exons which suggested a single rather than multiple products. Searching ESTs against the TIGR THCS was made to establish the presence of virtual mRNAs, if any and in fact, two assemblies representing the 5′and 3′ ends, respectively were found. Further, a bridging EST was found to join these two assemblies. A CAP3 assembler was used to construct an extended vitual mRNA of 2720bp. In other words, predicted ORF was produced by assembling the appropriate assemblies and ESTs into a virtual mRNA. This virtual mRNA was then translated into a protein of 474 amino acids using the translation tool and thus a putative full length protein sequence was generated.

Once the ORF has been predicted and putative protein has been found the next question is what to do to verify this putative novel protein *in silico*. As first step cross check for reading frame consistency and species orthologues by performing TBLASTN against all ESTs. In other words, the continuity of ORF is checked by translation searching for the unknown ORF against all ESTs. This will show the coverage of ORF by human ESTs as well as potential splice variants. Deletions in ESTs could represent splice variants. The next step involves matching of virtual mRNA back against GP using the BLAT search at UCSC. This delineates matching of 15 exons from putative 5'UTR to 3'UTR. In other words, it reveals 15 exons with the gene reading in the opposite orientation to its neighboring genes. Thus virtual mRNA lies very close to both neighboring genes thereby suggesting it to be a full length transcript.

The final step in analyzing a noverl gene is to assignment of function to the identified ORF. Two steps involved in this assignment of function are protein database search and motif analysis. One can see the sequence similarity scores of the novel ORF against the NCBI non-redundant protein database. In the present case there was sequence similarity over C-terminal section but the hits included the same proteins aasigned as UniGene homologies by Ensemble. Domain analysis using InterPro recognized two domains. The proline-rich domain was defined from a Prosite profile and the zinc finger was defined by both a Prosite Profile and SMART domain.

24.31   VALIDATION OF STRUCTURES OF GENES

Products of the sequencing of a genome is a complete, accurate catalogue of genes and their products, primarily mRNA transcripts and their cognate proteins. Such a catalogue can not be constructed by computational annotation alone, it requires validation on a genome-wide scale. Shoemaker et al. (2001) used 'exon' and 'tiling arrays' fabricated by ink-jet oligonucleotide synthesis and they devised experimental approach to validate and refine computational gene predictions and define full-length mRNA transcripts on the basis of co-regulated expression of their exons. Thus there are two approaches to determine the transcription profile of a cell type. These methods can provide more accurate gene numbers and allow the detection of mRNA splice variants and identification of the tissue and disease-specific conditions under which genes are expressed.

24.32   EXON ARRAYS

A high throughput microarray based experimental method to validate predicted exons, group the exons into genes by co-regulated expression and define full length mRNA transcripts is the method which involves the design and fabrication of '**exon arrays**' consisting of long (50-60bases) oligonucleotide probes derived from predicted exons, followed by hybridization with fluorescently labeled cDNAs derived from specific cell lines or normal or diseased tissues. Absolute intensities (measuring cellular abundances) and intensity ratios (measuring differential expression regulation) from hybridized cDNAs are used to identify those probes which represent authentic exons under the conditions tested. In addition, the expression data can define gene boundaries because adjacent exons that are co-regulated across many conditions are likely to be from the same transcript. Exon-based gene validation arrays can be limited by the fact that gene prediction programs perform best on 'internal exons' and not very well on initial and terminal exons or exons that correspond to the 5' or 3' UTRs of mRNAs. Oligonucleotide tiling arrays of overlapping probes can effectively address this challenge as they are constructed without any a priori knowledge of the possible exon content of a genomic sequence.

For a higher resolution view of gene structure 'tiling arrays' is used in which overlapping oligonucleotides are designed to blanket an entire genomic region of interest. This approach can potentially reveal exons not identified by current gene prediction algorithms and provide information about alternative splicing.

24.33 GENERATION OF cDNA SET

cDNA set comprising full length transcripts obtained by merging the Fantom-3 cDNA set with mouse cDNAs from GenBank and clustering the cDNAs into transcriptional units in which members share sequence transcribed from the same strand. Duplication (redundancy) in cDNA set might result from a number of factors including errors made when samples are gridded, internal initiation of reverse transcription, incomplete or variable splicing and differences in polyadenylation site usage which may account for about 19% of true 3' end variability in human. To increase the likelihood of discovering genes, selection should be biased towards clones with novel 3' end sequences.

Transcriptional Units(TUs)

TU refers to a segment of the genome from which transcripts are generated. It can be defined by the identification of a cluster of transcripts that contain a common core of genetic information(in some cases a protein-coding region). The existence of a TU is inferred from the identification of mRNAs through full-length cDNA isolation and sequencing. A single cDNA sequence may define a TU. Further, multiple cDNA transcripts sharing DNA sequence are termed as a single TU. For each TU the 5' boundary is defined as the most distal transcription start site(TSS) and the 3' boundary the most extreme poly(A) sequence. TUs are DNA strand specific and are typically bound by promoters at one end and termination sequences at the other. Thus alternatively spliced transcripts, alternative 5' start and 3' ends and variants arising from recombination are subsumed within a single TU even though they may generate protein or RNA products with very different functions. TUs on opposite strands are counted separately even if they overlap spatially in the genome. Thus antisense transcripts are considered to be made from separate TUs.

24.34 TILLING ARRAY TECHNOLOGY

Functional analysis of a genome requires accurate gene structure information and a complete gene inventory. Tiling array is a new type of microarray that can be used to survey genomic transcriptional activities and transcription factor binding sites at high resolution. In other words, tiling array is a microarray that interrogates genome with high density probes. In a typical tiling array, probes are distributed along chromosomes approximately evenly at a density of one probe per 10-100bp. When hybridized with RNA or chromatin immunoprecipitation (ChIP) samples, the array will detect genomic sequences that show transcriptional or transcriptional factor binding patterns of interest. Owing to high density of the probes a whole genome can be surveyed in an unbiased manner at a high resolution. There is need to do tiling array experiment under a number of differential developmental stages and identify genomic loci with specific temporal or spatial transcriptional or transcriptional factor binding patterns.

A dual experimental strategy was used to verify and correct the initial genome sequence annotation of the reference plant Arabidopsis. Sequencing full-length cDNAs and hybridizations using RNA populations from various tissues to a set of high-density oligonucleotide arrays spanning the entire genome allowed the accurate annotation of thousands of gene structure (Yamada et al., 2003). The initial identification of the transcriptional units in Arabidopsis genome was carried out largely by *ab initio* gene predictions, sequence homology, sequence motif analysis and other non-experimental methods. To identify transcription units in Arabidopsis genome a custom high-density ologonucleotide arrays that tile the entire genome and RNA samples prepared from a diverse set of tissues and treatments to ensure broad representation of transcriptional activity were used and validation of tiling chip detected transcription units was carried out through RT-PCR, amplification, cloning and sequencing.

Transcriptionally active regions of human genome have been mapped by a combination of the alignment of cDNAs to genomic sequences and the annotation of genomic sequences to predict coding regions (Rubin et al., 2000; Caron et al., 2001). But there is a need to develop empirical map of the transcriptionally active regions of the human genome at the nucleotide level and to relate the map to the sequence annotations derived from the above mentioned two general approaches. Annotation of genome guides in choosing which genomic regions to evaluate to determine the transcription profile of a cell type (exon array). Kapranov et al. (2002) used an empirical approach to create a collection of transcript maps using uniformly spaced probes (25-nt oligomer) that interrogate either every base or on average every 35bps of the sequences of chromosome in a systematic manner. Labelled double stranded cDNAs made form cytoplasmic polyadenylated RNA from 11 different tumor and fetal cell lines were hybridized to high density oligonucleotide arrays of two types. The first array consisted of a prefect complement (PM) and mismatched (MM) complement oligonucleotide probe for each base (DGCR) array which interrogated 362,901 contiguous nucleotides of D1George syndrome minimal critical region. The second array consisted of probe pairs (1,011,768) synthesized on a three array set interrogated approximately 35 million non-repetitive base pairs of chromosome 21 and 22. Whether or not a probe pair detected a RNA target is determined using a range of threshold values for the ratio (R) of PM to MM and for the difference (D) of PM-MM values. By pairing the R and D values for each experiment it is possible to estimate i. false positives, specificity and sensitivity rates on the basis of spiked bacterial RNA transcripts containing specific deletions and ii. sensitivity (only for DGCR) based on exon sequence detected by RT-PCR. Because of overlap of interrogating probes used in the design of DGCR array it was possible to join positive probes separated at most by a certain distance (max gap) and to then reject resultant regions whose length was less than a particular threshold (minimum), thus building maps with contiguious runs (contigs) of RNA.

24.35 FORMALDEHYDE ASSISTED ISOLATION OF REGULATORY ELEMENTS (FAIR)

It is a method to assay open chromatin using formaldehyde crosslinking followed by detection of the products using a genomic tiling array.

Advantages of this approach—The advantages of this approach are as follows.

1. Identification of new regions of transcription not yet observed by previous experimentation or sequence analyses.

2. Detection of RNA transcripts that have little or no coding capacity

3. Identification of alternative RNA isoforms of previously annotated genes

Many of the detected transcripts may be structurally connected to other partially characterized RNAs (i.e. ESTs). The estimate of relative abundance is very low based on the required levels of RT-PCR and PCR amplification required to detect them in the RNA and cDNA libraries and the time of exposure needed to detect those observed by Nothern hybridization. Sequences for many cDNAs are discarded as possible heteronuclear RNA contaminants because of low coding potential. The table 24.1 given below shows the application of tiling array technology in molecular genetics.

Table 24.1   Showing objective and the corresponding technology.

Transcription	Tiling array, integrated annotation
5′ends of transcripts	Tag sequencing
Histone modifications	Tiling array
Chromatin structure	QT-PCR, tiling array
Sequence specific factors	Tiling array, tag sequencing, promoter assays
Replication	Tiling array
Polymorphism	Resequencing, Copy number variation
Comparative sequence analysis	Genomic sequencing, multiple sequence alignments

24.36 GENOME-WIDE STUDY OF SPLICING BY SPLICING MICROARRAY

Protein coding information in eukaryotes is fragmented into exons which must be recognized and joined by the process of RNA splicing. Genome-wide analysis of splicing involves accurate parallel analyses of alternative splicing in higher organisms. Splicing takes place in the nucleus within spliceosome-a ribonucleoprotein complex, which transfers information within transcripts of eukaryotic genome to create sequences not found in the DNA. By its nature and position in the gene expression pathway splicing expands the possible interpretation of genomic information and does so under developmental and environmental influence. Understanding of the process of splicing is derived from studies on relatively few introns. How the process of splicing is integrated into genome function and evolution? In comparison to higher eukaryotes the single celled yeast contains relatively few spliceosomal introns and most have been correctly annotated. Microarray or fibre optic array probes can monitor splicing events and the idea of using probes positioned at exon-exon junctions to monitor splicing was suggested by Black (2000). Microarray can be used to monitor splicing at every exon-exon junction in more than 10,000 multi exon human genes in 52 tissues and cell lines to understand the biological roles and regulation of AS across different tissues and stages of development. It has provided evidence and tissue distribution of thousands of known and novel AS events. Microarray containing about 125,000 different 36-nucleotide junction probes have been used to monitor the exon-exon connections. Clark et al. (2002) designed microarray to distinguish spliced from unspliced RNA for each intron containing yeast gene. They measured the genome-wide effects on splicing caused by loss of 18 different mRNA processing factors. About 40 to 60% of human genes produce alternatively spliced transcripts. In a growing number of cases alternatively spliced mRNAs produced proteins of distinct or even antagonistic function. Improved expression profiling technologies must resolve changes in AS not simply by estimation exon representation but by providing direct evidence for exon joining. Oligonucleotides are designed to detect the splice junction (specific to spliced RNA and not found in the genome), intron, present in unspliced RNA and the second exon, common to spliced and unspliced RNA for each intron containing gene. The arrays thus contain three ologonucleotide probes for each intron containing genes as well as probes for control intronless gene. Intron probes detect unspliced RNA and lariats. Splice junction probes detect spliced mRNA and exon probes detect both spliced and unspliced RNAs. The validation of microarray data is done using RT-PCR. Splicing predictions from the array data are used to guide RT-PCR and sequencing validation efforts to specific transcript locations. This guided approach provides evidence of tissue specific AS in thousands of genes and has allowed to identify and sequence-verify splice variants not represented currently by ESTs or mRNAs. The oligonucleotides are printed on glass slides to create splicing-sensitive microarrays. The ability to distinguish differently spliced forms of RNA by using oligonucleotide microarrays has opened the way for expression profiling that accounts for alternative splicing and splicing regulations in higher cells. Oligonucleotide arrays designed to detect specific splicing products will be key to accurate parallel analysis of AS in higher organisms.

There are other approaches to study AS events but have limitations (Modrek and Lee, 2000). The full length cDNA sequence approach supplies gold standard transcript definition but is intensive and expensive. The EST approach provides evidence of a vast number of alternative isoforms but systematic study of AS using ESTs are hampered by protocol differences, transcript end bias and library coverage limitations.

24.37 DIGITAL POLONY EXON PROFILING (DPEP)

Alternative splicing in eukaryotic pre-mRNA is a powerful means of regulating gene expression and enhancing proteome diversity. Although dozens of *cis*-regulatory elements and transcription factors affecting alternative splicing have been identified but the mechanism of alternative splicing is still not understood. DPEP is a single molecule-based technology for studying complex alternative pre-mRNA splicing (Zhu et al., 2003). It allows to monitor the combinatorial diversity of exon inclusion in individual transcripts. A mini sequencing strategy provides single nucleotide resolution and the digital nature of the technology permits quantitation of individual splicing events. DPEP can be used to investigate the role of the physiological and pathological roles of alternatively spliced mRNAs as well as the mechanisms by which these mRNAs are produced.

The approach of profiling complex alternative splicing via the polymerase colony (polony) technology allows solid-phase parallel amplification of thousands to millions of DNA or RNA molecules such that each template gives rise to an individual colony of amplification products. One strand of each amplicon is covalently linked to an acrylamide matrix and serves as a template for probe hybridization and/or single base extensions (SEBs). Detection of multiple gene-specific sequences in a *cis* is achieved by combinations of spectrally distinct fluorophores and/or repeated cycles of probing. In the case of alternative splicing, combinatorial patterns of exon inclusion or exclusion can be unambiguously determined across multiple polonies in parallel. Furthermore, because each polony arises from a single molecule, the digital nature (Voselstein and Kinzler, 1999) yields a sensitive and accurate means of quantifying individual mRNA isoforms in one or more pools of interest. Though the current throughput of polony-based exon amplification is limited as compared to microarray-based splicing arrays (Clark et al., 2002; Yeakley et al., 2002), potential exists to obtain higher levels of multiplexing via quantum-dot labeled probes and/or fluorescent *in situ* sequencing. Polonies have also been used for genotyping, long range haplotyping and fluorescent *in situ* hybridization.

24.38 FUNCTIONAL GENOMICS TECHNIQUES

The different functional genomics techniques are:
1. Genome-wide expression profiling (using DNA micro–arrays and DNA chips (oligonucleotide chips), SAGE and GFP fusion proteins).
2. Two- hybrid screens for detecting protein-protein interaction and its variants
3. Transgenic strategies for analyzing gene function includes insertional mutagenesis (use of transposons or transgenes (Ac-Ds element), retrotransposons, MITES (Miniature inverted repeat) and T-DNA), gene tagging, enhancer or gene trap, etc.

Photolithography—In another strategy DNA is directly synthesized on the solid surface using **photolithography**. This technique makes use of nucleotide precursors that are activated by light and thus joins one nucleotide to the next. A computer is programmed with information regarding the nucleotide sequences to be synthesized at each point on a solid surface. The surface is washed successively with solutions containing one type of activated nucleotide (A*, G*, C*, T*). The activated nucleotides are blocked so that only one can be added to a chain in each cycle. The screen covering the solid surface is opened over the areas programmed to receive a particular nucleotide and a flash of light joins the nucleotide to the polymer in the uncovered areas. This cycle continues until the required sequences are built up on each spot on the surface. Many polymers with the same sequence are produced on each spot but surfaces also can have thousands of spots with different sequences. The DNA chip thus constructed is probed with mRNAs or cDNAs with fluorescently labeled deoxynucleotides from a particular cell type or cell culture which hybridize with the complementary sequences on the microarray and thus identify the genes being expressed.

The total mRNA is isolated from cells at two different stages of development and converted into cDNAs with fluorescently labeled nucleotides in two different colors. The mixtures of the fluorescent cDNAs are then used to probe the microarray and thus we can know the relative abundance of sequences at two developmental stages.

Photolithography is a technique for the light dependent engraving of the specific pattern on a solid surface used in printing process. The solid support ('plate') is coated with light sensitive emulsion and overlaid by a photographic film. The coated film is illuminated and the images of the film are reproduced on the plate. This technique is employed in **DNA chip technology**. In case of its use in the synthesis of oligonucleotide modifications of the usual phosphoramidite reagents are used, i.e. DMT group which protects the 5'-OH, is replaced by a phospholabile protective group. The synthesis of oligonucleotide precedes by photolithographically, deprotecting all the areas that will receive a common nucleoside and coupling this nucleoside by exposing the entire chip to an appropriate phosphramidite. This is achieved by so called masks made from chromium/glass which contains holes at positions where deprotection is required.

Electronic hybridization—The use of electronics is to move probe molecules to specific sites on the surface of the DNA chip where they concentrate. This electronic concentration of the probe at the target DNA on the chip promotes the rate of hybridization. Additionally, electronic stringency control (ie. the reversal of polarity or charge on the chip surface towards negative) lowers the amount of non-specifically bound DNA and reduces the background signals.

Gene array (gene chip)—Gene chip or microarray are wafers of silicon, glass or plastic that when mounted with microscopic DNA probes allow to study when and how gene expresses themselves. It refers to the ordered alignment of different gene sequences or parts of such sequences (i.e. in the form of oligonucleotides) or ORFs immobilized on supports of minute dimensions (e.g. nylone membrane, glass or quartz slide, plastic chip). A gene array may contain all gene sequences of an organism. For example, the 4290 protein-encoding genes of *E. coli* can be put on a mere space of 20 x 10cm or less. Such gene array can be used for high throughput expression profiling. In this technique total RNA from two organisms (cells, tissues, organs) to be compared is isolated, reverse transcribed into cDNA in the presence of one radioactively or fluorescently labeled nucleotides. After that the two cDNAs are separately hybridized to two identical gene arrays. After autoradiography phosphorimaging or fluorescence detection, the expression profiless of the two samples can be compared and up-or down-regulated genes can be identified.

A DNA array in DNA chip may contain 100,000 probe oligomers which is larger than the total number of genes even in higher organism. The spot size may be as small as ~ 150μ in diameter. The grid is typically a few centimeters across. To measure expression patterns the oligomeric probes are cDNAs or fragments of cDNA, reflecting the mRNAs for different genes. Oligomers of length ~ 50-80bp are required. For genotype analysis genomic DNA fragments of length 500-5000bp are used.

Affymetrix Gene Chip arrays (Lockhart et al., 1996; Marshall, 1999) are currently the most widely used microarrays. Short specific oligonucleotide probes are tethered and immobilized on the surface of Affymetrix arrays. Target cRNA is fluorescently labeled and hybridized to the array. A 2D image is then generated with each probe being identified by a position and an intensity. Each probe is 25bases long and each gene is represented by 11-20 probe pairs referred to as a probe set. Each probe pair comprises a perfect match (PM) probe and a mismatch (MM) probe. The PM probe and MM probe have almost the same base sequences except that the middle base of the MM probe is changed to the complementary one of that in the PM probe. The aim of the one base mismatch in the MM probe is to measure non-specific hybridization which occurs when the target binds to a probe that does not have the perfect complementary sequence. Therefore, by design, the MM probe acts as a background measurement for its corresponding PM probe. Probe-level analysis is the process of estimating the expression level of each gene from intensities of PM and MM probes in the corresponding probe-set. Thus this analysis provides the summary of the expression value for each gene. The aim of further analyses is to detect differentially expressed genes and to find co-regulated genes.

Microarray suffers from inherently noisy information (therefore requiring many replicates) and often limited by cost and availability of materials. Sheer volume of information is obtained from this analysis. Finally, expression level change of a gene may be corollary to a change in another gene and may not be the direct cause of the cellular phenotype. Additional information is required to place those genes in context.

24.39 COMPARATIVE GENOMICS

Sequencing of many genomes will provide database that can be used to assign gene function by genome comparison. Having identified a new sequence the comparison with sequence database is the simplest way to obtain (essentially biochemical) function. Computerized analyses are generally not sufficient to define gene function with a high level of confidence and so experimental confirmation is required in most cases. Indirect information on cellular or developmental function can be obtained from spatial and temporal expression patterns, for example, the presence of mRNA and/or protein in different cell types, during development, during pathogen infection, or in different environments. The subcellular localization and post-translational modifications of proteins can be informative as well. Knocking out or over expressing the gene permits the gene sequence to be linked to a phenotype from which a cellular role or a role in development may be deduced. Finally, Comparison of the fitness of plants carrying mutations or natural variants of the gene with wild-type in different environments will provide information on adaptive function of the gene. The number of genes can be predicted using sequence similarity, exon-microarray data and ORF size.

Computer prediction and alignment programs (see Roy, 2009) can identify many genes and predict some aspect of gene structure but it does not say where a gene starts or stops and when it should be turned on and off. Computation analyses can classify genes into families based on conserved motifs in the gene sequences but such functional insights are limited to some classification of protein's biochemical or cellular function based on motifs. Transcription factors and enzymes can be recognized for their motifs but not where and when they will be expressed and who their partners are in the encoded genome.

Problem with gene (ORF) finding approach—One possible approach to targeted gene finding is to use the knowledge of experimentally verified target genes in closely related species and utilize gene conservation across different species to identify putative target genes. This approach works by focusing on conservation of ORFs. ORF centric gene finding method may not be effective in discovering target genes that express under specific conditions. There is no direct correlation between gene function and gene expression although gene functions may be indicative of gene's responsiveness to certain stimuli. Gene expression is controlled mainly at transcription level where the binding between TFs and *cis*-regulatory DNA sequences (or *cis*-elements) in the upstream region of the genes play an important role. Therefore, if some *cis*-elements are known to be directly involved in gene transcription regulation in responding to specific stimuli then use the *cis*-elements to identify genes of interest. A large number of TFs and their binding *cis*-elements have been identified over years. Most of the TFs and their corresponding *cis*-binding elements in the yeast have been identified (Harbison et al., 2004). TRANSFAC- a database of *cis*-elements in many species have been established (Matys et al., 2001). Plant-specific *cis*-elements PLACE (Higo et al., 1999) and plantCARE (Lescot et al., 2002) are available.

24.40 GENE EXPRESSION PROCESS

During gene expression process, some substrings of the genome called genes are decoded to produce proteins. In order to start the gene expression process, a molecule called transcription factor (TF) binds to a binding site(BS), represented by a short substring in the promoter region of the gene. Genes seldom works alone and so one kind of TF may bind to the binding sites of several genes thereby allowing the genes to be expressed together. Such binding sites should then have the same length and similar patterns. In other words, these typical sites have a similar short DNA sequence pattern which is simply referred to as **motif**. Promoter determines whether or not a gene is expressed and how much mRNA and consequently-protein is produced. A promoter is a sequence that initiates and regulates the transcription of a gene. Protein binding sites in a promoter represent the most crucial element and the corresponding proteins are called transcription factors (TFs). TFBSs are usually short (~ 5-15bp) and degenerate (Stormo, 2000), making it difficult to define experimentally or computationally. There is large variety of TFs in the cells. Currently more than 1400 human TFs are known and a total of ~ 1850 to 3000 is estimated (Venter et al., 2001). Prediction of functional TFBSs is an important step in promoter analysis. Whether a TF has a global role (affecting all genes) or a specific role (affecting only some). Finding the common pattern, motif, of the binding sites from a set of sequences representing the promoter regions is an important problem for understanding how gene expression works. In other words, finding motif in DNA sequences plays an important role in deciphering transcriptional regulatory mechanisms and drug target identification. The functional sites are constrained to contain motifs since their changes will disrupt regulation which is detrimental to organism.

One way to describe TFBS is by nucleotide or position weight matrices (Stormo, 2000). A weight matrix pattern definition is superior to a simple IUPAC consensus sequence as it represents the complete nucleotide distribution for each single position. It also allows the quantification of the similarity between the nucleotide matrix and a potential TFBS detected in the sequence. TFBSs only carry the potential to bind to their corresponding proteins. However, they can occur every where in the genome and are by no means restricted to the regulatory regions. Sites outside regulatory regions are known to bind their TFs (Kodadek, 1998) and it is this context that differentiates a functional binding sites affecting gene regulation from a mere physical binding site (Elkon et al., 2003). TFBS prediction programs like **MatInspector** can infer the binding potential although not the functionality of a site. Functionality can ultimately be proven only by a lab experiment with defined settings particularly since potential binding sites in a promoter can be functional in certain cells, tissues or developmental stages and nonfunctional under different conditions. What we have seen above that the regulatory sequences usually contain many sites capable of binding TFs and the selection, which TFBS are functional for transcriptional control, depends on the biological context. Experimental expression analysis using expression vector, therefore, may be affected by additional TFBS in the vector as the number and combination of relevant TFBS may be influenced by the experimental conditions. So in such experiment it is essential to remove all potential TFBS from the sequence of the vector. This is achieved by creating point mutations since the deletion of complete binding sites or even single nucleotides can change the distance between functional binding sites and thus influence the interactions of binding proteins. But then there is a problem. Deleting one binding site might remove additional overlapping binding site from the sequence or might result in generation of a new binding site. A program called **Sequence Shaper** has been developed to make this kind of sequence. MatInspector can find most true positives TFBS matches and reduces the amount of false positive matches in a promoter region. Not all sites found are necessarily functional in the particular biological context. A first step in examining functionality is a comparative promoter analyses. Promoters sharing a common function, for example, promoters responsive to interferon or a set of actin promoters from different species can be compared. A binding site that occurs in most promoters particularly at a similar relative position within promoters is a presumably evolutionary conserved and this represents supporting evidence that the site may be functional. MatInspector can find the potential binding sites of various activators and repressors that bind to specific DNA regulatory sequences.

The concurrent expression of genes as observed in expression array analyses, in particular, is in part orchestrated by sets of common regulatory elements. Such an anlysis is complicated by the facts that co-expressed genes in an expression array experiment are not all necessarily co-regulated as different regulation mechanisms can lead to the same expression pattern. Effects like a secondary cascade of transcription activation during the time course can divide a co-expressed cluster of genes in subsets regarding co-regulation. Thus careful selection of gene clusters is a crucial step in successful promoter analysis (Catharius et al., 2005).

Transcription and regulation are co-ordinated actions beyond the traditional promoter sequences. Transcription factors previously thought to primarily bind promoters bind more generally and those which do not bind to promoters are equally likely to bind downstream of a transcription start site (TSS) as upstream. Many elements that previously were classified as distal enhancers are, in fact, close to one of the newly identified TSS. Only 35% of sites showing evidence of binding by multiple TFs are actually distal to a TSS.

24.41 TF-BINDING SITES

Binding sites for specific proteins are most important among regulatory elements. They consist of about 10 to 30 nucleotides, not all of which are equally important for protein binding. Individual protein binding sites may vary in part of their sequence even if they bind to the same protein. There are nucleotides which are in contact with the protein in a sequence-specific manner ('**recognition exons**') which usually represent the best conserved areas of a binding site. Different nucleotides are involved in more non-specific contacts to the DNA backbone (i.e. not sequence specific as they do not involve the bases A, G, C, or T) and there are internal spacers (introns) which are not in contact with the protein at all. All in all PBSs show enough sequence conservation to allow for detection of candidates by a variety of sequence similarity-based approaches (see Roy, 2009). Potential binding sites can be found almost anywhere in the genome and are not restricted to regulatory regions. A number of binding sites outside regulatory regions are also known to bind their respective binding proteins (Kodadek, 1998). Several functionally similar types of these stretches of DNA are already known and will be referred to as **elements**. These elements are neither restricted to regulatory regions nor individually sufficient for regulatory function of a promoter or enhancer. The function of complete regulatory region is composed of functions of the individual elements either in an additive manner (independent) or by synergistic effects (modules).

The transcription of genes is controlled by interaction between TFs and their BSs and so identifying and characterizing the BSs of a TF can provide a bettwe understanding of the function of the TF. The combinatorial presence and absence of transcription factor binding sites to, to a large degree, responsible for the complexity of gene regulation in virtually every organism (Wingender et al., 2000, 2001; Pickert et al., 1998; Kel-Margoulis et al., 2003). Identification of TFBSs can be through experimental techniques such as foot printing experiments or chromatin immunoprecipitation experiments.

24.42 EUKARYOTIC POLYMERASE II PROMOTERS

They are currently the best studied regulatory regions. The TF binding sites within promoters (and likewise most other regulatory sequences) do not show any obvious general patterns with respect to location and orientation within the promoter sequences. TF-BSs can be found virtually every where in promoters but in individual promoters possible locations are much more restricted. The function of TF-BS often depends on the relative location and especially on the sequence context of the binding site. The context of a TF-site is one of the major determinants of its role in transcription control. However, the context is not merely a few nucleotides around the binding site which could be more like extension of the BS rather than the context. More important is the context of other TFBSs located some distance that are often grouped together and such functional groups have been described in many cases.

In many cases a specific promoter function (e.g. a tissue-specific silencer) will require more than two sites simultaneously. Such groups of promoter subunits consisting of several TFBS that carry specific function independent of the promoter, will be referred as **promoter modules**. Within a molecular promoter module both sequential order and distance can be crucial for function which shows that these modules may be critical determinants of a promoter rather than individual binding sites. However, promoters can contain several modules that may use overlapping sets of binding sites. Therefore the conserved context of a particular binding site can not be determined from the primary sequence without additional information about the modular structure.

A motif is usually represented by a string (sequence) or a matrix. When a motif is represented by a 4 x l probability matrix, M, where l is the length of the binding sites, the ith column of M represents the occurrence probabilities of nucleotide A, C, T and G at the ith position of a binding site. Since biological signals are subject to mutations and usually do not appear exactly, they typically use probability weight matrix to represent motifs (for detail see Roy, 2009). According to traits of motif, motif discovery problem is to find a pattern in sample of sequences whose length is l and in every sample sequence there is a pattern which has no more than d mismatches with this motif pattern (Ma et al., 2006).

Although many real biological motifs can be better represented by a matrix, most existing algorithms can not guarantee finding the optimal matrix-represented matrix for a given set of sequences and those algorithms that can take a very long time to do so when l is large. When a length –l string P is used to represent the motif all binding sites are length –l strings similar to P with at most d point substitutions. In other words, Hamming distance between the motif and each binding site is at most d. Since there are a finite number of length l-string (4^l possible strings) many algorithms can guarantee finding the best string motif. The drawback is that some real biological motifs can not be represented by strings (Chin and Leung, 2006).

A sequence may contain zero or more variants of the motif which is represented by a length-l sequence called pattern, consisting of symbols{ A, C, T, G, N}. A variant of a pattern P is a substring exactly the same as P except that each wild card symbol N is replaced by A, C, T Or G where d is the number of wild card symbols in P. Patterns with d values are compared with their z-scores (the number of standard deviations by which the number of variants of a pattern in the input sequences exceeds its expected number). Patterns with higher z-scores are more likely to be the correct motif and the optimal motif is the pattern with the highest z-score.

Transcription factors are DNA-binding proteins at the terminal of signal transduction networks and in genomic sequences, a TF-binding site (motif) is a set of *cis*-regulatory elements that preserve a certain nucleotide composition, play a role in transcriptional regulation. Each TF recognizes a specific binding site composed of similar substrings, referred to as *cis*-regulatory variants. Recently, such subtle variations were hypothesized to also play a key role in transcriptional control (Cowles et al., 2002; Tamay et al., 2004). It is generally assumed that *cis*-regulatory variants are hard to be detected only by sequence analysis but rather require extensive experimental studies (Kawada and Sakakibara, 2006). Two assumptions are made under transcriptional regulation. The first is that the expression of a gene is governed by the binding sites of specific transcription factors on its promoter and the second is that the expression of a gene is a function of the concentration of specific transcription factors around its promoter. Approaches modeling the first associate gene expression level with putative binding motif on their promoter sequences whereas there are approaches which model gene expression levels from the expression levels of other genes, i.e. TFs and other regulators.

24.43 METHODS OF IDENTIFICATION OF TF BINDING SITES

Although there are many methods of identification of TF binding sites, it is still a challenging and unsolved problem.

1. **Computational methods**—The computational methods of detecting motif involves examining only upstream sequences of clustered and presumably co-regulated groups of genes or bound genes by the same TF and searching for statistically over-represented motif among them (for more see Roy, 2009).

2. **ChIP (Chromatin immunoprecipitation)**—Chromatin immunoprecipitation microarray experiments elucidates *in vivo* physical interactions between transcription factors and their chromosomal targets on the genome (Harbinson et al., 2004; Lee et al., 2002). Transcription factors are expected to interact primarily with promoters. To identify the targets of sequence-specific transcription factors SBF and MBF which activate gene expression during G1/S transition of the cell cycle in yeast chromatin immunoprecipitation and microarray hybridization were combined (Iyer et al., 2001). Proteins were cross–linked with formaldehyde to their target sites *in vivo*. DNA that was specifically cross-linked to either of the TFs was purified by immunoprecipitation using an antibody against either the native protein or an epitope tag that was fused to the protein. PCR analysis of immunoprecipitated DNA confirmed the specific association of TFs with several target promoters. After reversal of the cross links, immunoprecipitated DNA was amplified and fluorescently labeled with Cy5 fluorophore. Also, a separate control DNA sample was fluorescently labeled with the Cy3 fluorophore. Genomic target loci were identified by comparative hybridization of the immunoprecipitated and controlled DNA probes to a DNA microarray. The ratio of Cy5/Cy3 fluorescent intensities measured at each DNA element in the microarray provided a measure of the extent of binding of the TF to the corresponding genomic loci (Figure 24.1). In summary, in the ChIPchip experiments DNA is crosslinked *in vivo* to proteins at sites of DNA-protein interaction and sheared to 500bp-2kb fragments. The DNA-protein complexes are precipitated by antibodies specific to the TF of interest. The precipitated protein-bound DNA fragments are PCR amplified, fluorescently labeled and hybridized to microarrays containing every promoter (sometimes also every ORF) in the genome. DNA fragments which are consistently enriched by **ChIP-chip** over repeated experiments are identified as positive sequences containing the protein-DNA interacting loci at 1kb resolution. When compared with the gene expression data, the ChIP-chip data provide much more accurate information about the genome-wide location of *in vivo* TF-DNA interactions which enables to assign definitive class to some promoter sequences with high confidence. For detail see chapter 25.

This technique provides useful information about binding of a specific protein complex to DNA. The ChIP data provides information about not only TF-binding but also TF-DNA unbinding. True motifs appear only in the upstream sequences of the target genes, controlled and bound by the TF and do not appear in those of the unbound genes and so a discriminative approach is used to find true motifs that distinguish the upstream sequences between bound and unbound genes. Decision tree learning method is used for extracting sequence motifs given the positive and negative samples and motif is defines as informative substring (consensus sequence) which can correctly classify genes into proper class

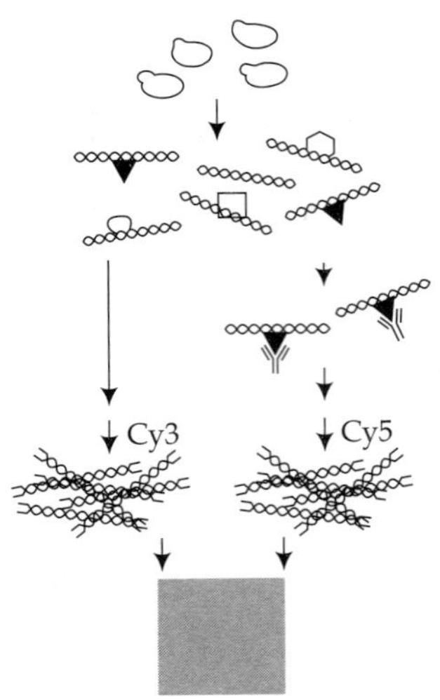

FIGURE 24.1 Chip-chip example: genome-wide localization analysis in yeast.

(positive/negative) based on their upstream sequences. This method takes into account unbound upstream sequence (negative samples) and bound sequences (positive samples). It defines motifs as 'discriminative' substrings which correctly distinguish between the upstream sequences of positive samples from those of negative ones. It uses a discriminative machine learning tool for detecting motif. Current models can not distinguish between the consensus sequences and their functional variants. In contrast with most existing methods Kawada and Sakakibara (2006) searched for main motifs and their functional variants by focusing on the subtle differences among substrings rather than allowing and unifying them.

3. **ChIP Seq**—How many TF is deployed across the entire genome for a given cell type? Direct physical interactions between TFs or cofactors and the chromosome can be detected by ChIP as described above. In ChIP experiments an immune reagent specific for a DNA binding factor is used to enrich target DNA sites to which the factor was bound. The enriched DNA sites are then identified and quantified. For the giga base size genomes of vertebrates it has been difficult to make ChIP measurements that combine high accuracy, whole genome completeness and high binding sites resolution. The data quality and depth issue dictate whether primary gene network structure can be inferred with reasonable certainty and comprehensiveness and how effectively the data can be used to discover binding site motifs by computational methods. For these purposes statistical robustness, sampling depth across the genome, absolute signal and signal –to- noise ratio must be good enough to detect nearly all *in vivo* binding sites for a regulator with minimal inclusion of false positives. Signal: noise ratio is the ratio of desired information (signal) to undesired background (noise). The signal: noise ratio thus depends on the signal level which is variable and the noise level which is more or less constant. A further challenge in genomes large or small is to map factor binding sites with high positional resolution. In addition to making computational discovery of binding motifs feasible, this dictates the quality of regulatory site annotation relative to other gene anatomy landmarks such as transcription start site, enahncers, introns and exons and conserved noncoding features. If high quality protein-DNA

interactome measurements can be performed regularly then *in vivo* protein-DNA interactions connect each TF with its direct targets to form a gene network scaffold. Johnson et al. (2007) developed a large scale chromatin immunoprecipitation assay (ChIP Seq) based on ultra high throughput DNA sequencing. They used a ultrahigh throughput DNA sequencing to gain sampling power and applied size selection on immuno-enriched DNA to enhance positional resolution. This assay differs from other large scale ChIP methods such as ChIP Array, also called ChIP-chip, ChIPSAGE (SACO) (Impey et al., 2004) or ChIPPet (Wei et al., 2006) in design, data produced and cost. Unlike SACO or ChIPPet it involves no plasmid library construction. Unlike microarray assays the vast majority of single copy sites in the genome is accessible for ChIPSeq assay rather than a subset selected to be array features. In addition ChIPSeq counts sequences and so avoids constraints imposed by array hybridization chemistry such as base composition constraints related to Tm, the temperature at which 50% of double stranded DNA or DNA-RNA hybrids is denatured, cross hybridization and structural interference. Finally, ChIPSeq is feasible for any sequenced genome rather than being restricted to species for which whole genome tiling arrays have been produced.

4. Serial analysis of chromatin occupancy (SACO)

The CHIP-on-chip approaches made possible the large scale mapping of the TF binding to chromatin. In those approaches the DNA fragments are selected by the CHIP are hybridized on custom-made DNA microarray (see chapter 25 for detail). In case of mammalian genomes such microarrays contain only sequences of proximal promoters (typically within 1kb of the transcription start site) of the currently annotated genes, an experimental choice which is inherently biased. High density microarrays which provide coverage at a high resolution of the non-repetitive sequences of a target portion of the human genome have been developed (Cawley et al., 2004) which allow unbiased genome–wide location analysis for selected TFs but this analysis has revealed that only a fraction of these sites are located within the proximal promoters. However, performing a complete genome-wide analysis of TF binding using tiled array would be very constly in case of sizeable genomes such as the mammalian because of high construction cost of such arrays, the SACO is an alternative technique for investigating the genome-wide mapping of TF of interest. SACO is also termed GMAT, STAGE and SEBE. In contrast to CHIP-on-chip assays that require building an array of promoter sequences which can bind to a given TF, this technique allows the unbiased genome-wide identification of all genomic targets for the TF. SACO has the ability to combine the specificity of CHIP with the sensitivity of SAGE. Steps involved in SACO are as follows.

1. This first step involves a CHIP-step in which the TF is cross-linked *in vivo* to its genomic targets, the DNA is sheared by sonication, isolated using a highly specific antibody for the TF, released from the protein-DNA complex by reverse cross-linking and purified.

2. The second step is a SAGE-like step in which the DNA fragments are amplified by ligation-mediated PCR and digested with the Nla III endonuclease-a four cutter enzyme which recognizes the sequence CATG. The resulting DNA fragments are ligated to linkers containing a recognition site for the MmeI endonuclease and then

cleaved with MmeI in order to release 21bp DNA fragments (which are referred to as tags) containing an NlaIII site. These tags are end-to-end ligated into 'ditags', concatenated and subcloned into sequencing vectors thus generating the SACO library.

The SACO experiment thus yields a collection of sequenced inserts that contain a large number of 21bp tags. Each tag 'labels' a genomic fragment of about 1kb (depending on the sonication resolution) that contains the tag and one or several sites to which the TF bound. Further processing of data is based on the assumption that the 21bp sequence of the tag can specifically identify a chromosomal location in an entire genome and therefore leads to the identification of the genes and genomic regions targeted by the TF. Marinsecu et al. (2006) have developed a START software-an automated tool for analysis of data generated by SACO.

Sequence tag analysis of genomic enrichment (STAGE)—It is a method similar to ChIP-chip for detecting protein factor binding regions but using extensive short sequence determination rather than genomic tiling arrays.

All these three methods (ChIP-chip, ChIP-PET and STAGE) use chromatin immunoprecipitation with specific antibodies to enrich for DNA in physical contact with the targeted epitope. This enriched DNA is then analyzed using either microarrays (ChIPchip) or high throughput sequencing (ChIP-PET and STAGE).

24.44 QUANTITATIVE MODELING OF REGULATORY NETWORK

Although CHIP has uncovered much information about the architecture or connectivity of the networks, any quantitative model would require the knowledge of both the concentration of TF proteins at a given time and the intensity with which they promote or repress transcription of their target genes. However, there are problems associated with these. First, measuring protein concentrations is a difficult task and little help can be gleaned from the knowledge of TF gene expression levels. Secondly, TFs are often post-transcriptionally regulated and have low and noisy expression level. Thirdly, the effect of a TF has on a target gene depends greatly on the experimental conditions. All these make experimental estimation of the strength of regulatory relationship a difficult task (Sanguinetti et al., 2006). A problem with CHIP data is the occurrence of false positives.

DNA microarray data and CHIP-chip binding analysis often form the basis of transcriptional regulatory analysis. Trancriptional analyses often use CHIP-chip data for defining transcriptional connectivities which are used to analyze gene expression data under the assumption that CHIP-chip-derived connectivities are biologically meaningful. Under this assumption CHIP-chip based network connectivities should explain more of the gene expression data than the random networks. Ideally these complementary data sources should provide sufficient approximations of gene transcription (expression data) and the transcriptional regulatory network (binding data). However, the ability of these data sources to provide such functionality is often adversely affected by a number of factors described above. The experimental noise associated with DNA microarray technology affects its ability to properly portray transcription while for CHIP-chip data, in addition to noise, environmental dependence in binding and uncorrelation between binding and regulation cause problems. These issues may impart uncertainity to transcriptional analyses. Brynildsen et al. (2006) used a combinatorial model of transcriptional regulation while attempting to describe gene expression from ten different environments and demonstrated that the currently available CHIP-chip data perform as well as random networks with the same connectivity density. This result shows that the approximations of transcription as expression data and the transcription network as CHIP-chip binding data are in certain instances insufficient and may generate misleading inferences from transcription analyses. So identify the agreement between gene expression and CHIP-chip binding data in order to ensure the accuracy of these approximations. This suggestion is based on the premise that inaccuracies are less likely to be present when separate data sources point to the same conclusion.

24.45 INTEGRATED ANALYSIS

DNA microarray is one of the most powerful techniques developed to survey the transcriptional profile of the entire genome. Microarray-based gene expression experiments provide information about gene, gene products and their function. Microarray data are generated for whole genomes, multiple genomes both within and across species. Many methods have been proposed for extracting biological meaning from microarray data including normalization and meta analysis, clustering, signature experiments, detection of differential expression. There is a need for intergrated analysis of microarray data because of difference in technology, protocols and experimental conditions. So the integration system must be robust. When examining among diverse biological datasets (microarray experiments from differing experimental conditions) it is critical to consider functional specificity, i.e. which biological processes are active in which experiments. For example, if datasets are from *S. cerevisiae* in experimental conditions inducing sporulation then it will inform us about the network related to meiosis-related genes. Integration can enable broader understanding of gene regulation in the context of specific pathways and can allow the discovery of co-expression relationship too weak to be detected in individual experiments.

24.46 GENE EXPRESSION PROFILES AS FINGERPRINTS

Gene expression profiling provides one approach to study cellular processes at the gene level.In other words, it has ben applied to elucidate the mechanisms underlying a biological pathway. Similar to DNA fingerprints one can have complex profiles consisting of thousands of individual observations which can serve as transcriptional fingerprints. Further, **expression map** (or transcription map) can be constructed here which refers to a graphical description of the precise locations of expressed genes (gene map) in different cell types of an organism. For construction of an expression map, the mRNAs of each cell type are isolated, reverse transcribed into cDNAs which are then sequenced and primer pairs-specific for each cDNA are then synthesized in the presence of fluoresecent nucleotide ([F] (dUTP)). The various cDNAs may either be assigned to specific chromosomes by previous flow cytometry and *in situ* hybridization or by linkage analysis. cDNA normalization greatly increases the efficiency of transcriptional analysis and functional screenings by equalization of the concentration of the different transcripts in a cDNA population.

Transcriptome map—It refers to a misleading term for a compilation of co-regulated genes.

Applications of transcriptional fingerprints

1. The fingerprints can be used for classification purposes or as tests for relatedness, in a similar fashion to the way in which DNA fingerprints are used in paternity testing.

2. The differences in gene expression are responsible for both morphological and phenotypic differences as well as an indicative of cellular responses to environmental stimuli and perturbations. It is a means to catalogue the biological responses to a large number of diverse perturbations. Thus transcriptome is highly dynamic and changes rapidly in response to perturbations or even during normal cellular events such as DNA replication and cell division. Transcriptional fingerprints can be used to determine the drug target. If a drug interacts with and inactivates a specific protein (or specific gene) the phenotype of the drug treated cell should be similar to the phenotype of a cell in which the gene encoding the protein has been genetically inactivated by mutation. Thus a comparison of expression profile of a drug treated cell and a mutated cell can be made and specific mutants can be matched to specific drugs. For example, the gene product of his3 gene was identified as the target of 3-aminotriazole.

3. Gene expression profiles can be used to classify drugs and their modes of action. Functional similarity and specificity of different purine analogues have been determined by comparing the genomes-wide effects on treated yeast, murines and human cells. Thus in humans it can help determine the causes and consequences of diseases, how drug and drug candidate works in cells and organism and what gene products might have therapeutic uses themselves or may be appropriate targets for therapeutic intervention.

4. Changes in multigene expression can provide clues about regulatory mechanisms and broader cellular functions and biochemical pathways.

24.47 OTHER GENE EXPRESSION METHODS

The expressed genes are usually isolated by a differential screening approach in which a cDNA library is screened with labeled single stranded cDNA which is obtained from converting the mRNA extracted from tissues after +/- treatments. Variations in this theme include subtractive hybridization and differential display. Besides arrarys methods there are other ways to measure mRNA abundance, gene expression and changes in gene expression. For measuring gene expression at the level of mRNA, northern blots, RT-PCR, nuclease protection, cDNA sequencing, clone hybridization, differential display, subtractive hybridization, cDNA fingerprinting, and SAGE have all been put to use to measure the expression levels of specific genes, characterize global expression profiles or to screen for significant differences in mRNA abundance. Large-scale gene expression assays may be performed using RT-PCR, SAGE or DNA-chip technology. RT-PCR is a technique for amplifying a specific region of a transcript. RT-PCR can be scaled up to cover the same number of genes as SAGE and DNA–chips by using robotics and capillary electrophoresis arrays for separating PCR products.

There are many approaches for measurement of gene expression at a single gene level. Standard methods such as northern blots, western blots or RT-PCR are simply used in a more targeted fashion to complement the broader measurements and to follow up on the genes, pathways and mechanisms implicated by the array results. Because the incidence of false positives can be made significantly lower, it is necessary to independently confirm every change for the results to be valid and trustworthy especially if conclusions are based on changes in set of genes rather than individual genes.

24.48   RNA (NOTHERN) BLOT ANALYSIS

Gene expression can be monitored with RNA blot analysis. In this analysis designated amounts of mRNA from wild-type and variant (say HAT4-transgenic plants) are spotted onto nylon membranes and probed with cDNAs. Analysis of RNA blots can be used to determine the size of transcript (mRNA), differences in transcript size or abundance of mRNA produced by genes. RNA blotting is a gel blotting technique in which RNA molecules are separated according to size by agarose or or PAGE and are transferred directly to a nitrocellulose filter or other matrices by electric or capillary forces. The single stranded nucleic acids may be fixed to the nictrocellulose filter by backing and are thus immobilized. Hybridization of specific, radioactively or non-radioactively labeled, single stranded probes to the immobilized RNA molecules allows the detection of individual RNAs out of the complex RNA population. The **non-combinatorial array based methods** involve the serial spotting of multiple clones or cDNAs onto nylon membranes or modified microscopic slide. These methods are inherently parallel for the analysis of sequence information and they complement the probe-based arrays in their ability to use previously non-sequenced biological materials for the expression analysis. However, hybridization to large cDNA or PCR products can be thermodynamically more stable than to a series of shorter discriminating oligonucleotides and can yield confusing cross-hybridization signals when examining closely related genes, gene families and other variants. In combination synthetic and mechanical technique build a powerful platform for studying gene expression and offers improvement over traditional cDNA library sequencing methods considering time and cost.

24.49   EXPRESSION MEASUREMENTS FROM SMALL AMOUNTS OF RNA-IDENTIFICATION OF GENES ASSOCIATED WITH DEVELOPMENT AND DIFFERENTIATION

Higher organisms contain about 25-40,000 different genes of which only a small fraction, perhaps 15% are expressed in any individual cells. It is the choice of which genes are expressed that determines all life processes such as development and differentiation, homeostasis, response to stress, cell cycle regulation, aging and even programmed cell death. Altered gene expression lies in the heart of the regulatory mechanism which control cell biology. In other words, differential gene expression underlies many fundamental biological processes such as embryo- and embryogenesis, cell and tissue differentiation, long term plasticity and cellular response to various stimuli and it is essential to identify and clone such differentially expressed genes for studying these processes. Comparisons of gene expression in different cell types provide the underlying information we need to analyze the biological processes that control over lives. The objective would be to identify differently expressed genes among the pproximately15, 000 individual mRNA species.

Most array experiments are done using RNA obtained from a million or more cells and getting such a large quantity of RNA is not a problem. But in case where it is essential to use fewer cells as when using a small organ from a fly or worm, sorted cells that express a rare marker or laser-capture microdissected tumor cells, efficient and reproducible mRNA amplification method is needed. There are two approaches to efficient and reproducible mRNA amplification. The first approach is a PCR-based approach whereas the second approach uses multiple rounds of linear amplification based on cDNA synthesis and a template directed *in vitro* transcription reaction.

PCR approach — The general strategy is to amplify partial cDNA sequences from subsets of mRNAs by reverse transcription and PCR (Prashar and Weismann, 1996). These short sequences are then displayed on a sequencing gel. Pairs of primers are selected so that each will amplify DNA from about 50 to 100 mRNAs because this number is optimal for display on the gel. Selection of 3′ primers takes advantage of the polyadenylate (poly A) tail present in most eukaryotic mRNAs to anchor the primer at the 3′ end, plus two additional 3′ basesA primer such as $5'\text{-}T_{11}$ CA would allow anchored annealing to mRNA containing TG located just upstream of their poly (A) tails. By probability this primer will recognize 1/12 of the total mRNA population as there are 12 different combinations of the last two 3′ bases, omitting T as the penultimate base. Any reverse transcribed cDNA species would be amplified by PCR if the distance at which a second primer anneals is smaller than 2 to 3 kb from the beginning of the poly (A) tail (an average molecular size of mRNA is 1.2kb). Ideally this annealing position should be within 500bp because cDNAs up to 500bp can be resolved by size on DNA sequencing gel. For a 5′ primer of arbitrary base sequence, annealing positions to cDNAs should be randomly distributed in distance from poly (A) tail. Therefore, the amplified products from various mRNAs will differ in size. After these PCR products have been labeled with $\alpha\text{-}^{35}$ S/-labelled deoxyadenosine triphosphate (dATP) they would be displayed by autoradiography as a ladder on a sequencing gel. The 5′ primer should, in theory, be short, 6 to 7 bp for it to anneal fairly frequently near the end of a cDNA strand. Whether such short primers can give specific DNA amplification by PCR? Although arbitrary primers 8 to 10 nucleotides in length have been used for DNA polymorphism analysis by PCR the standard PCR method uses primers of 20 or more nucleotides in length. In summary, it is a method to separate and clone individual mRNAs by means of PCR reaction. The key element is to use a set of oligonucleotide primers, one being anchored to the polyadenylate tail of a subset of mRNAs, the other being short and arbitrary in sequence so that it anneals at different positions relative to the first primer. The mRNA subpopulations defined by these primer pairs are amplified after reverse transcription and resolved on a DNA sequencing gel. When multiple primer sets are used, reproducible patterns of amplified cDNA fragments are obtained which showed strong dependence on sequence-specificity of either primer.

Early methods developed to identify and clone such genes were primarily based on the principle of subtractive hybridization. The limitations with these methods are that they can analyze only a fraction of the overall changes in gene expression. They require large amounts of RNA and are lengthy and laborious. Liang and Pardee developed a gel based technique which facilitates rapid and extensive analysis of differentially expressed mRNAs. The limitations with this technique are that there is lack of quantitative correlation with mRNA abundance, a significant incidence of false positives, variable reproducibility of the display patterns and there is under representation and redundancy of mRNA signals. All these make it difficult to fully evaluate differential gene expression.

Amplification of cDNAs at low primer annealing temperature of 40°C, a non-stringent PCR condition, is considered a major limitation of current gel display protocol. Adaptations of the original protocol have been reported to overcome some of these limitations such as the use of 1-base anchored oligo (dT) primer for increased representation of mRNAs and the use of long composite primers to achieve reproducible patterns under more stringent PCR conditions. However, all these modifications continue to evolve annealing of arbitrary primers at 40°C for cDNA amplification in the first few or all PCR cycles.

Global identification of gene—Global identification of gene can be made through Affymetrix 2-or genome tilling array and qPCR analysis is done to validate array result.

24.50 ALTERNATIVE APPROACH FOR cDNA DISPLAY ON GELS

Display patterns are generated when restriction enzyme-digested double stranded cDNA is ligated to an adaptor which mediates selective PCR amplification of 3′end fragments of cDNAs under high stringency PCR conditions instead of non-stringent arbitrary cDNA amplification. A diversity of pattern is generated by choosing different sets of restriction enzymes and anchored oligo (dT) primers with a heel. Since all cDNAs in a sample acquire a common heel from oligo (dT) primer during synthesis, most cDNAs molecules in a subset [determined by the anchor nucleotides of the oligo (dT)] can be displayed by choosing a combination of restriction enzymes, thus significantly reducing under representation or redundant representation of mRNAs. This approach provides near-quantitative information about the levels of gene expression, can resolve hidden differences in the display gel (hidden differences such as bands that differ in their sequence but co-migrate on a gel), produces a single band for each mRNA species and produces bands of predictable size for known genes sequences (bands corresponding to known cDNAs move to predictable positions in the gel and thus a powerful approach to correlate gel patterns with cDNA databases). This method produces consistently reproducible display patterns. This approach of monitoring changes in gene expression by selective amplification and display of 3′end restriction fragments of double stranded cDNAs can detect almost all mRNAs in a sample.

Thus we see that a combination of functional genomic mapping approaches can be used for the identification of genes. High throughput methods such as transcriptional profiling, protein interaction mapping and large scale phenotypic analysis have been applied individually to worms with considerable success. Although no single HT method can unequivocally define gene function, combining the data from any these complementary approaches is likely to provide greater insight. Combining data from complementary large scale approaches establishes a new paradigm in the field of functional genomics that makes it possible to accelerate functional discovery.

Classical approach—The classical approach to isolation of differentially expressed genes is the differential screening of a cDNA library with labeled probes from two or more different mRNA samples. Subtractive hybridization can be applied to enrich for sequences which are unique for one cell type.

Recently a method for RNA characterization by an arbitrarily primed reverse transcription-coupled PCR (AP-RT-PCR) has been developed. Patterns of gene expression are produced by low-stringency PCR with short arbitrarily selected primers followed by analysis of products of high resolution PAGE.

Gene expression fingerprinting (GEF)—It is based on the creation of nonoverlapping sets of 3′ terminal cDNA restriction fragments which are resolved by high resolution PAGE. This procedure gives rise to highly informative, fingerprint like patterns which can be used for comparative analysis of ensembles of expressed genes as well as for cloning of differentially expressed genes. This is based on ligation-mediated PCR and reaction is not sensitive to amplification conditions. The major disadvantage is that only abundant species are displayed as the number of amplified fragments applied to a single lane of a gel is about 2000.

24.51 MOLECULAR INDEXING

Class II S restriction enzymes cleave DNA at precise location outside their recognition sites and produce overhangs of unknown sequences. Molecular indexing refers to a series of techniques designed to characterize DNA fragments by these unknown sequences (Kato, 1996). It was applied to the description of the total mRNA population of a 3′ end cDNA fragment generated by class IIS restriction enzyme. This method is based on finding that *E. coli* ligase discriminates three nucleotides adjacent to the joining site. Fragments are discriminated by a library of 64 adaptors for all possible overhangs and selected fragments are PCR amplified using an adaptor-primer and anchored oligo (dT) primer. They are separated and displayed by a denaturing PAGE. Comparing electropherograms from various sources of RNA, differentially expressed genes can be identified easily. This method has advantages over display techniques based on arbitrarily primed PCR. In particular, this method can display most genes with low redundancy. However, amplified fragments correspond to the 3′ end of mRNA and there is not much chance of them containing coding regions. Kato described an alternative protocol for amplifying fragments from upstream regions. Original indexing procedures or differential display which are intended to amplify 3′ end cDNA fragments but amplified fragments are not restricted to 3′ end regions and have an increased likelihood of containing coding regions. Kato intended to amplify 5′end cDNA fragments-cDNAs digested by a class IIS restriction enzyme which produce a four nucleotide 5′ overhang.

24.52 TRANSCRIPTION BASED AMPLIFICATION SYSTEM (TAS)

A transcription based amplification system has been developed by Kwoh, (1989). The two steps involved in TAS are cDNA and RNA transcription. These are performed using an RNA template, reverse transcriptase and a chimaeric primer which is complementary to the RNA sequence and also contains an RNA polymerase binding site (T7 RNA polymerase). Up to

100 copies of RNA are produced and passed on to the RNA transcription step which increases the copy number of the template for cDNA synthesis in the next cycle of TAS. Thus unlike the standard PCR, few cycles are required to generate millions of copies of the target molecules.

Long accurate PCR (LA-PCR)—It is a variant of PCR which uses AMV reverse transcriptase in combination with a specific oligo (dT)-adaptor primer for the synthesis of first strand cDNA from a poly (A)$^+$ RNA (taking place at 42-60°C) and a Taq polymerase for a second strand synthesis. LA-PCR permits amplification of cDNA of up to 12.2 kb.

24.53 CATALOGUING OF TRANSCRIPTOME

The transcriptome can be catalogued based on 1. assembly of sequences from large scale EST sequencing projects 2. SAGE and 3. large scale sequencing of ORF sequence tags (Camargo et al., 2001).

Types of RNA transcripts—RNA transcripts are involved in many cellular functions either directly as biologically active molecules or indirectly by encoding other active molecules. There are different types of RNA transcripts.

1. Sets of RNA transcripts (e.g. mRNAs) which are encoded by distinct loci with each usually has a single biological role (for example, encoding a specific protein). This is the conventional view of genome organization but this simple picture is now looking more complex with the identification of many other types of transcripts.

2. Small nucleolar RNA s and microRNAs-   These RNAs are often encoded by regions that intercalate with protein coding genes. The discrepancy between the levels of observable mRNAs and large structural RNAs compared with the total RNAs in a cell suggests numerous RNA species yet to be classified.

3. Presence of RNA transcripts having specific role in chromatin maintenance and other regulatory control.

Transcript maps—There are three methods of identifying transcripts originating from genomic regions.

1. Hybridization of RNA (either total or poly (A)–selected) to unbiased tiling arrays
2. Tag sequencing of cap-selected RNA at the 5′ or joint 5′/3′ ends
3. Integrated annotation of available cDNA and EST sequence involving computation, manual and either experimental approaches.

Regions of transcript identified by an unbiased tiling array asssay is termed TxFrag (fragment of transcript). It represents a genomic region found to be present in a transcript. The regions identified by tag sequencing of cap-selected RNAs are called **CAGE** (a short sequence from the 5′end of a transcript) or **PET** (a short sequence containing both the 5′ and 3′ ends of a transcript) tags. Mapping of CAGE tags identifies the likely promoters and transcription start site of many genes identified by tilling array also called transfrags (Kampa et al., 2004). For this CAGE libraries are constructed, one is constructed with random primers and other with oligo d (T) primers. Mapping of the ends of transcripts can be used to identify the genomic span of the primary transcript. The region identified by integrated annotation is called GENCODE transcripts. The genecode annotations are categorized both by likely function (mainly the presence of an ORF) and by classification evidence (transcripts based solely on ESTs) This classification is not strongly correlated with expression levels.

Transcript fusion—The traditional view that many genes have one or more alternative transcripts (as a result of alternative splicing) that code for alternative proteins seems to be changing. The findings by ENCODE consortium suggest that a given gene may encode both multiple protein products and produce other transcripts that include sequences from both strands and from neighboring loci (often without encoding a different protein). A combination of RACE and tiling arrays ((Kapranov et al., 2005) can be used to analyze the diversity of transcripts. The transcript fragments detected using RACE followed by hybridization to tiling arrays are referred to as RxFrags. Analysis of 5′ ends of the transcripts derived from different tissues of a number of 399 protein coding loci showed that most loci had at least one RxFrag which often extends a considerable distance beyond the 5′ end of the locus. Thus the RACE-detected ends are different then the previously annotated TSS of each locus. The average distance of the extension is between 50kb and 100kb with many extensions (>20%) being more than 200kb. This extension represents the inclusion of at least one exon from an upstream gene which is consistent with the presence of overlapping genes in the human genome. To characterize further the 5′ RexFrag extensions, RT-PCR was performed followed by cloning and sequencing of 5′ RxFrags. Hybridization of the RT-PCR products to tiling arrays confirmed the transcript connectivity and sequencing clones confirmed transcript extensions. Thus detection of RxFrag extensions coupled with evidence of considerable intronic transcription shows that the protein coding loci are transcriptionally complex than previously thought. A 330kb interval of human chromosome 21 contained four annotated genes: DONSON, CRYL1, ITSN1 and ATP50, and a fusion transcript expressed in small intestine and consists of at least 3 exons from the ATP50 gene and at least 2 exons from the DONSON gene. A PET tag showed the termini of a transcript consistent with the RT-PCR product.

24.54 FUNCTIONAL ANALYSIS OF PSEUDOGENES

To analyze the function of pseudogenes various computational analyses can be carried out. To start with tiling array analysis can reveal whether or not the transcript contains a TxFrag. But the possible cross-hybridization between the pseudogenes and their corresponding parental genes might confound such analysis. To assess better the extent of pseudogene transcription, pseudogenes are examined for expression using RACE/tiling array analysis and transcripts are detected. Additional evidence for the transcription can come from their proximity (within 100bp of a pseudogene) to CAGE tags, PETs or cDNA/EST.

A pesudogene is a gene copy that does not produce a functional, full length protein. In other words, pseudogenes are defective copies of functional genes. They are narly as abundant as functional genes. The human genome is estimated to contain up to 20, 000 pseudogenes. Pseudogenes are often cast as evolutionary relics and a nuisance to genomic analysis. They arise by processes which are needed to create whole gene families such as those involved in immunity and smell. Hirotsune et al. (2003) demonstrated a specific regulatory role of an expressed pseudogene which point to the functional significance of noncoding RNA. They found the expression of the Makorin 1 gene to be controlled by one of its pseudogene copies, Makorin 1-p1. There are two ways in which this might happen (Lee, 2003). 1. **An RNA-mediated mechanism-** Here mRNA copies of the pseudogene and gene compete for a destabilizing protein that binds a crucial 700-nt region near the beginning of the mRNA. This destabilizing protein might be an RNA-digesting enzyme 2. **A DNA-mediated mechanism-** Here the regulatory elements in the 700-nt region of the pseudogene and gene compete for transcriptional repressor.

Identification of primary transcripts—The extent of primary (that is, unspliced) transcript can be examined using three technologies such as integrated annotation from GENCODE, PET tags and RxFrag extensions which will assess the presence of a nucleotide of a genomic sequence in a primary transcript. The result showed 93% of the bases of the ENCODE region overlapped transcripts identified by these technologies. The presence of PETs or RxFrags in a transcript which define the terminal ends of a transcript imply that the entire intervening DNA is transcribed and then processed. Again it shows broad amount of transcription across the human genome. But then other mechanisms such as *trans*-splicing or polymerase jumping not likely to be present in human would also produce these long termini.

24.55 CATALOGUING THE REGULATORY ELEMENTS

A simple view of transcriptional regulation involves five types of *cis*-acting regulatory sequences- promoters, enhancers, silencers, insulators and locus control regions (Maston et al., 2006). These regulatory elements are described in detail in chapter 16. Regulatory sequences reside outside the gene and they turn on and off. Regulatory sites occur virtually anywhere in the vast areas between protein coding regions (intergenic regions) and unlike non-functional regions they are conserved. It is essential to identify the regulatory elements that control the expression of each transcript and to understand how the function of these elements is co-ordinated to execute the complex cellular process. Overall, transcriptional regulation involves the interplay of multiple components whereby the availability of specific transcription factors and the accessibility of specific genomic regions determine whether a transcript is generated. The current overly simplified view of transcriptional regulation is that the consensus sequences of transcription factors binding sites (typically 6 to 10 bases) have very little information content and are present numerous times in the genome with great majority of these not participating in transcriptional regulation. At this point a number of questions can then be asked. Does chromatin structure then determine whether such a sequence has a regulatory role? Are there inter-factor interactions which integrate the signals from the multiple sites? How are signals from different distal regulatory elements coupled without affecting all neighboring genes?

Cataloguing the transcription start sites—It is certainly insufficient to simply take the sequence upstream from the start codon as 5′ UTR can often span additional 5′ exons in higher eukaryotes (Reese et al., 2000). Cataloguing of TSSs in ENCODE region was made on the basis of 5′ends of GENCODE annotated transcripts and the combined results of 5′ CAGE and PET tagging (the 5′ end capture technologies). The results suggested the presence of multiple transcription start sites (TSSs)within a single small segment (up to ~ 200bases) in many cases. This was due to some promoters containing TSSs with many very close precise initiation sites. TSSs can be grouped into three categories-known (present at the end of GENCODE-defined transcripts), novel (supported by other evidence) and unsupported.

DNaseI hypersensitive sites (DHSs)—DHSs and TSSs both reflect genomic regions thought to be enriched for regulatory information and many DHSs reside at or near TSSs. Sequence-specific transcription factors show a marked increase in binding across the broad region that encompasses each TSS. This increase is symmetric with binding equally likely upstream or downstream of a TSS. Many of the histone marks and Pol signals are now clearly asymmetrical with a persistent level of PolII into the genic regions. TSSs near CpG islands show a broader distribution of histone marks than those not near CpG islands. Finally, finding of some transcription factors such as E2F2, E2F4 and MYC, is extensive in case of active genes and is lower (or absent) in case of inactive genes.

RFBRs—It refers to regions with enriched binding of regulatory factors. RFBRSs can be identified using ChIP-chip data. The distribution of RFBRs is random and correlates with the position of TSSs. Further, factors for which binding sites are most enriched at the 5′ends of genes include histone modifications, TAFI and RNA Pol II with a hypo-phosphorylated carboxy terminal domain- confirming previous expectations.

Thus regulatory information around both TSS and DHSs in ENCODE genome has indicated presence of a large number of TSSs, almost 10-fold than the number of actual genes. The large number of TSSs might explain the extensive transcription described above. In the absence of this information on TSSs, many of the regulatory clustered would have been classified as residing distal to promoters. In other words, there is abundance of promoter-proximal regulatory elements but besides these the studies indicated a considerable number of putative distal regulatory elements.

Relationship between transcription and replication—On a large scale, early DNA replication in S phase is broadly correlated with gene density and transcriptional activity. However, this relationship is not universal as some actively transcribed genes replicate late and vice-versa. This relationship between transcription and replication is observed only when the signal of transcription is averaged over a large window (>100kb) which suggest that large scale chromosomal architecture may be more important than the activity of specific genes.

24.56 IDENTIFICATION OF EXPRESSED GENES

The expressed gene can be detected by detemining partial sequences of cDNA called 'expressed sequence tags' (ESTs).

24.56.1 Generation of ESTs

First, the mRNA is extracted from the material being studies and used as template by reverse transcriptase for cDNA synthesis. Then the cDNA is cloned into a suitable vector to produce a cDNA library. Random clones are selected and sequenced. In this approach single –pass sequences of 300 to 500bp are determined from one or both ends of randomly chosen cDNA clones. It is useful that the cDNAs in the library are cloned directionally so that the ESTs are isolated specifically from either the 5′or the 3′-end of the cDNA. 3′-ESTs often represent the 3′-untranslated region of the mRNA and are used to separate members of gene families that have similar coding sequences whereas 5′-ESTs generally represent coding sequences and give a better idea of the type of gene being expressed. EST analysis is used to find previously undiscovered tissue-specific genes or to tag as many expressed genes as possible. It is also used to examine the relative abundance of expressed genes. cDNA libraries used for the first two types of studies required often to be normalized so that highly expressed genes and rare genes are represented equally in the library or subtracted to reduce or eliminate the number of highly abundant clones (see subtractive hybridization). But an alternative to normalizing or subtracting libraries is to sequence a small number of ESTs from as many different tissues types and treatments as possible as different types of genes are highly expressed under different conditions. If the starting material for mRNA extraction is limiting then cDNA mini-libraries suitable for EST analysis are generated by arbitrarily primed RT-PCR ((Neto et al., 1997).

24.56.2 Uses of ESTs

There are many applications of ESTs as given below.

1. Obtaining the full length cDNAs which can be used in mutagenesis, transgenic and expression studies to analyze gene function

2. Designing PCR primers or hybridization probes for cloning full length cDNA

3. **Genome analysis and mapping**—EST analysis can be used to identify expressed genes. Now it is not possible to predict coding sequences in genomic DNA sequences reliably from sequence information alone (see Roy, 2009). Even when complete genomic sequences are available EST can be used to verify putative coding sequences and confirming intron and exon boundaries. Further, they can be used to distinguish pseudogenes from real genes and in identifying alternatively spliced transcripts which could never be predicted from the genomic sequence alone. ESTs can be converted into markers, 'sequence tagged sites' STSs which can be used for genomic mapping. An STS is a set of PCR primers which identifies a single gene and STS primers are designed from the EST sequence. An STS can be mapped to a specific region of the genome and if DNA contigs are available for the genome then the STS can specify a genomic clone. Mapping of ESTs to the genomic clones will provide a map of the expressed genes for each clone and thus can be used to identify **candidate genes** for inherited diseases.

4. **Study of gene expression**—Gene expression in different cells, tissues, organs or organisms under normal and various treatment conditions can be studied by surveying ESTs isolated from libraries constructed from different mRNA sources. In case of non-normalized and unsubtracted libraries for isolating ESTs the frequency of a particular EST indicates the relative abundance of the tagged gene. Finally, EST database can be searched for finding tissue-specific genes.

5. **Rapid identification of new genes and gene families**—Traditionally gene sequences are obtained by a number of method such as by isolating a protein, by identifying a mutant and cloning the mutated gene by complementation or by positional cloning (see chapter 16). Compared with these methods, ESTs can be used to rapidly identify new gene sequences. EST database can also be used to rapidly find genes similar to a gene of interest. Methods such as PCR using degenerate primers or hybridization at low stringency also enable to identify similar genes but these methods are often not as successful as database searches. Databases can be searched by using either nucleic

acid or protein sequences but protein comparison searches are often more fruitful as protein sequences are better conserved. Additionally databases can be used to search for similar genes in different organisms simultaneously thereby facilitating comparative studies. Once similar genes have been identified they can be translated into protein sequences and the databases are searched again to find the sequences most similar to this second set of genes. Thus large families of related genes can be rapidly identified by EST database analysis.

6. **Cloning specific genes**—EST can be used to identify and clone human genes related to diseases. EST database has been searched to identify human cDNAs homologous to previously cloned Drosophila genes that have an interesting developmental phenotype (Banfi et al., 1996). The human sequences were then mapped by several methods and the positions of the cDNAs were compared with the map positions of human disease genes. Some of the cDNAs mapped to regions containing human disease genes which cause symptoms similar to the defects in the corresponding Drosophila mutant genes. These cDNAs thus become the candidate genes for the disease genes. Further studies can be undertaken to determine whether the cDNAs and the disease genes are one and the same. Human ESTs related to known genes in a variety of model organisms can also be identified. In one case ESTs are mapped to both human and mouse and thus such genes can provide candidates for human disease genes.

In any organism where mutation has been mapped to a certain genomic region or to a genomic clone, ESTs mapping to the same region will identify candidate genes. In more unusual organisms where few genes have been previously isolated, EST analysis provides a rapidly generated survey of the types of genes expressed by the organism and is used to isolate novel genes from pathogenic organisms. In plants, Arabidopsis is often used as a model system for gene isolation and EST database searches provide a rapid means of isolating similar genes from many crop plants. STSs from ESTs representing important plant genes could be used as markers in selection program in plant breeding.

Although ESTs are overlapping and have high error rate but they are sufficiently accurate to unambiguously identify the corresponding gene in most cases. Any collection of ESTs cloned into a cloning vector constitutes the EST library. EST has numerous uses from genetic mapping to analyzing gene expression. EST database of the National Center of Biotechnology Information lists more than 39,000 Arabidopsis ESTs and more than one-half of the total set of genes (about 21,000) is already represented by an EST within 1.9 Mb of contiguous genomic sequence (Bevan et al., 1998) and about one-half of the predicted genes can be assigned to a functional category based on similarity with known proteins. It is difficult to know exactly how many different genes are represented by these ESTs as cDNA clones frequently are truncated at the 5′ end and sequencing of 5′ ends of cDNA clones often yields non-overlapping sequences corresponding to the same mRNA. EST databases have proven to be a very useful resource for finding genes and for interspecies sequence comparison and have provided markers for genetic and physical mapping and clones for expression analysis. The relative abundance of ESTs in libraries prepared from different organs and plants in different physiological conditions also provides information expression patterns for more abundant transcripts (Bouchez and Hofte, 1998). Thus what is required is to carry on EST sequencing programs in different plant species which will result in discovery of new genes.

EST array—It refers to an ordered alignment of different ESTs on supports of minute dimensions (e.g. nylon membranes, glass, quartz slides, silicon chips). EST array permits the simultaneous detection of thousands of expressed genes in a particular cell, tissue, organ or organism at a given time by hybridization of fluorochrome labeled cDNA preparations on the array and any hybridization event between an EST and cDNA is then detected by fluorescence.

EST-AFLP—It is a variant of conventional AFLP technique that allows to scan the 3' and/or 5' flanking regions of a gene for sequence polymorphism.

Promoter-proximal sequence tag (PST)—A method has been developed to tag promoter proximal sequences in mouse embryonic stem cells using a gene trap retrovirus shuttle vector. PSTs are ESTs derived from genomic DNA rather than cDNA and they are used similarly to screen sequence databases and to make STSs for mapping. PST has an additional advantage also. As each PST represents a specific ES cells line harboring a disrupted gene and mutant ES cells are used to generate mutant mouse strains it allows rapid progression from sequence analysis of a gene to analysis of its function.

24.56.3   Analysis of EST

ESTs are the resource for gene sequence data. They are important in gene discovery and identification and they are crucial for the discovery and identification of alternative splicing. ESTs are produced by one short sequencing of cDNA produced by cloning mRNA. ESTs are prone to sequencing error and vector sequence contamination and so methods for EST analyses need to take this into account. In order to reconstruct the original mRNA, EST analysis normally entails a **clustering stage**-attempting to group ESTs according to originating gene followed by an assembly or **consensus stage**- aiming to determine for each cluster one or more consensus sequences to which the sequences can be globally aligned. Ideally, one consensus sequence is produced for every mRNA isoform of the gene but in practice low coverage regions and high error rates make this goal difficult to obtain. As alternative transcripts are very common and so we must expect ESTs in a cluster to contain fragments, not only from a single correct mRNA sequence but from several equally correct related sequences. So an important challenge when processing EST sequence is the reconstruction of mRNA by assembly EST clusters into consensus sequence. Sequence assemblers such as Phrap, TIGR assembler (Sutton et al., 1995) and CAP3 (Huang and Madan, 1999) are used for constructing consensus sequences of ESTs. A complete EST analysis process incorporates many stages from base calling, quality and vector clipping, and repeat masking through sequence clustering and assembly, to analyses and detection of transcriptional features like SNPs and splice variants. The classical approach to sequence assembly use the 'overlap-layout-consensus' approach, i.e. they use pair-wise sequence alignments to find a best fit and build contigs by merging sequences that overlap (see chapter 28). Also, graph-based algorithm for generating EST consensus sequence has been developed (Malde et al., 2005). Since the EST in a cluster corresponding to a gene originate from multiple but related mRNAs sequences the graph resulting from applying this approach will partly be determined by the splice structure of the gene- a splice graph. Eulerian paths are computable in linear time. Sequence assembly is based on finding EPs in a graph based on work on sequencing by hybridization. It involves breaking down the sequencing data into fixed length, overlapping fragments or k-tuples. Computing assembly from the overlapping graph is N.P. complete (the Hamilton Path problem). A directed graph is constructed where each sequence is a vertex and edges represent overlaps between sequences, i.e. there is an edge from S1 to S2 and only if a suffix of S1 matches a prefix of S2. A correct assembly of the set of sequences is then constructed by finding a HP through this graph.

Tentative consensus sequence — It refers to the unique virtual transcript derived from comprehensive ESTs database by clustering of the sequences and assembly of cluster elements at high stringency (i.e. after removal of low quality, misclustered, chimeric sequences). TCs are generally longer than individual ESTs which comprise them so that TC can be used more efficiently for functional annotation.

Tentative human consensus (THC) — It refers to a consensus sequence for each putative protein derived from potential protein-coding regions deposited in databases. Sequences are first grouped together if they contain at least 40 bases with greater than 95% identity. Then these groups are assembled to generate a THC and discordant sequences are eliminated.

cDNA quality control — ESTs and full length cDNA sequences are used to confirm or revise computational annotations of genomic sequences. Because ESTs and full length cDNAs are generated by single pass sequencing, errors are frequent. Substitutions, insertions and deletions can alter reading frames or introduce premature termination codons. In addition, bacterial insertion sequences (ISs), ribosomal RNA and chimeric cDNA can contaiminate cDNA libraries. So there is a need to judge the quality of original cDNA data set. Hayden et al. (2005) developed cDNA quality control (cQC) to evaluate the quality of cDNA sequences and to provide corrected cDNA sequences. cQC is a program written in Perl (Practical Extraction and Report Language).

24.57 GENOME SEQUENCING

There are advantages of complete genome sequencing. As the rare transcripts and transcripts of genes that are induced under specific conditions (e.g. biotic and abiotic stresses) are not represented in EST databases and thus in order to have access to entire set of genes it is necessary to determine complete genomic sequence of different plant species. This sequence will provide information on the global structure of the genome including the relative order of genes on the chromosomes which is extremely useful for positional cloning strategies.

Positioning of introns — How to distinguish coding regions from non-coding intergenic sequences and introns? Comparisons with EST and cDNA sequences and sequence similarity to known coding sequences can be used to determine intron positions.

Intron-exon mapping — The localization of introns and exons within the coding region of a eukaryotic gene is done with the help of S1 mapping or heteroduplex mapping.

Splice site detection, Prediction of coding sequence from genomic sequence — For genes that do not match sequences in the databases the coding sequences need to be predicted from the genomic sequence. In case of human genome programs such as GRAIL have been developed to predict the beginnings and ends of genes and intron position with high reliability (Uberbacher and Mural, 1991). In plants splicing signals are still poorly defined and a better comprehension of splicing mechanism is required to improve gene prediction algorithms in future.

24.58 ISOLATION OF mRNA, ITS QUANTIFICATION AND AMPLIFICATION

mRNA is isolated by a guanidinium thiocyanate/CsCl microscale procedure. In case of isolation of messenger by affinity paper chromatography the poly (U) chains of more than 100 nucleotides in length are bound to the diazo-thiophenyl paper. This paper is thus used to isolate polyadenylated mRNA that binds to the poly (U) via hydrogen bonding. mRNA is also isolated and purified by chromatography discussed in chapter 30.

Enzyme linked fluorescence (ELF) signal amplification—It is a technique for the detection of specific mRNAs in cells or tissues section by *in situ* hybridization of a biotin-labeled DNA or RNA probes to the mRNA target. This technique works with a streptavidin alkaline phosphatase conjugate and a water soluble weakly fluorescent substrate whose phosphorylation by alkaline phosphatase leads to an insoluble fluorescent precipitate at the intracellular site of the mRNA which can be detected by fluorescence microscopy.

Global mRNA amplification—This technique refers to amplification of all mRNA of an mRNA population and produces sufficient amounts of low abundance mRNA prior to subsequent expression analyses. The amplification procedure used is called '**Eberwine procedure**' which is a technique for the linear amplification of an mRNA by T7RNA polymerase. In this technique the isolated mRNAs are hybridized to an oligo (dT) primers which contains T7 RNA polymerase promoter sequences. The oligo (dT) primed mRNA is then reverse transcribed into a single stranded cDNA. The ss cDNA is then converted into ds cDNAs by DNA polymerase I. Since each cDNA contains T7 promoter, large quantities of amplified RNA can be produced. This promoter drives cDNA synthesis by *in vitro* transcription.

Poly (A)-PCR—Poly (A)-PCR works with the generation of cDNAs by reverse transcriptase and subsequent amplification by PCR using sequence dependent primers. It is a variant of RT-PCR for global amplification of mRNA from samples containing only little mRNA or low abundance mRNA at lower detection limit (single cell, biopsies, needle aspirates) that capitalizes on the limiting size of the first strand cDNA to 300-700bases and therefore avoids a bias against the long transcripts during the amplification process.

RT-MPCR—It is a variant of conventional PCR in which multiple target mRNAs are simultaneously amplified in the same reaction tube using multiple primer pairs.

RT-PCR—For information on RT-PCR see chapter 31.

Anchored PCR—It is a variant of PCR which permits amplification of the cDNA sequences derived from polyadenylated mRNA whose 5′ sequences are not known. In such situation the conventional PCR can not be employed as only one specific primer, namely, oligo (dT) can be annealed to the poly (A) tail of the mRNA. However, if the 5′ end of the cDNA is extended by a homopolymeric sequence then a primer complementary to the sequence (anchor primer) can be designed and anchored to the 5′ end of the cDNA sequence. In the absence of both the 3′ and 5′ anchor primers the cDNA can be amplified by PCR.

The various RNA-PCR techniques thus allow amplification of cDNAs derived from small amounts of purified mRNA, tRNA, rRNA and viral RNA and detection of specific RNA at very low copy number and is therefore used in the analysis of gene expression at RNA level. This technique can be used for studying post-transcriptional modifications like alternate splicing. RNA-PCR can be combined with *in vitro* translation.

Quantitative real time PCR — It is used to measure the change in expression relative to that in unperturbed cells. qPCR data are filtered out to eliminate aberrant or inefficient reactions. The mean expression changes from each transcript in each experiment are calculated and only those changes that are greater then their standard errors are accepted as significant and used for further analysis.

Competitive reverse transcriptase PCR- (CRTPCR) — It is a variant of conventional PCR in which two templates of equal or similar length and common primer recognition sequences are simultaneously amplified. It is a technique used for the quantification of mRNAs of a cell, based on the addition of a known amount of PCR amplifiable competitor RNA to an mRNA sample and the reverse transcriptase PCR catalysed amplification of both the competitor and the target RNAS in the same reaction. The competitor RNA differs from the endogenous target mRNA by a few nucleotides and serves as exogenous standard in CRTPCR. Competitor RNA is designed in such a way that it is more resistant to the nuclease attack than normal RNAs and it does so by modification of modified nucleotides. Since the target RNA uses the same dNTPs-primer pair and DNA polymerase for amplification, a competition for these compounds follows. Now if the competitor is amplified at the same rate as the target RNA then the ratio of the amplified products will reflect the initial concentration of both. Since the quantity of competitor is known the amount of the target RNA in the sample can be estimated. The PCR product resulting from the amplification of competitor will differ from that of the target RNA by a small deletion, or insertion. The endogenous target can be cloned and modified by a deletion (that is about 10% of the target length). Thus it can be easily separated from endogenous target by agarose or PAGE.

Competitive/quantitiative PCR (CQPCR) — This technique is a variant of QPCR and used for exact quantification of mRNAs. In this technique the target mRNA and an exogenous competitor RNA (sharing partial homology with the target gene and recognised by the same primers) are simultaneously reverse transcribed and amplified. The competitor RNA is modified such that it neither contains an internal restriction site or that the amplification product differs in length from the mRNA amplification product. Thus as a consequence of sequence homology between the target mRNA and CRNA both are amplified with the same efficiency to produce a final concentration ratio reflecting the ratio prior to the reverse transcription. The amplification products contain mixture of heteroduplex in addition to the target and competitor homoduplexes which can be separated by denaturing high performance liquid chromatography. Each peak is then quantified to deduce the original amount of the target mRNA.

Adaptor tagged competitive PCR (ATAC-PCR) — It is a variant of the conventional QC-PCR used for the high through-put expression analysis of single genes.

Long –distance PCR (LD-PCR) — It has been designed to amplify DNA fragments up to 40kb as opposed to 6-8kb in conventional PCR, using Taq polymerase only. It requires the combination of two thermostable DNA polymerase, for example, *Thermus thermophilus* (Tth) DNA polymerase (with no proof reading activity) and Pfu (*Pyrococcus furiosus*) DNA polymerase (3'- 5' exonuclease activity and thus having proof reading activity). The specificity of Tth polymerase can be increased by inclusion of a monoclonal antibody raised against this enzyme which completely blocks the polymerase activity. At the start of thermal cycling process the enzyme-antibody complex dissociates from and renders the antibody ineffective. This step reduces artifacts of inaccurate amplification. The longer amplification product in LD-PCR results from the greater thermodynamic stability of one of the DNA polymerases at elevated temperatures.

24.59 RNA PROFILING

RNA profiling refers to the isolation, separation and visualization of preferably all RNAs of a cell, tissue, organ or organism.

24.59.1 Subtractive Hybridization/Differential Hybridization

It is important to identify and analyze genes whose expression levels and patterns are different in different cell types, tissues, developmental stages or particular conditions. Differential and subtractive hybridization methods have frequently been used to detect and isolate differentially expressed genes. Differential hybridization is easy and reliable but only effective for mRNA expressed abundantly in one of the two sample. Subtractive hybridization is effective in the detection and concentration of rare mRNA species but is rather empirical and has poor reproducibility. Subtractive hybridization is a procedure that increases the effective concentration of induced sequences expressed in an experimental RNA population (Target) but not in control population (driver) (Sargent, 1987). It has been successfully applied to clone mRNA sequences which are more abundant in one mRNA population than in another or to clone a DNA that is deleted in a mutant genome (Sangerstroem et al., 1997). This technique is used for the detection of sequences expressed in only one of the two cell types. In other words, it detects differences between the RNA in different cells, tissues, organisms or sexes under normal conditions, or during different growth phases, after various treatments, for example, hormone application, heat shock or in diseased (or mutant versus healthy (or wild type)) cells. It can also detect DNA differences between different genomes or between cell types where deletions or certain types of genetic rearrangements have occurred. In this technique cDNAs from mRNAs of one cell type, say A are hybridized to mRNAs from another cell type, say B. Only those sequences that are expressed in both cell types will form cDNA:mRNA hybrids which are then separated from single stranded mRNAs and cDNAs by hydroxyapatite chromatography. The fraction of single stranded sequences is then treated with alkali (which destroys the RNA) and contains the cDNAs from mRNAs expressed in cell type A only.

24.59.2   Steps Involved in Subtractive Hybridization

The five steps involved in subtractive hybridization are as follows.

1. Selection of material for isolating tester and driver nucleic acids
2. Production of tester and driver
3. Hybridization
4. Removal of tester-driver hybrids and excess driver (subtraction)
5. Isolation of complete sequence of target nucleic acid

The tester contains the target nucleic acid, the DNA or RNA differences that one wants to identify and the driver lacks the target sequences. The two nucleic acid populations are hybridized in the ratio of 10 driver:1 tester. The ratio of the driver to tester, the overall concentration of driver, the temperature and the length of hybridization are chosen based on the complexity of the driver and tester, the abundance class of the target nucleic acids and the length of the driver and tester sequences. As there is large excess of driver molecules, the tester sequences are more likely to form driver-tester hybrids than the double stranded tester. The sequences common only to both driver and tester hybridize, thus leaving the remaining tester sequences either single stranded or forming tester-tester pairs. The driver-tester, double stranded driver and any other single stranded driver are subsequently removed thus leaving only tester molecules. Multiple rounds of subtractive hybridization are usually carried out in order to identify the truly tester specific nucleic acid sequences. While selecting driver and tester it must be kept in mind that the less complex the source of tester and driver the more sequences they have in common.

Selection of driver/tester nucleic acid — Although in principle both tester and driver samples can be either DNA or RNA but in practice the tester is DNA as it is present in low concentration and DNA is more stable than RNA and the driver is RNA as the excess driver RNA can be easily removed enzymatically or by alkali treatment. In the basic protocol the RNA from tester source is reverse transcribed into cDNA, hybridized to poly A driver RNA, tester-driver hybrids removed, excess fresh driver added and the hybridization step is repeated once. The remaining target cDNA is either cloned or used to make probe. This basic protocol is employed in case of less complex nucleic acids. If the nucleic acid is complex or little starting material (tissue) is available then in that condition multiple rounds of hybridization-subtraction are required and there is a need to use a library or a PCR based technique. Testers and drivers are prepared from cDNA libraries as phagemids or as library inserts amplified by PCR or *in vitro* transcription. Alternatively, cDNA from tester and driver sources is ligated to different primers, amplified by PCR and hybridized. The different steps are repeated as required.

Methods used for subtraction — There are many different methods of subtraction depending upon the nature of tester and driver. If the driver is RNA then **hydroxyapatite chromatography** is employed for binding double-stranded driver and driver-tester hybrids and single stranded driver RNA is removed **chemically** (chemical crosslinking) or **enzymatically** leaving only single stranded cDNA tester after the subtraction. In case the tester is a single stranded phagemid library and the driver is a first-strand cDNA then in that condition the driver-tester hybrids is digested with a frequent cutter restriction enzymes and the hybridization mixture is used to infect bacteria. The single stranded tester phagemids infect the bacteria and thus get isolated. In some cDNA subfractions, cDNA covalently linked to a latex particle is used as driver and hybrid is removed by simple centrifugation. A common procedure is to employ **biotin-streptavidin** binding to separate nucleic acids. Streptavidin binds to biotinylated driver sequences and phenol extraction is used to remove the streptavidin protein and the bound driver-tester hybrids. Streptavidin can also be attached to beads or to column and used to remove excess driver and driver-tester hybrids.

The monitoring of effectiveness of subtraction is done through the use of radiolabeled tester and determination of the levels of single stranded tester decreases after subtraction. Alternatively enrichment for target sequences is monitored. If genes common to the driver and tester and one or more genes specific to tester are known then one can determine whether the tester specific gene is becoming more abundant compared with the common genes after each round of hybridization-subtraction.

Isolation of target sequences — Although after one or more hybridization and subtraction steps the resulting tester nucleic acid is expected to be greatly enriched for target sequence but it is still possible that rare sequences common to both the driver and the tester remain and in many cases the sequences isolated are only partial gene sequences. The remaining tester sequences are isolated and analyzed in a variety of ways. Tester can be made into an enriched library and probed with driver and tester sequences to isolate tester specific clones or the tester can be labeled and used to probe tester and driver libraries and to isolate full-length clones. The isolated tester sequences are further analyzed by Northern blotting, *in situ* hybridization or PCR techniques to determine whether or not the sequences are truly tester-specific.

Efficiency of subtractive hybridization depends on i. removal of tester/driver hybrids ii. driver: tester ratio iii. DNA concentration and salt concentration iv. enrichment strategies employed.

24.60 ALTERNATIVES TO STANDARDS SUBTRACTIVE HYBRIDIZATION TECHNIQUES

The different alternative to the standard subtractive hybridization technique are as follows.

1. Positive selection
2. Representational Difference analysis
3. Suppression subtractive Hybridization
4. Differential display
5. SAGE
6. Microarray

Positive selection — In the standard subtractive hybridization technique the hybridization of tester and driver is still carried out but rather then removing unwanted driver-tester and driver sequences by subtraction during step 4, double stranded tester sequences are positively selected for selective cloning or selective amplification. Although there are a number of methods to carry out positive selection but a simple method is to digest tester with a restriction enzyme generating cohesive ends while using sonication to shear the driver

DNA randomly. After hybridization, DNA ligase and vector DNA are added. The double stranded tester is only cloned into the vector and then it can be used to transform bacteria.

Identification and isolation of differentially expressed transcripts is generally achieved by differential display and related techniques, representational difference analysis, enzymatic degradation subtraction, linker capture subtraction and techniques involving physical removal of common sequence.

Representational difference Analysis (RDA)—RDA is PCR based technique used to detect and clone small sequence differences between the sequences of two or more genomes or DNA populations (Lisitsyn, Lisitsyn and Wigler, 1993).In other words, it is a technique to identify unique DNA sequences out of two complex and highly related genomes. It detects genome losses, rearrangements and amplifications of the cell genomes as well as pathogenic organisms. Further, it can be used to study cDNA libraries. In theory RDA eliminates fragments present in both populations, leaving only the difference. The elimination requires a vast excess of driver DNA to compete out cell sequences present in both representations. First, the complexity of both the target (tester) and driver genomes is reduced by digestion of genomic DNA with restriction enzymes recognizing 6bp recognition sequences. A high proportion of the resulting fragments do not fall into the amplifiable range of 0.1-10kb and therefore the complexity of the representation is reduced to about 2-10% of the complete genome. The olgonucleotide adaptors are ligated to the ends of the restriction fragments and adaptor-complementary primers are used to amplify the fragments using PCR. Only low molecular size fragments of <1kb are effectively amplified and represent a representation. After producing both representations their adaptors are removed by cleavage and only tester fragments are ligated to new adaptors at their 5′ ends. The target representation is mixed with excess of driver , melted and reannealed. The termini of formed duplexes are filled in with DNA polymerase and PCR is used to amplify the entire reaction. Self-annealed target (tester) duplexes are exponentially amplified as they contain two 5′ adaptors and thus could be filled in at both 3′ ends. Heteroduplexes are only linearly amplified and driver DNA is not amplified at all. Single stranded molecules are destroyed by mung bean nuclease and the cycles of hybridization and amplification are repeated. Reiteration of this process, i.e. cleavage of PCR products and ligation of new 5′ adaptors, leads to exponentially increasing enrichment of the target which is then cloned and sequenced. RDA has been applied to enrich for genomic fragments that differ in size or representation and to clone differentially expressed cDNAs. RDA does not solve the problem of the wide differences in abundance of individual mRNA species. Consequently multiple rounds of subtraction are still required. The mRNA differential display and RNA fingerprinting by arbitrary primed PCR (Welsh et al., 1992) are potentially faster methods for identifying differentially expressed genes. However, these methods have high rate of false positives biased for high copy number mRNA and might be inappropriate in experiments in which only a few genes are expected to vary. RDA may also be used for isolating probes linked to sites of genomic rearrangements whether occurring spontaneously and resulting in genetic disorders or cancer. The drawback with the current RDA technique is the requirement of pure driver DNA and relatively pure tester DNA.

cDNA-RDA (representational difference analysis)—It is a variant of RDA and allows to detect absolute differences between two or more populations of differentially expressed low abundance **mRNAs (Hubank and Schatz, 1994).** It is an effective method to identify fragments of differentially expressed genes. This technique was used to identify genes which are expressed in only a very small fraction of the cells from which Tester was derived. This implies that genes with very low levels of expression should also be amenable to isolation, provided they are not expressed in the cells used to prepare the Driver. Therefore, using cDNA-RDA it should be possible to isolate genes from defined chromosomal fragments present in somatic cell hybrids by comparing these hybrids with the parental cell lines from which they are derived (Groot and van de Oost, 1998).

Suppression subtractive hybridization—It combines a high subtraction efficiency with an equalized representation of differentially expressed sequences. This method is based on a specific form of PCR which permits exponential amplification of cDNAs which differ in abundance whereas amplification of sequences of identical abundance in two populations are suppressed. It is a method for generating differentially expressed (regulated) or tissue specific cDNA probes and libraries (Diatchenko et al., 1996). It requires a few micrograms of poly $(A)^+$ RNA from the two cell populations but such quantity of RNAs may be difficult to obtain. It is a variant of suppression PCR which allows to selectively amplify differentially expressed target cDNAs and simultaneously suppresses non-target DNA amplification. This technique is based on the selective and efficient suppression of the amplification of undesirable sequences, if long inverted terminal repeats are attached to the cDNAs. Further, it combines the normalization and subtraction steps in a single procedure. The normalization step equalizes the abundance of cDNAs within a target population and subtraction step excludes the common sequences between tester and driver populations. In this technique the tester (containing differentially expressed cDNAs) and driver cDNAs (containing these cDNAs at very low levels only) are first restricted with a 4-base cutter generating blunt ends (e.g. HaeII or RsaI). The tester cDNAs fragments are then divided into two samples (pop 1 and 2) and ligated to two different adaptors (adapt 1 and 2), resulting in two tester populations, resulting in two tester populations. The ends of the adaptors do not contain phosphate groups so that only the longer strand of each adaptor can be covalently attached to the 5 termini of cDNA. Then two successive hybridizations are employed. First, an excess of driver is added to each tester population, the sample denatured and allowed to anneal. The single stranded cDNA tester fraction (A) is normalized because the reannealing process producing homohybrid cDNAs (B) is much faster for more abundant cDNAs (second order kinetics of hybridization). Also the single stranded cDNAs in the tester fraction A are enriched for cDNAs from differentially expressed genes. Second, the two samples from the first hybridization are mixed. Only the remaining normalized and subtracted. Single stranded tester cDNAs can reassociate to form B, C and new E hybrids. Addition of a second portion of denatured D further enriched fraction E which is different from all other fractions as its cDNAs are linked to different adaptors at their 5'-termini (one from sample 1, the other from sample 2). These two different adaptor sequences allow preferential amplification of the subtracted normalized fraction E using PCR techniques and primers P1 and P2 directed against the terminal part of adaptor 1 and 2, respectively. The selective amplification requires the fill-in of the 3' –sticky ends with DNA polymerase. Exponential amplification can occur only with E type molecules. The B type molecules contain inverted repeats at their termini and form stable fold-back structures after each PCR step which are no templates for exponential PCR as intramolecular annealing of longer adaptor sequences is favored over the intermolecular annealing of the shorter primer (which is the basis of the suppression effect). Type A and D molecules do not contain primer binding sequences and type C molecules are amplified at a linear rate only. In a model system the suppression subtractive hybridization technique enriched for rare sequences over 1,000-fold in one round of subtractive hybridization. This technique is applicable to many molecular-genetic and positional cloning studies for the identification of disease, developmental, tissue-specific or other differentially expressed genes.

Magnetic-assisted subtraction technique—This is another technique for the detection of sequences expressed in only one of the two types of cell types. In this procedure total RNA is separately isolated from both cell types and each RNA is separately chromatographed over oligo $(dT)_n$ fixed to magnetic beads. The poly $(A)^+$ RNA which includes most mRNA will bound to the oligo (dT) is separated from the poly $(A)^-$ by magnetic force and washing. After that the poly $(A)^+$ RNA from the cell type, say A is converted into cDNA using reverse transcriptase. The resulting mRNA-cDNA hybrid is denatured such that the cDNA remains attached to the paramagnetic beads (driver cDNA). The same procedure is repeated with cell type, say B in which the cDNA (tracer cDNA) is produced from mRNA. All cDNAs that are present in equal amounts in both cell types are removed using a 25-fold excess of driver cDNA and thus only those cDNA will remain which are expressed in only cell type B.

Module shuffled primer PCR (MSP-PCR)—It is a variant of PCR which employs MSPs and drives the analysis of several to multiple genes in one single reaction tube. This technique is used for comparative gene expression profiling in different cell, tissue or organ.

Differential display reverse transcriptase-PCR (DDRT-PCR)—It is an RNA finger printing technique (Liang and Pardee, 1992). It is a technique used for the estimation of the number of expressed genes in different cell types and for detecting differences in expression by a **differential RNA display.** This technique allows the identification and molecular cloning of genes differentially expressed at two states in a given tissue. The differential RNA display refers to the visualization of all or a subset of all mRNA molecules of a given cell at a given time by techniques such as **DDRT-PCR.** In this technique a complete set of mRNAs from a particular cell type is used as template for reverse transcriptase and cDNA are synthesized using either oligodTVN (V = A, C, G, N = any deoxynucleotide triphosphate or simply oligo dT (12-18) primers). The cDNAs are then amplified using either primers of arbitrary sequence or specially designed **amplimers** (amplification primers) as reverse primers. The amplified fragments are then separated by denaturing a native agarose or polyacrylamide sequencing gels. The native gels are used to reduce complexity. The bands are then detected using either autoradiography if radioactive dATP (32p or 33p-dATP) is used to label the amplified fragments during the synthesis or by staining with EtBr. The highly resolved banding patterns from different cell types permit to visualize cDNAs that are specific for one but not for another cell type. Since this technique produces 'false positives', differently expressed cDNAs have to be reconfirmed by reamplification and northern analysis. **FDDRT-PCR-** This technique is similar to DDRT-PCR but it uses oligodTVN or simple oligdT as upstream primers and a primer of arbitrary sequence labeled with a fluorochrome (e.g. rhodamine) as downstream primer for amplification of specific mRNA and expresses differentially expressed cDNAs. The use of fluorescent primer avoids radioactivity, increases sensitivity and allows high throughput.

Compared with subtractive and differential hybridization methods the differential display is advantageous with respect to sensitivity and time requirement. However, PCR mediated amplification with arbitrary primers often gives false positives bands that can not detect signals in RNA blotting analysis. Furthermore, delicate control of amplification cycles and/or annealing temperature is necessary to detect differences in gene expression levels among samples.

RLGS—Hayashizaki et al., developed a novel method designated as restriction landmark gene scanning (RLGS) for the systematic analyses of genomic DNA. The principle of RLGS is based on using restriction enzyme sites as landmarks and high resolution two-dimensional gel electrophoresis. This method enables simultaneous visualization and quantitative determination of >1000 distinct genomic loci as gel spots. Thus RLGS is a very powerful technique which has been applied to genetic mapping, systematic detection of methyltable loci and search for aberrations in cancer DNA. Techniques to clone the target DNA fragments from RLGS gel spots have also been established.

Harukazu Suzuki et al. (1996) developed a new method for cDNA analysis in which labeled cDNA species of uniform length are prepared for each mRNA species and applied to the two-dimensional display system in RLGS. This restriction landmark cDNA scanning (RLCS) enables quantitative and simultaneous analyses of many cDNA species.

Ordered differential display—Recently many variants of an approach generally termed as mRNA differential display have been developed. The key procedure for obtaining discrete band patterns for comparisons in the most widely used strategy is amplification with arbitrary primers. However, such 'shotgun' comparison when the compared cDNA fragments are picked up from the initial pool at random (by means of amplification with arbitrary primers or any other means), is rather ineffective when more or less exhaustive investigation of the sample must be accompanied. There is an alternative approach called ordered differential display (Matz et al., 1997) which does not involve arbitrary priming for pattern generation and provides a possibility of a thorough step-by-step comparison of all mRNAs. To date, although some of the approaches related to differential display which have been proposed could be used for systematic investigation, these methods are rather complicated or may not have enough sensitivity. ODD is an extension of a technique employing display of 3'end restriction fragments of cDNAs. It utilizes more reliable energy for obtaining the representative pools of such fragments (based on PCR suppression effects) and propose an alternative way of their display which makes possible the systematic analysis of original 'shotgun' comparison. ODD is capable of producing stable patterns and allows direct sequencing of isolated band after their re-amplification.

A method for display of the 3'end restriction fragments of cDNAs is proposed, extending the idea reported recently. First, representative pools of such fragments are selectively amplified using PCR suppression effects. Then simplified subsets of these fragments suitable for comparison by PAGE are amplified by adaptor specific primers extended by two randomly picked bases at their extended 3' ends. By testing all possible combinations of extended primers the whole mRNA pool may be systematically investigated.

Equalization of cDNAs, subtractive hybridization and differential display (ESD)—It is a technique combining three procedures for the isolation of differentially expressed genes. In this technique mRNA is prepared from tissue, say A and B, respectively and converted to double stranded cDNA. The cDNA is digested with restriction endonuclease, AluI. The resulting cDNA fragments are ligated to specific adaptor sequences at both ends and are used as 'tracer'. Similarly, cDNAs from tissue B are ligated to a different adaptor and are used as 'driver'. The tracer and driver cDNAs are then amplified by PCR and the tracer is equalized (using dissociation-reassociation and hydroxyapatite gel chromatography to separate single stranded from souble stranded molecules). The single strand fraction in which the different cDNAs species are more or less equally represented is then amplified and converted to double stranded cDNAs by PCR with a B specific primer. The equalized tracer is then used in the subtractive hybridization. Small aliquots are then used for display i.e., amplified through PCR in the presence of a radioactive labeled nucleoside triphosphate. The PCR products are separated by sequencing gel electrophoresis and detected by autoradiography.

24.61 APPLICATIONS OF HIGH THROUGHPUT RNAi SCREENS IN FUNCTIONAL GENOMICS

Gene silencing via sense or antisense suppression has been a popular method for studying gene function. However, this method requires that several independent transgenic lines be generated for every gene. Further, essential genes can not be down regulated in this way as suppression would lead to dominant lethal phenotypes which can not be maintained. In mammals, many of the reverse genetic approaches are laborious and extensive and silencing genes with RNAi makes it possible to analyze many gene networks and offers a high throughput approach. With the availability of genome sequence a well established RNAi based screening technology can rapidly and systematically characterize gene function at a genome-wide scale to find new components. RNAi refers to the ability of double stranded RNA to interfere with the production of corresponding gene product. It can knock down the expression of any given gene. RNAi was first applied by microinjecting dsRNA into worm. Further, feeding worms dsRNA produced in bacteria also reduced gene expression. Microinjecting dsRNAs corresponding to 98% of *C. elegans* genes into adult worms offer insights into cells biological functions of a large set of genes. More specific functional studies can be carried out on a gene-by-gene basis. For example, determining the subcellular localization of proteins required for a distinct process will provide a straight forward first step in determining whether such proteins are directly responsible for cellular event or whether they regulate it from a distance. However, non-reproducibility of some phenotypes suggests that other gene requirements remain unidentified. RNAi-based phenotypes do not always recapitulate the 'null phenotype' the classical mutations can produce. A crucial aim upon completion of whole genome sequences is the functional analyses of all predicted genes. More than half of the genes in *C. elegans* have a human homologue. To screen most of the predicted genes in *C. elegans* by RNAi, a library of bacterial strains, each capable of expressing double stranded RNA designed to correspond (homologous) to a single gene was constructed (Kamath et al., 2003). Using this library wild-type C.elegans hermaphrodite was screened to identify genes for which RNAi reproducibly results in sterility, embryonic or larval lethality, show post-embryonic growth or a post-embryonic defect. The reusuable RNA library of bacterial clones will facilitate systematic analyses of the connections among gene sequences, chromosomal location and gene function.

The genes can be discovered by **loss-of-function genetic screens**. Application of a high-throughput RNAi screens of double stranded RNAs in culture cells can be employed to characterize the function of nearly all (91%) predicted Drosophila genes (Boutros et al., 2004) in cell growth and viability. The identification of gene function by cell-based RNAi screens requires generation of a **double stranded RNA library** targeting nearly all genes in the Drosophila genome. Primer pairs are designed to amplify gene –specific fragment which are then used for synthesis of double stranded RNAs. RNAi screens consist of a double stranded RNA library targeting nearly all genes in a genome. Treatment of cultured cells with the ds RNAs will lead to the depletion of the corresponding transcript and generation of specific and penetrant phenotype and thus providing an efficient approach for systematic loss of function phenotype analyses. Also, a quantitative assay of cell number that correlated the reduction of signal to dying cells as demonstrated by RNAi of the D-IAPI inhibitor of apoptosis.

Forward genetic screens — Geneticists have traditionally sought to gain insights into complex biological processes through forward genetic screens. Mutations are generated at random, phenotypes of interest are scored and the mutated gene is subsequently identified. Although this approach has been successful but limited by inherent biases in mutagenesis techniques, the large number of mutants that must be analyzed and the considerable efforts needed to identify the relevant genetic lesions. Moreover, most genes have multiple functions and a gene's function in one tissue can preclude its recovery in screens focused on function in other tissues. This is particularly true for genes that are essential in the early development of the organism (Dietzl et al., 2007).

Forward genetics refers to strategy to isolate a gene after its knock out (loss-of-function) by the insertion of, for example, transposon with concomitant change of the phenotype. The transposon sequence is then used as a tag to allow the isolation of the corresponding gene. Signal propagation through this pathway is probably regulated by a large network of moderate, context-specific proteins. The genes encoding these proteins may not have been discovered through traditional screens owing, in particular to the requirement of a visible phenotype. The inhibition of gene function by RNAi couples with availability of annotated genome sequences now enables systematic surveys of gene function by **reverse genetics**. One by one, the function of almost every predicted gene can be predicted and phenotypic consequences observed. Any phenotype is immediately linked to a specific DNA sequence. This method has been successfully used in genome-wide screens by applying double stranded RNAs to *Drosophila melanogaster* or mammalian cells in cultures.These cell-based assay systems enable detailed studies of biological processes but not the complex biology of whole organisms. Genome-wide RNAi screens can identify novel components of signal transduction pathways (Dasgupta et al., 2005). In other words, RNAi can be employed in dissecting complex biochemical signaling cascades. The output from traditional transcriptional pathway reporters may be biased by cell type, integrate unknown additional pathways and not linearly reflect endogenous signaling. To obtain a global view of RTK/ ERK signaling they performed unbiased RNAi, genome-wide high throughput screens. They developed a phosphospecific antibody reporter of proximal, endogenous pathway activity. Monoclonal antibodies recognizing the dually phosphorylated active form of mammalian ERK1/2 (dp ERK) have been characterized to report the dynamic activation of single Drosophila ERK isoform, Rolled, downstream of multiple RTKs. HT RNAi screens in cultured cells followed by functional analyses in model organisms prove to be rapid means of identifying regulators of signaling pathways implicated with development and disease. Only 25% of all known genes in Drosophila are associated with a readily obvious phenotype. Gene-to-phenotype analyses encompass the manipulation of genes through transgenic and knockout technologies to ask questions of gene function. The investigator has preconceived idea of the function of gene and therefore the expected phenotype. Thus gene-to-phenotype approach generally provides a confirmation of hypotheses.In the phenotype-to-gene approach, the classical genetics approach, there is no priori knowledge of the cause of that phenotype. This approach can lead to gene and pathway identification.

RNAi library in human cells—RNAi can be used to perform genetic screens for key regulators of cancer cell proliferation and survival. In mammalian cells, an RNAi response can be triggered by a 21-bp siRNAs which can cause strong but transient inhibition of gene expression. By contrast, vector expressed shRNAs can suppress gene expression over prolonged periods. As interfering RNAs that target such genes would be toxic, a retroviral vector is used for the inducible expression of shRNA (small hairpin RNA). In cells engineered to express the bacterial tetracycline repressor, this vector does not express shRNA until doxycyclin is added and can inducibly knockdown the expression of endogenous target genes by 50-70%. So create a library of shRNAs targeting human 2500 genes with 3-6 shRNAs per gene. In other words, construct a set of retroviral vectors encoding distinct shRNAs which target different human gene for expression. Construction of 3-6 different shRNA vectors against each gene increases the likelihood of obtaining a significant inhibition of gene expression. The oligonucleotides specifying the shRNAs are annealed and cloned in a high throughput fashion into pRetroSuper (pRS), a retroviral vector that contains the shRNA expression cassette (Berns et al., 2004). This vector based shRNA library can be used for functional genetic screens in both short term and long term assays using DNA transfection or retroviral transduction. For each gene transcript, three to six 19 nucleotide sequences are designed and hairpin derivatives (59-mer oligonucleotides) are cloned into pRS. Three to six vectors targeting one gene are combined in a single well of a 96-well plate. From each 96-well plate DNA was pooled and high-titre polyclonal virus was produced and used to infect cells.

siRNA bar code screens—The RNAi screen described above is time consuming in that individual colonies of cells must be isolated and hairpin vectors recovered and tested in second round selection. siRNA bar code screens are used to rapidly screen complex shRNA vector library in a polyclonal format. In other words, it can be used to rapidly identifying individual siRNA vectors associated with a specific phenotypes. This approach was pioneered in yeast (Shoemaker et al., 1996) and takes advantage of the fact that each hairpin vector contains a unique gene- specific molecular identifier: the 19-mer targeting sequence. This molecular bar code can be used to follow the relative abundance of individual vectors in a large population using DNA microarrays that contain the bar code oligonucleotides. Thus each shRNA construct is tagged with a different 60bp bar code to allow to monitor the abundance of each shRNA in the cell population. One can go for either local delivery or systemic delivery of siRNAs.

Bjorklund et al. (2006) analyzed the effect of loss of function of 70% Drosophila genes (= 90% of genes conserved in human) on cell cycle progression of S 2 cells using flow cytometry which allows direct determination of the fraction of cells in different phases of the cell cycle. Clear phenotypes including arrest in G1, G2/M and S, cytokinesis and DNA replication defects and apoptosis and/or cell death were observed when targeting known regulators of these processes. The cell cycle is a robust system in which compensatory mechanisms control its overall length (Reis and Edgar, 2004). The objective was to identify genes that control cell size, cytokinesis, cell death and/or apoptosis and the G1 and G2/M phases of the cell cycle. Classfication of genes into pathways by unsupervised hierarchical clustering on the basis of these phenotypes show that in addition to classical regulatory mechanisms such as Myc/Max, Cyclin/Cdk and E2F, cell cycle progression in S2 cells is controlled by vesicular and nuclear transport proteins.

Regulatory genenetwork of cell cycle of fission yeast—The gene products of four genes $cdc2^+$, $cdc25^+$, $wee1^+$ and $nim\ 1^+$ act in a regulatory gene network which determines the cell cycle timing of mitosis in fisson yeast. Cell size at mitosis in *S. pombe* is determined by expression of $cdc25^+$ and $nim\ 1^+$ inducer genes and inhibition of gene $wee\ 1^+$ which between them regulate the M-phase protein kinase p^{34cdc2}

Systematic reverse genetic screen—Systematic reverse genetic screens are possible with the availability of complete genome sequences and the introduction of RNA-mediated gene interference. Large scale RNAi-based surveys of gene function *in vivo* have thus so far been limited to the nematode, *C. elegans* and *Schmidtea mediterranea*. In these organisms RNAi is systemic and so gene interference can not easily be restricted to a specific cell type. In *Drosophila melanogaster*, RNAi is cell autonomous and can be triggered by the expression of a long double stranded 'hairpin' RNA from a transgene containing a gene fragment cloned as inverted repeat. Using the binary GAL4/UAS expression system such RNA transgenes can be used flexibly to target gene inactivation to potentially any desired cell type at any stage of animal's lifespan. If a genome-wide library of transgenic RNAi strains is available then it would be possible to conduct systematic RNA screens targeted to specific cell types in the intact animals (Dietzl et al., 2007). Genome-wide library of RNAi transgenes will enable the conditional inactivation of gene function in specific tissues of the intact organism. RNAi transgenes consist of short gene fragments cloned as inverted repeats.

Use of double stranded RNA probes—Double stranded-mediated interference (RNAi) of gene expression is a widely used method facilitating reverse genetic studies. dsRNA probes ranging from 200to 1000bp are derived from sequences that are present in the mature transcript but not from those in promoter or intergenic regions, are able to interfere with gene expression. Individual DNA fragments approximating 700bp in length containing coding sequences for the proteins to be knocked out are amplified using PCR. Each primer used in the PCR contains a 5′ T7RNA polymerase binding site followed by sequences specific for the target genes. The PCR products are purified and the purified PCR products are used as templates by using a MEGASCRIPT T7 transcription kit to produce dsRNA. If complementary strands of a specific RNA are combined to produce dsRNA, there is a potent and specific interference of protein expression. However, some variability of efficacy has been noted for different dsRNA probes that target the same gene. Only a few copies of dsRNA per cell are necessary to eliminate protein production suggesting that the mechanism is catalytic in nature and that it does not function by titrating endogenous mRNA as proposed for traditional antisense RNA techniques. The targeted gene does not appear to be mutated nor does transcription of the gene cease in the presence of dsRNA. However, the mRNA produced by the RNAi-targetted gene is absent from the cytoplasm and reduced in the nucleus. dsRNA exerts its effect during or following RNA processing but before final translation.

Use of chemically synthesized siRNA — Use of chemically synthesized siRNAs has helped to define gene function in vertebrate cells. Human cells are killed by long dsRNAs.

Advantages of RNAi — Use of RNAi in cell culture can be made to dissect signal transduction pathways. It has advantage over other methods requiring the introduction of DNA into cells. Transfection experiments can lead to nonphysiological concentrations of the recombinant protein often making the interpretation of result difficult. Secondly, transfection experiments often result in up-take of the DNA into significantly less than 100% of the cells but with RNAi all of the cells effectively take up the dsRNA because 95-99% of the protein disappears from the cultured cells. The use of RNAi requires only the production of dsRNA from a PCR product that has T7 RNA polymerase binding sites at each end. The results of protein 'knockout' experiment can be obtained within 2-3 days and so is a quick method in comparison to selective gene 'knock outs' in mammalian cells. RNAi helps in elucidation of the mechanism that allows dsRNA to inhibit target protein synthesis. It rapidly accelerates characterization of biochemical pathways. This procedure is similar to having inhibitors for specific proteins in a wide range of signal transduction cascades.

The observation of genetic interactions is the key to the definition of cellular networks. Gene silencing by RNAi in mammalian cells using siRNAS and shRNAs has become a valuable genetic tool. RNAi is a powerful technique with which to perform loss-of-function genetic screens in lower organisms and can greatly facilitate the identification of components of cellular signaling pathways. In invertebrates, RNAi has been harnessed as a tool to reduce endogenous gene expression through programming organisms or cells with homologous double stranded RNA. This approach has been used in large scale, genome-wide screens. With the advent of RNAi in mammals and refinement of techniques to trigger gene silencing, expression from any human or mouse transcripts can , in principle, be inhibited using siRNA and shRNAs.The new tools such as development of expression vectors to direct synthesis of shRNAs that act as siRNAs-like molecule to stably suppress gene expression, development of RNAi library and description of siRNA bar-code screens to rapidly identify individual siRNA vectors associated with a specific phenotype, will grealt facilitate large-scale loss-of-function genetic screens in mammalian cells. The large scale RNAi libraries can be used in specific, genetic applications in mammals and will become a valuable resource for gene analysis and gene discovery (Berns, 2004; Paddison et al., 2004). RNAi has enabled genetic approaches in both cultured mammalian cells and intact animals. Large scale screens of siRNA and shRNA collections have generally adopted a one-by-one approach interrogating phenotypes in a well-based format. This requires both considerable infrastructure and a substantial investment for each cell line to be screened. Alternatively, shRNA collections can be screened by assaying enrichment from pools but this limits the range of phenotypes that can be addressed. Whitehurst et al. (2007) devised a massive parallel study for screening shRNA collection for stable loss-of-function phenotypes. They combined a HTP cell-based one well/one gene screening platform with a genome-wide synthetic library of chemically synthesized siRNAs for systematic interrogation of the molecular underpinnings of cancer cell chemoresponsiveness.

24.62 CONDITIONAL KNOCK OUT/GENE TARGETTING

When null mutations produced by homologous recombination are lethal to embryo, then the role of gene product can not be studied at subsequent stages of development. Conditional gene targeting refers to the silencing (knocking out) of a specific gene under certain conditions (e.g. if a specific developmental stage is reached). **Conditional mutation** refers to any mutation which is only expressed under specific conditions (e.g high temperature).In other words, mutantions are predominantly conditional, if mutant protein only loses its function under certain conditions usually at a slightly raise temperature. Conditional knockouts can be produced by several techniques. Methods for creating conditional mutations permit control over the site and/or time of mutation *in vivo* and are based on site-specific recombination 34-bp catalyzed by a yeast or bacteriophage DNA recombinase. The locus of interest is modified by gene targeting such that the *loxP* sites in the same orientation are placed on either side of the selection cassette and at the 3' end of the gene to be mutated. For example, an essential exon of a target gene in one organism (e.g. mouse) can be experimentally flanked by two 34bp *lox P* (locus of crossingover) element (gene targeting) called *lox P* sites. The bacterial *cre* (which causes recombination) recombinase can catalyze the recombination of the DNA between the *lox P* sites. This process can lead to circularization or inversion, also deletion of the sequence between the *lox P* sites. Now the target gene remains intact as long as *cre* recombinase is expressed. Normally this phage P1 –derived enzyme is not present in eukaryotes. The gene encoding this enzyme is now transformed into a second mouse and the *lox P* and *cre* mice homozygous for the respective transgene are crossed so that the P sites and the *cre* recombinase are now present. Therefore, the *lox P* flanked exon is deleted, i.e. the target gene is silenced. If the deletion should occur at a distinct time only then the inducible promoters for the *cre* gene can be used. In other words, the ES cells are then transfected transiently with a *Cre* expression cassette, resulting in recombination between *lox P* sites and deletion of the intervening sequences. Cells in which selection cassette alone has been deleted are identified and used to generate mice that carry a functional gene flanked by *loxP* site. Homozygous offspring of those mice are crossed with mice which are heterozygous for the mutated gene and that also carry the *cre* recombinase gene driven by a tissue-, cell-, stage-specific promoter. Recombination and deletion mutation of the *loxP* flanked gene is restricted to the tissues of stage of development where/when the *cre* recombinase is expressed and permits direct investigation of the function of gene at these positions or stages.

Cre-lox **technology**—In addition to conditional mutations, this technique can be applied to a wide variety of targeted mutation strategies such as replacement of a gene from one species with the same gene from another species and insertion of a different gene into a particular genomic location (gene 'knock in'). It has also been used for the creation of subtle mutations to enable controlled activation of gene expression by insertion of a *lox-P* flanked stop sequence between a gene and its expression signals to produce complex genomic models for chromosomal rearrangements, translocations and deletions and to model diseases caused by mutations in somatic cells such as cancers and aging.

Functional analysis of mammalian gene relies, in part, on targeted mutations generated by homologous recombination in mice. Manipulation of transgene can be accomplished *in vivo* by the use of site-specific *Cre/loxP* recombination system of bacteriophage P1. *Cre* recombinase (causes recombination) (Sauer and Henderson, 1989) is a 38kDa typeI topoisomerase from bacteriophage P1 that catalyzes site specific recombination of DNA between two *loxP* sites which depends on the relative orientation of the two *loxP* sites. The enzyme recognizes the *loxP* site and introduces cuts in its 8bp core sequence. *loxP* (locus of crossingover) is a 34bp DNA sequence element of a bacteriophage P1, consists of two inverted repeats of 13bp each separated by a central asymmetric 8bp core sequence which functions as address site for *cre* recombinase. First four *cre* recombinase molecules bind to two adjacent *lox P* sites and form a DNA-protein complex. Within this complex the two core regions are cut and subsequently reciprocal strand exchange occurs leading to Holliday junctions.This recombination process yields different products depending on the position and relative orientation of the two *lox P* sequences. The polarity of the elements is determined by the core sequence and the polarities of the two adjacent elements determine the type of recombination. Tandemly repeated *loxP* sites (identical polarity) on one strand cause *cre* recombinase catalyzed excision of the intervening DNA sequence and its circularization, leaving a single *loxP* site behind in the genome.If the two adjacent *loxP* sites are in opposite orientation (opposite polarity) the enzyme will cause an inversion of the intervening DNA sequence. And if the two *loxP* sequences with identical polarity are localized on two different DNA strands (eg. on different chromosomes or a plasmid and a chromosome) then a translocation of one *loxP* flanking region onto other results. The integrated DNA resides between two directly repeated *loxP* sites (Figure 24.2).

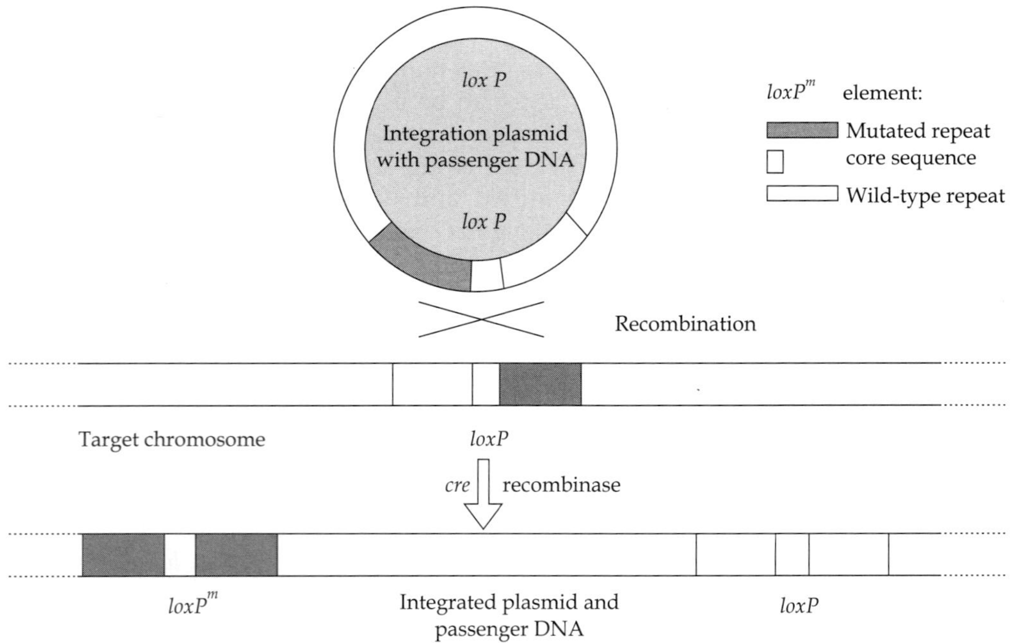

FIGURE 24.2   Showing Cre-lox technique.

The core recombinase efficiently directs both excision and insertion of DNA segments flanked by *loxP* sites in both bacterial and eukaryotic cells. Recombination occurs when two directly oriented *loxP* sites on a DNA substrate interacts with core recombinase. This interaction results in the excision of the interacting DNA molecule between the two *loxP* sites leaving a single *loxP* site in the genome. A strategy for adipose-specific inactivation of loxP-floxed gene segment has been developed. Transgenic mice have been established which express *Cre* recombinase under the control of adipose specific aP2 enhancer/promoter. Crossing of the aP2/*Cre* mice with any6 loxP-floxed gene will facilitate its functional analysis in adipose tissue.

The *cre/lox P* system is used for the precise excision of undesirable genes (or other sequences) in transgenic organisms. For example, selectable marker genes are essential for the selection of transformants in, for example, plant transformation experiments. Once a transformed palnt is identified the marker gene in its genome is not required. Its removal can be achieved with the *cre/lox* system by flanking the marker gene with directly repeated *loxP* sites in the targeting construct and the subsequent excision of the intervening marker gene by the transient expression of a co-transfected *cre* gene. Conversely, prior placement of a *loxP* site in the target genome allows subsequent targeting of this site by *cre* recomvbinase which directs single copy site specific insertion of a *loxP* carrying vector (with a gene, gene targeting).

Transient expression of the *cre* gene in a target cell is sufficient for a desirable limited recombination since continuous presence of the enzyme would induce a second round of recombination events which would remove the integrated vector (and the gene). Purified *cre* recombinase protein can also be directly introduced into cultured cells by , for example, lipofection and catalyzes the site specific chromosomal integration of a cotransfected *loxP* targeting vector and precise excision of genomic DNA flanked by directly repeated *loxP* sites.

Flp/FRT **system**—It is the yeast derived *Flp/FRT* system. Like *Cre/Lox* system it is used for modulating gene activities via recombination (site specific recombination). Both these systems target their specific 34bp recognition site *loxP* and *FRT*, respectively and catalyze a recombination event within these sites. Both systems work in a heterologous environment as no cofactors are required for this recombination and after the temperature optimum of Flp (O',Gorman et al., 1991) has been adjusted to 37°C. Both recombinase function efficiently in mice. Both systems allow DNA segments to be inverted or deleted in *cis* or to be connected in *trans*. Thus besides conditional gene (in) activation, this system has got several other applications such as chromosome engineering, recycling of selectable markers.

It is the most generally applied approach whereby genes are activated or silenced by deleting DNA fragments located between directly repeated recombination sites. In most cases, *loxP/FRT* sites are placed in introns and so the respective gene will generally show wild type characteristics until the gene or gene parts are deleted. Thus when a mouse containing a 'floxed' or 'flrted' gene is crossed with an animal expressing the respective recombinase under the control of a tissue specific promoter, recombination will occur only when the promoter driving *cre* or *flp* gene becomes active within the developmental program of the mouse.

Although both recombinase systems show an exquisite specificity, there are reports of degenerate *loxP* sites in mammalian genomes thus leading to undesirable effect. Therefore, high constitutive levels of *Cre* or *FLP* should be avoided (Gossen and Bujard, 2002).

Transcriptional control system based on elements of the tetracycline (TC) resistance operon (tet operon) of *E.coli*

There are two basic strategies to allow to conditionally modulate individual gene activities in eukaryotes in a highly specific, temporally defined and cell-type restricted fashion and the resulting phenotypic changes have provided insights into biological mechanisms and processes. The first strategy is site specific recombination- targeting genes directly by site specific recombination , resulting in activation, inactivation or alteration of the gene of interest. The second strategy aims at quantitatively and reversibly controlling a gene's function, generally leaving the endogenous gene itself untouched. Reversibly interfering with a gene's activity, for example, at the level of transcription adds another level of sophistication to the study of gene function *in vivo*. The ability to reversibly perturb a system permits not only comparison between the normal and perturbed state but also analysis of the system reactions after interference has ceased and during repeated cycles of gene activation/reactivation. This tetracycline response operon has the unusual specificity of interaction between the Tet repressor (TetR) and its specific DNA binding site, the tet operator (tetO) and between Tet repressor and its inducer, particularly Dox and the well studied chemical and physiological properties of the inducing agents, the various tetracyclines.

24.63 TRANSCRIPTION/EXPRESSION PROFILING

It refers to the determination of all expressed genes of a cell, a tissue, an organ or organism at a given time. Cells express a different range of genes at various stages in their development and functioning. This characteristic range of gene expression is the **expression profile** of the cell. The process of transcription profiling produces an expression profile. The expression profile changes continuously and is dependent on the developmental stage and the environment. It is a technique for simultaneous detection of the expression of thousands of genes whose cDNAs are immobilized on a positively charged nylon membrane to which the labeled cDNA from the target tissue (s) are hybridized. This technique starts with isolation of poly $(A)^+$-mRNAs from tissues, say A and B and are reversed transcribed separately into cDNAs using α-^{32}PdATP or α-^{32}PdCTP and enzyme, reverse transcriptase. These radiolabeled cDNA probes are then separately hybridized to two nylon membranes which are loaded with identical cDNAs. Each of these cDNAs is about 200-500bp in length and lacking the poly (A) tail and any repetitive DNA. All cDNAs are fixed on the nylon membranes in functional arrays, i.e. cDNAs from one organ in one and from another organ in another quadrant of the membrane. After hybridization and washing the hybridization pattern is visualized by autoradiography and quantified by phosphoimaging. Expression profile thus generated allows the detection of differential expression of thousands of genes in one single experiment.

Uses of genome-wide expression analysis — The genome-wide analysis of gene expression can provide many useful answers. When and where a gene product (RNA and/or protein) is expressed can provide important clues to its biological function. Expression profiles will help in characterizing the defects in developmental mutants. There are different approaches for monitoring mRNA levels. The various techniques described below can be used to describe in a systematic way mRNA levels in different tissues during the course of development or in different environmental conditions. The global gene expression analysis involves comprehensive search for all genes of a genome (which are expressed), their isolation, characterization and sequencing to construct a genome wide expression profile. The global gene expression can be determined by SAGE or MPSS. The global expression studies can be used to classify genes on the basis of their spatial and kinetic expression patterns. For example, studying gene expression pattern during shift from anaerobic to aerobic metabolism in yeast, classifying genes based on their induction or repression kinetics. New regulatory elements may be identified by comparing the regulatory sequences of genes of the same class. Finally, the kinetics of changes in gene expression in combination with expression profiles of mutants for known regulatory genes will allow study of expression networks. Other techniques such as systematic **whole-mount in situ hybridization** studies on various organs are essential for obtaining more precise information on cell-type-specific expression (see chapter 31 for detail on Whole-mount-ISH). Similar information can also be obtained by 'enhance' or 'gene trap' lines and are discussed in chapter 16.

24.64 METHODS FOR OBTAINING GENOME-WIDE mRNA EXPRESSION PATTERN

The global gene expression refers to comprehensive search for all genes of a genome that are expressed and their isolation, characterization and sequencing to establish a genome-wide expression profile. There are three recently devised methods for obtaining genome-wide mRNA expression data. The **differential gene expression (DGE) technology** is employed for the genome-wide detection and analysis of all expressed genes of a cell, tissue, organ or organism at a specific time and all changes occurring during lapse of time or after various natural or induced biotic or abiotic stress. In the closed differential gene expression techniques such as macroarray or microarray techniques only those genes are probed that are spotted on an array and probed with radioactively or fluorochrome labeled target sequences (usually cDNAs)whereas in case of open differential gene expression techniques such as cDNA-AFLP, SAGE, differential display, RT-PCR or TOGA, all differentially expressed genes can be profiled. In case of closed DGE the genes on chips are known in each case and genes not represented in the closed system 'chip' are not assessed. The open DGE technology does not require any a priori knowledge of the transcriptome so that the field of discovery is 'open'.

SAGE (Serial Analysis of Gene Expression)- SAGE (Serial analysis of gene expression)-
It is a technique for simultaneous detection, identification and quantification of virtually all genes expressed in a given cell at a given time. Further, it allows identification of unknown genes. SAGE is based on isolation of a short 9-14bp tag from a defined location within a transcript that contains unique and sufficient information to identify this transcript. SAGE is an extension of EST and is based on very short cDNA sequence tags from the 3' part of mRNA molecules, concatenated and cloned before sequencing (Velculescu et al., 1995).

SAGE is an extension of EST sequencing. SAGE is a technique that allows a rapid, detailed analysis of thousands of transcripts simultaneously. In other words, this method allows extremely rapid identification of thousands of genes and is used to identify new genes or to analyze relative levels (quantitative analyses) of gene expression. SAGE is based on the fact that 9-bp of sequence located at its 3′-end is all the sequence information required to identify a gene unambiguously. SAGE is based on the isolation of unique sequence tags from individual transcripts and concatenation of tags serially into long DNA molecules. Rapid sequencing of concatement clones reveals individual tags and enables quantitation and identification of cellular transcript. Thus, SAGE technology is based on three principles. First, a short nucleotide sequence tag (9 to 10 bp) contains sufficient information to uniquely identify a transcript provided it is isolated from a defined position within the transcript. Considering a random sequence of 9 nucleotides, there are 4^9 possible combination of 4 bases, or 262 144 sequences which is approximately five-fold the estimated number of genes in guman genome. Thus the first step in SAGE involves producing a 9-bp cDNA tag for each of the mRNAs in a population. Secondly, concatenation of short sequence tags allows the analysis of transcripts in a serial manner by the sequencing of multiple tags within a single clone. Here many unrelated tags are concatenated, the concatenated tags are cloned and random clones are sequenced. In other words, sequence tags can be linked together to form a long serial molecules (concatemers) which can be sequenced and cloned. Thirdly, counting of the number of times a particular tag is observed provides the expression level of the corresponding transcript. These sequences provide a spectrum of the genes expressed in the tissue and indicate their relative abundance. Thus many genes can be analyzed at once as only 9-bp of each gene is sequenced and many tags are sequenced in a single reaction. However, to be useful for most purposes full length genes corresponding to the tags must be subsequently identified and isolated. An inventory of transcripts is established based on very short cDNA sequence tags (9-11 bp) from the 3′ part of mRNA molecules, concatenated and cloned before sequencing (Velculescu et al., 1995). Double stranded DNA is first synthesized from mRNA by means of a biotynylated oligo (dT) primer. The cDNA is then cleaved with a restriction endonuclease (anchoring enzyme, AE) that would be expected to cleave most transcripts at least once. The restriction endonucleases with 4-bp recognition sequence (CATG) are generally used for the cleavage. The most 3′ portion of the cleaved cDNA is then isolated by binding to streptavidin beads. This process provides a unique site on each transcript that corresponds to the restriction site located closet to the poly (A) tail. The cDNA is then divided in half and ligated via the anchoring restriction site to one of the two linkers containing a type IIS restriction site (taging enzyme, TE). Type IIS restriction endonucleases cleave at a defined distance up to 20bp away from their asymmetric recognition sites. The linkers are designed in such a way that cleavage of the ligation products with the tagging enzyme results in release of the linker with a short piece of cDNA (Figure 24.3). After blunt ends are created the two pools of released tags are ligated to each other and the ligated tag then serves as templates for PCR amplification with primer specific to each linker. Thus the resulting amplification products contains two tags (one ditag) linked tail to tail and flanked by sites for the anchoring enzyme and in the final sequencing template this results in 4bp of punctuation per ditag. Cleavage of PCR products with anchoring enzyme will allow isolation of ditags that could then be concatenated by ligation, cloned and sequenced. These tags have proven sufficiently long to unambiguously identify corresponding genes in databases. Expression patterns for the different genes are reflected by the relative abundance of individual tags. SAGE patterns have been studied in human and yeast. A prerequisite for the identification of the tags is the availability of large sequence databases for the species under study.

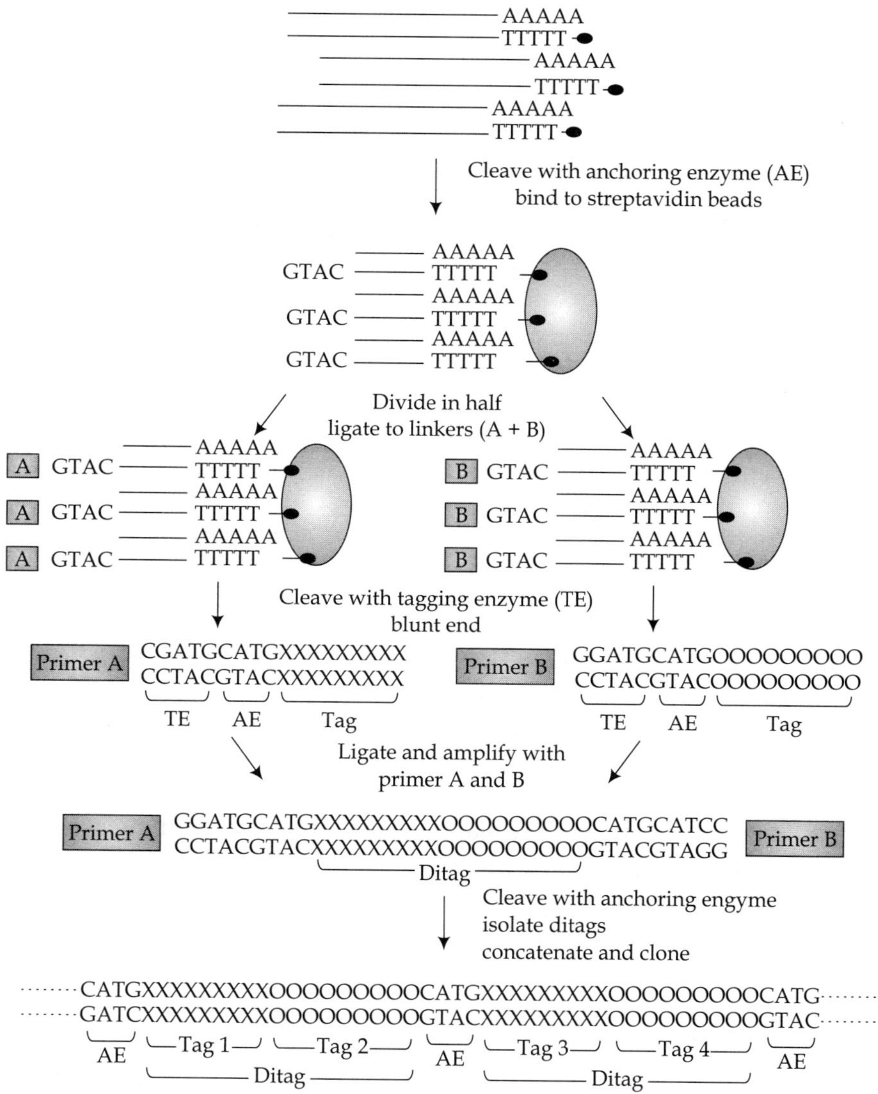

FIGURE 24.3 Showing SAGE procedure.

The concatemers contain approximately 30-40 tags. The number of tags that is obtained in a SAGE experiment ranges from 10,000 to 100, 000.

Processing of SAGE data—Three steps are involved. First, a list of tags is compiled from the concatemer sequences. Secondly, the SAGE tags are identified and finally the expression levels can be compared statistically. The extraction of tags from the concatemer sequence is straight forward since each concatemer consists of ditags which are separated by the CATG sequence. Each ditag contains one tag in the 5'-3' (sense) direction and a second tag in the 3'-5' (complementary-reverse) orientation. The ditags are extracted from the concatemers and duplicate ditags are removed because they are the most likely experimental artifacts. The length of the resulting ditags must be between 20 and 24bp. Shorter and longer ditags are discarded as experimental artifacts. Subsequently from each extracted ditag the sense and complementary-reverse (which is converted to a sense tag) tags are extracted and added to the list of SAGE tags. The number of times a tag occurs in the list directly reflects the expression level of the corresponding transcript.

The association between tag and transcript from which the tag is extracted is lost because of experimental procedure. Consequently after compiling the tag list (gene expression profile), each tag in the list has to be identified by matching it against a tag-to-gene map. This tag-to-gene mapping database must first be compiled by electronically extracting a tag from each mRNA/EST sequence in the GenBank database (by ePCR) and subsequently storing the annotated tag in the tag-to-gene mapping. Software programs and databases support the identification of the mRNAs corresponding to the tags in a SAGE library. However, this step is prone to errors and tag assignment requires manual verification. Wrong tags mainly arise from sequencing errors in ESTs and from errors in their 5' and 3' orientations.

Mapping of gene expression profile—In order to obtain a reliable mapping of gene expression profile to chromosomes, it is important to have a tag-to-gene mapping in which false positive tags are removed from the database. To improve the quality of SAGE analysis the Academic Medical center (AMC) tag-to-gene mapping process was constructed which excludes all possible false positive. This mapping is an improvement over the CGAP SAGE map tag to-gene mapping which contains many false positive tags. The AMC tag-to-gene mapping involves the following four steps.

1. Identification of the 3' end of cDNA clones and the electronic extraction of tags.
2. Removal of erroneous tags which result due to EST sequence error in the 10-bp tag.
3. Removal of erroneous tags that result because of EST sequence error in the CATG sequence.
4. Identification of antisense tags. Antisense tag is a tag with antisense orientation. Antisense tags refers to overlapping genes encoded on opposite DNA strand. Only the sense (5'-3') and complementary-reverse sequence orientations (3'-5') are considered in the electronic tag extraction procedure to construct the AMC tag-to-gene map.

Comparison of two SAGE libraries—The null hypothesis is that the differences between libraries result from random sampling. In other words, there is no difference in tag numbers between the two libraries. So, one will have to reject the null hypothesis that the observed counts in both libraries are equal before stating a pair-wise comparison of specific tags in two libraries. The unreliable tags include linker tags and redundant tags. The SAGE technique may provide tags derived from linker oligonucleotides used in library construction.

Redundant tags refer to some tags found for more than three UniGene clusters. Redundant tags may be generated by coincidental limited homologies between genes. Other redundant tags are derived from genes with a CATG close to the poly (A) tail. This generates tags with reduced sequence variability as most of the tag consist of an A stretch.

Transcriptome map—It provides an insight into the higher order- organization and regulation of expression in a genome. It presents gene expression profiles for any chromosomal region in normal and pathological lines (in case of human) and a search for genes that are overexpressed or silenced can be made. The transcriptome map provides three different ways to present gene expression profiles obtained with SAGE. Expression profiles can be given for all SAGE tags that could be linked to RH map- an extended view. Different tags may correspond to a single gene as they occur as a result of differential splicing or polyadenylation of the gene. Finally, we can have expression levels of all genes on a particular chromosome displayed- a whole chromosome view in case of a human chromosome (see Figure 24.4). From the whole chromosome view it is clear that there is a strong clustering of highly expressed genes in specific domains which are named **regions of increased gene expression** (RIDGEs) (Caron et al., 2001). The RIDGEs are observed on most chromosome. Analysis of RIDGEs suggests that many of them have high gene density. Correlation between gene expression and density of mapped genes is found for most RIDGEs. Typically RIDGEs contain 6 to 30 mapped genes per centiRay compared to 1 to 2 mapped genes per centiRay for weakly transcribed genes.

Difference between SAGE and DNA microarray—DNA microarray like SAGE enables the simultaneous measurement of the expression of thousands of genes in a single tissue. SAGE measures expression levels that directly reflect the fraction of mRNA in the cell, i.e. SAGE produces 'absolute' expression levels. In contrast DNA microarray techniques measure expression levels relative to a control condition. In DNA microarray experiments one only measures the genes for which the array contains the probes while in SAGE, one measures all mRNAs in the same experiment. So SAGE is very suitable for discovering new genes although low abundance transcripts are only likely to appear in large SAGE libraries while DNA microarray is suitable for quickly screening cells or tissues for the expression of a pre-selected set of genes.

The disadvantage is SAGE is that the extracted mRNA tags need to be identified *in silico*. In case of DNA microarray it is already known which probes (genes) are on the array.

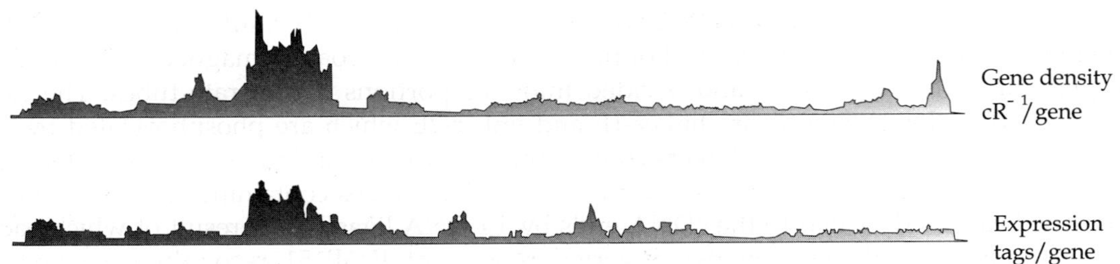

FIGURE 24.4   Showing regions of increased gene expression (RIDGEs). The upper figure shows the gene density and the lower as a moving median (adapted from Caron et al., 2001).

Further, construction of a SAGE library requires much more effort than carrying out a DNA microarray experiment.

This technique is powerful but not very convenient for the comparison of many different samples and for the study of the rarer transcripts. Although SAGE may prove quite useful in rapid surveys of gene expression differences between tissues or between normal developmental and diseased state, ESTs still have advantages over SAGE. The production and concatenation of the tags are somewhat cumbersome and the small amount of sequence information in the tags preludes many of the uses that the ESTs have.

SAGE class—It refers to any one of the four groups of SAGE tags which differ in their locations relative to known ORF, for example, within an ORF, within 500bp 3′ of an ORF, none of the two and on the strand opposite to an ORF.

Generation of longer cDNA fragments from SAGE tags for gene identification—It is a technique for the conversion of the conventionally short (9-13bp) tag sequences of the SAGE procedure into the corresponding 3′ cDNA fragments covering about 100 bases, based on a SAGE tag complementary (sense) primer and a single base-anchored ologonucleotide (dT) primer (antisense) to amplify the cDNA with Pfu polymerase in a conventional PCR. This technique expands the resolving power of SAGE and permits to discriminate between different cDNAs carrying the same SAGE tag sequence.

Variants of SAGE

Super SAGE—It allows the genome wide and quantitative gene expression profiling of cells, tissues, organs and organisms. It is a variant of SAGE and follows the SAGE protocol but involves the type III restriction endonuclease, EcoP151 that cleaves the cDNA template most distantly from its recognition site. Thus the resulting tags are 26 bp in length and so much more longer then tags from traditional SAGE (13bp) or long SAGE (19-21bp). The benefits of the super SAGE are two folds. First, the information content of super SAGE tag of 26bp is higher than the conventional tags and so allows to identify a gene directly from databases. Secondly, the ends of linker tag fragments produced by super SAGE are blunt-ended which ensure random association of the tags to form ditags. The super SAGE has the additional advantage of discovering host and pathogen messages simultaneously from the same infected material. The procedure starts with isolation of mRNA, its reverse transcription into single stranded cDNA using a reverse transcription primer with the following sequence.

5′- CTGATCTAGACCTACCGGATCC**CAGCAGT**- -3′

containing the 5′-CAGCAG-3′ recognition site for II type restriction enzyme,EcoP151. The ss cDNA is then converted into ds cDNA which is then digested with NlaIII. The 3′-end of the generated fragments of cDNAs is bound to streptavidin coated magnetic beads. The streptavidin cDNA is washed and divided into two portions in separate tubes. The two linkers used in this technique are linker-1E and linker-2E which are phosphorylated by T4 polynucleotide kinase and both linkers contain the EcoP151 recognition sequence. After that linker-1E and linker-2E, respectively are added to the two tubes containing cDNA bound to magnetic beads and ligated to the cDNA ends by T4-DNA ligase. As a result of which each cDNA fragment is flanked by two inverted repeats of EcoP151 recognition sequence. EcoP151 now recognizes the asymmetric hexameric sequence 5′-CAGCAG-3′ and cleaves the cDNA 25nt (in one strand) and 27nt (in another strand) downstream of the recognition site with 5′ overhangs of two bases. Two unmethylated and inversely oriented recognition sites in head-to-head orientation (5′-CAGCAG-N $_{(i)}$CTGCTG-3′) are essential for efficient cleavage. The linker ligated cDNA on the magnetic beads is then digested with EcoP151 at 37°C for 90 min. The generated cDNA fragments are then separated by PAGE and the ca 69bp′linker tag′ fragment (linker; 42bp, tag: ca27bp) is visualized by FITC (fluoroscein isothiocyanate, a fluorescent dye) fluorescence under UV light and collected from the gel. After that 'Linker-1E-tag and 'Linker-2E tag' fragments are mixed and their ends blunt-ended by filling in with DNA KOD polymerase (*Thermococcus kodakaraensis*) and subsequently ligated to each other. The resulting ditags are then amplified by PCR using biotinylated primer (ditag primer 1E: biotin-5′-CTAGGCTTAATACAGCAGCA-3′ and ditag primer-2e: 5′-polymerase (*Thermococcus kodakaraensis*) and subsequently ligated to each other. The resulting ditags are then amplified by PCR using biotinylated primer (ditag primer 1E: biotin-5′-CTAGGCTTAATACAGCAGCA-3′ and ditag primer-2e: 5′-TTCTAACGATGTACGC-AGCAGCA-3′) -3′) and the ditag PCR products are digested with restriction enzyme, NlaIII. The digested fragments are then separated by PAGE and the fragment ca54bp is isolated from the gel. This fragment is concatenated by ligation. The concatemers of sizes greater than 500bp are selected by PAGE which are then isolated and cloned into a plasmid vector. This vector is then used to transform *E.coli* cells by electroporation. The transformed *E.coli* is then selected by plating on LB (containing 100 µg/ml ampicillin, 100g/µl-Xgal and 0.1 mM-IPTG). Plasmid inserts are then amplified by colony PCR and sequences are analysed by SAGE200 software program for extraction of the 22-bp tags adjacent to CATG. The resulting 26-27bp sequence from each cDNA is called '**Super SAGE**' tag. The advantage of using EcoP151 over conventional enzymes is the longer tag which allows better identification of the underlying cDNA (or gene) by annotation.

A. Micro SAGE — It is a variant of the original SAGE used for the global analysis of gene expression patterns and requires only minute amounts of material (e.g. bioptic material or microdissections). Micro SAGE is run in a single streptavidin coated PCR tube (to which the RNA or cDNA remains immobilized) from RNA solution to the release of tags and thus avoids step-by-step losses. Also, reamplification of the excised ditags is reduced to only 8-15 cycles. In between different steps enzymes from the previous reactions are removed by heat inactivation and disposal so that after washing the reaction buffer and all ingradients for the next steps can be easily added. Micro SAGE also uses total RNA rather than polyadenylated RNA because the poly $(A)^{+}$- fraction is directly bound to the streptavidin coated wall of the tube via a biotinylated oligo (dT) primer which also serves as primer in subsequent cDNA anlysis.

B.   Mini-SAGE — It is a variant of SAGE and like microSAGE it is used for the analysis of global gene expression and uses only one microgram pf total RNA. Mini-SAGE uses a **phase lock gel** for the extraction of RNA from the original sample. In this technique the amount of linker oligonucleotide used in the ligation reaction is reduced so the interference with SAGE ditag amplification is minimized and which in turn increases the yield of SAGE ditags. It uses a single tube to perform different steps such as mRNA isolation, reverse transcription of mRNA into cDNA with a biotin-labeled oligo (dT) primer, digestion of cDNA, binding of digested biotin-labeled 3′-terminal cDNA fragments to stereptavidin –coupled magnetic beads, ligation of linker oligonucleotide containing recognition sites for a tagging enzyme to the bound cDNA fragments and release of cDNA tags.

C.   Long-SAGE—It is a variant of SAGE which generates 21bp tags derived from the 3′ ends of mRNA rather than 14bp tag in the original SAGE and is used for the quantification of transcript abundance in the RNA population of cell, tissue, organ or organism. SAGE was initially developed for gene expression profiling but is more powerful in transcript detection than EST sequencing. In a significant number of cases, a short sequence of 10 bases can be assigned to a single gene as a SAGE tag, immediately following the 3′ most Nla III recognition site (i.e. 5′-CATG-3′) in a transcript sequence. Long SAGE theoretically can be uniquely assigned to a single genomic position (Saha, et al., 2002). It might also assist in the correct identification of the genomic locus corresponding to a certain transcript. Long SAGE data can significantly improve the genome annotation by identifying novel genes and alternative transcripts (Wahl, et al., 2005). The goal of genome annotation is to identify sequence features that have a biological role in the organism. In this technique RNA is first extracted from the target cells and mRNA is then isolated. The mRNA preparation is then treated as per the conventional SAGE procedure but with the following changes.

1. After digestion of the cDNAs with NlaIII the linkers containing an MmeI recognition site are ligated to the 3′-ends of the cDNAs. Linker tag molecules are then released from cDNA through type IIS restriction enzyme, MmeI.

2. The resulting tags are then directly ligated with DNA ligase. Finally, tag concatemers are sequenced and the longer tags are analysed and matched to genomic sequence data. Matching of tags to genomic sequences results in localization of genes from which the tags ultimately are derived.

C. SAGE-lite—It is a variant of SAGE and used for global analysis of gene expression pattern and uses very small total amount of RNA (< 50ng). It is, therefore, used for expression of rare specimen, bioptic probes and microdissection material.

D. SAR (small amplified RNA)-SAGE—It is also a variant of conventional SAGE and uses very small amount of starting material (laser captured microdissected tissue or needle aspirates) and capitalizes on a linear amplification step of small mRNA fragment containing SAGE tags. In this method a low quantity (50ng or less) of total RNA is first converted to cDNA and the resulting dsRNAs are restricted with NlaIII restriction enzyme. After that an adaptor containing the T7 RNA polymerase promoter, upstream of a 16bp spacer and a CATG- terminus is ligated to the sticky ends of fragments of cDNAs attached to the oligo (dT) paramagnetic beads. Transcription is then catalysed by T7 RNA polymerase and extends from the last Nla III site of the transcripts to the poly (A) tail. Several rounds of transcription are performed successively with the same cDNA preparation. The resulting mRNA fragments are then separated from the beads and further processed as in conventional microSAGE protocol.

SAGE can quantitatively identify all transcripts expressed in a tissue or cell line. It is based on the extraction of a 10-base pair tag from a fixed position in each transcript and the sequencing of the thousands of these tags. Software programs and databases support the identification of the mRNAs corresponding to the tags in a SAGE library (Caron et al., 2001). However, this step is prone to errors and tag assignment requires manual verification. Wrong tags mainly stem from sequence errors in ESTs and from error in their 5′ and 3′ orientations. NCBI SAGE map database has electronically extracted tags from mRNAs and ESTs in UniGene clusters.

2. **Massively parallel signature sequencing (MPSS)** —It is a high throughput technique for sequencing millions of cDNA conjugated to oligonucleotide tags on the surface of 5μm diameter microbeads that avoids separate cDNA isolation, template processing and robotic procedure. 32-mer capture oligonucleotides are attached to the surface of separate microbeads by combinational synthesis such that each microbead has a unique tag for its complementary cDNA. Then mRNA is reverse transcribed into cDNA using oligo (dT) primers. The cDNA is restricted at both ends with a restriction endonuclease, DpnI. The complements of the capture oligonucleotides are attached to the poly (A) tail of each cDNA and the construct is cloned into an appropriate vector containing PCR handles which serve as binding sites for PCR based amplification of the tagged cDNA. The cDNA is then amplified with a fluorochrome labeled primer, denatured and the single stranded address tag containing fragments annealed to the surface of microbeads containing address tag sequences as hybridization anchors and then ligated (*in vitro* cloning). Each microbead shows about100,000 identical copies of a particular cDNA. The fluorescent microbeads (all containing a cDNA) are then separated from non-fluorescent ones (not containing a cDNA) by fluorescence activated cell shorter (FACS). Each single microbead in the library harbors multiple copies of a cDNA derived from different mRNAs. If a particular mRNA is highly abundant in the original sample then its sequence is represented on a large number of microbeads and vice-versa. In the original version of MPSS 16-20 bases at the free ends of the cloned templates on each microbead are sequenced (signature sequences).

In this procedure millions of templates containing microbeads are assembled first in a densely packed planar array at the bottom of a flow cell such that they remain fixed as sequencing reagents are pumped through the cell and their fluorescence can be monitored by imaging. Then the fluorophore at the end of cDNA is removed and the sequence at the end of the cDNA is determined in repetitive cycles of ligation of a short adaptor carrying restriction recognition site for a type II S restriction endonuclease, BbvI which binds with the adaptor and cuts the cDNA remotely and thus generating 4nt overhangs. After that a collection of 10-24 specially encoded adaptors are ligated to the overhangs and the coated tails interrogated by successive hybridization of 16 different fluorescent decoder oligonucleotide. This process is repeated many times in order to determine the signature of cDNA on the surface of each bead in the flow cell. The abundance of each in the original mRNA in the original sample is estimated by counting the number of clones with identical signatures.

Small RNAs have been identified using traditional cloning approaches (cDNA cloning approaches) which has limited how many could be characterized. Now MPSS technology has been used to identify over 1.5 million small RNAs from *A. thaliana*. MPSS sequences hundreds of thousands of molecules per reaction and provides quantitative information. Small RNAs are isolated by size fractionation on a polyacrylamide gel, RNA adaptors are sequentially ligated to the 5′ and 3′ ends and reverse transcriptase generated the first strand of cDNA which is amplified and then used as the template for MPSS (Brenner et al., 2000). For more on MPSS see chapter 12.

Polony multiplex analysis of gene expression—Although several techniques including microarrays and SAGE are available for studying large scale transcriptional changes, these technologies have limitations in that they preclude comprehensive interrogation of gene expression. Microarray-base approaches have limited ability to detect low abundance RNAs owing to cross-hybridization. Although SAGE provides the rigor of digital quantification and addresses some of the limitations associated with hybridization-based platforms, however, the cost of SAGE experiments limit sampling depth which diminishes sensitivity for interrogation of rare transcripts. To obtain comprehensive mRNA profiles that include rare and potentially important transcripts at as low as <1 copy per cell, Kim et al. (2007) developed a technique called **PMAGE**. This technique allows accurate quantitative assessment of mRNA expression because individual cDNA molecules are directly subjected to sequencing without antecedent library amplification, concatenation or subcloning. The individual cDNA template molecules are clonally amplified onto each polony bead in millions of paralle, compartmentalized droplets formed in a water-in-oil emulsion. Polony sequence-by-ligation (SBL) is used to provide an accurate, inexpensive and multiplexed platform for high throughput tag sequencing. SBL takes advantage of high discriminatory power of DNA ligase to iteratively label each bead with a fluorophore encoding the identity of a base within the template. Microscopy based detection of fluorescence ligation events allows the assessment of as many as 5 millions cDNA molecules per run and digital quantification of tag data accommodates rigorous statistical analyses over a broad range of mRNA expression. PMAGE incorporates an improved ligation based method to sequence 14 nucleotide tags derived from individual mRNA molecules.

Tag sequencing techniques—Tag sequencing is a pre-requisite for MPSS or SAGE. These are a series of high through-put RNA profiling techniques that are based on the sequencing of short nucleotide stretches (tags) for identifying a specific mRNA and thereby different mRNAs from each other.

2. **Hybridization techniques**—An alternative approach for monitoring mRNA levels is based on hybridization techniques.

Reverse northern technique—In this technique DNA fragments or oligonucleotides corresponding to different genes of cDNAs are immobilized on a solid support and hybridized to probes prepared from total mRNA pools extracted from cells, tissues, or whole organisms and converted to cDNA. The hybridization signal for each individual spot can be quantified automatically which in principle reflects the relative abundance of the corresponding mRNA in the total mRNA pool. The value of this approach lies in its propensity for miniaturization, allowing huge numbers of gene fragments to be analyzed simultaneously.

Variants of this technique—Different systems have been developed depending upon the source of target DNA and the nature of the solid support and the detection system.

A. High density DNA array—This is the simplest system (high density DNA array) which uses nylon filters in combination with radioactive probes, all purpose gridding robot and a radioactivity detection system such as a Phosphorimager. cDNAs or open reading frames (ORFs) identified in genomic sequences are PCR amplified and printed onto a solid support (microscope slide or nylon filters) using a gridding robot. Single stranded cDNA probes are synthesized from total mRNA populations using reverse transcriptase in the presence of labeled nucleotides and hybridized to the DNA arrays. Hybridization signals are detected using a two-dimensional radioactivity detector such as a Phosphorimager or for fluorescent probes, a modified confocal laser scanning microscope. The signals are quantified and processed using software. Software for the detection and quantification of signals is commercially available. The intensity of the hybridization signal is proportional to the abundance of the corresponding mRNA in the pool used to synthesize the probe. This technique can be used detect low-abundance mRNAs (down to 1:10,000 of the total mRNA population). A single 18-x18-mm microscope slide can contain the entire set of more than 6400 yeast ORFs.

Phosphoroimaging—It is a technique for the detection of radioactive isotopes. A polyester plate coated with fine crystals of photostimulatable phosphor ($BaFBr:Eu^{2+}$) works as imaging plate. This plate absorbs the energy emitted by the respective isotopes. In this technique the sample (a nylon membrane) covered with a Saran Wrap is exposed on the imaging plate inside a cassette. After the exposure the imaging plate is scanned with a laser beam and thus emits luminescence (proportional to the radiation intensity) which is collected into a photomultiplier tube and gets converted into electric signal. The imaging plate is reused once the data are erased by exposure of the plate to light.

B. DNA microarrays—Any microscale solid support (e.g. nylon membrane, nitrocellulose, glass, quartz, silicon or other synthetic materials) on which either DNA fragments, cDNAs, oligonucleotides, genes-ORF, peptides or proteins (e.g. antibodies) usually in nanolitre quantities are spotted in an ordered pattern (called array) using robotic devices at extremely high density is called **microarray**. Such microarrays are increasingly used for high throughput, expression profiling. We can have thus antibody chip array, gene chip array, EST array and exon array. A more sophisticated system uses DNA microarrays printed at high density on pretreated glass slides (Schena et al., 1995; De Risi et al., 1997). This system allows the use of the fluorescent probes and hybridization signals are detected by a confocal laser microscope. Two probes labeled with different fluorochromes are mixed and hybridized and the two colors are quantified simultaneously. This system allows simultaneous detection of hybridization signals of two probes in a single hybridization experiment. This system has a higher sensitivity (mRNA of abundance down to 1:100,000 can be reliably detected) than the above described system. The drawbacks with this technique are its cost and the requirement of a robot and scanner. Also, the arrays can not be reused which further escalates the cost. **Cross-talk-** It refers to undesirable and erroneous detection of a signal (e.g., fluorescent light of a specific wavelength) on a microarray by a channel which is specified for another wavelength. For example, when two fluorophores, say cyanin3 and cyanin 5, are used as labels for hybridization on the same chip and the emission of each fluorochrome is measured by a separate channel of detector then spill-over from one fluorochrome to the channel of other fluorochrome occurs because of their emission spectra extending over a relatively large range. Cross-talk can be minimized by taking a series of measures such as choice of fluorochromes, choice of lasers, excitation and emission filters.

C. DNA chips—It is miniaturized solid support onto which so called target molecules are spotted at a low, medium or high density or onto which nanochannels are microfabricated. Biochips are being used to follow changes in gene expression in response to abiotic stress. Using gene chips or microarrays representative gene can be analysed. DNA microarrays sometimes called **DNA chips** are used for rapid and simultaneous screening of thousands of genes. Microarrays consist of a number of probes that bind mRNA for specific genes. By determining how much mRNA binds to each probe the relative abundance of mRNA from each gene can be assessed. DNA segments, a few dozens to hundreds nucleotide in length, from known genes are amplified by PCR and placed on a solid surface. DNA chip is another technology for studying the mRNA levels. **Genome array/chip**- It refers to any microarray onto which a complete genome is spotted as, for example, genomic fragments. Genome arrays are available for E.coli, Arabidopsis and are expected for almost any organism whose genome has been sequenced. Genome arrays are employed for studying genome-wide gene expression and detection of co-regulated genes. **Universal array/microarray**- Universal array or one-chip-for-all refers to any gene array which contains all possible genes of the living world in the form of 10-16nucleotides long, chemically synthesized oligonucleotides. Universality requires one million spots per 10-mer per array or 64 million spots per array for 13-mers. Universal arrays permit to globally determine the gene expression patterns of any organism. **Modular chip/array/microarray**- It refers to any microarray which consists of several arrays separated by, for example, microfluid hybridization chambers. Each module can be separately used for specific experiments.

Most microarray experiments focus on identifying patterns of gene expression in a particular system, for example, comparing tumor tissue with normal tissue, analyzing gene expression responses to a stress stimulus or comparing expression patterns in a particular tissue over time. Conceptually differences between the various states should be reflected in changes in particular cellular pathways. Microarray should allow to identify the products of new genes and their contribution to these pathways. It should also identify cellular processes where most of the genes associated with a particular biological function are up-or down-regulated in a similar way. However, some genes known to be involved in a particular pathway are invariably missed whereas other apparently unrelated genes show expression profiles which are strikingly similar to bona fide pathway components- subject to independent validation using RT-PCR. The reason for this kind of deviation is that microarray studies fail to sufficiently sample the biological variability with a system.

Genes with conserved biological functions in different species (orthologs) would likely to retain similar patterns of expression while other associations occurring by chance would be filtered out by analysis of such a large multispecies data set. '**Metagene**' group, the orthologs are genes which have retained their functions through evolutionary history, are identified using sequence homology. Orthologs are identified by performing an all-against-all BLAST between every pair of protein sequences from each of the organism. A metagene is defined as a set of genes across multiple organisms whose protein sequences are one another's best reciprocal hit.

Photolithography—It is based on a technique to synthesize large amounts of different oligonucleotides *in situ* on a glass support using light directed, solid phase, combinatorial chemistry developed by Affymetrix (Santa Clara, CA). Oligonucleotides corresponding to the genes of interest are hybridized with fluorescent probes and signals are detected using techniques similar to the ones used for microarrays. In one experiment 260,000 25 –mers corresponding to a nearly complete set of yeast genes were synthesized *in situ* on four 1.28-cm^2 grids (Wodicka et al., 1997). Each ORF predicted from the genomic sequence was represented by about 20 oligonucleotides to cancel out difference in hybridization behaviour of different oligonucleotides and cross hybridization with related sequences. Further, the hybridization value for each oligonucleotide was corrected for the hybridization value for a second, negative control oligonucleotide with an identical sequence except for a single base difference. This technique shows a very high sensitivity and reliably detected a transcript present at one copy in 10 to 20 cells.

Production of cDNA arrays—The cDNA expression array refers to an ordered alignment of different complementary cDNAs or fragments of cDNAs or cDNA complementary oligonucleotides immobilized on a support. Such arrays may contain tens of thousands of different cDNAs on a surface (1x1cm or less) and are used to determine differential gene expression patterns. cDNA arrays can be produced by a number of methods. One particular method uses PCR amplified partial sequences of cDNAs. The reverse transcriptase PCR primers are designed from known cDNA sequences and used to amplify the corresponding cDNAs such that the amplified products are 200-600bp in length. The amplified products (amplicons) are cloned and are partially sequenced. The cloned fragments are again amplified, normalized and immobilized on positively charged nylone membrane. The sequence homologies among different cDNA amplicons are kept at a minimum. The hybridization probes are derived from total RNA or polyadenylated RNA of different specimen such as different tissue, organs or cells, reverse transcribed and labeled using oligo (dT), random or specific primers and hybridized to the arrays. Thus large scale up or down regulation of functionally related genes or gene classes can be determined.

Two color hybridization strategy is often used with cDNA microarrays. cDNA from two different conditions is labeled with two different fluorescent dyes (usually Cy3 and Cy5) and the two samples are co-hybridized to an array. After washing the array is scanned at two different wavelengths to detect the relative transcript abundance for each condition. It is based on the thinking that small set of genes are important to a process. **Transcript imaging**-It refers to the visualization of all transcripts, i.e., mRNAs of a cell at a specific time. A complete imaging can only be achieved by high-throughput techniques, for example, cDNA microarray screening.

Types of chips—Chips containing every gene can be produced allowing genome-wide expression analysis. This will help in elucidating the metabolic and disease pathways of the cell under a variety of developmental and environmental perturbations and have immediate applications in toxicological studies and pharmaceutical development. Chips containing the complete ORFs from the yeast genome have been developed. **Custom chips** refer to a series of chips with hundreds to thousands of full-length genes or segments obtained from various databases. **Standard chips** have been developed containing more than 65, 00 genes from the public databases.

Expression array (RNA chip)—It is a high density cDNA chip onto which multiple cDNAs, short fragments of cDNAs (see ESTs) or gene fragments are fixed which allows to determine the expression of a series of genes simultaneously. For an expression array, labeled cDNA from a target tissue is hybridized to the expression array and the hybridization patterns can be directly converted into information about the expressed genes in the sample–**cDNA map**-It refers to a graphical depiction of the order of expressed sequences–cDNAs (resembling exons) along a stretch of DNA, for example, a BAC clone , a chromosome. **Single nucleotide polymorphism (SNP) mapping chip**- It is used to identify the common polymorphism contained within the mapped STS collection (Lander, 1996). **Screening chip**- It will allow databasing of large number of polymorphisms. **Mapping chips/marker chips** - Mapping chips contain molecular markers and have a number of applications including studies of linkage, association and loss of heterozygosity measurements. To determine the gene involved in Berardinelli-Seip syndrome, the DNA from members of affected families is subjected to linkage analysis and homozygosity mapping with a genome wide panel containing ~400 microsatellite markers of known genetic location with an average spacing of ~10cM. In this technique, a fixed panel of primers specific for amplification and analysis of each marker is used to compare whole DNA of affected individuals with that of unaffected relatives. The measurements reveal the lengths of the repeats associated with each microsatellite. For every microsatellite, each observed length is an allele. Identifying microsatellite markers that are closely linked to the phenotype localizes the desired gene. The measurements are done efficiently and in parallel using commercial primer sets and instrumentation. Finer probing, mapping with additional markers localized the gene to a region of about 2.5Mb on chromosome 11. Microchips are particularly useful for detecting mutations and polymorphisms particularly SNPs. **SNP chips-** It will uncover how the SNP is associated with disease.

Methods for preparing labeled material for gene expression—There are different methods for preparing labeled material for gene expression. The RNA can be labeled directly using psoralen-biotin derivative or by ligation to an RNA molecule carrying biotin. Labeled nucleotides can be incorporated into cDNA during or after reverse transcription of polyadenylayed RNA or a cDNA can be generated which carries a T7 promoter at its 5′end. In the last case the double stranded cDNA serves as template for reverse transcription reaction in which labeled nucleotides are incorporated into CRNA. Commonly used labels include the fluorescein, Cy3 (or Cy5) or nonfluorescent biotin which is subsequently labeled by staining with a fluorescent streptavidin conjugates.

Total gene expression analysis (TOGA)—It is a technique for the automated high through-put analysis of the expression of nearly all genes in a given cell, tissue or organ. This technique is based on the fact that almost all mRNAs can be identified by an 8 nucleotide sequence and the distance of this sequence from the poly (A) tail. In this technique, poly (A)$^{+}$-mRNA is isolated first and double stranded cDNA issynthesized by reverse transcriptase using a pool of equimolar Not I-containing 5 biotinylated anchor primers, degenerate in their 3′ ultimate positions, for example, 5′ T$_{18}$VNN[V= A,C OR G; N=A,C,G or T]. One primer of this primer mixture initiates synthesis at a fixed position at the 3′ end of all copies of each mRNA species (defining a 3′ end point for each species) in the sample. Then the cDNAs are cleaved with MspI (recognition site: 5′-CCGG-3′), the 3′ fragments are isolated by streptavidin bead capture and released from the beads by NotI digestion. NotI cleaves an 8 nucleotide sequence within the anchor primer but rarely within the mRNA-derived part of the cDNAs. The resulting NotI-MspI fragments are then directly cloned into a ClaI-NotI-cleaved expression vector in an antisense orientation to its T3 RNA polymerase promoter and the constructs are transformed into an *E.coli*. The plasmids are then isolated. The insert-containing vectors are linearized with MspI which cleaves at several sites within the vector but not in the cDNA inserts or the T3 promoter and antisense cRNA transcripts of the cloned inserts are produced with T3 RNA polymerase. These transcripts contain known vector sequences (tags) touching the MspI and NotI sites. These cRNAs after removal of the plasmid DNA template with RNase free DNase serves as substates for reverse transcriptase using a primer complementary to the vector sequences. The the resulting cDNA is amplified with a primer extending across the non-reconstituted MspI/Cla I site with either A,C,G or T and a universal 3′ primer in a standard PCR. A subsequent PCR with a fluorescent 3′ primer and each of the 256 possible 5′ primers extending 4 bases into the inserts each one in a separate reaction) generates products which are separated on denaturing sequencing gels and the bands are detected by Laser-induced fluorescence. Thus, each final PCR product carries an indentity tag, a combination of an 8 nucleotide sequence (in case of MspI: $CCGGN_1N_2N_3N_4$) and its distance from the 3' end of the mRNA also known as vector-induced sequence added during TOGA processing.

cDNA-AFLP—Hybridization-based methods for genome-wide expression analysis such as microarrays require sequence information about the whole transcripts of the target organism. In the absence of such information differential display methods (Vos et al., 1995) such as c DNA-AFLP and its variants like HiCEP (Fukumura et al., 2003) are among the few available choices. Although sequence information is absolutely essential for cDNA-AFLP experimental design it is extremely useful for designing the experiment (sequences of genes cloned from the organism, EST collections or a genome of closely related organism). Thus cDNA-AFLP is a flexible tool which can be used even when genomic sequence information is not available yet all the available sequence information can be utilized. This technique can be used at the genome-wide level to identify previously unknown genes with interesting expression patterns.

In the differential display methods, the cDNA mixture synthesized from an mRNA sample is divided into small subsets called pools and cDNA in each pool are PCR amplified and separated using gel (or capillary) electrophoresis. When the above procedure is carried out for samples collected from different conditions, differences in the intensities of corresponding gel bands reflect the relative differences in the expression levels of genes. In c DNA-AFLP the division into subsets is a result of using restriction enzymes and selective PCR. The cDNAs are digested with two different restriction enzymes and adaptors are attached to the specific ends of the resulting fragments and the fragments are amplified using primers extended with additional selective nucleotides. Thus, for each selective primer pair only the fragments whose ends match the primer extensions get amplified and these fragments form a pool. Finally, the fragments in each pool are separated by electrophoresis. If the sequences of some transcripts are available from organism then their location on the gels can be predicted and thus some bands can be readily identified and their expression pattern recorded. On the other hand, an interesting yet unidentified band can be sequenced and in this way novel genes relevant for biological process being studied can be found.

It is a technique for monitoring the steady-state levels of a large number of mRNAs. In this technique mRNAs are isolated and reverse transcribed into cDNAs. After that the cDNAs are first restricted with a rare cutter enzyme, Bst YI and then with a frequent cutter, Mse I. The restriction fragments generated are then ligated to Bst YI and MseI adaptors for selective amplification in an AFLP procedure. This technique generates patterns resembling the complex AFLP pattern with genomic DNA.

Improvement over c DNA-AFLP—Only one pair of enzymes does not in practice produce a fragment for every c DNA molecule that could be amplified and detected by electrophoresis. The fragments generated from a particular cDNA can be too long or too short to be revealed by the electrophoretic setup or if either of the enzymes does not have a restriction site in the cDNA molecule, all fragments lack one or both of the two specific ends required for selective PCR. In addition, several cDNAs having the same nucleotides next to the restriction sites can produce fragments with the same length in which case these fragments end up in the same gel band and consequently do not give any useful information (Kivioja, et al., 2005).

The c DNA-AFLP has been improved so that only one fragment at most is obtained from each transcript. After digestion with the first restriction enzyme, the 3′ fragments of the cDNAs are captured. Only these are then digested with the second enzyme. This leads to at most one fragment per cDNA which has the specific ends of both enzymes. Obtaining at most one fragment from each transcript reduces the redundancy of the fragment pools and by reducing the total number of fragments, it also reduces the number of selected nucleotides needed for reasonable separation of bands on the gel. So concentrate on this variant of c DNA-AFLP-3′ variants.

Introduced AFLP (iAFLP) — It is a high throughput technique for the specific, selective and simultaneous expression profiling of thousands of genes. In this technique, mRNAs are first isolated from target cells or tissues, reverse transcribed into cDNAs. The cDNAs thus generated are cleaved with MboI and so called iAFLP adaptors are ligated to the resulting fragments using T_4 DNA ligase. The iAFLP adaptors carry common sequences at both termini and sequences of varying length internally. The polymorphic part of the adaptor consists of degenerate sequences of A,C and T residues to neutralize sequence specific effects of adaptors on the amplification rates. Then gene-specific primers and iAFLP adaptor specific primers are labeled with fluorochromes, are used to amplify the corresponding cDNA in a standard PCR. The amplified fragments are separated in a denaturing 10% PAG and detected by autoradiography or fluorography (in case of fluorochrome labeling). This technique allows to quantitate the absence of any transcript in the sample and several samples can be run simultaneously by labeling the adaptors with different fluorochromes.

24.65   NOT ALL mRNAs ENCODE PROTEINS

Transcription is not limited to protein coding genes. Recent studies have shown that most mRNAs do not encode proteins which thus explain the discrepancy between the small number of protein coding genes found in vertebrate genomes and the much larger and ever increasing number of polyadenylated transcripts identified by tag-sampling or microarray-based methods. Exploring the role and diversity of these numerous noncoding RNAs (ncRNASs) now constitutes a main challenge in transcriptome research (Claverie, 2004). Non-coding RNAs include structural RNAs (e.g. tRNA, rRNAs and snRNAs) and more recently discovered regulatory RNAs (e.g. miRNA) (for detail see chapter 23). Tilling array data provide evidence for transcription of all of these ncRNAs. Many known ncRNAs are characterized by a well –defined RNA secondary structure. The prediction of structured ncRNAs as well as functional structures in mRNA can be predicted using *de novo* ncRNA prediction algorithms including EvoFold and RNAs using multiple sequence alignments (for detail on these prediction see Roy, 2009) (Washietl et al., 2007).

Large scale studies of transcriptomes have used both cDNA cloning approaches and the interrogation of genome tilling arrays (Cheng et al.2005; Berton et al., 2004). In case of human 30,000-150,000 genes were predicted initially. After that bioinformatics analyses predicted around 35, 000 genes and now 25,000 genes have been predicted. The largest estimate of number of genes was based on the number of polyadenylated transcript 3′ ends identified through single pass sequencing of cDNA libraries. SAGE generated many transcripts, many of which do not correspond to annotated genes. A large fraction of the genome appears to give rise to polyadenylated transcripts that do not code for proteins and so there is discrepancy between the number of genes (protein coding) and the number of transcripts. Noncoding polyadenylated mRNA contributes to a large fraction of the EST sequences (and SAGE tags) which subsequently clustered as singletons. Initial analyses of the transcriptome were based on hybridization with probes derived from pre-defined or predicted gene sequences and so they did not reveal unexpected transcripts. When tiling arrays were introduced allowing the interrogation of genome sequences for corresponding transcripts at fixed interval irrespective of predicted gene location (a tiling array with 5′ nucleotide resolution which mapped the transcription activity along 10 human chromosomes) revealed that an average of 10% of the genome compared with the 1 to 2% represented by bona fide exons, corresponds to polyadenylated transcripts of which more than half do not overlap with known gene locations. Further studies with mouse genome showed transcription of 62% genome and identification of over 181, 000 independent transcripts of which half consisted of noncoding RNA. 70% of the mapped transcription units overlapped to some extent with a transcript from the opposite strand. The noncoding transcripts originate from the intergenic regions, introns or antisense strands.

miRNAs — miRNAs play an important regulatory function in eukaryotic gene expression via mRNA degradation or translation inhibition. An miRNA down regulates the translation of target mRNA through base pairing to these target mRNAs. In animals miRNAs tend to bind to the 3′ UTR of their target to repress translation. The pairing between miRNAs and their target usually includes short bulges and/or mismatches. In contrast, plant miRNAs bind to protein-coding region of the target mRNAs with 3 or fewer mismatches and induce target mRNA degradation or repress mRNA translation. miRNAs and their associated proteins appear to be one of the more abundant ribonucleoprotein complexes in the cell. A single organism may have hundreds of distinct miRNAs, some of which are expressed in stage, tissue or cell type specific patterns. Nonetheless, miRNAs whose expression is restricted to nonabundant cell types or specific environmental conditions could still be missed in cloning efforts. They are the smallest functional ncRNAs of animals including human, mouse and plants and are synthesized from a longer precursor (pre-miRNA) forming a hair-pin structure which contains the mature miRNA in either of its arm. Mature miRNAs range between 17 and 29 nucleotide in length and the majority of them are about 21-25 nucleotide long and have been found in a wide range of eukaryotes. The precursor miRNA is about 100 nucleotides for animals. miRNA sequences are thus quite short and miRNA unlike mRNA can be compared only at the nucleotide level and not at translated amino acid level. miRNAs tend to have highly constrained tissue and time –specific expression patterns and degradation products from mRNAs and other endogenous noncoding RNAs coexist with miRNAs and sometimes dominant in small RNA molecule samples extracted from cells (see chapter 23 for detail). The characteristic of animal miRNA is that their genes are often organized in tandem and are closely clustered. Thus one will have to distinguish the possible transcription of miRNA into two categories.

1. Co-transcribed miRNAs, i.e. miRNAs located in the introns of annotated host genes. The miRNAs in this case share the same ± 1000 up/downstream of the host gene.
2. Independently transcribed miRNA. These are not far away from the annotated genes. The transcribed miRNA can be further categorized into two: a). clustered

miRNAs (-1000 upstream of the first miRNA precursor in the cluster and + 1000 downstream of the last miRNA precursor in the cluster) and b). nonclustered miRNA (± 1000up/downstream of the miRNA precursor).

The second category is based on the observation of a prominent characteristic of animal miRNAs that their genes are often organized in tandem and are closely clustered. As mature miRNAs are 17-25 nucleotide long and its pre-miRNA is about 100 nucleotides in animals so distinguishing weakly conserved genes from random 'hits' is more difficult when searching for miRNAs than for protein-coding genes. Moreover, even in cases where there are large RNA families, sequence conservation is often at the secondary structure level, i.e. what is conserved are base pairing rather than the individual base sequence. Consequently, sequence alignment alone may fail to identify miRNAs that diverged too far apart within in their primary sequence while retaining their base-paired structure. For predicting miRNAs by computation method, one needs to define sequence and structure characteristics that differentiate known miRNA sequences from random genomic sequences and use these characteristics as constraints to screen intergenic regions/whole genomes (introns excluding these protein coding genes) in the target genome sequences for candidate miRNAs. ncRNAs lack in their primary sequence common statistical signals that could be exploited for reliable detection algorithms.

Results from comparative genomics study involving various species of Drosophila has demonstrated the evidence of miRNA processing from both arms of a miRNA hairpin and from both DNA strands of a miRNA locus in some cases (Stark et al., 2007)) leading to as many as four functional miRNA per locus. As miRNA/miRNA pairs are expressed from a single precursor and thus co-regulated whereas sense/antisense pairs are expressed from distinct promoters, the use of both arms or both strands provides general building blocks for the higher-level miRNA mediated regulation.

24.66 TILED OLIGINUCLEOTIDE

It refers to any one of a series of overlapping oligonucleotides which altogether span a genomic region, e.g. a mutation (for example, a SNP). Tiled oligonucleotides can be used to produce a tiling path across the region and neighboring sequences.

24.67 IMAGING OF GENE EXPRESSION

Gene expression can be studied using techniques such as Nothern and Western blotting, PCR and more recently by DNA microarrays and MS. However, these techniques are not sensitive enough to allow single-cell analysis of genes that are expressed at low levels. Furthermore, these ensembled averaged methods often mask stochastic gene expression events. Single molecule experiments providing imaging of gene expression at a single molecule level in living cells have been made possible because of two developments. First, at transcriptional level, single mRNA molecules can be detected and tracked in a living cell using multiple copies of a fluorescent mRNA binding proteins. At translation level the expression of single protein molecules can be tracked using a fast maturing and membrane tagging yellow fluorescent proteins (YFP). On genomic DNA a particular gene only exists in one (or a few) copy switching on and off stochastically to regulate biological functions (Xie, 2006).

24.68 GENE REDUNDANCY

It refers to the presence of many copies of a single gene in cell. Gene redundancy can be brought about by (i) a high degree of ploidy of the entire genome (ii) the presence of polytene chromosomes with lateral multiplicity of genes (iii) extra copies of parts of the genome (gene amplification) and (iv) linear multiplicity of genes in the chromosomes (gene reiteration). Multiple copies may be inherited or result from selective duplication during development. In theory, it is possible to directly measure the extent of redundancy for those genes whose corresponding mRNA molecules can be isolated. Hemoglobin and immunoglobulin (antibody) mRNA species are the predominant mRNA species within their respective cells. A direct titration can be made of the number of hemoglobin (immunoglobulin) genes capable of forming DNA-RNA hybrid molecules. Hybridization experiments can be conducted with DNA from those cells synthesizing hemoglobin (immunoglobulin) as well as with DNA from cells which are not do not produce hemoglobin (immunoglobulin). The results with DNA from nonhemoglobin (nonimmunoglobulin) producing cells will provide the answer. A positive answer will imply redundancy of chromosomally located genes. But if many identical copies are found only in the corresponding differentiated cells, it can be explained as the selective multiplication of a single chromosomal gene to give rise to a large number of extrachromosomal DNA fragments (Watson, 1970). In plants, each of the hormones (auxin, cytokinin and giberrellin) is pleiotropic in effects. Several hormones may affect the same response.Plant hormones are synthesized at several different sites in plants and not in any one specific gland or tissues. Each hormone brings about a variety of growth and morphogenetic responses. Thus there is an apparent redundancy in control of the same response.

Strains with mutations in either of the two yeast genes encoding hydroxy-methyl glutamyl CoA reductase grow as wild on rich or minimal medium but their distinct growth disadvantages are evident when they are co-cultivated with the wild type strain in liquid medium. It may therefore be possible to recognize and distinguish between genes that superficially appear to be functionally redundant.

24.69 FUNCTIONAL ANALYSIS

The functional analyses of gene include **gene trapping, gene targeting and chemical mutagenesis of ES cells** *in vitro*.

Forward genetics/Reverse genetics approach—There are various methods for determining functions of genes on a genome-wide scale.Sequence alignment-based comparisons are used to identify homologs between and within organisms. Trancriptional profiling is used to determine gene expression patterns whereas Y2H and other interaction analyses help identify pathways, networks and protein complexes. All these methods infer gene function but not in the context of the living organism. Determing such roles in the context has been the main strength of traditional forward genetics strategies where mutagenesis is typically followed by a phenotypic screen which ultimately leads to finding genes and determining its identity by sequencing it. Genetics relies on the study of variants either found in natural populations or induced by mutagenesis. The classical analysis of gene function involves identifying an organism with an abnormal set of physical and behaviour characteristics and then isolating the underlying mutant genes. The analysis of the inheritance of this variation in mapping populations provides information on the number of genetic factors determining the observed variation and their relative position on the chromosome. The variation can be discrete or discontinuous as is generally the case with simple traits (qualitative traits under control of one or a few genes) including mutations, or continuous as with most complex traits (especially those of agronomic importance such yield and yield components) which are mostly controlled by large number of genes (polygenes). **Forward genetic screens** provide insight into complex biological processes. Mutations are generated at random, phenotypes of interest are scored and the mutated genes are subsequently identified. The limitation of its application is limited by inherent biases in mutagenesis techniques, requirement of analysis of large number of mutants and considerable efforts for identification of relevant genetic lesions. Moreover, most genes have multiple functions and a gene's function in one tissue can preclude its recovery in screen focused on functions in other tissues. This is particularly true for genes that are essential in the early development of the organism. This '**forward genetics** approach is limited to rapidly reproducing organisms.

Once mutations that affect the process of interest have been identified they can be studied by standard genetic tools. The general methods are as follows.

1. Genetic complementation test can be used to estimate the number of different genes involved in the process.
2. The phenotypes of individuals carrying mutations in more than one of the genes can be studied to infer hierarchies of gene function.
3. Measuring the rates of meiotic recombination between a mutation and other genes can be used to map the relative chromosomal position of the mutated gene.

Other genetic tools—These approaches largely rely on the producrion of mosaic through chromosomal loss or by induced mitotic recombination and these are uniquely applied to Drosophila.

Reverse genetics approaches begin knocking out a specific gene to identify its function. Reverse genetics refers to the specific alteration of a pre-determined site within a gene or DNA sequence followed by analysis of the functional change in the expression of that gene which occurs as a consequence (Weissman, 1978) (Figure 25.5). In the genomics era we have the set of gene sequences for many organisms and the task is to determine the phenotype, i.e. to perform reverse genetics.

24.69.1 Reverse Genetics Strategies

There are two categories depending on whether the mutagenesis is targeted specifically to the locus of interest or is performed throughout the genome, followed by screening for a molar region ((Henikoff and Comai, 2003).

1. **Targetted mutagenesis**—Specific targeting of loci can be achieved using i. antisense RNA suppression; RNAi-based PTGS ii. Homologous recombination iii. Chimeric RNA/DNA oligonucleotides. They key features of antisense RNA suppression, RNAi-based PTGS are that they are dominant, effective for gene families, yield a phenotypic range, unpredictable outcome and require transformation.In case of RNAi-based PTGS the throughput is limited by the need to engineer a construct for each gene of interest and to individually transform

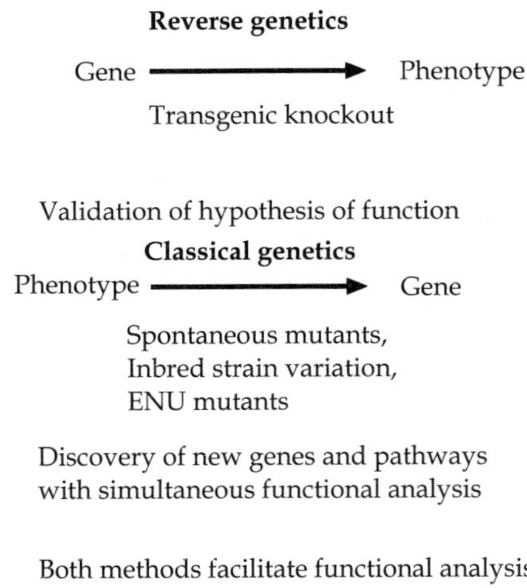

FIGURE 24.5   Showing two methods of functional analysis.

plants with each one. The antisense suppression methods are less reliable (Schuch, 1991). Although HR works in *E.coli* and yeast (Struhl et al., 1979) but is difficult or infeasible in multicellular eukaryotes which have less active HR systems. Further, HR would require transformation. Use of chimeric RNA/DNA hybrid oligonucleotides to induce base changes, insertions or deletions has been reported (Rice et al., 2000) but its application in plants has so for not been demonstrated.

2. Genome-wide mutagenesis—It includes insertional mutagenesis, fast neutron mutagenesis and size screening and targeting induced local lesions in genomes (TILING). In case of insertional mutagenesis, *in silico* screening for knockouts has been achieved and transformation and/or tissue culture step usually required. The features of fast neutron mutagenesis are that it is nontransgenic and can knockout gene blocks (Li et al., 2001). The features of TILING include nontransgenic, yielding allelic series, achievement of high throughput and suitability for small genes.

Chemical, radiation and transposon mutagenesis offer the most versatile methods for assessing gene function. Transposon and T-DNA (portion of the Ti (tumor inducing) plasmid that can be transferred to plant cells) insertions provide additional advantages of tagging the target gene molecularly and in many cases genetically via reporter and selectable marker genes carried by the insertions (Martienssen, 1998). These features can be combined with conventional mutagenesis to provide a comprehensive strategy for probing plant gene function. Classic chemical/physical mutagenesis methods reach saturation relatively easily (i.e. a high probability of recovering a mutation in every gene in the genome) but not with transposon mutagenesis. Access to the mutated gene is obtained using positional cloning strategies (i.e. cloning the gene based on its position on the genetic map). This cloning strategy is facilitated in case of species for which dense genetic maps with many visible and molecular markers exist and for which an almost complete physical map has been constructed using collection of overlapping fragments cloned in YACs or BACs. Although total sequencing of genome will provide the ultimate physical map but genomic sequencing along with these tools will greatly accelerate the positional cloning strategies. Further new mapping strategies based on 'mapping chips' will accelerate the map construction as well.

24.69.2 Probing Plant Gene Function and Expression With Transposons

Transposable elements are powerful mutagens for functional genomics. Transposable elements provides and convenient and flexible means to disrupt genes and thus allow to assess their functions. By engineering transposons to carry reporter genes and regulatory signals, the expression of target genes can be monitored and to some extent manipulated.

24.69.3 Insertional Mutagenesis

It is a powerful method for determining the function of plant genes identified in systematic genome and expressed sequence tag sequencing projects. Insertional mutagenesis using T-DNA or TE has been performed on a large scale and has the potential to provide a vast catalog of knockout mutations, however, a substantial fraction of genes will be missed due to inadequate sampling or insertion-site bias or because the probability of being mutagenized is proportional to the size of a gene. There are two types of approach with the transposons.

1. Use of plants with high copy numbers of transposons per genome
2. Use of plants with a single copy of transposon per genome.

In both cases the objective is to recover at least one informative insertion per gene. High copy number approach has the advantage that a relatively small population of plants would be required for complete genome coverage. The single copy approach offers a number of advantages. First, it can be used to integrate reporter genes into the genomes thereby permitting gene expression to be observed at each integration. The multicopy reporter gene insertions would give rise to a mixed pattern of expression. Second, single copy transposon can also be used to integrate genetic and physical markers into the genome and thus can be used in map-based genomic strategies. Thirdly, single copy transposon insertions can be sequenced directly and thus reducing the number of manipulations needed to isolate insertions in every gene. Multi –copy transposons can be used to heavily mutagenize plants with large genomes and recover insertions into genes of interest. On the other hand, single copy transposons allow the use of enhancer and gene trap reporter genes to monitor the patterns of gene expression as well as gene disruption. Furthermore, libraries of single-copy insertions can be sequenced systematically and screened for mutant phenotypes. These libraries represent the most economical method for systematic function search in plant genomes. In case of Arabidopsis with small genome the comparison of gene trap sequences with genomic and expressed sequence databases will permit determination of location of each insertion with nucleotide precision and thereby allowing their use as tools in positional cloning as well as wide scale gene disruption. By studying reporter gene expression patterns in viable heterozygotes the role of essential genes can be assessed in later development even if insertions are homozygous or haploid lethal. Further, by combining insertions in homologous genes the function of redundant genes can also be assessed.

Site- selected transposon mutagenesis — The insertional mutagenesis provides a more quick method of cloning a mutated gene. The transposons or the T-DNA which are able to insert at random within chromosomes, can be used as mutagens to create loss of function mutations in plants. Insertion of a mobile DNA element within a gene disrupts gene activity at the transcription level, translational level or both. Inserts can be directly selected to use a selectable marker (e.g. conferring kanamycin or Basta resistance harbored by the insertion sequence (transposon or T-DNA). The presence of a gene/enhancer trap with a reporter cassette allows expression of the gene at the site of insertion to be monitored.

Steps involved in site –selected mutagenesis

The steps involved in site-selected mutagenesis are as follows.

First, transposons are used to generate populations of model organisms (libraries) which can be searched for individuals with transposon insertions in any given gene. In other words, a genomic library from a transposon induced mutant individual that has been selected by its mutant phenotype is constructed. In the mutagenesis strategy the key parameters are

   i. The level of saturation, i.e. the probability of having at least one insertion in any gene which depends on the number of independent insertions in the population

   ii. The randomness of insertion of the element and

   iii. The number of insertions per line.

In order to have 95% probability of finding an insertion into any gene in organisms like Drosophila, Arabidopsis or *C. elegans* which have similar genome sizes and gene densities, a library needs to have around 120,000 insertions (Kaiser and Goodwin, 1990). The level of saturation depends on the size and organization of the genomes. Even so, most insertions are into non-coding regions and have no phenotypic effect. Secondary insertions or deletions must be generated at the locus by remobilizing the transposon to disrupt gene function.

Second, searches are carried out by amplifying DNA from these populations by using PCR with specific primers for transposons and from the gene. In other words, a molecular probe homologous to the mutagenizing transposon is used to screen the library for the positive clone. As the sequence of the inserted element is known the gene in which it is inserted can be easily recovered using various cloning or PCR-based strategies such as inverse PCR or tail PCR (Liu and Whittier, 1995) (Figure 25.6).

Thirdly, the positive pools (indicating presence of insertions) are rescreened by sib selection to identify individuals that have the desired insertion. The positive clones are then isolated and characterized by sequencing so that the mutated gene can be delimited.

In case of maize, the use of Robertson's *mutator* system of transposons *(Mu)* as insertion mutagens led to identification of three new null alleles. Transposition is facilitated by the MuDR autonomous transposon that encodes the enzymes required for transposition of other Robertson's Mutator transposon elements. The maize gene hcf106 has a Mu1 element inserted in the promoter region. This method can be used to isolate insertions in any DNA sequence in any genome, even those that might be lethal or those that might have no phenotypic effect. As the germinal excision is very rare in Mutator lines so new alleles are stable. Mutator insertion sites in maize can be sequenced in the systematic way. The individual PCR products from multiple elements are obtained by one of a number of anchored PCR technique (see chapter 21) and are resolved by gel electrophoresis. Individual bands are excised and the products are purified for sequencing. There are a number of problems associated with this technique. First, the somatic transposition which is not

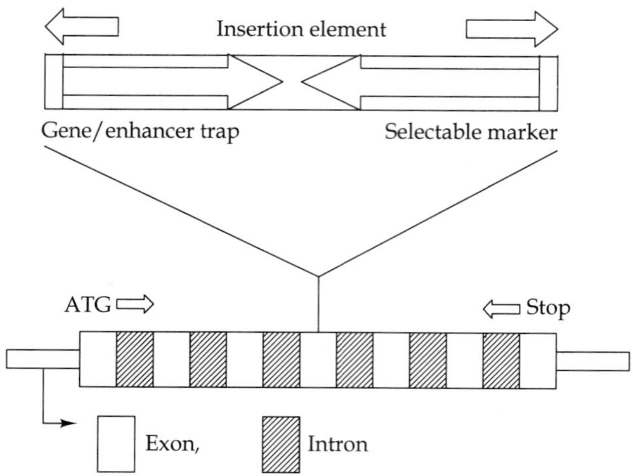

FIGURE 24.6 **Shows insertional (Transposon/T-DNA). Insertion of mobile element within a gene disrupts gene activity at the transcription level, translation level or both. Gene/enhancer trap with a reporter cassett allows monitoring of expression of the gene at the site of insertion. PCR is used the mutagenized population using primers specific for the gene and the insertion element. The insertion site can be recovered by standard PCR and by IPCR, T-PCR (After Bouchez and Hofte, 1998).**

transmitted germinally will also generate PCR products. Second, as each plant carries several hundred insertions and typically a handful of visible mutations, so sorting through mutations to determine which phenotype is caused by which insertion takes several generations and multiple PCR reactions. Finally, the disadvantage with this method is that many or perhaps most insertions fall in introns and other flanking regions (noncoding regions)and so confer weak or undetectable phenotypes and so this method can be used to generate deletion and secondary insertion alleles in order to stabilize and enhance any mutant phenotype. These regions are difficult to recognize in the absence of whole-genome sequencing. Amplification of cDNA by rapid amplification of cDNA end and PCR might be preferable in this case.

Requirement of special environment

To obtain information on the function of a gene requires creation of a loss-of-function mutation and study the phenotype of the resulting mutant. To reveal the phenotypic differences, it may be necessary to identify the correct environmental conditions for the expression of the phenotype. For example, the potassium-channel mutant AKT1 can be identified only at low potassium concentrations (Hirsch et al., 1998). Further, some mutations may have very subtle effects that are not detectable in standard experimental conditions and multigenerational populations studies have to be devised to reveal phenotypic differences. Finally, large-scale genome sequencing has shown that eukaryotic genomes are quite redundant with many genes duplicated in families and the functional analysis of such gene families requires the construction of lines carrying multiple mutations in different family members.

In vitro **transposition**—This technique can be used for large scale sequencing, production of gene knock-outs, gene knock-ins or truncated genes, study of structure-function relationships or generally the transfer of any DNA sequence into any target DNA. In this technique a stabilized transposome (generated *in vitro* in the absence of Mg^{2+}) which is a transient synaptic complex between a transposase (an enzyme which catalizes the excision, transfer and insertion of TE which carries the gene) and a transposon sequence containing a selectable marker gene, stably integrate the transposon into the target DNA. The transposed cells are identified on the basis of their resistance phenotypes. The transfer into target cells is through electroporation and its activation by cellular Mg^{2+}.

24.69.4   Site-directed Gene Targeting

This is a technique of the insertion of a gene into a specific chromosomal site using **homologous recombination.**Here gene disruption is by replacement of a wild type gene by a mutant version by homologous recombination. In this technique a foreign gene (e.g. a reporter gene) can be cloned into another gene (e.g. an alcohol dehydrogenase gene) *in vitro* so that it is flanked by sequences derived from the later. The foreign DNA sequence (the gene) is then transferred into chromosomal region containing the cellular counter part of the targeting gene (alcohol dehydrogenase gene in this case) through homologous recombination. In yeast, targeted gene disruption is the primary tool for this purpose because of high levels of homologous recombination in haploids and to some extent diploid yeast which makes gene disruption an exquisitely precise and efficient process. As embryonic stem (ES) cells can still enter the germ line after genetic manipulation in culture, homologous recombination between a native target chromosomal gene and exogenous DNA can be used in culture to modify specifically the target locus. In this technique of targeted disruption of gene X, a gene X-replacement vector that contaions an insertion of the neo^r gene in an exon of gene X and a linked HSV-tk gene, shows pairing with a chromosomal copy of gene X. Homologous recombination between the targeting vector and genomic X DNA results in the disruption of one copy of gene X and the loss of HSV-tk sequences. Such cells will be X⁻, neo^r and HSV-tk⁻ and will be resistant to both G418 and GANC. But because non-homologous insertion of exogenous DNA into the chromosome occurs through the ends of the linearized DNA, the HSV-tk gene remains linked to the neo^r gene and such cells will be X⁺, neo^r and HSV-tk⁺ and therefore resistant to G418 but sensitive to GANC (Mansour et al., 1988). This technique has been used to correct the mutant HPTRT gene in the ES cell line. For using this technique, a cloned fragment of the gene must be available and the intron-exon boundaries within that fragment must be defined. In case of Arabidopsis there has been some success in gene replacement but this method is still laborious involving the production of hundreds or thousands of transgenic plants for every gene assayed (Kempin et al., 1997) and thus not considered feasible on a large-scale.

24.69.5   Enhancer and Gene Trap Mutagenesis

Insertion mutagens can be engineered with reporter cassettes (enhancer and gene traps) which will report on the expression of the chromosomal gene at the site of insertion (Martienssen, 1998). Gene trap refers to a system which makes it possible to identify genes that give rise to a phenotypic effect when they are switched off. In this system a DNA construct is integrated into the genome of a cell line and integration should be as randomly as possible. The success of gene trap requires that the location where the DNA construct is integrated to be in the intron of a gene and if the intention is to switch off the function of the gene product, then the integration position should ideally be in one of the upstream introns. The integration can then have the effect of preventing the exons of this gene that are located downstream from being translated into protein. Instead a fusion protein results which is made up of N-terminus of the gene concerned and a reporter protein. **Gene traps** and **enhancer traps** are reporter genes that are not normally expressed unless they are integrated near or within a chromosomal gene. Enhancer traps are equipped with a minimal promoter that can respond to nearby enhancer whereas gene traps are equipped with a splice acceptor (see chapter 16 for detail) so that integration within introns leads to read through transcription and splicing. In both cases the expression of the reporter gene closely mimics that of the chromosomal gene. Enhancer trap and gene trap lines, each with a unique reporter gene have been established in Drosophila and in mouse. In case of plants, Arabidopsis, a system for gene trap and enhancer trap transposon mutagenesis has been devised in which Dissociation transposons (Ds) were engineered to carry a uidA[α-glucuronidase (GUS)] reporter gene and an NPT II kanamycin resistance gene. In the DsE (enhancer trap) construct (a laboratory term for a recombinant DNA molecule) the reporter gene is preceded by the -46 region of the CaMV35S promoter which has been shown to have no detectable transcriptional activity unless it is in the vicinity of an enhancer. In case of DsG (gene trap) construct the reporter gene is preceded by a triple splice acceptor and by a short intron so that insertion into chromosomal introns leads to reporter gene expression via alternate splicing in each reading frame. Additional splice donor sites in the end of Ds means that the reporter gene will also be expressed when DsG element is inserted into an exon. The elements are mobilized by crosses to transgenic plants carrying Activator transposase gene (Ac), transposase driven by the 35S promoter from CaMV. This results in high frequency of transposition. Transposable elements are selected by using a positive marker within the Ds element (the NPTII gene) and a negative marker adjacent to it (the iaaH gene). Positive – negative selection on naphthalene acetamide and kanamycin results in seedlings having retained the Ds element but lost the donor site from where it came. As selection depends on recombination between the transposed element and the donor site, the transposed element is always unlinked to the donor locus. 90% of the gene trap lines with carrying a single transposed element at random locations in the genome was found in Arabidopsis whereas 5% lines had multiple elements and the rest 5% had insertions that disrupted the negative marker gene on the T-DNA. Each line is designated as an ET (enhancer trap) or GT (gene trap) depending on the transposon. Mutagenized populations can be screened for lines expressing reporter gene in specific cell types or in specific environmental conditions. During the selection of transpositions DNA is prepared from each individual plants and then subjected to amplification by using TRAIL (thermal asymmetric interlaced)PCR. This procedure uses semi-nested primer from within the transposon and arbitrary degenerate primers to amplify genomic sequences flanking each insertion. By using a combination of different primers it is possible to amplify~ 95% of all the insertion sites by following a hierarchical tiered procedure by using a minimal number of PCR. These products are then sequenced directly without further resolution of gels. Now by comparing the sequence to known genomic and transcribed sequences, insertions into genes and their orientation can be identified.

24.69.6 Activation Tagging

It is based on the use of an insertion element carrying a strong enhancer or promoter, directing transcription into the region flanking the insertion helps in isolation of **gain of function** mutations in which ectopic activation of a flanking gene promotes a mutant phenotype (Wilson et al., 1996). In other words, if an insertion element carrying a strong outward facing promoter integrates adjacent to an endogenous gene, that gene will be activated by the promoter. Such mutations are generally dominant or semi-dominant.Unlike other insertion vectors which cause loss of function by interrupting a gene an activation tag causes gain of function through overexpression or ectopic expression. Ectopic means 'out of place' or ex topos (in latin-greek). **Ectopic expression** refers to the expression of a gene outside its normal location (domain) in a genome. An ectopic gene refers to a DNA sequence, gene or enzyme being present or active at an abnormal location. For example, all transgenes underly an ectopic expression at their insertion sites in the transformant's genome. Activation tagging is a modification of T-DNA tagging. Multiple copies of enhancer elements from the strong constitutive promoter of CaMV 35S are incorporated near the right border of a T-DNA of the T-DNA tagging vector pPCVICE-n 4HPT. This construct, if inserted near a gene, may cause enhanced transcription (overexpression) of that gene resulting in a dominant mutation. The new phenotype is recognized because of dominant mutation and tagging allows cloning of the gene (Hayashi et al., 1992). Kakimoto (1996) described the isolation of cytokinin-independent mutants of Arabidopsis, generated by activation transferred DNA (T-DNA) tagging and use of these mutants to idenfity a gene involved in cytokinin signal transduction.

24.69.7 Genetic Footprinting

A number of techniques are available for determining the biological function of yeast genes. These all involve a gene disruption strategy. In the gene disruption strategy, a mutation is created *in vitro* and introduced into the genome. The mutant strain's fitness is then tested under various physiological conditions. This approach is highly effective for analyzing individual genes but not for genome-wide functional analysis. Genetic footprinting is a genomic strategy for determining a gene's function given its sequence. It has been designed to allow to study the roles of thousands of genes together. In the genetic foot printing approach (Smith et al., 1995) insertional mutagenesis and selection are performed *en masse* in a large population of cell, in a manner that allows the effects of mutations in any DNA sequence under any particular selection, to be determined retrospectively using the PCR. Specifically, transposition of a marked Ty1 transposable element is introduced in a large population of cells, generating Ty1 insertional mutations at diverse sites. The mutagenized population is then divided into representative samples, each of which is subjected to one of a large set of selections or fractionations. DNA is prepared from selected cells. The recovery of cells carrying Ty1 insertions at a particular site following a particular selection can be determined retrospectively by using an aliquot of DNA from the selected cells as the template for PCR amplification.

A primer specific to the sequence under investigation and a second primer specific to the Ty1 element are used such that exponentially amplified products represent cells in which the sequence of interest is disrupted by a Ty1 insertional mutation. A role for a particular sequence under a particular set of selective conditions is inferred from depletion of the corresponding PCR product bands: the genetic '**foot print**'. This depletion reflects the selective depletion of cells bearing Ty1 insertions at that sequence.

Ty1 retrotransposon can insert at diverse sites in the nuclear genome although certain sites are strongly favored. Ty1 transposition into the CAN1, LYS2 and URA3 loci in yeast indicates a strong preference for insertion into noncoding regions, especially 5′ of the gene but Ty1 insertion into these coding regions has also been observed. The endogenous Ty1 elements present in most *S. cerevisiae* isolates, transpose infrequently, this rate can be increased 20-to-100-fold by overexpression of Ty1 to yield five or more insertions per genome.

24.69.8   Genome-wide Approaches to Gene Deletion

A powerful way to determine gene function is the phenotypic analysis of mutant missing the gene. Several genome-wide approaches have been proposed including genetic foot printing and random mutagenesis. While genetic foot printing has the advantage that all genes can be tested for their contribution to fitness under a particular growth condition relatively rapidly, it has the disadvantage that mutant strains can not be recovered. In addition, testing each additional condition is time consuming as the first. Random mutagenesis is relatively quick but the subsequent matching of phenotypes to genes is slower. In addition, with random approaches a certain fraction of genes may be missed even with over sampling. These limitations can not be overcome by deleting each gene in the genome in a directed fashion and by marking each yeast gene with a molecular bar code that allows the phenotypes of the mutant strains to be assayed in parallel.

The precise deletion of yeast gene can be accomplished using a PCR-mediated gene disruption strategy that exploits the high rate of homologous recombination in yeast. For this method, short regions of yeast sequence (~50bps) identical to those found upstream and downstream of a targeted gene are replaced at each end of the selectable marker gene via PCR. The resulting PCR product when introduced into yeast cells, can replace the targeted gene by HR. For most genes, more than 95% of the resulting yeast transformants carried the correct deletion. In addition, this method can be modified so as to introduce two molecular bar codes (UPTAG and DOWNTAG) into the deletion strain. The bar codes or tags are unique 20-bp oligomer (20-mer) sequences that serve as strain identifiers. These bar codes allow large number of deletion strains to be pooled and analyzed in parallel in competitive growth assays. This direct, simultaneous, competitive assay of fitness increases the sensitivity, accuracy and speed with which growth defects can be detected relative to the conventional methods (Winzeler et al., 1999). Using this high throughput strategy they constructed a total of 6925 S. cerevisiae strains, each with a precise deletion in one of 2026 ORFs. The phenotypes of more than 500 deletion strains were assayed in parallel.

Advantages and disadvantages of transposon/T-DNA based approaches—The main advantage of transposon based approaches is their relative ease in generating very large populations of insertions and their ability to use the partiality of many transposable elements to transpose to linked sites which makes it possible to remobilize the elements for insertion in the vicinity of the starting insertion site. But then T-DNA insertion mutagenesis results in fewer insertions (1-2 loci per line) and so difficult to achieve saturation. The insertions are stable, easy to maintain and do not show strong insertional biases.

24.69.9   Positional Cloning

The insertion library constructed using transposon (e.g. Ds) can be used for positional cloning. Once a collection of 1-2,000 insertions (Arabidopsis is thought to have 1-2000 genes) is mapped then it is possible to use them to map any new mutation at a very fine scale. Most new mutations are mapped to a 10-20-cM interval by using molecular markers. But with transposon as marker such a 10-20-cM interval will have approximately 20 transposons mapped within it. If each of the insertions carries a dominant genetic marker, the kanamycin resistance gene then any new mutant line can be crossed to these insertion lines and kanamycin resistant seedlings harbouring the mutation can be selected in F_2. The resistant progeny in the F_2 would have been produced as a result of recombination between the mutation and the Ds element. This technique thus allows the new mutation to be mapped to an interval between two transposons whose physical positions in the genome are known. The recombination breakpoints will allow positional cloning of the new mutation and the transposons themselves can then be used to tag the mutation through a short range transposition.

Remobilization—Transposon mutagen can be remobilized by re-introduction of the relevant transposase unlike T-DNA and other types of insertional mutagens. This will result in reversion of mutation which will confirm that it was caused by the transposon. Further, remobilization of transposons can also be used to generate mosaics. That is homozygous mutants that carry Ac will have somatic sectors that have lost the Ds transposon and so the gene function has been restored. Remobilization technique thus can be used for determining the site of action of a given gene as well as its expression pattern. Further, when a transposon is integrated not within the gene but close to it then can be used as a launching pad for local mutagenesis. The Ac/Ds transposon along with most eukaryotic invert-repeat transposons have a preference for integration within a few centomorgans of their starting location after transposition (Bancroft and Dean, 1993; James et al., 1995).

Lethal insertions—Essential genes are involved in cell division cycle (cell growth and division) control. However, adult function of essential genes can not be assessed if homozygous mutant do not survive to maturity. Gene traps and enhancer traps can give some indication of adult functions as reporter gene is expressed in viable heterozygotes and thus have the ability to identify essential genes. Lethal mutations caused by transposons can be confirmed easily by transposon-induced reversion. Each transposant (kanamycin resistance gene contained within the Ds element) with decreased seed set is testcrossed to plants carrying Ac. The F_1 will show presence of occasional branches with full seed set. This mosaicism indicates somatic excision of the Ds element and it is a very strong indication that the lethal phenotype is caused by insertion of Ds element.

24.69.10   Generation of Mutants in Essential Genes

The function of a gene can be known by studying its mutant alleles. The generation of mutant yeast strains from a cloned, non-essential yeast gene, is straight forward. To remove the wild type gene product, an essential step for the analysis of recessive allele, DNA at the wild type can be deleted entirely from the genome. A collection of mutant alleles can then be made and introduced into host cell using replicating plasmid vectors. The generation of mutants in an essential yeast gene poses a problem in removing the wild type allele since deletion or inactivation of the gene results in an inviable genotype. Two methods have been developed to overcome this problem and thus allow the selective mutagenesis of essential yeast genes:

1. **Integration–disruption method and**
2. **Plasmid shuffling**

The fist method, developed by Shortle et al. (1982;1984) uses homologous recombination to target *in vitro* mutagenized DNA to the chosen wild type gene locus. This procedure precisely replaces the resident chromosomal copy of any cloned gene. Homologous recombination of transforming DNA can also be used to create null alleles by gene disruption of cloned yeast gene. By mutagenizing and targeting truncated form of the gene, one can introduce mutations at the gene locus and destroy the wild-type gene in one step.There are a number of drawbacks with this strategy (Sikorski and Boeke, 1991). There is the requirement for a truncated form of gene. Entire coding sequence can not be mutagenized in one experiment. Problems can arise due to repetitive nature of the mutated locus and the fact that the wild gene is not deleted. There is problem of genetic instability because recombination can revert the locus to wild type. Finally, a large fraction of the conditional mutants thus far generated by integration have been shown to map to loci other than the mutagenized gene.

The second method uses replicating yeast episomes as a means of exchanging the wild-type gene for mutant copies. The basic scheme for the exchange is known as plasmid shuffling. In this strategy one copy of the gene of interest is inactivated in a diploid and a wild-type copy is propagated in the cell on an episome. This permits generation of a haploid strain with a chromosomal null allele. Mutagenized copies of the gene are then introduced into this cell on a second episome and exchanged (or shuffled) with wild-type version. Unlike integrative-disruptive procedure the entire coding sequence can be mutagenized in one experiment. Also, if the chromosomal copy of the gene has been inactivated by deletion which removes all of the sequences of the gene, the plasmid borne mutant alleles can not revert to wild type by homologous recombination.When necessary to examine the phenotype of a large number of *in vitro* generated constructs, plasmid shffle strategy is used.

Study of roles of TFs in plants and identification of TFs—Overexpression and antisense technology can be employed to study the role of TFs in plants. In case of overexpression, a gene is expressed from a high level constitutive or tissue-specific promoter in transgenic plants. Overexpression can either produce plants that accumulate high levels of TFs or knockout plants via inactivation of the transgene and/or the endogenous gene by co-suppression. Antisense technology in which an RNA is expressed that is complementary to a target mRNA, is employed to suppress the expression of endogenous TF gene. There are limitations with these technologies (Schwechheimer et al., 1998) and the alternative technique which allows the identification of mutant in specific genes employs PCR to screen large populations of plants containing T-DNA or transposon insertions.**Inducible gene expression**- This technique can be used to avoid the problem associated with overexpression. It allows temporal, spatial and quantitative control of gene expression in a mutant for the TF in a heterologous host plant or tissue. In order to prevent interference from endogenous plant genes inducible systems are based on non-plant components. They do not exert pleiotropic effects, are stable for a defined period of time and gene activation and expression is quick after induction. For studying plant TFs, post-translational induction makes use of animal steroid-inducible receptors. The steroid-binding domain of the glucocorticoid receptor is fused to a plant TF. This results in accumulation of the TF in the inactive unliganded state. Unliganded hormone-binding domain represses nuclear localization, DNA-binding and perhaps other activities of TF. After induction by a steroid , represson is relieved and active protein can rapidly enter the nucleus and exerts its TF function. Using inducible expression of TFse in combination with differential display should make large contribution to the identification of target genes activated by a TF.

24.70 IDENTIFICATION OF PLOIDY REGULATED GENES

Changes in the number of chromosome sets occur during the sexual cycle, during metazoan development and during tumor progression. Organisms with a sexual cycle double their ploidy upon fertilization and reduce their ploidy by half at meiosis. In the development of almost all plants and animals, specialized polyploidy and polytene cell types arise through endocycles, cell cycles lacking cell division. Cells of different ploidy typically show very different morphological and physiological characteristics. Thus the objective would be to look for genes whose expression relative to the total expression, increase or decrease in proportion to plody using oligonucleotide-probe microarray and the mRNA levels of all genes during exponential growth is monitored. Galitski et al. (19990 showed polidy dependent gene expression in eleven isogenic *S.cerevisiae* strains. They set two criteria to identify polidy regulated genes: 1. There should be significant correlation with an idealized expression pattern either directly or inversely proportional to ploidy 2. Minima are set for relative and absolute changes in expression for each gene across all strains.

24.71 SENSITIVITY OF GENE EXPRESSION TO MUTATIONS

Not all genes are equally sensitive to the effects of random spontaneous mutations. Landry et al. (2007) identified structural properties (presence of a TATA-box and *trans*-mutational target sizes) which greatly influence a gene's potential to undergo regulatory change. These determinants provide a mechanistic basis to serve as a foundation for more-releastic models for gene expression evolution. There are three main factors influencing the probability that a mutation affects the expression level of a genes): (i) The number of other genes that influence the expression of the focal gene (*trans*-mutational target size (ii) the number of regulatory elements controlling the expression of the gene (*cis*-mutational target size) and (iii) the distribution of the effect of mutations on expression.The *trans*-mutational target size of gene is composed of the number of genes in the genome which affect the expression level of the focal gene, weighted by their influence and their own mutational parameter. Larger *trans*-mutational target sizes have been shown to result in higher sensitivities of gene expression to mutation. The *cis*-mutational target size of a gene scales with the number and sizes of TFBSs, either directly through the number of nucleotides in the sites or indirectly through the number and variey of regulatory molecules binding to these sites. Eukaryotic genes differ in the composition of their *cis*-regulatory targets. About $1/5^{th}$ of the yeast genes contain a TATA box which modifies several aspects of their transcriptional regulation. TATA-box genes have large number *cis-* and *trans*-mutational target sizes relative to TATA-less genes.They determined the number of BSs per promoter and found genes with a large number of TFBSs to be more sensitive to spontaneous mutations affecting the level of gene expression. They thus found that sensitivity of gene expression to mutation increases with both increasing *trans*-mutational target size and the presence of a TATA box. Genes with greater sensitivity to mutations are also more sensitive to systematic environmental perturbations and stochastic noise. TATA-box containing genes are more likely to be subtelomeric, highly regulated by nucleosome and chromatin regulators and associated with higher rates of gene expression divergence among species and adaptation during experimental evolution.

24.72 MEASURES OF BIOLOGICAL COMPLEXITY

Biological diversity can be better explained by considering network of TFs (TFs are DNA-binding proteins which switch target gene off and on) and the genes they regulate rather than by simply counting the number of genes they regulate or the number of interactions between them. The number of genes in *C.elegans*, Drosophila, Arabidopsis (plant) and human (see chapter 16) shows that there must be other measures of complexity than the mere number of genes. A global analysis of transcriptional regulation in *E.coli* reveals that on average each TF regulates 3 genes and that each gene is under control of two TFs. With a limited number of genes, vertebrates manage to code for two highly complex subsystems- the immune system and the nervous system. Both can store large amounts of information based on fixed set of rules. These rules reside in residue variation generating mechanisms (such as reshuffling of immunoglobulin genes) and internal selective filters. Reshuffling of immunoglobulin genes generates enormous variety of antibodies. An internal selective filter than recognizes cells producing antibodies against self antigens, selects them out and destroys them. Of the two forms of genome complexity, one measured by gene number and the other by the connectivity of gene-regulation network, the complexity of organisms (in terms of morphology and behavior) correlates better with the second measure (Szathmary et al., 2001). Connectivity of gene regulation networks in eukaryotes is likely to be greater than that in bacteria, the prokaryotes.Clustering coefficient could be used to define relatively autonomous groups of developmental genes. The number of these groups (developmental modules) could then in turn provide a measure of developmental complexity. Table 24.2 shows different indices for measuring biological complexity.

24.73 SYNTHETIC BIOLOGY

Synthetic biology deals with the creation of biological systems from scratch or from components that had other functions (Adami, 2007). Synthetic biology is one approach of showing module function (see chapter 17). Here the objective would be to attempt to build or reconstruct functional modules and test them. This approach has already been used to construct and analyze artificial chromosomes made by assembling defined DNA elements (see chapter 10) and cellular oscillators made from networks of transcriptional regulatory proteins. The ultimate objective of the synthetic biology will be to design an organism to

Table 24.2   Genetic network and biological complexity.

Index	Scale	Relevance
No. of modes, N	Global	No. of relevant genes in a genetic network
No. of links, L	-do-	No. of gene interactions
Connectivity, C= 2L/[N (N-1)]	-do-	Realized fraction of all possible interactions
In -degree, Din	Local	No. of genes affecting a particular gene
Out-degree, Dou	-do-	No. of genes affected by a particular gene
Degree, D	-do-	No. of genes directly interacting with a particular gene
Av. Degree, Dav	Global	Average number of gene interactions per gene
Heterogeneity (the standard deviation of degrees)	-do-	Evenness of link distribution among genes
Clustering coefficient (the av. connectivity of subnetworks containing each node's neighbor, CC)	-do-	Appearance of tightly connected regulatory subnetworks
Average distance, Dav = [Sdij]/[N (N-1)]	-do-	No. of communication steps between two randomly chosen genes
Arc connectivity	-do-	Minimal no. of gene interactions whose deletions results in a disconnected network
Node connectivity	Global	Minimal no. of genes whose deletion results in a disconnected network

perform particular tasks. This will be done by designing an organism (genome) with the help of a computer, pressing the print botton to have the necessary DNA (genome) made and then to put that DNA into a cell to produce a custom-made creature- a whole genome engineering approach for practical ends. The aim here is to produce organism redesigned from scratch. Genome of one bacterium has been successfully replaced with that of another, transforming one species into another. In this genome transplantation, an organism's genome has been replaced with a wholly synthetic one made by DNA-synthesis technology. In the stripped-down genome, minimal bacterial genome of 381 genes identified in *Mycoplasma genitalium* has been used. This stripped genome would provide a chasis on which organisms with new functions can be designed by combining with genes from other organisms. Synthetic biology thus aims at making possible many new functions not by genome transplant but by fusing existing ones (Ball, 2007). Synthetic genome has been produced. Gibson et al. (2008) has built (reconstructed) bacterium genome *Mycoplasma genitalium* from DNA sequence information and raw chemicals. First, information on 582,970-bp DNA sequence of the genome was obtained from the database. This sequence was divided into shorter sections or cassettes of DNA up to ~7000 bp long and then those cassettes were constructed.The construction of cassette started with the synthesis of specific oligonucleotides, short DNA fragments up to hundred bp long. The subsets of oligonucleotides were combined to produce the cassettes. Thus the aim is to construct a genome encoding self-reproducing organism. Ribosome involved in protein synthesis has also been synthesized.

25

Gene Function and Interaction

25.1 GENE FUNCTION

Genes encode proteins but not all genes encode protein. Amino acid sequences of proteins are encoded in the linear structures of genes. In other words, there is exact point-by-point relationship between the order of amino acids along the polypeptide chain and the order of corresponding codons along the polynucleotide chain of the nucleic acid (Crick, 1963), i.e. there is colinearity of gene structure and protein structure. Although one gene-one enzyme hypothesis envisages single gene encoding single enzyme (or single polypeptide) but this single polypeptide can have either single enzymic function or dual or multiple enzymic functions or can synthesize polyproteins.

I. **Single genes–single enzyme function- Allelic complementation**—Allelic complementation occurs between different pairs of mutant alleles of the same gene. Two mutants can be said to be allelic if they are due to mutation in the same gene. Considering one locus with 2 alleles system (A, a) with A dominant over a, an individual with AA genotype produces a protein and the recessive (the mutant) aa also produces a protein. Although the heterozygote Aa may give wild phenotype at phenotypic level, biochemical analysis generally shows relatively low activity and/or reduced stability of the enzyme product. The general explanation for allelic complementation is that it results not in truly wild type product but rather in a hybrid protein formed from different mutant derivatives of the same polypeptide chain.

There are two general mechanisms of complementation between protein products of two mutant alleles.

A. **Conformational correction within a dimeric or oligomeric protein**—When the normal enzyme is a dimmer or oligomer composed of two or several identical polypeptide chains then when two different mutant polypeptides are synthesized a certain proportion of mixed enzyme molecules will be produced which would be active. Suppose there is one monomer being inactive because of loss of an essential amino acid side chain and the other monomer being inactive because of a faulty conformation. When the two monomers are packed together the first monomer will induce the correct conformation of the second monomer thus releasing its potential activity (Fincham, 1966, 1999).

B. **Reconstruction of a functional domain**—In this mechanism there is reconstruction of a functional protein by packing together separated domains of what is normally a single polypeptide chain. It is based on the principle of specific mutual fit between different domains of a globular protein. Complementation of this type has been observed between mutants of *E. coli* lac Z (encoding β-galactosidase) with deletions or chain termination causing non-overlapping deficiencies (Goldberg, 1969). This is possible if the two mutants are defective in different domains an adequately functional though not normal enzyme may be pieced together from their respective good domains.

II. **Single gene-single peptide-dual or multiple enzyme functions**—One of the best examples comes from the yeast mutants, called ura2, single gene encoding a single polypeptide with two independently functional domains, one responsible for *CPSase* (carbamoyl-phosphate synthetase) activity and the other determining ACTase (aspartic acid-carbamoyl transferase) activity. *CPase* catalyses the synthesis of CP and ATCase is involved in the transfer of the carbamoyl group from CP to aspartic acid to produce carbamoyl-aspartate, an intermediate in pyramidine biosynthesis.

III. **Single genes-polyproteins**—The best example is retrovirus which transcribes a long primary transcript which is translated into a long single polypeptide chain which is subsequently cleaved into separate proteins. The long transcript is processed in different ways to give different mRNAs representing different sets of genes.

Different products from the same transcript—A gene subjected to different modes of splicing can generate a family of different but overlapping mRNAs. In other words a single gene can encode a whole family of proteins depending on the mode of intron splicing. This type of genes has large number of exons, some of them extremely short and separated by much longer introns. Such genes have been identified in mouse, rat and human genomes. Differential splicing of mRNAs encoded by the single neural-cell adhesion molecule (N-CAM) gene lead to N-CAM molecules either with different modes of attachment to the cell membrane or different cytoplasmic domains (Cunnigham et al., 1987). The three N-CAM polypeptides include one large cytoplasmic domain polypeptide, one small cytoplasmic domain polypeptide and one small surface domain polypeptide and thus three types of mRNA are produced. The N-CAM gene contains at least 19 exons spanning more than 50kbs. 14 exons are common to the three major polypeptides and the individual chains thus arise by alternative splicing of the remaining five. Exon 1 includes the amino terminus of all three polypeptides. Exons 1 to 14 are shared by all three polypeptides. Exons 16, 17 and 19 appear in the mRNAs for the first two types of polypeptide but not for the third type. Exon 15 occurs only in the third type of mRNAs and exon 18 only in the first type of mRNA. Different mRNAs from the same transcript can also be generated as in case of RNA splicing in sexual development in Drosophila. If the ribonucleoprotein spliceosome complex fails to recognize one of these rather minimal landmarks (discussed above) of the introns and fastens instead on the acceptor sequence of the next intron downstream and thus splicing out two introns along with the intervening exon. Errors of this kind could result in two or more alternative RNAs encoding different polypeptide chains with perhaps different functions.

Genes nested within other gene's introns — The example of a gene nested within the intron of another gene is the *Gart* gene of Drosophila. As shown in Figure 25.1 it is transcribed and translated from left to right and encodes a trifunctional enzyme determining three enzyme activities (GARS, AIRS and GART) in the pathway of purine biosynthesis and thus more than one processed transcript is generated from the same gene. Gart gene has an intron-exon structure and in the Figure 25.1 Gart introns and exons are shown as stippled and empty bars, respectively. It has two polyadenylation/termination sites, T1 and T2. Because of the partially effective transcription termination sequence (polyadenylation site) present in the fourth intron a proportion of the transcripts ends in a truncated mRNA encoding only the GARS function but allows some transcripts to proceed into the AIRS and GART. Finally, the very long first intron of this Gart gene contains a 'nested' gene which encodes cuticle protein which is oriented in a direction opposite to that of Gart. This 'nested' gene itself contains a small intron and is transcribed and translated from right to left. The gart gene uses one DNA strand as template and the cuticle protein gene uses the other DNA strand (Henikoff et al., 1986).

Single transcripts-multiple proteins-polar effects on translation — The best example comes from the *E. coli* lac operon. In bacteria, *E. coli* the genes of related functions are often clustered into operons, that is in single units of transcription containing 2 or more ORFs. Within an operon the translation of one gene is not necessarily independent of another. Chain termination mutations in one gene often have marked depressing effect on the levels of translation of genes further downstream and the earlier the chain termination in the translation of the upstream gene the stronger the effect on downstream gene expression. The termination codon of one gene is followed closely by the initiation codon of the next. It is an

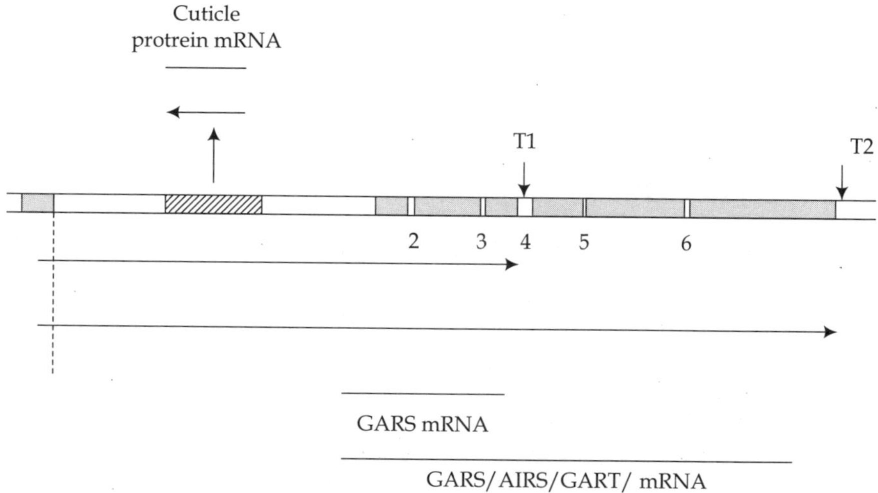

FIGURE 25.1   Showing an example of a gene 'nested' within the intron of another gene.

example of **polar mutation** in which the production of the particular enzyme for which the non-mutated gene encodes is prevented and the functioning of the operator-distal genes of the operon is relatively decreased (mRNA distal to the mutation remains free of ribosomes and is highly susceptible to endonucleolytic (ribonuclease) attack). The more upstream the mutation the longer the tract of exposed mRNA and the lower the level of surviving downstream messenger.

Study of gene function

Hybrid arrested release translation (HART) — It is a technique used for the identification of a protein encoded by a cloned DNA. This technique is based on the fact that an mRNA will not direct synthesis of a protein in a cell –free translation system when it is hybridized to its corresponding DNA complement. In this technique a crude preparation of mRNA encoding hundreds of distinct polypeptides is hybridized with a single stranded cloned cDNA. Those mRNAs which are homologous to the cloned cDNA will anneal it and in a subsequent *in vitro* translation system will not support the polypeptide synthesis. As 35S-methionine is added to the *in vitro* translation reaction mixture the newly synthesized radioactive polypeptides can be visualized after SDS PAGE by autoradiography. A comparison between the proteins generated from the unhybridized mRNA with those from the hybridized mRNA will allow the identification of protein encoded by the cloned cDNA. Full transcriptional activity of the unhybridized mRNA is restored when the mRNA is dissociated from the cloned cDNA by brief heating before its translation in a cell-free translation system.

Hybrid release translation (HRT) — It is a technique used for the isolation of specific mRNA from highly complex RNA mixtures. In this technique a specific DNA or cDNA from which the corresponding mRNA has to be isolated is covalently linked to a nitrocellulose filter. After that the RNA mixture is loaded onto the filter where only RNAs complementary to the cDNA or DNA will be bound. All non-bound RNA are washed through extensive washing and specific RNA is eluted from DNA-RNA hybrid by hot low salt elution buffer or by buffers containing formamide and used for characterization as described in case of HART. This technique is especially useful for isolation of rare mRNAs.

25.2 GENE INTERACTION

Protein rarely acts alone or very few proteins function in isolation and rather they interact with other proteins to perform particular function. Initially it was thought that the function of a protein, say A is its action on substrate (S) to form product (P) (see figure 25.2) but now the function of A is in the context of its interactions with other proteins. Most proteins function by interacting with other proteins in pairs or as components of larger complexes. It is important to know how genetic information results in concerted action of gene products in time and space to generate function. Genes never acts in isolation. Every aspect of the phenotype depends on a multitude of genes interacting in complex ways. Interactions of proteins with DNA and RNA are the heart of gene expression regulation. There are some proteins for which binding nucleic acids seem to be the primary function *in vivo*, while other proteins have dual functions of which one is the capacity to bind nucleic acid. Some of these proteins are involved in gene regulation while function of nucleic acid binding in others remains unknown. Several proteins that bind two or more different nucleic acids are

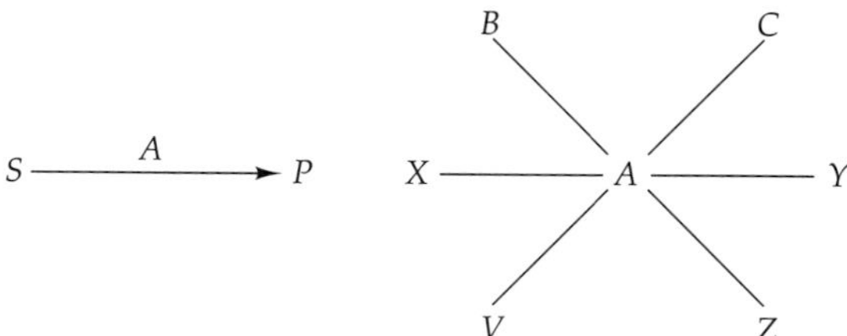

FIGURE 25.2 Showing traditional view of protein function (one protein – one function) and post-genomic view (one protein-many functions).

involved in gene regulation. There are other protein-nucleic acid interactions that are yet to be identified. For most proteins RNA ligands can be selected that bind with nanomolar affinities. Among these proteins are many not thought of as an RNA or DNA binders.

How a genome specifies the properties of an organism, can be known by elucidating interactions among its genes. The various interactions include protein-protein physical, gene-gene and protein gene interactions. Network constructions have focuses on either protein-protein interactions or a specific biological processes (Boulton et al., 2002) and thus they represent a subset of all genetic interactions.

25.3   MOLECULAR MECHANISM OF GENE INTERACTION

Interactions between genes or gene products may be between
1. DNA and protein
2. RNA and protein and
3. Protein and protein.

25.3.1   DNA-Protein Interaction

Genetic information on DNA is deciphered through protein-DNA interactions. So it is essential to know how proteins interact with DNA in order to understand living organism on a molecular basis. DNA-binding protein is a protein which recognizes specific DNA sequences (either single stranded or double stranded DNA) which are referred to as address sites or recognition sequences and bind there via electrostatic forces (binding site), mostly to bases of the major groove. In *E.coli* a defined sequence of at least 12 base pairs is required to specify functional binding sites. Many proteins bind to specific sites on the genome to regulate genome expression and maintenance. Transcriptional activators bind to specific promoter sequences. Further, distinct DNA binding proteins are also associated with origins of DNA replication, centromeres, telomeres, and other sites where they regulate chromosome replication, condensation, cohesion and other aspects of genome maintenance. These proteins can be classified according to their function into structural (e.g. histones), enzymatic (e.g. DNA polymerase) and regulatory proteins (activator, repressor, transcription factors). In other words, DNA-protein interactions may serve structural functions or regulatory functions (e.g. in complexes with DNA-modifying enzymes, DNA-dependent DNA and RNA polymerases and transcription factors). The binding of proteins (regulatory proteins) to the DNA of promoters, operators and enhancers is one of the main keys to the understanding of gene regulation. There are examples of several different proteins binding to the same regulatory element and interacting with each other as well as with the DNA. Transcriptional activators recruit chromatin modifying complexes and the transcription apparatus to initiate RNA synthesis. Reprogramming of gene expression that occurs as the cell progresses through cell cycle or when cells sense changes in their environment is effected in part by changes in the DNA binding status of transcriptional activators. There are genes encoding the RNA binding proteins of the spliceosomes which have universal function in the splicing-out of introns. The DNA binding sites for regulatory proteins are often inverted repeats of a short DNA sequence/a palindrome at which multiple (usually 2) subunits of a regulatory protein bind. For interaction with bases in the major groove of DNA a protein requires a relatively small structure that can stably protrude from protein surface. Binding domains of regulatory proteins tend to be small (60 to 90AA residues) and the structural motifs within these domain that are actually in contact with the DNA are small still. Several families of TFs have been identified in prokaryotes and eukaryotes based on their DNA-binding motifs and are as follows. These protein recognition domains form the actual molecular platforms on which the protein components of the complementary recognition surfaces are positioned in space. Readout of these hydrogen bond-based recognition interactions could be indirect (involving specific water molecules as intermediates) as well as direct and could be facilitated by sequence-specific distortion of the DNA, protein or both to bring appropriate charges into register and generally to improve the physical (and thermodynamic) complementarity of the interacting protein and nucleic acid surfaces.

1. **Helix-turn-helix**
2. **Leucine zipper**
3. **Basic helix-loop-helix**
4. **Zinc finger**
5. **Homeodomain**
6. **MYB proteins**
7. **MADS box- containing proteins**
8. **Other TFs** include APETALA2 (AP2) family, GRAS family and those with a $\beta\alpha\alpha$ motif in the DNA binding domain.

Figure 25.3 (A-F) show DNA binding motif of helix-turn-helix, lencine-zipper, helix-loop-helix, MYB and Zinc finger containing protein, respectively.

Besides these domains, there are specific protein-protein interaction domains that can form homo- and hetero dimmers of protein subunits that are both varied enough and extensive enough to recognize a spectrum of specific DNA target sites.

Helix-turn-helix — The lac repressor has this kind of DNA binding motif. It occurs largely in prokaryotes and some eukaryotes. It comprises about 20 AAs in two short α-helical segment, each 7 to 9 AA residues long and separated by a β turn. One of the two α helix is called a recognition helix that interacts with DNA in a sequence specific way. This α – helix is stacked on the other segments of protein structure so that it protrudes from the protein surface. The recognition helix is positioned in or nearly in the major groove of DNA.

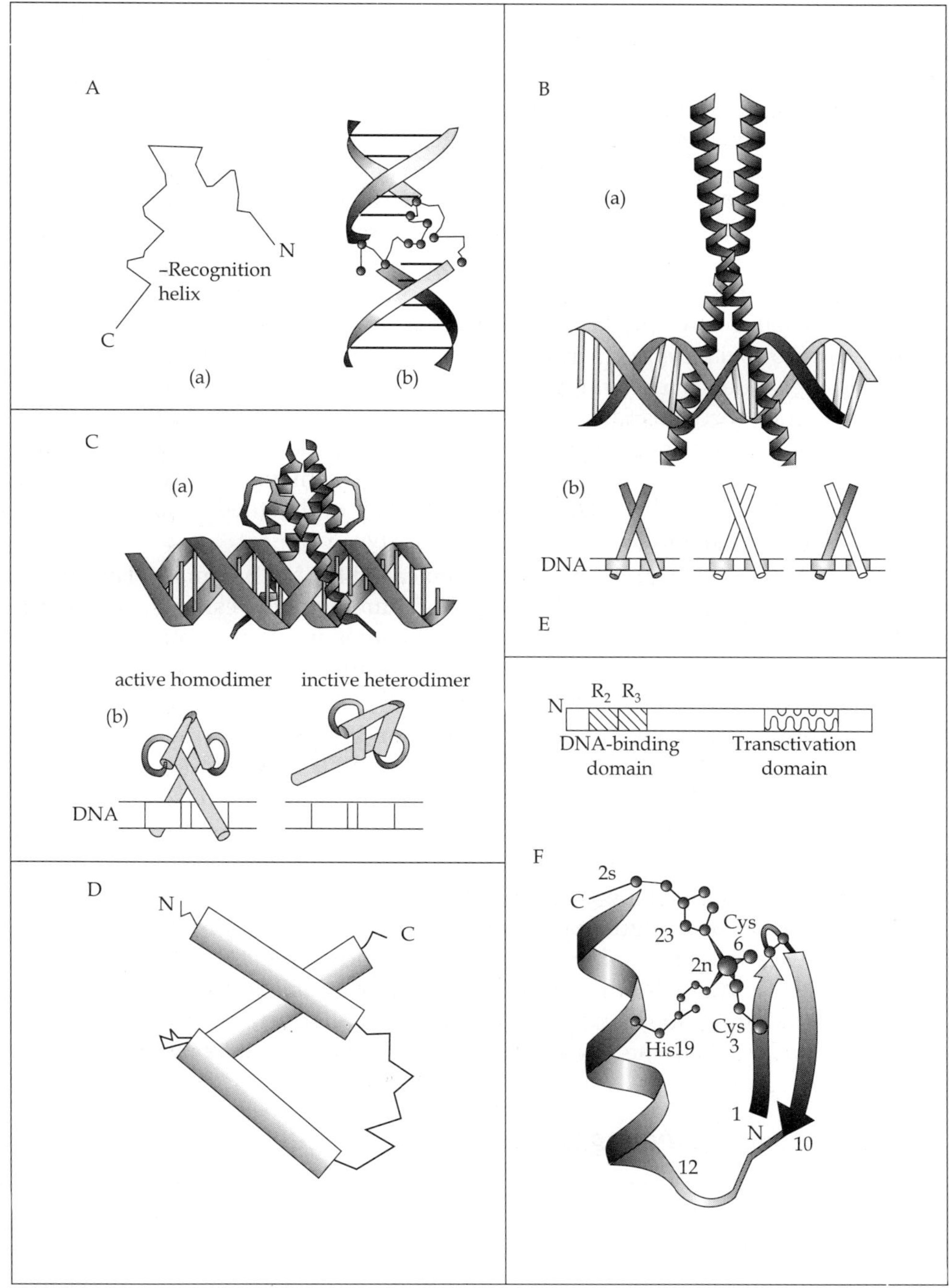

FIGURE 25.3(A-F) Showing DNA binding motif of A. helix-turn-helix containing protein. B. leucine-zipper protein (a. homodimer and b. heterodimer). C. helix-turn-helix D. homeodomain protein E. MYB protein and F. zinc finger containing protein (adapted from Alberts et al., 1994).

Zinc finger—Most eukaryotic DNA binding proteins contain Zinc fingers. Also, there are few examples of Zinc finger motif in prokaryotes. Zinc-finger was discovered during biochemical analysis of transcription factor, TF111A which regulates the 5S ribosomal RNA genes in *Xenopus laevis*. Zinc-fingers are more widely used to recognize RNA. As the interaction of single zinc finger with DNA is weak so many DNA binding proteins have multiple zinc fingers interacting simultaneously with DNA. The Zinc finger motif consists of a sequence of about 30 AA residues containing 2 Cysteins, 2 Histidines and 3 hydrophobic residues which are all at conserved positions. It forms an independent folded domain, stabilized by Zn^{2+} which can be used repeatedly in a modular fashion to achieve sequence – specific recognition of DNA. Thus half of a stretch of about 25 amino acids form an α helix whereas the other half forms two antiparallel β sheets, one of the β sheets together with α helix holds a zinc atom via Cys and His residues. Some Zinc fingers contain AA residues that are important in sequence discrimination whereas others appear to bind DNA non-specifically. Zinc-finger proteins of the classical Cys2 His2 type are the most frequently used class of transcription factor and account for about 3% of genes in the human genome. The domains all have the same structural framework but achieve chemical distinctiveness through variations in key residues. DNA binding is well understood but molecular basis of recognition of RNA by zinc-fingers has remained elusive.

MYB proteins—They have a DNA-binding domain called the MYB domain. This domain has two or three imperfect repeats (R1, R2, R3) of 51-53 amino acid residues, each containing tryptophan residue at conserved positions. MYB domains in plants generally contain only two repeats R2 and R3 whereas there are three repeats with 3 α helices in each in chicken and human. These proteins are well represented in Arabidopsis, petunia and maize.

MADS box protein—It has been derived from four initially identified members, yeast MCM1, Arabidopsis AGAMOUS and Antirrhinum DEFICIENS and the human serum response factor (SRF). The motif consists of 56 amino acid and is divided into basic hydrophilic region involved in DNA recognition and an acidic hydrophobic region which is important in flower development in plants.

Homeodomain—It is another type of DNA binding domain of a class of protein involved connected with translational regulators in eukaryotes. Its protein product includes a 60 – residues sequence called homeodomain (conserved domain). The DNA sequence that encodes this domain is called **homeobox**. Homeodomain contains three α helices. Helices 1 and 2 are separated by a loop and pack into antiparallel arrangement. Helix 3 is the recognition helix and binds to the DNA sequence. The helix-turn-helix motif is embedded in the homeodomain.

25.3.2 Methods for detecting protein-DNA interaction

Protein-DNA interactio+n plays an important role in transcriptional regulation. Transcription factors are isolated using DNA-protein interaction. There are four core methods for detecting DNA-protein interaction.

1. Gel shift
2. DNA-foot printing
3. *In vivo* cross linking and immunoprecipitation
4. Methylation inference assay

For individual sequences, these interactions have been studied using DNase foot printing and gel-shift assays whereas for unraveling protein-DNA interactions on a global scale functional genomic technology- the genome-wide application of chromatin-immunoprecipitation developed by Rick Young is employed.

1. **Gel shift**—This technique is based on the principle that in comparison with protein molecules DNA molecules are much smaller and therefore have much higher mobility in polyacrylamide gel (Fried and Crothers, 1981). Under favorable conditions unbound DNA can be distinguished from DNA associated with protein molecules because of relative mobility. Recently, several enhanced methods such as capillary electrophoretic mobility shift assay (Fraga et al., 2002) has been proposed to improve the performance of this approach.

2. **DNA-foot printing**—It is a method of identifying specific DNA sequences in DNA duplex where DNA-affine proteins are bound. This technique is based on the fact that such sequences are not accessible to endonucleases if proteins are bound to them. In this technique a 5′ end labeled double stranded target DNA segment is partially degraded by DNase in both the presence and absence of the putative binding protein. Degraded fragments are visualized by electrophoresis and autoradiography. The binding site on the DNA will be protected by the binding protein from the DNase degradation. Compared with gel shift methods, DNA foot printing not only confirms the interaction between the DNA and the binding protein but it can also elucidate the specific binding site of the protein (Seguin and Hammer, 1987; Galas and Schmitz, 1978).

3. *In vivo* **cross-linking and immunoprecipitation**—In this technique the binding protein is first covalently linked to DNA *in situ* using any of a variety of common cross-linking agents. Among these, UV light and formaldehyde have been widely used. After cross linking, chromosomal DNA is sheared, the protein precipitated using a specific antibody and bound DNA fragments co-precipitated. Reversal of cross-links releases bound DNA and so fragments can be identified by PCR and electrophoresis. This method is also called **Chromatin immunoprecipitation (ChIP)** (Kuo and Allis, 1999; Simpson, 1999).

4. **Methylation interference assay**—All three(1,2 and 4) methods suffer from limitations in that the DNA-protein interaction occurs *in vitro* and thus they show what factors can bind to the DNA rather than whether such factors actually bind to the DNA *in vivo* where a particular factor may be sequestrated in the cytoplasm or where its′ binding may be affected by the association of DNA with other proteins.

It is a technique used for localization of specific proteins or their modified forms in chromatin. In this technique chromatin is isolated, fragmented by micrococcal nuclease and the resulting nucleosomes mixed with an antibody raised against the protein in question (e.g. an acetylated histone). Immuno-conjugates are then immobilized on protein Agarose beads. The nonbound (or nonacetylated) nucleosomes are washed off and the proteins and the DNA from the antibody-bound, unbound and input chromatins are comparatively analyzed. Alternatively, formaldehyde can be used to covalently cross-link proteins to DNA. Formaldehyde reacts with lysine and arginine side chains of proteins and the purines and pyrimidine moieties of DNA. The DNA is then sheared into small fragments and antibodies against target proteins are used to purify 'cross-linked' DNA. The target sequence is then amplified by PCR and sequenced.

To investigate the molecular events that might accompany an increased accessibility of chromatin, the level of histone acetylation at the S and Ip region is measured by ChIP assay with antiacetylated histone H3 (AcH30 antiserum). The regulation of locus accessibility during recombination events at the immunoglobulin locus is thought to be mediated in part by specific promoters and enhancers which modulate the transcription and histone acetylation of recombination target sites (Nambu et al., 2003). **Acetylation mapping** refers to the determination of the number and precise location of acetyl groups in various histones of chromatin at a given time using chromatin immunoprecipitation. Acetylation refers to a post-translational modification of proteins, i.e. introduction of an acetyl residue. For example, histones are acetylated and consequently bind less strongly to DNA in nucleosome.

Genomic tiling path microarray / Genomic tiling array — In this technique large fragments of ~450-950bp covering $2\text{-}4 \times 10^6$ bp of a specific genomic region, are immobilized on a polylysine-coated glass slide microarray. These genomic fragments are generated by PCR amplification of the genomic DNA, producing overlapping fragments with 21-33 nucleotide long primers and span the region of interest. The selected protein (or proteins) are then applied to the tiling microarray and the DNA-protein interaction (s) are detected by appropriate techniques (eg. by fluorochromes or DamID chromatin profiling techniques) and plotted along the DNA. The resulting high resolution DNA-protein interaction map allows to identify DNA regions with low, medium or high binding capacity.

Yeast one hybrid system — It is a technique used for the *in vitro* isolation of genes encoding proteins that bind to a target DNA (**bait**) sequence (DNA-binding protein) (Li et al, 1993; Inouye et al., 1994). This technique is based on the fact that many eukaryotic transcription factors consist of a target specific sequence called **DNA-binding domain** (DBD) and a target independent **activation domain** (AD). The DBD consists of 1-174 amino acid residues, located at the N-terminus of TFs and that binds to specific sequence of DNA, the UAS. AD is a specific 30-100 amino acid domain of TFs located at the C-terminus, rich is acidic acids and is necessary for the transcriptional activity of the target gene. A cDNA encoding a potential DNA binding protein is fused to a sequence encoding an AD. This complex of DBD and AD drives the expression of a reporter gene (His3 or LacZ). In this technique a cassette containing tandem copies of a DNA target sequence for DNA binding protein is first inserted into a multiple cloning sites immediately upstream of the reporter-promoter in a yeast integration vector. The linearized vector is then transformed into competent yeast cells. The vector integrates at specific sites of the genome. The transformants are then selected by their URA phenotype and the transformants are called **reporter strain**. Then an activator domain fusion library containing candidate cDNA clones is transformed into the yeast reporter strain. In case of reporter gene, His3, the expression of His3 is enhanced after an AD-DNA binding protein hybrid interacts with the target DNA sequence. Thus the His^+ clones contains cDNA (gene) for putative DNA binding protein.

25.4 METHODS FOR DETERMINING THE BINDING SITES OF TFs ON A GENOME-WIDE SCALE

Differences in related individuals are generally attributed to changes in gene composition and/or changes in their regulation. Divergence of regulatory information had relied on the analysis of conserved sequences (computational approach) in putative promoter regions. However, the limitation with these approaches is that as TFBSs are often short and degenerate thereby making computational detection difficult. Further, requiring the conservation of motifs across species precludes the detection of sequences that are evolutionary divergent. The detection of binding sites with CHIP-chip analysis offers the ability to globally map transcription factor binding locations experimentally rather than computationally.

CHIP-chip (Chromatin immunoprecipitation and microarray chip technique) — This technique is used for the determination of the binding sites of TFs on a genome wide scale (Lee et al., 2002; Iyer et al., 2001). This method combines the CHIP technique with DNA microarray technology. Thousands of DNA fragments purified by the CHIP method are identified simultaneously by microarray experiments. Using CHIP-chip Lee et al. (2002) created a yeast regulatory network consisting of 106 TFs and 2363 target genes.

In this technique the bound and unbound sequence fragments are labeled with red and green dye, respectively and are then simultaneously hybridized onto an array (CHIP-chip). The relative intensities from the two channels (R/G ratios) provide a quantitative estimate for the binding affinities of a TF to all sequence regions of interest *in vivo*. Generally the measured affinities depend on the cellular conditions in which the binding of the TF is tested. Such changes can be because of differences in protein concentrations and DNA accessibility. So, a complementary approach of protein binding microarrays has been developed by Martha Bulykad and her collaborators (Mukherjee et al., 2004). This technique allows to quantify the relative affinities of TF to accessible double stranded DNA *in vitro* in terms of R/G ratio.

DNA adenine methyltransferase identification (DamID) — It is another technique to map DNA-protein interaction on genome-wide scale. The use of cross linking reagents can produce artifacts in CHIP-chip experiments. To overcome this problem, van Steensel and Hanikoff (2000, 2001) introduced a new technique called Dam ID. In this technique the DNA binding protein of interest is generally fused with *E.coli* DNA adenine methyltransferase (Dam). Dam methylates the N^6 position of adenine in the sequence 5′-GATC-3′ which occurs on average every 200-300bps in the fly genome. Upon *in vivo* binding of the protein to its target DNA sites, DNA around the target sites is preferentially methylated by the tethered Dam. Subsequently, genomic DNA is digested into small fragments by DpnI. DNA fragments without methylated GATCs are removed by DpnII digestion. The remaining methylated fragments are amplified by selective PCR and quantified by microarray analysis. Using Dam ID technique and genomic tiling path microarray, Sun et al. (2003) mapped protein-DNA interactions at high resolution along large genomic DNA segments of Drosophila.

25.5 GENOME WIDE LOCATION ANALYSIS FOR YEAST TRANSCRIPTION REGULATORS

In this procedure, transcriptional regulators were tagged by introducing the coding sequence for a c-myc epitope tag into the normal genomic locus for each regulator. Yeast strains were constructed so that each of the transcription factors contained a myc epitope tag. To increase the likelihood that tagged factors are expressed at all physiological levels they introduced epitope tag coding sequence into the genomic sequences encoding the COOH terminus of each regulator. Confirmation of the appropriate insertion of the tag and expression of the tagged protein was made by PCR and immunoblot analyses. In this way they constructed 106 strains of yeast. All 106 strains contained a single epitope tagged regulator whose expression could be detected in rich growth conditions. Chromatin immunoprecipitation was performed on each of these 106 strains. Promoter regions enriched through the chip procedure were identified by hybridization to microarrays containing a genome-wide set of yeast promoter regions. It allows to monitor the DNA x protein interaction across the entire yeast genome.

Ren et al. (2000) used a modified CHIP procedure to study DNA x protein interactions at a smaller number of specific DNA sites with DNA microarrays, i.e they used this to monitor binding of gene-specific transcription activators in yeast. Cells are fixed with formaldehyde, harvested and disrupted by sonification. The DNA extracts cross-linked to a protein of interest were enriched by immunoprecipitation with a specific antibody. After reversal of the cross-link, the enriched DNA was amplified and labeled with a fluorescent dye (cy5) with the use of ligation-mediated PCR. A sample of DNA that was not enriched by immunoprecipitation was subjected to LM-PCR in the presence of a different fluorophore (cy3) and both immunoprecipitated-enriched and unenriched pools of DNA were hybridized to a single DNA microarray containing yeast intergenic sequences. Relative intensities of spots (on microarray) were used as the basis for an error model that assigns a probe score to binding interactions. They identified sites bound by the transcription activator Gal4 in the yeast genome. Gal4 activates genes necessary for galactose metabolism. They identified genes whose promoter regions were bound by myc-tagged Gal4 and whose expressions were induced at least two-fold by galactose. They also used microarray and conventional CHIP techniques to show the genome-wide location of Gal4 protein. In the conventional CHIP procedure strains with (+) or without (–) a myc-tagged Gal4 protein were grown in galactose and amplification of the unenriched DNA and immunoprecipitated-enriched DNA was carried out.

To identify the *cis*-regulatory sequences that are likely to serve as recognition sites for transcriptional regulators, merge the data from genome-wide location analysis, phylogenetically conserved sequences and prior knowledge. Comparative genomics has been used to identify potential *cis*-regulatory sequences within yeast genome on the basis of phylogenetic conservation but this information alone does not reveal if or when transcriptional factors occupy these binding sites. There is a need for construction of yeast's transcriptional factors regulatory code by identifying the sequence elements that are bound by regulators under various conditions and that are conserved among Saccharomyces species. Environment-specific use of regulatory elements predicts mechanistic models for the function of a large population of yeast's transcriptional regulators (Harbinson et al., 2004). In yeast all 141 transcriptional factors are reported to have DNA binding and transcriptional activity.

25.6   IDENTIFICATION OF GENES AND REGULATORY ELEMENTS

Assembly of motifs into network structures- Identify group of genes that are both co-ordinately bound and co-ordinately expressed. Suppose there is found a set of genes G with set of regulators, S with a P value threshold of 0.001. Now find a large subset of genes in G that are similarly expressed over the entire set of expression data. These genes establish a **core expression profile**. If the expression profile of a gene deviates significantly from this core profile, then genes are dropped from the G. The remainder of the genome is scanned for genes with expression profiles that are similar to the core profile. Genes with a significant match in expression profile are then examined to see if the set of regulators S are bound. At this step the probability of a gene being bound by a set of regulators is used instead of the individual probabilities of that gene being bound by each of the individual regulators. As one is assaying the combined probability of the set of regulators being bound and is relying on similarity of expression patterns, one can relax the P value for individual binding events and thus recapture information which is lost as a result of the use of an arbitrary P value threshold. This process is repeated until all combinations of genes bound by regulators have been considered. The resulting sets of regulators and genes are essentially multi-input motif refined for common expression.

Motifs can be used as building blocks to construct large network structures. The network of transcriptional regulators that control other transcriptional regulators is highly connected suggesting that the network substructures for cellular functions such as cell cycle and development are themselves co-ordinated at the transcription level. Lee et al. identified network motifs which provide specific regulatory capacities.

Lerman et al., (2007) described a quick and general multiscale method for normalization of microarray data and identification of enriched targets without assuming any prior distribution. Its utility is greatest when the data is difficult to model statistically, for instance, when they contain unavoidable distortion and asymmetry. ChIP-on-chip data suffer from these problems. In ChIP-on-chip experiments, the enriched target DNA segments deviate in only one direction. That is, the M values (log ratio of input to IP signals) of enriched sites are mostly negative. Thus the statistical distribution of the corresponding M values is asymmetric, hard to model and thus difficult to estimate. Moreover, in such data the whole M values are frequently skewed, their observed distribution varies locally and the dependence of their local means on the A values is nonlinear. Further, one can encounter cDNA array data with asymmetric distribution of expression values in case of sex-biased genes in Drosophila. ChIP-on-chip experiments combine microarrays ('chips') with Chromatin immunoprecipitation (ChIP) assays for identifying the genomic loci bound by a given TF. The immunoprecipitated sample represents gene fragments bound to the transcription factor and is compared with a sample not subjected to immunoprecipitation and thus representing all genes equally ('input sample). The two samples are labeled with different fluorescent dyes and are co-hybridized onto a DNA microarray representing all gene promoters of the particular species studied. Those spots which show a significant increase in fluorescence in IP sample relative to the input sample are termed 'enriched spots' and are considered to represent the target genes bound by the TF in the cell nucleus.

25.7 Dam ID CHROMATIN PROFILING

It is a variant of Dam ID technique used for the high resolution of *in vivo* binding site mapping of proteins to a defined region of chromatin or genomic DNA. The gene encoding a chromatin protein or more generally a DNA-binding protein, is fused to *E.coli* Dam gene encoding Dam methyltransferase. The construct is then transfected into target cells and expressed *in vivo* as a fusion protein consisting of the full length Dam protein and the DNA-binding domain of a chromosomal protein. The DNA-binding domain of the chromosomal protein binds to its cognate sequence in the genome and guides the tethered Dam methyltransferase with it. Subsequently, the adenine residue is then digested by DpnI, size fractionated, subsequently isolated by sucrose density gradient centrifugation and labeled with a fluorochrome, for example, cyanine 3 using random priming technique. The corresponding control, the DNA-binding protein without Dam methylase, is treated the same way and labeled with cyanine 5. The combined samples are co-hybridized to, for example, a genomic tiling path microarray and binding sequences for the protein in question is identified by their cy3/cy5 fluorescence ratios which are corrected for unspecific binding of Dam and also local differences in chromatin accessibility. The identified sequences can then be mapped on, for example, chromosome map and thus this technique is also called **chromatin profiling**.

Prediction of double stranded binding site—PreDs (a server?) generates the molecular surface of the query protein and calculates the electrostatic potential and the local and global average curvatures on the molecule surface and judge if each vertex on the molecule surface is likely to appear at a DNA binding site or not. The prediction is based on the difference of the observed relative frequencies of the curvatures and electrostatic potentials between double stranded DNA binding sites and non-binding sites.

25.8 PREDICTING PROTEIN-PROTEIN INTERACTIONS FROM CORRELATION BETWEEN GENOMIC FEATURES

Two proteins are more likely to interact if the following genomic features are correlated.

1. **Correlated mRNA expressions**—Interacting proteins tend to have correlated expression profiles. Protein abundance can be directly and roughly measured by the presence or absence of the corresponding mRNA transcripts, though large difference can exist between mRNA level and protein abundance (Jensen et al., 2002). There are reports of a significant correlation of mRNA transcript levels among proteins that interact. This correlation is prominent for proteins in permanent complexes but less observed for those involved in transient complexes.

2. **Correlation between phenotypes of knock out mutants**—Phenotype of the knock out mutants can serve as another potential indicator suggesting whether two proteins are subunits of the same complex. The genetic deletion of different subunits of the same complex may disturb the function of a complex in the same way, thus producing a similar phenotype. Synthetic lethal interactions are generally enriched in genes that encode members of the same complex. More generally, if proteins function in related cellular processes, they have an increased probability of being in the same complex.

3. Co-localization of proteins — For interaction to occur, the proteins must localize to the same subcellular compartment at the same time. Thus co-localization serves as a useful predictor for protein-protein interaction.

Although two proteins are more likely to interact if the above mentioned genomic features are correlated but these indicators are rarely strong enough to directly predict protein-protein interaction.

Sequence alone is not sufficient to account for the specificity of Gal4 binding *in vivo* and additional factors such as chromatin structure contribute to specificity *in vivo*. Genome wide identification of DNA and RNA regulatory elements has provided insights into the combinatorial nature of regulatory networks. However, genome-wide rules relating the position of RE elements to the differential activity have not been defined. Bioinformatics has identified several *cis*-acting RNA motifs that regulate splicing but the *in vivo* relationship between the position of these motifs in pre-mRNA and the activity of RNA-binding proteins that recognize them is limited.

25.9 PROTEIN-PROTEIN INTERACTION

Molecular networks guide the biochemistry of a living cell. Its metabolic and signaling pathways are shaped by the network of interacting proteins whose production, in turn, is controlled by genetic regulatory network. For both interaction and regulatory networks, links between highly connected proteins are systematically suppressed whereas those between a highly connected and low-connected pairs of proteins are favored. This effect decreases the likelihood of cross-talk between different functional modules of the cell and increases the overall robustness of a network by localizing effects of deleterious perturbations. Direct physical interactions between pairs of proteins from one such network serves as a backbone for functional and structural relationships among its node and defines pathways for the propagation of various signals such as phosphorylation and allosteric regulation of proteins. Information about specific binding of proteins to each other has grown as a result of HTY2H experiments described in chapter 26. The production and degradation of proteins participating in the interaction network is controlled by genetic regulatory network of the cell formed by all pairs of proteins in which the first protein directly regulates the abundance of second protein. The majority of known cases of such regulation occurs at the level of transcription in which a TF positively or negatively regulates the RNA transcription of the controlled protein. The large scale structure of both these networks is characterized by high degree of interconnectedness where most pairs of nodes are linked to each other by at least one path (Maslov and Sneppen, 2002).

Interactomics refers to the whole repertoire of techniques employed for the isolation, purification and characterization of proteins and identification and molecular description of the interaction (s) with other proteins of the cells. Protein –protein interaction refers to the covalent and usually transient interaction (s) between two or more proteins necessary for the execution of a function or functions that each protein itself can not perform. Protein-protein interactions are the basis of cellular life and lead to the aggregation of a protein machines which enable the transfer of signals through signal transduction chains and guarantee high efficiency in metabolic pathway. The magnitude of P-P interactions can be represented by a

P-P interaction map which refers construction of the network of interactions set (preferably all) protein of a given cell at a given time point. Protein-protein interactions are commonly modeled by graph where nodes represent protein and edges represent physical interactions. Modeling and understanding the structure of these large networks are an important problem in bioinformatics. These require new mathematical and computational advances. Protein-protein interaction networks have been shown to have scale-free degree distribution.

Protein –protein interactions occur within a great variety of multi-protein complexes (ribosomes, spliceosomes, membrane structures, DNA and RNA polymerases), transcription and translation initiation complexes among many others. Within each complex the function of each component is more or less dependent on others. Furthermore, many proteins are structurally and functionally modified through the enzymic activities of other proteins. Phosphorylation and dephosphorylation, catalysed by protein kinases and phosphatases are the general modes of control of protein function. The importance of acetylation especially of histone proteins is being thought in gene regulation (Fincham, 1994). Clues to the molecular mechanism of interaction between genes have often been obtained from their sequences. DNA sequences may reveal ORFs encoding protein with DNA binding motifs or with affinities to known protein kinases. Regulatory proteins have domains for DNA binding, protein-protein interaction with RNA polymerase, other regulatory proteins or subunits of the same regulatory protein. Transcription factors (TF) function as gene activators often bind as dimmers to the DNA. The structural motifs that mediate protein-protein interaction tend to fall into one of the two categories. 1. Leucine Zipper and 2. basic helix-loop-helix. **Leucine Zipper**- It is an amphipathic α-helix with a series of hydrophobic AA residues concentrated on one side. Hydrophobic surface forms the area of contact between the two perptides of a dimmer. There is occurrence of Leu residue at every 7th position forming a straight line along the hydrophobic surface. They line up side by side as the interacting α-helices coil around each other. Regulatory proteins with LZs have a separate DNA binding domain with a high concentration of basic amino acid (Lys or Arg) residues which interacts with negatively charged phosphate of the DNA backbone. **Basic helix-loop-helix**- This type of motif occurs in regulatory proteins. Proteins share a conserved region of about 50AA residues which is important in both DNA binding and protein dimerization. This region can form two short amphipathic α-leices linked by a loop of variable length and is distinct from helix-turn-helix motif associated with DNA binding. The HLH motifs of two polypeptides interact to form dimmers. In these regulatory proteins DNA binding is mediated by an adjacent short AA sequence rich in basic residues similar to LZs. The additional three domains for protein-protein interaction that have been characterized in eukaryotes are glutamine rich, proline rich and acidic domains. The large-scale approaches to measure, detect and analyze protein-protein interactions include biochemistry such as co-immunoprecipitation or cross linking, molecular biology such as two hybrid system or phage display and genetics such as unlinked non-complementing mutant detection (see chapter 26 for detail).

25.9.1 Methods for Identifying Protein-Protein Interaction

Protein-protein interactions between two proteins, say X and Y, have generally been studied using biochemical techniques such as cross-linking, co-immunoprecipitation and co-fractionation by chromatography. Usually two-hybrid screening procedures (see chapter 26) are employed to systematically screen for such interactions.

Cross-linking—Interaction between X and Y is demonstrated where cells or cell lysates are exposed to a cross-linking agent and immunoprecipitation of protein X results in the co-precipitation of protein Y. Protein Y is released by cleavage of the cross-link.

Co-immunoprecipitation—Interaction between proteins X and Y is demonstrated by the addition of (usually monoclonal) antibodies against X to a cell lysate. Precipitation of the antibody-X complex results in the co-precipitation of protein Y.

Affinity chromatography—Interaction between X and Y is indicated by the capture of X on some kind of affinity matrix, e.g. a Sepharose column when a cell lysate is passed through. Protein Y also remains attached to the column by virtue of its interaction with protein X whereas the non-interacting protein are washed through. Affinity capture may be achieved using antibodies against X. Alternatively protein X may be expressed as a fusion with an epitope tag or a molecule such as GST which binds to glutathione-coated Sepharose beads. In a related technique termed a far-western blot (Blackwood and Eiseman, 1991), protein X can be immobilized on a membrane in a fashion similar to the western blot and used to screen for interacting proteins.

FRET—(Fluorescence resonance energy transfer)/BRET (Bioluminescence resonance energy transfer) - Interaction between X and Y is detected when energy is transferred from an excited donor fluorophore to a nearby acceptor fluorophore- a phenomenon called **FRET**. FRET occurs only when the two fluorophores are up to 10nm apart and can be detected by the charge in the emission wave length of the acceptor fluorophore (for more on FRET see chapter 31). It can be carried out if X and Y are conjugated with fluorophores such as Cy3 and Cy5. Alternatively, they can be expressed as fusions with different fluorescent proteins, e.g. enhance cyan fluorescent protein (donor) and enhanced yellow fluorescent protein (acceptor) in which case the technique is called **BRET**. The advantages with FRET/BRET analysis are that it can be carried *in vivo* and that transient as well as stable interactions can be detected (Day, 1998; Mahajan et al., 1998).

25.10   CONDITIONAL LETHAL MUTATIONS

It is also a means to identify protein-protein interactions. A mutation which renders a protein inactive at high or low temperature, for example, may be suppressed by a compensatory mutation in the gene encoding a protein that interacts with it. To conclude that two proteins interact, the suppression must be allele-specific, i.e. not caused by general suppression mechanism (as in case of tRNA non-sense suppressors). In addition, other events such as gene duplication suppress conditional lethals. Considering two proteins X and Y, interaction between X and Y is detected when a mutation in gene Y compensates for a mutation in gene X. One explanation for this is that a conformational change in X prevents interaction but this compensated by a complementary change in Y that restores interaction (Hartman and Roth, 1973). A variation in this theme is where overproduction of one protein leads to a mutant phenotype which can be compensated by the overproduction of a second (interacting) protein (Rin, 1991).

Dominant negatives—Multimerization of X can be demonstrated where overproduction of a mutant form of protein causes loss of function despite the presence of wild-type protein subunits. The phenomenon has been explained as that the mutant subunits sequester all the wild-type subunits into non-function multimers (Herskowitz, 1987).

25.11 SYNTHETIC LETHAL MUTATIONS

Genetic interactions between proteins are also identified by synthetic lethal mutations which are mutations in each of the two genes, neither of which is lethal alone but which are lethal when brought together in the same cell. On explanation for this is that individual mutations in X or Y cause changes that nevertheless preserve interactions but the combined effect of both mutations prevents the interaction (Koshland et al., 1985; Huffaker et al., 1987). One method to identify genetic interactions is by modifier screens (eg. synthetic lethal screens). **Synthetic lethal screens** are employed to identify genetic interactions between proteins. Small scale synthetic lethal screens have been used to identify genes involved in many cellular processes. However, this process requires easily detectable phenotypes. In case of metazoan, biological processes often involve phenotypes too complex to score in large scale screens and thus candidate genes are often tested. Relative to randomly paired genes, functionally interacting genes are more likely to have similar expression patterns and phenotypes, thus statistically combining those genetic features might lead to predictions of functional interactions.

Synthetic genetic array analysis — Recently Tong et al. (2001) introduced a systematic method to construct large- scale double array termed **synthetic genetic array** (SGA) in which double mutants were created by crossing a query mutation to an array of roughly 4700 deletion mutants and nonviable double mutants meiotic progeny were identified. SGA has generated a genetic network of 291 interactions among 204 genes. In other words, SGA analysis is an approach that automates the isolation of yeast double mutants and enables large-scale mapping of genetic interactions. In a typical SGA screen, a mutation in a query gene of interest is crossed to an array of viable gene deletion mutants to generate an output array of double mutants which can then be scored for specific phenotypes. **Synthetic lethal** or sick interactions in which combination of mutations in two genes causes cell death or reduced fitness, respectively, are of particular interest as they can identify genes whose products buffer one another and impinge on the same essential biological process. Synthetic lethal relationships may occur for genes acting in a single biochemical pathway or for genes within two distinct pathways if one process functionally compensates for or buffer the defects in the other.

Epistatic miniarray profile (EMAP)

Genetic interactions can provide functional information that is largely invisible to protein-protein interaction. Synthetic genetic array and diploid-based synthetic lethality analyses on microarray (dSLAM) approaches have enabled systematic identification of synthetic sickness/lethal relationships in *S. cerevisiae* in which pairs of gene deletions are far more deleterious than either of the individual deletions. Collins et al. (2007) exploited SGA strategy for generating double mutants to develop an approach called E-MAP. An E-MAP comprises quantitative measurements of genetic interactions between pairs of mutations within a defined subset of genes linked to one or more biological processes. E-MAPs are created by systematically generating yeast strains carrying each pair of mutations and measuring their growth rates. Genetic interactions are determined by comparing the observed fitness of the double mutants to an empirically determined typical fitness that would be expected on the basis of growth defects associated with each mutation. The E-MAP reveals that physical interactions fall into two classes distinguished by whether or not the individual protein acts coherently to carry out a common function.

Protein-based signal transduction systems are assembled through P-P interactions. These interactions are often specified by structurally defined 'domains' in one protein that bind to complementary, short, linear amino acid sequence motifs in another.

25.12   MECHANISMS CONTRIBUTING TO BINDING SPECIFICITY

Binding occurs only when a particular tyrosine, serine or threonine amino acid in the partner motif has been enzymatically tagged with a phosphate group. The phosphate contributes a large fraction of the total energy required for motif-domain binding. Additional amino acid flanking the phosphorylated residues fine tune the interaction, discriminating between specific and non-specific partners. But what about other moduler domains that recognize more promiscuous motifs with considerably lower affinity? How specificity is obtained for SH3 domains which recognize the core sequence motif proline-X-X- proline where X is any amino acid. This specificity results from evolutionary negative selection against non-specific interactions (Zarrinpar et al., 2003). A general feature is the presence of structural domains in one proteins and complementary motifs in their binding partners (ligands).

25.13   RNA-BINDING PROTEIN

It refers to any protein with an RNA recognition motif which allows the protein to bind to specific sites of an RNA. RNA recognition motif is a more or less conserved amino acid sequence motif of protein binding to specific address sites in RNAs. Human proteins, AUF1 (RRM2), 2UP1, PABP and related proteins contain phenylalanine at specific positions which in concert with flanking amino acid residues bind specific adenine moieties on the target RNA. Most of the RNA-protein interactions, however, are non-base specific stacking and van der Waals interactions rather than base or sequence-selective contacts. In one example from plants a glycine-rich stretch in a protein with an RRM binds to the 5′ –AAAATATCT-3′ in the promoter of the grp 7 gene specifying circadian rhythmus in *A. thaliana*.

RNA-protein interactions occur during RNA processing, translation and RNA virus assembly. Genes encoding RNA-binding proteins of spliceosomes are involved in intron splicing. Some intron- splicing is selective, thus making it possible to obtain different mature mRNAs from the same primary transcript. RNA recognition differs quite substantially from that of DNA with a greater percentage of protein-nucleic acid interactions made to the RNA base edge and sugar than to the phosphate backbone (Allers and Shano, 2001). The differences in RNA and DNA recognition thus highlight the importance of appropriate bioinformatics analysis.

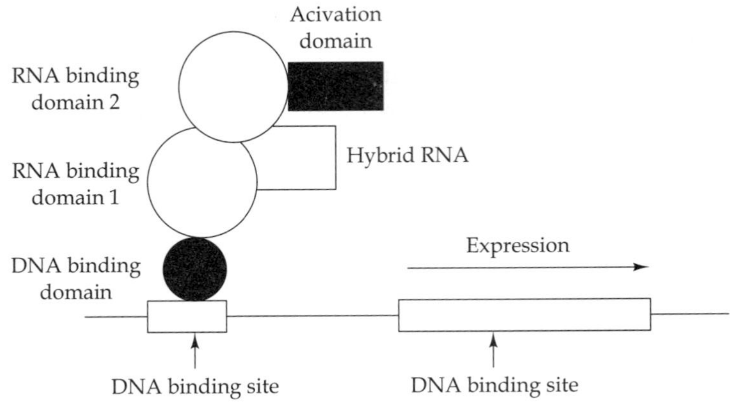

FIGURE 25.4 Showing yeast three-hybrid system for detecting RNA-protein interaction.

25.13.1 Methods for Studying RNA-Protein Interaction

1. **Three hybrid systems**—Three hybrid systems are used for detecting protein-RNA interactions *in vivo* (Sen Gupta et al., 1996). It is a variant of conventional three hybrid system designed for the detection of RNA-protein interaction (s) *in vivo* in which a hybrid RNA molecule (RNA **bait**) links two hybrid proteins and thus there are three components in this technique. The hybrid protein A contains an RNA-binding domain 1 fused to a DNA-binding domain (DB) and the hybrid protein B carries a different RNA-binding domain 2 fused to a transcriptional AD. Since the hybrid RNA contains recognition sequences for both RNA-binding proteins, it combines or bridges both and thereby activates a reporter gene which will be transcribed and whose activity can be easily monitored (Figure 25.4). In a specific case, hybrid protein is composed of a Lex A DNA-binding protein fused to the coat protein of a bacteriophage MS2 and hybrid protein B consists of the GAL4 protein transcription AD fused to the RNA-binding domain of a protein whose RNA-binding characteristics are to be detected (**prey**). The hybrid RNA in turns contains two MSRNA-binding sites, one of which is recognized by the MS2 coat protein of the protein A complex. The interaction between RNA and RNA-binding domain of the prey brings AD into close proximity to the promoter of the reporter gene thereby inducing its transcription in the *S. cerevisiae*.

The DNA sequences of all these components are cloned into separate expression vectors and co-transferred into a suitable yeast strain which carries two integrated reporter genes (HIS3 and LacZ), downstream of LexA binding sites. These binding sites may be present as multimers, for example, 4-mer in H1S3 and 8-mer in LacZ promoter. It also owns the gene encoding a Lex A DNA-binding domain. The MS2 coat protein fusion which recognizes and binds to the Lex A binding site upstream of the promoter drives the reporter genes. If the prey protein and bait RNA interact, they assemble at the Lex A binding site and thus the reporter gene is activated.

This hybrid system permits to detect even transient or weak interaction (s) between RNA and protein (s). It further, allows the identification of new RNA-binding proteins, isolation of different RNAs that bind to the same protein, design of synthetic RNA ligands with selective affinity for specific proteins and synthesis of inhibitors of RNA-protein interactions.

2. **Translational repression assay procedure**—It is a technique for the study of RNA-protein interactions and the effect of pharmacological compounds on the RNA-protein interactions **in vivo**, the cloning of RNA-binding proteins and the characterization of RNA sequence or protein domains essential for binding. This technique is based on the reduced translation of a reporter-mRNA to an indicator protein, if a protein binds to its cognate binding site, artificially introduced into the 5' UTR of the mRNA. The protein-mRNA complex inhibits the stable association of the small ribosomal subunit and thereby represses the translation of the message.

   Yeast cells are transformed with1. plasmid for the expression of an RNA-binding protein or a cDNA expression library, driven by a galactose-inducible promoter and 2. a plasmid for the expression of a reporter gene driven by a constitutively active promoter. Preferred reporter gene is the GFP or its derivatives. The binding site for the RNA-binding protein of interest is cloned into the 5' UTR of the mRNA encoding the reporter protein. Expression of the reporter is turned off in glucose medium (the promoter driving the expression of the RNA-binding protein is silenced), the phenotype is GFP$^+$. After induction of RNA-binding protein expression by replacement of glucose to galactose in the growth medium the translation of reporter mRNA is expressed, the phenotype of the cells is now GFP.

3. **Chromatin cross-linking with immune precipitation (CLIP)**—It is a technique for the detection of interaction (s) between proteins and cognate sequences within promoters *in vivo*. The different components are first crossed linked with formaldehyde *in situ* and the chromatin isolated by standard techniques. Then the cross-linked partners are isolated by immuno-precipitation with specific antibodies and analyzed. CLIP allows to probe promoter occupancy. Ule et al. (2003) identified a specific protein RNA targets or transcripts using this technique. In human RNA-binding proteins are associated with a large number of disorders. In this technique the tissue is directly irradiated with U.V-B light which forms covalent bonds between protein and RNA that are in direct contact, the relatively low efficiency of this reaction is compensated for by PCR amplification. Covalent binding allows rigorous purification schemes for obtaining highly purified protein-RNA complexes including stringent washing of immunoprecipitates, boiling in SDS, separating complexes on SDS-PAGE and transferring them to nitro-cellulose which retains protein-RNA complexes but not free RNA. Also, covalent cross-linking allows to partially digest the RNA while retaining the core element involved in protein-binding such that short (60 to 100 nt) RNA tags can be purified, thus allowing both the identification of the bound RNA species and the location of the binding site. Protein was removed with kinase K and RNA cloned with the use of linker ligation and RT-PCR. After identification of candidate RNAs the target is verified with the use of several tests such as sequence comparison with the known protein binding sites, demonstration of direct RNA-protein interaction and quantification of changes in alternative splicing in novel null mouse brain.

25.14 EPISTASIS

In human despite major progress much of the observed phenotypic variability can not be explained by mutations at a single locus leading to the exploration of oligogenic models of disease transmission in which multiple loci exert a synergistic effect to modify the penetrance and/or expressivity of disease trait. Epistasis can be defined in two ways. In Mendelian genetics, epistasis can be said to be occurring if the effect of a mutation in one gene is modified by the mutation in another gene. In other words, epistasis means that the phenotypic consequence of a mutation depends on the genetic background (genetic sequence) in which it occurs. The classical approach for deducing connection between gene functions is through demonstrating that the effect of a mutation in one gene is modified by mutation in another gene. In other words, epistasis is determined by the analysis of double mutants. It can also be said to occur if the **double- mutant** phenotype is not the sum of two single- mutants effects. If a mutation at one locus (say A) completely suppress the effect of the other locus (say B), that mutation (A gene) can be said to be **epistatic** and the double mutant would resemble the epistatic single mutant. The mutation or gene whose expression is suppressed or influenced (B gene) is called **hypostatic**. The various types of epistasis is shown in the figure 25.5. Epistasis can be defined as the dependence of the expression of an allele of one gene upon a second gene. In quantitative/population genetics, epistasis (or inter allelic interaction) occurs when the genotype at one locus influences the magnitude of allelic

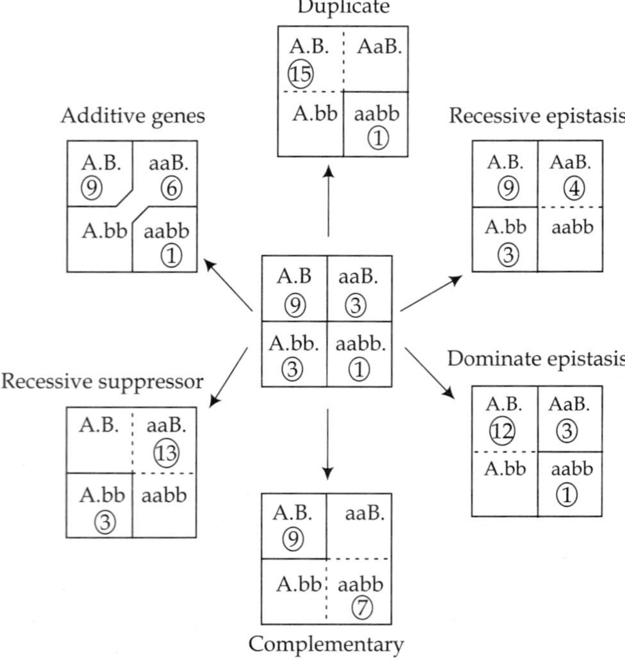

FIGURE 25.5 Showing different types of epistasis and phenotypic ratios in F_2 which deviates from 9:3:3:1.

effects at other locus. Here epistasis refers to the genotypic value contributed by gene interaction between loci. The aggregate genotypic value of an individual (G) may deviate from the sum of the genotypic values attributable to the first locus (G_A and the second locus (G_B). The deviation or non-additive component (I_{AXB}) is ascribed as epistasis or interactions between two loci. In the presence of epistasis, the aggregate genotypic value (G) would be as $G = G_A + G_B + I_{AXB}$. For detail on epistasis see books of Falconer (1960) and Mather and Jinks (1971), respectively.

How epistasis works?

There are several ways in which epistasis can work. One gene may provide the essential condition (in terms of metabolic intermediate or an activator essential for the transcription of another gene) for the activity of another. In such cases, loss of function of the first gene will make the second gene useless and the single gene will be just as disabled as the double mutant. A second possibility is that the first gene inhibits the function of the second by encoding a specific transcriptional repressor. In this condition the epistasis can work in the other way round. A null mutation (or reduced function mutation) in the controlled gene will make the controller gene inconsequential. Thus any gene under control of another whether by activation or repression can in principle mutate so as to work independently of control. Such liberating mutations will be more or less epistatic to mutations in the controlling gene, depending upon what else the latter has to do. The properties of the mutants –whether they have decreased or increased gene product and whether they are recessive or dominant, are required for the interpretation of epistatic relationships. Null mutations in an activator gene decrease whereas in a repressor gene, increase the output of the gene under control. Null mutations are usually recessive and mutations that free genes from control are usually dominant.

Epistatic analysis works only if two genes are unlinked and mutant alleles produce distinct phenotypes. Further, it is meaningful for genes that regulate discrete steps in a related event or pathway. If two genes regulate the same step then the phenotype may be intermediate in character or may reflect the dominant gene. If two genes affect different steps in the same pathway then it will be important to know the order in which they are expressed. Those later expressed are considered to be epistatic to those earlier expressed. Suppose there are A,B,C,D,E genes to be encoding proteins A,B,C,D,E and regulating sequential steps in a developmental pattern. Further, suppose that mutants d and e are known and they are recessive mutations conferring distinct or opposite phenotypes. To determine whether gene D acts ahead of gene E, the progeny of a cross between two plants heterozygous for alleles d and e at both genes is analyzed (figure 25.6).

If one mutant phenotype, say d⁻ is found in significantly large numbers than e⁻, it is concluded that gene D is epistatic to gene E. In the figure above the phenotype d⁻ outnumbers e⁻ by 1/16.

If two mutations possess a similar phenotype, a technique called 'transgression development analysis' is used to determine whether the products of the two wild genes act in the same pathway or in separate pathways that converge. This is accomplished by crossing the two mutants. If the phenotype of the double mutants shows an additive phenotype, it is likely that the products act in separate pathways that converge. For mutants with distinct phenotypes if one phenotype remains unaffected, i.e., one mutation completely masks the other, they are likely to occur in the same pathway.

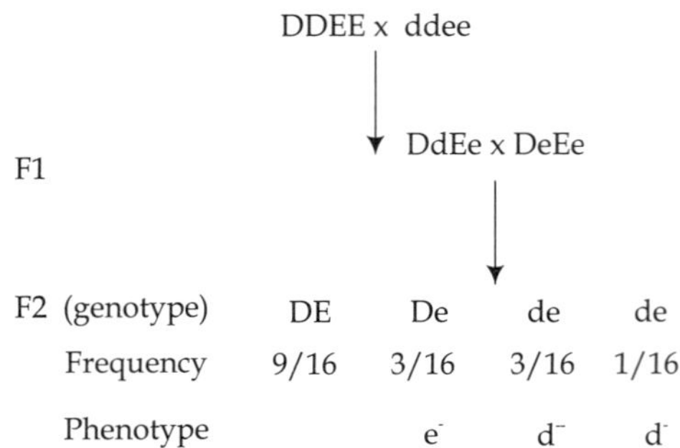

FIGURE 25.6   Showing genotypes, their frequencis and phenotypes in F_2 generation.

Types of epistasis—There are two types of epistasis and both kinds of epistasis are important. With sign epistasis the genotype at one locus may reverse the direction of allelic effects at another gene. In some cases the mechanisms of epistasis are known from functional studies (Johanson, 2000). At the level of individual genes, sign epsitasis influencing juvenile growth rate is caused by genetic polymorphism at serine/threonine protein kinase and patterns of nucleotide polymorphisms suggest that balancing selection maintains elevated level of genetic variation.

25.15   METHOD OF IDENTIFYING GENETIC INTERACTIONS

Functionally interacting genes are more likely to have similar expression patterns and phenotypes. Interactions among genes include protein-protein physical interactions as well as gene-gene and protein-gene interactions (Zhong and Sternberg, 2006). Genetic interactions can be identified by modifiers screens (e.g. a **synthetic lethal screens**), requires easily detectable phenotypes. A solid statistical model is required for producing reliable predictions. Two types of methods, Unions and Bayesian networks (Marcotte et al., 1999) are available. Bayesian network models are preferable to unions as they weight data according to their reliability but current methods often assume that datasets are independent (naïve Bayesian network). Logistic regression is a classical method for predicting binary counts (interaction vs no interaction). Yeast two hybrid data may not be as accurate as results of small-scale interactions studies. Further, a yeast two hybrid data describe physical rather than genetic interactions which includes both direct and indirect interactions.

Application of study of epistasis—When mutational and epistatic studies indicate that simple linear sequence of events is functioning in a specific process, biologists draw a pathway, connecting steps with arrows and assigning different steps to different genes (see chapter 17). When the results demand it, branch points and converging pathways are drawn. Such two-dimensional modeling is useful for presenting the results as well as interpretation but at some point, it would be better to mathematically abstract the processes.

An epistatic relationship between two gene functions is often used to infer their relative order within a genetic pathway. However, interpretation of such analysis depends on the type of genetic pathways. In case of biosynthetic pathways in which a series of genes modify a common intermediate substrate, epistatic genes are placed upstream of hypostatic genes. In case of pathways in which a pathway is a linear series of genes or gene products, the output depends solely on the activity of the terminal member of the pathway and other members can be thought of as a series of on-off switches that determine the active or inactive state of the output gene (Sakonju and Busturia, 1999). In contrast, in case of regulatory pathways where a member gene regulates the expression or activity of a downstream gene or gene product, epistatic genes are generally placed downstream. For example, a regulatory gene pathway determines the dorsoventral body axis of the embryo in *Drosophila melanogaster*. Most loss-of-function mutations which affect this pathway dorsalized the embryo whereas a gain-of-function mutant allele of Toll gene (Toll^D) ventralizes the embryo. Which of the two alleles is epistatic can be determined by studying the double mutant embryos. If the double-mutant shows a dorsalized embryo then the tester allele is epistatic over Toll^D and placed downstream of the Toll whereas if the double mutant shows a ventralized phenotype, the tester gene is placed upstream. Thus genes in the pathway can be placed upstream or downstream relative to the Toll gene. Regulatory pathways are not simple and there are situations in which epistatic genes act upstream (Avery and Wasserman, 1992).

25.16 BIOMOLECULAR NETWORK

Cells can be viewed as a complex network of interacting proteins, nucleic acids and other biomolecules and many fundamental cellular processes involve interactions among proteins and other biomolecules. Therefore, identifying interaction is a step toward defining protein function as clues about the function of an unknown protein can be known by studying its interaction with other proteins of known function. Proteins mediating the functions of organisms are strictly determined by the structure and activity of the genes that encode them. Data from gene knockout experiments and molecular analysis of individual genes reflect combinatorial regulation as well as redundancy of gene function (Crossin, 1994; Shashtri, 1994). These characteristics imply complex network. The interactions of biological macromolecules and the flow of regulatory information that controls development, bahviour and homoestasis can be considered a genetic network (Loomis and Sternberg, 1995). Genetic network consists of two subsystems- the proximal genetic network operating through *cis* (control regions of DNA) and *trans* (gene products which regulate *cis* regions) elements and the distal genetic network involving protein x protein interaction and protein-signalling factor interactions governing intra- and extracellular communication (gain determined by genes encoding the participating proteins). A biomolecular network can be viewed as a collection of nodes, some of which are connected by links. The nodes represent biomolecules whereas the links represent interactions. In other words, the nodes in such network are genes or their RNA and protein products and the connections are the regulatory and physical interactions among the RNAs, proteins and *cis*-regulatory DNA sequences of each gene. Genes are now being discovered and their primary sequences determined. The challenge now is to link the genes and their products into functional pathways, circuits and networks. There are many classes of molecular networks, each with different types of nodes and links (Xia et al., 2004).

1. **Protein-protein physical interaction networks**—In this type of network nodes represent proteins and links represent direct physical contacts between proteins. In

addition to direct interaction, two proteins can interact indirectly through other proteins when they belong to the same protein complex.

2. **Protein-protein genetic interaction networks**—Two genes can be said to interact genetically if a mutation in one gene either suppresses or enhances the phenotype of a mutation in its partner gene (Tong et al., 2001). Some workers refer the term 'genetic interaction' to a pair of so-called synthetic lethal genes, in which cell death occurs when this pair of genes is deleted simultaneously though neither deletion alone is lethal. Synthetic lethal relationships may occur between functionally redundant genes and therefore can be used to determine the function of unknown gene.

3. **Expression network**—In the expression network, genes that are co-expressed are considered connected. Genes linked in an expression network are not necessarily co-regulated as unrelated genes can sometimes show correlated expression simply by chance. The structure of an expression network can vary greatly across different experiments and even within the same experiment. Networks produced by different clustering algorithms are often distinct. Genes are clustered together into groups either by manual examination of data or by using statistical methods such as self-organizing maps, k-tuple means clustering or hierarchical clustering.

4. **Regulatory networks**—Protein-DNA interactions are important and common class of interactions. Most DNA-binding proteins are transcription factors that regulate the expression of target genes. Combinatorial use of TFs further complicates simple interactions of target genes for a given TF. A regulatory network thus consists of TFs and their targets with a specific directionality to the connection between a TF and its target. TF can either up or down regulate expression of their target genes. Downstream of the signal transduction pathways a complex arrays of gene regulation network takes place. The transcriptional regulatory networks mix heterogeneous physical interactions (protein-protein, protein-DNA and protein-RNA) and genetic interactions (activation, inhibition, etc). Gene regulation networks are, however, still more studied at a higher level of abstraction.

5. **Metabolic networks**—These networks describe the biochemical reactions within metabolic pathways in the cell. Here nodes represent metabolic substrate/products and links represent metabolic reaction. Microbial metabolic pathways required to isolate and subtly perturb specific genes before intricacies of the feedback loops and alternate controls became clear.

6. **Signalling networks**—They represent signal transduction pathways through protein-protein interactions and protein-small molecule interaction. Here nodes represent protein or small molecules and links represent signal transduction event. Signal transduction pathways are particular examples of internal cell pathways describing the cascades of molecular interactions from the reception of an extracellular signal (e.g. binding of cytokinine to its receptor) to the activator of transcription factors triggering the transcription of specific gene. The signal transduction networks are generally described in terms of physical interactions between proteins (e.g. binding or phosphorylation, etc).

Signalling networks are thus protein networks that describe post-translational modifications, degradation, subcellular localization or physical interactions of proteins.

There are different definitions of gene regulatory networks depending on the nature of nodes(i.e. DNAs, RNAs, proteins or a mixture) and the links(regulatory relationships).

Analyses of regulatory networks (such as those involving signal transduction and transcriptional regulation cascades) illustrate combinatorial action.

Networks can be divided into two categories: directed and undirected. Physical interaction, genetic interaction and expression networks are **undirected** which implies no directionality or causality in the interactions. Regulatory, metabolic and signaling networks are **directed networks** implying directionality in their linkages. A node in the directed network may have an incoming degree and an outgoing degree which are completely independent. The incoming degree of node refers to the number of edges pointing toward this node whereas its outgoing degree refers to the number of edges pointing out of this node. Clustering coefficient can not be calculated for directed networks.

Biological processes have been added during evolution to adapt to gene expression and function to different needs. Genetic networks are the product (and material) of evolution and the evolutionary divergence of particular networks can be used to understand their properties. Comparison of different networks within the same organism will throw light on the basic organization principles of the cells and cross species comparison of the networks will tell us how these networks evolve.

25.17 COMPARISON OF NETWORKS (EXPRESSION AND REGULATORY NETWORKS)

Yu et al. (2002) compared the regulatory networks and expression network in *S.cerevisiae* and showed that co-regulated genes were generally co-expressed and correlation in expression profiles was highest for genes targeted by multiple TFs. Furthermore, co-regulated gene pairs tended to share similar functions and there were subdivisions within individual network motifs that separate the regulation of genes of distinct functions. The expression profiles of TFs and their target genes displayed more complex relationships than simple correlations with regulatory response of the target genes often being delayed.

Mapping and understanding of interaction networks will show how a cell actually operates in time and space. Interaction networks tell us whether or not two molecules interact and also how strongly and how quickly they interact. It will also tell us how functional elements interact with and regulate each other. In addition to mapping out all possible networks it is important to know under which conditions (cellular state, environment type and protein modification type) each interaction is present in a cell. Exploring the protein-protein interaction map reveals biological pathways and allows prediction of protein function. Protein interaction maps are now available for four eukaryotic organisms such as budding yeast (Iteo et al., 2001; Utez et al., 2000), worm (C.elegans) (Li et al., 2004), Drosophila (Formstecher et al., 2005; Giot et al., 2003) and human (Stelze et al., 2005; Rual et al., 2005). The sizes of the networks are given in the table 25.1.

Protein-protein interaction maps thus form a large intricate networks leading to a vision of cell biology as an integrated system. Protein-protein interactions network are fundamental to understanding of cellular processes. They are being used for assignment of function to uncharacterized proteins and searching for remote similarities between proteins. Although

Table 25.1 Showing different organisms and the size of their interactions.

Organism	Size (No. of interactions/graph edges)
Yeast	~6000
Worm	5000
Fly	20, 000
Human	5500

there is a need for extracting and revealing functional information they contain but the sizes and complexity of these networks make their analysis a difficult task. Baudot et al (2006) developed a method called PRODISTIN (protein distance based on interaction) which allows a functional classification of the proteins according to the identity of their interacting partners (Brun et al., 2003). The central idea in this interaction-based functional clustering is to compare interaction partners for all protein pairs assuming that the more two proteins share interacting partners, the more they should be functionally related. Applying this to the yeast protein-protein interaction network they showed that the method (i) clusters proteins participating in the same cellular processes in the same functional classes, (ii) predicts function for unknown proteins and (iii) is statistically valid.

The focus now is on identifying new components defining the regulatory inputs and outputs of each node and delineating the physiologically relevant pathways. In case of analysis of nodes, the objective now is to know how many inputs and outputs potentially exist. For example, analysis of individual protein might reveal three distinct binding sites for other proteins or analysis of the cis-regulatory regions of a gene might reveal some twenty different proteins specifically interacting with these sequences. New tools for identifying interacting components including genetic enhancer and suppressor screens, two-hybrid library screening and affinity chromatography will help in elucidating the network.

Genetic screens for mutations that alter the phenotype of other mutations can identify novel interactions and components but they do not guarantee the direction of the connections. Biochemical detection methods also identify novel interactions but do not guarantee their functional relevance. But when used together with independent tests, they add to the wealth of connections that complement the result from genome sequencing projects.

Mutational analysis allows the function of each gene to be examined by determining the effect of its elimination or alteration on the organism's phenotype. Genes can be altered one at a time to assess their individual contribution or genomes can be generated with multiple changes to understand the relations of the specific genes and their products. When the loss of a specific gene results in a clear physiological consequence it indicates the normal role of the gene in a well defined process. However, site-directed mutants in multicellular organisms often have phenotypes that differ little from wild-type, thereby indicating partial functional redundancy or complete irrelevance of the gene in the process under study. **Characteristics of networks-** Some networks stay to counter and respond to differing pressures and requirement and breaking any given connection will not bring down the whole structure. Some networks underlie homeostasis and others drive spatial and temporal changes during ontogeny.

25.18   STRUCTURE-BASED PREDICTION OF PROTEIN FUNCTION

Predicting the function of proteins that the genes encode is a challenge. When proteins are unrelated to others (orthologues) of known activity, bioinformatics inference for function becomes problematic. It will be therefore useful to interrogate protein structure for function directly. Structure-based prediction has been used with some success for inhibition design, substrate prediction has proven difficult. The function of an enzyme of unknown activity Tm 0936 from *Thermotoga maritime* has been predicted by docking high-energy intermediate forms of thousands of candidate metabolites (Hermann et al., 2007). Complexes of macromolecules invariably reveal a striking degree of spatial and electrostatic complemenatrity between the contacting surfaces and thus there is a need of modeling such interfaces. Molecular surfaces are fitted to each other by a new solution to the problem of docking a ligand into the active site of a protein molecule. The aim is to dock a ligand into the active site of a protein molecule in such a way that optimizes complementarity, steric hindrance and electrostatic energy. The procedure constructs patterns of points on the surfaces and superimposes them upon each other using a least square best fit algorithm. This brings surfaces into contact and provides a direct measure of their local complementarity.

Docking is divided into two tasks: searching and filtering. Searching consists in matching patterns all over the ligand surface against patterns in the protein active site to generate a large set of transformations (co-ordinate translations and rotations), each of which docks the ligand into the active site in different orientation. Filtering refers to the process of evaluating these docking and keeping the best one (Bacon and Moult, 1992). Patterns are inscribed on molecular surfaces as sequences of points in 3-dimensional space. They are superimposed on one another, docking the ligand into the protein active site, using the algorithm of McLachlan (1979) which was originally designed for superimposing and comparing sequences of C^α co-ordinates and thereby to bring the surfaces together. However, there are limitations with this technique. Besides problems of sampling and scoring in docking, substrate prediction confronts many other challenges including many possible substrates to consider and many reactions that an enzyme might catalyze. Furthermore, enzymes preferably recognize transition states over the ground state structures that usually represented in docking.

26

Proteomics

26.1 PROTEOME

Information on mRNA levels does not give a complete picture of the way gene expression is regulated. Our aim is to identify proteins that are up or down regulated. Protein expression data are much more informative but are much more difficult to obtain in a parallel fashion. Proteins have a lower throughput, are less sensitive. Hydrophobic and large proteins usually do not enter the second dimension of the gel; difficult to visualize all but the most abundant proteins and difficult to define normal protein expression patterns that can be compared with the disease state. Proteome approach has been used in yeast to study gene function through the generation of knock out or overexpression mutants and for the analysis of changes in protein profiles on two dimensional gels. mRNA levels are more informative about cell state and the activity of gene and for most genes changes in mRNA abundance are related to changes in protein abundance and so go for methods of array-based mRNA expression profiling. Protein expression analysis will be most useful under the following conditions.

1. Analyzing samples that do not contain mRNA such as some body fluids.
2. Where protein abundance does not correlated with mRNA abundance
3. Where critical changes involve post-translational modifications of proteins such as glycosylation or phosphorylation (which are important for protein function) rather than changes in protein abundance
4. Where an overview of most abundant proteins in a specialized source is itself of importance
5. Where a two-dimensional gel allows a relatively comprehensive overview of a sample proteome such as that of a microbe.

But the analysis of proteins is expected to provide more direct understanding of function and regulation than analysis of genes because proteins are one step closer to function than are genes. The applications of proteomic approaches include

(i) identification of proteins, their isoforms (isoform refers to any duplicated gene or the protein encoded by a duplicate gene) and their prevalence in each tissue,

(ii) characterizing the biochemical and cellular functions of each protein and

(iii) the analysis of protein regulation and its relation to other regulatory networks (Bertone and Snyder, 2005).

Thus the objective in proteomics is to understand the expression, function and regulation of the entire set of proteins encoded by an organism. This information will help in understanding how complex processes occur at the molecular level, how they differ in various cell types and how they are changed in biotic or abiotic stress conditions. **Classical proteomics** refers to identifying and quantifying the expression levels of proteins localized in specific protein complexes. **Functional proteomics** refers to the identification of interaction and cell processes.

26.2 METHODS FOR MONITORING PROTEINS

Methods for monitoring protein levels either directly or indirectly include western blot, two-dimensional gels, methods based on protein or peptide chromatographic separation and mass spectrometric detection, methods that use specific protein-fusion reporter constructs and colorometric readouts and methods based on characterization of actively translated polysomal mRNA. Low throughput technologies such as co-immunoprecipitation, far-western blots, pull downs are used for studies on individual proteins. However, study of interactions at the proteome level requires high throughput assays. Functional proteomics requires the set of technologies used to generate interaction data on a large scale.

26.3 CELLULAR EXPRESSION PATTERNS/PROTEIN PROFILING

Proteome-wide characterization permits the construction of global maps of differentially expressed proteins. By comparing several sets of expression patterns under different conditions (e.g. wild-type vs mutant or healthy vs diseased) or at different time stages one can deduce a cluster of co-regulated proteins which could be interpreted as a protein expression 'network'. This network can be used for elucidation of cell pathways, characterization of cell types or identification of pathogenic agents. This network is complementary to gene regulation networks produced by transcriptome techniques. If protein networks give information about co-regulation of proteins and their response to specific conditions, they are not completely informative about the biochemical function of gene products. The function of a protein is defined by the role it takes in cell pathways and interactions in which it participates with other cell components (e.g. DNA, RNA, protein, metabolites or lipids) and interaction networks deal with only proteins.

Cellular proteins can be isolated and sequenced. A number of methods are available for profiling protein expression which include 2-DGE or liquid chromatography followed by tandem MS. Mass spectrometry is the preferred method for in-depth assessment of protein production and function as it can extract a large amount of quantitative information from such samples. High resolution methods such as Fourier transform mass spectrometry should allow the detection of several thousand proteins (see chapters 30 and 31 for details on these methods). Two-dimensional gel electrophoresis can be used for separation of up to 1000 different proteins and Mass spectrometry can then be used to partially sequence individual protein and thus assign each to a gene. Two dimensional PAGE is at present the method of choice to study the abundance and posttranslational modification of several hundred proteins in parallel (Humphery-Smith and Blackstock, 1997). Resolution and reproducibility of the separation system has now greatly improved and algorithms for automatic spot quantification have developed. Furthermore, protein spots on gels are now more easily identified with improved method for N-terminal and internal microsequencing with the use of MS techniques. These techniques can determine very accurately the molecular mass of fragments released upon protease treatment of minute amount of protein excised from gels. Further, comparison with theoretical mass predicted from sequences in protein databases allows for the attribution of a sequence to the protein fragments provided that a complete, organism-specific protein database is available. The presence or absence of a particular protein in samples from different tissues, from similar tissues at different stages of development or from tissues treated in ways that simulate a variety of biological conditions such as drought or flooding can help in defining protein function. At present this technique allows monitoring of only a few thousand abundant proteins. The other techniques based on MS may be developed in future for studying total protein complements.

26.3.1 Subcellular fractionation

Although current approaches allow the detection of thousands of protein but what is required is an approach which can truly profile tens of thousands of different protein isoforms present in each cell. And so what is also required is to obtain information on intracellular localization of a subset of proteins as in case of plasma membrane protein or protein from other organelles and even multisubunit protein complexes. In other words, one method for detecting more proteins is the subcellular fractionation. This approach will reduce the complexity of protein extracts while rare proteins can be enriched and thus more readily detected. Further, functional information can be obtained by determining the subcellular compartments where individual proteins are localized. In other words, the determination of subcellular location of each gene product will provide information regarding the cellular function of many proteins. Thus nuclear, chloroplast, amyloplast, plasma membrane, peroxisome, endoplasmic reticulum, cell wall and mitochondrial proteins should be characterized and have been characterized in Arabidopsis.

Recalcitrant proteins — Attention should be paid to isolation of classes of proteins such as membrane and other hydrophobic proteins which are recalcitrant to isolation and analysis. Membrane proteins although comprising 30% of the typical proteome, tendency to aggregate and precipitate in solution confounds their analysis.

26.3.2 Protein chips

One method to obtain biochemical function of each protein is to systematically express proteins in recombinant form and subject them to biochemical analyses. The microarray format provides a robust and convenient platform for the simultaneous analysis of thousands of individual protein samples and facilitates the design of sophisticated and reproducible biochemical experiments under highly specific conditions.

We have seen in chapter 24 that DNA chips refer to microarrays which consist of short DNA sequences immobilized on a surface. By determining which spots bind to mRNA extracts from a sample one can obtain snapshot of the activity of thousands of genes simultaneously. Similarly, there is a need of protein chips that will carry tens of thousands of protein 'capture' molecules. For DNA chips designing capture molecules and developing readout systems are relatively easy. Strands of DNA can bind tightly and specifically to mRNAs with a complementary sequence. Further, if mRNAs in a sample are tagged with a fluorescent dye, it is simple to determine where on the array they are bound. In case of protein chip specific capture molecules must be designed for all possible proteins encoded by the genome including the modified forms of proteins. Capture molecules which will bind with high affinity to one protein alone is extremely difficult. Capture molecules need to bind with high affinity as hormones, growth factors and intracellular signaling proteins are present in only at very low concentrations. Capture molecules must be able to identify target proteins from nanomolecular (10^{-9} molar) to picomolar (10^{-12} molar) solutions.

The application of microarray technology to proteomics yielded the protein chips. Recombinant proteins can be expressed and purified in a large scale format. These proteins are pooled into wells and assayed for functions such as enzymatic activity. This approach has been termed '**Biochemical Genetics**'. It is an *in vitro* technique for protein functional assay on a large scale. Protein chip technology can be directly applied to protein interaction networks as large number of immobilized proteins can be probed with labeled substrate in a single experiment. It is used for studying the proteome of a cell. Proteins can also be immobilized on a solid surface. When an array of antibodies to particular proteins are immobilized on the solid surface and a sample of proteins is added then if the protein that binds any of the antibodies is present in the sample and can be detected by a solid form of the ELISA assay. In fact, such an approach is essentially a large-scale version of ELISAs. In one version protein chip is probed with fluorescent labeled proteins from two different cell lysates. Cell lysates are labeled by different fluorophores and mixed such that the color acts as a readouts for the change in abundance of the protein bound to the antibody. This system needs specific and well characterized antibodies and a number of technical problems have to be sorted out before it could be used in proteome analysis. In another modifications, peptides, protein fragments or proteins may also be immobilized onto chips and samples (e.g. phage library or plasma serum) applied onto the chip followed by detection of binding. One approach using protein chips couples the above techniques with a direct MALADI readout of the bound material. Thousands of proteins are spotted in an ordered array onto the glass slide or other solid support to allow visualization of protein-protein, for example, protein-antibody or protein-ligand interaction (s). Glass slide is first treated with an aldehyde-containing silane which reacts with primary amines of protein to establish a Schiff's base linkage (covalent coupling). Alternatively, the glass surface can be coated with poly-L-lysine (non-covalent coupling). The purified proteins, each in a nano droplets (containing billions of protein copies with concentration of 100ug/ml or less) are spotted onto the glass slide which is additionally coated with bovine serum albumin (BSA) to prevent denaturation of the spotted protein. The spotted proteins are then covalently bound to the glass surface. For peptides or very small proteins, a molecular layer of BSA is first attached to the glass which is then activated with N,N'-disuccinimicarbonate to yield BSA-N—hydroxy succinimide and quenched with glycine. The peptides are then displayed on the top of the BSA monolayer which readily exposes them for interaction (s) with target molecules. The ligand is fluorescently labeled and exposed to the protein array and binding of ligand to a protein (or to many proteins) is detected by fluorescence. Using different fluorophores with non-overlapping excitation and emission spectra expands the number of simultaneously detectable interactions. Kumar and snyder 2000) developed a microarray technology in which purified, active proteins from almost entire yeast proteome are printed onto a microscopic slide at high density such that thousands of protein interactions (and other protein functions) can be assayed simultaneously.

Arenkov et al. (2000) created a protein antibody-based protein microchip, containing 0.2nl spots of gel substrate in which proteins were immobilized. This platform allowed electrophoresis to be used to enhance mixing of substrates.

A protein chip can be prepared in several ways. The surface can be immobilized with recombinant proteins or their domains such as the bacterially expressed GST-fusion proteins and then cell lysates containing putative interacting partners are applied to the chip followed by washing to remove the unbound material. The bound proteins can then be eluted and identified by MS. Alternatively, instead of cell lysates, a phage cDNA display library can be applied to the chip followed by washing and amplification steps to isolate individual interacting phage particles. The inserts in these phage particles can then be sequenced to determine the identity of interacting partners.

Combinatorial protein array—It is a variant of protein chip onto which thousands or even millions of stable protein complexes rather than individual proteins are arrayed. **Antibody array/microarray/chip**—The classical capture molecules for proteins are antibodies. Hundreds of monoclonal antibodies directed against proteins of known functions are spotted on a membrane or other solid support in an ordered array, each at a pre-determined position. The antibodies on such an array bind to their complementary antigens (and linked to secondary proteins). In this technique antibodies are first bound to the chip by conventional chemistry as in case of protein chip. Then protein lysates prepared from one organism, say A and B, respectively. The A proteins are labeled with, for example, cyanin 3 and the B proteins are labeled with cyanin 5 and they are incubated with the array. After the binding of an antigen by the corresponding antibody the former is detected by a fluorescence scanner. The antigen can also be detected by, for example, immunoblotting and its identity is deduced from the position of the resulting signal. In this way, antibodies arrays allow to profile thousands of proteins simultaneously. It can also be used to screen for variations of protein levels, protein-protein interactions, and protein expression patterns in various tissues or can be employed for studies of post-translational protein modifications. Alternatively monoclonal antibodies can be bound onto the surface of microarray chips to which peptides or proteins from cell extract are bound. Similarly, **combinatorial antibody arrays** with more than 100,000 antibodies can be created through combination from fewer than 800 component heavy and light chain parts. Such arrays can be used to probe protein-protein interactions. A combinatorial library refers to a large collection of molecules which are chemically related but designed to display highly diverse combinations of chemical reactivities and physical structures. So far antibodies are the only capture molecules that have high specificity and sensitivity.

The advent of protein-based microarrays allows the global observation of biochemical activities on a scale where hundreds or thousands of proteins can be simultaneously screened for protein-protein, protein-nucleic acid and small-molecule interactions as well as posttranslational modifications. In many cases, proteins are likely to have multiple cellular roles and thus the ability to systematically analyze them in a high throughput manner should reveal biochemical properties for proteins not previously appreciated.

Problems in protein array—The principal challenges in protein array are (i) creation of a comprehensive expression clone library, (ii) high-throughput protein production including expression, isolation and purification and (iii) adaptation of DNA microarray technology to accommodate protein substrates. The greatest obstacle in developing functional protein microarrays is the challenge number (i) from which a large number of distinct proteins samples can be produced. Further, the substrate materials associated with conventional protein assays are not often compatible with robotic arrayers and thus can not provide the sensitivity or dynamic range expected from microarray experiments or contribute high fluorescence background resulting in low signal-to-noise ratios.

Investigating protein function—There are three main ways of investigating protein function:

1. Sequence and structural comparison with genes and proteins of known function
2. Determining when and where a gene is expressed (determination of subcellular location of each gene product)
3. Investigating the interactions of a protein with other proteins.

Homology searches can be used to show likely functions of genes but the phenotypic role of individual gene family members is usually obscure without assigning gene function by genetic analysis.

The function of a protein can be defined by determining what it binds to. The association of a protein of unknown function with a protein of known function can provide useful information about its function. Understanding of interactions is important in order to appreciate the mechanisms by which cellular pathways function and interlink. There are different techniques for investigating protein-protein interactions. The use of global two-hybrid interactions or affinity purification followed by mass spectroscopy in other organisms such as yeast has provided considerable information regarding interacting partners.

26.3.3 Approaches to protein purification and defining protein-protein interaction

There are two approaches to study of protein and protein-protein interaction. *In vitro* and *in vivo* techniques have been employed to interacting proteins. In comparison to Y2H (described below), *in vivo* pull-down methods detect protein complexes rather than binary protein-protein physical interactions. The 'pull down' methods refer to those methods in which the isolation of specific compounds from complex mixtures is through their high affinity binding to immobilized capture molecules (that are bound to magnetic beads or stationary phases of affinity column). Further, *in vivo* techniques use either a native or modified protein bait protein to identify and precipitate interacting partners. Most experiments studying protein-protein interactions through pull down techniques consist of three parts:

1. Bait presentation
2. Affinity purification and
3. Analysis of the recovered complex (Boulton et al., 2002).

Gavin et al. (2002) and Ho et al. (2000) took a different approach which is particularly effective at identifying protein complexes which contain three or more components. With the advent of ultrasensitive mass-spectrometric protein identification methods, it is now possible to identify directly protein complexes on a proteome–wide scale. To date, generation of large-scale protein-protein interaction maps has relied on the Y2H system which detects binary interactions through the activation of reporter gene expression. The different steps involved in the **co-precipitation/mass spectrometry approach** can be carried out on proteome-wide scale are as follows:

1. An affinity tag is first attached to target protein (the bait).
2. Bait proteins are systematically precipitated along with many associated proteins on an affinity column. This approach is particularly effective at identifying protein complexes that contain three or more components.
3. Purified protein complexes are resolved by one-dimensional SDS-PAGE, a technique which involves running an electric charge through complexes on a gel so that proteins become separated according to mass.
4. Proteins are excised from the gel, digested with the trypsin and analyzed by MS.
5. Database search algorithms are then used to identify specific proteins from their mass spectra.

They attached tags to hundreds of different proteins (to create 'bait' proteins). Then they introduced DNA encoding these bait proteins into yeast cells allowing the modified proteins to be expressed in the cells and to form physical complexes with other proteins. Then using the tag each bait protein was pulled out often fishing out the entire complex with it (and hence the term **bait**). The proteins extracted with the tagged bait were identified using MS. This approach thus uses tandem affinity purification (TAP) and MS in a large scale approach to characterize multiprotein complexes. Gene-specific cassettes containing the TAP tag generated by PCR were inserted by homologous recombination at the 3′ end of the genes. After growing the cells to mod—log phase assemblies are purified from total cell lysates by TAP. This technique combines a first high affinity purification, mild elution using a site-specific protease and a second affinity purification to obtain protein complexes with high efficiency and specificity. The purified protein assemblies were separated by denaturing gel electrophoresis. Individual bands were digested by trypsin and analyzed by MALDI-TOFMS and sequences are inferred from mass and sequences are identified by comparing with a database of predicted proteins encoded by the organism's genome by search algorithm (see Roy, 2009). If mass alone can not predict the exact sequence, fragmentation methods (nano-electrospray tandem mass spectrometry) can be used to produce stretches of up to 16 amino acids of sequence from femtomolar amounts of protein fragment. Ho et al. used high throughput mass spectrometric protein complex identification method. They chose a set of bait proteins representatives of a variety of different functional classes. Further, one step immuno-affinity purification based on the Flag epitope tag was used to capture bait proteins which were transiently overexpressed from the heterologous GAL1 or tet promoters. Proteins from different individuals were resolved by SDS-PAGE, visualized by Coomassie stain, excised from the gel and subjected to trypsin digestion before mass spectrometric analysis. As the isolation procedure often yielded multiple proteins from single excised bands which can not be resolved by peptide-mass fingerprinting alone so tandem mass spectrometry

(MS/MS) fragmentation was used to identify unambiguously proteins in each gel slice. TAP tagging has proved invaluable for purification of complexes from different cellular compartments. It allows identification of low abundance proteins that could not be detected by approaches involving expression proteomics. Further, it allows purification of very large complexes.

Drawbacks of this approach—It results in production of a significant number of false-positive interactions and simultaneously it fails to identify many known interactions. Like most other large-scale studies these results are also imperfect and so it is essential to merge data from different sources to obtain accurate understanding of protein networks. The limitation with mass spectrometric analysis of protein interactions is that the protein complexes first need to be isolated by physical methods such as electrophoresis. Further, many proteins will not be detected because, for example, they are too scarce, large, small, acidic, or alkaline to be studied by two-dimensional gel analysis or the interaction is too weak and transient to survive affinity purification.

1. **Use of gene/protein tag**—One of the generic ways of identifying the interaction partners of a new protein is to tag it with an epitope. The protein can then be over-expressed in cells and together with its interaction partners immunoprecipitated by an antibody against the epitope. Expression libraries contain many types of tags or fusion domains which facilitate library screening and downstream manipulation of individual clones. It requires full length cDNA clone of a gene. There are at least four types of protein fusion moieties commonly available.

A. **Epitope tags**—ETs are typically short amino acid segments fused to either the N-or C-terminus of the cloned protein. **Flag sequence** refers to any short nucleotide sequence inserted into a gene which does not change the information of this gene but allows to identify the expressed gene product (fusion protein) very easily. In other words, epitope refers to chemically defined antigenic determinant. ET provides ready-made immunoreactivity to the cloned gene product through antibodies. **Epitope tagging** (Munro and Pelham, 1984) is the process of fusing a set of amino acid residues that are recognized as an **antigenic determinant** to a protein of interest. Antibodies interact directly with an antigen through a small portion of the antigen-antigenic determinant or epitope. This technique is used for the detection, isolation, purification and characterization of a specific protein, determination of its intracellular localization and identification of other protein that interacts with it. **Epitope tag** is a short oligonucleotide sequence encoding an epitope domain (a 3-14 amino acid long peptide) of an antigen. Any individual antibody binds only to a particular structure within the antigen is called its antigenic determinant or **epitope**. In this approach if a candidate gene (encoding a protein of interest) cloned is fused to an **epitope** tag (an antigenic determinant) either at 5′ or 3′ end and the resulting gene construct is then introduced into an epitope tagging vector and transformed into a target organism by suitable gene transfer technique (Figure 26.1). The expression of the candidate gene and its epitope tag leads to the generation of a fusion protein and can be separated using the antibody (monoclonal antibody) which will bind to the epitope. And if that tagged gene is expressed in a cell and interacts with other proteins then that interacting protein can also be isolated and identified. When the crude extract of the cells expressing the tagged protein is passed through a column of immobilized

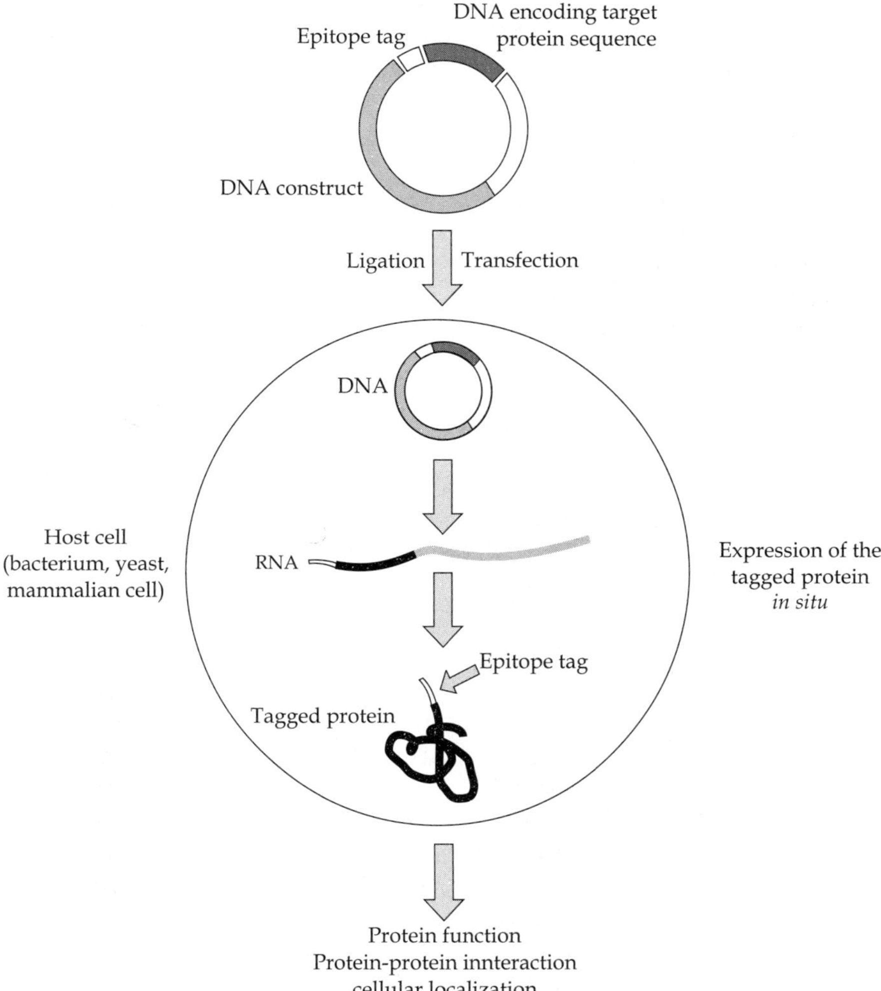

FIGURE 26.1 **Showing epitope tagging procedure.**

antibody (immuno affinity chromatography) then the tagged protein will bind to the antibody. Sometimes the protein associated with the tagged protein is also retained on the column. Finally, the tag is cleaved from the protein using specific protease.

Site of epitope tagging—The amino (N) or carboxyl (C) terminus of the protein is typically chosen as a tagging site because the ends of proteins are more likely to be accessible to the antibody and to be susceptible to modifications without affecting function. Given on next page are the two epitopes, 9E10 and 12CA5 derived from a portion of the human c-myc gene and influenza hemagglutinin protein, respectively. Antibody 12CA5 recognizes an epitope, 12CA5 whereas antibody 9E10 recognizes 9E10 epitope.

9E10 epitope GAACAAAA<u>GCTTATTT</u>CTCAAGAAGACTTG (Hind III)

12CA5 epitope TACCCATACGAC<u>GTCCCA</u>GACTACGCT (AaII)

These epitope sequences are marked via the inclusion of a restriction site. The underlined sequences indicate restriction enzyme recognition sites. Other epitope tags with their size, sequence and antibody are given below in the Table (26.1) (Jarvick and Telmer, 1998).

Table 26.1 Showing various epitopes, their size, sequence and antibody.

Epitopes	Size (number of amino acids)	Sequence	Antibody	
AU1	6	DTYRI	AU1	
AU5	6	TDFYLK	AU5	
B tag	6	QYPALT	D11, F10	Without highly charged
c-myc	10	EQKLISEEDL	9E10	Highly charged, popular for epitope tagging
FLAG	8	DYKDDDDK	M1, M2, M5	Highly charged, popular for epitope tagging
Glu-Glu	6	EYMPME OR	GLU-GLU EEMPME	
HA	9	YPYDVPDYA	12CA5, 3F10, HA.11	Highly charged, popular for epitope tagging
His6	6	HHHHHH	6-His, 6xHis, HIS-11	
HSV	11	QPELAPEDPED	HSV.TAG	
HTTPHH	6	HTTPHH	PHHTT	Without highly charged
IRS	5	RYIRS	IRSI	
KT3	6	PPEPET	KT3	
Protein C	12	EDQVDPRLIDGK	HPC4	
T7	11	MASMTGGQQMG	17.Tag	Without highly charged
V5	14	GKJPIPNPLLGLDST	V5	
VSV-G	6	MNRLGK	P5D4	

Approaches to epitope tagging—The two standard approaches to tagging a cloned gene are: (i) an epitope encoded oligonucleotide is inserted into the coding sequence or (ii) the coding sequence is inserted into an expression vector which already carries the epitope tag. Addition of the epitope coding sequence to the protein coding sequence is accomplished by oligonucleotide-mediated site directed mutagenesis of the corresponding gene. Alternatively, the additional amino acids can be introduced via the PCR. **Site-directed mutagenesis**-The gene to be altered is cloned into either an M13 vector or a yeast shuttle vector bearing a single stranded bacteriophage origin of replication. An oligonucleotide is synthesized that contains epitope coding sequence flanked by 15-17 nucleotides that are complementary to DNA on each side of the insertion site. The single stranded DNA is produced and the oligonucleotide is annealed to the single stranded DNA form of vector. The oligonucleotide serves as a primer for *in vitro* synthesis of the complementary DNA strand. Selection against the template strand allows efficient recovery of mutant. For example, the template strand can be made sensitive to nucleases either by incorporation of uracil during replication in a dut¯urg¯ strain 10 or by specific nicking *in vitro*.

Advantage of epitope tagging approach over traditional approach (antibody generated directly against protein of interest) — 1. Tagged protein can be monitored with a well characterized monoclonal antibody whose immunoprecipitation capabilities and spectrum of cross-reactivity with non-tagged proteins are already known. 2. Proteins that associate with a tagged protein (protein-protein interaction) can be identified in a manner not possible with antibodies generated against the protein of interest. Such associated proteins will be absent in the immunoprecipitates from extracts prepared from the cell lacking tagged protein. 3. The use of independently raised antibodies can not provide a comparable negative control. 4. The intracellular location of epitope tagged proteins can be identified in immunofluorescence experiments. 5. Epitope tagging approach may be particularly useful for discriminating among similar gene products (Kolodziej and Young, 1991). The tag does not add new biological activity to the protein and thus the epitope tagged protein retains to a large extent normal structure and function.

The limitation with epitope tagging approach is the potentially deleterious effect that the additional amino acids may have on protein function. However, the ability of the altered protein to function *in vivo* is readily resolved in yeast by testing complementation of a null allele by the modified gene.

Multi copy tag sequence — The sensitivity of epitope tagging can be increased through the use of tandem copies of the tag instead of just one. If any tag sequence can be reiterated up to 10 copies per tag then its detection in the corresponding protein becomes easier as it produces stronger signals, for example , on western blots. Larger tags are tolerated as in most cases they are in external loop or at termini where they do not greatly perturb the rest of protein.But then the number of multi copy tag has to be optimized otherwise it may interfere with the function of the tagged protein. **Multi epitope imaging**-It is a technique of whole cell protein sequencing. This technique is used for the simultaneous detection and imaging of many different proteins in a tissue section. This technique is based on sequential *in situ* interactions between fluorescence labeled specific antibodies and their epitopes. This technique further shows the locations of those proteins within a cell and their changes in different phases of life cycle or after environmental changes.

Epitope screening — It is a technique which is used to search expression library for specific protein (the antigen) using a monoclonal antibody raised against the antigen. In this technique the clones constituting the expression library are plated and different colonies are allowed to grow. The colonies are then transferred onto a nitrocellulose or nylon based filter membrane which is then treated with a solution containing the specific antibody. The colonies expressing the corresponding protein are then localized by antibody-antigen interaction and the antigen-antibody complex is finally visualized by fluorescence labeled anti-antibody. **Monoclonal antibody** refers to a population of immunoglobulins originating from one single clone of plasma cells and therefore consisting of structurally and functionally identical antibodies. In the organism the monoclonal antibodies are produced by tumor cells of the immune systems (myelomas). However, it is experimentally possible to fuse such myeloma cells with antibody producing plasma cells (spleen cell or B lymphocytes). These hybrid cells (**hybridomas**) proliferate into a clone which can be maintained in tissue culture and can be used as a source of monoclonal antibodies. The word 'Hybridoma' is generally used to describe continuously growing cell line which is a hybrid between a malignant cell and a normal cell. Myeloma cells would provide the immortality required for continuous culture. **Monoclonal antibodies** are important tools for the characterization and identification of proteins. If it is used in conjunction with a cDNA library cloned in an expression vector, it can be used to find the gene from which the protein was translated. The plant antibody refers to monoclonal antibody which is synthesized by transgenic plants. Genes for the heavy and light chain peptide of I_gG antibodies, have been transferred into separate tobacco mesophyll cells by Agrobacterium mediated gene transfer which were regenerated to complete plants expressing the foreign gene. Conventional crossing of these two transgenic plants leads to a plant harboring both gene encoding the k-light chain and the gene for the γ-heavy chain. The plant is able to synthesize a complete I_gG antibody. Immunoglobulin refers to group of molecules consisting of k or λ light chains and μ, δ, γ, ε or α heavy chains. All antibodies are immunoglobulins but not all immunoglobulins have known antibody properties. There are five classes of antibody such as IgM, IgD, IgG, IgA and IgE. The classification of antibody structure is based on the basis of the constant portion of the heavy chain. The epitope can also be detected by **polyclonal antibody**. Polyclonal antibody refers to an antibody (immunoglobulin) synthesized by different clones of B lymphocytes. In response to an antigen (any substance which elicits a specific immune response), an organism usually generates a heterogeneous mixture of specific antibodies with differing binding specificities and affinities. Thus every natural induced antiserum is polyclonal. For example, there are two types of lymphocytes: B cells and T cells. B cells express immunoglobulins at cell surface whereas T cells produces T Cell Receptor (TCR). Further, for some epitopes only one antibody is available but for others there are multiple choices including monoclonals and polyclonals. Antibodies recognize the epitope whether it is internal to the protein or at a terminus but in some cases recognition is dependent on localization, e.g. the tag is only recognized when it is at the extreme C terminus.

Epitope tagged proteins are detected using western blotting, immunoprecipitation, ELISA, immunofluorescence, immunoelectron microscopy, surface plasmon resonance, etc.

B. **Affinity tags**—One way of studying protein-protein interaction is to purify the entire multiprotein complex by affinity based methods and this can be achieved in a variety of ways such as GST-fusion proteins, antibodies, peptides, DNA, RNA or a small molecule binding specifically to a cellular target. Each protein is expressed in different amounts and has different properties. Thus to apply the same purification protocol across the range of proteins researchers have engineered proteins with generic tags that will bind to an affinity ligand. Widely used tags include a small peptide of histidine residues (6X His), a calmodulin-binding peptide, the streptavidin friendly biotin, the cellulose binding tag, the maltose-binding protein (Nus A) and glutathione S-transferase (GST). Although a variety of protein tags are available one common protein tag is a 6-His tag which comprises a string of six histidine residues. A gene contiguous with a His tag will produce protein which will have extra His sequences either at its carboxyl end or amino terminus. Histidine tags are incorporated into the protein coding gene by primers that carry six 5'-CAT-3' triplets. This tagged protein can be separated by chromatography on column with immobilized nickel to which the

His sequences will bind tightly. So 6-His fusion protein can be purified from other cellular proteins by affinity chromatography. But the problem with this method of protein purification and thus in turn study of gene function is that the additional nucleotide sequence can interfere with the function of gene or the additional amino acid residues in an epitope tag or Histine tag can affect protein activity.

C. **Domain tags**—Domain tags are independent units of structure and function which maintain or enhance certain activities of the proteins from which they were derived. Common examples include

(i) **Glutathionine-S transferase (GST)**—Making fusion proteins such as GST-fusions is another generic way to obtain interaction partners. In other words, interacting proteins can be isolated using GST tag. The multiprotein complex associates with the 'bait' which is immobilized on a solid support. After washing away the proteins that interact non-specifically, the protein complex is eluted, separated by gel electrophoresis and analyzed by MS. Thus components of an entire multiprotein complex can be identified. GST enables facile purification over glutathionine-agarose resins. The glutathione S-transferase (GST) gene fusion system consists of a vector which contains the carboxy terminus of the GST gene from *Schstosoma japonicum* driven by the tac promoter, a multiple cloning site and an antibiotic resistance marker (e.g. Amp^r). Any foreign DNA cloned into the polylinker is easily expressed as fusion protein containing GST as an affinity tag tail. This fusion protein can be isolated and purified on a glutathione Sepharose 4B affinity chromatography column. Compared to conventional approaches this method allows a rapid and sensitive assignment of catalytic function to ORFs and is generally applicable for detection of any type of activity. The disadvantages of this approach are that it requires several steps and that prevalent enzymes or activities such as kinases and phosphatases must be screened in smaller pools. The advantage with this technique is that proteins that are non-specifically bound to the matrix or the tag itself are not eluted. The eluted proteins are resolved by one or two-dimensional gel electrophoresis and compared with GST alone. The bands or spots corresponding to proteins specifically bound to the tagged proteins are excised and analyzed by mass spectrometry. S14 protein is immobilized onto agarose beads using a GST-tag. Nuclear cell extracts are incubated with the beads and the beads are washed extensively. Thrombin is used to cleave between the GST and the S14 protein which results in elution of all proteins that are specifically bound to S14.

(ii) **Green fluorescent protein (GFP)**—It is an autofluorescent protein which can be monitored in both quantitative and localization assay. When a target gene is fused with a gene for **green fluorescent protein**, this reporter construct generates a fusion protein which is highly fluorescent and thus can be observed under microscope. This technique thus allows to monitor the location of fusion protein and its movement in a cell. This strategy has been used to characterize multiprotein spliceosomal complex. Expression of a GFP-tagged version of a protein, SPF45 was identified from the gel. HeLa cells were transiently transfected with a plasmid encoding SPF45 which was tagged. The inserts from both the plasmids are then sequenced to obtain a pair of interacting genes.

    D. Cleavage sequences—Cleavage sequences are commonly introduced between fusion tags and the cloned genes so that tags can be removed following affinity purification. Cleavage elements are mostly proteinase recognition sequences although self-cleaving protein domains called **inteins** (discussed in chapter on protein) are becoming increasingly available.

26.4 YEAST-TWO-HYBRID (Y2H) SYSTEM

It is a genetic approach for detection of even relatively weak and transient protein-protein interactions *in vivo* and the identification of genes encoding interaction proteins based on dual modular composition of many eukaryotic transcriptional activators. It can measure interactions between two known proteins or polypeptides and identify domains and structural features of protein in these interactions. It can also search for unknown partners (prey) of a given protein (bait) and thus can isolate new proteins.

The Y2H is based on the principle that a protein domain that binds specifically to DNA sequences (BD) is fused to a polypeptide termed '**bait**' and a domain that recruits the transcription machinery (AD) is fused to a polypeptide termed '**prey**'. The basis of the assay is that transcription of a reporter gene will occur only if the bait and the prey polypeptides interact together. In this approach the detection of protein-protein interaction is based on the properties of the Gal4 protein (Gal4p) which activates transcription of genes (Gal1, Gal2, PGM2, Gal7, Gal10 and MEL1) scattered over different chromosomes of yeast. The metabolism of galactose in which galactose is converted to galactose-6-phosphate is regulated by genes, Gal3, Gal4 and Gal80 of which Gal 4 plays a central role. Gal4p has two domains-DNA binding domain, DBD (which binds to a specific DNA sequence) and activation domain,AD (which activates the RNA polymerase that synthesizes mRNA from an adjacent reporter gene). These two domains are stable when separated but activation of the RNA polymerase requires interaction between the two domains and the interaction is associated with the expression of the reporter gene. The two strains of yeast-strain 1 with Gal4p-binding domain and strain 2 with Gal4p-activating domain can be constructed. When the two strains are crossed and the mated mixture is plated on a medium which selects only the yeast with the expressed reporter gene then the surviving colonies of yeast can be said to have that interacting fusion protein pairs and the sequencing of the fusion proteins will reveal which proteins are interacting. In other words, once a putative interaction is detected, the ORF is identified simply by sequencing of the relevant clones. It is a generic method. The development of 2H method was based on the three findings: 1. modular nature of the transcription factors 2. phenomenon of 'recruitment' i.e. complexion of the DNA binding domain with a transcription activation domain and 3. that transcription could be activated by binary protein complexes in which DNA binding and activation domains resided on separate polypeptide chains. Fields and Song (1989) measured the interaction between two yeast proteins involved in regulating the SUC2 gene, Snf1 and Snf4 by expressing them as chimeras. One chimera contained the DBD of Gal4 at the amino terminus of Snf1 and the other contained the AD fromGal4 at the amino terminus of Snf4. Interaction between the two chimeric proteins brought the AD to a lacZ reporter containing Gal4 binding sites and was detected when β-galactosidase encoded by lacZ caused a colony to turn blue on X-Gal indicator plates (Figure 26.2). Here Snf1 and Snf4 are not transcription factors and thus this finding uncoupled the study of interaction from the requirement that either underivatised partner affect transcription. The use of transcription activation as a phenotype can be made to identify cDNA libraries protein that modifies transcription activity of the chimeric proteins.

DNA-binding domain (DBDs)—Protein sequences from databases have shown proteins with sequence motifs that bind specific sites on DNA. These proteins (and their DNA-binding domains) are often rotationally symmetric dimmers or tetramers. Gal4 and Gcn4 are dimmers. In eukaryotes, these typically bind to 10-to-20-bp sites (recognition sites) within the promoter or enhancer region of a gene. **Activation domains (ADs)**-Many transcription factors contain at least one AD. ADs are thought to function by interacting directly or indirectly through intermediary proteins with RNA Pol II or III associated proteins in the vicinity of transcription start site. Thus there is modular organization of eukaryotic gene regulatory proteins, i.e. transcription factors (or transcription activators) are frequently **modular** with at least two separable domains: one that binds to DNA (BD) and another that activates transcription (AD). Activation domain refers to a specific 30-100 amino acid domain of TFs located at the C-terminus and rich in acidic amino acid that can form amphipathic α-helical structures and is essential for the transcriptional activation of the target gene. For example, the yeast TF GAL4 harbors such an AD which can be discriminated into two regions. I-residues 148-236, II-768-881 residues, either of which activates transcription when fused to DBD (residue 1-147) -a specific domain of TFs located at the N-terminus that bind to specific sequences of DNA (upstream activating sequences, UAS). The activity of region I is proportional to its content of acidic residues. Principally three different features of ADs can be discriminated, an acidic, negatively charged domain (in GAL4, GCN4), a glutamine rich domain (eg., HAPI, HAPII, GAL11, OCT-1, OCT-2, JUN, AP-2, SRF, Sp1) and a proline rich domain (eg. CTF/NF-1, AP-2, JUN, OCT-2, SRF). All these regions establish contacts to other proteins.

Use of additional reporters—2H system is modified to use additional reporter whose expression is required for yeast growth. Colonies in which both reporters are active are characterized. 1. To show that activation of the reporters is due to the activation tagged protein (sometimes called the prey or fish protein) and not to a yeast mutation) and this is done by reintroducing the library plasmid into the original strain that expresses the bait and recapitulating activation of the reporters. 2. Library encoded protein interacts specifically with the protein of interest and not with, for example, the LexA or Gal4 moiety of the bait or with other unrelated bait proteins and this is tested by introducing the library plasmid into strains expressing other bait fusions and showing that cDNA expression results in activation of reporters only in strains containing the original bait.

Two-bait systems—In this two-reporter system transcription of selectable and counter selectable reporters is directed by different baits. Such a system is useful in isolating proteins that simultaneously contact the different baits (bridging proteins) and those that contact one bait but not the other (discriminatory).Because the different baits can be closely related and in fact can be allelic variants of the same protein, this system facilitates the isolation of proteins which interact with either a wild-type or a mutant type variant of a protein and of peptide aptamers which interact with either wild-type or with mutant allelic forms (White,

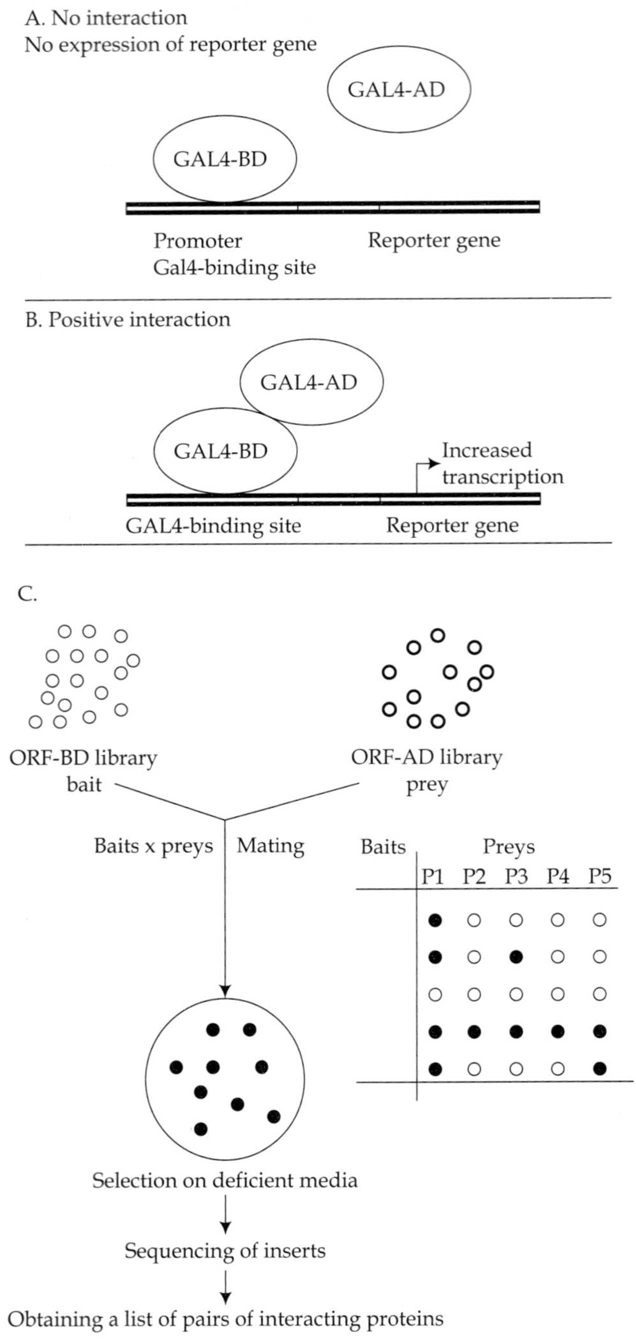

FIGURE 26.2   Showing yeast two-hybrid strategies. Figures A and B show principle of the yeast two-hybrid assay. Figure C shows the matrix approach uses the same collection of proteins used as bait and prey, for obtaining a list of interacting proteins. The matrix approach leads to the identification of autoactivators (B4 for example) and sticky prey proteins (P1 for example).

1996). Two-bait system like other systems in which a third protein is expressed, can detect interactions that depend on bridging or modification by a third protein. In this system different baits are bound to DNA upstream of different reporters. These systems have been used to identify proteins that interact with different domains of a protein (Snf1) (Jaing and Carlson (1996)), different alleles of a protein (Ras) (Xu et al., 1994) and identify mutant proteins that differentially bind to two known interactors of a wild-type protein (Ste5) (Inouye et al., 1997).

26.5 DIFFERENT 2H SYSTEMS

Different 2H systems differ in their components (reporters, expression vectors). There are three systems most common in use. All three systems use lacZ reporters drived from a GAL1-lacZ fusion that has the yeast GAL1 promoter lacking its upstream transcriptional control region and a small portion of the GAL1 coding region fused in –frame with lacZ. Then there is a second reporter gene in all three systems whose transcription restores a nutritional prototropy. Yeast strain bearing chromosomal HIS3 or LEU2 reporters with binding sites for either Gal4 or LexA. These selectable reporters allow detection of even weak interactions. **Bait/prey expression vectors**-These systems differ in their bait vectors. Baits are expressed from plasmids as moieties fused to C-terminal to either a LexA or Gal4 moiety. Both contain sequences that promote dimerization of the fused protein and thus binding to the reporters. Gal4 amino terminus contains a nuclear localization signal while LexA does not. Thus the nuclear concentration is lower for many LexA bait and thus many LexA fusions are not produced in sufficient amounts to efficiently occupy the reporters leading to loss of sensitivity. Finally, these systems differ in their prey systems. One used a relatively weak AD B42 encoded by *E. coli* (Gyuris et al. 1993) and another used Gal4 ADII derived from the C-terminus of protein (Elledge et al.) and the third used a more powerful AD derived from herpes simplex virus VP16 (Brent and Finley Jr, 1997).

Mammalian 2H system—Two different 2H systems are used in mammals. One system uses the yeast Gal4 DNA binding domain and VP16 activation domain and the reporter gene used is chloramphenicol acetyl transferase (CAT) and like yeast it relies on the reconstitution of active TF from two different chimeric proteins. Another system uses SV40 T antigen as reporter and replication of a plasmid that requires the large T antigen for replication as a readout.

26.6 OTHER PROTEIN INTERACTION METHODS

Besides yeast 2H methods other methods for studying protein interactions are: modified 2H methods, other reconstitution methods, fluorescent resonance energy transfer, protein mass spectrometry and evanescent wave methods (Mendelsohn and Brent, 1999). These methods reveal not only the partners of particular protein but also show how tightly the interacting protein touch one another and where and when those contacts occur in living cells.

26.7 OTHER TRANSCRIPTION-BASED INTERACTION DETECTION METHODS

1. **Use of a third protein**—Some protein-protein interactions can not be detected with standard 2H methods as they require a third molecule. The detection of binary interaction thus depends on one or more proteins. For example, affinity of two proteins that contact each other directly may be enhanced by the expression of a third protein which contacts both. Thus interaction between two proteins depends on a third protein. In other words, the two proteins do not make direct contact but interact via a third, bridging protein. The bridging proteins can either be stable or transient (particularly when interaction depends on post-translational modifications which are relatively uncommon in yeast).

2. **Use of small molecules**—The interaction between two proteins can depend on a third, non-protein, ligand such as hormone receptor (thyroid hormone receptor). An expression library can be screened for proteins or ligands enhance the interaction.

   Other reconstitution methods—Transcriptional activation is not the only biochemical activity which can be reconstituted by the protein-protein interaction.

1. Interacting proteins reconstitute guanine exchange factor (GEF) or Ras by bringing together the catalytic domain with the required membrane localized domain which then complements localization yeast carrying a temperature-sensitive mutation in yeast GEF. This system is suited for studying interaction between cytoplasmic and membrane-proximal proteins as well as proteins that activate transcription. This system can be employed only in appropriately engineered yeast cells and are prone to false positives.

2. **Protein reconstitution**—Interacting protein fragments can sometimes reconstitute a **split protein's** function. In this system a mutated NH2 terminal fragment of ubiquitin and a COOH-terminal fragment fused to a bait moiety(Johansson and Varsharsky, 1994), could upon interaction reconstitute ubiquitin whose cleavage from the COOH terminal fragment could be detected by protein immunoblotting. **Split-ubiquitin two hybrid system**- It is a variant of yeast two hybrid system which allows to identify interactions between integral membrane proteins and to screen for small molecules inhibiting such interactions. Ubiquitin is a ubiquitiously occurring phylogenetically highly conserved globular and compact 85kDa acidic eukaryotic protein of 76 amino acids which can be conjugated to a variety of cellular and nuclear proteins. Drastic increases in ubiquitin and ubiquitin-conjugated proteins are observed during a variety of human disorders such as Alzheimer's and Parkinson's diseases. In this technique the gene encoding one membrane protein is fused to a sequence coding for the N-terminal half of ubiquitin (Nub) and a separate fusion is made between the sequences encoding the C-terminal half of ubiquitin (Cub) and an other membrane protein. Both these constructs are transformed in yeast cells and expressed. When both membrane proteins interact, the two halves of the ubiquitin molecule come into close proximity to form a quasi-native ubiquitin. This system is therefore based on ubiquitin which is attached to the N-terminus of proteins to tag them for proteasomal degradation. Ubiquitin tagged proteins are recognized by ubiquitin specific proteases

(UBPs) resulting in cleavage and subsequent degradation of the attached proteins. Ubiquitin can be split into an N-terminal (Nub) and a C-terminal (Cub) half. The two parts retain an affinity for each other and spontaneously reassemble into the socalled split-ubiquitin which is recpognized by UBPs. These in turn cleave the peptide bound adjacent to the C-terminal amino acid residues of the Cub (i.e. glycine). Now if a reporter protein is fused to the C-terminus of the Cub then it will be cleaved upon assembly of the Nub and Cub regions. A point mutation in the N-terminal half of ubiquitin (NubG) completely abolishes the affinity of both halves for each other.

Stagljar et al. (1998) described a transcription based selectable version of this ubiquitin split protein sensor in which upon interaction ubiquitin reconstitution and cleavage, a transcription factor activator is released to activate a nuclear localized reporter.

3. **Enzyme reconstitution** — Interacting proteins can reconstitute enzymatic activity (**split enzyme's function**) and the reconstituted enzymatic activity is the scored phenotype. Interaction-mediated reassembly of enzymatic activity has been described for β-galactosidase (β-gal) in *E.coli*, DHFR (mouse dihydrofolate reductase) and CysA adenyl cyclase in *Bordetella pertusses*. The β-gal experiments bring together NH2 terminal and COOH-terminal fragments of a protein via fused interacting partners. Karimova et al. (1998) showed reconstitution of adenylyl cyclase in *E.coli*. cAMP produced by this enzyme activated CAP and in turn enabled transcription from the lac or mal operons.

26.8 APPROACHES FOR LARGE SCALE THROUGHPUT STUDIES

Our ultimate objective is to identify the interaction among a large number of proteins (even all of the proteins in a genome, i.e., complete set of interactions among the proteins of a cell). The different experimental methods available for this purpose can be grouped into two main categories: small-scale (low-throughput) and large-scale (high throughput). Given a set of proteins, small-scale techniques such as co-immunoprecipitation determine the interaction between one pair of proteins at a time whereas large-scale techniques such asY2H and TAP tagging allow identification of large number of interacting pairs simultaneously in a single experiment. The advantages with LTP experiments is that they allow much more precise identification of the interacting partners than HTP procedures which are more error prone.

Yeast two hybrid is the main large-scale technology to build protein interaction maps. Although genomic technologies have been developed that focus predominantly on the use of DNA array approaches to measure, for example, the expression of large sets of genes, the comparable strategies for protein analysis are not available. This is possible because haploids exist in yeast in two mating types that mate with one another and form diploids when they are physically juxtaposed. Plasmids encoding potentially interacting proteins can be introduced into the same nucleus. Thus the experiment starts with mating of haploid strains of opposite mating types containing the potential interactors and determination of the transcription phenotypes of the reporters in the resulting diploids. The following three approaches are used involving genomic two hybrid analysis.

1. **Matrix approach** (one vs one)-This approach is used to systematically test pairs of proteins for an interaction phenotype and a positive can indicate that these particular pairs of proteins interact. It thus uses a collection of predefined ORFs, usually full length proteins as both bait and prey for interaction assays. Combinations of bait and prey can be assessed individually or after pooling cells expressing different bait and prey proteins. The limitation with matrix approach is that it tests only predefined proteins.

2. **Array approach**-(one vs all)-It involves construction of an array of hybrid proteins-a protein array of activation-domain hybrids (array screens). In this approach the interactions of a single DB fusion protein are examined against a pool of AD fusion proteins and thus the whole proteome coverage can be achieved against a single DB fusion. In other words, yeast clones containing ORFs as fusions to DNA or ADs are arrayed into a grid and the ORFs to be tested as reciprocal fusions, are screened against the entire grid to identify interacting clones. In short, yeast cells can be transformed with individual ORF activation domain fusions. These cells can be grown in an array format on plates or filters such that each element of the array contains a yeast clone with a unique ORF. Such as array can be probed in mating assay with yeast cells containing a single ORF DNA binding domain fusion, one at a time. The nutritional selection ensures that only the yeast cell containing interacting partners survive. The interacting clones can be re-screened to reduce false positives or be sequenced directly.

3. **Library or pooling method** (all vs all)-In this strategy two different strains containing two different cDNA libraries are prepared. In one case the ORFs are expressed as GAL4-BD fusions and in other case they are expressed as GAL4-AD fusions. The two strains are then mated and diploids selected on deficient media where only the yeast cells expressing interacting proteins survive. Thus this approach involves yeast strain expressing different DBfusion mass mated with strains expressing AD hybrids. By this approach it is possible to test every protein in the organism against every other protein. In other words, one set of ORFs are first pooled to generate a library and then reciprocal ORF-fusions are mated with the library one by one or several at a time. This fragment (or polypeptide) library screening strategy (Finlat and Brent, 1994) uses exhaustive libraries to screen for the identification of new protein interacting partners. When applied to functionally related proteins, it results in connection of uncharacterized proteins to specific pathways. It can be applied to whole cellular interactomes. Screening numerous randomly generated fragments contained in the libraries also allow the determination of interacting domains defined experimentally as the common sequence shared by selected overlapping prey fragments (Rain et al., 2001)

It is also feasible to use large scale two-hybrid screens to explore interaction relevant to a specific pathway or biological process. The yeast activation domain array and library also allows the detection *in vivo* of nucleic acid-protein and small molecule-protein interactions through the use of hybrid molecules in a fashion similar to that for protein-protein interactions (Utez et al., 2000). For 2H array a set of yeast colonies was derived from about 6,000 individual transformants. The transformation process inserts one of the yeast ORFs into a Gal4 transcription activation domain vector to create a hybrid protein. To enable quick, large scale transformation, those ORFs were generated as a set of PCR products with 70bp sequences at their 5′ and 3′ ends that precisely matched sequences in the activation domain vector pOAD. These sequences allowed highly efficient recombination between the ORFs and the linearized vector. After transformation of a yeast 2H reporter strain, the colonies from each transformation plate were pooled to constitute a single array element with entire array contained on sixteen microarray plates of 384 colonies. Further, a set 192 DNA binding domain hybrids were simultaneously generated by transformation into a strain of opposite mating type of PCR products with a Gal4 DNA binding domain vector. To screen for protein interaction a transformant containing one of the DNA binding domain hybrids is mated to all of the transformants of the array and selection of diploids was made using markers carried on the 2hybrid plasmid. The diploids were then transferred to selective plates deficient in histidine and positive colonies for the 2hybrid reporter HIS3 gene were identified by their positions on the array.

The high throughput screens of an activation domain library was made by pooling transformants containing the roughly 6,000 potential ORFs fused to the Gal4 activation domain. The hybrid proteins were derived similarly to those in the 2hybrid array with each of the potential ORFs cloned separately into a GAL4 DNA binding domain vector in addition to the Gal4 activation domain vector. The same PCR products and recipient plasmids were used to generate collections of transformants, each consisting of 64bar coded 96-well plates. Transformants from the Gal4 activation domain collection were then pooled to form activation domain library. To screen for protein interaction, each DNA binding domain hybrid transformant was mated in duplicate to the activation domain library. Mating mixes were transferred to selective plates to select diploid cell that expressed interacting pairs and activated both reporter genes (URA3 and lacZ).

Data collection with Y2H—The two-hybrid strategy is shown in Figure 26.2. The matrix approach (first column) uses the same collection of proteins as bait (B1-B5) and prey (P1-P5).The results can be presented in a matrix where **bait autoactivators** (for example, B4) and **sticky prey protein** (P1, for example) interact with many proteins are identified and discarded. An autoactivator bait can be used in the screening process with more stringent selective conditions. Sticky prey proteins are identified as fragments of protein which are often selected regardless of the bait protein (Wojcik and Hamburger, 2003). The final result can be summarized as a list of interactions which can be heterodimer (B2-P3) or homodimer (B5-P5). The library screening approach identifies for each interacting prey protein the domain of interaction with a given bait.

Data processing—Data requires post-processing when a partner is screened against a library and has selected a target/prey. The prey must be sequenced and identified using BLAST. In case of Y2H using fragment libraries, the functionally interacting domains can be precisely mapped on proteins. The common sequence shared by the selected overlapping prey fragments defines the smallest docking site selected by the bait.

The main advantage of these methods is that they can be performed with a high throughput and in an automated fashion. High throughput methods performed in an array format have enabled large scale projects for the characterization, protein-protein interactions and the biochemical analysis of protein function. These studies provide only potential interactions that have to be confirmed or eliminated by further experimentation. Systematic mapping of P-P interactions using Y2H system using low levels of expression of both Gal 4 DNA-binding domain (DB) and Gal4 activation domain (AD), hybrid proteins (DB-X and AD-Y or DB-ORF and AD-ORF), three different Y2H inducible reporter gene and a plasmid-shuffling counter selection to eliminate systematically *de novo* autoactivators has been reported (Ruial et al., 2005).

Confirmation of validity of an interaction—There are different techniques for confirming the validity of an interaction. Confirmation of validity of an interaction can also come from the 2H experiments themselves. The proposed interaction should also be observed by a different technique such as co-immunoprecipitation of the putative interactors from the appropriate cell or tissues. But co-immunoprecipitation experiments are conducted *in vitro* experiments and thus are subjected to artifacts. The strongest such criterion is suppression of an interaction defect. It can also come from demonstration that interaction is specific and makes biological sense. Another validity of an interaction is affinity. Interaction affinity, normally is quantified as the equilibrium dissociation constant or Kd which can be crudely estimated from interaction phenotypes based on stoichiometry of the interacting components.

Many important and specific interactions from enzyme-substrate interactions to cooperative interactions between gene regulatory proteins are weak and that correlation between the strength of an interaction and its significance is thus imperfect at least.

26.9 INTERACTOME ANALYSIS

It refers to the study of interactions between the biological molecules on a global scale. All proteins interact with other proteins and thus form regulated complexes and pathways in cells. There are two fundamental principles that can account for how did interacting components that contribute to a single function evolve (Kuriyan and Eisenberg, 2007). The interactions between proteins increase with their concentration and the localization of proteins in a cell markedly increases effective concentrations. Therefore, the effect of a mutation which results in two proteins interacting is increased by colocalization of these proteins and thus can lead to the formation of protein networks and to the regulation of protein function by allosteric mechanisms.

For studying how protein-protein interaction networks relate to multicellular functions, one has to map a large fraction of the interactome network. High throughput mapping of protein interactions allow global survey of protein interaction of organisms. In other words, for understanding biological processes it is important to consider protein functions in the context of complex molecular networks and the study of such networks requires the availability of proteome-wide protein-protein interaction or interactome maps or protein networks. The **protein-protein interaction map** refers to the construction of the complex network of interactions between (preferably all) proteins of a given cell at a given point of time. Biological pathways and networks are built upon the identification of protein interactions. Traditionally information about protein-protein interactions is collected from small scale screening and the accuracy of each interaction is validated with multiple experiments. The interactome network can be studied with high throughput yeast two hybrid screens and protein chip technology. Since protein networks reflect the functional grouping of these interacting or co-ordinalely induced/suppressed genes, the roles of the subunits of co-expressed genes may be resolved using combined data. This may be done by evaluating the topological features of the sets of genes identified by microarray experiments. Topological features of the protein networks have been demonstrated to reflect the functionality of the interacting genes. For example, essential genes in yeast tend to be well connected and globally centered in the protein network. Centrality of genes is associated with the essential functions of the genes. Essential genes are those when mutated are lethal, tend to be well connected. Genes that are not essential but provide vital function in toxin metabolism also have high number of edges associated with the nodes albeit less well connected than that of essential genes (Wachi and Wu, 2005). Furthermore, globally centered interactions are more likely to be well conserved and serve as an evolutionary backbone for the network. The upregulated genes are rich in global hubs. Global hubs are well connected nodes that are also located near the center of the protein network. The same claim is lacking in the down regulated genes. Topological and biological features of this interactome networks as well as the integration with phenome and transcriptome data sets will lead to numerous biological hypotheses. The yeast, *Saccharis cerevisiae* has been used to develop a eukaryotic unicellular interactome map. The current version of Worm interactome (WI5) map contains ~5500 interactions. Figure 26.3 shows the procedure for construction of protein-protein interaction map.

Y2H is widely used in protein-protein physical interaction assays (Fields and Song, 1989). The Y2H system des not demand any protein purification, isolation or manipulation. The proteins to be tested are expressed by the yeast cells and a result is easily seen by *in vivo* reporter gene assay. 2hybrid methods are suited to dissect the genetic pathway and their extensions will give insight into pathways that have not yet been discovered. Proteome signature refers to the specific patterns of the various subproteomes of a cell at a specific time and a particular developmental stage as detected by 2-D PAGE of the individual subproteome and subsequent staining of the proteome spots. **Proteome signatures** (or more precisely characteristic indicator proteins) are characteristic for a specific condition of the cell (e.g. after stress, during mitotic change, after a knock-out of a gene).

Drawbacks with Y2H

The drawbacks with the two-hybrid system are as follows (Legrain et al., 2001; Zu et al., 2003).

a. They are not comprehensive. Most well characterized proteins interact with 5-7 other proteins. Thus the estimates of the number of interactions expected in yeast are around 30,000-40,000, significantly higher than 4500 interactions identified so far.

b. They identify a large number of false positives. The HTP technology generates false positives or false negatives depending on assay conditions. False negative interactions are biological interactions which are missed because of incorrect folding, inadequate subcellular localization, lack of specific post-translational modifications, etc. The matrix approach is prone to generate a high level of false negatives because only two assays are performed for each interaction (bait vs prey and reciprocal) whereas the fragment library approach allows testing of millions of potential interactions simultaneously. Searching for potential interactions in case of screening a random

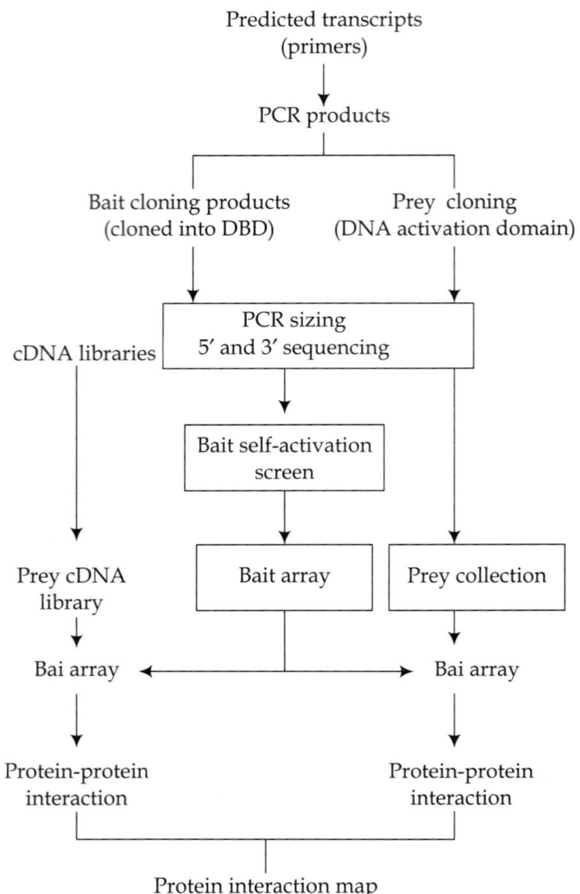

FIGURE 26.3 **Construction of protein-protein interaction map (after Giot et al., 2003).**

fragment library increases the probability of selecting non-significant interacting polypeptides because of the following reasons. Some bait proteins might have a predisposition to activate the transcription of reporter genes without specific interaction with any prey proteins. Further, some chimeric prey protein termed **sticky proteins** may similarly be non-specificly selected by many independent bait proteins. Thus discarding **autoactivators** or sticky proteins will lead to reduction in the rate of false positives although it may also mean a slightly increased number of false negatives.

c. Interaction occurs in the nucleus and uses transcriptional readout. Consequently, the interactions of many membrane proteins and transcription factors can not be measured.

Two-hybrid approach is especially powerful when analyzing smaller genomes because most of the genes are easily known from the genome sequence and because the method has the potential to uncover all the possible combinations quickly and relatively easily. More than 4000 interactions have been identified from high-throughput-yeast two hybrid screens (Li et al., 2004).

Reverse yeast two-hybrid—It is a variant of the standard two-hybrid system, designed for the detection of protein-protein interaction (s) in *S.cerevisiae*. It is based on negative selection, i.e. expression of interacting hybrid proteins increases the expression of a counter selectable marker, toxic under particular conditions. Under these conditions any failure of the proteins to interact provides a selective advantage (for example, survival). Therefore, mutants in which protein-protein interaction is not possible, can be identified from non-growing yeast colonies in which interaction occurs. Similarly, compounds preventing the interactions can be selected from a large collection in the same way. Interruption of a specific protein interaction is an expected consequence of many missense loss-of-function mutations. The most widely used toxic reporter gene is URA3 encoding orotidine-5′-phosphate decarboxylase (part of uracil biosynthetic pathway). It can be used for identification of compounds and peptides that disrupt protein-protein interactions. It is used for developing drugs that have activities *in vivo* as opposed to drug screens that are conventionally done *in vitro*.

The two-hybrid systems that can identify **disruption of protein interactions** (reverse 2H-system) have been implemented in the following three ways.

I. The positive selectable reporter is replaced by a counterselectable reporter such as UAR3, LYS2 or CYH2. Using this technique Vidal et al. (1996) isolated point mutants of the positive cell cycle regulator E2F that do not interact with DP1.

II. Two reporters are coupled such that interacting proteins induce expression of DNA binding protein, Tn10 tetracycline repressor (TetR) which then represses transcription of a second TetRop-HIS3 gene so that only cells in which the original protein-interaction is disrupted, grows in the absence of histidine (Shih et al., 1996).

III. Use of peptide aptamers that potentially disrupt protein interactions in cells engineered to express two interacting proteins at lower concentrations and identification of interactions by diminution of the positive signals (Geyer et al., 1999)

The reverse systems are useful in allowing selection of mutations in one or the other hybrid protein that abolish interactions. They may also be useful in identifying reagents that disrupt specific interactions.

1. **Interaction trap**—It is an improved version of the two-hybrid system for the *in vivo* detection of interactions between proteins. In this technique a gene coding a known protein ('**bait**') is cloned into the so called bait vector and expressed as a fusion protein to a DNA-binding protein (e.g Lex A from *E. coli*). The gene encoding the potentially interacting protein ('**prey**') or a cDNA clone acting as source for interacting proteins is cloned into a second vector called prey vector and expressed as a fusion protein to a transcription activation domain. Both vectors are transformed into a yeast, S. cerevisiae host strain (**reporter strain**). The reporter strain contains one or two copies of the target sequence for the DNA binding protein (in this case Lex A-binding site) at upstream of a reporter gene (lacZ or His3) which is integrated into its genome. When the bait and the prey proteins interact the protein complex formed will activate the transcription of the reporter gene. In case of reporter gene, His3 the interaction is detected by the growth of the reporter strain in the absence of histidine whereas in

case of reporter gene, lacZ the interaction is detected by β-galactosidase expression. Thus the DNA fragment (gene) encoding the interactive protein can be isolated and further characterized. **Three hybrid system**-The three hybrid system detects interaction (s) in ternary protein complex. It is based on conventional transcriptional activation domain fusion vector and a modified DNA –binding domain fusion vector with the two separate cloning sites, one for the prey-encoding gene, the other for the third protein. The third protein is conditionally expressed from a methionine promoter (active only in the absence of amino acid). The three hybrid system is fully dependent on the function of the third protein which either links the two other, normally not reactive proteins or stabilizes a possible weak interaction between both proteins, enzymatically modifies one or both proteins as , e.g., phosphokinase or inhibits protein-protein interaction (s). The three hybrid system then also allows to identify protein (s) that inhibit protein-protein interaction. Prey-Prey is a part of hybrid protein component of yeastTH system encoded by a hybrid gene consisting of fusion of a cDNA or genomic DNA fragments (the prey *per se* whose interaction with the so called bait has to be tested). If the prey and bait interacts the activation domain of the prey construct comes into close proximity with the DNA –binding site which induces transcription of a reporter gene. The bait refers to a part of hybrid protein component yeast TH system encoded by a hybrid gene consisting of fusion of the coding region of a DNA-binding protein and a DNA segment encoding the bait *per se* that interacts with the so-called prey.

Advantages and disadvantages of *in vitro* and *in vivo* techniques—*In vitro* techniques permit control over ionic concentration, temperature and other factors. Interaction probed by YR2H technique which must occur in nucleus is unlikely to recover only significant partners and false-positive results might be obtained if the interacting proteins are localized to different cell compartments, express at different times or otherwise inaccessible to binding under normal conditions. In the *in vitro* techniques such as protein chip assays results can be visualized for individual interacting partners. Advantage with *in vivo* techniques is that a near-native interaction can be probed provided that tagging and bait expression do not interfere with the replication of endogenous levels of protein activity-proper folding, post-translational modification and accessibility of biologically relevant partners are generally assumed still. But abundance of proteins and solutes in the cell yield misleading results. In the pooling techniques over or underrepresentation in bait/prey pools can influence results and positives must be identified by sequencing or bar code analyses.

Advantages of *in vivo* pull down techniques—The advantages include the relative ease of analyzing complete complexes and the use of native, processed and post-translationally modified proteins as a reagent to target potential interactions in the natural environments and at normal abundance levels. If suitable antibody exists to the native protein then endogenous proteins can be assayed. Since sufficient antibodies do not exist to attack unmodified proteins with requisite specificity and affinity then more general techniques such as tagging are used for large scale assay. Generic tagging involves addition of a sequence onto the gene of interest, encoding a tag which is recognized by a convenient antibody.

26.10 LARGE–SCALE APPROACHES TO ASSIGN FUNCTION TO GENE PRODUCTS

Several large scale approaches to assign cellular function to new gene products include the following.

1. Monitoring of mRNA expression (chips and SAGE)
2. Loss of function approaches combined with subcellular localization of screens
3. RNA-mediated interference
4. Gene trap in mice
5. Computer *in silico* methods (protein fusions, gene neighboring, structural predictions)
6. Extensive two-hybrid screens
7. Protein chip analysis.

26.11 TANDEM AFFINITY PURIFICATION (TAP) SYSTEM

It is a recent tagging strategy which facilitates recovery of highly pure protein preparations. It is a technique for quantitation of proteins in crude cellular extracts or in purified samples or in extreme cases in a whole proteome. In this technique, a sequence encoding two protein A modules is inserted into the target gene that it tags the C-terminals of the encoded protein. If this gene is expressed, i.e. the encoded protein is synthesized then its TAP tag can be bound to the immunoglobulin G with high affinity. Usually the cell extract is transferred to a polyvinylidine difluoride membrane which is washed and blocked (to minimize unspecific background). Then the target is detected by a one-step immunoaffinity procedure with an anti-peroxidase IgG conjugate and chemiluminescence. TAP and MS in a large-scale approach is used to characterize multiprotein complexes. This system consists of calmodulin-binding domain and the protein-A I_g-binding domain separated by the tobacco etch virus (TEV) protease target sequence (Rignaut et al., 1999). The bait protein is recovered with an immunoglobulin-bound solid support and after washing, it is released from the support by protease cleaving. After initial purification, the recovered sample is passed over a calmodulin column, pending elution with ethylene glycol-bis (β-aminoethyl ether)-N, N, N, N′-tetra acetic acid or other Ca^{2+} chelators. This two-stage purification ensures low background noise and corresponding high sample purity but then it risk loosing weak interacting partners or complex components due to harsh purification procedures. After the bait/interactor complexes is purified, components of the complex can be identified by MS. MALADI-TOF, ESI and tandem MS/MS have enhanced the accuracy while permitting ionization (and therefore characterization) of larger biomolecules. In general MS proteomics experiments comprise five stages (Aebersold and Mann, 2003). The first three stages involve purification typically ending in one-dimensional gel electrophoresis, tryptic digestion to generate short peptides and HPLC separation of the tryptic digest. The final two stages are tandem MS assays. The high accuracy of MS spectra, combined with knowledge of the genomic sequence in question allows rapid and accurate identification of the proteins involved in the recovered complex.

26.12 AFFINITY TAGGING AND MASS SPECTROSCOPY

Affinity purification is used to identify protein-protein interactions or validate their presence. It is used in a large-scale approach to characterize multiprotein complexes in S. cerevisiae. Eukaryotic proteome can be studies as a network of protein complexes at the level of organization beyond binary interactions. The endogenous protein coding genes can be fused to the coding sequences of a tandem affinity purification (TAP) tag and the resulting tagged proteins are then purified along with their associated partners and subsequently separated by gel electrophoresis. The co-purifying proteins are identified by MS. Of the various high throughput experimental methods used to identify protein-protein interactions, TAP of affinity-tagged proteins expressed from their natural chromosomal locations followed by mass spectrometry has provided the best coverage and accuracy. In another approach the inducible FLAG epitope-tagged fusions are constructed, proteins are overexpressed and purified along with their associated proteins. As the overexpressed proteins are likely to interact with more proteins and thus they may also associate with proteins with which they do not normally bind, they yield false positives. A FLAG epitope is an epitope which contains within it a recognition site for protease enterokinase. In the absence of genome-wide screens where many complexes are retrieved repeatedly through a 'reverse purification' process, assignment of a component to a particular complex relied heavily on experimental strategy and arbitrary thresholds.

26.13 ANALYSIS OF BIOCHEMICAL ACTIVITIES

Information about protein function can be derived from the analyses of biochemical activities. Major task in analysis of biochemical activities is purification and identification of polypeptide or polypeptides responsible for the activity. Purification is a prerequisite for cloning of genes and subsequent biochemical and genetic study. Alternative to purification is expression cloning-the introduction of cDNA pools into various host cells followed by screening for activity and identifying the responsible DNA. Expression cloning is limited to proteins that are easily detectable in the background of host cell proteins.

26.14 GENOMIC METHOD FOR IDENTIFYING YEAST GENES ENCODING BIOCHEMICAL ACTIVITIES

The biochemical genomics strategy has advantages over conventional purification or expression cloning as it is quick and sensitive. Identification of an ORF associated activity takes less time (days) whereas purification or expression cloning takes months or years. Further, activities can be detected in the purified GST-ORF pools but it simply can not be detected in extracts or cells in case of purification or expression cloning.

GST-ORF are individual expressed at higher levels and are largely free of extract proteins after purification. Activities can be measured for hours without competing activities that destroy the substrate, the product or the enzymes. The steps involved in this strategy are as follows (Martzen et al., 1999).

1. Construct an array of yeast strains (6144) with each carrying a plasmid expressing a different GST-ORF fusion under control of Pcup1 promoter. In other words, each strain contains a different yeast ORF fused to GST.

2. To identify genes encoding a particular biochemical activities, strains are grown in defined pools (64 pools of 96 fusion each) and GST-ORFs are purified.

3. Assay the pools for a particular activity and active pools are deconvoluted to identify the strain and ORF responsible for the activity.

4. The ORF-associated activities included a cyclic phosphodiesterase, an Appr-1'-p processing activity and a cytochrome c methyl transferase.

In principle, the biochemical activities of proteins can be systematically probed by producing proteins in a high throughput manner and analyzing the functions of hundreds or thousands of protein samples in parallel using protein microarrays. Major limitations in screening an entire proteome array have been the ability to generate the necessary expression clones and also the expression and purification of proteins in a high throughput fashion. Zhu et al. (2001) constructed a yeast proteome microarray containing ~80% yeast proteins and screened it for a number of biochemical activities. First, they constructed a collection of 5800 yeast ORFs cloned into a high copy number expression vector using recombination cloning. The yeast proteins are fused to glutathionine S transferase polyhistidine (GST-Hisx6) at their NH2-termini and expressed in yeast using inducible GAL1 promoter. The yeast expression strains contain individual plasmids in which the correct ORFs have been shown to be fused in frame to GST by DNA sequencing. The proteins were expressed in yeast to help ensure that the proteins were modified and folded properly. Using a 96-well format, 1152 samples were purified at once from yeast extracts using glutathionine-agarose beads. 0.1% Triton was added to lysis buffer and washed to ensure that the purified proteins were free from lipids. In addition to conventional use arrays could be used in two other ways: 1.To determine the range of potential substrate proteins for any protein-modifying enzyme (such as a protein kinase) before genetic or biochemical tests to establish authentic substrate 2. To identify genes encoding proteins that bind any particular macromolecule, ligand or drug.

In case of using proteome chip, they printed 6566 preparations representing 5800 different yeast proteins in duplicate onto glass slide using a microarrayer. Initially they used aldehyde-treated microscopic slides in which fusion proteins attach to the surface through primary amines at their NH2-termini or other residues of the protein. But in the subsequent experiments proteins were spotted onto nickel coated slides in which the fusion proteins attach via their Hisx6 tags and presumably uniformly orient away from the surface. This proteome chip yielded superior signals. The proteome chips were tested by probing for several protein-protein interactions and protein-lipid interactions. In case of testing for protein-protein interactions the yeast proteome was probed with biotinylated calmodulin in the presence of Ca. Calmodulin is a highly conserved calcium binding protein involved in many calcium regulated cellular processes. The bound biotinylated protein was detected using Cy3 labeled streptavidin.

26.15   PROTEIN MICROARRAYS

It can be used to globally analyze the activities of proteins including their binding to proteins, nucleic acids, lipids, carbohydrates and small molecules, antibodies and drugs. It can also be used to identify *in vivo* post-translational modifications such as glycosylation or phosphorylation using lectins or antibodies. In this technique sets of proteins of interest or an entire proteome is overexpressed, purified, distributed in an addressable array format and then assayed. There are several types of protein array formats:

1. Nanowells
2. Solid surface supports such as glass slides
3. Hydrogels (the thick absorbent surfaces).

Nanowell arrays contain wells of 1mm or less in diameter and are made of a plastic such as polydimethylsiloxane by using a mold or alternatively wells can be etched in glass. The substrates are first immobilized on the surface of individual nanowells and individual proteins are incubated with the substrates. Finally, the nanowell chips are analyzed using a Phosphoimager for identifying protein-substrate interactions. Mac Beath and Schreiber (2000) used microarray technology and robotics to spot nanoliter volumes of protein onto aldehyde coated glass sides. The abundance of lysine residues in most proteins, combined with a reactive N-terminal amine, allows protein to become covalently linked to the slide surface in a number of possible orientations.

The more common functional protein microarray format is the glass microscope slides. Proteins are attached to the surface through either direct covalent methods, linkers, or affinity tags and the bound proteins are then assayed for binding or enzymatic activities. This technique requires to produce all proteins in functional forms which further requires high quality and comprehensive expression libraries and methods that yield a large number of proteins. This has been made possible by the development of **proteome chip**. This microarray is composed of more than 5800 individually cloned, overexpressed and purified proteins. The proteome chip is then probed using binding proteins. The interacting proteins are then identified and their addresses on the array known using a laser scanner. In other words, fabricate microarrays of the purified domains which are then be queried with each fluorescent peptides. In order to reduce experimental variation and to expedite the processing of hundreds of arrays, produce protein microarrays in microtitre plates. Samples are printed in duplicate and a small amount of cyanine 5-labelled albumin is introduced into each sample to facilitate image analysis. Initially the proteome chip is used to identify protein-protein interactions by screening for binding targets of interaction protein but then it can be used to identify the binding targets of secondary messengers, i.e. the chip is probed with proteins incorporated in liposomes. Proteome microarray can be used to screen protein-drug interactions and detect post-translational modifications. Zhu et al. (2000) created the first whole-proteome chip-a glass slide. Gene products were attached to nickel-coated glass slides via 6His tags. It contained over 80% of known yeast ORF gene products. Protein-protein studies can be carried out by probing biotinylated calmodulin in the presence of calcium. Calmodulin binding partners were visualized by probing with Cy3 labeled streptavidin. This demonstrated that biotinylated constructs of virtually any protein could be used to probe the proteome chip thereby visualizing protein-protein interactions. They uncovered several known calmodulin interactors and found significant number of novel potential interaction partners, many of which shared a motif thought to be important for calmodulin binding.

26.16 EVANESCENT WAVE METHODS-SURFACE PLASMON RESONANCE

Protein microarrays can also be used to analyze the kinetics of protein-protein interactions via real-time detection methods. **Surface plasmon resonance** (Schuck, 1997) is a versatile detection tool to analyze the kinetics of protein-ligand interactions over a wide range of weights, affinities and binding sites in 'surface plasmon resonance' devices in which there is a thin layer of metal at the interface between the glass and the aqueous medium. **Ligands** refers to proteins, peptides, antibodies, antigens, allergens and small molecules which can be immobilized in high density on modified surfaces to form functional and analytical protein microarrays. A planar wave guide has been used as the detection method to develop an antibody array biosensor for studying the kinetics of antigen-antibody interactions. **SPR** is an electron charge density phenomenon (excitation) which occurs at the boundary between a metal (gold, silver) and a dielectric medium (water, air) and permits to detect interactions (binding or dissociation) between a ligand (e.g. an antigen) and a biomolecule (e.g.an antibody) and to measure real-time reaction kinetics between two or more molecules. If a layer of bait protein is bound near the metallic surface and a solution containing an interacting protein flows past this surface, the resonance angle changes as the protein in solution binds the bait. The rate of increase in mass is the association rate. When a solution without interacting protein flows past, the mass at the surface gradually decreases as the interacting protein dissociates and is washed away. The rate of that decrease is the dissociation rate. These rates give the interaction affinity. This technique makes use of a 50nm thin metal layer (e.g. gold) in which surface plasmons, called plasmonic surface polaritons (localized regions of oscillating electrons) can be excited by impinging light from a monochromatic laser. At a defined wave length and angle of incidence of light and favorable refractive index of the material, the light energy is absorbed by the plasmons and therefore, no light is reflected from the surface. There exists an angle of incident light (a resonance angle) at which some of the energy in the evanescent wave is dissipated into the electron cloud (palsmon) in the metal. This angle depends on the local refractive index which in turn depends on the mass of proteins bound to the interface. A photodetector monitoring the intensity of the reflected light will therefore show minimal reflection density. The sensor surface is coated with either a hydrophilic polymer modified with carboxyl or amino groups or polystyrene into which the ligands are embedded by covalent coupling to the support , usually through a flexible spacer. When the analyte (e.g. an antibody) is added it interacts with the immobilized ligand (e.g. an antigen) and changes the refractive index at the gold surface and thus results in shifting the angle of intensity minimum. Such optical biosensors do not require labeled analytes. **SPR biosensor**-It is a **biochip** used for the real-time detection of interaction (s) between two unlabeled molecules (DNA and protein) based on SPR. One interacting partner is bound to the chip by covalent bonds, adsorption or spacer molecules. The other interacting partners which may be single molecule species but also include crude cell extracts, supernatants, or column fractions, are then applied to the sensor surface. SPR is used to detect minimal mass change as a consequence of binding or subsequent dissociation resulting in sensorgram. After regeneration of the chip by removal of ligands, it may be used for another round of SPR measurements.

SPR biosensors can be applied for DNA-DNA, DNA-RNA, DNA-protein, RNA-protein, protein-protein and DNA-, RNA-or protein-ligand interaction (s).

26.17 PEPTIDE MICROARRAY/PEPTIDE CHIP

It is a variant of protein microarray which can be used as substrates for enzymatic activities and as ligands when probing with proteins or other molecules. In this technique a part of a polypeptide chain is immobilized on the glass surface. Immobilized 9-mer peptide (thought to be important for the phosphorylation of the enzyme, tyrosine kinase) substrates arrayed at high density on a gold-coated glass surface was used to characterize the substrate specificity of this enzyme using SPR. Peptide microarray harboring > 1000 peptide species was used in the study of substrate specificity of a human protein kinase Nek6. Thus consensus sequences determined *in vitro* can be used as a means of identifying physiologically relevant substrates for any enzyme. Finally, **carbohydrate microarrays** have now recently been introduced. Wang et al. used carbohydrate-based microarray for analyzing the different types of anticarbohydrate antibodies in human and mammalian sera.

A summary of the different technologies for interaction proteomics is given in the table 26.2 (Zhu et al., 2003). All these approach involve over-expressing the protein coding genes and screening the expressed protein for biochemical activities

26.18 GENETIC SUPPRESSION VS 2H METHODS

Given a mutation in a gene, one could then isolate unlinked mutations which suppress the effect of the first mutation. From these observations one could infer that these mutations might lies in proteins that functioned in the same pathway as the first gene. For example, gain-of-function mutations that suppressed recessive mutations might function downstream in the same pathway while the recessive mutations that suppressed other recessive mutations might define genes whose products touched one another. In both cases identification of second-site suppressor identified genes that were conceptually linked. There is analogy between 2H experiments and suppressor genetics but linkage is physical rather than conceptual. Although the fact that proteins touch one another, does not mean that they function in the same process, proteins that touch one another frequently do. Thus like suppressor analysis interactor analysis can identify members of genetic pathways. Further, like suppressor analysis, positions and identities of nodes in those pathways would sometimes give clues to the function of pathway members and the pathway as a whole. But unlike suppressor analysis, it can not reveal causal linkage, only strong physical one, so proteins that acts through third proteins or small molecule intermediates or that interact with one another more loosely than detection threshold, are not revealed. Further, unlike reciprocal temperature shift experiments on partners identified by suppressor analysis, 2H experiments can not reveal the order with which components act.

Table 26.2 Showing various approaches for proteome analysis, their application and advantage.

Approach	Application	Advantage	Disadvantages
Yeast two-hybrid	Protein-protein interactions, protein-DNA interactions	High throughput and systematic to reveal protein interaction	No control over interaction conditions. Interactions are usually in the nucleus
Affinity tagging/MS	Dissecting protein complexes	*In vivo* interactions involving multiple partners	May miss weak or transient interactions, difficult to detect false positives
Antibody array	Protein profiling, protein detection, clinical diagnostics	Very sensitive and low amount of sample, potential in biomarker and drug development	Use restricted by the number and quality of available antibodies. Semi-quantitative protein detection
Functional protein array lipid, protein-small molecule, enzyme-substrateinteractions as well as drug discovery and posttranslational modifications	Protein-protein, protein-biochemical activities of proteins and high throughput drug and drug drug discovery and target discovery	Potentials for analyzing	*In vitro* assays
Peptide array	Enzyme-substrate interaction and drug discovery	Sensitive and straight forward way to identify epitopes	Expensive, *in vitro* assays
Carbohydrate array	Carbohydrate-mediated molecular recognition and infection response	Sensitive way to study carbohydrate-mediated molecular events	*In vitro* assays, difficult to obtain carbohydrate anti- in pure forms
Small molecule array	Protein-small molecule interaction, drug discovery, enzyme specificity profiling	Requires minimum amount of sample and high sensitivity.	*In vitro* assays, requiring to improve throughput to cover 10^6 molecules in a normal combinatorial library.

26.19 QUANTIFICATION OF THE AFFINITIES OF MOLECULAR INTERACTIONS

Quantifying the affinities of molecular interactions is a technical challenge. First, any particular molecular interaction is determined by a large number of variables and so obtaining equilibrium dissociation constants requires one to perform dozens of assays and as the concentrations of various components are systematically varied, increasing the number of measurements needed. Secondly, many molecular interactions are transient and show nanomolar to micromolar affinities and thus lead to rapid loss of bound material or little bound material in the first place. These factors are much more problematic for high throughput methods such as Y2H and TAP-MS where transient interactions are frequently missed. Protein-protein and protein-DNA binding microarrays are especially susceptible because of stringent wash requirements, causing rapid loss of weakly bound material. But as network topologies are becoming well characterized information about the elements comprising these networks have remained minimal and in the realm of low throughput biology. In order to model and predict the behavior of these complex systems, the underlying interactions between elements will have to be quantitatively characterized. In a network of interacting proteins most interactions are of necessity low affinity and findings from biochemical measurements of affinities has shown that many individual steps are dependent on coincidence of multiple interactions.

26.20 PROTEIN ENGINEERING

Protein engineering refers to the modification of the physico-chemical or biological properties, i.e. reaction kinetics, substrate affinity, substrate specificity, effector sensitivity, thermostability or lability, intracellular location of a naturally occurring protein with the ultimate objective of improving the quality of protein. One of the technique used in protein engineering is *in vitro* mutagenesis which allows to change the coding capacity of a gene at a defined position, for example, within a region encoding the active centre of the protein. In consequence the engineered protein adopts different properties which are useful in biotechnology. It is another approach to probe protein function.

Combinatorial biosynthesis—It is a term used for the recombination of domains in enzymes in order to produce useful secondary metabolites. In this technique hybrid genes are produced through genetic engineering techniques which consist of a new combination of domains originating from different genes. Further, specific domains can be exchanged for others. Domains can be added or removed or mutations in these domains can be introduced. These terms also emcompasses the expression of foreign genes encoding modified enzymes in the target organism and generally aims at producing novel secondary metabolites.

26.21 DRUG DISCOVERY

Molecules which bind a target protein can be used to develop diagnostic tests for that protein or they can be used for probing the function of that protein. Molecules which inhibit a target protein have potential to be used directly for therapeutic applications. Small molecules, the most common stating point for potential drug do not readily bind to the large, flat contact surfaces that are normally involved in protein-protein interactions. Mutational analysis has indicated that a small subset of the amino acid residues at a protein-protein interaction interface contributes most of the free energy for binding. Small molecules designed to bind strongly to such 'hotspots' can successfully inhibit protein-protein interactions and so these interfaces might be more tractable than has been believed (Wells and McClendon, 2007). The contact surfaces involved in protein-protein interactions are large (~1500-3000Å) compared with those involved in protein-small molecule interactions (~300-1000Å). Furthermore, the contact surfaces of the proteins that interact with other proteins are generally flat and often lack the grooves and pockets present at the surfaces of the proteins that bind to small molecules. There are two types of protein binding agents: macromolecules such as protein or nucleic acids and small molecules.

Inhibitors-Proteins—Both monoclonal and polyclonal antibodies have been used as affinity reagents to probe and inhibit protein function. But recently methods such as **phage antibody-display, ribosome display and mRNA display** have been developed to speed up the process of drug discovery. These approaches require the construction of large repertoires of folded domains with potential binding activities which are then selected by multiple rounds of affinity purification. The binding affinity of the resulting candidate clones can be further improved through subsequent mutagenesis and selection strategies.

26.21.1 Phage-Display Technology

It is a technique used for the presentation of distinct peptides or proteins on bacterial surfaces which uses bacteriophages such as M13, fd, f1 as carriers for these display molecules and permits to identify peptides or proteins with desirable binding properties. In other words, it is a method where bacteriophage particles are made to express either a peptide or protein of interest fused to a capsid or coat protein. Phage display is used for the establishment of libraries of peptides or proteins (eg., enzymes) variants or oligopeptide inhibitors for various target molecules, for the isolation of enzyme variants in a better or modified affinity for the substrate and charged catalytic properties or for the detection of enzyme variants with increased stability. In other words, it can be used to screen for peptide epitopes, peptide ligands, enzyme substrate or single side chain antibody fragments. Although combinatorial peptide libraries have generally been used in most phage display based studies, more large scale protein interaction studies can now be carried out if the products of cDNA libraries are displayed on phage particles.

Phage display technology can be used to screen a library of polypeptides for interaction with a target protein. Each polypeptide is expressed on the surface of a bacteriophage particle as a fusion with a phage coat protein. This provides a physical link between the expressed polypeptide and its encoding gene. The phage displayed polypeptide can be selected by binding to a target using affinity chromatography and further characterization by amplification and sequencing of the corresponding gene located within the phage particle. This technique is particularly suited for screening libraries of random polypeptides variants such as antibody fragment and can be combined to the complementary Y2H technique in order to obtain more relevant result (Wojcik and Hamburger, 2003).

Production of large libraries of antibodies using a procedure called phage display — The steps involved in the production of large libraries of antibodies are a follows (Vaughan et al., 1996).

1. Cloning of antibody gene from WBC of healthy individuals and inserting them into bacteriophages which infect *E. coli*.
2. Phages reproduce in cultures of *E.coli*.
3. Infected cells eventually rupture releasing phages into the growth medium
4. Removal of bacterial cells through centrifusion leaving a population of phages, each of which carries on its surface the antibody encoded by the inserted gene.
5. Antibodies which bind to a particular target are then fished or 'panned' from this population typically using a plastic surface on which the protein in question is immobilized.

It can be performed with high throughput and like Y2H, is simple. Depending on the particular class of proteins being studied such as cytoplasmic vs cell surface proteins, this method may be superior or inferior to the Y2H system as the interactions take place in solution as opposed to the nucleus of the yeast cell. Further, this method is applicable in principle to transfer factors which are not amenable to the Y2H system (Pandey and Snyder, 2000).Cambridge Antibody Technology (CAT) is producing antibody since 1990 and has identified specific antibodies that bind with high affinity to individual targets. It has produced antibody against the tumor necrosis factor α (Mahler et al., 1997). There is a need for construction of bigger libraries as bigger the library greater the chance of finding high affinity antibodies. The limitation with phage display is that the cumbersome step of fermenting bacterial cultures makes it difficult to expand the libraries further.

26.21.2 Ribosome display

It is an *in vitro* technique used for the exposure of a nascent, correctly folded protein on the surfaces of ribosomes in cell-free lysates in which the exposed protein is physically coupled to its mRNA. The exposed protein can be used for molecular selection (e.g. binding to surface-immobilized ligands) and the corresponding mRNA of potential candidates can be directly used for sequencing. In this technique phage is substituted with a ribosome, the cellular machinery which translates mRNA into protein. Normally, an mRNA molecule passes through the ribosome like ticker-tape and is released along with the newly synthesized protein molecule when a 'stop codon' is reached. In Aptein's technology stop codons are eliminated so that the completed antibody and its mRNA remain bound together on the ribosome. It is a cell free system and so is more amenable to automation. With this technique it is possible to construct libraries that are orders of magnitude higher than those created by phage display.

The drawbacks with antibodies are that they tend to be denatured, loose structure when heated or exposed to other stresses such as changes in pH and that antibodies and antibodies technology are covered by patents. So the other alternatives are as follows.

26.21.3 Combinatorial Engineering of Protein Scaffolds

A scaffold is a domain of a large protein which like an antibody can be made to bind to an enormous range of proteins by subtly changing its sequence of amino acids. The scaffolds are more stable than antibodies to heat or other stresses and are sufficiently small that they or the genetic sequences to encode them, can be synthesized chemically. The scaffolds will allow development of chips that can analyze thousands of proteins simultaneously.

Two companies-Phylos and Affibody are involved in the combinatorial engineering of protein scaffolds. Phylos uses a 100-amino acid domain from a structural protein called human fibronectin for its scaffold. Like the binding sites of antibodies this domain consists of a rigid unit that supports loops of varying lengths. Small changes in amino acid sequence change the shape of these loops and hence the structures to which the domain will bind. By substituting different amino acids in the loops held by a scaffold trillions or more variants can be constructed and libraries of these can then be searched to identify any that bind to a particular protein. Phylos has constructed vast libraries of scaffold variants called 'trinectins' using mRNA-display system.

Affibody uses as its scaffold a domain of Staphylococcus protein A, a bacterial surface protein which consists of 58 amino acid residues. It normally interacts via a binding surface made up of 13 of its amino acid with immunoglobulin G molecules, the main class of antibodies in the blood. By substitution these 13 amino acids either singly or in combination, Affibody scaffolds can be made to bind to a variety of proteins in theory. It is possible to create 10^{16} different affibodies. Affibody has developed libraries containing up to 10^8 variants and now looking for a cell free alternative to boost libraries. They are trying to develop system which does not involve passage of phages through a bacterial system and thus more suited to automation. The problems with *in vitro* techniques are that first, there is difficulty of generating sufficiently large libraries and secondly, selected molecules might cross-react with other proteins. Considering these limitations, it is now thought to develop *in vivo* screening for high affinity binders in which antibodies are derived using mice engineered to have a human immune system. Each mouse is immunized with multiple antigen proteins and after 40 days its blood plasma is screened against a protein chip with the same antigens attached.

26.21.4 mRNA Display

It is a technique used for *in vitro* discovery and directed molecular evolution of new peptides and proteins from combinatorial libraries in which the mRNA molecules are covalently bound to the peptide or protein they encode *in vitro*. It is a cell free system similar to CAT ribosome display except that mRNA remains bound to the protein produced and the ribosomes are washed away (Roberts, 1999).Individual mRNAs associated with individual proteins can be identified by reverse transcribing and amplified by PCR. The mRNA tags can thus, in principle, identify those proteins encoded by a large pool of mRNAs whose interaction is not blocked by association with the large, negative cloud of mRNA and that can interact under the dilute conditions of this assay. mRNA display in which proteins are covalently linked to their encoding mRNAs, is used to select for functional proteins from an *in vitro* translated protein library of > 10^2 independent sequences without the constraints imposed by any *in vivo* step. This technique has been used to evolve new peptides and proteins that can bind a specific ligand, from both random-sequence libraries and libraries based on a known protein fold. Seeling and Szostak (2007) isolated RNA ligases from a library that is based on zinc finger scaffold, followed by *in vitro* directed evolution to further optimize these enzymes.

26.22 READOUT SYSTEMS

General protein stains can be used in systems in which the capture molecules are also proteins. Labeling all the proteins in a sample with fluorescent tag as with mRNAs detected by conventional DNA microarrays is not a viable option as different proteins take up the tags to different extent. Currently most systems rely on adding labeled antibodies to the proteins after they have been captured on the chips. But these 'sand wich assays' are hard to use as (i) it is difficult to get sufficiently large number of antibodies in solution and (ii) antibodies may bind to nontarget proteins. To overcome these problems one approach is to label the capture molecules so that they will signal when they are bound to their target protein. The FLAME system is based on fluorescence resonance energy transfer (FRET). In this system the affibody capture molecules are engineered to include two tags, one of which fluoresces but it is suppressed by the second if it is close by. When an antibody binds to its target the two tags move away from one another and fluorescence occurs. Other approaches rely on detecting the physiochemical changes which occur when a capture molecule binds to its target molecule and include a method called surface Plasma Resonance (SPR) which detects differences in refractive index at the surface of a capture molecule (Nelson et al., 2000). Using this system one can screen cross-reactivity of antibodies, specificity and high affinity in one go.

Creation of a combinatorial library of macromolecules — The simplest way to generate a combinatorial library of nucleic acids is to prepare random-sequence DNA molecules using an automated DNA synthesizer. Modern automated DNA synthesizers allow the synthesis of up to about 120mers although the yield of amplifiable DNA molecules deteriorates significantly for lengths above 100. Thus allowing for primer binding sites of 15-20 nucleotides at each end of the DNA, the maximum number of random nucleotides that can be included within a synthetic DNA molecule is about 80. Ideally the four deoxynucleosidephsphoramidites should be pre-mixed at a ratio that compensates for their differential coupling efficiency ensuring their equal representation in the synthesized library. RNA molecules can be transcribed from random sequence DNA templates which results in 10-to 1000 fold amplification due to the ability of RNA polymerase to generate multiple RNA transcripts per DNA template. Even when generating a library of DNA molecules one typically carries out a few cycles of PCR amplification thus replacing the synthetic DNA with biochemically prepared DNA that does not contain residual protecting groups or other chemical lesions.

The number of possible sequences for a nucleic acid of length n is 4^n ($\sim 10^n$). Thus a combinatorial library of one copy each of all possible 25 mers would contain about 10^{15} molecules or ~ 1nmol of material. This is close to the limit of what can be produced using conventional methods especially if one wishes to have an average 10-100 copies of each sequence so that allowing sampling statistics, almost all sequences are represented in the population. It may not, however, be possible to achieve certain catalytic function with nucleic acid molecules that contain only 25 residues. Thus combinatorial libraries of larger random sequence molecules often are prepared with the realization that such libraries contain only a very sparse sampling of all possible sequences. One proceeds with the assumption that there are many possible sequences with desired activity so that even a sparse sampling will contain at least a few of them.

26.23 NUCLEIC ACIDS

In stead of proteins as capture molecules one can go for nucleic acids. Oligonucleotides can also be selected to bind proteins. Nucleic acids which bind a wide variety of molecules have been selected using a method called **SELEX** (systematic evolution of ligands by exponential enrichment) (Wilson and Szostak, 1999). SELEX is a technique for the development of high-affinity RNA ligands (RNA aptamers) which recognize and bind specific amino acid domains of a target peptide or protein or low molecular weight ligands. In other words, SELEX is a method for rapid experimental identification of nucleic acid sequences (usually RNA) that have particular catalytic or ligand-binding properties. **Aptamer** refers to any synthetic single stranded, 30-50 nucleotides long DNA or RNA oligonucleotide which folds into a distinct three-dimensional configuration and thereby recognizing target molecules with affinities and specificities comparable to monoclonal antibodies. In other words, aptamers are short strings of DNA or RNA that constitute specific binding partners for a range of proteins. Aptamers can be synthesized chemically and chips could be produced using the same high throughput already perfected for DNA microarrays. The limitation with this technique is that aptamers have not been shown to bind specifically to individual proteins when confronted with a complex mixture. SomaLogic has constructed a library of 10^{15} DNA molecules in which thymidine is replaced with bromodeoxyuracil.

In this technique a large repertoire of different single stranded RNA or DNA sequences, completely randomized at specific positions are produced by *in vitro* transcription of (usually synthetic) single stranded DNA templates varying at eight positions critical for binding. In other words, a DNA pool is chemically synthesized with a region of random or mutagenized sequence flanked on each side by constant sequence and with a T7RNA polymerase promoter at each 5′end (Figure 26.4). This DNA is amplified by a few cycles of PCR and subsequently transcribed *in vitro* to make the RNA pool. Theoretically, 65,536 individual RNA species can be created in this way. The transcripts are then mixed with target proteins and RNA-protein complexes are retained on nitrocellulose filters. In other words, RNA pools can be partitioned based on whether they bind to the same chosen target compound by passing them through an affinity column derivatized with the target. After removing the non-bound RNA and protein the candidate RNA are eluted from the filters, reverse transcribed into cDNA using AMV reverse transcriptase and cDNAs are amplified by Taq DNA polymerase in a PCR. One of the primers contains a T7 RNA polymerase promoter. The double stranded DNA amplification products are then transcribed *in vitro* and the resulting RNAs are used in the second round of selection. With successive rounds of selection the ratio of active: inactive sequences increases. After 5-10 rounds, the pool become dominated by the once rare molecule which can bind the target ligand. In other words, multiple rounds (5-10) of this selection procedure enrich the RNA population for RNA molecules with high affinity

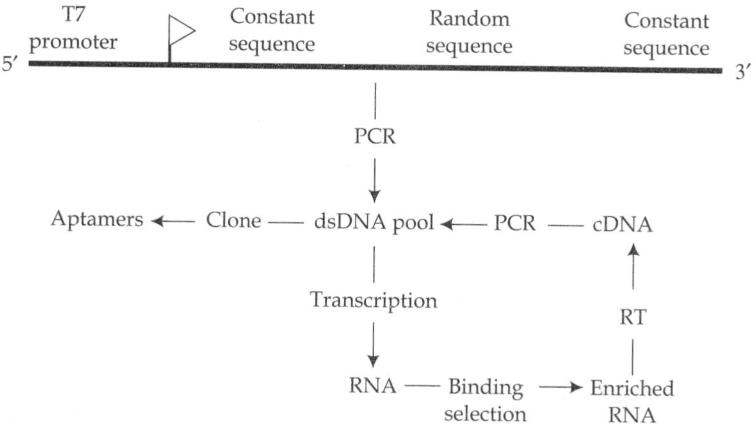

FIGURE 26.4   Showing a typical aptamer pool and structure and scheme for selection process (After Wilson and Szostak, 1999).

for the protein which can be isolated, cloned and sequenced. The practical limit for the complexity of an RNA mixture in SELEX is about 10^{15} different sequences which allows for the complete randomization of 25 nucleotides ($4^{25} = 10^{15}$). If longer RNAs are used then RNA pool used to start the search does not include all possible sequences.

This technique can also be simplified to find either double stranded or single stranded sequences that bind to targets or it can be modified to allow for the selection of catalytic RNA or DNA species. In other words, SELEX can analyze the optimal binding sequences and their sequence context for any DNA-binding proteins such as transcriptional factors, repressors, replication proteins and translational repressors at ribosome binding sites. The interactions of small molecules, for example, amino acids, nucleotides with specific RNAs can be studied by binding the effectors to solid supports and partition those RNAs which interact specifically with these substrates. SELEX thus allows the isolation of protein targets such as bacteriophage T4 DNA polymerase and HIV reverse transcriptase or nucleotides with affinity for low molecular weight targets such as ATP, tryptophan, arginine, theophyllin, caffeine and others. In addition to its use in exploring the function of RNA SELEX has an important practical use in identifying short RNAs with pharmaceutical applications. Finding an aptamer which binds specifically to every potential therapeutic target may be impossible but the capacity of SELEX to rapidly select and amplify a specific oligonucleotide from a pool of sequences makes this approach promising for the generation of new therapies.

26.23.1 Genomic SELEX

It is an extension of SELEX. An increasing number of proteins are being identified which regulate gene expression by binding specific nucleic acids *in vivo*. A method termed genomic SELEX facilitates the rapid identification of networks of protein-nucleic acid interactions by identifying within the genomic sequences of an organism the highest affinity sites for any protein of the organism (Ellington and Szostak, 1990). As with its progenitor, SELEX of random sequence nucleic acids genomic SELEX involves iterative binding, partitioning and amplification of nucleic acids. The two methods differ in that the variable region of nucleic acid library for genomic SELEX is derived from the genome of an organism. Genomic SELEX is an extension of SELEX. In SELEX the nucleic acids that bind tightly to a portion of interest are identified through successive rounds of binding, partitioning and amplification. In SELEX as originally developed the library contains $10^{14\text{-}15}$ random sequences. PCR amplification requires that the nucleic acid sequences of interest be flanked by fixed sequence primer annealing sites. A T7 promoter is included in one of the primer annealing sites so that the library can be expressed as RNA. In genomic SELEX the library contains sequences derived from genome of the organism of interest flanked by a fixed regions which allow PCR amplification and transcription.

In vitro selection or SELEX allows rare functional RNA or DNA molecules to be isolated from pools of over 10^{15} different sequences. By selecting high affinity and specificity nucleic acid ligands for proteins, promising new therapeutic and diagnostic reagents have been identified. Selection experiments have been carried out for identifying ribozymes that catalyze a variety of chemical transformations including RNA cleavage, ligation and synthesis as well as alkylation and acyl-transfer reactions and N$-$glycosidic peptide bond formation. The ligands that emerge from *in vitro* selections are called 'aptamers'. SELEX makes use of large populations of random RNA or DNA sequences as the raw material for the selection of rare functional molecules.

In vitro selection has been used to identify aptamers to targets covering wide range of sizes including small ions, small molecules, peptides, single proteins, organelle, viruses and even entire cells. By generating aptamers to proteins which do not normally interact with RNA, selections have been used to make reagents for interfering with biological functions of specific proteins.

RNA is capable of recognizing a large variety of small molecules including other nucleotides and free nucleobases, amino acids, cofactors, basic antibiotics and transition-state analogs. It can recognize both planar and nonplanar compounds having overall – (negative) or + (positive) charges. Even molecules largely hydrophobic such as valine or tryptophane can be recognized. Many of these RNA aptamers show impressive specificity. RNA secondary structure can for a precise binding pocket for its selected ligand. Most random RNA sequences are highly structured because of favorable binding energy for Watson-Crick and G-U Wobble base pairs and from the statistical likelihood that many such matches will form in a random string. A random pool of RNA is a collection of relatively stable 3-D shapes. The number of shapes is immense and at least one of them will likely to bind a chosen target or accelerate a particular reaction.

The first RNA aptamer targeted to small biomolecule was directed against ATP. Aptamer was isolated by passing a random RNA pool over a column of immobilized ATP, washing away unbound sequences and then eluting with free ATP. After six cycles of selection and amplification ATP-binding sequences began to dominate the pool. Cloning and sequencing revealed a large number of highly divergent and thus independently derived sequences. Inspection of these sequences showed that they all possessed a common 11-nucleotide motif that always occurred within the same secondary structure context (a hair-pin loop with an internal asymmetric purine-rich bulge). Synthesis of this minimal structure confirmed that this domain was indeed responsible for the observed ATP binding with a K_D of about 1 uM. Aptamers exist in a random sequence at a frequency of about one in 10^{11}.

Single stranded DNA is also capable of recognizing a variety of small molecules including organic dyes, ATP, porphyrins and arginine. Most DNA aptamers do not function if they are converted into RNA and vice-versa. Different ligands present distinct levels of difficulty for recognition. 3-D structure solutions determined by NMR spectroscopy have revealed the bases for ligand recognition in several cases.

26.23.2 Photoaptamer System

In this system one end of each aptamer is covalently bound to the chip surface. Target proteins in a sample are captured by their individual aptamers and the chip is then exposed to UV light which causes the bromodeoxyuracil to cross-link with the captured proteins. Unbound proteins can then be washed away and the remainder can be identified using a general protein stain. One of two requirements of a protein chip is to create a reliable and rapid readout system and this problem is solved with photoaptamers.

Concentrations of different proteins in a sample can vary over several orders of magnitude and thus detecting them all may require samples to be split, diluted to different extents and analyzed repeatedly on a duplicate chips or alternatively splitting capture molecules for higher and lower concentration proteins between different chips. The cost of protein chips will increase with the number of proteins on each chip and so CAT has opted for a low-density chips, a chip that could follow a few dozens proteins in a cell signaling pathways could be a great way to unravel the whole pathway. But then there is a need for chips carrying tens of thousands of protein capture molecules.

26.24 SMALL MOLECULES

Small molecules, natural or synthetic, are a means to dissect protein functions and regulatory mechanisms. Small molecules can be identified using a microarray format. Small compounds from a combinatorial chemistry library can be immobilized to form a high density small molecule microarray which can be probed with fluorescently labeled target proteins to identify new ligands for these proteins. Library of small molecules can be constructed by joining them to a peptide nucleic acid (PNA) tag. These PNA tags provide them the address for the structure of the corresponding small molecules as well as immobilize them at specific sites on the chip. Information about small molecules can come from the analysis of gene expression profiles and chemical compounds inhibiting the function of a particular gene can be used to form **small molecule microarray**. This strategy has the potential to generate several tailored probes for every protein of interest in an entire proteome and thus will facilitate the development of pharmaceutical agents.

Immunoprecipitation is a technique for isolating a peptide or protein. It uses a specific antibody raised against this peptide or protein to precipitate it out of the complex mixture of compounds as, for example, cellular extract. The precipitated antibody-antigen complex can then be resolved and the protein is analyzed.

Combining protein interaction technologies—Using a number of protein interaction technologies in a series, one can select an enzyme/protein and then detect its partner with which it interacts, isolate the protein complexes and then finally, characterize the constituent proteins. For example, one can use split enzyme approach to select enzyme and then characterize *in vivo* the proteins that interact with a particular bait and in parallel using that bait as an affinity tag to isolate protein complexes whose constituents can be characterized by mass spectrometry. Using biochemical, structural and proteomic data, the major protein interactions can be organized as a pathway protein interactome.

26.25 ANALYSIS OF MULTIPLE SETS OF DATA

In order to identify and characterize components of biological pathways and to elucidate the response of these pathways it is essential to integrate multiple sets of data and verify using alternative methods rather than drawing conclusion based on single set of data. High-throughput yeast two-hybrid experiments, affinity protein complex purification analyses, correlated mRNA expression profiles, genetic interactions and *in silico* predictions have all helped in identifying and characterizing components of biological pathways as well as elucidating the response of these pathways. No single set of any high throughput data is definitive and integration of multiple datasets and verification using alternative methods is always required before drawing any firm conclusion. For example, in the study of relationship of protein-protein interactions with the expression of mRNAs encoding the components, Jensen et al. reported that the mRNA levels of the subunits in the same protein complexes showed significant coexpression patterns over a time course. By contrast, protein interactions identified by the genome-wide Y2H approach showed only a weak relationship with gene expression. therefore, protein-protein interactions identified by the Y2H approach should be independently confirmed using other methods such as protein chip and/or affinity/MS approaches or by more traditional methods such as co-immunoprecipitation. Multiple sets of data can be generated by gene expression profile studies. A reference database of expression profiles corresponding to diverse mutations and chemical treatments in an organism is constructed. By pattern matching to the database even subtle difference in profiles can be revealed. This approach resulted in predicting the function of eight uncharacterized gene in *S. cerevisiae* which was later confirmed experimentally (Hughes et al,). This approach can be applied to characterize pharmacological perturbations and thus this *in silico* (**computed**) approach has potential in drug discovery and drug target identification. **Virtual screening** *(in silico* screening)-It is a technique to identify compounds with desirable properties such as binding to a target receptor protein, by computational analyses of real or virtual chemical or genetic databases. Screening of millions of candidate molecules with potential structural affinity to biological targets based on 3D structure of the target, results in a collection of lead compounds from which high affinity ligands can in turn be rationally designed using the software programs like GRID, Flex, LUDI and others.

Multiple sets of data can also be generated by different types of high-throughput methods. By integrating expression profiles, MS analysis, and known protein-protein interaction information in the database Ideker et al. (2001b) built, tested and refined the existing model and suggested new hypotheses about the regulation of galactose utilization and physical interactions between galactose and a variety of other metabolic pathways. Global set of data generated with high-throughput approaches can be integrated to evaluate the quality and improve the confidence of individual data sets. Integration of the expression profiles and two hybrid interactions has functionally annotated more than 300 previously uncharacterized genes (Kemmeren et al.,). Further more, such *in silico* predictions were validated experimentally.

Examination of the protein interactions by different techniques such as yeast two-hybrid system, affinity protein complex, purification analysis, correlated mRNA expression profiles, genetic interactions and *in silico* prediction has shown that only about 2400 of about 80,000 purported interactions between yeast proteins are supported by more than one method. Thus each technique produced a unique coverage of interactions in terms of gene categories which suggests that these methods have their specific strengths and weaknesses. For example, most proteins interaction data sets are heavily biased toward proteins of high abundance and toward particular cellular localization of interacting proteins. In addition, the degree of evolutionary novelty of proteins plays a role in causing biased interaction coverage. Further, comparison of data with a reference set of trusted interactions in order to assess accuracy and coverage of large interaction data showed that the highest accuracy was achieved for interactions supported by more than one method. Therefore, in order to increase coverage and accuracy for protein interactions as many complementary methods as possible should be employed.

26.26 PROTEIN LOCALIZATION

Analysis of the function of a particular gene product typically involves determining

    I. the expression profile of the gene

    II. the subcellular location of the protein and

    III. the phenotype of a null strain lacking the protein.

In other words, for each gene, three analyses such as analyses for characterization of gene's expression pattern and subcellular distribution of the encoded protein and the phenotypic analysis of a disruption mutant are to be carried out. Besides all the above analyses conditional alleles of the gene are often created as a means of generating additional information about *in vivo* function of a protein. Ross-Macdonald et al. (1994) developed a multifunctional, transposon-based system that simultaneous generates constructs for all the above analyses.

There are many programs for predicting localization of amino acid sequence in eukaryotic cells. PSORT, SSROTII and Signal protein family of programs and others use sorting signal information for prediction (Horton et al., 2006). However, majority of prediction programs developed use amino acid content in some form. These methods exploit the long stranding observation that amino acid content correlates strongly with localization state (Nishikawa and Ooi, 1982).

26.26.1 Mini transposon Tagging

The transposons create insertion mutations in the target gene allowing phenotype analysis. The transposons can be reduced by *Cre-lox* single stranded recombination to a smaller element that leaves an epitope tag inserted in the encoded protein. The epitope tag also has the potential to create conditional alleles. Mutagenesis of cloned yeast genes using min-transposon (mTn) constructs can circumvent several steps. By a simple manipulation of *E.coli* strains, mTns containing an expression reporter construct can be inserted at multiple independent sites in a cloned gene. The mutagenized genes may then be reintroduced into the yeast genome by homologous recombination. Yeast strains resulting from such transposon mutagenesis have been used to analyze coding regions and disruption phenotypes, structure-function relationships, differential gene expression and protein localization. Epitope tagging is another useful technique. By engineering an epitope recognized by a commercially available antibody into a protein of interest the time and cost of generating specific antibodies and associated reagents is avoided. They constructed a set of mTns which incorporate all of these features. With a single cloning step followed by a simple mutagenesis procedure using one of these mTns, it is possible to generate multiple constructs which will permit rapid analysis of expression pattern, disruption phenotype and protein localization for a given gene. The mTns contains the coding regions for either β-galactosidase (lacZ; β-gal) or the *Aequorea victoria* green fluorescent protein (GFP). Depending on the transposon used this yeast gene is fused to a coding region for β-gal or GFP. Gene expression can therefore be monitored by chemical or fluorescence assays. In –frame fusions between a yeast coding region and the transposon can be identified by β-gal activity or fluorescence. The transposon insertion creates a truncation of the mutagenized gene providing disruption alleles for the phenotype analysis. The inserted transposon can then be reduced by *Cre*-mediated site-specific recombination to an element of less than 300bp that encode several tandem copies of an epitope which can be used for immunodetection. Further, while a protein containing the epitope tag is often functional, conditional mutations may also be produced by this method.

Transposon strategy has been employed for protein localization. In order to monitor the relative levels of protein expression a transposon tagging strategy using a mini-transposon (mTn) is used to generate a library of random insertions in yeast DNA in an *E. coli* plasmid. The mTn contains a promoter less and 5′-truncated lacZ gene near one end of the transposon and a coding sequence for three copies of an epitope tag at the other end. In other words, a transposon contains an *E.coli* lac Z gene lacking its ATG translation initiation codon and promoter adjacent to a lox site at one end of the transposon and a coding sequence for three copies of a hemagluttinin epitope tag adjacent to another lox site at the other end of the transposon. The transposition is mediated in *E.coli* and the mutagenized DNA is then shuttled into yeast. The mutagenized yeast DNA is prepared in a 96 well format, digested with the restriction endonuclease Not1 to free the yeast DNA for the plasmid and individually transformed into a diploid yeast strain. The insertion allele replaces one of the chromosomal copies. When the lacZ gene is inserted in-frame within an yeast ORF, its transcription and translation can be visualized in yeast cells using assays for β-galactosidase and such colonies will turn blue. A large portion of the inserted cassette is then excised at the *lox* sites via *Cre*-mediated recombination, leaving behind the ORF with a short in-frame epitope tag coding sequence. The subcellular location of the protein is then determined by indirect immunofluorescence. Using this random transposon tagging strategy as well as an approach in which ORFs were fused directly to an epitope tag, Kumar et al. localized approximately 55% of the proteome and described the first '**localizome**'-the subcellular localization of most proteins of an organism.

Limitations — The shortcomings with this approach are that first, the library is not complete and secondly, as the tagged proteins are expressed from their native promoters, their localization information is biased toward abundant proteins. Further, as proteins in the transposon tagging approach are visualized through immunostaining on the fixed cells, the dynamics of protein localization and transportation can not be analyzed.

In order to develop a real-time detection method, proteins were tagged with GFP in a genomic library and tagged plasmid library was transformed into the *S. pombe* cells. Out of 6954 transformants showing GFP, 728 showed distinct localization patterns. When plasmids from 728 strains were recovered and analyzed, 250 genes were found to have GFP tags in – frame. In mammalian system, systematic GFP tagging of cDNA clones have been accomplished. Simpson et al. cloned 107 novel human cDNA to produce both N-and C-terminal fusions to GFP and about 100 proteins showed a clear pattern of intracellular localization. On the basis of their sequence homology they predicted the locations of 47% of those novel cDNAs which was in good agreement with the experimental results.

26.26.2 Large Scale Analysis of Protein Localization

The previous large scale analysis of protein localization includes transposon-mediated random epitope tagging and plasmid-based over expression of epitope tagged proteins. However, epitope tagging of partial ORFs can interrupt important localization signals and overexpression of proteins may saturate intracellular transport mechanisms leading to abnormal subcellular localization. To overcome this problem Huh et al. (2003) generated a yeast strain collection expressing full length proteins, tagged at the carboxyl terminal end with GFP, free from their endogenous promoters by inserting the coding sequence of *Aequorea victoria* GFP inframe immediately preceding the stop codon of each ORF. With this strategy wild-type levels and patterns of protein expression are minimally perturbed. GFP signals can be monitored in living cells without disrupting cellular integrity.

Construction and analysis of a GFP-tagged library — Each ORF at its chromosomal location is tagged through oligonucleotide-directed homologous recombination. For each of the annotated ORFs, a pair of oligonucleotide is generated that has homology to the desired chromosomal insertion site at the 5′ end of each primer and homology to a vector containing the GFP tag at the 3′ end. These primers are then used to amplify the GFP tag and an auxotrophic marker from a plasmid template and and the resulting PCR fragments are then transformed into a haploid yeast strain. Transformants are assayed by genomic PCR with one primer specific for the GFP tag and a second specific for each ORF to determine whether the cassettes is integrated at the appropriate locus. Thus they grew a total of 6029 strains with chromosomally GFP-tagged ORFs in medium and analyzed by fluorescence microscopy. Micrographs of each GFP-tagged strain, lacking ORFidentifiers are independently evaluated by two scores and initially classified into one or more of the 12 subcellular localization categories. Refining of the categories was done by performing a series of co-localization experiments. Haploid reference strains carrying monomeric red fluorescent protein (mRFP) fusions to proteins whose localization has been characterized previously were mated to approximately 700 GFP strain that were not assigned definitive localization by GFP microscopy alone and the resulting diploid cells were analyzed by fluorescence microscopy. On the basis of this analysis proteins were assigned to additional 11 localization categories.

26.27   GLOBAL ANALYSIS OF PROTEIN EXPRESSION

The use of terms such as protein expression and protein expression profiling should be avoided as only genes can be expressed (into mRNAs which in turn are translated into proteins). mRNA transcript expression patterns are similar for groups of functionally related genes, Further, mRNA abundance is also similar within certain cellular compartments. However, transcriptional co-regulation has not been directly compared to subcellular protein localization on a proteome wide scale (Ghaemmaghami et al., 2003).

Individually tagging each of the annotated ORFs with an epitope tag will result in fusion proteins that are expressed under the control of their natural promoters. The fusion library allows the immunodetection and immunopurification of the entire yeast proteome using a single antibody enabling the development of a range of high throughput assays. They synthesized 6,234 pairs of ORF specific oligonucleotide primers for the construction of epitope tagged yeast fusion libraries. Each of the oligonucletide pairs have shared 3′ ends that allow for PCR amplification of a common insertion cassette as well as gene-specific 5′ ends that allow for the precise introduction via homologous recombination of the amplified insertion cassettes as a perfect in-frame fusion at the carboxyl end of the coding region of each gene. The insertion cassettes contained the coding region for a modified version of tandem affinity protein purification (TAP) tag which consists of a calmodulin binding peptide, a TEV cleavage site and two IgG binding domains of *Staphylococcus aureus* protein A as well as a selectable marker. Western blot analysis using an antibody which specifically recognizes the TAP tag, showed that the large majority (>95%) of detected fusion proteins migrate predominantly as a single band of approximately expected molecular mass. Further, tagging does not hinder their regulated proteolysis by the ubiquitin/proteasome degradation system and that tag itself is rapidly destroyed during the targeted degradation of the fusion protein.

Finding functions of small set of proteins—The standard approach for finding functions of small set of proteins has been to combine classical genetics (isolation and phenotypic analyses of mutants) and biochemistry (analyses of protein activity *in vitro*) with recombinant genetics (manipulating cloned DNA segments and reintroducing it at homologous sites in yeast genome).

26.28 COMPUTATIONAL DETECTION OF FUNCTIONAL LINKAGES

Protein-protein interaction can be inferred on the basis of comparative genomics, detailed sequence and structural analysis, correlation of protein functional genomic features and existence of conserved interactions in other organisms. There are techniques which supply functional information for many proteins simultaneously. These methods detect functional linkages between proteins. If the function of one of the proteins is known, then the linked proteins act in the same pathway or complex as the first protein. The extension of the two-hybrid screen to a genome-wide assay has detected over 1000 putative protein-protein interactions. Putative (or probable) protein refers to any protein whose amino acid sequence has only limited similarity to already characterized proteins (as, for example, checked by comparison via SWISS PROT database). Another technique is the analysis of correlated mRNA expression levels. By correlating those mRNAs whose expression levels have changed-one can establish functional linkages between the proteins encoded by the correlated mRNAs. Genes with similar expression behaviour (for example, increasing and decreasing together under similar circumstances) are likely to be related functionally. In this way genes without previous functional assignments can be given tentative assignments or assigned a role in biological process based on the known function of genes in the same cluster (that is the concept of '**guilt by association**). In other words, two interacting proteins should share the same function.

There are three computational methods for the detection of functional linkages:

1. Phylogenetic profile
2. Analyzing fusion patterns of protein domains
3. Gene neighbor methods.

1. **Phylogenetic profile**—If the two proteins have the same phylogenetic profile, i.e. the same pattern of presence or absence in all surveyed genomes it can be inferred that the two proteins have a functional link (Pellegrini et al., 1999). Considering n fully sequenced genomes, there will be up to 2^n phylogenetic profile. Any two proteins having identical or similar phylogenetic profiles are likely to be engaged in a common pathway or complex. In other words, proteins with similar phylogenetic trees are

more likely to interact with one another. 'Homology method' is widely used to extend the knowledge of protein function from one protein to its cousins which are presumably descended from the same common ancestral protein. Functional assignments by homology methods usually involve identification of some molecular function of the protein and not the cellular function.

2. **Domain fusion or Rosetta stone**—Frequently, separate proteins, say A and B in one organism are expressed as a fusion protein in some other species and when two proteins are expressed as a fused protein, the two domains of A and B are almost certainly linked in function. In other words, they participate in the same structural complex, in the same biological pathway, in the same biological process or some times to physically interact. This type of functional linkage can not be detected by 'Homology search' as A and B have unrelated sequences and also because the fused protein has similarity to both A and B, it is termed as a **Rosetta stone sequence** (Marcotte et al., 1999; Enright et al., 1999).

3. **Gene neighbor methods**—In several genomes if the genes that encode two proteins are neighbors on the same chromosome, the proteins tend to be functionally linked (Wu and Maniatis, 1999). These are based on the observations that functionally related proteins in bacteria tend to cluster along the chromosome to form operons. It is a powerful method in uncovering functional linkages in prokaryotes where operons are common. It also shows promise for analyzing interacting proteins in eukaryotes as operon like cluster structures have been observed. Considering proteins, one will have to carry out the analysis of patterns of location of proteins on chromosome and identify protein pairs that are adjacent along the chromosome. Protein pairs are likely to share similar functions if such chromosomal proximity is conserved across multiple genomes. Further, conserved gene order can also be used as an indicator for functional interaction. Overbeek et al. (1999) extended this analysis by building synteny groups, i.e. gene clusters across organisms in order to infer functional links. They defined a gene cluster as a set of genes located in the same strand and in which the maximal intergenic distance is 300bps. If two genes say XA and XB in a given cluster of genome A have orthologues XA and XB in a cluster of genome B, then they can be said to be functionally linked.

Different methods of prediction of protein function is summarized in Figure 26.5.

26.29 PREDICTION OF PROTEIN-PROTEIN INTERACTIONS BASED ON SEQUENCE AND STRUCTURAL ANALYSIS

The two prediction methods are based on the hypothesis that interacting protein tend to coevolve. In the first method the coevolution of interacting protein families is measured by the similarity of phyogenetic trees constructed from multiple sequence alignments of two protein families. Thus phylogenetic trees of all proteins are constructed and proteins with similar phylogenetic trees can be said to be interacting more with one another. In the second method, the coevolutionary signal is multiple sequence alignments if further analyzed in terms of correlated mutations and a protein pair can be said to be interacting if there is accumulation of correlated mutations between the interacting partners.

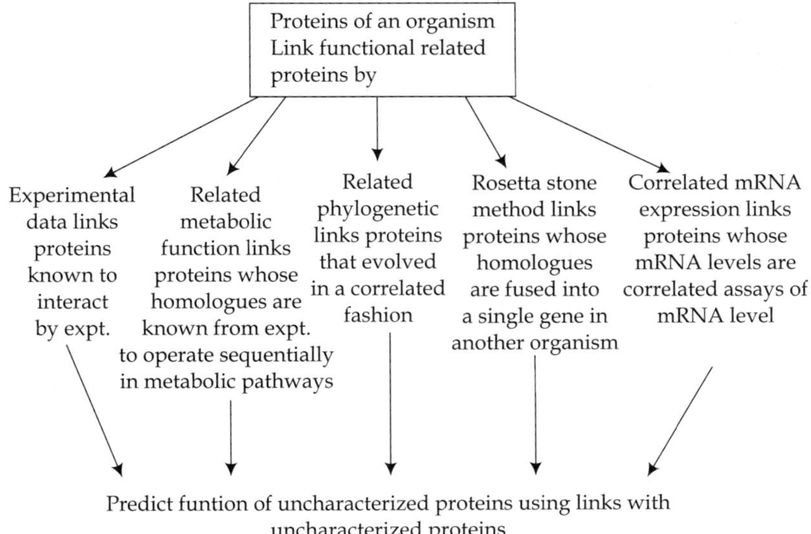

FIGURE 26.5 **Showing different methods of predicting protein function.**

Certain pairs of sequence motifs and structural families preferentially interact. To identify such pairs one will have to classify known protein interactions in terms of interaction between sequence motifs and structural families and pairs of sequence motifs and structural families which are overrepresented in the interaction dataset can then be identified. A new protein pair is likely to interact if it can be classified into one of these overrepresented sequence motifs or structural families.

Protein-protein interaction can be predicted using neural network (machine-learning techniques) based on sequence information and associated physicochemical properties such as charge, hydrophobicity and surface tension (Janin et al., 2003)

26.30 CONSTRUCTION OF REGULATORY NETWORKS

Besides providing information about protein function, protein interaction information can be used to build complex networks, protein-protein interaction networks (see chapter 17 for details on networks). Protein-protein interaction networks cluster proteins participating in the same cellular processes in the same functional classes. Functional classification of proteins can be made according to the identity of their interacting partners. Interaction-based functional clustering is done to compare interaction partners for all protein pairs assuming that the more two proteins share interacting partners, the more they should be functionally related. Construction of regulatory circuits is essential for improving plants. In the network, individual proteins appear as nodes and edges represent the physical relationship between proteins. Protein interaction data can be combined with other genetic information (from global analysis of gene expression, transcription factor binding and posttranslational modifications) , using existing functional classification of proteins to reconstruct biochemical pathways. These associations can be useful in mapping the relationships between individual proteins and as well as their shared participation in subnetworks. For example, it has been shown that proteins which interact with many partners and subsequently influence many biological processes, tend to be essential. In case of organisms whose genetic and proteomic constituents are ill defined, these techniques can be applied to infer functionality through evolutionary conservation of gene sequences and protein domains between species.

26.31 CHEMICAL PHENOTYPING/PROFILING

Chemical or metabolic profiling refers to isolation of all metabolites of a cell, their separation by, for example, liquid chromatography, capillary electrophoresis, gas chromatography or matrix-assisted laser disorption/ionization and identification by, for example, matching the mass of each compound with reference masses or using internal standards. Metabolic profiling can be used for constructing a **metabolic map**-an inventory of all metabolites of a cell at a given time or cataloguing of up-and down-regulated compounds as a result of intrinsic or environmental stimuli. Chemical profiling allows to monitor entire pathways simultaneously.

26.32 MODULAR ORGANIZATION OF THE CELL MACHINERY

There are procedures that partition proteins in complexes in two types: core components which are present in most isoforms and attachments present only in some of them.

26.33 NANOTECHNOLOGY

Nanotechnology-literally refers to 'dwarf technology' is the exploration and exploitation of the material world on the scale of billionths of a metre: a human hair is , by comparison, huge with a thickness of 80,000 nanometers. This technology will help in building things-materials, from bottom up, assembling them atom by atom, molecule by molecule. This would give total control over the end products which would have fascinating properties and functions. Among the most exciting applications of nanoscience is the ability to interact with living cells and even DNA-most promising kinds of drugs and health monitors as well as new insights into the basic machinery of life. Currently, one application is creating remarkable interfaces between inorganic and biological structures. Clusters of nanoparticles of gold have been introduced on a graphite surface into a liquid soup of biological proteins. The proteins become trapped onto these 'designer binding sites'. This immobilization of proteins is also an important step towards analyzing proteins using advanced atomic force microscopy. Thus there is a potential for watching how individual complexes of proteins are formed, especially medically important molecules such as cytokines-the signaling chemicals implicated in all manner of disease. The nanoscience thus meets the sharp end of genetics-genomics and proteomics. Asthma may well be due to nanoparticles which are way below the micron size and to detect this novel particles mass spectrometer is being developed for preparing nanosurfaces. This extraordinary technique of 'molecular dissection' is being pushed forward whereby atoms can be made to shoot off a surface by tuning the tip of a Scanning Tunnelling

Microscope. With the advent of this technology it is now possible to imagine doing chemistry in a tiny labs on a single chip-a number of microfluidic channels etched in silicon or glass and using vanishingly small amount of active substances. Microfluidics can help looking at the reaction of drugs to single cells and their genes, a prelude to more effective gene therapy (see chapter 8) and to tiny distributed implants that will monitor the whole body and alert individual to any thing that needs attention.

Nanotechnology involves the manufacture and manipulation of the macromolecules at the molecular or atomic level. It includes a series of techniques designed to study molecules at the nanometer level, billionths of a metre (e.g.Scanning tunneling microscopy, STM, Scanning tunneling current and AFM (atomic force microscopy)), scanning the repulsive atomic forces between a sample and a probe and thereby producing high resolution surface topographies of proteins and nucleic acids. Nascent polypeptide/DNA or RNA-Nascent polypeptide refers to a chain of amino acids linked through peptide bonds which is being formed (*in statu nascendi*) and still attached to the 50S (bacteria) or 60S (eukaryotes) ribosomes unit through a tRNA. Nascent DNA or RNA refers to a chain of nucleotides linked together by phosphodiester bonds which is being formed (*in statu nascendi*).

27

Oligos-Chemical Genomics

27.1 OLIGOS

An oligodeoxynucleotide (oligo) of more than 15 to 17 nucleotides would have a unique sequence relative to the entire human genome (Stein and Cheng, 1993). In principle, a suitable oligo should be able to interfere (hybridize) in a sequence–specific manner with a specific RNA molecule *in vivo* and thereby inhibit its subsequent use. In most cases the target is mRNA which can not be translated into protein. A property of complementarity, specificity is conferred by Watson-Crick base pair formation, if an appropriate target can be identified. All that is required is knowledge of gene' nucleotide sequences so that an antisense oligonucleotide with the complementary sequence can be synthesized. Further, as the gene expression is blocked the consequent observed effects will help in elucidating the physiological role of the gene and its product. In other words, antisense oligos can selectively down regulate gene expression. Both antisense DNA and RNA oligonucleotides are used to study cellular processes. Antisense offers a good alternative to gene knock out techniques considering cost, time and resource requirement. An antisense gene refers to any gene introduced into a target organism and transcribed into an antisense RNA. Thus an antisense DNA (double stranded DNA) encodes an antisense RNA. It is widely used in gene function determination, drug targets and pathway studies (Taylor et al., 1991). Unfortunately not all antisense oligos are effective in inhibiting protein synthesis. Typically less than 20% of AOs are effective.

27.2 MECHANISM OF INTERFERENCE

There are three ways in which an antisense oligo can interfere with nucleotide sequence.

1. An oligo can form double stranded RNA molecule with sense mRNA which can not be processed and/or exported to the cytoplasm (at least in mammals) will be preferentially degraded (as in Drosophila) or arrest translation by blocking ribosomal sites.

2. Oligos complementary to genomic DNA can interact with it by means of Hoogsteen base pairing in the major groove and thus inhibit transcription in tissue culture by inducing triple helix formation. In other words, antisense RNA binds to genomic DNA sequences to form triple helix structure which interferes with the binding of DNA-affine protein (e.g. transcription factors).

3. Antisense RNA can enter transcription bubble (where single stranded DNA is available for RNA polymerase) bind to their cognate sequences and reduce processivity of the transcriptional complex. In bacteria translation can be inhibited by the interference of anitsense RNA with the mRNA binding to the ribosome.

27.3 CLASSES OF OLIGOS

There are four classes of oligos: 1. O, phosphodiester or normal DNA 2. OET (ethyl), phosphotriester 3. Me (methyl) phosphonate 4. S, phosphorothioate. Class 4 differs from class 1 oligo in that the oxygen in class 1 is replaced by sulfur. Of the four classes which class 4 oligo is best understood and has produced the broadest range of activities. The stability of oligonucleotide refers to their capacity to withstand degradation or hydrolysis by nucleases. Oligos which can be used as therapeutic agents should have the following characteristics.

1. It can be synthesized easily and in bulk
2. It must be stable *in vivo*
3. It must be able to enter the target cell
4. It must be retained by the target cell
5. It must interact not interact with other cellular targets
6. It should not interact in a non-sequence-specific manner with other macromolecules.

Phosphorothioate oligos meet some but not all of these criteria. Most interesting modifications to date are those that alter the sugar moiety and the backbone. Modifications such as 2′ methoxyethoxy enhance affinity for RNA potency *in vivo*. It provides a remarkable increase in stability.

27.4 SITES OF ANTISENSE ACTIVITY

Selection of sites in a RNA molecule at which optimal antisense activity may be induced is dependent on the terminating mechanism and influenced by the chemical class of the oligonucleotides. Each RNA shows a unique pattern of sites of sensitivity. It is important to define potentially optimal sites in mRNA species. To a large extent this must be determined empirically for each RNA target and every new class of oligonucleotides.

Usually the antisense RNAs contain one to three stem-loop structures. The corresponding sense RNAs frequently are longer, possess the complementary stem-loops and additional structures. The loops determine the specificity of pairing between antisense and sense RNA and the stems are responsible for the stability of the antisense RNA. Oligos targeted to donor-acceptor sites for splicing pre-mRNA inhibit HIV replication. Regions that are important for viral function are poly (A) addition signal sequence or the cap site, the R-segment or the splice acceptor site for the tat gene sequence.

Accessibility to target sites is important in determining the efficiency of AOs. Several methods for accessible sites identification have been developed. Of which mRNA accessible site tagging (MAST) (Zhanj et al., 2003) has greater advantage over RNaseH mapping, gel shift, oligonucleitde array and random reverse transcriptase priming in throughput, cost and complexity.

27.5 CHEMICAL GENOMICS/PROTEOMICS

Chemical genomics refers to development of novel, selective, small cell permeable and biologically active ligands or selection of already existing compound that binds to the cellular proteins and affect their function (s) either negatively or positively. The resulting changes in protein function permits to study whole metabolic pathways. Ligands are small molecular weight molecules which bind to a specific complementary site of another molecule. For example, a substrate whose 3-dimensional configuration permits its binding to the catalytically active site of an enzyme molecule is a termed a ligand. Smaller weight cellular compounds (ligands) interact with macromolecules such as DNA, RNA, proteins or peptides. The study of interactions of ligands (usually organic) with peptides or proteins is referred to as **chemical proteomics**. It allows a high throughput screening of whole libraries of ligand molecules for their binding potential to proteins in a study to detect drug candidates simultaneously. Chemical genetics aims at developing small ligands (chemical probes) which affect protein function (s) *in vivo* and thus perturbing complex intracellular processes in order to understand complete biological process and to control cellular pathways. Thus chemical genetics and chemical genomics yield information regarding complex cellular processes and the underlying proteins.

The use of chemicals to interrogate molecular processes provides a novel avenue for the rapid and effective dissection of biological mechanisms and gene networks in ways not feasible with conventional methods such as mutation-based approaches (Raikhel and Pirrung, 2005). In other words, chemical genomics facilitates the identification of new pathways and networks. The **chemical genomics** approach uses small molecules to modify or disrupt the functions of specific genes/proteins (Stockwell, 2000; Dobson, 2004; Lipinski and Hopkins, 2004) in contrast to classic genetics in which mutations are used to disrupt gene function. The underlying theory is that the functions of most proteins can be altered by the binding of a chemical which can be found by screening large libraries for compounds that specifically affect a measurable process.

Chemical genomics involves (i) library assembly/synthesis or the creation of chemical diverse libraries of compounds (ii)screening or the identification of compounds that affect a biological process of interest (iii) target identification or the discovery of the protein targets of active compounds and (iv) target function and network discovery. Chemical genomics can address loss-of-function, lethality and gene redundancy and allows instantaneous, reversible, tunable and conditional control of a phenotype and thus has many advantages over traditional genetic approaches. The aim over here is to discover chemicals that specifically disrupt a process or function of a protein and once these chemical are identified, one can combine their use with genetic screens to identify genes involved in the same process (functional genomics analyses).

27.6 SMALL MOLECULES

More than 10 million pure compounds are available. The potential diversity (defined as the number of unique chemical structures) of compounds composed of C, H, O, S, P and the halogens of molecular weight <1000 likely exceeds 10^{60}. Bioactive chemicals have been used by plant physiologists to unravel several mechanisms including inhibitor of GA biosynthesis, inhibitors of ethylene action, inhibitors of auxin transport, cytoskeleton-disrupting drugs and inhibitors of GDP-GTP exchange proteins but the problem has the complexities associated with understanding the action mode of the compounds at molecular level. What is now required is to work out the specificity of selected compounds and possible uptake and metabolism of the compounds as these will affect the discovery and analysis of the bioactive compounds.

Applications — Chemical genetics approaches have been used to dissect pathways in single – celled systems such as bacteria, yeast and mammalian cell cultures. The challenge in mammalian systems is the identification of target proteins because of the lack of an efficient molecular genetic/genomic approach. However, chemical genomics approaches when integrated with genomics/proteomics tools can be applied in plant systems (eg. Arabidopsis) can effectively dissect a complex gene network affected by chemicals of interest, rapidly identify their gene targets and efficiently investigate their mode of action. It can thus address the issues of overlapping gene function in gene families, lethal loci and control of dosage- and tissue-specific applications.

Critical to application of chemical genomics approach is the ability to screen for mutants resistant or hypersensitive to chemicals and to identify responsible genes using genomics based methods including a completely sequenced genome, whole-genome microarrays, a large collection of knockout and activation tagged mutations, a rich array of mutants and ease of cloning a gene by map-based cloning aided by DNA microarrays. Mutations conferring resistance/hypersensitivity can either affect the target of the chemical, upstream and downstream components of its target. Mutations in a component upstream/downstream of the target protein will not only provide insights into the function of the target genes but also give a new high throughput method for identifying new genes in the corresponding network. In other words, chemical genomics will enhance the power of existing genomic tools including the large collection of T-DNA insertion lines from which one can recover mutants with altered chemical sensitivity that otherwise may not show a phenotype.

28

Genome Comparison

28.1 APPLICATION OF COMPARATIVE GENOMICS

The study of genes (genomics) and proteins (proteomics) on the scale of whole cells and organisms and their comparison (comparison of genome composition by DNA and proteome composition by AA) will provide information on evolutionary relationship between different species. Comparative mapping studies between related species using common set of molecular markers will reveal co-linearity over short chromosomal segments or whole chromosome and thus comparative genome analysis will provide information on whether or not there is an overall **syntenic** relationship (sharing of conserved gene order) between chromosomes and divergence at the level of SNP and insertion/deletion. Sequencing many genomes will provide database that can be used to assign gene functions by genome comparisons- a field of study called **comparative genomics**. The mere presence of combinations of genes in particular genomes can indicate at protein function. When the two genes always appear together in a genome it suggests that the proteins they encode may be functionally related and thus comparative genomics can be used to identify functionally related genes.

Comparative genomics is a powerful methodology for extracting biological information directly from DNA sequence. Comparison of few (two-four) closely related genomes have been successful in the discovery of protein-coding genes, RNA genes, miRNA genes and catalogues of regulatory elements. In principle, the resolution and discovery power of these studies should increase with the number of genomes thereby enabling the discovery of all conserved functional elements. However, it will pose challenges arising from sequencing, assembly or alignment artifacts and from movement or loss of motif instances in individual species. Comparative genomics involving genomes of twelve different species of Drosophila has shown that protein-coding regions show highly constrained codon substitution frequencies and insertions and deletions that are heavily biased to be multiple of three. RNA genes and structures tolerate substitutions which preserve base pairing. MicroRNA hairpins show a characteristic conservation profile with high conservation in the stem and mutations in loop region. Regulatory motifs are marked by high levels of genome-wide conservation (Stark et al., 2007) and post-transcriptional motifs show strand-biased conservation. Long protein coding exons (>300nts) are recovered at very high rates even with few closely related species but more species and larger distances are important for recovering short exons, miRNAs and regulatory elements. Annotations of protein-coding genes remain difficult in metazoans because of short exons and complex gene structures with abundant alternate splicing.

28.2 *DE NOVO* IDENTIFICATION OF PROTEIN CODING SEQUENCES/ORFs

De novo identification of complete set of protein coding sequences remains imperfect. Gene identification is typically based on a comparison of genomic sequence to known mRNA transcripts in other organisms. The basic approach is to identify ORFs that are too long to have occurred by chance. However, stop codons occur at a frequency of only ~1 in 20 in random sequence. Thus ORFs of $\geq$ 60 amino acids will occur frequently by chance (~5% under a Poisson model) and even ORFs of $\geq$ 150 amino acids will appear by chance in a large genome (approximately 0.05%). Yeast genes typically contain few and larger exons but nonetheless there are too many exons encoding $\leq$ 150 amino acid and thus it will be difficult to discriminate those from randomly occurring ORFs. Higher eukaryotes contain many small exons. The typical size of exon is ~125bp for internal coding exons in mammals. So the objective is to identify ORF theoretically encoding proteins of $\geq$ 100 amino acids and that do not overlap a half of their length. In comparative genomic analysis one tests the ORF seen in one species by observing whether the orthologous sequence in related species also encode an ORF. True protein- coding ORFs will be under strong selective pressure to preserve ORF whereas spurious ORFs will accumulate frameshifts and stop codons. Comparative genome analysis not only improves the recognition of true ORFs, it also yields much more accurate definitions of gene structure including translation start, translation stop and intron boundaries. Start of translation is defined as the first in-frame ATG codon. However, actual start of translation could lie3′ to this point or (if sequencing errors or mutations have obscured an earlier in-frame ATG codon) 5′ to this point. The stop codons seem to show more variability in position than start codon.

Donor, branching point and acceptor splice sites were strongly conserved with respect to both location and sequence. Moreover, exon boundaries closely demarcated the domains of sequence conservation as measured by both nucleotide identity and absence of indels.

Rapid evolving genes include acquisition or loss of entire genes, the rapid divergence of nucleotide sequence or the presence of large insertions (numerous insertions and deletions). In case of rapid protein change nucleotide changes in protein-coding regions are silent or affect individual amino acid. Additional mechanism of rapid protein change involves closely spaced compensatory indels that affect the translation of small contiguous amino acid stretches. They also include the loss and gain of stop codons (by a nucleotide substitution or a frameshifting indel) which may result in the rapid change of protein segments or the translation of previously non-coding regions. Such events were frequently observed near telomeric regions and may affect silenced genes or recently inactivated pseudogenes. Slow evolving genes are the mating type genes ($MAT_{\alpha 1}$, $MAT_{\alpha 2}$ $MAT_{\alpha 3}$) of S. cerevisiae. They show 100% conservation at the amino acid level over its entire length and 100% conservation at the nucleotide level as well and thus differs sharply from the typical pattern seen for protein coding genes which show relaxed constraints in third positions of codons. $MAT_{\alpha2}$ may function not by encoding a protein but rather by encoding an antisense RNA or a DNA site. There are two independent evolutionary signatures unique to protein coding regions. The first is the **reading frame conservation** (RFC) which observes the tendency of the nucleotide insertions and deletions to preserve the coding reading frame. The RFC test is used to classify each ORF as biologically meaningful or meaningless on the basis of the proportion of ORF over which reading frame is locally conserved in each of the species The second is **codon substitution** frequencies which observe mutational biases towards synonymous codon substitutions and conservative amino acid changes similar to the nonsynonymous/ synonymous substitution ratio KA/Ks.

28.3 GENOME-WIDE IDENTIFICATION OF REGULATORY ELEMENTS

Cross-species conservation has previously been used to identify putative genes or regulatory elements in small genomic regions. Regulatory elements are typically short (6-15bp) sequences, tolerate some degree of sequence variation and follow few known rules. Direct identification of regulatory elements is more challenging than for genes. The experimental manipulation involves systematic mutation of individual promoter regions. This method is laborious and unfit for genome-wide analysis. The second method involves computational analysis of single genomes. It has been successfully used to identify regulatory elements associated with known sets of genes. These approaches do not have sufficient power to allow comprehensive direct identification of regulatory elements. The third approach involves comparative genomics. This method offers various approaches for finding regulatory elements and the simplest approach is to perform cross-species sequence alignment to find **phylogenetic footprints**-the regions of unusually high conservation. Conserved sequence motifs reside upstream of the genes with similar functional annotations or similar expression patterns or those bound by the same transcription factor. Several of intergenic regions have highly conserved sequences immediately upstream of the translational start site that are longer and more conserved than 6-to-10 nucleotide length expected for transcriptional regulatory elements. Genes which have in their promoters a conserved TATA box have a broader and higher peak of conservation which suggests that promoters contain more regulatory sequences than average promoter or that they are slowly evolving than average promoter. Further, lower sequence identity at 75 to 100bp upstream of the ATG codon suggests that there may be spatial restriction on regulatory sequences that prevents them from acting close to transcriptional start site. The intergenic regions are relatively short (average ~500bp) in *S. cerevisiae*. Intergenic sequences of the *sensu stricto* species are similar enough that most orthologous sequences can be accurately identified and aligned with CLUSTALW. Orthologous intergenic sequences of distantly related species could only be identified by aligning the sequences of their associated predicted proteins using BLASTX. The distribution of sequence identity within intergenic regions is not uniform. There is relative uniformity of sequence identity in intergenic regions downstream of genes which harbour no promoter elements. The best approach for direct identification of regulatory elements is to cluster genes into functionally related subsets (for example, genes with a common biochemical function or co-ordinated transcription) and then search for common sequence motif in the vicinity of the genes. The limitation with this approach is that it requires prior information about gene function which is often not available and never comprehensive and is inherently limited in its ability to extract information from a single genome.

Comparative genomis will define promoters and other islands of intergenic conservation. One can use the conservation properties of known regulatory motifs. Binding site for TF, Gal4 whose sequence motif is CGGn (11)CCG which contains 11 unspecified bases. Gal4 regulates genes including GAL1 and GAL10 genes and is involved in galactose metabolism. Gal4 motifs occur three times in this intergenic region and the fourth binding site for Gal4 differs this consensus by one nucleotide in S. cerevisiae. Gal4 motif occurs 96 times in intergenic regions and 415 times in genic (protein-coding) regions. Intergenic region motifs Gal4 have a conservation rate that is ~5-fold higher than equivalent random motifs. Intergenic motifs are more frequently conserved than genic occurrences. By contrast, random motifs are less frequently conserved in the intergenic regions than in genic regions. Finally, intergenic region motif shows a higher conservation rate in divergent compared with the convergent intergenic regions (those that lie upstream compared with downstream of both flanking genes). Comparative genomics and species-specific experimental studies provide complementary approaches to biological signal discovery. Comparative studies help pinpoint evolutionary selected functional elements across diverse conditions whereas experimental studies reveal stage and tissue-specific information as well as species-specific sites.

28.4   METHODOLOGY FOR GENOME-WIDE DISCOVERY OF MOTIFS

There can be *de novo* discovery of regulatory motifs directly from the genome. Methods for prediction of functional sequence motifs use stringent criteria for sequence motif definition (i.e. searching for n-mers). Most regulatory sequences are turning up as conserved n-mers. Motif conservation score (MCS) is calculated on the basis of the conservation rate of the motif in intergenic region. MCS is measured in standard deviation above the rate of comparable control motifs (Kellis et al., 2003). Most of the known motifs show strong conservation with 60% having MCS = 4 which is substantially higher than expected by chance. Thus the first step would be to identify conserved '**mini motifs**' first and then use them to construct full motif. Mini motifs are sequences of the form $XYZ_{n\ (0-21)}UVW$ and thus consist of two triplets of specified bases interrupted by a fixed number (from 0 to 21) of unspecified bases, for example, TAGGAT, ATAnnGGC to the Gal4 motif itself ($CGG_{n\ (11)}CCG$). Conserved mini motifs are then defined according to the following three criteria. In each case the conservation rates are normalized to appropriate random controls.

1. Intergenic conservation (CC1) when the mini motif shows a significantly high conservation rate in intergenic region.
2. Intergenic-genic conservation (CC2) when the mini motif shows significantly higher conservation in intergenic regions than in genic regions.
3. Upstream-downstream conservation (CC3) when the mini motif shows significantly different conservation rates when it occurs upstream compared with downstream of a gene.

28.5 DISTINCTION BETWEEN DIFFERENT TYPES OF FUNCTIONAL ELEMENTS

Measures of nucleotide conservation alone do not distinguish between different types of functional elements. Many functional elements which tolerate abundant 'silent' mutations such as protein coding exons and many regulatory motifs might not be detected on the basis of strong nucleotide conservation. Multiple closely related genomes provide neutral divergence for recognition of functional regions in stretches of highly conserved nucleotides.

28.6 CHIP-BOUND MOTIFS AND CONSERVED MOTIFS

It is important to determine whether CHIP-bound motifs that lack conservation are biologically meaningful. Stark et al. (2007) reported that non-conserved CHIP-bound sites may be of decreased biological significance. This is based on the observation that motif instances that are both CHIP-bound and conserved show strongest functional enrichment in muscle genes for Mef2 whereas motif instances derived by CHIP alone show substantially reduced enrichment levels. CHIP-bound motifs that were not conserved showed decreased enrichment in muscles/mesoderm development where the factor are known to act suggesting that the potential lineage-specific roles may lie outside the regulator's conserved function. In other words, in case of TFBSs the functionally meaningful sites for TFs lie outside CHIP-bound regions.

28.7 RELATIONSHIP BETWEEN CHROMOSOMAL GENE ORDER AND FUNCTION

Genes are not randomly distributed on the chromosomes and that genes which are physically close to each other lead to represent groups of genes with similar functional relationship even if they are not contiguous. These groups of genes are called '**gene clusters**' or 'gene teams' in two or more genomes. Identifying conserved gene clusters is important for many biological problems such as comparative genome mapping, studying transcriptional neighbourhoods, predicting gene functions and other problems. Certeel (2002) introduced the concept of a '**gene team**' which is a set of orthologous genes which appear in two or more species possibly in different order yet with distance between the genes in the team for each chromosome is always within a certain threshold,

If a newly discovered gene is related by sequence homologies to a gene previously known in different or the same species, the function of the unknown gene can be entirely or partly defined by that relationship. When two genes share sequence similarities (nucleotide sequence in DNA or amino acid sequence in the proteins they encode their sequences) are said to be homologous and the proteins they encode are **homologs**. Two genes from different species but possessing a clear sequence and functional relationship to each other are called **orthologous** and their protein products are **orthologs**. In other words two homologous genes found in different species are said to be orthologous. Thus if the function of a gene is known in one species , this information can be used to assign gene function to the ortholog found in other species. Genes with sequence and functional similarity to each other within a single species are called **paralogous** and their protein products are **paralogs**. Sometimes even the order of genes on a chromosome is conserved over large segments of the genome of closely related species. The conserved gene order is called **synteny**. It provides additional evidence for an orthologous relationship between genes at identical locations within the related segments. Further, certain AA sequences associated with particular structural motifs may be identified within a protein. The presence of a structural motif may suggest that it catalyzes ATP hydrolysis, binds to DNA or forms a complex with Zn ions and thus helping to define function. But at present sequences can not be associated with particular structural motifs. There is thus a need for studying structure of proteins and protein domains (structural proteomics) on large scale in order to assign function based on structural relationships. The effort will be to define the extent of variation in structural motifs. The function of a newly discovered protein is found if its structural folds relate to motifs with known function in the database. The homoelogous relationship (synteny) between genomes of various crops have been worked out. An **annotated genome** includes in addition to the DNA sequence itself, a description of the likely function of each gene product derived from comparisons with other genomic sequences and known protein function.

Cross-species sequence comparison is a powerful method for identifying functional regions in a genome. Evolutionary conservation has guided in the discovery of many novel genes and regulatory elements. The evolutionary conserved regions are a hunting ground for transcriptional regulatory signals which determine when, where and in what quantities genes are expressed. Besides these, genetic variation in these elements may be responsible for individual variability of gene expression which can therefore define susceptibility to disease (Stranger et al., 2005). Although sequences for coding proteins are strongly conserved across species, they encompass a small portion of a vertebrate genome. Some fraction of non-coding sequences is also conserved in the phylogeny of vertebrates.

Gramineae family—Comparison of genetic map of wheat, rye and barley shows that apart from a number of gross chromosome rearrangements such as the position of nucleolus organizer regions, the order of loci is very similar reflecting a general evolutionary conservation of linkage groups structure. There is presence of extensive homoeology in several of the genomes of wheat, rice and maize. Many wheat chromosomes contained homoeologous genes and genomic DNA fragments in an order similar to that found in rice. Gene content and order are highly conserved at both the map and megabase level between different species within the grass family but the amount and organization of repetitive sequences have diverged. **Macrosynteny**- The conserved order of large genomic blocks in the megabase range in the genomes of related (but also unrelated) species can be detected by techniques such as chromosomal *in situ* suppression hybridization and FISH. The comparative genomics has revealed that cereal genomes are composed of similar genomic building blocks (linkage blocks). **Microsynteny**- The conserved order, sequence and orientation of gene, conserved gene repertoire and conserved gene spacing (similar length of intergenic regions) in the range of about 100-500kb in the genome of closely related species can be detected. Microsynteny analysis using rice YAC revealed similarities in marker orders between rice, barley or wheat. Although gene sequences and their map orders are highly conserved but neither are size, sequence or composition of the DNA in between genes and thus most of the genetic diversity between species lies in the intergenic regions. Gene location in one species can be used to predict presence and location of orthologous loci in other species based on comparative mapping which accelerates map based cloning of orthologous genes. **Solanaceae family**- Potato genomes mapped with tomato random genomic and cDNA clones have shown that 1. nucleotide sequences are well conserved between potato and tomato and the karyotypes of these two related genera are very similar at the diploid level. 2. five major paracentric-like inversions constitute the major difference between them and 3. homosequential alignments of DNA markers between potato and tomato were recognised. The genome size estimated from nuclear DNA content of diploid by flow cytometry has shown 1.9pg in case of potato and 2.0pg in case of tomato. The genome size varies between 700cM-1600cM. Tomato markers hybridized with the potato genome and potato markers hybridized to tomato genome show similar (G+ C)%, 37.4 in potato and 41.4% in tomato. Only small structural differences could be detected between potato and tomato.

28.8 COMPARATIVE GENE MAPPING

It is a technique through which known coding sequences from one species are used to screen and map related coding sequences in other species. In this technique all available sequences for a particular gene from different species are first aligned and highly homologous regions are identified. The primers are then designed (e.g for adjacent exons) and used to amplify the intervening introns to detect sequence polymorphism. Comparative genomic based approach is used to identify gene regulatory sequences primarily on a gene-by-gene basis. It can further be used for identification of sequence signature. Identification of core promoter elements based on their defined position immediately upstream if each gene and their nearly universal activation by RNA polymerase II. Identification of distant gene regulatory sequences (enhancers) that direct precise spatial and temporal patterns of expression has been limited despite their established roles in development, phenotypic diversity and human disease.

28.9 COMPARATIVE GENOME MAPPING

The comparative genome mapping aims at the detection of conservation of the overall karyotype, the chromosomal architecture (synteny) and the gene order in genomes of related species. The comparative genetics uses conserved gene sequences (e.g. exons) from a gene of one species to detect similar or identical sequences in other species. RFLP probes from rice can be used to map corresponding sequences on wheat genome especially since they are sufficiently homologous to cross-hybridize. Thus construction of genetic linkage map in species of unknown genomic composition can be done by using probes from a related species.

Model plant species—The model plant species are *Arabidopsis thaliana* for dicots and rice (*Oryza sativa*) for monocots. Arabidopsis has one of the smallest genomes observed in higher plants with very low levels of repetitive DNA. Rice has one of the smallest genomes among the monocots. Rice has genome size of 430Mb, extensive genetic map with 3267 markers with relative ease of transformation and synteny with other cereal crops. Average exon size is comparable to that of Arabidopsis but average intron size is about 3.6 times larger. Average transcriptome size is similar (120kb) in both species. The G + C content of coding and noncoding regions in rice are higher than in Arabidopsis. Rice coding regions are especially rich in G+C. This characteristic is reflected by the biased usage of G/C at the third position of codons within the predicted gene. Buoyant density studies have shown that rice genes are localized in (G+C) rich islands that occupy 24% of genome (Sasaki et al., 2002). Collinearity between dicot and monocot species suggests extensive genome reshuffling which has occurred since their divergence. The comparison between rice subspecies as well as varieties shows two principal sequence differences including SNPs and Indels which might be useful for dissecting the molecular basis of the phenotypic differences that are often associated with different subspecific or varietal lines.

Types and relative numbers of genes in rice look very similar to those found in Arabidopsis. Comparative genetic maps within grass family show extensive regions of conserved gene content and order. About one-third genes found in these two species are not found in any fungal or animal genome sequence so far. These include many thousands of genes involved in photosynthesis and photo-morphogenesis. Plants contain many genes particularly those involved in basic intermediary metabolism or gene replication, repair and expression that are very similar to those in animals and fungi. Gene families are common in all eukaryotes and both rice and Arabidopsis have got higher copy number of many gene families than have been observed in fungi or animals. A large percentage of these genes are on unlinked chromosomes, among collinear clusters of other duplicated genes which suggesting evolution of rice and Arabidopsis involving polyploidization and/or segmental duplication. Many of these genes encode proteins with very different functions in plants and animals.

More than 80% of the genes that have been annotated in Arabidopsis are also found in rice. Further, more than 45% of the predicted rice genes do not have homologs in Arabidopsis. Most of these 'extra' rice genes are hypothetical genes. Most of the rice genes that are not found in Arabidopsis are artifacts of annotation. Comparisons of intraspecific whole genome sequences in rice or Arabidopsis can reveal SNP genomic regions with markedly low or high levels, possible indicators of positive or balancing selection, both of which are signatures of adaptive evolution (Nelson, 2001). The mark of selection on candidate loci identified in this way can then be verified by sampling more individuals within the species. The occurrence of intraspecific variation for phenotypic traits of interest also allows the identification of genes responsible for these traits for associations among naturally occurring genome-wide polymorphisms using linkage disequilibrium mapping- a technique whose viability is SNP.

Comparison of genome of different species has shown that the predicted gene number in *C. elegans* is higher than in Drosophila. Arabidopsis (weed) has about as many genes as humans have. Human have an average number of genes for a higher eukaryote. This result is surprising as it was through that the humans, the most evolved species could have the highest number of genes. But the fact is that the polyploids will have many more genes than humans. The estimates of the number of genes in different species are conservative because **hypothetical genes** (genes that were less than 300bp in predicted size) were not considered and further in the current annotation process great amounts of hand annotation were performed. Hypothetical genes are those genes that are not found in EST database.

28.10 COMPARISON BETWEEN SPECIES

Comparisons of genomes of different species have shown the following.

1. Human genome has about 25,000-40,000 genes which is only about twice the number needed to make a fruit fly, worm or plants.

2. There is high degree of alternative splicing in human than in other species which shows that more proteins are encoded per gene in human than in other species.

3. Physical and behavioral differences between species are not related in a simple way to gene number.

4. There are four times as many genes in some gene families in human genome as compared to fruit fly.

5. Although 90% of the domains (discrete structural units of proteins) found in human proteins are also present in fruit fly, worm proteins, they have been shuffled to create nearly twice as many different arrangements in human. This shows evolution of new domains in vertebrates.

6. 60% of human predicted proteins have sequence similarity to proteins from other species.

7. Just over 40% of the human predicted proteins show similarity with fruit fly or worm proteins.

8. 61% of fruit fly proteins, 43% of worm proteins and 46% of yeast proteins have sequence similarity to predicted human proteins.

9. 1/3 of yeast, fruit fly, worm and human proteins do not show strong similarity to known protein from other species and thus these might have acquired species-specific functions. Alternatively ORFs that encode these proteins are maintained in a new way, one that is independent of the precise amino acid sequence and thus is free to evolve rapidly. The rapidly evolving proteins are less likely to have essential functions and also less likely to be conserved during evolution.

All cells have at least one mechanism called **nonsense-mediated decay of mRNA**, for detecting imperfect ORFs, irrespective of the amino acid sequence that they encode.

28.11 COMPARISON OF DISTANTLY RELATED GENOMES

Comparison of distantly related species is important from evolution point of view. DNA segments that have a function are more likely to retain their sequence during evolution than non-functional segments. So DNA segments that are conserved between species are likely to have important functions. The ideal species for comparison are those whose form, physiology and behavior are as similar as possible but whose genomes have evolved sufficiently that nonfunctional sequences have had time to diverge. In practice, there may not be one ideal species as different genes and regulatory sites evolve at different rates.

28.12 COMPARISON OF CLOSELY RELATED GENOMES

The comparison of closely related genomes is done for determining the function of individual DNA segments. One of the uses of sequence comparison is to determine the structure of genes (exons and introns). High degree of alternative splicing in vertebrate makes this comparative approach particularly important. Gene finding algorithms can not easily predict the existence of alternate forms of an mRNA without experimental information and this information is difficult to obtain in case of rare mRNAs. Comparison of closely related genomes can also help in identifying gene control regions. Conserved sequences, in fact, correspond to functional control elements in individual genes. The proteins that control gene expression often detect sequence features that elude the best computer algorithms and may use information from contacts with other proteins that is difficult to model.

Differences in related individuals are generally attributed to changes in gene composition and/or changes in their regulation. The examination of the divergence of regulatory information has depended on the analyses of conserved sequences in putative promoter regions. These approaches are limited as TFBSs are often short and degenerate, thus making computational detection difficult. Further, the assumption of the conservation of motifs across species precludes the detection of sequences that are evolutionary divergent. TFBSs have diverged substantially faster than ortholog content. The gene regulation resulting from TF binding is likely to be a major cause of divergence between related species (Borneman et al., 2007). The detection of binding sites with CHIP-chip analyses offers the ability to globally map TFBSs experimentally rather than computationally.

28.13 COMPUTATIONAL GENOMICS

It refers to development and application of software programs in the evaluation and management of sequence data of different genomes. It aims at closing the sequence gap and relates the sequence to function, i.e. identification of splice sites, start codon, stop codon and determination of the function or possible function of the encoded protein, avoiding misinterpretation and false signal background. *In silico* **gene-** It refers to an ORF detected by evaluation of primary sequence data of an organism using software programs. The software program is also used to screen genomic sequences in the database for novel genes or potential splice variants- a technique called *in silico* **gene hunting**. *In silico* **mapping-** It is a variant of comparative mapping used for the construction of physical map of unsequenced genome or part of it using the available sequence database and *in silico* **map** refers to a physical map produced using *in silico* mapping. For example, the sequence databases of rice genome (which is almost completely sequenced) can be screened for an interesting gene, the gene located on a BAC clone and this clone or its orthologue (i.e. a homologous clone from another species) can be used to construct a physical map of the region around the same gene in rye, *H. vulgare* genome. *In silico* mapping exploits the microsynteny and/or macrosynteny between related (and also less related) genomes. *In silico* cloning – It refers to application of software packages to design and virtually perform a complete cloning procedure. Such computer programs also allow to document any step in recombinant DNA experiments. *In silico* transcriptomics- It refers to screening of cDNA and/or expressed sequence tag data for genes conditioning a particular cellular state (e.g disease, stress, inflammation , tumor or injury). *In silico* **proteomics**- It refers to study of protein data bases for comparing primary sequences, secondary and tertiary structures, to extract information about functional or structural domains, to catalogue protein families and cluster and to relate any proteome to already finished proteome.

28.14   COMPARATIVE GENOMIC COMPARISON

After sequencing of genome one would like to identify functional elements encoded in a genome (Kellis et al., 2003). First of all, one would like to identify the genes encoding protein sequences. *De novo* identification of complete set of protein-coding sequence remains imperfect. Results from *de novo* gene prediction are not reliable and so gene identification is typically based on a comparison of genomic sequences to known mRNA or to genes in other organisms.

Identification of functional regions in genomes can be carried out by searching for conservation among genome sequences as functional regions are thought to be under stabilizing selection and should be preferentially conserved over evolutionary time. This approach has been use successfully in the annotation of animal and yeast genomes (Miller et al., 2004) but has not yet been used extensively in plants as only two complete and quite distant plants genomes are currently available. Both coding and regulatory regions can be identified by locating genomic areas under purifying selection although regulatory regions tend to be less conserved.

28.15   GENOME EVOLUTION

With the availability of whole-genome sequences as well as other genomic resources such as microarray methods, EST libraries, high throughput resequencing technologies, the comparative method will draw inferences about the evolution of genome structure and function. Comparisons of whole genome sequences at both intraspecific and interspecific levels will address several key issues surrounding the evolution of genome architectures. These include the extents and rates of change in genome structure and size, patterns of large scale genome duplications (including polyploidization), the dynamics of the origins and extinction of genes, role of selection acting on large-scale variation in genome structure and organization, the evolutionary forces that determine transposable element activity and number and the functional consequences of these mobile elements and the extent and impact of epigenetic markings on genome and organismal evolution (Caicedo and Purugganan, 2005).

The genome evolution at large scale involves the rapid structural evolution in the telomeric regions which can also be observed at the gene level. One-to-one matches are used as orthologous landmarks to define the large scale alignment which showed strong conservation of synteny across the species. First, the gene families show significant changes in number, order and orientation. Secondly, there is reciprocal translocation, inversion, segmental duplication. Reciprocal translocation can be identified by pulse-field gel electrophoresis. Sequences at the chromosomal break points suggested possible mechanisms which underlie the rearrangements. Inversions are flanked by tRNA genes in opposite transcriptional orientation and usually the same isoacceptor type. Reciprocal translocations occurred between Ty elements in seven cases and between similar pairs of ribosomal protein genes in two cases. Finally, one segmental duplication involves 'donor' and 'recipient' regions that are descendants of an ancient duplication in the yeast genome.

One can also study the genome evolution at the nucleotide level by generating local nucleotide level alignments around each orthologous ORFs.

Comparative genome analysis of a handful of related species has the power to identify genes (species specific genes, rapid evolving genes, slow evolving genes), define gene structure, highlight rapid and slow evolutionary change, recognize regulatory elements and reveal combinatorial control of gene regulation.

Causes for change in amount of coding DNA during evolution

During evolution the DNA content and in particular the amount of coding DNA has increased. This additional DNA (the new genetic material) has come not as a result of *de novo* synthesis of DNA but because of duplication of pre-existing DNA followed by sequence divergence, i.e. duplication followed by a frameshift mutation. Duplication can occur in many ways: (i) unequal crossingover (ii) transposition and (iii) polyploidy. Single genes, chromosomal regions and even entire genomes can undergo duplication. Unequal crossingover can lead to the generation of tandemly arranged copies of the same DNA sequence. It can also give rise to repeated segments within a gene. Thus repeats (duplications) can be of whole genes or parts of genes. Once crossingover has generated a tandem repeat, further duplication events are more likely as chromosome can pair 'out of register' (either in meiosis or between sister strands in mitosis). Although unequal crossing-over provides a mechanism for concerted evolution of tandemly repeated units, it can not explain the concerted evolution of a set of similar units spread around the genome as if pairing and crossingover take place between such units then the result would be a major chromosomal rearrangement (in fact, a translocation, if elements were on different chromosomes) which would lead to lowered fertility. Gene duplication is a major source of new genes and functions. The duplicated genes are free to evolve new functions as the original function is maintained by the other copy. However, studies have revealed that most new genes do not have novel functions and instead, paralogous gene pairs are often 'subfunctionalized' with two or more functions being partially or completely subdivided between the genes after gene duplications.

28.16   CONSEQUENCES OF GENOMIC DUPLICATION

Consequences are genomic duplication are as follows.

1. Loss of all or part of duplicated sequences through deletion or degeneration causing non-functionality. This will lead to rapid speciation- gene loss in response to whole genome duplication.

2. One or both duplicated sequences might acquire new functions. They might undergo mutations leading to new characteristics and functions, i.e. duplicated genes with divergent functions. Gene mutation results in abnormal levels or functions of their protein products.

3. New functional sequences arise mainly through differential degeneration of the two copies so that the newly evolved functions mutually complement the original pre-duplication function. This is common in multifunctional genes and can occur in one of the three ways. The two copies can evolve different functions which may be overlapping or non-overlapping, permitting complementation and maintenance of the original function as well as expansion of functionality. Alternatively in bi-functional or multifunctional genes in which different functions are encoded by different sequence modules within the same gene, duplicated copies can reciprocally loose or change function in different modules. This could also result in the coverage of original function and its possible extension. Finally, regulatory differentiation may occur permitting differential spatial or temporal expression of one or both copies. New function can arise through a combination of these processes. From the study of evolution of one pair of duplicated genes (GAL1 and GAL3) in *S. cerevisiae*, Hittinger and Carroll (2007) concluded that they diverged following a combination of differentiation and degeneracy in different modules. In case of duplication of genes (Hst1p and Sir2p in *S.cerevisiae*) encoding the histone deacetylase enzymes, Hickman et al. (2007) concluded that although they have different functions, they use the same biochemical process for their activity. The functional differentiation of these two enzymes is due to their distinct protein interaction domains which allow the selection of different protein partners and complex associations. Consequently, Hst1p is involved in the repression of gene expressed in meiotic cell division and Sir2p functions in a process called silent mating type duplication.

Comparative genomic analyses of several sequences from species of yeast, *S.cerevisiae*, have shown the following general principles underlying the evolution of duplicated genes (Louis, 2007).

1. A common outcome is loss of one of the two copies of the duplicated gene by deletion or degeneration.

2. The two copies of a gene can diverse in sequence either maintaining the ability to complement or substitute for each other or diverging completely to have non-overlapping functions.

3. Differential functional modules of the two copies are lost or diverge so that together the products of the two copies perform the original function but each is determining a single aspect of the function.

4. The regulatory regions of the two copies can diverge, permitting spatial and temporal expansion of functionality as well as enhanced expression level.

New functions rarely appear to arise from changes in biochemical properties but are frequently associated with regulatory changes. Divergence rate of DNA sequences encoding genes is generally slower than non-coding sequences.

Transposition leads to replication of a DNA sequence and this copy is inserted at a new site in the chromosome. This sometimes involves transcription to RNA and reversed transcription to DNA before insertion (see chapter 22 for detail).

28.17 SELECTION OF SUITABLE SET OF RELATED SPECIES

The optimal choice of species depends on multiple considerations largely related to the evolutionary tree connecting the species. In evolutionary tree the branch length denotes the number of substitution per site in genic and intergenic sequence. Pair-wise distances are calculated suing the Kimura two parameter model and tree are constructed using the UPGMA method (Roy, 2009). The relative rate of closet species is computed suing the neighbor-joining method.

1. The branch length, t, between species should be short enough to allow orthologous sequence to be readily aligned.

2. The total branch length of the tree should be large enough that non-functional sites will have undergone substantially more drift than functional sites thereby providing an adequate degree of signal to noise enrichment.

3. The species should represent as narrow a taxon as possible as we seek to identify genomic elements common to the species, it can explain only aspects of biology shared across the taxon.

Signal-to-noise ratio is high in yeast genome as protein coding regions comprise approximately 70% of the genome and regulatory elements comprise perhaps about 15% of the intergenic regions. But human genome has much lower signal-to-noise ratio and the corresponding regions account for only ~2% and ~3%, respectively. Enrichment can be made in case of human genome by filtering out the repeat sequences that comprise half of the human genome, i.e. separate euchromatin from heterochromatin. Further enrichment can be obtained by increasing the greater number of species.

Combinational control — There is presence of a limited number of genome-wide regulatory motif but there is large number of transcriptionally regulated processes. This suggests regulation to be dependent on combinatorial control. There should then be a search for motifs that occur in the same intergenic regions much more frequently than would be expected by chance. In a single genome few significant correlations are found (functional instances the motifs are overwhelmed by a much larger number of random occurrences). Cross-species conservation greatly reduces this random noise and reveals biologically meaningful correlations.

Proteins, the products of genes are embedded in a large scale interaction networks that represent integrated functional units at the molecular genetic level. Thus to understand the evolution of function, it becomes essential to understand the evolutionary dynamics of molecular genetic networks. Expression profiling with microarrays combined with EST and genome sequence data allow to identify the interacting candidate genes of genetic networks and examine how the patterns of interactions change across species.

29

In Vivo and *in Vitro* Systems for Transcription and Translation

29.1 NEED FOR CELL FREE SYSTEMS

Functional analysis of cloned DNA fragment (gene (s)) involves characterization of its gene products (RNA or proteins) and regulation of their synthesis in appropriate *in vivo* (within the living organism)and *in vitro* (in the test tube)systems. The gene products, RNAs transcribed from the fragment can be characterized and their regulatory sequences (operator-promoter) identified. The assays of the gene products depend on the functioning of the gene product (e.g. enzyme activity) or its structure (e.g. immuno-detection). As we know that the objective in cloning is to isolate a particular piece of DNA or mRNA copy in a pure state, the question is how to identify which particular recombinant DNA clone in the large population of clones contains the desired gene. There are two ways of identification of recombinant DNA. Identification of cloned DNA by 'indirect expression', i.e. by study of expression of the mRNA after selection by DNA-RNA hybridization in a cell-free system (HAT, HART) and by 'direct expression' of protein product. The method of direct expression is particularly useful for low-abundance mRNAs for the protein products of which no sequence data are available but against which antibodies have been raised. In both cases the gene product, protein, can be identified immunologically, or electrophoretically or in particular case by its biological activity. As the screening of large number of clones by these methods is difficult, one can only select mRNAs of particular size range using sucrose density gradients. If antibody is available then one can attempt to enrich the mRNA by immunological isolation of polysomes containing nascent peptides of protein. The particular recombinant DNA can also be identified by random sequencing to search for cDNA clones corresponding to abundant proteins of known sequence (Constanzo et al., 1983).

There are a number of systems available for studying the translation of cloned gene in eukaryotic cells but all these systems involve use of intact cells. The intact cell system of microinjection into frog oocytes (or eggs) has the advantage that translation may continue for several days and thus allows transient transcription of a cloned gene but cell-free systems has greater utility in non-specialized laboratory. The large size of the nucleus in the oocyte facilitates this operation. This technique requires special equipment and considerable practice. Now more generally accessible is the technique of calcium phosphate co-precipitation of DNA onto the tissue culture cells which take up the precipitate by a process called **phagocytosis**.

The **cell free system** is a general term used for a mixture of cytoplasmic components required for *in vitro* translation. This system is free from intact cells, membranes and nucleic acids. Cell-free extracts are required for *in vitro* transcription as well as *in vitro* translation. *In vitro* translation is a method to translate isolated and purified mRNAs into their corresponding proteins *in vitro*. The mRNA preparations are mixed with cell extracts. In case of prokaryotic mRNA translation the cell extracts from *E. coli* or *Bacillus stearothermophilus* are used whereas in case of eukaryotic mRNA translation cell extracts from rabbit reticulocyte lysates (Pelham and Jackson, 1976), *Xenopus laevis* oocytes or wheat germ (Roberts and Paterson, 1973; Anderson et al., 1973) are used.

29.2   CELL-FREE EXTRACT

Cell-free extracts are used as starting material for *in vitro* experiments (e.g *in vitro* translation) or for purification of macromolecules such as proteins, nucleic acids. It is usually a buffered solution containing all soluble components of cells that have been obtained after rupturing by physical (e.g sonication), chemical (e.g. osmotic shock) or enzymatic means. The subcellular particles (e.g plastids, mitochondria, nuclei, dicotyosomes) as well as membranes are removed by centrifugation. The two most widely used systems for cell –free translation are derived from specialized cells and are described below.

29.2.1   Wheat Germ System

It is a complete *in vitro* translation system obtained from wheat embryo (Marcu and Dudock, 1974). It includes ribosomes, a full complement of tRNAs, aminoacyl tRNA-synthetases, amino acids (20 amino acids with usually one of them radioactively labeled , for example, 35S methionine) , enzymes, initiation, elongation and termination factors, K^+ and Mg^{2+}, ATP and GTP and an energy regenerationg system (e.g. creatine phosphokinase). This system is used to translate heterologous mRNAs into proteins *in vitro*. Since the wheat germ is practically free from endogenous mRNA any translation activity is fully dependent on exogenous mRNAs. In other words, wheat germ system may translate weakly initiating mRNAs more efficiently. However, large mRNAs, coding for proteins of more than 80kDa are only inefficiently translated as their synthesis is frequently not completed. The biggest problem with this system is its tendency to produce incomplete translation products due to its premature termination.

29.2.2   Rabbit Reticulocyte Lysate

It is also a complete *in vitro* translation system made from rabbit reticulocyte (Pelham and Jackson, 1976), includes ribosomes, a full complement of tRNAs, aminoacyl tRNA-synthetases, initiation, elongation and termination factors. This system is treated with micrococcal nuclease which destroys endogenous mRNAs. Thus the translational activity of the lysate is fully dependent on the exogenously added mRNAs. In addition to compounds essential for *in vitro* translation such as K^+ AMD Mg^{2+} ions, nucleotides like ATP and GTP and energy-generating system (like creatine phosphokinase) hemin can be added to ensure that the *in vitro* translation of exogenous mRNA proceeds to completion. The lysate is usually aliquoted and stored over liquid nitrogen.

The choice between the two systems will depend on the type of mRNAs to be translated. The main disadvantage with reticulocyte system is that it requires removal of endogenous mRNA. Although endogenous mRNA can be removed by incubation with a Ca^{2+}-dependent endonuclease which is subsequently inactivated by EGTA, a chelator of Ca^{2+} significant amounts of mRNA fragments including sequences which contain initiation sites remain. And so the added mRNAs will have to compete with these fragments for ribosome binding sites and thus weakly initiating mRNA species may not be translated efficiently (Jagus, 1987). The advantage with this system is that the requirement of higher quantity of reticulocyte can be met as one rabbit will yield 20-30ml of reticulocyte lysate.

29.2.3 Prokaryotic Cell-Free System

The prokaryotic cell-free system for translation was originally developed by Nirenberg and Matthaei (1961) for their work on genetic code. The ribosomes, tRNAs and soluble factors required for translation of exogenous mRNAs are present in the supernatant obtained after centrifugation of the disrupted cells.

29.2.4 Use of Human-Rodent Hybrid Cells

Human-rodent cell hybrids have played an important role in the analysis of human chromosome. Mammalian cells in culture can be induced to fuse and thus either human–mouse or human-hamster cell hybrids can be obtained. Cell fusion is followed by nuclear fusion and the hybrid cells thus formed contains initially diploid sets of chromosomes of both species. But as growth proceeds human chromosomes tend to get lost and thus subclones containing one or a few human chromosomes can be isolated and the identity of the surviving human chromosome can be established by chromosome banding. Further, they can be used as a source of chromosome specific probes. This hybrid cell technique can be used for isolating human segmental chromosome involved in disease condition, NF. This technique can also be used for isolating fragments of human chromosomes. Human cells or better hybrid cells with just one human chromosome can be heavily irradiated with X-ray to break the chromosome into a number of fragments before fusion with rodent cells. Few fragments of human chromosome may persist autonomously but some might be integrated into the rodent chromosomes. The presence of a small fragment of human chromosome can be detected by the probe specific for human *Alu* element, if not by chromosome banding. Further, as the different cell lines contain different and overlapping segments of the same chromosome, they can be used for mapping-**radiation mapping** study.

29.3 *IN VITRO* TRANSCRIPTION

In order to study the mechanisms of transcription as well as to identify the transcription factors and DNA sequences that control and regulate gene expression, cell free systems which accurately and specifically transcribe exogenously added DNA are required. *In vitro* systems have been developed in which accurate transcription of all three types of RNA polymerases-RNA polymerase I, II and III can be obtained. There are two basic approaches that have been successful. One approach is to prepare concentrated cell lysates which contain not only the factors required for transcription but also sufficient amounts of RNA polymerase so that addition of purified enzyme is not needed. *In vitro* transcription (*in vitro* RNA synthesis, cell free transcription) is a method to transcribe cloned genes into their corresponding transcripts using specially prepared whole-cell extracts from He La cells or Drosophila embryos and specific transcription vectors (expression vectors) that contain endogenous RNA polymerase II. This whole-cell extract initiates mRNA synthesis from exogenous templates *in vitro* (Manley et al., 1980, 1987).

In another approach purified RNA polymerase is supplemented with cell extracts which contain factors required for accurate transcription. This system is used primarily for RNA polymerase II-mediated transcription

29.4 COUPLED TRANSCRIPTION-TRANSLATION SYSTEM

It is a combination of expression vector and *in vitro* translation system (rabbit reticulocyte lysate or wheat germ system) which permits transcription of a cloned gene with simultaneous translation of the resulting mRNA into protein. Here the expression vector contains a polylinker flanked by an SP6 and T7 RNA polymerase promoter in opposite direction, a selective marker gene, *ori*, origin of replication (derived from *E. coli*) and a leader sequence for higher level of translation of any foreign sequences linked to the leader. A hybrid promoter is more effective then either of the promoters.

In vitro transcription-translation systems or transcriptional and translational fusions are often required to prove that an ORF in DNA actually encodes a protein.

29.5 *IN VIVO* TRANSCRIPTION-TRANSLATION SYSTEM

29.5.1 Mini Cells

If the host chromosome codes for a product that is similar or identical to that produced by the cloned fragment then it is essential to use mutant host cell that no longer synthesizes this product or that synthesizes a distinguishable version of it. If no assay exists for a product encoded by the cloned fragment then its detection in viable cells is rendered difficult due to high level of chromosomal gene products. In such situation it is necessary to suppress the expression of chromosomal genes and this can be accomplished by the use of 'mini cells'. Specific mutant (min A min B, the double mutant) strains of bacteria, *E. coli* or *Bacillus subtilis* produce two types of cells, one spherically shaped small cells (called **mini cells**') and the other normal wild type cells, continuously during their growth. Mini cells mutant are defective in cell division functions and septum, instead of forming in the middle of a growing cell, thereby dividing the cell into two equal halves and ensuring equal division of the chromosomes, forms close to one of the cell poles producing a large but otherwise normal cell and a 'mini cell'. The small 'mini cells' are easily separated from their normal sized cells by sucrose density gradient centrifugation. The mini cells contain plasmids (if the parental cell contains plasmids) but not the chromosomal DNA and are capable of synthesizing RNA and proteins and thus possess most of the metabolic potential of the normal cell. These mini cells thus can be used to detect the expression of plasmid borne genes and further proteins encoded by the genes can be characterized (Roozen, Fenwick and Curtiss, 1971). Mini cells are also useful for the identification of gene products encoded by DNA fragments cloned in λ vectors. In this case mini cells carrying a plasmid that encodes the λ repressor serves to depress the synthesis of λ proteins which otherwise are actively produced in the mini cells (Reeve, 1977).

29.5.2 Maxi Cells

Maxi cells are cells of a bacterial strain that is highly sensitive to UV irradiation. The irradiation with UV light of bacteria, *E.coli* or *Bacillus subtilis* (redA,uvrA) leads to degradation of the chromosomal DNA and thus cessation of the chromosomal DNA synthesis. But the plasmids are not damaged through UV irradiation and therefore continue to survive, replicate and express the genes. These '**maxi cells**' then can be used to study cloned genes in a system free from chromosomal background (Sancar, Hack and Rupp, 1979). As the 'maxi cells' can be produced more easily than 'mini cells', it is a rather more attractive system for the identification of products of actively expressed genes.

Although *in vitro* systems are more time consuming but they are more sensitive and informative as they enable the investigator to modify the synthesis conditions and to add or subtract components. The analysis of the transcriptional units of a cloned DNA fragment may be made using RNA polymerase holoenzyme (an enzyme that consists of two or more subunits and is active only when all the subunits are assembled, for example, DNA dependent RNA polymerase consists of core enzyme and a sigma factor) and radioactive ribonucleotides and either supercoiled plasmid or isolated fragment as template. There are reports that supercoiled DNA is required for the efficient transcription from some promoters (Levine and Rupp, 1978). The transcripts made *in vitro* or *in vivo* can be sized by polyacrlamide gel electrophoresis and their template regions located by hybridization to electrophoretically separated restriction endonuclease generated sub-fragments of the cloned DNA fragment which have been transferred to the nitrocellulose filter. Cell-free coupled transcription-translation systems (Zubay, Chambers and Cheong, 1970; Gold and Schweiger, 1971) have been successfully used to synthesize and analyse the regulation of plasmid – encoded proteins *in vitro*. Finally, comparison of proteins produced *in vitro* and *in vivo* can be made to investigate the possible processing of proteins which are localized elsewhere.

High throughput protein purification—One widely used method of purifying hundreds of proteins in parallel involves their expression in a heterologous system. Once a target protein is identified the corresponding cDNA is cloned into an expression vector for producing the protein in *E. coli* or any other organism. The purification of the expressed recombinant protein from bacteria involves cell lysis, incubation of the lysate with an affinity resins, washes and elution. Additional purification steps such as ion exchange and size exclusion chromatography (see chapter 30) are required for protein crystallography.

The limitation with the HT protein production is the expression of enough properly folded protein in bacterial cells. Proteins expressed in *E.coli* can accumulate as inactive, misfolded aggregates called **inclusion bodies**. Proteins in this form can be purified under denaturing conditions but need to be folded into their native conformations. Only a fraction of proteins can be overproduced in *E.coli* in sufficient quantity and without the formation of inclusion body aggregates or the proteolytic degradation of expressed proteins.

Alternative expression systems — The alternative expression systems include cell cultures from eukaryotic organisms such as insect cells and cell free, *in vitro* protein expression. Researchers have developed procedures for expressing proteins in insect cells or cell free extracts to overcome some of these problems.

Cell free systems — Markley and colleagues developed a pipe line using wheat germ cell free protein translation as a way to produce proteins for nuclear magnetic resonance structure determination. CFSs simplify the purification efforts as only the protein of interest is expressed and labeled. CFS requires smaller volumes, avoiding lengthy concentration steps and it lends itself to the labeling, a requirement for NMR. One can harvest twice as many folded proteins using cell free system than with expression in *E.coli*. The limitation with this system is about 79% of human protein will be expressed , about one half of these will produce protein in sufficient quantities for structural studies (10-20mg) and half of those will remain stably folded for NMR studies. Structural genomics or structural proteomics aims at providing three-dimensional information for all proteins. This information can be used to ascribe function to a protein and to reveal or invalidate drug targets.

It is essential to know where, when and to what extent a protein is expressed. Determination of the protein component of a given cell or tissue uses a two dimensional electrophoresis or liquid chromatography to resolve the proteins in the samples and mass spectrometry to then identify the individual proteins. But then it is difficult to identify rare proteins as the cell extracts are dominated by a few very abundant proteins. Thus the most abundant proteins from a sample are removed. Use of antibodies bound to inert beads which are then packed in liquid chromatography column or for smaller samples, spin columns is made. The enriched material is collected and fractionated. The proteins are divided in two dimensions. First in a two dimension gel samples are separated by isoelectric focusing and size and then injected into a liquid chromatography column.

30

Chromatography and Electrophoresis

30.1 CHROMATOGRAPHY

Study of macromolecules such as protein, nucleic acid, polysaccharides and lipids (not classified as micromolecule) involves the isolation and separation of compounds from the crude extract. When the cells are break open the macromolecules –proteins are released into solution which is called a **crude extract**. The crude extract is then subjected to treatments to separate proteins on the basis of size or charge into different fractions and this process is called **fractionation**. **Chromatography** and **electrophoresis** are two most valuable techniques for resolving mixtures of molecules with closely related chemical properties. These techniques separate molecules on the basis of their difference in mass, size, shape, charge and absorption effect. There are three components involved in the different techniques. The first component is the mixture to be resolved. The second is the stationary phase in the form of a column or film. The third component is a mobile phase which is a liquid in most cases and it produces a medium for the molecules to move through. The mobile phase is gaseous and acts as a carrier of the molecules to be separated which are in the form of vapor as in case of gas chromatography. The molecules move through the medium under the influence of a hydrodynamic or electrical force and are distributed between the two phases on the basis of differences in absorption, ion exchange, solubility or size depending upon the particular method applied. The stationary phase is solid and may play an active part as in case of absorption or ion exchange chromatography or may be present simply to support the aqueous phase and be relatively inert as in case of partition chromatography and zone electrophoresis. Chromatography is a process in which complex mixtures of molecules are separated by many repeated partitioning between a flowing (mobile) phase and a stationary phase. It differs from electrophoresis in use of electric field. No electric field is applied in chromatography.

Dialysis—But first the most inexpensive method of purification such as dialysis should be used. Dialysis is a procedure used for gross separation of large and small molecules and proteins are separated because of its larger size. The partially purified extract is filled in a bag or tube of a semi-permeable membrane. When it is suspended in a buffered solution of an appropriate pH ionic strength the membrane allows the exchange of salts and other smaller molecules and buffer but not proteins. The dialysis thus retains large molecule, the proteins within the bag or tube whereas the other solutes in the protein preparation pass freely through the membrane and come to an equilibrium with the solution outside the membrane. It might be used for removing ammonium sulfate from the protein preparation. It is advisable to desalt the sample before subjecting to chromatography.

Column chromatography — The column chromatography separates proteins on the basis of charge, size, binding affinity and other properties of protein. The column is a solid porous material made up of some form of plastic and constitutes the stationary phase. The buffer solution which flows through the solid matrix constitutes the mobile phase. The solution (mixture), protein preparation is supplied from the reservoir at the top and the effluent is constantly replaced and additional solution mixture is added. The protein solution layered on the top of the column percolates through the solid matrix and forms bands within the mobile phase. As proteins move through the column they are retarded to different degrees by their interaction with the matrix. The overall protein band thus widens as it moves through the column. Individual types of protein gradually separate from each other, forming bands within the broader protein. The rate at which the protein solution can flow through the column usually decreases with the length of the column. Separation (resolution) improves as the length of the column increases. Individual proteins migrate faster or more slowly through the column depending upon their properties (size, charge, binding affinity, etc.).

High performance liquid chromatography — It is an improvement over column chromatography. HPLC makes use of high pressure pump that speed up the movement of protein molecules down the column as well as high quality chromatographic material that can tolerate the crushing force of the pressurized flow. It reduces the transit time on the column and thus can limit diffusional spreading of protein band and thus greatly improves resolution. It can also be used to determine the types and amount of monosaccharides.

Ion Exchange Chromatography — It is used for purification of proteins. It exploits the differences in the sign and magnitude of net electrical charge of protein at a given pH. In IEC separation of compounds is done through the use of insoluble matrix containing labile ions capable of exchanging with ions in the surrounding media. There are two types of IEC.

1. Cation exchange chromatography
2. Anion exchange chromatography.

The insoluble solid matrix, made up of synthetic polymer, has negatively charged groups (anionic group) bounded to it in CEC whereas in AEC the solid matrix is bound with positively charged groups (cationic group). The solid matrix in CEC has affinity for positive ions and thus proteins with a net positive charge will migrate through the column more slowly in comparison with a protein with net negative charge. In case of CEC proteins with a more negative net charge will move faster and elute earlier. The affinity of each protein for the charged groups on the column is affected by the pH (which determines the ionization state of the molecule) and concentration of the competing free salt ions in the surrounding solutions. The resolution is optimized by gradually changing the pH and/or salt concentration of the mobile phase and thus with the formation of a pH or salt gradient. It can be used to separate oligonucleotides in the mixtures.

Size exclusion chromatography—SEC (or gel filtration) separates proteins on the basis of differences in size. The column matrix is made up of a crossed-linked polymer (gel) with pores of selected size. Protein solution (mixture of large and small molecules) is placed on the top of the column and they pass down the column, larger molecules will migrate faster in comparison to smaller protein molecules as they are too large to enter the pores in the beads and hence take a more direct short route through the column. The smaller proteins molecules enter the pores in the beads, become trapped and are slowed and follow a longer path. Eventually, complete separation occurs with the large molecules leaving the column first and the smaller ones last. It is opposite to method of molecular sieving in which the small ones emerge first while the large ones are held back. It can be used to separate oligosaccharides from the mixtures of oligosaccharides.

Affinity chromatography—In this technique the separation of protein is based on their differences in binding specificities. The macromolecules retained on the column are those that have been specifically bound to a ligand cross- linked to the beads. The term ligand refers to a group or a molecule that binds to a macromolecule such as protein or nucleic acid. Macromolecules that do not bind to the ligand are washed through the column. The bound protein is eluted (washed out of the column) by a solution containing free ligand. Lectin affinity chromatography can be used to separate oligosaccharides from the mixture.

In case of purification of proteins, early fractionation of protein should be based on differences in protein solubility which is a function of pH, temperature, salt conc. and other factors. The solubility of a protein is generally lowered at higher salt conc.- an effect called **salting out**. Thus use of right salt conc. will selectively precipitate (separate) some proteins while others remain in the solution. The ammonium salt, $(NH4)_2$ SO4, which is highly soluble in water can be used for this purpose. So it would be better to start with *salting out* procedure (the most inexpensive procedure) and then use different methods in sequence. The sample size become small after each purification step and then finally use the most sophisticated and expensive chromatographic procedure. Initially the sample size is large and if chromatography is used then it will require enough amount of chromatographic material and that is why it is not used in early stages of purification of proteins. Figure 30.1 shows separation and purification of transcription factor Sp1, a protein through the use of Sepaharose affinity chromatography column.

Paper chromatography—The macromolecules that are readily soluble in organic solvent but less soluble in water can be separated by adsorption chromatography whereas ionizable water soluble compounds are best separated by ion exchange chromatography and compounds that are soluble both in water and organic solvent can be readily separated by partitioning chromatography. Before using paper chromatography the protein sample is desalted as the presence of excess salt will result in a poor chromatogram with spreading of spots and changes in their Rf values (the ratio of distance traveled by a compound to that moved by the solvent). The apparatus consists of a sheet of Whatman No. 1 paper supported on a frame so that one edge is in contact with a trough filled with the solvent the choice of which depends on the mixture investigated (Consden, Gorden and Martin, 1941). The sample is then applied to the paper with a micro-pipette. The mixture is spotted on to the paper and dried. After that the solvent is allowed to flow along the sheet either by gravity (descending chromatography) or capillary attraction (ascending chromatography) for almost a day. The

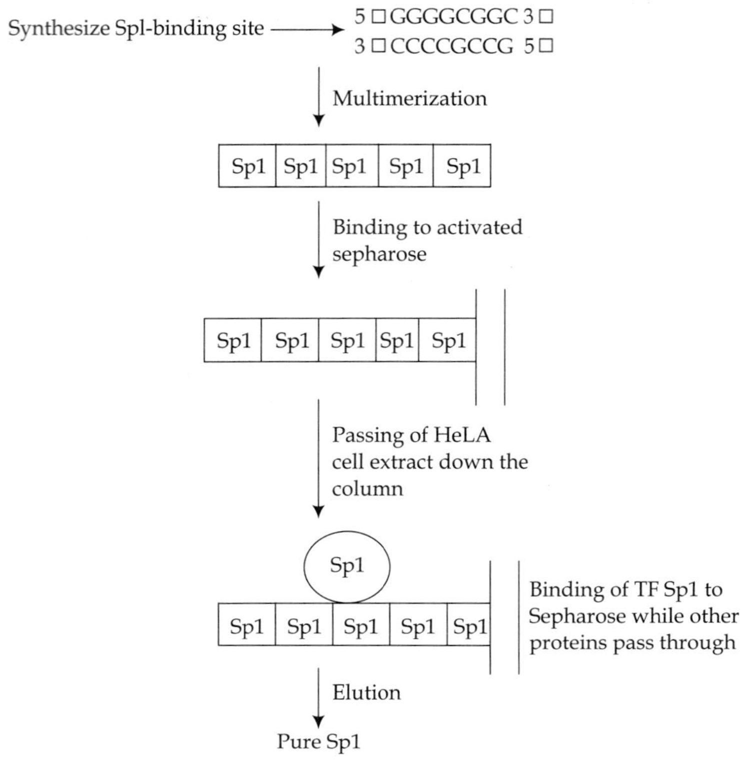

FIGURE 30.1   Showing purification of transcription factor Sp1 on an affinity column using coupling of multiple copies of Sp1 binding site on a Sepharose support (Kadonaga and Tjian, 1986).

solvent front is marked and after drying the paper, the positions of the compound present in the mixture is visualized by staining. It can be used for separation of sugars. One can go using two dimensional paper chromatography for separation of a mixture of amino acids. Ascending chromatography is more frequently employed and it has the advantage that separation can be carried out in the two dimensions. In this two dimensional chromatography the mixture is separated in the first solvent which should be volatile and then after drying the paper is turned through 90° and separation is carried out in the second solvent. The compounds (e.g. different AAs) can be identified by comparing their positions with a map of known compounds developed under the same conditions. Figure 30.2 shows the fingerprint (the separated peptide fragments) pattern in case of normal and sickel cell anemia hemoglobin. The polypeptide chains are broken by trypsin and the peptide fragments are separated on paper by electrophoresis followed by chromatography at right angles to the direction of electrophoresis.

Thin layer chromatography- In this technique the separation of compounds is made on a thin layer of supporting medium. It is very similar to paper chromatography but has the added advantages that a variety of supporting media and mixtures can be used. Fluorescent dye can be added to the medium in order to identify the spots. Separation can be by

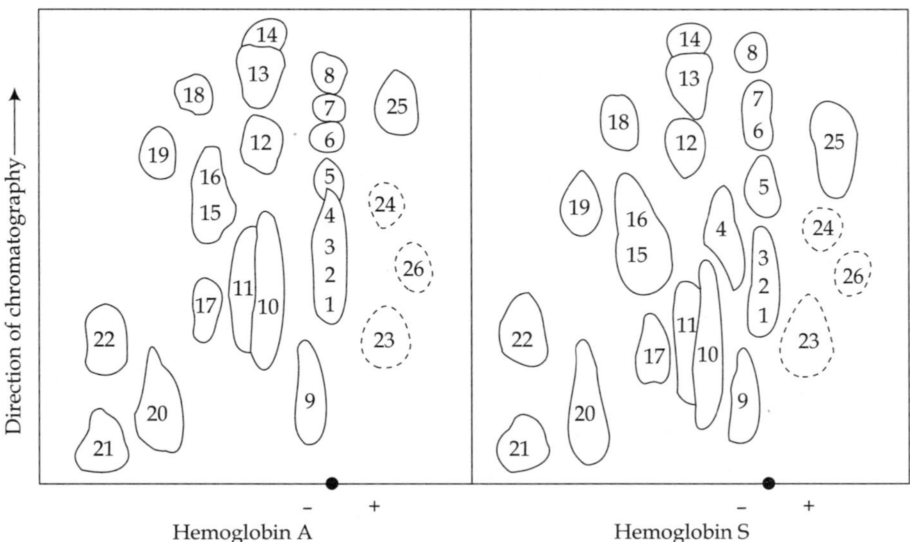

FIGURE 30.2 Showing the "fingerprint" patttern of normal hemoglobin, A, and of sickle cell anemia hemoglobin, S.

adsorption, ion exchange, partitioning chromatography or gel filtration depending upon the nature of medium used. Separation is very quick and completed within an hour. As the spots are very compact so it is possible to detect compounds at a low conc. Than on paper chromatography. Further, separated compounds can be detected by corrosive sprays and elevated temperatures with some thin layer materials which is not possible with paper chromatography.

In case of lipid separation TLC on a silicic acid employs the same principle as the adsorption chromatography. A thin layer of silica gel is spread onto a glass plate to which the lipids adhere. A sample of lipids is applied near one edge of the plate which is dipped in a shallow container of an organic solvent or solvent mixture and this whole system is enclosed within a chamber saturated with the solvent vapor. As the solvent rises on the plate by capillary action it carries with it the lipids. Neutral or less polar lipids move faster and occupies the highest position whereas the polar lipids bind to silicic acid and thus move slowly and occupy the lowest position. The lipids can be detected after separation by spraying the plate with a dye (rhodamine) that fluoresces when associated with lipids or by exposing the plate to iodine flumes which reacts with unsaturated fatty acids and develops a yellow brown colour.

Adsorption chromatography—In this technique the column is packed with adsorbents (Tswett, 1903). Some of the common adsorbents used are alumina, silica gel, calcium phosphate, charcoal and sucrose. The adsorbent properties of some of these materials can be modified by pre-treatment with acid or by heating. To start with the column is filled with solvent and then slurr of adsorbent is added slowly. The adsorbent is allowed to settle and the excess solvent is allowed to run off and this process is repeated until column is the height required and after that the column is thoroughly washed with solvent and the level of liquid is allowed to fall to just above the surface of the adsorbent. When the aim is to separate lipids of different polarity an insoluble material such as silica gel (a form of silicic acid, Si (OH)4) is packed into a glass column and the lipid mixture is applied at the top. Polar lipids bind lightly to silica gel while the neutral lipids don't and thus pass directly through the column and are the ones that emerge in the first chloroform wash. The polar lipids are then eluted on order of increasing polarity by washing the column with solvents of progressively higher polarity. Uncharged but polar lipids are eluted with acetone and very polar or charged lipids (e.g. glycerophospholipids) are eluted with the methanol.

Gas liquid chromatography — In GLC the mobile phase is gaseous and acts as a carrier for molecules which are in vapour form and need to be separated and the solid phase consists of the column of solid adsorbent. GLC can be used to resolve mixtures of volatile lipid derivatives. It separates volatile components of mixture on the basis of their relative tendencies to dissolve in the inert material packed in the column and to volatilize and move through the column carried by an inert gas such as helium. Some lipids are volatile whereas others are not and in such cases thses are treated with chemicals which transform them in such a form that it become volatile. For example, phospholipids are not volatile and for fatty acid analysis it is first heated with methanol/HCL or methanol/NaOH mixture which converts them into their methyl esters which when heated produces volatile compounds. When these fatty methyl acyl esters are loaded in the column and the column is heated to volatilize the compounds, the soluble acy methyl ester is dissolved in the adsorbent and the less soluble lipids are carried away the stream of helium and emerge first from the column. The order of separation depends on the nature of solid adsorbent and the boiling point of the components of lipid mixture. It can also be used to identify and quantify the monosaccharide unit to yield overall composition of the polymer.

30.2 MASS SPECTROMETRY AND NUCLEAR MAGNETIC RESONANCE

Both these techniques can be used to determine the three dimensional structure of macromolecules- carbohydrate, nucleic acid and small to average sized protein.

Mass spectrometry — Mass spectrometry is extremely useful in proteome research. It accurately measures the molecular mass of a protein which is one of the critical parameters for its identification. **Principle-** The principle underlying the MS is that the macromolecules to be analysed are first ionized in a vacuum and when this newly charged molecules are introduced into an electric/or magnetic field their paths through the field are a function of their mass-to-charge ratio, m/z, from which mass (M) of an analyte can be derived.

Suppose that a protein acquires a number of protons when it is injected into the gas phase and which in ESIMS creats a spectrum of species with different peaks (with different mass-to-charge ratios). Each peak differs from the neighboring peak by a charge difference of 1 and a mass difference of 1 (1 proton). Thus mass of the protein can be determined from any two neighboring peaks.

The m/z for one peak will be, $(m/z)2 = (M + n2\ X)/n2$ — — — - (1) and for second neighboring peak, $(m/z)1 = (M + (n2+1)\ X)/(n2+1)$ — — — — — — (2), where M is the mass of protein, X is the mass of the added group (protons in this case) and n2 is the number of charges. From these two equations, (1) and (2), n_2 will be determined first and then finally M will be determined as $n_2 = ((m/z)-1)/((m/z)2-(m/z)1)$ and $M = n_2 ((m/z)2- X))$. There are two variants of MS. 1. Matrix assisted laser desorption/ionization ms (**MALDI MS**) and 2. Electrospray ionization ms (**ESI MS**). In MALDI MS the proteins are placed in a light absorbing matrix and after that the proteins are ionized by a short pulse of laser light. These ionized proteins thus move into the vacuum system. In ESI MS the macromolecules in solution are directly forced from the liquid to gaseous phase. The solution of macromolecules is passed through a charged needle kept at a very high electric potential which disperses the solution into a very fine mist of charged microdroplets. The solvent surrounding the macromolecules rapidly evaporates and put the charged macromolecular ions into a gas phase. MS can be used for cataloguing cellular proteins separated on a two dimensional gel.

Proteome research—MS is ideal for sequencing short stretches (20 to 30 AA residues) of polynucleotide and thus can identify the unknown protein. The sequence information is determined using a technique called **Tandem MS** or MS/MS. After proteolytic hydrolysis a protein solution is injected into a MS-1. The different peptides are sorted so that only one type is selected for further analysis. This selected peptide is further fragmented into a chamber between two MSs and m/z is measured in MS-2 (which is attached to a detector) for each fragment. Two types of ions (b or y) depending upon whether the charge is retained on amino terminal side or carboxyl terminal side, are generated during second fragmentation and thus it produces a spectrum of peaks representing the peptide fragments (AAs). Each peak represents one AA residue and the different peaks differ by the mass of a particular AA in the original peptide. MS technique has now been computerized which provides more accurate estimates of mass.

Nuclear Magnetic Resonance—NMR is a manifestation of nuclear spin angular momentum shown by atomic nuclei. The atoms that possess this kind of spin that gives rise to NMR signal are H, 13C, 15N, 19F and 31P. The nuclear spin generates a magnetic dipole. When a single type of macromolecule in solution is subjected to a strong stable field the magnetic dipoles are aligned in the field in one of the two parallel (low energy) or antiparallel (high energy) orientations. When a short pulse (~10 us) of electromagnetic energy is applied at right angles to the nuclei aligned in the magnetic field, nuclei absorb some energy and thus switch to higher energy state and thus an absorption spectrum results which contains information regarding the identity of nuclei and their immediate chemical environment.

As one dimensional NMR produces a two complex spectrum to be analysed, one can switch over to two dimensional NMR technique which will allow measurement of distance dependent coupling of nuclear spins in atoms through space (a method called, NOESY, Nuclear Overhausen) or coupling of nuclear spins in atoms connected by co-valent bonds (a method called TOSCY, Total correlation spectroscopy). The NOE provides information regarding the distance between individual atoms whereas TOSCY identifies which NOE signal reflects atoms that are linked by co-valent bonds and thus these two techniques together will provide a complete three dimensional structure and thus each atom giving rise to signal is identified. The NMR technique is also computerized and it generates a family of closely related structures that represents the range of conformations consistent with the NOE distance constraints.

30.3 X-RAY DIFFRACTION/CRYSTALLOGRAPHY

The spacing of different kinds of atoms in a complex organic molecule such as protein, nucleic acid in the form of crystal lattice can be determined by measuring the locations and intensities of spots produced on a photographic film by a beam of X-ray of a given wave length after diffraction by the electrons of the atoms. Each atom in the molecule makes a contribution to each spot and thus a regular array of spots called reflection is generated and an electron –density map of protein/nucleic acid is reconstructed from the overall diffraction pattern of spots. The complex molecule shows a repeating pattern as in case of protein and nucleic acid and numerous atoms in the molecule produce thousands of spots which must be analyzed by a computer attached to a detector.

Principle — When the light from a point source falls on an object it gets diffracted and when these scatters waves are recombined using a series of lenses it produces an enlarged image of the object. The resolving power of this system depends on the wave length of the light. The wave length of the visible light ranges from 400 to 700nm and so object smaller than half of the wave length of the incident light can not be resolved. Thus to resolve macromolecules like protein, nucleic acid one must use X-ray with wave length in the range of 0.7 to 1.5 +Å (0.07 to 0.15nm). As there are no lenses to recombine the diffracted X-rays to produce an image , the X-ray diffraction pattern obtained is used to construct image using mathematical technique called a '**Fourior transform**'.

The amount of information obtained through the X-ray diffraction depends on the degree of structural order in the sample. The sample requires a highly ordered protein or nucleic acid crystals. Thus once a crystal is obtained it is placed between the X-ray source and a detector which is connected to computer which then builds a model for the structure that is consistent with the electron-density map.

X-ray crystallography can be applied successfully to problems too large to be structurally analysed by NMR. Crystal derived structure almost always represents a functional conformation of protein obtained by NMR which can also determine the three dimensional structure of macromolecules.

30.4 ELECTROPHORESIS

It is a chromatographic technique used for separation of mixtures of macromolecules (ionic compounds)- proteins or nucleic acid , based on movement of charged macromolecules in response to an electric field. The visualization of electrophoretically separated macromolecules (proteins, RNA, DNA) is done by using more or less specific dyes such as Coomassie stains (a triphenyl methane anionic dye) for proteins, EtBr for dsDNA, double stranded regions in ssDNA or RNA and Stains-all (a chemical, (1-ethyl-2-3 (1-ethylnaphtho-1,2-d-thiazolene-2-ylidine)-2-methylpropenylnaphtho-1,2-d-thiazolium bromide) for DNA (blue), RNA (bluish purple and proteins (pink or red).

Principles underlying electrophoresis — Charged particles in solution migrate to electrodes of opposite charge in response to an electric field. This principle is used in electrophoresis to separate molecules differing in charges. The elctrophoretic mobility of a protein , μ depends on velocity of particle (V) and electrical potential (E) as $\mu = V/E = z/f$ where z is the net charge of the molecule and f is frictional resistance.. When the proteins pass through the gel, the gel slows the migration approximately in proportion to their charge to mass ratio. Migration is further affected by shape of the protein. Thus the electrophoretic mobility of a protein depends on its charge and mass. The rate of movement of a protein is determined primarily by the electric charge on the protein which in turn depends on the protein's amino acid sequence. As the different forms of a protein or different proteins differ in amino acid sequence they will have different charges (Changes in coding sequence under many circumstances, not all circumstances, will lead to changes in the primary structure of protein) and thus they will move with at different rates through the gel.

One of the advantages of gel electrophoresis over chromatography is the possibility of varying the gel concentration continuously so as to achieve any effective pore size within a continuum. The average pore size in a gel can range from a fraction of a nanometer to 50 to 100 nm (corresponding to the diameter of a virus) and up to nearly a micron. Another important advantage is its adaptability to gradients of pore size and to the simultaneous analysis of samples containing large number of components. The gel electrophoresis of multiple samples can be carried out simultaneously on multiple channels of vertically or horizontally oriented gel slabs, using either a single gel concentration of concentration gradient. A vertical gel generally provides better resolution but it puts more strain on the gel. **Separation of proteins**—The elctrophoteric apparatus consists of three components. 1. Gel 2. Buffer tank 3. Electrode. Electrophoretic separation of complex mixtures of proteins can be accomplished in several types of support media including starch, polyacrylamide and agarose gels and acetate membranes. Starch and polyacrylamide gels are widely used to study enzyme polymorphisms because of their higher resolving power. Gel comprises a network of pore through which the molecules must travel and they act as a molecular sieve.

30.4.1 Starch Gel Electrophoresis

Starch gel electrophoresis (Smithies, 1955) continues to be preferred for most studies involving the analysis of a large number of individuals for several to many different enzymes despite the greater resolving power of the PAGE. It is preferred over PAGE because of low cost and thick malleable gels. Starch gels after electrophoresis can be sliced a prescribed number (8-10 slices) of times and stained simultaneously for independent enzyme activities and thus phenotypes at up to 16 to 30 loci could be determined from each of up to 60 samples in a single run whereas polyacrylamide gels are stained for single enzyme. In starch gel electrophoresis carbohydrate is formed into a gel which is placed on the electrode tank in buffer solution attached to an electric current. The gel is covered with a chilled water pack which keeps the gel cool during the electrophoresis. The protein samples are inserted into slots at one end (negative side) of the gel which is moved through the gel by an electric force to the positive side. Samples are applied to gel as extracts absorbed onto wicks that are typically cut or punched from the filter paper (often Whatmann 3MM although other grades can be used). The gel slot is opened and the samples wicks are inserted. Wicks are often removed from the gel after approximately 15 minutes of electrophoresis but resolution is of equivalent quality if the cathodal and anodal sides of the origin are pressed together by inserting a 3MM thick beveled section of acrylic between the cathodal end of the mold and the cathodal end of the gel. Control samples representing reproducible genotypes run on all gels throughout a study and wicks containing bromophenol blue or food coloring as **tracker dyes** are often interspersed with sample wicks. A constant voltage or current is required for electrophoresis. A field strength from 2 to 8 V/am length is suitable for most separation at room temperature. At 250 volts a slightly alkaline 12% (W/V) starch gel will typically yield acceptable migration after 3-4 hours. The strength and duration of the electric field must be determined empirically for each specific isozyme system, buffer combination and pore size. After the run, electric current is turned off and the gel is treated with enzyme specific stain to reveal mobility differences. Bands are usually sharpest if electrophoresis is performed at 4 ° C. The gel slices are cut by pulling a tightened wire or nylon thread horizontally through the gel slab on an acrylic slicing bed. The figure 30.3 shows the banding patterns in case of lactic acid dehydrogenase (LDH) isozymes. The tracking dyes are organic compounds which allow the visualization of the running progress during electrophoretic run (gel electrophoresis).In most electrophoretic systems both bromophenol blue and xylene cyanol are used. Bromophenol blue migrates with DNA fragments of about 10-100bp in length whereas bromocresol green, methyl green and xylene cyanol migrate with DNA fragments of about 5kb in length.

Preparation of gel—The quality of starch gel involves desirable gel handling properties, adequate isozyme separation and clarity. Starch gel is prepared by one of the two methods. In the first method micro-oven is used which gives the greatest reprodubility of results. Approximately three-fourth of the buffer required for a single gel is heated in a flask in a microwave oven until boiling while the remaining one-fourth buffer is mixed with the starch (and sucrose) in another flask at room temperature and mixed to form an even suspension. The boiling buffer from the microwave oven is then poured into the starch suspension, vigorously mixed for few seconds and then the flask is returned to the oven where it is heated again until it begins to boil and finally the solution clarifies. The solution is then degassed with an aspirator to remove air bubbles and after that it is poured into acrylic gel mold. Where microwave oven is not available the solution of starch and buffer (all of the

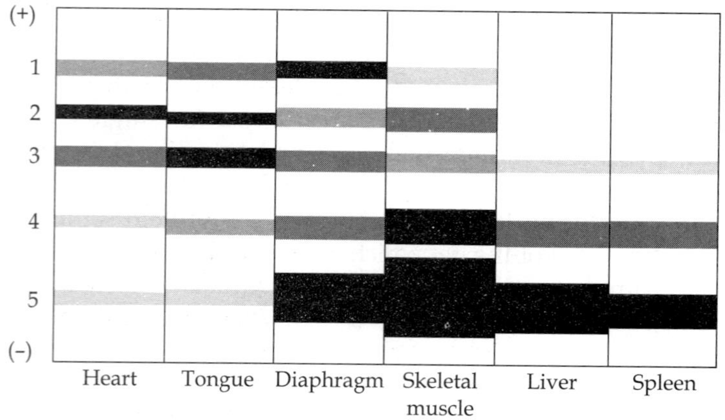

FIGURE 30.3 Showing electrophoretic patterns of LDH isozymes from adult mouse tissues. (From Markert, C.L., and Ursprung, H. 1962. Develop. Biol. 5:363-381.)

buffer) is heated over a flame or hot plate. The suspension is continuously and vigorously swirled until the suspension is free of lumps and a clear boiling solution is obtained. Clear agarose solution are allowed to cool to 60° C before the gel is cast. The gel is treated with 0.5% photoflo to prevent sticking of the gel. The gel molds are available in a variety of sizes and they generally range in volumes from 300 to 600 ml. The thickness of the gel can be increased to comprehensive screening runs to test for large number of enzymes. Further, uneven gel thickness will cause variations in enzyme migration rates in different regions of the gel. Before loading of samples the gels are cooled to approximately 4° C in a refrigerator or on trays with ice and then finally covered with polyethylene film or Saran Wrap in order to prevent dehydration and left over night before loading for electrophoresis. Once the gel is cool the polyethylene covering is trimmed so that it covers only the surface of the gel. After that the wrap is pulled back from the cathodal edge of the gel and a slit is cut perpendicular to the gel surface at approximately 3 cm from the edge of the gel which serves as the origin. A polyacrlamide or agarose slab gel can be placed either horizontally (flat bed gel or **horizontal gel**) or vertically (**vertical gel**) during electrophoresis.

Buffer—The tank is filled with electrode buffer. A buffer solution is one that resists pH change on the addition of either acid or alkali. Such solutions are required in many biochemical experiments where the pH needs to be accurately controlled.The contact between the gel and electrode buffer may be either direct or though cloth or sponge wicks. Types of gel molds are available which provide direct contact between gel and electrode buffer. A few gel and electrode buffer types, each with several modifications are being used for resolution of a broad spectrum of enzymes are : Histidine-citrate, pH 5.7 (Stuber et al., 1977), Morpholine-citrate, pH 6.1 (Clayton and Tretiak, 1972), Tris-citrate, pH7.0 (Meizel and Markert, 1967), Citrate/histidine pH 7.0 (Flides and Harris, 1966), Sodium- borate, pH 8.0/ Tris-citrate, pH8.6 (Poulik, 1957), Lithium-borate, pH 8.3/Tris-citrate, pH 8.3 (Aston and Braden, 1961) and Tris-borate, pH 8.6 (Markert and Faulhaber, 1965).

Staining—After electrophoresis the gel is removed from the electrophoretic apparatus and gel slices are cut. These gel slices are immersed into the staining solution. The substrate and other reagents diffuse into the gel where they are acted upon by the enzyme present in the gel. The enzyme detection (appearance of a band) is primarily based upon precipitation of soluble indicator dyes which become insoluble and colored in zones of enzyme activity. The various types of dyes being used are tetrazolium salt, diazonium salts or various redox dyes. Natural or artificial fluorochromes and U.V. light are used for detection of some enzymes.

The results from electrophoresis depend on the cooking method applied in preparation of starch gel, gel and electrode buffer, electrophoretic apparatus of which the most critical is the electrode buffer.

30.4.2 Polyacrylamide Gel Electrophoresis

It is an important technique for separation and characterization of proteins. Acrylamide is a synthetic hydrophilic substance. It polymerises in the presence of free radicals and bis-acrylamide to for a gel.The gel is thus made up of cross-linked polymer. The gel substance is stable and inert and the gel shows uniform porosity and transparency.The most valuable property of PAGE is that the stringency of molecular sieving can be varied by altering the acylamide concentration and thus increasing flexibility in the range of protein molecular weights or dimensions that are separable (Leaback, 1976). The effective diameter of the pores through which the proteins migrate is proportional to the concentration of acrylamide monomer. Acceptible resolution is usually obtained between 6 and 12% acrylamide and may actually be maximized in a discontinuous step orientation (6 – 9 – 12%), top layer having low acrylamide conc. whereas lower sections having higher acrylamide concentration. The gel is cast in a sealed glass and generally run in a vertical orientation. The samples are cleared and layered directly on the top of gel (comb or wells or depressions) as opposed to the use of crude extract and wicks in case of starch gel electrophoresis.

In this type of electrophoresis only substances migrating anodally can be visualized. All extractions and electrophoresis should be conducted at temperature < 10 °C (usually 2 to 4 ° C) to avoid thermal denaturation. The proteins move into the gel when an electric field is applied. In PAGE the migration of a protein in the gel is a function of its size and shape. Proteins are detected after electrophoresis by treating the gel with a stain such as Coomassive blue or silver stain which binds to the protein. Each band on the gel represents a different protein (or protein subunits). Further, smaller proteins move through the gel more rapidly than larger proteins and thus smaller proteins are found near the bottom of the gel whereas the larger proteins are found near the top of the gel. Polyacrlamide gel percentages for resolution of proteins of different molecular weights are given in the Table 30.1 below.

Table 30.1 Showing relationship polyacrylamide gel percentage and molecular weight of protein separated.

Polyacylamide gel (%)	Protein size range (KDa)
8	40-200
10	21-100
12	10-40

30.4.3 SDS Page

This is an electrophoretic method employed for estimation of the degree of a particular protein preparation purity and its molecular weight. It is a powerful procedure for separating different polypeptides. It makes use of detergent- sodium dodecyl sulfate (SDS) which denatures the protein. SDS binds to most protein in amounts roughly proportional to the molecular weight of the protein. Roughly one molecule of SDS binds with every two AA residues. The bound SDS contributes a large net negative charge to protein, rendering protein's own charge insignificant and thus each protein has a similar charge to mass ratio. Further, as the SDS denatures the proteins, most proteins assume a similar shape and so SDS PAGE separates proteins on the basis of mass (molecular weight) with smaller peptides migrating rapidly whereas larger peptides moving slowly. After the electrophoresis the proteins are visualized by adding dye such as Coomassive blue. With this SDS PAGE one can monitor the progress of protein purification procedure as the number of bands visible on the gel will decrease after each fractionation step.

When the objective is to estimate the molecular weight of an unknown protein (Lane 2), standard proteins (marker protein) of known molecular weights are also subjected to electrophoresis (Lane 1). There is a linear relationship between the log M of marker proteins and their relative migration and the molecular weight of an unknown protein can be read from the graph by knowing the position of the unknown on the linear slope.

Many proteins have subunit structures and have two or more different subunits. These subunits (individual polypeptides) will generally be separated by SDS treatment and a separate band will appear for each subunit.

30.4.4 Two-dimensional Gel Electrophoresis

It is a tool for obtaining global picture of the expression levels of a proteome under various conditions. It separates proteins on the basis of two of their characteristics: isoelectric point and molecular weight. It combines isoelectric focusing, a procedure of determining the isoelectric point (pI) of a protein and SDS electrophoresis and are used in sequence. Isoelectric focusing separates proteins on the basis of their isoelectric point. A stable pH gradient is established in the gel by adding a mixture of low molecular weight organic acid and bases (ampholytes) onto the gel which is distributed in an electric field generated across the gel. The pH is high near the top of the gel whereas it is lower near the bottom of the gel. When the protein mixture is placed in a well on the gel and the electric field is applied the proteins migrate until each reaches equivalent to its pI (when pH =pI, the net charge of a protein is zero) and thus proteins with different pIs are distributed differently through out the gel. The protein with high pI is near the top of the gel and with low pI is near the bottom of the gel. In the two dimensional electrophoresis proteins are first separated by isoelectric focusing in a cylindrical gel and after completion of separation this cylindrical gel is placed horizontally on a second slab shaped gel (SDS (sodium dodecyl sulfate) polyacrylamide gel) which separates proteins according to their molecular weights. Spots are detected using color stains, fluorescent dyes or radioactive labels.This procedure is more sensitive than either IF or SDS PAGE. TDE separates either proteins of different pI with similar molecular weight or proteins with similar pI but different molecular weight. The horizontal separation shows differences in pI whereas the vertical separation shows differences in molecular weight. Although thousands of protein can be resolved in a single slab gel the 2-D gel electrophoresis has the disadvantages that it is cumbersome, has poor dynamic range and is biased towards abundant and soluble proteins. Also, 2-D gel electrophoresis alone can not provide the identity of the proteins which have been resolved.

Two dimensional PAGE is a powerful tool for analysis and detection of proteins from complex biological systems (O'Farrell, 1975). This system allows simultaneous determination of molecular weight and approximate isoelectric points of proteins. Since these two parameters are unrelated, it is possible to obtain an almost uniform distribution of protein spots across a two-dimensional gel. More than 1000 proteins can be resolved with this system but it is capable of resolving a maximum of 5,000 proteins. A protein containing as little as one disintegration per minute of either 14C or 35S can be detected by autoradiography and a protein species representing as little as 10^{-4} to 10^{-5} of 1% of the total protein can be detected and quantified. The reproducibility of the separation is sufficient to permit each spot on one separation to be matched with a spot on a different separation. It provides a method for estimation of the number of proteins made by any biological system. This system can resolve proteins differing in a single charge and consequently can be used in the analysis of *in vivo* modifications resulting in a change in charge. Finally, proteins whose charge is changed by missense mutations can be identified.

30.4.5 Capillary electrophoresis

Capillary electrophoresis involves electrophoretic separations in narrow bore plastic tubes, glass capillaries and thin films between parallel plates in which charged molecules are separated in the presence of an electric field. The advantages of capillary electrophoresis include high precision, ease of automation, high sample throughput, use of low sample volumes and minimal sample pretreatment(Megeed et al., 2007)

30.5 WESTERN BLOTTING

The separated polypeptides in the polyacrylamide gel from PAGE can also be transferred to a nitrocellulose membrane and individual proteins can be detected using specific antibodies. The transfer of proteins from acrylamide gel to nitrocellulose membrane is called **western blotting**. Proteins separated by PAGE is transferred electrophoretically (using an electric current) to nitrocellulose. The nitrocellulose membrane with immobilized proteins in then placed in an antibody solution which binds with specific protein (s). The non-bounded antibodies are washed off the membrane. The presence of the primary (or initial) antibody is detected by placing the membrane in a solution containing the secondary antibody. This secondary antibody is linked to either radioactive isotope and thus permitting autoradiography or an enzyme that produces a visible (colored) product when a proper substrate is added. This secondary antibody-enzyme (or labeled radioactive isotope) binds to primary antibody. Thus this immunoblot assay is a powerful tool for identifying and characterizing specific gene products. The colored product forms only along the bands containing the proteins of interest. Thus through this technique a very small amount of protein in a sample can be detected and its molecular weight can be approximated. Thus it can be used for identifying and quantifying a specific protein in a complex sample.

This western blot thus uses the technique called **ELISA** (enzyme linked immunosorbent assay) which is used for rapid screening and quantification of the antigen (protein). ELISA is used for the detection of specific proteins using two antibodies, a primary antibody and a secondary antibody. The primary antibody binds to the protein under investigation (antigen) and is in turn is recognized by a secondary antibody linked to an indicator enzyme whose activity can be quantified. The activity measured is proportional to the quantity of primary antibody and consequently of the protein. The different steps involved in ELISA are as follows.

1. The wells in a usually 96-well polystyrene plate are coated with proteins in samples (antigens)
2. The surface is washed with non-specific protein-often casein (obtained from dry non-fat milk powder) to block the protein in subsequent steps and also from adsorbing to these surfaces. Also, block the unoccupied sites on the surface with non-specific proteins.
3. The surface is then treated with a solution containing primary antibody. Here the antibody binds the specific antigen.

4. The unbound antibody is washed off and again the surface is again treated with a solution containing secondary antibody which is linked to enzyme.

5. The antibody-enzyme complex binds primary antibody. Again, the unbound secondary antibody is washed off and the substrate of the antibody-enzyme is added.

6. The enzyme linked to secondary antibody which binds primary antibody which in turn binds specific antigen catalyzes the reaction with the substrate and form a colored product which indicates the presence of specific antigen. The intensity of color is proportional to the concentration of the protein of interest in the sample. The second antibody is joined to a colorimetric or fluorescent or chemiluminescent label.

ELISA allows accurate determination of protein levels.

Study of nucleic acid—Gel electrophoresis which is discussed above is also used for separation of DNA molecules and estimation of their sizes. The gel is composed of agarose or polyacrlamide. As the DNA molecules have constant charge per unit mass they separate in agarose and acrylamide gel almost entirely on the basis of size or conformation. The DNA fragments differing in size move through an electric field at different rates. DNA is negatively charged and thus it moves towards positive electrode.

30.6 AGAROSE GEL ELECTROPHORESIS

An agarose gel is a complex network of polymeric molecules whose average pore size depends on the buffer composition and types and concentration of agarose used. It is used for separating larger DNA molecules. Each DNA sample with the help of micropipette is loaded in the sample wells of the gel placed in an electrophoretic chamber. A comb with teeth of the desired size is used to form the wells in the gel. The comb is located at about 0.5 mm above the glass plate so that agarose can run under the comb. The agarose gel is prepared by pouring melted agarose on the surface of a glass plate with comb in the above mentioned position. The comb is removed after the agarose is solidified and then it is placed in the electrophoresis chamber. Agarose gel is made from complex polysaccharide agarose buffered at somewhat alkaline pH. Increasing concentration of agarose will reduce the pore size which in turn will retard fragments more drastically than the smaller ones. Horizontal gel electrophoresis system should be used for agarose concentrations lower than 0.5% whereas higher agarose concentrations allow the use of conventional vertical slab gel cast between glass plates. An appropriate running buffer solution is filled in the chamber. Prior to loading the samples DNA solutions are mixed with gel-loading buffer which contains sucrose, glycerol or Ficoll so that its density is greater than that of the running buffer. The loading buffer usually contains a dye such as bromophenol blue in order to monitor the rate of migration of molecules. After covering the buffer chamber the electric current is applied across the gel. The voltage or amperage must be desired one. The electrophoresis of DNA is usually carried out at 5-10 V/cm. The DNA fragments in the samples move through the gel in response to the electric field. Each fragment migrates a total distance proportional to the logarithm of its size (Maniatis et al., 1975). But this rule is limited to only a range of molecular weights. Thus it has been advised to use DNA fragments of known molecular weights as standard molecular weight marker and run them in a separate lane (slot) of the same gel. After the indicator dye, bromophenol blue, has migrated to the end of the gel the gel is removed from the chamber and placed in a solution of ethidium bromide. Ethidium bromide binds to DNA and fluoresces when illuminated with U.V. light. Thus if a sample contains a limited number of DNA fragments, the position of each fragment size can be located in the gel by viewing the gel under U.V. light. The conc. of each fragment size produces a bright fluorescent band across the gel. In other words, each band represents a collection of DNA molecules of the same size class. Longer DNA molecule will be collected near the top of the gel whereas the smaller DNA molecule will be collected near the bottom of the gel. The gel thus slows down the movement of larger molecules through sieve in comparison to small molecules of DNA. If a sample containing DNA molecules of known sizes (standards) is simultaneously run then the sizes of the unknown DNA molecules can also be estimated. The percentage of agarose can be increased to separate small sized DNA fragments but for very small species (up to 1 kb) PAGE can be used.

Low melting point (LMP) agarose—It is a specific agarose which melts at 65°C, remains fluid at 37°C and solidifies below 25°C. Since the resolving characteristics of both LMP agarose and standard agarose are similar the former is ideally applicable for the recovery of DNA fragments after gel electrophoresis.

30.7 POLYACRYLAMIDE GEL ELECTROPHORESIS

Agarose gel is used for separating larger DNA molecules whereas polyacrylamide gel is used for separating smaller DNA molecules. In practice the composition of the gel determines the size of the DNA molecules that can be separated. A agarose gel composed of 0.5 % agarose which is 0.5cm thick and thus have relatively large pores can be used for molecules on the size range 1 to 30 kb and thus molecules differing by 1 to 2 kb can be distinguished whereas PAGE composed of 40% PAG which is 0.3mm thick and has very small pores can be used to separate molecules in the size range 1 to 300bp and molecules differing by just one nucleotide can be distinguished. Thus by varying the concentration of agarose (agar or polyacrylamide) (Table 30.2) and electrophoretic conditions fragments differing in by thousand bases or just one base can be separated. The polyacrylamide gel matrix slows the mobility of larger DNA fragments so very long electrphoretic run times are required to separate larger fragments than 1000bp. The molecular weight range that can be separated using conventional agarose gel electrophoresis is limited to sizes smaller than 50 kb of DNA and molecules greater than 50 kb will show the same mobility and thus can not be separated

Table 30.2 Showing relationship between acrylamid percentage and the size of DNA fragments sparated.

Acrylamide (%)	Size of the ds DNA fragments separated
3.5	100-1000
5.0	80-5000
8.0	60-400
12.0	40-200
15.0	25-150
20.0	6-100

by conventional gel electrophoresis. DNA fragments larger than 50 kb can not be resolved efficiently by agarose or PAGE because electrophoretic resolution decreases above this molecular weight and larger DNA fragments show anomalously high electrophoretic mobility and this behaviour may be due to condensation and orientation effect.

In conventional gel electrophoresis, the mobility of DNA decreases as the size of the molecule increases until a limiting mobility, independent of size, is reached. Above that size electrophoresis offers no resolution.

Eckhardt gel electrophoresis — It is a technique for the rapid analysis of recombinant plasmid DNA. In this technique either the bacteria containing the recombinant plasmid are lysed in the slots of an agarose gel or alternatively the plasmids are isolated and then loaded directly on the gel. After electrophoresis the plasmids containing the insert can be discriminated from the recircularized plasmids by their decreased electrophoretic mobility.

Visualization of DNA molecules in gel — The DNA in agarose and polyacrylamide gels are stained with ethidium bromide (EtBr) and the bands are observed under U.V. light. Bands of DNA (as little as 0.05ug of DNA in one band) stained with EtBr can be detected as visible fluorescence when the gel is illuminated in U.V. light. Although the light of 254nm wave length gives maximum sensitivity it will damage the DNA and thus the intact DNA molecules can not be recovered from gels. So the light of 366nm wave length is used and in practice long wave red light is commonly filtered out by red interference filters. The bands are clearly visible only when significant DNA is present in bands. If less than 25 ng of DNA is present per band then staining with ETBR will be of no help and in that case one can go for autoradiography of radioactively labeled DNA. If the DNA molecules are labeled with radioactive marker then the DNA molecules can be visualized by placing an X –ray sensitive photographic film over the gel. Thus the photographic film is exposed to the radioactive DNA and thus reveals the banding pattern. A little as 2ng of DNA per band can be visualized by autoradiography of the labeled DNA. The intercalation of EtBr into a portion of DNA double helix increases the spacing between successive base pairs, distorts the regular phosphate backbone and reduces the pitch of the helix.

Silver staining — It is a technique for the visualization of proteins and nucleic acids in polyacrylamide gels in which silver ions react with the macromolecules at pH>10, form complexes and are subsequently reduced to elementary silver. The silver is deposited at the sites of reduction and can be visualized easily.

Denaturating gel — It refers to a gel electrophoresis matrix (gel) containing compounds for the denaturation of macromolecules such as proteins, RNAs, DNAs which are to be separated. The electrophoretic migration of such denatured molecules is almost exclusively dependent on their molecular weight (molecular length) whereas the secondary, tertiary or quarternary structures have no impact on rate of migration under these conditions. Most proteins adopt random coil conformations when denatured in the presence of high concentrations of urea or with β-mercaptoethanol and SDS (see SDS PAGE). RNAs are denatured by methyl mercuric hydroxide (CH_3H_gOH) or glyoxal (HOOCCOOH) to form completely single stranded molecule. DNA can be denatured during electrophoresis by high pH (>10) and methyl mercuric hydroxide or glyoxal (alkaline gel, DGGE).

Sequencing gel — It is an ultra thin high resolution gel used for the fractionation of single stranded (denatured) DNA fragments according to their length. It allows separation of fragments that differ in length by only one single nucleotide. It contains 6-20% polyacrylamide and 7 M urea to minimize the effects of DNA secondary structures on the electrophoretic mobility. It is also used for determining the sequences of bases in DNA molecule in DNA sequencing.

30.8 PULSED FIELD GEL ELECTROPHORESIS (PFGE)/ORTHOGONAL FIELD ALTERATION GEL ELECTROPHORESIS

In the conventional gel electrophoresis the electric field is applied along the length of the gel and the DNA molecules migrate in a straight line towards the anode (positive) pole (Figure 30.4). The different sized molecules are separated in the gel because they migrate at different rates but DNA molecules larger than 50 kb can not be resolved as the difference in migration rates between larger DNA molecules becomes smaller and smaller. This problem with standard gel electrophoresis can be overcome by PFGE. Schwartz and Cantor devised PFGE during 1982-84 for separating DNA fragments ranging from 35 to 2000kbp. In PFGE electric field, instead of being applied directly along the length of the gel as in case of standard gel electrophoresis, alternates between two pairs of electrodes. Each pair of electrodes is set at an angle of 45° to the length of the gel and the result is the generation of pulse field. The gel concentration is lower than in conventional gel electrophoresis so that to allow migration of large DNA molecules. It is a variant of gel electrophoresis that allows the separation of large DNA fragments up to several megabase pairs (up to 3 million bps or 3Mbp) (Ganal et al., 1989). DNA molecules are forced to change the direction on which they are moving by exposing them to alternating electric field (Figure 30.5). The time each field is on is called the pulse time. Shift between alternating fields is instantaneous and pulse time is tuned to the size molecules that need to be resolved. The speed at which a molecule changes direction is molecular weight dependent whereas the ordinary electrophoretic mobility of large DNA molecule is independent of molecular weight. When an electric field is applied the DNA molecule migrates for a short time in the direction of the current while elongating in the same direction and when the electric field changes the DNA molecule changes its conformation and reorient before migrating in the direction of the altered field. Thus with every change in field direction each DNA molecule has to realign though 90° before its migration can continue. The time needed to reorient is related to molecular length, the larger

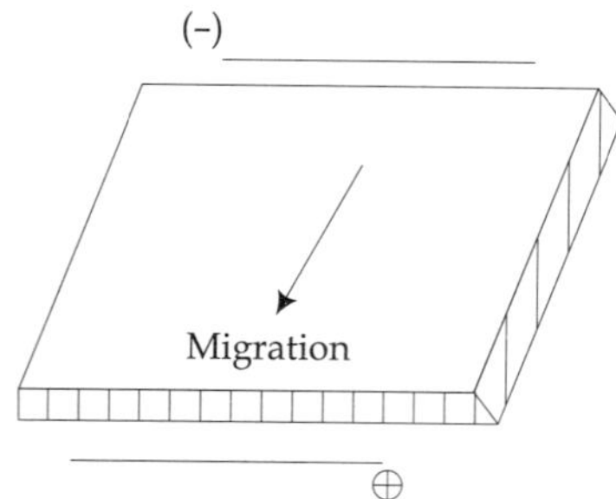

FIGURE 30.4 Showing direction of migration of DNA molecules in Agarose gel electrophoresis.

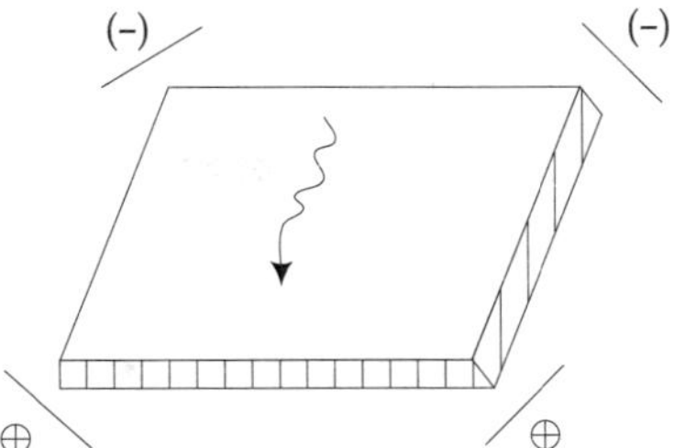

FIGURE 30.5   Showing direction of migration of DNA molecules in Pulsed-field gel electrophoresis.

the DNA molecule the more time it takes to reorient and thus larger DNA molecules spend more time switching direction and migrate more slowly in comparison to smaller DNA molecules. The net movement of the DNA molecule in the gel is still from one end to the other in a more or less in a straight line. Optimum separation is obtained in case of those DNA molecules which spend most of a pulse time reorienting and very little time moving because movement is not dependent on molecular size. Resolution thus is based on retardation differences rather than movement of DNA. Variations in pulse time, field strength, shape, temperature and gel concentration will have effects on resolution. Sized range resolved in PFG is roughly as the product of square of the field strength and pulse time. The field strength is calculated as the ratio of total voltage applied to the shortest measurable distance between anode and cathode electrodes. The effective PFG running conditions are 3-10 V/cm, temperature 12-15° and pulse time 100 seconds which is suitable for resolving the DNA of size 0.05 Mbp to 1 Mbp. The amount of sample needed depends on the pulse time, electrode configuration and number of bands into which the DNA will be resolved. For best resolution low amounts of DNA are used. Longer run time (20hrs) results in resolution whereas shorter run time results in lesser resolution, straight bands and is extremely useful in quickly evaluating an experimental result. The comb should be placed perpendicular to the surface of the gel and always at the same place for higher reproducibility. Moving anode downwards the center will result in less curvature and less resolution whereas moving the electrode out further towards the corner will increase the curvature and band sharpening effects. To maximize resolution use short run time, use the centre gel lane and move the anode more towards the centre of the side of the box.

Estimation of molecular sizes is done by comparing migration distances of samples and standards in different gel lanes. Mobility (or size) comparisons between gel lanes may be made by dropping a diagonal from the well to the band of interest. One may ignore the sinuosities at the top of the gel, however, the curvature at bottom can not be ignored and to minimize the curvature DNA molecules are run to about 5 cm above the bottom of the gel and the run times are usually 40 - 48 hrs for the best resolution. With PFG DNA molecules 30 to 2000 kb in length can be separated. PFG can be used for separating large DNA fragments or whole chromosomes of yeast, *Saccharomyces cerevisae* (2n = 16), several filamentous fungi and protozoans such as the malaria parasite, *Plasmodium fulciparum*. It can also be used to determine the total size of genome in case of bacterial chromosome (e.g. *R.meliloti*) and location of genes in case of yeast (*S. cerevisae*). The separated chromosomes are stained with ethidium bromide and thus are visualized under U.V. light. Further, size of the DNA molecule can be determined by comparing them to standards of known size. Also, the DNA from the gel can be transferred to a nitrocellulose membrane using the Southern transfer method which can be hybridized to a probe to identify/locate a sequence (gene). As DNA breaks during extraction so precautions must be taken to prevent the DNA sample from breaking while being prepared for PFGE.

Various types of PFGE are chosen depending upon the sizes of fragments to separate. The reliability of this method depends on the identification of complete set of fragments generated by the restriction treatments. The main drawbacks of PFGE are: (i) nonambiguous resolution of the digestion fragments may be hindered if the fragments are too numerous or if two or more have very similar lengths and (ii) fragments too small to be detected or so small that they would have eluted from the gel before it was examined may be generated.

30.9 CONTOUR CLAMPED HOMOGENEOUS ELECTRIC FIELD ELECTROPHORESIS (CHEF)

It is a variant of PFGE described above and separates DNA fragments as large as 2 Megabase in a homogeneous electric field generated by hexagonal arrangement of multiple electrodes (e.g. 24 electrodes). The electrodes of opposite sides are alternatively activated. The electric current flows in an angle of 120° against the direction of migration of the DNA. The permanently changing direction of the electric current induces permanent changes in the conformation and orientation of the DNA fragments. The bigger sized fragments take longer time for the reorientation than shorter fragments and thus get separated. PFGE and related techniques such as CHEF and FIGE can be used to purify DNA from individual chromosomes from the gel and thus a series of chromosomal libraries can be prepared. Further, chromosomal DNA molecules can be immobilized on a nitrocellulose or nylone membrane by Southern transfer and chromosome carrying a cloned gene can be detected by hybridization analysis.

30.10 DENATURING GRADIENT GEL ELECTROPHORESIS (DGGE)

Two DNA fragments differing by changes as small as a single base substitution, deletion, insertion or mismatch can be physically separated from each other by a method called DGGE. It involves use of acrylamide gel electrophoresis containing linear gradient of DNA denaturants such as formamide and urea or temperature.

Theory

In this process we are exploiting sequence dependent melting differences. Even the two very similar DNA fragments often have different melting temperature and so they can be separated electrophoretically. Change in base sequences as small as single base substitution will change the degree of stacking significantly enough to change the TM by over 1°C. In other words, even a 1bp change in a DNA fragment of 0.5kb or more is often enough to make a perceptible difference to the stability of the double stranded structure. Within a DNA fragment there can be blocks of sequences called melting domains (MDs). They will melt co-operatively at discrete temperature called melting temperature, Tm for a given domain. MDs are between 25 to several hundred base pairs in length. Two adjacent domains may differ in Tm by several degrees and have fairly sharp boundaries. A DNA fragment of 100-1000 bps length can have 2to 5 MDs. MDs of restriction fragments from most naturally occurring source are between 65 and 80°C. The order of bases on a strand determines the degree of stacking, contributing more to the stabilization of DNA than their hydrogen bond. Stacking interactions between adjacent bases on the same strand of DNA stabilize the DNA in helical form, contribute more to the determination of thermodynamic stability of the double helix and hence the degree of stability used in base pairing of opposite strands. There is thus an alteration of Tm. The base sequence affects the TM of a domain and the conformation of a DNA fragment affects its migration in a gel matrix driven by an electric field. The rate of migration depends on its molecular weight (at a linear rate). The DNA enters the gel as double stranded and there the migration rate is dependent on its molecular weight.   The point in the gel at which denaturant concentration and bath temperature equals Tm, a domain (block of segment) becomes single stranded and now the mobility of this DNA fragment is retarded because of branched chain structure which becomes entangled in the pores of the gel. The degree of mobility retardation is roughly a function of length of melted and un-melted DNA regions, greater the melted region, higher the mobility retardation. The position in the gel at which DNA fragment begins to slow is equivalent to Tm of the lowest MD of the fragment. Two DNA fragments differing in Tm will begin slowing down at different positions in the gel and will be separated from each other at the end of the run. As the temperature or denaturant concentration is gradually raised the double stranded fragment goes through a distinct pattern of melting behavior. A-T melt more easily than G-C base pairs and mismatches destabilize neighbouring base pairs to different extents depending on the nature of mismatch. A branched DNA fragment is required for separation of the wild type and mutant fragment where the mutation falls in the melted regions of the fragment. Base changes in all but higher temperature MDs are accessible to analysis by DGGE. Base changes in the last domain to melt are generally not detectable by the gel system because as this domain melts, DNA is completely separated resulting in loss of sequence dependent mobility. This problem can be solved by attaching a high temperature melting domain, the **GC lamp**, next to the DNA fragments with domains of lower temperature, thus rendering the entire attached fragment accessible to the analysis. GC-clamp is a synthetic oligodeoxynucleotide with a high GC content (>70%) which is inserted into a cloning vector as a stabilizing sequence. It prevents the melting of the region where they are located.

Mismatches in a DNA fragment tend to cause enough destacking of bases and thus a large decrease in stability occurs which makes it easier to melt the DNA fragment. Destabilization by as much as 6°C can be due to a single base mismatch. 100 % of all possible changes in lower temperature MD of DNA fragment have a detectable shift in mobility. On average over half of the length of DNA fragments in the size of 100-1000bp fall in lower tem. MD so over 50% of all possible base changes in theses fragments can be detected by DGGE. The denaturing reagents used are urea and/or formamide.

Field-inversion gel electrophoresis (FIGE)—It is also called inversed field gel electrophoresis or reversed field gel electrophoresis. It is a technique for the enhancing the resolution of DNA molecules in the range of 15 to more than700kb. It employs periodic inversion of a uniform electric field with concomitant cycles of forward and reverse migration of the molecules.

Rotating gel electrophoresis (RGE)—It is a technique for separating large DNA molecules in the size range of 50kb to grater than 7000kb using a single homogeneous electric field but changing the orientation of the field in relation to the gel by periodically rotating the gel.

Temperature gradient gel electrophoresis (TGGE)—It is an electrophoretic technique used for the analysis of conformational, transitions and sequence variation of DNA and RNA and protein-nucleic acid interaction. This technique combines the gel electrophoresis (separation of macromolecules on the basis of size, charge and conformation) with a superimposed temperature gradient perpendicular to the electric field (separation by differing thermal stabilities of the different macromolecules). **Temperature sweep gel electrophoresis**- It is a variant of TGGE technique used for the detection of point mutations in DNA. A decrease in electrophoretic mobility of DNA fragment occurs if localized regions begin to melt. The melting point of such regions is changed by a point mutation so that base substitution in these first denatured regions can be detected.

30.11 SOUTHERN BLOT

Blotting refers to the immobilization of sample of nucleic acid on a solid support, generally nylon or nitrocellulose membrane. E. M. Southern in 1975 devised a procedure called 'Southern blot' for identifying the location of gene and other sequences on restriction fragments separated by gel electrophoresis. Migration rates of DNA molecules were inversely proportional to the logarithms of the molecular weights. He showed that fragment length or molecular weight when plotted against the reciprocal of mobility yielded a straight line over a wide range than the semi-logarithmic plot. The different steps involved in southern blotting are as follows.

1. DNA molecules separated by gel electrophoresis are transferred to nitrocellulose or nylone membrane. The transfer of DNA to membrane is called 'Southern blot'

2. DNA is denatured either prior to or during the transfer by placing the gel in an alkaline solution such as NaOH. The double stranded DNA thus gets separated.

3. The membrane, dried by simply drying (baking) or U.V. light, induced cross-linking to the filter. The DNA molecules get immobilized. The cross-linking can be induced by chemical as well.

4. Radioactive DNA probe (containing the gene or sequence of interest) is then hybridized or annealed with immobilized DNA on the membrane.

5. The probe will anneal and form double helix only with the complementary DNA strand. The non-annealed probe is washed off the membrane.

6. After washing the membrane it is exposed to X-ray film (autoradiography) which will detect the radioactivity in the bound probe.

7. After autoradiogram is developed the dark bands show the position (s) of the DNA sequences that were hybridized with the probe and thus the gene (s) and other sequences on the DNA molecules are located.

The southern blot technique requires 1. relatively fresh DNA samples and 2. larger amount of DNA.

Blotting (blot transfer)—It refers to any procedure that transfers electrophoretically separated DNAs, DNA fragments, RNAs, RNA fragments or proteins from a separation gel (agarose, polyacrylamide) to a paper or membrane matrix (nitrocellulose, nylon membranes). Binding of nucleic acid to nitrocellulose is noncovalent. There are three blotting variants.

1. **Capillary blotting** which transfers molecules through capillary forces
2. **Electro blotting** that transfers molecules by in an electric field
3. **Vacuum blotting-** It is a variant of capillary transfer (Southern transfer) used for blotting of the DNA fragments from 1kb up to the size of the whole chromosomes from an agarose gel to nitrocellulose or nylon matrices employing controlled vacuum.

Study of RNA

The principles underlying separation of DNA, RNA and proteins are largely the same but they involve different protocols. As the RNA molecules are more sensitive to degradation by RNase so one must wear gloves at all times during analysis in order to prevent contaminations of solution with RNase on one's finger. In addition, all glass wares and chemical agents must be baked in order to make them RNase free. Further, as the RNA molecules have secondary structure so they must be denatured during electrophoresis if one wants to separate them on the basis of size. Denaturation of RNA molecules is done by adding formaldehyde (or other chemical agents) to loading and running buffer. After electrophoresis the RNA molecules will get separated.

30.12 NOTHERN BLOT

The RNA molecules separated by an electrophoresis containing agarose gel under denaturating conditions are transferred to a nylone membrane by a procedure called 'Nothern blotting'. After transfer of RNA molecules to membrane, these molecules are hybridized to either RNA or DNA probes just as with Southern blots and autoradiography is done (Alwine et al., 1977). Nothern blot hybridization is used for studying gene expression. It will provide information regarding whether a particular gene is transcribed in all tissues or only in certain specific tissues. Further, it can be used to study the temporal pattern of expression of individual gene (s) i.e. at what stage during growth and development a particular gene is expressed. Finally, northern blot hybridization provides information regarding the levels of RNA transcripts present in the tissue but it does not say anything about its causes. The quantity of a RNA transcript depends on the rate of transcription or rate of transcript turn over. Only the sophisticated analysis can distinguish between these causes (discussed in chapter 24).

Disadvantages with nitrocellulose solid support—Although nitrocellulose remains the solid support of choice for both DNA and RNA, there are disadvantages with nitrocellulose solid support (Wahl et al., 1987). Firstly, unlike nylon, it is extremely friable particularly when dry and secondly, it binds fragments smaller than 200-300 bases poorly. The first problem is solved by keeping filter membrane moist at all but the fixation step as drying or baking is essential for permanent binding to nitrocellulose. To overcome the second problem use of diazotized cellulose, a solid support can be used. A solid support with reactive groups which make possible covalent binding of nucleic acids can be used or nylon filters which require UV irradiation to cross-link either DNAor RNA to the solid support can be employed.

South-Western blotting—It is a method for the rapid characterization of both a DNA-binding protein and its specific binding site on the genomic DNA.

30.13 AUTORADIOGRAPHY

It is a method of detecting radioactively labeled molecules (DNA, RNA, protein) by placing a photosensitive emulsion (consisting of silver halide crystals suspended in clear phase composed mainly of gelatin) or a film onto the surface of a radioactive specimen and the **autoradiograph** refers to the photographic documentation of the positions of the radioactively labeled molecules in tissue sections or on filters or on gels. The radioactive emissions produce black spots (areas) on the developed film (autoradiograph). Autoradiography is best suited to weak to medium strength β-emitting radionuclides such as 3H, 14C , 35S and not to highly energetic β-particles such as those from 32P or γ-rays emitted from isotope, 125I.

Fluorography—It is a special version of autoradiography in which the detection of radioactively labeled molecules are enhanced by appropriate scintillatiors, for example, 2, 5-diphenyloxazol, PPO or sodium salicylate. In this technique the specimen is soaked with the scintillator and then exposed to an appropriate film and the **fluorogram** is an autoradiograph (autoradiogram) in which the signals generated by the radioactively labeled materials were enhanced by scintillators. Fluorography is mostly used for improving the detection of weak β-emitters.

30.14 ELECTRON MICROSCOPY

Electron microscopy is used to visualize the nucleic acid molecules and thus can be used to measure the size of DNA molecules. The electron microscopy technique is now being used in the study of DNA-RNA hybrids. The most graphic way of showing whether or not the genomic DNA sequences are identical to cDNA sequences is by electron microscopy. And there it is used for detection of introns in a gene by **heteroduplex** formation. The DNA is a right handed double helix in solution was confirmed by electron microscopic study.

Solid phase methods—Solid phase methods have proved useful for separation, synthesis and diagnostic detection of biomolecules (Uhlen, 1989). The solid phase approach produces reproducible reaction with high yields and allows solution to be changed rapidly and thus suitable for automation. The solid phase may consist of the walls of test tube or microtitre wells, polymer particles such as agarose or silica, packed in column. Magnetic particles can be used as a solid support for the separation of non-classified cell suspensions or lysates without the need for centrifugation or filtration. If magnetic beads are nonporous then adsorption and desorption of biomolecules occur at the surface providing reaction kinetics similar to those found in free solution. Magnetic particles were made earlier by the polymerization of acrylamide and agarose with paramagnetic materials and were heterogeneous in size and magnetic contents but now hydrophilic beads have been developed which are identical in size and density and amount of magnetized material. Magnetic bead was originally developed for immunoassays for immunomagnetic separation using immobilized monoclonal and polyclonal antibodies where the magnetic beads were directed to targets such as cells, organelles or macromolecules. The speed of separation was linked to specificity of antibodies.

Specific DNA fragments can be linked (or coupled) to the beads and can be used as probes or templates and binding can be through a reactive amino or thiol group on a synthesized oligonucleotide. In a more versatile system, magnetic particles are covalently bound to streptavidin for the directed immobilization of both double stranded and single stranded biotinylated DNA.The non-covalent biotin-streptavidin interaction allows DNA manipulation reactions such as strand melting, elution and hybridization (using alkali, temperature or formamide) to be performed without interfering with the binding of the DNA to the beads.

A combination of three microsystems, miniaturizing the sample preparation (less of the potentially infectious specimen is necessary), reaction chamber (a faster reaction possible using nanoliter to femtoliter volumes of reagents), detection apparatus (less of reacted sample is needed for measurement) in one microfabricated device will create a single micro total analysis system. In other words, there would be a reduction in sample volume, a reduction in reagent volume, high throughput of the same, reduction of contamination and increase ease of use of thorough automation which is necessary to reduce the cost of diagnostic test and increase of the speed in obtaining results. Surface chemistry plays significant role in any reaction performed inside the microfabricated device. The surface to volume ratio in a conventional PCR reaction tube is about $1.5mm^2/\mu l$ which increases to about 5-fold in case of a capillary tube and 13-fold for a PCR chip. An oxidized silicon (Sio_2) surface gave consistent amplifications comparable with reactions performed in a conventional PCR-tube ((Shoffner et al., 1996)

Introduction of biotin into DNA — There are different approaches for introducing biotin into one of the strand of DNA. In one approach, synthesis of oligonucleotide containing a chemically introduced biotin molecules is made and then directly the oligonucleotide is immobilized onto the streptavidin coated bead. In the more flexible approach, endonuclease restriction of a purified plasmid is performed, followed by fill-in reaction with DNA polymerase using one or several biotinylated oligonucleotides. PCR can be performed using Taq polymerase and a biotinylated oligonucleotide.

Molecular applications — The use of magnetic beads in combination with biotin-streptavidin system has allowed the direct solid phase sequencing of both genomic and plasmid DNA. It allows the *in vitro* amplification and sequencing reactions to be performed under optimal conditions as the buffers and enzymes can be changed easily. Direct sequencing of plasmid DNA can be made starting from single bacterial colony without the need for previous template purification. Magnetic system can be substituted for conventional affinity chromatography for purification of DNA–binding proteins. Beads containing single stranded DNA can be used for separation and isolation of RNA and single stranded DNA. It is also possible to isolate specific RNA molecules using cloned or synthesized DNA fragments as probes and RNA molecules recovered can be subsequently used for solid phase cDNA synthesis. Specific single stranded DNA fragments can be recovered from a complex mixture of fragments. DNA fragments obtained from plasmid, lambda, cosmid or genomic DNA could be specifically recovered by magnetic beads and subsequently amplified by PCR. Thus in this way chromosome walking and other operations could be achieved without the need for subcloning. Finally magnetic particles can be used for separation steps during the analysis of specific DNA or RNA sequences. Specific hybridization complexes can be purified using a magnetic approach called target cycling/background reduction in which hybrids are captured from solution with a complementary sequence attached to magnetic particles.

31

Molecular Cytogenetics and Genetic Engineering Techniques

For understanding the structure and function of the chromosome and for karyotype analysis it is essential to isolate, classify and purify the chromosomes and this is where the various methods such as cytogenetic, cytochemical and instrumental methods play an important role. Purification of individual chromosomes is important for several reasons. Enriched or pure chromosome fractions are analyzed to provide information on the structure of DNA or protein, to transfer genetic information to whole cells or to map genes by *in vitro* hybridization. In general conventional techniques have not been able to provide chromosomes of sufficient purity for high resolution biological or biochemical studies. Chromosome analysis and purification by flow methods will contribute greatly to this end. The different flow cytometric techniques for separating different cells organelles, separating nuclei, chromosomes, protoplasts and mitochondria, discussed this chapter are as follows.

1. Flow cytometry
2. Flow sorting (FACS)
3. DNA cytophotometry,

   Immunocytochemistry,

   Image analysis.

Of particular relevance are then techniques that allow chromosome isolation and analysis through flow- sorting. The procedure, initially developed for studying mammalian chromosomes, involves fundamental steps such as induction of cell cycle synchrony, accumulation of cells in metaphase, isolation of chromosomes from the cell and flow-sorting of chromosomes which have been optimized for a successful application to plants (Lucretti and Dolezel, 1995).

31.1 FLOW CYTOMETRY

Flow cytometry is a method for the isolation of cells and chromosomes that are labeled with fluorescent dyes (e.g. acridine orange, DAPI, EtBr). The cell suspension is forced into a flow chamber where it is hydrodynamically focused and accelerated through an orifice of 50-100 μm in diameter so that only one cell at a time can pass. The cells then traverse a light beam (for nonfluorescent cells) or a laser beam (for fluorescent cells or chromosomes) at higher speed (100-1000 cells per second). The cells in the flow stream either scatter the light or the fluorochromes used to stain the cells emit fluorescence light after laser excitation. In both cases the optical sensors can detect and quantify the signals and tens of thousands of cells can thus be analysed within minutes. Special equipment permits selection of subpopulation of cells (Fluorescence activated cell sorter, FACS). FACS is a technique for the separation of different cells types from a mixture of cells by loading the cells with a fluorochrome and detection of different cells by differences in laser induced fluorescence. FACS can be applied for the sorting of sperm cells carrying either X or Y.

Flow cytometry of isolated chromosomes is an approach that provides rapid measurement of individual chromosomes. In this approach chromosomes are stained in aqueous suspension with an appropriate fluorochrome and are forced to flow at high speed through a narrow laser beam that excites the stain. The emitted fluorescence is measured photometrically and the data are presented in the form of frequency distribution of chromosome fluorescence. The peaks of this frequency distribution are due to individual chromosomes or groups of chromosomes of similar fluorescence. The peak mean is proportional to chromosome fluorescence and the peak area is proportional to the chromosome frequency of occurrence. Thus the frequency distribution serves as a karyotype. The Figure 31.1 shows the fluorescence distribution of human chromosomes (Carrano et al., 1979) measured by high resolution flow cytometry accomplished on the LLL dual laser cytometer utilizing a Spectra Physics 171-05 argon ion laser operating at 351-364nm with a flow rate of about 1000 chromosomes per second. The data are collected into a pulse-height analyzer and

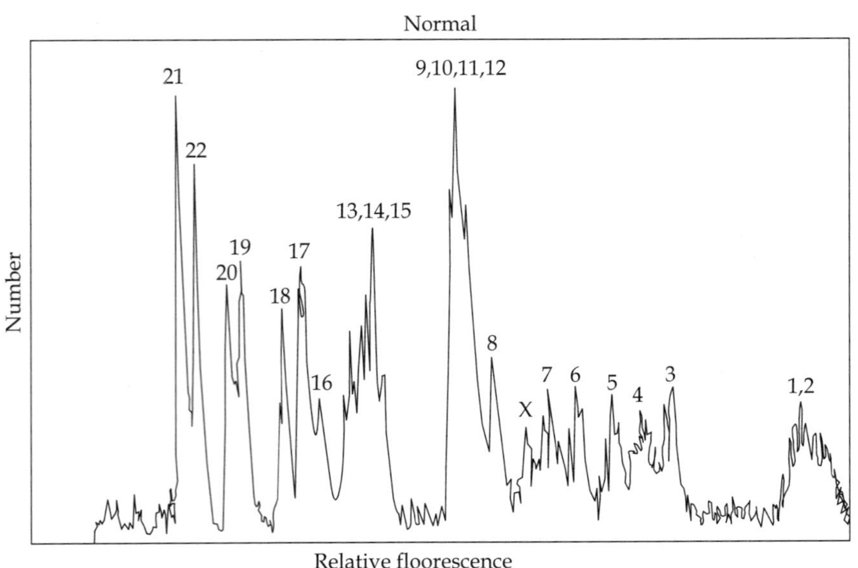

FIGURE 31.1 **Showing sorting of human chromosomes by relative flurescence flow cytometry.**

subjected to an iterative least-squares method that fits a sum of Gaussian distributions to each peaks plus a power function to the underlying continuum. The continuum is the area under the Gaussian distribution and attributes to fluorescence debris, clumping and non-specific staining. The computer generated line is drawn through the data. The relative fluorescence is normalized to the mean of peak A = 1.0. Flow cytometric analysis of suspensions of isolated chromosomes enables classification of chromosomes in the form of histograms (also called **flow karyotypes**) based on biochemical features such as their relative DNA, base and protein content and various morphological parameters. The inherent higher difficulty in flow sorting and flow karyotyping in plants, due to similarity of DNA content, base composition and chromosome morphology among the individual chromosomes of most plant genomes are overcome by use of genetically reconstructed karyotypes where chromosomes are made morphologically distinct by known translocations and by combined use of certain fluorochromes for chromosome staining (Dolezi and Lucretti, 1995) as well as by specific fluorescent labeling attainable, for instance, with primed *in situ* (PRINS) labeling of specific DNA probes (Pich et al., 1995; Gualberti et al., 1996). For high resolution studies on chromosome structure and morphology, flow sorting of single –type chromosome fractions has been shown to represent a powerful tool for construction of chromosome specific DNA libraries and physical localization of single or low copy DNA sequences (functional genes).

31.2 FLOW SORTING

The conventional methods for purifying chromosomes rely upon velocity or isopycnic sedimentation, zonal centrifugation or selective filtration. Flow sorting can be used to separate chromosomes (or cells) on the basis of their staining properties. Thus different chromosomes comprising the karyotype can be separated with **fluorescence activated chromosome sorter (FACS)**. Similarly, bacterial cells comprising fluorescent and non-fluorescent cells can be separated with fluorescence activated cell sorter (FACS). There are two methods to identify the chromosomes associated with each peak in the flow distribution: 1. Flow sorting and 2. DNA photocytometry. Flow sorting is performed with the LLL flow sorter also equipped with a Spectra Physics 171-05 argon ion laser. By flow sorting on the basis of ethidium bromide fluorescence human chromosomes can be separated. The sorted chromosomes can then be stained with quinacrine dihydrochloride in case of human chromosomes for banding analysis.

Applications of FCM—FCM can be used for estimating genome size (also the specific AT and GC content), analyzing the cell cycle (determining the proportions of cells in cell cycle phases such as G1, S and G2/M) and ploidy level of cells and its change (s) during tissue culture, determining changes in genome size during evolution and cell differentiation, measurement of the DNA content of chromosomes (karyotyping-flow karyotyping), sorting of chromosomes and production of chromosome enriched libraries. Isolation of individual chromosomes will simplify the process of analyses and sequencing of the plant genome. Further, transformation of protoplast can be made with specific single chromosome when transfer of linked characters is envisaged. FCM can be used for characterization of cell surface receptor populations particularly in lymphoid cells as well as analyses of cellular

RNA and protein contents and of specific enzyme activities (flow cytoenzymology). In case of animal systems flow sorting has most frequently been employed for the isolation of rare cells from the populations based on specific fluorescent markers identified through flow cytometry and for isolation of individual chromosomes (Muirhead et al., 1987). In case of plant system direct isolation of rare mutant or variant types including those possessing enhanced capabilities for embryogenesis (based perhaps on cell size) and those showing altered photosynthetic activities. It can also be used for isolation of mutant or variant gametophytes such as pollen at the microscope level and the isolation of transformed protoplast *in vivo*.

Dyes — All flow cytometric analyses of nuclear or cellular DNA content and the cell cycle require the use of fluorochromatic dyes that bind to the cellular DNA and emit a fluorescence signal which is linearly proportional to the amount of DNA that is present. Three classes of dyes are employed in animal cell system for the measurement of DNA content.

I class — It comprises those dyes that intercalate double stranded nucleic acids and include Ethidium bromide (EtBr) and propidium iodide (PI). Signal directly reflects the amount of nucleic acid present within the cell. As double stranded regions of RNA can also intercalate EtBe or PI, a pre- treatment with ribonuclease (RNase) is essential if cell cycle information is required.

II class — It a acridine orange (AO) and it dichromatically distinguishes between RNA (red fluorescence) from DNA (green fluorescence) which allows simultaneous flow analysis of DNA and RNA levels within cells.

III class — It includes those fluorochromatic dyes that specifically bind to DNA- DAPI (4', 6-diamideno-2-phenyl indole), Hoechst 33258 and 33342 and mithramycin (MI), chromomycin A3 and olivomycin. The first three have AT and the remainder has GC preferences.

31.3 CHROMSOME STRETCHING

It is a technique for the generation of highly elongated (stretched) metaphase chromosomes. The unfixed hypotonically treated metaphase chromosomes are subjected to shearing by centrifugation which results in their extension, about 5-20 times the normal length. As the chromatin is less condensed in stretched chromosomes they are used as hybridization targets in technique, for example, FISH.

31.4 CHROMOSOME FIBER STRETCHING

It is a variant of chromosome stretching technique which capitalizes on the isolation of metaphase chromosome by chromosome sorting (Flow cytometry) onto glass slides. After drying the sorted chromosomes, proteinase K is added for 5-8 min. and then ethanol: acetic acid (3: 1) fixative is used to stretch the chromosomes. Stretching probably occurs at the buffer-ethanol/acetic acid interface. The product is dried and fixed by baking (1´ at 65°C). Fiber stretching is different for different chromosomes and achieves a 15-100 fold extension as compared to the normal metaphase chromosomes. The extended fibers then can be hybridized to labeled probes as in case of FISH. Baking refers to fixation of nucleic acids on either nitrocellulose filters or microscopic sides which is achieved by incubation of filters (at 80°C for 2hrs) or slides. In order to obtain elongated human chromosomes in metaphase spreads prepared from human lymphocytes, the cultured cell can be treated with Colcemid (0,1 ug/ml) for 10 to 30 min. and hypotonic treatment (0.075M KCL is carried out for 12 to 18 min. followed by standard methanol-acetic acid fixation) (Yunis, 1976).

31.5   MOLECULAR COMBING

It is a method of stretching cloned or native DNA molecules on a microscope. In this method the termini of solubilized DNA are bound to the silnated surface of a glass slide. The DNA solution is then covered with an untreated coverslip such that the drop spreads uniformly and the solution is evaporated. This method leads to the straightening of the bound DNA molecules to which the labeled probes can easily be hybridized in comparison to the condensed chromosomes. The DNA combing refers to the preparation of extended fibres from metaphase chromosomes for *in situ* hybridization with fluorochromes labeled probes (representing genes) and thereby identifying the genes.

31.6   CHROMOSOME MICRODISSECTION

The isolated individual chromosome can be microdissected using atomic force microscopy or laser microdissection. The microdissected pieces still harbour at least 10^6 bp and are used for the construction of subgenomic libraries and thereby use in BAC or map based cloning.

31.7   DNA CYTOPHOTOMETRY

The DNA stain content of chromosomes in metaphase cells can be measured by using a scanning cytophotometer. The human chromosomes are stained with gallocyanin chrome alum under specific conditions and the stained chromosomes are scanned on a flying spot microscope. The scans are digitized and processed by computer. The end results are the measurements of the integrated optical density (DNA stain content) for each chromosome.

31.8   IMAGE ANALYSIS

Computerized image analysis is carried out for karyotype anlaysis, particularly in case of species having chromosomes of small sized and similar arm ratio (Venora and co-workers,1991-1995). This technique can also be used to quantify heterochromatic regions of C-banded chromosomes.

31.9   MICRODENSITOMETER (SCANNER)

It is an instrument used for continuous measurement of light transmitted through a chromatogram, electropherogram or a developed film (e.g X-ray autoradiograph) and the resulting densitogram can be quantified. This technique can be used for estimation of genome size.

31.10 MATRIX ASSISTED LASER DESCRIPTION IONIZATION MASS SPECTROMETER (MALADI)

It is an instrument for determining the mass of a gene (generally a DNA sequence). Basically this **gene balance** is a MS. The DNA sample is first embedded into a matrix which is evaporated by a short laser pulse and this leads to release of DNA molecules into a gas phase where they are ionized by collision with matrix molecules. These ionized molecules are then accelerated into a field free channel. DNA molecules with different base sequences, i.e. different mass, reach a detector at different times which in turn allow to estimate their precise mass. This gene balance can be used to estimate the different masses of alleles. Similarly, MALADI-PSD-MS allows to determine the masses of peptide fragments generated by ionization of isolated proteins.

31.11 MOLECULAR CYTOGENETICS TECHNIQUES

Cytogenetics deals with studies of organization, function, transmission and recombination of the genetic materials originally using microscopy, scanning, or transmission electromicroscopy of chromosomes during mitosis or meiosis (see Roy, 2009) but more recently by the molecular techniques described below. High-resolution cytogenetics analysis is carried out in cases (such as detection or microdeletion/duplication) in which a particular chromosomal region needs to be studied in greater detail. The different molecular **hybridization techniques** are discussed below. Hybridization experiments are carried out to detect sequence homology between two different nucleic acid molecules and thus aims at visualization of specific sequence and thereby detection of gene. Hybridization refers to the formation of duplex molecules from complementary single strands and thus we can see DNA-DNA, DNA-RNA or RNA-RNA hybrids. Usually a single strand DNA is labeled either radioactively or non-radioactively and used as probe which may anneal to homologous sequence in the other single stranded nucleic acid molecules. The formation of stable double strand depends on the stringency of the reaction conditions. The specific reaction conditions in the hybridization experiment allow formation of duplexes from single strand molecules with complete or a high degree of base complementarity only. Stringency depends on the hybridization and washing temperatures and on salt concentration of the washing buffer and increasing stringency is obtained by the use of progressively diluted salt solutions. The most stringent conditions will allow the formation of stable hybrids between perfectly complementary sequences only whereas low stringency permits the formation of duplexes from single stranded molecules with a certain degree of base mismatches. The resulting hybrid molecules can be detected by autoradiography or various methods depending upon the kind of label used. Hybridization may be carried out either within the cells or tissues sections (*in situ* hybridization, squash dot hybridization) or with electrophoretically separated DNA or RNA which is transferred to a filter by blotting.

1. *in situ* **hybridization (ISH)**- The first ISH studies were developed in the late 1960s and 1970s (Gall and Pardue, 1969; John et al., 1969) as a controlled variant of chromosome banding (see Cytogenetics by Roy (2009)) in which the molecular structure of the labeled bands or regions was at least potentially known. The spatial

distribution of a target nucleic acid in a tissue, cell or chromosome can be visualized in ISH techniques. This technique is widely used for physical mapping in plant genomes of low copy and highly repeated genes and sequences on squashed mitotic and meiotic chromosome preparations. It is a method to identify specific DNA sequences (gene loci) on intact chromosomes (usually metaphase spread) or RNA sequences in a cell by hybridization with radioactively labeled complementary nucleic acid probes, frequently synthetic oligocucleotides. The material is usually squashed on microscopic sides, the DNA denatured and then hybridized to the tritium labeled probe. After that a photographic emulsion is layered onto the preparation and the location of hybrids is visualized through autoradiographs. In the past radioactive ISH (using 3_H (tritium)-DNA or 125 I-RNA, tritium, also C^{14}) labeled probes) was mostly used. Hybridzation can also be performed with non-radioactively labeled probe (biotinylation of nucleic acid). The steps involved in various ISH techniques are as follows (Figure 31.2).

1. Fixing of the sample to preserve its structure
2. Limited proteinase digestion in order to provide access to the hybridization probe
3. Addition of hybridizing probe
4. Washing to remove unhybrized probe
5. Detection of hybrids

There are thus two variants of ISH depending upon the types (radioisotope or fluorescence)of probes used for detection of hybrids.

   A. **Radioactive *in situ* hybridization, RISH.**

   B. **Fluorescent *in situ* hybridization, FISH.**

In case of former the probes used are labeled with radioactive isotopes whereas in the later fluorochrome labeled probes are used. RISH is rarely used because radiation free

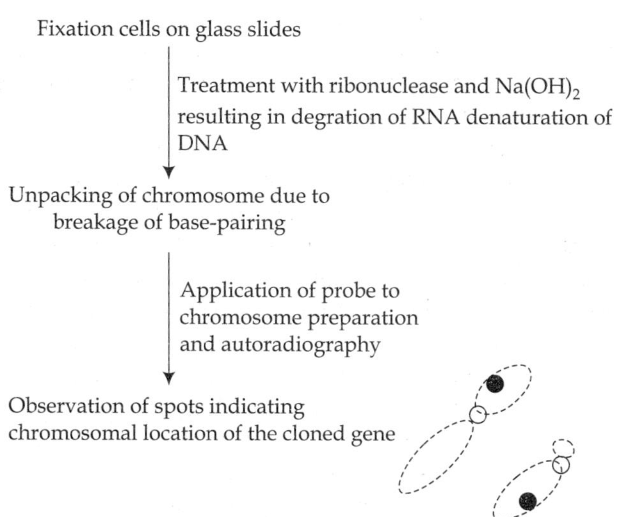

FIGURE 31.2   Showing steps in *in situ* hybridization.

fluorochromes can be more elegantly employed in FISH described below. The precision of the fluorescent probes is limited only by the resolving power of the light microscope. Probing by fluorescent method is better than probing by radioactivity probe as the later always gives a scatter of silver grains in the radiosensitive film around the labeled locus. The radioisotopes required are the ones that emit very short range radiation and not like 32_P is very energetic and would blacken the film over a distance far exceeding the dimensions of the chromosomes. Through this method chromosomal localization of zein genes in maize, legumin genes in *Vicia faba*, gliadin genes in Dsypyrum and several repeated DNA sequences in species such as *Vicia faba*, *Allium*, *Phaseolus coccineus* have been reported. ISH has been widely used for the determining the exact location of cloned genes on polytene chromosomes from dipterae especially Drosophila. Although the scatter of silver grains in the exposed film is no too great in relation to the size of the polytene chromosome band of Drosophila, this method is far from ideal for ordinary chromosomes. ISH has also been used to localize the transcripts of specific genes in cells and tissues sections.

2. **Fluorescence *in situ* hybridization (FISH)**- The substitution in the 1980 of flurorescence for radioactivity in FISH was an advance over ISH (Bauman et al., 1980). The FISH technique is quicker, safer and cheaper and at the same time the results can be polychromatic and have higher resolution. Recently employment of nonisotopic ISH (NISH) and in particular, FISH has greatly enhanced the potentiality of such a molecular cytogenetic technique particularly because of the possibility of visualizing simultaneously and/successively several different probes on the same specimen, thus maximizing the amount of information that can be obtained from a given chromosome preparation. Through the use of this technique the study of structure, organization and diversity of chromosomes of various crop species and their related wild species is now being conducted. This technique in combination with meiotic cytogenetic analysis can identify different chromosomes of wheat x rye hybrid plants and hence their pairing behaviour, particularly that of each individual rye chromosome with those of the wheat chromosome. In formation of this kind is fundamental in interspecific gene transfer based on homoeologous recombination. The relative cytogenetic affinity between the donor and recipient chromosomes, to the limited size of the introgressed segment can be proved by means of FISH analysis. Further, this analysis greatly facilitates characterization and selection of desired genotypes (Ceoloni et al., 1998). FISH is an important technique of interphase genetics which involves localization of specific sequences on interphase chromosomes by hybridization of fluorochrome or radioactively labeled probes (e.g. BAC clones, genes , YAC clones) to interphase nuclei in which chromatin is less condensed and is therefore more accessible and detection of hybridization events by autoradiography or fluorography. In other words, it is a technique for localizing probes to different chromosomes. It enables assignment of probes to chromosome bands and probes can be ordered for physical mapping project using multicolor labeling. It is possible to hybridize different types of labeled DNA (using biotin, degeoxygenin or directly labeled with a fluorochrome) to metaphase chromosomes using either (i). large genomic clones (YAC, P1, cosmid, Phage) after suppression of repetitive sequences (ii). a few kb of unique single copy DNA or cDNA sequence (iii). PCR products or (iv). DNA synthesized directly *in situ* by oligonucleotide priming. This technique is particularly

useful for studying chromosomal rearrangements. Rearrangements such as chromosome duplications or translocation of a segment of one chromosome to another can be typed relatively more quickly by FISH than by conventional techniques. Conventional techniques such as G-banding are routinely used to detect chromosome rearrangements and aneuploids. Chromosome abnormalities can also be detected using FISH with chromosome-specific paints. The application of the 24 different chromosome paints to metaphase chromosomes on a single slide would detect majority of chromosomal abnormalities (Larin et al., 1994). Chromosome-specific centromeric repeats and chromosome specific DNA libraries are frequently used as probes for FISH because of their utility in revealing chromosome aneuploidy or aberrations in interphase cells and tissues as well as identification of marker chromosomes unrecognizable by conventional banding methods. Three sets of distinguishable fluorophores emitting in the green (fluorescecin), red (rhodamine or Texas Red) and blue (AMCA or Cascade Blue) have been used for FISH to date.

FISH is thus a technique of sequence specific detection of DNA via hybridization. In other words, it is an *in situ* DNA detection technique in which fluorescent dyes are conjugated to DNA probes (Pinkle et al., 1986). FISH is an invasive technique which can not be applied to living cells. FISH is used as an adjunct to routine cytogenetics to achieve a higher sensitivity, specificity and resolution of chromosome aberration than is possible by banding analysis. For detecting a specific DNA sequence in chromosomes in nuclei the samples are usually fixed and denatured before hybridization to the DNA probe which contains the specific sequence. In order to make the DNA probe permeable to the nucleoplasm, the cell wall and cytosol are at least partly removed from plant samples by either mechanical (including flow sorting) or chemical (such as enzymatic digestion) methods before denaturing the DNA. Fluorescent microscope can precisely distinguish many of the fluorescent dyes. To date, most colors of fluorescent dye combinations in a single cell that have been achieved are 27 for human chromosomes (Lam et al., 2004). More dye combinations are theoretically possible by combining more color variants and DNA probes.

The technique of FISH can be used to study chromosome organization in plants using telomere and centromere probes containing repetitive sequences. In wheat the Ph1 locus is shown to be important for correct pairing and recombination. FISH reveals that the Ph1 locus reduces non-homologue association at their centromeres (Martinez-perez et al., 2001). Interphase chromosomes in Arabidopsis are organized as well defined chromocenters from which euchromatic loops emnate (Fransz et al., 2002).FISH can be combined with other staining technique to obtain better results. Combined chromosome specific FISH using BACs neighboring perincentromeric regions or NORs and DAPI staining of chromocenters showed that these condensed repetitive regions can interact at a high frequency in interphase cells.

FISH-a molecular cytogenetics technique enables one to interrogate regions of human genome too small to be visualized by conventional cytogenetics techniques. It has become a standard approach for commonly known microdeletions and microduplications. In this technique a DNA probe is labeled with a nucleotide that is conjugated with a fluorescent molecule or a reporter molecule (termed direct labeling or indirect labeling, respectively). After hybridization to metaphase chromosome or interphase cell and detection of signals, the probes can be visualized in the nucleus with the help of fluorescence microscopy. A variety of probes including YAC, BAC, Cosmids and plasmids have been used to identify microdeletions and microduplications. Analysis of microdeletions is carried out in metaphase cells whereas microduplication analysis involving FISH is carried out on interphase nuclei (Trask, 1991).

31.12 INTERPHASE CYTOGENETICS

Here one can analyze single genes or specific chromosomal regions in interphase nuclei. Genomic features on metaphase chromosomes can be detected in interphase nucleic by **non-isotopic** *in situ* **hybridization (NISH)**. Complex probe sets such as pools of clones from a single chromosome library have been used under suppressive hybridization conditions to decorate or paint individual chromosomes in a highly specific manner. Some of the strategies employed in the following interphase cytogenetic studies are shown below schematically (Lichter and ward, 1990).

In each lane two metaphase chromosomes of a particular karyotype are shown as white and shaded chromosomes. Two different signals as produced by NISH are shown as white and black circles. Signals in interphase nuclei are shown on the right side of each lane. Here see the number or locations of the signals. In Figures 31.3 (a-f), **a** shows detection of trisomic chromosomal material, **b** shows detection of subgenomic deletions or loss of a single gene, figures **c and d** show identification of a specific translocation which brings sequences of different chromosomes together, **e and f** show characterization of a chromosomal break point or of break point sequences. In the normal cell (e) two signals are seen whereas in case of translocation (f) one single pair and two separate signals are seen. Correspondingly, a single clone spanning the break point will give three versus the expected two signals in cells with translocated chromosomes. If the interphase cells are in G2-phase of the cell cycle all signals are visible as doublets versus the singlets shown in this scheme for the G1-phase nuclei. Further, viral integration sites can be enumerated by NISH.

Gene mapping—NISH can also be applied in gene mapping studies. There are reports of visualization of probes $\geq$ 1kb long using either conventional microscopy or digital imaging microscopy. In case of probes of higher complexity (i.e. containing interspersed repetitive sequences) suppressive hybridization can be employed and thus one can obtain highly specific delineation of the genome regions from which the probes have been derived. Suppressive hybridization has been used to map sequences cloned on YACs, cosmids, phage and plasmid vectors. With cosmids probes about 90% of the target sequences in cell population can be delineated by fluorescence and on metaphase chromosomes spread more than 80% shows specific signals on both chromatids of both chromosome homologs. As only signals present in parallel on both chromatids are counted as true signals in case of metaphase spread the precision mapping is greatly reduced. The use of smaller probe sizes reduces the number of informative metaphases to 20-25%, thereby increasing analysis time but both sensitivity and throughput are superior to isotopic detection methods. Hybridization signals from two probes can be spatially resolved on metaphase chromosomes when the probes are only several hundred kilobases apart but a minimum of 1-2-Mbp separation is required to permit their physical order to be established. To further improve the spatial resolution of gene mapping, co-hybridization with closely spaced probes from a

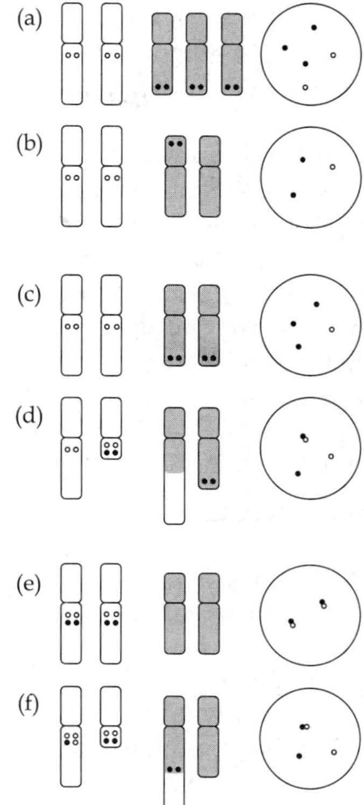

FIGURE 31.3 (a-f) Showing genomic features on metaphase chromosome to be detected in interphase nuclei by NISH. a showing detection of trisomic, b showing detection of submicroscopic deletion or loss of a single gene. C and D showing dentification of a specific translocation and e and detection of a chromosomal break or of break point sequences (adapted from Lichter and ward, 1990).

single genomic region have been performed on interphase nuclei. The combination of metaphase and interphase nuclear mapping particularly using multiple probes offers opportunity to physically order genomic DNA segments with a resolution presently achieved by gel electrophoretic methods and it thus provides a new bridge to interrelate physical and genetic linkage information. Further, other chromosome specific probes often clones of the alphoid DNA repeat family are being used to detect numerical chromosome changes in nuclei of cells in specimens from other inherited diseases.

In FISH, generally two detection techniques can be used.

I. **Direct detection** — This involves DNA probes that are nick-translated with dUTP conjugated to a fluorochrome such as fluorescien-tetramathylrhodamine or Texas red via an allylamine linker and fluorescence can be directly visualized by fluorescent microscopy.

II. **Indirect detection** — The indirect detection employs hapten (refers to any molecule which can react with an appropriate antibody to form a precipitate)-modified DNA probes. The different haptens (e.g. biotin, digoxigenin or also dinitrophenol) are introduced via fluorescent antibody techniques. The DNA probe is ligated to an antigenic protein that binds tightly to a specific antibody. The antibody itself is coupled to a dye that fluoresces intensely in U.V. light. The chromosomal DNA is then hybridized to the antigen-bound probe which in turn is detected by the fluorescent antibody. The current method of choice is **bioluminescence** or **chemiluminescence**. Bioluminescence refers to the emission of photons (light) by living organism (e.g. bacteria, diatoms, fungi, insects, fish, jellyfish and worms). Bacterial bioluminescence is generated in bacteria related to the genus *Vibrio*. Intracellular bioluminescence is produced in specialized cells which are frequently arranged in regular organs in fishes whereas extracellular bioluminescence originates from an organism but is usually secreted into a slime surrounding the producer. In case of chemiluminescence a biotin label is detected by avidin/streptavidin coupled to alkali phosphatase or horseradish peroxidase whereas digoxigenin label is recognized and bound by specific antibody linked to alkaline phosphatase or horse-radish peroxidase. Biotin is a low molecular weight vitamin which binds to proteins, hormones or DNAs. Avidin or streptavidin is used for the detection of biotin in biotinylated nucleic acids.

Fluorochrome, fluorescent dye or fluorophore is a chemical substance which emits fluorescent light after appropriate excitation. It reacts more or less specifically with a particular cell component and permits its detection, for example, the fluorescence detection of DNA with DAPI. Fluorochromes, some of them in their isothiocynate derivatives, e.g. xanthylium salts, fluorescein isothiocyante and rhodamine B isothiocyante or heteroxanthylium salts (oxazines), dipyrromethines, cyanine or stable bezopyran hemicyanins may be used for non-radioactive labeling of nucleic acids or antibodies. The fluorescent dyes are numerous with many spectrally distinct chromophores spanning the optimal range of U.V. to near infrared.

Variants of ISH—There are many variants of ISH depending on the type of DNA sequences, types of sequences (DNA/RNA), number of sequences (or gene loci) detected simultaneously, number of chromosomes, number of genomes, etc which are described as follows.

1. **Anti-sense fluorescent in situ hybridization (AS-FISH)**—It can be used to discriminate between the transcribed anti-sense (strand of a DNA duplex serving as template for the synthesis of RNA molecule and has base sequence complementary to the transcript) and the non-transcribed (sense strand) sequence. It uses antisense oligonucleotide coupled to a fluorochrome for detection of the coding strand of DNA to which it hybridizes.

2. **Chromogenic in situ hybridization (chromogenic ISH)**—In this technique the identification of specific DNA sequences on intact chromosome is through hybridization of non-radioactively labeled probes such as biotin, digoxygenin, dinitrophenol or fluorescein to target DNA and detection of the hybridization events by a specific antibody raised against the label and coupled to an enzyme such as alkaline phosphatase, horseradish peroxidase. These enzymes convert a colourless substrate to a colored product which is deposited at the site of hybridization.

   Enzyme conjugated antibody (enzyme linked antibody)—When an antibody is covalently linked to an enzyme it is called enzyme conjugated antibody. Such enzyme conjugated antibodies are used in the DNA detection systems for the visualization of the binding of the non-radioactively labeled DNA probe. The conjugated antibody has been raised against biotin or digoxygenin. The enzyme may also be linked to anti-antibodies for the detection of primary antibodies. Such conjugates are used in techniques such as ELISA, Western blotting.

3. **Chromosomal in situ suppression hybridization (CISS hybridization)** — Chromosome specific labeling is achieved by suppression hybridization. CISS hybridization permits direct visualization of the presence or absence of a genomic DNA sequence in a given deletion or duplication or its position relative to a translocation break point in a rapid and straightforward manner. It is a variant of non-radioactive ISH and hybridization is with complex biotinylated probes, for example, λ, cosmids or YAC libraries of a specific chromosme. As such chromosome libraries also contain ubiquitous, highly repetitive sequences, any hybridization to a chromosome preparation would detect sequences on all chromosomes. To prevent this unspecific hybridization the probe is denatured in the presence of an excess amount of genomic DNA and then pre-incubated at 37°C. The cot conditions chosen permit complete reannealing of highly repetitive sequences of the probe. Thus there hybridization to the chromosomal preparation is blocked (suppressed). Subsequent ISH with residual denatured probe and the detection of hybrids with the avidine-horse-radish peroxidase yield highly specific hybridization patterns and allows the selective visualization of specific chromosomal regions **(chromosome painting)**. The use of deletions, duplications and translocations will be value in mapping chromosomal regions of interest.

4. **Filter hybridization** — In this technique one of the nucleic acid components of the hybridization is immobilized on a membrane (originally nitrocellulose but more recently charged nylon membranes have also been used).

   It includes a variety of methods to detect specific denatured, for example, single stranded DNA or RNA sequences that are immobilized on a filter (nitrocellulose) by using a radioactively labeled RNA or single stranded DNA probe. The probe will anneal to homologous sequences and the hybrid thus formed will remain on the filter whereas the non-bound probe is washed off.

Colony hybridization — It is an *in situ* procedure for the detection of a particular DNA segment within a population of transformed bacteria containing with a large number of different recombinant sequences (bacterial libraries). In this technique the bacterial colonies to be screened are replica plated from a master plate onto nitrocellulose filter and lysed. After that the liberated DNA is denatured and fixed to the filter by baking. A radioactively labeled probe, complementary to the sequence under investigation is then hybridized to the filer bound DNA and the position of the colony containing the recombinant DNA looked for is localized by autoradiography. The desirable clones can then easily be selected from the master plate (Figure 31.4).

Grunstein and Hogness (1975) devised a method of transferring duplicate copies of colonies from a 'master' agar plate to nitrocellulose filter (replica plating) on which the plasmid DNA could be immobilized after lysis of cells. Now if the sequence of the protein or part thereof is known then it is possible to screen the colonies with a radiolabelled, 32P–

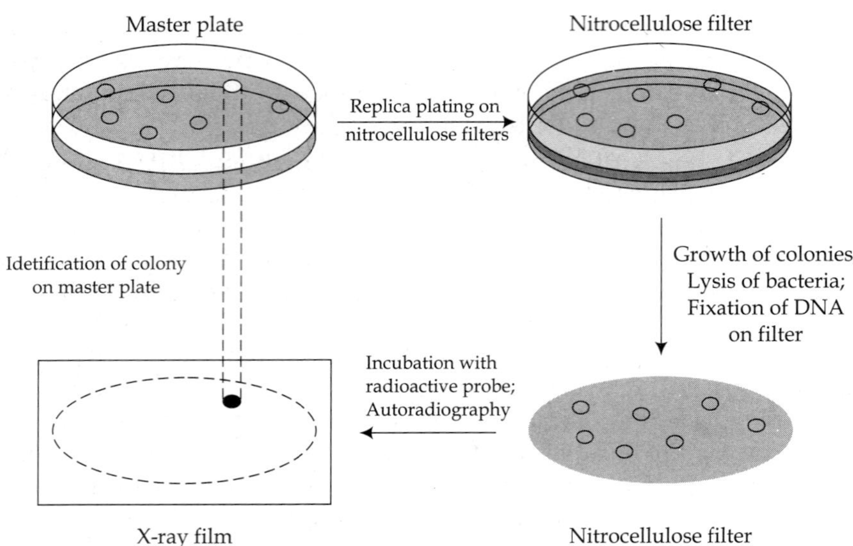

FIGURE 31.4 **Schematic representation of the principle of colony hybridisation for the identification of recombinant bacterial clones (Grunstein and wallis, 1979).**

oligonucleotide probe corresponding to the nucleotide sequence predicted by a part of the amino acid sequence (Montgomery et al., 1978).

Replica plating—It is a method to reproduce identical patterns of bacterial colonies using a cylindrical block covered with velveteen which is pressed onto a Petri dish(master plate) containing the colonies. About 20% of the bacteria will attach to the block and may then be transferred to secondary plate containing suitable growth media. A single pad usually gives 5 to 8 replicas.

Plaque hybridization—It is an *in situ* gene screening procedure used for the detection of a particular DNA sequence within a population of transformed bacteria harbouring large amounts of different cloned sequences (phage libraries). The detection is through hybridization *in situ* with a radioactively labeled RNA or DNA probes with complementarility with the sequence sought. In this technique the nitrocellulose filter is placed on an agar plate containing bacteriophage plaques. The unbound recombinant phage DNA is bound to the filter (plaque lift, phage lift), denatured and fixed to the filter by baking. The radioactive probe is then hybridized to the filter bound DNA and the position of the plaque containing the complementary sequence is located by autoradiography. Finally, the desirable plaques are then easily recovered from the master plate. Plaque/colony hybridization and filter hybridization are conventional methods for the isolation and recovery of cDNA clones and used for screening of cDNA libraries. PCR-based methods are also for arraying and screening of cDNA libraries.

5. **Solution hybridization**—In this technique annealing of complementary single strands of nucleic acids to double stranded helical molecules (DNA-DNA,DNA-RNA,RNA-RNA) takes place in solution. Solution hybridization or liquid hybridization is made in a technique called **Reverse Southern hybridization**. This technique is used to

detect specific DNA sequences by hybridizing them in solution to a suitable radioactively labeled probe. The thymidine residues of this probe are chemically modified by a psoralen derivative which permits the U.V. light induced cross-linking of the probe-target DNA hybrids. Subsequent electrophoresis in a denaturing gel allows the separation of cross-linked hybrid molecules from single strands. The hybrids can then be visualized by autoradiography. In the conventional Southern blotting procedure the DNA fragments are separated by gel electrophoresis and transferred to a memebrane prior to hybridization.

6. **Dot blotting**—It refers to the accurate and rapid measurement of the abundance of specific RNA or DNA sequences. In this technique the samples are dotted directly onto the filters without previous separation by electrophoresis. Filters are then treated as for Southern or northern hybridization. This technique of molecular cloning is used for rapid screening of transformants, transfectants, cell cultures, blood cells and even solid tissue samples (Costanzi and Gillespie, 1987).

Dot blot technique—It is a quick hybridization protocol for the detection and quantification of specific nucleic acid sequences (DNA, RNA). Hybridization of immobilized nucleic acids (DNA, RNA) on paper support is not restricted to material transferred from agarose gels. The recombinant DNA in bacterial colonies or bacteriophage plaques can be detected after transfer from petri dishes or for multiple samples of total cellular RNA can be applied directly to the nitrocellulose and can get detected. In this technique the aqueous nucleic acid sample is applied directly to a hybridization membrane (nitrocellulose or nylon membrane) as a dot blot of liquid and the sample filtered through the membrane by means of suction. This is usually achieved by the use of 'dot blot apparatus'. Subsequent hybridization with the probe of interest is carried out essentially for Southern blotting (DNA-DNA) or Nothern blotting (DNA-RNA). This technique can be used to determine the relative amounts of a particular DNA or RNA in a mixture by using control samples which contain known concentrations of the target sequence and employing densitometry to quantitate the relative hybridization signals from the dot blots (van Helden and Olliver, 1981).

Cytoplasmic dot hybridization—It is a variant of dot blot hybridization which uses whole cells as source of DNA or RNA. These cells are then transferred onto appropriate membranes, fixed and denatured and hybridized to radioactive or non-radioactive probes.

Chromosome painting—By extending the FISH technique entire chromosome can be specifically decorated-a technique called chromosome painting (Schrock et al. (1996); Zink et al., 1998). In this technique each chromosome is isolated by flow sorting or microdissection. The isolated chromosomes are then used as templates for PCR with degenerated primers and fluorescent derivatives of 2'-deoxycytosine 5' triphosphate (dCTP) to produce DNA probes. A single fluorescent derivative of dCTP or a mixture of dCTP derivatives with distinct fluorescent labels can be used to create chromosome specific probes for different colors. In other words, chromosome painting probes (chromosome specific composite libraries) generated by PCR from the flow sorted chromosomes are directly labeled with nucleotides conjugated to different fluorophores such as Cy2, Spectrum Green, Cy3, Texas Red and Cy5 or combinations there of. The probes are hybridized to target samples such as permeablized, fixed nuclei or chromosome spread. Typically, unlabeled DNA fragments containing disperse repeat sequences from the target chromosomes are included to increase specificity of the probes by suppressing cross-chromosome hybridization due to the presence and abundance of these repetitive DNA sequences all over the genome. In other words, repetitive sequences in the composite libraries are bloked by the addition of an excess of unlabeled DNA enriched for repetitive sequences (Cot-1 DNA). This technique is widely used in mammals, birds, insects to characterize chromosomes.

Reverse chromosome painting—It is a variant of chromosome painting which uses probes from aberrant chromosomes for ISH to metaphase chromosome spreads and detection of hybridization events by FISH. Abnormal chromosomes are characterized by highlighting normal chromosomes. Aberrant chromosomes are first isolated by Flow cytometry. Then they are amplified and labeled by degenerate oligonucleotide primed PCR and used for ISH under suppression conditions (avoidance of undesirable background signals). Reverse chromosome painting is used to identify marker chromosomes, to analyze chromosome break points, and to characterize deletions and insertions.

7. **Multicolor *in situ* hybridization**—It is a method for identifying several specific sequences of intact chromosomes simultaneously by hybridization with different probes such that each of which is labeled with a specific fluorochrome, for example, probe A with fluoresceine isothiocyanate, probe B with rhodamine B isothiocyanate, probe C with coumarin derivative), each of which detects a specific chromosomal site. The simultaneous use of several differently labeled probes in one single experiment produces multicolor chromosome picture. This cytogenetic technique can be used for differential staining of chromosome region specific areas as color bands and thus can be used for high resolution multicolor banding. The bands are generated by FISH of metaphase spreads of chromosomes with dissection libraries labeled with different fluorochromes. The different fluorescence intensities along the different chromosomes allow to assign different pseudocolors to specific chromosome regions.

8. **Multiplex fluorescent *in situ* hybridization (M-FISH)**—It is a technique for the simultaneous detection and discrimination of all different chromosomes in a metaphase spread by different colorization. Chromosome specific libraries are first labeled with distinct combination of fluorochromes. Then the different specifically labeled chromosome libraries are hybridized onto spreads of metaphase chromosomes (or cell nuclei) and the individual fluorochromes detected by epifluorescence microscopy (using all filters to excite all the fluorochromes in the sample) coupled to a cooled charge-coupled device (CCD)camera. The different fluorescence signals permit unequivocally assignment of a specific fluorogram to a specific chromosome. Combinatorial labeling of probes (i.e., with two or more different reporters) increases the number of target sequences which can be detected simultaneously by FISH (Ried et al., 1992). The ability to visualize multiple probes simultaneously should streamline the screening of specimens for chromosomal aneuploidies and/or chromosomal rearrangements. In addition, by incorporating one or more appropriate reference clones (e.g., centromere repeats or unique sequence genes) in the experiments, the assignment of gene dosage (loss of heterozygosity, aneuploidy and mosaic) or defining boundaries of chromosomal deletions should be more definitive. The generation of physical mapping data using either metaphase or interphase should be facilitated with combinatorial fluorescence as would studies on understanding of intranuclear topography of gene and chromosomes.

9. **Whole–mount *in situ* hybridization** (Whole-mount-ISH)—This technique is used to identify specific sequences in cells, tissues or organs of a whole plant or animal or cross-sections of them (e.g. embryo) by hybridization with the radioactively labeled probes. Usually the organism is prepared for ISH by dehybration-rehydration, hydrogen peroxide treatment to decrease background, proteinase k digestion to increase the accessibility of mRNAs for the probe and refixation.

10. **Zoo fluorescent ISH (Zoo-FISH)**—It is used for localization of fluorescence labeled probes, for example, BAC clones, genes on panels of metaphase chromosomes from different yet related animals.

    Q-FISH (Quantitative-FISH)—It is a variant of conventional FISH and used for quantitative estimation of telomeric lengths and employs a peptide nucleic acid (PNA) probes labeled with fluorochromosomes (e.g cyanin3) rather than the DNA or RNA probes in the traditional method. The PNA possesses an uncharged backbone and can hybridize to the target DNA under an extremely low ionic condition which prevents target DNA renaturation. Under these conditions fluorescence intensity correlates linearly with the number of bound fluorophores and therefore permits a quantitative estimate of telomeric repeat numbers.

11. **FISH BAC (or BAC-FISH)**—It is a variant of FISH technique. This method involves hybridization of fluorescently labeled bacterial artificial chromosome clones onto metaphase chromosome spreads and visualization of hybridization events by fluorescent microscopy. FISH-BAC allows localization of genes known to reside on a particular BAC onto specific chromosomes, detection of translocations, inversions, deletions or other chromosomal rearrangements.

12. **Primed *in situ* labeling (PRINS)**—It is a sensitive variant of the ISH technique , used for detecting specific DNA sequences in metaphase chromosomes. In this technique metaphase chromosomes spreads are prepared, synthetic oligodeoxynucleotides or short DNA fragments are hybridized to the chromosomes *in situ* and used as primers for DNA polymerase (e.g. *Thermus aquaticus* DNA polymerase) catalyzed extension in the presence of biotynylated or digoxigenin –labeled nucleotides using the chromosomal DNA as the template. The newly synthesized strand is visualized with fluorescence (e.g FITC) labeled –avidin or anti-digoxygenin- Fab fragments. This technique can be used for detection of DNA sequences (**DNA-PRINS**) as well as for visualization of RNA *in situ* (**RNA-PRINS**). The RNA-PRINS employs oligodeoxynu-cleotides as primers for reverse transcription (RTase)catalyzed extension with labeled nucleoside triphosphate using the RNA (e.g mRNA) as template.

13. **Genomic *in situ* hybridization (GISH) and its applications**—GISH is considered a further advance over FISH because in the former whole chromosomes become labeled and this is particularly advantageous in studies such as genome identification and meiotic analyses. Further, GISH can be applied to cells at interphase as well as those undergoing division. However, FISH is still the preferred technique in the investigations of NOR sequences, telomere sequences or specific repetitive sequences. In some cases where a repetitive sequence is well dispersed throughout the whole genome, FISH using such a sequence as a probe produces results similar to that of GISH with whole target genome being labeled. GISH was originally developed for

animal hybrid cell-lines (Pinkel et al., 1986) and first used on plants in 1987 (Schwarzacher et al., 1989) and they coined the term GISH). It is primarily the highly repetitive DNA rather than the low copy DNA (unique sequence DNA, the gene) that responds to probing by GISH. So the best results are obtained with species which have high repetitive DNA which is dispersed more or less evenly along the length of the chromosomes. GISH is a comparative, not absolute, approach. According to the experimental conditions imposed, virtually any total genomic DNA probe could be used to produce positive labeling with any target total genomic DNA or conversely not to produce positive labeling (even with identical DNA).But as it is extremely difficult to reproduce exact experimental conditions, GISH is not a technique which lends itself to probing a range of target diploid taxa from related diploid species to establish relationships but rather one that probes a target cell which contains two or more different genomes. Thus what is important is to provide the experimental conditions that allow the probe to label one of the genomes in the target cell but not the other. This is accomplished by two methods. Firstly, the stringency of the experimental conditions (which include temperature, time, reagent strength and degree of DNA fragmentation) can be varied whereby the degree of similarity of the two DNA sequences that would allow reannealing (and therefore labeling) can be raised or lowered. Secondly, the technique of blocking can be used whereby the labeled total genomic DNA probe is mixed with an excess of unlabeled total DNA (the block) from the nontarget genome in the target cells. The block hybridizes with sequences in common between the block and probe and thus mainly genome-specific sequences in the target genome remain as sites for hybridization with the labeled probe. The degree of excess of block over probe can be a critical factor. For rather distantly related probe-target pairs (e.g. *Triticum/Secale, Solanum/Lycopersicon, Leymus/Hordeum, Festuca/Vulipa, Festuca/Lolium, Gasterial/Aloe*) often no blocking at all is required although some workers have used blocks in ratios higher than 100:1. In other words, different genera can be readily distinguished often without the need for DNA-blocking but blocking is necessary when differentiating between species within a genus (e.g. Hordeum, Solanum, Nicotiana). Thus GISH can be used with or without a block. GISH is a technique which enables one to label differentially genomes or parts of genomes within cells. This technique is a variant of FISH. In this method the total genomic DNA of one species is used as probe together with unlabeled or differently labeled total genomic DNA of the other species, the later 'blocking homologous sequences shared by the two genomes (Ceoloni et al., 1998; Galasso et al., 1997). In this technique specific sequences of intact chromosomes are identified by hybridization of metaphase chromosome spreads with radioactively labeled or fluorochrome-conjugated genomic probes (e.g. from a genomic library) (Leitch et al., 1994). Thus chromosomes in several interspecific hybrid plants that carry chromosomes from different species such as the hybrid of wheat and rye, barley and rye or mung bean x urd bean can be painted. For painting alien chromosomes, one or both parental chromosomes are extracted and used as probes. GISH has been mainly used to characterize the hybrid status of progenies from a cross of two distinct parental species. Studies have shown segregation of the two parental genomes into distinct and separate domains within the nuclei of the hybrid. Further, when a strong concentric arrangement of the two genomes is observed in hybrids there the phenotype of the hybrid resembled the parent that contributed to the outer whorl genome. Thus nuclear location may determine gene activity.

The applications of GISH can be divided into four categories (Stace and Bailey, 1999). I. Chromosome disposition in interphase and mitotic cells II. Identification of genomes III. Recognition of parts of genomes (e.g. alien chromosomes, intergenic translocations or recombinations, B-chromosomes IV. Meiotic studies (what is pairing with what?). Although GISH is in theory and in practice equally as applicable to meiotic as to interphase or mitotic problems but few studies have so far been reported probably because optimum preparations are less easily obtained. Stages from pachytene onwards are suitable for study by GISH. The existence of separate chromosomal or genomic domains which is supposed to control transcription and gene expression, preferential loss of chromosomes wide sexual or somatic hybrids and pairing at meiosis can be studied. GISH can be used to label not only whole of one genome but it has the potential to paint the different genomes in different colors (Bennett, 1995). All three genomes (A, B and D) of the hexaploid wheat (*Triticum aestivum*) can be distinguished using simultaneous probes from their putative ancestors. The use of different dyes produced tricolored target chromosome spreads (Mukai et al., 1993). GISH is equally effective in identifying odd chromosomes or segments of chromosomes in foreign cells. Fragments as small as 50-100Mbp can be detected by GISH (Bennett, 1995) which is about five times higher than the minimum size detectable by ISH. Alien chromosomes in addition and substitution lines can be detected. Translocations or recombinations in cells containing two or more genomes can be identified. The application of GISH to somatic cells provides information on pairing, crossingover and recombination in sexual hybrids and on translocations in sexual and somatic hybrids. The remarkable success of GISH in probing cereal and other grass genomes is due to the facts that most grasses have relatively large C-values (i.e. large amounts of repetitive DNA) and that these repetitive sequences are dispersed along the chromosome length. It is the highly repetitive DNA that forms most of the species-specific DNA in the chromosomes and the gene sequences being relatively uniform in related species (Anamthawat-Jonsson and Heslop-Harrison, 1995). These repetitive sequences vary in amount (lowest in taxa with low C-values) and distribution. For example, the repetitive DNA is less abundant in *Glycine* and highly localized in *Festuca rubra*. The synteny (gene order) has been maintained across large chromosome sections of at least three of the five major subfamilies of grasses and that most quick genome evolution occurs in the repetitive DNA. Two classes of highly repetitive DNA are found: one mostly in paracentromeric and telomeric regions and the other dispersed along the chromosome length. The class one of the highly repetitive DNA predominates in *F. rubra* whereas class second predominates in *Secale cereale*. **Partial labeling of genomes**-The normal expection in GISH is that a complete genome is either labeled or not. If a whole genome is obviously not labeled then it usually indicates the presence of partial genomes (e.g. alien chromosomes or other segments) or the occurrence of sexual recombination or somatic translocation. Here it must be kept in mind that it is not possible to distinguish between sexual recombination and somatic translocation in a later generation of a sexual hybrids unless chromosome homologies have been established.

14. **Fiber-FISH**—It is used for detection and quantification of target sequences (such as genes or gene families) that work with hybridization of different probes lebelled with different fluorochromes to chromsome fibers. The fibers are generated either by molecular combing or dynamic molecular combing.

15. **Extended fiber *in situ* hybridization (EDF-ISH)**—It is a variant of ISH which permits an extreme resolution of fluorochrome signals in FISH. It is a technique for visualizing specific sequences (e.g. telomeres, rDNA genes) on spread DNA fibers protruding from a nuclear halo (i.e. a proteinous-nuclear scaffold).

    Competition hybridization—It is a variant of hybridization reaction in which increasing amounts of both a radioactive labeled and unradioactive labeled RNA (competitorRNA) are added to a fixed amount of unlabeled DNA. If both RNAs are totally unrelated then hybridization of the labeled RNA to the DNA is driven to saturation (control). But if the two RNAs possess homologies then the unlabeled competitor RNA will compete with the labeled RNA for homologous target sequences on the DNA and depending on the degree of homology the amount of the labeled RNA in the DNA-RNA hybrid will be reduced. Thus competition hybridization permits to estimate the degree of base homology between the two RNA molecules.

31.13   RNASE PROTECTION ASSAY

It is a technique used for detection, quantification and characterization of specific mRNA molecule out of complex mixtures of total RNAs. This assay is very important for mapping of transcription start and termination sites, determination of intron-exon structures and discovery of mutations (deletions and insertions). In this assay the total RNA is isolated and hybridized to an excess of specific labeled antisense RNA probe. This probe is produced by cloning of a cDNA fragment into a plasmid in between SP6, -T7 and T3 DNA dependent RNA polymerase promoters. Addition of the respective RNA polymerases allows to synthesize high specific activity radiolabeled antisense RNA probes. After hybridization of these probes RNase A and RNase T1 are added which digest single stranded RNAs and free antisense probes but do not attack RNA-DNA hybrids. After inactivation of the RNases the RNase protected hybrids are electrophoresed in the denaturing polyacrlamide gels. These gels are subsequently dries and exposed to X-ray films. Finally, the quantity and lengths of these hybrids are determined by autoradiography or phosphoimaging. The absolute quantity of mRNA can be obtained by comparing the hybridization signal intensity with signal strengths of calibration curve generated with synthetic sense strand target RNA. In case of S1 nuclease assay S1 nuclease is added after hybridization of the probes to total RNAs. The S1 nuclease digests single stranded RNAs and free antisense probes but only marginally attacks the RNA-DNA hybrids.

31.14   IMAGING AND VISUALIZATION RESEARCH TOOLS

Images are a means to communicate. They convey information which words could not articulate. Imaging has evolved much further than simple snapshots and with visualization, is now used as a research technique across a whole range of disciplines. Imaging via the technology of microscopes enables visualization of things too tiny to be visible by naked eye. Thus it can be used to study structure and function of genes and genomes. Besides it can be used for understanding the structure of protein. The three dimensional (3D) imaging of proteins involved in various fatal diseases in human is vital to better understand their mechanisms. Study of structure of protein domains is essential as they will reveal mechanisms for the development of drugs.

Atomic force microscopic imaging (AFM) — AFM along with STM (scanning tunneling microscopy) and NSOM (near –filed scanning optical microscopy) is an advanced microscopy technique of scanning probe microscopy. It is a technique for visualization of protein-DNA complexes, identification of gene start sites, counting of gene number, gene location and their orientation on a defined stretches of DNA (e.g. a BCA clone). In this technique the proteins of such protein-DNA complexes are detected by specific antibodies which result in increase of the mass of complex. AFM is a technique for higher resolution imaging of surfaces without the use of optical lenses or electron beams. It is being employed for the analyses of biological surfaces, localization of proteins, e.g., restriction endonucleases,or transcription factors on cognate regions in DNA and the generation of ultrahigh density chips-nanochips. In this technique the surface of the object is scanned with an ultrafine tip of 10μm length and a radius of less than 10nm which is attached to the lower side of a gold-coated silicium or silicium nitride microcantilever. The tip is dragged across the surface, driven by piezoelectric forces. Any irregularity on the surface results in a bend of the thin flexible cantilever which is detected by a scanning laser beam whose reflected light is directed into a multisegment photodiode. The intensity of the measured photon flow is proportional to the deviation of the cantilever from its relaxed position. There are two imaging techniques. In the **contact modus** the tip is directly placed on the surface of the object whereas in the **resonance modus** the tip is oscillating and the surface irregularities dampen the oscillation. The computer program then translates the photon flow into a surface profile and finally a three dimensional topographic image of the surface is computed.

31.15   POLYMERASE CHAIN REACTION (PCR)

PCR is powerful technique for amplifying and detecting specific of DNA sequences from a large pool of irrevelant DNA sequences. This technique requires some knowledge of DNA sequences flanking the segment that one is interested in amplifying. The sequence in genomic DNA amplified by PCR techniques is called '**amplicon**'. This term is also used for the sequence produced by PCR. With PCR (Mullis, 1983) specific regions of DNA within a genome can be amplified by as much as million fold and they can be visualized without the use of radioactive or any type of labeled probe (ethidium bromide). Two-deoxyolig- onucleotide primers, 20-30 bp in length and composed of nucleotide sequences at the termini or ends of a DNA segment are synthesized. The sequence of denaturation of DNA, annealing (annealing of primers to their complementary sequences) and extension reaction or polymerization (extension of the annealed primers with DNA polymerase) constitute one cycle of reaction. The product of a reaction cycle serves as templates for subsequent cycles of reaction and thus it is called a polymerase chain reaction (PCR).The DNA sample is heat denatured first and after that the primer anneals (or hybridizes) each DNA strand at specific sites which subsequently are extended by DNA polymerase, Taq polymerase. The two oligonucleotide primers thus flank the DNA segments to be amplified. The DNA synthesis proceeds across the region between the primers and thus resulting in doubling the amount of DNA fragment. In other words, one cycle of PCR results in two replicas of each double template molecule originally present in the mixture (Figure 31.5). The repeated cycles of denaturation-annealing and extension results in exponential (or geometric amplification) accumulation of the specific target fragment. The fragment length is the distance between 5′ termini of the primers on the target sequence. Any region of duplex DNA up to ~5kb can be amplified using two oligonucleotide primers. The Klenow fragment of *E. coli* DNA polymerase I which is normally used for polymerization reaction is not used in PCR because it is thermolabile. The *Taq polymerase* obtained from bacteria, *Thermus aquaticus*. Taq polymerase is heat resistant and survives many cycles of heating, annealing and polymerization and thus does not need to be replaced each cycle. *T. aquaticus* grows in hot springs and the bacteria survive at 90° C. Amplification works well for fragments up to 2000bp and several fragments can be amplified simultaneously in the same reaction. The reaction mixtures in PCR consist of DNA sample, Taq polymerase, buffer, dNTPs and a pair of primers. The buffer for PCR with Taq polymerase contains 50Mm Kcl, 10mM Tri-HCl, pH 8.4, 2.5Mm $MgCl_2$ and 100 µg/ml gelatin. Usually a 10X stock solution is prepared and stored at -20° C. The concentration of each dNTP (dATP, dCTP, dTTP and dGTP) is determined by neutralizing these with NaOH. 2mM of each triphosphate is used and the resulting 8 mM dNTP solution is stored at -20° C. The concentration of Taq polymerase is 2 units (1-4) per 100 µl reaction volume. Higher concentration than 1-4 units may result in greater production of non-specific PCR products and reduces the yield of desired gene fragment. The temperature ranges for denaturation, annealing and extension reaction are as follows: Denaturation (90-95, average 93°C but usually 94°C), Annealing (40-60°C but usually around 37°C) and Extension (70-75°C, average 72°C). The temperature at annealing

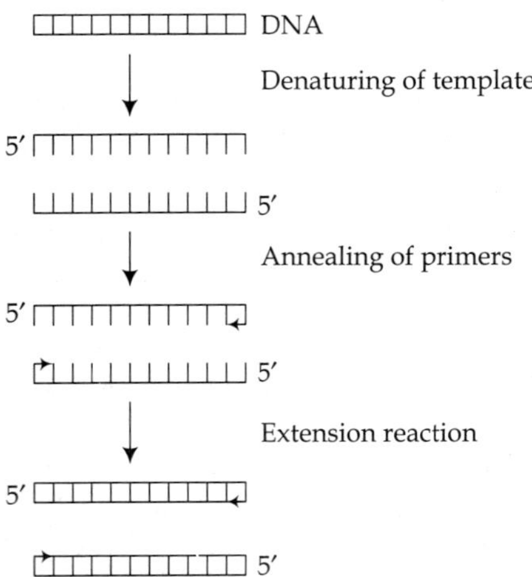

FIGURE 31.5   Steps in a standard PCR.

depends on 1. the length of the primers and 2. the GC content of the primers. For primers of only 12-15 bases annealing should be done at 40°C whereas the annealing temperature is~50-60° C for a 20 bp oligonucleotide primer with ~50% GC. At extension temperature of 72° C there is maximum activity of Taq polymerase. A look at the temperature ranges of denaturation, annealing and extension reaction shows that denaturation step is followed by cooling or lowering of temperature and annealing is followed by heating. As the extension reaction takes place at relatively higher temperature the secondary structure in the DNA template that often interferes with *in vitro* DNA elongation of DNA polymerase is abolished. The cycle of heating, cooling and extension (replication) is repeated 20 to 30 times over few hours and thus after 20 cycles there will be> 10 molecules of DNA (number of copies) but in fact the efficiency of PCR is usually between 20 to 50 %. After 30 cycles of PCR amplification the target sequence accounts for 99.9% of the total DNA and is thus sufficiently concentrated and that for most purposes additional purification is unnecessary. The time of incubation at 70-75° C during extension depends on the length of the target gene being amplified. The time taken from denatuartion to annealing or from annealing to extension depends on the type of equipments used for heating and cooling and the ramps. Temperature should be at least 90° C for strand separation and to have a margin of assurance one can go to 93°C. Typical values for 100 ul reaction volumes in 1.5 microcentrifuge tube with water bath at 72-93°C, 93-55°C and 55-72° C should be 1 min., 1min. and 2 minutes, respectively.

If this cycle of denaturation, annealing and extension is repeated up to 50 times than it will generate up to 1 billion-fold identical copies of the DNA template. This is the reason that PCR is becoming more important in quality control(QC) procedures where there is a need to, for example, measure the exact load of bacterial DNA in food sample.

Primers—Primer is a short oligomer complementary to the template to which an enzyme adds additional monomeric subunits. In fact, primers are single stranded flanking sequences. Primers are typically short stretch of synthetic DNA, ~20 bases in length. It provides a free OH (hydroxyl) group at the 3′ end to which a new nucleotide unit is added. In practice, primers longer than 30 bases are rarely used. Short primers are used when limited information is available about the DNA fragment and when there is problem of primer stability at the extension temperature. When DNA sequence to be amplified is of higher complexity long primers are used. **Amplimer** refers to the primer of a PCR. The primer should have the following characteristics (Saiki et al., 1988).

1. Selecting primers with an average GC content of ~ 50%

2. Avoid primers with substantial secondary structures

3. The primers should be checked against each other for complementarity.

Sequences not complementary to the template can be added at the 5′ end of the primers. These sequences become incorporated into the double stranded PCR products and provides a means of introducing restriction sites or regulatory elements (e.g. promoters) at the end of the amplified target sequence. The primers including a restriction endonuclease site will facilitate the subsequent cloning of the amplified DNA. The structural gene can be put under different promoters and cloned to study its expression and functional products. Also, the two structural genes can be fused and its product can be studied. Primers are necessary for

   i. cDNA synthesis

  ii. sequencing and

 iii. sequence amplification in PCR.

Types of primers—A priori knowledge of the nucleotide sequences flanking the loci is an absolute requirement for the design of primers. **Random primer** refers to an oligonucleotide primer that has a four fold-degeneracy at each position, i.e. each of the four bases (A,C,G or T) may equally well occupy a position. For example, 5′-NNNNN-3′ where N can be any of the four bases A,C,G or T. **Arbitrary primer** is an oligonucleotide primer that has a specific but arbitrary chosen sequence (i.e. a single base at each position hence either A or C or G or T). Primers of arbitrary sequence are used to amplify sequences of genomes using the conventional PCR technologies.

Non-targetted amplification methods—These methods use arbitrarily chosen sequences as primers. Any given oligonucleotide will pair with templates which are not strictly homologous under relaxed stringency. PCR leads to multilocus amplification instead of targeting unique locus. AP-PCR and RAPD are based on this principle. Because of the weak stability of the priming reaction the RAPD method lacks reproducibility and many factors can compromise the results. The definition of mono-locus specific primers, more stable for analysis of large sample is impossible. There are also primers used for priming specific DNA sequences, for example, **T primer** and **P primer**. P primer is a primer of arbitrary sequence which serves as a forward primer in the sequence differential display reverse transcription PCR in combination with a reverse oligo (dT) primer targeting at the poly (A) tail of eukaryotic mRNA ('T primer'). Sequence specific primer is one which permits amplification of specific sequence out of a complex genome whereas a gene specific primer is one which is complementary to a part of a gene (or its cDNA) and serves as a primer for amplification of this gene by PCR technique. A **unidirectional primer** is one which contains a homopolymeric (dT) tail at its 3′ ends and one or more restriction sites in close proximity. These are designed to prime cDNA synthesis from poly (A)$^+$-mRNA while generating a specific restriction site at the 3′ end of the cDNA. **Anchored primers**-Primers are specifically designed oligodeoxynucleotides that contain a central region of varying length (e.g. 20nt) and additionally one , two or more bases at the 3′- or 5′ end. The central region has high sequence homology to a specific locus in the genomic DNA whereas anchor (s) extend (s) the specificity into the adjacent 3′ or 5′ flanking sequences. The employment of such primers in conventional PCR will reduce the number of amplified products from a genomic template and thus reduces the complexity of amplified fragment patterns.

Fluorescent primer—It refers to any oligonucleotide that has been labeled by one or more fluorochromes and used as primer in PCR reaction.

Primers are used for different purposes other than amplification. The primer DNA (DNA primer) which is a single stranded DNA fragment is used by DNA polymerase III for DNA replication. Similarly, RNA primer (primer RNA), a short oligonucleotide of about 10-15nt in length is synthesized by RNA primase, serves as a primer in the DNA polymerase catalyzed synthesis of an Okazaki fragment resulting in replication of lagging strand of DNA in DNA replication.

The PCR can be used with very small amount of DNA sample. The PCR can amplify as little as one DNA molecule and thus the DNA obtained even from a single hair, a drop of blood or a small semen can be subjected to PCR. Further, DNA samples obtained from months or years old material can be analysed. Thus DNA from fossils (human remains or plant or plant parts buried) can be analysed.

The amplified PCR product is subjected to agarose gel electrophoresis and the difference in the products can be detected by different bands in the agarose gel. The size of the products can be determined by PAGE. The Taq polymerase has allowed PCR to be completely computerized. The reagents are mixed and filled in the reaction vessel which is inserted into a programmable **thermal cycler** and the machine is programmed to set the DNA denaturation temperature , primer annealing temperature, primer extension tome, number of cycles and other parameters.

Efficiency of PCR is measured by the number of amplified molecules produced and it depends on the length of the primers used and the temperature conditions and time allowed during the reaction cycle. **Primers** are the key to the success of PCR. If the right type of primers is used then it will result in amplification of a single DNA fragment but if wrong type of primers is used then there will be either no amplification at all or wrong fragment and/or many fragments will be amplified. The correct or incorrect type depends on the nucleotide sequence and the length of base pairs constituting the primers. Primers are short oligomers, complementary to the template to which DNA polymerase adds additional monomeric subunits. So nucleotide sequence of primers must correspond with the sequence flanking the target region on the template molecule. In practice, each of the two primers must ,of course, be complementary (if not identical) to its template strand for hybridization to occur and 3' ends of the primers should point towards one another.

Ideally the length of DNA fragment to be amplified is < 1 kb and it should not be greater than 3 kb. Although with standard PCR technique fragments up to 10 kb can be amplified but the longer fragments will be less efficiently amplified and thus result in loss of reproducibility of result. With modifications in techniques amplification of fragments up to 40 kb is possible.

The length of primers could be short or long. Too small primers will hybridize to non-target sites and give undesired amplification products (many molecules of too many DNA fragments). For example, a primer of 8 base pair will be able to anneal once every 65536bp (4^8) giving approximately 46000 positions in a genome of 3,000,000 kb and thus this pair of primers will not give a single specific amplification product. But when a pair of primers of 17 bp is used then it will pair once every 17179869184 bp which is 5 times greater than the size of human genome and there is thus the possibility of hybridization of the 17-mer primer at just one specific site and thus the PCR would give a single specific product. The length of primers also affects the rate at which they hybridize to the template DNA. The longer the primer the slower is the rate of hybridization. Longer primers will not be able to hybridize completely to the template in the time allowed during the reaction cycle and thus will result in less efficient amplification.

Too high annealing temperature will lead to no hybridization between primer and template DNA whereas too low temperature will produce mismatched hybrids which will lead to increase in hybridization sites and thus amplification of non-target DNA fragments. There is thus need of ideal annealing temperature in the PCR. The ideal temperature is 1-2° C below the melting temperature (Tm) of the primer template- hybrid. The base composition of a DNA fragment (primer) affects Tm. Optimal annealing and washing conditions for nucleic acid hybridization depends on the knowledge of GC content of the reactant. The melting temperature, Tm, is calculated as TM= 4 x (G+C) + (A+T) in which G+C is the number of G and C nucleotides and A+T is the number of A and T nucleotides in the primer sequence. Thus, Tm for each primer is calculated and accordingly the annealing temperature is adjusted. Further, as two primers are used in PCR so they should be designed in such a way that they have identical Tms. Primers are designed by copying a fragment of cDNA clone from a cDNA library. The annealing temperature in degree centigrade (TA) is calculated by the following formula (Wu et al., 1991).

$$T_A = 1.46 (A + T + 2C + 2G) + 22.$$

In case of RAPD primers there are 1,048,576 possible combinations of ten bases for primers. However, the majority of RAPD primers rather than having random nucleotide contents, are constrained to have 60-70% GC (24,576 possible combinations). It is known that GC content is not evenly distributed in plant genomes (Li and Graur, 1991) and so the RAPD technique may preferentially screen GC-rich regions. The high degree of polymorphism at CpG sites (methylated regions of eukaryotic genomes) may result in overestimates of nucleotide diversity of very similar taxa (Harris, 1999).

Fidelity of DNA polymerase — Any DNA or RNA dependent DNA polymerase erroneously incorporates wrong bases into the polymerization product. There is absence of 3'-5' exonuclease activity (proof reading activity) in Taq polymerase and so if it misincorporates a dNTP, subsequent extension of the strand either proceeds either slowly or stops completely. To overcome this problem another thermostable polymerase with proof reading capability is added. Similarly, AMV reverse transcriptase introduces1 error/17000 bases whereas Moloney murine leukemia virus (MuLV) introduces 1 error per 30,000 bases and so the latter is the desired reverse transcriptase for cDNA synthesis.

Limitation with PCR — The two limitations are (i). fidelity of the final product and (ii). the size of the product span that can be amplified (Barnes, 1994). The first problem is partially addressed by the replacement of Taq DNA polymerase by Pfu. PCR amplification rapidly becomes inefficient or non-existent as the length of the target span exceeds 5-6kb. The Taq polymerase stops at the site of depurination which arises because of high temperature. Single stranded DNA purinates faster than double stranded DNA. Depurination may well limit long distance PCR. So length and duration of the heat denaturation step of PCR be kept to a minimum. The short denaturation time found to be optimal (estimated 5 seconds or the reaction itself at 95°C) was surprisingly effective for the amplification of 35kb. Thus in case of RAPD, factors influencing its reproducibility include: primer and primer concentration (ii) Taq polymerase source (iii) magnesium ion concentration (iv) template concentration (v) thermocycler and (vi) temperature profile (Devos and Gale, 1992; Williams et al., 1993).

Hot-start PCR — It is generally recognized as the most reliable way to perform PCR (Taylor and Logan, 1995). A hot-start PCR procedure prevents primer extension before the PCR reagents attain a high temperature. This ensures adequate stringency in the first thermal cycle. The simplest way of conducting hot-start PCR is to use anti-Taq DNA polymerase antibodies capable of inhibiting Taq DNA polymerase activity until they become denatured at high temperature.

Applications of PCR—PCR is a selective amplification technique. The PCR technology can be applied in the various following areas. It can be used for detection of known nucleotide substitution associated with sickle cell anemia, diagnosis of hereditary and infectious diseases, routine DNA amplification, site directed mutagenesis, cloning of genomic DNAs and cDNAs, DNA sequencing, DNA fingerprinting and in forensic science. It can be used for the detection of allelic polymorphisms of DNA sequences unique to rare in a population as well as the cross-species isolation of homologous genes.

Genomic cloning/marker generation—PCR allows to identify and isolate one or few copies a specific target sequence from a complex background. It allows rapid cloning of genes even when the exact sequence of the gene is not known. This is helpful when restriction sites in the sequence of interest are too frequent or limiting and DNA must be moved from one vector to another or DNA sequences must be deleted or juxtaposed. With PCR gene can be synthesized *in vitro*. Large quantities of DNA template can be amplified directly from the trace amounts of genomic DNA using primers designed from pre-determined sequence data from the native gene. The DNA synthesized from PCR can be sequenced directly. By incorporating suitable restriction sites into the primer sequences the PCR product can be cloned in an appropriately cleaved vector to facilitate further manipulation. Amplifying and sequencing DNA segments whose flanking sequences are unknown by a process called reverse PCR which is described below. DNA polymorphism can be studied by using repeat specific primers (Alu sequences in human and microsatellite in plants). PCR methods have been used to map STSs and ESTs. PCR has produced markers for fine mapping and integrating genomic and genetic maps. PCR-based techniques are used to map sequences specific to cloned DNA such as BAC, YAC, cosmid, contigs or segments of chromosomes (using radiation hybrid mapping techniques).

Genome sequencing—Modified PCR techniques have proved ideal for generating DNA template for sequencing and have been directly applied to bacterial colonies and phage plaques and thereby obviating any DNA preparation.

***In vitro* mutagenesis**—As the oligonucleotide primers have the ability to bind to the template DNA even when they are not an exact sequence complement, base substitution, addition, deletion within the sequence defined by the primers can be incorporated and the function of the modified form can be compared with the native molecule. In other words, mutation can be created or engineered *in vitro* and gene function can be studied (see chapter 21).The PCR technique has been used to construct mutations *in vitro* and to amplify single copy sequences within complex DNA mixtures for facile cloning and analysis (Saiki et al., 1985).

RNA analysis/Analysis of gene expression—RNA molecules can also be analyzed by PCR by first converting them to cDNA by reverse transcription using reverse transcriptase. The resulting DNA transcript can then be amplified by standard PCR technique. Through this technique then even rare mRNA molecules can be detected and monitored. If only a part of the amino acid sequence of protein is known, these data can be used to synthesize degenerate primers which can amplify a segment of cDNA. This in turn can be used as a probe to screen a cDNA of library for the desired full length clone. Alternatively, if only one end of a sequence is known then cDNA generated using the known primer can be tailed with a homoplymer, poly (G) sequence to which an 'anchor primer' that includes a poly (C) stretch can bind and act as an unknown PCR primer. This strategy has been used to amplify T cell receptor transcript, the 5'-end of which encode the variable (and therefore unknown) region of the receptor. Various PCR-based methods have been developed to identify differentially expressed genes (see **subtractive hybridization**) and two of the most commonly used techniques are representational difference analysis and differential display (see chapter 24 for detail). PCR is also used to analyse the relative levels of gene expression in different tissues or after different treatments (see chapter 24).

Forensic application—Analysis of biological material such as hairs, blood, saliva, semen stains can be analysed.

Molecular palaeontology—Direct sequencing of genomes will provide insight into evolutionary history.

Medical uses—PCR technology can be used in medical research, genetic counseling and clinical investigation. PCR-based tests have been developed for a number of genetic diseases such as Haemoglobin and its disorders, Phenylketonuria, Duchenne muscular dystrophy, Cystic fibrosis. Further, PCR is ideal for the detection of pathogens such as HIV-1, herpes, hepatitis and cytomegatovirus as it is possible to amplify a specific sequence present at a low concentration within a complex of DNA mixtures. It can be used for amplification of DNA sequences flanking the crossover sites of a characteristic chromosomal translocation.

In situ **PCR-** *In situ* PCR is useful for visualizing the location of rare transcripts which are difficult to visualize by normal *in situ* hybridization.

31.16 VARIANTS OF PCR

The PCR technology includes a myriad of techniques used to amplify a specific DNA segment and to modify the amplified sequence simultaneously, for example, the application of PCR-add on primers, the introduction of mutations in PCR mutagenesis or the *in vitro* recombination of two specific DNA fragments in recombinant PCR. There are many variants of PCR depending upon the types of sequences we are interested in amplifying and the types and numbers of primers used in the polymerase chain reaction.

AP-PCR—It is a modification of the conventional PCR (Welsh and McClelland, 1990) that allows the amplification of the target sequences with arbitrarily chosen primers (any oligonucleotide primer that has a specific but arbitrarily chosen sequence, i.e. a single base at each position and hence either A or C or G or T) of 18-32 bp in length and high primer-template ratio (1-500) but without prior knowledge of sequences of the target genome. This technique is used for establishing fingerprints of polymorphism between bacteria and fungal strains or different plant varieties.

Inverse PCR—This technique allows amplification of unknown DNA sequences flanking a core region of known sequences. In this technique DNA containing the core region is digested with appropriate restriction endonucleases that cut outside the core region and generate a fragment of suitable size for PCR amplification. This fragment is then circularized using T4 DNA ligase under conditions favouring the formation of monomeric circles. Primers for PCR, homologous to the ends of the core region and with 3' ends pointing away from each other (inverse orientation) are annealed so that the PCR elongation proceeds across the unknown region of the circular molecule rather than across the known core separating the primers (Figure 31.6). For the amplification the circle as such can be used but linearlization may enhance the chain reaction. This technique is used for the study of 5′ or 3′ flanking regions of the coding sequences or transposons and for isolating and characterizing sequences that are contiguous to a DNA fragment (or chromosome segment) of known sequence (a technique called '**chromosome crawling**').

Anchored PCR—For information on anchored PCR see chapter 24.

Allele-specific PCR—It is a variant of conventional PCR that permits the amplification of specific alleles or DNA sequences variants. For example, a single base difference which discriminates both alleles, can be detected by using two primers, one possessing a 3′ end specific for allele A and the other, a 3′ specific for allele a. A third primer is designed to bind to sequences downstream of the alleleic polymorphic sites which are identical for both genomes. **Cascade –PCR-** A variant of PCR designed to reduce the complexicity of AFLP patterns from giant genomes such as genomes of maize, wheat or lilies, by a series of PCR steps. Each of these PCR steps uses a more selective amplification primer then the previous one. The first round of PCR cycles employs a primer with one base 3′ overhang. The second round of PCR cycles uses the same primer but a 2 base 3′ overhang and finally the third round of PCR cycles uses the same primer but now with a 3 base 3′ overhang for extension.

Suppression-PCR—A variant of PCR used for isolation of promoters and regulatory elements, for walking from the sequences of cloned cDNAs for exact determination of exon-intron boundaries of split genes and up or down stream walking from sequence tagged sites (STS)s. In this technique the genomic DNA is separately digested with a series of restriction endonucleases with 6bp recognition sequences that generate blunt ends, e.g. DraI, EcoRV, PuvII, Sca I and Ssp I. Then a special adaptor is ligated to the blunt ends of restriction fragments using T4 DNA ligase. The adaptor possesses a blunt end to facilitate ligation to the

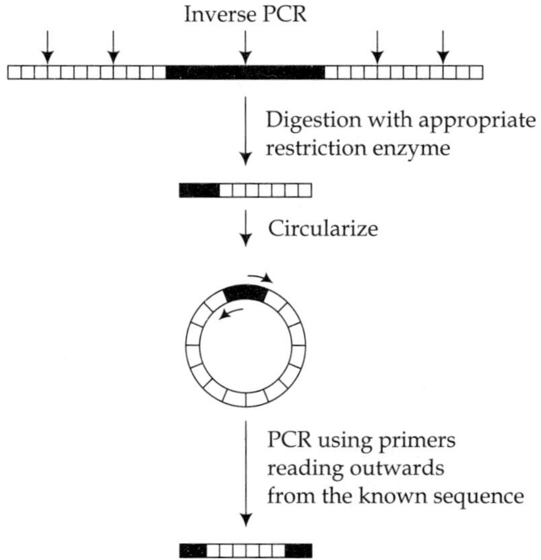

FIGURE 31.6   Showing inverse PCR procedure.

blunt ended genomic restriction fragments, carries two rare restriction enzyme sites, for example, Not I and Srf I/SmaI, for generating staggered or blunt ends, respectively, for cloning into suitable cut vectors and an amino group on the 3' end of the lower strand. This group blocks the extension of the lower adaptor strand and prevents the formation of a primer binding site (unless a defined distal gene specific primer extends a DNA strand opposite to the upper strand of a adaptor). After adaptor ligation, an aliquot of the DNA fragments serves as template for a PCR with an adaptor primer and a gene specific primer using a mixture of two long distance thermostable DNA polymerases, namely, *Thermus thermophillous* DNA polymerase and vent TM DNA polymerase. The adaptor primer is shorter than adaptor and is capable of double stranded adaptor sequences at both ends (non-specific DNA synthesis), ends of the individual strands will form pan handle structures due to the presence of the inverted terminal repeats. These panhandles are more stable than primer-template hybrids and suppress exponential amplification. If a distal gene specific primer extends a DNA strand through the adaptor then the extension product contains the adaptor sequence on one end only and thus can not form a panhandle structure and thus in this case amplification proceeds normally. Finally, the amplification products are separated by agarose gel electrophoresis and EtBr fluorescence.

PCR-RFLP—It detects restriction endonuclease site variation in two or more genomes. In other words, any variation (s) in the length of amplified fragments from genomic DNA of two or more individuals of a species can be known using PCR technique. The amplified fragments are subsequently cut with a specific restriction endonuclease.

e-PCR—(exclusive PCR)- A variant of PCR which amplifies preferentially non-hairpin forming single stranded DNA molecule.

TAIL–PCR—(Thermally asymmetric PCR)- A variant of conventional PCR in which temperature (primer annealing temperature) programme switches between low and high. This variant of PCR thus permits to use two primers of different sequences (e.g a RAPD primer and a microsatellite primer) to amplify genomic regions flanked by a microsatellite and the address site for RAPD primer.

Asymmetric PCR—In this PCR technique unequal amounts of the two primers are used. During the first 20 or so cycles double stranded DNA is produced in an exponential fashion but beyond this the primer at lower concentration becomes depleted and subsequent cycles generate predominantly single stranded DNA by extension from the non-limiting primer. The single stranded DNA accumulates linearly and is more amenable to Sanger sequencing techniques than the corresponding double stranded DNA molecules.

Those PCR techniques that employ one primer only are as follows.

1. AP-PCR-
2. DNAAF
3. RAPD

Patched circle PCR (PC-PCR)—It can be used for site directed mutagenesis (see chapter 21).

Vectorette PCR—V-PCR is used for the amplification, cloning and sequencing of the unknown DNA sequences adjacent to a known sequence. It is a variant of PCR used for chromosome walking in a specific direction, characterization of sequences flanking inserted DNA (e.g. transgenes, transposons or viruses) or amplification of terminal sequences of BAC clones. In this technique, one or several vectorette libraries are constructed through digestion of the target DNA with only one or a variety of restriction enzymes and ligation of an excess of vectorettes to the restriction fragments generated, catalyzed by T4 DNA ligase. Vectorettes are synthetic oligonucleotide duplexes consisting of two complementary strands containing an internal mismatched sequence and either a blunt or stick end complementary to the overhang generated by the chosen restriction endonuclease (s). Then a locus-specific primer is used in a first PCR cycle to amplify a new complementary strand to which vectorette primer, identical to the mismatched sequence of the vectorette's bottom strand, can anneal in the second PCR cycle. Now in the subsequent cycle, priming occurs from both the locus-specific and vectorette primer thus include sequences from then unknown region. Priming from other vectorette-flanked fragments that do not contain the known sequence, is therefore not possible. The amplified fragments can either be cloned or sequenced with a sequencing primer identical to the 3' end of the vectorette's botton strand. After sequencing a primer can be designed which allows further walking into flanking unknown DNA using a new aliquot of the vectorette library. Thus this technique is employed for chromosome walking in a specific direction, characterization of sequences flanking inserted DNA or amplification of terminal sequences of bacterial or yeast artificial chromosome clones. **Nested PCR-** It is a modification of the standard PCR which improves the yield of specific target sequences. In the standard PCR genomic DNA is denatured and annealed with an excess of two oligonucleotide amplimers which bind to sequences up- and downstream of the target DNA. These amplimers are then extended using thermostable DNA polymerases. The DNA is gain denatured, annealed to the same oligonucleotides and extended in second cycle and this process is repeated some 20-30 times. Since the polymerase reads beyond the target DNA, a population of fragments arises, the lengths of which exceed that of the target DNA. In order to restrict the PCR to the target DNA, a second set of amplimers (nested oligos) is annealed to sequences within the target DNA. This results in amplification of only target DNA.

Recombinant–PCR—It is a variant of conventional PCR and it permits the *in vitro* recombination of two different DNA fragments, for example, a promoter of gene A and coding sequence of gene B, using two antiparallel PCR add on primers that carry complementary 5' termini. These primers are employed to amplify different sequences separately. And after amplification the different sequences are annealed by their complementary ends and thereby generating a fusion product of promoter of gene A and gene B.

Motif-primed PCR—It is a variant of the conventional PCR and uses primers complementary to conserved DNA sequence motifs (e.g. parts of promoters, consensus sequences for DNA binding proteins or regulatory domains of gene families) important for gene function and regulation.

Quantitative-PCR—Q-PCR or real-time PCR or TaqMan PCR is a technique used for the detection of the accumulation of the amplification products during conventional PCRs and their quantification. DNA damage and repair is important in both mutagenesis and cancer therapy. Southern blotting based methods have proved extremely valuable at the level of individual genes. PCR based techniques are equally sensitive, more rapid and uses less DNA compared with Southern blotting and allow mapping of damage and repair at the sub-gene level (~500bp). Q-PCR is based on the fact that many DNA lesions can block the Taq polymerase and thereby result in a decrease in amplification of a damaged DNA segment compared to the amplification of the same segment in a non-damaged template. Q-PCR assays thus has the potential to measure the lesion frequency and subsequent repair of any DNA damaging agent which blocks Taq polymerase (or other polymerases used in PCR) in any DNA segment for which flanking sequences are known. Strand specific-QPCR (ssQPCR) allows the strand specific measurement of DNA damage and repair in mammalian cells (Bingham et al., 1996). The various techniques of Q-PCR can be grouped in two broad categories:

1. **Intercalator-based methods** — These methods involve intercalating dyes such as EtBr in each amplification reaction, irradiating the sample with U.V light and detecting the resulting fluorescence light with a computer aided, cooled charge-coupled device (CCD) camera. By plotting the fluorescence increase vs. cycle number amplification plots are generated which allows to quantify the products. The limitation with this technique is that both specific and non-specific products generate fluorescence signals which make quantification obsolete.

2. **Probe based methods** — 5′ nuclease PCR and similar probe based quantification protocols permit to detect only specific amplification products in real time. This technique exploits the 5′ nuclease activity of Taq polymerase to cleave the probe-target hybrids during amplification when the enzyme extends from an upstream primer into the region of the probe. This cleavage is can be visualized by increased fluorescence, if the oligonucleotide probe contains both a reporter fluorochrome at its 5′ end and a quencher dye at its 3′ end. The close proximity of both fluorochromes (1.5-6nm) results in a Forster type fluorescence energy transfer, leading to the suppression of the reporter (quenching) which is relaxed when the probe is hydrolyzed. Quenching refers to the reduction or complete inactivation of an excited state of a distinct molecule, for example, a fluorochrome, by a **quencher** and it does this either by energy transfer, electron transfer or by chemical reaction.

Another version of Q-PCR is the competitor PCR in which a synthetic DNA or RNA is used as internal standard (competitor amplicon) that contains the same primer binding sites and has the same amplification efficiency as the target but has a different size to discriminate it from the target. Probe based PCR are reproducible.

Qualitative vs quantitative PCR — In a standard PCR the end point analysis of PCR products is not quantitative as the efficiency of amplification reaction varies during the course of PCR(the exponential amplification continues only up to 30 PCR cycles) and approaches a plateau phase at later cycles and thus it can not be used to estimate the initial amount of DNA at the start of the PCR cycle. Thus quantitative PCR which detects small amounts of DNA and provides answers in yes or no like qualitative PCR , also allows to estimate the initial copy number of the DNA sequence that is amplified and detected. The quantitative PCR thus either controls the amplification efficiency throughout the reaction or enables the measurement of PCR product during the exponential phase of PCR.

Real-time — PCR- Real -time PCR and RT-PCR are powerful technologies for PCR-based quantification of nucleic acids. Real-time PCR allows direct monitoring and quantification of the generated PCR product during the course of the amplification reaction without the need to further manipulate the PCR after PCR. This has been made possible because of two developments. (i) Detection of PCR products through the use of fluorescent dyes (ii)Development of machine for reading of fluorescent signal in real-time during the course of reaction in each PCR cycle. The copy number of the initial template is determined by the PCR cycle at which the PCR product is detected. Fluorescence is measured in each PCR cycle and fluorescence intensity is directly proportional to the amount of PCR product generated at a given point in time(Loeffert, 2007).There are two main methods of detection of PCR products in real-time PCR. Detection is through either by fluorescent dyes such as SYBR® Green I which bind to double stranded PCR product(but not to single stranded PCR primers)(Figure 31.7) or by oligonucleotide probes that are emit light upon binding to specific PCR products. The latter depends on the use of dual-labelled probes that harbour a fluorescent dye such as6-carboxyfluorescein(FAMTM) and a quencher. Light is emitted only when the quencher and fluorophore are separated and this separation occurs either by the probe undergoing a conformational change as is the case in Molecular Beacons (Figure 31.8) (Tyagi and Kramer, 1996) or by hydrolysis of the probe due to 5'-3' exonuclease activity of Taq DNA polymerase (Figure 31.9) (TaqMan probes, Livak et al., 1995). The other less frequently used types of probe or fluorescently modified primer include FRET hybridization probe(Figure 31.10).

Competitive oligonucleotide PCR (COPCR)—It is a variant of PCR which uses three oligodeoxyribonucleotide primers (two forward primers and one common reverse primer) to detect single base pair changes in a specific DNA target molecule. One forward primer is specific for the wild-type whereas the other is for the mutant sequence.

PCR add on primer—In this technique of PCR the synthetic primer carries a recognition site for a restriction endonuclease. Such add on primer with 5' overhanging termini are annealed to the target sequence. The amplified products thus contain the desired restriction site which can be easily cloned into appropriately cloned vector.

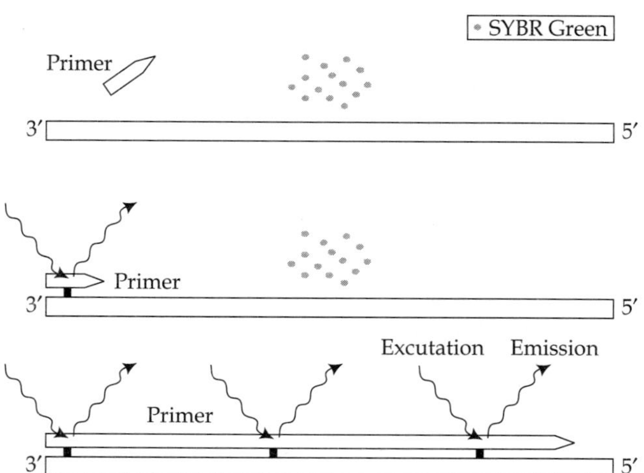

FIGURE 31.7   Principle of SYBR Green I based detection of PCR products in real-time PCR (Adapted from Loeffert, 2007).

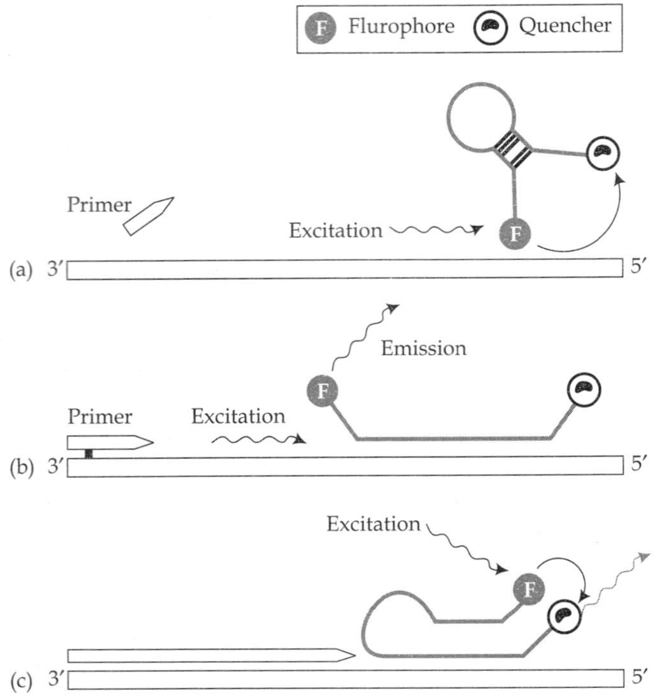

FIGURE 31.8 Showing the principle of Molecular Beacons in real-time PCR. (a) When not bound to its target sequence, the Molecular Beacon forms a hairpin structure. The proximity of the fluorescent reporter with the quencher prevents the reporter with fluorescing. (b) During teh PCR annealing step, the Molecular Beacon probe hybridizes to its target sequence. This seprates the fluorescent reporter and quencher, resulting in fluorescent signal. (c) During the extension step of PCR, the probe is displaced from the target sequence, bringing the fluorophore and quencher into closer proximity, resulting in quenching of fluoroscence. (Adapted from Loeffert, 2007).

Linker adaptor PCR—It is a variant of conventional PCR and used for amplification of genomic restriction fragments of unknown sequences. In this technique the restriction fragments generated by restriction endonucleases are ligated to double strand oligonucleotides (linker adaptor). Adaptors are attached to the ends of genomic restriction fragments. Then primers (specific) complementary to the adaptors are used to amplify the sequences between adaptors.

Whole genome–PCR—It is a variant of PCR which is used to select and amplify specific genome fragments. In this technique the genomic DNA is either sonicated or cleaved with the restriction endinuclease, MboI (recognition sequence, 5′-GATC-3′). The generated fragments are then blunt ended with Klenow fragment of E.coli DNA polymerase I. The blunt ended fragments are then ligated to linkers (catch linkers). The linker linked DNA fragments are then cleaved with another restriction endonuclease, XhoI (recognition sequence, 5′-CTCGAG-3′). The desired fragments are then selected using radioactive probes such as proteins if DNA protein interactions are to be studies. Finally, primers complementary to linkers are then used to amplify the selected sequence using PCR.

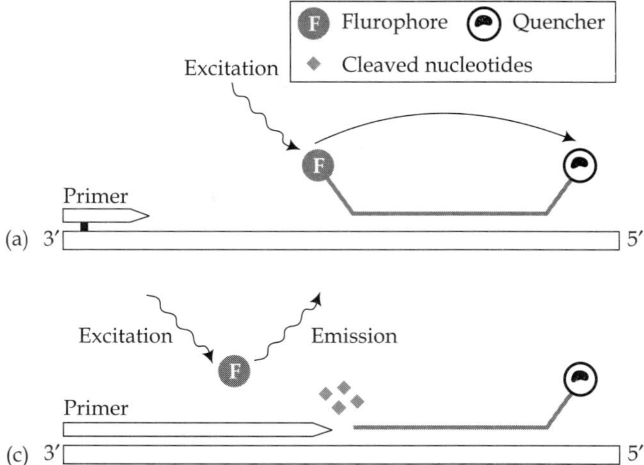

FIGURE 31.9   Showing the principle of TaqMan probes in real-time PCR. (a) Both thye TaqMan probe and PCR primers anneal to the target sequence during the PCR annealing step. The proximity of the fluorescent reporter from fluorescing. (b) During the combined PCR annealing/extension step, Taq DNA polymerase extends the primer. When the enzyme reaches the probe, its 5′-3′ exonuclease activity cleaves the fluorescent reporter from the probe. The fluoresecnt signal from the free reporter is measured. (Adapted from Loeffer, 2007).

Degenerate oligonucleotide-primed PCR- It is a variant of standard PCR which uses degenerate primers to amplify a representative portion of a genome including highly, moderately and low repetitive and unique sequences. A typical DOP primer has a tripartite structure, a central cassette of six degenerated nucleotides and a 3′ and 5′-flanking part, each composed of a specific sequence. Usually a two-step PCR is employed with the first cycles at low temperatures (e.g 30°C, allowing non-specific binding of the short 3′end sequence and the adjacent central cassette of the degenerated nucleotides of the primer to the template and non-specific amplification during the initial cycles) and subsequent cycles at high annealing temperatures (e.g. 62°C) permitting specific amplification by binding of all nucleotides of the primers to the target to increase the stringency. All sequences synthesized in the first step are now exponentially amplified resulting in accumulation of target sequences. Further, a restriction site, for example, XhoI in the DOP primer is used for cloning of the DOP-PCR product into a suitable vector.

AluI-PCR— It is a variant of PCR in which primers complementary to AluI sequences are used to amplify the regions flanked by the two neighboring AluI islands and thus through this technique DNA polymorphisms in this inter AluI region are detected.

Interspersed repetitive sequence-PCR— In this technique primers complementary to repetitive sequence elements, for example, LINES, SINES and especially AluI family are used to amplify the genomic region between such elements. IRS-PCR products allows to characterize the genomic DNA in between the IRS elements (e.g by DNA sequencing) or can be used as probe for ISH of chromosomes.

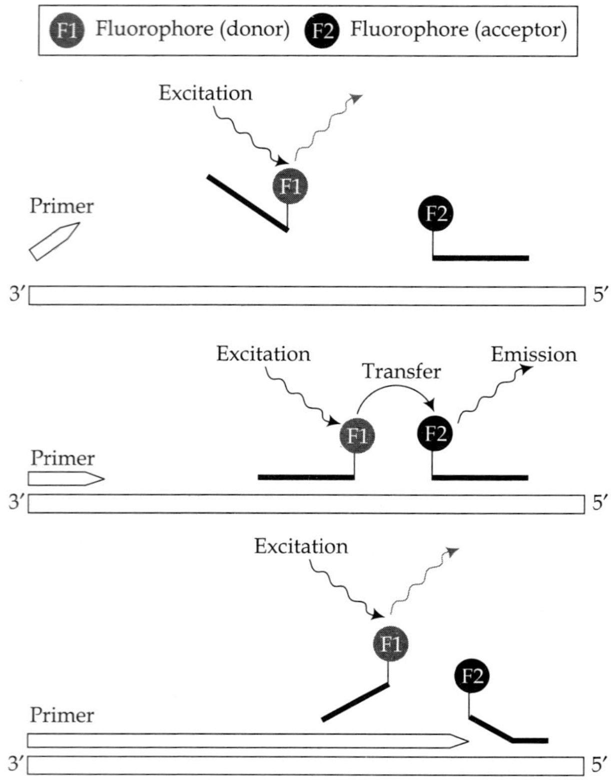

FIGURE 31.10   Showing the principle of FRET probes in real-time PCR. (a) When not bound to their target sequence, no fluorescent signal from the acceptro fluoropohore is detected. (b) During teh PCR annealing step, both FRET probes hybridize to the target sequence. This brings the donor and acceptor fluorophore into close proximity, allowing energy transfer between the fluorophores and resulting in fluoroscent signal from the acceptor fluorophore that is detected. (c) During the extension step of PCR, the probes are displaced from the target sequence and the acceptor fluorophore is no longer able to generate a fluoroscent signal. (Adapted from Loeffert, 2007).

Colony-PCR — It is a PCR technique used for the amplification of inserts cloned into vector plasmid taken directly from the bacterial host cells. In this technique the bacterial colonies are lysed at 95°C for 10 minutes and the resulting debris centrifuged and the supernatant is used as template for amplification using PCR with vector primer. Colony PCR thus serves as an easy and rapid screening procedure for cloned inserts.

Sexual PCR — Advances in recombination technology involves fragmenting the gene in random manner and then allowing the fragments to serve as primers for each other to reassemble shuffled versions of the gene (Stemmer, 1994). This technique does not require knowledge of gene sequence and thus can be applied to entire populations of nucleic acid molecules. It has long been possible to shuffle segments of a gene or gene family by taking advantage of the restriction sites or primer binding sites that flank those segments. Through either restriction digestion or primer extension at internal sites the gene can be broken into component segments which then can be reassembled in various combinations by either ligation or polymerase-mediated overlap extension. Sexual PCR has been widely used in the directed evolution of proteins. Recombination appears to be more important for proteins than nucleic acids as proteins have an inherent domain structure and are composed of 20 different amino acids which prevent complete representation of mutant beyond the three-error class.

On-chip PCR (solid phase PCR) — It is a variant of standard PCR which allows to perform all the cycling processes on a glass or silicon chip. In summary, primers are first spotted onto silanized galss chip, covalently linked through their 5' phosphates by , for example, the EDC-methylimidazole method, blocked with succinic anhydride in diemethylformamide and then hybridized to a mixture of target DNAs. The target then hybridizes to its complementary sequence on the primer and an added DNA polymerase extends the primer and synthesizes a chip bound double stranded DNA. The a second PCR cycle denaturates this double stranded DNA and now a second fluorophore-conjugated primer anneals to the previously synthesized strand and the DNA polymerase completes the synthesis of a fluorescently labeled double stranded DNA. That can be detected with laser beam. It is favorable for on-chip PCR to double the amount of primers, d NTPs and DNA polymerase and to extend the PCR time by 100% (except for the synthesis step). On-chip PCR permits to amplify fragments of up to 7kb in length.

In this on-chip PCR technique the primer extension reaction occurs on a chip-bound target molecule and so this configuration allows to use high stringency for the hybridization reaction between primer and target.

RT-PCR- It is an *in vitro* RNA amplification technique which uses retroviral reverse transcriptase or thermostable *Thermus thermophilus* (Tth) DNA polymerase to produce a cDNA on the RNA template. This cDNA is then amplified using the standard PCR techniques. The Tth reverse transcriptase catalyzes both the reverse transcription in the presence of Mncl2 and the amplification of the resulting cDNA in the presence of Mgcl2. Furthermore, its catalytic activity is impaired by increasing the reaction temperature to destabilize complex secondary structures of the RNA for stringent primer annealing. RNA-PCR permits the amplification of cDNA derived from small amounts of the purified mRNA, tRNA, rRNA and viral RNAs and identification of specific RNAs at a very low copy number and is therefore used in the analysis of gene expression at the RNA level (for example, for the study of post transcriptional modifications such as alternate splicing, also for the analysis of mRNA populations of very small cell populations, ideally even of single cell). RNA-PCR can also be combined with *in vitro* translation.

RT-PCR can be carried out in either one- step or two- step. In two-step RT-PCR, the RNA is reverse transcribed into cDNA in the first step using oligo-dT primers, random primers or less often gene-specific primers. The second step involves addition of an aliquot of the reverse transcription reaction to the PCR. The use of random primers or oligo- dT primers allows the analysis of multiple transcripts from a single cDNA sample). In case of one-step RT-PCR both reverse transcription and PCR take place in one tube sequentially. The use of gene-specific primers in one-step RT-PCR allows rapid processing of many samples simultaneously. Q-RT-PCR is a popular method for single gene-transcriptional analysis.

31.17 FLUORESCENCE PROBES FOR IMAGING IN CELL BIOLOGY

1. **Fluorophores**—These are small organic fluorophores (<1kD) used for covalent labeling of macromolecules. Fluorephores could directly recognize organelles, nucleic acids and certain important ion in living cells. Hundreds of such dyes are commercially available (Haugland, 2005). These dyes lack specificity for any particular protein and most applications use antibodies in fixed and premeabilized cells. **Fluorescent antibody techniques**- It is a technique to identify a specific protein within a cell using sections or the section of a specific tissue. There are two methods to localize the protein. In the direct method an antibody is used against the target protein. This antibody may be coupled to a fluorochrome which binds directly to the protein (antigen). The antigen-antibody-fluorochrome complex can then be visualized by fluorescence microscopy. In another approach an unlabeled primary antibody binds to the target protein. Then a secondary antibody (called anti-antibody) tagged with a fluorochrome (fluorochrome conjugated antibody) is directed against the primary antibody.

2. **Fluoresecent proteins**—Fluoresecent proteins have enabled non-invasive imaging in living cells and organisms of reporter gene expression, protein trafficking and many dynamic biochemical signals. The utility of fluorescent proteins has come from the discovery (Shimomura et al., 1962), gene cloning (Prasher et al., 1992) and heterologous expression of the green fluorescent protein (GFP) from jelly fish, *Acquorea victoria* (Tsien, 1998). GFP reporter gene differs from GUS, LUC, CAT in that the later reporters usually involve killed (or fixed) and sectioned material whereas the GFP can be used to report *in vivo* or lving cells. GFP emits green fluorescence when excited by U.V. light. Expression of GFP alone or in most genetic fusions with other proteins results in visible fluorescence without requiring any cofactors other than O_2 because the chromophore is generated by spontaneous cyclization and oxidation of three amino acids buried at the heart of 2.4-4nm beta barrels. Green fluorescent protein is just one member of a large family of homologous fluorescent protein. They come in different colors from variations in chromophore covalent structure and noncovalent environment. Mutations have generated fluorescent proteins that are photoswitchable from dark to bright or from one color to another color. Such photoswitching can be reversible or irreversible and is useful for monitoring protein fusion, trafficking and age.

   Principle and method of application—When two fluorophores with overlapping emission/absorption spectra are within ~100A of one another and their transition dipoles are approximately oriented, stimulation of the higher energy donor fluorophore excites the lower energy acceptor fluorophore, causing it to emit photons. This phenomenon is called **fluorescence resonance energy transfer** (FRET). In such experiments, one protein is fused to a FRET donor and the other to a FRET acceptor. The proteins are expressed inside cells and their interaction is monitored by fluorescence microscopy or light microscopy. For example, FRET between BFP-Bc12 and GFP-Bax revealed that Bc12 (a cell death inhibitor) and Bax (a cell death potentiator) interacted inside mitochondria. FRET can, in principle, register protein-protein interaction in any cellular compartment. When two fluorescent moieties are more than 100A apart, FRET can not occur even though the proteins interact. However, fusion proteins with extremely long hinge regions that bear a moiety from a protein that homodimerizes weakly could generate a strong FRET signal upon interaction. **Limitations with FRET-** The limitations with the FRET are: 1. low signal-to-noise ratios of the two available mutant GFP pairs usable for FRET 2.rapid photobleaching of GFP and mutant GFP and 3.working of FRET only over distances up to ~100A. Fluorophores are buried 12Å within the GFP monomer. Xu etal. (1998) have proposed **bioluminescence resonance energy transfer (BRET)** for detection of protein interactions by using Renilla luciferase to transfer energy to yellow FP, a mutant GFP. Excitation by bioluminescence eliminates photobleaching and autofluorescence associated with fluorescent excitation in FRET but requires that the assayed cells be exposed to the luciferase substrate coelenterazine. BRET is overall less sensitive than FRET and is subjected to poorly understood distance constraints (Xu et al., 1999). Photobleaching refers to irreversible light catalyzed degradation of any fluorochrome. It determines the half time of the fluorochrome and thereby the utility of any particular fluorochrome for FISH.

3. **Quantum dots** — QDs are nanocrystals which fluoresce at sharp and discrete wavelengths depending on their size, have high extinction coefficients (10 to 100) times those of small fluorophores and fluorescent proteins and have good quantum yields. What was crucial for biological applications was the development of coatings that make QDs water soluble, prevent quenching by water and allow conjugation to protein-targetting molecules such as antibodies and streptavidin (Michalet et al., 2005). QD's use is restricted to permeabilized cells or extracellular or endocytosed proteins as the large size of QDs efficiently conjugated to biomolecules (~10 to 30nm) prevents traversal of intact membranes. QDs typically contain a CdSe or CdTe core and ZnS shell. QDs are photostable and this photostability of QDs allows repeated imaging of single molecules and their size and electron density permits correlated electron microscopy. Gold, siliver nanodots formed within hydrophobic dendrites are also highly efficient and wave length tunable but their application is awaited.

QDs fluoresce up to 100 times longer than conventional fluorochromes and are used for the sensitive non-isotopic detection of biomolecules. QD is a 1-5nm nanocrystal with a Cadmium-selenium core and a zinc sulfide cover which can be excited by UV light and emits fluorescence light over long periods of time depending on the crystal sizes, over several months. QDs are chemically synthesized at high temperatures and normally possess a hydrophobic coat that can be replaced by polyacrylate which allows coupling to antibodies or modified by negatively charged dihydroliponic acid (DHLA). DHLA permits specific binding of specific antibodies or other proteins. Also, the pores of polystyrene microbeads can be loaded with QDs of different sizes and emission spectra, by coupling them to short oligonucleotide sequences. These can be hybridized to specific target sequences and the hybridization events can be detected by long term fluorescence light of different wavelengths.

4. **Immunolabelling** — The most wide spread technique to detect endogenous proteins using fluorescence is labeling with a primary antibody followed by amplification with secondary antibody conjugated to small organic dyes, a phycobiliprotein (derived from cyanobacteria) or a QD. Alternatively, primary antibodies can be directly

conjugated to fluorophores or to biotin which is then detected using streptavidin. Direct conjugation is especially useful when injecting antibodies into live cells or to increase spectral diversity when analyzing multiproteins. When high quality antibodies to target proteins are not available, the target can be recombinantly expressed with an epitope tag although it is no longer completely endogenous. As the accuracy of the protein recognition depends on the specificity of the primary antibodies, this should be validated using parallel methods. The disadvanges of immunofluorescence are that it is usually restricted to permeabilized cells or extracellular or endocytosed proteins and multivalency of these probes might lead to oligomerization of target proteins in living cells. In standard immunolabelling technique the size of the fluorophore-targetting complex typically exceeds 200kD and thus might interfere with multiprotein recognition in protein complexes.

5. **Genetic tagging**—The limitation with GFP can be overcome by the use of chemical fluorescent labeling of proteins inside cells. UAG suppressor or tRNAs designed to force incorporation of a fluorescent amino acid derivative into a protein whose coding sequence is modified to include UAG (Turcatti et al., 1996). Efforts are on to use new nucleotides to extend the genetic code beyond its current 64 codons may even enable suspected interacting proteins to be tagged with unique codons that direct incorporation of FRET capable fluorophores. *In vivo* labeling requires chemical synthesis of fluoreophore charged tRNAs and their microinjection into living cells. Another approach is expressing of proteins containing a compact arsenic binding domain which binds a cell permeable fluorescein-arsenic derivatives (Griffin et al., 1998). A key advantage of genetically encoded fluorescent proteins is that they can be covalently fused to protein of interest so that targeting is precise. Transfection and transgenic techniques often make exogenous DNA easier than dyes to deliver to cells or organisms. The limitations include the need for ectopic expression, the significant size of fluorescent proteins, the possibility that the fusion may interfere with the function of the protein of interest and the restriction to fluorescence as the only useful property. Therefore, several hybrid systems such as FRET (receptor) direct, FRET (Fluorescence resonance energy transfer) (sensor) indirect, BIFC (Bimolecular fluorescence complementation), Photoactivation, FRAP (Fluoresence recovery after photobleaching), Pulse-charge, Photo-oxidation for EM and CALI (Chromophore activated light inactivation), are described by which small molecules can be covalently targeted to genetically specified proteins inside or on the surface of the living cells either by spontaneous attachment or enzymatic ligation (Giepmans et al., 2006).

The most developed technique is the tetracysteine biarsenal system (Griffin and Adams, 1998) which requires modification of the target protein by a 12 residue peptide sequence that includes four cysteines which binds membrane-permeable biarsenical molecules notably the green and red dyes "FlAsH" and "ReAsH" with picomolar affinity. Tetracysteine-biarsenicals also enable manipulations not readily possible with fluorescent proteins such affinity purification, fluorophore associated light inactivation, co-translational detection of protein synthesis, pulse-chase labeling and correlative electron microscopy localization. However, it gives higher background fluorescence and poorer contrast than fluorescent proteins and thus have not yet been demonstrated in intact transgenic animals, require the cysteines to be reduced during labeling and do not permit two different proteins in the same compartment to be simultaneously labeled with different probes. CALI is used for manipulation of protein activity. It inactivates protein of interest with much better spatio-temporal control than possible with genetic knock outs or RNAi. Photo-activation locally changes the spectrum of the FP allowing movement of the marked vs unmarked protein to be visualized. In the Pulse-charge, a tetracysteine-tagged protein is labeled with FIAsH (green). After removal of free FIAsH and biosynthesis of new copies the later are detected with ReAsH (red dots). FRET can detect dynamic protein-protein interactions in living cells of ectopically expressed F—tagged proteins.

Although genetic tagging is an unprecedented tool in live cell protein imaging, other methods should be used simultaneously (small organic dyes (antibody targeted),QD (antibody targeted), FP) to address whether tags perturbs endogenous protein fusion (Table 31.1). These include fluorophores conjugated to well characterized antibodies, often specific for given post-translational modifications as well as genetic knock-in experiments to test whether the modified protein is an adequate *in vivo* replacement for its endogenous parent. Conversely, immunofluorescence labeling of FP-tagged target proteins test the specificity of antibodies. Immunodetection using bright and stable QDs lowers the detection limit, increases the feasibility of multiprotein detection and can be applied in correlated EM studies and in vitro assays that currently rely on enzymatic activity such as Western blots and ELISA. Fluorescence technique is central in the analysis of protein networks and system biology.

Table 31.1   Showing comparison of different tools in living cell imaging.

Types of study	Fluorophore, small organic dyes (antibody-targeted)	QDs (antibody – targeted)	FPs	Genetic tags with small dyes
Endogenous proteins	++	+	_	_
Clinical specimens	++	+	_	_
Animals	Ex vivo	Ex vivo	Transgene	Transgene, Ex vivo
Primary tissues	++	+	Transgene/virus	Transgene/virus
Living cells in cultures	Surface	Surface	++	+
Multiple proteins simultaneously	++	++	++	_
Interactions	+/_	+/_	++	Combination with Fp
Turnover/synthesis	_	_	+	+
Activation state	Photopsecific	Photospecific	FRET sensors	Combination with FP
CALI	+	_	+	++
EM	+/_	++	+/_	+
In gel fluorescence	_	_	+	+
Western blot	_	+	_	_
Protein microarrays	++	+	_	_
Advantages	Diversity of properties	Bright and photostable	Living cells and specificity	Living cells and small size
Limitations	Targeting in living cells	Targeting and penetration	Ectopic expression	Ectopic expression and background staining.

References

Adams, M.D., Celinker, S.E., Holt, R.A. et al. 2000. The genome sequence of *Drosophila melanogaster*. Science, Vol. 287, 2185-2195.

Adams, M.D., Killey, J.M., Gocayne, J.D., Dubnick, M., Polymeropoulos, M.H., Xiao, H., Merril, C.R., Wu, A., Olde, B., Moreno, R.F., Kerlavage, A.R.McCombie, W.R. and Venter, J.C. 1991. Complementary DNA sequencing: Expressed Sequence Tags and Human Genome Project. Science, Vol. 252, 1651-1656

Adams, R.L.P.; Knowler, J.T. and Leader, D.P. 1986. The Biochemistry of Nucleic Acids. Chapman and Hall. London.

Aderem, A. 2005. System Biology: its practice and challenges. Cell, 121: 511-513

Adhya, S. et al. 1991. A laboratory guide to *in vitro* studies of protein-DNA interactions. Basel, Birkhaüser.

Agarwala, R., Applegate, D.L., Maglott, D., Schuler, G.D. and Schaffer, A.A. 2000. A fast and scalable radiation hybrid map construction and integration strategy. Genome Res., 10: 350-364

Akutsu, T. et al., 2006. On the complexity of finding control strategies for Boolean network. In Proc. of the 4th Asia-Pacific Bioinformatics conference(eds. Jiang et al). Imperial College Press, London. Pp. 99-108

Akutsu, T., Miyano, S. and Kuhara, S. 2000. Inferring qualitative relations in genetic networks and metabolic pathways. Bioinformatics, 16: 727-734

Albert, R., Jeong, H. and Barabasi, A.L. 2000. Nature, 378-382

Allison, D.B. 1997. Transmission-disequilibrium tests for quantitative traits. Am. J. Hum. Genet., 60: 676-690

Alon, U. 2003. Biological networks: the tinkerer as an engineer. Science, 301:1866-67

Altshuler, D. et al. 2000. A SNP map of human genome generated by reduced representation shotgun sequencing. Nature, 407: 513-16

Amaral, L.A., Diaz-Guilera, A. et al. 2004. Emergence of complex dynamics in a simple model of signaling networks. Proc. Natl. Acad. Sci., U.S.A. 101: 15551-15555

Aravin, A.A. et al. 2007. Developmentally regulated piRNA clusters implicate MILI in transposon control. Science, 316: 744-47

Baker, B.S. 1989. Sex in flies: the splice of life. Nature, 259: 471-9

Balakirev, E.S. and Ayala, F.J. 2003. Pseudogenes: Are they junk or functional DNA? Ann. Rev. Genet., 37: 123-152

Ball, P. 2007. Designs for life. Nature, 448: 32-33

Barnes, M.R. 2003. Predictive functional analysis of polymorphisms: An overview. In Bioinformatics for Geneticists(eds.) M.R.Barnes and I.C. Gray. John Wiley & Sons. Ltd.

Bartolomei, S.M. and Tilghman, S.M.1997. Genomic imprinting in mammals. Ann. Rev. Genet., 31: 493-526

Bass, B.L. 2002. RNA editing by adenosine deaminases that act on RNA. Ann. Rev. Biochem., 71: 817-846

Basten, C., Weir, B.S. and Zeng, Z.B. 1997. QTL cartographer: A Reference Manual and Tutorial for QTL Mapping. Deptt. Of Statistics, North Carolina State University, Raleigh, NC.

Beasley, E.M., Myers, R.M., Cox, D.R. and Lazzeroni, L.C. 1999. Statistical refinement of primer design parameters. In Michael, D.H.G., Innis, A and Sninsky, J.J.(eds), PCR Applications. Academic press, New York., pp. 55-71

Beckmann, J.S. and Soller, M. 1983. Restriction fragment length polymorphisms in genetic improvement: methodologies, mapping and costs. Theor. Appl. Genet., 67, 35-43

Bender, J. 2004. DNA methylation and epigenetics. Ann. Rev. Plant Biol., 55: 41-68

Bender, W., Akam, M., Karch, F. et al. 1983. Moecular genetics of the bithorax complex in Drosophila melanogaster. Science, 221: 23-9

Benfey, P.N. 2005. Vision statement for plant physiology. Developmental networks. Plant physiology., Vol. 138, pp. 548-549

Bennett, S.T. and Todd, J.A.1996. Human type 1 diabetes and the insulin gene: principles of mapping polygenes. Ann Rev. Genet., 30: 343-370

Bennetzen, J. 2002. Opening the door to comparative plant biology. Science, 296: 60-63

Benzer, S. 1959. On the topology of gene fine structure. Proc. NATL. Acad. Sci., U.S.A., 45:1607-20

Berger S.L. and Kimel, A.R.(eds.) 1987. Methods in Enzymology, Guide to Molecular Cloning Techniques. Vol. 152, Academic Press.

Berger, S.L. 2007. The complex language of chromatin regulation during transcription. Nature, 447: 407-12

Berget, S.M. 1995. Exon recognition in vertebrate splicing. J.Biol. Chem., 270: 2411-14

Bernardi, G. 2000. Isochores and the evolutionary genomics of vertebrates. Gene, 241: 3-17

Bertone, P and Snyder, M. 2005. Vision statement for plant physiology., Prospects and challenges in proteomics. Plant Physiology., Vol. 138, pp. 560-562

Berzer, J., Suzyki, T. et al. 2001. Genetic mapping with SNP markers in Drosophila. Nature Genet., 29: 475-481

Biémont, C. and Vieira, C. 2006. Junk DNA as an evolutionary force. Nature, 443, 521-525

Bird, A. 2007. Perceptions of epigenetics. Nature, 447: 396-398

Bird, A.P. 1986. CPPG-rich islands and the function of DNA methylation. Nature, 321: 209-13

Birnbaum, K. et al. 2003. A gene expression map of the Arabidopsis root. Science, 302: 1956-60

Björklund, M. et al. 2006. Identification of pathways regulating cell size and cell cycle progression by RNAi. Nature, 439: 1009-1013

Black, D.L. 2003. Mechanisms of alternative pre-messenger RNA splicing. Ann. Rev. Biochem., 72: 291-336

Blackwood, E.M. and Kadonaga, J.T. 1998. Going the distance: a current view of enhancer action. Science, 281:61-63

Bluthgen, N., Kiebasa, S.M.and Herzel, H. 2005. Inferring combinatorial regulation of transcription in silico. Nucleic Acids Res., 33(1): 272-279

Bode, J., Benham, C., Knopp, A. and Meilke, C. 2000. Transcriptional augmentation: modulation of gene expression by scaffold/ matrix- attached regions(S/MARelements). Crit. Rev. Eukaryot Gene Expr., 10: 73-90

Bolivar, F. et al. 1977. Construction and characterization of new cloning vectors. II. A multipurpose cloning system. Gene, 2, 95-113-p BR322

Bouchez, D. and Höfte, H. 1998. Functional genomics in plants. Plant Physiol., 118: 725-732

Boulikas, T. 1995. Chromatin domains and prediction of MAR sequences. Int. Rev.Cytol., 162A: 279-388

Boyer, H.W. and Nicosia, S. 1978. Genetic engineering. Elsevier/ North Holland.

Brenner, S. 2000. Inverse genetics. Curr. Biol., 10: R649

Briggs, S.P. and Singer,T. 2005. Vision statement for plant physiology. Genetic networks. Plant physiology, Vol.138, pp. 5420544

Britton, R.J. and Kohne, D.E. 1968. Repeated sequences in DNA. Science, 161: 529-31

Brooks, J.E. 1987. Properties and uses of restriction endonucleases. Methods in enzymology, Vol. 152: 113-129

Brown, C.D., Johnson, D.S. and Sidow, A. 2007. Functional architecture and evolution of transcriptional elements that drive gene copexpression. Science, 317: 1557-60

Brown, J.W.S. and Simpson, C.G. 1998. Splice site selection in plant pre-mRNA splicing. Ann. Rev. Plant. Physiol. Plant Mol. Biol., 49: 77-126

Brown, T.A. 1994. DNA sequencing. IRL press, Oxford

Buchan, J.R. and Parker, R. 2007. Two faces of miRNA. Science, 1877-78

Buchman, T.G. 2002. The community of the self. Nature, 420: 246-251

Buck, M. J. and Lieb, J.D. 2004. ChIP-chip: considerations for the design, analysis and application of genome wide chromatin immunoprecipitation experiments. Genomics, 83: 349-60

Bult, C.J., White, O. et al. 1996. Complete genome sequence of the Methanogenic Archaeon, Methanococcus jannaschii. Science, 273: 1058-1073

Burge, C., Tuschi and Sharp, P.A. 1999. Splicing of precursors to mRNA by spliceosome. In R.F. Gestland, T.R. Cech and J.F. Atkins(eds), The RNA world, 2nd edn. Cold Spring Harbor Laboratory press, New York.

Burke, D.T. et al. 1987. Cloning of large segments of exogenous DNA into yeast by means of artificial chromosome vectors. Science, 236: 806-812

Butte, A.J., Tamayo, P., Slonim, D., Golub, T.R. and Kohne, I.S. 2000. Discovering functional relationships between RNA expression and chemotherapeutic susceptibility using relevance networks. Proc. Natl. Acad. Sci., U.S.A., 97: 12182-12186

Caicedo, A.L. and Purugganan, M.D. 2005. Vision statement for plant physiology. Comparative plant genomics. Frontiers and prospects. Plant physiology, Vol. 138, pp. 545-547

Carey, J. 1991. Gel retardation. Methods Enzymol, 208: 103-17

Carle, G.E. and Olson, M. V. 1985. An electrophoretic karyotype for yeast. Proc.Natl. Acad. Sci., U.S.A., 82, 3756-3760-an application of OFAGE.

Carninci, P. et al. 2005. The transcriptional landscape of the mammalian genome. Science, 309: 1559-63

Caron, H., van Schaik, B., van der Mee, M. et al. 2001. The human transcriptome map: clustering of highly expressed genes in chromosomal domains. Science, 291: 1289-1292

Carrano, A.V., Gray, J.W., Langlois, R.G., Burkhart-Schultz, K.J. and Van Dilla, M.A. 1979. Measurement and purification of human chromosomes by flow cytometry and sorting. Proc. Natl. Acad. Sci., U.S.A., 76, PP. 1382-1384.

Carrington, J.C. 2005. Vision statement for plant Physiology. Small RNAs and Arabidopsis. A fast forward look. Plant Physiology, Vol. 138, PP. 565-566

Cattenach, B. 1974. Position-effect variegation in the mouse. Genet Res., Camb., 23: 291-306

Catteneo, R. 1991. Different types of editing. Ann. Rev, Genet., 25:71-88.

Chan, S.W. et al. 2004. RNA silencing genes control de novo DNA methylation. Science, 303: 1336

Charnay, P. et al. 1980. Biosynthesis of hepatitis B virus surface antigen in *Escherichia coli*. Nature, 286: 893- 895.

Chee, M., Yang, R. et al. 1996. Accessing genetic information with high density DNA arrays. Science, 274: 610-626

Chesler, E.J., Rodriguez-Zas, S.L.and Mogil, J.S.2001. *In silico* mapping of mouse quantitative trail loci. Science, 294: 2423

Chin, F.Y.L. and Leung, H.C.M. 2006. An efficient algorithm for starting motif discovery.In Proc. of the 4th Asia-Pacific Bioinformatics conference(eds. Jiang et al). Imperial College Press, London. Pp. 79-88

Chumakov, I.M., Rigault, P. et al. 1995. A YAC contig map of the human genome. Nature, 377(Suppl.) 175-297

Churchill, G.A and Doerge, R.W. 1994. Empirical threshold values for quantitative trait mapping. Genetics, 138:963-971

Clauset, A. et al. 2008. Hierarchical structure and prediction of missing links in networks. Nature, 453: 98-101

Coen, E.S. and Meyerowitz, E.M. 1991. The war of whorls: genetic interactions controlling flower development. Nature, 353: 31-7

Colbert et al. (2001). High-throughput screening for induced point mutations. Plant Physiology, 126: 480-4.

Collins, A. and Morton, N.E. 1998. Mapping a disease locus by allelic association. Proc. Natl. Acad. Sci., U.S.A., 95: 1741-1745

Collins, A., Lonjou, C. and Morton, N.E. 1999. Genetic epidemiology of single-nucleotide polymorphisms. Proc. Natl. Acad. Sci., U.S.A., 96: 15173-15177

Court, D.L., Sawitzke, J.A. and Thomason, L.C. 2002. Genetic engineering using homologous recombination. Ann, Rev. Genet., 36: 361-388

Cox, D.R., Burmeister, M. et al. 1990. Radiation hybrid mapping: a somatic cell genetic method for constructing high-resolution maps of mammalian chromosomes. Science, 250: 245-250

Cox, M.M. and Phillips, G.N.(Jr.)(eds.). 2007. Handbook of Proteins Structure, Function and methods. John Wiley and Sons, Ltd., England.

Cramer, P. 2007. Extending the message. Nature, 448: 142-143

Creighton, T.E. 1999. Encyclopedia of Molecular Biology. Vols 1-4. John Wiley & Sons, INC. New York

Cunningham, B.A., Himperly, J.J., Murray, B.A. et al. 1987. Neural cell adhesion molecule: structure, immunoglobulin-like domains, cell surface modulation and alternative RNA splicing. Science, 236: 799-806.

Cutler, S. and McCourt, P. 2005. Vision statement for plant physiology. Dude, where's my phenotype? Dealing with redundancy in signaling networks. Plant Physiology, Vol. 138, pp. 558-559.

Daly, M.J., Rioux, J.D., Schaffner, S.F., Hudson, T.J. and Lander, E.S. 2001. High-resolution haplotype structure in the human genome. Nature Genet., 29: 229-232

Darvasi, A. 2001. *In silico* mapping of mouse quantitative trait loci. Science, 294: 2423

Dawson, E., Abecasis, G.R. et al., 2002. A first-generation linkage disequilibrium map of human chromosome 22. Nature, 418: 544-548

De Angelis, M.H., Flaswinkle, H. et al., 2000. Genome-wide, large scale production of mutant mice by ENU mutagenesis. Nature Gene, 25: 444-447.

Debrauwere, H., Gendrel, C.G., Llechat, S. and Dutreix, M. 1997. Differences and similarities between various tandem repeat sequences: minisatellites and microsatellites. Biochimie., 79: 577-86

Deininger, P.L. 1989. SINES: short interspersed repeated DNA elements in higher eukaryotes. In Berg, P.E. and Howe, M.H.(eds.) Mobile DNA, pp. 53-109, American Soc. Microbiology, Washington, DC.

Deloukas, P., Schuler, G.D. et al. 1998. A physical map of 30, 000 human genes. Science, 282: 744-746

Demerec, M. 1940. Genetic behavior of euchromatic segments inserted into heterochromatin. Genetics, 25: 618-27

DeRisi, J.L., Iyer, V.R. and Brown, P.O. 1997. Exploring the metabolic and genetic control of gene expression on genome scale. Science, 278: 680-686

Derynk, R. et al. 1980. Expression of human fibroblast interferon gene in *Escherichia coli*. Nature, 287: 193-197.

Devlin, B. and Risch, N. 1995. A comparison of linkage disequilibrium measures for fine-scale mapping. Genomics, 29: 311-322

Dib, C., Faure, S. et al. 1996. A comprehensive genetic map of the human genome based on 5,264 microsatellites. Nature, 380: 152-154

Dietzl, G. et al. 2007. A genome-wide transgenic RNA library for conditional gene inactivation in Drosophila. Nature, 448: 151-56

Dignam, J.D. et al. 1983. Eukaryotic gene transfer with purified components. Methods in Enzymology, 101: 582-598

Doherty, E.A and Doudwa, J.A. 2000. Ribozymes structure and mechanisms Ann. Rev. Biochem., 69: 597-615

Downs, J.A., Nussenzweig, M.C. and Nussenzweig, A. 2007. Chromatin dynamics and preservation of genetic information. Nature, 447: 951-958

Drosophila 12 Genomes Consortium. 2007. Evolution of genes and genomes on the Drosophila phylogeny. Nature, 450: 203-218

Dryja, T.P., Hahn, L.B., Cowley, G.S., McGee, T.L. and Berson, E.L. 1991. Mutation spectrum of the rhodopsin gene among patients with autosomal dominant retinitis pigmentosa. Proc. Natl. Acad. Sci., U.S.A., 88: 9370-3

Dugaiczyk, A., Woo, S.L.C., Colbert, D.A. et al., 1979. The ovalbumin gene: cloning and molecular organization of the entire natural gene. Proc. Natl. Acad. Sci., U.S.A., 76: 2253-7.

Duncan, I.W. 2002. Transvection effects in Drosophila. Ann. Rev. Genet., 36: 521-556

Dunham, I., Shimizu, N., Roe, B.A. et al. 1999. The DNA sequence of human chromosome 22. Nature, 402: 489-495

Eberwine, J. et al. 1992. Analysis of gene expression in single line neuron. Proc. Natl. Acad. Sci., U.S.A., 89: 3010-3014

Eisen, M.B., Spellman, P.T., Brown, P.O., Botstein, D. 1998. Cluster analysis and display of genome-wide expression patterns. Proc. Natl. Acad. Sci., U.S.A., 95: 14863-14868

Eisenberg, D., Marcotte, E.M., Xenarios, I. and Yeates, T.O. 2000. Protein function in the post-genomic era. Nature, 405: 823-826

Emtage, J.S. et al. 1980. Influenza antigenic determinants are expressed from haemagglutinin genes cloned in Escherichia coli. Nature, 283: 171- 174.

Endy, D. and Brent, R. 2001. Modeling cellular behavior. Nature, 409: 391-395

Engels, W.R. 1989. P elements in Drosophila. In Berg, P.E. and Howe, M.H.(eds.) Mobile DNA, pp. 53-109, American Soc. Microbiology, Washington, DC.

Enright, A.J., Iliopoulos, I., Kyrpides, N.C., Ouzounis, C.A.1999. Protein interaction maps for complete genomes based on gene fusion events. Nature, 402: 86-90

Erlich, H.A., Gelfand, D. and Sninsky, J.J. 1991. Recent advances in polymerase chain reaction. Science, Vol. 252. 1643-1651.

Fagard, and Vaucheret, H. 2000. (Trans) gene silencing in plants: How many mechanisms? Ann. Rev. Plant Physiol. Plant Mol. Biol., 51: 167-94

Feng, Qi. et al., 2002. Sequence and analysis of rice chromosome 4. Nature, Vol. 420, pp. 316-420

Fields, S. and Song, O. 1989. A novel genetic system to detect protein-protein interactions. Nature, 340: 245-246

Fincham, J.R.S. 1966. Genetic Complementation. W.A.Benjamin, New York.

Fincham, J.R.S. 1994. Genetic Analysis, Principles, Scope and Objectives. Blackwell Science. London.

Fincham, J.R.S. 1999. Gene structure. Encyclopedia of Molecular Biology. John Wiley & Sons. New York. p.982

Finley, R.L. Jr, and Brent, R. 1994. Interaction mating reveals binary and ternary connections between Drosophila cell cycle regulators. Proc. Natl. Acad. Sci., U.S.A., 91: 12980-12984

Finnagan, E.J., Genger, R.K., Peacock, W.J. and Dennis, E.S. 1998. DNA-methylation in plants. Ann. Rev. Plant Physiol. Plant Mol. Biol., 49: 223-248

Fleischmann, R.D., Adams, M.D. et al. 1995. Whole-genome random sequencing and assembly of *Haemophilus influenzae* Rd. Science, 269: 496-512

Flint-Garcia, S.A., Thornberry, J.M. and Buckler, IV, E.S. 2003. Structure of linkage disequilibrium in plants. Ann. Rev. Plant Biol., 54: 357-374

Fraser, C.M., Gocayne, J.D. et al. 1995. The minimal gene complement of *Mycoplasma genitalium*. Science, 270: 397-403

Fridman, E., Pleban, T. and Zamir, D. 2000. A recombination hotspot delimits a wild-species quantitative trait locus for tomato sugar content to 484-bp within an invertase gene. Proc. Natl. Acad. Sci., U.S.A., 97:4718-4723

Gabriel, S.B., Schaffner, S.F. et al. 2002. The structure of haplotype blocks in the human genome. Science, 296: 2225-2229

Gallant, T. D.; Maier, D. and Stoerer. 1980. On finding minimal length superstrings. J. Comp. Syst. Sci., 20: 50-58

Gatz, C. 1997. Chemical control of gene expression. Ann. Rev. Plant Physiol. Plant Mol. Biol., 48: 89-108

Gavin, A-C, Bosche, M., et al. 2002. Functional analysis of yeast proteome by systematic analysis of protein complexes. Nature, 415: 141-47

Gavin, A.C. et al. 2006. Proteome survey reveals modularity of the yeast cell machinery. Nature, 440: 631-636

Gerasimova, T.I. and Corces, V.G. 2001. Chromatin insulators and boundaries. Effects on transcription and nuclear organization. Ann. Rev. Genet., 35: 193-208

Gershoni, J.M. 1999. Blotting. Encyclopedia of Molecular Biology. John Wiley & Sons. New York. p. 301

Giaever, G., Chu, A.M. et al. 2002. Functional profiling of the *Saccharomyces cerevisiae* genome. Nature, 418: 387-391

Gimelbrant, A., Hutchinson, J.N., Thompson, B.R. and Chess, A. 2007. Wide spread monoallelic expression on human autosomes. Science, 318: 1136-1140

Giot, L. et al. 2003. A protein interaction map of Drosophila melanogaster. Science, 302: 1727-1736

Goeddel, D.V. et al. 1979. Expression in Escherichia coli of chemically synthesized genes for human insulin. Proc. Natl. Acad. Sci., U.S.A, 76: 106-110.

Goeddel, D.V. et al. 1980. Human leukocyte interferon produced by *E.coli* is biologically active. Nature, 287: 411-416.

Goff, S.A., Ricke, D. et al., 2002. A draft sequence of the rice genome(*Oryza sativa*) I. ssp. Japonica). Science, 296: 92-100

Goffeau, A., Barrel, B.G. et al 1996. Life with 6000 genes. Science, 274: 546-567

Goldbeter, A. 2002. Computational approaches to cellular rhythms. Nature, 420: 238-245

Goldstein, D.B. and Schlotterer, C.(eds.) 1999. Microsatellites- evolution and applications. Oxford university Press, oxford, U.K.

Golub, T.R., Slonim, D.K. et al. 1999. Molecular classification of cancer: class discovery and class prediction by gene expression monitoring. Science, 286: 531-537

Gott, J.M. and Emeson, R.B. 2000. Functions and mechanisms of RNA editing. Ann. Rev. Genet., 34:449-532

Graber, J.H., Cantor, C.R., Mohr, S.C. and Smith, 1999. In silico detection of control signals: mRNA 3′ end processing sequences in diverse species. Proc. Natl. Acad. Sci., U.S.A.,96: 14055-14060

Gresham, D. et al. 2006. Genome-wide detection of polymorphisms at nucleotide resolution with a single DNA microarray. Science, 1932-36

Grewal, S.I. and Elgin, S.C.R. 2007. Transcription and RNAi in the formation of heterochromatin. Nature, 447: 399-406

Grupe, A., Germer, S. et al., 2001. In silico mapping of complex disease-related traits in mice. Science, 292: 1915-1918

Gunawardane, L. et al. 2007. A slicer-mediated mechanism for repeat-associated siRNA 5′ end formation in Drosophila. Science, 315: 1587-1590

Gutierres, R.A., Shasha, D.E. and Coruzzi, G.M. 2005. Vision statement for plant physiology. Systems biology for virtual plant. Plant Physiology, Vol. 138, pp. 550-554

Haig, D. 2004. Genomic imprinting and kinship: How good is the evidence? Ann. Rev. Genet., 38: 5530586

Haley, C.S. and Knott, S.A. 1992. A simple regression method for mapping quantitative trait loci in line crosses using flanking markers. J. Hered., 69: 315-324

Hammond-Kosack, M.C.U.and Bevan, M.W. 1993. A practical guide to ligation-mediated PCR footprinting and in vivo DNA analysis using plant tissues. Plant Mol. Biol. Rep., 11: 249-72

Handerson, I.R. and Jacobsen, S.E. 2007. Epigenetic inheritance in plants. Nature, 447: 418-424

Harbison, C., Gordon, D., Lee, T. et al. 2004. Transcriptional regulatory code of a eukaryotic genome. Nature, 431: 99-104

Hartwell, L.H. et al. 1999. From molecular to modular cell biology. Nature, 402: C47-52

Hasty, J. et al. 2002. Engineered gene circuits. Nature, 420: 224-230

Hattori, M. and Tailor, T.D. 2001. Part three in the book of genes. Nature, 414: 854-55

Hattori, M., Fujiyama, A. et al. 2000. The DNA sequence of human chromosome 21. Nature, 405: 311-19

Heid, C.A., Stevens, J., Livak, K.J. and Williams, P.M. 1996. Real time quantitative PCR. Genome methods., 986-994

Hendrick, P.W. 1987. Gametic disequilibrium measures.: proceed with caution. Genetics, 117: 331-341

Henikoff, S. and Comai, L. 2003. Single-nucleotide mutations for plant functional genomics. Ann. Rev. Plant Biol., 54: 375-402

Henikoff, S. et al. 1997. Gene Families: The taxonomy of protein paralogs and chimeras. Science, 278: 609-614

Henikoff, S., Keen, M.A., Fechtel, K. and Fristrom, J.W. 1986. Gene within a gene: nested Drosophila genes encode unrelated proteins on opposite DNA strands. Cell, 44:33-42.

Hernalsteens, J.-P. et al. 1980. The Agrobacterium tumefaciens Ti plasmid as a host vector system for introducing foreign DNA in plant cells. Nature, 287: 654-656.

Hieter, P. and Boguski, M. 1997. Functional genomics: It's all how you read it. Science, Vol. 278, 601-2

Higuchi, M., Antonarakis, S.E., Kasch, L. et al. 1991. Molecular characterization of mild-to-moderate haemophilia A: detection of the mutation in 25 of 29 pateints by denaturing gradient gel electrophoresis. Proc. Natl. Acad. Sci., U.S.A., 88: 8307-11

Hill. W.G. and Weir, B.S. 1994. Maximum –likelihood estimation of gene location by linkage disequilibrium. Ann. J. Hum. Genet., 54: 705-714

HMSO. 1980. 'BIOTECHNOLOGY' The Report of a Joint Working Party.

Ho, Y., Gruhler, A. et al. 2002. Systematic identification of protein complexes in S. cerevisiae by mass spectrometry. Nature, 415: 180-83

Hogenesch, J.B., Chin, K.A. et al. 2001. A comparison of the Celera and Ensemble predicted gene sets reveals little overlap in novel genes. Cell, 106: 413-15

Hollaender, A. 1977. Genetic Engineering for Nitrogen Fixation. Plenum.

Holliday, R. and Pugh, J.E. 1975. DNA modification mechanisms and gene activity during development. Science, 187: 226-32

Holm, L. and Sander, C. 1996. Mapping the protein universe. Science, 273: 595-602

Horton, P. et al. 2006. Protein subcellular localization prediction with wolf PSORT. In Proc. of the 4th Asia-Pacific Bioinformatics conference(eds. Jiang et al). Imperial College Press, London. Pp. 39-48

Hoskin, R.A. et al. 2007. Sequence finishing and mapping of *Drosophila melanogaster* heterochromatin. Science, 316: 1625-1628

Houry, W.A., Frishman, D., Eckersorn, C., Lottspeich, F. and Hartl, F.U. 1999. Nature, 402: 147-154

Hseih, J. and Fire, A. 2000. Recognition and silencing of repeated DNA. Ann. Rev. Genet., 34: 187-204

Huyen, M.A. and Bork, P. 1998. Measuring genome evolution. Proc. Natl. Acad. Sci., U.S.A., 95: 5849-56

Hwang, H-W, Wentzel, E.A. and Mendell, J.T. 2007. A hexanucleotide element directs microRNA nuclear import. Science.

Ideker, T. et al. 2001b. Integrated genomic and proteomic analyses of a systematically perturbed metabolic network. Science, 292: 929-34

Ideker, T., Galitski, T. and Hood, L. 2001. A new approach to decoding life: system biology. Ann. Rev. Genomics Hum Genet., 2: 343-72

International Human Genome Sequencing Consortium. 2001. Initial sequencing and analysis of the human genome. Nature, 409: 860-92

International Human Genome Sequencing Consortium. 2004. Finishing the euchromatic sequence of the human genome. Nature, 431: 931-94

International SNP Map Working Group. 2001. A map of human genome sequence variation containing 1.42 million single nucleotide polymorphisms. Nature, 409: 928-33

Itakura, K. et al. 1977. Expression in *Escherichia coli* of a chemically synthesized gene for hormone somatostatin. Science, 198: 1056-1063.

Ito, T. et al. 2001. A comprehensive two-hybrid analysis to explore the yeast protein interaction. Proc. Natl. Acad. Sci., U.S.A., 98: 4569-74

Ito, T., Tashiro, K., Muta, S., Ozawa, R., Chiba, T. et al. 2000. Toward a protein-protein interaction map of the budding yeast: a comprehensive system to examine two-hybrid interactions in all possible combination between the yeast proteins. Proc. Natl Acad. Sci., U.S.A., 97: 1143-1147

Jarvik, J.W.and Telmer, C.A. 1998. Epitope tagging. Ann. Rev. Genet., 32: 601-618

Jensen, R.C. 1993. Interval mapping of multiple quantitative trait loci. Genetics, 135: 205-211

Jensen, R.C. and Stam, P. 1994. High resolution of quantitative traits into multiple loci via interval mapping. Genetics, 136: 1447-1455

Jeong, H., Mason, S.P., Barabasi, A-L., Oltvai, Z.N. 2001. Lethality and centrality in protein networks. Nature, 411: 41-42

Jeong, H., Tombor, B., Albert, R., Oltvai, Z.N. and Barabasi, A-L. 2000.The large-scale organization of metabolic networks. Nature, 407: 651-654

Johnson, G.C., Esposito, L. et al. 2001. Haplotype tagging for the identification of common disease genes. Nature Genet., 29: 233-237

Jukes, T.H., Osawa, S., Muto, A. and Lehman, N. 1987. Evolution of anticodons: variations in the genetic code. Cold Spring Harbor Symposia on Quantitative Biology, 52: 769-76

Kahl, Günter. 2004. The Dictionary of Gene Technology. Vols. 1 and 2. Wiley-VCH Verlag GmbH & Co. KGaA

Karlin, S.; Campbell, A.M. and Mrazek, J. 1998. Comparative DNA analysis across diverse genomes. Ann. Rev. Genet., 32: 185-225

Katsanis, N., Worley, K.C. and Lupski, J.R. 2001. An evaluation of the draft human genome sequence. Nature Genet., 29: 88-91

Kawada, Y. and Sakakibara, Y. 2006. Discriminative detection of cis-acting regulatory variation from location data. In Proc. of the 4th Asia-Pacific Bioinformatics conference(eds. Jiang et al). Imperial College Press, London. Pp. 89-98

Kawahara, Y. et al. 2007. Redirection of silencing targets by adenosine to inosine editing of miRNAs. Science, 315: 1137-1140

Kawai, J., Shinagawa, A. et al. 2001. Functional annotation of a full length mouse cDNA collection. Nature, 409: 685-90

Kendrew, J. Sir(ed.). 1994. The Encyclopedia of Molecular Biology. Blackwell Science.

Kent, W.J. and Haussler, D. 2001. Assembly of the working draft of the human genome with GigAssembler. Genome Res., 11: 1541-48

Kim et al. 2007. Polony multiplex analysis of gene expression(PMAGE) in mouse hypertrophic cardiomyopathy.

Kim, T.H. et al. 2005. A high-resolution map of active promoters in the human genome. Nature, 436: 876-880

Kitano, H. 2002. Computational systems biology. Nature, 420: 206-210

Koonin, E.V. et al. 2002. The structure of the protein universe and genome evolution. Nature, 420: 218-223

Korbel, J.O. et al. 2007. Paired end mapping reveals extensive structural variation in the human genome. Science, 318: 420-426

Kornberg, R.D. 1974. Chromatin structure: A repeating unit of histones and DNA. Science, 184: 868-870

Kozak, M. 1991. Structural features in eukaryotic mRNAs that modulate the initiation of translation. J. Biol. Chem., 266(30): 19867-19870

Krammer, A. 1996. The structure and function of proteins involved in mammalian pre-mRNA splicing. Ann. Rev. Biochem., 65: 367-409

Krogan, N.J. et al. 2006. Global landscape of protein complexes in the yeast *S. cerevisiae*. Nature, 440: 637-643

Kumar, A. and Bennetzen, J. 1999. Plant retrotransposons. Ann. Rev. Genet., 33: 479-532

Kupper, H. et al. 1981. Cloning of cDNA of major antigen of foot and mouth disease virus and expression in E.coli. Nature, 289: 555-559.

Kwok, P.Y. 2001. Methods for genotyping single nucleotide polymorphisms. Ann Rev Genomics Hum. Genet., 2: 235-258

Lander, E.S. and Botstein, D. 1989. Mapping Mendelian factors underlying quantitative traits using RFLP linkage maps. Genetics, 121: 185-99

Lander, E.S., Linton, L.M., Birren, B. et al. 2001. Initial sequencing and analysis of the human genome. Nature, 409: 860-921

Landry, C.R. et al. 2007. Genetic properties influencing the evolvability of gene expression. Science, 317: 118-121

Lane, D. 1999. F plasmid. Encyclopedia of Molecular Biology. John Wiley & Sons, New York. p.891

Latchman, D.S. 2008. Eukaryotic Transcription Factors. Academic Press.

Lee, T., Rinaldi, N., Robert, F. et al. 2002. Transcriptional regulatory networks in Saccharomyces cerevisiae. Science, 298: 799-804

Lehman, L., Brueckner, F. and Cramer, P.2007. Molecular basis of RNA-dependent RNA polymerase II activity. Nature, 450: 445-449

Leitch, A.R., Schwarzacher, T., Jackson, D. and Leitch, I.j. 1994. *In situ* hybridization: a practical guide, BIOS Scientific Publishers.

Leslie, M. 2004. Protein Matchmaking. Science, 305: 1381

Li, C.W.; Chang, W.C. and Chen, B.S.2006. System identification and robustness analysis of the circadian regulatory network via microarray data in Arabidopsis thaliana. In Proc. of the 4th Asia-Pacific Bioinformatics conference(eds. Jiang et al). Imperial College Press, London. PP. 27-38

Li, S. et al. 2004. A map of interactome network of the metazoan C. *elegans*. Science, 303: 540-43

Liang, P. and Pardee, A.B. 1992. Differential display of eukaryotic mRNA by means of polymerase chain reaction. Science, 257: 967-971

Lichter, P and Ward, D.C. 1990. Is non-isotopic *in situ* hybridization finally coming of age? Nature, Vol. 345, pp.93-95

Lichter, P. et al. 1990. High-resolution mapping of human chromosome 11 by *in situ* hybridization with cosmid clones. Science, 247:64-69

Liu, Y.G. and Whittier, R. 1995. Thermal asymmetric interlaced PCR: automable amplification and sequencing of insert end fragments from P1 and YAC clones for chromosome walking. Genomes.

Livak, K.J. et al., 1995. Oligonucleotides with fluorescent dyes at opposite ends provide a quenched probe system useful for detecting PCR products and nucleic acid hybridization. PCR Methods Appl., 4: 357-362.

Lockhart, D.J and Winzeler, E.A. 2000. Genomics, gene expression and DNA arrays. Nature, 405(6788): 827-836

Lockhart, D.J., Dong, H., Byrne, M.C. et al. 1996. Expression monitoring by hybridization to high-density oligonucleotide arrays. Nature Biotechnol., 14:1675-1680

Loeffert, D. 2007. Polymerase Chain Reaction(PCR): An Analytical Tool in Bioprocessing. In:Ganapathy Subramanian(ed.) Bioseparation and Bioprocessing. A Handbook. Vol. 2. PP. 693-718. WILEY-VCH.

Lönning, W. E. and Saedler, H. 2002. Chromosome rearrangements and transposable elements. Ann. Rev. Genet., 36:389-416

Looger, L.L., Lalonde, S. and Frommer, W.B. 2005. Vision statement for plant physiology. Genetically encoded FRET sensors for visualizing metabolites with subcellular resolution in living cells. Plant Physiology., Vol. 138, pp. 555-557

Ma, P.C.H. and Chan, K.C.C. 2006. Inference of gene regulatory network from microarray data. In Proc. of the 4th Asia-Pacific Bioinformatics conference(eds. Jiang et al). Imperial College Press, London. pp17-26

Ma, Y. et al. 2006. EDAM: an efficient clique discovery algorithm with frequency transformation for finding motifs. In Proc. of the 4th Asia-Pacific Bioinformatics conference(eds. Jiang et al). Imperial College Press, London. PP. 119-128

MacBeath, G. and Schreiber, S.L. 2000. Science, 289: 1760-1763

Mackay, T.F.C. 2001. The genetic architecture of quantitative traits. Ann. Rev. Genet., 35:303-339

Mainly, J.L. et al. 1983. In vitro transcription: whole-cell extract. Methods in Enzymology, 101: 568-582

Maniatis,T.; Fritsch, E.F. and Sambrook. 1982. Molecular cloning: A laboratory Manual. Cold Spring Harbor Laboratory, New York

Manly, K.F. and Olson, J.M. 1999. Overview of QTL mapping software and introduction to map manager QT. Mamm. Genome, 10:327-334

Manly, K.F., Codmore, R.H. and Meer, J.M. 2001. Map Manager QTX, cross-platform software for genetic mapping. Mamm. Genome., 12: 930-932

Mann, M., Hendrickson, R.C. and Pandey, A. 2001. Analyses of protein and proteomes by mass spectrometry. Ann Rev. Biochem., 70: 437-474

Mansour, S.L., Thomas, K.R. and Capecchi, M.R. 1988. Disruption of the proto-oncogene int-2 in mouse embryo derived cells: a general strategy for targeting mutations to non-selectable genes. Nature, 336: 348-52

Marcotte, E.M., Pellegrini, M., Ng, H.L., Rice, D.W., Yeates, T.O and Eisenberg, D. 1999. Detecting protein function and protein-protein interactions from genome sequences. Science, 285: 751-753

Marcotte, E.M., Pellegrini, M., Thompson, M.J., Yeates, T.O and Eisenberg, D. 1999. A combined algorithm for genome wide prediction of protein function. Nature, 402: 83-86

Margulies et al. 2005. Genome sequencing in microfabricated high-density picolitre reactors. Nature., 437: 376-380

Marth, G.T., Korf, I. et al. 2001. A general approach to single nucleotide polymorphism discovery. Nature Genet., 23: 453-456

Martienssen, R.A. 1998. Functional genomics: probing plant gene function and expression with transposons. Proc. Natl. Acad. Sci., U.S.A., 95: 8145-50

Martienssen, R.A. 1998. Functional genomics: Probing plant gene function and expression with transposons. Proc. Natl. Acad. Sci., U.S.A., Vol. 95, pp. 2021-2026

Martin, C., Carpenter, R., Sommer, H., Saedler, H. and Coen, E.S. 1985. Molecular analysis of instability in flower pigmentation in Antirrhinum majus, following isolation of the pallida locus by transposon tagging. EMBO J.4:1625-30

Martin, G.B. et al. 1993. Map-based cloning of a protein kinase gene conferring disease resistance in tomato. Science, 262: 1432-36

Martinez-Laborda, A. Gonzalez-Reyes, A. and Morata, G. 1992. Trans regulation in the Ultrabithorax gene of Drosophila: alterations in the promoter enhance transvection. EMBO J., 11: 3545-62.

Martzen, M.R. et al. 1999. A biochemical genomics approach for identifying genes by the activity of their products. Science, 286: 1153-1155

Maxam, A. and Gilbert, W. 1977. A new method of sequencing DNA. Proc. Natl., Acad. Sci., U.S.A., 74, 560-4

McAdams, H.H. and Shapiro, L. 1995. Circuit simulation of genetic networks. Science, 269: 650-656

McCraith, S.,Holtzman, T., Moss, B. and Fields, S. 2000. Proc. Natl. Acad. Sci., U.S.A., 97: 4879-4884

Megeed, Z. et al. 2007.Electrokinetic separations. In : Subramanian, G.(ed.). Bioseparation and Bioprocessing. A Handbook. Vol. 2. Weley –VCH.pp. 719-734.

Members of the Complex Trait Consortium. 2003. The nature and identification of quantitative trait loci: a community's view. Nat. Rev. Genet., 4: 911-916

Mendelsohn, A.R.and Brent, R. 1999. Protein interaction methods-toward an endgame. Science, 284: 1948-50

Micales, J.A., Bonde, M.R. and Peterson, G.L. 1986. The use of isozyme analysis in fungal taxonomy and genetics. Mycotaxon, Vol. XXVII, pp. 405-449.

Michalatos-Beloin, S., Tishkoff, S.A., Bentley, K.L. et al.1996. Molecular haplotyping of genetic markers 10kb apart by allele-specific long-range PCR. Nucleic Acids Res., 24: 4841-4843

Mikkelsen, T.S., Manching, K. et al. 2007. Genome-wide maps of chromatin state in pluripotent and lineage committed cells. Nature, 448: 553-560

Miller, J.E. 1978. A proposal for the genetic manipulation of plant metabolism. J. Theor. Biol., 74: 153-154.

Ming, R. et al. 2008. The Draft genome of the transgenic tropical fruit tree papaya (*Carica papaya* Linneaus). Nature, 452: 991-996

Mirkin, S.M. 2007. Expandable DNA repeats and human disease. Nature, 447: 932-940

Mockler, T.C. et al. 2005. Applications of DNA tilling arrays for whole-genome analysis. Genomics, 85: 1-15

Moore, G. 2000. Cereal chromosome structure, evolution and pairing. Ann. Rev. Plant Physiol. Plant Mol. Biol., 51: 195-222

Moore, K.J. and Nagle, D.L. 2000. Complex trait analysis in the mouse: The strengths, the limitations and the promise yet to come. Ann. Rev. Genet., 34: 653-686

Moore, P.B. Structural motifs in RNA. 1999. Ann. Rev. Biochem., 67: 287-300

Morton, N.E. 1955. Sequential tests for detection of linkage. Am. J. Hum. Genet., 7: 277-318

Mott, R., Talbot, C.J., Turri, M.G., Collins, A.C. and Flint, J.2000. A method for fine mapping quantitative trait loci in outbred animal stocks. Proc. natl. Acad. Sci., U.S.A., 97: 12649-12654

Mouse Genome sequencing Consortium. 2002. Initial sequencing and Comparative analysis of the Mouse genome. Nature

Mullis, K.B. 1990. The unusual origin of the polymerase chain reaction. Sci. Am., 262:56-65

Mullis, K.B. and Faloora, F. 1987. Specific synthesis of DNA in vitro via a polymerase catalyzed chain reaction. Methods Enzymol., 155: 335-50

Myer, P. 1996. Homology-dependent gene silencing in plants. Ann. Rev. Plant Physiol. Plant Mol. Biol., 47: 23-48

Myers, E.W. et al., 2000. A whole-genome assembly of Drosophila. Science, Vol. 287, 2196-2216.

Myers, E.W., Sutton, G.G. et al. 2000. A whole-genome assembly of Drosophila. Science, 287: 2196-2204

Nelson L.D. and Cox, M.M. 2005.Lehninger principles of Biochemistry. 4th edition

Newman, J.R., Wolf, E. and Kim, P.S. 2000. Proc. Natl. Acad. Sci., U.S.A., 97: 13203-13208

Nilsson, S., Helou, K. et al. 2001. Rat-mouse and rat-human comparative maps based on gene homology and high-resolution zoo-FISH. Genomics, 74: 287-298

Nolan, P.M., Peters, J. et al. 2000. A systematic genome-wide, phenotype-driven mutagenesis program for gene function studies in mouse. Nature Genet., 25: 440-443

Noordewier, M.O. and Warren, P.V. 2001. Gene expression microarrays and the integration of biological knowledge. Trends Biotechnol., 19: 412-415

Norris, R., Norris, F.Christiansen, L. and Fil , N. 1983. Efficient site-directed mutagenesis by simultaneous use of two primers. Nucleic Acids Res., 11: 5103-12.

Novina, C.D. and Sharp, P.A. 2004. The RNAi revolution. Nature., Vol. 430, 161-164

Nowacki, M. et al. 2008. RNA-mediated epigenetic programming of a genome rearrangement pathway. Nature, 451: 153-158

Ochman, H., Lawrence, J.G. and Groisman, E.A. 2000. Lateral gene transfer and the nature of bacterial innovation. Nature, 405(6784): 299-304

Old, R.W. and Primrose, S.B. 1980. Principles of Gene Manipulation. Blackwells.

Olins, D.L. and Olins, D.E. 1974. Spheroid chromatin unit(v Bodies). Science, 183: 330-332

Olivier, M., Agarwal, A. et al. 2001. A high-resolution radiation map of the human genome draft sequence. Science, 291: 1298-1302

Olson, M., Hood, L., Cantor, C. and Botstein, D.1989. A common language for physical mapping of the human genome. Science, Vol. 245, 1434-35

Osoegawa, K., Tateno, M. et al. 2000. Bacterial artificial chromosome libraries for mouse sequencing and functional analysis. Genome Res.,10: 116-128

Overbeek, R., Fonstein, M., D'Souza, M., Pusch, G.D. and Maltsev, N. 1999. The use of gene clusters to infer functional coupling. Proc. Natl. Acad. Sci., U.S.A., 96: 2896-2901

Pak, J. and Fire, A. 2007. Distinct populations of primary and secondary effectors during RNAi in C. elegans. Science, 315: 241-244

Pandey, A. and Mann, M. 2000. Proteomics to study genes and genomes. Nature, 405: 837-846

Patil, N., Berno, A.J. et al. 2001. Blocks of limited haplotype diversity revealed by high-resolution scanning of human chromosome 21. Science, U.S.A., 294: 1719-1723

Pellegrini, M., Marcotte, E.M., Thompson, M.J., Eisenberg, D. and Yeates, T.O. 1999. Assigning protein functions by comparative genome analysis: protein phylogenetic profiles. Proc. Natl. Acad. Sci., U.S.A., 96: 4285-4288

Pennisi, E. 2000. Ideas fly at Gene-Finding Jamboree. Science, Vol. 287, pp. 2182-84

Perbal, B. 1999. S1 nuclease. Encyclopedia of Molecular Biology. John Wiley & Sons. New York, p.2263

Peri, S and Pandey, A. 2001. A reassessment of the translation initiation codon in vertebrates. Trends Genet., 17: 685-687

Poelwijk, F.J. et al. 2007. Empirical fitness landscapes reveal accessible evolutionary paths. Nature, 445: 383-386

Primrose, S.B., Ţwyman, R.M. and Old, R.W.2001. Principles of Gene Manipulation. Blackwell Science Ltd, Australia

Quellette, M.M. and Wright, W.E. 1995. Use of reiterative selection for defining protein-nucleic acid interactions. Curr. Opin. Biotechnol., 6: 65-72

Raikhel, N. and Pirrung, M. 2005. Vision statement for plant physiology. Adding precision tools to the plant biologists' toolbox with chemical genomics. Plant physiology., Vol. 138. pp. 563-564

Rain. J.C., Selig, L., de Reuse, H., Battaglia, V., Reverdy, C. et al. 2001. The protein-protein interaction map of Helicobacter pylori. Nature, 409: 211-216

Ray, Wu. (ed.). 1979. Methods in Enzymology, Recombinant DNA. Vol. 68, Academic Press.

Reich, D. et al. 2001. Linkage disequilibrium in the human genome. Nature., 411: 199-204

Reidy, M.C., Timm, E.A. Jr and Stewart, C.C. 1995. Quantitative RT-PCR for measuring gene expression. Biotechniques, 18: 70-74, 76

Reik, W. 2007. Stability and flexibility of epigenetic gene regulation in mammalian development. Nature, 447: 425-432

Reinert, K. and Huson, D. 2007. Bioinformatics support for genome-sequencing projects. In: Bioinformatics-From Genomes to Therapies(ED.) T. Lengauer. Vol. 1: Molecular sequence and Structure. Wiley-VCH.pp.25-55.

Ren, B. et al. 2000. Genome-wide location and function of DNA binding proteins. Science, 290: 2306-9.

Ried, T., Baldini, A., Rand, T.C. and Ward, D.C. 1992. Simultaneous visualization of seven different DNA probes by in situ hybridization using combinatorial fluorescence and digital imaging microscopy. Proc. Natl. Acad. Sci., U.S.A., 89: 1388-92

Risch, N. 2000. Searching for genetic determinants in the new millennium. Nature, 405: 847-856

Roberts, L. 1989. New game plan for genome mapping. Science, Vol. 245,1438-1440

Roberts, R.C., Bobrow, M. and Bentley, D.R.. 1992. Point mutations in dystrophin gene. Proc. Natl. Acad. Sci., U.S.A., 89: 2331-5

Rommens, R.C., Ianuzzi, M.C., Kerem, B.S. et al. 1989. Identification of the cystic fibrosis gene. Chromosome walking and jumping. Science, 245: 1059-65

Ronaghi et al. 1996. Real-time DNA sequencing using detection of pyrophosphate release. Ann. Biochem., 242: 84-89

Ronaghi, M., Uhlen, M. and Nyren, P.A. A sequencing method based on real-time pyrophosphate. Science, 281: 363-365

Ross-Macdonald, P. et al. 1999. Large-scale analysis of the yeast genome by transposon tagging and gene disruption. Nature, 402: 413-418

Ross-Macdonald, P., Sheen, A., Roeder, G.S. and Snyder, M.A. 1997. A multipurpose transposon for analyzing protein production, localization and function in Saccharomyces cerevisiae. Proc. Natl. Acad. Sci., U.S.A., 94: 190-195

Rowen, L.; Mahairas, G. and Hood, L. 1997. Sequencing the human genome. Science, 278: 605-607

Roy, D. 2000. Plant Breeding-Analysis and Exploitation of Variation. Narosa, New Delhi and Alpha Science, U.K.

Roy, D. 2009 Cytogenetics. Narosa, New Delhi and Alpha Science, U.K.

Roy, D. 2009. Bioinformatics, Narosa, New Delhi and Alpha Science, U.K.

Rual, J.F. et al. 2005. Towards a proteome-scale map o protein-protein interaction network. Nature, 437: 1173-178

Rubin, G.M. and Spradling, A.C. 1982. Genetic transformation of Drosophila with transposable element vectors. Science, 218: 348-53.

Rubin, G.M., Hong, L., Brokstein, P., Evans-Holm, M., Frise, E., Stapleton, M. and Harvey, D. 2000. A Drosophila Complementary DNA resource. Science, Vol. 287, 2222-2224

Ruby, S.W. and Abelson, J. 1991. Pre-mRNA splicing in yeast. Trends genet., 7: 79-85

Sachidananadam, R., Weissman, D. et al. 2001. A map of human genome sequence variation containing 1.42 million single nucleotide polymorphisms. Nature, 409: 928-933

Saiki, R.K. et al. 1988. Primer-directed enzymatic amplification of DNA with a thermostable DNA polymerase. Science, 239: 487-91

Saleh-A. et al. 2006. A genome-wide search for ribozymes reveals an HDV-like sequence in the human CPEB3 gene. Science, 1788-92

Sambrook, J., Fritsch, E.F. and Maniatis, T. 1989. Molecular Cloning: a Laboratory Manual. Cold Spring harbor Laboratory, New York.

Sanger F. et al. 1980. Cloning in single-stranded bacteriophage as an aid to rapid DNA sequencing. J.Mol.Biol., 143, 161-78-M13 vectors.

Sanger, F. et al. 1977. DNA sequencing with chain-terminating inhibitors. Proc. Natl. Acad. Sci., U.S.A., 74, 5463-7

Sanger, F., Coulson, A.R., Hong, G.F., Hill, D.F. and Peterson, G.B. 1982. Nucleotide sequence of bacteriophage lambda DNA. J.Mol. Biol., 162: 729-773

Sasaki, T., et al. 2002. The genome sequence and structure of rice chromosome 1. Nature, Vol. 420, pp. 312-316

Schena, M., Shalon, D., Davis, R.W. and Brown, P.O. 1995. Quantitative monitoring of gene expression patterns with a complementary DNA microarray. Science, 270: 467

Schena, M., Shalon, D., Heller, R., Chai, A., Brown, P.O. and Davis, R.W. 1996. Parallel genome analysis: microarray- based expression monitoring of 1000 genes. Proc. Natl. Acad. Sci., U.S.A., 93: 10614-10619

Schizyya, H., Biren, B. Kim, U.j., Mancino, V., Slepak, T., Tachiiri, Y. and Simon, M. 1992. Cloning and stable maintenance of 300-kilobase-pair fragments of human DNA in Escherichia coli using an F-factor-based vector. Proc. Natl. Acad. Sci., U.S.A 89: 8794-7

Schlabach, M. et al. 2007. Cancer proliferation gene discovery through functional genomics. Science, 319: 620-624

Schmid, E.M. and McMohan, T. 2007. Integrating molecular and network biology to decode endocytosis. Nature, 448: 883-888

Schmidt, R. and Dean, C. 1995. Plant Genomes: A current molecular description. Ann. Rev. Plant Physiol. Plant Mol. Biol., 46: 395-418

Schmidt, R., West, J., Love, K., Lenehan, Z., Lister, C., Thompson, H., Bouchez, D. and Dean, C. 1995. Physical map and organization of *Arabidopsis thaliana* chromosome 4. Science, Vol. 270, 480-485.

Schuler, G.D. 1997. Sequence mapping by electronic PCR. Genome Res., 7: 541-550

Schwechheimer, C.Zourelidou, M. and Bevan, M.W. 1998. Plant transcription factor studies. Ann. Rev. Plant Physiol. Plant Mol. Biol., 49: 127-150

Scott, H.S. and Chrast, R. 2001. Global transcript expression profiling by Serial Analysis of Gene Expression(SAGE). Genet. Eng., 23: 201-219

Searls, D.B. 2002. The language of genes. Nature, 420: 211-217

Sebak, O.K. and Laskin, A.I. 1979. Genetics of Industrial Microorganisms. ASM.

Segal, E. et al. 2006. A genomic code for nucleosome positioning. Nature, 442: 772-778

Segal, E. et al. 2008. Predicting expression patterns for regulatory sequence in Drosophila segmentation. Nature, 451: 535-540

SenGupta, D.J. et al. 1996. A three-hybrid system to detect RNA-protein interactions in vivo. Proc. Natl. Acad. Sci., U.S.A., 93: 8496-8501

Service, R.F. 2006. Gene sequencing. The race for the $1000 genome. Science, 311: 1544-46

Setlow, J.K. and Hollaender, A. 1979, 1980. Genetic Engineering: Principles and Methods. Vol.1 and Vol. 2. Plenum.

Shaffer, L.G. and Lupski, J.R. 2000. Molecular mechanisms for constitutional chromosomal rearrangements in humans. Ann Rev. Genet., 34: 297-329

Sham, P.C. and Curtis, D. 1995. An extended transmission/ disequilibrium test(TDT) for multiple allele marker loci. Ann. Hum Genet., 59: 323-336

Sharp, P.A. 1994. Split genes and RNA splicing. Cell, 77: 805-815

Shen, L.X., Basilion, J.P. and Stantoon, V.P. Jr.(1999). Single-nucleotide polymorphisms can cause different structural folds of mRNA. Proc. Natl. Acad. Sci., U.S.A., 96: 7871-7876

Shiraki et al. 2003. Proc. Natl. Acad. Sci., U.S.A., 100: 15776

Sijen, T. et al. 2007. Secondary siRNAs result from unprimed RNA synthesis and form a distinct class. Science, 315: 244-246

Sinden, R.R. et al. 2005. DNA twists and flips. Nature, 437: 1097-98

Singer, J.B. et al. 2004. Genetic dissection of complex traits with chromosome substitution strains of mice. Science, Vol. 304, 445-448

Smale, S.T. and Kadonaga, J.T. 2003. The RNA polymerase II core promoter. Ann. Rev. Biochem., 72: 449-480

Smith, V., Chou, K.N., lashkari, D., Botstein, D. and Brown, P.o. 1996. Functional analysis of the genes of yeast chromosome V by genetic footprinting. Science, 274: 2069-2074

Snel, B., Bork, P. and Huynen, M.A. 2002. The identification of functional modules from the genomic association of genes. Proc. Natl. Acad. Sci., U.S.A., 99: 5890-5895

Soderlund, C., Humphray, S., Dunham, A. and French, L.2000. Contigs built with fingerprints, markers and FPC V4.7. Genome Res., 10: 1772-87

Solter, D. 1988. Differential imprinting and expression of maternal and paternal genomes. Ann. Rev. Genet., 22: 127-146

Sontheimer, E. and Steitz, J.A. 1993. The U5 and U6 small nuclear RNAs as active site components of spliceosome. Science, 262: 1989-1996

Southern, E.M. 1975. Detection of specific sequences among DNA fragments by gel electrophoresis. J. Mol. Biol., 98, 503-507- Southern hybridization

Spector, D.L. 2003. The dynamics of chromosome organization and gene regulation. Ann. Rev. Biochem., 72: 573-608

Spencer, H.G.2000. Population genetics and evolution of genomic imprinting. Ann. Rev. Genet., 34: 457-478

Spielman, R. et al. 1993. Transmission test for linkage disequilibrium: the insulin gene region and insulin-independent diabetes mellitus. Am. J. Hum. Genet., 52: 506-516

Stace, C.A. and Bailey, J.P. 1999. The value of genomic in situ hybridization(GISH) in plant taxonomic and evolutionary studies. In Molecular systematics and plant evolution(eds. P.M. Hollingsworth, R.M. Bateman and R.J. Gornall), Taylor and Francis, London, pp. 199-210

Stark, A. et al. 2007. Discovery of functional elements in 12 Drosophila genomes using evolutionary signatures. Nature, 450: 219-232

Steen, R.G., Kwitek,-Black, A.E. et al. 1999. A high- density integrated genetic linkage and radiation hybrid map of the laboratory rat. Genome Res., 9, AP1-8, INSERT.

Stein, L. 2001. Genome annotation: from sequence to biology. Nature Rev. Genet., 2:493-503

Sternberg, N.1990. Bacteriophage P1 cloning system for the isolation, amplification, and recovery of DNA fragments as large as 100 kilobase pairs. Proc. Natl. Acad. Sci., U.S.A., 87 pp. 103-107

Stranger, B.E. et al. 2007. Relative impact of nucleotide and copy number variation on gene expression phenotypes. Science, 315: 848-853

Strausberg,R.L., Feingold, E.A., Klausner, R.D. and Collins, F.C. 2000. The mammalian gene collection. Science, 286: 455-457

Strobel, S.A. 1999. Ribozyme/ catalytic RNA. Encyclopedia of Molecular Biology. John Wiley & Sons. New York. p. 2202

Sudarson, et al. 2006. Tandem riboswitches architectures exhibit complex gene control function. Science, 314: 300-304

Sugiura, M. 1997. Plant *in vitro* transcription systems. Ann. Rev. Plant Physiol. Plant Mol. Biol., 48: 383-398

Sun, J.S. and Manely, J.L.1995. A novel U2-U6snRNA structure is necessary for mammalian mRNA slicing. Genes and Dev., 9: 843-854

Surrani, M.A. 1991. Genomic imprinting: developmental significance and molecular mechanism. Curr. Opin. Genet. Dev., 1: 241-6

Szathmary, E., Jordan, F and Pal, C. 2001. Can we explain biological diversity. Science, 292: 1315-17

Tamayo, P. et al. 1999. Interpreting patterns of gene expression with self-organizing maps: methods and application to hematopoietic differentiation. Proc. Natl. Acad. Sci., U.S.A. ,96: 2907-2912

Tatusov, R.L. 1997. A genomic perspective on protein families. Science, 278: 631-637

Temple, L.K.F., McLeod, R.S., Gallinger, S. and Wright, J.G. 2001. Defining disease in genomics era. Science, 293: 807-808

The Arabidopsis Genome Initiative. 2000. Analysis of the genome sequence of the flowering plant Arabidopsis thaliana. Nature, 408: 796-815

The *C. elegans* Sequencing Consortium. 1998. Genome sequence of the nematode C. elegans: A platform for investigating biology. Science, Vol. 282, 2012-2018.

The ENCODE Project Consortium. 2007. Identification and analysis of functional elements in 1% of the human genome by the ENCODE pilot project. Nature, Vol. 447, 799-

The French-Italian Public Consortium for Grapevine Genome Characterization. 2007. Nature, 449: 463-467

The International Haplotype Map Consortium. 2005. Nature, 437: 1299

The RIKEN exploration Research Group PhaseII Team and The FANTOM Consortium. 2001. Functional annotation of a full-length mouse cDNA collection. Nature, 409: 685-690

Thomas, J.W.T., Blakesley, R.W. et al. 2003. Comparative analyses of multi-species sequences from targeted genomic regions. Nature, 424: 788-793

Tijsterman, M., Ketting, R.F., Hudson, A.M. and Cooley, L. 2002. The genetics of RNA silencing. Ann. Rev. Genet., 36:489-520

Timmis, K.N. 1981. Gene Manipulation *In vitro*. In : SGM Symposium. 31. S.W.Glover and D.A.Hopwood(eds.).

Timmis, K.N. and Puhler, A. 1979. Plasmids of Medical, Environmental and Commercial Importance. Elsevier/ North Holland.

Tompa, M.; N. Li., T.L Bailey et al. 2005. Assessing computational tools for the discovery of transcription factor binding sites. Nat. Biotechnol., 23: 137-144

Tong, A.H. et al. 2001. Systematic genetic analysis with ordered arrays of yeast deletion mutants. Science, 294: 236402368

Tong, A.H., Lesage, D. et al. 2001. Global mapping of the yeast genetic interaction network. Science, 303: 808-813

Trifonov, E.N. 1999. Sequence codes. Encyclopedia of Molecular Biology. John Wiley & Sons. New York. p.2324

Tyagi, S. and Kramer, F.R. 1996. Molecular beacons: probes that fluoresce upon hybridization. Nat. Biotechnol., 14: 303-308.

Utez, , P., Giot, L. et al. 2000. A comprehensive analysis of protein-protein interactions in Saccharomyces cerevisiae. Nature, 403: 623-627

Vagin, V.V. et al. 2006. A distinct small RNA pathway silences selfish genetic elements in the germ line. Science, 313: 320-324

Van Ooijen, J.W., Maliepaard, C. 1996a. MapQTL version 3.0: software for the calculation of QTL positions on genetic maps. Plant genome IV.

Van Ooijen, J.W., Maliepaard, C. 1996a. MapQTL version 3.0: software for the calculation of QTL positions on genetic maps. CPRO-DLLO: Wageningen.

Varani, G. etal. 1999. RNA structure. Encyclopedia of Molecular Biology. John Wiley & Sons. New York. p.2232

Varmus, H. and Brown, P. 1989. Retroviruses. In Berg, P.E. and Howe, M.H.(eds.) Mobile DNA, pp. 53-109, American Soc. Microbiology, Washington, DC.

Vassylycv, D.G. et al. 2007. Structural basis for transcription elongation by bacterial RNA polymerase. Nature, 448: 157-162

Vasudevan, S., Tong, Y. and Steitz, A. 2007. Switching from repression to activation: microRNAs can up-regulate translation. Science, 318: 1931-34

Velculescu, V.E., Zhang, L., Vogelstein, B., and Kinzler, K.W. 1995. Serial analysis of gene expression. Science, 270, 484-487

Venter, J.C., Smith, H.O. and Hood, L.1996. A new strategy for genome sequencing. Nature, Vol. 381, 364-367.

Venter, J.C.; Adams, M.D., Myers, E.W. et al. 2001. The sequence of human genome. Science., 291: 1304-1351

Vieux, E.F., Kwok, P.Y., Miller, R.D. 2002. Primer design for PCR and sequencing in high-throughput analysis of SNPs. Biotechniques.

Vollmer, S.J. and Yanofsky, C. 1986. Efficient cloning of genes in Neurospora crassa. Proc. Natl. Acad. Sci., U.S.A., 83: 4867-73

Von Mering, C. et al. 2002. Comparative assessment of large-scale datasets of protein-protein interactions. Nature, 417: 399-403

Vukmirovic, O.G. and Tilghman, S.M. 2000. Exploring genome space. Nature, 405: 820-822

Wahle, E. and Keller, W. 1992. The biochemistry of 3-end cleavage and polyadenylation of messenger RNA precursors. Ann. Rev., Biochem., 61: 419-440

Wahle, E. and Keller, W. 1992. The biochemistry of the 3-end cleavage and polyadenylation of messenger RNA precursors. Ann. Rev. Biochem, 61: 419-440

Walbot, V. 2000. A green chapter in the book of life. Nature, Vol. 408, pp.794

Walhout, A.J. and Vidal, M.A. 1999. A genetic strategy to eliminate self-activator bait prior to high throughput yeast two –hybrid screens. Genome Res., 9: 1128-34

Walhout, A.J., Sordella, R. et al. 2000. Protein-protein interaction mapping in C. elegans using proteins involved in vulva development. Science, 287: 116-122

Wang, D.G., Fan, J.-B. et al. 1998. Large-scale identification, mapping and genotyping of single-nucleotide polymorphisms in the human genome. Science, 280: 1077-1081

Wang. A.M., Doyle, M.V. and Mark, D.F. 1989. Quantitation of mRNA by the polymerase chain reaction. Proc. Natl. Acad. Sci., U.S.A., 86: 9717-21

Wassarman, D.A. and Steitz, J.A. 1992. interactions of small nuclear RNA's with precursor messenger RNA during in vitro splicing. Science, 257: 1918-1925

Wasserman, W.W., Palumbo, M., Thompson, W., Fickett, J.W. and Lawrence, C.E. 2000.Human-mouse genome comparisons to locate regulatory sites. Nature Genet., 26: 225-228

Watanabe, T.K. Bihoreau, M.T. et al. 1999. A radiation hybrid map of the rat genome containing 5,255 markers. Nature Genet., 22: 27-36

Weber, J.L. and Myers, E.W. 1997. Human whole-genome shotgun sequencing. Genome Res., 7: 401-9

Weigel, D and Nordborg, M. 2005. Vision statement for plant physiology. Natural variation in Arabidopsis. How do we find the causal genes? Plant Physiology, Vol. 138, pp. 567-568

Weir, B.S. 1996. Genetic Data Analysis II. Sinauer Associates Inc Publishers: Sunderland, MA, U.S.A.

Werner, T. 1999. Identification and characterization of promoters in eukaryotic DNA sequences. Mammalian Genome, 10: 168-175

Whitelegge, J.P., le Coutre, J. et al. 1999. Toward the bilayer proteome, electrospray ionization mass spectroscopy of large, intact transmembrane protein. Proc. Natl. Acad. Sci., U.S.A., 96: 10695-10698

Wicks, S.R., Yeh, R.T. et al. 2001. Rapid gene mapping in Caenorhabditis elegans using a high density polymorphism map. Nature Genet., 28: 160-164

Will, G.L. and Luhrmann, R. 1999. RNA splicing. Encyclopedia of Molecular Biology. John Wiley & Sons. New York. p. 2230

Williams, J.G.K., Kubelik, A.R., Lival, K.J., Rafalski, J.A. and Tingey, S.V. .DNA polymorphisms amplified by arbitrary primers are useful genetic markers. Nucleic Acid Research., Vol. 18, No. 22, 6531-6535

Williams, N. 1997. Bioinformatics: how to get databases talking the same language. Science, 275(5298): 301-302

Wilson, A.K. 1999. Expressed sequence tag. Encyclopedia of Molecular Biology. John Wiley & Sons. New York. p.879

Wilson, A.K. 1999. Polymerase chain reaction. Encyclopedia of Molecular Biology. John Wiley & Sons. New York. p. 1903

Wilson, A.K. 1999. Subtractive hybridization. Encyclopedia of Molecular Biology. John Wiley & Sons. New York. p. 2458

Wilson, D.S. and Szostak, J.W. 1999. In vitro selection of functional nucleic acids. Ann Rev. Biochem., 68: 611-647

Windass, J.D. et al. 1980. Improved conversion of methanol to single –cell protein by Methylophilus methylotrophus. Nature, 287: 396-401.

Winnacker, W.-L. 1987. From Genes to Clones. VCH

Wittwer, C.T. et al., 1997. The LightCycler: a microvolume multisample fluorimeter with rapid temperature control. Biotechniques, 22: 176-181.

Wu, L. and Russell, P. 1993. Nim 1 kinase promotes mitosis by inactivating Wee 1 tyrosine kinase. Nature, 363: 738-41

Xia, Y. et al. 2004. Analyzing cellular biochemistry in terms of molecular networks. Ann. Rev. Biochem., 73, 1051-1087

Xie, X. and Ott, J. 1993. Testing linkage disequilibrium between a disease gene and marker loci. Am. J. Hum. Genet., 53:1107

Xie, X. et al. 2005. Systematic discovery of regulatory motifs in human promoters and 3′UTR by comparison of several mammals. Nature, 434: 338-345

Yamada, K., Lim, J. et al. 2003. Empirical analysis of transcriptional activity in the Arabidopsis genome. Science, 302: 842-46

Yamashita, R., Suzuku, Y., Sugano, S. and Nakai, K. 2005. Genome-wide analysis reveals strong correlation between CpG islands with nearby transcription start sites of genes and their tissue specificity. Gene, 350: 129-36

Yao, M.-C. 2008. RNA rules. Nature, 451: 131-32

Yarus, M. 2007. Bring them back alive. Nature, 438: 40

Yin, H. and Lin, H. 2007. An epigenetic activation role of Piwi and a Piwi-associated piRNA in Drosophila melanogaster. Nature, 450: 304-308

Yu, A., Zhao, C. et al. 2001. Comparison of human genetic and sequence-based physical maps. Nature, 409: 951-953

Yu, J., Hu, S.N. et al. 2002. A draft sequence of the rice genome(*Oryza sativa* I. ssp. Indica). Science, 296: 79-92

Yuh, C.-H., Bolouri, H. and Davidson, E.H. 1998. Genomic Cis-regulatory logic: experimental and computational analysis of a sea urchin gene. Science, 279: 1896-1902

Zamore, P.D. 2007. Genomic defence with a slice of pi. Nature, 446: 864-865

Zarembinski, T.I., Hung. L.W. et al. 1998. Structure-based assignment of the biochemical function of a hypothetical protein: A test case of structural genomics. Proc. Natl. Acad. Sci., U.S.A., 95: 15189-15193

Zeng, Z.-B. 1993. Theoretical basis for separation of multiple linked gene effects in mapping quantitative trait loci. Proc. Natl. Acad. Sci., U.S.A., 90: 10972-10976

Zeng, Z.-B. 1994. Precison mapping of quantitative trait loci. Genetics, 136: 1457-1468

Zeng, Z.-B.,Kao, C.H. and Basten, C.J.1999. Estimating the genetic architecture of quantitative traits. Genet. Res., 74: 279-289

Zhang, H.; Zhang, R. and Liang, P. 1996. Differential screening of gene expression difference enriched by differential display. Nucleic Acids Res., 24(No. 12): 2454-2455

Zhang, L., Zhou, W., Velculescu, V.E. et al. 1997. Gene expression profiles in normal and cancer cells. Science, 276: 1268-1272

Zhang, Q. et al. 1999. Messenger RNA translation state: the second dimension of high-throughput expression screening. Proc. Natl. Acad. Sci., U.S.A., 96: 10632-36

Zhong, W and Sternberg, P.W. 2006. Genome-wide prediction of *C. elegans* genetic interactions. Science, 311: 1481-84

Zhu, H., Bilgin, M. and Snyder, M. 2003. Proteomics. Ann. Rev. Biochem., 72, 783-812

Zhu, T. and Wang, X. 2000. Large-scale profiling of the Arabidopsis transcriptome. Plant Physiol., 124: 1472-76

Zhu,H., Bilgin, M. et al. 2004. Global analysis of protein acitivities using proteome chips. Science, 293: 2401-05

Index